Advances in Intelligent Systems and Computing

Volume 856

Series editor

Janusz Kacprzyk, Polish Academy of Sciences, Warsaw, Poland
e-mail: kacprzyk@ibspan.waw.pl

The series "Advances in Intelligent Systems and Computing" contains publications on theory, applications, and design methods of Intelligent Systems and Intelligent Computing. Virtually all disciplines such as engineering, natural sciences, computer and information science, ICT, economics, business, e-commerce, environment, healthcare, life science are covered. The list of topics spans all the areas of modern intelligent systems and computing such as: computational intelligence, soft computing including neural networks, fuzzy systems, evolutionary computing and the fusion of these paradigms, social intelligence, ambient intelligence, computational neuroscience, artificial life, virtual worlds and society, cognitive science and systems, Perception and Vision, DNA and immune based systems, self-organizing and adaptive systems, e-Learning and teaching, human-centered and human-centric computing, recommender systems, intelligent control, robotics and mechatronics including human-machine teaming, knowledge-based paradigms, learning paradigms, machine ethics, intelligent data analysis, knowledge management, intelligent agents, intelligent decision making and support, intelligent network security, trust management, interactive entertainment, Web intelligence and multimedia.

The publications within "Advances in Intelligent Systems and Computing" are primarily proceedings of important conferences, symposia and congresses. They cover significant recent developments in the field, both of a foundational and applicable character. An important characteristic feature of the series is the short publication time and world-wide distribution. This permits a rapid and broad dissemination of research results.

More information about this series at http://www.springer.com/series/11156

Kevin Deng · Zhengtao Yu
Srikanta Patnaik · John Wang
Editors

Recent Developments in Mechatronics and Intelligent Robotics

Proceedings of International Conference on Mechatronics and Intelligent Robotics (ICMIR2018)

Set 1

Springer

Editors
Kevin Deng
School of Transportation Science and Engineering
Beihang University
Beijing, China

Srikanta Patnaik
Department of Computer Science and Engineering
Soa University
Bhubaneswar, Odisha, India

Zhengtao Yu
School of Information Engineering and Automation
Kunming University of Science and Technology
Kunming, China

John Wang
Department of Information Management and Business Analytics
Montclair State University
Montclair, NJ, USA

ISSN 2194-5357 ISSN 2194-5365 (electronic)
Advances in Intelligent Systems and Computing
ISBN 978-3-030-00213-8 ISBN 978-3-030-00214-5 (eBook)
https://doi.org/10.1007/978-3-030-00214-5

Library of Congress Control Number: 2018953052

This Springer imprint is published by the registered company Springer Nature Switzerland AG
The registered company address is: Gewerbestrasse 11, 6330 Cham, Switzerland

Preface

I, Prof. Srikanta Patnaik, Programme Chair of the Conference, welcome all the participants to the International Conference on Mechatronics and Intelligent Robotics (ICMIR2018) held at Kunming, China, during 19–20 May 2018. This annual conference is being organized each year by Interscience Research Network, an international professional body, in association with International Journal of Computational Vision and Robotics, and International Journal of Information and Communication Technology, published by Inderscience Publishing House. I must welcome this year's General Chairs, Prof. Dr. Zhengtao Yu, Dean, School of Information Engineering and Automation, Kunming University of Science and Technology, China, and Prof. Feng Qiao, Shenyang Jianzhu University, Shenyang, China, for their generous contribution to make ICMIR2018 a success.

Like every edition, ICMIR2018 was academically very rich, and we had two eminent professors as keynote speakers, namely Prof. John Wang, Department of Information Management and Business Analytics, School of Business, Montclair State University, USA, and Dr. Nilanjan Dey, Department of Information Technology, Techno India College of Technology, Kolkata, India.

There has been a paradigm shift in this domain after the introduction of IoT and Web-enabled systems. This area covers a wide range of applications by now, such as IoT-enabled manufacturing, Industry 4.0, robot-assisted manufacturing; advanced mechanisms and robotics; systems modelling and analysis; instrumentation and device control; automation systems; intelligent sensing and control; medical robotics; and autonomous and complex systems. New technologies are constantly emerging, which are enabling applications in various domains and services.

Mechatronics and Intelligent Robotics is no longer a functional area within the department of mechanical or electronics, but is an integral part of the manufacturing function of any organization. In recent times, Mechatronics and Intelligent Robotics is probably the single most important facilitator of the manufacturing process. The strong influence of Mechatronics and Intelligent Robotics is now the deciding factor in shaping the process of globalization, particularly in the productive, manufacturing and commercial spheres, and is also widely recognized. Creating economic opportunities and contributing to monotony reduction is another thrust area for the emerging epoch of Mechatronics and Intelligent Robotics.

This edition of ICMIR covers the following areas but is not limited to: Intelligent mechatronics, robotics and biomimetics; Novel and unconventional mechatronics systems; Modelling and control of mechatronics systems; Elements, structures, mechanisms of micro- and nano-systems; Sensors, wireless sensor networks and multi-sensor data fusion; Biomedical and rehabilitation engineering, prosthetics and artificial organs, AI, neural networks and fuzzy logic in mechatronics and robotics; Industrial automation, process control and networked control systems; Telerobotics, human–computer interaction and human–robot interaction; Artificial intelligence; Bio-inspired robotics; Control algorithms and control systems; Design theories and principles; Evolutional robotics; Field robotics; Force sensors, accelerometers and other measuring devices; Healthcare robotics; Human–robot interaction; Kinematics and dynamics analysis; Manufacturing robotics; Mathematical and computational methodologies in robotics; Medical robotics; Parallel robots and manipulators; Robotic cognition and emotion; Robotic perception and decision; Sensor integration, fusion and perception; Social robotics. This volume covers various articles covering the recent developments in the area of Mechatronics and Intelligent Robotics categorized into six tracks, such as follows:

1. Intelligent Systems
2. Robotics
3. Intelligent Sensors and Actuators
4. Mechatronics
5. Computational Vision
6. Machine Learning and Soft Computing

Srikanta Patnaik
Programme Chair: ICMIR2018

Acknowledgement

Like every year, this edition of ICMIR2018 was also attended by more than 130 participants, and 155 papers were shortlisted and published in this proceedings. The papers covered in this proceeding are the result of the efforts of the researchers working in broad domains of Mechatronics and Intelligent Robotics. We are thankful to the authors and paper contributors of this volume.

We are thankful to the editor in chief and the anonymous review committee members of the Springer series on *Advances in Intelligent Systems and Computing*, for their support to bring out the proceedings of ICMIR2018.

We are thankful to our friends, namely Prof. John Wang, School of Business, Montclair State University, USA, and Dr. Nilanjan Dey, Techno India College of Technology, Kolkata, India, for their keynote addresses. We are also thankful to the experts and reviewers who have worked for this volume despite the veil of their anonymity.

We should not forget to inform you that the next edition of the conference, i.e. 3rd International Conference on Mechatronics and Intelligent Robotics (ICMIR-2019), shall be held at Kunming, China, in association with Kunming University of Science and Technology, Kunming, China, during the last week of April 2019.

It was really a nice experience to interact with the scholars and researchers who came from various parts of China and outside China to participate in ICMIR2018. In addition to the academic part, the participants must have enjoyed their stay, participation and presentation during the conference and sightseeing trip at Kunming.

I am sure that the readers will be benefited from this volume of AISC series volume on "Recent Developments in Mechatronics and Intelligent Robotics: Proceedings of International Conference on Mechatronics and Intelligent Robotics (ICMIR2018)".

Contents

Mechatronics

Machine Learning Soft Computing

Intelligent Systems

Effective Prediction of Microblog User Retweet Behaviors Based on Markov Random Field

Ren Guoheng(✉), Wang Wei, and Chen Liyong

School of Network Engineering, Zhoukou Normal University, Zhoukou, China
renguoheng@126.com

Abstract. Prediction of user retweet behavior is of great significance to the topic detection and diffusion range control. Based on the data of Sina microblog, we propose an effective prediction framework based on Markov Random Field for finding the global optimal solution to improve the prediction accuracy. Under the proposed method, we systematically analyze the various factors affecting the retweet behavior and the retweet constraints between users, and then construct the prediction model to perform the prediction. Experimental analysis shows that the prediction accuracy can be improved significantly compared to traditional methods.

Keywords: Microblog data · Retweeting behavior prediction
Markov random field · Logistic regression

1 Introduction

With the development of mobile internet technology, social networks have springing up in people's daily life. Micro-blog has greatly accelerated the speed of information transmission in the network. When a microblog is forwarded by different users, the audience will increase in the geometric level, which leads to the uneasy controllability of the dissemination and diffusion of information. Therefore, predicting user retweet behaviors can more accurately detect the spread and trend of a micro-blog. Based on the prediction results, the information dissemination can be effectively intervened and the scope of information transmission can be controlled.

Information diffusion model [1–3] in social network is widely used in studying the transmission of internet public opinion, micro-blog marketing and other fields. So, the accuracy precision of retweet behavior is an important basis for the construction of information diffusion model. In fact, in the process of retweet behaviors prediction, except microblog contents and user attributes, the social network structure frequently have significant impact on improving the prediction accuracy. In this case, traditional existing methods have some disadvantages as follows: (1) entirely ignoring the influence of the social network structure on predicting user's behaviors; (2) only taking some numerical values (e.g., the number of followers) related with the social network structure as certain as the component feature to predict user retweet behaviors; (3) fail to appropriately model the behavior constraints between users (e.g., ignoring the types of the relationships between users). As a result, these methods frequently fail to achieve satisfactory results.

K. Deng et al. (Eds.): ICMIR 2018, AISC 856, pp. 3–11, 2019.
https://doi.org/10.1007/978-3-030-00214-5_1

With a view to solving the above problems, a generic framework based on Markov Random Field is proposed for predicting user retweeting behaviors. We comprehensively analyze the influences on retweeting behavior caused by various factors, including user characteristics, microblog contents, retweeting behavior constraints between users, group retweet priors, so the proposed method can globally perform the prediction and have an expected results.

2 Problem Statement

Given a microblog social network $G = <U, E>$, where $U = \{u_i\}(i = 1, \ldots, n)$ and $E = \{u_i, u_j\}$ $(i, j = 1, \ldots, n)$ respectively denote the set of users and a set of social relationships between users. For the new posted microblog m, let $y_i \in \{0, 1\}$ denote the retweet behavior label of user u_i (i.e., $y_i = 1$ indicates that user u_i retweets T and $y_i = 0$ indicates that the user discard T). Then, our goal is how to effectively determine the corresponding set of retweet behavior labels $Y = \{y_i\}(i = 1, \ldots, n)$ for the set of users $U = \{u_i\}(i = 1, \ldots, n)$.

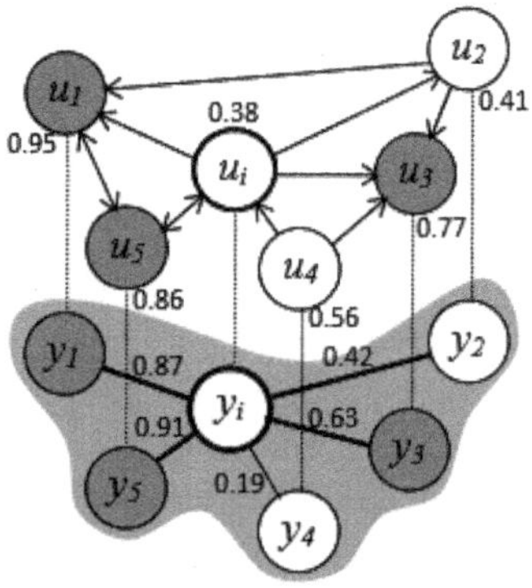

Fig. 1. User retweet behavior prediction under the MRF framework

In practice, for user u_i, his/her behavior (i.e., retweet or discard T) is not only determined by user attributes (e.g., gender, hobby) and microblog contents (e.g., the similarity between the current microblog and all the history microblogs), but also the behavior constraints between himself/herself and his/her neighboring users (i.e., its followees and followers). As shown in Fig. 1, if the followees of user u_i, i.e., u_1, u_3 and u_5 have retweeted m (gray circles), user u_i is also likely to reweet m. This is mainly caused by the retweet behaviors of his/her followees. Apparently, the more number of followees who retweet m, the higher probability that user u_i retweets m. For example, if user u_2 also retweet m, user u_i is more likely to retweet m.

In additions, after u_i etweets *m* the behaviors of its followers, i.e., u_2 nd u_4 also will be accordingly influenced. Further, these influences derive from not only the directly related users (e.g., u_3 and u_i are directly related to u_4 but also the indirectly related users (e.g., u_1 and u_5 are indirectly related to u_4). That is, the retweet behaviors of a user usually have been globally restricted by more users (including directly and indirectly related to it) in a social network (e.g., the light grey region). Thus, in order to better predict the retweet behavior of a user, the constraints between user and the retweet priors caused by more user behaviors should be comprehensively considered instead of the only influence derived from the directly related users.

3 Proposed Retweet Behavior Prediction Model Under Markov Random Field

In this paper, we formulate the problem of predicting user retweet behaviors as an energy minimization problem under the MRF framework, the proposed energy function is defined as:

$$E(Y) = \sum\nolimits_{i=1}^{N} \left(D_T(y_i, u_i) + \lambda \cdot \sum\nolimits_{j \in N(i)} \psi_{i,j} \cdot \delta\left(y_i \neq y_j\right) \right) \tag{1}$$

where two energy terms respectively encode the local user retweet behaviors and retweet behavior constraints between users. N(i) is the set of users directly related to user u_i (including followers and followees). λ is the weight constant.

3.1 Local User Retweet Behavior

The data term $D_T(y_i, u_i)$ encodes the cost that user u_i retweets T and defined as:

$$D_T(y_i, u_i) = |y_i - P(u_i, T)| \tag{2}$$

In Eq. (2), $y_i \in \{0, 1\}$ is the retweet label for user u_i, $P(u_i, T)$ is the probability obtained according to only the features comprised of u_i's attributes and T's contents, and is thus called as the local user retweet behavior.

3.1.1 User Retweet Behavior Basic Features

User attributes and microblog contents are commonly used in predicting user retweet behaviors, and also are adopted in the paper. The basic features are extracted from released microblog, as is shown Table 1. In our experiments, our method appears insensitive to the feature extraction because of the incorporating the retweet behavior constraints and group retweet priors.

The importance of each user attribute feature is different in the prediction. The meaning and function of microblog user attribute features have been introduced in more detail in the same type of paper, we will no longer describe them in this paper.

In the process of retweeting microblog, users tend to forward a microblog that is interested. The probability distribution of documents combined from history microblogs and microblog on the predetermined topics (such as politics, social security, etc.) is obtained by LDA (Latent Dirichlet Allocation) model [7]. The similarity of the topic is obtained by the cosine distance [5]. The cosine similarity is defined as:

$$L(d_i, T) = \frac{LDA(d_i) \cdot LDA(T)}{||LDA(d_i)||\,||LDA(T)||} \tag{3}$$

In order to facilitate the experimental analysis, we adopt the formula (4) to normalize the type value of the features shown in Table 1.

$$f' = \frac{f - f_{min}}{f_{max} - f_{min}} \tag{4}$$

Where f is the original feature, f' is the normalized feature, f_{min} and f_{max} respectively represent the minimum and maximum values of the current feature with respect to all users.

Table 1. User retweet behavior features

No.	Type	Features
1	User attributes	Number of followers
2		Number of followees
3		Certification
4		Number of tweets
5		Number of retweets
6		Retweet activity
7	Microblog contents	Topic similarity
8		Times of retweets
9		Length
10		URL tag
11		@ tag

3.1.2 Local Retweet Probability

We utilize the resulting features to obtain the local retweet probability $P(u_i, T)$ based on the logical regression model. Given the normalized feature vector X_i^{ut} of user u_i, the $P(u_i, T)$ is defined as:

$$P(u_i, T) = \sigma\left(w^T X_i^{ut}\right) = \frac{1}{1 + exp(-w^T X_i^{ut})} \tag{5}$$

Where w is feature weight vector, the value is got by specific risk function minimization of gradient descent algorithm.

$$w^* = \underset{w}{argmin} \sum_{i=1}^{N} \left(\mathcal{L}(y_i, P(u_i, T)) + \lambda_3 \cdot ||w||_2^2\right) \tag{6}$$

In Eq. (6), the loss function $\mathcal{L}(\cdot)$ uses cross entropy loss function, and N represents the sample number, $||\cdot||_2$ is the canonical regularization term of L_2, and λ_3 is the parameter to control the regularization intensity.

3.2 User Retweet Behavior Constraints

In Eq. (1), the lower order energy term measures user u_i retweet behavior affected by the followers of user u_i and paying attention to user u_j retweet behavior, the punishment $\psi_{i,j}$ determined according to the retweet similarity features between the users.

$$\psi_{i,j} = exp\left(P\left(u_i, u_j\right)/\sigma\right) \tag{7}$$

In Eq. (7), the punishment is controlled by the parameter σ (σ is set to 5 in the experiment), $P\left(u_i, u_j\right)$ is determined according to retweet similarity features (such as interest preference, common concern, etc.) between users.

3.2.1 Topic Preferences

If users have similar topic preferences, they will have a consistent retweet behavior for the same microblog (both retweeting or not retweeting). We measure the topic preferences similarity between user u_i and user u_j by adopting Eq. (3).

3.2.2 Follow Each Other and Common Concerns

User u_i and user u_j will be have larger probability on the same retweeting behaviors if they follow each other. In this paper, the corresponding features value of follow each other is 1, otherwise, it's 0. Furthermore, the users are more likely to forward the same microblog when the number of users with common concern is larger. The features value is obtained by Eq. (8):

$$S_{ij} = \frac{U_i \cap U_j}{U_i \cup U_j} \tag{8}$$

In Eq. (8), U_i represents the user set of u_i follows in all users.

3.2.3 Retweet Each Other and Common Retweet

The more microblogs retweet each other between user u_i and user u_j, the more similar topic preference, they are more likely to have the same forwarding behavior for the same microblog. The features value is calculated by Eq. (9):

$$R_{ij} = \max(T_{ij}/T_i, T_{ji}/T_i) \tag{9}$$

In Eq. (9), T_{ij} represents the number of the forwarded microblogs by user u_j in all the tweeting microblogs of user u_i, T_i represents the total number of the retweeted microblogs by user u_i.

Moreover, the number of microblogs common retweet bitterly measures the common interest and retweeting tendency, the value is calculated by Eq. (10):

$$M_{ij} = \frac{T_i \cap T_j}{T_i \cup T_j} \tag{10}$$

3.3 Group Retweet Priors

When users retweet microblog, they usually find followees and followers who have been retweet microblog in a local area that is centered around him, the constraints between user's retweet behaviors will have a certain impact on the current users' subjective intention of retweeting microblog. the higher-order energy term of measuring group retweet priors is defined as:

$$\varphi_c(f_c) = \begin{cases} \frac{\rho}{Q} \cdot \lambda_{\max} & \rho \le Q \\ \lambda_{\max} & otherwise \end{cases} \tag{11}$$

Where, the constant λ_{max} is the punishment of group retweet behavior, and ρ is the average of the retweet behaviors similarity probability P (u_i, u_j) in the network G_i^τ, Q represents the proportion of all users that have the local retweet probability of microblog T smaller than the specified threshold ϵ (experimental setting is 0.5).

4 Experimental Results

In this paper, three experiments have been carried out on the public data set [7] to verify the effectiveness of the prediction algorithm (ERBP, Energy-based Retweet Behavior Predicting). The data set contains the base information of 1787443 Sina users and the social-network structure. For a larger social network, which is divided into several sub social networks based on algorithm [8], and then we complete the user retweet behavior prediction in each sub network. The prediction result of the original social network is the average of all the prediction results of the sub social network.

For the convenience of experimental analysis, the local retweet probability is called LERBP (Local ERBP), and the user retweet behavior model only containing data items and lower order energy terms is called PERBP (Pairwise ERBP). We adopt the evaluation index of method [4] to measure the performance of the proposed method.

4.1 Influence of the Retweet Behavior Constraints and the Group Retweet Priors on the Prediction Accuracy

To analyze the impact of retweet behavior constraints and group retweet priors on the user retweet behavior, we observe the variation tendency of prediction accuracy by changing parameter values of energy function. We focus on investigating the influence of the retweet behavior constraint on the precision based on the LERBP. As shown in Fig. 2, with the weight λ_1 increase, the precision first rise and then drop. Similarly to the weight λ_1, with the weight λ_2 increases, the trend of prediction accuracy increased first and then decreased. As a whole, the algorithm has a better performance when the weight λ_1 and weight λ_2 re set to 0.6 and 0.3 respectively.

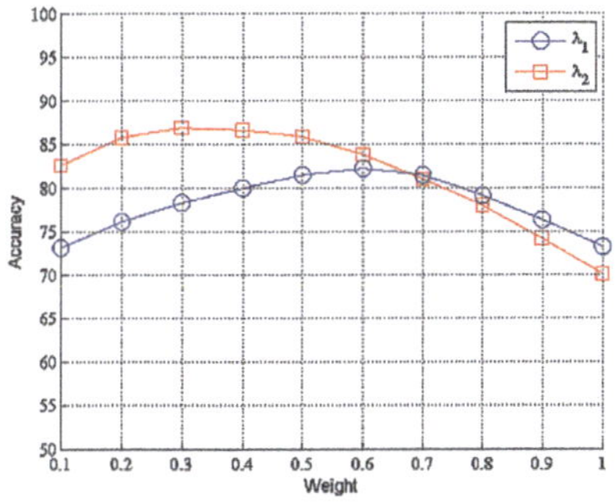

Fig. 2. Impacts of retweet behavior constraints and group retweet priors

For further verification, an analysis of the prediction results based on sub-networks containing different number of users is made. As shown in Table 2, the prediction accuracy has not changed greatly with the increase of user number (the number of users is less than 60 K).

Table 2. Prediction results over different sub networks ($\lambda_1 = 0.6$, $\lambda_2 = 0.3$)

Results	4,679	9,098	13,976	29,301	36,884	50,034
Recall	0.8536	0.8766	0.8623	0.8479	0.8637	0.8522
Precision	0.8345	0.8801	0.8409	0.8701	0.8525	0.8797
F1-score	0.8439	0.8783	0.8515	0.8589	0.8581	0.8657

4.2 User Retweet Behaviors Prediction

As shown in Table 3, due to the lack of taking social relations between users, the prediction accuracy of SVM (Support Vector Machine) and LERBP is low. In contrast, the prediction accuracy of method [5] and method [6] is improved to a certain extent for taking into account the impact of the retweet behavior constraints and social relationships, but neither of the two methods can get higher precision because of not globally measuring the constraints relationship.

Based on the Markov Random Field, PERBP preferably analyzes retweet behavior features and retweet behavior constraints, not only conducive to highlight the personalized forwarding, but also conducive to describe the common features of retweet behavior of more users. PERBP obtained the prediction results of global optimization. Based on PERBP, ERBP further fuses the group retweet priors, and more accurately reflects the essential characteristics of user retweet behavior. The corresponding prediction precision of ERBP is further improved.

Table 3. User retweet behavior prediction ($\lambda_1 = 0.6$, $\lambda_2 = 0.3$)

	SVM	[6]	[5]	LERBP	PERBP	ERBP
Recall	0.6076	0.7354	0.6121	0.5380	0.8687	0.8976
Precision	0.5881	0.6623	0.8628	0.5091	0.8254	0.8693
F1-score	0.5977	0.6969	0.7161	0.5232	0.8465	0.8832

4.3 Microblog Retweet Path Prediction

The microblog retweeting path is defined as the path containing the most users who retweet it, and the length of the retweeting path is defined as the number of all users in this path. Further, the prediction of retweeting path is deemed successful only when the retweet behavior of all users is accurately predicted.

ERBP obtains relatively better prediction results taking into account both the retweet behavior constraints and group retweet priors compared to the method [4] simply using basic features, the experiment result as shown in Table 4.

Table 4. Microblog retweet path prediction ($\lambda_1 = 0.6$, $\lambda_2 = 0.3$)

Path length	Total path	Number of path correctly predicted		Precision	
2	20435	[4]	ERBP	[4]	ERBP
3	5337	7951	12308	0.3891	0.6023
4	446	132	239	0.2954	0.5349
5	186	47	87	0.2509	0.4688
6	93	15	39	0.1576	0.4196

5 Conclusion

The paper proposes a generic framework based on Markov Random Field to find the global optimal solution to preform retweet behaviors prediction in a social network. The framework effectively incorporates the influence of basic features and individual retweet characteristics and group retweet priors on user retweet behavior. The method proposed in the paper achieves higher precision compared with other method in the experiment. At present, the shortcomings and improvements of the proposed method lie in: when the size of social network is large, it is divided into multiple sub networks by the community identification; however, the size controllability of these sub social networks and the solution efficiency of energy model need to be further explored. Thus, how to construct a prediction model more effective and easily solved is a problem that is worthy to further study.

Acknowledgments. This work is supported by the Science and Technology Development Project of Henan province (No. 172102310727), the College Science Research Project of Henan province (No. 17A520018, 17A520019).

References

1. Pezzoni, F., An, J., Passarella, A., et al.: Why Do I Retweet It? An Information Propagation Model for Microblogs. Social Informatics, pp. 360–369. Springer, Berlin (2013)
2. Yang, J., Leskovec, J.: Modeling information diffusion in implicit networks. In: IEEE International Conference on Data Mining. IEEE Computer Society, pp. 599–608 (2010)
3. Huang, D., Zhou, J., Yang, F., et al.: Understanding retweeting behaviors in twitter. J. Comput. Inf. Syst. **11**(13), 4625–4634 (2015)
4. Cao, J., Wu, J., Shi, W., et al.: Sina microblog information diffusion analysis and prediction. J. Comput. **37**(4), 779–790 (2014)
5. Tang, X., Miao, Q., Quan, Y., et al.: Predicting individual retweet behavior by user similarity. Knowl. Based Syst. **89(C)**, 681–688 (2015)
6. Peng, H.K., Zhu, J., Piao, D. et al.: Retweet modeling using conditional random fields. Data Mining Workshops, pp. 336–343 (2011)
7. Zhang, J., Tang, J., Li, J., et al.: Who influenced you? Predicting retweet via social influence locality. ACM Trans. Knowl. Discov. Data **9**(3), 1–26 (2015)
8. Kohli, P., Ladický, L., Torr, P.H.S.: Robust higher order potentials for enforcing label consistency. Int. J. Comput. Vision **82**(3), 302–324 (2009)

Denoising Point Cloud Based on the Connexity of Spatial Grid

Fu Siyong and Wu Lushen(✉)

School of Mechanical and Electrical Engineering, Nanchang University, Nanchang 330031, China
wulushen@163.com

Abstract. To ensure no loss of feature information of model and rapidly eliminate the high density noise clusters of point cloud. A kind of de-noising algorithm based on the connexity of spatial grid was proposed. Firstly, the concept of connected region in the binary image was extended to 3D point cloud model, divided the point cloud model into uniform space hexahedron, created the concept of adjacent lattice cell, then determined the pixel value of cell grid according to the density of the point in it, and the connected area was established by matching the pixel values of the adjacent units, judged whether the lattice cell was a noise unit according the size of connected region, so that the large scale noise was filtered. Finally, used the bilateral filtering to smoothing the small-scale noise. The experimental results show the proposed method could filter out the noise while preserving the details feature of model, the maximum error of point cloud is 1.7245E−2 mm, and standard deviation is 0.4012E−2 mm after filtering, the computation time is 53 s far less than that of the clustering algorithm; under the interference of different intensity noise, the proposed algorithm can achieve better de-noising effects and retain the high-frequency feature information of point cloud.

Keywords: Point cloud smoothing · Bilateral filtering · Connected region

1 Introduction

3D digital technology has been widely used in many fields, such as VR/AR Game, film and television stunt, virtual reality, industrial design and many other fields [1, 2]. The existing 3D scanner can easily acquire data on the surface of the object, but because of the influence of the object characteristics, scanning environment, lighting and disturbing factors, the collected point cloud data is unavoidable with noise. These noises have great interference to the subsequent processing of point cloud, such as the feature extraction of the point cloud, the registration of the surface and the reconfiguration of model.

In recent years, many scholars have done a lot of research on the filtering methods of three dimensional point clouds. Taubin [2] extended the Laplace filtering algorithm to the three dimensional point cloud model, but the algorithm would lead to the model shrinkage and the loss of the subtle features. The bilateral filtering method was proposed by Jones [3], it can ensure that the features of the model can be well maintained

K. Deng et al. (Eds.): ICMIR 2018, AISC 856, pp. 12–19, 2019.
https://doi.org/10.1007/978-3-030-00214-5_2

while filtering the noise. However, when there was large scale noise in the model, the factor of the filter and the feature retention factors could not be adaptively adjusted for different regions, then the model was easy to be overly smooth. Fleishman [4] used the statistical methods to identify outliers, then smoothed the point cloud by the moving least squares method. Gu [5] classed the point cloud by the curvature, then used the mean filter and the bilateral filter method to filtered it according to the different classes points. The algorithm was simple, but the accuracy of the calculation value of the point cloud curvature was affected by the noise, it could not deal the high density point cloud with noise. The improved anisotropic diffusion method was applied to smooth the face point cloud [6], and achieved a good results, but the type of noise processed was only small scale noise. A method based on the fuzzy C means clustering algorithm and the bilateral filtering was proposed [7], the clustering algorithm was used to filter the large scale noise, and the bilateral filtering was used to filter the small scale noise and the effect was better. The core idea of literature [7–10] was to use different filtering methods based on the scale of the noise, the large scale noise filtered by clustering method, and the small scale noise was filtered out by the bilateral filtering, but the large scale noise could not be completely filtered out by the clustering method, and it took a long time. Zhu [11] used hexahedron to segment point cloud data, and used the density of each point cloud to analyze whether the points in the grid was noise point. When there is a high density noise cluster in the model, it could not identify the noise unit.

To sum up, the mainstream filtering method was to adopt different methods based on different types of noise, the large scale noise was filter out by the cluster method, but it could deal with the high density noise cluster, and it would take a long time. So, we proposed a method based on the connexity of spatial grid, the point cloud model was decomposed into a number of uniform grids, and, the noise grids were identified based on the connectivity of the grid and the size of the connected area.

2 Filtering Method for Large Scale Noise

2.1 The Idea of Algorithm

The surface of the model is a continuous surface, and the normal point cloud data should be closely distributed over the surface of the model, while the large scale noise is generally far away from the surface of the model [12, 13]. In order to filter the large scale noise, a denoising method based on the connectivity of spatial grid was proposed. Firstly, the point cloud model was divided into uniform space hexahedron, as shown in Fig. 1, because the large scale noise was far away from the surface of model, and was isolated, but the normal data points were distributed continuously, so we could identify the large scale noise based on this feature. Then, the density information of the point cloud data in the grid cell was calculated, and the pixel value of each unit was determined according to the density of points in the unit, when the density was less than a certain threshold, the pixel value was 0. After analyzed the connectivity of adjacent units and calculated the size of connected area. Finally, the noise unit was identified according to the pixel value of the cell body and the size of the connected area of the unit body.

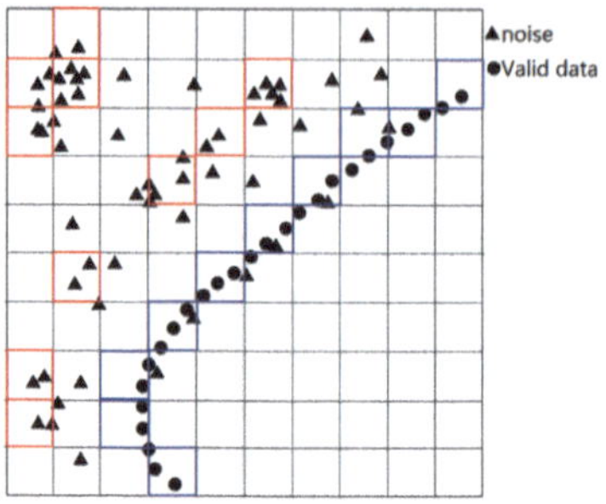

Fig. 1. A plane map of a space grid

2.2 Spatial Grid Segmentation of Point Clouds

According to the density information of the point cloud data, the minimum distance d_x, d_y, d_z of the X, Y, Z three directions were set. That is to say, if the point cloud model is divided by the distance of d_x, d_y, d_z, then each grid only contains one point at most in theory [21]. Considering the practicability of the algorithm and the overall running speed, the distance between the smallest grids was magnified by X_s, Y_s, Z_s times, where $X_s, Y_s, Z_s \in [2, 5]$, in this paper, we set $X_s = Y_s = Z_s = 3$. The steps of the specific segmentation algorithm are as follows:

(1) Firstly, find the most valued $X_{max}, X_{min}, Y_{max}, Y_{min}, Z_{max}, Z_{min}$ in the three directions of X, Y, Z, then the point cloud model could be surrounded by a rectangular body, and the 8 vertex of the rectangle were $(X_{min}, Y_{min}, Z_{min})$, $(X_{min}, Y_{max}, Z_{min})$, $(X_{min}, Y_{min}, Z_{max})$, $(X_{min}, Y_{max}, Z_{max})$, $(X_{max}, Y_{min}, Z_{min})$, $(X_{max}, Y_{max}, Z_{min})$, $(X_{max}, Y_{min}, Z_{max})$, $(X_{max}, Y_{min}, Z_{min})$.

(2) According to the size of the minimum scale D_x, D_y, D_z, the number of segments divided by the three directions of X, Y, Z can be determined.

$$\begin{cases} N_x = \lceil (X_{max} - X_{min})/D_x \rceil \\ N_y = \lceil (Y_{max} - Y_{min})/D_y \rceil \\ N_z = \lceil (Z_{max} - Z_{min})/D_z \rceil \end{cases} \quad (1)$$

Where $\lceil \bullet \rceil$ was the operation to round, then the point cloud model was divided into $N_x \times N_y \times N_z$ space grid, and we call each space grid as a unit.

(3) The index number $C(u, v, w)$ of each unit was established. We set the index number of the unit containing the point $(X_{min}, Y_{min}, Z_{min})$ to $C(1, 1, 1)$. Then, the data of the point cloud in an arbitrary unit body $C(u, v, w)$ should be satisfied

$$\begin{cases} x = \{x | X_{min} \leq x < X_{min} + u * D_x,\ u \leq N_x\} \\ y = \{y | Y_{min} \leq x < Y_{min} + v * D_y,\ v \leq N_y\} \\ z = \{z | Z_{min} \leq x < Z_{min} + w * D_z,\ w \leq N_z\} \end{cases} \quad (2)$$

Through the above form, all points in the point cloud model could be divided into each unit body, and any unit body contains 26 adjacent units, as shown in the following figure (Fig. 2).

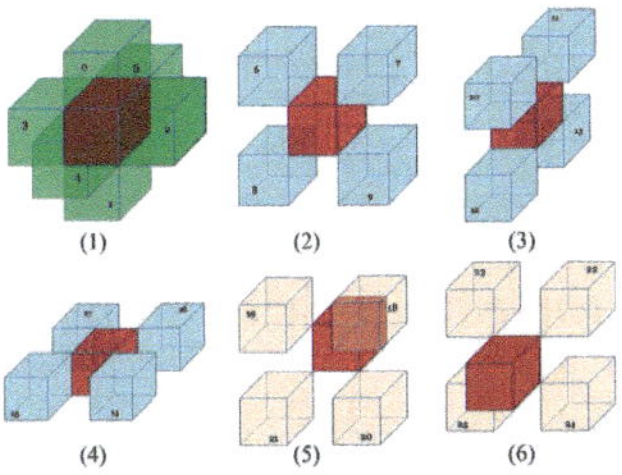

Fig. 2. 26 adjacent unit bodies

2.3 Filtering of Large Scale Noise

The pixels with the same pixel value at any two adjacent positions in the two value grayscale image are called the two pixels connected. We replaced the "the value of pixel in the image" to "the number of point cloud data points in the unit body".

Definition 1. If the number of point cloud data in an arbitrary unit was greater than the threshold value N_t, then let $C(u, v, w) = 0$, instead, let $C(u, v, w) = 1$. The value of N_t was determined by the density of the original point cloud and the minimum gap between the grid.

Definition 2. The value of the $C(u, v, w)$ was called the pixel value of the unit body, and the pixel values of any two adjacent units are equal and 1, then the two unit body is considered to be connected.

Definition 3. If a plurality of continuous unit connected, then the number of connected units as the connecting length.

The specific steps of noise determination based on the unit connectivity analysis are as follows:

Step1: Each unit was traversed, the number of the point in the cell unit was saved in N_c, and the value of the $C(u, v, w)$ was computed. If $C(u, v, w) = 0$, then, it was considered that the points in the cell unit were noise and can be deleted directly.

Step2: The unit cel was cut into the N_y slice in the direction of Y, and the sized of the connected area of the unit was calculated. Let $v = k$, the loop traverses the index u and w, then found the connected area of the unit and calculated its number N_l according this definitions, if $N_l < NL_{threshold}$, then set $C(u, v, w) = 0$, it meant that the pixel value of the unit cell was cleared, where $NL_{threshold}$ was dependent on the maximum width of the noise cluster.

Step3. The loop runs ***Step2*** N_y times, found the connected area in all the slices, and updated the value of $C(u, v, w)$.

The recognition of the connected domain of the unit body in the above step was only based on the adjacent surface area, the cell body of the pixel $C(u, v, w) = 0$ might be a pseudo noise unit. Therefore, in order to enhance the reliability of the algorithm, we could identify whether the pixel of adjacent edges and adjacent vertices unit was the real and effective data, that was, go to step ***Step4.***

Step4. Whether the pixel values of adjacent edges, adjacent vertices units of the unit of the $C(u, v, w) = 0$'s unit were 1, If it was true, the data of the unit was considered to be real, otherwise the data of the unit was deleted.

3 Bilateral Filtering for Small Scale Noise

The anisotropic denoising algorithm based on bilateral filtering was very effective for small scale noise, and could preserve the sharp features of the model while filtering noise.

$$p_k^{'} = p_k + \alpha \times n \tag{4}$$

Where p_k was the original data point, $p_k^{'}$ a filtered data point, and n was normal vector of point p_k, α was a factor of bilateral filter, it could be expressed as

$$\alpha = \frac{\sum_{j=1}^{m} W_c(\|p_k - p_j\|) W_s(\|\langle p_k - p_j, n\rangle\|)\langle p_k - p_j, n\rangle}{\sum_{j=1}^{m} W_c(\|p_k - p_j\|) W_s(\|\langle p_k - p_j, n\rangle\|)} \tag{5}$$

Where $W_c = e^{-x^2/2\sigma_c^2}$ was the filter weight function, $W_s = e^{-y^2/2\sigma_s^2}$ was the feature preserving weighting function, the σ_c was the distance influence factor and generally set to the radius of its neighborhood [22].

The concrete steps of the algorithm:

Step1. Found the m neighborhood points for each data point p_k.

Step2. Calculated W_c and W_s.

Step3. Calculated the bilateral filter factor α by the formula (5).

Step4. Calculated the value of $p_k^{'}$ by the formula (4)

Step5. Loop calculation until all data points were updated after the program is finished.

4 The Analysis of Experiment and Result

4.1 Robustness Verification of the Algorithm

In order to verify the robustness of the algorithm, used C++ language to realize point cloud smoothing algorithm on the VS2010 platform, and called the OpenGL library function to display the point cloud. selected the standard Bunny model is. The standard Bunny model was adopted in the experimental model, it had 35974 effective points and adding different intensities of noise, as shown in Fig. 3, set $rate = 0.3$, $NL_{threshold} = 28$,

Fig. 3. Bunny model after adding different intensity noise; (a) 3040 noise point (b) 8054 noise point (c) 12928 noise point

Fig. 4. The denoised model by the proposed method; (a) 3040 noise point (b) 8054 noise point (c) 12928 noise point

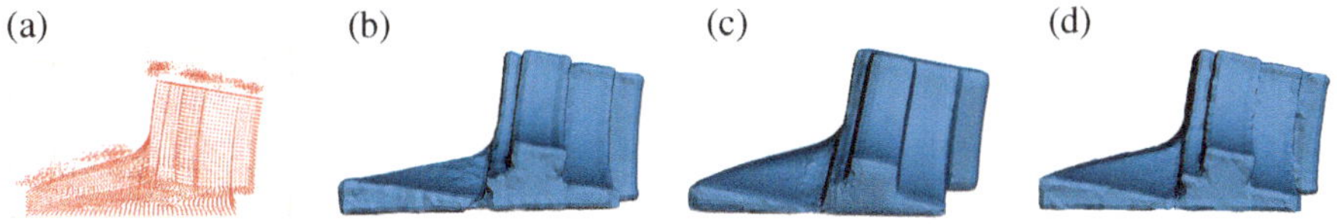

Fig. 5. Triangulation model after de-noising with different methods; (a) model after adding noise (b) FCM (c) multi-scale density analysis (d) the proposed method

$\sigma_s = \frac{1}{5}\delta_c$, the denoised model by the proposed method was shown in Fig. 4. While filtering noise, the detail features of the model were well preserved. However, with the increase of the number of large scale noise points, the details of the model are smooth after filtering (Fig. 5).

4.2 Performance Analysis of the Algorithm

In order to further verify the performance of this algorithm, we added 5 dB Gauss noise and 1245 random distribution of large scale noise points in the fandisk model, as shown in Fig. 6, and the proposed method was compared with FCM algorithm and the method based on density analysis.

The clustering algorithm and the proposed method could better preserve the details of the model, but the time-consuming 53 *s* of the proposed method was far less than

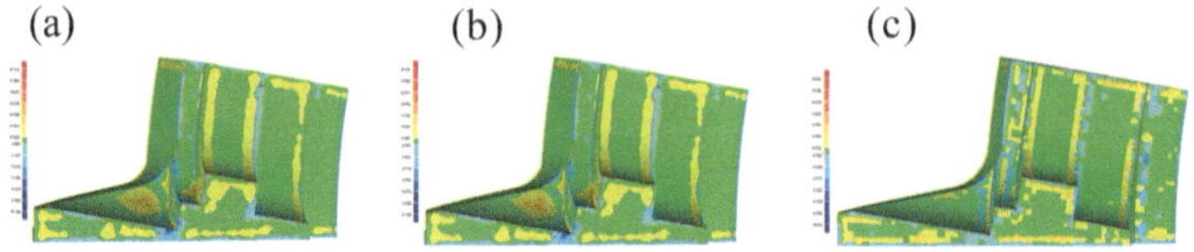

Fig. 6. The deviation chromatogram of the model from the original model after the de-noising of different methods; (a) FCM (b) multi-scale density analysis (c) the proposed method

the 103 *s* of clustering algorithm, the maximum error of the proposed method was 1.7245E−2, and average error was 0.2145E−2, was less of that the FCM and the method based on density analysis.

5 Summary

A point cloud denoising algorithm based on spatial grid connectivity analysis was proposed in this paper. Firstly, the concept of connected area in the two valued image was extended to point cloud space grid, and the pixel value of the cell body was constructed according to the density information of the point cloud data in the unit. Then the connected area of the unit was established by matching the pixel value of the adjacent unit body. Finally, according to the size of the connected region and the density information of the adjacent unit body density, it could be used to determine whether the unit was noise. The experimental results show that the proposed algorithm has good robustness. The filtering results of fandisk model show that the standard deviation of filtered algorithm was 0.4012E−2 mm, less than that of FCM and multi-scale density analysis. it showed that the proposed method could preserve the detail features of the model while filtering large scale noise.

Acknowledgments. The authors would like to thank all the reviewers for their valuable comments.

References

1. Wang, X., Cai, J., Wu, Z., et al.: Normal estimation and normal orientation for point cloud model based on improved local surface fitting. J. Comput. Aided Des. Comput. Gr. **27**(4), 614–620 (2015)
2. Taubin, G.: A signal processing approach to fair surface design. In: Conference on Computer Graphics and Interactive Techniques, pp. 351–358. ACM (1995)
3. Jones, T.R., Durand, F., Desbrun, M.: Non-iterative, feature-preserving mesh smoothing. ACM Trans. Gr. **22**(3), 943–949 (2004)
4. Fleishman, S., Cohen-Or, D.: Robust moving least-squares fitting with sharp features. ACM Trans. Gr. **24**(3), 544–552 (2005)
5. Gu, X.: A filtering algorithm for scattered point cloud based on curvature features classification. J. Inf. Comput. Sci. **12**(2), 525–532 (2015)

6. Wibowo, S.A., Kim, S.: Three-dimensional face point cloud smoothing based on modified anisotropic diffusion method. Int. J. Fuzzy L. Intell. Syst. **14**(2), 84–90 (2014)
7. Wang, L., Yuan, B.: Robust fuzzy C-means and bilateral point clouds denoising. J. Beijing Jiaotong Univ. **32**(2), 18–21 (2008)
8. Li, J., Zhang, C., Fan, H.: Point cloud denoising algorithm based on swarm intelligent. Comput. Integr. Manuf. Syst. **17**(5), 935–945 (2011)
9. Guo, J., Chen, X., Lu, J.-M., et al.: Points cloud self-adaptive denoising algorithm based on density k-means and improved bilateral fitering. Transducer Microsyst. Technol. **35**(7), 147–149 (2016)
10. Xu, L., Huang, X., Li, G.: Point clouds denoise algorithm based on fuzzy c-means and mean filter. Mach. Build. Autom. **45**(4), 5–7 (2016)
11. Zhu, J., Hu, X., Zhang, Z., et al.: hierachical outlier detection for point cloud data using a density analysis method. Acta Geodaetica Cartogr. Sin. **44**(3), 282–291 (2015)
12. Ramiya, A.M., Nidamanuri, R.R., Ramakrishnan, K.: A supervoxel-based spectro-spatial approach for 3D urban point cloud labelling. Int. J. Remote Sens. **37**(17), 4172–4200 (2016)
13. Li, W.L., Xie, H., Zhang, G., et al.: Adaptive bilateral smoothing for a point-sampled blade surface. IEEE/ASME Trans. Mechatron. **21**(6), 2805–2816 (2016)

A New Active Learning Semi-supervised Community Detection Algorithm in Complex Networks

Jinjin Wang[1(✉)] and Mingwei Leng[2]

[1] State Key Laboratory of Intense Pulsed Radiation Simulation and Effect (Northwest Institute of Nuclear Technology), Xi'an 710024, China
wangjinjin@nint.ac.cn

[2] School of Education Science and Technology, Northwest Minzu University, Lanzhou 730000, China

Abstract. Community detection is an important method for analyzing community structure of complex networks. Semi-supervised community detection algorithms use prior information to improve the performance of community detection. But how to obtain these prior information is a challenging problem. In this paper, we present a new active learning semi-supervised community detection algorithm based on node relevance and similarities measures. Firstly, it calculates similarity for each neighbored nodes. Then it using our combine rule and node relevance measure to select informative nodes in active learning step. Secondly, it expands community using these selected nodes with our hierarchical expanding rule in semi-supervised strategy. Finally, the algorithm is demonstrated with four real-world networks and a artificial network benchmark. Extensive experiments show that it effectively improves community detection results and has lower computational complexity compared with some community detection algorithms.

Keywords: Artificial intelligence · Active learning · Complex network Community detection

1 Introduction

Community detection has been a hot topic in the fields of physics and computer science in last decade. The size, shape or diameter of the community in real-world complex networks is unknown. Therefore, community detection has been studied as graph partition problem; it is a NP-hard problem. Nodes in a same group or community are more similar and densely connected than the outside nodes. Nodes in different communities may be connected with sparser nodes, and they are dissimilar. A community in a network is a group of nodes which probably share common properties or play similar functional roles. Community detection in complex networks helps us to understand the network structure and analyze the network characters. To find more efficient and accurate community detecting algorithms is arousing great interest in last decade. There are many algorithms for community detection have been proposed in recent years. The popular methods are: modularity methods [1–3], hierarchical methods

K. Deng et al. (Eds.): ICMIR 2018, AISC 856, pp. 20–32, 2019.
https://doi.org/10.1007/978-3-030-00214-5_3

[1, 3–5] and label propagation [6–8] methods. Modularity algorithm was first proposed by Newman [1]. It measured by modularity Q for detecting community structure in networks, which is referred to as NM. Q measures the community quality of a network. Many algorithms perform community detection by maximization of modularity [1]. Raghavan [6] first introduced a label propagation algorithm (LPA) based on the diffusion of information in networks. The main idea of LPA is simple. Girvan and Newman [3] introduced a hierarchical divisive algorithm (called GN) which splits network into communities by removing edges from the network iteratively. GN algorithm uses shortest-path measure as betweenness. The measure is recalculated after each removal. EBC [9] uses the edge betweenness in another paper instead of the shortest-path betweenness in GN.

But modularity algorithm, label algorithm and hierarchical algorithm is insufficient in some cases as. These algorithms only make use of the network topology information, which completely ignore the background information of the network. However, we may know some prior information that could be useful in detecting the community structures. Incorporating prior knowledge into the process of detecting community may be the most efficient method for improving the performance of community detection algorithms [10]. Active learning semi-supervised is a new community detection method. It was first introduced into semi-supervised community detection algorithm by Leng et al. in 2013 [11]. This kind of methods is mainly research on how to select useful nodes using active learning and then detect community structure in semi-supervised method. Cheng et al. [12] proposes an active semi-supervised algorithm based on must-link and cannot-link Constraints. Leng et al. [10] proposed a new active semi-supervised algorithm based on asymmetric similarity measure recently. It includes an asymmetric similarity method for nodes similarities and a fast active semi-supervised community detection method.

In our paper, we present a new active learning semi-supervised community detection algorithm based on node relevance and similarities measures. Our new active learning semi-supervised community detection algorithm called ASCD. It has two steps. Extensive experiments show that it effectively improves community detection results and has lower computational complexity compared with some community detection algorithms. The rest of the paper is organized as follows. Section 2 gives definitions and notations we designed. In Sect. 3, we introduce our method in detail. In Sect. 4, we demonstrate our method with some standard networks and compare it with some other community detection algorithms. Sect. 5 gives the conclusion.

2 Some Definitions We Designed

In the network $G(V, E)$, $C = \{C_i | i = 1, 2,\ldots, c\}$ is the community structure, C_i denotes community i in network G, $L = \{L_i | i = 1, 2,\ldots, c\}$ is the set of community labels, L_i is the label of C_i, l_v denotes the label of node v.

2.1 Node Relevance

Each node plays different roles in a network. Some of them are more important than others in the given situation. We have paid attention to this difference. In the network G (V, E), we design the relevance of a node. It denotes a node's cohesion to others. It is in Eq. (1).

$$R(v) = \sum_{w \in N(v)} link(C_v, w) \tag{1}$$

For each node $v \in G$, $w \in N(v)$, w is one of the neighbors of node$v$. The value of $link(C_v, w)$ is 1 when node w is combined to C_v by node v. Otherwise, it equals 0. It will be introduced carefully in active learning step.

2.2 The Most Relevance Node of V

In $G(V, E)$, for each node $v \in G$, $w \in N(v)$, We initialize n_{vw} as similarity of node v and w. It shows in Eq. (2).We treat each single node as a community. We design *single-node community similarity measure*, which is given in Eq. (3).

$$n_{vw} = 1 - \frac{|(N(v) \backslash N(w)) \cup (N(w) \backslash N(v))|}{|N(v)| + |N(w)|} \tag{2}$$

$$S_{vw} = \frac{2 * n_{vw}}{\sum_{v' \in N(v)} n_{vv'} + \sum_{w' \in N(w)} n_{ww'}} \tag{3}$$

For $m \in N(v)$, if S_{vm} is the biggest, we define m as the most relevance node of v.

2.3 The New Label of Unlabeled Node in Expanding

In $G(V, E)$, for each node $v \in G$, $w \in N(v)$,We initialize n_{vw} as similarity of node v and w. It shows in Eq. (2).We treat each single node as a community. We design *single-node community similarity measure*, which is given in Eq. (3).

$$l_v = \underset{L_i \in L(v), L_i \neq \phi}{\arg\max} \sum_{w \in N(v)} n_{vw} \delta(L_i, l_w) \tag{4}$$

$$\delta(L_i, l_w) = \begin{cases} 1, & L_i = l_w \\ 0, & L_i \neq l_w \end{cases} \tag{5}$$

2.4 The Agglomeration Force of a Node

In the network *G*(*V*, *E*), the agglomeration force of *v* is defined below which is A_v in Eq. (6).

$$A_v = \sum_{w \in N(v)} n_{vw} \tag{6}$$

3 ASCD Algorithm

In this section, we introduce our algorithm ASCD. ASCD is an active learning semi-supervised algorithm. Several measures defined to be used in active learning and semi-supervised step. ASCD can prevent single-node communities and monster communities from being created, especially when the inner community link densities are close. It also can get the correct community numbers in most cases.

Our community detection algorithm consists of two steps: (1) it selects a few nodes from community network after active learning. We have designed a node agglomeration evaluation standard. Our method could guarantee each community has a node be selected in most cases which helps improving the accuracy in semi-supervised step. (2) We design semi-supervised strategy to detect community. The step uses nodes selected from step (1) as labeled nodes. Semi-supervised strategy uses these labeled nodes to detect the other unlabeled nodes. Finally, we detect the community structure.

3.1 Active Learning Choosing Nodes

In this step, we select representative nodes from complex community network using active learning method. We initialize S_{vw} as similarity of single-node community *v* and *w* which are adjacent. We sort S_{vw} in descending order. The descending order is put in a list L_s. Each item of L_s records *v*, *w*, and S_{vw}. We sequentially get each item from L_s to merge community *v* and *w*. It contains three cases. In the combination rule, we have considered the priority of S_{vw}. If *v* and *w* are still two single-node communities, we merge *v* and *w* into a community. If *v* has not been merged with other community and *w* has already been merged before, we add *v* to the community which *w* belongs to, vice versa. If *v* and *w* both are not two single-node communities which means they have already been merged with other communities before, it will not be merged this time. We can find that *v* and *w* are not most relevance nodes each other according to Sect. 2.2. In the combination rule, we break ties uniformly regularly. After this step, there are no single-node communities in the network now. Then the network *G*(*V*, *E*) has the structure $C = \{C_i|\ i = 1,2,\ldots, k\}$.We select node *m* which has the biggest value of *R*(*m*) in each C_i. Node *m*is the core node in C_i. We get the active learning set $LabeledNodes = \{m_i|i = 1,2,\ldots, k\}$.

The algorithm for active learning is given in Algorithm 1.

Algorithm 1. Active learning choosing nodes

Input: Network $G(V, E)$

for each edge(v, w) $\in E$ **do**

$R(v) = R(w) = 0$;

$l_v = v.Id$;

$l_w = w.Id$;

S[v][w] $= S_{vw}$;

L_s .append((l_v , l_w ,S[v][w]));

end for

L_s = SortDescendingOrder(L_s);

for each item(l_v , l_w ,S[v][w]) in L_s **do**

if v and w both are not expanded **then**

$l_v = l_w$;

end if

if v or w is expanded **then**

if v is expanded **then**

$l_w = l_v$;

$R(v)$++;

else $l_v = l_w$;

$R(w)$++;

end if

end if

end for

for each $C_i \in C$ **do**

l_m = MaximalRelevance(C_i);

if $R(m) \neq 0$ **then**

LabeledNodes .append(l_m);

end if

end for

for each $v \in V$ **do**

if $v \in$ *LabeledNodes* **then**

query l_v;

else l_v =null;

end if

end for

Output: *LabeledNodes*

Complexity Analysis: The algorithm needs $O(m\ \Delta^2)$ time in lines 1–7, where Δ is the network average degree and m is sum of the edges in the network. It takes $O(n\log n)$ time in line 8. And it also needs $O(m)$ time in lines 9–21. The worst time complexity of lines 22–27 is $O(n/2)$. Lines 28–33 take $O(n)$ time. So the time complexity of algorithm 1is $O(n\log n + m\ \Delta^2)$.

3.2 Semi-supervised Agglomeration

After active learning step, we have prior information in this step. We semi-supervised expand our community using these labeled nodes. Av denotes the agglomeration force of v. First, we calculate Av of all unlabeled nodes. The values are put in a list in descending order. We'll use this list to expand unlabeled nodes.

For unlabeled nodes, if some of their neighbors' labels are known, we'll expand them using the ruler in Sect. 2.3. If the node has been expanded, it'll be removed from the expanding list. In this step, we'll not change the labeled nodes' labels we get from oracle. While the expanding list is empty, the expanding step is ending. It is shown in Algorithm 2.

Complexity Analysis: The algorithm takes $O(n)$ time in lines 1–8. It needs $O(n\log n)$ time in line 9. We calculate S_{vw}'s time complexity in Algorithm 1 which needs $O(\Delta)$ time in lines 12–20. And in the while circle, it executes this calculation on each unlabeled nodes. So lines 10–28 needs $O(m)$ time. The complexity of Algorithm 2 is $O(n\log n)$.

Taking the complexity of Algorithm 1 and 2 into account, we can deduce that the total time complexity of ASCD is $O(n\log n + m\ \Delta^2)$. In the sparse network, ASCD's time complexity is $O(n\log n)$ as Δ is a constant.

Algorithm 2:Semi-supervised agglomeration

Input: Network $G(V, E)$, $LabeledNodes$, $C = \{C_i \mid i = 1, 2, \ldots, k\}$

MaxSimilarity = 0;
MaxSimilarityLabel = null;
for each $C_i \in C$ **do**
  Sim[L_i] = 0;
end for
for each $v \in V \setminus LabeledNodes$ **do**
LS .append((l_v , A_v));
end for
Rank LS in descending order by A_v ;
while LS is not null **do**
for each (l_v , A_v) $\in LS$ **do**
for each $w \in N(v)$ **do**
if l_w ≠null **then**
    Sim[l_w] += S_{vw} ;
if Sim[l_w] ≥ MaxSimilarity **then**
    MaxSimilarity = Sim[l_w];
    MaxSimilarityLabel = l_w ;
end if
end if
end for
if MaxSimilarityLabel ≠null **then**
l_v = MaxSimilarityLabel ;
LS .remove((l_v , A_v));
   MaxSimilarity = 0;
   MaxSimilarityLabel = null;
end if
end for
end while

Output: $G(V, E, C)$

4 Results and Discussion

4.1 The Networks

In this section, we test our algorithm ASCD both with several real-world networks and some commonly used artificial networks. These networks are: a network of friendships between members of a university karate club [13]; a network of frequent associations between dolphins living near Doubtful Sound, New Zealand [14, 15]; an artificial graph proposed by Girvan and Newman [16]; a risk map network is a map network of a popular chessboard game [17]; a collaboration network is a scientists' collaboration network in Santa Fe institute [16, 18]. As most of the researchers did, we uniformly treat all networks as undirected and unweighted networks and exclude all self-loop edges.

The karate club network is widely used as a benchmark for detecting communities. It is a social network of friendships between 34 members in a karate club at a US university in 1970s. Because of a disagreement between the club's administrator and the club's instructor, the karate club split into two groups.

The dolphin network is a community of 62 bottlenose dolphins living in Doubtful Sound, New Zealand. The network is usually splits into two large communities and one of the large community splits into three smaller communities in the paper.

Girvan and Newman also proposed a popular benchmark. It is a large set of artificial, computer generated graphs. The graph has 128 vertices divided into four communities of 32 vertices each.

The risk map network is a map network of a popular chessboard game. The network has six communities. Each community represents one continent in the game. The network has 118 vertices.

The collaboration network is a scientists' collaboration network in Santa Fe institute. There are 118 scientists in the institute. It includes four groups which are structure of RNA, statistical physics, agent-based models and mathematical ecology. In a same group, scientists do similar research. These networks are shown in Figs. 1, 2, 3, 4 and 5.

4.2 The Results and the Accuracy of ASCD in These Networks

In the artificial graph, our algorithm ASCD selects nodes set {11, 14, 116, 21, 25, 32, 40, 50, 53, 57, 59, 6, 79, 81, 84, 87, 88, 96, 110, 111, 115, 117, 120} in active learning. The artificial graph's core nodes are not obviously. 17.9% of the nodes were selected. Figure 6 shows the results of the community structure detected by ASCD. All of nodes are divided into correct communities. The accuracy of ASCD in the artificial network is 100%.

In the karate network, our active learning semi-supervised algorithm selects nodes set {2, 7, 25, 30, 33}. 14.7% of the nodes are selected in active learning step. Each

Fig. 1. The karate network

Fig. 2. The bottlenose dolphin network

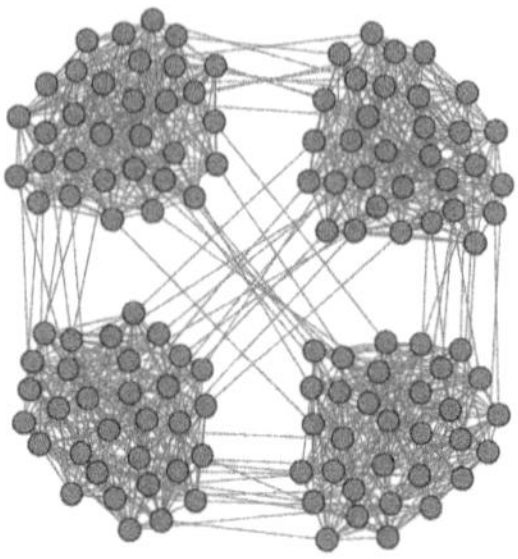

Fig. 3. The artificial graph by Girvan and Newman

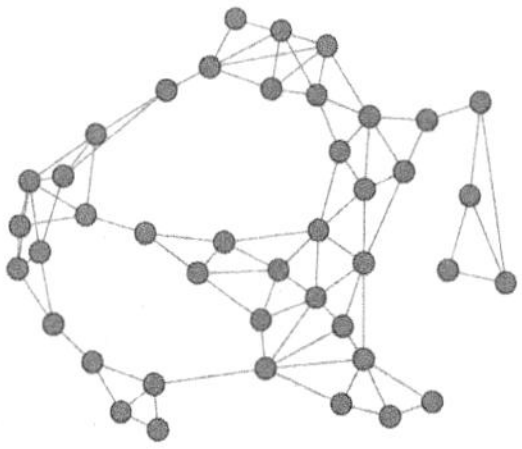

Fig. 4. The risk map network

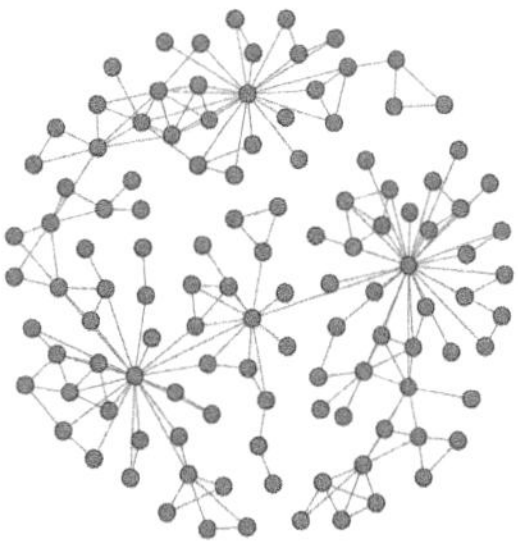

Fig. 5. The collaboration network

community has nodes to be selected are the basic of the correct community detection in the semi-supervised step. Figure 7 shows the results of the community structure detected by ASCD. All of nodes are divided into correct communities. The accuracy of ASCD in the karate network is 100%.

In the dolphin network, nodes set {Number1, SN100, TR77, Jonah, SN63, Scabs, Trigger, Patchback, Web} is selected. The nodes are selected at 14.5%. The result is shown in Fig. 8 All of nodes are divided into correct communities contrasting to [15]. The accuracy of ASCD in the karate network is 100%.

In the risk map network, the active learning step selects nodes set {2, 15, 21, 22, 25, 30, 34, 37, 40}. ASCD select nodes at 21.4%. The result is shown in Fig. 9. Three nodes are misclassified by our method contrasting to the result of the paper [17]. The accuracy is 92.8%. ASCD doesn't have good community detection on the network.

In the collaboration network, the nodes set is {3, 6, 7, 8, 24, 30, 35, 39, 59, 60, 65, 71, 92, 102, 104, 114}. 13.5% of the nodes were selected. Figure 10 shows the results of the community structure detected by ASCD. Only one node 83 is misclassified by our method contrasting to the result of the paper [15]. 83 only has two edges which connects two large communities. Its neighbors are 82 and 102. As the expanding order

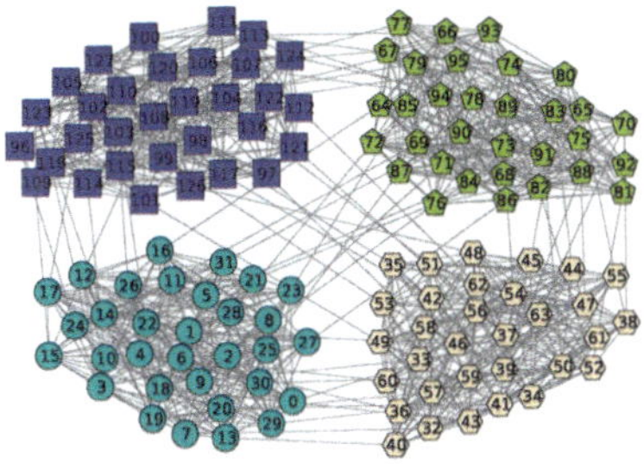

Fig. 6. The artificial graph

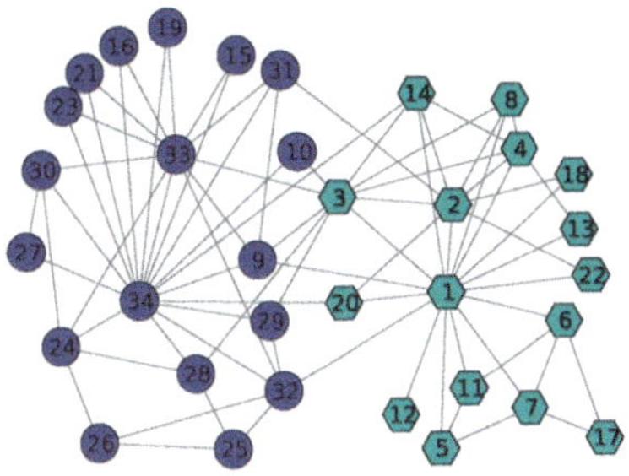

Fig. 7. The karate network

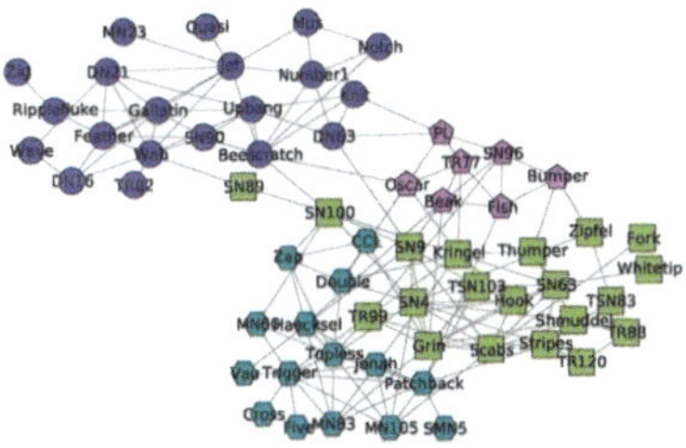

Fig. 8. The dolphin network

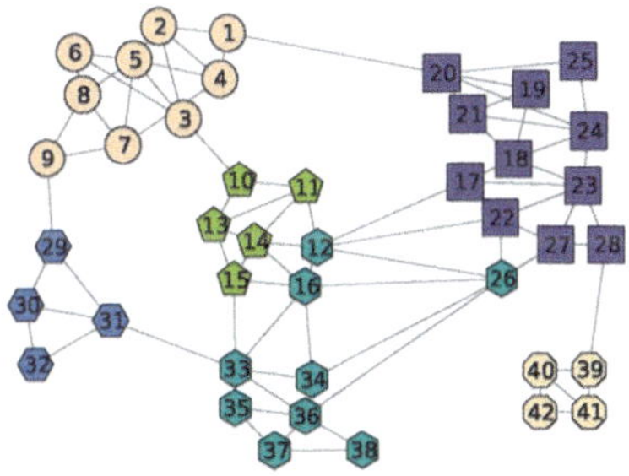

Fig. 9. The risk map network

in semi-supervised step, 83 is expanded into the community which 102 belongs to. The accuracy of ASCD is 99.2%.

4.3 Comparing with Other Algorithms

We apply each algorithm one hundred times on each network. NM is a representative hierarchical agglomeration algorithm in [1]. EBC uses the edge betweenness in another paper [9] instead of the shortest-path betweenness in GN. EBC is a hierarchical divisive

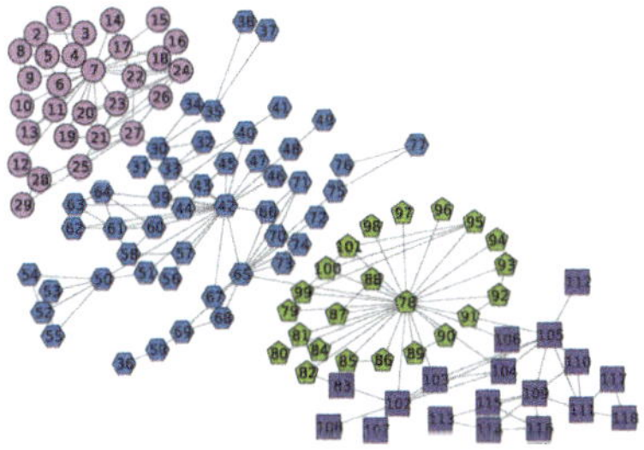

Fig. 10. The collaboration network

algorithm. LPA is a label propagation algorithm in [6]. It is a near linear time algorithm. The complexity of ASCD is O($n\log n$) in sparse network. But NM needs O(n^2) in sparse network. EBC's run time is even O(n^3). In Table 1, the quality of the detected communities is measured by the modularity Qin [1]. It shows the maximal modularity, the average modularity, Nc is the community count of NM, EBC.

In Table 1, due to NM, EBC are hierarchical agglomeration algorithms, we choose the N_c of which they all have largest Q. As our ASCD is stable, it has the same community structure in each run. And it has larger Q_{avg} in these networks. As LPA algorithm is not stable, it may have different community structure in different runs. We don't give the N_c of LPA.

We also compare the accuracy in Table 2. ASCD has the largest accuracy in each network. ASCD has more clear community structure by overall consideration.

5 Conclusion

In this paper, we have introduced a new community detection algorithm ASCD. ASCD is an active learning semi-supervised algorithm. ASCD treats each node as a single-node community first. We combine these single-node communities using our combine rule. Then we select representative nodes with our node relevance measure. This step

Table 1. Comparisons among some algorithms with modularity Q

Network	Algorithm						
	ASCD		NM		EBC		LPA
	N_c	Q_{avg}	N_c	Q_{avg}	N_c	Q_{avg}	Q_{avg}
Karate	2	0.371	2	**0.372**	2	0.360	0.344
Dolphin	4	**0.526**	4	0.490	4	0.458	0.469
Collaboration	4	0.677	4	0.696	4	**0.697**	0.630
Risk map	6	0.622	6	**0.624**	6	0.616	0.597
Artificial graph	4	**0.689**	4	**0.689**	4	**0.689**	**0.689**

Table 2. Comparisons among some algorithms with accuracy

Network	Algorithm			
	N_c	NM (accuracy)	EBC (accuracy)	ASCD (accuracy)
Karate	2	0.970	0.970	**1.000**
Dolphin	4	0.870	0.742	**1.000**
Collaboration	4	0.720	0.712	**0.992**
Risk map	6	**0.928**	0.952	**0.928**
Artificial graph	4	**1.000**	**1.000**	**1.000**

called active learning. We use these selected nodes to guide our community detection in semi-supervised strategy. Experiments show that ASCD has a better accuracy than algorithm NM and EBC. ASCD also has the highest modularity values in dolphin and artificial graph. Combine Q and accuracy evaluate criterions, ASCD performs well. ASCD has only one community structure in each run. It is a stable algorithm.

Our contribution aims at solving current shortcomings of low accuracy and long run time of algorithms. Finally, ASCD prevents both the creation of single-node communities and monster communities. It benefits from our proposed active learning step and semi-supervised strategy. ASCD is a new active learning semi-supervised community detection algorithm which has a good performance. Its entire run time in the worst case is $O(n\log n)$.

References

1. Newman, M.E.J.: Fast algorithm for detecting community structure in networks. Phys. Rev. E **69**, 066133 (2004)
2. Clauset, A., Newman, M.E.J., Moore, C.: Finding community structure in very large networks. Phys. Rev. E **70**, 066111 (2004)
3. Newman, M.E.J., Girvan, M.: Finding and evaluating community structure in networks. Phys. Rev. E **69**, 026113 (2004)
4. Lee, M.-J., Lee, J., Park, J.Y., Choi, R.H., Chung, C.: QUBE: a Quick algorithm for Updating Betweenness centrality. In: Proceedings of the 21st International Conference on World Wide Web, pp. 351–361 (2012)
5. Weng, W., Zhu, S., Xu, H.: Hierarchical community detection algorithm based on local similarity. J. Dig. Inf. Manag. **12**(4), 274–280 (2014)
6. Raghavan, U.N., Albert, R., Kumara, S.: Near linear time algorithm to detect community structures in large-scale networks. Phys. Rev. E **76**, 036106 (2007)
7. Herbiet, G.J., Bouvry, P.: SHARC: community-based partitioning for mobile ad hoc networks using neighborhood similarity. In: IEEE WoWMoM 2010, Montreal, pp. 1–9 (2010)
8. Liu, X., Murata, T.: Advanced modularity-specialized label propagation algorithm for detecting communities in networks. Physica A **389**(7), 1493–1500 (2010)
9. Brandes, U.: On variants of shortest-path betweenness centrality and their generic computation. Soc. Netw. **30**(2), 136–145 (2008)
10. Leng, M.W., Huang, L., Li, L.J., Zhou, H.H., Cheng, J.J., Chen, X.Y.: Active semisupervised community detection based on asymmetric similarity measure. Int. J. Mod. Phys. B **29**(13), 7821–7828 (2015)
11. Leng, M., Yao, Y., Cheng, J., Lv, W., Chen, X.: Active semisupervised community detection algorithm with label propagation. In: DASFAA2013, vol. 2, pp. 324–338 (2013)
12. Cheng, J.J., Leng, M.W., Li, L.J., Zhou, H.H., Chen, X.Y.: Active semi-supervised community detection based on must-link and cannot-link constraints. PLoS One **9**, 10 (2014)
13. Zachary, W.W.: An information flow model for conflict and fission in small groups. J. Anthropol. Res. **33**, 452–473 (1977)
14. Lusseau, D., Schneider, K., Boisseau, O.J., Haase, P., Slooten, E., Dawson, S.M.: The bottlenose dolphincommunity of doubtful sound features a large proportion of long-lasting associations. Behav. Ecol. Sociobiol. **54**, 396–405 (2003)

15. Li, K., Gong, X.F., Guan, S.G., Lai, C.H.: Efficient algorithm based on neighborhood overlap for communityidentification in complex networks. Physica A **391**(4), 1788–1796 (2012)
16. Girvan, M., Newman, M.E.J.: Community structure in social and biological networks. Proc. Natl. Acad. Sci. **99**(12), 7821–7826 (2002)
17. Steinhaeuser, K., Chawla, N.V.: Identifying and evaluating community structure in complex networks. Pattern Recognit. **31**, 413–421 (2010)
18. Lim, K.H., Datta, A.: A seed-centric community detection algorithm based on an expanding ring search. Proc. First Australas. Web Conf. **144**, 21–25 (2013)

Modeling and Numerical Simulation of Drawing Process of Hemisphere-Shaped Part Based on Electromagnetic Blank Holding System

Hao Li[1], Qiang Wang[1](✉), and Fang He[2]

[1] School of Mechanical Engineering, University of Jinan, Jinan, Shandong, China
me_wangq@ujn.edu.cn
[2] School of Electrical Engineering, University of Jinan, Jinan, Shandong, China

Abstract. A novel Electromagnetic Blank Holding System (EBHS) has been proposed in this paper. Traditional mechanical and hydraulic force has been replaced by Electromagnetic Force (EMF) in EBHS. Mathematical expression of the relationship between Blank Holder Force (BHF) and EMF has been given according to established force model. Besides, EMF numerical simulation has been performed by ANSYS/Multiphysics to obtain the distribution of EMF in EBHS. Additionally, three kinds of Stroke vs BHF curves have been proposed and hemisphere-shaped part drawing numerical simulation has been carried out by LS-DYNA basded on the three curves. The simulation results illustrate that the application of EBHS can effectively restrain wrinkling and over-thinning during hemisphere-shaped part drawing.

Keywords: Hemisphere-shaped part · Drawing process · BHF
EMF

1 Introduction

Hemisphere-shaped part is always applied in the field of manufacturing of lamps and household electrical appliances. The forming process of hemisphere-shaped part is sheet metal drawing. During the drawing process, tangential force in flange region of blank is always larger than the force in other directions. So wrinkling often occurs in the flange region of sheet metal. To restrain occurrence of wrinkling, applying BHF on flange region is always a good approach. But if the BHF is too large, the drawing part will be fractured. Therefore, applying reasonable BHF to flange region is essential to acquire high quality hemisphere-shaped part.

Traditional Blank Holding systems can be divided into mechanical and hydraulic Blank Holding systems. Mechanical Blank Holding devices usually contain spring, rubber pad, nitrogen cylinder and so on. The acquirement of BHF in hydraulic Blank Holding system is through controlling oil hydraulic circuit. Although traditional Blank Holding devices can reduce drawing defects during sheet metal drawing, they can only offer a BHF curve with rising tendency, which is unhelpful to the drawing quality of

K. Deng et al. (Eds.): ICMIR 2018, AISC 856, pp. 33–39, 2019.
https://doi.org/10.1007/978-3-030-00214-5_4

hemisphere-shaped part. In recent years, many scholars and experts home and abroad have paid their attention to the optimization of BHF curve and the research of variable-BHF novel devices. As a novel Blank Holding technology, EBHS has been more and more popular in the filed of sheet metal drawing. Satoshi et al. investigated the optimization of pulsating and variable BHF for identification of formability window for cylindrical part drawing [1]. Kitayama et al. optimized variable BHF trajectory by sequential approximate optimization with RBF network [2]. Seo et al. carried out corresponding experiments about EBHS and verified the advantage of EBHS [3]. Lai et al. studied EBHS based on electromagnetic forming and proposed corresponding devices [4]. Lin et al. proposed a PID control strategy to find out the most optimal BHF curve during sheet metal forming [5]. Li et al. proposed a novel design method of EBHS and researched the effect of radial electromagnetic force acting on the flange region [6]. They found that the radial electromagnetic force is helpful to sheet metal forming.

2 Design of EBHS

2.1 Modeling of EBHS

EBHS mainly contains drawing system and control system. In Fig. 1, it can be seen that the drawing system mainly contains drawing elements and Blank Holding elements. Drawing elements includes punch, die and sheet metal. The Blank Holding elements is consisted of Blank Holder, attractive element, fixed element and coils. Control system of EBHS contains coil system and power system. Designed EBHS drawing tool is shown in Fig. 2.

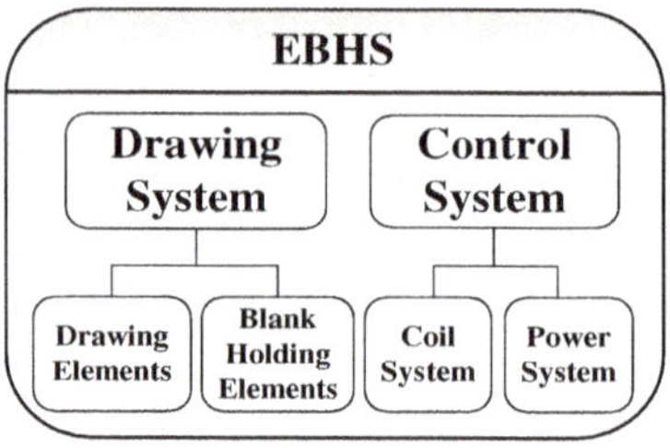

Fig. 1. Sketch of EBHS

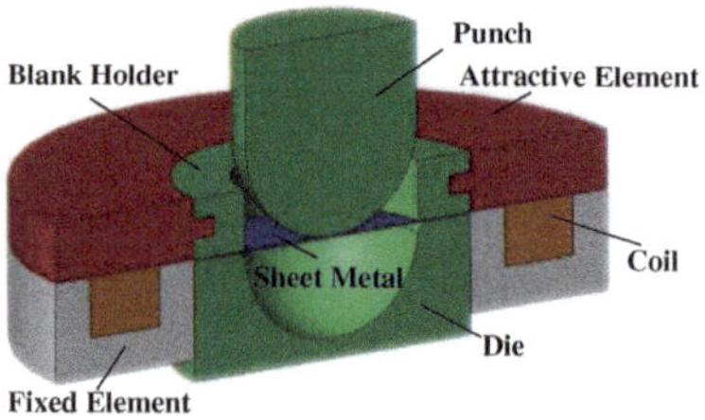

Fig. 2. Schematic diagram of EBHS

2.2 Principle of EBHS

As is described in introduction, BHF is replaced by EMF in EBHS. When sheet metal drawing is beginning, DC current is charged into the coil system simultaneously. Nearly at the same time, it will produce EMF in the gap between attractive element and fixed element. Then, EMF will transferred into BHF through designed mechanical structure. Besides, the amplitude of BHF can be flexibly changed through adjusting the amplitude of DC current. The working principle schematic of EBHS has been given in Fig. 3.

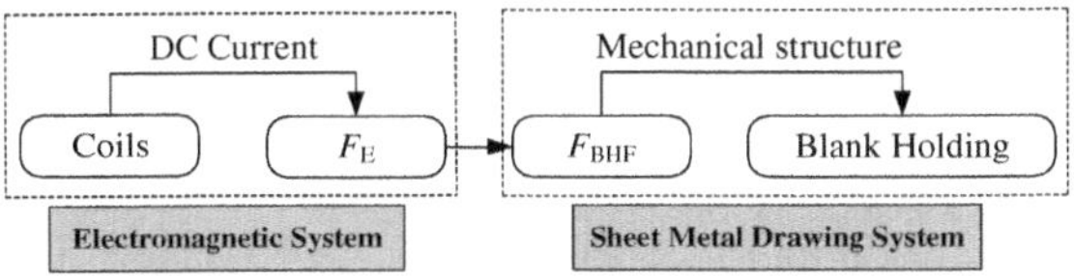

Fig. 3. Principle Schematic of EBHS

3 Calculation of BHF and EMF

3.1 Theoretical Calculation of BHF and EMF

Based on the established EBHS 3D model, calculation of BHF has been performed. Combining with designed structure size of EBHS, the BHF theoretical calculation formula 1 is derived and the corresponding BHF calculation result is also obtained, the value of which is 17226 N.

For accurately express the relationship between Holding force and EMF, Blank Holder and attractive element are regarded as an integral element in this paper. The force relationship of EBHS is shown in Fig. 4. According to Newton Second Law, the forces acting on the assemble element should satisfy the equivalence of force and moment in all directions. Therefore, the sum of force and moment of the two forces ought to be zero and concrete mathematical expressions have been shown in formula 2 and formula 3. Besides, the value of Holding force and BHF ought to be equal in value according to Newton First Law. According to formula 2, formula 3 and the calculation result of BHF, the relationship expression of BHF and EMF is acquired, which is shown in formula 4 and the value of EMF can be inferred as 8326 N.

$$F_{\mathrm{BHF}} = \frac{\pi}{4}\left(D^2 - D_1^2\right) \cdot P, P = 2.75\,\mathrm{MPa} \tag{1}$$

$$F_{\mathrm{E}} + F_{\mathrm{H}} = 0, F_{\mathrm{BHF}} = F_{\mathrm{H}} \tag{2}$$

$$F_{\mathrm{H}} \cdot L_1 + F_{\mathrm{E}} \cdot L_2 = 0 \tag{3}$$

$$F_{\mathrm{BHF}} = -\frac{F_{\mathrm{E}} L_2}{L_1} \tag{4}$$

F_{BHF} - Blank Holder Force, F_{E} - Electromagnetic Force, F_{H} - Holding Force, L_1 - Distance between original point and F_{H}, L_2 - Distance between original point and F_{E}.

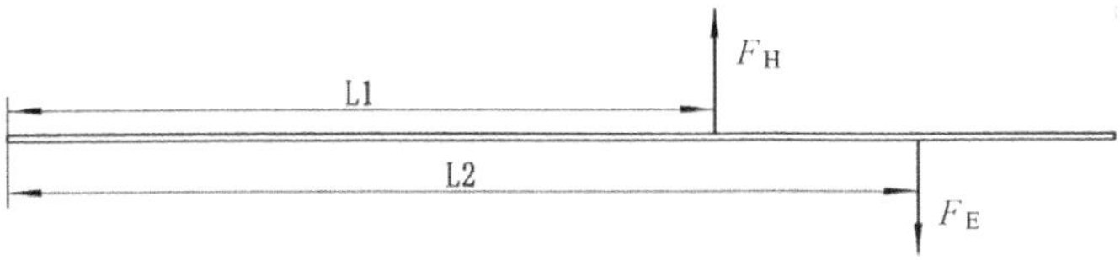

Fig. 4. Force relationship of EBHS

3.2 Numerical Simulation of EMF

To verify whether the EMF offered by EBHS can satisfy the requirement of Blank Holding of hemisphere-shaped part, numerical simulation of EMF in EBHS has been performed by ANSYS/Multiphysics. Firstly, 2D EBHS model is established as is shown in Fig. 5. In addition, the materials of die and coils are chosen die steel and copper while the permeability of which is 1. The material of attractive element is chosen Q235 and its corresponding permeability is 4000.

Through numerical simulation of EMF, the distribution of EMF, magnetic flux density distribution and magnetic flux density vector distribution have been obtained. The corresponding simulation results are shown in Figs. 6, 7 and 8. According to the simulation result, it can be seen that the maximum EMF is mainly distributed in the gap between attractive element and fixed element. Besides, the maximum EMF is up to 14554 N, which is larger than the calculated value. Additionally, the maximum magnetic flux density and corresponding vector also occur in the same region, which illustrates that the designed EBHS structure is reasonable.

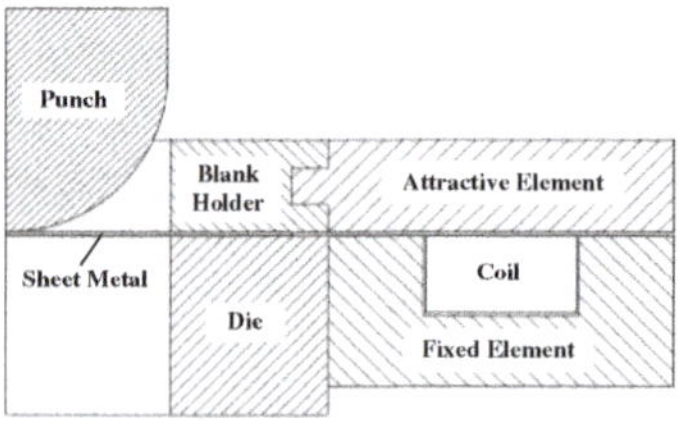

Fig. 5. 2D model of EBHS

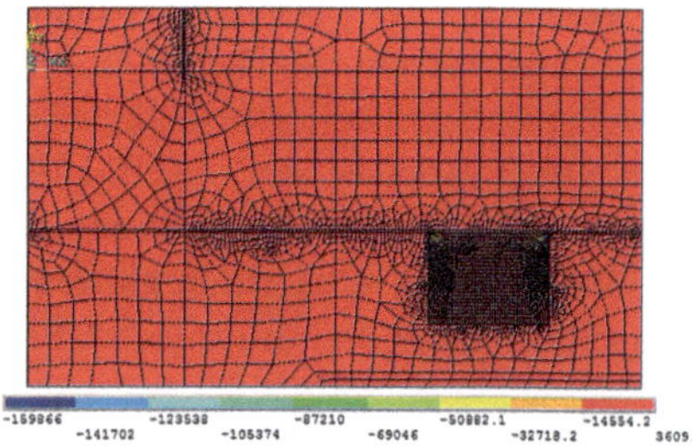

Fig. 6. EMF distribution

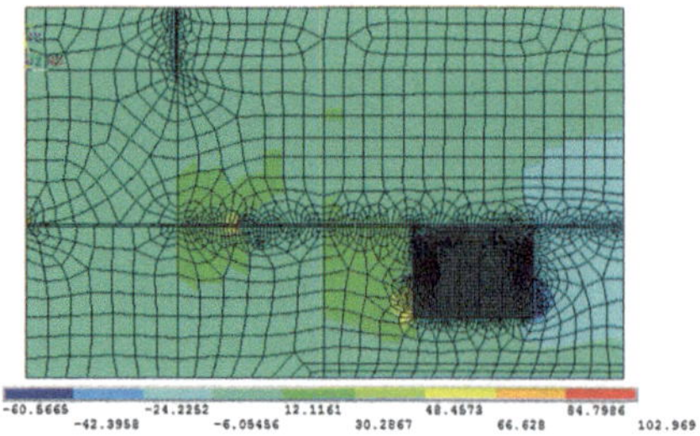

Fig. 7. Magnetic flux density distribution

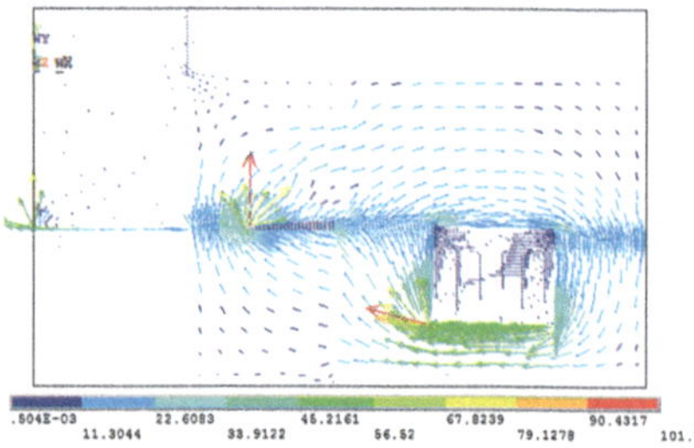

Fig. 8. Magnetic flux density vector distribution

4 Numerical Simulation of Hemisphere-Shaped Part Drawing

4.1 Modeling of Drawing Process of Hemisphere-Shaped Part

The size of designed hemisphere-shaped part is given in Fig. 9 and the FE model of hemisphere-shaped part drawing is shown in Fig. 10. The parameters' settlement of drawing system is also been accomplished in LS-DYNA. The friction coefficient between die and sheet metal is set as 0.1 and the drawing height is set as 50 mm.

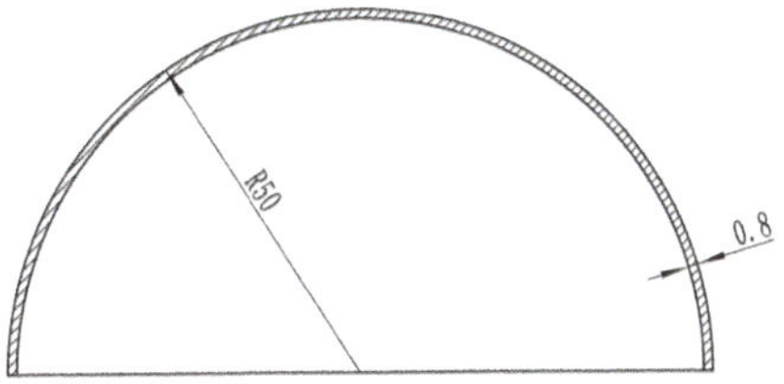

Fig. 9. The size of hemisphere-shaped part

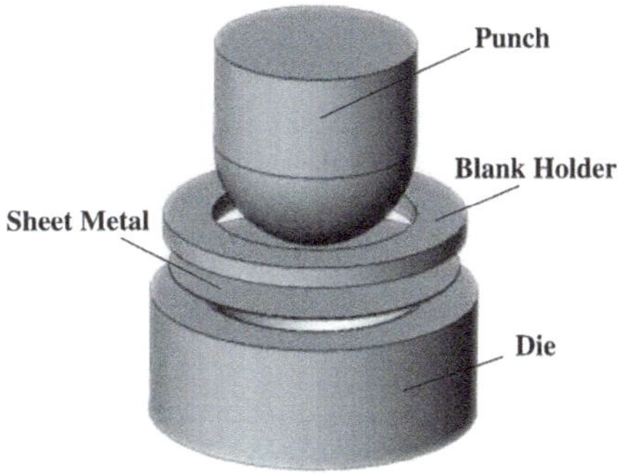

Fig. 10. FE model of drawing

4.2 Drawing Effectiveness of Traditional and Electromagnetic Blank Holder System

To compare the drawing effectiveness with the use of traditional Blank Holder system and EBHS, three BHF curves has been proposed as is shown in Fig. 11.

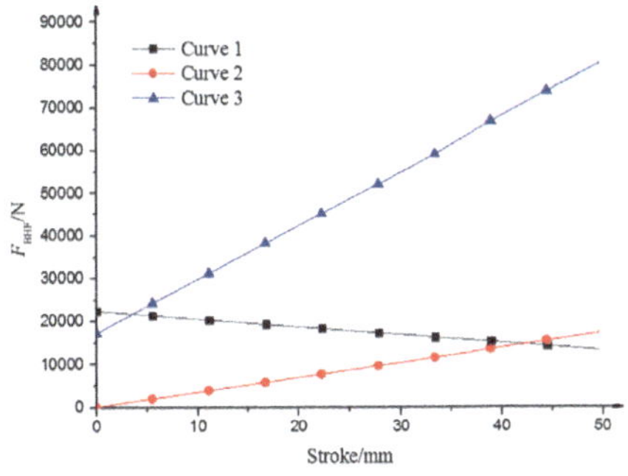

Fig. 11. Stroke vs F_{BHF} curve

Curve 1 represents BHF offered by EBHS and the corresponding tendency is decreasing. Furthermore, the initial value of Curve 1 is 1.3 times of the calculated value and the final value of the curve is 70% of the calculated value. Curve 2 and Curve 3 represent the BHF created by traditional Blank Holder devices, the tendency of which are all increasing. The difference between the two curves is that the original value of Curve 2 is zero and the initial value of Curve 3 is the calculated value.

Based on above proposed Stroke vs F_{BHF} curves, numerical simulation of hemisphere-shaped part drawing has been carried out by LS-DYNA. From Fig. 12, it can be easily seen that the flange region will occur serious wrinkling with the use of the Blank Holding device without applying initial BHF (Curve 2). Figure 13 shows the thinning rate contrary of Curve 1 and Curve 3 at the end of hemisphere-shaped part drawing. According to the simulation results, since hemisphere-shaped part drawing with increasing tendency (Curve 3) can effectively restrain wrinkling in flange region, it will induce over-thinning in the bottom of hemisphere-shaped part. In addition, the value of thinning rate in the bottom of the part has reached to 18.0%, which will easily cause the region fractured. Contrarily, If taking use of Curve 1 (producing by EBHS), the thinning rate in bottom region of the hemisphere-shaped part will be decreased to 10.6%, which is 38.78% lower than the former thinning rate.

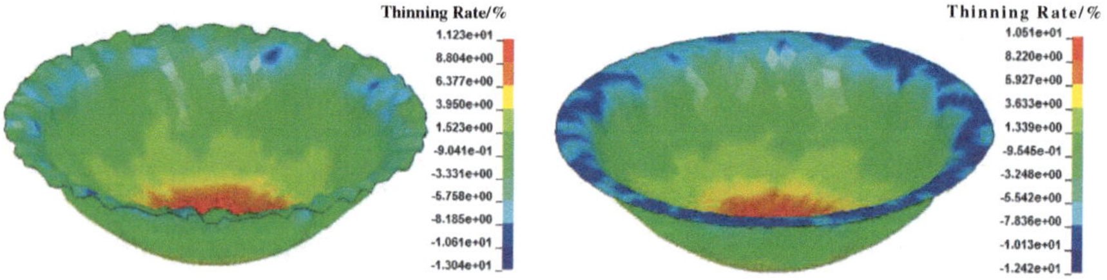

(a) Curve 2 (b) Curve 1

Fig. 12. Comparison of wrinkling and thinning rate (S = 37.5 mm)

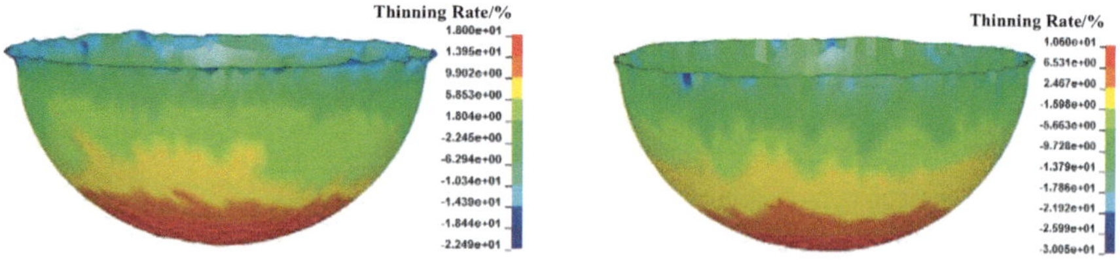

(a) Curve 3 (a) Curve 1

Fig. 13. Comparison of thinning rate contrary at the end of drawing (S = 50 mm)

From above numerical simulation results of hemisphere-shaped part drawing, it can be seen that the novel EBHS can not only effectively restrain wrinkling in the flange region of blank, but also can restrain the over-thinning of hemisphere-shaped part, which extremely improves the drawing quality of hemisphere-shaped part.

5 Conclusions

(1) A novel EBHS is proposed in this paper, traditional BHF is replaced by EMF in the system.
(2) Mathematical expression describing the relationship between BHF and EMF has been established.
(3) Through EMF numerical simulation, it can be found that the EMF produced by EBHS can satisfy the requirement of BHF.
(4) By comparing the drawing results based on EBHS and traditional Blank Holder systems, it is verified that the application of EBHS can effectively restrain wrinkling and over-thinning during hemisphere-shaped part drawing.

References

1. Satoshi, K.: Numerical investigation and optimization of pulsating and variable blank holder force for identification of formability window for deep drawing of cylindrical cup. Int. J. Adv. Manuf. Technol. **82**, 583–593 (2016)
2. Kitayama, S., Kita, K., Yamazaki, K.: Optimization of variable blank holder force trajectory by sequential approximate optimization with RBF network. Adv. Manuf. Technol. **61**, 1067–1083 (2012)
3. Seo, Y.R.: Electromagnetic blank restrainer in sheet metal forming processes. Int. J. Mech. Sci. **50**, 743–751 (2008)
4. Lai, Z.P., Cao, Q.L., Han, X.T., Xiong, Q., Deng, F.X., Zhang, X., Chen, Q., Li, L.: Design, implementation, and testing of a pulsed electromagnetic blank holder system. IEEE Trans. Appl. Supercond. **26**(4), 1–5 (2016)
5. Lin, Z.Q., Wang, W.R., Chen, G.L.: A new strategy to optimize variable blank holder force towards improving the forming limits of aluminum sheet metal forming. J. Mater. Process. Technol. **183**(2–3), 339–346 (2007)
6. Lai, Z.P., Cao, Q.L., Zhang, B., Han, X.T., Zhou, Z.Y., Li, L.: Radial Lorentz force augmented deep drawing for large drawing ratio using a novel dual-coil electromagnetic forming system. J. Mater. Process. Technol. **222**, 13–20 (2015)

Modelling and Simulating for Marine Exhaust Gas Boiler

Fu Yu[1], Yu Yongfeng[1], Wang Honghai[1], Yang Ming[1], Zhu Donghua[2], and Lv Bo[3(✉)]

[1] Jiujiang Haitian Equipment Manufacture Co., Ltd, Jiujiang, Jiangxi, China
[2] Wuhan Branch of China Classification Society, Wuhan, Hubei, China
[3] College of Marine Engineering, Dalian Maritime University, Dalian, Liaoning, China
1036942534@qq.com

Abstract. Based on the established dynamic mathematical model, the MATLAB/Simulink is used to analyze the dynamic characteristics of the boiler, and the model of the four zones, namely, the first convection zone, the second convection zone, the metal zone and the steam zone, are used to establish the mathematical model. Steam pressure, water volume, steam temperature and drum water level were simulated, observed and analyzed of experimental simulation results. As can be seen, the dynamic operation of the ship's exhaust boiler is in good condition and stable.

Keywords: Exhaust gas boiler · MATLAB and simulink · Drum water level Modular modeling

1 Introduction

The marine exhaust gas boiler is an important equipment of waste heat utilization. In the ship sailing, the exhaust boiler using the waste heat of exhaust gas from main engine to produce a lot of low pressure steam to meet the heating and daily life, in energy-saving emission reduction achieved fruitful. Therefore, the study of the dynamic performance of exhaust boiler is particularly important.

Boilers are controlled by means of two control loops. Control loop for controlling the pressure on the boiler and the water level in the boiler. These controllers are independent I/O (input and output) or internal variables are not exchanged between the control loops [1]. The most important is the dynamic performance of the boilers. The requirements for dynamic performance are increasing inherent contradictions: natural circulation, and disposable boilers need to be used to absorb shrinkage and expansion during dynamic operation. Boilers can absorb large reservoirs in the boiler. On the other hand, the thickness of the material in the pressurized vessel is approximately proportional to the diameter, i.e., the higher pressure, the larger material thickness. But the allowable temperature decreases as the wall thickness increases and the gradient of the pressurized vessel decreases [2].

The ship exhaust gas boiler is a complex, multivariable nonlinear control system. However, the steam boiler is controlled by the background theory based on the SISO

K. Deng et al. (Eds.): ICMIR 2018, AISC 856, pp. 40–48, 2019.
https://doi.org/10.1007/978-3-030-00214-5_5

(single input single output) process controller. This is the general steam boiler. The vapor pressure is controlled by the PID, in some cases were controlled by the feed-forward complement of the steam flow. The water level is controlled by a so-called single, two or three-element controller. The water level feedback controller is usually a pure proportional action.

2 Mathematical Model of Exhaust Boiler

The Fig. 1 is the model of the exhaust boiler body.

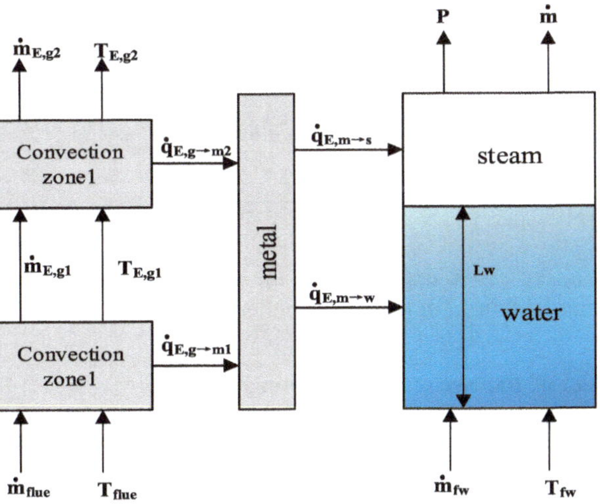

Fig. 1. Schematic diagram of the division of the exhaust boiler module

2.1 The Model of the First Convection Zone

The amount of flue gas flowing into the convection zone 1 is equal to the amount of flue gas flowing out of the marine engine:

$$\dot{q}_{E,g1} = \dot{q}_{flue} \tag{1}$$

$$\dot{q}_{flue} - \dot{q}_{E,g1} - \dot{q}_{E,g\to m1} = \frac{dU_{E,g1}}{dt} \tag{2}$$

Where:

$\dot{q}_{flue}$ is the unit time into the first convection zone of flue gas energy, kJ/s; $\dot{q}_{E,g1}$ is to the unit time outflow the first convection zone flue gas energy, kJ/s; $\dot{q}_{E,g\to m1}$ is the unit time of the flue gas and boiler metal wall convection heat transfer, kJ/s; $U_{E,g1}$ is the first convection zone flue gas energy, kJ/s.

The marine engine smoke flow into convective zone heat calculation:

$$\dot{q}_{flue} = \dot{m}_{flue} h_{flue} = \dot{m}_{flue} c_{p,flue}(T_{flue} - T_{ref}) \tag{3}$$

Smoke flow outflow convection a zone heat calculation:

$$\dot{q}_{E,g1} = \dot{m}_{E,g1} h_{E,g1} = \dot{m}_{E,g1} c_{p,E,g1}(T_{E,g1} - T_{ref}) \tag{4}$$

Calculation of convective heat transfer of flue gas and boiler metal wall:

$$\dot{q}_{E,g\to,m1} = A_{E,g1}\alpha_{E,g1}(\overline{T}_{E,g1} - T_{E,m}) \tag{5}$$

Calculation of flue gas energy in the first convection zone:

$$U_{E,g1} = M_{E,g1}h_{E,g1} = V_{E,g1}\rho_{E,g1}c_{p,E,g1}(\overline{T}_{E,g1} - T_{ref}) \tag{6}$$

$$\overline{T}_{E,g1} = \frac{T_{E,g1} + T_{flue}}{2} \tag{7}$$

Among them, the above expression in the subscript indicates the meaning of the exhaust gas boiler, and in front of the gas boiler distinction, the following is the same.

2.2 The Model of the Second Convection Zone

The amount of flue gas flowing into the convection area is equal to the amount of flue gas flowing out:

$$\dot{m}_{E,g2} = \dot{m}_{E,g1} \tag{8}$$

$$\dot{q}_{E,g1} - \dot{q}_{E,g2} - \dot{q}_{E,g\to m2} = \frac{dU_{E,g2}}{dt} \tag{9}$$

Where:

$\dot{q}_{E,g1}$ is the unit time into the second convection zone of flue gas energy, kJ/s; $\dot{q}_{E,g2}$ is to the unit time outflow the second convection zone flue gas energy, kJ/s; $\dot{q}_{E,g\to m2}$ is the unit time of the flue gas and boiler metal wall convection heat transfer, kJ/s; $U_{E,g2}$ is the second convection zone flue gas energy, kJ/s.

Calculation of heat flow in flue gas outflow convection Zone:

$$\dot{q}_{E,g2} = \dot{m}_{E,g2} h_{E,g2} = \dot{m}_{E,g2} c_{p,E,g2}(T_{E,g2} - T_{ref}) \tag{10}$$

Calculation of convective heat transfer of flue gas and boiler metal wall:

$$\dot{q}_{E,g\rightarrow,m2} = A_{E,g2}\alpha_{E,g2}(\overline{T}_{E,g2} - T_{E,m}) \tag{11}$$

Calculation of flue gas energy in the second convection zone:

$$U_{E,g2} = M_{E,g2}h_{E,g2} = V_{E,g2}\rho_{E,g2}c_{p,E,g2}(\overline{T}_{E,g2} - T_{ref}) \tag{12}$$

$$\overline{T}_{E,g2} = \frac{T_{E,g2} + T_{E,g1}}{2} \tag{13}$$

2.3 The Model of the Metal Zone

The heat transfer model of the metal wall of the boiler mainly considers the heat transfer effect of the metal wall on the heat transfer in the soft water area, and the energy conservation equation of the metal pipe wall:

$$\dot{q}_{E,g1\rightarrow m1} + \dot{q}_{E,g\rightarrow m2} - \dot{q}_{E,m\rightarrow w} - \dot{q}_{E,m\rightarrow s} = \frac{dU_{E,m}}{dt} \tag{14}$$

Where:

$\dot{q}_{E,m\rightarrow w}$ is the unit time metal pipe wall to pass to the energy of the boiler water, kJ/s; $\dot{q}_{E,m\rightarrow s}$ is the unit time the metal pipe wall passes to the steam energy, kJ/s.

Calculation of energy of metal pipe wall to boiler water:

$$\dot{q}_{E,m\rightarrow w} = A_{E,g1}\alpha_{E,g1,w}(T_{E,m} - T_{E,w}) \tag{15}$$

Metal pipe wall to the steam energy calculation:

$$\dot{q}_{E,m\rightarrow s} = A_{E,g2}\alpha_{E,g2,s}(T_{E,m} - T_{E,s}) \tag{16}$$

Calculation of thermal energy of metal tube wall:

$$U_{E,m} = M_{E,m}c_{p,E,m}(T_{E,m} - T_{ref}) \tag{17}$$

Where:

$c_{p,E,m}$ is the specific heat capacity of the boiler metal tube, Value is 450 kJ/(kg·K);

2.4 The Model of the Steam and Water Zone

The main parameters of the steam and water zone model are the boiler water and steam pressure [3]. Figure 2 shows the steam or water zone dynamic process diagram.

The mass conservation and energy conservation of the steam and water zone are analyzed as follows:

$$\frac{d}{dt}(\rho_s(V_s + V_b) + \rho_w V_w) = \dot{m}_{fw} - \dot{m}_s \tag{18}$$

Where:

ρ_s, ρ_w respectively, saturated steam, saturated water density, kJ/m^3; $\dot{m}_{fw}$, $\dot{m}_s$ respectively, for the water quality flow and steam mass flow, kg/s; V_s, V_b, V_w respectively, the volume of steam, underwater bubbles and water space, m^3. Due to the volume of the entire boiler area $V_t = V_w + V_s + V_b$, so the formula (18) can be written:

$$(V_t - V_w)\frac{d\rho_s}{dP_s} + V_w\frac{d\rho_s}{dP_s}\frac{dP_s}{dt} + (\rho_w - \rho_s)\frac{dV_w}{dt} = \dot{m}_{fw} - \dot{m}_s \quad (19)$$

Where:
P_s is the saturated steam pressure, Pa.

$$\frac{d(U_w + U_b + U_s)}{dt} = \dot{q}_{E,m\to w} + \dot{q}_{E,m\to s} + \dot{q}_{fw} - \dot{q}_s \quad (20)$$

Where:
U_w, U_b, U_s are the internal energies of water, bubbles and steam, respectively, J. Because the internal energy relationship of the working fluid is:

$$\begin{cases} U = M \cdot u = \rho \cdot V \cdot (h - \upsilon \cdot p) \\ \upsilon \cdot p = 1 \end{cases} \quad (21)$$

Where:
h is the specific enthalpy of working fluid, J/kg; υ, m^3/kg.
Feedwater and steam energy calculated energy:

$$\dot{q}_{fw} = h_{fw}\,\dot{m}_{fw} \quad (22)$$

$$\dot{q}_s = h_s\,\dot{m}_s \quad (23)$$

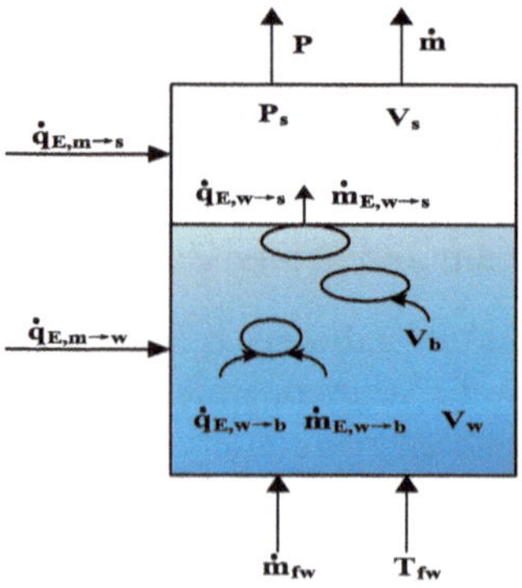

Fig. 2. Principle of the soda zone model

Using the chain differential principle, the formula (20) is further developed:

$$\left(\rho_w V_w \frac{dh_w}{dP_s} + h_w V_w \frac{d\rho_w}{dP_s} + \rho_s (V_t - V_w) \frac{dh_s}{dP_s} + h_s (V_t - V_w) \frac{d\rho_s}{dP_s} - V_t\right) \frac{dP_s}{dt} \\ + (h_w \rho_w - h_s \rho_s) \frac{dV_w}{dt} = \overset{\bullet}{q}_{E,m\to w} + \overset{\bullet}{q}_{E,m\to s} + h_{fw} \overset{\bullet}{m}_{fw} - h_s \overset{\bullet}{m}_s \tag{24}$$

To further describe the relationship between boiler water level and the underwater bubbles, a detailed model of underwater bubbles needs to be described.

For exhaust boilers, the water level can be expressed as:

$$L_w = (V_w + V_b)/A_w \tag{25}$$

It can be seen from Eq. (25) that the boiler water level is directly related to the volume of bubbles under the water surface.

The quality conservation and energy conservation of the area under the water surface are analyzed:

$$\frac{d(\rho_s V_b)}{dt} = \overset{\bullet}{m}_{w\to b} - \overset{\bullet}{m}_{b\to s} \tag{26}$$

$$\frac{d(\rho_w V_w)}{dt} = \overset{\bullet}{m}_{fw} - \overset{\bullet}{m}_{w\to b} \tag{27}$$

Since the mass flow of water into the bubble and the mass flow from the bubble to the steam is uncertain, it is difficult to describe it in quantitative mathematical relationships. It can be described for empirical formula:

$$\overset{\bullet}{m}_{b\to s} = \gamma \frac{V_b}{V_w} + \beta \overset{\bullet}{m}_{w\to b} \tag{28}$$

From the formula of (25), (26), (27) and (28), we can get the formula (29).

$$\left(V_b \frac{d\rho_s}{dP_s} + (1-\beta) V_w \frac{d\rho_w}{dP_s}\right) \frac{dP_s}{dt} + (1-\beta)\rho_w \frac{dV_w}{dt} + \rho_s \frac{dV_b}{dt} = (1-\beta) \overset{\bullet}{m}_{fw} - \gamma \frac{V_b}{V_w} \tag{29}$$

From the formula of (29) we can see that the change is the main factor causes a phenomenon called shrink-and-swell.

3 Simulation

The exhaust boiler of the various components of the system partition, in accordance with the process of working fluid and the relationship between the overlapping together. We can get the boiler simulation model of the module diagram, as shown in Fig. 3.

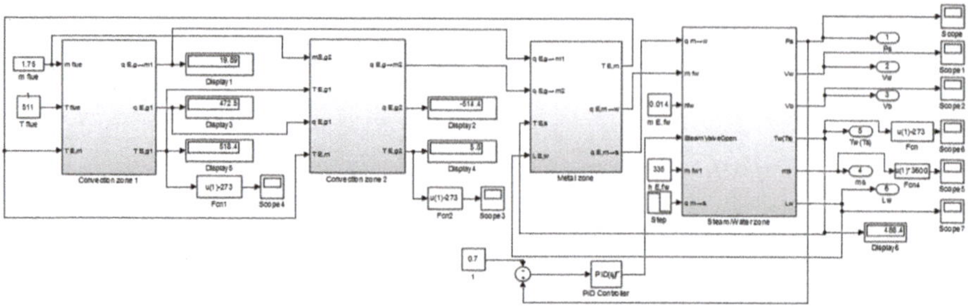

Fig. 3. Modular model of exhaust boiler

To simulate exhaust boiler steam pressure, steam temperature and water level.

Before the simulation, setting the simulation parameters and the simulation time is 0 to 2500 s, the simulation algorithm for the fixed step size, step size is 0.1. when time is 1800 s, adding water flow disturbance. The PID parameters for controlling the steam pressure are:

The simulation results as shown in Fig. 4.

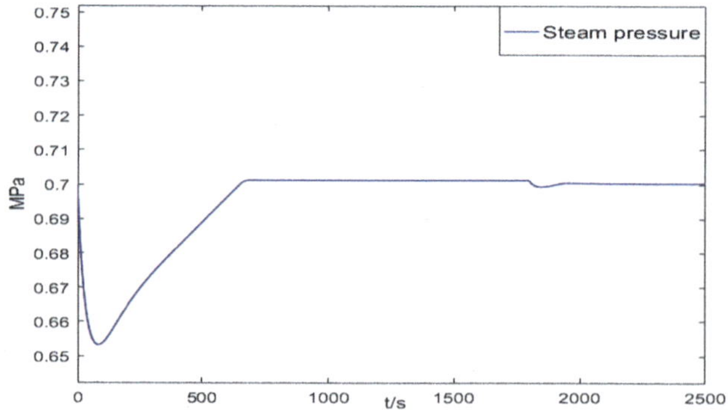

Fig. 4. Simulation for steam pressure

To simulate the boiler water level. Set the simulation step for the fixed step length of 0.1, the total simulation time is 2500 s. when time is 1800 s, adding water flow disturbance. The simulation results as shown in Fig. 5.

To simulate the volume of boiler water. Set the simulation step for the fixed step length of 0.1, the total simulation time is 2500 s. When time is 1800 s, adding water flow disturbance, control the volume of boiler water stability. The simulation results as shown in Fig. 6.

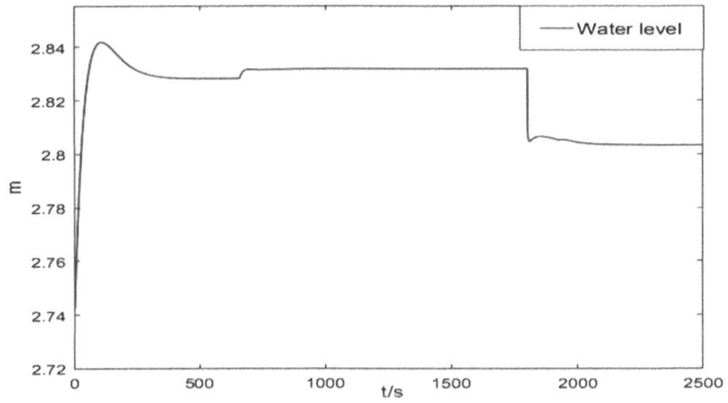

Fig. 5. Simulation for water level

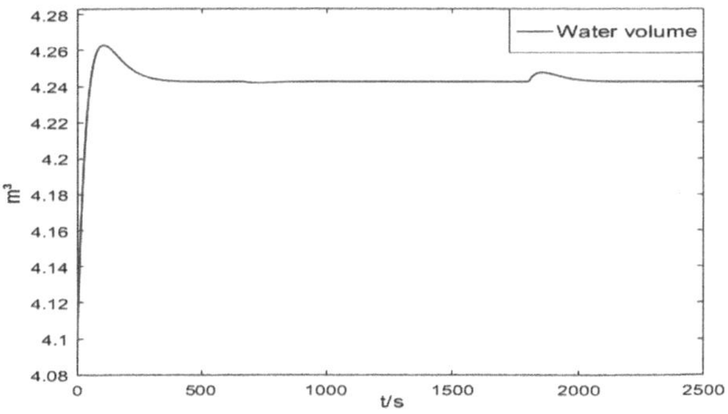

Fig. 6. Simulation for the volume of boiler water

4 Conclusion

Based on the research of ship exhaust boiler, a model of simulated boiler has been designed according to the principle and characteristics of the actual boiler system. The model is composed of four sub-modules in the entire system of the boiler, for each sub-module, the corresponding algebraic equation is established. Subsequently, these models are integrated into the global model of the system. The simulation results are compared with those of the actual boiler, which show that the dynamic performance is in accordance with the actual performance and it turns out to be good, and the expected effect is achieved.

Acknowledgments. The study was supported by "High Technology Ship Research Program of the Ministry of Industry and Information Technology of the People's Republic of China" ([2016 NO.25]). The authors also gratefully acknowledge the helpful comments and suggestions of the reviewers, which have improved the presentation.

References

1. Dukelow, S.G.: The Control of Boilers, 2nd edn. Instrument Society of America, Research Triangle Park (1991)
2. Sørensen, K., Condra, T.: Modeling and simulating fire tube boiler performance. In: Proceedings of SIMS 2003-44th Conference on Simulation and Modeling, pp. 9–12 (2003)
3. Solberg, B., Karstensen, C.M.S., Andersen, P., Pedersen, P.S., Hvistendahl, P.U.: Model-based control of a bottom fired marine boiler. In: IFAC Proceedings Volumes, vol. 38, no. 1, pp. 314–319 (2005)

Congestion Aware Multipath Routing Algorithm Based on Coding-Aware in Wireless Mesh Networks

YuKun Yao, Juan Li(✉), and Yi Zhang

Chongqing University of Posts and Telecommunications, Chongqing, China
juan.li.cqupt@foxmail.com

Abstract. The load balanced coding-aware multipath routing algorithm may have the problems of high load node forwarding routing request message leading to a waste of bandwidth resources during routing discovery and the coding gain only considers the number of coded flows that can't make the path actual coding gain optimal. To solve these problems, a coding-aware and congestion aware multipath routing algorithm in wireless mesh networks (CACAMA) by proposed in this paper. By introducing two mechanisms of congestion aware based coding-aware routing metrics and congestion aware mechanisms, the CACAMA algorithm has obvious advantages in reducing packet loss rate and improving network throughput. Simulation results show that, compared with LCMR, the performance of CACAMA algorithm has been improved in terms of packet loss rate, network throughput and end to end delay.

Keywords: Wireless mesh networks · Coding-aware · Congestion aware Multipath routing

1 Introduction

Wireless mesh network is a multi-point to multi-point distributed network. It is considered to be a new solution to solve the bottleneck problem of the last mile of wireless access [1, 2]. Because of the radio characteristics of the wireless media, there are some problems in the wireless mesh network, such as channel fading and interference. Network coding [3] allows intermediate nodes to code and forward the received information, increase the information volume of single transmission, and significantly improve network throughput. The broadcast characteristics of wireless media provide favorable conditions for the implementation of network coding and decoding. The application of network coding to wireless mesh network has become a research hotspot.

Katti designed an original wireless network forwarding framework called COPE [4], which combined network coding theory and practical requirements. The following two limitations exist in COPE: The first is that the coding structure of the COPE is limited to the two hop range, and the second is the coding opportunity in COPE depends on the established routing. In order to solve the problem in the COPE, the concept of coding-aware proposed by Ni et al. [5]. The main idea of coding-aware is to detect the coding opportunities in the route discovery process, and the routing

K. Deng et al. (Eds.): ICMIR 2018, AISC 856, pp. 49–56, 2019.
https://doi.org/10.1007/978-3-030-00214-5_6

algorithm based on coding awareness has become a hot topic of research [6–8]. The work in [9] proposed Distributed Coding-Aware Routing (DCAR), and defined the coding condition of two raw data streams in the cross node, and proposed a coding-aware routing metric (CRM) that can reflect the performance between paths with coding opportunities and non-coding opportunities.

The work in [10] proposed a wireless mesh network coding-aware wireless routing protocol to support COPE applications called OCR. The OCR protocol evaluation path performance through consumption function, which considers both coding gain and path hop count. The network throughput can be greatly improved by OCR, compared to the traditional routing protocol. Reference [11] considered a coding-aware routing metric with load balancing (CRM-LB) based on CRM, which takes into account the degree of communication density and network congestion of all nodes on the path P. And proposed a wireless mesh network multicast routing algorithm (CLR) based on CRM-LB, the CLR algorithm comprehensively considers the increasing encoding opportunity and the balanced network load, and can better support network coding and balance the network load.

Compared with single path routing algorithm, multipath routing algorithm can make full use of network resources, enhance the fault tolerance of network, and facilitate the load balancing of network. A multi-path routing mechanism called CAMP based on network coding is proposed in [12]. The mechanism allows data packets to be transmitted dynamically on multiple paths, sharing a traffic load on multiple paths and getting higher throughput. Xing et al. proposed a multi path routing algorithm based on load balanced coding (LCMR) [13], in LCMR algorithm, high load nodes or congested nodes need to forward new route request messages in the process of route discovery and not adjust the proportion of traffic assignment during data transmission, resulting in increased routing overhead, packet loss rate and end-to-end delay.

2 Problem Description

We found the following problems exist in LCMR routing algorithm: Firstly, the coding gain in the routing metric LCRM only considers the number of coded streams and can't optimize the actual coding gain of the path. The coding opportunities that exist in the process of routing discovery are extremely likely to disappear in the process of data transmission, therefore the current path of greater coding gain in the whole process of data transmission in the actual encoding gain is not large. It is impossible to choose the optimal path of the actual coding gain, which can not make full use of the coding opportunities in the network to give full play to the advantages of network coding. Secondly, the load of the whole path measured by the load of all nodes on the path. It can't be ensured that there is no high load node or congested node in the path where the sum of the loads is small, which ignores the impact of a single node load in the path on the entire path performance. The paths contain node with higher congestion probability can not excluded in advance. If the paths contain high load nodes or congested nodes, the end-to-end delay will increase, and the high load node or the congestion node forward the new routing request message, which will lead to the redundancy of routing request message and increase the routing cost.

3 CACAMA Routing Algorithm

3.1 The Routing Metric CACM

In order to discuss the routing metric CACM more intuitively, the coding conditions are described first. Then, the parameter encoding factor and the load interference factor and the expected transmission time are explained in detail.

Before analyzing the network coding condition, let us define symbols. f indicates a data flow, $a \in f$ denotes node a belongs to the routing of data flow f while the source node is s and the destination node is d. $N(a)$ means the single-hop neighbor set of node a. $For(a, f)$ indicates the forwarding nodes set of node a in the routing of data flow f. $Bac(a, f)$ indicates the backward nodes set of node a in the routing of data flow f. For example, in Fig. 1, $For(c, f_1) = \{a, b\}$, $Bac(c, f_1) = \{h\}$, $For(c, f_2) = \{d\}$, $Bac(c, f_2) = \{i, j\}$. When two flows intersect in a node and satisfy the network coding conditions, the packets of flow can then be encoded.

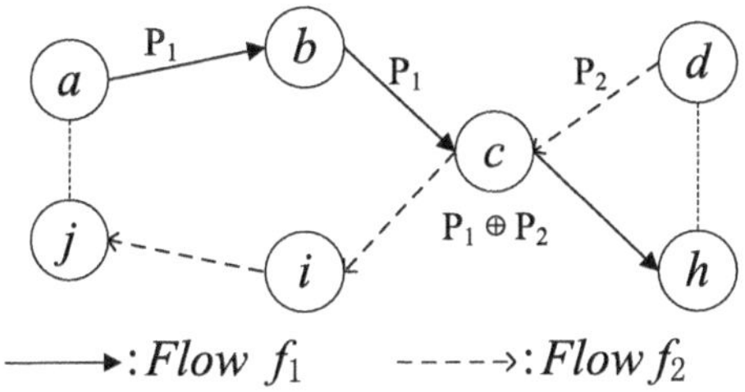

Fig. 1. Multi-hop coding-aware

For the flow f_1 and f_2 which intersect at node c, if the following conditions are satisfied, network coding is feasible [9].

1. Existing node $d_1 \in Bac(c, f_1)$ while $d_1 \in N(s_2) \wedge s_2 \in For(c, f_2)$ or $d_1 \in For(c, f_2)$.
2. Existing node $d_2 \in Bac(c, f_2)$ while $d_2 \in N(s_1) \wedge s_1 \in For(c, f_1)$ or $d_2 \in For(c, f_1)$.

As in Fig. 1, there are two flows $a \rightarrow b \rightarrow c \rightarrow h$ and $d \rightarrow c \rightarrow i \rightarrow j$ intersect at node c. Node c codes the packets from these two flows and broadcast the coded packets to both nodes h and i. When the two-way delivery rate of the link $l_{c,i}$ between the node c and the node i is $\rho_{c,i}$.

In the traditional data exchange, the node c successfully completes the exchange of the original data packet p_1 and p_2 respectively. The total expected transmission times are as follows

$$N_T = \frac{1}{\rho_{b,c}} + \frac{1}{\rho_{c,h}} + \frac{1}{\rho_{d,c}} + \frac{1}{\rho_{c,i}} \quad (1)$$

According to the coding condition, there is a coding opportunity in node c. In node c codes $p_1 \oplus p_2$, and broadcasts it, node h and j can decode the coded packet and achieve the expected packet. The total expected transmission times are as follows

$$N_c = N_T - \frac{1}{\rho_{c,i} + \rho_{d,c} - \rho_{c,i}\rho_{d,c}} \tag{2}$$

Compared with the traditional data exchange method, the number of expected transmissions reduced after the network coding at node c is defined as follows

$$G = \frac{1}{\rho_{c,i} + \rho_{d,c} - \rho_{c,i}\rho_{d,c}} \tag{3}$$

The data stream f_1 coded with the new data stream f_2 may complete transmission during the transmission of the data stream f_2, result in the coding opportunities disappear in the node c. Therefore, the number of expected transmission times reduced by the actual use of network coding will be lower than the calculated value in formula (3). It is necessary to consider the problem of data matching between the coded streams, and the formula for calculating the data matching factor is presented as follows

$$\lambda = \frac{1}{q_{old}} \tag{4}$$

Accordingly, the data matching factor λ, which denotes the reciprocal of the data quantity of old flow q_{old}. The encoding factor is used to reflect the actual coding gain of the node, and the encoding factor of the coded node C is presented as follows

$$GF(\mathrm{c}) = \frac{\lambda}{G_c} \tag{5}$$

The load degree of the node has a direct impact on queuing delay. When the load degree of the node is high, the queuing delay of the new arrival data packet is large, and the load degree of the node is measured by the cache occupancy rate. The cache occupancy of the node c is defined as follows

$$O(c) = \frac{q(c)}{Q(c)} \tag{6}$$

where $O(c)$ is denotes the ratio between the length of node c cache queue occupied $q(c)$, and the length of node c total cache queue $Q(c)$.

The load interference factor L (c) is used to reflect the load degree of the node c and its neighbor interference nodes, and the expression is as follows

$$L(c) = O(c) + I(c) \tag{7}$$

where $I(c)$ is load intensity of the neighbor interference nodes of node c, which can reflect the load situation of the region. The calculation formula of $I(c)$ is as follows

$$I(c) = \frac{1}{N(c)}\sum_{w \in S(c)} \frac{q(k)}{Q(k)} + \left(\frac{1}{N(c)}\sum_{w \in S(c)}\left(\frac{q(k)}{Q(k)} - \frac{1}{N(c)}\sum_{w \in S(c)}\frac{q(k)}{Q(k)}\right)^2\right)^{\frac{1}{2}} \quad (8)$$

where N (c) is the number of neighbor interference nodes that cause interference to node c, S (c) is a set of neighbor interference nodes that cause interference to node c, q (k) is the length of neighbor interference node k cache queue occupied, and Q (k) is the length of neighbor interference node k total cache queue.

The congestion aware coding-aware routing metrics take into account coding opportunities, path load and interference conditions, and link quality. The CACM metric of node c is as follows

$$CACM(c) = GF(c) + ETT(c) + L(c) \quad (9)$$

where *ETT* is expected transmission time, which means the time it takes to successfully transmit a data packet on a link.

The CACM value of the path P is defined as follows

$$CACM_P = \sum_{c \in P} CACM(c) \quad (10)$$

The coding-aware congestion aware multipath routing algorithm CACAMA gives priority to the path that is smaller in the metric value CACM. According to the routing metric CACM of each path, the source node selects the optimal m paths. Then, the source node calculates the rate of flow allocation for each path. If the selected m route route metric values are M_P^1, M_P^2, …, M_P^k, …, M_P^m, the flow distribution ratio of the path k $(k \in m)$ is as follows

$$f_k = \frac{\frac{1}{M_P^k}}{\sum_{j=1}^{m} \frac{1}{M_P^j}} \quad (11)$$

3.2 Congestion Aware Mechanism

When there are high load nodes or congested nodes in the path, the overall performance of the path will be affected by a single node. In multipath routing, when the source node selects the path, it should avoid choosing a path that contains a high load node or a congestion node. Congestion aware mechanisms can exclude the paths including high load nodes or congestion nodes in the route discovery process, it reduces the routing overhead and the redundancy of the routing request messages.

When the source node s needs to send the data, it initiates the route discovery by broadcasting the Route Request (RREQ) message. After a relay node receives a new routing request message, it calculates the cache occupancy rate and judge the load situation. In order to facilitate the discussion of node load conditions, two thresholds of

T_{min} and T_{max} are set for the cache occupancy rate, and the specific steps of the congestion perception mechanism are as follows

Step 1. The relay node calculates the link cache occupancy after received RREQ, and compares the cache occupancy rate O with T_{min} and T_{max}.

1. If $O < T_{min}$, the node is in a lighter load state and the probability of congestion is lower, and the relay node should continue forward RREQ message.
2. If $O > T_{max}$, the node is in a high load state and has a higher probability of congestion, and execute step 3. Otherwise, step 2 is executed.

Step 2. The relay node compare packet reception rate R_r and packet transmission rate R_s.

1. If $T_{min} < O < T_{max}$ *and* $R_s > R_r$, the node is in a moderate load state and the probability of congestion is lower, and the relay node should continue forward RREQ message.
2. If $T_{min} < O < T_{max}$ *and* $R_s < R_r$, the node is in a moderate load state and the probability of congestion is higher, execute step 3.

Step 3. Discarded the routing request message RREQ, give up the forwarding of the routing request message.

In the route discovery process, the relay node that is in the congestion state or is in the state of congestion no longer continues to forward the new routing request message. It is possible to exclude paths that contain high load nodes or congestion nodes in the routing discovery phase. The congestion detection mechanism of load sensing can effectively reduce the routing overhead and avoid network congestion.

4 Simulation Results and Analysis

To verify the performance of CACAMA, we utilize OPNET to simulate and analyze the results. The network model of the simulation experiment is a WMN scenario in which 40 nodes are randomly distributed in a rectangular area of 1000×1000 m^2. We use 802.11b and UDP traffic sources. The transmission range of each wireless node is set to 250 and the carrier sensing range is set to 550, increasing network load in a way that gradually increases the number of data streams. The routing algorithms CACAMA, COPE and LCRM are simulated respectively, according to the simulation results, the performance comparison and analysis of three routing algorithms based on simulation results.

Accordingly, Fig. 2 presents the comparison of the average throughput in three algorithms. The average network throughput of CACAMA algorithm under different network loads is higher than that of LCMR and COPE. The main reason is that CACAMA avoids the problem of network congestion caused by data flow convergence in the coding-aware routing, which can effectively improve the average throughput of the network.

Figure 3 show that, as the network load increased, the average end-to-end delay of the three routing algorithms increased. However, under different network loads, the average end-to-end delay of CACAMA algorithm is lower than LCMR and COPE.

Because CACAMA is more inclined to choose the path of light load, it can reduce the queuing time delay of data packets.

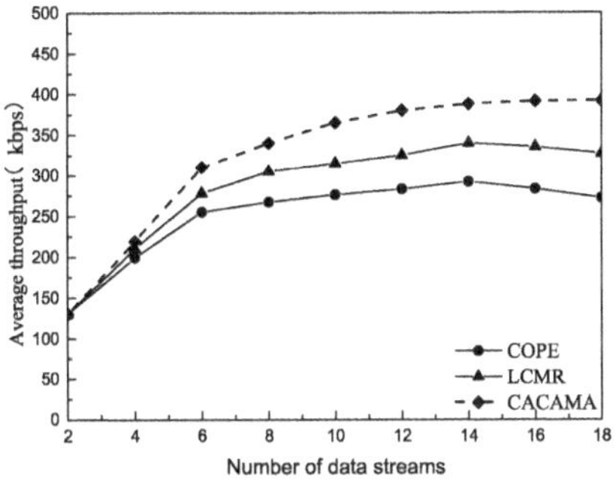

Fig. 2. Average throughput under different number of data streams

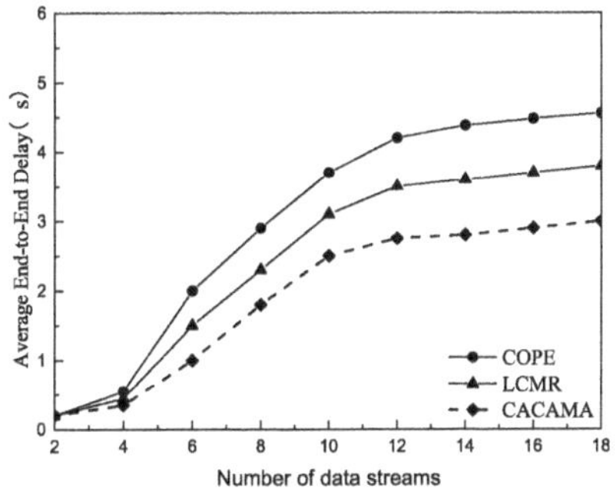

Fig. 3. Average end-to-end delay under different number of data streams

As is shown in Fig. 4, the packet delivery rate of the 3 routing algorithms decreases with the increase of network load. Because CACAMA uses the mechanism of congestion sensing in route discovery, it can exclude the high load nodes that are in a state

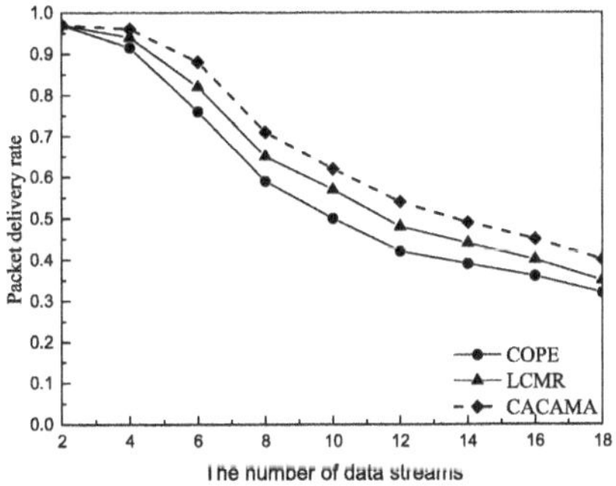

Fig. 4. Packet delivery rate under different number of data streams

of congestion or congestion soon. Therefore, the path chosen by source node does not contain congestion nodes, which can reduce data packet loss due to congestion. CACAMA can maintain high packet delivery rate under high load.

5 Conclusion

In this paper, a multi-path routing algorithm (CACAMA) for coding-aware load balancing in wireless mesh networks is proposed, and the routing metric CACM consider path actual coding gain and load and interference is basis on routing metric LCRM. It can exclude the congestion sensing mechanism of high load nodes and congestion nodes in the stage of routing discovery. Simulation results show that CACAMA has significant advantages in improving network throughput, reduce network end-to-end delay and improving packet delivery rate.

References

1. Akyildiz, I.F., Wang, X., Wang, W.: Wireless mesh networks: a survey. Comput. Netw. **47** (4), 445–487 (2005)
2. Akyildiz, I.F., Wang, X.: A survey on wireless mesh networks. IEEE Commun. Mag. **43**(9), S23–S30 (2005)
3. Ahlswede, R., Cai, N., Li, S.Y.R., et al.: Network information flow. IEEE Trans. Inf. Theory **46**(4), 1204–1216 (2000)
4. Katti, S., Rahul, H., Hu, W., et al.: XORs in the air: practical wireless network coding. IEEE/ACM Trans. Netw. **16**(3), 497–510 (2008)
5. Ni, B., Santhapuri, N., Zhong, Z., et al.: Routing with opportunistically coded exchanges in wireless mesh networks. In: IEEE Workshop on Wireless Mesh Networks, 2006. Wimesh 2006, pp. 157–159. IEEE (2006)
6. Prashanthi, V., Babu, D.S., Rao, C.V.G.: Network coding-based communication in wireless ad-hoc networks. In: International Conference on Communications and Signal Processing, pp. 1040–1044. IEEE (2014)
7. Yazdanpanah, M., Assi, C., Sebbah, S., et al.: Does network coding combined with interference cancellation bring any gain to a wireless network? IEEE/ACM Trans. Netw. **23** (5), 1485–1500 (2015)
8. Kok, G.X., Chow, C.O., Ishii, H.: Improving network coding in wireless ad hoc networks. Ad Hoc Netw. **33**(C), 16–34 (2015)
9. Le, J., Lui, J.C.S., Chiu, D.M.: DCAR: distributed coding-aware routing in wireless networks. IEEE Trans. Mob. Comput. **9**(4), 596–608 (2010)
10. Fan, K., Li, L.X., Long, D.Y.: Study of on-demand COPE-aware routing protocol in wireless mesh networks. J. Commun. **30**(01), 128–134 (2009)
11. Shen, X.J., Chen, Z.G., Liu, L.: Load balancing multicast routing based on network coding in wireless mesh network. J. Commun. **36**(4), 89–95 (2015)
12. Chen, H.G., Li, H.X., Han, S., et al.: Network coding-aware multipath routing in multi-hop wireless networks. J. Softw. **21**(08), 1908–1919 (2010)
13. Xing, S., Wang, R., Huang, H., et al.: Load balanced coding aware multipath routing for wireless mesh networks. Chin. J. Electron. **24**(1), 8–12 (2015)

FPGA Implementation of Rate Compatible Non-binary LDPC Codes

Deng Zhibin[1], Meng Jiahui[2(✉)], Wang Ziyong[1], Zhang Liang[2], and Gao Jingpeng[2]

[1] Wuhan Maritime Communication Research Institute, Wuhan 430000, China
[2] College of Information and Communication Engineering, Harbin Engineering University, Harbin 150001, China
mengjiahui@hrbeu.edu.cn

Abstract. In order to solve the problem of the requirements of communication on UAN, an improved construction of Rate-compatible Low-Density Parity-Codes (RC-LDPC) based on progress edge growth (PEG) is proposed. The algorithm uses high bit rate to low bit rate compatible with the way to achieve multi-rate, after constructing the master code with high bit rate, we continue to use an improved PEG algorithm to increase the decoding performance of the master code by increasing the large ring in the matrix. The simulation shows that this code covered multi-rates, and the performance of this code is better than other traditional LDPC codes with single rate. For this code, based on exclusive OR (XOR) gate array and random access memory (RAM) combined with the coding architecture, reduce the encoder for the on-chip resource occupancy, to achieve multi-code rate multi-code length switch. The encoder was implemented on the chip of Cyclone IV with Verilog HDL language. The reports show that the encoder reduce the application of the cyclic shift memory and the occupancy of the chip resource at the same time.

Keywords: Rate-compatible Low-Density Parity-Codes
Progress edge growth · Field-programmable gate array · Shift register

1 Introduction

In 1962, Gallager proposed low density parity codes (LDPC) in his doctoral dissertation [1]. In 1996, scholars Mackay and Neal rediscovered the LDPC code proposed by Gallager, and found that its performance is close to the Shannon limit, making LDPC code fast known as the research focus in the field of channel coding [2]. At the end of 1980s, Hagenauer scholars proposed rate compatible punctured convolutional code (RCPC). After encoding, the coed words in the convolutional code are subjected to different puncturing matrices to obtain different codewords [3]. RCPC code to solve the problem of variable bit rate at the same time, through the punctuation matrix to adjust the rate of bit error correction code for the development of a new idea. As proposed in 2000, rate compatible punctured turbo (RCPT) code [4]. In 2002, the rate compatible low density parity check (RC-LDPC) code proposed by the scholar Li

K. Deng et al. (Eds.): ICMIR 2018, AISC 856, pp. 57–64, 2019.
https://doi.org/10.1007/978-3-030-00214-5_7

proved that the LDPC code is only 1 dB from the Shannon limit in the hybrid automatic repeat quest (HARQ) system from the point of view of throughput [5]. Since 2004, Ha [6, 7] and Tian [8] scholars have studied the RC-LDPC codes. Under a moderate rate mother code condition, only one pair of coders can achieve arbitrary rate of change at low bit rate to high bit rate, almost without increasing the complexity of the system.

LDPC code is applied in space, optical fiber, personal communication system, asymmetric digital subscriber line (ADSL) and magnetic recording equipment for its excellent performance, concise form and good application prospect. Therefore, the LDPC code and the RC-LDPC codes play an important role in the wireless mobile communication system. In wireless mobile communication system, most of the channels are time-varying. In order to achieve greater throughput, communication systems can adaptively change the code length and code rate of error correction codes according to the channel conditions. LDPC code can adaptively change the corresponding bit rate, change the dynamic range of bit rate, adapt to the changing channel environment, ensure bit error rate and improve transmission reliability.

The traditional method of variable code rate is to select different coders when selecting different code rates, but at the same time increase the system complexity. In order to improve this shortcoming, a coding method for obtaining a different bit rate is proposed for a mother code by deleting, extending, and shortening. The method only needs to design a pair of coders based on the mother code, and the other newly generated sub-codes of the various bit rates can be successfully decode by the decoder of the mother code. A series of variable bit rate LDPC codes generated by this method are called code RC-LDPC codes.

At present, according to the structure characteristics of RC-LDPC code, it can be divided into three types: Punctured, Extended, Shortened. The advantages and disadvantages of the above three types of RC-LDPC codes and the scope of application are summarized, as shown in Table 1.

Table 1. Comparison of different types of RC-LDPC codes

Type	Advantages	Disadvantages	Rate-compatible
RCP-LDPC	Only one checksum matrix is stored	The punctured node will still participate in the encoding	Compatibility of low bit rate mother code to high bit rate
RCE-LDPC	Only when the bit rate is reduced will only add the new check node to code, reduce the waste	Different check matrixes may be stored for different code rates	Compatibility of high bit rate mother code to low bit rate
RCS-LDPC	Only one checksum matrix is stored	The information may be pre-segmented	Compatibility of high bit rate mother code to low bit rate

2 Construction and Performance Analysis of Linear Time Coded RC-LDPC Codes

In order to facilitate hardware implementation, we use multi bit rate from high bitrate to low bit rate compatible, which is different from RCS-LDPC code. After constructing high bit rate mother code, we use an improved PEG algorithm [10, 11]. The algorithm gradually adds the edge of the bit node to the checkpoint node in the Tanner graph, so that the ring in the Tanner graph is always kept in a large state.

It is assumed that the degree sequence of the bit node is $D_v = \left\{d_{v_0}, d_{v_1}, \ldots, d_{v_{(n-1)}}\right\}$, where d_{v_j} is the degree of the bit node v_j. The degree sequence of the check node is $D_c = \left\{d_{c_0}, d_{c_1}, \ldots, d_{c_{(m-1)}}\right\}$, where d_{c_j} is the degree of the check node c_j. E is a set of edges, $E = E_{v_0} \cup E_{v_1} \cup \cdots \cup E_{v_{(n-1)}}$, where E_{v_j} contains all the edges to connect to v_j, and the side of k that connects v_j is expressed as $E_{v_j}^k$, $0 \le k \le d_{v_j} - 1$. Define a set of check nodes contained in a tree of depth l extended with a bit node v_j as a root node as $N_{v_j}^l$, with a complement of $\overline{N}_{v_j}^l$, then $N_{v_j}^l + \overline{N}_{v_j}^l = S_c$.

Based on the above definition of symbols, the specific steps of the PEG algorithm are given.

The first step, the use of PEG algorithm to construct a high rate of the mother matrix;

The second step, the short ring in the mother matrix is search for, and the number of short rings in each series is counted, arranged in descending order of the number of rings, and the column labels included in the short ring are stored;

The third step, using the PEG algorithm, adds new edges to the short loops with different included labels so that they from a lager loop;

The fourth step, search the newborn short ring matrix, repeat the first step until the rate of the newborn matrix code rate up to the request;

The fifth step, the number of short rings is further reduced by shifting the order of the newly constructed regions.

As shown in Fig. 1, an improved PEG algorithm is given.

In order to better study the performance of PEG-RC-LDPC constructed in this paper, its information sequence length is 36 bit, code rate covers 7/8, 3/4, 2/3, 1/2. Taking the 3/4 code rate as the mother matrix, an extended structure was constructed and compared with QC-LDPC codes of 1/2 code rate performance and several similar parameters. The parameters are set as follows: the encoder uses BPSK modulation, the channel is AWGN channel, the mean value is zero variance σ_N^2, the maximum iteration number is 25 times. The system simulation block diagram is shown in Fig. 2.

As can be seen from Fig. 3, from the perspective of code rate, the performance of the low rate matrix generated by the PEG-RC-LDPC construction method is slightly better than that of the QC-LDPC code under the same parameter conditions.

From Fig. 4, we can see that from the perspective of code length, LDPC decoding is superior to convolutional code in the same parameter condition and decoding algorithm using BP decoding algorithm, and the longer the code length is, the better the performance is. The performance of the improved PEG-RC-LDPC algorithm presented in this paper is superior to the traditional QC-LDPC algorithm under the same condition.

for($j = 0, j \le n-1, j++$)
{for($k = 0, k \le d_{v_j -1} - 1, k++$)
{if($k = 0$)
{ $E^0_{v_j} \leftarrow \text{edge}\left(c_i, v_j\right)$,where $E^0_{v_j}$ is the first edge incident to v_j, and c_i is a check node having the lowest check-node degree under the current graph setting $E_{v_0} \cup E_{v_1} \cup \ldots \cup E_{v_{(j-1)}}$.}
else
{expanding a tree from symbol node v_j up to depth 1 under the current graph setting such that $\overline{N}^{l}_{v_j} \ne \varnothing$ but $\overline{N}^{l+1}_{v_j} \ne \varnothing$,or the cardinality of $N^l_{v_j}$ stops increasing but is less than m, then $E^k_{v_j} \leftarrow \text{edge}\left(c_i, v_j\right)$, where $E^k_{v_j}$ is the kth edge incident to v_j ,and c_j is a check node picked from the set $\overline{N}^{l}_{v_j}$ having the lowest check-node degree .}}}
end

Fig. 1. An improved PEG algorithm

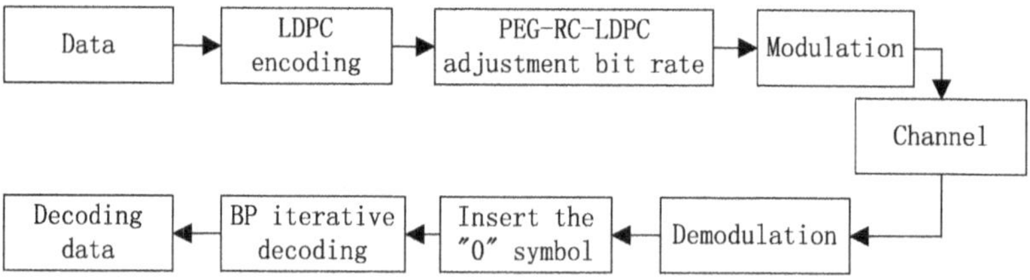

Fig. 2. The system simulation block diagram

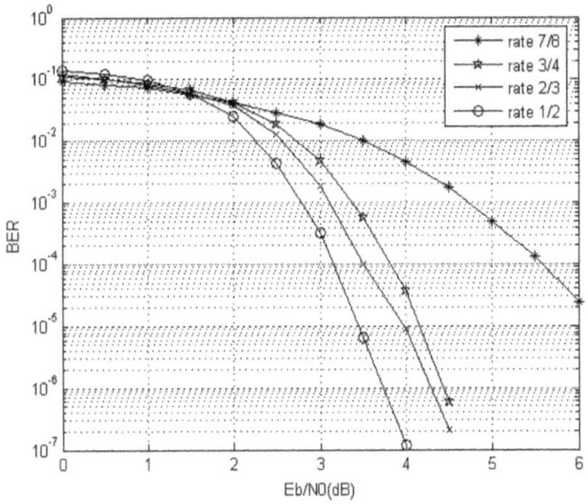

Fig. 3. Performance of information frame length 288 bits curve family constructed by PEG-RC-LDPC algorithm

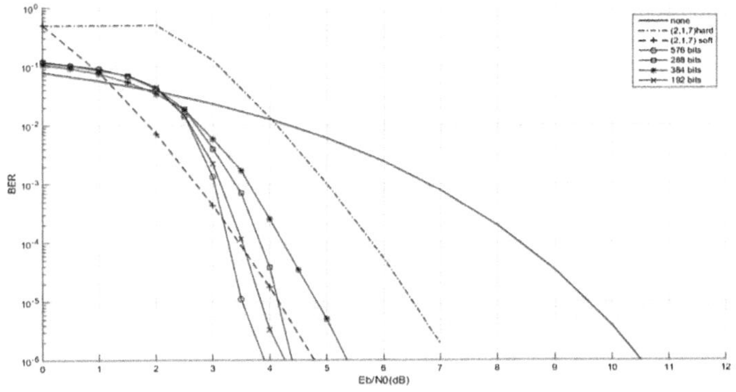

Fig. 4. Comparison between PEG-RC-LDPC algorithm and QC-LDPC algorithm

3 Hardware Implementation and Performance Analysis of RC-LDPC Encoder

In this paper, the Quasi-cyclic LDPC code (QC-LDPC) is used to encode. The encoding of QC-LDPC codes can be completed in the linear time with shift register, and it is convenient for partial parallel decoding. Therefore, it has more advantages in implementation. By generating cyclic matrix, we can get the generating matrix of cyclic row form. Based on the generation matrix of QC-LDPC code cyclic determinant form, a variety of encoders can be designed with a simple shift register [12].

The check matrix $\boldsymbol{H}$ of LDPC code is $m \times n$ dimension and full rank, an information sequence s of length k, a check matrix p after encoding, and a generator matrix $\boldsymbol{G}$ of $k \times n$ are input. The check matrix $H = [H_1 H_2]$, using the orthogonality $HG^T = 0$ between $\boldsymbol{H}$ and $\boldsymbol{G}$, which can be obtained as follows:

$$HG^T = [H_1 H_2] \cdot \begin{bmatrix} s^T \\ p^T \end{bmatrix} = H_1 s^T + H_2 p^T = 0 \tag{1}$$

$$pH_2^T = sH_1^T \tag{2}$$

sH_1^T is recorded as q, and the information sequence q and check sequence p are rewritten to $q = (q_1, q_2, \ldots, q_m)$ and $p = (p_1, p_2, \ldots, p_m)$ according to the length of the cyclic block, then the formula (2) is as follows:

$$(p_1, p_2, \ldots, p_m) \cdot \left(h_{21}^T, h_{22}^T, \ldots, h_{2m}^T\right) = (q_1, q_2, \ldots, q_m) \tag{3}$$

$$\begin{aligned} p_1 &= q_1 \oplus p_1 \oplus ph_{21}^T \\ p_2 &= q_2 \oplus p_2 \oplus ph_{22}^T \\ &\vdots \\ p_m &= q_m \oplus p_m \oplus ph_{2m}^T \end{aligned} \tag{4}$$

H_2 is a lower trigonometric simulation with a main diagonal line of 1 and full rank. In the formula (4), the encoder can be converted into a pre (i − 1) term module 2 accumulator on the check sequence, so the encoding process of the encoder is divided into two stages:

The information sequence s is multiplied by H_1^T, and the intermediate sequence q is calculated and stored.

Using QC-LDPC code check matrix, the cyclic block size is $b \times b$. According to the coding characteristic of QC-LDPC, the structure of the cyclic block is also determined when the offset $a_{i,j}$ of the first row or the first column of any non-zero cyclic block $A_{i,j}$ in the parity check matrix is determined. This article uses a fixed column offset, $0 \leq a_{i,j} \leq b$, so that $s = (s_1, s_2, \ldots, s_t)$, $H_1 = (h_{11}, h_{12}, \ldots, h_{1t})$. From the previous derivation of $q = sH_1^T$, we continue to deduce the $q^T = H_1 s^T$ and expand to:

$$s_1 \times h_{11} \oplus s_2 \times h_{12} \oplus \cdots \oplus s_t \times h_{1t} = q^T \tag{5}$$

By formula (5), we can get the value of q^T by summation of every column and corresponding information bits in check matrix H_1, so that the XOR processing module can be completed through a RAM module in the first stage. At the same time, according to the length of s^T, control and match it, and calculate the number of columns in check matrix H_1, we can get multiple information sequence length encoders, and achieve multiple code lengths.

The results of the middle sequence q and $p_i \oplus ph_{2i}^T$ are calculated, that is to find the check sequence p.

Let the i line j column element in H_2^T be $b_{i,j}$, $1 \leq i \leq m$, $1 \leq j \leq m$, then $h_{2i}^T = (b_{i,1}, b_{i,2}, \ldots, b_{i,i-1}, 1, 0, \ldots, 0)^T$, so the formula (4) is rewritten as:

$$p_i^T = \begin{cases} q_i^T, i = 0 \\ q_i^T \oplus b_{i,1} \times p_1^T \oplus b_{i,2} \times p_2^T \oplus \cdots \oplus b_{i,i-1} \times p_{i-1}^T, 0 < i \leq m \end{cases} \tag{6}$$

It is known by formula (6) that p_m is only related to q_m and $(q_1, q_2, \cdots, q_{m-1})$, so the calculated p_m overlay data on $(q_1, q_2, \cdots, q_{m-1})$, which has no effect on the generation of the checkout sequence. Therefore, the XOR module calls the same RAM area in the above two stages to reduce the occupancy of some resources. When generating check sequence P, according to the length of check column required for corresponding bit rate, the computation sequence of XOR processing module can be finished ahead of time, and then the check sequence with high bit rate can be generated, and multiple bit rate can be implemented.

In order to further reduce the consumption of encoder for on-chip resources and reduce the resources occupied by XOR array and matrix storage, the encoder adopts a coding architecture based on XOR gate array and RAM.

Using the Verilog HDL language in Quartus II 13.1, using the Cyclone IV series chip. The code length of the information sequence for 24 bit and 36 bit is designed, which covers the encoder of the 1/2, 2/3, and 3/4 code rates. Because of the large number of calls to Block RAM resources, it greatly reduces the occupation of logical resource classes and effectively reduces the proportion of resources (Table 2).

Table 2. The coder core coding unit resource occupancy table for the information sequence length 36 bit

Resource type	Number	The proportion of /%
Total combinational functions	587	<1%
Dedicated logic registers	167	<1%

4 Conclusion

In this paper, an improved RC-LDPC code construction method based on PEG algorithm is proposed. By this method, an RC-LDPC code with a lower triangular structure and compatible with multi-code rate and easy hardware implementation is constructed and guaranteed to be good performance. In this article, an encoder architecture is proposed by calling RAM. By calling ROM and RAM for the encoding output, greatly reducing the shift register and XOR gate array for the occupation of resources. The generating matrix of the check matrix does not need to be obtained, and only the matrix corresponding to the check sequence in the check matrix needs to be transposed. As a result, the hardware complexity is reduced and the coding efficiency is also improved.

References

1. Gallager, R.: Low Destiny Parity-Check Codes. Doctoral dissertation, MIT Press (1963)
2. Mackay, D.J., Neal, R.M.: Near Shannon limit performance of low density parity check codes. Electron. Lett. **32**(18), 1645–1646 (1996)
3. Hagenauer, J.: Rate-compatible punctured convolutional codes (RCPC codes) and their applications. IEEE Trans. Commun. **36**(4), 389–400 (1998)
4. Rowitch, N., Milstein, L.B.: On the performance of hybrid FEC/ARQ systems using rate compatible punctured turbo (RCPT) codes. IEEE Trans. Commun. **48**(6), 948–959 (2000)
5. Li, J., Narayanan, K.: Rate-compatible low density parity check codes for capacity approaching ARQ scheme in packet data communications. In: International Conference on Communication, Internet, and Information Technology (CIIT) (2002)
6. Ha, J., Kim, J.: Rate-compatible puncturing of low-density parity-check codes. IEEE Trans. Inf. Theory **50**(11), 2824–2836 (2004)
7. Ha, J., Kim, J.: Rate-compatible puncturing of low-density parity-check codes with short block lengths. IEEE Trans. Inf. Theory **52**(2), 728–738 (2006)
8. Klinc, D., Ha, J.: On rate-adaptability of nonbinary LDPC codes. In: 5th International Symposium on Turbo Codes and Related Topics, pp. 231–236 (2008)
9. Yazdani, M.R., Banthashemi, A.H.: On construction of rate-compatible low-density Parity-check codes. IEEE Commun. Lett. **8**(3), 159–161 (2004)
10. Zhang, J.: Research on the construction of quasi cyclic LDPC code based on PEG algorithm. Electron. Dev. **35**(6), 647–651 (2012)
11. Xiao, H., Banthshemi, A.H.: Improved progressive-edge-growth (PEG) construction of irregular LDPC codes. In: Global Telecommunications Conference, 2004. GLOBECOM 2004, vol. 1, pp. 489–492. IEEE (2004)
12. Li, Z., Chen, L., Zeng, L., et al.: Efficient encoding of quasi-cyclic low-density parity-check codes. IEEE Trans. Commun. **54**(1), 71–81 (2006)

Rnk New Algorithm Based on N-Cut Graph Segmentation

Xiaomin Xie, Fan Zhang(✉), Xingwang Wang,
Yong Zeng, and Changkai Li

Institute of Electronic Communication Engineering,
Anhui Xinhua University, Hefei 230088, China
357106305@qq.com, salas0527@163.com

Abstract. In this paper, we propose a new Rnk segmentation algorithm based on N-cut criterion, which is more reasonable than common segmentation criteria. The algorithm can improve the existing segmentation by changing the vertex pairs of N-cut values continuously, and finally get the optimal segmentation. A hashing technique is proposed to improve the efficiency of finding the optimal exchange vertex pairs. When the graph is dense matrix, the improvement effect is especially obvious. The experimental results of random graph segmentation show that the proposed algorithm is more reasonable and faster than the traditional KL algorithm.

Keywords: Partitioning of right graph · N-cut criterion
Rnk segmentation algorithm · Hashing technique

1 Introduction

Graph segmentation, also known as graph aggregation problem, is a method of dividing graphs into several sub-modules according to a certain standard. Graph segmentation problem has been proved to be NP-complete problem, and most of the existing studies seek to obtain suboptimal solutions within a tolerable range. In practical application, heuristic algorithm is often used to solve this kind of problem. In recent years, there are many methods to study graph segmentation problems, such as iterative methods [1, 3] to improve the segmentation effect continuously, greedy segmentation method [4] based on graph "potential energy", segmentation method [5] based on partition strategy, graph-based clustering method [6] and graph segmentation method [2, 7, 8] based on maximum flow minimum cut theory and minimum cut tree, and so on. Clustering method and minimum cut tree method are both the problem of transforming the segmentation problem into the problem of finding the equivalent optimal eigenvalue and linear programming. At the same time, there are many studies on the criteria for evaluating segmentation effect, such as minimum cost, minimum diameter, k-center, k-median criteria, etc. In the new algorithm proposed in this paper, more reasonable segmentation evaluation N-cut criteria [9–11] are used, and the idea of traditional KL algorithm is used to continuously improve the random initial segmentation until the optimal solution is obtained.

K. Deng et al. (Eds.): ICMIR 2018, AISC 856, pp. 65–71, 2019.
https://doi.org/10.1007/978-3-030-00214-5_8

2 Analysis of Graph Segmentation Evaluation Criteria

The graph $G = (V, E)$ is divided into subsets A and B $(A \cup B = V, A \cap B = \emptyset)$, first of all, there is a standard to evaluate the quality of segmentation. A good segmentation should make the coupling degree of vertices in the sub-graph after segmentation the highest relative to other segmentation, and the coupling degree between sub-graphs is the lowest. There are many general evaluation criteria for segmentation quality, such as minimum radius, minimum cost, etc., but these evaluation criteria are not the most reasonable.

KL algorithm is an algorithm to optimize graph segmentation under the minimum cut criterion, even if cut(A,B) $= \sum_{u \in A, v \in B} w(u, v)$ is minimized, the criterion only takes into account the minimum degree of coupling between sub-graphs, and ignores the node coupling in the module. According to the objective function, although the goal of minimum cutting cost can be achieved, but contrary to the standard of optimal segmentation, may be formed in the sub-module relationship independent group, contrary to the requirement of close relationship within the same sub-graph, so that the overall cannot reach the optimal segmentation. New segmentation criteria are proposed. For convenience, the following definitions are made:

$$\text{The cutting cost of the figure is: } \mathrm{cut(A, B)} = \sum_{u \in A, v \in B} w(u, v) \tag{1}$$

$$\text{The cost of subgraph } A \text{ is: } \mathrm{assoc(A, V)} = \sum_{u \in A, t \in V} w(u, t) \tag{2}$$

Define the N-cut criteria for evaluating the advantages and disadvantages of partitioned subgraph (A, B) as follows:

$$\mathrm{Ncut(A,B)} = \frac{\mathrm{cut(A,B)}}{\mathrm{assoc(A,V)}} + \frac{\mathrm{cut(A,B)}}{\mathrm{assoc(B,V)}} \tag{3}$$

Criteria for evaluate that effect of secant a on the degree of coupling between vertex of subgraph are also defined:

$$\mathrm{Nassoc(A,B)} = \frac{\mathrm{assoc(A,A)}}{\mathrm{assoc(A,V)}} + \frac{\mathrm{assoc(B,B)}}{\mathrm{assoc(B,V)}} \tag{4}$$

Equations 3 and 4 both reflect the extent to which segmentation affects vertex coupling. Ncut(A,B) lower value indicates that vertices with high coupling are assigned to one subgraph and vertices with low coupling are assigned to different subgraph. A larger value of Nassoc(A,B) indicates that a vertex with a high degree of couple is assigned to that same subgraph. Then a more ideal segmentation should be to find the segmentation that makes Ncut(A,B) minimum and Nassoc(A,B) maximum among all the partitions of the graph.

Since Ncut(A,B) = 2 – Nassoc(A,B), it can be seen that the two criteria are equivalent. The Ncut(A,B) standard considers not only the coupling between vertices in the same module in the graph, but also the coupling between vertices in different modules in the graph. It is an ideal standard for evaluating the segmentation quality. In this paper, Ncut(A,B) is used as the standard for evaluating the segmentation cost. Although standard Ncut(A,B) may incur higher cut costs than minimum cut criteria, the overall cost is still lower than the resulting subgraph size.

3 Definition of Problem

The general module partitioning problem can be modeled as a partitioning problem for undirected weighted join graph $G = (V, E)$, where V represents a set of nodes, E represents a set of possible interconnections between each node, and for each edge $(u, v) \in E$ in the graph, a weight $w(u, v)$ represents a join cost and a weight function $w : E \to R$. For set S, the number of nodes contained in the set is defined as the potential of the subset, and it is noted that the potential of set S is $|S|$. Defines the state in which the graph is divided into (A, B) as the cut state of the graph.

The above graph segmentation problem can be expressed as:

Input: undirected weighted join graph $G = (V, E)$ after modeling.

Output: divides the graph into subsets A and B that satisfy the area balance condition ($A \cup B = V$, $A \cap B = \varnothing$).

In order to control that number of node contained in sets A and B, the balance factor $r(r = {}^{|A|}/_{|V|})$ of the control area is set, here a is taken as 0.5, i.e. an aliquot.

Objective: to minimize the cost $\text{Ncut(A,B)} = \frac{\text{cut(A,B)}}{\text{assoc(A,V)}} + \frac{\text{cut(A,B)}}{\text{assoc(B,V)}}$ (N-cut criterion) of cut (A, B) of a graph.

4 Description of Rnk Algorithm

4.1 Introduction of KL Algorithm

In view of the algorithm proposed in this paper, using the basic idea of KL algorithm and related terms, KL algorithm is briefly introduced in this section. First define the relevant formula, assuming that the vertex set of fig. G is divided into subsets P1, P2, for any vertex $u \in V(G)$, the outer cost E_u and the inner cost I_u of vertex u are defined respectively as follows:

$$E_u = \sum_{v \in p_i} w(u, v), \quad i = 1 \text{ or } 2,\ u \notin p_i,\ (u, v) \in E(G)$$

$$I_u = \sum_{v \in p_i} w(u, v), \quad i = 1 \text{ or } 2,\ u \in p_i,\ (u, v) \in E(G)$$

And defines the gain $D_u = E_u - I_u$ of vertex u, which represents a change in the gain of moving vertex u to another set, assuming $u \in$ P1, $v \in$ P2, then defines the amount of gain change that exchange u, v brings to the whole

$$\text{Gain}_{(u,v)} = D_u + D_v - 2 * w(u, v) \tag{5}$$

Suppose $D_K^{'}$ is defined as the D- domain of the leading node k of the u after the exchange vertex pair (u, v), then

$$D_K^{'} = D_k + 2 * w(v, k) - 2 * w(u, k) \tag{6}$$

4.2 A New Rnk Algorithm

From the discussion in the second section, it can be seen that the method of optimizing graph segmentation according to minimum cut cost as objective function is not ideal. In this paper, a new algorithm of graph segmentation based on N-cut criterion is proposed.

Assuming that the two vertices $u \in A$, $v \in B$ in the graph are exchanged u and v, the gain change is $g_{(u,v)}$, and the corresponding new cut cost is $\text{Ncut}_{\text{new}}(\text{A}, \text{B})$. And a new assoc(A,V) value, expressed as $\text{assoc(A,V)}_{\text{new}}$, is obtained.

$$\text{Make } D_{(u,v)} = E_u - I_v - w(u, v) \tag{7}$$

Theorem 1: after exchange, get:

$$\text{cut(A,B)}_{\text{new}} = \text{cut(A,B)}_{\text{old}} - g_{(u,v)} \tag{8}$$

$$\text{assoc(A,V)}_{\text{new}} = \text{assoc(A,V)}_{\text{old}} - D_{(u,v)} \tag{9}$$

$$\text{Ncut}_{\text{new}}(\text{A}, \text{B}) = \frac{\text{cut(A,B)}_{\text{old}} - g_{(u,v)}}{\text{assoc(A,V)}_{\text{old}} - D_{(u,v)}} + \frac{\text{cut(A,B)}_{\text{old}} - g_{(u,v)}}{\text{assoc(B,V)}_{\text{old}} - D_{(u,v)}} \tag{10}$$

If $\text{Ncut}_{(a_k,b_k)}\text{(A,B)} \Leftarrow \text{Ncut(A,B)}$ indicates that Ncut(A, B) has reached an optimal value. By improving the Ncut(A,B) value of the segmentation, the algorithm finally achieves the optimal segmentation on the whole. Rnk algorithm flow chart is as follows (Fig. 1).

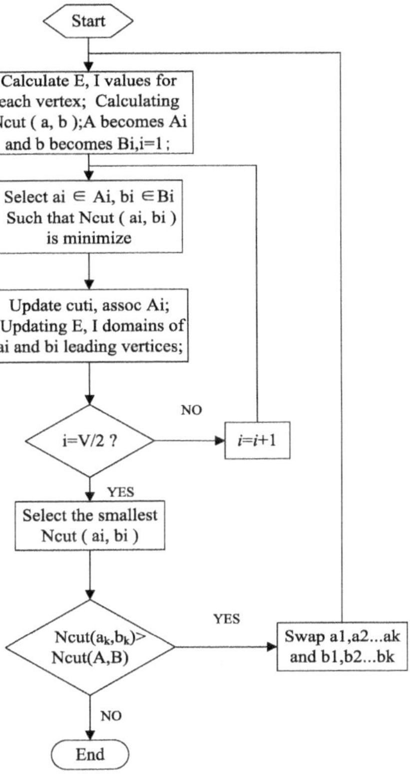

Fig. 1. Rnk algorithm flow chart

5 Experimental Results and Analysis

Table 1 shows the results of two algorithms for random graph partitioning. KL, Rnk (the algorithm in this paper) are implemented using C language and compared with each other. VC is used to compile the program, the running environment is in the CPU frequency of 2.7 GHz, 512 MB of memory, operating system for windows XP PC execution. The initial partition of each graph is random and then optimize using two algorithm.

From the above experiments, we can know that Rnk algorithm does not necessarily get the minimum cut cost, but under the more reasonable Ncut evaluation criteria, the algorithm can get a better segmentation than KL algorithm. Because the algorithm may fall into local optimal solution, Rnk algorithm does not get optimal solution compared with KL in some cases. The Rnk algorithm and KL algorithm get the same optimal solution when the optimal node found by KL algorithm and the optimal node found by this algorithm are the same.

Table 1. Output results and data comparison

Random graph	Cutting cost			N-cut		
	KL algorithm	RNK algorithm	Improvement ratio (%)	KL algorithm	RNK algorithm	Improvement ratio (%)
V = 35, N = 260	152.95	152.95	0	1.267	1.267	0
V = 75, N = 580	695.55	678.58	2.88	1.408	1.287	1.97
V = 110, N = 496	130.78	129.88	0.69	0.827	0.823	0.61
V = 110, N = 995	340.10	345.51	−1.59	1.004	1.016	−1.18
V = 110, N = 2449	1672.56	1668.92	0.32	1.243	1.242	0.092
V = 210, N = 995	258.67	255.77	1.12	0.813	0.806	0.87
V = 210, N = 12058	4636.85	4622.85	0.52	1.419	1.417	0.26
V = 210, N = 18026	7430.24	7430.24	0	1.264	1.264	0
V = 410, N = 4604	1355.6	1279.05	5.65	1.010	0.968	4.34
V = 410, N = 25916	10395.3	10426.4	1.96	1.214	1.209	1.54
V = 410, N = 41913	18219.6	18215.8	0.021	1.251	1.250	0.8
V = 820, N = 8363	1882.77	1845.84	1.96	0.921	0.907	1.54
V = 820, N = 58560	35612.4	35690.0	−0.22	1.235	1.237	−0.16
V = 820, N = 137922	59207.00	59098.8	0.18	1.264	1.262	0.16

6 Concluding Remarks

In this paper, an improved graph segmentation algorithm based on KL algorithm is proposed by using a more reasonable criterion for evaluating the graph segmentation effect than the minimum cut. Aiming at the problem that the algorithm is time-consuming to find the optimal vertex pair, a hash technique is proposed to improve the search efficiency. Experiments show that this method can obtain better segmentation than KL algorithm. For the problem that the algorithm may fall into local optimal solution, further research and analysis are needed and improved.

Fund Project. Special Funding Project of China Postdoctoral Science Foundation (2014T70967); Natural Science Research Key Project of Anhui Province Higher School (KJ2017A630); Quality Engineering Project of Anhui Provincial (2016jxtd055); Key Construction Discipline Project at College Level of Anhui Xinhua University (zdxk201702); Institute Project at College Level of Anhui Xinhua University (yjs201706).

References

1. Kernighan, B.W., Lin, S.: An efficient heuristic procedure for partitioning graphs. Bell Syst. Tech. J. **1970**(49), 291–307 (1970)
2. Shi, J., Malik, J.: Normalized cuts and image segmentation. IEEE Trans. Pattern Anal. Mach. Intell. **22**(8), 888–905 (2000)
3. Ren, X., Malik, J.: Learning a classification model for segmentation. In: Proceedings of the 9th IEEE International Conference on Computer Vision, pp. 10–17 (2003)
4. Molinari, F., Zeng, G., Suri, J.S.: Greedy technique and its validation for fusion of two segmentation paradigms leading to an accurate IMT measure in plaqued carotid arterial ultrasound. J. Vasc. Ultrasound **34**(2), 63–73 (2010)
5. Ottaviano, G., Venturini R.: Partitioned Elias-Fano indexes. In: Proceedings of the 37th International ACM SIGIR Conference on Research and Development in Information Retrieval, pp. 273–282 (2014)
6. Van Dongen, S.M.: Graph clustering by flow simulation. Ph.D. thesis, University of Utrecht (2000)
7. Felzenszwalb, P.F., Huttenlocher, D.P.: Efficient graph-based image segmentation. Int. J. Comput. Vis. **59**(2), 167–181 (2004)
8. Han, X., Fu, Y., Zhang, H.: A fast two-step marker-controlled watershed image segmentation method. In: Proceedings of 2012 IEEE International Conference on Mechatronics and Automation, Chengdu, China, 5–8 August 2012
9. Felzenszwalb, P.F., Huttenlocher, D.P.: Efficient graph-based image segmentation. Int. J. Comput. Vis. **59**(2), 167–181 (2004)
10. Husain, R.A., Zayed, A.S., Ahmed, W.M., Elhaji, H.S.: Image segmentation with improved watershed algorithm using radial bases function neural networks. In: 2015 16th International Conference on Sciences and Techniques of Automatic Control and Computer Engineering, pp. 121–126 (2015)
11. Browet, A., De Vleeschouwer, C., Jacques, L., Mathiah, N., Saykali, B., Migeotte, I.: Cell segmentation with random ferns and graph-cuts. In: 2016 IEEE International Conference on Image Processing, pp. 4145–4149 (2016)

Different Ways of Research on Pricing Done Math Model Base on Big Data

Chenghui Yang[1,2,3,4(✉)], Xing Peng[5], Zhongfu Luo[5], and Zeyi Liu[5]

[1] College of Electrical Engineering, Northwest Minzu University, No. 1, Northwest Xincun, Lanzhou 730030, Gansu, China
690393099@qq.com

[2] China National Institute of Information Research, No. 1, Northwest Xincun, Lanzhou 730030, Gansu, China

[3] School of Automation and Electrical Engineering, Lanzhou Jiaotong University, No. 188, AnNing West Rode, Lanzhou 730070, Gansu, China

[4] Gansu Provincial Engineering Research Center for Artificial Intelligence and Graphics and Image Processing, No. 188, AnNing West Rode, Lanzhou 730070, Gansu, China

[5] School of Civil Engineering, Northwest Minzu University, No. 1, Northwest Xincun, Lanzhou 730030, Gansu, China

Abstract. The use of MATLAB, EXCEL and other software for the classification of data presentment, the completion of the task, through the comparison between the region and the price to find out the reasons for the task is not completed. Combined with the workflow scheduling framework, we take the solution of releasing crowdsourced packages to APP members by combining the tasks together, and combining with the operation flow of crowdsourcing mode, and using the reverse-oriented pricing method to get a new pricing method. Determine the optimal solution, come to the unfinished task area new pricing, improve the completion of the task rate, solve the "photo shoot money" APP task pricing problem.

Keywords: Operation process · Crowdsourcing · Utility network model MATLAB

1 Introduction

"Photo Earning Money" is a self-service service based on the mobile Internet. Users download the APP, register as a member of the APP, then take a photo-taking task on the APP, and earn a nominal reward for the task after completing the established task. This self-service crowd sourcing platform based on the mobile Internet provides businesses with a variety of commercial inspections and information gathering, which can greatly save the investigation cost compared with the traditional market research methods.

In this paper, through the establishment of a model and specific algorithms, crowd sourcing strategy, combined with the development of system platform to solve a new task pricing.

K. Deng et al. (Eds.): ICMIR 2018, AISC 856, pp. 72–82, 2019.
https://doi.org/10.1007/978-3-030-00214-5_9

Based on the obtained results of clustering analysis model, decision tree and linear fitting, reduction method and price game theory, this paper establishes a new pricing model by combining some advantages and characteristics of this paper. The one hand, so that many APP members can easily earn a nominal fee and access to a large number of members, on the other hand effectively ensure the authenticity and reliability of the survey data to shorten the investigation cycle.

Therefore, APP has become the core of the platform operation, and the task of APP pricing is its core elements. If the pricing is unreasonable, some tasks will be unattended, which led to the completion of the task failed. You need to find out the pricing rules, establish a model, the use of the optimal program, the final unfinished mission area to obtain new pricing, improve the completion of the task rate.

2 Modeling Preparation

It is better to face the rationality of pricing provided by members and effectively complete the utility grid model, as shown in Fig. 1. The utility grid is made up of several self-made task providers SP. Each SP has a set of program operation mechanism to ensure the maximum benefit: Under the analysis of SP analysis module, draw the price function between the member and the provider to determine the number of the required member objects and the number of tasks in different locations; the task pricing module Members completed the task volume analysis module to get the demand data for each task, the maximum profit of SP as the goal, to define the remuneration of each task, the decision-making is reasonable. If there is a mismatch then undo or re-configure [1].

Market driven scientific workflow scheduling framework MSWSF, by the scientific workflow management system SWMS, task providers SP and the market. SP in the catalog SCD release tasks and task attributes [2]. The QOS feedback from SWMS is fed back to SP through QOS as reference for task quality and adjustment of remuneration. Task Level Agreements SLAs define the types of tasks, qualifications and payments required by SPs and SWMSs. SWMS further monitors the volume and quality of market-ready tasks by monitoring WFEE. There are analytical module, data prediction module, task discovery module, task reservation module, actuator, data transmitter, QOS monitor and other components. From the framework of market-driven scientific workflow scheduling, task providers can search the task location, membership distribution [16], the number of tasks and the number of members through task retrieval to make the best judgment for further reasonable pricing. As shown in Fig. 2.

3 Crowd Sourcing Model

Through the crowd sourcing approach to the traditional space can not be separated tasks, the effective organization of a large number of members who are willing to earn remuneration, to complete the task, to overcome the traffic, personnel scarcity, low prices in the form of portfolio sales, the overall package. Crowd sourcing can be divided into multiple tasks in the same area to be collectively integrated into one member, or multiple members in one area can complete a complex task.

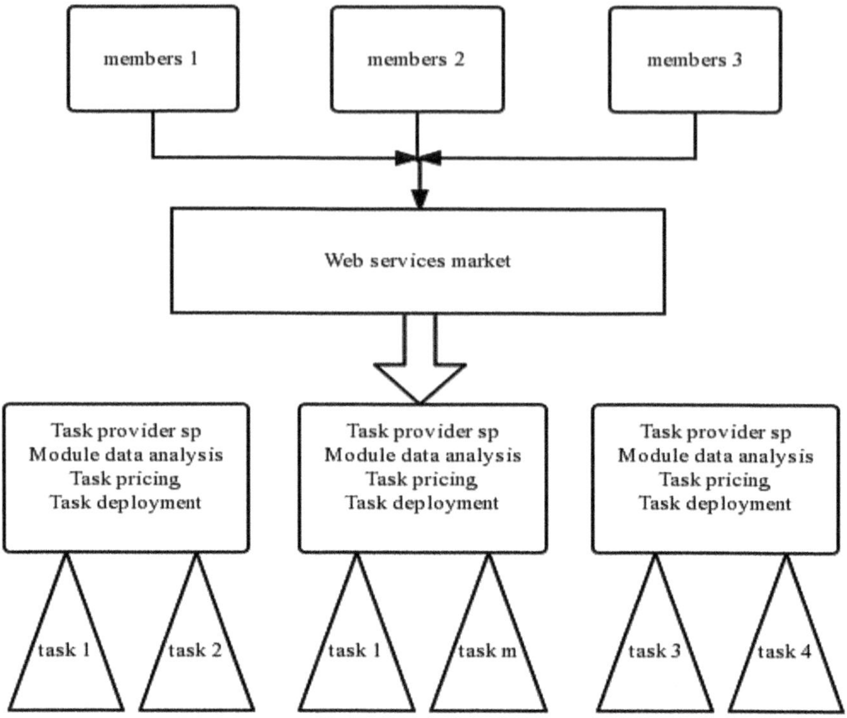

Fig. 1. Utility network architecture

According to the target task can be integrated in the operating mode, can be divided into open source code and non-open source code tasks.

i. The process of open source code task operation: SP build platform, in one area to provide the task, members go to complete the task. The works received by the Open Source Code Mandate complement each other, increasing their value as the number of collected works increases, i.e. [3], the value of all works integrated into one whole increases. The operation flow is relatively simple, the mission proponent releases the mission; the APP member uploads the photo achievement to the system platform set up by the mission proponent, and can obtain the corresponding remuneration as long as the SP complies with the requirements.
ii. Non-open-source code task operation process: SP release tasks, collect photos of selected photos and in many APP member program to select the highest quality works. Non-open source code works won the competition between the alternative, known as the membership map process [17].

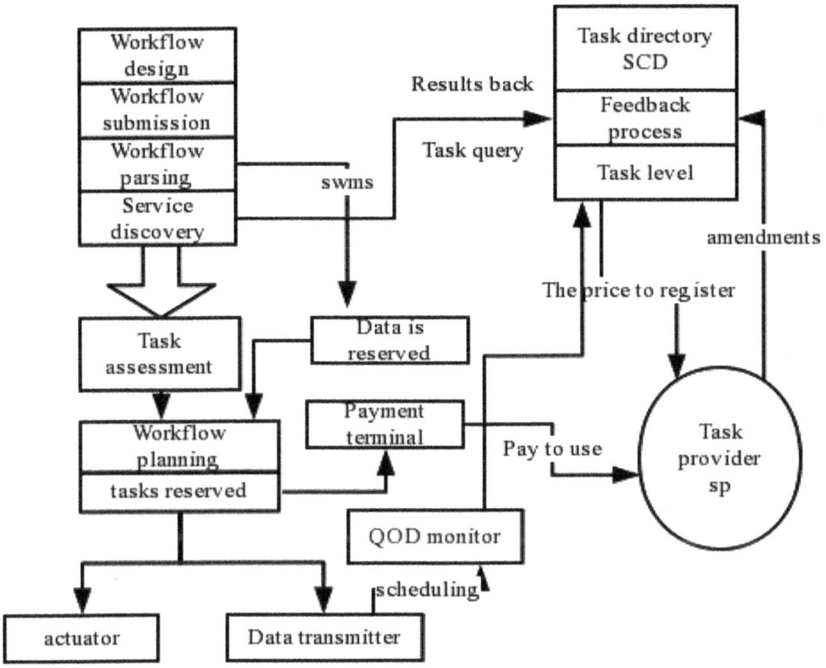

Fig. 2. Scientific workflow schuduling framework driven byworkflow adjustment market MSWSF

Membership map requires membership platform for each APP member to open up personal space, membership platform can derive through member map profitability, such as transaction volume commission, advertising or auction ranking, the completion of the task of the credibility of the value of pricing. First of all, APP members set up personal space on the platform; then, SP searches for the required APP members; then the two parties get in touch and decide whether to cooperate; before the cooperation, the members negotiate pricing with the SP to establish cooperation intention [4]; Platform prepaid remuneration. Members in the million sad task to submit photos to the SP portfolio, through the acceptance and meet the requirements of the SP, the third party will pay the remuneration, if not satisfied with the place, it will return to the task distribution and bidding system, SP will increase the price of this task in the system, to attract outstanding members to complete the task, continue to cycle the previous process, until there is a satisfactory SP photos, this task can be ended. Finally, the completion of the transaction SP and member system platform for evaluation [5], to further reflect the importance of credibility. Process shown in Fig. 3.

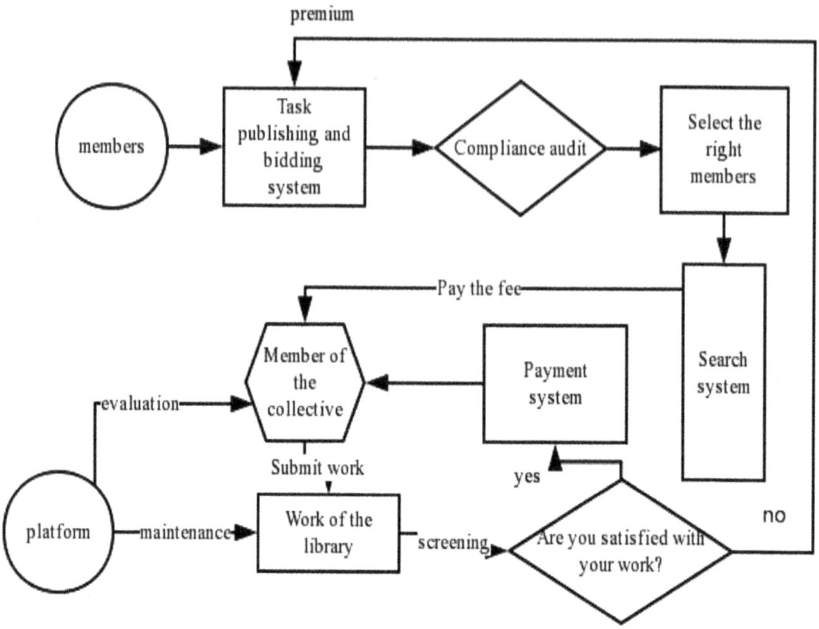

Fig. 3. Member map flow chart

4 Balanced Pricing

Crowd sourcing needs to consider participation in the number of members and tasks pricing, the optimal number and the cost of the end.

From the marginal analysis of economics, the additional expenditure is equal to the additional income when the critical point, the economic individual to obtain the maximum profit [6]. MSWSF framework, SP income refers to the APP member to complete their tasks to get the sum of remuneration, spending refers to the SP needs to provide the task. Suppose the demand function is d = f (p), p, d is the task compensation and task demand. If the SP deploys n tasks S1, S2, …, Sn, revenue $B = \sum_{i=1}^{n} p_i d_i$, Pi is Si need to pay remuneration, di is the demand of Si task [15]. The task of deployment takes a certain amount of membership, but the total amount is limited. Considering the limited number of m members, the revenue is constrained by $\sum_{i=1}^{n} d_i v_i^j = c_j$, j = 1,2 … m, v_i^j is the total consumption of the jth APP member rj per unit time. This paper is first selected $d_i = f(p_i) = (a_i - p_i)/b_i$ as the demand function of task Si, the parameters ai, bi values can be obtained by linear regression [7]. Therefore, to simplify B = g (p1, p2, …, pn), the constraint is a function of pi cj = uj (p1, p2, …, pn) with j = 1,2, …, m and the solution target is maxB. Use Lagrange multiplier method to solve as follows

$$\max B = g(p_1, p_2, \ldots, p_n) + \sum_{j=1}^{m} \lambda_j (c_j - u_j(p_1, p_2, \ldots, p_n)) \tag{1}$$

The first-order condition of B's extremum is then:

$$B_{\lambda_j} = c_j - u_j(p_1, p_2, \ldots, p_n) = 0 \quad \mathrm{j} = 1, 2, \ldots, \mathrm{m} \tag{2}$$

$$B_{\mathrm{p}_i} = g_{p_i} - \sum_{j=1}^{m} \lambda_j u_{p_i} = 0 \quad \mathrm{i} = 1, 2, \ldots, \mathrm{n} \tag{3}$$

$$g_{p_i} = \frac{^{\partial} g(p_1, p_2, \ldots p_n)}{^{\partial} p_i} = \frac{(a_i - 2p_i)}{b_i} \tag{4}$$

$$u_{j_{p_i}} = \frac{^{\partial} \mathrm{u}_j(p_1, p_2, \ldots p_n)}{^{\partial} p_i} = \frac{v_i^j}{b_i} \tag{5}$$

Formula (4) and formula (5) into Eq. (3) obtained:

$$a_i - 2p_i = \sum_{j=1}^{m} \lambda_j v_i^k$$

Simplified (2) to get

$$\sum_{i=1}^{n} v_i^j (a_i - p_i)/b_i = c_j \Rightarrow \sum_{i=1}^{n} M_i^j \cdot p_i = N_j$$

j = 1, 2, …, m

$$N_j = \sum_{j=1}^{m} a_i v_i^j / b_i - c_j, \mathrm{M}_i^j = \frac{v_i^j}{b_i} \tag{6}$$

If m $\geq$ n, and the augmented matrix determinant is not 0, pi can be directly obtained from (7); otherwise, m + n equations with m + n variables are obtained by the simultaneous Eqs. (6) and (7). Solving pi with linear algebra theory and MATLAB [8]. gpi is the increment of revenue for each increment of pi, ujpi is the increment of APP member consumption for each increment of pi, and (6) (7) is the equilibrium point equation of constrained optimization problem, which indicates that for every increment in each direction When the marginal benefits brought by the units are equal, the total benefit B allocated by the total number of APP members is the best.

APP membership and demanders is the mapping between the cost optimization mapping strategy, with the lowest cost to complete the best camera tasks. Turning this problem into a computational model, the scientific workflow can be abstracted as a directed acyclic graph G = (T, E), T is a task set, and E is a data-dependent set. $c \times \sum_{k=1}^{|E|} |d_k^{rs}$ is data transmission costs, c is the unit data transmission price ($\backslash$ Mb), d_k^{rs} is the task of Tr and Ts data transfer, scientific workflow execution within a certain period of time D [9]. Therefore, the formalization of scientific workflow scheduling is:

$$
\begin{aligned}
& Min\left(\sum_{j=1}^{|T|} p_i^j t_i^j + c \times \sum^{|E|} d_k^{rs}\right) \\
& \text{s.t} \quad \sum_{j=1}^{n} t_i^j + \sum_{k=1}^{m} \frac{d_k^{rs}}{B^{RS}} \leq D
\end{aligned}
\tag{7}
$$

General scheduling algorithm framework:
Procedure S che(G, D, set)
//G is scientific workflow, S is set of services, D is deadlime

Step 1: for j = 1 to |T|
Step 2: Setj $\Leftarrow$ get available n_j services for task T_j
Step 3: sort S_i^j, i = 1, 2, … n_j in Setj in non-descending order of $p_i^j t_i^j$ value
Step 4: select $S_{k_j}^j$ whose $p_i^j t_i^j$ is min in Setj for T_j and record in rb
Step 5: end for
Step 6: Flag = true; ct $\Leftarrow$ critical tasks;
Step 7: while(Flag){heu S che(G.S.D. ct.rb)}

5 Adjustment of Pricing

Based on the location of the more concentrated areas, the user will choose crowdsourcing release APP members to receive the task, members can be crowdsourcing approach reverse price-fixing method to get the gratuity. First, we need to accurately calculate the total task cost C, and then add a certain target profit π, the number of photos Q, get the service price $P = \frac{C+\pi}{Q}$. However, it is difficult to accurately calculate the task cost, especially the personalized task product is difficult to scale [10]. Consider also the task price elasticity Ed [14]. Assuming that the pre-transfer price is P, then the pricing formula is $P = \frac{1}{2}\left(3 - \frac{1}{|E|}\right)$, Suppose the demand function of the task result is: $P = \alpha + \beta\chi$, α, β $\geq$ 0, Pc is the price of other tasks, β is the price of previous tasks . The weight of the influence, θ represents the weight of the recent task price changes, γ represents the weight of the impact of the average price of crowdsourcing members on demand, from which we can draw the conclusion that long-term task price strategy is [11]

$$\lim_{n\to\infty} P_n = \frac{\left\{\begin{array}{l} \alpha-\gamma+\theta+C\beta+\dfrac{C\theta}{P_C}+ \\ \sqrt{\begin{array}{l}\left(\alpha-\gamma+\theta+C\beta+\frac{C\theta}{P_C}\right)^2 \\ +8C\gamma\left(\beta+\frac{\theta}{P_C}\right)\end{array}} \end{array}\right\}}{2+\left(\beta+\frac{\theta}{P_c}\right)} \tag{8}$$

The value of the task product depends more on the perceived value and reputation value of the task product on the cost of the task. The higher the photo quality, the higher the satisfaction, the higher the pricing, and the task outcome pricing will be different from other crowdsourcing Member pricing changes and changes.

In this rule, the average quotation of outsourcing industry is positively correlated with the SP price. The task difficulty, the task duration, the similar tasks of the member system platform and the membership number of the member system platform have a positive influence on the task publisher bidding. The mode strategy is shown in Fig. 4 Show [12].

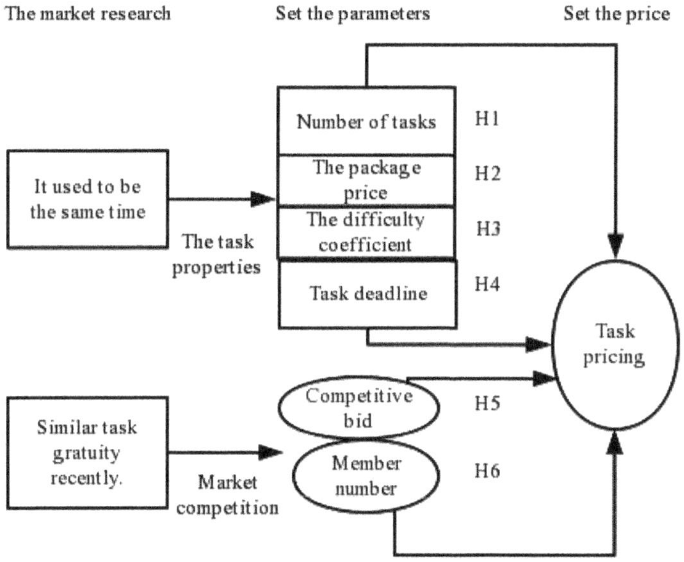

Fig. 4. Study model

By adopting the crowd sourcing model and reverse-directed pricing method to obtain the remuneration, the completion of the task of completing the settlement and the efficiency of completion are improved. The SP pays the remuneration after passing the inspection of the member's quality and gives the member a rating to obtain the credit value. In the future tasks, the use of high credit value to accept more tasks, SP is also easy to choose from members to complete the task, improve the completion rate.

6 Innovation Points

This paper mainly focuses on the task pricing game between task providers and members under the framework of workflow scheduling. The task pricing business negotiation is the highlight and the biggest innovation point of the whole pricing.

i. On the basis of the crowd sourcing mode, the task proponent can discuss with the members, and the task proponent and task proponent, members and members can also cooperate with each other to become a collective. In a region with higher cost, a member can Complete multiple tasks, empathy, a task can also be completed by multiple members.
ii. For a task to complete the quality of the task proposed by the satisfaction of their choice of works, are not satisfied with the payment of remuneration, the members also have the task platform to retrieve their willingness to complete the task.
iii. The whole pricing algorithm, taking into account differences in economic traffic population in different regions, according to different locations have different pricing, the current data processing in the mature context, accurate application of data simulation, combined with the characteristics of each region to establish an algorithm Simulation, considering the cost optimization strategy, the use of reverse-oriented price method, a new pricing method, the task pricing more flexible and humane.
iv. The model has its own corresponding system, the system design based on the current upgrade, monitoring the completion of the task, the task is completed, there will be providers to evaluate their credibility will be involved.

This paper mainly focuses on the task pricing game between the task provider and the member under the workflow scheduling framework, and the task pricing business negotiation is the highlight point of the whole pricing and the biggest innovation point.

7 Conclusion

In order to improve the reasonable pricing and task completion rate, the mutual benefits between the members and the SP are maximized. In terms of the difficulty, quantity and pricing of the tasks, both parties can hold discussions and the members can compete with each other to obtain high-quality photos. This model can be used to study the balance of the information taking and sharing of economy and the balance between resource utilization rate and resource node income. It is convenient to promote and has a wide range of application prospects.

Project Funding. 1. 2017 "13th five-year plan" education scientific planning key project of gansu province "advanced education targeted poverty alleviation research" (Item no.: GS[2017]GHBZ034);2. Northwest national university 2017 central university project fund. (Item no.: 31920170079);3. Research and practice on the course process reform of digital electronic technology in northwest nationalities university in 2017. (Item no.: 2017XJJG-22);4. In 2017, the special fund for basic scientific research operation of the central university of northwest nationalities university. (Item no.: 31920170141);5. In 2016, the youth team of the central university project fund of northwest nationalities university. (Item no.: 31920160003);6. Chang Jiang scholars and innovation team development plan funding. (IRT_16R36)7. 2018 Northwest University for Nationalities opened the laboratory project, Central University (SYSKF-2018002;SYSKF-2018007;SYSKF-2018025;SYSKF-2018030)8. Supported by Program of 2018 Scientific research project of colleges and universities of Gansu province department of education, "advanced education targeted poverty alleviation research": (Item no.: 2018B-026)

References

1. Wu, R., Sun, P., Sun, Y.: Distributed task plan dynamic adjustment model and algorithm. Syst. Eng. Electron. Technol. **39**(02), 322–328 (2017)
2. Kong, Y.: Research on the Marketing Strategy of Membership System in Z Company. Donghua University, Shanghai (2016)
3. Sun, X.: Research on Task Allocation in Public Packet Environment. Yangzhou University, Yangzhou (2016)
4. Liu, Y.: Discuss the application of network membership marketing. Oper. Manag. (32):274 (2013)
5. Liu, X.: Research on the Influencing Factors of the Bidding Behavior of Task Publishers in Crowdsourcing. University of Chongqing, Chongqing (2012)
6. Ma, Y., Gong, B., Zou, L.: Scientific workflow scheduling strategy based on balanced pricing and cost gradient. Acta Electron. Sinica **38**(10), 2416–2421 (2010)
7. Hong-tao Z.: Information helps retail industry to perfect membership marketing. Modern Ind. (22):50–53 2008
8. Yu, J., Buyya, A.: A taxonomy of workflow management systems For grid computing. J. Grid Comput. **3**(3–4), 171–200 (2005)
9. Yu, J., Buyya, R., Tham, C. K.: Cost-based scheduling of scientific workflow applications on utility grids. In: Proceedings of the 1st IEEE International Conference on e-Science and Grid Computing, pp. 140–147. IEEE, Melbourne (2005)
10. Wu, J., Deng, L., Hu, Z.: Price negotiation and control in grid computing economic model. J. N. Polytech. Univ. **26**(4), 497–502 (2008)
11. Li, L., Liu, Y., Ma, X.: Grid resource allocation based on combined two-way auction. Acta Electron. Sinica **37**(1), 165–169 (2009)

12. Li, H., Zhong, Y., Joan, L.: A banking based grid recourse alloca-tion scheduling. In: Proceedings of the 3rd International Conference on Grid and Pervasive Computing. Kunming: IEEE, pp. 239–244 (2008)
13. Chen, H., Wang, Y.: Cost-time optimization scheduling algorithm for grid directed acyclic graph. Acta Electron. Sinica **33**(8), 1375–1380 (2005)
14. Chenghui, Y., Enen, R., Dang, J.: Analysis research of control method model on automobile brake test rig. Przegląd Elektrotechniczny **8**(5), 375–378 (2012)
15. Yang, C., Ren, E., Jianwu, D.: A new automotive brake test control analysis of mathematical models. Energy Educ. Sci. Technol. Part A: Energy Sci. Res. **31**(1), 143–149 (2013)

Information Security Requirements of Cloud Computing Information System

Haohao Song(✉)

The Third Research Institute of Ministry of Public Security, Shanghai, China
songhh@mctc.org.cn

Abstract. With the development of cloud computing technology, the deployment and application of cloud computing information system is more and more widely, and the information security problem of cloud computing information system is becoming increasingly prominent, and some problems even seriously affect the social and national security. However, cloud computing information system has its particularity, and also makes it different from traditional information systems in the face of security threats, security issues and the need to consider the security measures. Traditional information security products cannot be applied to the field of cloud computing, so there has been a dedicated cloud computing information security products. This paper analyzes the difference between cloud computing information system and the traditional information system, introduces the current several typical cloud computing information security products, and analyzes the information security requirements of cloud computing information system.

Keywords: Cloud computing · Information security
Information security requirements

1 Introduction

Cloud computing is a style of computing in which dynamically scalable and often virtualized resources are provided as a service over the Internet.

The National Institute of Standards and Technology (NIST) definition: cloud computing is a model for enabling ubiquitous, convenient, on-demand network access to a shared pool of configurable computing resources (e.g., networks, servers, storage, applications and services) that can be rapidly provisioned and released with minimal management effort or service provider interaction. The application scope of services of various cloud computing is expanding, and the influence is immeasurable.

In order to improve the security of information system using cloud computing (referred to as: cloud computing system), is now generally considered from two aspects: one is to improve the self-security of the cloud computing system, security architecture started from the design stage, and the implementation it in aspect of research, deployment and development of cloud computing system; one is by additional information security means (such as the deployment of security products in the cloud computing system) for repairing security vulnerabilities of the system itself to a certain extent. Due to the requirements of cloud computing system for high stability

K. Deng et al. (Eds.): ICMIR 2018, AISC 856, pp. 83–89, 2019.
https://doi.org/10.1007/978-3-030-00214-5_10

and high availability, it is the easiest way to realize and accept for the enhancing security of cloud computing by means of additional information security guarantee.

Cloud computing information system is essentially a kind of information system, so the security problem of cloud computing information system is the combination of information system security and cloud computing characteristics. National Security Agency (NSA) in America has developed several versions of Information Technical Assurance Framework (IATF) [1], which has become an authoritative framework for information security assurance. IATF divides information security into four parts: Protection, Detection, Response and Recovery. Firstly, various measures are adopted to protect the objects that need to be protected, and then the security objects are tracked and detected by means of corresponding detection methods to keep abreast of their security status. If the security status of the security object is changed, especially from security to insecurity, emergency measures should be taken immediately to respond to it until the security state of the security object is restored. These four parts complement each other and are indispensable, forming a recognized PDRR model, as shown in Fig. 1.

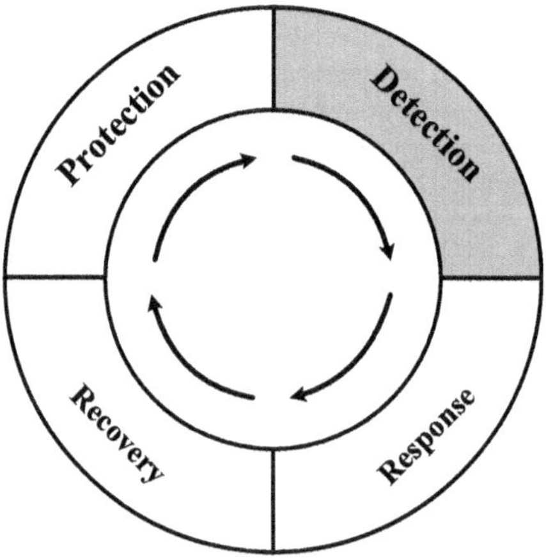

Fig. 1. PDRR model

This paper analyzes from the PDDR model of the "detection" aspects to test and evaluate information security products dedicated to the protection of cloud computing information system—cloud computing information security products, and establish test evaluation system of cloud computing information security products, provides professional theoretical support and practical guidance for cloud computing information security product evaluation. On the one hand help the suppliers to enhance security capabilities of their products, on the other hand, provide support for the security for the user to buy cloud computing information security products. Through the comprehensive analysis of the evaluation results, the paper provides the real and comprehensive security situation analysis report for the relevant authorities, and promotes the development of information security in the field of cloud computing.

2 Cloud Computing Information Security Products

Due to the characteristics of cloud computing information system, cloud computing information security products suitable for cloud computing information system need to focus on the characteristics of cloud computing information system. The use of relatively typical cloud computing and information security products include cloud security operating system, desktop cloud security management system, security cloud storage systems, cloud computing security defense products and so on [2].

2.1 Security Cloud Operating System

The cloud operating system is the cloud platform integrated management system based on the infrastructure hardware resources such as server, storage, network, the basic software such as middleware and database, managing of mass foundational software and hardware resources. It has more security features, including the remaining information protection function, memory isolation, user data protection, and anti-denial of service attacks and so on.

2.2 Security Desktop Cloud System

Security desktop cloud system virtualizes the desktop, centralizes the virtual desktop on remote server, to achieve the isolation of the desktop, on-demand distribution and rapid delivery. It has more security features, including the session control, virtual desktop restore initial state, patch management and anti-malware software loading and so on.

2.3 Security Cloud Storage System

Security cloud storage system refers to the product that through the cluster, grid technology or distributed file system. It has more security features, including key management, data storage isolation integrity detection function, and remote transmission security and so on.

2.4 Cloud Security Comprehensive Defense Product

Cloud security comprehensive defense product is the integrated defense product based on the cloud computing, elastic expansion, mainly on the cloud computing platform and applications of cloud computing services and upper business, preventing malicious attacks from external cloud platform, Inside virtual machine or virtual machine. It has the security functions, including the linkage and response with the cloud computing platform and between functions, anti-attack ability and so on.

3 Information Security Requirements of Cloud Computing Information System

3.1 Security Threat

Before summarizing the information security requirements of cloud computing information systems, we need to confirm the security threats faced by cloud computing technology. Let's analyze it from four levels [3–8].

(1) Certification Level
Cloud Computing supports authentication and access to large numbers of users, user authentication and access management must be fully automated, it is needed to improve authentication access management experience, and simplify the user authentication process of cloud, such as all business unified SSO and authorization management in the cloud. Supporting the mobility and distributed network computing is the important characteristics of cloud computing, which increases the difficulty of user authentication management, in order to achieve the user whenever and wherever possible can access cloud computing resources, the access authentication will be accepted from different positions and different client login. If the entrance of a city is attacked break by an intruder, like, breaking the door, which will run with the city. Once the intruder enters the internal cloud computing system, it is bound to master the internal resources, do some damage or theft of bad things, causing serious losses to cloud computing users. The intruder usually uses the rented virtual machine to attack the virtualization management platform by using the operating system or the homepage loophole to illegally intercept the user data, illegally obtain the user password, and illegally obtain the sensitive information. So, the door to cloud computing is very important, or all the data presented to illegal intruders.

(2) Data Level
Cloud computing deals with massive data, and these data uses the technology of distributed computing, that is a lot of data calculation that is accomplished by computing resources in all common, so there is a large amount of intermediate data need to be transferred by the network, the process is not protected, there is a big security risk. Not only the data processed, but the data stored in the data center all are threated in security. These data can be stored encrypted, users can encrypt data through the client, then the data stored in the cloud, user data encryption key stored in the client, and the cloud platform cannot get the key to decrypt the data. Also, when the user is signed out from the cloud, the user's data space can be directly released to other users, and if these data are not emptied in time, other users can access to the original user's private information, data leakage may exist. Cloud computing is the lack of ability to identify the data content, data often get direct calculation, the lack of inspection and verification mechanisms is that it tends to make some invalid data or falsified data mixed in, which on the one hand may affect the calculation results, on the other hand also takes up a lot of computing resources, impact the effect of cloud computing.

(3) Virtualization Level
Cloud computing use virtualization technology incisively and vividly, it can be said that without virtualization technology, cloud computing will lose the meaning of existence. However, virtualization technology amplifies security threats and exposes the system to the outside world. The virtual machine is dynamically created and migrated, and the security measures of the virtual machine must be automatically created and migrated automatically. The virtual machine itself can migrate anywhere in the two layer network. It is difficult to protect the virtual machine, especially in the process of migration. The virtual machine does not automatically create no security measures or security measures easily leading to key that is used to access and management of virtual machine is stolen, the corresponding service is attacked, weak passwords or no password account is stolen, virtualization increases the security threat, and there is no good way to protect. As we all know, Hypervisor is the core technology of virtualization, which can capture CPU instructions, act as an intermediary for instructions to access hardware controllers and peripherals, coordinate all CPU resource allocation, and run on the highest priority that is higher than the privileges of the operating system. Once Hypervisor is attacked, all of the virtual machines on the Hypervisor will not have any security, directly exposed to the attack, which will bring great hidden dangers to the system.

(4) Network Level
Cloud computing relies on the network to complete the distributed computing. The essence of cloud computing is the use of the network, centralizing computing resources at different locations, and then through the collaborative software, let all the computing resources to work together to complete some calculation function. This process in cloud computing, requires a large amount of data transmission through the network. In the transmission process, data privacy and integrity are threatened, while the traditional calculation does not involve, the traditional computing data directly into a specific server to complete the calculation, as long as the server has security protection, which can guarantee the basic calculation process interference. In the cloud computing process, to consider network security factors, it is suggested that quantum communication, in order to solve the network security problems in the process of data transmission. If in the cloud computing process, data is quantum cryptography, that will greatly enhance security capabilities, reduce security threats. Cloud computing must be based on the network that can be accessed at any time, so that users can easily access the cloud computing resources through the network, which makes the cloud computing resources need distributed deployment routing; more complex domain name configuration is, more vulnerable to network attacks cloud computing is. For IaaS, DDoS attacks are not only from the outside network, but also from the internal network. Including user data leakage caused by improper isolation measures, users suffer other malicious users attack from the same physical environment, and so on. The traditional network attacks is amplified in the cloud environment, so the security protection scheme is need for cloud computing network environment.

3.2 Information Security Requirements

Based on the security threats of cloud computing technology, the corresponding information security requirements of cloud computing information system is put forward from 4 levels of authentication, data, virtualization and network.

(1) Certification Level
For the user authentication level of cloud computing information system, because users access to system resources through the Internet authentication, so the user should be authenticated by two factor (to avoid only static password authentication), authentication information should be transmitted through the Internet not plaintext. For static password, it should have password complexity and regular replacement requirements; continuous identification failure should adopt corresponding measures to prevent violent speculation.

For the administrator authentication level of cloud computing information system, it should be not that opening the administrator access background function of cloud computing information system in Internet, it should only allow internal network access. At the same time, it should strengthen the means and security measures of its certification.

(2) Data Level
In the massive data of cloud computing information system, the most important is the user's personal information and important business data. The two part of the information is also proposed in the network security law needing to focus on the protection. In the storage and transmission of these data, the encryption measures with higher strength should be adopted, and the integrity check mechanism should be added.
(3) Virtualization Level
Virtualization technology of cloud computing information system should ensure the absolute isolation of virtual resources, take effective measures to prevent the escape of virtual machines and so on, and ensure the security of using virtualization technology.
(4) Network Level
Cloud computing information system should take effective measures to prevent the most frequent DDoS attacks and WEB vulnerability attacks, take isolation measures to prevent the user data leakage and the user service interruption, and provide log records for related attacks, and to trace and block the source of the attack in the whole network.

4 Conclusion

Due to the rapid development of cloud computing information security and related products, the security assessment of cloud computing belongs to a new field. The difference between cloud computing information system and the traditional information system is analyzed, the current several typical characteristics of cloud computing information security products is introduced, and the information security requirements of cloud computing information system is analyzed.

Acknowledgments. The research work in the paper is supported by the Public Security Theory and Soft Science Research Project of MPS (Grant No. 2017LLYJGASS018).

References

1. Information Assurance Technical Framework, 3rd edn. NSA (2000). http://www.iatf.net
2. Song, H.: The development and application of information security products based on cloud computing. In: Proceedings of the International Conference on Mechatronics and Intelligent Robotics, ICMIR 2017, vol. 2, pp. 223–228 (2018)
3. MPS Information Classified Security Protection Evaluation Center. Information security Classified evaluater training course. Electronic Industry Press (2011)
4. Feng, D., Zhang, M., Zhang, Y.: Cloud computing security research. J. Softw. **22**(1), 71–83 (2011)
5. Zhang, C., Jing, Y., et al.: Cloud Computing Security Architecture. Science Press, Henderson (2014)
6. Fang, J., Hao, W., Bai, S.: Virtualization security issues in cloud computing. Telecommun. Sci. **4**, 135–140 (2012)
7. Brunette, G., Mogull, R.: Cloud security alliance. Security Guidance for Critical Areas of Focus in Cloud Computing v2.1 (2009)
8. Liu, F., Tong, J., Mao, J., Bohn, R., Messina, J., Badger, L., Leaf, D.: Cloud Computing Reference Architecture. National Institute of Standards and Technology

Application of Mining and Analysis on Score Data

Shuhao Yu[1(✉)], Shenglong Zhu[2], and Chengqi Zhu[3]

[1] School of Electronic and Information Engineering, West Anhui University, Lu'an 237012, China
yush@wxc.edu.cn
[2] Anhui Electric Power Science and Research, Hefei 230022, China
[3] School of Foreign Languages, West Anhui University, Lu'an 237012, China

Abstract. Data mining is one of the key technologies in big data and one of the domains with the rapid development in education. The data mining techniques are curtly discussed on the basis of analyzing the score data in universities at the present time. Take the university student score as an example, the terms of settlement of the score analysis is suggested according to data mining techniques. The application of data mining technology for the analysis of the exploration of factors related to student scores and the student scores can offer suggestions for curriculum design, teaching and education administration.

Keywords: Data mining · Big data · Education big data · Score data

1 Introduction

In recent years, big data as a new trend in the development of information technology has penetrated into all walks of life. An explosion of education information and related data processing technology of innovation become the focus as a new vision. How to make use of information and analysis of these data, such as how affect the exchange of information, knowledge transfer and learning effect, affect the decision-making of teaching and learning mode optimization, has become the current education workers and consistent focus on aspects of learners. The in-depth application of the two key technologies, such as learning analysis and data mining brings new opportunities for education.

With the development of big data and development of education big data processing and analysis has become a driving force to promote the development of education reform and cause the attention of the government and the administrative department of education. In October 2012, the U.S. department of education released the "Enhancing Teaching and Learning through Educational Data Mining and Learning Analytics". Through education big data analysis and mining, the "big data" education application and data mining will be promoted, which will lead to education development.

K. Deng et al. (Eds.): ICMIR 2018, AISC 856, pp. 90–96, 2019.
https://doi.org/10.1007/978-3-030-00214-5_11

2 Data Mining

Data mining refers to a system that mines interesting knowledge from a large amount of data stored in databases, data warehouses or other information repositories. In recent years, in order to promote the application of data mining in practice, many researchers have done a lot of research on the architecture of data mining system [1]. By and large, data mining extracts implicit information from large, incomplete, noisy, fuzzy or random big data of practical application. The process of data mining can obtain the potential and useful information which people don't know in advance.

Data mining is an iterative procedure of knowledge discovery based on the description of the mining results, analysis and evaluation, optimizing the data mining model and algorithm for mining. Data pre-processing plays an important role for data mining. This process includes data cleaning, data transformation, data reduction and data evaluations [2]. The process of data mining is shown in Fig. 1.

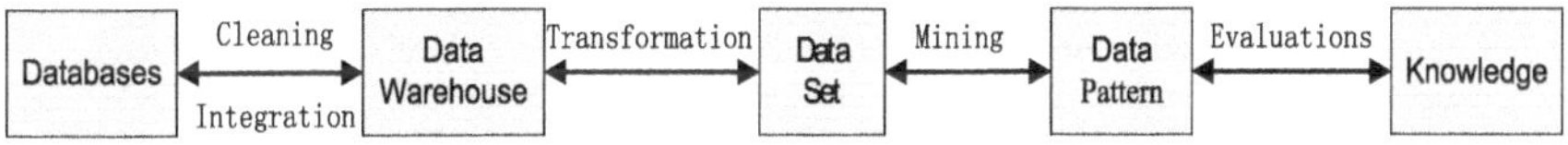

Fig. 1. Database knowledge discovery process

Association rule mining is finding the relationship among data items in massive data. The support degree and confidence level of association rules reflect the usefulness and certainty of the discovered rules respectively [3]. In general, the user can define two thresholds, the minimum support threshold and the minimum confidence threshold. When the rules support and confidence to meet the two thresholds, that this rule is effective, otherwise is invalid [4].

3 Data Mining

After the above data processing, the relatively clean data is obtained, and the data will be processed further. This analysis system mainly consists of user management, system maintenance, database consolidation, data analysis and several modules. The contents of the test paper are sorted by chapters, then the data association analysis is conducted to find the relationship among the knowledge points in each chapter.

In the current education system, the student's performance depends on internal evaluation and final assessment. The internal evaluation is based on students' classroom test, seminar, homework, teacher's assignment, general ability, attendance rate and laboratory work performance in education activity. The final examination is a student's score at the end of the semester. Each student must obtain the minimum mark, pass the semester and the final exam.

3.1 Decision-Making Tree

In decision tree, each fork node tree denotes a quantity of selections, each leaf node denotes a selection. Decision trees are often used to get information about decision making. The decision tree begins from a root node that contains user action. From this node, each user node is recursively divided according to the learning algorithm. The end solution is a decision tree that each branch denotes a reasonable decision. Three widely used decision tree learning algorithms as following: ID3, C4.5 and ASSISTANT. The basic idea of the ID3 algorithm is to test each attribute to construct the decision tree in each tree node by using the greedy search through a given set. A basic algorithm for inducing a decision tree list as Fig. 2.

```
(1)  create a node N;
(2)  if tuples in D are all of the same class, C then
(3)       return N as a leaf node labeled with the class C;
(4)  if attribute_list is empty then
(5)       return N as a leaf node labeled with the majority class in D; // majority voting
(6)  apply Attribute_selection_method(D, attribute_list) to find "best" splitting_criterion;
(7)  label node N with splitting_criterion;
(8)  if splitting_attribute is discrete-valued and
          multiway splits allowed then // not restricted to binary trees
(9)       attribute_list ← attribute_list − splitting_attribute; // remove splitting_attribute
(10) for each outcome j of splitting_criterion
     // partition the tuples and grow subtrees for each partition
(11)      let D_j be the set of data tuples in D satisfying outcome j; // a partition
(12)      if D_j is empty then
(13)           attach a leaf labeled with the majority class in D to node N;
(14)      else attach the node returned by Generate_decision_tree(D_j, attribute_list) to node N;
     endfor
(15) return N;
```

Fig. 2. Basic algorithm of a decision tree

3.2 Measure Impurities

The pure (or heterogeneity) of the measurement is determined by a table of attributes and classes that are included. If it involves only one class, we say pure or consistent, otherwise we say that the table is not pure or heterogeneous. There are some indicators to measure the impurity of the measurement. The most famous measure of impurity is Gini index, classification error and entropy.

$$Entropy(D) = -\sum_{i=1}^{m} p_i log_2(p_i) \tag{1}$$

The Gini index measures the impurity of D, it defines as following:

$$Gini(D) = 1 - \sum_{i=1}^{m} p_i^2 \tag{2}$$

Where Pi is the probability that a tuple in D remains with class C_i. it is computed by $|C_{i,D}|/|D|$.

3.3 Division Standard

The best attribute of nodes in the tree can be determined by information gain.C4.5, a successor of ID3. It adopts a kind of normalization by using a "split information" value which can be defined as following:

$$Split\ Information(S, A) = -\sum\nolimits_{i=1}^{n} \frac{|S_i|}{|S|} \log_2 \frac{|S_i|}{|S|} \quad (3)$$

The term $\frac{|S_i|}{|S|}$ serves on the weight of the *j*th partition. Split Information (*S*, *A*) is the anticipated information needed to categorize a tuple from S based on the partitioning by *A*.

$$Gain(S, A) = Entropy(S) = -\sum\nolimits_{V \in Values(A)} \frac{|S_V|}{|S|} Entropy(S_V) \quad (4)$$

The first item in the gain equation is the entropy of the initial set S. The second item is the anticipated value of the entropy for the set S using the property A partition. The expected value of entropy in the second item is the sum of the entropy of each subset. By weighting the instance of Gain(S, A), that is $\frac{|S_V|}{|S|}$, then you get the decrease of the expected entropy because you give me the value of the property A.

4 Score Mining System

Correct analysis of student achievement is the key to guarantee the teaching work smoothly. Together with the teachers' basic information analysis, the system can better grasp the difficulty degree of examination paper and reveals some rules of "teaching" and "learning" phenomenon. This can provide the better guidance for teaching and learning. And it also can provide the basis for education teaching planning and decision-making to improve the teaching effect and results. In view of the need of teaching management, the score data mining technology is applied to design student achievement mining system. The contents include reliability and validity analysis of students' achievements. The relevance of knowledge points in each chapter. The influence of teacher's title, age and other factors on teaching and students' professional courses on examination results.

4.1 System Function Design

As previously mentioned, to design the system, we analysis the feasibility and requirement of the score mining system firstly. The system mainly realizes the four functions of information initialization, data acquisition, data preprocessing, data analysis and data mining. The score analysis flow chart shows as Fig. 3.

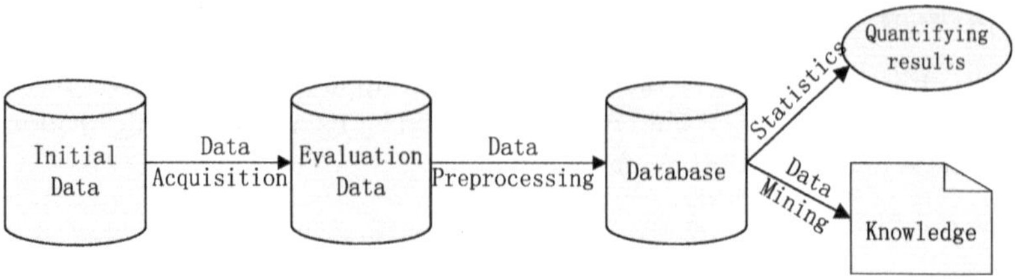

Fig. 3. Score analysis flow chart

4.2 System Initialization

System initialization is the basis of system operation. The main objective is to accurately and timely complete the data collection, verification, modification and other operations necessary for the evaluation of the system. In addition, it need to consider that data mining analysis subsystem required for the processing of all kinds of data and information, especially when some data to evaluate the data acquisition is unnecessary because it is very important in data mining analysis. For example, in the analysis of data mining system, we need to analyze the teacher's teaching effect with teachers' age, educational background, title etc. If relevant it requires the system have the complete eligible teachers file data and in the evaluation of information collection. Therefore, the data initialization module needs to comprehensively consider the requirements of subsequent modules and initialize the data information. The main tasks that need to be completed for information initialization are:

(1) Identify objects and targets. According to the students at ordinary times the basic situation. Such as learning interest, learning knowledge in front of the situation, the teacher's teaching effect of practice after class, etc. Meanwhile, it need to consider the final exam grades, academic title on teacher student performance analysis and the influence degree and wish to gain with the analysis of the results to guide the teaching work in the future.
(2) Correct integration of data from different data sources. The mining data for evaluation of the system require the class, teachers and students as well as the realization of data mining. It need detailed files are exist in different application systems. This process needs to integrate the data from different systems to the operation of this system in the database.
(3) After the completion of data integration, all kinds of information tables needed by the integrated data table. These tables include as following: teacher information table, class information table, class teacher's appointment information table, etc. To guarantee the accuracy of the data of the absolute, the system uses automatically generated by the machine and checks the method of combining the artificial which first using the machine to automatically generate all kinds of data table. Then it will need to check the units of information points on the Web server that carried out by relevant staff carefully checking.
(4) The identification number is generated when students evaluate. To achieve anonymous evaluation, the system generates a unique identification number for each student to display in the evaluation data. At the same time, the number is put

into the student information database to determine the real source of each teacher's evaluation data.

(5) Collect information from peer experts. Due to the limitation of time and conditions, we need to set up a peer specialist auxiliary acquisition interface of classroom teaching quality evaluation information. So, it need to facilitate the administrator to this part of the evaluation data submitted to the corresponding database.
(6) Evaluation criteria setting. The evaluation index editing interface is set up, and users can flexibly implement the modification and setting of evaluation indexes and weights.

4.3 Data Acquisition

Participation in the operation of the data acquisition is the most part of the mining system. The user here is mainly divided into two categories: teachers and students. The goal is to achieve large-scale evaluation of the full and accurate data acquisition [5]. System application way of data collection have online sampling offline human input. The system can use computer data conversion function directly open source database for data acquisition and can also be transmitted through the network file for data collection [6]. The collected content includes teacher data, student situation data, examination result data and so on.

4.4 Data Preprocessing

Data preprocessing is to prepare data for data mining analysis module. The preprocessing of data is mainly:

(1) Data integration. Using database technology to combine data from multiple data sources which mainly includes the student assessment information, student information, teacher information, course arrangement and student achievement information, etc. In this process, some data need to be calculated and transformed to facilitate data mining system to analyze the required data information.
(2) Data cleansing. Data cleaning usually includes work such as inconsistent data processing, noise data smoothing, filling missing data values and data transformation. The operation of data cleaning can find that some missing attribute value of the interest which can adopt the method of artificial fill. The principle of fill is based on the records of other attribute values as a filter and screening in the database. Then it chooses the most attribute values fill the vacancy. In the process of teacher assessment, the evaluation criterion is divided into five grades: excellent, good, middle, passing and poor, and the value of database can reduce to 10, 8, 6, 4 and 2 accordingly.
(3) Processing of duplicate records and conflict records. Prior to data mining, there is a possibility of duplication and conflict in the analysis data selected by the user which needs to be handled in advance. Through the above data preprocessing operation, the system has a complete and clean data output suitable for mining to ensure that subsequent data mining can produce useful rules.

5 Conclusion

Data mining is a complex data processing process and also a creative thinking process. In this paper, by analyzing the relation between the mining students course influence degree, as a reference in turn can also dig the course itself. We introduce the technology of data mining and design the mining system for students score system. Mining and analyzing the students score and other valuable information such as curriculum and course grade can guide teachers in the teaching reform, professional talent training scheme designing and improve teaching quality.

Acknowledgements. This research was supported by the National Natural Science Foundation of China (NSFC) (No. 61702375) and the Universities Excellent Young Talents Foundation of Anhui Province (No. gxyqZD2016249).

References

1. Han, J., Kamber, M.: Data Mining: Concepts and Techniques (2007)
2. Gao, Y.: Progress of data mining in China. J. Nanjing Univ. Nat. Sci. **4**, 351–353 (2011)
3. Ye, F.: Basic research on data mining technology. Manuf. Autom. **1**, 204–207 (2011)
4. Cui, X.: Application of association rules for miningapriori algorithm to analysis of students performance. J. Hebei North Univ. Nat. Sci. Ed. **1**, 44–47 (2011)
5. Wei, Y., Zhang, C.: An exploration of data warehouse and data mining in teaching practice. High. Educ. Forum **1**, 94–96,99 (2011)
6. Wang, X.: Research and application of association rules in data mining technology. Coal Technol. **8**, 205–207 (2011)

Improved CUSUM Model Based on Commodity Sales Forecasting

Zhang Yi(✉), Long Hua, Tian Yu-jie, Shao Yu-bin, and Du Qing-zhi

Faculty of Information Engineering and Automation,
Kunming University of Science and Technology,
Kunming 650000, Yunnan, China
1353369570@qq.com

Abstract. CUSUM algorithm is one of the core methods which is commonly used in early prediction of events both at home and abroad. It has been shown in many years of practice that the forecasting method is too simple and the timeliness of the analysis is not ideal. In this paper, the improved EW_CUSUM algorithm came into being. First, calculate the amount of expected data under different baseline windows according to the amount of historical data generated. Then, calculate the baseline level based on the expected data amount. Finally, add the prediction statistics under two baselines. The required forecasting statistics will be helpful to figure out the high incidence of the situations. This paper validates the effectiveness of this method in commodity sales forecasting by using this method. This method has successfully improved the ability of forecasting events, advanced the time of the forecast and enhanced the timeliness of forecasting.

Keywords: CUSUM algorithm · Prediction · Timeliness

1 Introduction

The core theory of corporate marketing is to study the marketing activities, discover its rules and improve products and services. The forecast of commodity sales refers to the combination of computer technology and market decision-making through the specific analysis and forecasting method, in order to estimate the sales status in the future. Its emergence provides the data support for the enterprises, which is going to make decisions and solve the industry problems. Enterprises would be able to arrange production according to the number of sales forecasts, and also to avoid product demand or inventory backlog. From this we can see that the accuracy of sales forecasting is particularly important for the formulation of marketing plans, safety stock and even the supply chain of the entire product [1, 2].

There are different methods in sales forecasting model, such as time series analysis, regression and correlation analysis, neural network etc. Each model has its own advantages and disadvantages, such as good at extracting the information of the original data from different perspectives, reflecting the characteristics of a certain aspect of the data [3]. In the analysis process of CUSUM model prediction, researchers have proposed various improved method. Reference [4] improved the threshold of the

K. Deng et al. (Eds.): ICMIR 2018, AISC 856, pp. 97–103, 2019.
https://doi.org/10.1007/978-3-030-00214-5_12

classic CUSUM model warning. References [5, 6] improve its false alarm rate and omission rate by using statistical Non-parametric change detection method when facing with the CUSUM model parameter problems. It has been found that the CUSUM model accumulates some historical observations directly and seeks the expectation, which will lead to the issue of excessive warning [7–9]. In this paper, we proposes a new way that we should process the historical data of the classical CUSUM model when it comes to estimate the daily occurrence of an event and make relevant statistics. The expected warning value obtained will be more reasonable and rigorous, and the warning effect will be more sensitive and effective.

2 Algorithm and Analysis of CUSUM

2.1 Source of Data

In this paper, we use the daily sales volume of JiaHua Flower Cake (from 2016 to 2017), a large supermarket in our city, as a research sample.

2.2 CUSUM Algorithm

The CUSUM model was born in Cambridge, England, in 1945 [10, 11]. Its central idea is a control chart that highlights the shift between observations and horizontal values. It can magnify the volatility of observed data so that weak anomalies can be captured quickly and sensitively.

Suppose that in a period of time, there are m time points observe that the daily occurrence data of certain types of events $X_1, X_2, \ldots, X_i, \ldots X_m$ are independent and obey the distribution of $N(\mu_1, \sigma^2)$. The model focuses on the average value over the entire time period m. Assuming that the average value of the events occurring after the time n < m is offset, the occurrence of the events $X_{n+1}, X_{n+2}, \ldots$ would follow the distribution $N(\mu_2, \sigma^2)$.

If $\mu_1 < \mu_2$, then the hypothesis test at the significant level has the following results:

Set the original hypothesis H_0: the entire incident occurred in a stable state

Alternative hypothesis H_1: The number of occurrences is abnormal

The test under the assumed conditions meet the statistic:

$$S = \frac{X_i - \mu_1}{\sigma} \sim N(0, 1) \tag{1}$$

When the observed data values would meet

$$|X_i| > X_{1-\frac{\sigma}{2}} \tag{2}$$

At this point, the whole process would be considered abnormal, where $X_{1-\frac{\sigma}{2}}$ is the distribution $N(0, 1)$'s $1 - \frac{\sigma}{2}$ quantiles. In other words, $X_{1-\frac{\sigma}{2}}$ is the critical value of distribution $N(0, 1)$.

2.3 The Classic CUSUM Model

If the study subjects are compliance of $X \sim N(\mu, \sigma)$, the predictive statistics of the forecasting system are:

$$S_0 = 0 \tag{3}$$

$$z_i = (X_i - \mu_i)/\sigma \tag{4}$$

$$S_i = \max(0, S_{i-1} + z_i - k) \tag{5}$$

The initial value is $S_0 = 0$, while k is the reference value, which is one of the pending parameters in the CUSUM model. S_i is the required prediction statistic, S_{i-1} represents the statistical prediction value of the previous stage. z_i is a deviation of the current observation from the historical baseline. X_i is the current observation, and μ_i is the mean of the historical baseline. σ_i is the standard deviation of the historical baseline. In the CUSUM model, the historical baseline is selected according to the characteristics of the object. Generally, one way is to use the recently historical data as the baseline, and the other is a long baseline based on contemporaneous history data over the past few years and a small amount of observational data, which is swinging around near the history data. Once the prediction threshold h (another unknown parameter in the CUSUM model) in predictive analysis is set and $S_i > h$, it means that the prediction statistic exceeds the prediction value. Under this circumstance, the prediction system sends a prediction signal to get the department prepared.

2.4 The Inadequate of Classical CUSUM Model

When using the CUSUM model to predict the events in the time dimension, the observed occurrences on the day 'i' is greatly affected by recent data while the impact of the occurrence of the longer ones on the day 'i' are weaker. But in the CUSUM prediction model, the analysis and processing used under the same baseline window for the daily occurrence, usually does not take any factors into consider, and directly calculate the level under the baseline, which result in that the statistical deviation is rough, which makes the whole prediction model lack of validity.

3 Improved EW_CUSUM Model

There are m time points $X_1, X_2, \ldots, X_t, \ldots X_m$, $(10 < t < m, t \subset N^+, m \subset N^+)$ to observe the daily data occurrence of certain events and they are independent of each other and obey distribution $N(\mu, \sigma^2)$. Next, use the parameters g (Parameters that define the position of the baseline window, $g = 1, 2, 3\ldots, 7$) to select the required short baseline window.

Baseline window A: Set $g = 0$, the baseline window start from the day t-1 to the day t-7.

Baseline window B: Set $g = 2$, the baseline window start from the day t-2 to the day t-9.

According to the corresponding baseline window A, B, take a daily weight of raw data, which determines the level of impact on the expected amount of event data requested on day t. When the object compliance with $X \sim N(\mu, \delta)$, the t-day's expected amount of event data under corresponding baseline window is achievable. Formula is as follows:

$$y'_t = (1 - \lambda) \sum_{k=1}^{7} \lambda^{\mathrm{k}} X_{t-k-g} \tag{6}$$

where λ is the weight parameter, $\lambda \subset (0, 1)$, k is the size of the baseline window, and parameter g is the control of start and end position of the baseline window.

The daily expected changes in the corresponding baseline windows A and B will be obtained after the historical data processing. $y'_{t(A)}$ is the expected daily occurrences in the window A, $y'_{t(B)}$ is the expected daily occurrences of the events in the baseline window B, through which can calculate the average of the corresponding historical baseline $\mu'_{t(A)}, \mu'_{\mathrm{t}(B)}$ and standard deviation $\delta'_{t(A)}, \delta'_{t(B)}$. Thus, the prediction statistics in the predictive system are as follows:

$$S'_0 = 0 \tag{7}$$

$$z'_t = \frac{X_t - \mu'_{t(A)}}{\sigma'_{t(A)}} + \frac{X_t - \mu' t_{(B)}}{\sigma'_{t(B)}} \tag{8}$$

$$S'_t = \max(0, S'_{t-1} + z'_t - k) \tag{9}$$

The initial value $S'_0 = 0$, k is the reference value and also one of the pending parameters in the CUSUM model. S'_{t} is the required prediction statistics. S'_{t-1} represents the statistical prediction of the previous phase. z'_t is the deviation of the current observation from the historical one. X_t is the i-day observation which can get the required forecasting statistics under different baseline windows. Finally, compare the forecasting statistics with the threshold h to find out whether there is any deviation. If answer is yes, give corresponding measures.

4 Experimental Analysis and Verification

By getting the daily sales of JiaHua flowers cake from 2016 to 2017 in a supermarket as sample, shown in Fig. 1. experiments establish on Tomcat server using JAVA and SQL database. First, take the analysis of sales in May, June and July as an example, and then verify the prediction of sudden increase in sales during three months. Next, compare the predicted results of the CUSUM model with the C_CUSUM model using the timeliness index (where the timeliness /day = the predicted indication date of the event - the start of the event date). Finally, the supermarket would be able to provide guidance for the purchase of goods through the sales of the sudden increase in sales forecast.

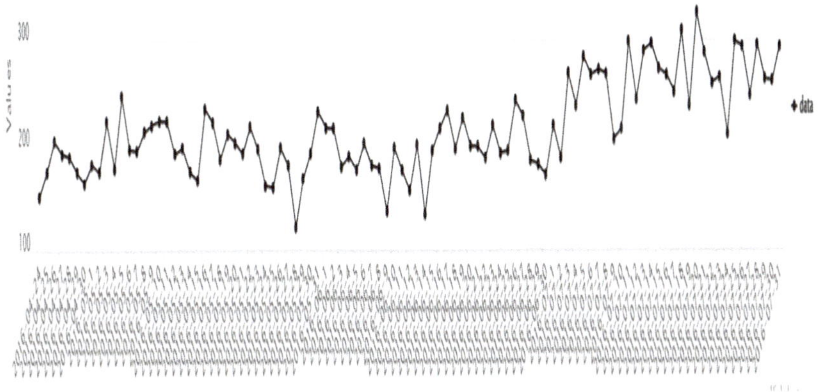

Fig. 1. The daily sales of JiaHua flowers cake from 2016 to 2017 in a supermarket

The daily sales volume of jia hua Flowers Cake in May, June and July was the experimental samples. Prospective experiments were performed on the data using the CUSUM model and C_CUSUM respectively. In the model, we choose different combination test for threshold h (h = 1, 2, 3), the parameter k (k = 0.5, 1.0, 1.5) and weight parameter λ (λ = 0.3, 4). Finally, the parameters of the model are determined as h = 2.5, k = 1, λ = 0.3. The experimental results shown in Figure 2 obviously shows that the prediction and analysis ability of EW_CUSUM model is obviously higher than that of CUSUM model when the threshold h = 2, and also verifies the effectiveness of EW_CUSUM model in detecting abnormal events.

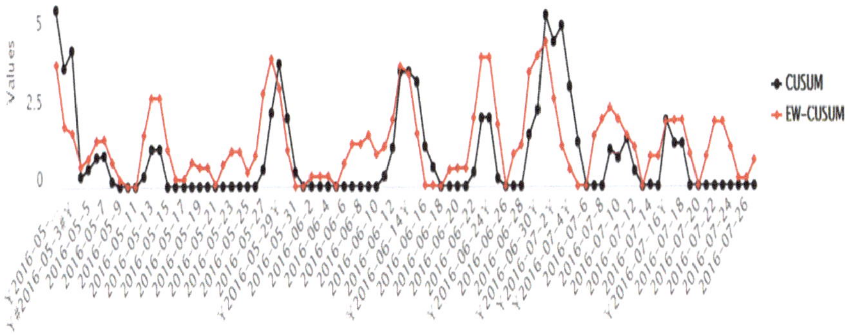

Fig. 2. The experimental results

The results of the experimental data analysis for May, June and July are shown in Table 1. The supermarket has a total of eight high-profile sales events, of which the CUSUM model results show that there are five predictions for the situation and seven for the EW_CUSUM model. By comparing the timeliness index (the time interval

Table 1. The results of the experimental data analysis

Event code	CUSUM model timeliness (day)	EW_CUSUM model timeliness (day)
1	3	4
2	–	2
3	2	3
4	3	3
5	–	2
6	3	5
7	–	0

between the start date of the event and the early warning date) of the abnormal prediction of each event, the timeliness of the analysis and prediction of the EW_CUSUM model is fully verified to be superior to that of the CUSUM model.

5 Conclusion

In this paper, base on JiaHua's Sales history data in a supermarket,we use EW_CUSUM model to obtain the expected data generated by per day under different baseline windows. Then add the two forecasting statistics under the baseline to get the required forecasting statistics. Finally, make an early forecast for the supermarket's JiaHua sales, and then give the departments guidance on the planning of this product. The experimental result also verifies the validity of the EW_CUSUM model and predictions are advanced. It also improved the timeliness of the whole model's predictive ability compared with the CUSUM model. As a result, this model gained more time for the enterprise to produce marketing plans. However, this model only conducts independent predictive analysis on the time dimension. The factors, which affect the sales of commodities, is not comprehensive. In addition, the sales of the products will be greatly affected by external factors such as holidays, seasonal choices and shop activities. Therefore, When developing a marketing plan, companies need to take this seriously. Moreover, this model does not consider the influence of spatial correlation across regions. If we introduce spatial influencing factors and influence factors of commodity sales in this model, we believe there will be new breakthroughs.

References

1. Xu, Y.: Research and Implementation of Sales Forecast of Enterprise Marketing System Based on Hadoop. Zhe Jiang University of Technology, Hangzhou (2015)
2. Li, Y.-L.: Design and implementation of sales forecast system in retail enterprises. China Manag. Infor. **18**(01), 64–66 (2015)
3. Sun, J.: Research on Algorithms of Sales Prediction and Personalized Recommendation Based on Retail Stores. Tianjin Normal University, Tiandjin (2016)
4. Jo, A., Riaz, M.: Mixed multivariate EWMA-CUSUM control charts for an improved process monitoring. Commun. Stat. **46**(14), 6980–6993 (2016)

5. Va, S., Papagalou, F.: Application of anomaly detection algorithms for detecting SYN flooding attacks. Global
6. Bu, Y., Zhang, H., Wang, R.: Network abnormal traffic detection based on improved CUSUM algorithm. Appl. Res. Comput. (2), 500–501 (2009)
7. Zhou, D., Yang, W., Lan, Y.J., et al.: Effect of scale and duration of outbreak on the effectiveness of CUSUM early warning model. Chin. J. Epidemiol. (6), 617–621 (2012)
8. Deng, F., Han, W., Zhu, Q., et al.: Improved CUSUM network traffic anomaly detection. Shandong Agric. Univ. J.: Nat. Sci. Ver. **3**, 356–359 (2014)
9. Sun, M., Song, X.: Cumulative score control chart based on improved cumulative and control chart. Stat. Decis. (18), 37–40 (2017)
10. Zaman, B., Abbas, N., Riaz, M.: Mixed CUSUM-EWMA chart for monitoring process dispersion. Int. J. Adv. Manuf. Technol. **86**(9–12), 3025–3039 (2016)
11. Xie, M., Goh, T.N., Lu, X.S.: A comparative study of CCC and CUSUM charts. Q. Reliab. Eng. Int. **14**(5), 315–339 (2015)
12. Cho, H.: Change-point detection in panel data via double CUSUM statistic. Electron. J. Stat. **10**(2), 2000–2038 (2016)
13. Sanusi, R.A., Riaz, M., Abbas N.: Using FIR to improve CUSUM charts for monitoring process dispersion. Q. Reliab. Eng. Int. **33** (2017)
14. Weib, C.H., Testik, M.C.: Residuals-based CUSUM charts for poisson INAR(1) processes. J. Q. Technol. A Q. J. Methods Appl. Relat. Top. **47**(1), 30–42 (2017)
15. Withers, C.S., Saralees, N.: A general class of cusum statistics. Prob. Eng. Inf. Sci. **29**(3), 361–384 (2015)
16. Zhao, H., Shu, L., Jiang, W.: An adaptive CUSUM chart for monitoring poisson rates with increasing population sizes. Eur. J. Ind. Eng. **9**(5), 592–715 (2015)

A Smooth Transition Algorithm Between Track Segments in CNC Machining

Jian-xin Xiao[1(✉)], Hui Zhang[2], Yong Zhang[2], and Guolin Li[1]

[1] Department of Electronic Engineering, Tsinghua University, Beijing 100084, China
xjxl5@mail.tsinghua.edu.cn
[2] Department of Mechanical Engineering, Tsinghua University, Beijing 100084, China

Abstract. The complex surface is the most common surface form in the numerical control process. In the numerical control system, the complex surface is often approached by a continuous straight segment, but the speed discontinuity between adjacent straight segments seriously affects the surface processing quality, thus often being used between track segments. Inserting a straight line or curve improves the speed continuity. This paper proposes an improved circular arc path planning algorithm based on the allowable error in the existing arc transfer algorithm, which can improve the transfer trajectory while ensuring the accuracy of the transfer trajectory. At the same time, based on the flexible acceleration and deceleration control method, a seven-section S-shaped acceleration/deceleration curve with continuous acceleration and jerk bounded is used in this paper to implement tool path speed planning and interpolation. The experimental results show that the proposed algorithm can significantly improve the transition speed between the track segments and improve the processing efficiency of complex curved surfaces under the same processing accuracy.

Keywords: Partial transfer method · Arc transition · Route planning
Feedrate scheduling

1 Introduction

In the aerospace, IT, woodworking furniture and other industries, high-speed and high-efficiency CNC machining of non-ferrous metals, composite materials, and non-metal low-density materials has a wide range of demands, with increasingly higher requirements being imposed on its processing efficiency and quality. With the rapid development of electronic technology and information technology, computer-aided design and manufacturing (CAD/CAM) systems can dissect part contour trajectories into linear G01-expressed straight-line segment trajectories, and are controlled by CNC [1]. In the numerical control machining, there are two ways to deal with the straight line segment: (1) Globally interpolate or fit the straight line trajectory segment; (2) Insert a straight line or curve between adjacent linear trajectory segments to achieve continuous speed transfer.

K. Deng et al. (Eds.): ICMIR 2018, AISC 856, pp. 104–109, 2019.
https://doi.org/10.1007/978-3-030-00214-5_13

Global Interpolation or Fitting [2] is a numerical control system that uses a smooth curve to globally approximate a straight trajectory, which transforms from a large number of consecutive straight segments to a global curve form. The characteristics of curve interpolation or fitting lie in the smooth and continuous machined surface as well as the improved machining efficiency. But it also has its limitations: First of all, in the process of global interpolation or fitting, the error is more difficult to be controlled; secondly, in the process of converting to a curve, iterative calculations are needed, which require large amounts of calculations, long time, and high requirements for real-time systems.

The method of realizing the trajectory transition between adjacent trajectory segments by using a straight line, an arc, or a spline curve [3–8] is more suitable for real-time system implementation because of simple calculation. This paper proposes a path planning algorithm that makes full use of the allowable errors of processing and improves arc transfer rate based on the arc transfer algorithm. It also considers the combination of path planning and speed planning. The path planning is based on the increase of allowable error. Radius is to improve transfer speed and propose a solution for continuous transfer of multiple long straight lines. Velocity planning adopts S-shaped velocity planning method, and deduces a speed planning solution model. Finally, it compares with existing algorithms to verify the proposed transfer algorithm and find that the speed control algorithm is feasible and effective.

2 Arc Transfer Path Planning Algorithm

2.1 Arc Transit Path Planning Model

Set the straight line trajectory segment L_1 and L_2 intersect at point A. In the actual processing of the numerical control system, the machining direction is $L_1 \rightarrow L_2$. Literature [4] has proposed a method of inserting a small arc curve at the corners of adjacent tracks to increase the transition speed so that the speed at the corner does not drop to zero. Figure 1 is a transition model based on arc transition transition smoothing algorithm. Insert a circle centered at O' in the adjacent corners, $O'B'$ arcs $B'C'$ and AO' intersecting with arc $B'C'$ at E points, and calculate the arc according to the geometric relation. Radius: $R' = e\tan\theta/(\sec\theta - \tan\theta)$, of which θ is half of the angle between adjacent linear trajectories, e is the distance between the intersection point A and the point E on the arc, that is, the allowable error of the transition section. Calculate the parameters of the B', C' and O' arcs based on the conditions of the linear trajectories L_1, L_2 parameters, the arc radius R', and the coordinates of the A points.

According to the literature [4], the transfer speed solution formula infers that: based on the arc transition transition smoothing algorithm, the transition speed between the track segments is proportional to the arc radius. In the numerical control part machining process, within the allowable tolerance range of $\pm e$, in accordance with this, this paper proposes an improved algorithm: the use of the transition allowable error $\pm e$, increase the insertion arc radius, and improve the transfer speed. As shown in Fig. 1, insert an arc PQ with a circle center O and a radius $OP = OQ$, where $L_1^{'}$ and $L_2^{'}$ are two lines parallel to the straight line trajectories L_1 and L_2, respectively. The distance between

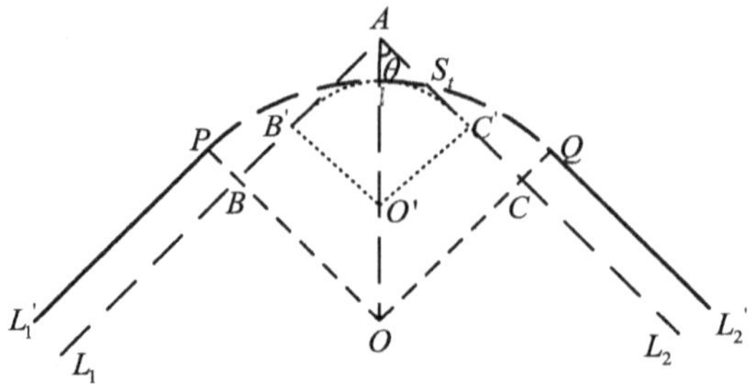

Fig. 1. Arc transition transition algorithm model

lines $L_1^{'}$, $L_2^{'}$ and lines L_1, L_2 respectively is the error constraint value. e, P and Q are the inserting points and the tangent points of $L_1^{'}$, $L_2^{'}$ respectively. The radii OP and OQ intersect with the command trajectories L_1 and L_2 at B and C, then the arc radius $R = OB + BP$ is inserted. The actual machining path after the optimization is based on the algorithm. It is $L_1^{'} \rightarrow PQ \rightarrow L_2^{'}$.

According to the optimization algorithm, the path planning is performed on the linear track segments L_1 and L_2, and the setting $e_{\max}$ is the error at the maximum interpolation speed of the arc transition between the track segments, and the radius of the inserted arc between the adjacent track segments can be obtained: $R = (2 * e_{\max} * \cos\theta/(1 - \sin\theta)) * \tan\theta + e_{\max}$. After comparison, the radius of the arc inserted between adjacent track segments is at least twice that of the method proposed by [4].

2.2 Speed Planning Algorithm

Taking into account the machine's own mechanical properties, the allowable acceleration of the machine tool itself is bounded, but when machining curve sections have a small radius of curvature, the actual acceleration may exceed the maximum acceleration allowed by the machine tool. Due to the small radius of the arc inserted in this algorithm, it is necessary to consider the acceleration limit.

Set the transfer speed to V, and the acceleration allowed by the system is $a_{\max}$. According to the acceleration constraint, the transfer speed can be obtained as: $V \leq \sqrt{a_{\max} R}$.

3 Experimental Verification

This section experimentally verifies the arc transition algorithm and speed control algorithm between the trajectories. The CNC machining parameters are set to: the maximum machining speed of the machine tool: $V_m = 100$ mm/s, the maximum allowable acceleration of the shaft: $A_m = 3000$ mm/ms^2, the default jerk: $J_m = 40000$ mm/ms^3, the interpolation period of $e = 0.1$ mm to facilitate theoretical

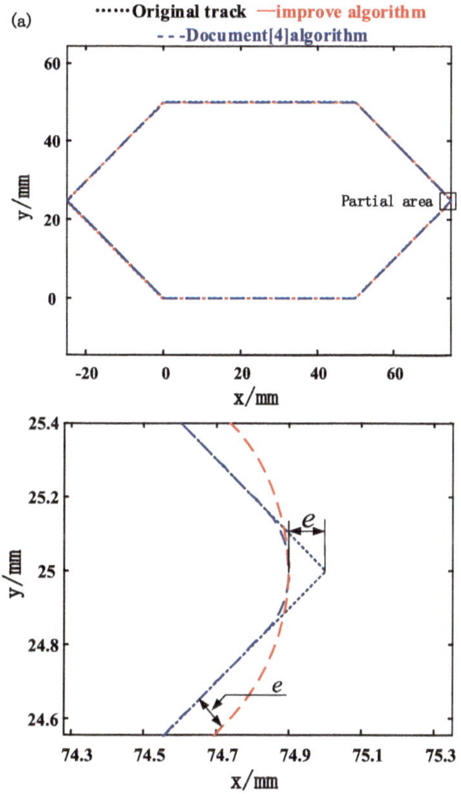

Fig. 2. Simulation experiment test track (a) Overall track (b) Local track

analysis and set the allowable error at the connection point between the track sections. Figure 1 shows the simulation results (Fig. 2).

Based on the continuous acceleration control algorithm proposed in Sect. 2, the speed planning and trajectory interpolation process is completed by two different circular arc transition methods. In addition, the actual processing speed, acceleration, jerk, and other trajectory information are calculated based on the interpolation results. As shown in Fig. 3, the actual processing speed, acceleration, and jerk are all within the given constraints, and the effectiveness of the proposed algorithm is verified. In terms of transfer speed, the arc transfer algorithm proposed in this paper is between the track segments. The speed is obviously higher than that in [4]. In terms of machining efficiency, the arc switching algorithm processing time is 2.637 s. Compared with the processing time of 2.904 s by method [4], the machining efficiency has been improved by about 9.2%.

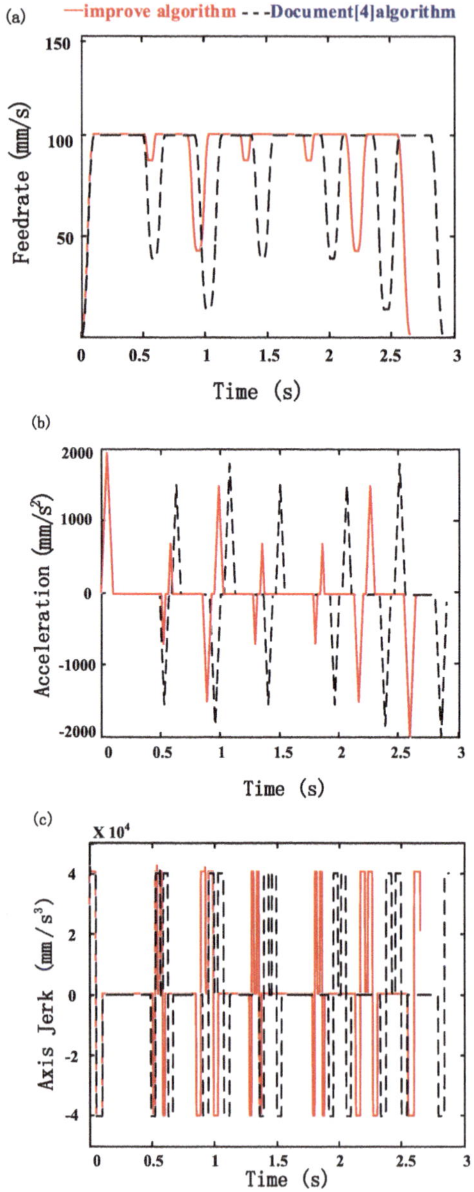

Fig. 3. Trajectory information (a) Speed curve (b) Acceleration curve (c) Axis Jerk curve

4 Conclusion

This paper proposes a small straight-line trajectory routing algorithm based on circular arc curves, which incorporates transfer errors into the transfer process, thus realizing non-zero transfer speed control. The algorithm can obtain more efficient arc transfer processes under the condition of acceleration constraints. The most common 7-section S-shaped acceleration and deceleration curves are used to achieve linear acceleration and deceleration control of the arc-arc trajectory, so as to ensure that the speed and acceleration curves are continuous, and the jerk bounded speed planning control. Compared with the traditional algorithm switching algorithm, the proposed algorithm significantly improves the processing efficiency for about 9.2%.

Acknowledgments. This work has been supported by China's national key technical support program (NO. 2015BAI01B16).

References

1. Erkorkmaz, K., Altintas, Y.: High speed CNC system design. Part II: modeling and Identification of Feed Drives. Int. J. Mach. Tools Manuf. **41**(10), 1487–1509 (2001)
2. Zhao, J., Zhu, B., Liu, F., et al.: Smooth interpolation of short lines in five-axis CNC machining. J. Mech. Eng. **52**(12), 1–8 (2016)
3. Ye, P., Zhao, S.: Study on control algorithm for micro-line continuous interpolation. China Mech. Eng. **15**, 38–40 (2004)
4. Lv, Q., Zhang, H., Yang, K., et al.: Study on the method of increasing turning velocity during CNC continuous machining. Technol. Test **07**, 79–83 (2008)
5. He, J., You, P., Wang, H.: A micro-line transition algorithm based on ferguson spline for high speed machining (HSM). Chin. J. Mech. Eng. **17**, 2085–2089 (2008)
6. Zhang, X., Yu, D., Yang, D., et al.: Corner curve transition interpolation algorithm for high speed machining of micro-line segment. J. Mech. Eng. **2**(19), 183–191 (2010)
7. Huang, J., Song, A., Tao, J., et al.: Smoothing transfer for the corner of adjacent NC processing segment. J. Shanghai Jiaotong Univ. **47**(05), 734–739 (2013)
8. Bi, Q.Z., Jin, Y.Q., Wang, Y.H., et al.: RETRACTED: an analytical curvature-continuous Bézier transition algorithm for high-speed machining of a linear tool path. Int. J. Mach. Tools Manuf. **57**, 55–65 (2012)

The Practice of the College Students' Network Security Quality Education

Shaoyun Guan(✉), Hui Zhang, Guibin Cao, Xinglan Wang, and Xuena Han

Harbin University of Commerce, Harbin, China
gsyj91@163.com

Abstract. With the increasingly development of network, various information security incidents are happening frequently. College students are becoming the main victims and high-risk groups of all kinds of security incidents due to their lack of social experience and their weakness of information security awareness. In this paper, a stepwise network and information security education mode are constructed, then a teaching design for college students' information security quality is expounded from three aspects: teaching objectives, contents and methods, at last this paper discusses the relationship between teachers and students in teaching practice.

Keywords: College students · Information security · Education

1 Introduction

Whether we want to see it or not, college students are becoming the main victims of various network attacks and information security incidents in recent years. It brings bad influence on individuals, families, society and country. In May 2016, Wei Zhexi, a student at Xi'an University of Electronic Science and Technology, died from a medical advertisement in Baidu; in August, many college students, such as Xu Yuyu and Song Yining, lost their young lives because of fraud online. Many students and families suffered from the frequent incidents of "Internet credits" and "bare notes". In May 2017 global blue blackmail virus hit colleges and universities hardest, with the graduation thesis being encrypted and the risk of failing to graduate on time. There are also all kinds of countless things that caused loss and damage, such as fake part-time jobs, Internet frauds, phishing, information disclosure, network infiltration, network attacks and so on. The Internet is a double-edged sword for college students. If it is used properly, it is a treasure house of Alibaba or a "open sesame", and there are endless wealth, and unexpected opportunities. Otherwise, it's a Pandora's box. It's a disaster. Not only will it influence the daily life of the individuals, but also it may even threaten national security. Just as president Xi pointed out, "there is no national security without network security". "Not realizing that risk is the biggest risk". Education is the best firewall for network and information security threats and events. Through the network and information security quality education, the students can master the necessary safety awareness, skills and basic knowledge. They can understand and comply with the

K. Deng et al. (Eds.): ICMIR 2018, AISC 856, pp. 110–114, 2019.
https://doi.org/10.1007/978-3-030-00214-5_14

network moral norms of the space. It helps to avoid network security events and maintain campus harmony, social stability and national security.

2 To Construct a Stepwise Network and Information Security Education Mode

Network and information security education is a complex system project involving computer, communication, law and management. It can't be built overnight. It is a good strategy to build a stepwise teaching mode. Coordinated work of various departments in universities can realize the multi-level popularization of information security quality education effectively.

Stepwise network security education model is shown in Fig. 1

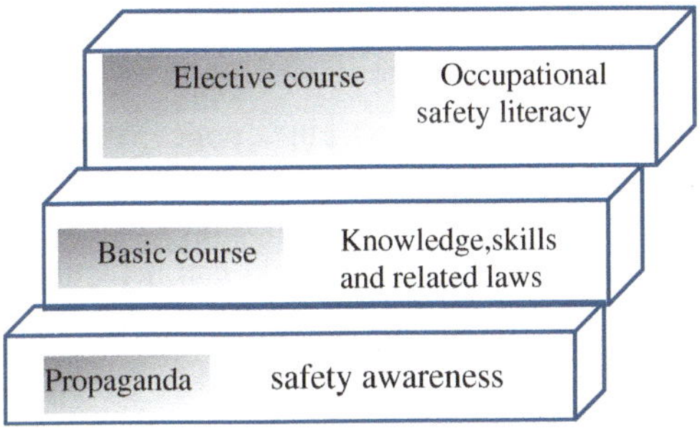

Fig. 1. The stepwise security education model

2.1 To Improve Safety Awareness in a Timely Manner in Many Ways

College students' consciousness of network security should be strengthened Through a variety of forms such as lectures, posters, brochures, the campus network, TV, newspaper and so on. They should know the various risks in network application so as to avoid them. The awareness education of the network security for college students must be timely. Now cyberspace environment is not optimistic and network crimes are still rampant. Personal network leakage is so serious that the crimes are more and more accurate. It is a fatal weakness for college students. It's better to warn the students of the high-risk and high-probability safety accidents. For many network security incidents, people usually say "never too late, never too late". But if we can build a barrier of security awareness earlier, then we can avoid the loss of money, the harm of physical and even the cost of life.

2.2 To Master Systematic Knowledge by Learning Public Basic Courses

Network and information security education can't just stay in consciousness, It is systematic and informative, and it should not be fragmented and marginalized. President Xi noted that "network security and informationization are the twin wings of a bird and two wheels of a driving force." However, in the current education system in our country, the informational education has always been a "wheel". With the popularization of the informationization, we should popularize the informationization since primary school. But network and information security education is almost missing in education system. It is so difficult for students to face the more and more complex network security situation, because of their lack of security awareness and their lack of knowledge and skills reserve. Network security education should be incorporated into the public education scheme as well as other traditional campus safety education. Students should be taught the basic knowledge and skills of network security by learning public basic courses (such as computer, law, etc.), and students should obey relevant laws and moral norms. What's more, Internet and information security education is more complex, because threats change rapidly with the development of various technologies and applications.

2.3 To Strengthen Occupational Safety Awareness by Learning Elective Courses

With the promotion of China's "Internet plus" policy, college students should also master the knowledge and skills of network security related to their major and industry. In future positions, talent will not be the short board of the "barrel effect". So colleges and universities should set public elective courses and network security + professional curriculum to meet the basic requirements of professionals.

3 The Teaching Design of Network Security Education

3.1 The Teaching Objective Should Meet the Requirements of the Times

The goal of teaching includes not only the professional knowledge but also the requirements of the students' network security literacy corresponding to the country and society. That is to say that the teaching objectives are more in line with the requirements of the times. In formulating the teaching goal, we should consult the requirements of the publicity week of network security in recent years and we should also consult the requirements of the communist youth league in network security in order to educate the students who have higher safety consciousness, better network literacy and more protection skills. With the certain knowledge of network security, college students can play important roles in building a safe network situation in modern life.

3.2 Teaching Content Should Be Targeted and Practical

Network security education is different from fire prevention, burglary prevention and earthquake prevention it involves much more knowledge and stronger professional technical. And its technology updates more quickly. All sorts of threats happen in endlessly, So it is a course that requires lifelong learning for college freshmen, who are lack of professional knowledge and social experience. The design of its teaching content should conform to the rules of learning and the students' knowledge background and focus on students. And it includes four subjects: the safety awareness, skills, codes of ethics and basic knowledge. The purpose is to make students identify potential risk and vulnerability so as to defense the known security risks and prevent the unknown security risks. Meanwhile, we should teach students obey the rules and laws of cyberspace, and not to harm others. Before teaching this part, we should investigate the security situation of the network to ensure the practicality of the content.

3.3 Teaching Methods Should Stimulate Students' Interest

The students' learning interest has an important influence on the teaching effect. Some hot spot network security events can arouse students' interest easily. so that students can understand the essential knowledge of the network security. We should select those hot cases that are new and typical and that have strong influence and high visibility. For example, when introducing storage security, we can introduce the following incidents: the photo incident of Edison Chen in 2008, the indecent photos of the apple phone of the Hollywood star in 2014, and the Baidu incident in 2016. Students are so interested in these incidents that they will focus on the nature of these events – the various safety precautions and techniques for storing security. It is inevitable that network security education will involve cryptography and some theoretical knowledge that is not easy to present. Students are tired of them. With some spy dramas such as "Lurk", "The wind" and "Cybercrime Investigation", it is easier for students to understand the visualized knowledge. In addition, Students should be provided with experimental operation platform to simulate network security attack and defense process, which can motivate students to master the skills of prevention and treatment of practice.

4 Conclusion

During the university's network security education, teachers' professional quality cannot be ignored. Teachers are the basic guarantee that complete teaching work, and they play leading roles in the whole teaching work. In the limited teaching plan, what to teach and how to teach? So teachers should have professional knowledge background, accept new things actively and focus on the development of information network security. They should also have the consciousness of safety warning and the strong sense of responsibility.

In addition, in the process of education, students should have a dialectical understanding of the present and future of network security. We should neither refuse new things and technologies because of various security incidents, nor forget about potential threats. It is necessary to let the students understand that there is no absolute security in cyberspace. Threats and vulnerabilities will persist with the progress of technology. The "two sides" of the technological development process cannot be eliminated. It can only rise in a spiral of attack and defense. We must prevent actively and dispose quickly to minimize the risk of harm. Building your own "firewall" is the most effective way to avoid network and information security events.

Acknowledgments. This thesis is a stage achievement of the project (No. 17EDE315) of the Heilongjiang Province Society Scientific Fund and the project (No. SJGY20170054) of the Heilongjiang Province Education Fund Many people have contributed to it. This thesis will never be accomplished without all the invaluable contributions selflessly from my group members. It is my greatest pleasure and honor working with my group. I'm deeply grateful to them.

References

1. Li, H.: The enlightenment of American information security education and training. Comput. Educ. (2015)
2. Tan, Y.: International comparison and enlightenment of network security education for college students. Electron. Gov. (2017)
3. Zhang, B.: Research on embedded teaching of information literacy education. Mod. Intell. (2015)
4. Liu, X-P., Song J-H.: Development of Domestic Information Security Training. Inf. Secur. Technol. 74–77 (2010)
5. Information on. http://www.moe.gov.cn/jyb_xxgk/moe_1777/moe_1778/201511/t20151127_221423.html

Study on Rapid Layering Construction Technology of Panoramic Model for Electrical Equipment

Zhimin He[1,2](✉), Lin Peng[1], Haiyun Han[1], Min Xu[1], Gang Wang[1], Hai Yu[1], Xingchuan Bao[1], Zhansheng Hou[1], He Wang[1], Liang Zhu[1], and Zehao Zhang[1]

[1] Global Energy Interconnection Research Institute, Nanri Road 8, Nanjing 210003, China
825097034@qq.com

[2] State Grid Key Laboratory of Information and Network Security, Beiqijia, Beijing 100000, China

Abstract. Computer fast scan modeling for converter valves starts with 3D modeling technology based on computer geometry modeling, visual modeling technology based on depth of field sensors, automatic layered reconstruction technology of internal and external parts models of electrical equipment, developing the rapid layering construction technology research of a panoramic model for electrical equipment. Taking the converter valve as the object, the traditional 3D modeling, fast scanning modeling and model layering are determined, and a computer-identifiable model is created using fast depth-of-field scanning technology, and a refined 3D model of the traditional design stage is hierarchically displayed. Realize the all-round display of the interior and exterior parts of the power equipment model on the job site.

Keywords: Rapid layering · 3D modeling · Fast scan modeling

1 Introduction

1.1 Background

With the acceleration of smart grid construction, large-scale electrical equipment such as DC converter valves are constantly emerging. Due to the complex structure of equipment, multiple types of monitoring data, and abstraction of background monitoring pages, it is difficult for on-site personnel to accurately and intuitively understand the internal and external structures of the equipment and the real-time running status of each part. The virtual 3D model of the traditional power equipment in the design and manufacturing stage is difficult to be used for physical equipment computer identification, and does not show the internal structure of the equipment during operation. The large-scale power equipment lacks a panoramic model that can be used for computer recognition and display of internal and external hierarchical structures. At the same time, with the rapid development of panoramic modeling technology and augmented reality technology, it can provide new and intuitive intelligent solutions for on-site

K. Deng et al. (Eds.): ICMIR 2018, AISC 856, pp. 115–122, 2019.
https://doi.org/10.1007/978-3-030-00214-5_15

operation and maintenance of large-scale equipment. Therefore, it is necessary to study the panoramic scanning modeling technology and model layering technology of large-scale electric equipment, and to use augmented reality technology to display the monitoring data of the equipment during operation and the panoramic model, so that the on-site operators can more intuitively understand the inside and outside of the equipment. In the case of construction, the corresponding component operation status is quickly obtained, and the on-site operation personnel are assisted to more accurately and comprehensively check the operation status of the entire device and the components. Through the key technologies mastered in this study, technical assistance is provided for the further development of intelligent operation and maintenance systems for large-scale power equipment based on panoramic models, and the efficiency of on-site inspection operations of large-scale power equipment and the accuracy of on-site equipment diagnosis are improved.

1.2 Necessity of Subject

The technical achievements of this project will lay a solid foundation for the future development of the panoramic augmented reality inspection operation system. Through the subsequent research and development of the practical system, it can be applied to the operation and maintenance of DC converter valves, and the realization of panoramic layered overlay display The internal and external 3D model of the flow valve combines the identified equipment components with its corresponding monitoring data, and is superimposed and displayed around the equipment components, so that the on-site operators can more intuitively understand the internal and external configuration of the equipment and the overall equipment and components operation. State, improve the efficiency of on-site operation, and assist in the on-site diagnosis of the status of the converter valve, improve the accuracy of field device fault diagnosis, and extend the service life of equipment, with obvious direct and indirect benefits.

2 Domestic and Abroad Research Status

2.1 Abroad Research Status

The mature 3D modeling software on the market is mainly foreign, mainly including the Solid Works software introduced by Dassault Corporation in 1995. It is one of the world's largest installed and best-use software. PRO/E software is a comprehensive 3D software integrating CAD/CAM/CAE functions. It plays an important role in the field of 3D modeling software and it is the new standard in the field of mechanical CAD/CAM/CAE in today's world. The recognition and promotion of the industry is one of the most successful CAD/CAM/CAE software. 3DS MAX is a PC-based 3D animation rendering and production software developed by Discreet. Autodesk Maya is the world's top 3D animation software produced by the United States Autodesk Corporation. It has high production efficiency and strong rendering sense. It is a movie-level high-end production software. With the development of these modeling software, the corresponding 3D modeling technology has gradually increased, but also makes people's lives more and more abundant.

The panoramic scanning modeling technology mainly refers to the scanning modeling of computer 3D images. The early 3D image scanning modeling technology is mainly the study of matching and stitching algorithms. The earliest research can be traced back to 1975, after 1990s, the development of automated imaging technology. Rapidly, a series of high-resolution panoramic imaging systems such as PAL cameras emerged. Due to phase-dependent frequency domain methods, there is a high demand for rotation and deformation of images. Successive scholars have turned to the spatial domain based on optical aberration and feature matching. Solve the problem. With the development of computer graphics technology, more and more research institutes and companies have joined the field, making great progress in the field of global scene environment modeling. The University of Washington and Microsoft Labs have developed a real-time visual SLAM system to establish 3D points. Cloud map.

3D model layering techniques and methods have become hot topics in the rapid processing of formed data. Some scholars at home and abroad have conducted some research on the direct layering of 3D CAD models. Foreign countries started earlier in the research of model layering algorithms, and the research is deeper. Jamieson et al. (1995) used C language to study the direct layering of 3D CAD models on Unisolids solid modeling kernel Parasolid, and used the calculation layer area. The method of deviation adaptively determines the layer thickness. Zhao et al. (2000) used the same layer thickness calculation method and used Auto CAD Run Time Extension (ARX) to study the adaptive layering method of CAD models. Ma et al. used a similar adaptive layer thickness calculation method and studied a selective fill scanning method based on Mechanical Desk Top (1999) and Unigraphics (2004), respectively.

2.2 Domestic Research Status

The development of China's 3D software CAD/CAE began in the late 1970s and its development has been rapid. In the development of CAD/CAM software on the microcomputer platform, China and foreign countries have similar starting points. All of them use Visual C, OpenGL and other tools for software development. Many domestic universities, software companies and enterprises have developed advanced and have their own characteristics on this basis. CAD/CAE software or modules that are in line with Chinese users' habits, some of which have been promoted and used. For example, CAXA 3D modeling software developed by Beijing Digital Technology Co., Ltd. also occupies a large share in the low-end market. The metal plastic forming 3D finite element simulation system developed by Shanghai Jiaotong University's Zhongmo Company has reached the international advanced level in its rigid (sticky) plastic finite element analyzer and dynamic boundary processing technology. Chinese Academy of Sciences CASS Software Group and Beijing Kaisi Bohong Application Engineering Co., Ltd. have developed their own proprietary PICAD system and series of software. However, the domestic 3D modeling software is still at a low level compared with other foreign industries, mainly occupying the domestic low-end industry market, and still has a long way to go in terms of core technology and market share.

The level of domestic 3D image scanning modeling algorithms is slightly lower than that of foreign countries. However, after years of development, it has also made remarkable achievements. Dr. Liu Yanli of Central South University proposed a VO-EIF SLAM method based on visual track deduction and extended information filtering, and finally completed 3D synchronous positioning and map construction based on color information and depth information. The CAD&CG laboratory of Zhejiang University used hierarchical layering and pre-blanking technology to realize the virtual building environment roaming.

In China, there has been considerable progress in the hierarchical algorithm of the model. Chen Xubing et al. (2000) studied the direct slicing of the 3D CAD model using the macro language as a secondary development tool based on the Power SHAPE modeling software. Yang Jia et al. (2002) used Object ARX's AcBr library to study the direct delamination method of boundary-expressed 3D CAD models. According to Longgang et al. (2009), based on the principle that a smooth curve can be enveloped by its upper tangent line, a slice thickness calculation method based on a fixed-length tangent line is proposed.

3 Key Technology Research

After investigation and analysis, this scheme is aimed at the 3D modeling of the converter valve, taking the reference to the existing equipment design drawings and the use of modeling software for hand book drawing and point cloud modeling. Reference to the existing equipment design drawings and desk accounts using modeling software hand-drawn, with the characteristics of exempt from the industry acquisition, cost base; The point cloud modeling is the most direct and common 3D model to achieve accurate modeling of equipment, This method can not only make up for blind spots that are incomplete due to incomplete drawings and cannot be modeled by equipment design and ledgers. At the same time, the device models modeled by point clouds, called true 3D models, can satisfy some applications that rely on high-precision model data demand.

Using traditional modeling techniques and model layered reconstruction techniques, using a converter valve as a target, a 3D layered model is constructed, and models such as structural layer, electrical layer (water and electricity, etc.), and operation and assembly between layers are divided. Layer construction. The use of rapid depth-of-field scanning technology to establish a computer-readable main component depth of field cloud model for computer automatic object recognition and matching, to achieve model recognition, registration overlay and tracking functions, as shown in Fig. 1.

3.1 Traditional 3D Modeling

The idea of 3D modeling method of converter valve based on two-dimensional drawings is that the 3D model of the 3D model of the converter valve is quickly converted into two-dimensional and symbolic product drawings using the existing converter valve design, construction, and completion phases. Common technical

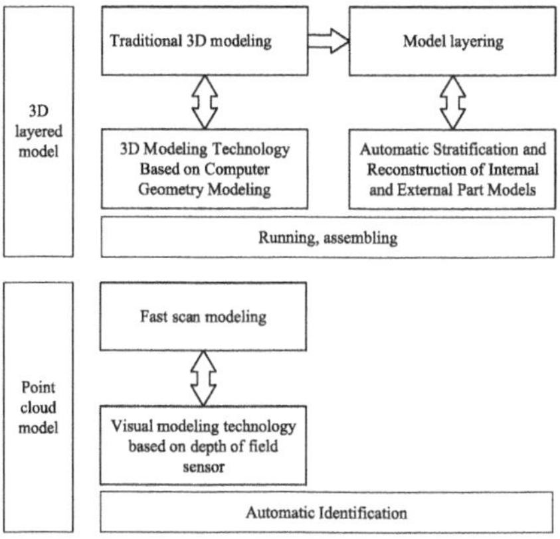

Fig. 1. Implementation

means. It is characterized by the use of 3D modeling software or parameter drivers to construct 3D grid devices and their scenes after designing in 2D CAD and other platforms.

Using the basic idea of building a static component library and then constructing a dynamic converter valve, the method and steps for converting the two-dimensional graphic data of the DC converter valve into 3D model data are as follows:

(1) Build a component library
First, collect data, including the valve parameter table, converter valve mosaic, existing equipment design and other relevant information. Draw a two-dimensional graphic of the converter valve model, which consists of a rectangle, a cylinder, a line, and an S-curve. Because there are a large number of duplicate components in the model, you can draw one component first and then reuse it multiple times. 3D parametric modeling and attribute information configuration of the converter valve component library are established, and a corresponding component library is built.
(2) Converter valve splicing
Through the installation configuration document of the converter valve, the connection mode and the angle of the converter valve are obtained, and a two-dimensional graphic of the converter valve model is drawn according to the description sequence. Secondly, the line segments parallel to the internal components of the converter valve and the connecting lines are made by scatter, and the two-dimensional figures after the scatter processing are saved in the DXF format.
(3) Data processing
First, the DXF file is automatically processed to extract the graphic information and normalized. Second, the two-dimensional graphic coordinate data is converted into node coordinates, and the sub-grid is divided into different groups to facilitate the calculation of the parameter assignment. Again, the program will directly generate the data files needed for the calculation and save it in txt format. Finally, a 3D data model can be generated by calling thc txt data file generated by the program and adding boundary conditions, initial conditions, and rock mechanics parameters.

3.2 Fast Scan Modeling

The visual modeling system based on depth of field sensor is composed of hardware and software. Through the calibration of sensors, the conversion formulas of depth data and point cloud data are obtained. After analyzing the sensor accuracy, it is determined that the ICP point cloud splicing algorithm and the MC surface modeling algorithm are used as the main algorithm. The scanned 3D model can be used for subsequent model identification and achieve the function of registration overlay and tracking.

The main hardware components of the visual modeling system based on the depth of field sensor include a depth of field sensor and a PC. When the measurement is performed, the operator holds the depth sensor and photographs the part around the part to obtain the depth data under different shooting angles. At the same time, these deep data are transmitted from the USB interface to the host computer and processed in the host computer. After the measurement is completed, the host computer can generate a standard 3D model file format.

Visual modeling is mainly divided into depth-of-field sensor point cloud image acquisition, point cloud image processing, model database construction, and real-time identification algorithms, as shown in Fig. 2.

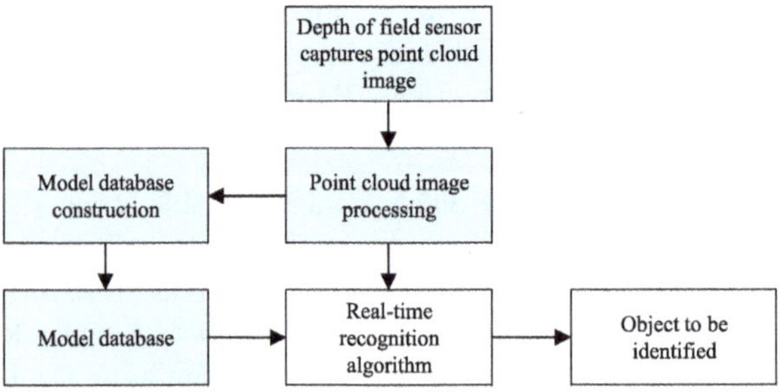

Fig. 2. The main software process

The cloud image of the current scene captured by the depth sensor contains noise effects, passes through filtering and other preprocessing stages, passes into the point cloud image processing stage, and uses the European cluster segmentation method to extract the point cloud containing the recognized object. Clustering, which enters the lower channel, is a real-time identification phase, and entering the left channel is building a model database.

Point cloud image processing includes preprocessing, data registration, data fusion, data correction, and the like. After a series of registration and splicing processes, a point cloud model whose main content is 3D coordinates is generated. The models required in this project are standard 3D surface model formats such as OBJ, STL, and so on. 3D surface reconstruction using Marching cubes algorithm. Through the above steps, the complete physical model can be obtained.

3.3 Model Layering

Combined with the 3D model of power equipment, the internal and external component models are automatically layered to fully display the internal structural features of the equipment. Taking a DC converter valve as an example, it is necessary to perform layered construction of models such as operation and assembly between structural layers, electrical layers (water and electricity, etc.), and various layers of components. Since the traditional model layering technology requires conversion of the model format (such as STL, etc.), it will bring about the loss of model accuracy, leading to the problem of decreasing the quality of details after layering. Therefore, it is necessary to use the 3D model direct tiering technology to avoid the loss of accuracy due to the conversion of the model format.

The 3D model direct hierarchical reconstruction process is as follows:

(1) The establishment of three-dimensional CAD model

Establishing CAD 3D models can be built directly with existing CAD software, such as Pro/E, I-DEAS, AutoCAD, UG, etc.; 2D patterns of existing products can also be converted to form 3D models, or lasers can be applied to product entities. Scanning, CT tomography, point cloud data, and then reverse engineering method to construct a three-dimensional model.

(2) Virtual slice processing

In the CAD software system, select the appropriate machining direction according to the characteristics of the workpiece, cut the CAD 3D model with a series of planes at a certain interval in the machining height direction, and then extract the outline information of the section, usually the interval is 0.05–0.5 mm. The commonly used is 0.1 mm. The smaller the interval, the higher the forming accuracy, but the longer the forming time, the lower the efficiency; on the contrary, the accuracy is low, but the efficiency is high.

(3) Scan path planning

Inside the cross-section profile after the slicing process, an optimal scanning path is sought. Typical are: parallel scan path, contour scan path, fractal scan path, etc.

(4) Hierarchical model

After slicing the outline of the CAD model and the well-defined scan path, under the control of the computer, the scanning motion is performed according to the contour information of each section, the model is divided layer by layer, and then the layers are separated from each other, and a layered prototype product is finally obtained.

Object ARX is used to directly layer 3D solid model in 3D modeling software. Layering directly on the basis of 3D model data can accurately extract model section profile data at a specified height, avoiding the STL format conversion. Profile data accuracy loss problem.

4 Conclusion

Through in-depth research on the rapid layered construction technology of panoramic models of electrical equipment, breakthroughs in project technology bottlenecks, mastered the core technologies for the panoramic modeling of electrical equipment.

Combining depth-of-field scanning modeling technology with traditional 3D modeling technology, a computer-readable model is created using fast depth-of-field scanning technology, and then the feature matching degree calculation is performed with the existing model to complete device identification.

Using the direct layering technique for 3D models, without the aid of an intermediary file conversion method, the 3D model of large-scale power equipment is used to shape the geometrical characteristics of the slice scanning contours in different distributions of the tops of the slices, and accurately extract the 3D model at a specified height. The profile data of the cross-section, at the same time, the different internal structures such as the physical layer and the electrical layer are hierarchically displayed in real time, so as to solve the loss problem of the model geometric description accuracy caused by the data format conversion.

Acknowledgments. This work was financially supported by the science and technology project to State Grid Corporation "Research on the interactive technology of intelligent transportation and maintenance information fusion based on panoramic layered electrical equipment model". I would like to express my heartfelt gratitude to my colleagues and friends, they gave me a lot of useful advices during the process of studying this topic research and also provided enthusiastic help in the process of typesetting and writing thesis! At the same time, I want to thank all the scholars who are quoted in this paper. Due to my limited academic level, there are some disadvantages to writing a paper, and I will solicit the criticism and corrections from experts and scholars.

References

Bi, S., Zhang, G., Hou, R., Liang, J.: Comparing research on 3D modeling technology and its implement methods. J. Wuhan Univ. Technol. **32**(16), 26–30 (2010)

Fan, L., Tang, J.: Expectation and review on overhead transmission lines 3D modeling methods. South. Energy Constr. **4**(2), 120–125 (2017)

Zhao, C.: Research and application of substation visualization based on 3D panorama and rapid modeling technology. Master's degree Dissertation, Shandong University (2013)

Li, Y.: Research on key techniques of 3D surface reconstruction based on depth camera. Doctoral Dissertation, Zhejiang University (2015)

He, Q., Cheng, H., Yang, X.: A review of 3D model process technology oriented to 3D printing. Manuf. Technol. Mach. Tools **6**, 54–61 (2016)

Bearing Compound Fault Diagnosis Based on Morphological Filtering and Independent Component Analysis

Feng Liu[1(✉)], Lian-jiang Xu[1], and Yun-hai Yan[2]

[1] College of Mechanical and Electrical Engineering, Yunnan Open University, Kunming 650500, China
L744492418@163.com

[2] Yunnan Tobacco Machinery Co., Ltd., Kunming 650106, China

Abstract. The state monitoring and fault diagnosis of rolling bearings are of great significance. The fault signal of rolling bearing contains periodic impact components, and the impact frequency reflects the location information of bearing fault. In order to effectively diagnose the early and weak fault of rolling bearing, the method of morphological filtering was used to pretreat the fault signal of rolling bearing in this paper. Then, the independent component analysis (ICA) was applied to separate the signal after morphological filtering. The results of experimental study on the compound fault signals of rolling bearing inner and outer ring show that this method can effectively identify and separate the fault characteristics of rolling bearing.

Keywords: Rolling bearing · Fault diagnosis
Independent component analysis · Morphological filtering

1 Introduction

Rolling bearings are the most common parts of rotating machinery, and its state has great influence on the working conditions of the machine. According to the statistics, 30 percent of the faults of rotating machinery are caused by bearing faults [1]. How to extract sufficient quantity, real and objective information from various mechanical state signals is the key to the success of fault diagnosis [2]. During the operation of rolling bearing, the acquisition signal is disturbed by the vibration of a large number of non-monitoring components, so that the effective signal can be drowned. Moreover, the fault signal is very weak, often causing the fault information to be drowned in the background noise and interference, thus making the signal feature extraction and fault diagnosis become very difficult [3, 4].

Independent component analysis based on high order statistics can effectively find the independent component in the mixed signal and has gained wide attention in the field of mechanical fault signal processing [5]. However, ICA takes the independence of the result of the separation as the criterion, so that it can separate the other irrelevant signals or noises as independent sources of information. In the case of more disturbance

K. Deng et al. (Eds.): ICMIR 2018, AISC 856, pp. 123–129, 2019.
https://doi.org/10.1007/978-3-030-00214-5_16

source, it is not conducive to making the correct judgment. Therefore, pretreatment can be done before the fault signal analysis.

2 The Principle of Morphological Filtering

Mathematical morphology is a tool of signal processing and analyzing based on set theory, and morphological transformation is a nonlinear transformation based on mathematical morphology.

The basic idea of mathematical morphology is based on the geometrical features of signals. A structural element is designed to be calculated with the analysis signal to achieve the purpose of extracting useful information, maintaining the characteristic details of the signal and suppressing the noise. The basic operation of mathematical morphology includes corrosion, expansion, opening and closing.

Because of the anti-expansibility and the expansion of morphology, the signal has statistical deviation in the process of morphological filtering that has a bad effect on noise suppression. In order to eliminate this influence, open-closed and closed-open combination filters are constructed as:

$$Y(n) = [OC\ (f\ (n)) + CO\ (f\ (n))]/2$$

3 Independent Component Analysis

Before the mixed signals been blind separated by FastICA, it should be pretreated that such as centralization and whitening. Centralization is one method that used to make sure the observation signal's mean equal to 0. Whitening is another method that translates the observation signal into irrelevant unit variance.

A linear structure element has been chosen for independent component analysis that its amplitude is 0 and length is 10. Through the open-closed and closed-open combination filters, which the compound fault signals can be filtered by morphological filtering. Then, the filtered signals will be analyzed though the independent component.

4 Acquisition of Bearing Fault Signal

The signal acquisition system of bearing fault diagnosis composed by NI9234 data acquisition card, PCB acceleration sensor, eddy current sensor, and computer. Through the VI that used for multi channels data acquisition that designed by the LabVIEW that it is a kind of virtual instrument software. It can be used for collecting the fault bearing's acceleration and speed when the fault bearing is working.

The interface of the data acquisition system is shown in Fig. 1.

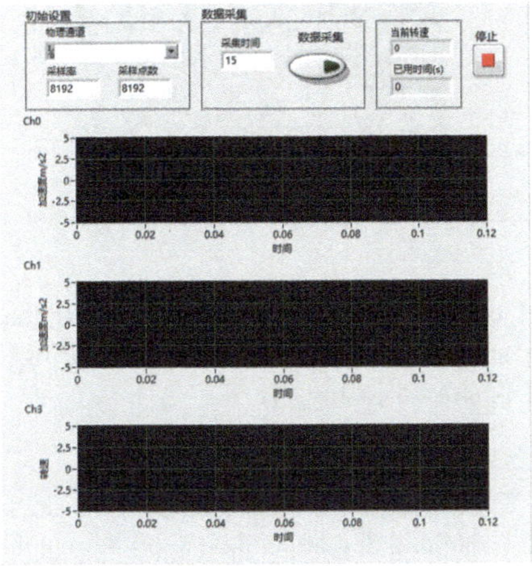

Fig. 1. Data acquisition system interface

5 Experimental Verification

In order to verify the practicality and effectiveness of the method, the rolling bearing which its inner and outer ring has composite fault that has been chosen for the experiments. The parameters of the signal acquisition system that the speed of 800 r/min, the sampling frequency is 8192 Hz, the sampling points for N = 8192 that will be used for the experiments. And the parameters of the rolling bearing that pitch circle diameter D = 39 mm, roller diameter d = 7.5 mm, rolling body number Z = 12, contact angle alpha = 0 deg will be used in the experiments too. According to the parameters, the rolling bearing's faults are that outer ring's fault is 64.61 Hz, the inner ring's fault is 95.38 Hz, and the cage fault is 5.30 Hz.

The experimental device is shown in Fig. 2, and the vibration acceleration installation position is shown in Fig. 3.

Fig. 2. Experimental device

Fig. 3. Acceleration installation position

A series number of experimental data were collected that including the normal rolling bearing. In the same external condition, the rolling bearing's time domain waveform of the signal is shown in Fig. 4.

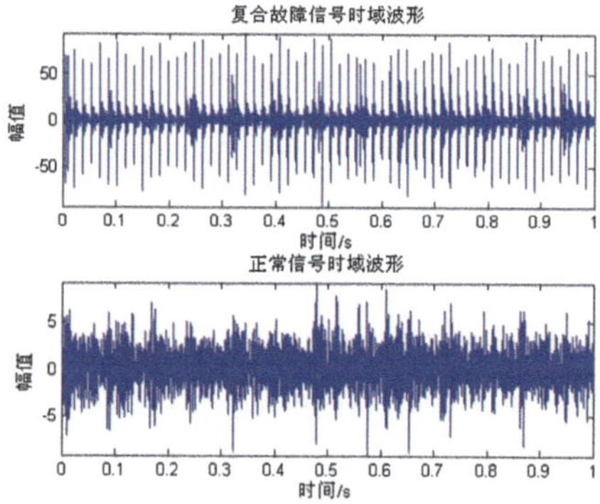

Fig. 4. Time domain waveform of signal acquisition

The method that the Fourier transform which direct used in locating the fault characteristics that the fault signal features cannot be shown by the method, it shows in Fig. 5.

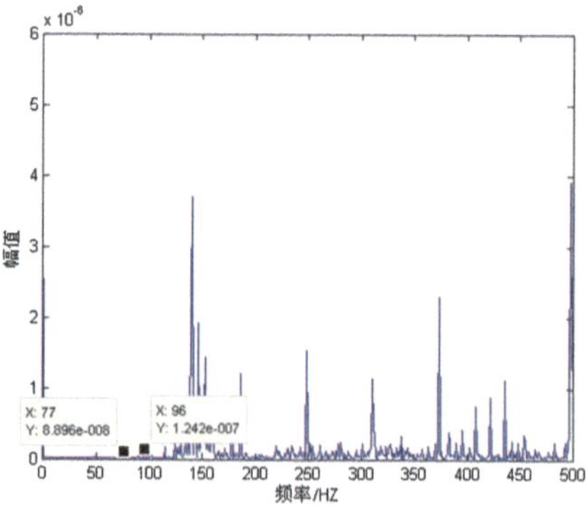

Fig. 5. Envelope demodulation spectrum of fault signal

The method that the FastICA which direct used in isolating the fault signal features without pretreatment that the result is shown in Fig. 6, but it can't show the fault signal features either.

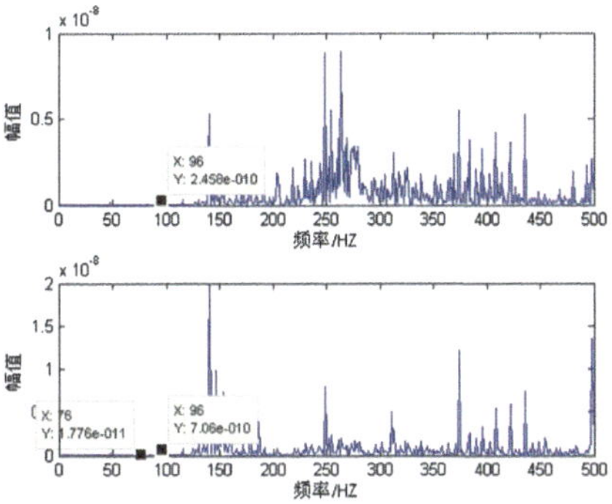

Fig. 6. Envelope demodulation spectrum of FastICA separation

The composite fault signal that has been separated by the methods that called morphological filtering and independent component analysis, and the demodulation spectrum is shown in Fig. 7.

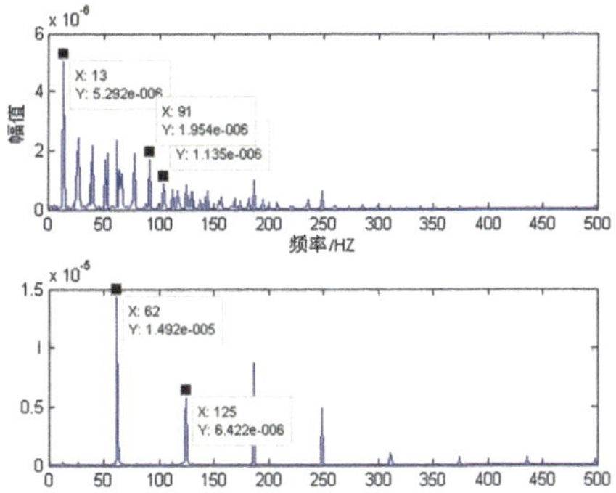

Fig. 7. Envelope demodulation spectrum of MF-FastICA separation

Though Fig. 7 first figure it can be known that there are base frequency lines of 13 and 19 Hz, their frequency doubling is the bearing inner fault frequency that is 95.38 Hz, it belongs to the complex inner fault features, although it has some errors. From the second figure it can be known that the base frequency lines of 62 Hz and its frequency doubling, it means that the method can effectively separate the complex fault signals.

6 Fault State Identification of Rolling Bearing

From the Fig. 4 it can be known that there is a lot of difference between the fault signal and normal signal of bearings. And the fault signals that separated by the morphological filtering and independent component have the same situation. In order to get a bigger difference, the feature that will be expanded 10 times than normal time, and in the clustering of feature values with the means of function, then it shown in Fig. 8. And its further improved clustering graph shown in Fig. 9. Judging from the center point of a number of known classes, then judging whether the detected data is normal or fault.

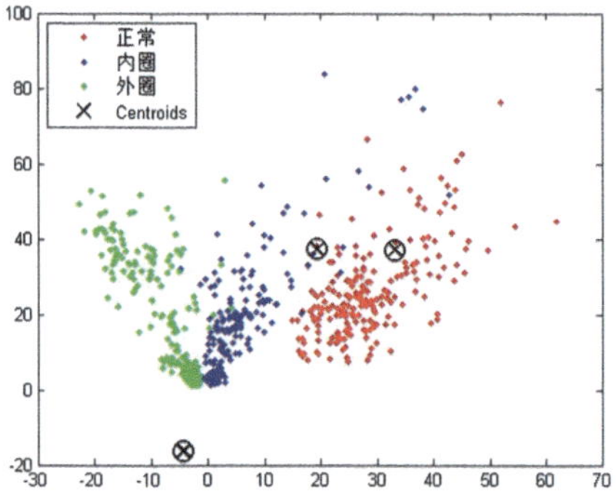

Fig. 8. Mean peak to peak value

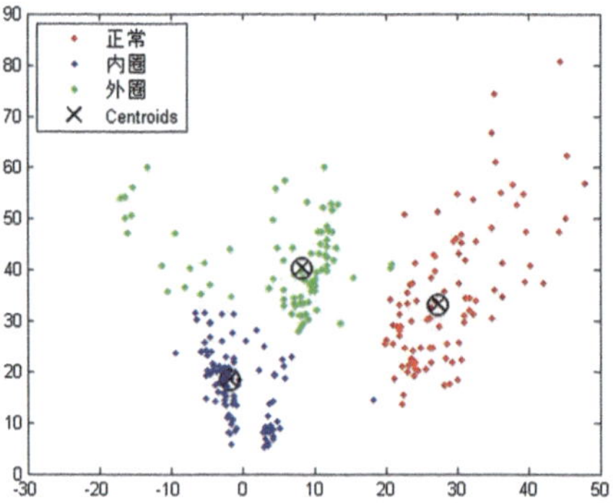

Fig. 9. Improved clustering graph

7 Conclusion

The method of rolling bearing analysis that based on the morphological filtering and independent component analysis has been shown in this paper. The experimental results show that this method can effectively isolate the fault signals. In the end, the features of fault signals have been extracted that used to judge their fault types.

References

1. Zhong, B., Huang, R.: Mechanical fault diagnosis. Machinery Industry Press, Beijing (2007)
2. He, Z., Cheng, J., Wang, T., et al.: Theory and Application of Mechanical Fault Diagnosis. Higher Education Press, Beijing (2010)
3. Lv, Y., Xiong, S., Ling, X., et al.: Fault diagnosis of rolling bearing based on wavelet packet and EMD processing. J. Taiyuan Univ. Technol. **41**(2), 178–182 (2010)
4. Hu, A., Tang, G., An, L.: Noise reduction method of rotating machinery vibration signal based on mathematical morphology. J. Mech. Eng. **42**(4), 127–130 (2006)
5. Xu, Y., Zhang, Q., He, Z.: Independent component analysis and its application in fault diagnosis. Vibration and shock **23**(2), 104–107 (2004)
6. Li, Y., Wu, X., Liu, X., He, W.: Blind separation of bearing faults based on morphological filtering and independent component analysis. Electron. Meas. Technol. **33**(9), 9 (2010)

Research on Social Responsibility of Artificial Intelligence Based on ISO 26000

Wei-wei Zhao(✉)

China National Institute of Standardization, Beijing, China
Zhaoww@cnis.gov.cn

Abstract. Due to the limitations of artificial intelligence (AI) technology, the policies and regulations are lagged, the social responsibility of artificial intelligence has a series of problems, such as human rights, environmental problems. Based on ISO 26000, which is the seven core issues of social responsibility: organizational governance, human rights, labor, environment, fair operating practice, consumer issues, community participation and development. We describe the problems of artificial intelligence in these seven aspects in this paper. Then we analyzed and corresponding countermeasures which are discussed. Finally, we can draw a conclusion that artificial intelligence should pay attention to the performance of social responsibility.

Keywords: ISO 26000 · Social responsibility · Artificial intelligence

1 Introduction

As a new science and technology, artificial intelligence is mainly concerned with the theory, method and application of research and development for simulating and expanding human intelligence. Artificial intelligence has become an important global research area. It not only brings all kinds of technological advances and quality of life, but also brings all kinds of "creative destruction", problems and challenges, the uproar surrounding the US election fake news events and the adjustments made by the Facebook. This is mainly from the technical ethical order, that is, the level of "art" to deal with, and less "morality" reflection and organizational norms at the institutional level. In the face of these challenges, social responsibility may be the key. Based on ISO 26000, social responsibility means "responsibility of an organization for the impacts of its decisions and activities on society and the environment, through transparent and ethical behaviour". The pursuit of sustainability development has become the common goal of the world.

2 Background

More and more people are beginning to pay attention to the social responsibility of AI technology. Many countries have made relevant policies. For example, the White House Office of Science and Technology Policy (OSTP) of American released two reports on artificial intelligence in 2016, "Preparing for the Future of artificial

K. Deng et al. (Eds.): ICMIR 2018, AISC 856, pp. 130–137, 2019.
https://doi.org/10.1007/978-3-030-00214-5_17

intelligence". "The National Artificial Intelligence Research and Development Strategic Plan"; China published the "New Generation of artificial Intelligence Development Plan" in 2017, which proposed "establishing artificial intelligence laws, regulations, ethical norms and policy systems to develop artificial intelligence security assessment and control capabilities." Isaac Asimov set out his famous three laws of robotics in 'I, Robot,'. It has become the default research and development criterion in artificial intelligence technology. Richard O. Mason, an expert in management information science, put forward the famous "PAPA" theory in his four Ethical issues in the Information Age. That is, the right to information privacy, the right to accurate information, the right to information property, the right to access information resources. Early in 2017, Hawking and Musk supported the 23 basic principles of the Asilomar conference on scientific research. These principles are closely related to social responsibility. Up to now, the research on social responsibility management of AI technology is not systematic and comprehensive. With the development of technology, the problem of social responsibility management brought by AI technology will be more and more prominent.

3 Measures

Based on ISO 26000, there are seven core themes of social responsibility: organizational governance, human rights, labor practices, environment, fair operation practices, consumer issues, community participation and development. Figure 1 provides an overview of ISO 26000, and is intended to assist us in understanding the structure of this paper. From the overall analysis of artificial intelligence in these seven aspects of the existing problems, according to these problems to carry out the corresponding countermeasures. Finally, it is concluded the aspects that artificial intelligence assumes social responsibility should pay attention to.

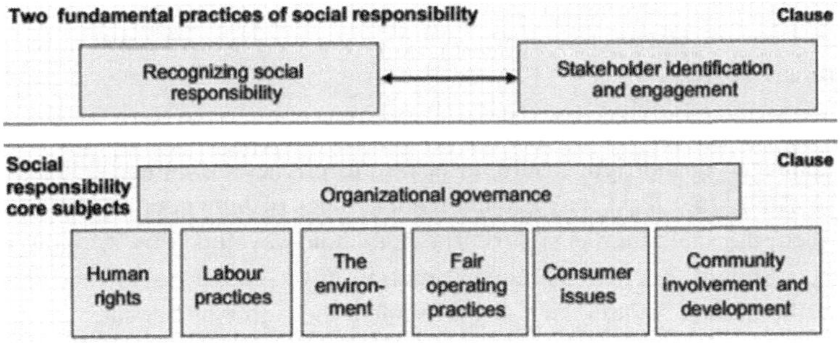

Fig. 1. Social responsibility core subjects bases on ISO 26000.

4 Social Responsibility Core Themes of Artificial Intelligence

4.1 Organizational Governance

Organizational governance is a core subject in ISO 26000, in terms of AI, which platform governance is the key part. The development of artificial intelligence has made the traditional industry boundaries increasingly blurred, and the platform is an important stage for all artificial intelligence. Buyers and sellers form the demand-side and supply-side of the market. The platform facilitates small and medium-sized suppliers and demanders, it make the market more efficient, and enhances the bargaining power of small users, Improve platform consumer welfare, especially for farmers, disabled people and other vulnerable groups. But it also brings some drawbacks, such as winner-take-all, hitchhiking, and diminishing marginal benefits.

Using artificial intelligence can help organizations to analyze and judge "truth and false" from the aspects of user identification, academic certificate, work certificate, etc. However, remote identity authentication and other measures can not completely prevent the disorder of false information from happening again. Although human beings have the vast majority of power in organizational governance, they have gradually lost their autonomy and discretion in their management. Humans will eventually create highly humanoid robots, and these robots will be involved in all kinds of productive and biological activities of human beings. Whether the machine has the ability to manage (effectiveness legitimacy) or whether the machine has the power of manage (power legitimacy), mankind will face a severe challenge. On the one hand, machines have the ability to manage human beings, Can robots lead humans throughout their governance activities? on the other hand, the management of human beings may indeed damage the subjectivity of human beings, because society may prefer the justice and unbiased nature of artificial intelligence. As well as the high degree of dependence formed by human beings and the inherent division and efficiency competition in society, it tends to accept that machines gradually participate in and even take over the management of human beings.

4.2 Human Rights

The development of AI, while liberating the labor force, may also violate human rights, such as endangering the right to life, to health, to privacy, to human freedom, and so on. At the same time, if AI has certain characteristics of human consciousness, So do we give them the same human rights? The humanoid developed by AI, how to define their rights? Should this sort of human emotion have human rights? and if so, what rights can they have? Would such a right violate the rights of human?

4.3 Labor Practices

With the continuous development of artificial intelligence, this highly intelligent body coupled with high mechanical tools, can replace nearly all human labor. Even the traditional industry must be provided by a human labor, and also replaced by the humanoid robot to achieve the goal. Is this a violation of the human labor rights. The artificial intelligence platform of IBM, with the help of some medical institutions, they can help doctors diagnose and treat cancer, at the same time, they undergo surgery. The mechanical arm are much more flexible and stable than humans. They can implement some difficult operation.

4.4 Environment

Electronic waste exacerbates environmental degradation and brings environmental radiation; with the rapid development of artificial intelligence technology, the garbage generated by the replacement of related products will be even more severe. Artificial intelligence technology plays an important role in the development of the universe. The resulting space waste also threatens the rational use of space by human beings and poses certain dangers to space navigation, which may aggravate the ecological crisis.

4.5 Fair Operation Practices

There is no clear legal provision on how to deal with the contradiction caused by artificial intelligence technology. If the developer has racial prejudice, gender bias and so on, the resulting product will also have algorithm discrimination. Improper development and use of artificial intelligence may cause devastating disasters to the entire human being, such as by terrorist groups, which all resulting in new inequalities. It may lead to the accumulation of wealth in the hands of a small number of people, resulting in new inequalities.

4.6 Consumer Issues

Regarding the use of artificial intelligence technology, the basic rights and obligations of ordinary consumers are not clearly regulated. There is no clear penalty rule. This is a liability gap. There is no consensus on who will be responsible, how to take responsibility, how to distribute responsibility, and so on. Is it the developer, the owner, and so on. Or determine artificial intelligence legal personality, how to obtain compensation for the injured, there is no clear conclusion.

4.7 Community Participation and Development

In addition to offline communities, there are online communities. Cyberspace is a real virtual existence, an independent world without physical space. Personal information includes private information that the party does not want to know or is inconvenient for others to know. In an information society, these scattered and extensive personal information may be collected and processed to form a so-called digital personality-a personality that is largely forced to be established. In the present and the future, when the mobile Internet, big data and machine intelligence are combined, we live in a "privacy free society." The consciousness of social responsibility will also be weakened with the expansion of this kind of space-time. They embody the equality of human beings, but it also makes human activity less binding.

5 Causes of Problems

5.1 Technical Limitations

Intelligent robots, however, lack the unique conscience of mankind. For example, people most use artificial intelligence technology in modern military warfare, in such a context, how to maximize the avoidance of civilian casualties and reduce the destruction of environment. The reason for this concern is essentially due to the limitations of artificial intelligence technology. Artificial intelligence technology was used in military warfare, aerospace development, everyday life at home, children's smart toys, and so on. These all bring convenience to human life, but if these machines go wrong, they will also bring about panic to humanity. This is ultimately a problem of social responsibility due to technical inadequacies.

5.2 Lag of Policies and Regulations

Artificial intelligence technology as the emerging development technology, and it is not a long time, which causes the related social responsibility problem that are gone beyond the scope of the existing law, such as the environment, the labor practice. These problems are more constrained by the social responsibility consciousness of researchers and developers. Since they are not mandatory that the effect is not very obvious. At present, many countries have made efforts. Japan has drafted the guidelines on the Safety of next Generation Robots, and South Korea has drafted the Robot Ethics Charter, although these countries have made efforts. But this policy is still relatively simple, need human being to improve constantly.

5.3 Human Being's Own Social Responsibility Consciousness Is Insufficient

The self-cultivation and belief of researchers and developers will greatly affect their social responsibility. If developers, designers and users have poor motives to treat artificial intelligence, such as for nuclear testing, inventing intelligent weapons for war, which will eventually destroy the harmony and stability of human society. If people use artificial intelligence technology to carry out illegal and criminal activities, it will pose a threat to the public interest, public safety. In addition, the random dissemination of false information about artificial intelligence will make people panic when people don't know enough about artificial intelligence. The scientific and cultural qualities of the developers and users of artificial intelligence technology can greatly affect the sustainable and healthy development of AI technology.

6 Suggestions

6.1 Organizational Governance

The focus of the future is on platform governance. Researchers and artificial intelligence developers should be nurtured a culture of mutual cooperation, trust and transparency. To integrate more social responsibility concepts into the initial stages of technology development to ensure the sustainable development of artificial intelligence technology. Strengthen the social responsibility of researchers and developers. In order to make sure the development direction of artificial intelligence technology is right, enhance the public awareness of social responsibility, correctly view of artificial intelligence technology, and create a good atmosphere for the development of artificial intelligence technology. The owner of the platform needs to govern both parties, mediate the behavior of the participants, and gain value in the process of meeting the needs of the stakeholders. Through the enactment of relevant laws, strictly monitor and regulate the development and use of new products of artificial intelligence, such as social responsibility laws.

6.2 Human Rights

The artificial intelligence system should be designed and operated in accordance with the concepts of human dignity, rights, freedoms and cultural diversity. Adhere to the most basic social ethics, eliminate social discrimination, Follow the principle of social equality. Prohibit the development of artificial intelligence for improper purposes. The developer's idea will influence the possible direction of an artificial intelligence. To some extent, the artificial intelligence technology will be promoted the public interest of society, If the loss of social benefits is too great compared with the damage caused by it, the technology should be abandoned in time. Since artificial intelligence systems can analyze and utilize data, people should have the right to access, manage and control the data they produce. The economic prosperity created by artificial intelligence should be widely shared for the benefit of all mankind. Humans should choose how and whether to represent artificial intelligence to make decisions to achieve human goals. The power achieved by controlling advanced artificial intelligence systems should respect and improve the social and civic advancement on which a healthy society is based. Artificial intelligence technology should benefit as many people as possible.

6.3 Labor Practices

Enterprises need to increase relevant training, enhance the competitiveness of their employees. On the one hand for themselves, but also for the sustainable development of enterprise. Artificial intelligence technology development might unintentionally reduce job opportunities, but also create employment opportunities. In the era of artificial intelligence, The International Federation of Robotics (IFR) has predicted that robots will create 3.6 jobs for every robot they produce in the future, adding to the total number of jobs that will be created in the future.

6.4 Environment

An arms race with lethal autonomous weapons should be avoided. A comprehensive, multilevel and effective social responsibility management system should be established to guide and monitor the development of artificial intelligence technology. Which will not damage the human social environment. Acceptance trial phase, sufficient time to ensure foolproof. In the application phase, there are complete usage instructions, matters needing attention.

6.5 Fair Operation Practices

This requires transnational, cross-cutting cooperation. For example, lawyers, human rights experts, government policy makers, associations, etc., if an artificial intelligence system causes damage, there should be a way to find out why. Whenever a self-developed system is involved, it should provide a persuasive explanation and be audited by a competent person in judicial decisions. To improve the transparency associated with big data and the application of artificial intelligence. Encourage to be publicly available for research results that do not involve user privacy and trade secrets. Encourage their information flow in particular with regard to the use of block chain technical management data. Try to break technology giant's monopoly on big data technically and institutionally as far as possible.

6.6 Consumer Issues

There is a contradiction between individual freedom and social responsibility. In the age of big data, the development of technology and the protection of privacy begin to contradict. With the help of artificial intelligence technology, the company collects the information of the user without limitation, which may expose the privacy of the individual. In the face of artificial intelligence technology potential threat to personal privacy, one is to enhance the awareness of self-protection. We are privacy whistle blowers in the current era of mobile interconnection and the future Internet of things, too many apps are installed on smart phones, too many words from social networks that are unsuitable for public use, and there are all sorts of electronics. From cameras with GPS to smart appliances connected to WIFI, private movements and life information are carefully recorded. In the era of artificial intelligence, personal privacy is mined by machines, so privacy protection must begin with itself. The second is to strengthen the responsibility of enterprises to protect the privacy of users. Enterprises must have the will and responsibility to protect the privacy of users, which will be the key to the effective protection of user privacy.

6.7 Community Participation and Development

Promote local employment, regional construction, cooperation between schools and enterprises, promote the development of educational undertakings, and build cooperative training bases between schools and enterprises, To provide a relatively undecided base for production practice and employment for local students, to train technical personnel for enterprises and communities, and to cooperate with enterprises to run classes. On the one hand, it will improve the production efficiency of enterprises, on the other hand, it will enrich the educational resources of colleges and universities. Promote the growth of local artificial intelligence talent. Pay attention to the network community. Regulate the developers, users and owners of artificial intelligence, do not harm the legitimate interests of the state, society and individuals.

7 Conclusion

It's good news for the artificial intelligence industry only to be regulated within the framework of social responsibility. But in the long run, the subversive of today will be subverted by the more savage tonight if without rules and rules. This paper discusses the social responsibility problem of artificial intelligence with some related suggestions. We believe with the development of technology and the continuous improvement of human cognitive ability, many problems will eventually be solved. The cooperation of users and social responsibility specialists seeks more benefits for the future life of mankind.

Acknowledgments. This work was funded by the National Key R&D Program of China under grant Nos. 2017YFF0207603 and 2016YFF0202502.

References

1. Baum, S.D.: Social choice ethics in artificial intelligence. Ai Soc. **3**, 1–12 (2017)
2. Bostrom, N.: Ethical issues in advanced artificial intelligence. Humans & in Artificial Intelligence, pp. 12–17 (2003)
3. Cormier, D., Gordon, I.M., Magnan, M.: Corporate ethical lapses: do markets and stakeholders care? Manag. Decis. **54**(10), 2485–2506 (2016)
4. Huang, M.H., Rust, R.T.: Artificial intelligence in service. J. Serv. Res. **1**, 1220302093 (2018)
5. ISO ISO 26000:2010 - Guidance on social responsibility (2010)
6. Li, X., Zhang, T.: An exploration on artificial intelligence application: from security, privacy and ethic perspective (2017)
7. Mason, R.O.: Four ethical issues of the information age. MIS Q. **10**(1), 5–12 (1986)
8. Wilson, H.J., Daugherty, P.R., Morini-Bianzino, N.: The jobs that artificial intelligence will create. Mit Sloan Manag. Rev. **58**(4), 14–16 (2017)

Transmission Efficiency Study of Planetary Gears Transmission Based on Directed Digraph and Screw Theories

Jianying Li[1], Zhiyong Xu[2], Tianjun Zhu[1](✉), Jie Liu[1], Qingchun Hu[3], Chaoming Cai[1], and Changfu Zong[1]

[1] School of Mechanical and Automotive Engineering, ZhaoQing University, Zhaoqing 516260, Guangdong, China
360103919@qq.com
[2] School of Electronics and Electrical Engineering, ZhaoQing University, Zhaoqing 516260, Guangdong, China
[3] School of Mechanical and Automotive Engineering, South China University of Technology, Guangzhou 510640, China

Abstract. The transmission efficiency formula of 2K-H planetary gears transmission based on the directed digraph and screw theories are deduced considering the influence of the gearbox shell and friction torques of the gear pairs, the formula are associated with the pitch radii of the of the sun gear and ring gear and the friction coefficient magnitude of the gear pairs. The transmission efficiency of 2K-H planetary gears transmission increases with the number of teeth of the sun gear and the ring gear increasing, but it decreases with the friction coefficient magnitude increasing, which the influence degrees are not the same.

Keywords: Directed digraph · Screw theory · Instantaneous virtual power Transmission efficiency

1 Introduction

Planetary gear has the advantages of small size, light weight, large load capacity and high transmission efficiency. It has been widely used in automobile, shipbuilding, aerospace and electromechanical products [1]. The scholars studied the transmission efficiency of planetary gears transmission using the graph theory. Ma and Xu [2] proposed a new algorithm for enumerating planetary gear trains based on the graph theory, and defined the vertex characteristics of the graph firstly. Chen [3] studied the sensitivity of two-stage planetary gears transmission using the graph theory. Based on the graph method, Pennestrì and Valentini [4] systematically analyzed the power flow and transmission efficiency of planetary gear trains, and then the method was introduced to analyze the motion and power flow of the bevel gear planetary gear [5]. Laus and Simas [6] applied the graph theory to analyze the transmission efficiency of compound planetary gear. Salgado and Del Castillo [7] analyzed jointly the efficiency and transmission ratio range of 4-, 5-, and 6-link planetary gear trains based on the graphical representation. Yutao and Di [8] established the dynamics model of planetary

K. Deng et al. (Eds.): ICMIR 2018, AISC 856, pp. 138–146, 2019.
https://doi.org/10.1007/978-3-030-00214-5_18

gear trains based on the bond graph theory and proposed three sub-models of the carrier-planet gear, ring-planet gear and sun-planet gear. Le [9] and Gomà Ayats [10] studied the complex planetary gears transmission using the super-graph theory. Based on the directed graph and screw theory, the formula of 2K-H planetary gears transmission efficiency is deduced considering the influence of the friction torques of the gear pairs, it is revealed that the influence of the geometric parameters and the friction coefficient magnitude of the gear pairs.

2 Directed Digraph and Screw Theories of 2K-H Planetary Gears Transmission

2.1 Directed Digraph Representation of 2K-H Planetary Gears Transmission [11, 12]

The diagram of 2K-H planetary gears transmission is shown in Fig. 1, in which 0, 1, 2, 3, 4 represent the gearbox shell, the sun gear, the planet gear, the ring gear and the carrier, respectively. A represents the rotary pair of the gearbox shell 0 with the sun gear 1, B represents the rotary pair of the gearbox shell 0 with the carrier 4, C represents the gear pair of the sun gear 1 with the planet gear 2, D represents the gear pair of the ring gear 3 with the planet gear 2, E represents the rotary pair of the planet gear 2 with the carrier 4. Figure 2 is the coupling graph G_C of Fig. 1, where the gearbox shell and the components are represented by the small solid black dots, respectively, the solid line is used as the rotary pair and the gear pair is attached 'cross' on the solid line.

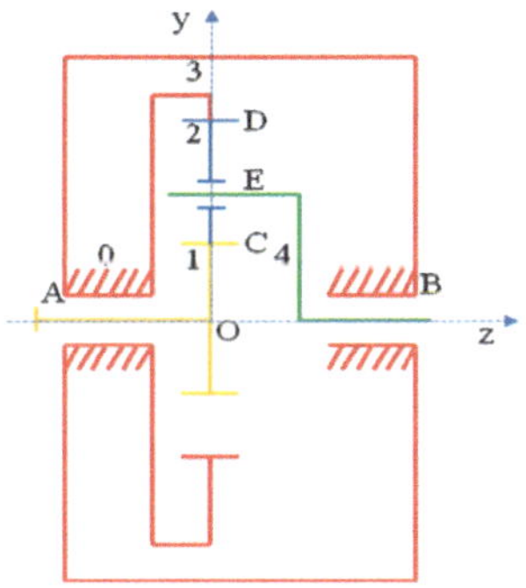

Fig. 1. 2K-H planetary gears transmission diagram

2.2 Screw Theory

Hunt [13] pointed out that any motion of a rigid body in space is a helical motion, that is, a rotation around an instantaneous helical axis and a concentric moving. Any space force system on the rigid body can be combined with the force line vector and the collinear force couple vector, which are described by the kinematic screw $\$^m$ and force

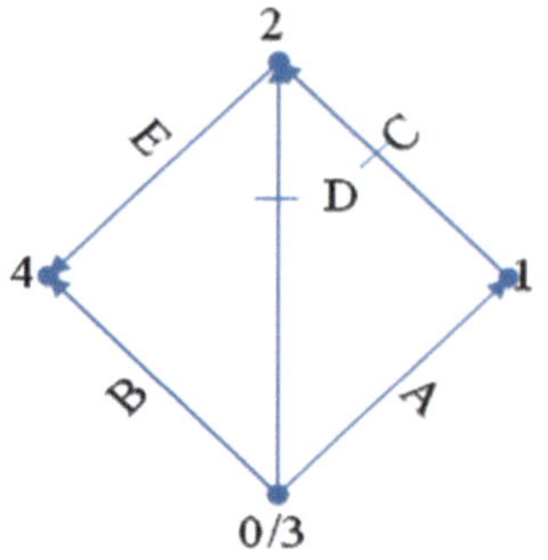

Fig. 2. 2K-H planetary gears transmission coupling graph G_C

screw $\$^a$, the scalar product of these two screws represents the instantaneous power produced by the kinematic screw $\$^m$ and force screw $\$^a$, as shown in Eq. (1)

$$P = \$^a \cdot \$^m = rR + sS + tT + uU + vV + wW \tag{1}$$

where $\$^a = \{R, S, T; U, V, W\}$, $\$^m = \{r, s, t; u, v, w\}$. $\boldsymbol{P} = \{R, S, T\}$ is the force couple vector, $\boldsymbol{Q} = \{U, V, W\}$ is the force line vector, $\boldsymbol{p} = \{r, s, t\}$ is the angular velocity vector, $\boldsymbol{q} = \{u, v, w\}$ is the linear velocity vector. It is pointed out that all the motion pairs can be represented by the screw in Ref. [14], and the rotary pair is the screw of zero pitch. The motion pair of 2K-H planetary gears transmission is divided into the gear pair and the rotary pair in 1.1, so the instantaneous power made by the force screw $\$^a$ on these the kinematic screw $\$^m$ is equal to the instantaneous power represented by Eq. (1).

3 Study on the Transmission Efficiency of 2K-H Planetary Gears Transmission

Assuming r_1, r_2 and r_3 represent the pitch radii of the sun gear, planet gear and ring gear, respectively, the coordinate system yOz is established in Fig. 1, there are following two assumptions in this paper:

(1) The length of each gear shaft of planetary gears transmission are not consider, that is, each gear pairs mesh forces is in the $z = 0$ plane.
(2) Each gear of the planetary gears transmission is thin spur gears, and the tooth contact points are on the y-axis.

3.1 Motion Analysis

The motion analysis of 2K-H planetary gears transmission shows that the direct coupling unit motion matrix $\hat{M}_D$ [6] is

$$\hat{M}_D = \begin{array}{c} \\ t \\ u \end{array}\!\!\begin{array}{c} \begin{array}{ccccc} t_A & t_B & t_C & t_D & t_E \end{array} \\ \begin{bmatrix} 1 & 1 & 1 & 1 & 1 \\ 0 & 0 & r_1 & r_3 & r_1+r_2 \end{bmatrix} \end{array} \tag{2}$$

The coupling graph G_C and the motion graph G_M of 2K-H planetary gears transmission studied in this paper are the same. Therefore, the circuit matrix **B** of G_C is equal to the circuit matrix $\mathbf{B_M}$ of G_M.

$$B = B_M = \begin{array}{c} \\ C \\ E \end{array}\!\!\begin{array}{c} \begin{array}{ccccc} A & B & C & D & E \end{array} \\ \begin{bmatrix} 1 & 0 & 1 & -1 & 0 \\ 0 & -1 & 0 & 1 & 1 \end{bmatrix} \end{array} \tag{3}$$

Using the representation of the network unit motion matrix $[\hat{M}_N]_{dl \times F}$ in Ref. [6], it can be obtained

$$\hat{M}_N = \begin{array}{c} \\ t \\ u \\ t \\ u \end{array}\!\!\begin{array}{c} \begin{array}{ccccc} t_A & t_B & t_C & t_D & t_E \end{array} \\ \begin{bmatrix} 1 & 0 & 1 & -1 & 0 \\ 0 & 0 & r_1 & -r_3 & 0 \\ 0 & -1 & 0 & 1 & 1 \\ 0 & 0 & 0 & r_3 & r_1+r_2 \end{bmatrix} \end{array} \tag{4}$$

Combining Eq. (4) and Davies equation in Ref. [15], the null space $\hat{M}_N$ can be obtained

$$\left[\hat{M}_N\right]_{dl \times F} [\psi]_F = [0]_{dl}$$

$$Null\left(\hat{M}_N\right) = \begin{bmatrix} 1 & \dfrac{r_1^2 + r_1(r_2 - r_3)}{(r_1 - r_3)(r_1 + r_2)} & \dfrac{r_3}{r_1 - r_3} & \dfrac{r_1}{r_1 - r_3} & \dfrac{-r_1 r_3}{(r_1 - r_3)(r_1 + r_2)} \end{bmatrix}^T \tag{5}$$

Arbitrarily selecting t_A as the basic variable, the motion spiral matrix $[M]_{2\times5}$ of 2K-H planetary gears transmission [6] shown in Fig. 1 is

$$[M]_{2\times5} = \hat{M}_D diag\left(Null\left(\hat{M}_N\right)\right) t_A$$

$$= \begin{array}{c} \\ t \\ u \end{array}\!\!\begin{array}{c} \begin{array}{ccccc} \$_A^m & \$_B^m & \$_C^m & \$_D^m & \$_E^m \end{array} \\ \begin{bmatrix} (r_1 - r_3)(r_1 + r_2) & r_1^2 + r_1(r_2 - r_3) & r_3(r_1 + r_2) & r_1(r_1 + r_2) & -r_1 r_3 \\ 0 & 0 & r_1 r_3(r_1 + r_2) & r_1 r_3(r_1 + r_2) & -r_1 r_3(r_1 + r_2) \end{bmatrix} \end{array} \frac{t_A}{(r_1 - r_3)(r_1 + r_2)} \tag{6}$$

3.2 Force Analysis

The force action of 2K-H planetary gears transmission is analyzed in two cases.

Case one: neglecting the influence of the friction torque of the gear pair of 2K-H planetary gears transmission, the direct coupling unit force action matrix $\hat{A}_D$ [6] is

$$\hat{A}_D = \begin{matrix} \\ T \\ U \end{matrix} \begin{matrix} T_A & U_A & T_B & U_B & U_C & U_D & U_E \\ \left[\begin{matrix} 1 \\ 0 \end{matrix}\right. & \begin{matrix} 0 \\ 1 \end{matrix} & \begin{matrix} 1 \\ 0 \end{matrix} & \begin{matrix} 0 \\ 1 \end{matrix} & \begin{matrix} -r_1 \\ 1 \end{matrix} & \begin{matrix} -r_3 \\ 1 \end{matrix} & \left.\begin{matrix} -(r_1+r_2) \\ 1 \end{matrix}\right] \end{matrix} \quad (7)$$

where T_A, T_B is the external torque magnitude, T_C, T_D is the friction torque magnitude of the gear pair, U_A, U_B and U_E is the force magnitude of the rotary pair, U_C, U_D is the force magnitude of the gear pair.

Selecting A, B, and D as the cut set of the force action graph G_A, then the cut set matrix $\mathbf{Q_A}$ of G_A is

$$Q_A = \begin{matrix} \\ \\ A \\ B \\ D \end{matrix} \begin{matrix} A & A & B & B & C & D & E \\ T_A & U_A & T_B & U_B & U_C & U_D & U_E \\ \left[\begin{matrix} 1 \\ 0 \\ 0 \end{matrix}\right. & \begin{matrix} 1 \\ 0 \\ 0 \end{matrix} & \begin{matrix} 0 \\ 1 \\ 0 \end{matrix} & \begin{matrix} 0 \\ 1 \\ 0 \end{matrix} & \begin{matrix} -1 \\ 0 \\ 1 \end{matrix} & \begin{matrix} 0 \\ 0 \\ 1 \end{matrix} & \left.\begin{matrix} 0 \\ 1 \\ -1 \end{matrix}\right] \end{matrix} \quad (8)$$

As the same way, the representation method of the network unit force action matrix $[\hat{A}_N]_{dk \times C}$ in Ref. [6], and Davies equation $[\hat{A}_N]_{dk \times C}[\phi]_C = [0]_{dk}$ in Ref. [15] are used for getting the null space of $\hat{A}_N$

$$Null(\hat{A}_N) = \begin{bmatrix} 1 & -\frac{1}{r_1} & \frac{r_1+r_2}{r_1} & -\frac{1}{r_1} & \frac{r_3-r_1-r_2}{r_1(r_3-r_1)} & \frac{r_2}{r_1(r_3-r_1)} & \frac{1}{r_1} \end{bmatrix}^T \quad (9)$$

Arbitrarily selecting T_A as the basic variable, the force spiral matrix $[A]_{2\times5}$ of 2K-H planetary gears transmission [6] shown in Fig. 1 is

$$[A]_{2\times5} = \hat{A}_D diag\left(Null\left(\hat{A}_N\right)\right) T_A$$

$$= \begin{matrix} T \\ U \end{matrix} \begin{bmatrix} \$_A^a & \$_B^a & \$_C^a & \$_D^a & \$_E^a \\ 1 & \frac{r_1+r_2}{r_1} & -\frac{r_3-r_1-r_2}{r_3-r_1} & -\frac{r_3 r_2}{r_1(r_3-r_1)} & -\frac{r_1+r_2}{r_1} \\ -\frac{1}{r_1} & -\frac{1}{r_1} & \frac{r_3-r_1-r_2}{r_1(r_3-r_1)} & \frac{r_2}{r_1(r_3-r_1)} & \frac{1}{r_1} \end{bmatrix} T_A \quad (10)$$

Case two: consider the influence of the friction torques of the gear pair T_C and T_D of 2K-H planetary gears transmissions [16], the analysis method is the same as above, arbitrarily selecting T_A as the basic variable, then the force spiral matrix $[A]_{2\times 5}$ of 2K-H planetary gears transmission shown in Fig. 1 is

$$[A]_{2\times 5} = \hat{A}_D diag\left(Null\left(D_A\right)\right)T_A$$

$$= \begin{matrix} T \\ U \end{matrix}\begin{bmatrix} \overset{\$_A^a}{1} & \overset{\$_B^a}{\dfrac{(r_1+r_2)(\zeta_C-\zeta_D+r_3-r_1)}{(\zeta_C-r_1)(r_3-r_1-r_2-\zeta_D)}} & \overset{\$_C^a}{1} & \overset{\$_D^a}{\dfrac{(\zeta_C+r_2)(\zeta_D-r_3)}{(\zeta_C-r_1)(r_3-r_1-r_2-\zeta_D)}} & \overset{\$_E^a}{-\dfrac{(r_1+r_2)(\zeta_C-\zeta_D+r_3-r_1)}{(\zeta_C-r_1)(r_3-r_1-r_2-\zeta_D)}} \\ \dfrac{1}{\zeta_C-r_1} & \dfrac{\zeta_D-\zeta_C+r_1-r_3}{(\zeta_C-r_1)(r_3-r_1-r_2-\zeta_D)} & \dfrac{1}{\zeta_C-r_1} & \dfrac{\zeta_C+r_2}{(\zeta_C-r_1)(r_3-r_1-r_2-\zeta_D)} & \dfrac{\zeta_C-\zeta_D+r_3-r_1}{(\zeta_C-r_1)(r_3-r_1-r_2-\zeta_D)} \end{bmatrix} T_A \tag{11}$$

where ζ_C, ζ_D is the frictional coefficient magnitude.

3.3 Force Analysis

Taking the carrier 4 as the analysis object, the order-reduction matrix A_4 of the coupling graph G_C of 2K-H planetary gears transmissions is

$$A_4 = \begin{matrix} \\ 1 \\ 2 \\ 0/3 \end{matrix}\begin{matrix} \begin{matrix} A & B & C & D & E \end{matrix} \\ \begin{bmatrix} -1 & 0 & 1 & 0 & 0 \\ 0 & 0 & -1 & -1 & 1 \\ 1 & 1 & 0 & 1 & 0 \end{bmatrix} \end{matrix} \tag{12}$$

Using the power flow calculation method in Ref. [6], the Moore-Penrose pseudo-matrices $A_4^{\dagger}$ and M_4 of A_4 are

$$A_4^{\dagger} = \begin{matrix} A \\ B \\ C \\ D \\ E \end{matrix}\begin{bmatrix} -4 & -1 & 1 \\ 4 & 3 & 5 \\ 4 & -1 & 1 \\ 0 & -2 & 2 \\ 4 & 5 & 3 \end{bmatrix}\frac{1}{8} \tag{13}$$

$$M_4 = \begin{matrix} t \\ u \end{matrix}\begin{bmatrix} \overset{{}^4\$_1^m}{-r_2r_3} & \overset{{}^4\$_2^m}{r_1r_3} & \overset{{}^4\$_{0/3}^m}{-\left(r_1^2+r_1r_2-r_1r_3\right)} \\ 0 & r_1r_3\left(r_1+r_2\right) & 0 \end{bmatrix}\frac{t_A}{\left(r_1-r_3\right)\left(r_1+r_2\right)} \tag{14}$$

Observing Figs. 1 and 2, and combining Eqs. (1) and (10), it can be seen that the powers of the sun gear 1 and the planet gear 2, the planet gear 2 and the ring gear 3 are transmitted through the gear pair C and D, P_{12}, P_{23} can be obtained by

$$P_{12} = \$_C^a\,{}^4\$_1^m = \$_C^a\,{}^4\$_2^m = -\frac{T_A t_A r_2 r_3 (r_3 - r_1 - r_2)}{(r_1 - r_3)^2 (r_1 + r_2)} \tag{15}$$

$$P_{23} = \$_D^a {}^4\$_2^m = \$_D^a {}^4\$_{0/3}^m = \frac{T_A t_A r_2 r_3 (r_3 - r_1 - r_2)}{(r_1 - r_3)^2 (r_1 + r_2)} \tag{16}$$

Seen from the coupling graph G_C of Fig. 2, it can be obtained $T_A t_A \rangle 0$, $r_3 - r_1 - r_2 \rangle 0$, then $P_{12} \langle 0$, so the actual power flow of 2K-H planetary gears transmissions is from the planet gear 2 to the sun gear 1.

3.4 Transmission Efficiency Study of 2K-H Planetary Gears Transmission

Seen from the coupling graph G_C of Fig. 2, it can be known that the rotation pair B is the input and the rotation pair A is the output. Combining Eq. (1) and the geometry relationship $r_1 + r_2 = r_3 - r_2$ of Fig. 1, the transmission efficiency formula of 2K-H planetary gears transmission studied in this paper can be obtained

$$\eta = \frac{-P_A}{P_B} = \frac{-\$_A^a \cdot \$_A^m}{\$_B^a \cdot \$_B^m} = \frac{(r_1 + r_2)(r_3 - r_1)\eta_{21}}{r_1 r_2 \eta_{21} + r_2 r_3 \eta_{23}} \tag{17}$$

where η_{21}, η_{23} is the transmission efficiency of the sun gear 1 and the planet gear 2, the planet gear 2 and the ring gear 3.

3.4.1 Influence of the Pitch Radii of the Sun Gear and Ring Gear

For studying the influence of the pitch radii of the sun gear 1 and ring gear 3 on the transmission efficiency of 2K-H planetary gears transmission, the number of teeth range of the sun gear 1 is taken as 20–40, the transmission ratio range of 2K-H planetary gears transmission is taken as 2.8–13, and the frictional coefficient magnitude ζ_C, ζ_D is 0.08. Seen from Fig. 3, the transmission efficiency of 2K-H planetary gears transmission increases with the number of teeth of the sun gear 1 and the ring gear 3 increasing.

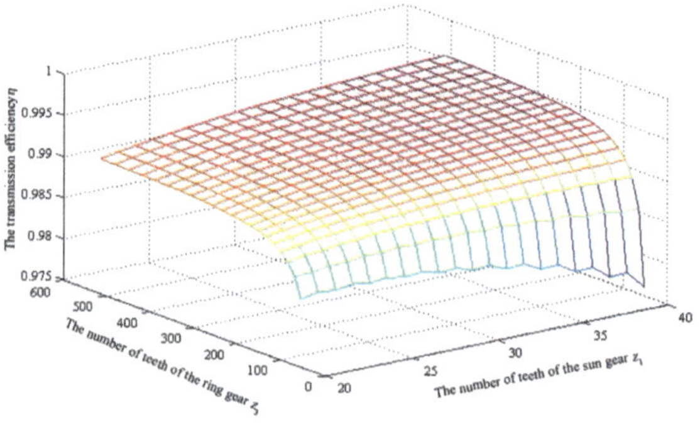

Fig. 3. Influence of the number of teeth of the sun gear and ring gear on the transmission efficiency of 2K-H planetary gears transmission

3.4.2 Influence of the Frictional Coefficient Magnitude

For studying the influence of the friction coefficient magnitude on the transmission efficiency of 2K-H planetary gears transmission, the number of teeth of the sun gear 1 and the ring gear 3 are taken as 20 and 56, respectively, and the frictional coefficient magnitude range ζ_C, ζ_D is taken as 0.06–0.10. Seen from Fig. 4, with the increase of the friction coefficient magnitude ζ_C, the transmission efficiency of 2K-H planetary gears transmission decreases obviously. The transmission efficiency of 2K-H planetary gears transmission decreases slightly with the increase of the friction coefficient magnitude ζ_D.

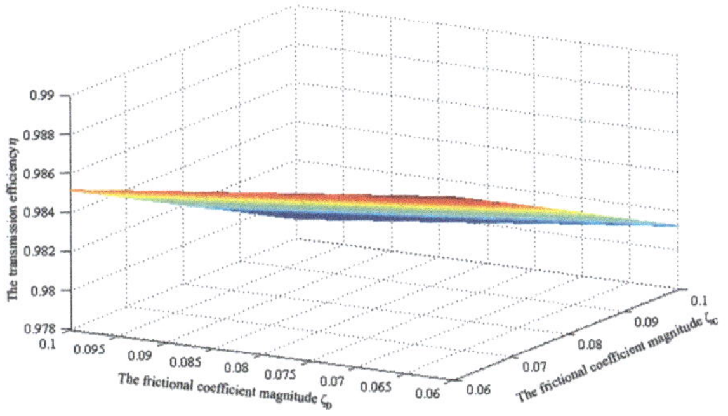

Fig. 4. Influence of the frictional coefficient magnitude on the transmission efficiency of 2K-H planetary gears transmission

4 Conclusion

(1) The transmission efficiency formula of 2K-H planetary gears transmission based on the directed digraph and screw theories are deduced considering the influence of the friction torques of the gear pair, the formula is associated with the pitch radii of the of the sun gear and ring gear, and the friction coefficient magnitude of the gear pair.

(2) The transmission efficiency of 2K-H planetary gears transmission increases with the number of teeth of the sun gear and the ring gear increasing, which decreases with the friction coefficient magnitude ζ_C and ζ_D increasing, the former decreases obviously, but the latter decreases slightly.

Acknowledgements. School of natural science fund project (201728); Practical teaching reform project (sjjx201627); Zhaoqing science and technology innovation guidance project (201704030201); Quality engineering and teaching reform project (zlgc201751); Guangdong province natural science fund (2014A030311045), Featured innovation projects of Guangdong province university (2016KTSCX154).

References

1. Rao, Z.: Planetary Gear Design. Chemical Industry Press, Beijing (2003)
2. Ma, M., Xu, X.: A novel algorithm for enumeration of the planetary gear train based on graph theory. Adv. Mater. Res. **199–200**, 392–399 (2011)
3. Chen, M.: Efficiency sensitivity of two-stage planetary gear transmission system based on graph theory. Appl. Mech. Mater. **215–216**, 1058–1061 (2012)
4. Pennestrì, E., Mariti, L., Valentini, P.P., et al.: Efficiency evaluation of gearboxes for parallel hybrid vehicles: theory and applications. Mech. Mach. Theory **49**, 157–176 (2012)
5. Delpio, G., Pennestrì, E., Valentini, P.P.: Kinematic and power-flow analysis of bevel gears planetary gear trains with gyroscopic complexity. Mech. Mach. Theory **70**, 523–537 (2013)
6. Laus, L.P., Simas, H., Martins, D.: Efficiency of gear trains determined using graph and screw theories. Mech. Mach. Theory **52**, 296–325 (2012)
7. Salgado, D.R., Del Castillo, J.M.: Analysis of the transmission ratio and efficiency ranges of the four-, five-, and six-link planetary gear trains. Mech. Mach. Theory **73**, 218–243 (2014)
8. Yutao, L., Di, T.: Dynamics modeling of planetary gear set considering meshing stiffness based on bond graph. Procedia Eng. **24**, 850–855 (2011)
9. Le, X., Zhang, Y., Liu, H., et al.: The research on general modeling methods of the complicated transmission based on hypergraph theory. Appl. Mech. Mater. **86**, 704–708 (2011)
10. Ayats, J.R.G., Diego Ayala, U., Canela, J.M., et al.: Hypergraphs for the analysis of complex mechanisms comprising planetary gear trains and other variable or fixed transmissions. Mech. Mach. Theory **51**, 217–229 (2012)
11. Bang Jensen, J., Gutin, G.: Directed Graph Theory, Algorithms and Applications (trans: Yao, B., Zhang, Z.). Science Press, Beijing (2009)
12. Zuanwu, L.: Application of Graph Theory. National Defense University Press, Changsha (2006)
13. Hunt, K.H.: Kinematic Geometry of Mechanisms. Oxford University Press, Oxford (1978)
14. Huang, Z., Zhao, Y., Zhao, T.: Advanced Space Agencies. Higher Education Press, Beijing (2006)
15. Davies, T.H.: Freedom and constraint in coupling networks. Proc. Inst. Mech. Eng. Part C J. Mech. Eng. Sci. **220**(7), 989–1010 (2006)
16. Armstrong Helouvry, B., Dupont, P., Wit, C.C.D.: A survey of models, analysis tools and compensation methods for the control of machines with friction. Automatica **30**(7), 1083–1138 (1994)

Study on the Construction of Lightning Protection and Safety Supervision System in FuJian

Lei Zhang[1], JinQuan Zeng[2(✉)], Min Cai[3], Teng Lv[4], and Qingli Lai[5]

[1] FuJian Provincial Meteorological Service, Fuzhou 350001, China
[2] Meteorological Disasters Defending Technique Centre of Fujian Province, Fuzhou 350001, China
187240935@qq.com
[3] Fuzhou Meteorological Bureau, Fuzhou 350008, China
[4] Fujian Province Meteorological Institute, Fuzhou 350001, China
[5] Fujian Center of Public Awareness and Education for Meteorological Science, Fuzhou 350008, China

Abstract. FuJian city has built lightning protection supervision system in 2013, and the system has basic information collection, daily safety management, lightning protection business data integration and the full information management functions of the ultimate decision-making analysis. The system not only has save time, reducing cost and improving administrative efficiency, but also has special significance for the accident disposal of lightning disaster. The subsequent catties out analysis and processing for the large amount of stored data in the system, and we will get some useful knowledge and can help the improvement of the municipal lightning protection business process. This paper mainly introduces the system realization technology and present application advantages.

Keywords: Lightning protection business · Risk assessment
Security management system · App

1 Introduction

After the statistics found, lightning protection business in recent years has outstanding achievement in FuJian city, it needs for lightning detection unit up to more than 3300 in each year, acceptance unit up to 2500 and the completion of lightning protection number up to more than 500 [1]. As a department of the mine industry is mainly responsible for the current new project of mine safety administrative license, audit new renovation project lightning protection engineering design drawings, the completion of the project acceptance, lightning disaster risk assessment, lightning accident investigation, detection supervision units lightning protection and other social management functions. A lot of lightning protection and integrity risk propose the renewal and higher requirements for lightning protection and safety management [2, 3]. In order to further do strengthen the supervision of lightning protection, people should increase the intensity of mine safety protection supervision work, strengthen mine safety production

K. Deng et al. (Eds.): ICMIR 2018, AISC 856, pp. 147–152, 2019.
https://doi.org/10.1007/978-3-030-00214-5_19

track inspection, reference FuJian city management platform and hazardous chemicals monitoring platform and other successful experiences, at the same times people also refer to increasingly sophisticated information systems and advanced mobile Internet technology to build the mine safety supervision platform in FuJian city, it can realize real-time supervision for the unit mine safety and bring into the city's key regulatory units, to effectively improve the real-time monitoring level of mine safety.

2 System Design Scheme

2.1 System Design Target

The design goal of lightning protection supervision platform is to be able to become a real-time information query platform, the lightning protection information of city key units can quickly query in real time giving a visual interface of the GIS system, to meet the needs of management mine safety supervision; for the industry provided interactive platform, inspectors by handheld mobile terminal will real-time test unit information transmission platform, to meet the needs of information platform real time update; for the safety supervision platform, the system will automatically prompt unit testing time, annotation refuse inspection unit, unqualified, rectification unit, to meet the required for tracking supervision inspection [4, 5]. According to the business management process of lightning protection center, it can achieve the core of the safety and supervision, to build real time, anywhere and integrated regulatory information system. Through strengthening the security monitoring and management, strengthen the management and control of the field, system can realize the centralized safety control, so as to improve the lightning protection ability and response ability of the lightning protection center.

2.2 System Overall Structure

According to the business management process of lightning protection center, the security supervision platform is the core of the security supervision, to build the real-time, anywhere and integrated supervision information system, and to achieve mobile information reporting and personnel management as well as centralized security monitoring and dynamic tracking [6]. Through strengthening the security monitoring and management, we can strengthen the field management and control to realize the centralized safety control, so as to improve the safety ability and response ability of lightning protection center.

The system user is divided into business personnel of using mobile App secure information acquisition system and management personnel of using PC security management system [7–9]. According to each user's operating authority, users will enter the respective operating interface in each login por. The overall frame diagram of system is shown in Fig. 1.

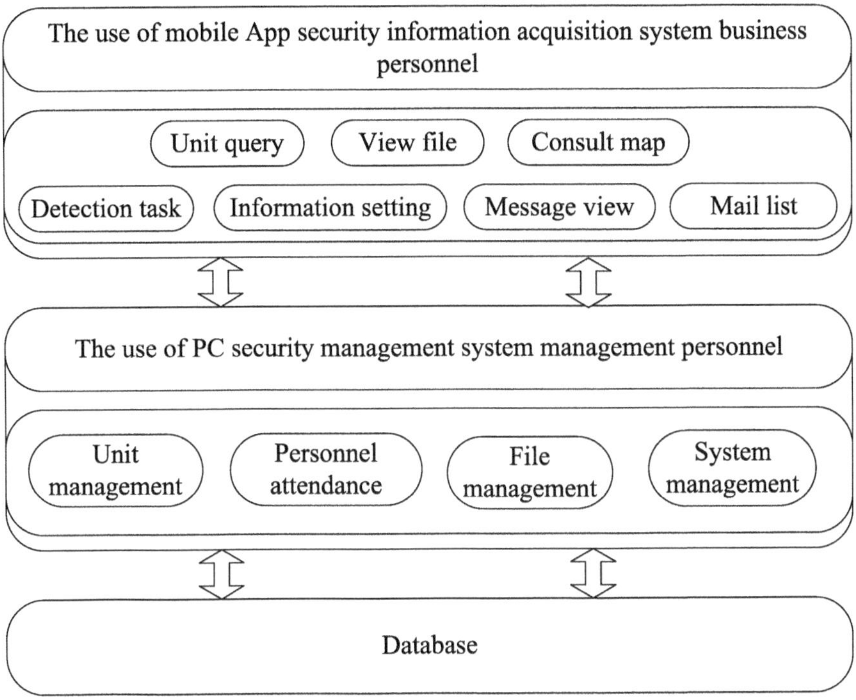

Fig. 1. The overall structure diagram of the system

3 Build Business Model

In the process of business model design, lightning protection and disaster reduction will delete the elements in work or secondary position, so that the corresponding business model of mine safety supervision platform in FuJian city is much more simple, clear and easy to deal with than the real affairs [10]. The business model of mine safety supervision platform in FuJian city mainly includes security check processing process, new units' excavation treatment process, daily inspection business process, inspection personnel management process and audit process.

In the lightning protection daily test business process as an example, when management promotions through logging in safety management system of FuJian mine safety supervision platform based on PC, people can make the test plan according to the relevant requirements of lightning protection and disaster reduction department, after FuJian mine safety supervision platform included plans, it will send information to the responsible unit area's inspection personnel; when the inspection personnel using a mobile phone logins in security information collection system of lightning supervision safety platform based on mobile phone App, it will be subject to inspection plan, it is one by one to test the indicators of supervision units for lightning protection and disaster reduction in accordance with the relevant requirements of the test plan within the specified time, and the use of mobile phone APP records each detected value, they respectively have inspection personnel and units responsible after completion test to carry on signature confirmation [11]. As the security management system managers based on PC, they can real time check the data uploaded by the inspection personnel, the data will be imported in the old system to analyze, and then the old system generates reports to send management personnel and inspection personnel. The process chart of daily inspection business is shown in Fig. 2.

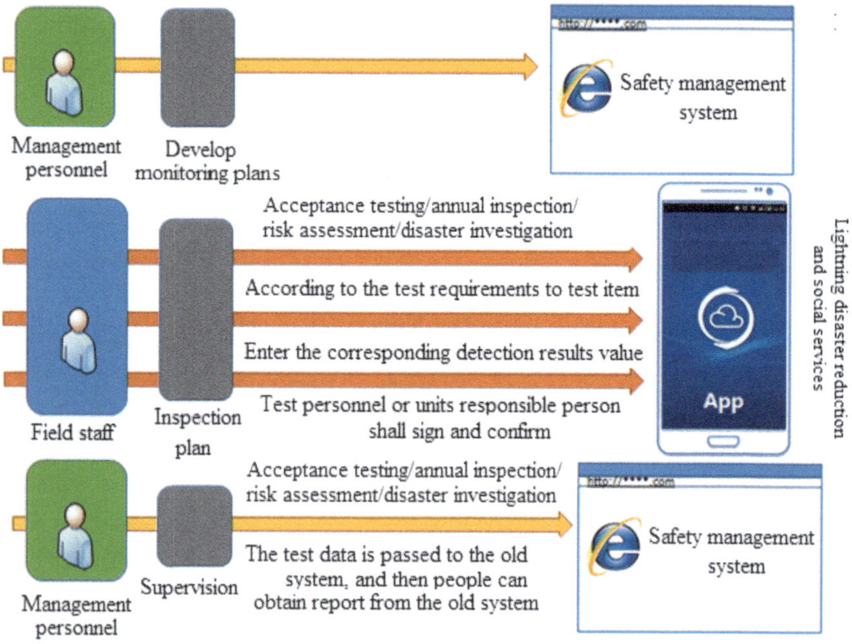

Fig. 2. The process chart of daily inspection business

4 System Function and Knowledge Discovery

According to the demand analysis of the early stage, the system can be divided into five functional modules. The system overall function diagram is shown in Fig. 3.

In this paper, a large number of data systems are analyzed by Apriori algorithm, and then some meaningful knowledge information is obtained. The most important mining frequent itemsets flow chart of using Apriori algorithm is shown in Fig. 4.

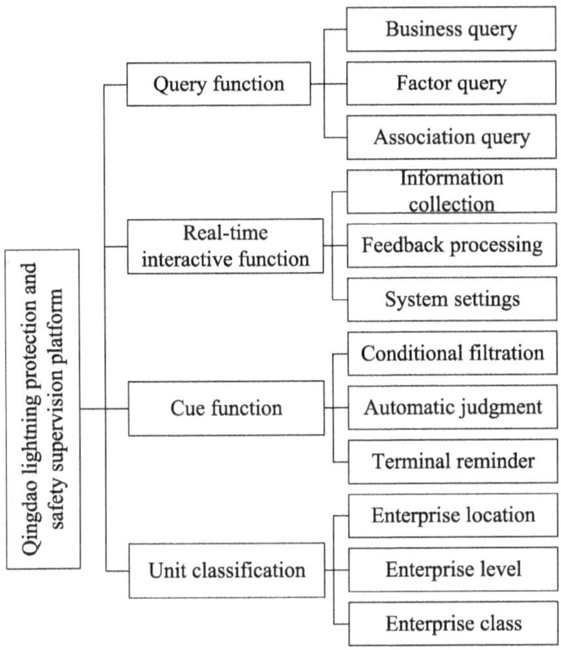

Fig. 3. The overall function diagram of lightning protection and safety supervision platform in FuJian city

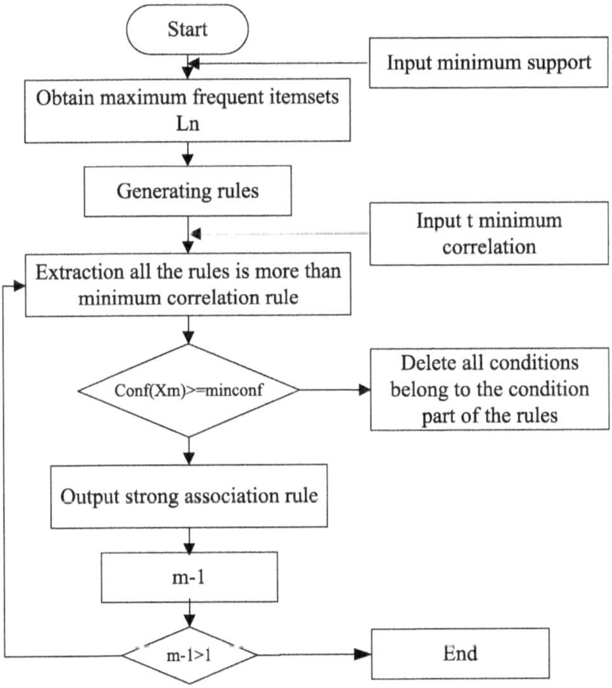

Fig. 4. The flow chart of Apriori frequent itemsets algorithm

5 Conclusions

In order to further improve the mine safety supervision work, increase the mine safety production supervision and strengthen mine safety production track inspection, we have the aid of increasingly sophisticated information systems and advanced mobile Internet technology to build mine safety supervision platform in FuJian city, at the same times we can be the city's key regulatory units into a platform, to realize real-time supervision of the unit mine safety and effectively improve the mine safety real-time monitoring level.

Acknowledgments. Supported by the research on the key technology of intelligent protection against lightning disaster in sensitive industry (No. 2018K05).

References

1. F.X. L.: Database Principle and Technology, pp. 312–330. Electronics Industry Press, Beijing (2013)
2. B. T.: Research on the Office Automation System of Local Government in County Level, pp. 3–16. Kunming University of Science and Technology, Kunming (2014)
3. C.J. Z.: Design of small and medium enterprises office automation system. Shandong Institute of Computer Science and Information Technology and Information Technology Seminar in 2005, pp. 1112–1117 (2005)
4. Y.H. W.: Web Application Framework and Its Case Based on MVC Design Model, pp. 1–9. Jilin University, Jilin (2014)
5. Jacobson, I., Booch, G., Rumbaugh, J.: UML User Guide, pp. 11–110. Mechanical Industry Press, Beijing (2014)
6. Jorgensen, P.C.: Software Testing, pp. 145–150. Mechanical Industry Press, Beijing (2004)
7. Zahran, S.: Software Process Improvement, pp. 19–39. Mechanical Industry Press, Beijing (2013)
8. Yang, F.Q.: Software reuse and relevant technology. Comput. Sci. **26**(5), 112–114 (2014)
9. Yang, F.Q.: Software reuse and relevant technology. Comput. Sci. Chinese with English Abstract [8] **26**(5), 1–4 (1999)
10. Zhang, D.Q.: The impact of cloud computing on the development of e-commerce in the future. Acad. Exch. **4**, 135–138 (2014)
11. Gray, A.: The Elements of C++ Style, pp. 560–589. Cambridge University Press, Cambridge (2013)

An Error Compensation Method Based on Millimeter Wave Radar

Chen Cao[1](✉) and Yu-Chen Liu[2]

[1] Research Institute of Highway Ministry of Transport, Haidian District Xitucheng Road No. 8, Beijing 100088, China
c.cao@rioh.cn

[2] Beijing GOTEC ITS Technology Co., Ltd., Haidian District Xitucheng Road No. 8, Beijing 100088, China

Abstract. In this paper, mainly studies the effective target determination method of millimeter-wave radar (MMW) used in autonomous vehicles, also focuses on data acquisition and error compensation for millimeter wave radar by using Delphi ESR radar and USBCANII converter, and the millimeter wave radar error correction method is obtained by analyzing radar error by radar positioning test.

Keywords: Error compensation · Millimeter wave radar
Radar positioning test

1 Introduction

In the vehicle driving environment, the measured value of millimeter-wave radar may be derived from the target vehicle and may also be derived from other disturbances. Because of the instability of the millimeter wave radar's own work and the inhomogeneity of the echo energy, false targets may appear. At the same time, the positioning error of the millimeter wave radar itself may lead to inaccurate positioning.

An error compensation method for millimeter-wave radar is sought to help the vehicle-mounted millimeter-wave radar to locate the target accurately.

2 The MMW Radar Composition and Structure

The MMW radar sends electromagnetic waves around and calculates the distance, Angle and speed of the detected object by analyzing the reflection signal. The wavelength of electromagnetic waves emitted by millimeter-wave radar is millimeter-wave (1–10 mm) and the frequency is 24 GHz–300 GHz, and the range of small millimeter wave radar for commercial use is usually 100 m–200 m. Millimeter-wave radar has many advantages, such as strong detection ability, good target discrimination performance, strong ability to eliminate false alarm, good night performance, and suitable for rain and snow, etc. With the development of millimeter wave radar technology, millimeter-wave radar is not only good in performance, perfect in function, volume and weight. At the same time as the product innovation, each big brand of MMW radar

K. Deng et al. (Eds.): ICMIR 2018, AISC 856, pp. 153–159, 2019.
https://doi.org/10.1007/978-3-030-00214-5_20

prices are falling, so by more and more attention in many areas, especially in the intelligent autopilot vehicle, and is at a relatively frontier field of unmanned aerial vehicle. Because of the research requires MMW radar multi-target detection, Mid-range detection and long-range detection capacity, The ESR (electronically scanned millimeter-wave radar) produced by Delphi company was selected as a sample. The Delphi ESR as shown in Fig. 1.

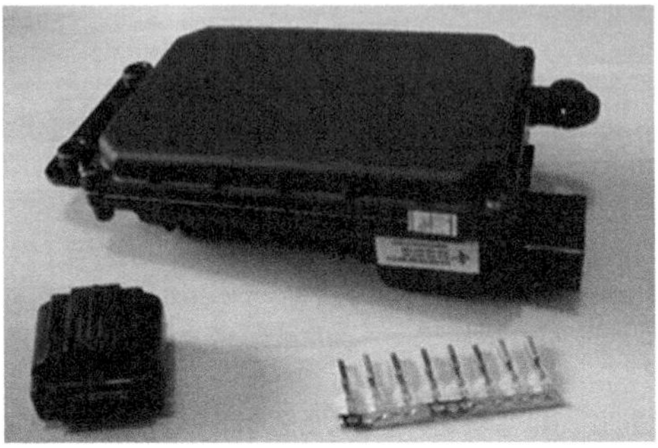

Fig. 1. Delphi ESR radar

Delphi's multimode ESR provides wide coverage at mid-range and high-resolution long range coverage using a single radar. Wide, mid-range coverage not only allows vehicles cutting in from adjacent lanes to be detected but also identifies vehicles and pedestrians across the width of the equipped vehicle. Long-range coverage provides accurate range and speed data with powerful object discrimination that can identify up to 64 targets in the vehicle's path, it's performance parameter is shown in Table 1.

Table 1. Delphi ESR radar performance data

Delphi ESR radar data		Long-range distance	Mid-range distance
Feature	Band	76–77 GHz	
	Size	130 × 90 × 39 mm (w × h × d)	
Refresh rate		50 ms	50 ms
Coverage area	Distance	1–175 m	1–60 m
	Relative speed	−100 to +25 m/s	−100 to +25 m/s
	Horizontal view Angle	±10°	±45°
Accuracy	Distance	±0.5 m	±0.25 m
	Relative speed	±0.12 m/s	±0.12 m/s
	Angle	±0.5°	±1°
Multi-target discrimination	Distance	2.5 m	1.3 m
	Relative speed	0.25 m/s	0.25 m/s
	Angle	3.5°	12°

3 Data Acquisition

Because radar transmits data through can port, the USBCANII is selected in this research. As shown in the Fig. 2, CAN data is collected, and CAN data is converted into serial port and sent to the computer. Found in the experiment, after data conversion, using USBCANII software development kit for acquisition of radar data frame is not complete, so that each target data acquisition of data is not the same time, which lead to the radar to detect partial obstruction cannot be output as target of leak in pairs. Therefore, the following measures are taken to ensure the integrity of the data before the interpretation of the radar data protocol. First, the VCI_ GetReceiveNum function is called to determine whether the radar data in the cache region reaches the threshold T (T = 70) before the data collection of each call to the VCI Receive function. Secondly, the data received is spliced. Usually incomplete data is not lost for data, but two frames of data exist in two different frames. Mainly through to the radar data within the frame of long tail frame head frame and the frame to check, find the frame head of the frame head after frameinfo1 copied into the array of data in a frame of data arrival will frame before the end of the data copied to array within frameinfo1 joining together, so it can realize the integrity of the data frame. After the guarantee data integrity of the radar data according to specified radar resolution protocol to radar data by data parsing hexadecimal agreement for decimal contains multiple objective Angle, distance and speed information of decimal data. In order to improve the operation efficiency of the software, the optimized measures such as displacement operation and multiplication and division are adopted, and the program of the optimized collection program is not able to calculate the speed of 1 ms on the onboard computer.

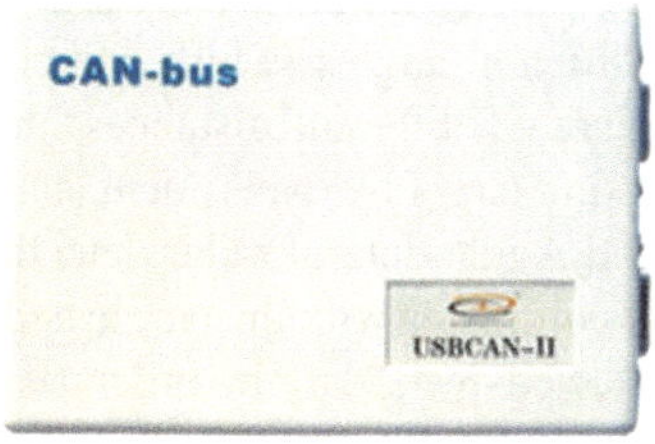

Fig. 2. USBCAN II

4 Error Compensation

In radar positioning experiment, there is a certain angular deviation between the position of radar output target and the target real value, as shown in Fig. 3. Figure 3(b) red box for radar to detect the target location, and with blue box manual calibration for target true location, the angular deviation in schematic diagram in Fig. 3(a), because of the higher accuracy of radar target positioning is better for the detection accuracy of the later target. So designed the experimental analysis of radar error and error compensation.

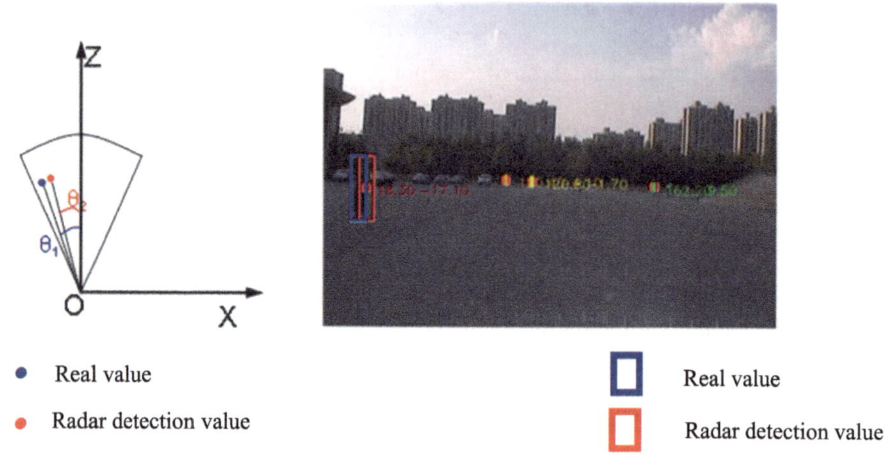

(a) Objective Angle deviation schematic diagram

(b) Actual position and radar detection position comparison diagram

Fig. 3. The angle error of radar

In order to prevent the influence of the background objects on the radar, an open space of 100 m × 40 m was selected for the experiment, as shown in Fig. 4(a). In 10 m to 78 m and −16° to 16° between every 4 m and every 2° to determine a coordinate point, a total of 17 × 18 points, as the real value of these coordinates, and draw the corresponding marked points on the ground. As shown in Fig. 4(b), the above diagram shows the marking lines of different angles (white lines in the figure). The following picture shows the marking lines of different distances, thus forming each coordinate point. Human as a radar detecting target by experiment station in each coordinate point, we can record the distance and Angle data of radar detection and the real distance and Angle data of corresponding coordinates with the Angle and distance values of the target in different positions are detected by radar. In order to ensure that the data of the compensation function is not identical with the test data, two experiments were conducted on the above coordinate points, and the radar detection value and the real value were recorded respectively, and a total of 306 × 2 sets of data were obtained.

The distance error and Angle error of the radar are analyzed by the radar detection value and the real value of 306 sets, and the distance error and Angle error of the radar are calculated according to the formula (1).

$$\begin{aligned} E_{\mathrm{D}} &= D_{radar} - D_{groundtruth} \\ E_{A} &= A_{radar} - A_{groundtruth} \end{aligned} \tag{1}$$

In the above equation, $D_{groundtruth}$ is the real value of distance, D_{radar} is the distance value of radar detection and E_{D} is the distance error. $A_{groundtruth}$ is the real value of the Angle, A_{radar} is the Angle value of the radar detection and E_{A} is the Angle error.

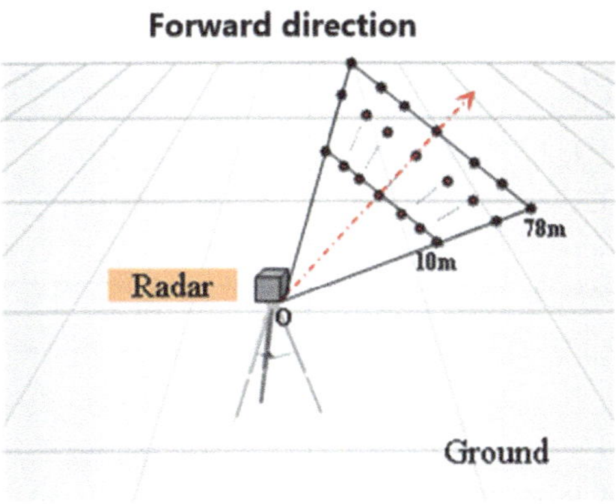

(a) Ground structure network coordinate point

(b) Actual coordinate point

Fig. 4. Radar positioning experiment schematic diagram

The distance error analysis results are shown in Fig. 5. The distance error is between −8.15 cm and 11.55 cm, and the error has little effect on the target in the image. Therefore, it mainly analyzes the Angle error of radar.

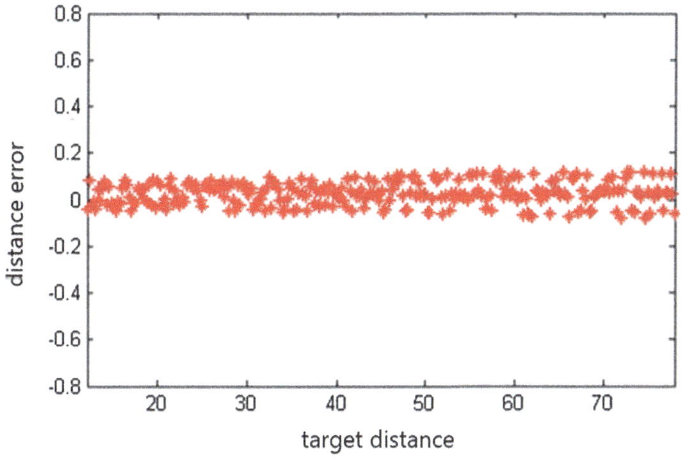

Fig. 5. The relationship between radar distance and target distance diagram

The correction method of radar Angle error is to construct the correction function of Angle error by using least square method for Angle error. In order to prevent the left Angle error of radar and the right Angle error, the Angle error of the left and right angles is analyzed respectively, and the corresponding Angle error function is shown in Eq. (2).

$$
\begin{aligned}
E_{angel} &= 0.001029A_{radar}^2 + 0.004303A_{radar} + 0.654002 \quad (A_{radar} < 0) \\
E_{angel} &= -0.007778A_{radar}^2 + 0.090453A_{radar} + 0.269875 \quad (A_{radar} > 0)
\end{aligned} \tag{2}
$$

In this equation, A_{radar} is radar detection Angle, E_{angel} is corresponding error. So, the error correction function is defined as:

$$F_{angle} = -E_{angel} \tag{3}$$

The angle error correction functions is:

$$A_F = A_{radar} - E_{angel} \tag{4}$$

In order to test the correction effect of compensation function, take the rest of the 306 groups of radar detection value and real value test error compensation function of correction effect, and compared with fixed before the error, the experimental results as shown in Table 2.

Table 2. The before-and-after comparison of radar angle error by moid

	Left-hand angle error/°	Right-hand angle error/°
	Average value	Average value
Before modified	0.444	0.152
After modified	0.047	0.031

5 Conclusion

By comparing the error data of the millimeter wave radar positioning test, it is found that the error accuracy of the modified algorithm is much higher than that of the original output data. Because this article USES the Delphi ESR and CAN - BUS II are widely used in automotive radar at the present stage, the error compensation method of the millimeter wave radar is written into the radar processor to reduce the positioning error of the target.

References

1. Su, H.-Y., Zhu, H.-C.: Study on phase error calibration method for stepped frequency radar. Syst. Eng. Electron. **30**(2), 253–256 (2008)
2. Shi, X.: Application and development of millimetre-wave radars. Telecommun. Eng. **46**(1), 1–9 (2006)
3. Gao, Z.-H., Wang, J., Tong, J., et al.: Target motion state estimation for vehicle-borne millimeter-wave radar. J. Jilin Univ. (Eng. Technol. Ed.) **44**(6), 1537–1544 (2014)

4. Huang, W.: Design and Implementation of the Obstacle Detection System Based on Radar and Machine Vision. Wuhan University of Technology, Wuhan (2010)
5. Ausherman, D.A., Kozma, A., Walker, J.L., et al.: Developments in radar imaging. IEEE Trans. Aerosp. Electron. Syst. AES **20**(4), 363–400 (2008)
6. Kuroda, H., Okai, F., Takano, K.: A study on estimating the aiming angle error of millimeter wave radar for automobile. IEEJ Trans. Ind. Appl. **125**(3), 286–292 (2005)
7. Peng, D.L., Zeng, D., Huang, X.P.: A method of multi-sensor and multi-target tracking and fusion based on double-threshold technique. Int. J. Sens. Wirel. Commun. Control **2**, 1 (2012)
8. Zhang, Y., Xiao-Jun, H.U.: Study on the method of correction for radar azimuth angle refraction error. Mod. Radar **36**, 43 (2008)

Research on Road Transportation Safety Management

Xuan Dong(✉)

Key Laboratory of Operation Safety Technology on Transport Vehicles Ministry of Transport, Research Institute of Highway Ministry of Transport, Courtyard 8#, Xitucheng Road, Haidian District, Beijing 100088, People's Republic of China
x.dong@rioh.cn

Abstract. The safety management of road transportation is not only an objective requirement and an inevitable choice to promote the scientific development and healthy development of the road transport industry, but also a difficult and focal point in the management of the road transport industry. Through the statistical analysis of road traffic accident cases, the characteristics of road traffic accident law in China are summed up in this paper, and the outstanding problems existing in the current road transportation safety management are analyzed. On the basis of road traffic safety management, the future road safety management ideas in China are put forward.

Keywords: Road transport · Security management · Accident rule

1 Introduction

With the rapid development of China's economy and society, the volume of passenger transport and turnover, the volume of freight and turnover, and the amount of operating vehicles are increasing year by year, and the demand for social transportation is vigorous. From the perspective of development, the scale of road transportation industry will continue to grow, and its contribution to social and economic development will also increase. Security is the premise and foundation of development. From the point of view of guaranteeing and improving the people's livelihood and serving the social and economic construction, the workload of road transportation safety management will be more and more large, and the demand for the road transportation is becoming more and more high. The importance of road transportation safety management is more prominent. In recent years, the road transport industry has made great achievements in safety management. However, compared with the developed countries in the world, the level of road transport safety has a certain distance, especially the serious accidents still occur, and the security situation is still very severe.

K. Deng et al. (Eds.): ICMIR 2018, AISC 856, pp. 160–166, 2019.
https://doi.org/10.1007/978-3-030-00214-5_21

2 Characteristics of Road Traffic Accident Law

Through statistical analysis from 2009 to 2017, the road traffic accident management of more than three people in China has been killed at a time. This article summarizes the law of road traffic accidents.

(1) The Accident occurred most in the first three quarters
The first quarter was affected by the peak flow of Spring Festival passenger transport and the bad weather such as rain and snow. In the two quarter and the three quarter, large scale production, plus the three quarter in the summer season, is also a season of accidents. Statistics show that in 2014, the first three quarters of the death of 3 people and more traffic accidents accounted for 82.7% of the year.
(2) February, March and July and August are high incidence of accidents
In recent years, the monthly distribution of road traffic accidents occurred in 3 or more people died in 2, March, July and August. In February and March, due to the surge of Spring Festival passenger traffic and bad weather conditions, it is easy to cause road traffic accidents. In the summer of July and August, transportation demand was strong, and affected by the hot weather, the driver's fatigue driving and mood irritability and other factors had caused more accidents.
(3) There is still more time for the accident in the daytime
From the whole day's time, there are two periods of road traffic accidents, which are from 9 a.m. to 12 p.m. and 14 p.m. to 17 p.m. However, most of the major road traffic accidents related to freight cars have occurred in the near future, most of which have occurred in the middle of the night, indicating that the fatigue driving of the truck drivers has not been effectively solved.
(4) Collisions are still the most important forms of the accident
According to the statistical analysis in recent years, collision is the main form of traffic accidents in the road transportation industry, in which the proportion of frontal collision is higher in the collision accident. The main causes of the collision accidents are speeding, illegal driving, rain and snow, vehicle out of control and poor sight.
(5) The driver factor is still the main cause of the accident
In accordance with the situation in the world, traffic accidents caused by driver factors still occupy the main proportion of the total number of accidents, such as illegal overspeed, illegal overloading, illegal driving, illegal overtaking, fatigue driving and improper operation.

3 Main Problems in the Safety Management of Road Transportation

(1) The responsibility of the safety of the transportation enterprise is not implemented
One is that understanding is not in place. It is mainly manifested in the lack of a correct understanding of the main responsibilities of enterprises' safety in production, the relationship between security and development, and the relationship

between economic benefits and safety input. A lot of transportation enterprises are just sticking to their mouths and hanging on the wall for production safety work. Enterprises are the main body of production and management, the main body of security obligations, and the direct right of production safety. Enterprises are the main body of production and operation, the enterprises directly control the factors of production and operation, the enterprises play a decisive role in production and operation, and the enterprises have the direct responsibility for the safety accidents.

The two is that the system is not implemented. In terms of safety in production, the industry management departments have formulated rules and regulations, but transport enterprises have not been well implemented. The main performance is that the enterprise safety responsibility performance is not in place, the management regulations, standards and systems of safety production are not in place, which leads to the lack of human resources in safety production, the prevention measures of safety production management, the safety hidden trouble clearance work, the implementation of daily safety production work, and the basic work of enterprise safety. Solid, safety education, safety training and safety technology guarantee have different degrees of deficiency.

Three is the lack of effective management means. The key to road transport safety production is to solve the problems on the road. Accidents happen on the road. As a matter of fact, enterprises are in a state of "invisible, unable to hear, and unable to manage" after the operation of vehicles. If there is no support of scientific and technological means, we can not effectively monitor the situation on the road, and transport enterprises have no effective means to implement the main responsibility of safety.

(2) Great hidden danger in rural passenger transport
With the in-depth implementation of the "village and village communication" project, the safety problem of rural passenger transport gradually appears, becoming an important aspect of the traffic accident increment. There are many factors that affect the safety of passenger transport in rural areas, which are mainly involved in the following aspects:

First, the safety consciousness of rural traffic participants is weak. With the rapid development of rural roads, the rural traffic conditions have been greatly improved, but the safety awareness of rural passenger traffic drivers and other traffic participants did not go up. Most of the traffic participants in rural areas are low education, lack of awareness of traffic safety, illegal transportation, such as overloading transportation and vehicle "sick" transportation.

Two, the lack of security facilities is more prominent. Because rural roads are many, wide and large, the focus of early stage rural road construction is to solve the basic travel problem of farmers. Due to many factors such as funds, after the construction of rural highway main body, the safety and warning signs and other facilities are inadequate.

The three is that the equipment of the rural passenger vehicle is relatively backward. Under the influence of insufficient funds and low economic benefits, compared with long-distance bus, the technical conditions of rural road bus are relatively poor, vehicle maintenance is not in place, and the monitoring ability of vehicle running is insufficient, which restricts the service level and security ability of rural road passenger transport.

Four, the main body of rural passenger transport is small and scattered, and the ability of security and risk prevention is relatively poor. Rural passenger transport operators are generally "one car, one household" business mode, few scale enterprises.

4 Effect of Road Transportation Safety Management

After 2012, the management departments at all levels of the road transport industry have done a lot of work on the safety management of road transportation, which greatly improved the improvement of road transport safety. The main work in road traffic safety management is as follows:

(1) Strengthening the responsibility of safe production in transportation enterprises
In accordance with the relevant provisions of the road traffic safety law and the road transport regulations, combining the functions and characteristics of the transportation departments, the management responsibility of the safety source of the "three customs and one supervision" is carried out, and the laws and characteristics of the road transportation production safety are deeply grasped, and the basic management work is firmly grasped and the responsibility system is set up. Efforts should be made to promote the institutionalization and standardization of production safety.

First, carry out special rectification work for road transport enterprises. The transportation department organizes the transportation departments at all levels to examine the operating qualifications of the transportation enterprises, strictly control the technical status of the operating vehicles, examine the qualifications of the employees of the transportation enterprises, improve the construction of the safety management system of the transportation enterprises, and actively guide the enterprises to install the dynamic monitoring equipment, such as satellite positioning and loading, and so on. The safety management equipment of the road transportation enterprises has been obviously improved, the dynamic supervision ability of the vehicles and personnel has been steadily improved, and the scientific level of the safety production management of the transportation enterprises has been gradually improved.

Two, we should speed up the pace of legalization of road transport safety. According to the road traffic safety law, a series of regulations such as "road passenger transport and passenger station management regulations" and "road passenger transport enterprise safety management standard" are formulated and implemented.

Three, we should improve the credibility evaluation mechanism of road transport quality. In order to establish and improve the competition mechanism and market exit mechanism of the survival of the fittest, promote the safe production, good faith operation and high quality service of the transportation enterprises, make the production safety of the enterprise as an important index, and take the heavy traffic liability accident as "one vote veto". Through the establishment of the quality credit rating system of the enterprise, and the organic combination of the assessment results and the business rights of the passenger passenger class line, a more perfect transportation market integrity system has been established.

(2) Give full play to the important supporting role of science and technology in road transport safety work.

First, integrate the industry dynamic supervision resources, and build a nationwide key vehicle operation network control system. The Ministry of transportation has integrated the dynamic monitoring resources of the industry, and the construction of the key operating vehicle network joint control system. It provides effective technical means for the joint supervision of trans regional and cross departments of operating vehicles, and provides security support for major domestic activities. A dynamic public information exchange platform for key business vehicles has been set up to achieve continuous transmission and exchange of vehicle dynamic data between provincial platforms.

The two is to set standards in time and lay the technical foundation for the unified and standardized operation of the system. Because of the lack of unified standards, the functional differences between each system platform are large, and the vehicle terminal is varied. It is simply a simple information exchange. It can not carry out effective supervision and assessment, and provide a higher level of service. The Ministry of transport has formulated a series of standards for the satellite positioning system of road transport vehicles, and has formed a relatively complete standard system.

The three is to do a good job of demonstration and application. The installation and use of satellite positioning devices are commonly focused on installation and light usage. It is more important for the Ministry of transport to put forward the idea that installation is not an end. In 2012, the Ministry of transportation and transportation combined with the Beidou demonstration and application project to do all the preparations for the implementation of the project of "key transport process monitoring and management service demonstration system project" in the field of road transportation. Through demonstration application projects, the application level of two aspects should be improved. (1) improve the application level of transportation enterprises. Through the formulation of the corresponding management methods, standardize the enterprise monitoring terminal, supervise the transportation enterprises to carry out the responsibility of the monitoring subject; (2) improve the application level of the management department, and enhance the ability of joint supervision by establishing inter departmental information sharing and linkage law enforcement mechanism.

5 Ideas of Road Transportation Safety Management in the Future

From the aspects of strengthening the driver management, the safety management of the transportation enterprises, the dynamic supervision of the road transportation, and improving the safety supervision system of the rural road passenger transport, this paper looks forward to the future road transportation safety management work in our country.

(1) Strengthen the driver management

We will continue to promote the quality education project of the driver. The first is to strengthen the cooperation with the public security department, reform the existing examination mode, highlight the training of safety skills, introduce the notice on strengthening the training of motorists and examination management, jointly convene a meeting, and deploy the related work; two is to study and formulate standard teaching standards and promote the teaching management letter. To further strengthen the implementation of the training syllabus for motor vehicle drivers; three, starting with improving the safety awareness and operational skills of the operating drivers, by revising the examination syllabus, setting up the contents of the training and examination of the operating drivers, strictly testing procedures, strengthening the assessment of operating drivers, and establishing Road transport drivers. The driver's continuing education system will carry out the warning education through case analysis, improve the operating driver's integrity assessment mechanism and establish a long-term mechanism for the management of the employees.

(2) Strengthening the safety management of transportation enterprises

The key is to urge the road transport enterprises to implement the main responsibility of safety production. The first is to further implement the notice of the State Council on Further Strengthening the work on the safety of production of enterprises (National [2010] 23), to study and formulate the "standard for the safety production management of road transportation enterprises", to clarify the main responsibility of the transportation enterprises, to refine the safety management institutions, responsibilities and various safety management systems of the transportation enterprises. The two is to study and formulate "road transport enterprise safety assessment method", the introduction of third party organizations to evaluate the safety of road transportation enterprises, and establish a long-term mechanism to guide the continuous improvement of enterprise safety management.

(3) Strengthening the dynamic supervision of road transportation

The use of modern information technology to strengthen the monitoring of key business vehicles is an effective means to ensure road transport safety. The effective carrier of carrying out the responsibility of the safety main body of the transportation enterprise has solved the problem that the former transportation enterprises "can not see, hear and manage" after the operating vehicles go on the road. One is to supervise and urge the relevant departments to urge the "notice on strengthening the dynamic supervision of road transport vehicles" jointly issued by the four ministries and commissions at all levels; two is to establish a long-term operating mechanism of the networked joint control system; three is to carry out the demonstration project in depth; four is to actively communicate with the public security, safety supervision and other regulatory departments and establish the Department. The door coordination and cooperation mechanism.

(4)
Improving the safety supervision system of rural road passenger transport

First, to develop the social management of road traffic safety in rural areas by carrying out the "safe and unimpeded county city building activities", and to form a social work mechanism for road traffic safety with government leadership, cooperation, social linkage, joint management and comprehensive governance, fully play the role of the grass-roots government and establish a sound "county management". Rural road safety work mechanism implemented by township package and village implementation. The two is to cooperate with the public security department to do a good job of propaganda and education on safety knowledge, promote the "civilized traffic action plan", promote the socialization and institutionalization of traffic safety and education, strengthen the construction of the rural propaganda network, and deepen the propaganda of "entering the countryside, entering the enterprise, entering the community, entering the campus and entering the construction site" by strengthening the construction of the rural propaganda network. The three is to vigorously promote the integration of urban and rural passenger transport. We should further improve the policy support, financial support and tax and fee preferential measures for rural passenger transport, and improve the rural passenger transport infrastructure, service quality and safety management level.

Acknowledgments. This work was supported by a Central Public Welfare Scientific Research Institute of basic research funds (2017-9020).

References

1. Standing Committee of the National People's Congress. Road Traffic Safety Law of the People's Republic of China (2011)
2. State Council. People's Republic of China road traffic regulations (2004)
3. Ministry of communications. Regulations on road passenger transport and passenger station management (2005)
4. Ministry of communications, Ministry of public security, State Administration of work safety supervision and administration, safety management standard for road passenger transport enterprises (2012)

Diagnosis of Intermittent Fault for Gear System Using WPT-GHMM

Xiaoyu Fang, Jianfeng Qu(✉), Yi Chai, Ting Zhong, and Xinhua Yan

College of Automation, Chongqing University, Chongqing 400030, China
qujianfeng@cqu.edu.cn

Abstract. With the periodic rotation of gear, the characteristic of intermittent fault in gear system is periodic with fixed amplitude. The aim at this study is to identify the three states of the gear system, i.e., normal state (NS), intermittent fault state (IFS) and permanent fault state (PFS). Based on Wavelet Packet Transform (WPT) and Gaussian Mixture Hidden Markov Model (GHMM), an algorithm used in the gear intermittent fault detection and classification is put forward in this paper. The WPT method is adopted to decompose the vibration signal of gear into different individual frequency bands and extract the energy feature vector. Then input the observation sequence composed of the energy feature vectors into GHMM for training and identification. According to the experimental results, this method can be effectively applied to gear intermittent fault diagnosis.

Keywords: Intermittent fault diagnosis · Gear system · WPT
Energy feature vector · GHMM

1 Introduction

Gear as one of the power transmission devices, which directly affect the performance of mechanical equipment [1]. Before the permanent fault occurs, the system usually has experienced the IFS. The gear device is cyclical rotation driven by the motor, so that most of these intermittent failures are periodic. It usually takes a long time to generate wear and crack in gear, so it is considered that the wear and fracture degree of the equipment is invariable [2]. In addition, it will be eventually transformed into permanent fault due to the cumulative effect of intermittent faults in gear system.

Recently, the study of intermittent fault diagnosis has received extensive concern from the researchers. Guo et al. propose an intermittent fault diagnosis approach based on EMD-HMM to restrain false alarm for Built-in test [3]. Obeid et al. adopt an adapted wavelet transform to identify early intermittent interturn fault [4]. Correcher et al. present two ways of modeling intermittent failure dynamics (probabilistic model and temporal model) to diagnose intermittent fault, and tested them with experimental data [5]. Considering the complexity of intermittent faults, the traditional data based fault diagnosis methods have many problems in application, and it is difficult to construct an analytical model that can accurately describe the intermittent fault system.

For fault recognition, the traditional method mainly deals with the static classification process, which is not suitable for dynamic signal processing, and the training

K. Deng et al. (Eds.): ICMIR 2018, AISC 856, pp. 167–173, 2019.
https://doi.org/10.1007/978-3-030-00214-5_22

requires a large number of samples. Intermittent fault samples are difficult to obtain, in the case of limited number of samples, the optimization process is easy to be trapped in the local extremum [6]. Hidden Markov model(HMM) is a dynamic time series pattern recognition method, which overcomes the shortcomings of traditional fault diagnosis methods with only one-time point [7].

In this paper, a method based on the WPT-GHMM is demonstrated. In this method, reliable model can be trained by a small amount of samples. According to the principle of pattern matching, the most similar model to the detected signal is identified as the recognition result. Discrete hidden Markov model(DHMM) may cause substantial loss of effective information during quantization coding. So it is more advantageous to use GHMM with continuous observed probability density functions than DHMM with discrete symbol observations during fault diagnosis [8].

This remainder of this paper is organized as follows: Sect. 2 presents the introduction of WPT and GHMM, Sect. 3 shows the case study performed to validate the method and Sect. 4 concludes the paper.

2 Methodology

The intermittent fault of the gear system can be regarded as a periodic fault with fixed amplitude. For this characteristic of intermittent fault, we use a method of WPT-GHMM for fault diagnosis in this study. The intermittent fault diagnosis framework for the proposed methodology is shown in Fig. 1. Gear fault diagnosis includes fault feature extraction and fault classification. Before feature extraction, we collect the vibration signals of three states (NS, IFS and PFS) and pretreat the signals by Z-score standardization. Then, according to time series, we divide the pretreated vibration signals into several frames. In feature extraction, the energy feature vectors are extracted by WPT. In fault classification, the Tri-State GHMM models are trained by Baum-Welch algorithm. After training, the observation sequence is inputted into the trained GHMM for identifying the different fault types.

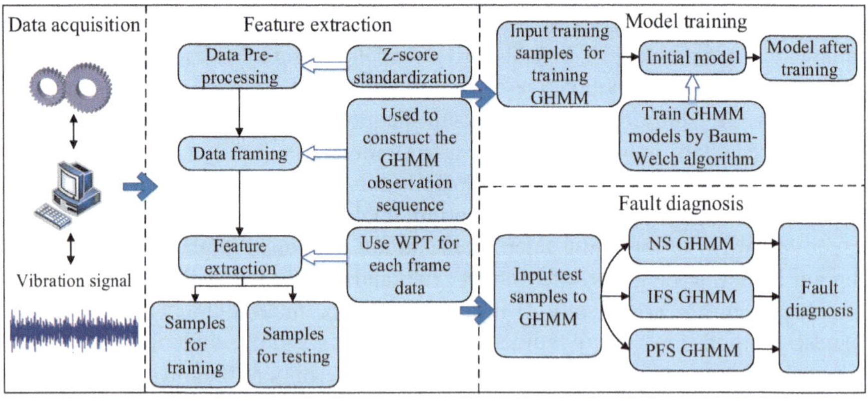

Fig. 1. Block diagram of the method

2.1 Feature Extraction Based on WPT

At present, there are many ways to extract the features of gear fault, such as Fast Fourier Transform, Wavelet transform, Wavelet packet transform, Cepstrum and Wigner distribution. Gear vibration signal is non-stationary and non-linear, some traditional methods cannot ideally extract the fault feature [1].

WPT is a time-frequency analysis method which is suitable for non-stationary signals. Compared with wavelet transform, WPT overcome the shortcomings of wavelet transform, so the whole signal frequency range can be uniformly decomposed. It improves the overall time-frequency resolution.

The WPT algorithm is as follows [9]:

$$\begin{cases} d_{j+1,2n} = \sum_{m} h(m-2k) d_{j,n} \\ d_{j+1,2n+1} = \sum_{m} g(m-2k) d_{j,n} \end{cases} \tag{1}$$

Where $d_{j,n}$ denotes the wavelet coefficients at the j level, n sub-band, $d_{j+1,2n}$ and $d_{j+1,2n+1}$ denotes the wavelet coefficients at the $j+1$ level, $2n$ and $2n+1$ sub-bands, m is the number of the wavelet coefficients.

2.2 Energy Feature Extraction

When the gear is intermittent or permanent fault state, the energy of vibration signal will change in some frequency bands. Intermittent fault is a state before the permanent fault occurs. It occurs at the initial stage of failure, and its energy feature is also different from permanent faults. Therefore, the mapping relationship between the energy information of each frequency band of the vibration signal and the fault information can be established, and then the type of the fault can be identified.

After the kth frame signal is decomposed by the j-layer wavelet packet, its total energy is also decomposed into 2^j independent frequency ranges. Therefore, it can be used as the original signal feature information by measuring the signal energy distribution in various frequency bands. Let (j, i) be the ith node of the jth layer, and the corresponding reconstructed signal is denoted as $S_{j,i}^{(k)}$, its energy recorded as $E_{j,i}^{(k)}$, then the corresponding energy of node (j, i) is calculated as [10]:

$$E_{j,i}^{(k)} = \int \left| S_{j,i}^{(k)}(t) \right|^2 dt = \sum_{x-1}^{N} \left| S_{j,i}^{(k)}[x] \right|^2 \tag{2}$$

Where j is the number of wavelet packets, $i = [0, 1, \ldots, 2^j - 1]$ is the ith node of jth layer, and N is the length of the signal. So the extracted feature vector of kth frame is

$$T^{(k)} = \left\{ E_{j,0}^{(k)}, E_{j,1}^{(k)}, \ldots, E_{j,2^j-1}^{(k)} \right\} = \left\{ o_1^{(k)}, o_2^{(k)}, \ldots, o_{T_k}^{(k)} \right\} \tag{3}$$

Where $T^{(k)} = \left\{ \vec{o}_1^{(k)}, \vec{o}_2^{(k)}, \ldots, \vec{o}_{T_k}^{(k)} \right\}$ is the kth observation vector sequence.

2.3 State Classification Based on GHMM

HMM is a double random process, not only the transfer between states is random, but also the observation of each state is random. In practical applications, it is often encountered that the observation value is represented by a continuous signal. Although the data can be input to DHMM after discrete coding processing, the process may cause the loss of effective information features. Therefore, in this case, GHMM can solve the problem that information may be lost in quantization process [11].

The definition of GHMM is as follows:

To get GHMM which models statistical characteristics of such vector sequences, N states and M mixtures are selected as

$$S = \{s_1, s_2, \ldots, s_N\} \tag{4}$$

$$X = [x_{nm}]_{N \times M} \tag{5}$$

Where x_{nm} is kth mixture of state s_n, N is the number of state and M is the number of mixture. Then, GHMM can be expresses as

$$\lambda = (\pi, A, C, \mu, U) \tag{6}$$

Where π is initial state probability distribution, A is state transition probability distribution; C is mixture gain of Gaussian mixture model (GMM), μ is mean of GMM, U is covariance of GMM.

If the probability density function $b_{t,n}^{(k)}$ of observation vector $o_t^{(k)}$ in state s_n is the Gaussian density function, The probability density function is given as

$$b_{t,n}^{(k)} = \sum_{m=1}^{M} c_{nm} \times \frac{\exp\left(-0.5(\vec{o}_t^{(k)} - \vec{\mu}_{nm})^T \cdot U_{nm}^{-1}(\vec{o}_t^{(k)} - \vec{\mu}_{nm})\right)}{2\pi\sqrt{|U_{nm}|}} \tag{7}$$

Where $\vec{\mu}_{nm}$ is a D-dimensional mean vector and is a $D \times D$ dimensional covariance matrix at x_{nm}, c_{nm} is mth gain of Gaussian weighting in state s_n.

Based on the above definition, the HMM model of gear system is established in Fig. 2. Where a_{ij} is state transition probability, s_1 denotes NS, s_2 denotes IFS, s_3 PFS, $O^{(s_n)}$ is an observational sequence corresponding to state s_n.

The model uses the Tri-State Hidden Markov Chain with left and right type. The concrete steps for training GHMM are as follows:

(1) Divide the observation sequence $O^{(s_n)}$ into two parts, namely $O_1^{(s_n)}$, which represents the training samples, and $O_2^{(s_n)}$, which are the testing samples.
(2) The observation sequence $O_1^{(s_1)}$, $O_1^{(s_2)}$ and $O_1^{(s_3)}$ are respectively input into the initial model for training, and three GHMM models are obtained, they are NS GHMM, IFS GHMM and PFS GHMM. In this paper, the Baum-Welch algorithm is used to solve the sample training problem. The algorithm adjusts the parameter of model λ_n by continuous iteration to maximize the probability $P(O_1^{(s_n)}|\lambda_n)$.

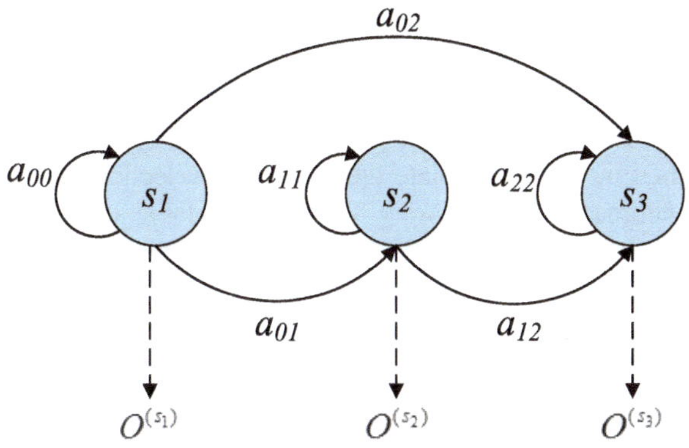

Fig. 2. HMM model of gear system

(3) Input the samples $O_2^{(s_n)}$ to the trained GHMMs respectively. Calculate the probability under different models by forward-backward algorithm. The model with the highest output probability corresponds to the recognition result.

3 Experiments and Analysis

The gear vibration data are obtained through rotating machinery vibration. The three types of the vibration signal are collected at 5120 sample/s and 1470 RPM. The intermittent fault is caused by the partial wear of gear. Due to the cumulative effect of intermittent fault, the partial wear of the gear will result in gear breakage. So the experiment uses partial wear data of gear as intermittent fault data and gear breakage data caused by intermittent fault as permanent fault data. The length of each original collected signal is about 100,000, and 100 samples for each vibration condition are extracted. The vibration signals of the three states are shown in Fig. 3.

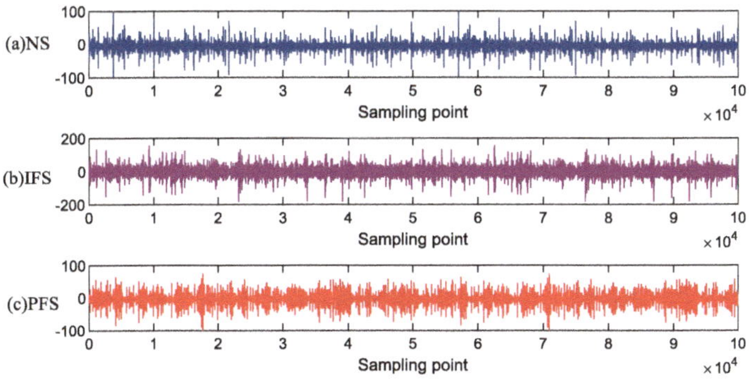

Fig. 3. The vibration signals of the three states

Taking 1000 points as a frame, the vibration signals of the three states after pretreatment are divided into 100 frames. The maximum frequency of the collected signal is about 5120 Hz, the lowest frequency is approximately 0 Hz, and the range of each band is about 640 Hz after three-layer wavelet packet decomposition.

At present, there is not a complete theory for the selection of the mother wavelet function, so sometimes it is necessary to find a suitable choice by experience or multiple attempts. In this paper, DB11 mother wavelet function is selected to decompose each frame with three-layers of wavelet packet. Then a set of eigenvectors is extracted for each frame to form a sequence of eigenvectors.

In general, the initial state of gear system is normal, then the initial state probability vector of GHMM is selected as follows:

$$\pi = [1, 0, 0] \tag{8}$$

From Fig. 2, the initial state transfer matrix is evenly selected as follows:

$$A = \begin{bmatrix} 0.5 & 0.5 & 0 \\ 0 & 0.5 & 0.5 \\ 0 & 0 & 1 \end{bmatrix} \tag{9}$$

For each Gaussian probability density function $b_{t,n}^{(k)}$, in order to reduce the number of model revaluation and increase the training speed, the k-means clustering algorithm is used to estimate the initial parameters. In this study, each GHMM has three states, and each state adopts three Gaussian mixtures. The GHMM parameters are estimated for each state by Baum-Welch algorithm. The IFS recognition results of GHMM are shown in Fig. 4. It can be seen that GHMM has good effect in identifying intermittent fault.

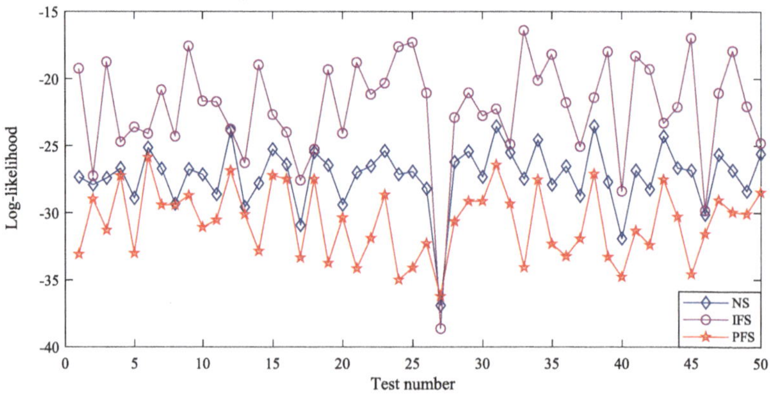

Fig. 4. IFS recognition results of GHMM

4 Conclusions

In this paper, a method based on WPT-GHMM for gear intermittent fault diagnosis is put forward. In this method, the vibration signal is decomposed by WPT and the energy feature vector is extracted on different individual frequency bands. Then the observation sequence is respectively input to the three models of GHMM for training and recognition. The effectiveness of the proposed intermittent fault diagnosis method is verified by using the gear data. The results show that the method has obvious advantages. In future works, we will predict the remaining life of the gear system based on intermittent fault information.

Acknowledgments. This work was supported by the Natural Science Foundation of Chongqing City, China (cstc2016jcyjA0504).

References

1. Zheng, S.L., Wang, X.Y., Nanchang Hangkong University: Gear fault pattern recognition based on the optimum wavelet packet decomposition and HMM. Fail. Anal. Prevent. **9**(6), 330–334 (2014)
2. Zhou, D.H., Shi, J.T., Xiao, H.E., et al.: Review of intermittent fault diagnosis techniques for dynamic systems. Acta Autom. Sin. **40**(2), 161–171 (2014)
3. Guo, M., Ni, S., Zhu, J.: Intermittent fault diagnosis for built-in test system based on EMD and HMM. J. Vib. Measur. Diagn. **32**(3), 467–470 (2012)
4. Obeid, N.H., Battiston, A., Boileau, T., et al.: Early intermittent inter-turn fault detection and localization for a permanent magnet synchronous motor of electrical vehicles using wavelet transform. IEEE Trans. Transp. Electr. **3**(3), 694–702 (2017)
5. Correcher, A., Garcia, E., Morant, F., et al.: Intermittent failure dynamics characterization. IEEE Trans. Reliab. **61**(3), 649–658 (2012)
6. Li, Q., Liang, S.Y., Yang, J.: Bearing fault pattern recognition using harmonic wavelet sample entropy and hidden markov model. J. Shanghai Jiaotong Univ. **50**(5), 723–729 (2016)
7. Zhu, J., Ge, Z., Song, Z.: HMM-driven robust probabilistic principal component analyzer for dynamic process fault classification. IEEE Trans. Ind. Electron. **62**(6), 3814–3821 (2015)
8. Luo, H., Lu, W., Wang, Y., et al.: A novel approach for analog fault diagnosis based on stochastic signal analysis and improved GHMM. Measurement **81**, 26–35 (2016)
9. Wang, Z., Zhang, Q., Xiong, J., et al.: Fault diagnosis of a rolling bearing using wavelet packet denoising and random forests. IEEE Sens. J. **17**(17), 5581–5588 (2017)
10. Wang, X.W., Zhang, Y.J., Hou, Y.X.: A novel fault line selection method of distribution network based on wavelet packet energy. Appl. Mech. Mater. **340**, 572–577 (2013)
11. Jiang, H., Zhang, J.J., Gao, W., et al.: Fault detection, identification, and location in smart grid based on data-driven computational methods. IEEE Trans. Smart Grid **5**(6), 2947–2956 (2014)

Diagnosis Intermittent Fault of Bearing Based on EEMD-HMM

Xiaoyan He, Jianfeng Qu(✉), Ting Zhong, and Xinhua Yan

College of Automation, Chongqing University, Chongqing 400030, China
qujianfeng@cqu.edu.cn

Abstract. Due to the randomness and short duration of the intermittent bearing faults, it is very hard to be found. Aiming at the mechanism that the bearings can be accumulate damage rapidly and cause permanent faults due to intermittent faults, this paper put forward a method that based on ensemble empirical mode decomposition (EEMD) and hidden Markov model (HMM). First, pre-preprocess the original bearing signal and then decompose these through EEMD. Second, several intrinsic mode functions (IMFs) will be got, then chose the first several IMFs with large energy for feature extraction, and the energy and energy entropy of each IMF are obtained as the observed value of the system state. At last, input the observed value into the trained HMM for decision making and obtain the maximum likelihood probability value as the recognition result. Through this operation, it can determine the current system status.

Keywords: Bearing · Ensemble empirical mode decomposition (EEMD) Intermittent fault · Hidden Markov model (HMM)

1 Introduction

At present, the diagnosis of bearing intermittent faults mostly focuses on the methods based on probability and artificial intelligence. However, the probability of fault and intermittent fault must to be known in advance if use the method of probabilistic, but these are almost impossible to be known in advance usually. Because the method of neural network discriminates the state of system from the state at a single moment and neglects the information before and after the intermittent fault, so it would be wrong easily.

For the intermittent fault, a great deal of research has been carried out by professionals both at home and abroad, and great progress has been made. Sedighi et al. [1] proposed a method that intermittent fault detection based on model; Soldani [2] proposed an approach for the detection and localization of intermittent faults in discrete events systems with partial observability; Zhou et al. [3] proposed a new intermittent fault diagnostic technique based on EMD and neural network methods. Deng et al. [4] has developed a discrete event systems approach to discriminating intermittent from permanent faults; Wu et al. [5] based on the relationship analysis between intermittent fault and false-alarm, build intermittent fault diagnosis model of power system based on HMM-SVM.

K. Deng et al. (Eds.): ICMIR 2018, AISC 856, pp. 174–182, 2019.
https://doi.org/10.1007/978-3-030-00214-5_23

For non-stationary time series signal, Wu and Huang [6] proposed a new signal processing method EEMD based on EMD, which overcomes the modal mixing problem of EMD decomposition. It can self-adaptively decompose a complicated signal into several intrinsic mode functions (IMFs) based on the local characteristic time scales of the signals. Now many researchers have done a lot of work and made great progress, EEMD has been widely applied for fault diagnosis [7–9].

HMM is a time series based statistical model for dynamic time series, especially for non-stationary, low repeatability and recurrence of the signal has a strong pattern recognition capability, and it has the advantages of high recognition efficiency and speed of the discrete input signal characteristics. As an important research direction in the field of signal processing, it is gradually extended to other aspects, especially in the field of fault diagnosis and achieved great results [10, 11].

The rest of this paper is organized as follows: Sect. 2 describes the framework of signal processing and identification. Section 3 a experiment is carried out to validate the method. The conclusions are given in Sect. 4.

2 Signal Processing and Identification

2.1 The Framework of Intermittent Fault Diagnosis

The EEMD can adaptively decompose the signal based on the information of the signal itself, and get a group of IMF from high frequency to low frequency. When the faults happened, the signal energy in different frequency bands will change, so according to calculate the energy entropy of each IMF to extract the signal's sensitive feature information, can it judge whether there is a fault and identify the fault's type. The algorithm flow chart of the whole fault diagnosis is shown in Fig. 1.

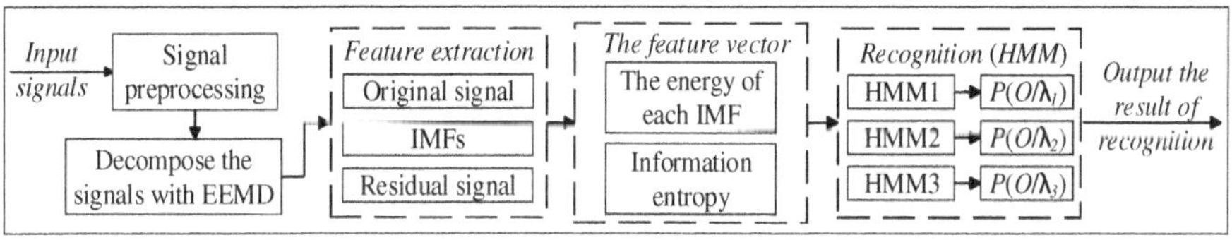

Fig. 1. Flow chart of fault diagnosis method based on EEMD-HMM

2.2 Signal Decomposition

EEMD is a method of fully adaptive data analysis for non-stationary and nonlinear signals, which overcomes the modal mixing problem of EMD decomposition. Because Gaussian white noise has uniform frequency distribution of statistical characteristics, so when the Gaussian white noise add to the signal, the signal is continuous on the different scales, which will reduces the degree of mode mixing.

Through this method, can calculate the ensemble means of the corresponding IMFs of the decompositions:

$$e_i(t) = \frac{1}{M}\sum_{k=1}^{M} c_{k,j}(t) \tag{1}$$

Where $e_i(t)$ is the i th IMF decomposed through EEMD, with $k = 1, 2, 3, \ldots, M$ and $j = 1, 2, 3, \ldots, N$, $c_{i,j}(t)(c_{i,1}, c_{i,2}, \ldots, c_{i,N})$ represents each IMF, which contains different frequencies.

2.3 Signal Feature Extraction

The IMFs decomposed by EEMD contain the components of the signal in different frequency bands from high to low, and the changes of the energy in each frequency band indicate the occurrence of the fault. Considering that the fault information is mainly concentrated in the high-frequency bands, that is, the first several IMFs with a large weight ratio, the energy of the first few IMF components is selected as the eigenvector to identify the working state and the fault type.

For a given signal $x(t)$, suppose the IMFs' component energy which decomposed by EEMD are $x_i(t)$, then the energy E_i of each $x_i(t)$ can be expressed as follows:

$$E_i = \int_{-\infty}^{\infty} x^2(t)dt = \int_{-\infty}^{\infty} [\sum_{i=1}^{n} x_i(t)]^2 dt \tag{2}$$

Where E_i is the i th IMF's energy, $i = 1, 2, 3, \ldots, n$ and n is the number of IMFs.

The first n IMFs components containing main fault information are selected to form the vector T and a new feature vector $T^{'}$:

$$T = [E_1, E_2, E_3, \ldots, E_n] \tag{3}$$

$$T^{'} = [E_1/E, E_2/E, E_3/E, \ldots, E_n/E] \tag{4}$$

Where n is the number of IMFs, E is the sum of all IMF's energy.

2.4 Fault Identification

HMM is a statistical model developed on the basis of Markov chain, which is a double random processing. The real state of the model can't be directly seen, only to conjecture the state and the characteristics through the observation. In summary, HMM can be expressed as follows:

$$\lambda = (A, B, \pi) \tag{5}$$

An HMM model λ can be composed of three basic elements. $\lambda = (A, B, \alpha)$, where A is the hidden state transition probability distribution matrix; B is the observation probability matrix, α is the initial probability distribution vector of the model. The model is defined the probability distribution of each hidden state at the initial moment.

There are three types of system states. Extract the feature vectors T' from training samples of normal, intermittent and fault state respectively. The algorithm adjusts the model parameter λ by continuous iteration to maximize the probability $P(O|\lambda)$. On the actual state monitoring, make the feature vector which through the EEMD decomposition method as the sequence of observations and put the observation vector into the already trained HMM model, then the forward-backward algorithm to calculate the probability of observation vector in different models $P(O|\lambda)$. Input the samples to be identified to the trained HMM model respectively. Calculate the probability $P(O|\overline{\lambda})$ of observation vector under different models by forward-backward algorithm. The state of the system is determined by the HMM model with the highest probability $P(O|\overline{\lambda})$.

3 Experiment and Result Analysis

3.1 Data Collection

In order to collect relevant data and to test the proposed methodology, an experimental platform has been established as Fig. 2. The platform used to test the bearings is consists of a motor, a torque transducer/encoder, a dynamometer, and a computer. Various subtle faults of the bearing are man-made to simulate real faults. Experiments all in the drive motor 100 N m load situation, collect vibration signals by the acceleration sensor which locate on the motor housing through the magnetic base. In these experiments, the signal sampling frequency is 12 kHz, and the motor speed is 2400 r/min.

Fig. 2. Test platform of bearing fault

Relevant data is collected for each experiment, each type of the data consists of 120,000 recordings (time series sequences) collected in a period of time. A total of 100 samples are collected from each type and each sample datum contains 1200 points.

3.2 Fault Feature Extraction

The time-domain graph of one of the sensor signals in a normal state, intermittent state and fault state are shown in Fig. 3.

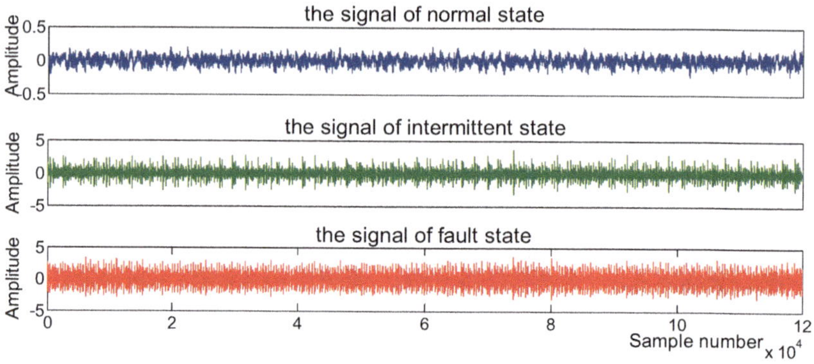

Fig. 3. The signals of normal, intermittent and fault state

EEMD is the first method to decompose the sensor signals into several IMFs, and the result is shown in Fig. 4. What shown in Fig. 5 are IMFs which decomposed through EMD. Compare with the group of figures obtained from EEMD, the group of figures obtained from EMD observed revealed a mode-mixing phenomenon, and the figures obtained from EEMD are better for the same signal.

In order to illustrate the effect of the proposed method is better, this paper presents a method that compare the two groups result of the experiments with EMD and EEMD. The experiment uses these two methods to process the signals from the same sensor, and then compares the results of the processing to obtain which one is the more intuitive effect method.

When the three types of signals are obtained, this paper select zero-mean normalization to preprocess these signals. The three kinds of preprocessed sensor signals are divided into 60 groups in the form of every 2000 points, each group of signals will get several IMFs through EEMD or EMD. Since the first six IMFs that processed by EEMD and EMD already contain 85%–99% of the original signal's information, so these IMFs components are selected as the target.

The IMFs components are arranged in descending order of frequency as $c_1(t), c_2(t), \cdots, c_6(t)$, and $T' = [E_1/E, E_2/E, E_3/E, \ldots, E_6/E]$ can be get according to Eqs. (2), (3) and (4).

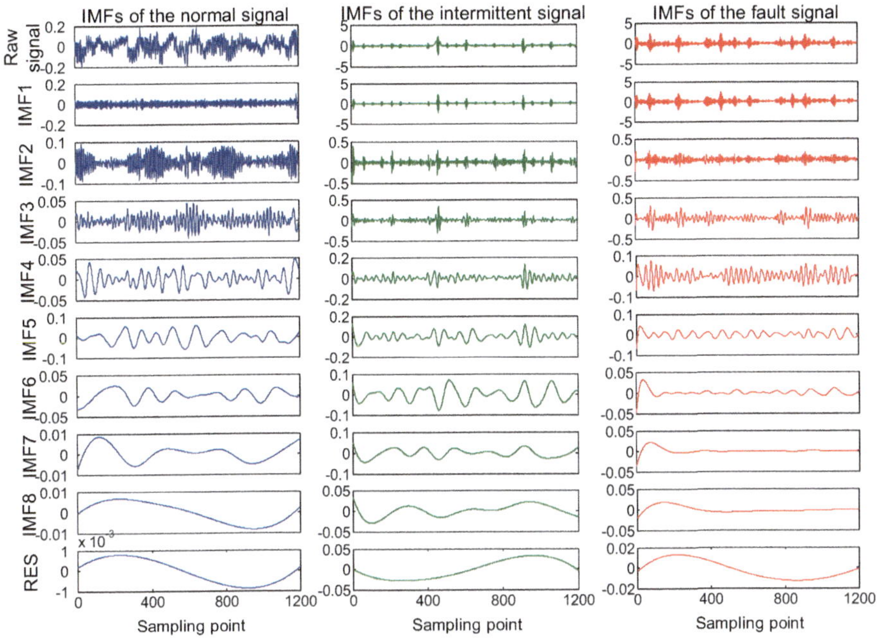

Fig. 4. IMFs of the sensor signal obtained through EEMD

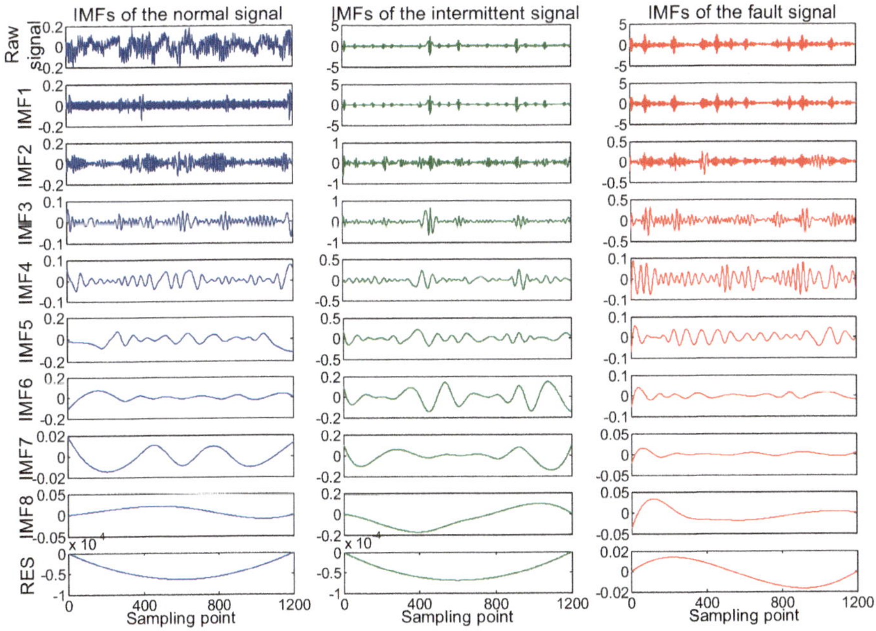

Fig. 5. MFs of the sensor signal obtained through EMD

3.3 Intermittent Fault Identification

The HMM model requires that the value entered into the matrix must be a positive integer, but the IMF energies obtained through the EEMD and EMD are unqualified, so the obtained energy matrix needs to be scalar quantized.

In order to compare the effects of EEMD-HMM and EMD-HMM, a series of experiments were conducted. The eigenvectors extracted from the sample sequence are scalar quantized to be the observed values of the system state. The normal state HMM (λ_1), intermittent state HMM (λ_2) and fault state HMM (λ_3) are obtained by using the groups of training samples, and during the processing, the result will convergence after about 25 times train generally.

For the 50 groups of normal state samples through by EMD and EEMD, input these samples' T' to the HMM models and the results is shown in Fig. 6. In this figure each curve shows the output probability when a sample is input to HMM models, and the model with the highest output probability represents the state recognition result. The same 50 groups of intermittent and fault state validation samples through by EMD-HMM and EEMD-HMM were tested, the results shown in Figs. 7 and 8.

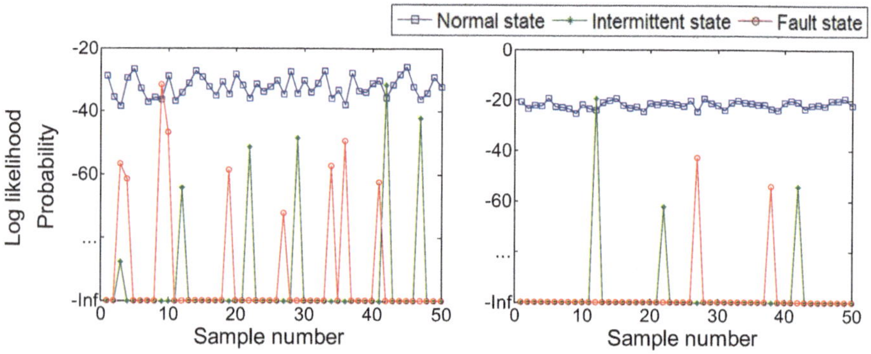

Fig. 6. The probabilities of normal state by EMD-HMM and EEMD-HMM

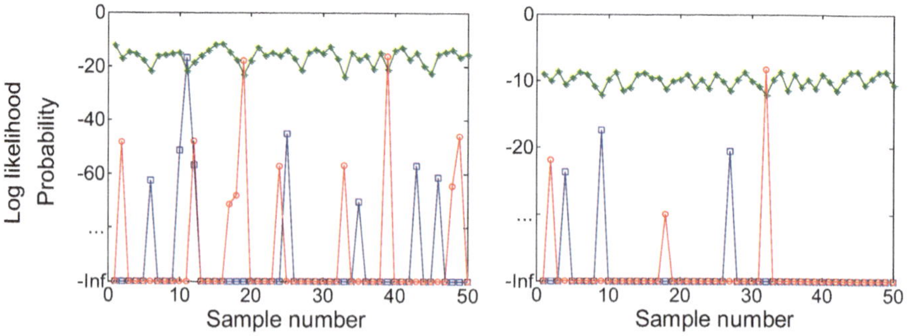

Fig. 7. The probabilities of intermittent state by EMD-HMM and EEMD-HMM

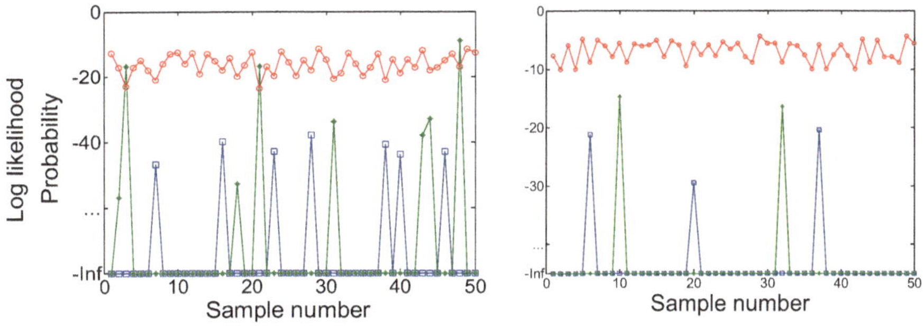

Fig. 8. The probabilities of fault state under EMD-HMM and EEMD-HMM

The test results show that most of the samples can be accurately classified in the 50 sets of test data of each state. The results of the experiment show that the average recognition of EMD-HMM is 94.7%, while the results of EEMD-HMM is 98.7%. It can be seen from the above figures that EEMD-HMM can achieve a better recognition of highly complex fault types and the classification accuracy can reach over 96%. Therefore, EEMD-HMM can achieve the purpose of improving the accuracy of pattern recognition and reducing the complexity of the model.

4 Conclusion

This paper presents a method of intermittent fault diagnosis of bearing based on EEMD and HMM. According to this method, multiple IMFs and one residual energy can be obtained by performing EEMD on the original data signal, and the energy and energy entropy of each IMF can be used as the eigenvector of bearing fault information. Then the intermittent fault diagnosis of bearing is carried out by using HMM which has high generalization ability to solve the problem of intermittent bearing detection. According to the experimental results, EEMD is an adaptive signal processing method with anti-aliasing effect and experiments also show that the performance of EEMD and HMM is superior to the combination of EMD and HMM.

Acknowledgments. This work was supported by the Natural Science Foundation of Chongqing City, China (cstc2016jcyjA0504).

References

1. Sedighi, T., Phillips, P., Foote, P.D.: Model-based intermittent fault detection. Procedia CIRP **11**, 68–73 (2013)
2. Soldani, S., Combacau, M., Thomas, J., et al.: Intermittent fault detection through message exchanges: a coherence based approach. Fault Detect. Superv. Saf. Tech. Process. **39**(13), 1479–1484 (2007)

3. Zhou, H., Liu, G.J., Qiu, J., et al.: Intermittent fault diagnosis under extreme vibration environment based on EMD and neural network. Key Eng. Mater. **584**(584), 97–101 (2014)
4. Deng, G., Jing, Q., Liu, G., et al.: A discrete event systems approach to discriminating intermittent from permanent faults. Chin. J. Aeronaut. (English version) **27**(2), 390–396 (2014)
5. Wu, X.J., Zhang, B.F.: Intermittent fault diagnosis method of power system based on HMM-SVM characteristics. Appl. Mech. Mater. **63–64**, 850–854 (2011)
6. Wu, Z., Huang, N.E.: Ensemble empirical mode decomposition: a noise-assisted data analysis method. Adv. Adapt. Data Anal. **1**(01), 1–41 (2005)
7. Liu, Z., Liu, Y., Shan, H., et al.: A fault diagnosis methodology for gear pump based on EEMD and Bayesian network. PLoS ONE **10**(5), e0125703 (2015)
8. Cheng, G., Chen, X., et al.: Study on planetary gear fault diagnosis based on entropy feature fusion of ensemble empirical mode decomposition. Measurement **91**, 140–154 (2016)
9. Zhang, M., Jian, T., Zhang, X., et al.: Intelligent diagnosis of short hydraulic signal based on improved EEMD and SVM with few low-dimensional training samples. Chin. J. Mech. Eng. **29**(2), 396–405 (2016)
10. Saini, R., Roy, P.P., Dogra, D.P.: A segmental HMM based trajectory classification using genetic algorithm. Expert Syst. Appl. **93** (2018)
11. Lyons, J., Paliwal, K.K., Dehzangi, A., et al.: Protein fold recognition using HMM-HMM alignment and dynamic programming. J. Theor. Biol. **393**, 67–74 (2016)

Using Support Vector Machine Optimized with ACO for Analysis of Multi-component of Spectral Data

Xiang Han(✉), Dong-xiang Zhang, and Jin-yan Yang

School of Chemistry and Chemical Engineering,
Beijing Institute of Technology, Beijing 100081, China
hanxianghx94@163.com

Abstract. This paper present a improved method based on the principle of soft sensor for analyzing overlapped spectra in the case of small samples. The method combines wavelet packet transform and support vector machine for improving the performance of noise reduction filtering, feature extraction as well as improving the prediction accuracy of the soft sensor model. Wavelet transform decomposes the original signal into wavelets of multiple frequency bands to filter out clutter other than the signal band. The feature vectors of the spectral signals are extracted and applied as inputs to the SVM. Support vector machine is applied for least squares regression of input and output data to solve the nonlinear problem of multi-component systems. Ant colony algorithm is applied for optimizing of training parameters. Proper parameters can improve the accuracy and generalization ability of the method. The multi-component overlapped spectra is analyzed by using the method, three kinds of ions of Cu(II), Co (II), Pb(II) the average relative errors are <6%. The result shows the system performed very well. This method offers an promising method for analysis of multi-component overlapped spectra.

Keywords: Support vector machine · Soft sensor · Ant colony algorithm Multi-component · Overlapped spectra

1 Introduction

In the extraction process of metal ions, the concentration of metal ions in the extraction process is measured to ensure the normal and stable extraction process. The traditional methods employ UV spectrometer for sampling and analysis of continuous determination of metal ions [1], however, the detection of a variety of metal ion concentration need for analysis of samples for metal ion separation due to the nonlinear of spectral overlap. The traditional method of multi-component metal ion analysis is complex and time consuming, which can not meet the requirements of real time and online testing in industry.

There are many researchers presented many methods that can analyze the multiple component directly with the separation of metal ions substituting metal ion separation. They employ artificial neural network to construct the detection system to analyze the overlap of the spectrum directly for measuring the concentration of metal ions [2]. Artificial neural network (ANN) is extremely widely used in the modeling of various

K. Deng et al. (Eds.): ICMIR 2018, AISC 856, pp. 183–190, 2019.
https://doi.org/10.1007/978-3-030-00214-5_24

systems in the field of detection and control [3]. It is constructed according to the weights of input and output data of many samples to approximate any nonlinear system.

In these methods, the artificial neural network achieves the analysis of metal ion concentration in the extraction solution of multiple components. However, the artificial neural network needs a large number of sample data for training. The training speed of this method is slow, the number of training samples required is large, and it is easy to fall into local minimum in training, it is difficult to seek balance in the generalization ability and precision. a lot of research has carried out about the optimal solution of balance position of generalization ability and precision.

Consequently, the purpose of this research is to develop a method with support vector machine to achieve high accuracy and good generalization ability in case of small sample. Wavelet transform is applied for denoising and extracting the energy characteristics of signals. Ant colony algorithm is applied for optimizing of training parameters. Support vector machine is based on the structural risk minimization (SRM) optimal principle which guarantees the learning machine has good generalization ability [4]. Our research employed Cu(II), Co(II), and Pb(II) as raw data for the UV spectra data of three ion mixed solutions. The method can availably solve the problems of bad generalization ability and precision. The experimental results show that the method has high detection accuracy and fast response time, and suitable for a variety of soft measurement modeling.

2 Optimization Method

2.1 Wavelet Packet Transform

The basic idea of wavelet transform is the multi-resolution analysis. wavelet packet transform (WPT) decomposes not only scale space V_j but also wavelet space W_j. WPT is a more precise signal analysis method to further decompose the high frequency which can not be achieved by multi-resolution analysis [5]. Therefore wavelet packet transform has been more extensively applied than wavelet transform.

The two scale equation of multi-resolution analysis is:

$$\begin{cases} \psi(t) = \sum_n h[n]\psi_{-1,n}(t) = \sqrt{2}\sum_n (h[n] \times \psi(2t - n)) \\ \phi(t) = \sum_n g[n]\psi_{-1,n}(t) = \sqrt{2}\sum_n (g[n] \times \psi(2t - n)) \end{cases} \tag{1}$$

Where the $\psi(t)$ is the wavelet function and $\varphi(t)$ is the basic scale function, h[n] and g[n] are orthogonal filter group.

By introducing $u_0(t) = \varphi(t)$ and $u_1(t) = \psi(t)$, the two scale equation can be transformed into the following equation:

$$\begin{cases} u_{2n}(t) = \sqrt{2}\sum_{k\in Z}(h[k] \times u_n(2t-k)) \\ u_{2n+1}(t) = \sqrt{2}\sum_{k\in Z}(g[k] \times u_n(2t-k)) \end{cases} \quad (2)$$

Where the two coefficients h[k], g[k] are orthogonal and the relationship of h[k] and g[k] is:

$$g(k) = (-1)^k \times h(1-k) \quad (3)$$

Noise is often the high frequency signal, so denoising can be achieved by threshold operation on the high frequency coefficients obtained by wavelet decomposition [6].

After preprocessing, the signal is decomposed by wavelet, the energy of each frequency section can be given by Eq. (4) after wavelet decomposition [7], where the total energy is given by Eq. (5)

$$E_J^i = \sum_{k,l}(g_{j,k}^l)^2 \quad (4)$$

$$E = \left(\sum_{j=0}^{J}|E_J^i|^2\right)^{\frac{1}{2}} \quad (5)$$

The energy feature samples are obtained by normalization, Eq. (6)

$$T = \left[\frac{E_J^0}{E}, \frac{E_J^1}{E}, \frac{E_J^2}{E}, \ldots \frac{E_J^n}{E}\right] \quad (6)$$

The wavelet transform is used to extract the feature vectors of the signals, and then the feature vectors are input into the SVM to classify the signals. The raw data is greatly compressed after the wavelet transform feature extraction, so it can effectively reduce the number of input layer units and greatly improve the computational efficiency.

2.2 Ant Colony Algorithm

Ant colony optimization algorithm is a optimization method inspired by the efficient feeding behavior of ants in the nature. It is originally used to solve the shortest path problem [8]. SVM involves three parameters: width of kernel function (σ), penalty parameter C, the parameter (ε) of the insensitive loss function. The basic idea of optimizing parameters of SVM by ACO is as follows:

1. Set C, σ and ε with several effective digits to make optimization variables X (standing for the three variables) have n-dimensional component, i.e., $X = \{x_1, x_2, \ldots, x_n\}$. Each component is equally divided into N panel point, then N × n panel points are presented in xOy plane.
2. When t = 0, m ants are located at the starting point, then the transition probability from L_{i-1} to L_i of every ant is calculated by Eq. (7):

$$P(x_i, y_{i,j}, t) = \left[\tau^{\alpha}(x_i, y_{i,j}, t) \times \eta^{\beta}(x_i, y_{i,j}, t)\right] / \left[\sum_{j=0}^{9} \tau^{\alpha}(x_i, y_{i,j}, t) \times \eta^{\beta}(x_i, y_{i,j}, t)\right] \quad (7)$$

   Where $\tau(x_i, y_i, j, t)$ is the remaining pheromone on $Knot(x_i, y_i, j)$ at t, $\eta(x_i, y_i, j, t)$ is the expectation value from $Knot(x_{i-1}, y_i, j)$ to $Knot(x_i, y_i, j)$ which is related to value of the optimal target function F on the above cycle. The factors α and β are relative to pheromone trail and heuristic information.
3. After n time units, all ants reach the terminal point. By calculating the objective functions, the obtained values F* are compared to determine the optical route of this cycle and record the related factors C, σ and ε. The optical route is used to solve the problem of optimal solution of the objective function.
4. Update pheromone of all panel points at this time according to Eq. (8):

$$\tau(x_i, y_{i,j}, t+12) = (1-\rho)\tau(x_i, y_{j,i}, t) + \Delta\tau(x_i, y_{j,i}) + \tau_{\max} \quad (8)$$

$$\Delta\tau_k(x_i, y_{i,j}) = \begin{cases} \frac{Q}{F_k} & \text{if ant k pass } Knot(x_i, y_{i,j}) \text{ in this cycle} \\ 0 & \text{other} \end{cases} \quad (9)$$

Where ρ is the evaporation rate of pheromone ($0 < \rho < 1$). Q is a replace constant and affects the convergence of algorithm. F_k is the optimum solution of objective function for ant_k on this cycle.

3 Results

For evaluating the performance of a multivariate calibration model, the mean square error of prediction (MSEP) and the square of correlation coefficient (R^2) are given by Eqs. (10) and (11), respectively:

$$MSEP = \sum_{i=1}^{l} (a_i - \hat{a}_i)^2 / l \quad (10)$$

Where a_i and $\hat{a}_i$ are the actual value and predicted value, respectively. The i is the number of samples. Where $\bar{a}_i$ is the mean value of actual concentration.

$$R^2 = \sum_{i=1}^{l} (\hat{a}_i - \bar{a}_i)^2 / \sum_{i=1}^{l} (a_i - \bar{a}_i)^2 \qquad (11)$$

The wavelet decomposition is used to preprocess the data, and ACO optimizes the kernel parameters of the SVM, and the SVM outputs the prediction results. The system flow is shown in Fig. 1.

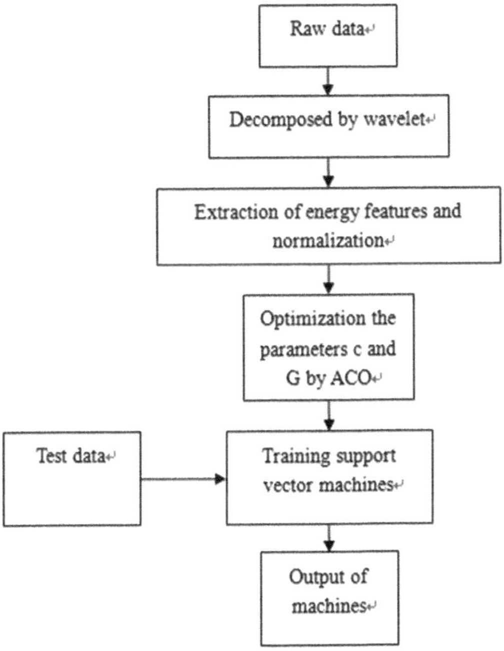

Fig. 1. Component analysis flow chart of multi-component system

3.1 Absorption Spectra of Various Ions in Solution

Figure 2 shows the data of of 8 μg/mL Cu(II), Co(II) and Pb(II) and mixture. It showed that there was a clear overlapping spectra of the above metal ions which prevented the simultaneous value of multiple ion concentrations. However, SVM could be used to solve this problem.

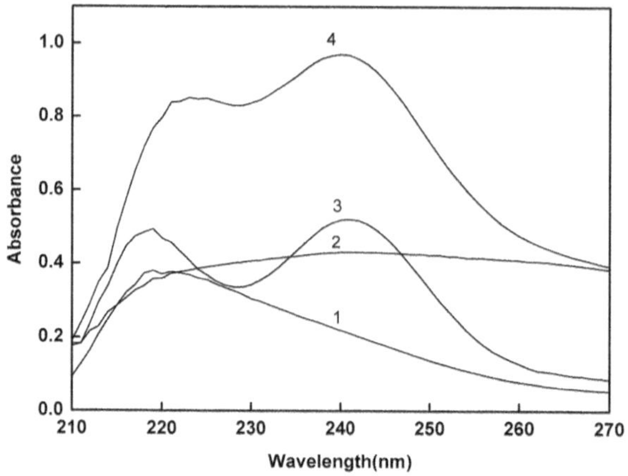

Fig. 2. Absorbance spectra of 1: Co(II), 2: Cu(II), 3: Pb(II), 4: their mixture

3.2 Ant Colony Algorithm Optimizing the Parameters of SVM

Generalization ability and accuracy are two important performance parameters of support vector machine. R^2 can be used as an index to evaluate the generalization ability and MESP as the evaluation index of accuracy.

In order to evaluate ACO method, genetic algorithm (GA) was applied in the study of optimizing parameters. The MSEP and R^2 for two methods and control group are given in the Table 1.

Table 1. MSEP and R^2 values for multi-component system by the two optimizing methods

Method	MSEP (10^{-2})			R^2		
	Cu^{2+}	Co^{2+}	Pb^{2+}	Cu^{2+}	Co^{2+}	Pb^{2+}
No pre-processing	6.28	5.89	6.63	0.78	0.49	0.74
ACO	0.09	0.37	0.12	0.99	0.96	0.99
GA	0.10	0.42	0.59	0.98	0.96	0.97

Table 2. Actual and calculated concentration and relative error of the unknown

Sample no	Relative error (%)		
	Cu^{2+}	Co^{2+}	Pb^{2+}
1	6.71	−7.95	3.68
2	−7.96	2.46	6.17
3	2.46	0.55	−1.92
4	−6.29	−6.29	2.09
5	−6.21	6.21	6.83
Mean relative error	5.93	4.70	4.14

SVM model with optimum parameters was established by learning training data. Table 2 shows that the concentrations and relative error of three metal ions were calculated by SVM. The relative errors of Cu(II), Co(II) and Pb(II) were 5.93%, 4.70% and 4.14%, respectively.

4 Discussion

The experimental results show that the model can effectively approximate the nonlinear and strongly coupled multi-component system, so as to realize the fast and accurate analysis of multi-components. We find that WPT can extract the characteristics of signals and effectively reduce the number of inputs, and the ant colony algorithm can optimize the prediction accuracy of SVM model. Our analysis model is based on the support vector machine combined with WPT to eliminate the noise and uses ant colony algorithm to optimize the learning parameters. Prior to this study, Gao also used the WPT denoising method [9], but they used the RBF neural network algorithm to build the model rather than SVM. In addition, some researchers also use SVM algorithm to solve nonlinear system problems.

However, the neural network model constructed requires a large number of raw data as training samples. Support vector function to obtain more accurate analysis results in small samples. There are substantial differences between previous research about SVM and our model which uses ant colony algorithm to optimize learning parameters of SVM. There are substantial differences between previous research about SVM and our model which uses ant colony algorithm to optimize learning parameters of SVM, so that the relative error of model prediction results could be reduced as soon as possible.

In the experiment, the model predicted concentrations are shown in the Table 2, the data in the table which the predicted ion concentration and metal ion concentration is very close to the actual model, the relative errors are relatively small, most can meet the detection requirements. However, here is little known about potential of machine learning for us to speculate about whether there is the better learning algorithm, so we choose to leave the question open for future investigations to resolve.

5 Conclusion

This paper proposed a method based on the data-driven principle, to analyze the spectrophotometry of multi-components for predicting the concentration of each component. The method combines ant colony algorithm, wavelet transform, and support vector machine to predict the concentration of each component of a multi-component system that is difficult to measure online by using spectral data which could be easily measured. The method used artificial intelligence and machine learning to establish a soft measurement model, and performed noise reduction filtering, feature extraction and least-squares regression on complex nonlinear multi-spectral compound spectral data, thus realizing on-line detection of component concentrations. From the analysis results of the multi-component composite spectrum experiment, wavelet transform can not only denoise, but also effectively reduce the number of inputs. From the analysis results of model prediction error and correlation, ant colony algorithm can effectively reduce the error and improve the correlation in the case of small sample. Therefore, it is feasible to use wavelet transform and support vector machines optimized by ant colony algorithm to realize multi-component analysis of spectral data.

Acknowledgments. The work has been supported by the International Science & Technology Cooperation Program of China (2014DFR61080).

References

1. Sun, X., Zhang, W., Fu, Z., et al.: Study on the testing method of metal ions in water by MPT-AES. Ind. Water Treat. **125**(8), 49–52 (2005)
2. Loele, G., De Luca, M., Dinç, E., et al.: Artificial neural network combined with principal component analysis for resolution of complex pharmaceutical formulations. Chem. Pharm. Bull. **59**(1), 35 (2011)
3. Khan, Z.H., Alin, T.S., Akter, A.: Price prediction of share market using artificial neural network (ANN). Int. J. Comput. Appl. **22**(2), 42 (2011)
4. Besalatpour, A., Hajabbasi, M., Ayoubi, S., et al.: Prediction of soil physical properties by optimized support vector machines. Int. Agrophys. **26**(2), 109–115 (2012)
5. Donoho, D.L.: De-noising by soft-thresholding. IEEE Trans. Inf. Theory **41**, 613–627 (1995)
6. Khare, V., Santhosh, J., Anand, S., et al.: Performance comparison of neural network training methods based on wavelet packet transform for classification of five mental tasks. J. Biomed. Sci. Eng. **3**(6), 612–617 (2010)
7. An, Z.P., Xiao, Z.H., Chen, Y.F., et al.: Analysis of wavelet packet energy spectrum and power spectrum and its application to fault diagnosis of hydropower units. J. Hydroelectr. Eng. **34**(6), 271–278 (2015)
8. Zhang, X., Chen, X., He, Z.: An ACO-based algorithm for parameter optimization of support vector machines. Expert Syst. Appl. **37**, 6618–6628 (2010)
9. Gao, L., Ren, S.: Combining orthogonal signal correction and wavelet packet transform with partial least squares to analyze overlapping spectra of three kinds of metal ions. Chemom. Intell. Lab. Syst. **100**, 57–65 (2010)

Online Update Joint Dictionary Learning Method for Hyperspectral Image Super Resolution

Mengen Xu(✉) and BaoLing Xie

Army Artillery and Air Defense Forces Academy of PLA, Hefei 230031, Anhui, China
2695315332@qq.com

Abstract. Aiming at solving the problems that the commonly used single-dictionary learning and offline-update dictionary learning are not precise enough for the sparse coding of input image signal, the quality of reconstruction is not high and the robustness is poor in current image super-resolution reconstruction, online update joint dictionary learning method is proposed. According to the reconstruction constraints and sparse prior, the least square method is used to learn and update the joint dictionary, which can effectively improve the precision of sparse coding. Experimental results show that the proposed method can improve the quality and robustness of reconstructed hyperspectral images when compared with the classical image super-resolution reconstruction algorithm, ensure that the computation time does not increase excessively.

Keywords: Hyperspectral image · Sparse coding · Online update
Joint dictionary learning

1 Introduction

Hyperspectral imaging is an important part of comprehensive earth observation since the 1980s. By combining imaging and spectroscopy, hyperspectral imaging technology is rich in spatial information, radiation information and spectral information. It has the characteristics of continuous spectrum and unified map [1]. Hyperspectral images have been widely used in geological prospecting, marine monitoring and battlefield reconnaissance due to their good spectral characteristics [2]. In addition, in medical imaging, spatial resolution spectroscopy obtained by hyperspectral imaging provides diagnostic information on histophysiology, morphology and composition [3]. Spatial resolution of remote sensing images is an important index to measure the quality of hyperspectral remote sensing images [4].

With the application of super-resolution (SR) in signal processing, remote sensing reconnaissance images, computer vision and medical imaging, more and more researches have been devoted to image SR technology. At present, the image SR problem based on sparse representation and dictionary learning has attracted the attention and in-depth study of many scholars at home and abroad. The dictionary size and learning method are the key points in the sparse representation-based SR process

K. Deng et al. (Eds.): ICMIR 2018, AISC 856, pp. 191–199, 2019.
https://doi.org/10.1007/978-3-030-00214-5_25

because the size of the dictionary directly determines the operation speed of the SR and the quality of the reconstructed image. The image super-resolution algorithm based on sparse representation of original image blocks proposed by Yang [5] is one of the successful algorithms to apply sparse representation and dictionary learning to image SR earlier, but there are still some problems. First, the dictionary update operation is not performed during the reconstruction process, resulting in the algorithm having good reconstruction results only for specific images. Second, the noise in the learning process of the dictionary is very sensitive and has poor robustness. Based on this, Lu [6] proposed an online robust dictionary learning algorithm to solve the problem of updating and robustness in Yang's dictionary learning algorithm. However, this algorithm is mainly used for single-dictionary learning. In the image SR process of reconstruction, the sparse coding of the input image signal is not accurate enough, resulting in low reconstruction image quality.

Aiming at the above problems in dictionary learning and SR reconstruction, this paper proposes a method of online update joint dictionary learning method. Sparse representation of the input image, combined with the dictionary learning theory, the updated high-low resolution dictionary online update, then use the updated dictionary to complete the image SR reconstruction.

2 Related Research Work

Image SR problem is an ill-posed inverse problem, which needs to set two constraints: The first is reconstruction constraint [7], refers to the recovered high-resolution image X should be consistent with the input low-resolution image Y in the image observation model. The second one is sparse prior [8]. Suppose that high-resolution image blocks can be sparsely represented by a determined overcomplete dictionary and their sparse representation can be recovered from low-resolution observed images.

(1) Reconstruction constraints: the low-resolution image Y is obtained by blurring and down sampling a high-resolution image X:

$$Y = SHX + v \tag{1}$$

where S is the down sampling operator, H is the blur filter, and v is the Gaussian noise with a mean of zero and a known variance.

(2) Sparse priori: a high-resolution image block is obtained by up-sampling the training image and then a high-resolution dictionary is learned to obtain a sparse linear combination of the image blocks x in the high-resolution image X in the dictionary:

$$x \approx \mathbf{D}_h \boldsymbol{\alpha}\ , \quad \boldsymbol{\alpha} \in \mathbb{R}^K, \|\boldsymbol{\alpha}\|_0 \ll K \tag{2}$$

Among them, α represents the sparse coefficient matrix. Only a very small number of non-0 elements in the matrix can be regularized with the L0 paradigm $\mathbf{R}(x) = \|\boldsymbol{\alpha}\|_0$. In the process of seeking α, due to the non-convexity of the function, direct calculation is difficult to get the optimal solution, so the L2 paradigm convex treatment, and then find the approximate solution:

$$\hat{\boldsymbol{\alpha}} = \mathrm{Argmin}_{\boldsymbol{\alpha}}\{||y - SH\mathbf{D}\boldsymbol{\alpha}||_2^2 + \lambda||\boldsymbol{\alpha}||_1\} \tag{3}$$

$\hat{\boldsymbol{\alpha}}$ represents the approximate optimal solution matrix, λ represents regularization coefficient.

3 Online Update Joint Dictionary Learning Method

3.1 Single Dictionary Learning

In image SR reconstruction, sparse coding is one of the important links. By learning sample sets $\mathcal{X} = \{\mathbf{x}_1, \ldots, \mathbf{x}_n\}$, we can get the dictionary:

$$\mathbf{D} = \arg\min_{\mathbf{D},\alpha} ||\mathbf{x} - \mathbf{D}\alpha||_2^2 + \lambda||\alpha||_1 \quad s.t. \ ||\mathbf{D}_i||_2^2 \le 1, i = 1, 2, \ldots, K \tag{4}$$

The previous term is a constraint term used to remove the dithering in the dictionary; the latter term represents the sparseness of the dictionary.

For variables $\mathbf{D}$ and α, formula (4) is a non-convex function, it is difficult to find the optimal solution simultaneously. However, by fixing either of the two variables, Eq. (4) becomes a convex function for the other variable, and its solution can be obtained. Therefore, you can iteratively solve the variable $\mathbf{D}$ and α:

(1) Initialize the dictionary by using a Gaussian random matrix;
(2) Fixed dictionary $\mathbf{D}$, using linear programming, according to formula (5) to achieve the update α;

$$\alpha = \arg\min_{\alpha} ||\mathbf{x} - \mathbf{D}\alpha||_2^2 + \lambda||\alpha||_1 \tag{5}$$

(3) Fixed α, according to Eq. (6) update $\mathbf{D}$ by quadratically constrained quadratic programming;

$$\mathbf{D} = \arg\min_{\mathbf{D}} ||\mathbf{x} - \mathbf{D}\alpha||_2^2 \quad s.t. \ ||\mathbf{D}_i||_2^2 \le 1, i = 1, 2, \ldots, K \tag{6}$$

4. Iterate steps (2) and (3) until the function converges.

3.2 Joint Dictionary Learning

Given a pair of training image blocks $P = \{\mathbf{X}^h, \mathbf{Y}^l\}$, $\mathbf{X}^h = \{\mathbf{x}_1, \ldots, \mathbf{x}_n\}$ represents a set of up-sampled high-resolution image blocks, $\mathbf{Y}^l = \{\mathbf{y}_1, \ldots, \mathbf{y}_n\}$ represents the corresponding low-resolution image blocks. The goal of joint dictionary learning is to perform dictionary learning from high-low resolution image blocks so that the sparse representation coefficients of high-resolution image blocks are the same as their corresponding low-resolution image blocks. The sparse coding problems of high-low resolution dictionary are shown as follows:

$$\mathbf{D}_h = \underset{\{\mathbf{D}_h,\alpha\}}{\arg\min} \left\| \mathbf{X}^h - \mathbf{D}_h\alpha \right\|_2^2 + \lambda\|\alpha\|_1 \tag{7}$$

$$\mathbf{D}_l = \underset{\{\mathbf{D}_l,\alpha\}}{\arg\min} \left\| \mathbf{Y}^l - \mathbf{D}_l\alpha \right\|_2^2 + \lambda\|\alpha\|_1 \tag{8}$$

The high-low resolution dictionary has the same sparse matrix in sparse coding, and the joint objective function (9) can be obtained by combining formulas (7) and (8):

$$\min_{\{\mathbf{D}_h,\mathbf{D}_l,\alpha\}} \frac{1}{M}\left\|\mathbf{X}^h - \mathbf{D}_h\alpha\right\|_2^2 + \frac{1}{N}\left\|\mathbf{Y}^l - \mathbf{D}_l\alpha\right\|_2^2 + \lambda\left(\frac{1}{M} + \frac{1}{N}\right)\|\alpha\|_1 \tag{9}$$

N and M denote the dimension sizes of high-resolution and low-resolution image blocks respectively, $1/N$ and $1/M$ are used to balance the value terms in formulas (7) and (8), so that formula (9) can be rewritten as follows:

$$\begin{gathered}\min_{\{\mathbf{D}_h,\mathbf{D}_l,\alpha\}} \|\mathbf{X}_c - \mathbf{D}_c\alpha\|_2^2 + \hat{\lambda}\|\alpha\|_1 \\ \mathbf{X}_c = \begin{bmatrix} \frac{1}{\sqrt{M}}\mathbf{X}^h \\ \frac{1}{\sqrt{N}}\mathbf{Y}^l \end{bmatrix}, \mathbf{D}_c = \begin{bmatrix} \frac{1}{\sqrt{M}}\mathbf{D}_h \\ \frac{1}{\sqrt{N}}\mathbf{D}_l \end{bmatrix}, \hat{\lambda} = \lambda\left(\frac{1}{M} + \frac{1}{N}\right)\end{gathered} \tag{10}$$

Through the above changes, we can achieve single-dictionary learning strategy for joint dictionary learning and acquire the joint dictionary.

3.3 Online Update Dictionary Learning

The traditional method of solving Eq. (3) is iteratively updating $\mathbf{D}$ and α until it converges. This is an offline dictionary training method, and then applied to the image reconstruction process. Although the speed is fast, they take a long time and dictionaries can not be updated automatically, possibly resulting in poor reconstructed images. In this paper, we use an online robust dictionary learning strategy to solve this problem. Online dictionary update can update dictionary according to different LR images, and find the optimal combination of sparse representation to improve the resolution of reconstructed images and reduce the total time consuming.

In this paper, Iterative Weighted Least Squares algorithm is used to update the dictionary $\mathbf{D}$ and sparse coefficient matrix α. The algorithm steps are as follows:

Input: Image signal $\mathcal{X} = \{\mathbf{x}_1, \ldots, \mathbf{x}_n\}$, initial dictionary $\mathbf{D}_0$,λ and the minimum amount of data in one iteration h;

(1) Update the sparse coefficient matrix α according to formula (11);

$$\hat{\alpha}_j = \arg\min \alpha_j \|\mathbf{x}_i - \mathbf{D}\alpha_i\|_1^1 + \lambda\|\alpha_i\|_1 \tag{11}$$

(2) Update $\mathbf{D}_n$ according to $\mathbf{D}_{n-h}$ and $\{\hat{\alpha}_{n-h+1,...,}\hat{\alpha}_n\}$;
(3) Repeat steps (2) and (3) until the function converges or reaches the upper limit of the number of iterations;

Output: Updated dictionary $\mathbf{D}_n$ and sparse coefficient matrix α.

3.4 Online Update Joint Dictionary Learning

Firstly, the algorithm constructs a high-low resolution dictionary pair, learns the initial joint dictionary pairs adaptively for different input images, and obtains the perfect overcomplete dictionary pair. Then the sparse coefficient matrix of LR image is obtained, and then multiplies it with the HR dictionary to obtain the HR image blocks; finally completes the entire reconstruction process through image fusion.

By learning online dictionary, this method not only improves the sparse representation accuracy of images for different scenes, but also improves the efficiency of the algorithm. After the introduction above, combined with online dictionary learning and joint dictionary learning process, summed up the specific steps based on online update dictionary learning method is as follows.

Input: $\mathbf{X}^h = \{\mathbf{x}_1,...,\mathbf{x}_n\}, \mathbf{Y}^l = \{\mathbf{y}_1,...,\mathbf{y}_n\}, \hat{\lambda} = \lambda(\frac{1}{N}+\frac{1}{M})$, $\mathbf{D}_{h0}, \mathbf{D}_{l0}$, h.

for each $\{\mathbf{x}_{n-h+1},...,\mathbf{x}_n\}$ in $\mathbf{X}_c$ **do**

for $j = n-h+1 : n$ **do**

update $\min\limits_{\{\mathbf{D}_h,\mathbf{D}_l,\alpha\}} \|\mathbf{X}_c - \mathbf{D}_c\alpha\|_2^2 + \hat{\lambda}\|\alpha\|_1$;

end for

update $\mathbf{D}_{c(n)}$ by combing $\mathbf{D}_{c(n-h)}$ and $\{\alpha_{n-h+1,...,}\alpha_n\}$;

$n = n+1$;

end for

Output: $\mathbf{D}_{c(n)}$ and α.

4 Experiment Results and Analysis

4.1 Experimental Description

In order to test the effectiveness of the proposed method, the test images were taken from the public data and the collected experimental images. The proposed method was compared with the Sparse Coding Super Resolution (ScSR) method and the Sparse

Matrix Factorization (SMF) method [9]. In order to evaluate the reconstruction effect of this method, we compare and analyze it from the subjective and objective aspects. Subjectively from the reconstruction of visual effects and objectively from peak signal-to-noise ratio (PSNR) and structural similarity (SSIM) comparison of image reconstruction quality. This method of image processing experiments operating environment is: Lenovo ideapad700, Intel Core i5-6300HQ, CPU @ 2.30 GHz, 8 GB RAM, MATLAB R2014a.

4.2 Super-Resolution Reconstruction Results

In this paper, AVIRIS Dataset's "University Campus" image and the acquired "truck" hyperspectral image are used. The spectral range of "University Campus" image is 400 nm to 1000 nm and the spectral resolution is 20 nm. The spectral range of "truck" experimental image is 400 nm to 700 nm and the spectral resolution is 10 nm. In the experiment, Gaussian blur and down-sampled high-resolution reference images were processed separately to get the low spatial resolution image to be processed. The size of the Gaussian blur kernel is 8×8, the down-sampling factor of row and column is 3, the initial dictionary size of high-low resolution is 1024, the sparsity coefficient$\lambda$is 0.1, and the minimum batch data size is $h = 100$.

Experiment 1: The effectiveness of SR reconstruction algorithm based on online update joint dictionary learning.

Figure 1 compares the reconstruction of "University Campus" images with different SR algorithms. The figure shows a reconstructed image of a 600 nm band. Figure 1(a) is a high-resolution reference image. As shown in Fig. 1(b), the reconstructed image of the ScSR method is relatively fuzzy. Figure 1(c) is a reconstructed image of the SMF method, which restored many details compared with the ScSR and improved the edge sharpness of the image and the overall definition. Figure 1(d) is a reconstructed image of the method in this paper, which improves the sharpness of the edge and restores more detail information.

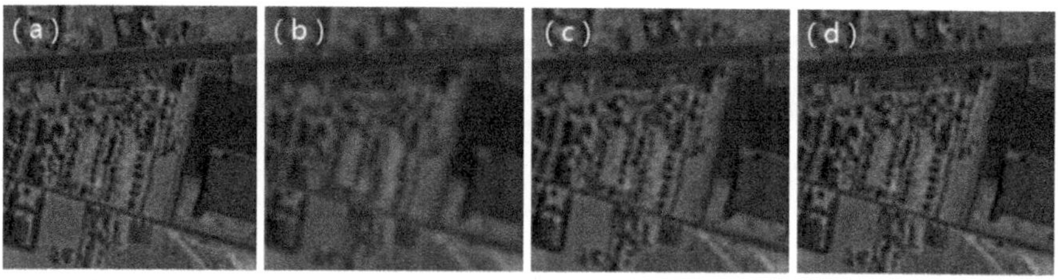

Fig. 1. Reconstruction effect of different SR methods of university image (a) high-resolution reference image; (b) ScSR method; (c) SMF method; (d) this method

Figure 2 compares the effect of the "truck" model image reconstruction in different SR methods, which shows the reconstructed image of a 540 nm band. The proposed method reconstructs the contour edge and detail information of the "truck" better and improves the image edge sharpness and the detail around the "truck" image.

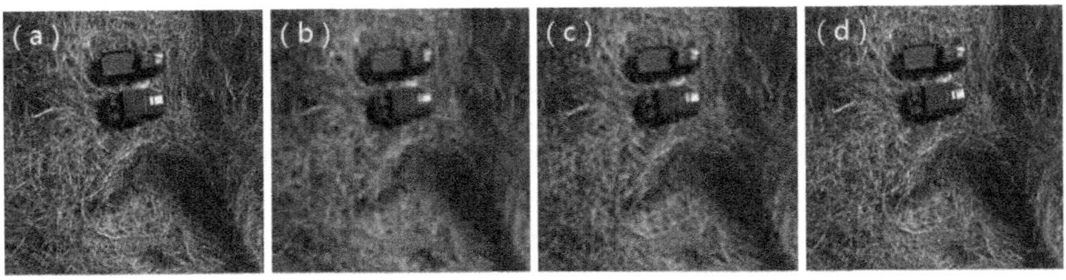

Fig. 2. Reconstruction effect of different SR methods of truck model (a) high-resolution reference image; (b) ScSR method; (c) SMF method; (d) this method

In order to further objectively evaluate the SR methods, the reconstruction effect of different SR methods were compared and analyzed by using PSNR and SSIM. At 3 times the reconstruction magnification, the evaluation results shown in Table 1. The experimental results show that the objective evaluation indexes of this method are superior to the other two methods.

Table 1. Comparison of PSNR (dB) and SSIM evaluation index of different methods

Image	ScSR	SMF	This method
Figure 1	27.33	28.13	**29.78**
	0.6823	0.7642	**0.8626**
Figure 2	25.35	26.03	**27.64**
	0.6415	0.7374	**0.7878**
Average	26.34	27.08	**28.71**
	0.6619	0.7508	**0.8252**

The proposed algorithm in this chapter is superior to the other two contrast algorithms in the subjective visual effects and the objective evaluation of reconstructed image quality.

Experiment 2: Robustness analysis of image super-resolution reconstruction algorithm based on online update joint dictionary learning.

In order to prove the anti-noise performance of this algorithm, the truck image is sequentially weighted with Gaussian noise whose variance is gradually increased from 0 to 15 and progressive value is 5. The PSNR and SSIM of the reconstructed algorithm with different degrees of Gaussian noise are analyzed. The specific experimental results of the two indicators are shown in Tables 2 and 3 respectively.

From Tables 2 and 3, we can see that the image quality obtained by the three reconstruction algorithms all decrease when the added Gaussian noise intensity gradually increases. In contrast, both PSNR and SSIM, the reconstruction results of the proposed algorithm under Gaussian noise with different intensities are still better than the results of ScSR and SMF method.

Table 2. PSNR (dB) of truck image reconstruction with different intensity noise

Method	Variance			
	0	5	10	15
ScSR	29.4752	28.4335	26.6258	25.1946
SMF	31.2534	30.4636	28.3862	27.0578
This method	**32.0242**	**31.2536**	**29.3694**	**28.0225**

Table 3. SSIM of truck image reconstruction with different intensity noise

Method	Variance			
	0	5	10	15
ScSR	0.8712	0.8152	0.7268	0.6066
SMF	0.8956	0.8675	0.7968	0.7048
This method	**0.9089**	**0.8865**	**0.8151**	**0.7445**

5 Conclusion

Aiming at the defects of commonly used offline dictionary and single dictionary learning method, an online update joint dictionary learning method is proposed to hyperspectral image super resolution. Through contrast experiments with the classical ScSR algorithm and SMF method, it is proved that the SR algorithm proposed in this paper is better both in subjective visual aspects and objective evaluation index, as well as the robustness of the reconstructed image against noise, the algorithm in this paper has obvious advantages, which verifies the validity.

References

1. Yi, C.: Research on Spatial-Spectral Resolution Enhancement of Hyperspectral Image, pp. 1–5. Northwestern Polytechnical University, Xian (2016)
2. Wu, L.D., Yao, Z.H., Ren, Z.W.: Techniques on processing of hyperspectral images in perception of battlefield environment. J. Equip. Acad. **28**(3), 1–7 (2017)
3. Liu, L.X., Li, M.Z., Zhao, Z.G., et al.: Recent advances of hyperspectral imaging application in biomedicine. Chin. J. Lasers **45**(4), 1–17 (2018)
4. Zhang, L.P., Li, J.Y.: Review and prospect of hyperspectral image sparse information processing. J. Remote Sens. **20**(5), 1091–1101 (2016)
5. Yang, J.C., Wright, J., Huang, T., et al.: Image super-resolution as sparse representation of raw image patches. In: IEEE Computer Society Conference on Computer Vision and Pattern Recognition, DBLP, pp. 1–8 (2008)
6. Lu, C., Shi, J., Jia, J.: Online robust dictionary learning. In: IEEE Conference on Computer Vision and Pattern Recognition, pp. 415–422 (2013)

7. Xu, G.M., Xue, M.G., Cui, H.C.: Robust single image super-resolution reconstruction model and algorithm based on over-complete dictionary. J. Comput. Aided Des. Comput. Graph **24** (12), 1599–1605 (2012)
8. Sulam, J., Elad, M.: Expected patch log likelihood with a sparse prior. In: Energy Minimization Methods in Computer Vision and Pattern Recognition, pp. 99–111 (2015)
9. Huang, B., Song, H., Cui, H., et al.: Spatial and spectral image fusion using sparse matrix factorization. IEEE Trans. Geosci. Remote Sens. **52**(3), 1693–1704 (2014)

Research on Logistics E-commerce Distribution Network Optimization Based on WebGIS

Chao Xie(✉)

School of Management, Wuhan Donghu University, Wuhan 430212, China
Xiechao188@163.com

Abstract. E-commerce logistics is developing vigorously, and logistics distribution is the key. The control of distribution cost depends on the optimization of distribution network. The core issue of WebGIS is to achieve geographic information sharing, so as to solve the problem of distribution planning. In this paper, a logistics distribution network optimization model is built on the basis of WebGIS, and the distribution path is obtained by selecting the distribution center, using the address matching search target and planning the detailed route, and the model is verified by an example.

Keywords: WebGIS · Logistics distribution · Network optimization

1 Introduction

With the prosperity of e-commerce economy, modern logistics has been developed unprecedentedly, and logistics distribution is the focus of e-commerce logistics. The efficiency of distribution depends to a great extent on the processing of information, including goods information and geographic information, how to deal with various kinds of information, and support the scientific scheduling and rational arrangement of logistics activities. The decisions of distribution centers and route optimization are regarded as the main points of the development of e-commerce logistics [1]. The development of WebGIS technology provides an opportunity for logistics services. Through WebGIS, geographic information can be shared, geographic data services are provided for users, and a strong geographic location data support for distribution center planning and line arrangement is provided. The optimization research of e-commerce logistics network based on WebGIS helps to improve the efficiency and accuracy of network optimization, and has certain practical significance [2].

2 Research Status of Logistics Distribution Network

2.1 Logistics Distribution Network Model

The general logistics distribution network model takes the logistics cost as the objective function, and the distribution of distribution tools owned by the distribution center does

K. Deng et al. (Eds.): ICMIR 2018, AISC 856, pp. 200–206, 2019.
https://doi.org/10.1007/978-3-030-00214-5_26

not exceed the upper limit. The vehicle does not exceed the rated distance, and the distribution time does not exceed the upper limit of the customer's requirement (part model as the objective function) and constructs an optimization model. Then, we use genetic algorithm, simulated annealing algorithm or ant colony algorithm to solve the problem, and the result is usually the order of delivery [3]. There are also some studies that consider more complex, larger and more efficient distribution ways: building distribution centers, dividing distribution areas, and multi point supply. In this model, the distribution center should be selected first, then the two level distribution path planning is carried out, and the distribution center and distribution path results are obtained. Most of the model is also optimized by the logistics cost.

2.2 Research on Optimization of Logistics Distribution Network Based on WebGIS

At present, there are many researches on the model of logistics distribution. There are many researches on distribution path, vehicle scheduling and distribution center selection, but there are few studies on the optimization of distribution network optimization. Some scholars build logistics network based on WebGIS based on the construction of logistics information system, network database and so on. It provides a research foundation for network optimization. Generally speaking, logistics distribution network optimization based on WebGIS consists of three parts: path analysis, address matching and resource allocation [4].

According to the advantages of WebGIS geographic information technology and the characteristics of its integration of logistics distribution network, this paper constructs a logistics e-business distribution network model based on WebGIS based on the network model of relevant scholars, and has done the corresponding example verification to provide reference for further research.

3 Construction of Distribution Network Optimization Model

The network optimization based on WebGIS mainly solves the problem of resource allocation among the nodes, which is divided into two steps: distribution point selection, distribution point to demand point path analysis and optimization.

3.1 Network Description

$G = (V, E, D)$ represents the distribution network of logistics e-commerce, $V = (v_1, v_1, \cdots, v_n)$ represents the node of the network, the $E = (e_1, e_1, \cdots, e_n)$ represents the edge of the network, and the D represents the distance between the nodes. As the edge of the network, it can be expressed as follows:

$$D = \begin{bmatrix} d_{11} & \cdots & d_{1n} \\ \vdots & d_{ij} & \vdots \\ d_{n1} & \cdots & d_{nn} \end{bmatrix} \tag{1}$$

3.2 The Selection of Distribution Points

3.2.1 Hypothetical Conditions

Firstly, there is a distribution node $V_1(i = 1, 2, \cdots, n)$ in the logistics e-business distribution network, one of which will be selected as the distribution center, as the stock and transfer center, and there are N customer demand nodes $x_i(i = 1, 2, \cdots, n)$ in the network.

Secondly, the requirement of the distribution center and the distribution of the customer node are the historical data, and the distribution path is selected as the actual demand, and the demand is generated by the customer. This paper assumes that the demand of the customer node remains unchanged in the distribution.

Thirdly, the delivery cost c_i of the customer node i is determined by the demand q_i, the distribution distance l_{ij} and the unit delivery cost α, $c_i = q_i \times l_{ij} \times \alpha$.

Fourth, the distribution of distribution points cannot exceed the upper limit of distribution. When multiple customer node planning paths find that multiple customer points need to be supplied by the same distribution point in excess of the upper limit of supply, it is carried out according to the principle of first arrival, and replanning the route for the unable to meet the node.

3.2.2 The Selection of Distribution Point Based on Cost

According to the first part, for the description of network, we use Floyd algorithm to solve every demand point to the nearest distribution point path by inverse, and we can use the following matrix representation.

$$L = \begin{bmatrix} l_{1(n+1)} & \cdots & l_{1(n+m)} \\ \vdots & l_{i(n+j)} & \vdots \\ l_{m(n+1)} & \cdots & l_{m(n+m)} \end{bmatrix} \tag{2}$$

The node 1 to m is the distribution center selected from n, $n+1$ to $n+m$ node is the customer demand point, $l_{i(n+j)}$ represents the shortest distance between the customer node and the distribution center $i(i \leq n)$. It is assumed that the customer demand node j history demand is h_{qj}, so the total cost of each distribution center is TC.

$$TC = \sum_{j=1}^{n} c_j = \alpha \sum_{j=1}^{n} l_j \times h_{qj} \tag{3}$$

Among them, $l_j = \min(l_{1j}, \cdots, l_{ij}, \cdots, l_{mj})$ is the nearest distribution center from customer demand node.

Therefore, we can use the following objective function to solve the optimal mobile distribution center.

$$\min TC = \min \alpha \times \sum_{j=1}^{n} l_j \times h_{qj} \tag{4}$$

$$s.t. \begin{cases} l_j = \min\,(l_{1j}, \cdots, l_{ij}, \cdots, l_{mj}) \\ i = 1, 2, \cdots, m,\, j = n+1, \cdots, n+m \end{cases}$$

The cost of all distribution centers is the total cost of the whole logistics network, with the total cost minimized as the objective function. The distribution node, which has the hourly cost, is selected as the distribution center. The general distribution center has the center position. It is close to the aggregate area of the customer demand node, and the traffic is advantageous, which helps to get the goods ready in advance. Quick completion of customer demand distribution. The distribution center is generally based on the historical data, the historical customer gathering area, the historical demand and so on, and the actual demand and customer aggregation tend to change with the change of the market. Therefore, the distribution center should also need regular or irregular change, considering the extra cost, this article will not take the examination for the time being. Think about this.

3.3 Distribution Path Planning

After the distribution center is determined, each distribution area needs to be delineated, and the specific customer demand distribution route is determined according to the storage situation of the goods and the distribution of vehicle distribution. Here we introduce the WebGIS network optimization and address matching technology to improve the efficiency of path finding.

3.3.1 WebGIS Address Matching

There is often a problem in the distribution of regional e-commerce logistics and distribution. The actual distribution route will be different from the original line. In order to avoid the replanning of the temporary route, improve the quality and efficiency of the route planning, according to the customer's single address and the network address, the distribution target is determined jointly. And use WebGIS geographic information database to view real-time geographic information, such as Baidu map and other geographic information services, to ensure accurate goals and real-time route conditions.

3.3.2 Distribution Route Planning

For the distribution center i, the upper limit of distribution is Q, the vehicle m vehicle is distributed and the maximum load of the vehicle k is P_k. The distribution range of the distribution center is expressed by the set, and the actual demand of the customer demand point is expressed by q_i. The demand point requires the longest time of arrival of the goods by t_i, and the average transportation speed of the vehicle is constant v.

Due to the demand for distribution speed of e-commerce customers, we choose the shortest distance and the fastest delivery as the goal.

According to the shortest path of the node to the distribution center, we assign the distribution scope of each distribution center as follows.

$$A_i = \{j|l_{ij} = \min\{l_{1j}, l_{2j}, \cdots, l_{ij}, l_{mj}\}\} \tag{5}$$

According to the principle of nearby distribution, we collect customers' points, get the distribution scope of each distribution center, and form the whole distribution network. According to the priority of customer satisfaction (in real time, order distribution according to the order time), along the way, the order of delivery is determined by the principle of concurrent distribution.

Firstly, calculate the most urgent distribution demand of each distribution center.

$$\Delta_i = \left\{i|\min\left(t_i - {}^{l_{ij}}/_{v}\right)\right\}, \quad \Delta_i \in A_i, j = 1, 2, \cdots, n \tag{6}$$

Secondly, give priority to the distribution of customers who have the smallest difference between waiting time and delivery time in distribution center, use WebGIS address matching technology to query routes, and distribute according to routes.

Thirdly, to judge whether the distribution path passes through the node, whether or not it has obtained the distribution and whether the demand is satisfied, then $n = n - 1$, turn step 1 into the next distribution, and if the final delivery is obtained, it is judged whether the supply can meet the needs of the target customers, that is, the difference between the maximum load P_k of the vehicle k and the demand of the target customers is full. If the demand on foot is met, if it can satisfy h customers, then supply $n = n - h$ and change step 1.

4 An Example of Logistics E-business Distribution

The development of e-commerce logistics has been developed unprecedentedly. In front of us, we built the distribution model. In order to verify the practicability of the model, the following is illustrated by a simple example.

There are 10 customer demand points in the e-business logistics distribution network in a certain area. 2 distribution centers are selected from 7 distribution points. Each distribution point has one vehicle and 6 tons vehicle, the average speed is 60 km/h. The historical demand of each demand point (the annual demand of the previous year is used to determine the distribution center) and the current needs. The distance between points is defined according to the form (Table 1).

Table 1. Nodal material demand list (demand:ton, time:day)

Demand/Node	V_1	V_2	V_3	V_4	V_5	V_6	V_7	V_8	V_9	V_{10}
Total demand of the previous year	46.8	45	43.2	72	61.2	64.8	57.6	46.8	64.8	82.8
Current demand	2.4	3.15	2.7	1.95	2.25	3	2.85	3.45	2.25	1.8
Time requirements	2.5	3.5	2	1	3	3	2	3.5	2	1.5

The shortest path algorithm (Floyd algorithm) is used to get the path matrix. According to the formula (4), the distribution center is calculated, and the result is the distribution center V_{13}, V_{14}.

The address matching technology is used to search the target address and the real-time route. According to the basis formula (6), the distribution range sets of each distribution center are calculated. The two distribution center sets are V_{13} responsible for customer requirement node $V_2, V_3, V_5, V_6, V_7, V_{10}$, and V_{14} is responsible for V_1, V_4, V_8, V_9. According to formula and distribution steps, the distribution route is as follows:

$$V_{13} \text{ route is } V_{13} \rightarrow V_3 \rightarrow V_8 \rightarrow V_1 \rightarrow V_{13} \rightarrow V_9;$$
$$V_{14} \text{ route is } V_{14} \rightarrow V_4 \rightarrow V_7 \rightarrow V_2 \rightarrow V_6 \rightarrow V_{14} \rightarrow V_5 \rightarrow V_{10}.$$

From the result of the distribution path, we can see that the distribution center V is selected, indicating that the two points are in the center of the customer demand point. The two distribution centers are transported by two rounds, on the one hand, the vehicle load is limited, on the other hand, the position of the customer demand point is more scattered. Furthermore, in this paper, the logistics distribution center is selected for the minimum cost of distribution network, and the distribution route is further planned. The distribution scheme obtained is lower than the normal distribution cost, and the customer satisfaction is higher (Table 2).

Table 2. Distance table between nodes (unit: kilometre)

	V_1	V_2	V_3	V_4	V_5	V_6	V_7	V_8	V_9	V_{10}	V_{11}	V_{12}	V_{13}	V_{14}	V_{15}
V_1	0	33	24	36	40	∞	∞	34	∞	∞	∞	∞	34	∞	∞
V_2	33	0	25	10	∞	16	37	∞	∞	18	0	16	16	19	∞
V_3	24	25	0	18	∞	27	∞	22	38	∞	16	∞	∞	∞	∞
V_4	36	10	18	0	16	16	19	∞	∞	27	16	∞	0	∞	∞
V_5	40	∞	∞	16	0	∞	∞	∞	∞	∞	19	∞	∞	0	∞
V_6	∞	16	27	16	∞	0	∞	∞	16	22	∞	∞	∞	∞	0
V_7	∞	37	∞	19	∞	∞	0	∞	∞	38	∞	∞	46	∞	∞
V_8	34	∞	22	∞	∞	∞	∞	0	∞	∞	40	41	∞	∞	∞
V_9	∞	∞	38	∞	∞	46	∞	∞	0	35	46	∞	∞	0	35
V_{10}	∞	∞	∞	40	41	∞	∞	∞	35	0	∞	∞	∞	35	0
V_{11}	∞	16	27	16	∞	19	∞	∞	0	∞	∞	∞	∞	46	∞
V_{12}	∞	37	∞	19	22	∞	∞	∞	∞	0	∞	0	∞	∞	∞
V_{13}	34	∞	22	∞	38	∞	∞	46	∞	∞	0	∞	0	∞	∞
V_{14}	∞	∞	38	∞	∞	40	41	∞	∞	∞	35	∞	∞	0	35
V_{15}	∞	∞	∞	40	41	∞	∞	∞	35	0	∞	∞	∞	35	0

5 Conclusion

The research of logistics distribution network is not a new topic. The research on logistics distribution path planning, distribution point location, vehicle scheduling problem is everywhere, but most of the research is based on unilateral consideration, without considering the distribution center, distribution path and many other considerations, especially in the geographic information system and Web technology fast. The development of logistics distribution network with WebGIS has become a new research breakthrough point. In this paper, the main points of WebGIS research are analyzed, and the optimization characteristics of logistics distribution network in WebGIS Research (such as path optimization, resource allocation, etc.) are explored, and a logistics distribution network model based on WebGIS is constructed, and the distribution center and customer satisfaction are chosen to plan and distribute the distribution path with the goal of cost. The two step is to study the network optimization, get the distribution center and the best delivery route, and solve the logistics distribution problem of e-commerce. However, considering the shallow application of WebGIS technology in this paper, the construction and technical analysis of the deeper logistics information and GIS integration platform need to be further studied. At the same time, the research of the distribution center as a fixed point and more close to the cost optimization of the mobile distribution center needs further development.

References

1. Fan, Z.: Research on logistics management bated on GIS and electronic commerce. J. Jiamusi Vocat. Inst. **12**, 474–475 (2016)
2. Baoan, W.: Analysis and design of grain logistics distribution and optimization system based on web GIS in e-commerce environment. Grain Distrib. Technol. **3**, 32–34 (2012)
3. Conger, Z., Hu, X.: Application of GIS technology in optimization of distribution routing under consideration of carbon emissions. J. Jianghan (Univ. Nat. Sci. Ed.) **3**, 268–273 (2017)
4. Ding, Q.: Study on the site selection of e-commerce distribution center based on GIS. Master's Degree Thesis of Chang'an University (2013)

Prediction of Aviation Material Demand Based on Poisson Distribution and Bayesian Method

Penghui Niu(✉), Wei Hu, and Zhen Wang

Army Aviation Institute of PLA, Beijing, China
niu_penghui@163.com

Abstract. During the initial aviation material recommended period, the fault rate of parts is constant, random probability of aviation material demand satisfied Poisson theory, the Poisson distribution can be carried out to predict air materiel demand; for material which has certain historical fault information, we can introduce Bayesian method, then calculate the demand for spare parts. The validity of the method is proved by examples.

Keywords: Prediction of aviation material demand · Poisson distribution Bayesian method

1 Introduction

Prediction is the speculation of future events or some variables based on the past and present conditions. The prediction of aviation material demand is to predict future demand and development rule of aviation material supply according to the past and present statistics of aviation material system. In particular, the prediction of the aviation material demand refers to the various estimates of the future demand for some accessories and equipment of the aviation material by scientific methods. The prediction of aviation material demand is an important subject in the research of aviation material supply system and an important link to guide the collecting of aviation material correctly. The prediction of aviation material demand is a complex work process, it need to not only master the change of security tasks, the statistics, product quality, market quotation and all kinds of information in the past and at present, but also give full play to the initiative of aviation material staff and estimate the state of future correctly. Aviation material demand prediction work must be planned and steps, do not repeat, no omission, orderly, improve the efficiency and quality of work. Generally, it can be divided into the following steps: the first step is to identify the prediction object; the second step is to collect and analyze the data; the third step is to choose the prediction method; the fourth step is to set up the prediction model; the fifth step is to predict; the sixth step is to analyze the prediction results.

Planned replacement required aviation material items and quantities can be obtained directly according to the maintenance task information. Now, we study unplanned replacement aviation material reserve quantity, practical experience indicates that when the material failure rate remains unchanged, the randomness demand satisfies Poisson distribution. The calculation method based on Poisson formula is

K. Deng et al. (Eds.): ICMIR 2018, AISC 856, pp. 207–213, 2019.
https://doi.org/10.1007/978-3-030-00214-5_27

recommended in the first order of the aviation material. This method is suitable for the turnover parts with larger randomness and smaller demand. Prediction of follow-up spare parts is mainly used in the prediction model based on the maintenance theory. The study introduces Bayesian method, according to the type of life distribution when aviation equipment is in use, using corresponding conjugate prior distribution, combined with the existing historical fault data, calculate the hyper parameters using the prior moment, and then to calculate the aviation material demand [1–9].

2 Prediction of Aviation Material Demand

2.1 Prediction Based on Poisson Distribution

(1) The stochastic probability of the aviation material demand
The failure rate of the aviation material components is constant in the initial period, and the stochastic probability of the aviation material demand satisfies the Poisson distribution theory, that is:

$$P\{R \leq m\} = e^{-D_{rst}} \times \sum\nolimits_{0}^{m} \frac{(D_{rst})^{m}}{ml}. \tag{1}$$

In this formula: P—Aviation material support rate; R—Aviation material dismantling demand; m—Recommended amount of aviation material; D_{rst}—Expected demand for replenishment period. In the actual case, the value of m is an integer. The value of P is determined by the requirements proposed by the customer. The minimum m value which satisfies a given P value is the recommended amount of the calculated aviation material.

(2) The calculation process of the recommendation of aviation material
(a) Calculate annual demand D_{ann}:

$$D_{ann} = \frac{FH \times FS \times QPA}{MTBUR} \tag{2}$$

In this formula: D_{ann}—Annual demand; FH—Annual flying hours; FS—Fleet size; QPA—Quantity of aircraft loading; MTBUR—Average unplanned dismantling interval time.

(b) Compare annual demand D_{ann} with minimum annual demand MAD: Analyzing the storage risk of parts by comparing the annual demand D_{ann} with the minimum annual demand MAD. If $D_{ann} \geq MAD$, recommend this aviation material; otherwise don't recommend this aviation material.

$$MAD = 1/Y. \tag{3}$$

In this formula: Y—Guarantee years, in years.

(c) $D_{ann} \geq MAD$, Calculate the expected demand for replenishment period, consumptive part:

$$D_{rst} = D_{ann} \times \left(\frac{LT + AT}{365}\right). \tag{4}$$

Rotables/repairable spare,

$$D_{rst} = D_{ann} \times \left[\left(\frac{MSPT + TT}{365}\right) \times \left(1 - \frac{SR}{1000}\right) + \left(\frac{SR}{1000}\right) \times \left(\frac{LT + AT}{365}\right)\right]. \tag{5}$$

In this formula: LT—Lead Time; AT—Administration Time; MSPT—Mean Shop Processing Time; TT—Transit Time; MSPT + TT = RTAT, Repair turnaround time; SR—Scrap Rate.

(d) Calculate the recommended number: that satisfies the requirements of the guarantee rate by iteration: put D_{rst} from c. and the preset m into (1), Refer to the Poisson distribution table, and increase the m value of the iteration until $P\{R \leq m\}$ is greater than or equal to the requirements of aviation materiel support rate, at this time m is the minimum recommended amount which satisfies the guarantee rate [10–12].

2.2 Prediction Based on Bayesian Method

(1) Determine the initial estimate λ_0 of the failure rate
According to the historical fault information of this kind of aviation materials, the average value of the failure rate is calculated as the initial estimate λ_0 in the prediction of the aviation material.

(2) Determine probability distribution density function of equipment fault
Suppose that the service life of a kind of aviation materials is exponentially distributed, and the empirical failure rate is λ_0. To satisfy the guarantee rate P, the probability distribution density function of the equipment failure is:

$$P(j \leq s) = \sum_{j=0}^{s} \frac{(n\lambda_0 t)^j}{j!} \exp(-n\lambda_0 t). \tag{6}$$

In this formula: n is the amount of installation for the project; t is the time of work for the project; s is the number of spare parts required for the project.

(3) Determine a priori distribution and a priori prediction distribution of Bayesian method
According to the probability distribution density function of equipment failure, the corresponding prior distribution or conjugate prior distribution is determined, and the priori prediction distribution of this kind of aviation material is determined. According to Bayes method, the conjugate prior distribution of mean value obeyed exponential distribution is Gamma distribution, its shape parameter is $\alpha(\alpha > 0)$, and scale parameter is $\beta(\beta > 0)$:

$$G(\lambda|\alpha, \beta) = \begin{cases} 0, \lambda \leq 0 \\ \frac{\beta^{\alpha}\lambda^{\alpha-1}}{\Gamma(\alpha)} e^{-\beta\lambda}, \lambda > 0 \end{cases} \tag{7}$$

If during the time t, we can sense r times fault, prior density function is changed into $G\left(\lambda|\acute{\alpha},\acute{\beta}\right)$, in which $\acute{\alpha} = \alpha + r$, $\acute{\beta} = \beta + t$. The prior prediction distribution of the k times fault in the prediction period L is as follows:

$$p(k|\alpha,\beta) = \int_0^{\infty} \frac{(-\lambda nL)^k \exp(\lambda nL)}{k!} \frac{\beta^\alpha \lambda^{\alpha-1} \exp(-\beta\lambda)}{\Gamma(\alpha)} \cdot d$$
$$= \frac{\Gamma(k+\alpha)}{k! \cdot \Gamma(\alpha)} \cdot \frac{\beta}{nL+\beta} \cdot \left(\frac{nL}{nL+\beta}\right)^k. \quad (8)$$

Prediction cycle:

$$L = \frac{\text{a period of time to be predicted}}{\text{the time period used to estimate the failure rate}}.$$

(4) Calculate super parameters

$$\alpha/\beta = \omega\lambda_0. \quad (9)$$

$$\int_0^{\delta\lambda_0} \frac{\beta^\alpha \lambda^{\alpha-1} \exp(-\lambda\beta)}{\Gamma(\alpha)} \cdot d\lambda = 0.95. \quad (10)$$

(a) Calculate parameter ω and δ:
Because of N historical fault data had been recorded in use, so compare N observed failure rate λ_1 with the initial failure rate estimator λ_0, calculating the ratio λ_1/λ_0, and let ω equal to the average value of the ratio. Calculate Security probability $P \times N$ and round number, order ratio of value from small to large, let δ equal to the ratio of $P \times N$.
(b) Calculate super parameter α, β:
Put ω, δ into (9) and (10), calculate super parameter α and β, the following two formulas are used to update the super parameters α and β:

$$\acute{\alpha} = \alpha + r, \quad (11)$$

$$\acute{\beta} = \beta + tn. \quad (12)$$

Among them, t is a unit of time, r is the number of failures in t, and n is the amount of the installation of the aviation material.

(5) Calculate spare parts demand in prediction cycle L

The formula for calculating the spare parts satisfying the guarantee probability P is

$$\left(\frac{\beta}{nL+\beta}\right)^{\alpha} + \sum_{j=1}^{S}\left[\left(\prod_{k=1}^{j}\left(\frac{\alpha+j}{k}-1\right)\right)\cdot\left(\frac{\beta}{nL+\beta}\right)^{\alpha}\cdot\left(\frac{nL}{nL+\beta}\right)^{j}\right] \ge P \quad (13)$$

Put α, β and n, L, P into (13), The requirement value of spare parts S which satisfies the requirements of the guarantee probability is calculated [13–16].

3 Application Example

3.1 Example 1

There are some aviation material spare parts in an installed state. The relation curve between demand and time is shown in Fig. 1. There are 10 spare parts, and the time of installation is as follows: 0, 200, 1250, 1300, 2400, 2500, 3650, 4850, 5000, 6200, unit is h, failure rate is 6×10^{-4}/h. The demand for the time point 7000 h is calculated. Through the operation of the program, the demand is 26, which is consistent with the actual situation.

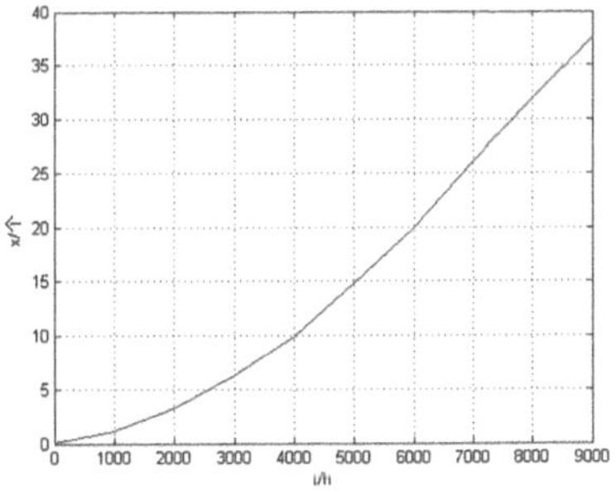

Fig. 1. The relation curve between demand and time.

3.2 Example 2

As an example, the above algorithm is verified by an example of the demand rate of a certain aircraft spare part on a flying regiment aircraft. Now we have known the actual demand rate of spare parts for 3 squadrons (Tables 1, 2 and 3). According to the above algorithm, we predicate demand in August, September, October and November for each squadron, the prediction results are shown in Tables 4, 5 and 6. We can see from the tables, the average prediction error is less or equal to 0.002, the relative error is small.

Table 1. The data of squadron 1

Month	3	4	5	6	7	8	9	10	11
Actual demand rate (quantity/days)	0.2	0.16	0.18	0.21	0.14	0.17	0.13	0.19	0.22

Table 2. The data of squadron 2

Month	3	4	5	6	7	8	9	10	11
Actual demand rate (quantity/days)	0.17	0.18	0.21	0.16	0.22	0.15	0.13	0.19	0.14

Table 3. The data of squadron 3

Month	3	4	5	6	7	8	9	10	11
Actual demand rate (quantity/days)	0.21	0.18	0.20	0.15	0.12	0.15	0.17	0.16	0.19

Table 4. The prediction results of squadron 1

Month	8	9	10	11
Actual value	0.17	0.13	0.19	0.22
Prediction results	0.169	0.133	0.192	0.218
Forecast error	0.001	0.003	0.002	0.002
Relative error	0.6%	2.3%	1.1%	0.9%

Table 5. The prediction results of squadron 2

Month	8	9	10	11
Actual value	0.15	0.13	0.19	0.14
Prediction results	0.147	0.132	0.188	0.141
Forecast error	0.003	0.002	0.002	0.001
Relative error	2%	1.5%	1.1%	0.7%

Table 6. The prediction results of squadron 3

Month	8	9	10	11
Actual value	0.15	0.17	0.16	0.19
Prediction results	0.147	0.168	0.161	0.189
Forecast error	0.003	0.002	0.001	0.001
Relative error	2%	1.2%	0.6%	0.5%

4 Conclusion

With regard to inventory management for initial spares recommended period, during this period, the fault rate of parts is constant, random probability of aviation material demand satisfies Poisson distribution theory, the Poisson distribution can be carried out to predict aviation materiel demand. With regard to materials which have certain historical fault information, Bayesian method can be used to calculate the demand for spare parts. The validity of the method is proved by examples.

References

1. Zhang, Y., Liang, J.: Overview and enlightenment of demand forecasting method research about aircraft spare parts. J. Civ. Aviat. Univ. China **32**(1), 92–96 (2014)
2. Sha, H., Cai, J., Lin, H.: Research on safety stock of air-material under the laws of the different needs. J. Logist. Sci. Tech. **11**, 47–49 (2011)
3. Kang, Z., Xu, K., Yang, C.: Application of ANFIS in aviation material spare parts requirement prognostication. J. Sichuan Ordnance **36**(1), 84–86 (2015)
4. Aronis, K.-P., Dekker, R.: Inventory control of spare parts using a Bayesian approach: a case study. Eur. J. Oper. Res. **154**, 730–739 (2003)
5. Yang, Y., Wang, D., Chang, S.: Research on forecasting the demand of aviation spares by simulating based on SPN and arena. Value Eng. **8**, 129–130 (2012)
6. Ruan, M., Li, Q., Huang, A.: Optimization of cooperative ordering project for consumable spare parts under (R, Q) inventory policy. Trans. Beijing Inst. Technol. **7**(33), 680–684 (2013)
7. Zhu, S., Li, S., Li, W., et al.: A spare parts providing model with the maximal availability. J. Air Force Eng. Univ. Nat. Sci. Ed. **2**(6), 22–24 (2005)
8. Liu, Z., Liu, Y., Wang, D.: Aeronautic spare parts supervise model based on operating availability and demand analysis. Aeronaut. Comput. Tech. **5**(37), 38–41 (2007)
9. Zhang, D., Ming, X., Zhao, C., et al.: Spare parts demand forecasting method based on BP neural networks and equipment characters. Mach. Des. Res. **1**(26), 72–76 (2010)
10. Luan, Y., Han, Q., Zhang, Y.: Study on storage control model of repairable aerial material based on Poisson distribution. J. Value Eng. 45–46 (2016)
11. Zhao, J., Xu, T., Ge, X., et al.: Consumption forecasting of missile spare parts based on wavelet transform and revised GM-ARMA model. J. Beijing Univ. Aeronaut. Astronaut. **4** (39), 553–558 (2013)
12. Gao, K., Xing, G., Sun, D., et al.: Application of grey correlation SVM in reserve consuming prediction of spare parts. Electron. Opt. Control **3**(19), 100–105 (2012)
13. Zhao, H., Wang, J., Zhu, G.: A predictive model of the following spare parts based on Bayes. J. Microelectron. Comput. **21**(9), 104–106 (2005)
14. Marcot, B.G.: Metrics for evaluating performance and uncertainty of Bayesian network models. Ecol. Model. **230**, 50–62 (2012)
15. Douail, N., Csaba, H.: Diagnosis support system based on clinical guidelines: comparison between case-based fuzzy cognitive maps and Bayesian networks. Comput. Methods Programs Biomed. **113**, 133–143 (2014)
16. Wolbrecht, E., Dambrosio, B., Paasch, R., et al.: Monitoring and diagnosis of a multistage manufacturing process using Bayesian networks. Artif. Intell. Eng. Des. Anal. Manuf. **1**(14), 53–67 (2000)

Implicit Neural Dynamics for Real-Time Matrix Inversion

Xuanjiao Lv[1(✉)], Yue Huang[1], Huiru Cao[1], Zhi Yang[2], and Zhiguo Tan[1]

[1] Nanfang College of Sun Yat-Sen University, Guangzhou 510970, China
lxj_258369@163.com
[2] School of Data and Computer Science, Sun Yat-Sen University, Guangzhou 510006, China

Abstract. This paper develops a new implicit neural dynamic to real-time find the inverse of time-invariant (or say static, constant) matrix. Such a neural model is proven to have higher convergence rate than other existing neural models, specifically, the gradient-based neural dynamic, Zhang neural dynamic. Additionally, for comparison and verification, a simulative example is conducted. The corresponding results further validate the effectiveness and superiority of the presented model for real-time matrix inversion.

Keywords: Implicit neural dynamic · Matrix inversion
Zhang neural dynamic · Convergence rate

1 Introduction

Let us consider the following defining equation of matrix inverse:

$$AX(t) = I, \ \ or \ X(t)A = I, \tag{1}$$

where $A \in R^{n \times n}$ is a time-invariant (or say static, constant) matrix, $I \in R^{n \times n}$ is the identity matrix, and $X(t) \in R^{n \times n}$ corresponds to the theoretical inverse of matrix A, which is the unknown matrix to be found. A lot of applications in scientific and engineering fields involve matrix inversion, such as robot kinematics [1], statistics [2, 3], visual surveillance [4], and so on.

Due to its significant role, great efforts have been contributed to the solution of matrix inverse problems. Recent research works [5–10] show that neural dynamic approach is a powerful tool to solve matrix-related problems owing to its advantages (e.g., parallel processing, distributed storage) over conventional iterative numerical algorithms. Therefore, many neural dynamic models are proposed and exploited. For example, a recurrent neural network, which can be recognized as a classic gradient-based neural dynamic (GND), was proposed for constant matrix inversion in [5]. Recently, Zhang et al. developed a novel neural dynamic model (termed ZND) [6, 7]. Such a model is intrinsically designed for time-varying matrix-related problems solving and thus quite different from GND in methodology. To pursue superior convergence rate, a hybrid recurrent implicit dynamic was developed by Chen [9] (termed Chen's

K. Deng et al. (Eds.): ICMIR 2018, AISC 856, pp. 214–219, 2019.
https://doi.org/10.1007/978-3-030-00214-5_28

model). It is known that convergence performance is always of cardinal significance for neural dynamics. Motivated by this point, an improved implicit neural dynamic is proposed for static matrix inversion in this paper.

2 Neural Dynamic Models Description

2.1 GND Model

GND is a classic hopfield-type neural network and can be employed to solve static matrix-involved problems. As for the matrix inverse problem, a scale function is first defined as $\rho(t) = ||AX(t) - I||_F^2/2$ for GND model, where $||\cdot||_F$ denotes the Frobenius norm. In order to drive $\rho(t)$ approach 0 (we thus obtain the corresponding $X(t)$ which is the theoretical inverse A^{-1}), the gradient descent direction of $\rho(t)$ is taken into consider, which yields the following GND model for matrix inversion [9, 10]:

$$\dot{X}(t) = -\kappa A^T AX(t) + \kappa A^T, \tag{2}$$

where $\dot{X}(t)$ is the time derivative of $X(t)$, $\kappa > 0$ is a tunable parameter.

2.2 ZND Model

Based on in-depth study of neural networks, Zhang et al. designed a novel neural dynamic model which aims at solving time-varying problems. According to the design idea [7], a matrix-valued error function is first defined as $\psi(t) = AX(t) - I \in R^{n\times n}$. Then, to make $\psi(t)$ tend to zero matrix, the following design formula is adopted [7, 8]:

$$\dot{\psi}(t) = -\kappa\psi(t).$$

By expanding the above equation, we have the ZND model for static matrix inversion:

$$A\dot{X}(t) = -\kappa AX(t) + \kappa I. \tag{3}$$

2.3 Proposed Implicit Neural Dynamic

In this paper, to pursue superior convergence performance, we propose following neural dynamic for real-time static matrix inversion:

$$A\dot{X}(t) = -\kappa(AA^T + A^T A + I)(AX(t) - I). \tag{4}$$

It is worth pointing out here that GND model (2) is described in explicit form of dynamic systems (i.e., $\dot{X}(t) = \ldots$), while ZND model (3), and the proposed model (4) are described in implicit form of dynamics systems (i.e., $A\dot{X}(t) = \ldots$). Implicit dynamics are usually associated with actual physical systems [11], and can be converted to explicit dynamics if necessary. In this sense, the proposed model (4) and ZND model (3) are much better than GND model (2).

3 Theoretical Analysis

In this section, we present and analyze the convergence performance of the neural dynamic models (2)–(4).

Lemma 1. Suppose static matrix $A \in R^{n\times n}$ in Eq. (1) is nonsingular, neural state matrix $X(t)$ of GND model (2) is globally exponentially convergent to exact inverse $X^* = A^{-1}$, no matter what initial value $X(0)$ is. In addition, the convergence rate is $\kappa\alpha$, where α is the minimum eigenvalue of A^TA.

Proof. See [10] for details and omit here due to space limitation.

Lemma 2. Suppose static matrix $A \in R^{n\times n}$ in Eq. (1) is nonsingular, neural state matrix $X(t)$ of ZND model (3) is globally exponentially convergent to exact inverse $X^* = A^{-1}$, no matter what initial value $X(0)$ is. In addition, the convergence rate is κ.

Proof. See [7] for details and omit here due to space limitation.

Theorem 1. Suppose static matrix $A \in R^{n\times n}$ in Eq. (1) is nonsingular, neural state matrix $X(t)$ of the proposed model (4) is globally exponentially convergent to exact inverse $X^* = A^{-1}$, no matter what initial value $X(0)$ is. In addition, the convergence rate is at least $\kappa(1+2\alpha)$, where α is the minimum eigenvalue of A^TA.

Proof. We first define solution error as $E(t) = X(t) - X^* = X(t) - A^{-1}$. $\dot{E}(t)$ is readily obtained as $\dot{E}(t) = \dot{X}(t)$. Consequently, the neural model (4) is equivalently formulated as

$$A\dot{E}(t) = -\kappa(AA^T + A^TA + I)AE(t). \tag{5}$$

Note that, because matrix A is nonsingular, the matrices A^TA and AA^T are all invertible and have same eigenvalues. Now, we use Lyapunov theory to investigate the convergence performance of model (5). Let us choose $\varsigma(t) = ||AE(t)||_F^2/2 \geq 0$ as Lyapunov function. Evidently, $\varsigma(t)$ is positive definite. The time derivative of $\varsigma(t)$ along the trajectory of neural dynamic system (4) can be computed as

$$\begin{aligned}\dot{\varsigma}(t) &= trace(E^T(t)A^TA\dot{E}(t))\\ &= trace(E^T(t)A^T(-\kappa(AA^T + A^TA + I))AE(t))\\ &= -\kappa trace(E^T(t)A^T(AA^T + A^TA + I)AE(t))\\ &\leq -\kappa(1+2a)||AE(t)||_F^2 \leq 0.\end{aligned} \tag{6}$$

In addition, $\dot{\varsigma}(t)$ is equal to 0 if and only if $E(t) = 0$. Therefore, $\dot{\varsigma}(t)$ is negative definite. By the Lyapunov theory [7, 9, 10], we conclude that the unique equilibrium point $E(t) = 0$ (i.e., $X(t) = X^*$) is of global and asymptotical stability.

Moreover, in view of $\varsigma(t) = ||AE(t)||_F^2/2$, inequality (6) can be further written as:

$$\dot{\varsigma}(t) \leq -2\kappa(1+2a)\varsigma(t).$$

Solving above differential inequality, we have $\varsigma(t) \leq \varsigma(0)\exp(-2\kappa(1+2a)t)$ with $\varsigma(0) = ||AE(0)||_F^2/2$, which leads to: $||AE(t)||_F \leq ||AE(0)||_F \exp(-\kappa(1+2a)t)$.

Define $M = ||AE(0)||_F/\sqrt{\alpha}$, in view of $\sqrt{\alpha}||E(t)||_F \leq ||AE(t)||_F$, we have:

$$||X(t) - X^*||_F = ||E(t)||_F \leq \frac{1}{\sqrt{\alpha}}||AE(t)||_F \leq M\exp(-\kappa(1+2a)t),$$

Therefore, we draw the conclusion that the neural model (4) is globally exponentially convergent to its equilibrium point $E(t) = 0$ with exponential convergence rate being at least $\kappa(1+2a)$. The proof is thus completed.

It is worth mentioning here that the convergence rate of the proposed neural mode (4) is higher than that of GND model (2), ZND model (3), and Chen's model [9] (with convergence rate being $\kappa(1+a)$). In this sense, our mode (4) is much superior to them.

4 Simulation Study

In this section, a computer simulation example is performed to further verify the theoretical results presented in Sect. 3.

Let us consider the following nonsingular matrix A and its exact inverse A^{-1}:

$$A = \begin{Bmatrix} 2 & 1 & -1 \\ 2 & 1 & 0 \\ 1 & -1 & 1 \end{Bmatrix}, \quad A^{-1} = X^* = \begin{Bmatrix} 1/3 & 0 & 1/3 \\ -2/3 & 1 & -2/3 \\ 1 & 1 & 0 \end{Bmatrix}.$$

For the above given matrix A, the minimum eigenvalue of A^TA is $\alpha = 0.27$. For verification, the exact inverse A^{-1} of the matrix A is also given above. Design parameter $\kappa = 100$. All the elements of initial neural state $X(0)$ are zero. Now, GND model (2), ZND model (3), Chen's model [9], and the proposed model (4) are all utilized to real-time solve the inverse of matrix A, with the corresponding simulation results shown in Figs. 1, 2 and 3.

From Figs. 1 and 2, we can see that all the entries of neural state $X(t)$ are convergent to their corresponding entries of A^{-1} for ZND model (3), Chen's model and the proposed model (4), which illustrates the effectiveness of all the neural dynamic models for real-time matrix inversion. More importantly, as observed from Fig. 3, for the GND model (2), ZND model (3), and Chen's model, the times of the residual error $||X(t) - X^*||_F$ to reach a small value $M\exp(-5)$ are about 0.185 s (i.e., $5/\kappa\alpha$), 0.05 s (i.e., $5/\kappa$), 0.04 s (i.e., $5/\kappa(1+\alpha)$), respectively. While for the proposed neural model (4),

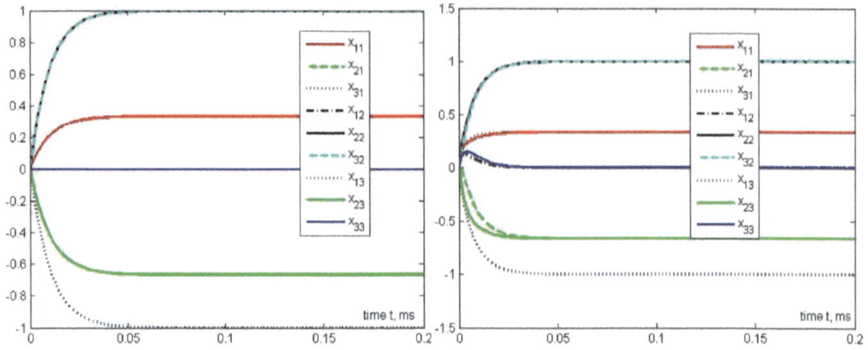

(a) ZND model (3) **(b)** Chen's model

Fig. 1. Neural state $X(t)$ of ZND model (3) and Chen's model for matrix inversion

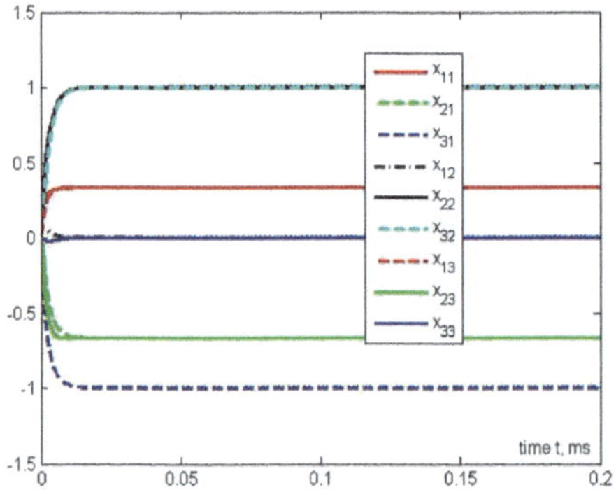

Fig. 2. Neural state $X(t)$ of our model (4)

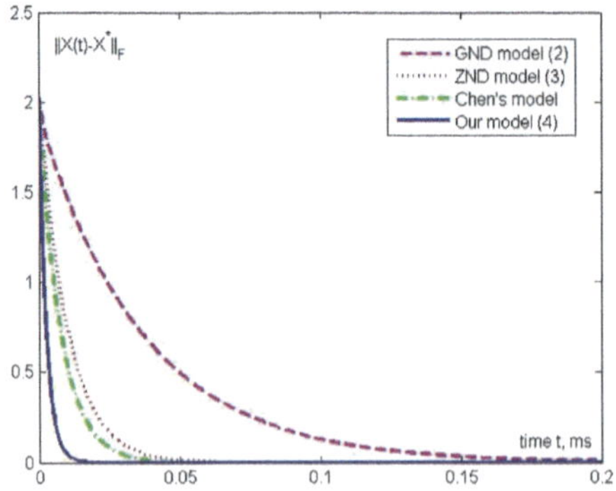

Fig. 3. Residual error $||X(t) - X^*||_F$ produced by models (2)–(4), and Chen's model

the time of the residual errors $||X(t) - X^*||_F$ to reach the same small value $M\exp(-5)$ is only about 0.03 s (i.e.,$5/\kappa(1+2\alpha)$)! These simulation results verify the theoretical analyses presented in Sect. 3, and also demonstrate the superiority of our model (4) in comparison to GND model (2), ZND model (3) and Chen's model for matrix inversion.

5 Conclusion

In this paper, an improved implicit neural dynamic model is proposed and investigated for constant matrix inversion. In addition, three theorems about convergence performance of GND, ZND and the proposed models are also presented. Both theoretical analysis and simulation results have demonstrated the effectiveness and superiority of the presented model for real-time matrix inversion.

References

1. Sturges, R.H.: Anolog matrix inversion (robot kinematics). IEEE J. Robot. Autom. **4**(2), 157–162 (1988)
2. Zhang, Y., Leithead, W.E.: Exploiting hessian matrix and trust region algorithm in hyper parameters estimation of Gaussian process. Appl. Math. Comput. **171**, 1264–1281 (2005)
3. Zhang, Y., Leithead, W.E., Leith, D.J.: Time-series Gaussian process regression based on toeplitz computation of $O(N^2)$ operations and $O(N)$-level storage. In: Proceedings of 44th IEEE Conference Decision Control, pp. 3711–3716 (2005)
4. Chen, K., Loy, C.C., Gong, S., Xiang, T.: Feature mining for localised crowd counting. In: Proceedings of British Machine Vision Conference, pp. 21.1–21.11 (2012)
5. Wang, J.: A recurrent neural network for real-time matrix inversion. Appl. Math. Comput. **55**, 89–100 (1993)
6. Zhang, Y., Chen, K., Tan, H.: Performance analysis of gradient neural network exploited for online time-varying matrix inversion. IEEE Trans. Autom. Control **54**, 1940–1945 (2009)
7. Zhang, Y., Ge, S.S.: Design and analysis of a general recurrent neural network model for time-varying matrix inversion. IEEE Trans. Neural Netw. **16**, 1477–1490 (2005)
8. Jin, L., Li, S., Liao, B., Zhang, Z.: Zeroing neural networks: a survey. Neurocomputing **267**, 597–604 (2017)
9. Chen, K.: Recurrent implicit dynamics for online matrix inversion. Appl. Math. Comput. **219**(20), 10218–10224 (2013)
10. Zhang, Y., Shi, Y., Chen, K., Wang, C.: Global exponential convergence and stability of gradient-based neural network for online matrix inversion. Appl. Math. Comput. **215**, 1301–1306 (2009)
11. Yi, C., Zhang, Y.: Analogue recurrent neural network for linear algebraic equation solving. Electron. Lett. **44**(18), 1078–1079 (2008)

Logistics Service Innovation Based on Demand

Gang Liu(✉)

School of Economy and Management, Tianjin Agricultural University, Jinjing Road 22 in Xiqing District, Tianjin 300384, China
liugang_tianjin@163.com

Abstract. The changes in environment have increased the pressure to logistics service innovation. In order to analyze how to improve the quality of logistics service through service innovation, the paper developes a conceptual model consists of combination of technology innovation, concept innovation and solution innovation, and takes Cainiao as a case to verify the model. We propose that logistics firms should set up new service idea oriented customer demand; pay close attention to advanced technology in different fields; and provide supply chain solutions for customers which integrating multiple logistics functions.

Keywords: Logistics · Service innovation · Demand · Cainiao

1 Introduction

Logistics is the core function in supply chain. For customers, the expectations for logistics service are getting higher. The changes in environment such as globalization, individualized demand have increased competition for cost and quality and thus the pressure to logistics service innovation. The need for logistics service innovation has been increased over past years. Excellent logistics service has become an important source of competitive advantage for customers. More and more logistics firms seek to provide sophisticated services through innovation in order to increase market share. The paper mainly analyzes the innovation management of logistics firms in order to explore how to provide better service for customers through service innovation.

2 Literature Review

General innovation management research is concentrate on the development of technical product. Logistics service innovation refers to a new, helpful idea, procedure, or practice in logistics operations that is different from a company's current practice (Grawe 2009). Many studies have analyzed logistics service innovation from perspective of resource-based and customer-value (Grawe et al. 2014). Wallenburg (2009) found a positive correlation between logistics service innovation and customer loyalty. Daugherty et al. (2011) discovered that organizational structure can affect innovation capability of logistics enterprises. Some key factors which lead to logistics service innovation were identified, such as knowledge acquisition, investment in infrastructure,

K. Deng et al. (Eds.): ICMIR 2018, AISC 856, pp. 220–224, 2019.
https://doi.org/10.1007/978-3-030-00214-5_29

relationship networks, supply chain learning management, technology, combination of resources across supply chain (Chapman et al. 2003; Flint et al. 2008; Wagner 2008). For logistics service innovation, new ideas generate from specific demand of customers or the needs of supply chain processes (Flint et al. 2005, 2008). The purpose of this paper is to develop a theoretical model of logistics service innovation based on demand.

3 Logistics Service Innovation Based on Demand

The development of new technology and globalization competition has inspired firms to seek new business models to overcome the challenge of fierce competition. Firms are paying more attention to their core competencies. Many companies choose to outsource logistics to professional logistics firms. As market competition intensifying, the expectation of firms for logistics service is getting higher. Simple logistics services have been difficult to meet the needs of customers. Firms demand for logistics service are becoming fragmented and individualized, index such as quick response, logistics flexibility, time compression are getting more concerned by customers. New customer needs will bring pressure to be innovative. Logistics companies should provide innovative services in order to respond to the variety of customer needs. Bilderbeek et al. (1998) developed a four dimensional model of service innovation, including new service concept, new client interface, new service delivery system/organisation, technological options. The paper builds a logistics service innovation model base on demand according to this model (Fig. 1).

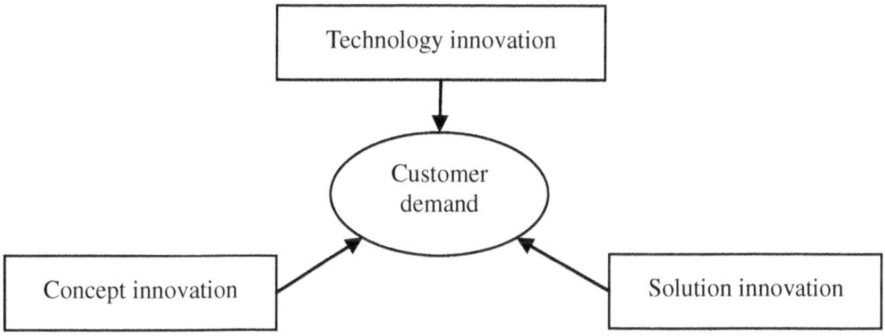

Fig. 1. Logistics service innovation model based on demand

3.1 Technology Innovation

Unlike other types of services, technology has been playing an important role in logistics service. Logistics services are more dependent on physical equipment and technology. There is a great link between technology and logistics service innovation. Some logistics innovations are directly technology-driven innovation, certain equipment has play an important role in the progress of logistics innovation. Novel technologies such as GPS or RFID have been used in creating new logistics service or promoting process innovations. Meanwhile information and automation technology are widely applied in logistics service, become the basis for service innovation. The examples of information and automation technology use in Table 1 illustrate that the direction of future logistics technology.

Table 1. Examples of future logistics technology

Logistics technology	Description
Warehouse technology	Warehouse technology is mainly used to mprove the efficiency of storage, which including robot and automatic sorting, wearable equipment, unmanned forklift, cargo recognition et al.
Transportation technology	The main direction of tansportation technology is pilotless, such as unmanned truck, pilotless plane
Terminal technology	Terminal technology innovation is mainly for improving convenience. Intelligent express cabinet is a typical terminal technology
Data technology	Data technology has become an important basis for logistics innovation, which mainly including internet of things, big data technology, artificial intelligence

3.2 Concept Innovation

Concept innovation means design the new idea of how to implement a solution to the problem. New concept or idea is deriving from dynamic changes in market demand. Concept innovation may be a completely novel idea about logistics service; new idea also may be from the existing service patterns in other fields. For example, HACCP is applied in the control for logistics process. Some examples of "concept innovation" in logistics service are showed in the following in order to further interpret it.

1. Consumers are often unable to receive their express delivery on time due to time reasons. Automatic express terminal is a new concept, a new way of thinking about express receiving that meets user need.
2. As many clients especially small enterprises are facing a problem of lacking of capital. The concept of logistics finance not only to support client financing, but also reducing the risk of financial institutions.
3. A single pattern of transportation is difficult to meet huge demand for logistics service. Intermodality is a new transportation idea that can improve the efficiency of logistics and reduce cost.

3.3 Solution Innovation

Solution innovation aims at meeting the personalized needs of the users. Logistics enterprises should provide professional service according to the characteristic of different goods or industry through solution innovation. A new solution may package new technology and new concept to create complete service product to end customers. Good solutions can promote customers outsourcing logistics function so that they can concentrate on their core business. Solution innovation can include the following types:

1. A new solution to certain logistics functions such as transportation or storage, for example improving the efficiency of transportation through GPS network design or planning method.
2. A new solution is comprise of multiple functions in order to promote the efficiency of supply chain, such as certain logistics service package integrating warehousing and distribution.
3. A new solution to certain industry which is based on the analysis of the characteristics of this industry, such as electrical industry solution or fast moving consumer goods solution.

4 Case Study

In order to illustrate the theoretical model of logistics innovation presented in the third part, we employ a case study of Cainiao. This firm, located in Zhejiang province in china, which focus on providing internet platform services for logistics network. Cainiao is committed to building a global logistics network with partners, in order to improve the efficiency of logistics and inventory turnover. The paper discusses the ways that Cainiao has taken to promote logisitics service innovation.

Firsty, Cainiao has set up E.T. logistics lab in order to introduce advanced technology into logistics industry. The lab mainly develop new intelligent machinery technologies instead of manual operation. The following technological achievements have been achieved in E.T.lab. (1) Terminal distribution robot. Terminal distribution robot is characteristic of intelligent planning, dynamic recognition, automatic evasion, environmental analysis et al., which can solve the problem of "last kilometre" in distribution. (2) Augmented reality (AR) technology. AR technology is used to realize intelligent sorting and navigation, which can visualize the operations in the warehouse. (3) Unmanned aerial vehicle (UAV) security system. UAV security system is mainly used to implement the task of security patrol in Logistics Park.

Secondly, Cainiao bring the idea of internet coordination which is characteristic of open and intelligent into logistics industry. Internet coordination instead of self-built logistics is better adapted to future logistics demand. The business logic established by Cainiao is to build platforms, so that different service providers, businesses and consumers can achieve efficient connection in supply chain, so as to enhance logistics efficiency and service quality and reduce logistics costs.

Thirdly, Cainiao provide professional logistics solutions for customers according to the characteristics of different industries, for example small home appliances, food and health products, 3C digital products, beauty industry, clothing industry et al. Indicators such as product shape, seasonality, return frequency, timeliness will be considered in the process of solutions design. Then according to customer demand, specialized and qualified logistics service providers can be selected through data analysis results. Professional logistics solution scan provide advantages of marketing competition for customers and increase their market share.

5 Conclusions

Innovation in the logistics context, such as concept, technology, solution, has not only benefited the logistics firms and advanced logistics industry, and also Logistics service innovation has been shown to add value to customers. High quality logistics services can make customers more focused on the development of core competencies, in some cases, logistics directly becomes the core competitiveness of customers. Firstly, logistics firms should set up new service idea oriented customer demand. Secondly, logistics firms should be sensitive to the development of new technologies in different fields, and applying new technology to improve logistics service or develop new service. Finally, logistics firms should provide supply chain solutions for customers which integrating multiple logistics functions.

Acknowledgements. This work was partially supported by the key base of the humanities and social sciences—the open fund project of the rural modernization research center in Tianjin, also supported by the innovation and entrepreneurship training program for college students in Tianjin (Project No: 201510061029), and supported by the investigation and research project of Tianjin Rural Work Committee (TJNWY2017001-02), china.

References

Grawe, S.J.: Logistics innovation: a literature-based conceptual framework. Int. J. Logist. Manag. **20**(3), 360–377 (2009)

Grawe, S.J., Autry, C.W., Daugherty, P.J.: Organizational implants and logistics service innovation: a relational social capital perspective. Transp. J. **53**(2), 180–210 (2014)

Wallenburg, C.: Innovation in logistics outsourcing relationships: proactive improvement by logistics service providers as a driver of customer loyalty. J. Supply Chain Manag. **45**(2), 75–93 (2009)

Daugherty, P.J., Chen, H., Ferrin, B.G.: Organizational structure and logistics service innovation. Int. J. Logist. Manag. **22**(1), 26–51 (2011)

Chapman, R.L., Soosay, C., Kandampully, J.: Innovation in logistic services and the new business model. Int. J. Phys. Distrib. Logist. Manag. **33**(7), 630–650 (2003)

Flint, D.J., Larsson, E., Gammelgaard, B.: Exploring processes for customer value insights, supply chain learning, and innovation: an international study. J. Bus. Logist. **29**(1), 257–281 (2008)

Wagner, S.M.: Innovation management in the German transportation industry. J. Bus. Logist. **29**(2), 215–232 (2008)

Flint, D.J., Larsson, E., Gammelgaard, B., Mentzer, J.T.: Logistics innovation: a customer value-oriented social process. J. Bus. Logist. **26**(1), 113–147 (2005)

Application of Online Payment Mode in University Charging Management

Hao Zhu[1(✉)], Yaohua Xie[2], Mengshu Hou[1], Kai Yan[1], and Tingwei Li[1]

[1] Information Center, University of Electronic Science and Technology of China, Chengdu 611731, China
zhuhao@uestc.edu.cn
[2] Yale University, New Haven, USA

Abstract. There are many payment items in a university that require students to pay. In the traditional payment mode, students have to queue up at the financial department to pay. This mode is very wasteful of students' time and energy. The workload of the staff of the financial department is very large, the work efficiency is low, and mistakes often occur. In order to solve this problem, this paper proposes an online payment mode. The online payment mode can adopt flexible solutions according to the specific conditions of the charging department. The online payment mode reconstructs and simplifies the payment process in the university. It comprehensively improves the efficiency of the university charging management.

Keywords: Paying by hand · Online payment system · Process reengineering System integration

1 Introduction

When the students are in university, they will need to pay for tuition, accommodation, examination registration fees, and other expenses. In the traditional payment mode, the students need to line up and pay the fees in the toll lobby of the financial department. They can pay the fees by credit card or cash. Due to the large number of students in the university, they often need to queue for a long time. Since the number of the financial staff is limited, the workload of each financial staff is very large. Due to the need for manual banknote counting and bill filling, the work efficiency is very low. With the development of the third-party payments, it has become possible to use the online payments in university charging management. In this paper, an online payment system [1] is proposed. In the context of the country's vigorous promotion of the "Internet+" [2], the online payment mode is a good solution, which can fundamentally improve the efficiency of university charging management.

1.1 Disadvantages of the Traditional Payment Mode

The university's traditional charging management mainly depends on manual completion. The payment, accounting, auditing, and statistics are all done manually. It is

K. Deng et al. (Eds.): ICMIR 2018, AISC 856, pp. 225–230, 2019.
https://doi.org/10.1007/978-3-030-00214-5_30

inefficient and the user experience is extremely poor. The reason is that there are many problems under the traditional payment mode.

The students need to pay insurance when they are in university. In the traditional mode, the monitor must collect the insurance fee of every student in his class, and then he can go to the financial department to pay the fee. In the process, if any student does not pay, it will cause the class to fail to complete the financial payment and the insurance policy will not be effective. At the same time, the financial department also needs to set up a special charging window and arrange for staff to charge. The efficiency is low.

2 Online Payment System

The drawbacks of the traditional payment mode are obvious. The reason for the problem is that we have not effectively used information technology to reconstruct and simplify the process. We have not stood in the perspective of the payers and the financial department staff to consider the issue. Then, we found that the online payment system can solve this problem.

The online payment system is not a traditional application system, but a payment platform. There is no need to build a separate payment system for each department of the university. The charging items of each department can be carried out online through the platform. The architecture, main functions, and applications of the online payment system are discussed in the paper. The architecture of the online payment system is shown in Fig. 1.

2.1 Basic Platform

The online payment system is integrated with the university's unified identity authentication platform. Paying users, charging department administrators, and financial department staff can all use the account and password of the unified identity authentication platform [3] to log in. It is very convenient and it also facilitates the authorization management of the system. At the same time, the online payment system is also integrated with the shared database platform [4].

2.2 Unified Payment Portal

The system centralizes all on-campus payment items in the same payment portal. The paying users do not need to visit the website of each charging department to find the payment items. The users can find the payment details related to themselves in the unified payment portal. On the other hand, the users can also check all their historical payment records in the unified payment portal.

2.3 Unified Payment Gateway

The online payment system generates online payment orders through the unified payment gateway [5]. The unified payment gateway can connect with the third-party

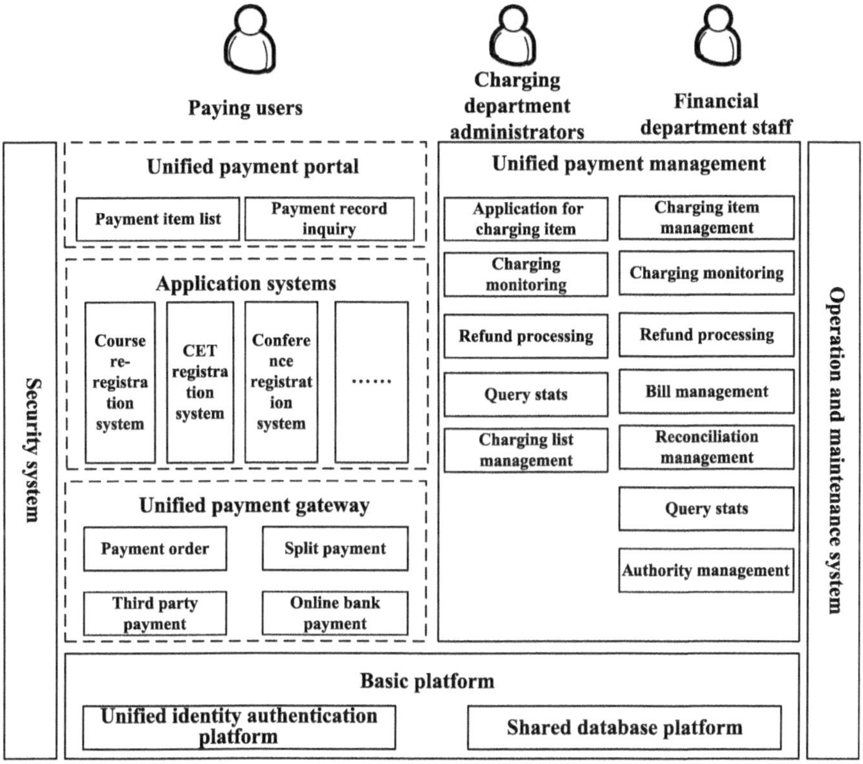

Fig. 1. Architecture of the online payment system

payment and the online bank. It can realize payment, refund and reconciliation of on-campus charging items. It also allows users to split payments.

2.4 Application Systems

The application systems of the charging departments can be integrated with the online payment system. When users pay, they will first submit relevant information in the application system of the charging department. Then they can complete the payment through the online payment system.

2.5 Unified Payment Management

The charging department administrators can monitor the charging process. They can check the current payment status of the charging item in real time. Historical payment records can also be queried at any time. It can quickly generate all kinds of statistical reports.

3 Application of Online Payment Mode in University Charging Management

Regardless of whether there is an application system in the charging department, online payment can be realized. It can cover existed charging items. There are no application systems for the charging items such as insurance charging. The charging process without application system is shown in Fig. 2.

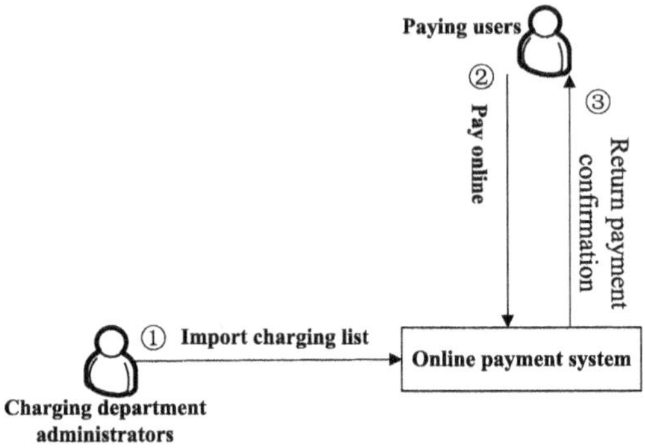

Fig. 2. Payment process without application system

① The charging department administrators import the charging list including the charging items, charging objects, and charging amounts into the online payment system.
② After the paying users log in to the online payment system, they can select the corresponding charging items for online payment.
③ The system returns payment confirmation information to the paying user.

In the traditional charging mode, the monitor must collect insurance fees for each student in his class before he can hand in the financial department. Any student who does not pay will affect the class. In the online payment mode, the student pays by himself online, and the student who has not pay only influences himself.

If there is an application system in the charging department, the system interface can be used for system integration to realize online payment. The data flow diagram is shown in Fig. 3.

The online payment order is transferred from the database of online payment system to the database of the application system. Then the payment result is returned to the database of the application system. The charging process with application system is shown in Fig. 4.

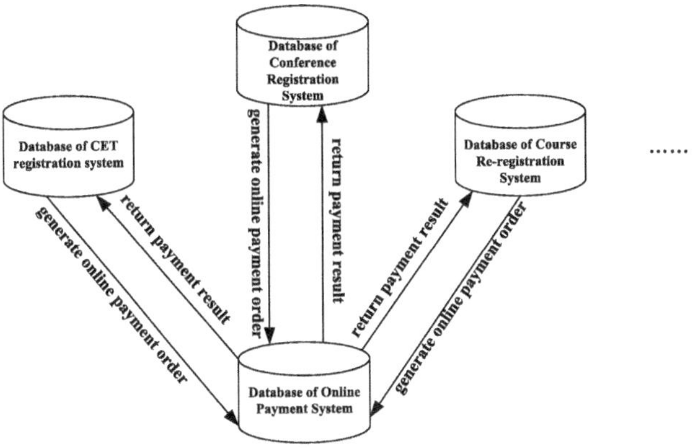

Fig. 3. Data flow diagram

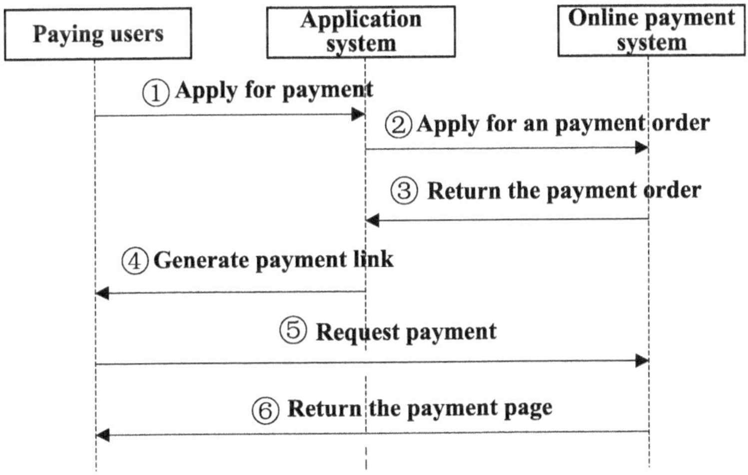

Fig. 4. Payment process with application system

① The paying users apply for payment in the application system of the charging department.
② The application system of the charging department applies to the online payment system for a payment order.
③ The online payment system returns the payment order to the application system of the charging department.
④ The application system of the charging department generates a payment link page on the user interface.
⑤ The paying users initiate a payment request to the online payment system.
⑥ The online payment system returns the payment page to the paying users.

With the improvement of the service level, the university provides freshmen with the freedom to choose bedding. Freshmen can choose their favorite bedding. The logistics management system is integrated with the online payment system. Freshmen can pay for the bedding online.

With the online payment system, students no longer have to go to the financial department to line up and pay. The financial department does not need to arrange staff to charge manually, and the workload is greatly reduced. At the same time, the system's log record is very detailed. All payment records can be queried in real time, and the university's financial management work efficiency is improved.

4 Conclusions

The online payment mode is realized through the online payment system. It changed the backward situation of the traditional charging mode. Its ultimate goal is to serve the users. It has improved the efficiency of the charging department, reduced the burden on payment, and improved the user satisfaction.

References

1. Wang, C.: B2C online payment system construction. Manuf. Autom. **33**(12), 184–186 (2010)
2. Wu, M., Liu, H., Ren, Y.: Internet plus campus: a new phase of "smart campus" construction of colleges. J. Distance Educ. **229**(4), 8–13 (2015)
3. Ma, R.: Research and implementation of SSO. Comput. Eng. Sci. **31**(2), 145–149 (2009)
4. Sun, Q., Chen, P., Huang, L., et al.: Research on sharing service platform building for affairs management data of university. China Educ. Technol. **350**, 69–74 (2016)
5. Li, F., Guan, L., Zhao, J., et al.: Mobile payment protocols based on digital certificate. Comput. Sci. **39**(11A), 19–23 (2012)

Digital Inheritance of the Traditional Dulong Culture in Yunnan Based on Gesture Interaction

Wu Xu[1], Yilin Li[1(✉)], Yiwen Huo[1], Jing Tao[1], Wei Guo[1], and Xinye Shao[2]

[1] School of Electrical and Information Engineering, Yunnan Minzu University, Kunming, China
670889576@qq.com
[2] Faculty of Information Engineering and Automation, Kunming University of Science and Technology, Kunming, China

Abstract. Dulong culture is an vivid and abundant part of the cultural heritage of minorities in China. The traditional protection methods are limited by time and space which can't retain cultural treasures in a comprehensive and long-term manner. In this paper, we collect minority cultural by scanning, create models by SketchUp, real-time render by 3dsMax to enhance the three-dimensionality of the graphics. The user can recognize the immersive experience in the virtual scene through the palm fingertips and feel the charm of traditional culture anytime and anywhere. It breaks normal forms of cultural communication, such as traditional museum exhibitions, and improves the inheritance of Chinese culture greatly.

Keywords: Virtual reality · 3D modeling · Gesture interaction

1 Introduction

The ethnic minority culture in Yunnan is a colorful and diverse part of the diversity of Chinese culture. The Dulong traditional culture is one of them. Dulong is one of the 8 ethnic groups in Yunnan Province. The Dulong River valley is their only colony. Its unique geography and natural environment, extensive agriculture based on fishing, hunting, and collecting activities gradually formed its unique national culture [1]. However, for a long period of time in the past, due to insufficient emphasis and weak protection techniques, customs such as festivals, rituals, and building houses gradually faded. The reshaping of the inheritance system faces enormous difficulties.

Nowadays, new media and Internet technologies have become important carriers for the protection of intangible cultural heritage. The traditional protection of images, pictures and texts could not complete the protection of the Dulong culture. Digital inheritance by virtual reality came into being. Digitization refers to the use of certain devices to convert information such as graphs, texts, sounds, and images into binary digits. Then, we could display Dulong culture on the network platform again through three-dimensional modeling, three-dimensional rendering of virtual scenes, and gesture

K. Deng et al. (Eds.): ICMIR 2018, AISC 856, pp. 231–239, 2019.
https://doi.org/10.1007/978-3-030-00214-5_31

interaction. Therefore, using digital technology can not only record the data saved in the traditional way with less storage space, but also make use of dynamic simulation to enable people to feel the charm of Dulong culture at anytime and anywhere. So as to expand their influence and promote the economic and cultural development of ethnic minority areas.

2 Digital Preservation of Dulong Traditional Cultural Relics

Firstly, high-definition digital photography, image scanning and other means are used to obtain information. Secondly, SketchUp is used to draw sketches to perform pre-model making, combined with fine-tuning and 3D rendering of 3dsMax models [2]. These model data are stored in a computer and digitized preservation of physical data such as residential buildings and bamboo ware is completed.

2.1 Data Acquisition and Processing

The virtual scene of the Dulong residential building is based on the real scene. The accuracy and reliability of the scene data are directly related to the quality of the scene construction. Scene data mainly includes topographic data and feature data.

(1) Topographic data
Virtual topographic data can rely on Google Earth satellite imagery. The high resolution of satellite imagery ensures the deep application of the ground environment and provides rich imagery information for the topographic features of virtual scenes.

(2) Feature data
The ground elevation data can be obtained by means of a total station to measure the horizontal distance and the inclination angle through the mathematical formula for calculating the elevation. For elevation data with lower accuracy requirements, Google Earth 2D images can be imported into SketchUp, and the solar azimuth angle can be calibrated. At the same time, the shadow of the 3D model of the feature can be adjusted to completely coincide with the shadow of the 2D image to measure the elevation. Unobserved features can be photographed by a multi-angle digital camera, and the estimated elevations are based on the ratio of reference objects. Get the scene data diagram as shown in Fig. 1.

2.2 Modeling Process

SketchUp is a three-dimensional design platform that directly faces the design process. Based on the triangular surface modeling system, a massive 3D model library can be directly invoked from a three-dimensional scene [3], and the wooden building model has high construction efficiency. Also, it can output the built model as a 3ds file for subsequent rendering calls.

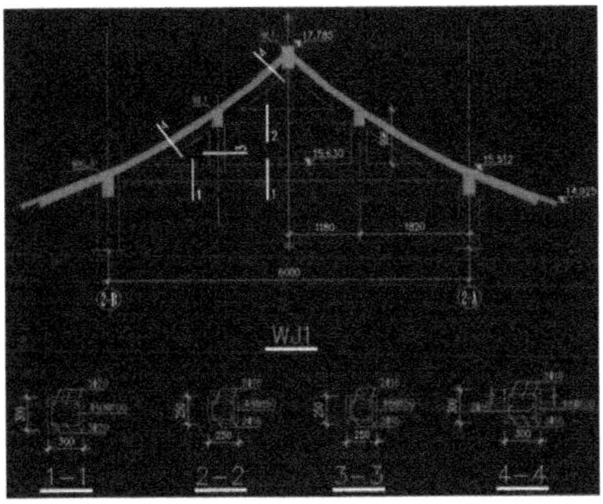

Fig. 1. Scene data of buildings

(1) Generation of building planes

Taking advantage of the constructive solid geometry (CSG) modeling concept, the building model mainly adopts the geometric modeling method, and the detailed structure enhances the realism of the model through real-time mapping. First, import Google Earth satellite images and display the map in two dimensions to provide accurate coordinates and references for scene modeling. The set view is the top view as the plane reference, drawing the outline with the SketchUp drawing tool and automatically closing the Dulong residential building plane. The plane outline outline should be consistent with the satellite Orthophoto Image.

(2) Create a pre-model
Mainly the wall and roof parts, the wall model is based on the ground plane, use the "push and pull" tool to pull up the ground plane to generate accurate wall. The inclined roof model can first move the building plane to the top of the wall, draw the ridge line, and then vertically move the roof line to the corresponding height to generate the roof. In the end, each component will be constructed and stitched to form a preliminary model of residential buildings.
(3) Using 3dsMax to refine the model structure
The 3dsMax stackable modeling step gives the model making flexibility. These advantages fully meet the structural requirements of Dulong ethnic dwelling buildings, fishing gear and other complex models. For example, when constructing fishing gear, 3dsMax is used to directly draw out each slender bamboo strip and assemble it according to its structure to realize the construction of a model. The kd-tree with segmentation strategy is particularly suitable for dense geometry and highly occluded environments [4]. The kd-tree build process is as follows:

```
kdNode* kdBuild (triangles T, voxel V)// Create recursion
{ if( Terminate ( T, V ))      // Split space
f( return new kdNode ( T ));
else
{ P = FindPlane ( T, V);      //Finding a suitable segmentation plane
( VL, VR) = Split V with p; //VL is the left half space & VR is the right half space
TL={ t∈T and ( t ∩ VL) ≠ϕ }; //Defining the left half space patch
TR={ t∈T and ( t ∩ VR) ≠ϕ}; //Defining the right half space patch
Return new kdNode ( p , kdBuild (TL, VL), kdBuild (TR, VR )); }
}
```

The production process is shown in Fig. 2, and the final results are shown in Fig. 3.

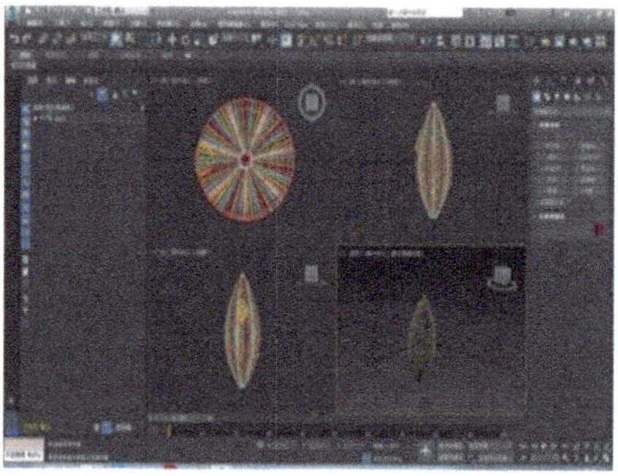

Fig. 2. Production process diagram

Fig. 3. Effect diagram

2.3 Virtual Scene Stereo Rendering

In order to achieve three-dimensional real-time image rendering, three-dimensional rendering to achieve real-time, the first solution is the continuity of the rendering. The number of three-dimensional scene rendering frames should be more than 30 frames per second. The processing speed of the graphics card in the simulation system, the selection and acceleration of the 3D rendering engine, and the complexity of the simulation scene all affect the rendering speed [5]. The realism and immersion of the virtual scene depend on the quality of the rendering of the picture. In order to improve the saturation of the picture, a higher resolution is needed. The rendering of the Dulong characters before and after is shown below. Before the rendering, the Dulong figure model was grayish and unable to see the figure, dress and other information. After the rendering, the lines of the clothes were clearly visible and the characters were vivid (Fig. 4).

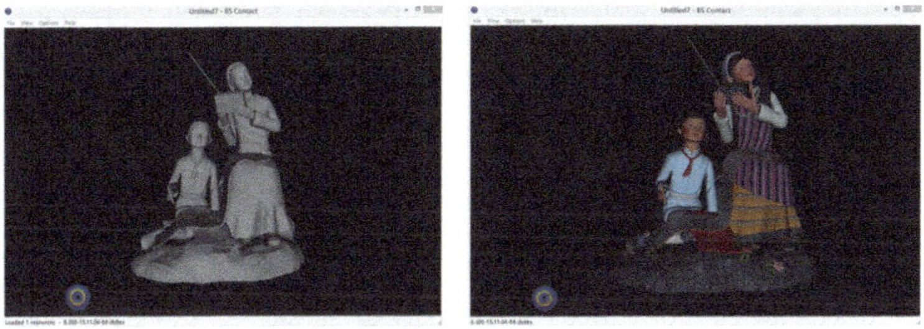

Fig. 4. Rendering contrast chart

3 Dulong Traditional Culture Interactive System

Gestures are one of the most common natural communication mediums for humans. Because of its natural, intuitive, easy to remember and other advantages in the human interface and virtual environment has a good interaction [6]. Data gloves and other equipment are expensive and inconvenient to wear and are not suitable for everyday use. The method based on computer vision is not comprehensive enough, and the algorithm is more complicated and it also brings obstacles to the recognition of gestures. Surface Electromyography (SEMG) is affected by the filtering action of the skin tissue. There are many variations in the surface myoelectric signal and the original myoelectric signal, which cannot provide more accurate data for the study. Therefore, this article uses a naked interaction.

Naked hand interaction is mainly divided into two major parts: gesture segmentation, gesture tracking [7]. Kinect is only for large gestures and is insensitive to fine movements of fingers and other parts. Leap Motion's fingertip recognition somatosensory controller's recognition accuracy can reach sub-millimeter level, and it can be compatible with various types of PCs and mobile phones. Therefore, by using Kinect to capture the three-dimensional information of the hand and Leap Motion to recognize the depth information of the hand at a short distance, detailed position information of the finger joint points can be obtained, and fingertips, fingers, palms, and other classification regions are obtained. Then, the Gaussian background subtraction is applied to the two-dimensional color image, and the hand region is extracted using the skin color feature. Then the finger area and palm extraction are performed on the hand area to classify the gestures.

3.1 Gesture Segmentation Based on GMBM

We use handoff regions based on mixed Gaussian background subtraction. The Gaussian Mixture Background Model (GMBM) is a method based on the background environment of the scene, which is based on the distribution of each pixel point in the video stream image frame in continuous time, and each pixel in the background image. Point, declare the K Muse distribution to represent the function form of the pixel, K is the number of peaks in the distribution of the pixel. A Gaussian mixture model consisting of three Gaussian probability density functions is defined as follows:

$$P(x) = a_1 g(x; \mu_1, \sigma_1^2) + a_2 g(x; \mu_2, \sigma_2^2) + \cdots + a_n g(x; \mu_n, \sigma_n^2) \tag{1}$$

Where p(x) represents the probability of pixel x, a_n represents the weight of the nth Gaussian model, and μ_n represents the mean value of pixel x in the nth Gaussian model. In establishing the Dulong dwelling house model, the distribution parameter model of each pixel is established through the progressive relationship of Gaussian distribution. The GMBM is a weighted sum of multiple single Muse Models. It can represent the multi-bump shape of a pixel over a period of time. The value of K is between 3 and 5. Taking the extraction of the foreground of Dulong ethnic dwellings as an example, a matching value is sought in the K single Gaussian distributions, and a pixel that satisfies the background of the environmental scene is obtained. To obtain the set S of edge points of the residential outlines, the elliptic curve is to be fitted by calculating the outline points of the image boundary. The least squares method is commonly used, and its algebraic form is as follows:

$$F(x) = Ax^2 + Bxy + CY^2 + Dx + Ey + F = 0 \tag{2}$$

3.2 Fingertip Palm Feature Extraction

After the above initialization is completed, feature extraction of the palm fingertips is performed. Only by obtaining an accurate palm center of gravity, can we obtain an accurate fingertip position. The positional relationship between the fingertip and palm is shown in the figure below (Fig. 5).

Using the shape of the palm (except for fingers) is similar to that of a circle, the position of the forehand palm can be detected more accurately by the Huffman circle transformation. The final palm position is obtained by the weighted average of the center of gravity and contour center of gravity:

$$P_c = P_{cog} * W_1 + P_{cht} * W_2 \tag{3}$$

Among them, P_{cog} is the palm of the hand, P_{cht} is the palm Hoffman's center, P_c is the final palm. W_1 is the weight of the palm center, W_2 is the weight of the center of the circle. The palm point is used as the reference point e. Two points n_1 and n_2 of equal pixel steps are selected clockwise and counterclockwise along the contour. The step length used in this article is 50 pixels. According to e, n_1, n_2, the fingertip outline circle and radius R can be determined, and the threshold h_0 is set. If the fingertip outline radius is smaller than h_0, the point can be judged as a fingertip point.

The definition of the interaction gesture template is given by the change of the fingertip feature of the continuous frame and the position of the palm: (1) Open state: The number of fingertips detected in the image frame is greater than four. (2) Fist state: The number of fingertips detected in the image frame is zero. (3) Tracking status:

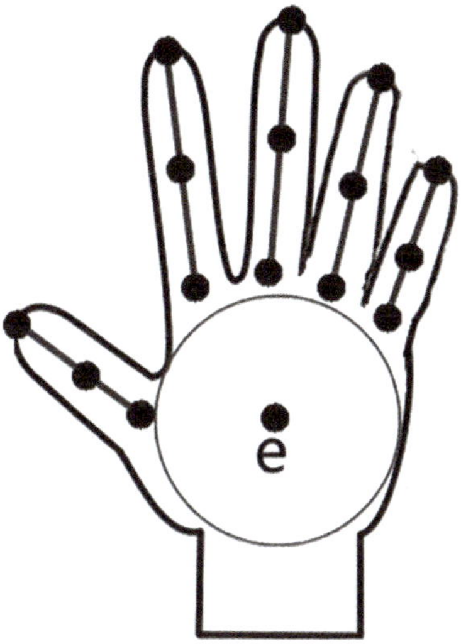

Fig. 5. Palm finger feature

Kinect image frame detected the fingertip and other skeletal points. The meaning of the following gesture actions is defined according to the gesture template:

(1) Mobile gesture: In the tracking state, the displacement of the palm coordinate between the two frames is directly mapped to the displacement of the screen coordinate cursor through the coordinate number, so that the movement is realized as shown in Fig. 6.
(2) Click gesture: It consists of continuous gesture frames in the tracking state. In the case of tracking the human hand, the change in position of the palm position coordinate between two consecutive frames is confined to a very small area, and the time exceeds 1.5 s, namely, it is determined as the click event of the screen position corresponding to the palm position.
(3) Pickup Gesture: It consists of two key frames: open state, fist state, two or more frames in succession when the hand is tracked, the number of fingertips of the hand changes from five to zero and the palm position within a certain range per second,

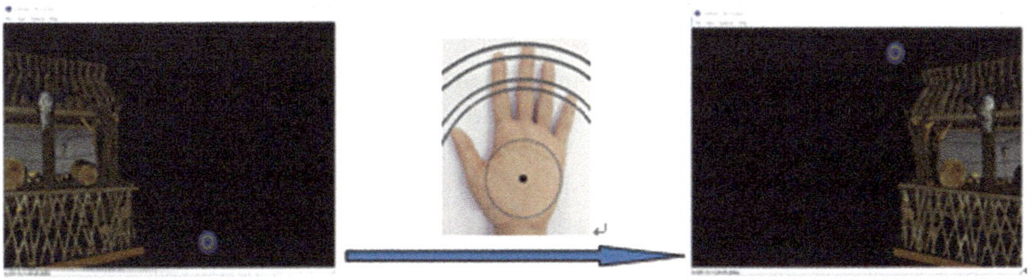

Fig. 6. Mobile gesture interaction diagram

when the number of fingertips changes in the sequence of 5 > 0 + 0, 5 + 4 + 0, etc., the number of fingertips in the middle of any descending sequence, it is determined as a pick-up gesture [7].

(4) Zoom gestures: In the state of two-handed fists, the positions of the palms of both hands change in opposite directions. The changed distance is converted into a scaling factor as shown in Fig. 7.

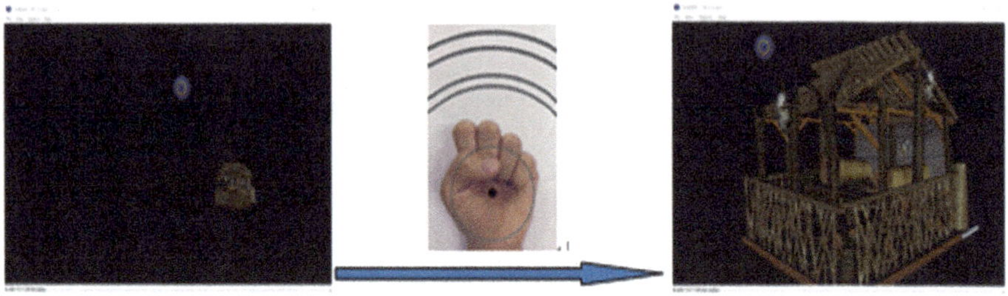

Fig. 7. Zoom gesture interaction diagram

The detection accuracy of the gesture template based on pixel classification is very high, and it can be used in human-computer interaction to achieve a good user experience. On this basis, many 3D human-computer interaction systems can be implemented, and the recognition algorithm can also recognize more gestures.

4 Summary

Ethnic minority culture plays an important role in Chinese traditional culture. In the background of the Internet+, this paper takes the traditional Dulong culture in Yunnan as the research object. Using three-dimensional modeling and three-dimensional rendering technology, Dulong traditional cultural resources are converted into digital storage forms and imported into virtual platforms. A Gaussian background subtraction method is used to extract hand areas for bare hand interactions. Thereby it improves the interactive process which rely on Data gloves, VR glasses and other devices in the past. Also it provides a reference for the protection and inheritance of other ethnic traditional cultures.

Acknowledgements. This research is supported by National Natural Science Foundation of China ("contact + non-contact" hybrid interface design strategy and its key human factors research. Grant: 61741206), and also supported by "The Key Laboratory of Ethnic Culture Resources Informatization".

References

1. Li, J.: The problems in the preservation of the dulong ethnic culture and the solutions. Acad. Explor. **10**(5), 35–44 (2009)
2. She, Z.: The production of virtual campus' 3D model based on google sketchup. Beijing Surv. Mapp. **3**, 78–84 (2017)
3. Lee, S., Yan, J.: The impact of 3D CAD interfaces on user ideation: a comparative analysis using SketchUp and Silhouette Modeler. Des. Stud. **2**(1), 53–55 (2014)
4. Meng, Y., Lu, C., Han, Y., et al.: Design method of fusion interface for virtual reality software and typical 3D model based on OpenGL. Comput. Eng. Appl. **53**, 7–13 (2017)
5. Xiang, M., Xu, W., Zhao, L., et al.: The research of real-time mapping in the restoration of lacquer based on X3D. Sci. J. Inf. Eng. **3**(6), 111–115 (2013)
6. Zhang, K.: Research on Depth Data Fused Human Computer Interaction Hand Gesture Recognition. Central China Normal University, Wuhan (2013)
7. Wang, K.: A Hand Gesture Interaction System for EAST Based on Kinect and Fingertip Recognition. University of Science and Technology of China, Wuhan (2017)
8. Yan, H.: Multi-Fingertips Tracking in 3D Space Using Kinect and its Application. Zhejiang University, Zhejiang (2013)

Research on the Digital Promotion and Development of the Achang Forging Skills in Yunnan

Wu Xu[1], Jing Tao[1(✉)], Yiwen Huo[1], Yilin Li[1], Chunjie Jin[1], and Xinye Shao[2]

[1] School of Electrical and Information Engineering, Yunnan Minzu University, Kunming, China
815032364@qq.com
[2] Faculty of Information Engineering and Automation, Kunming University of Science and Technology, Kunming, China

Abstract. With the strengthening of the globalization trend and the acceleration of the modernization process, people tend to think that traditional techniques and traditional culture are unable to integrate into social development, young people cannot recognize the intangible cultural heritage and the survival of traditional skills has been more and more impacted. In recent years, the rapid development of VR technology breaks the limitations of time and space due to its high digital accuracy, people can experience the charm of traditional culture more intuitively and solve the problem of people's identification of intangible cultural heritage. In this paper, we first use SolidWorks to model the tools needed for forging swords, and optimize the model to create the interactive three-dimensional scene under streaming media; then we store environmental information and models in the database, we can experience the traditional forging sword process personally by wearable devices, such as data gloves which can recognize people's action; finally, we use JavaScript technology as the core to develop PC-based and mobile-based clients, so that people can experience the process of forging the sword online. By researching and digitally protecting the forging skills of the sword, we will provide reference for the development and protection of other ethnic minority special techniques.

Keywords: Husa sword forging · Virtual reality · Image optimization JavaScript technology

1 Introduction

In recent years, the country has put forward the "precise poverty alleviation" strategy, including "accurate poverty alleviation" - cultural poverty alleviation, which focuses on cultural ideas. The Provincial Department of Culture of Yunnan Province has made poverty alleviation as a major livelihood project. It continuously promotes cultural poverty alleviation projects, the development of traditional handicrafts in Yunnan, the protection of non-manufactured production and the transformation of production and lifestyle. These measures not only inherit the unique cultures of the ethnic groups in

K. Deng et al. (Eds.): ICMIR 2018, AISC 856, pp. 240–248, 2019.
https://doi.org/10.1007/978-3-030-00214-5_32

Yunnan, but also promote employment and entrepreneurship in the process of inheritance, as a result, it has turned the country out of poverty.

Yunnan Province is located in the southwestern border of China, where many ethnic minorities live in. Achang is one of the earliest indigenous peoples in Yunnan, China. Its culture is rich and colorful, and the ironware it manufactures is extremely famous. It is famous for its "husa sword". People also engrave some symbol of good luck on the sword and equip with beautiful scabbards which are made of wood, leather, silver and other raw materials, which makes the sword more ornamental and represents "Sword culture" from the local ethnic characteristics of Yunnan [1]. In recent years, because of the impact of modern mechanized processing, fewer and fewer people have used hand-made cutting tools, and they are slowly disappearing. Simply relying on traditional methods of protection can no longer attract the attention of most people, and traditional hand-crafting techniques are urgently needed to protect.

VR technology is a modern information technology that has emerged in these two years. Nowadays, it is increasingly used in cultural fields such as the protection of world cultural heritage. Combining Achang's protection of forging skills with VR technology, we can store it in a computer network by a digital way and realize 3D digital archiving. Due to advantages of information technology like rapidity, stability and sustainable development, the development of information products or systems that are appropriate to the traditional cultural characteristics of ethnic minorities, and ultimately we will achieve the protection, inheritance and development of traditional ethnic culture.

1.1 Modeling Tool Selection and Optimization

1.1.1 The Choice of Modeling Tools

There are many modeling methods available now, such as C++, OpenGL graphics library modeling method and virtual modeling language VRML modeling method. The modeling software is 3DMAX, Pro/E, SolidWorks and so on. This article uses SolidWorks software + VRML language modeling, we import tool pictures into SolidWorks for object modeling and then form VRML(.wrl)files through a series of transformation, modification, input, output by VRML language. It reduces the workload of virtual modeling greatly and improves efficiency. The model of the husa sword and forged tool is shown in the figure below (Figs. 1, 2).

1.1.2 The Optimization of Model Images

Expect forged tool models, we also need to build a complex scene of forging sword, a large number of three-dimensional models will make the graphics run slowly, the virtual scene presented to the audience is not real enough. We need to improve the modeling technology, increase the fluency of the scene and the realism of the scene to improve the user's immersion. The techniques for optimizing the image in this paper are as follows.

(1) Level of detail technology

Level of Detail (LOD) technology can increase not only the speed and quality of model loading but also real-time performance, which enhance the fluency of the

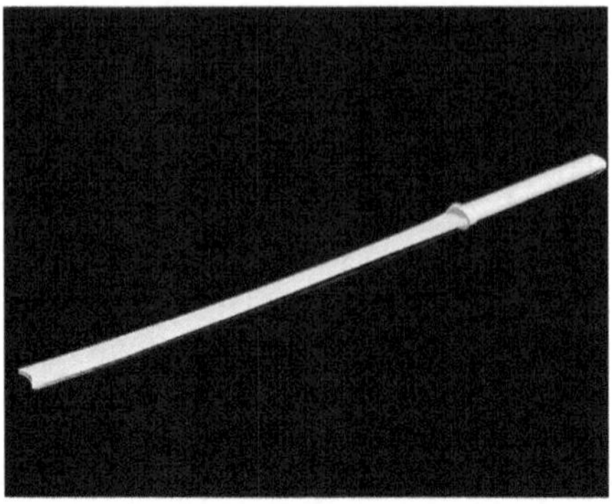

Fig. 1. Husa sword model

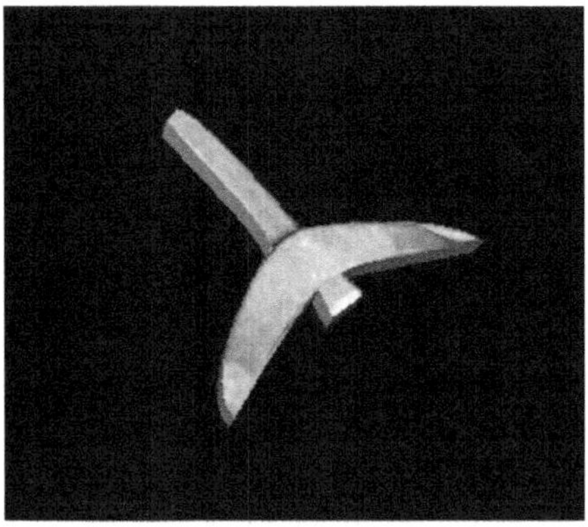

Fig. 2. Tool model for forging sword

environment loading. Meanwhile, the geometric complexity of scenes can be reduced by successive simplification of the surface details. Depending on the accuracy, different details are applied to the surface of the same original model. With the viewpoint transformation, the detail degree of the object model is changed to improve the display speed. We use a fine model when observing an object from near, and a coarser model when observing an object from a distance. Among them, the coarse model consists of relatively few polygons, while the fine model consists of a relatively large number of polygons [2]. The conversion diagram of LOD is shown in Fig. 3.

Among them, "Switch In" and "Switch Out" represent the distance of turn into and turn off, these two values determine the display range of LOD. The value of "Switch In" must be greater than that of "Switch Out". In addition, the combination of LOD technology and Morphing technology can enhance the smoothness of LOD transformation.

The Morphing transition is actually controlled by the Center, Switch In, Switch Out, and Transition which are shown in Fig. 4. We divide the range of transition into

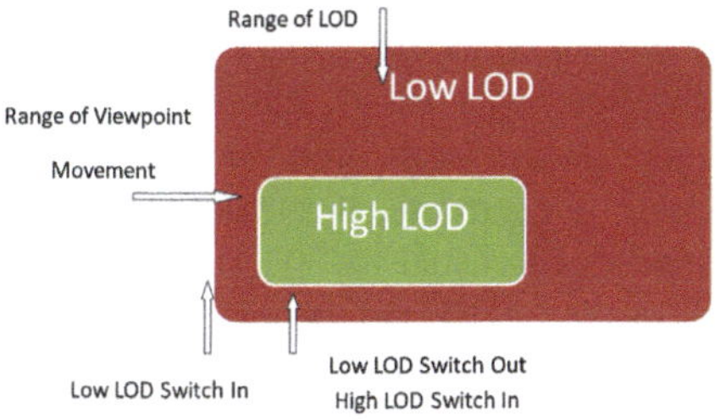

Fig. 3. Conversion diagram of LOD

several segments equally, and the transition vertices will be divided into several steps into the actual model vertices in the transition range [3]. When the viewpoint enters the visible range, the transition vertices of the model are displayed firstly. As the viewpoint closes to the center, the transition vertex is gradually transferred to the actual model vertices which show up until the actual model is surpassed, the shorter the transitional range, the faster the transition, and the smoother the picture we see.

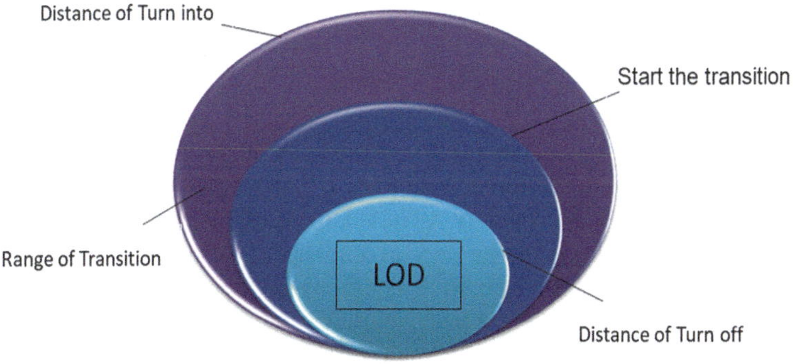

Fig. 4. LOD's morphing transition

(2) Texture mapping technology
In the process of modeling, in order to enhance the sense of reality, we will often achieve the overall effect at the expense of local details. In some complex models, Texture mapping technology is used to replace the model details with texture, the realistic texture can not only improve the level of detail and the reality of the model, but also do not increase the complexity of three-dimensional geometric modeling, and reduce the number of polygons in the model. The textures are two-dimensional images that are mapped to the surface of the three-dimensional image. The surface of the model built in the virtual reality system is very smooth and unrealistic. Modeling with models would be troublesome and varies of the patches, but the texture can be very simple to achieve the desired effect. The texture image is defined as the u, v coordinate plane, which is mapped to the three-dimensional

coordinate system. Then, we adjust color and shadow data of the image on the surface to eliminate the abnormal effect caused by the relationship between pixels and pixels. When the three-dimensional model is transformed and projected onto the screen, the mapped texture is also rotated and resized, as if it is drawn on the surface of the model [4]. Therefore, the texture technology can make the scene models or device models in virtual reality have surface texture and color shade, enhance the realism of the model, reduce the number of patches and resources of the model, and improve the loading efficiency (Fig. 5).

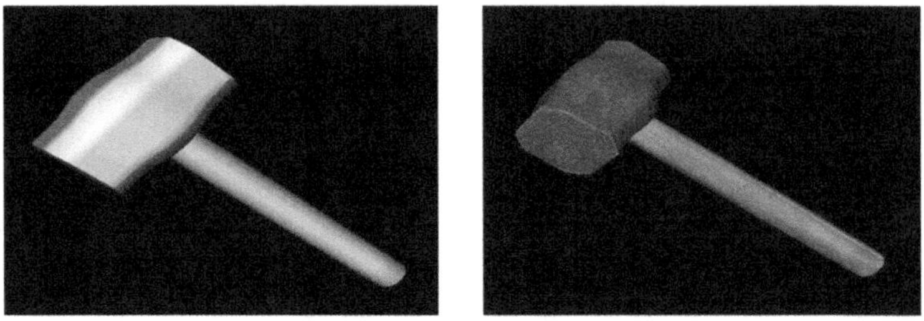

Fig. 5. Contrast diagram of texture mapping technology

(3) Instantiation technology

In the three-dimensional complex scene model, there will be multiple objects with the same geometry but different positions. If the normal copying method is used, the number of polygons will be doubled for each additional object. Using the instantiation technology, the number of similar objects can be increased while the number of polygons is not increased. The same geometry can share the same model data and be placed in different places through matrix transformation. In this case, we only need one geometry data storage space, which can greatly save memory space. The matrix can be expressed as:

$$T = \begin{bmatrix} a_{11} & a_{12} & a_{13} & a_{14} \\ a_{21} & a_{22} & a_{23} & a_{24} \\ a_{31} & a_{32} & a_{33} & a_{34} \\ a_{41} & a_{42} & a_{43} & a_{44} \end{bmatrix} \quad (1)$$

T can be divided into four matrices from the function of transformation, T produces proportional, rotational, and other geometric transformations; [a41 a42 a43] produces a translation transform; [a14 a24 a34] produces a projective transform; [a44] produces a global scale transform. Using the above four matrices can accomplish different transformations of the same data model.

(4) Degree of Freedom of Joint

Because of the high requirement of the model for the freedom of movement in virtual reality, we use DOF technique which is a node model to set the degrees of freedom for the dynamic behavior of objects. Using this technology can make the model object have the ability to move, the node can control all its child nodes to move or rotate according to the set range of freedom [5]. We can insert it into the database to increase the movement of the geometry, or create a local coordinate system that controls the movement of the geometry around the axis in it.

1.2 Multi-channel Interaction Technology in Environment Roaming

Modeling and dynamic display technology based on image is the direct construction of virtual scenes by using the actual scene images, which has the advantages of quickness and simplicity. The disadvantage is that the observation point and the observation direction are limited, and a complete interactive operation cannot be achieved. In order to enhance the immersion and natural interaction, we use multi-channel human-machine interface technology in the environmental roaming, which let people have a more real experience of forging process [6]. The virtual scene roaming interaction is suitable for adopting natural interaction channels such as gestures and stereoscopic display devices. The following describes the stereoscopic display technology and gesture interactive roaming used in the system.

1.2.1 Stereoscopic Display

Stereoscopic display is very important for realizing the immersion of virtual reality. Using stereo vision can be used to construct three-dimensional geometric information of scenes and models. The stereoscopically displayed image can be obtained through helmets, liquid crystal light valve glasses, etc. The stereoscopic vision model requires that the two viewpoints should be separately calculated to generate the left-eye and right-eye views, so we need two centers of projection. According to the relative position between the surface of projection, the human eye, and the surface of object, there may be a negative parallax and a positive parallax. In the following, we take negative parallax as an example and calculate based on the perspective principle:

Let the distance between the two eyes be $2x_0$. If the middle point of the line between the two eyes is defined as the origin of the coordinate system, the line segment is on the X axis, and the Z axis is perpendicular to the projection plane. Use the right hand system to create a three-dimensional coordinate system. The coordinates of the two eyes are $(x_0,0,0)$ and $(-x_0,0,0)$, and the coordinates of the object point are (x_1,y_1,z_1). The distance from the viewpoint to the object is $O_{distance}$, the distance from the projection plane to the viewpoint is $P_{distance}$. The coordinates of the projection point of the object on the projection plane are respectively: (x_{11},y_{11},z_2) and (x_{12},y_{12},z_2), where $z_1 = O_{distance}$, $z_2 = P_{distance}$, The diagram is shown below (Fig. 6).

When we calculate the coordinates of the subpoint, we use the right eye position as the coordinate origin for convenience. After calculation, the following formulas can be obtained:

$$x_{11} = [(x_0 + x_1)P_{distance}/O_{distance}] - x_0 \tag{2}$$

$$y_{11} = y_1 \times (P_{distance}/O_{distance}) \tag{3}$$

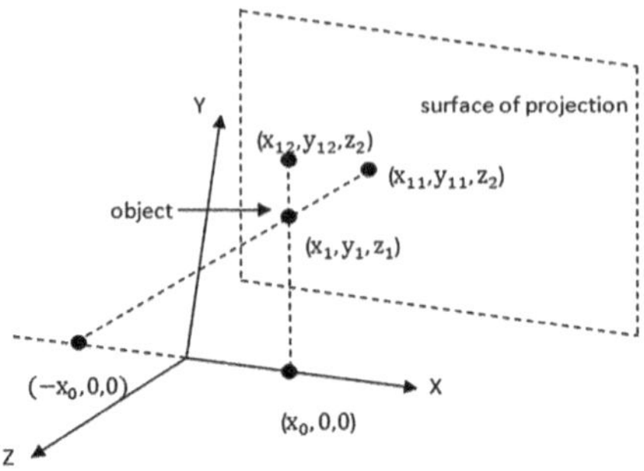

Fig. 6. Schematic diagram

In the same way, we can get x_{12}, y_{12}. When calculating the left picture, x_0 is positive; when calculating the right picture, x_0 is negative. After producing a stereoscopic image pair, each eye can only see one image of a stereoscopic image pair when viewing with a helmet or stereoscopic glasses, which can give a three-dimensional impression.

1.2.2 Gesture Interactive Roaming

Gesture interaction is a frequently-used interaction method in people's daily life, we can use gestures for virtual scene roaming. Gestures can complete interactive tasks such as: grabbing and shaking hands. Gesture input devices include: mouse, pens, and data gloves; gesture input based on mouse and pen is suitable for the development of miniaturization and personalization, gesture-based data glove-based input is currently the ideal choice, and we can completely rely on data gloves in virtual scenes freely. In the virtual scene, the model of the hand is used as the avatar of the user, and the user can freely specify the roaming route through the hand movement, manipulate the objects in the virtual environment.

During the process of scene roaming, the interaction tasks of viewpoints are roughly divided into seven basic tasks: forward, backward, upward, top, left, right, and return to the starting point; operations on objects in the scene can define corresponding operation gestures. Such as: grab objects, open the door and so on. Firstly, gesture interaction completes the simulation of the human hand, it is divided into two parts: the palm and the fingers. The thumb has two joints, and the rest of four fingers each have three joints. There is a corresponding constraint relationship between the palm and fingers, and the various joints of the fingers. For example, the movement of the palm can drive the movement of the fingers, the movement of the bottom joint in each finger can affect the movement of the top joint. Based on this constraint relationship, when modeling the entire hand, the entire hand model is used as a node in the scene. After the physical model of the hand is established, gesture synthesis is performed, which is performed by the computer according to the user's hand wearing the data gloves. Postures with certain meanings are generated and displayed graphically. The data of the data glove is used to drive the movement of the physics model. It can be described by a series of state transitions and constraints. The state transformation matrix is used to calculate the spatial position and current shape of the gesture [7].

2 The Promotion of the Mobile Terminal of the Sword Forging Skills

With the rapid development of new media technologies, people's lives are increasingly connected with computers, mobile phones and other electronic devices. However, most of the protection methods are still explained by the static viewing modes such as the museum's pictures, object exhibitions. There is little protection for real operation that allow people to perform real-world operations on electronic devices. It greatly reduces people's interest in it and directly affects the spread and development of this traditional art. We can create web pages or develop a Web App to display multi-faceted forging techniques so that people could experience it personally.

JavaScript is a general scripting language for web page creation. Its function is to respond to the user's operations on the web page, such as simple clicks and input. The task of JavaScript is to give the web page the ability to "dynamically interact". JavaScript is linked by browser to build interpreter in browser, so that HTML static web page can be changed into dynamic webpage. The J Query framework is a widely used JavaScript framework. The J Query framework has mature technology and rich functions. It is an important development framework for web site developers to develop web page dynamic features. With this framework, we can save developing time, ensure developing efficiency, and complete rich functions.

In addition to using web pages to build virtual reality development platforms, we can also promote them on mobile devices such as mobile phones. We can develop a Web App. The Web App client is developed by JavaScript and the server uses Java technology. Web App can achieve many functions that can be realized by Native App,

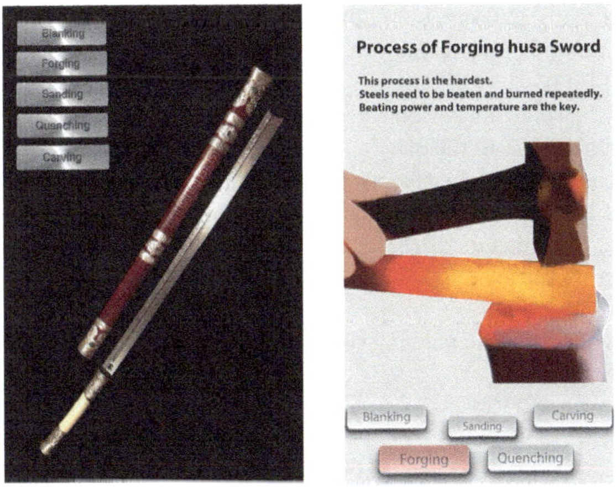

Fig. 7. Mobile phone terminal Web App application display

such as local file and data storage, audio and video playback functions. Web App can be easily cross-platform without user downloads, and it can be dynamically upgraded like traditional websites [8]. Web App has the characteristics of App, but more Web features. Web Apps can be linked to each other like traditional internet web pages, and jump from one web app directly to another web app. This is a very good choice either from the user experience level or from the data transmission between applications (Fig. 7).

3 Conclusion

In this paper, we have developed a digital display platform for forging technology of husa swords by VR technology. The image is optimized to increase the authenticity of the model, and gesture interaction with data glove is used to enhance user's immersion. We also display the forging techniques with PC and mobile phone terminal application system, expand the publicity channels for sword forging skills through display methods such as PC client and mobile application system, enhance the understanding of ordinary people and rejuvenate this ancient skill. so that this old skill will achieve the inheritance, protection, and innovation of the forging skills. It also provides a new reference for the protection of traditional culture and techniques of other ethnic minorities.

Acknowledgments. This research is supported by National Natural Science Foundation of China ("contact + non-contact" hybrid interface design strategy and its key human factors research. Grant: 61741206), and also supported by "The Key Laboratory of Ethnic Culture Resources Informatization".

References

1. Wang, J.: Research on the inheritance and protection of Achang Nationality's historical culture. J. Pu'er Coll. **32**(04), 57–59 (2016)
2. Gao, F.: Research on optimization of 3D Scene construction in virtual reality. Sci. Technol. Innov. Herald **14**(27), 113 (2017)
3. Ming, J., Yan, M.: Three dimensional geological surface creation based on morphing. Geogr. Geo-Inf. Sci. **30**(01), 37–40 (2014)
4. Su J.: Research and implementation of 3D real-time rendering technology—rendering framework and key rendering technology based on scene graph, School of Information Engineering and Automation (2015)
5. Zhao, C., Zhang, P., Zhang, Y.: DOF program design of vega prime. Program Lang. **21**, 27–30 (2013)
6. Xu, F.: Implementation of the roaming system for 3D virtual campus. Comput. Program Skills Maint. **09**, 91–92 (2015)
7. Yi, J., Cheng, J., Ku, X.: Review of gesture recognition based on vision. Comput. Sci. **43**(06), 103–108 (2016)
8. Yi, X., Li, Y.: Applications and differences of between JavaScript and Java in web development. Commun. Technol. **46**(06), 145–147 (2016)

Detection and Defense of SYN Flood Attack Based on Winpcap

Xiuyu Zhong(✉) and Yang Liu

School of Computer Science, Jiaying University, Guangdong, China
mzzxy@jyu.edu.cn

Abstract. With the development and application of computer network technology, the threat of network security is getting worse. SQL inpouring attack, ARP cheat, DNS cheat and various attacks of denial service occur frequently, such as SYN flood attack. This paper presented the design of detection and defense system of SYN flood attacks on Winpcap platform. The system includes three modules, such as sniffer module, analysis module and active defense module. Simulation results show that the detection rate of system is high, it can provide reference for tracking SYN flood attack.

Keywords: SYN flood attack · Winpcap · Sniffer · Detection Defense

1 Introduction

The report of China Internet network security in 2016 shows that China's web site security still faces complex and grim situation [1]. In view of the particularity of the network security field, high requirements of the government and schools for network security, and the urgent needs of companies for higher security requirements, so the purpose of the system development is loging and alerting the network attack, responsible person master the first scene in real time, the network administrator is not required to retrieve the system log, automatic defense of conventional DDoS flood attacks without on duty.

Many scholars have made research on SYN flood attack, Zhang presented a defensive method of SYN flood attack based on retransmission timeout mechanism [2]. Ding and Tian presented an adaptive threshold algorithm which could detect SYN flood attack quickly based on dual stack network firewall [3]. Aborujilah and Musa presented a method to detect TCP SYN flood attack by analysis its effects on CPU and network performance through performing several experiments with different parameters [4]. Gilang and Yusuf presented a method to detect TCP SYN flood DDoS attack using artificial immune systems [5]. Arshadi and Amir presented a novel approach for detecting SYN flooding attacks by investigating the entropy of SYN packet inter-arrival times as a measure of randomness [6].

K. Deng et al. (Eds.): ICMIR 2018, AISC 856, pp. 249–257, 2019.
https://doi.org/10.1007/978-3-030-00214-5_33

2 Brief Introduction of Winpcap

Winpcap(windows packet capture) is a system for capturing and analyzing packets under Windows platform, it includes a kernel-level packet filter, a low-level DLL (Packet.dll) and an advanced system-independent DLL (Wpcap.dll). The structure of Winpcap is shown in Figs. 1 [7]. Winpcap provides the following four functions to monitor packets.

(1) Capturing packets which are sent or received by hosts and exchanged between each other.
(2) filtering packets according to custom rules before sending packets to applications.
(3) Sending the original packets on the network.
(4) Collecting statistical information during network communication.

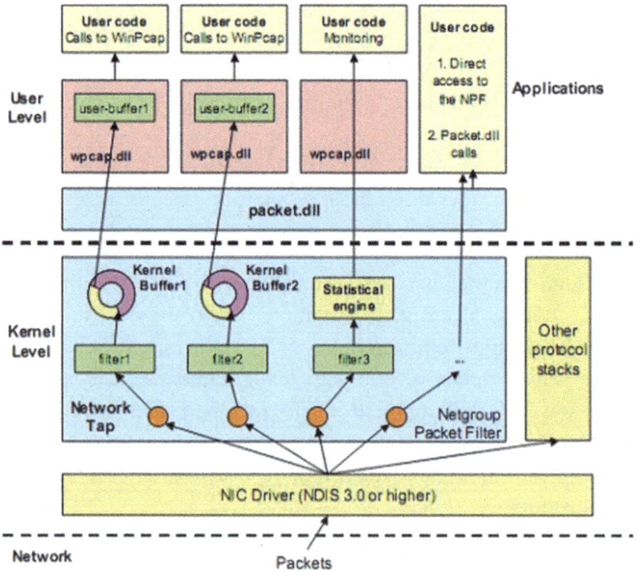

Fig. 1. Structure of Winpcap

3 System Design

3.1 High Level Design

The system adopts C/S architecture, includes three modules, such as the under sniffer module, analysis module and active defense module. Each module runs independently to reduce coupling. Each module can be run on different threads on processing data, so that it can take full advantage of the host machine's multi-core devices to speed up the process. The monitor is mainly to determine the packet type, extracts the information fields, and creates data finger. The network packets are captured and send on Winpcap, including the information of TCP service type and service agreement, IP packet protocol, ARP packet, EAP packet, EAP packet, etc.

The system is written by C++, Winpcap is used to capture packets on the bottom layer, packets are analyzed and attacks are alerted on the upper level, Plug-in intercepts various attacks to realize active defense.

In order to accurately realize the dynamic identification of attacks,the system builds a simulation attack victim based on Linux system, designs a Web program for query.

3.2 Data Structure Design

Various structures is defined from the bottom to upper according to the OSI model on data structure design, including structure packet, structure UDP head,structure IP and seal verification, structure physical frame head,structure Arp frame,structure ARP packet,structure IP header,structure TCP header, structure TCP pseudo header, structure UDP header,structure UDP pseudo header,structure ICMP header,structure ICMP packet. structure TCP pseudo header is used to compute TCP checksums in order to ensure the effectiveness of TCP performance, structure UDP pseudo header is used to compute checksums only. Furthermore, structure attack information is defined, including attack packet type, attack IP and attack characteristic key, specific definition is as follows.

```
typedef struct
{       unsigned char flag;
        unsigned int srcip;
        unsigned char code[33];
}Attack_Infor;
```

3.3 Low Level Design

(1) Sniffer Module
The module is responsible for the direct interaction with the network card, The operating system provides the driver, Winpcap provides the program interface of application layer. The sockets interface provided by Windows is not used because the socket listener packet can run on Windows Server only, The generic version of the Windows operating system kernel prohibits the user layer from sending and listening to the original packet. packets are monitored by starting a new thread, and the user interface is refreshed by setting a time callback, part code is as follows.

```
        WriteLog("Initialization interface success!");
        DWORD addr;
        monitor=new NetMonitor;
        monitor->pdlg=this;
        SetTimer(1,100,0);
        __asm
        {mov addr,offset NetMonitor::run;}
   CloseHandle(CreateThread(0,0,(LPTHREAD_START_ROUTINE)addr,monitor,0,
  0));
```

Before listening, various network adapters of the current computer can be obtained by enumerating system network adapters, and a network adapter is selected by default as a work network adapter. As far as possible, all the packets that are passed through the intranet or external network entries are copied to the local computer, so that the physical layer will not leak the packets.

(2) Analysis Module
The C++ main program is designed for analysis and sniffing,other plug-ins are written by Python, the user interface is designed with MFC so that the GUI will not become a system bottleneck,at the same time, the coupling of the system is reduced by the plug-in separation. The key data is analysed by pattern matching algorithm, classifying and counting the real-time packets. Some new attack characteristic database can be written by the third-party plug-ins, and then submitted to the plug-in for depth analysis after initial processing by anonymous pipes or network links. The main program implements the detection of most common network attack. The main code is as follows.

```
void NetMonitor::handledata( u_char* capBuff, bpf_u_int32 caplen )
{
if(caplen<sizeof(Ethernet_header))return;
      Ethernet_header* ether=(Ethernet_header*)capBuff;
      char smac[18]={0};
      char dmac[18]={0};
      sprintf(smac,"%02x:%02x:%02x:%02x:%02x:%02x",
               ether->srcmac[0],
               ether->srcmac[1],
               ether->srcmac[2],
               ether->srcmac[3],
               ether->srcmac[4],
               ether->srcmac[5]
            );
      sprintf(dmac,"%02x:%02x:%02x:%02x:%02x:%02x",
               ether->desmac[0],
               ether->desmac[1],
               ether->desmac[2],
               ether->desmac[3],
               ether->desmac[4],
               ether->desmac[5]
            );
      for(int i=0;i<pdlg->m_list.GetItemCount();i++)
      {
       if (pdlg->m_list.GetItemText(i,2).Compare(dmac)==0 ||
          pdlg->m_list.GetItemText(i,2).Compare(smac)==0)
          {
  ItemData* id=(ItemData*)pdlg->m_list.GetItemData(i);
  id->packet+=1;
  }
}
if (ntohs(ether->ethertype)==0x0806 && caplen>=sizeof(ARP_Packet))
{
```

```
                ARP_Packet* arp=(ARP_Packet*)capBuff;
                if(ntohs(arp->arpframe.flag)!=2)return;
                SOCKADDR_IN ip;
                ip.sin_addr.S_un.S_addr=arp->arpframe.send_prot_addr;
                ip.sin_family=AF_INET;
                char* ipstr=inet_ntoa(ip.sin_addr);
                char scount[20]={0};
                sprintf(scount,"%d",pdlg->m_list.GetItemCount());
                for(int i=0;i<pdlg->m_list.GetItemCount();i++)
                {
                  if (pdlg->m_list.GetItemText(i,1).Compare(ipstr)==0)
                              return;
                }
                pdlg->m_list.LockWindowUpdate();
                int i=pdlg->m_list.InsertItem(0,scount);
                pdlg->m_list.SetItemText(i,1,ipstr);
                pdlg->m_list.SetItemText(i,2,smac);
                pdlg->m_list.SetItemText(i,3,"0");
                pdlg->m_list.SetItemText(i,4,"0");
                pdlg->m_list.SetItemText(i,5,"0");
                ItemData*  id = new ItemData;
                id->packet=0;
                id->synpacket=0;
                pdlg->m_list.SetItemData(i,(DWORD)id);
                pdlg->m_list.UnlockWindowUpdate();
                pdlg->WriteLog("intranet protected machine was added to list ");
         }
          if (htons(ether->ethertype)==0x0800 &&
       caplen>=(sizeof(Ethernet_header)+sizeof(IP_HEADER)+sizeof(Tcp_Header)))
         {
         IP_HEADER* iphead=(IP_HEADER*)(capBuff+sizeof(Ethernet_header));
         if (iphead->proto!=0x6)
                 return;
Tcp_Header*
tcphead=(Tcp_Header*)(capBuff+sizeof(Ethernet_header)+sizeof(IP_HEADER));
         if(tcphead->flags!=0x2)
                 return;
         for(int i=0;i<pdlg->m_list.GetItemCount();i++)
         {
         if ( pdlg->m_list.GetItemText(i,2).Compare(dmac)==0||pdlg-
       >m_list.GetItemText(i,2).Compare(smac)==0 )
         {
           ItemData* id=(ItemData*)pdlg->m_list.GetItemData(i);
           id->synpacket+=1;
         }
         }
         }
         }
```

SYN flood attacks is a typical Denial-of-service attack, the purpose of the attacker is to deplete the server TCP connect resources, and stop responding to the normal TCP connection request. So, Linux gives a parameter called Tcp_syncookies to deal with it, when the SYN queue is full, TCP creates a special sequence number to send back with the source address port, destination address port, and timestamp, if it is a attacker, it does not respond, if it is a normal connection, the SYN cookie will be sent back, and then the server can build the connection through cookies,even if it's not in the SYN queue. But tcp_syncookies can't be used to handle normal load connection because synccookies is a compromised version of the TCP protocol, not strictly. For normal requests, it should adjust the three TCP parameters, the first one is Tcp_synack_retries which can be used to reduce the number of retries, the second is Tcp_-max_syn_backlog which increases the number of SYN connections, the third is tcp_abort_on_overflow which refuses to connect for overloading requestes. SYN must also comply with TCP checksum rules, or it will be discarded by network devices.

(3) Defense Module
In this module, if the connection reaches a preset threshold, the connection is directly alerted. TCP and RST packet are sent by plug-in,the information of dis-connect link or attacker's IP MAC are inserted into the firewall's iptables rules, even in the rules of the hardware firewall. Plug-ins can define their own behavior, the main program does not care about Plug-ins or sub-application implementation of what kind of interception or rule functions, so the maximum dynamic updates are allowed, even using scripting languages like Lua, python, and other scripting languages to implement the hot update of the plug-in subprogram.

The defense module can also record original log, provide onsite information for network administrators to track the source of the attack, and the generated logs can be stored automatically by date, and each log has a corresponding UTC.

The main program also has a built-in SQL injection detection mechanism, if the preset SQL injection statements are detected in the HTTP packet, the program's active defense and alarm mechanism will be triggered.

4 System TEST

Several attacks are simulated in a simulated small LAN. The scene is described as follows, there is a large number of TCP SYN handshake packets, the source IP address are randomly forged in the IP header of each packet, the destination IP address is the IP address of the attacked machine, the port in the TCP header is one of open ports of the target machine,the port 135.139.445 of Windows and the port 22 of Linux are common attack ports,In a large number of simulation attacks, the SYN flood attacke flood is detected in the intranet machine with IP 192.168.1.45, the attack IP is displayed in the log area and the source MAC address is logged in the background log, the MAC source address is an important reference, in practical applications, an attacker can be tracked based on the source MAC address and the machine's routing table snapshot to find out where the last hop came from. Most attackers do not use the real IP address to launch the attack, but use a fake source IP to implement the attack. In order to make the attack more efficient and the firewall disable, most attackers would randomize some range fields in most of the packet headers, make most of the data fingerprint technology ineffective. However, as long as the restriction on the exchange node of the internal system is not allowed to emit the source IP packets of the non-local subnet, this kind of internal attack can be eliminated. But for now, most network administrators do not have set up such rules, it gives potential attackers the convenience of attacks.

The system initialization interface is shown in Fig. 2, it displays the details of device area list and attack. The interface is designed with six columns, the first column represents the host sequence number, the second column represents the IP address, the third column represents the MAC address, the fourth column represents the number of processed packets, the fifth column represents SYN, the sixth column indicates the SYN rate. The system alert interface is shown in Fig. 3, the 6 columns meaning of the interface design is the same as that of Fig. 2. Figure 3 indicates that the host with IP address 192.168.1.45 is being SYN flood attacked.

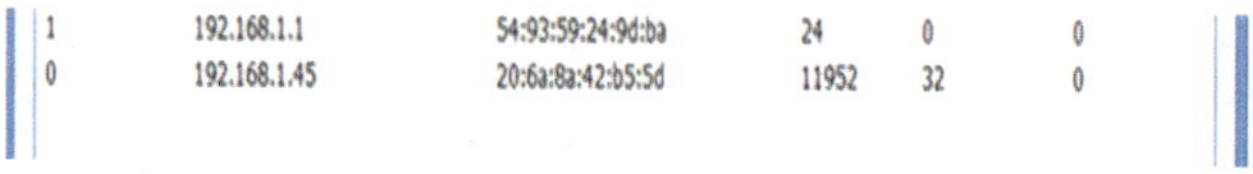

Fig. 2. System initial

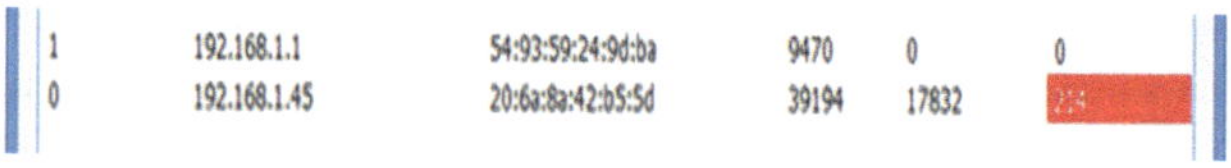

Fig. 3. Alarm display

5 Conclusion

This paper mainly describes the detection and defense of SYN flood attack based on Winpcap. The system can recorde and alert to network attack, let the responsible person grasp the scene in time. Network administrators are no longer required to retrieve system logs.The conventional DDoS, CC and other flood attacks can be defensed without duty, the detailes of SYN flood attack can be tracked.

Acknowledgments. The authors want to thank construction project of college students' practice education base in local universities of education ministry and the teaching quality and teaching reform project in universities of guangdong province for their general support for the research (with grant no. [2013]48 teacher's letter and [2016]233 high letter of yue jiao respectively).

References

1. The report of China Internet network Security in 2016, pp. 1–5, December 2016
2. Zhang, D.: A defensive method of SYN flood attack based on retransmission timeout mechanism. Electron. Sci. Technl. **04**(04), 87–90 (2017)

3. Ding, P., Tian, Z.: Detection and defense of SYN flood attacks based on dual stack network firewall. In: 2016 IEEE First International Conference on Data Science in Cyberspace, pp. 526–531 (2016)
4. Aborujilah, A., Musa, S.: Detecting TCP SYN based flooding attacks by analyzing CPU and network resources performance. In: 2014 3rd International Conference on Advanced Computer Science Applications and Technologies, pp. 157–161 (2014)
5. Gilang, R., Yusuf K.: Design of TCP SYN Flood DDoS attack detection using artificial immune systems. In: 2016 IEEE 6th International Conference on System Engineering and Technology, pp. 72–76 (2016)
6. Laleh, A., Amir, H.J.: Entropy based SYN flooding detection. In: 36th Annual IEEE Conference on Local Computer Networks, pp. 139–142 (2011)
7. An, X., Lu, X.: Packet capture and protocol analysis based on Winpcap. In: 2016 International conference on robots & intelligent system, pp. 272–275 (2016)

Application of Weighted Nuclear Norm Denoising Algorithm in Diffusion-Weighted Image

Sijie Li, Jianfeng He(✉), and Sanli Yi

School of Information Engineering and Automation,
Kunming University of Science and Technology, Kunming 650500, China
jfenghe@kmust.edu.cn

Abstract. Diffusion-weighted imaging is a noninvasive method for detecting the diffusion of water molecules in living tissues, which requires high accuracy of the data. Diffusion-weighted images have a high degree of self-similarity. And it is widely used in the nerve fiber tracking in human brain. According to the characteristics of the image, the weighted nuclear norm denoising algorithm is proposed for the diffusion-weighted images denoising. By processing similar blocks of the image, the image can be reconstructed. The algorithm is compared with some traditional algorithms and we propose a new method to describe the denoising effect. The experiment shows that the weighted nuclear norm denoising algorithm makes good use of the self-similarity of the diffusion-weighted images, and achieves the purpose of image denoising through the processing of similar blocks. Can obtain better results from the proposed algorithm, the accuracy and validity of DTI data are improved, and it is very helpful for the subsequent processing of the image.

Keywords: Denoising · Diffusion-weighted imaging
Weighted nuclear norm denoising algorithm · Nerve fiber tracking

1 Introduction

Diffusion tensor imaging (DTI) is the non-invasive technique to acquire the white matter structure of the brain. From DTI, the symptoms of acute cerebral infarction can be discovered as early as possible, and some problems related to the brain nerve can also be revealed. It is widely used in the nerve fiber tracking in human brain. However, noise will affect the accuracy of the DTI data and the subsequent processing, and the diffusion tensor images are calculated by the diffusion-weighted images(DWI). Therefore, it is necessary to remove the noise in DWI.

There are many researches on DWI denoising at home and abroad. High-order singular value decomposition to the DWI denoising, the algorithm significantly reduces stripe artifacts in homogeneous regions [1]. A robust variational method for simultaneously smoothing and estimating of DTI [2]. A similarity measure based on the local mean and on a modified structure-similarity is defined by Bao et al. [3]. An algorithm for DWI data enhancement based on structural adaptive smoothing in both voxel space and diffusion-gradient space introduced by Becker et al. [4]. In a series of papers,

K. Deng et al. (Eds.): ICMIR 2018, AISC 856, pp. 258–263, 2019.
https://doi.org/10.1007/978-3-030-00214-5_34

smoothing of the diffusion-weighted data is achieved via a weighted Total Variation (TV) norm minimization, which strives to smooth while retaining all relevant detail [5].

The above algorithms can effectively improve the quality of DWI, but these algorithms do not pay attention to the texture details in DWI. DWI have a high degree of self-similarity and rich feature details. The weighted nuclear norm denoising algorithm is proposed for the DWI denoising by exploiting the image's nonlocal self-similarity. Based on the above characteristics, in this paper we applied the algorithm to DWI denoising, and it can also save the texture details of the image.

2 Method

2.1 Diffusion Tensor Imaging

Diffusion tensor imaging technology can detect the diffusion of water molecules in the brain. The diffusion motion of water molecules in the human brain can be expressed by diffusion tensor, and the diffusion state can indirectly reflect the microstructure characteristics and change of the tissue. Each voxel in DTI consists of a diffusion tensor **D**. The diffusion tensor is a symmetric positive definite matrix.

Diffusion tensor data can be used to process many problems in the brain, such as tracking neural fibers. The DTI is calculated based on the Stejskal–Tanner formula for the DWI [6, 7]. So the DTI data calculated by the formula is unreliable. Therefore, it is important for DWI to remove noise.

2.2 Weighted Nuclear Norm Denoising Algorithm

The weighted nuclear norm denoising algorithm can also be called the weighted nuclear norm minimization (WNNM) proposed by Gu et al. [8]. The algorithm exploits the self-similarity of the image to denoising. The algorithm searches for the target block's nonlocal similar blocks across the image. The original image can be estimated from the similar block matrix while reducing the noise. For a noisy image, the model is as follows:

$$y = x + n \tag{1}$$

Where y is the observed noisy image, x is the original image without noise, and n is assumed to be additive Gaussian white noise with zero mean and variance σ^2. The noisy image is divided into several local blocks. The local block finds similar image blocks by block matching from the whole image, and by stacking all the similar blocks into a similar block matrix. So there has:

$$Y_j = X_j + N_j \tag{2}$$

Where $\boldsymbol{X_j}$ and $\boldsymbol{N_j}$ are the block matrices of original similar and noise, and j represents the j-th local block. The image can be denoised by shrinking the singular values of similar block matrices. But different singular values have different importance, so different singular values are assigned different weights in WNNM. By applying the WNNM to $\boldsymbol{Y}_j$ to estimate $\boldsymbol{X}_j$ and aggregating all the denoised blocks, the whole image can be estimated. By using the noise variance σ_n^2 to normalize the F-norm data fidelity term $\|\boldsymbol{Y}_j - \boldsymbol{X}_j\|_F^2$, there have the following energy function:

$$\widehat{X}_j = argmin_{X_j} \frac{1}{\sigma_n^2} \|Y_j - X_j\|_F^2 + \|X_j\|_{\omega,*} \tag{3}$$

Where $\widehat{\boldsymbol{X}}_j$ is an estimated block matrix, and $\|\boldsymbol{X}_j\|_{\omega,*}$ is a weighted nuclear norm matrix, which can be defined as:

$$\|X_j\|_{\omega,*} = \sum\nolimits_i |\omega_i \sigma_i(X_j)|_1 \tag{4}$$

In (4), $\sigma_i(\boldsymbol{X}_j)$ is the i-th singular value of matrix $\boldsymbol{X}_j$. ω_i is a weight assigned to $\sigma_i(\boldsymbol{X}_j)$. There have:

$$\omega_i = c\sqrt{n}/(\sigma_i(X_j) + \varepsilon) \tag{5}$$

Where n is the number of similar blocks, c is a constant and $c > 0$, ε is a constant that avoids the denominator to be zero. Since the singular values of $\boldsymbol{X}_j$ is unknown, so the initial $\sigma_i(\boldsymbol{X}_j)$ can be estimated as:

$$\widehat{\sigma}_i(X_j) = \sqrt{\max(\sigma_i^2(Y_j) - n\sigma_n^2, 0)} \tag{6}$$

Where σ_n^2 is the noise variance, $\sigma_i^2(\boldsymbol{Y}_j)$ is the i-th singular value of $\boldsymbol{Y}_j$. $\widehat{\boldsymbol{X}}_j$ in (3) can be obtained by $\widehat{\boldsymbol{X}}_j = US_\omega(\boldsymbol{\Sigma})V^T$, where $\boldsymbol{\Sigma}$ is a singular value matrix and $S_\omega(\boldsymbol{\Sigma})$ is the generalized soft-thresholding operator with weight vector ω, by using $[\boldsymbol{U}, \boldsymbol{\Sigma}, \boldsymbol{V}] = SVD(\boldsymbol{Yj})$ can be obtained matrices $\boldsymbol{U}$ and $\boldsymbol{V}$, the soft-thresholding operator is as follows:

$$S_\omega(\Sigma)_{ii} = max(\Sigma_{ii} - \omega_i, 0) \tag{7}$$

Where Σ_{ii} is each diagonal element in $\boldsymbol{\Sigma}$. By aggregating all blocks together, the image x can be reconstructed.

The WNNM algorithm exploit the self-similarity between the non-local blocks and the characteristics of singular values to process the image. The algorithm can maintain the local structure of the image. More texture detail information in the image can be preserved. The DWI image have a high degree of self-similarity. Therefore, WNNM is proposed to DWI denoising. This algorithm can save more texture details in DWI as much as possible, and improve the accuracy of the data.

3 Experiment

To quantitatively evaluate the performance of the algorithm, DWI were expressed by the tensor field. The diffusion weighted direction of the experiment was six.

Our real data comes from an ex vivo human brain, they were acquired in http://www.mathworks.com/matlabcentral/fileexchange/21130-dti-and-fiber-trackin. with a size of $128 \times 128 \times 58 \times 7$. The number of layers scanned is 58 layers, each layer has 7 images. Each image has a size of 128×128. The diffusion sensitivity coefficient $b = 1000\ \mathrm{s/mm^2}$.

In the experiment, we first remove the noise in DWI by the algorithms and the DTI data is calculated based on the denoised image. Then the classical Assignment by Continuous Tracking (FACT) algorithm [11] is used to fiber tracking of the DTI data. The FACT is the most commonly used algorithm in nerve fiber tracking. The tracking results are shown in Fig. 1.

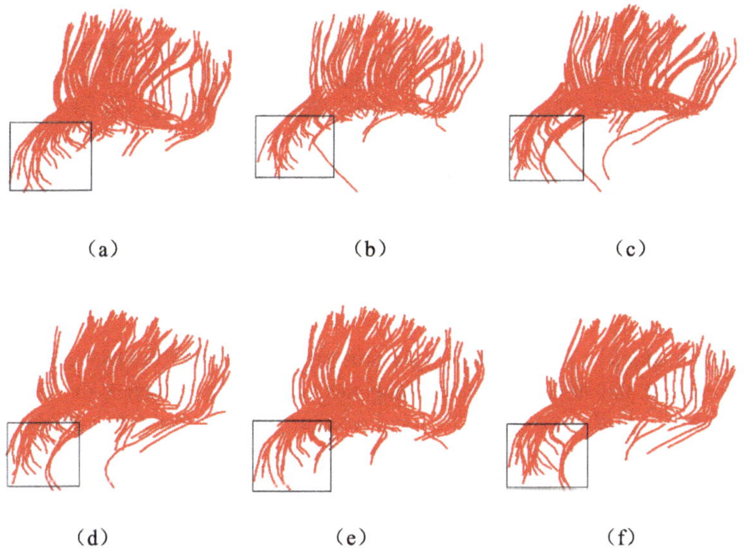

Fig. 1. Nerve fiber maps traced after denoising of DWI images by using various denoising algorithms (a) TV; (b) PM; (c) isotropic; (d) wiener; (e) texture detection; (f) WNNM

Figure 1(a)–(f) are the nerve fiber maps that obtained by fiber tracking. It can be seen from Fig. 1 that the density of fibers in the nerve fiber maps obtained by WNNM is significantly larger, and the fibers are generally longer and smoother.

For real data, it is necessary to judge the effect of nerve fiber tracking. The commonly method to illustrate is based on the number of nerve fibers and the length of the longest fiber. However, because of the influence of noise in DWI, sometimes the longer fibers are mistakenly divided into several shorter fibers. Therefore, judging the effect of denoising based on the number of fibers and the length of the longest fiber are not reliable. We propose the average fiber length to judge the effect. The longer the average fiber length, the smoother the tracked fiber, which can better illustrate the denoising effect. The average fiber length is calculated as follows:

$$ave_{length} = \frac{\sum_{i=1}^{n} length(i)}{n} \quad (8)$$

Where n is the number of fibers and *length* is the length of each fiber.

Table 1. Statistics of nerve fiber tracking

	TV	PM	Isotropic;	Wiener	Texture detection	WNNM
Fiber number	190	205	219	219	209	228
Max length	43	34	52	48	47	43
Average length	20.7	20.5024	21.1096	20.4292	20.4115	**21.3304**

Statistics of nerve fibers obtained after tracking are shown in Table 1.

In Table 1, we can see that the average length of the nerve fibers obtained by WNNM is longer. So the algorithm is more effective. The longest fiber obtained by isotropic algorithm, but according to Fig. 1(d) can be seen, the nerve fiber obtained by the algorithm has the wrong fibers because of the false tracking. Therefore, it is further explained that the longest fiber length is not reliable to judge the effect.

4 Conclusion

The weighted nuclear norm denoising algorithm mainly exploits the self-similarity of DWI to remove the noise. By processing the similar block matrix of the image, the optimal solution is obtained, and the local structure and texture details of the image are preserved well. The experimental results show that the proposed algorithm can not only remove the noise well, but also preserve more texture details in DWI.

Acknowledgements. This research is supported by the National Natural Science Foundation of China (Project No: 11265007).

References

1. Zhang, X., Peng, J., Xu, M., et al.: Denoise diffusion-weighted images using higher-order singular value decomposition. Neuroimage **156**, 128–145 (2017)
2. Liu, M., Vemuri, B.C., Deriche, R.: A robust variational approach for simultaneous smoothing and estimation of DTI. Neuroimage **67**(2), 33–41 (2013)
3. Bao, L., Robini, M., Liu, W., et al.: Structure-adaptive sparse denoising for diffusion-tensor MRI. Med. Image Anal. **17**(4), 442–457 (2013)

4. Becker, S.M.A., Tabelow, K., Voss, H.U., et al.: Position-orientation adaptive smoothing of diffusion weighted magnetic resonance data (POAS). Med. Image Anal. **16**(6), 1142–1155 (2012)
5. Mcgraw, T., Vemuri, B.C., Chen, Y., et al.: DT-MRI denoising and neuronal fiber tracking. Med. Image Anal. **8**(2), 95–111 (2004)
6. Basser, P.J., Mattiello, J., LeBihan, D.: MR diffusion tensor spectroscopy and imaging. Biophys. J. **66**(1), 259 (1994)
7. Stejskal, E.O., Tanner, J.E.: Spin diffusion measurements: spin echoes in the presence of a time-dependent field gradient. J. Chem. Phys. **42**(1), 288–292 (1965)
8. Gu, S., Zhang, L., Zuo, W., et al.: Weighted nuclear norm minimization with application to image denoising. In: Computer Vision and Pattern Recognition, pp. 2862–2869. IEEE (2014)
9. Martin-Fernandez, M., Muñoz-Moreno, E., Cammoun, L., et al.: Sequential anisotropic multichannel Wiener filtering with Rician bias correction applied to 3D regularization of DWI data. Med. Image Anal. **13**(6), 910 (2009)
10. Rafsanjani, H.K., Sedaaghi, M.H., Saryazdi, S.: An adaptive diffusion coefficient selection for image denoising[J]. Digit. Signal Proc. **64**, 75–82 (2017)
11. Kim, K.H., Ronen, I., Formisano, E., et al.: Robust fiber tracking method by vector selection criterion in diffusion tensor images. In: 26th Annual International Conference of the IEEE Engineering in Medicine and Biology Society, IEMBS2004, pp. 1080–1083. IEEE (2014)

A Computational Experiment-Based Study on the Evolution of Third-Party Brand Building Behavior

Duan Qi(✉) and Sun Qiu

China National Institute of Standardization, Beijing, China
duanqi@cnis.gov.cn

Abstract. Based on the learning mechanism of enterprises and the application of the theories and methods of computational experiment, this paper builds a simulation model of the multi-subject brand building behavior with the joint participation of the government, the public and the third-party institutions within the region, and studies the interactive operation between the enterprise and third-party institutions and the dynamic evolution law of brand building behavior of enterprises through the model and computational experiment. The study has two major findings: (1) The implementation of third-party operation in brand building means the introduction of market-oriented mechanism to brand building and the transformation of the direct responsibility of an enterprise into an indirect economic responsibility, which explains the mechanism of existence of third-party brand building institutions and proves the importance and necessity of third-party brand building; (2) The cooperation model in which the enterprise can make a subjective and dynamic decision on the costs of third-party brand building is more in line with the development needs of third-party brand building.

Keywords: Brand building · Third party · Evolutionary game theory

1 Introduction

Developed countries have always attached great importance to the study of brand strategies, and relied on third-party institutions, including well-known universities, consulting and advertising companies and periodical offices, to conduct research on management methods such as brand evaluation and brand cultivation based on the characteristics of development of domestic industries and enterprises and with an emphasis on the factors of stakeholder [1–3] and finance [4] in brand development. They have put forward a large number of quantitative models of brand cultivation, and established a complete set of "nation-industry-enterprise" brand strategies and a policy support system [5–7] on the basis of brand alliance.

With China's increasing attention on brand and the proposal of brand strategy, "promoting building by evaluation" has become the main idea in brand building, and created a trend in the theoretical study of brand evaluation. However, compared with the other major economies in the world, China's understanding of brand is not profound enough in terms of both theory and practice. China emphasizes quality

K. Deng et al. (Eds.): ICMIR 2018, AISC 856, pp. 264–273, 2019.
https://doi.org/10.1007/978-3-030-00214-5_35

improvement [8] on a macro level and popularity promotion [9] on a micro level, and has not yet established its brand cultivation and management system and mechanism which can grasp the overall situation, possess full content, cover all industries, highlight the whole and have progressive levels. The establishment of such system and mechanism must base on a thorough analysis of the evolution of the brand building behavior of enterprises. As the western model of brand building which seeks the assistance of third-party institutions has been introduced to China, in the current system of enterprise development, the main groups of enterprises, local governments and the public all behave different for the optimization of their own goals in different situations. Moreover, the participation of third-party institutions results in even greater complexity [10] of the whole dynamic system due to the delay of decision-making, the non-linearity, the negative feedback loop and the interaction between subjects. Therefore, in the context of the third-party institutions' participation in brand building market, this study adopts a method of computational experiment [11], builds a simulation model of multi-subject brand building behavior to simulate the operation of third-party institutions, and conducts a dynamic analysis of brand building behavior of enterprises. Through the changes of several important parameters, this study analyzes the operational efficiency of third-party institutions and the law of brand building behavior of enterprises and obtains positive implications on management from the law.

2 Model Building

2.1 Fundamental Assumptions and Background Setting

Suppose that there are a number of enterprises of varying industries, products and scales in the market, which all have the need for brand building. Accordingly, there is also the government and the public that have an influence on the enterprises. As the enterprises with the need for evaluation lack expertise in brand building, third-party institutions for evaluation and cultivation emerge in the market. Consequently, the interactive game behavior among the government, the public, enterprises and third-party institutions occurs.

This study adopts a method of computational experiment to simulate the evaluation behavior of enterprises with bounded rationality in different situations and summarize the evolution law of the enterprise behavior through analysis of its changes in each cycle. The fundamental assumptions are as follows: (1) In the regional market, there is one third-party evaluation and cultivation institution and a total of w enterprises with building need. For the convenience of observing the evolution law of the evaluation behavior of enterprises and the third-party institution, suppose that the number of various types of enterprises is fixed with neither increase nor decrease. (2) Despite that the third-party evaluation institution is supported by the government; it is essentially a profit-making institution with characteristics of enterprises, such as running for profit and taking full responsibility for its own profits and losses. (3) The collaborators of the third-party institution are mostly enterprises and the government; therefore, the influence of the public is small enough to be neglected. (4) The enterprises with the need for brand building in the market can be divided into five types, among which G1, G2 and

G3 respectively represent the impatient type, the urgent type and the eager type. All of the three types refer to the enterprises that need to conduct brand diagnosis through evaluation behavior, so as to make the enterprises carry out self-inspection and improve current business situation through brand building or transformation. However, the degree of urgency on the need for brand consulting and building varies among these three types. G4 represents the wait-and-see type, which refers to the enterprises who do not have a strong need for brand building and take a wait-and-see attitude. G5 represents the passive type; enterprises of this type basically have no subjective need for brand building and only conduct brand building when pushed by the government.

2.2 Descriptions of the Subjects in the System

(1) The government
The foremost function of the government is to reward the enterprises which actively conduct brand building. Let A_{it} be the rewards to an enterprise by the government, then

$$A_{it} = \begin{cases} \eta A_0, & A > A_0 \\ 0, & A \leq A_0 \end{cases} \tag{1}$$

In the formula, A_{it} represents the government's rewards to the enterprise i for its brand building in the year t; "$A_0 > 0$" means the enterprise gets the rewards for brand building. It needs to note that participating in the activity of brand evaluation or consulting is just an action of diagnosis of the current situation of the brand's various factors, and that only through constant improvement on the basis of evaluation results can the enterprise gain further building fruits. Therefore, participating in the activity of brand evaluation or consulting is not bound to get rewards from the government, and η is the probability of getting rewards. Finally, the government will publicize the evaluation results.

(2) The public
Public's brand awareness will have an influence on the price of the enterprise's products. Let $p_{i,t}$ be the product price of the enterprise i in the year t, then

$$p_{i,t} = p_{i,t-1}(1 - \varepsilon_{i,t} - BC_{i,t}) \tag{2}$$

In the formula, BC_t represents the public brand awareness of the enterprise i in the year t, and ε_{it} is an adjustment coefficient to adjust the positive or negative value of $\varepsilon_{it}BC_{it}$, that is

$$\varepsilon_{it} = \begin{cases} -1, & BV_{i,t} > BV_{i,t-1} \\ 1, & BV_{i,t} < BV_{i,t-1} \end{cases} \tag{3}$$

In the formula, $BV_{i,t}$ represents the brand value of the enterprise i in the year t.

(3) The enterprises to be evaluated
Since enterprises are profit-oriented economic organizations, it is most appropriate to start with profits in the study of the characteristics of their behaviors. For

manufacturing enterprises, their industrialized production and processing activities can be studied with the Cobb–Douglas production function proposed by C.W. Cobb and Paul H. Douglas which is used to represent the relationship between input and output. The function is as follows:

$$OV_{i,t} = L_{i,t}^{\alpha} K_{i,t}^{\beta} \tag{4}$$

In the function, $OV_{i,t}$ represents the total industrial output of the enterprise i in the year t, $L_{i,t}$ represents the labor input of the enterprise i in the year t, $K_{i,t}$ represents the capital input of the enterprise i in the year t, and α and β are constants.

In order to calculate the profit of the enterprise i in the year t, it is necessary to first figure out the production cost. To simplify the model, the following formula is used for calculation (C_1 and C_2 are constants):

$$C_{i,t} = C_1 K_{i,t} + C_2 L_{i,t} \tag{5}$$

For enterprises which emphasize brand building, it is necessary to invest on brand building every year in terms of capital investment. An enterprise's attitude towards brand building can influence the initial public brand awareness of the enterprise. The more attention an enterprise pays to brand cultivation and protection, the more initial input the enterprise will put into it. The model is as follows:

$$I_{i,1} = DCog_i \tag{6}$$

In the formula, $I_{i,1}$ represents the input in brand building of the enterprise i in the first year, D is a control parameter, and Cog_i represents the degree of attention the enterprise i pays to brand building, that is, the enterprise's awareness of brand building. However, from the second year onward, the enterprise's input in brand building $I_{i,t}(t \geq 2)$ will change as the external environment changes.

The value of brand value $BV_{i,t}$ has relations with the input in brand building $I_{i,t}$, and R is correlation coefficient. The specific expression in the model is as follows:

$$BV_{i,t} = \frac{RI_{i,t}}{1 + RI_{i,t}} \tag{7}$$

Then, it is necessary to calculate the enterprise's $E_{i,t}$ profit:

$$E_{i,t} = OV_{i,t} - C_{i,t} - I_{i,t} \tag{8}$$

(4) The third-party evaluation institution
Owing to the national brand building strategy, the government gives financial allocation S_t to the third-party evaluation institution, so as to promote the theoretical and practical development of brand evaluation, consulting and building of

manufacturing enterprises. The third-party institution's behaviors can be divided into the following three types.

First, its main behavior is to carry out brand consulting and diagnosis for the enterprises to be evaluated by means of contract or agreement. Suppose that the cost for the third-party evaluation institution to evaluate every enterprise is φ.

Second, its another main behavior is to study and develop the technology of brand evaluation, diagnosis and consulting. The necessary R&D input is IT_t; the value of primary input IT_1 is assigned by the system. In every cycle, $IT_t (t \geq 2)$ will change as the external environment changes. Suppose that there is a total of n enterprises to be evaluated in the year t, then the cost for the third-party evaluation institution is:

$$CT_t = n\varphi + IT_t \tag{9}$$

Finally, since the third-party evaluation institution is essentially in the form of enterprise, it is a profit-oriented economic organization. Suppose that the quotation for the third-party institution to evaluate an enterprise is θ, then the profit of the third-party institution is:

$$ET_t = n(\theta - \varphi) + IT_t + S_t \tag{10}$$

2.3 Parameter Setting

Suppose that the number of enterprises with the need for brand building is 300. Table 1 shows the parameters of the system and the primary assignment rule.

3 Design of the Computational Experiment

In order to better study the dynamic changes of the third-party institution's operation and the enterprises' building behaviors, this study constructs four experimental situations according to the sufficiency of the enterprise's input in brand building, the cooperation model between the third-party institution and the enterprise and the fluctuation of transaction amount, as shown in Table 2.

4 Results and Discussion

For the classification of brand building in the fundamental assumptions in Sect. 2.1, the classification standard is as follows. Suppose that the enterprise type is G_μ ($\mu \in \{1, 2, 3, 4, 5\}$), which is determined by ρ (the proportion of input in brand building, that is, the ratio of input in brand building to total industrial output; $\rho = \frac{I_{i,t}}{OV_{i,t}}$).

Table 1. The parameters of the system and the primary assignment rule

Parameter	Range	Meaning and assignment rule
η	0.3	The probability that the enterprise gets the reward. Set after repeated tests and assigned only for the purpose of highlighting the structure (with no other influence)
A_0	50	The probability that the enterprise gets the government reward for brand building. Set after repeated tests and assigned only for the purpose of highlighting the structure (with no other influence)
BC_t	0.01–0.10	Public brand awareness. Determined by individual characteristics and randomly distributed
Cog_i	33–100	The degree of attention the enterprise pays to brand building. Determined by individual characteristics and randomly distributed
D	0.3	The control parameter of primary input in brand building. Set after repeated tests and assigned only for the purpose of highlighting the structure (with no other influence)
$L_{i,t}$	20–1000	The number of employees in the enterprise. The range of middle and small-sized enterprises in the latest partition criterion for enterprises
$K_{i,t}$	300–40000	The operating income of the enterprise. The range of middle and small-sized enterprises in the latest partition criterion for enterprises
$p_{i,t}$	30–50	Product price. Determined by individual characteristics and randomly distributed
α	0.3	The elasticity coefficient of labor output. Estimated value according to previous literature
β	0.3	The elasticity coefficient of capital output. Estimated value according to previous literature
C_1 C_2	1.5, 2	The control parameter of production cost. Set after repeated tests and assigned only for the purpose of highlighting the structure (with no other influence)

4.1 Analysis of the Evolution Law of Enterprise Brand Building Behavior in Different Situations

The cooperation model between the enterprise and the third-party institution varies in different situations, which will result in the difference in the schedule of brand building. Therefore, a simulation experiment is conducted to obtain the changes in the numbers of enterprises of various types in different situations, as shown in Fig. 1.

In Fig. 1(a) and (b) respectively represents the specific conditions of Situation 1 and Situation 2. It can be inferred from Fig. 1(a) that, the final result of the development of enterprise brand building behavior is to make all enterprises in the region become G1 enterprises, from which the advantage of cooperation is shown and the importance and necessity of the existence of third-party institutions can be reflected.

Table 2. The situations of the computational experiment

Situation	Dynamic transaction amount	Static transaction amount
Insufficient input in brand building	Situation 1: The parameters and data change periodically with the historical operating conditions or external environment	Situation 2: The enterprise does not cooperate with the third-party institution, that is, $ET_t = 0$; the parameters and data of the enterprise to be evaluated change periodically with the historical operating conditions or external environment
Sufficient input in brand building	Situation 3: The consulting fee for brand building fluctuates randomly with the enterprise's current situation of brand building. $A_{i,t} = 0$; $\varepsilon_{i,t} = 1$	Situation 4: Owing to the signing of an agreement, the consulting fee for brand building is fixed in all cycles. $A_{i,t} = 0$; $\varepsilon_{i,t} = 1$; $I_{i,t}$ is a constant

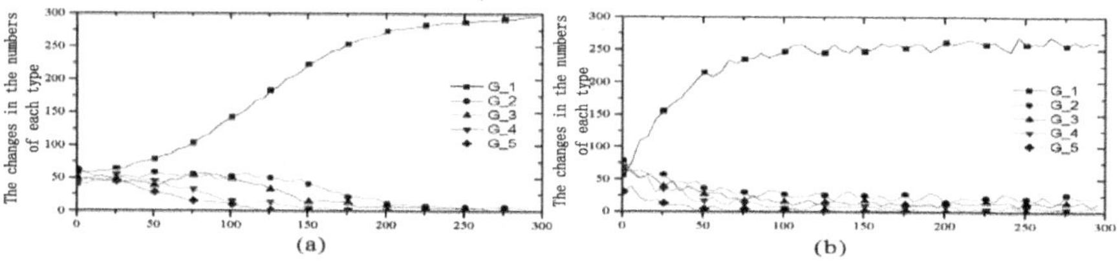

Fig. 1. The changes in the numbers of enterprises of various types in different situations

From Fig. 1(b), it can be seen that in the primary stage of Situation 2 (when the enterprise does not cooperate with the third-party institution), the model assumes that the numbers of various types of enterprises are roughly the same. However, the enterprises in the region will change their types over time according to factors such as government reward and public brand awareness. As for the overall trend, the number of G1 enterprises will gradually increase and the trend of brand building tends to be better, while the status of the types in the region is always in fluctuation. By comparing (a) and (b), it can be discovered that under the same conditions of government reward and public awareness, the schedule of brand building in Situation 1 obviously falls behind that in Situation 2. The reason for such phenomenon is that the enterprise in Situation 1 needs to game and cooperate with the third-party institution. Due to the concern for input in brand building and the desire for price concession of the third-party institution, the enterprise is not willing to cooperate with the third-party institution or rely less on the third-party institution. In Situation 2, however, the enterprise takes full responsibility for their profits and losses and can thus quickly discover advantages in brand building. Therefore, it can make relatively quick decision to carry out brand building.

4.2 Analysis of the Evolution Law of Operating Profits of the Third-Party Institution in Different Interaction Models

This study assumes that there are four types of interaction between the enterprise and the third-party institution, and constructs four situations accordingly. However, as the way of charging varies in different situations, the four situations can be compared by the third-party institution's profits in each situation. In Fig. 2(a), (b), (c) and (d) respectively correspond to Situation 1, 2, 3 and 4.

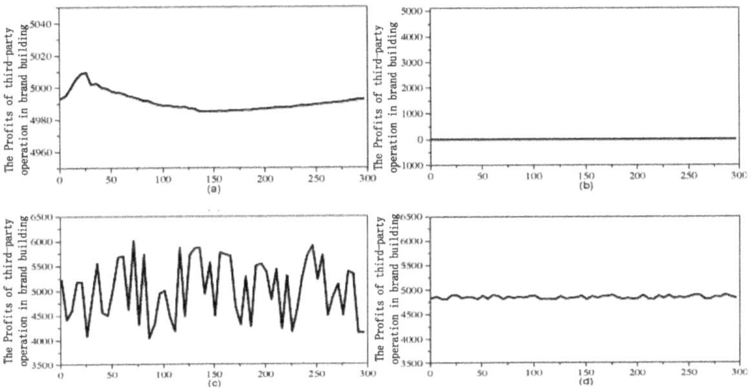

Fig. 2. The influence of the third-party institution's charge for brand building on its profits and market

As shown in Fig. 2(a), in Situation 1, the fluctuation of the third-party institution's profits has certain trend and tendency: after it enters the market, the stimulation to the enterprise by the newly-emerged institution and the government reward can quickly bring about a profit peak for the third-party institution in early operation; as the enterprises realizes that there is a conflict between the input in brand building and the temporary maximization of profits, it will gradually adjust its capital input in brand building; after a period of adaptation, a relatively stable cooperation model between the two sides is formed. In Fig. 2(b), since the enterprise does not cooperate with the third-party institution, the profits of third-party institution is constantly 0. In Situation 3 in Fig. 2(c), the profits of the third-party institution fluctuate with the enterprise's input in brand building in the cycle, and the fluctuation is not regular. In Situation 4 shown in Fig. 2(d), the profits of the third-party institution stay relatively flat, because of the strategic agreement on fixed consulting fee for brand building it signs with the enterprise. The little fluctuation is attributed to the difference in R&D costs in different years. To sum up, it can be concluded that the cooperation model in which the enterprise can make a subjective and dynamic decision on the input in brand building is more in line with the development needs of the third-party institution.

5 Conclusion

The introduction of third-party institutions in brand building is to transform the enterprise's direct responsibility for brand building into an indirect economic responsibility under the market mechanism. In such circumstance, the intermediary institution which is responsible for consulting and cultivation centralizes the capital input, serves as an operating entity, and becomes a pilot program for socialized paid service, management and operation. This study has the following contributions: (1) the diversity, complexity and systematization of the interaction between the third-party institution and the enterprise can be seen from the model analysis and experiment results, and the dynamic analysis should be conducted according to different situations; (2) this study analyzes the schedule of enterprise's brand building and the evolution law of enterprise behavior in different situations, and proves the importance and necessity of third-party institutions by means of dynamic simulation.

For the convenience of research, this study abstracts the attributes of enterprise and third-party institution, simplifies some rules and avoids some complicated factors such as the enterprise's industrial characteristics, the difference in profit cycle and market position and so on, which to some extent reduces the authenticity of the study. Besides, the public in the model only includes consumers; public organizations such as community and NGO are not included. In future research, the model should be improved in this respect, so as to better reflect the situation of the realistic society.

Acknowledgements. This research was financially supported by China Central Public-interest Scientific Institution Basal Research Fund (Grant No. 552018Y-5930).

References

1. Jones, R.: Finding sources of brand value developing a stakeholder model of brand equity. Brand Manag. **13**(1), 10–32 (2005)
2. Washburn, J.H., Plank, R.E.: Measuring brand equity: an evaluation of a consumer-based brand equity scale. J. Market. Theory Pract. **10**, 46–61 (2002)
3. Wangbing, H., Zhengming, H.: Review and prospect of brand equity research based on customer perspective. Bus. Dyn. **7**, 88 (2011)
4. Hutton, P.: The emerging importance of brand energy in the financial services sector. J. Financ. Serv. Market. **9**(4), 307–317 (2005)
5. Chengjun, W.: Characteristics of assessment industry in Canada and the United States and its enlightenment. State Owned Assets Manag. **10**, 52–55 (2001)
6. Xingyuan, W.: Research on Brand-Name Ecosystem Analysis Theory and Management Strategy: Exploration from the Perspective of Ecology, pp. 1–5. Economic Science Press, Beijing (2007)
7. Zhang, Y., Zhang, R., Liu, J.: Research on Brand Ecological Theory and Management Method, p. 38. China Economic Publishing House, Beijing (2013)
8. Quanlu, L., Chunmei, H.: Review on the research of brand equity and its evaluation model. Commer. Times **06**(6), 41–43 (2013)
9. Fanghua, W., Jie, C.: Review on the research of brand-based issues. Chin. J. Manag. **3**(5), 622–628 (2006)

10. Yurong, H., Jiebin, Q.: Research on the efficiency of China's investment. Stat. Decis. **22**, 126–128 (2014)
11. Zhaohan, S., Zhang, J., Jianguo, D.: Theory and Applications of Social Science Computational Experiments. Joint Publishing, Shanghai (2009)

A Study on Analyze Body Characteristics and Predict Human Size of Young Women from Northeastern China

Yao Tong[1], Yao Chong[2], Wang Jun[1], and Pan Li[1](✉)

[1] National Demonstration Center for Experimental Fashion Design and Engineering Education, School of Fashion, Dalian Polytechnic University, Qinggongyuan, Ganjingzi District, Dalian, Liaoning, China
434396675@qq.com

[2] Air Force Harbin Flight Academy, Nantong Street, Nangang District, Harbin, Heilongjiang, China

Abstract. The MTM manufacture mode is becoming more and more widely used in the clothing enterprise. How to obtain efficient and accurate data about the human body becomes the most important step in the MTM mode. In this paper, the body characteristics of young women aged 18–25 years old in Northeast China were studied, which used the non-contact 3D body measurement technology to get the date for 450 young women's body characteristics. The results show that their average height is 162.15 cm, and 15 characteristic values of the other body parts are given. Select 370 groups as training data, 80 groups as test data, and use the linear model and the neural network to predict the body characteristics on the basis of the incomplete data. To be more specific, based on the analysis of human body characteristics and the method of body type classification, select the data of 5 key body parts from 15 human features (height, weight, chest circumference, waist circumference, hip circumference) as input data and 10 parts data as output data, use the regression equation listed by linear regression and neural network algorithm to establish the human body data forecasting model to predict the output value, and then validate and optimize the result of the experiment. Experimental results show that the prediction error to the human body of the prediction mode is within the scope of the permit. The contrast test shows that the method is effective to predict human body data for clothing purpose, and with the increase of the training sample amount, forecasting accuracy is significantly improved.

Keywords: 3D body measurement · Linear regression · Neural networks
Young women of northeastern china · Human size prediction

1 Introduction

For garment industry, to master accurate body size is crucial. In our country, the current measurement on body size still complies with the national standard 1997 calculated twenty years ago. However, in the last twenty years, the living standard has been improved. The intake of nutrients and the change of living environment have affected

K. Deng et al. (Eds.): ICMIR 2018, AISC 856, pp. 274–280, 2019.
https://doi.org/10.1007/978-3-030-00214-5_36

people's physical characteristics to a certain degree. Therefore, human's physical characteristics varies a lot because of their different habits, customs, location, diet and other factors. Thus it is necessary to measure the human body in different areas [1].

There are two common methods to measure human body: The manual method and non-contact 3D body measurement. Non-contact measurement has been widely applied in the clothing and the human body research because of its precise, fast and comprehensive access to human body data information [2].

With the advent of the era of big data, using big data to forecast the unknown are applied in many ways. In industrial manufacturing, Mesut Kumru and Pinar Yildiz Kumru used neural network to predict metal parts manufacturing time, in order to improve the precision of the delivery time [3]. Since the human body is a complex curved surface, the clothing-oriented human body size also has its particularity. Some scholars used the big data to predict the human body from different angles. Zhenzhen huang adopted the method of gray model and fuzzy membership functions to estimate the body size parameters to predict secondary positions from a small size sample. The model saves much measuring time and significantly accelerates the design efficiency and improves the design quality [4]. Kaixuan Liu proposed a BP neural network model to predict all pattern making-related body dimensions by inputting few key human body dimensions [5]. Xiaoning Jing established the multidimensional information system to synthetically predict the shape and size of human body, in order to gain the body type, detailed measurement information and human body model of the sample close to the customer [6].

In this paper, the body characteristics of young women in Northeast China were studied and the characteristic of the body was analyzed. Also, Linear regression and neural network were used to predict human size of young women from northeastern China.

2 Description and Analysis on Body Characteristics of Young Women from Northeastern China

2.1 Sample Selection

The non-contact measurement with the Human Solution 3D body scanner were used to get the body dimensions. 18–25 years old young women were selected from the population in Liaoning, Heilongjiang, Jilin and Neimenggu provinces in the North-eastern China. They were primarily recruited from the colleges, universities, offices, factories and companies. Select 450 data about 15 parts of a human body. Since human body has some gender, age and geographical features, we have to restrict these factors in order to improve the size prediction accuracy.

2.2 Data Analysis

The descriptive statistics of the main 15 body parts including weight, body height, bust girth, waist girth, buttock girth, seventh vertebra, distance waist-knee, waist height, breast height, mid neck girth, under bust circumference, high hip girth, belly circumference, shoulder width, arm length were shown in Table 1. In the list of mean, max, min and standard deviation.

Table 1. The description of main body dimension unit: [cm]

No.	Body part	Mean	Max	Min	Std. deviation
1	Weight	53.8978	79.40	35.60	7.20253
2	Body height	162.1487	178.60	145.70	5.84426
3	Bust girth	85.2883	105.80	71.00	5.40573
4	Waist girth	71.1869	114.20	56.90	7.65046
5	Buttock girth	92.8627	111.10	79.60	4.84612
6	Seventh vertebra	137.0553	152.00	122.80	5.35445
7	Distance waist-knee	57.8758	66.80	51.40	2.64249
8	Waist height	100.1629	113.40	89.00	4.55633
9	Breast height	116.1656	132.70	104.40	5.02524
10	Mid neck girth	32.267	41.50	27.30	1.73887
11	Under bust circumference	74.7478	95.30	63.00	5.47535
12	High hip girth	80.6936	109.00	64.70	6.59316
13	Belly circumference	78.5253	107.50	62.20	6.76451
14	Shoulder width	38.7262	48.00	30.50	2.08212
15	Arm length	56.1224	63.70	49.00	2.78414

3 To Establish the Parametric Prediction Mode of Young Women from Northeastern China

3.1 Parametric Prediction Mode Based on Linear Regression

The basic train of thought using the linear regression method to predict the young women's body size is as follows: Choose the body parts which are easily and accurately to measure as input data, and then obtain the data on secondary human body dimensions by regression equation. The input data on key body parts are weight, body height, bust girth, waist girth, buttock girth, a total of five. The secondary human body dimensions to predict are seventh vertebra, waist-knee distance, waist height, breast height, mid neck girth, under bust circumference, high hip girth, belly circumference, shoulder width, arm length, a total of ten.

Randomly select 370 groups of data for correlation analysis, that is, to analyze the correlation between each input data (including weight, body height, bust girth, waist girth, buttock girth) and each predicted one (including seventh vertebra, waist-knee distance, waist height, breast height, mid neck girth, under bust circumference, high hip girth, belly circumference, shoulder width, and arm length). The results are as follows:

(1) Seventh vertebra, waist-knee distance, waist height and breast height are highly correlated with body height.

(2) Under bust circumference, high hip girth and belly circumference are highly correlated with weight, bust girth, waist girth and buttock girth. Mid neck girth presents significant correlation with weight, bust girth, waist girth, buttock girth.
(3) Arm length presents significant correlation with body height while weakly correlated with weight, bust girth, waist girth, buttock girth.
(4) Shoulder width presents weak correlation with weight, bust girth, waist girth and buttock girth.

According to the results of the correlation analysis, linear regression analysis is applied to 370 training data. The linear regression equation is obtained as follows, setting the body height = h (Tables 2 and 3).

Table 2. The linear regression equation of height direction

(y)	Regression equation
Seventh vertebra	y = 0.912h − 10.813
Waist-knee distance	y = 0.423h − 10.63
Waist height	y = 0.744h − 20.525
Breast height	y = 0.83h − 18.43
Arm length	y = 0.322h + 4.229

Table 3. The linear regression equation of surrounded degree

(y)	Regression equation
Mid neck girth	y = 0.043B + 0.06Z + 0.061W + 0.041H + 17.109
Underbust circumference	y = 0.41B + 0.248Z + 0.199W − 0.063H + 18.045
High hip girth	y = 0.165B + 0.073Z + 0.554W + 0.369H − 10.611
Belly circumference	y = 0.189B + 0.1Z + 0.361W + 0.492H − 14.373
Shoulder width	y = − 0.083B + 0.118Z − 0.003W + 0.081H + 32.148

3.2 Parametric Prediction Mode Base on Neural Networks

BP neural network is a kind of multi-layer feedforward neural network. The main characteristic of the network is to pass before the signal, the error back propagation. There were a total of 450 subjects involved in model building, in which 370 were for training the proposed model, 80 are for validating the model. The general scheme model is described in Fig. 1. Let X = (x1, x2, … xn) be the input item of the predicting model. In this experiment, we set n = 5; there are "weight", "body height", "bust girth", "waist girth", "buttock girth". Let Y = (y1, y2, … ym) be a set of m secondary body dimensions, Y is the output item of the predicting model. We set m = 10; there are "seventh vertebra", "distance waist-knee", "waist height", "breast height", "mid neck girth", "underbust circumference", "high hip girth", "belly circumference", "shoulder width", "arm length". As can be seen from the graph, the BP neural network can be regarded as a nonlinear function, and the input values and the predicted values are the independent variables and dependent variables of the function. Use function relation expression for: ym = f(x1, x2,… xn).

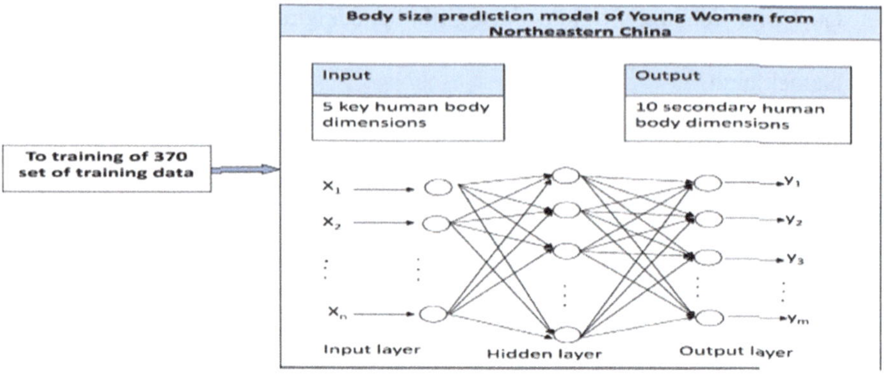

Fig. 1. Body size prediction model based on neural network

3.3 Results and Discussion

This experiment uses the linear regression and BP neural network to structure size prediction model for young women from northeastern China. Take seventh vertebra for example. The difference between predictive value and measured value of seventh vertebra is as follows: the left is based on linear regression while the right is based on BP neural network. As can be seen from the pictures, due to individual differences, the predictive values and measured values are not completely consistent, but the prediction error to the human body of the prediction mode is within the scope of the permit. Although the body data itself has certain instability, such as the body surrounded degree change due to breathing and the body height nuance arising from different standing postures, the two models control the difference within the scope of the permit. Therefore, we can say that the method is effective to predict human body data for clothing purpose (Fig. 2).

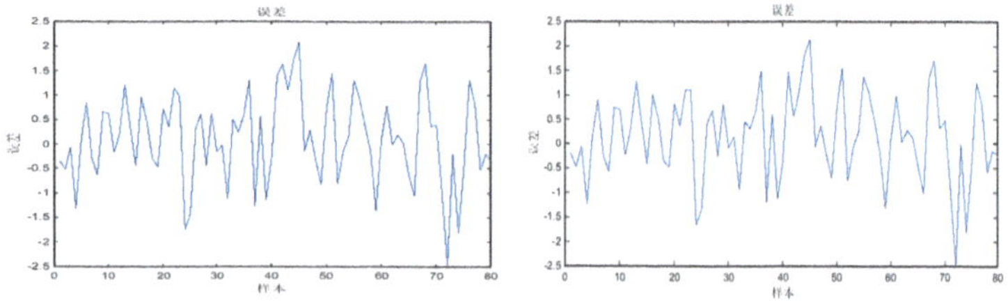

Fig. 2. The difference contrast between predictive value and measured value of seventh vertebra

Substitute the rest parts into the linear equations and neural network prediction model, and then calculate the root mean square error (RMSE) difference between the measured value and predicted value. RMSE refers to the square root of the sum of the squares due to error between the observed value and true value divided by the observation frequency n. It can well reflect the precision of the measurement. Compare the RMSE of linear regression model with neural network model, and the results are as follows (Table 4).

Table 4. RMSE comparison of linear regression model and neural network prediction model

Body part	The RMSE between outputs and targets based on linear regression model	The RMSE between outputs and targets based on neural network prediction model
Seventh vertebra	0.9100	0.9072
Distance waist-knee	0.9300	0.9465
Waist height	1.3992	1.4268
Breast height	1.5308	1.4903
Arm length	2.4857	2.5956
Mid neck girth	1.1443	1.1479
Under bust circumference	2.4302	2.5680
High hip girth	2.7652	1.7977
Belly circumference	1.9378	1.6994
Shoulder width	1.6304	1.6590

4 Conclusion

In this paper, the body characteristics of young women aged 18–25 in Northeast China were studied, which used the non-contact 3D body measurement technology to get the date for 450 young women's body characteristics. It used the linear model and the neural network to establish the body size predict model. After verification, both models are effective.

(1) 15 body size characteristics of Young Women from Northeastern China have been obtained
(2) Both linear regression model and neural network model are effective. Since linear regression model is simple, practical and easy to understand, it can be used to predict human body data for clothing purpose. Through the two models, it can be seen that the 10 secondary human body dimensions have a close linear relationship with the 5 key human body dimensions
(3) Test results show that the mathematical model can be applied to repair the incomplete data on human body

Acknowledgements. This work was financially supported y the fund of Liaoning education department (2016J010).

References

1. Vadood, M., Esfandarani, M.S., Johari, M.S.: Developing a new suit sizing system using neural network. J. Eng. Fibers Fabr. **10**(2), 108–112 (2015)
2. Wang, J.: The body feature study of young women from Northeastern China. Adv. Text. Eng. Mater. **8**, 543–546 (2014)
3. Kumrua, M., Kumrub, P.Y.: Using artificial neural networks to forecast operation times in metal industry. Int. J. Comput. Integr. Manuf. **27**(1), 48–59 (2014)
4. Huang, Z., Wang, X., Zhang, H.: Predicting body parameters from data of small sample size. J. Tianjin Polytech. Univ. **429**(2), 34–38 (2010)
5. Liu, K., Wang, J.: Using artificial intelligence to predict human body dimensions for pattern making. In: FLINS, pp. 996–1001 (2016)
6. Jing, X.: A Study on Multidimensional Information Synthetically Predictive the Somatotype and Size of Human Body System. Tianjin Polytechnic University, Tianjin (2013)

Empirical Analysis of the Relationship of Environmental Quality and Level of Economic Development

Zhang Bin and Jiang Peng(✉)

Institute of Agricultural Information and Economics BAAFS, Beijing 100097, China
jiangpeng0451@126.com

Abstract. This article choose the environmental performance index and per capita GDP as a measure, to study if between the world environmental quality and economic development level exists the relationship, through least squares method acquired the two trend curve, and ultimately concluded that existence of the relationship between environmental quality and level of economic development, the two changes will have an impact on each other.

Keywords: Environmental performance index · Economic development level Least squares method

1 Introduction

Environmental quality is the degree to which the overall environment, or some of its elements, in a given period of time is appropriate to the survival, reproduction and socio-economic development of living beings, especially human beings. The level of economic development is the scale, speed and level of a country's economic development. Gross national product, national income, per capita national income, economic growth and economic growth are the common indicators that reflect the level of economic development in a country. From the perspective of the world, the developed countries have higher level of economic development and better environmental quality, while developing countries have lower level of economic development and worse environmental quality. From a global perspective: Whether there is necessarily a certain connection between the level of economic development and environmental protection is still worth discussing. In this paper, we will quantitatively analyze the changes in the ranking of economic development and environmental quality in 124 countries and regions in the world and finally determine whether there is a necessary relationship between the level of economic development and the quality of the environment.

K. Deng et al. (Eds.): ICMIR 2018, AISC 856, pp. 281–289, 2019.
https://doi.org/10.1007/978-3-030-00214-5_37

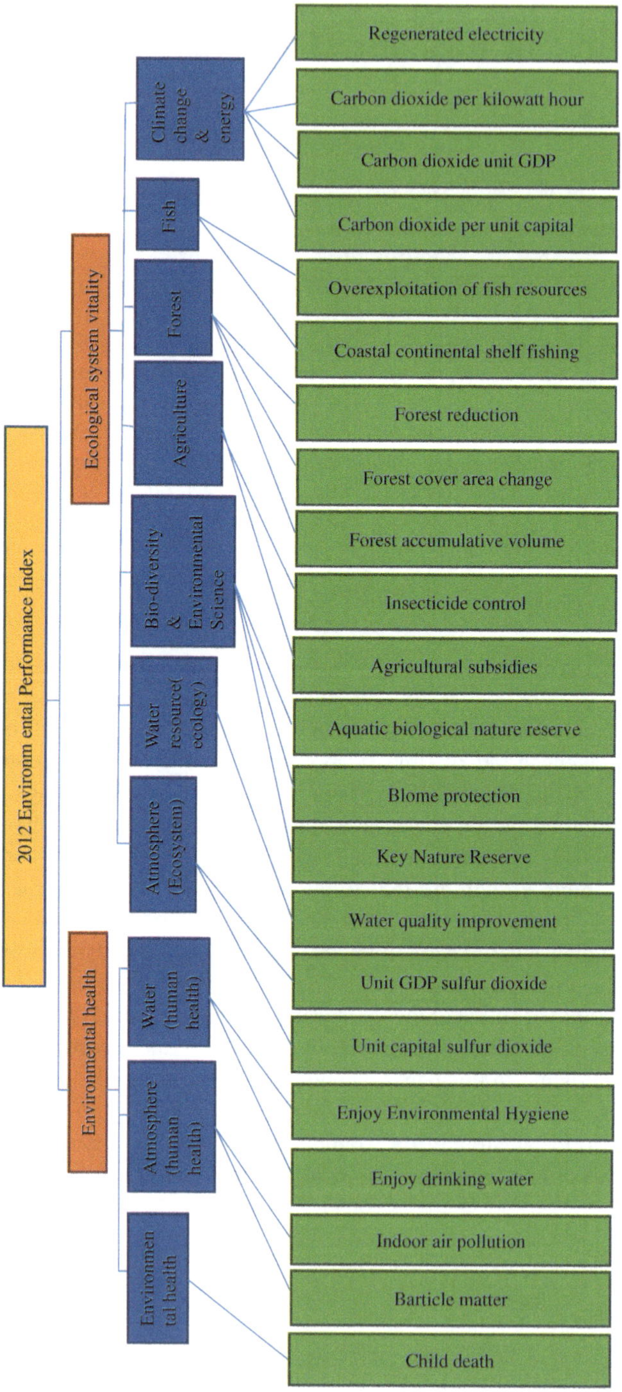
2012 Environm ental Performance Index
Ecological system vitality
Environmental health
Climate change & energy
Fish
Forest
Agriculture
Bio-diversity & Environmental Science
Water resource(ecology)
Atmosphere (Ecosystem)
Water (human health)
Atmosphere (human health)
Environmen tal health
Regenerated electricity
Carbon dioxide per kilowatt hour
Carbon dioxide unit GDP
Carbon dioxide per unit capital
Overexploitation of fish resources
Coastal continental shelf fishing
Forest reduction
Forest cover area change
Forest accumulative volume
Insecticide control
Agricultural subsidies
Aquatic biological nature reserve
Blome protection
Key Nature Reserve
Water quality improvement
Unit GDP sulfur dioxide
Unit capital sulfur dioxide
Enjoy Environmental Hygiene
Enjoy drinking water
Indoor air pollution
Barticle matter
Child death

2 Overview of Environmental Quality and Economic Development

2.1 Environmental Quality Overview

Environment is composed of various elements of natural environment and social environment. Therefore, the quality of environment includes the comprehensive quality of the environment and the quality of various environmental elements, such as the atmospheric quality, the water quality, the soil quality, the biological quality, the urban quality, production quality, cultural quality and so on. The advantages and disadvantages of all kinds of environmental factors are evaluated according to the aforementioned requirements of human beings. Therefore, the quality of the environment is linked with the evaluation of the environmental quality, that is, the determination of the specific quality of the environment requires the assessment of the quality of the environment; and the result of the evaluation indicates the quality of the environment. Therefore, environmental quality assessment is to measure the quality of the environment, environmental quality is the result of environmental quality assessment. In the research, development and utilization of the environment, human beings need to evaluate the environmental quality in order to determine the quality of the environment. There must be a standard for the evaluation so that a system of environmental quality standards is produced. Since the quality of the environment is evaluated on the basis of various human requirements, there are many kinds of environmental quality standards depending on different requirements.

Environmental Performance Index referred to as EPI (ENVIORNMENTAL PERFORMANCE INDEX). The EPI was developed by the Environmental Law and Policy Center of Yale University and the International Geoscience Information Network Center of Columbia University (CIESIN) and based on the "Environmental Sustainability Index" compiled for four years in a row from 2002 to 2005. The EPI was developed by the Center for Environmental Law and Policy at Yale University and CIESIN of Columbia University on the basis of the Environmental Sustainability Index compiled for four consecutive years from 2002 to 2005. At present, 2006EPI, 2008EPI, 2010EPI, 2011EPI and 2012EPI have been launched.

2006EPI contains 16 of the six policy categories and expanded to 25 in 2008EPI. 2010EPI is based on the top ten policy categories based on 25 indicators. The indicator system established by the EPI focuses on environmental sustainability and the current environmental performance of each country and collects data through a series of indicators identified by policy makers and experts as representing core pollution and natural resource management challenges. The categories do not have precise answers, but the indicators of their choice form a comprehensive set of indicators that reflect the current focus of the social environment challenge.

Specifically, EPI aim to help policymakers identify environmental issues, track trends in pollution control and natural resource management, determine which policies produce good results and which ones do not and provide them for cross-national and cross-sector performance Comparison of the benchmark, looking for "similar groups", thematic way to distinguish between advanced and backward, positioning the best example of action and successful policy model. The Environmental Performance Index (EPI) focuses on two basic environmental goals: (1) reducing the environmental stress on human health; and (2) enhancing ecosystem vitality and promoting sound management of natural resources.

2010EPI uses 25 environmental indicators and follows ten policy categories to rank environmental performance in 163 countries. The ten policy established categories are Environmental Burden of Disease, Impact of Water on Humanity, Human Impact of Air Pollution, Impact of Air Pollution on Ecosystems, Impact of Water on Ecosystems, Biodiversity and Habitats, Forests fisheries agriculture and climate change. The 25 indicators are mainly selected from a wide range of environmental policy documents; widely accepted by experts within each policy area; the Millennium Development Goals, the Intergovernmental Panel on Climate Change, Environmental Outlook 4 "and other dialogues reached consensus; environmental agreement about environmental policy debate; the expert's comments and so on. The index defines the policy objectives of environmental management and calculates the gap between the actual performance of each country and the target. As a quantitative tool for pollution control and natural resource management, it provides a powerful tool for policy improvement and provides a more solid analytical basis for environmental decision-making.

In this paper, the environmental performance index (EPI) is selected as a measure of environmental quality, while if the EPI index is high, the environmental quality is good. If the EPI index is small, the environmental quality is poor.

2.2 Overview of the Level of Economic Development

The level of economic development in a country or region can be measured in terms of its size (stock) and speed (increase). "Economies of scale" refers to the total amount of wealth that a country can produce in a given time frame, from basic household items to sophisticated means of production, to the wealth of various cultural and spiritual products. The most commonly used indicator of economic size is gross domestic product (GDP), which comprehensively represents the sum of the wealth (goods and services) produced by a country over a period of time. In addition, the measurement of economic size is divided into absolute size and relative size measurement. The absolute size measures the total amount of GDP a country or region has over a given period of time, regardless of how much of the country's GDP is created by the workforce. Relative size indicators are concerned with the relationship between a country's population (or workforce) and its total GDP. Among the relative size indicators, the most commonly used is the "per capita GDP" indicator. In terms of economic growth, the most commonly used indicator is the "annual growth rate of GDP".

In this paper, the object of study is the quality of the environment, while the environment is the object of human beings, therefore the quality of the environment with people's production and life are inseparable. Therefore, the per capita GDP (per capita GDP) is taken as a measure of the economic development level of a country.

3 Data

3.1 Data Sources

Environmental performance indicators of environmental quality from 2012 EPI rank rankings released by Yale University and Columbia University. The 2012 EPI indicator system is divided into two aspects, a total of 10 categories, 22 indicators.

The data of per capita GDP comes from the calculation of the per capita GDP of various countries and regions in the world in the 2010 World Economic Outlook of the International Monetary Fund (IMF).

3.2 Data Processing

The table above shows the environmental performance indicators of 124 countries and regions, which comprehensively reflects the environmental quality of each country.

The above table shows the per capita GDP of 124 countries. The ranking is based on the EPI values in Table 1 as the standard.

Table 1. Environmental performance indicators of countries in the world

Country	EPI	Country	EPI	Country	EPI	Country	EPI
Switzerland	76.69	Thailand	59.98	Botswana	53.74	Qatar	46.59
Latvia	70.37	Nicaragua	59.23	Côte d'Ivoire	53.55	Kyrgyzstan	46.33
Norway	69.92	Ireland	58.69	Zimbabwe	52.76	Ukraine	46.31
Luxembourg	69.20	Canada	58.41	Myanmar	52.72	Serbia	46.14
Costa Rica	69.03	Nepal	57.97	Ethiopia	52.71	Sudan	46.00
France	69.00	Panama	57.94	Honduras	52.54	Morocco	45.76
Austria	68.92	Gabon	57.91	Dominica	52.44	Russia	45.43
Italy	68.90	Portugal	57.64	Paraguay	52.40	Mongolia	45.37
Sweden	68.82	Philippines	57.40	Indonesia	52.29	Moldova	45.21
United Kingdom	68.82	Korea	57.20	El Salvador	52.08	Turkey	44.80
Germany	66.91	Cyprus	57.15	Guatemala	51.88	Oman	44.00
Slovakia	66.62	Hungary	57.12	Namibia	50.68	Azerbaijan	43.11
Iceland	66.28	Uruguay	57.06	Vietnam	50.64	Cameroon	42.97
New Zealand	66.05	Georgia	56.84	Benin	50.38	Syria	42.75
Netherlands	65.65	Australia	56.61	Peru	50.29	Iran	42.73
Lithuania	65.50	United States of America	56.59	Kenya	49.28	Bangladesh	42.55
Czech Republic	64.79	Argentina	56.48	Mexico	49.11	China	42.24
Finland	64.44	Singapore	56.36	Togo	48.66	Jordan	42.16
Croatia	64.16	Bulgaria	56.28	Algeria	48.56	Haiti	41.15

(*continued*)

Table 1. (*continued*)

Country	EPI	Country	EPI	Country	EPI	Country	EPI
Denmark	63.61	Estonia	56.09	Malta	48.51	Nigeria	40.14
Poland	63.47	Sri Lanka	55.72	Romania	48.34	Pakistan	39.56
Japan	63.36	Venezuela	55.62	Mozambique	47.82	Tajikistan	38.78
Belgium	63.02	Zambia	55.56	Angola	47.57	Eritrea	38.39
Malaysia	62.51	Chile	55.34	Ghana	47.50	India	36.23
Brunei Darussalam	62.49	Cambodia	55.29	Armenia	47.48	Kuwait	35.54
Colombia	62.33	Egypt	55.18	Lebanon	47.35	Yemen	35.49
Slovenia	62.25	Israel	54.64	Congo	47.18	South Africa	34.55
Brazil	60.90	Bolivia	54.57	Trinidad	47.04	Kazakhstan	32.94
Ecuador	60.55	Jamaica	54.36	Macedonia	46.96	Uzbekistan	32.24
Spain	60.31	Tanzania	54.26	Senegal	46.73	Turkmenistan	31.75
Greece	60.04	Belarus	53.88	Tunisia	46.66	Iraq	25.32

The data processing is based on the following idea, and the EPI index and GDP per capita are treated as comparable data with the same extreme value (the maximum value is 124) and the same polarity (the bigger the environmental quality is, the bigger the GDP is). In this paper, we need to process the environmental performance indicators of 124 countries in the above table. Divide the data in the table by the maximum of 124 data and multiply it by 124 to get the environmental index value with the maximum value of 124. In the treatment of per capita GDP, we should not consider the specific per capita GDP of each country, but only rank the per capita GDP of each country according to the rank of per capita GDP examined in Table 1.

4 Quantitative Analysis of the Relationship Between Environmental Quality and Economic Development

4.1 Methodology

The analysis of the relationship between environmental quality and economic development should not adopt the traditional LOGISTIC method, because the traditional Logistic analysis can determine whether the level of economic development is related to the environmental quality, but it is not graphical and trendy. Although it can produce results, we can not draw that what is the law of economic development in the end with the decline of environmental quality, or as the level of economic development to enhance the quality of the environment will produce what changes. Therefore, in this article the traditional LOGISTIC method does not been used.

In addition, according to IMF data, Luxembourg's per capita GDP is US $ 104,511.9, the per capita GDP of the Congo is only 171.5 U.S. dollars, a difference of 600 times. In this case, if the traditional method of using GDP as a comparative factor is still used compared with the quality of the environment, because extreme values are too large, even a 100% change in the minimum will not have a significant effect on the maximum value, resulting in a general relationship not showing a specific trend. Therefore, in this paper, GDP does not been used, according to the size of each country's GDP, we ranked them from the first place to the 124 and take it as a factor of economic development. This method can not only avoid the impact of terminal value, but also reflect the trend. The purpose of this study is not to get an exact figure, instead,

Table 2. GDP ranking of countries in the world

Country	Per capita GDP	Country	Per capita GDP	Country	Per capita GDP	Country	Per capita GDP
Switzerland	67559.57	Thailand	3939.63	Botswana	6406.85	Qatar	68871.71
Latvia	11607.17	Nicaragua	971.81	Côte d'Ivoire	1052.01	Kyrgyzstan	1304.25
Norway	79085	Ireland	51356.28	Zimbabwe	374.78	Ukraine	2542.11
Luxembourg	104511.9	Canada	39668.62	Myanmar	459.37	Serbia	5808.88
Costa Rica	6345.18	Nepal	451.94	Ethiopia	390.27	Sudan	1397.78
France	42747.22	Panama	7132.72	Honduras	1822.68	Morocco	2864.54
Austria	45989.22	Gabon	7468.3	Dominican Republic	5176.01	Russia	8693.8
Italy	35435.15	Portugal	21407.62	Paraguay	2336.89	Mongolia	1560.18
Sweden	43986.18	Philippines	1745.6	Indonesia	2329.45	Moldova	1514.48
United Kingdom	35334.32	Korea	17074.33	El Salvador	3623.27	Turkey	8723.41
Germany	40874.64	Cyprus	29619.5	Guatemala	2661.94	Oman	18013
Slovakia	16281.59	Hungary	12926.5	Namibia	4542.95	Azerbaijan	4807.12
Iceland	37976.65	Uruguay	9425.53	Vietnam	1059.94	Cameroon	1115.28
New Zealand	27259.12	Georgia	2448.37	Benin	711.25	Syria	2578.82
Netherlands	48222.92	Australia	45586.53	Peru	4356.03	Iran	4459.68
Lithuania	11171.87	United States	46380.91	Kenya	911.95	Bangladesh	573.79
Czech Republic	18557.14	Argentina	7725.66	Mexico	8134.75	China	3677.86
Finland	44491.54	Singapore	37293.42	Togo	421.84	Jordan	3828.61
Croatia	14242.99	Bulgaria	6223.4	Algeria	4026.89	Haiti	733.18
Denmark	56115.35	Estonia	14266.54	Malta	19111.16	Nigeria	1141.91
Poland	11287.92	Sri Lanka	2041.41	Romania	7542.47	Pakistan	1016.73
Japan	39731.04	Venezuela	11789	Mozambique	464.54	Tajikistan	766.82
Belgium	43533.25	Zambia	1086.05	Angola	3971.59	Eritrea	362.9
Malaysia	6896.74	Chile	9525.37	Ghana	671.33	India	1030.79
Brunei	26325.48	Cambodia	774.95	Armenia	2667.56	Kuwait	31482.03
Colombia	5087.43	Egypt	2450.39	Lebanon	8706.7	Yemen	1060.96
Slovenia	24417.46	Israel	26796.71	Congo	171.5	South Africa	5823.58
Brazil	8220.36	Bolivia	1723.51	Trinidad and Tobago	15580.79	Kazakhstan	7019.04
Ecuador	4059.26	Jamaica	4390.33	Macedonia	4482.02	Uzbekistan	1175.71
Spain	31946.3	Tanzania	550.54	Senegal	993.65	Turkmenistan	3242.08
Greece	29634.92	Belarus	5165.8	Tunisia	3851.57	Iraq	2107.91

it reflects the relationship between environmental quality and the level of economic development in general.

4.2 Calculation

This paper uses the MATLAB r2011a, first input the environmental performance index (Table 1) and the gross domestic product (Table 2) rank, then calculate environmental Performance Index, finally calculate the environmental performance trend curve and GDP ranking trend curve the least squares method.

4.3 Results

The above figure shows a comparison of the trend of the environmental performance Index and the per capita GDP under the least squares method, and the blue curve above is the change of environmental performance index, and the lower red curve is the per capita GDP change chart. It can be seen from the figure that with the decline of environmental quality, the level of economic development also shows a downward trend, indicating that there is a clear relationship between environmental quality and the level of economic development.

As shown in the above picture, as the level of economic development, environmental quality changes, the above blue curve for the environmental performance Index, the red curve below for GDP per capita, from this perspective we can also see that with the level of economic development decline in the quality of the environment showed a downward trend. However, due to the environmental quality can not be unlimited decline in the end, it shows a stable situation.

In conclusion, we can see from Figs. 1 and 2 that there is a relationship between the level of economic development and environmental quality in the worldwide.

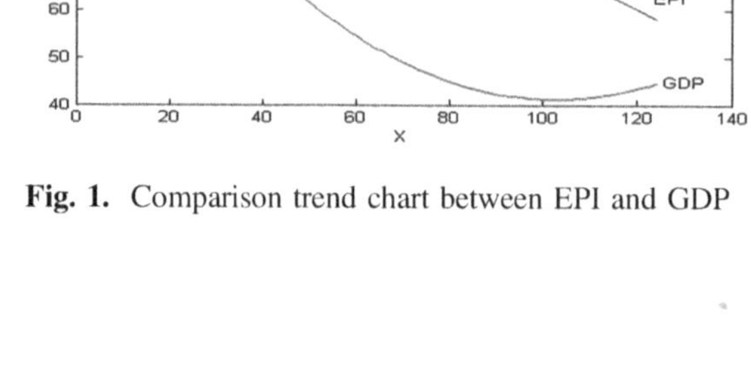

Fig. 1. Comparison trend chart between EPI and GDP

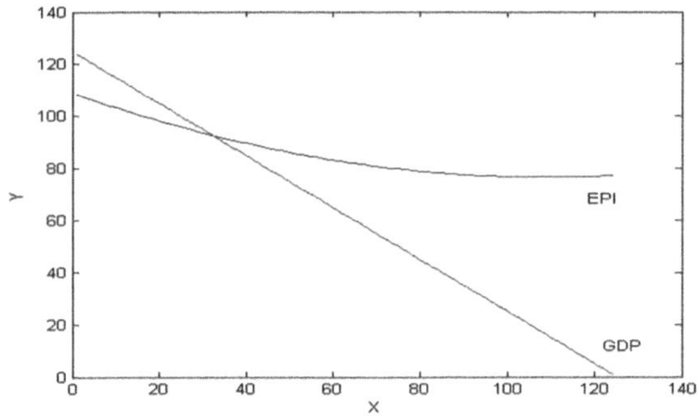

Fig. 2. Comparison trend chart between EPI and GDP

Fund Project: Funds for reform and development of public welfare institutions.

Research of the Redundancy Control Technique for Servo System

Xuan Wu(✉), Jia Wei Zheng, Hong Yu Liu, and Mei Hong Huang

Shanghai Institute of Spaceflight Control Technology, Shanghai, China
wuxuanhome@hotmail.com

Abstract. Servo system is an important part of the rocket control system. The control computer sends the instruction to servo system by 1553B bus, then operator drive the load move according to the order. This paper redesign the redundancy control technique of the servo system,improve the reliability during the occasion that one of the feedback has failure. The design is considered to be feasible with experimental verification.

Keywords: Redundancy · Reliability · Dynamic performance

1 Introduction

In most hydraulic servo system currently in service, fault absorption method has been used as the redundancy technology, the servo system consists of three redundant servo valve and three redundant feedback. However the method will not cause system failure but will cause system performance degradation.

This paper proposed a new redundancy control technique to improve the performance of the system for several types of failures without sacrificing system reliability. By taking three redundant votes of the feedback, this method can effectively improve performance of the system during the occasion that one of the feedback has failure, with the minimum hardware and software changes.

2 The Principle of the Servo Controller

The servo controller consists of a control electric circuit(RT remote controller), its bus/control module take up the address 20 to communicate with the bus controller, then calculates based on the date of the instruction and feedback. The collected analog signals has been converted, amplified by the analog convert module.

By 1553B bus, the control computer sends the yawing instruction to servo controller, inside which the CPU reads the value of the instruction and carries out the control algorithm with it and the displacement feedback voltage value converted from A/D chip, and then the control value will be obtained, which is converted to drive current by D/A chip and V/I circuit and outputted to the servo mechanism. And the CPU carries out one servo control operation per 1 ms, and refreshes the displacement

K. Deng et al. (Eds.): ICMIR 2018, AISC 856, pp. 290–296, 2019.
https://doi.org/10.1007/978-3-030-00214-5_38

feedback voltage value and the drive current once per 1 ms, as well as obtains an updated value of yawing instruction per 20 ms.

The servo controller carries out 3 times operations of three of the displacement feedback voltage values and the yawing instructions, and the 3 outputted drive currents are imputed to 3 motors of the servo valve, and the working performance of the servo system will be affected when one of the feedback is interrupted.

See Fig. 1 for the work flow diagram of the servo controller without the voting function.

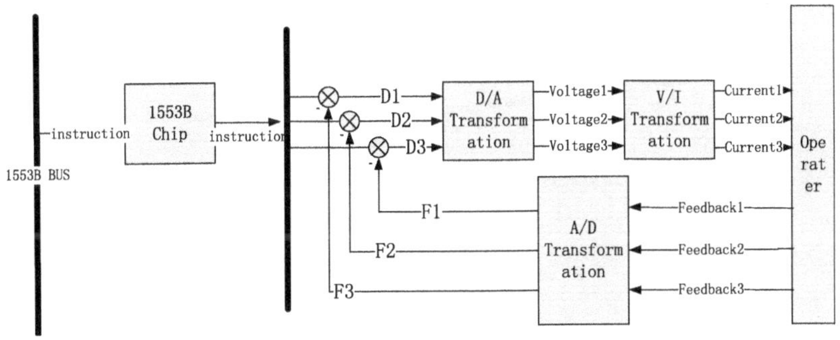

Fig. 1. Flow diagram of the servo controller without the voting function

It can be seen that fault absorption method is used in the redundancy Control technique, and if one of the displacement is interrupted, the left two can still work and the system will not stop working completely. But the system performance definitely decreases for one torque motor of the servo valve fails continuously.

2.1 Test Results of no Voting Status

The dynamic performance of the servo system is tested with the real engine load with a no voting controller (Table 1 and Fig. 2).

Table 1. Test results of no voting status

Controller state	One feedback failure state A		No failure state B	
Frequency/Hz	Amplitude/dB	Phase/°	Amplitude/dB	Phase/°
0.318	−0.027	−3.611	−0.032	−2.058
2.387	−0.013	−33.821	−0.088	−28.855
6.366	−0.346	/	−0.679	/
17.507	−22.716	/	−22.911	/
20.69	−22.5	/	−24.628	/

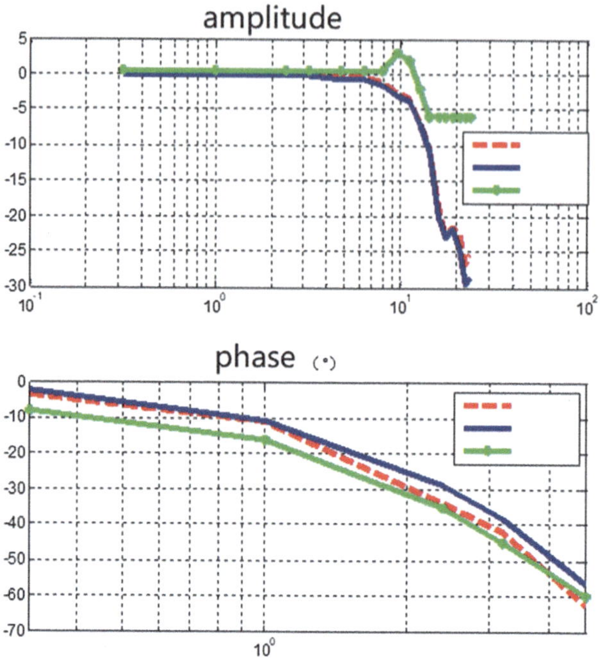

Fig. 2. Bode diagram of no voting status

2.2 Conclusion

From the comparison of the data above, it can be seen that both the values of amplitude and phase of the controller with no voting status have reached the standard; when one of the displacement is interrupted, the value of amplitude reaches the standard and similar to that of the normal condition, and the phase falls completely behind the normal state. The phase exceeds the system index at the point of 4.775, which indicates a reduction occurring in the rapidity of system response to controller with no voting status when one of the displacement is interrupted.

3 The Redundancy Voting Algorithm

3.1 Design of the Redundancy Voting Algorithm

Modify the controller software, then compare the difference value of two from the three values of the displacement feedback voltage and judge whether there is something wrong in the line displacement. If there is something wrong, the average value output of two values of the right displacement feedback voltage will be adopted to the control algorithm, and the outputted three values of drive current in the end are imputed to the three motors of the servo valve.

The main idea of the algorithm involves: two values will be remained from the three values of the feedback voltage, which means that the average value of the remained two values is taken as the final feedback value for the three ones after comparing them and removing the one with the biggest error (Fig. 3).

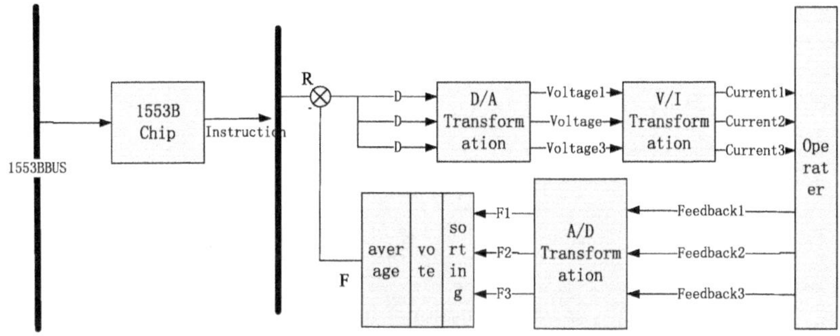

Fig. 3. Flow diagram of the servo controller with the voting function

The figure above is the block diagram of the redundancy algorithm. And F1, F2, and F3 inside refer to the three displacement feedback signals of the servo system; F refers to the displacement feedback result of the redundancy algorithm; D refers to the feedback error of the instruction and the displacement.

First of all, order the feedback 1, 2 and 3, suppose:

$$F1 \geq F2 \geq F3 \tag{1}$$

$$E_{12} = |F1 - F2|, \quad E_{23} = |F2 - F3| \tag{2}$$

And set the average value of the voting dimensional output as the outputted result of the redundancy algorithm. Namely

$$F1 \geq F2 \geq F3 \tag{3}$$

$$E_{12} = |F1 - F2|, \quad E_{23} = |F2 - F3| \tag{4}$$

When E12 > E23, the result of two from three is F2 and F3, so F1 is removed, and the displacement feedback result is F = (F2 + F3)/2;

When E12 < E2, the result of two from three is F1 and F2, so F3 is removed, and the displacement feedback result is F = (F1 + F2)/2;

When E12 = E13, the displacement feedback result is F2;

All the three outputs will be calculated with feedback results after [2] the voting, and the calculated value is only one, which will effectively enhance the system reliability of the one with wrong state feedback, and few changes in the software. Accordingly, a reprogramming is conducted on the control software of the controller for this improvement.

3.2 Test Verification and Analysis

In order to verify the validity and high-reliability of the redundant voting system, the system experiment of its dynamic characteristics is made in normal or one wrong state [3]. It will be conducted by the two kinds of controllers with voting and no voting function on the basis of controlling the load of the simulated system engine. The algorithm used in the two controllers is same in the experiment and only the feedback voting values are changed. The algorithm equation is as follows:

$$\begin{aligned} U_c(k) &= K_1 \cdot e(k) + K_2 \cdot e(k-1) + K_3 \cdot e(k-2) + K_4 \cdot e(k-3) \\ &+ K_5 \cdot e(k-4) + K_6 \cdot U_c(k-1) + K_7 \cdot U_c(k-2) \\ &+ K_8 \cdot U_c(k-3) + K_9 \cdot U_c(k-3) \end{aligned} \quad (5)$$

$$T = 0.001, \quad \text{sampling period.}$$

$$\text{e} = (\text{instruction - feedback}).$$

$$\text{Uc}: \ \text{D/A output.}$$

The experiment of dynamic characteristics with voting state is conducted on the real engine, and the results are as follows (Table 2 and Fig. 4):

Table 2. Test results of no voting status

Controller state	One feedback failure state A		No failure state B	
Frequency/Hz	Amplitude/dB	Phase/°	Frequency/Hz	Amplitude/dB
0.318	−0.098	−2.293	−0.091	−1.408
2.387	−0.115	−27.749	−0.181	−28.673
6.366	−0.645		−0.624	
17.507	−23.218		−23.428	
20.69	−25.084		−24.947	

3.3 Conclusion of the Test

Value of amplitude: it can be obtained from the comparison of the data above that under the condition of one wrong state, when the frequency characteristic amplitude value of the servo system with no redundancy voting are compared with each other, the value of amplitude with no voting is slightly higher during the frequency band of 2–8 Hz, and no resonance peaks of 0 occur at the two states, which indicates a reasonable design of the control rate of the system control algorithm.

Phase: both the yawing displacement and the line displacement with no voting state fall behind at the whole frequency band; while no falls are witnessed in the dynamic

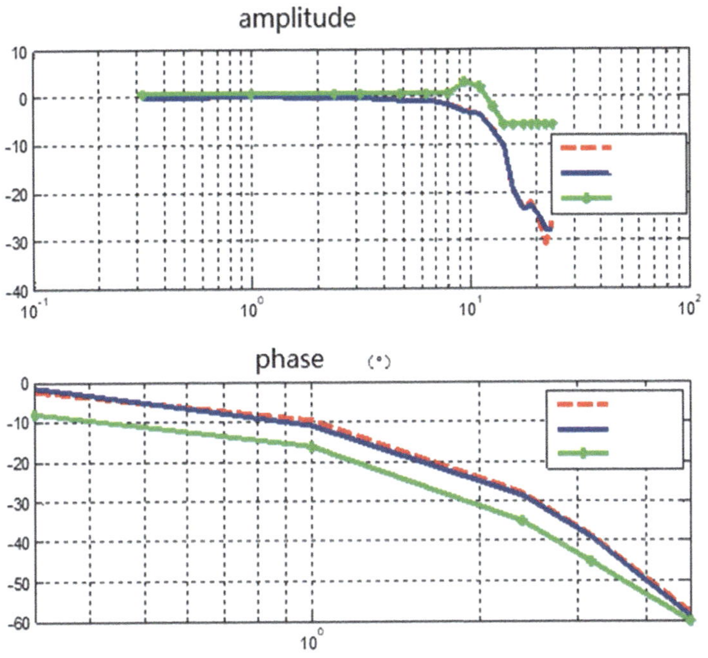

Fig. 4. Bode diagram of voting status

characteristics of the servo system with the redundancy voting function in all the three experiments, which indicates the redundancy voting function can effectively eliminate the performance reduction caused by the fault of one of the displacement feedback.

Under the condition of adding the redundancy voting function, both the amplitude and phase reach the standard at the normal and one wrong states, and their situations are pretty identical, which indicates that the redundancy voting function can effectively eliminate the performance reduction caused by the fault of one of the displacement feedback to make the system to meet the target requirements.

4 Conclusions

The adopted method of digital redundancy voting system can effectively eliminate the performance reduction caused by the fault of one of the displacement feedback to make the system to meet the target requirements. The experimental data can meet the improvement target, and the dynamic performance index under one wrong state of the servo system with voting function has been improved obviously and the target of a good system performance under one wrong state has been achieved, which indicates the feasibility of the redundancy voting system.

Acknowledgments. In writing this paper, I have benefited from the presence of my teachers and my colleagues. I hereby extend my thanks to them for their kind help.

References

1. Hui, G., Shun, C.: The exploration of the development trend of electro-hydraulic proportional technology. Mach. Tool Hydraul. (2005). 102 pages
2. Shousong, H.: Principle of Automatic Control, 4th Version. Science Press (2004). 253 pages
3. Zhou, P., He, Z.: MATLAB Numerical Analysis. Beijing: Machinery Industry Press (2009). 163 pages

Teaching and Learning in Design of Deduced Tree for Extensive Innovative Software

Bifeng Guo, Rui Fan(✉), Sheng Gao, Lihua Cai, and Chuan Yue

Faculty of Software Technology, Guangdong Ocean University, Zhanjiang, China
fanrui@gdou.edu.cn

Abstract. With the development of era, innovative thinking crowds into human's minds. Thus, innovative thinking becomes increasing essential in working field. In order to explore technology of artificial intelligence, we developed a team studying the law of human's minds. In combination with Extenics theory and mind mapping, we find a new way for innovative thinking that thoughts are printed into a mind mapping with different shapes and colors. Besides, we also analyze some way of innovation and design a software model to verify our way for innovation and study further.

Keywords: Deduced tree · Teaching and learning · Innovative software

1 Introduction

Students usually repeat the same homework and experiment in common courses. Most of them are lack of innovative thinking according to their tasks. It is vital to take some measures to improve the innovative abilities of the students. We have some innovation courses for them. In this way, on the one hand, it helps us collect new ideas for studying and explore new ways for innovative thinking. Moreover, we can explore the law of innovative thinking and make the problems solved by robots someday automatically. Namely, it is vital to achieve artificial intelligence of innovative thinking.

1.1 Teaching Methods of Extenics and Innovation

To foster the innovative abilities of students, we teach them a new Chinese principle called Extenics [1]. We arouse their interests by telling them an old Chinese historical story which is relevant to Extensive innovation thinking. Then, we teach them the whole procedure of solving contradictory problems in Extenics in combination with classical stories. After they have learned about the Extensive innovative methods, they are asked to find out the contradictory problems around them and manage to address the issues by the innovative way. Firstly, they should print the contradictory problems into a paper as a form of mind mapping. Then each of them is asked to submit the solutions of the problem in the innovative teaching courses.

By using Extensive innovative deduced tree for innovative work in courses, students extend their minds much. Compared with common innovative courses, they put up with more effective ways for innovative thinking. In contrast, students who use

K. Deng et al. (Eds.): ICMIR 2018, AISC 856, pp. 297–304, 2019.
https://doi.org/10.1007/978-3-030-00214-5_39

Extensive innovative deduced tree spend less time deducing contradictory problems. Consequently, it is confirmed in courses that Extensive innovative deduced tree is a better way for innovative thinking and we would like to set up a team with both professors and students for further research.

1.2 Introduction of Developed Team

From the innovative courses we know that some of the students are interested in Extenics theory or innovative thinking. We set up an innovation team and provide an environment for them to have innovative thinking. To further study, they learn some solutions of contradictory problems deduced by mathematical ways [2]. Up to now, they have done some work for innovative methods including developing innovative software. In theory, they have developed a software modeling of Extenics by Intelligent Service Components [3]. To simplify the process of calculating, they developed Extenics innovation software in Android operating system [4], which helps us save much time to deduce a problem. However, it can't clearly reflect the procedures of people's solving the problem. Self-adaptive software is developed to explore innovative thinking in a new way [5]. Later on, a new way is found to express the whole procedure of solving the problem and that is mind mapping [6]. Then they try to combine mind mapping with Extensive mathematical ways according to human's direction of thinking [7]. Besides, we also analyze other ways of innovative thinking [8]. According to the way they improved, more contradictory problems are addresses by this method. They solved a marketing problem with Extensive mind mapping [9]. They also solved a contradictory problem of hotel food management [10]. They are now developing web Extensive software offering more mechanisms for innovative thinking.

2 Insights of Extensive Innovative Software

2.1 Cognition Process of Innovative Thinking

After the team built, our cognition of innovative thinking has changed for times. At the beginning, we consider that people mind should be guild according to algorithm and formula. We try for times and found the problem that it is difficult to continue doing the same innovative thinking problem. Like programming codes, usually we forget what we have coded after a period of time. Thus, we think the thoughts should be recorded in some ways. From them on, we record the key points of innovative thinking in a paper every time we have an innovative thought. In this way, sometimes we fall in a mess and ignore some important details. By consulting other innovative ways, we find a better way to record what we are thinking which is called mind mapping. Mind mapping provides a record way in which the whole innovative thinking procedure can be recorded. However, some data, formulas and relationships are hard to describe in the mind mapping. To solve the problem, we decide to design an Ex-mind mapping tree to make it work better.

2.2 Reasons for Developing Deduced Tree

Extensible mind mapping means Extenics in combination with mind mapping. Namely, we solve the contradictory problem in a mind mapping where the thoughts of Extenics are printed. Today, innovative thinking of human races promotes the development of society and technology. Generally, an innovative thinking is created accidentally [11]. People often create the thinking in combination with both others' thinking and their own thinking. Once people don't carve the thought and it will fade out in the mind.

Mind mapping is beneficial to memorizing the innovative thinking much [12]. First of all, mind mapping shows the whole procedure of one thing. However, people might stop to do another thing while during the innovative thinking. Carving the key procedure to the mind mapping reminds us of the procedure and innovative thinking that is lost. Secondly, Mind mapping makes us concentrate on key points. With mind mapping, key words are shown in a picture making us think about the mentioned matters instead of the uselessness [13]. Thirdly, mind mapping triggers our right brain. Human's brain is made up of left brain and right brain. Left brain is used to deal with logic vocabulary and digit while right brain is used to deal with abstract thinking creativity and imagination. There are some kind of shapes, key words with colors and icons in the mind mapping so that the whole brain of people can be active, which is good for performing innovative thinking.

However, there are some disadvantages of mind mapping. Firstly, it can't deal with the problem with much information. Mind mapping extended as a form of tree and if there are some relationships among each leaf or node, it is hard to carve the relationships in the mind mapping, which is more complex than the normal ways. Secondly, learning to draw a mind mapping also need some time [14]. As for the simple problem, people might spend less time solving the problem in a normal way. On the other hand, people have to spend time in learning to draw a correct mind mapping or the drawn mind mapping is just in person and others are difficult to read the picture. Thirdly, mind mapping doesn't fit the problem needing thinking systematically. For example, when performing computer programming, mind mapping is hard to describe many conditions of logic. At the time, other way of describing the mind matters. Flow chart would fit programming rather than mind mapping.

Extenics in combination with mind mapping not only solves some problems of Extenics and mind mapping. Extenics provides with us mathematical formulas and ways to think about the contradictory problem. Extenics provides a thinking of decomposition problems and key point selection, which is vital to solve contradictory problems. Nevertheless, it doesn't state how to lead human's mind into a right direction to think about the contradictory problem, causing people might spend much time thinking about the uselessness. Besides, deducing a contradictory problem might be a long-time-work. In general, people spend some times solving the same problem. Once the innovative thinking stops, it is hard to pick up for the next time. In Extenics, some relationships might be a mess, such as the relationship between original problem and each kernel problem and relationship between comprehensive dependent function and each dependent function. The advantages of using it are shown as follows.

1. Extensible mind mapping makes those who learn Extenics know quickly to understand the ways and procedure of solving the problem because all of the Extensible mind mappings are similar. For instance, an original problem is divided into some sub-problems and a sub-problem can be confirmed a kernel problem.
2. It can deal with the information with much relationship. From the model shown in Fig. 1 we can see that the relationship between original problem and kernel problem is below the kernel problem. As for the relationship among goal base elements, conditional base elements and dependent function, we can deal with the problem by their carrying a form in which the relationships are shown clearly.

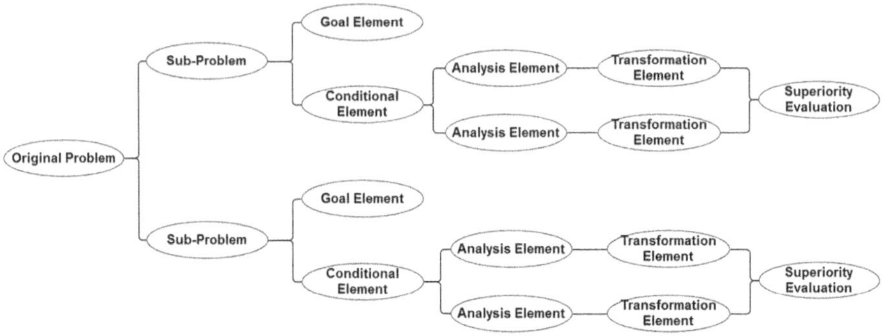

Fig. 1. Extenics innovative thinking in mind mapping

3. It triggers human's right brain while performing an innovative thinking. Most contents of Extenics are digits and logic that runs in human's left brain. It is a better way to use both left brain and right brain to improve the efficiency. Mind mapping offers us some shapes and color, triggering right brains of human.
4. It is easy to record fleeting innovative thinking. While deducing a difficulty contradictory problem, people might find it difficult to solve and stop to do other things. However, people might think it up at any time which is easy to fade out. Carving fleeting thought into the mind mapping might be a good way to record this innovative thought.

Extensible mind mapping doesn't solve all problems existing in Extenics. Extensible mind mapping provides a better environment for people and lead them to a right direction to solve the contradictory problem, which doesn't mean that the contradictory problem is bound to be solved. Besides, it doesn't solve some definitions of some contradictory problems such us whether the ladder can be put into a room. But these can be solved as the improvement of Extenics goes. Consequently, it is essential to use the method of Extensible mind mapping to solve contradictory problem.

3 Design and Application of Deduced Tree

In the mind mapping diagram (see Fig. 1), the whole procedure of innovative thinking with Extenics is shown. However, some relationships among base elements and problems can't be shown clearly.

To describe the relationships among element and dependent functions, we add an editable form to each base element and dependent functions (see Fig. 2).

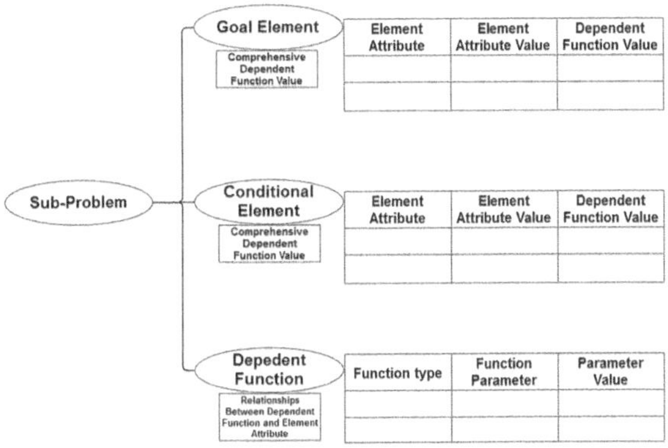

Fig. 2. Improved extensive mind mapping

From Fig. 2 we can see a sub-problem consists of goal base element, conditional base element and dependent function. For goal base element and conditional base element, its names of element attributes, values of each element attributes and value of dependent function. Each set of them is in the same line in correspondence. For dependent function, each element attribute should match an independent function. Each dependent function includes function type, function parameter and parameter value. After all element attributes and attribute value confirmed, the dependent function values are calculated by computer shown in the corresponding form. The comprehensive dependent function will be shown below itself. To make the relationships between each dependent functions and element attributes, an interface is design shown in Fig. 3.

From the figure above, we can see the comprehensive function can be set in the left interface and the compatibility can be set according to the demand shown in the right interface. After confirming, the relationships can be shown in the main deduced tree interface.

To express the procedure of Extenics analysis, another form of base elements are shown in Fig. 4.

From Fig. 4 we can see a conditional element is transformed into several analysis elements. The corresponding functions of each element attribute can be conveniently selected so that the procedure of Extenics analysis becomes more portable.

As for Extenics transformation and superiority evaluation, we design an anti-bracket for the selected element shown in Fig. 5.

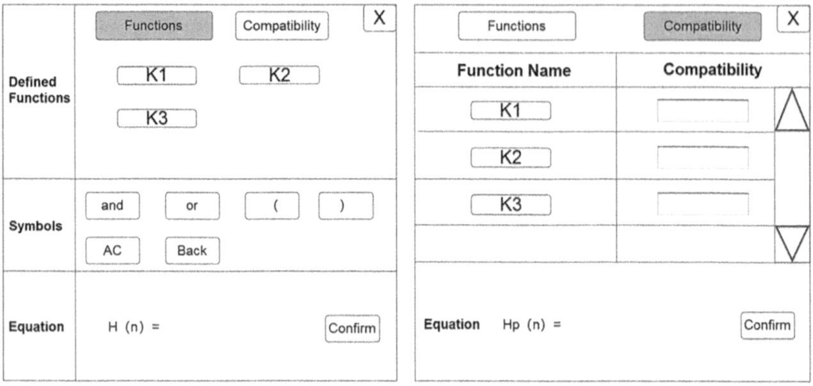

Fig. 3. Functions and compatibility confirm interface

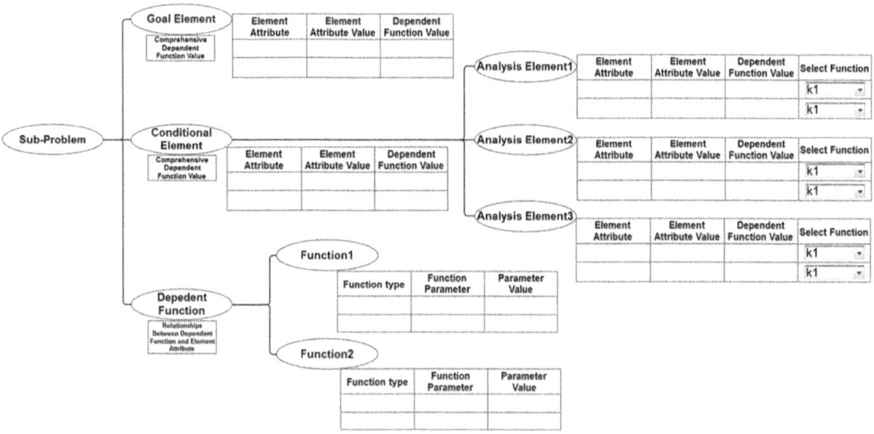

Fig. 4. Procedure of extenics analysis

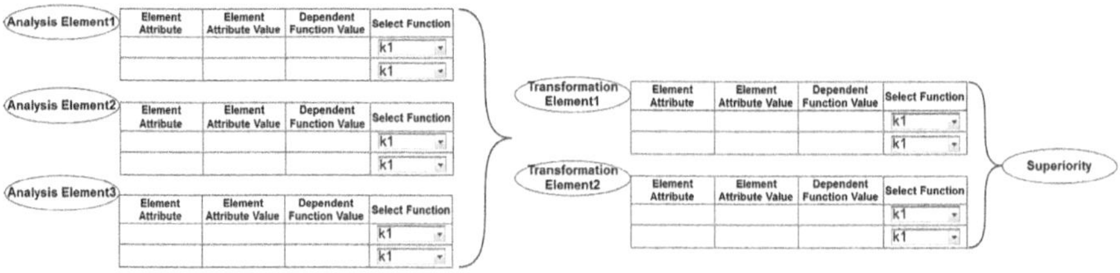

Fig. 5. Procedure of extenics transformation and superiority evaluation

We think an-bracket should be the best way to describe the relationship between them. In this ways above, the software for Extensive deduced tree can be developed and benefits for those who are interested in Extenics or innovative thinking.

Sum up these ways for innovative thinking, the whole Extensive deduced tree should be shown in Fig. 6. The whole procedure of innovative thinking can be seen in a mind mapping. Besides, the relationship between each data can be seen clearly in the form.

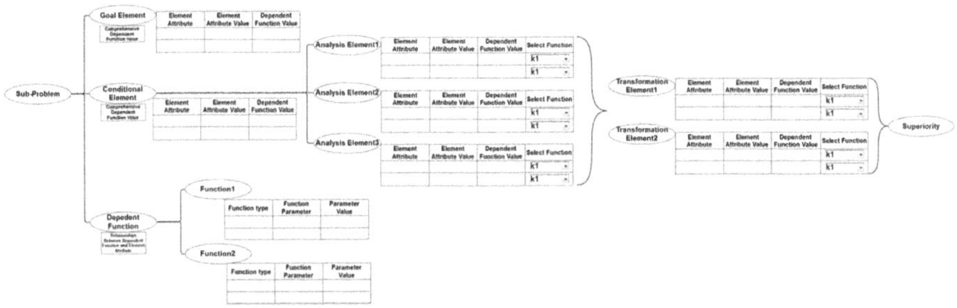

Fig. 6. Procedure of extensive innovative thinking

4 Evaluation and Future Work

The advantages of designing Extensive innovative deduced trees are shown as follows.

1. It can deal with the information with much relationship.
2. It triggers human's right brain while performing an innovative thinking.
3. It is easy to record fleeting innovative thinking.

However, there remains some shortcoming needing improving.

1. It can't deal with the problem with much information.
2. Mind mapping doesn't fit the problem needing thinking systematically.

As for advanced work of innovative mind thinking, we first manage to develop the software of Extensive deduced tree for innovative thinking uses according to the design above. Besides, we will improve the software by adding functions of studying Extenics and innovative thinking, discussing system and user management. After the software running well, we will collect some ways of innovative thinking and find out the law of human's innovative thoughts by big data processing. Finally, we will achieve artificial intelligence for innovative thinking by the explored law.

Acknowledgements. The research is supported by Guangdong Provincial Science and Technology Project (2014A040402010), Guangdong Province Innovation and Entrepreneurship Training Program for College Students (201710566036) and Guangdong Ocean University Excellent Courses Project for Software Engineering (JP2010007).

References

1. Yang, C., Cai, W., Tu.: Extension research, application and development. Syst. Sci. Math. **36**(09), 1507–1512 (2016)
2. Yang, C., Cai, W.: The intelligentized treatment of extenics and contradictions. Technol. Guide **32**(36), 15–20 (2014)
3. Fan, R.: Modeling extenics innovation software by intelligent service components. Open Cybern. Syst. J. **8**(1), 1–7 (2014)
4. Yan, S., Fan, R., Chen, Y., et al.: Research on web services-based extenics aided innovation system. Procedia Comput. Sci. **107**(5), 103–110 (2017)
5. Fan, R., Peng, Y., Chen, Y., Lei, G., Liu, X.: Adaptive software formalization method based on extenics. J. Intell. Syst. **10**(06), 901–911 (2015)
6. Deng, Q., Wu, Y., Weng, Y.: The understanding and practice of construction of system structure and extension of mind mapping method and application. Math. Pract. Knowing **45** (12), 94–99 (2015)
7. Zhu, C.K.: The thinking direction of original educational researching: remodeling the spirit of dialectical materialism. J. Northeast Normal Univ. (2004)
8. Liu, S.: How to cultivate the innovative thinking way of students in the teaching of animation key-frame design. In: Conference on Informatisation in Education, Management and Business, pp. 499–509 (2014)
9. Guo, B., Fan, R., Huang, C.W., et al.: The best marketing plan of photographing for money deduced by extensible mind mapping. In: 4th Annual International Conference on Wireless Communication and Sensor Network (WCSN 2017), vol. 17 no 13, p. 03003 (2018)
10. Ma, F., Fan, R., Huang, C., et al.: The solution of excess ingredients in hotels deduced by extensible mind mapping. In: 4th Annual International Conference on Wireless Communication and Sensor Network (WCSN 2017), vol. 17 no. 3, p. 03004 (2018)
11. Bingyao, F.: Innovative Thinking Mechanism. Jilin University (2014)
12. Zhang, X.: The advantages of mind mapping at ideological and political aspects. Teach. Co. **04**, 133–134 (2017)
13. Chen, Z., Lu, T.: A preliminary study on the evaluation criteria of the subject map of discipline. Jiangsu Educ. Res. **10**, 48–51 (2014)
14. Geng, Z., Liu, L., Hu.: Providence study mode in research of mind mapping based on fragmentation, contemporary education practice and teaching. Mod. Educ. Pract. Teach. Res. **07**, 126–127 + 161 (2017)

Intelligent Indoor Environment Monitoring System Based on IoT and Cloud Platform

Haoming Zhang(✉) and Ping Liu

Army Academy of Artillery and Air Defense, Hefei 230031, AnHui, China
710223917@qq.com

Abstract. In order to achieve intelligent, comprehensive and real-time monitoring of indoor environmental parameters, an indoor environment monitoring system based on IoT and cloud computing technology was designed. The system integrates IoT and cloud computing technology for indoor environment monitoring, and uses the cloud computing information technology platform to establish an evaluation model of indoor thermal comfort and air quality based on BP neural network. This is of great significance to the monitoring, statistics and analysis of the comfort of the residents in the residential environment. Experiments show that the system has low cost, low power consumption, simple operation, stable operation, high measurement accuracy and good practical value.

Keywords: Indoor environment · Monitoring · IoT · Cloud computing
Artificial intelligence

1 Introduction

With the development of society, people are increasingly aware of the importance of environmental issues. The quality of the indoor environment is now attracting people's attention. The indoor environment is the micro-environment of the building built by humans for the needs of life and production. In the whole life of a person, most of the time is spent indoors [1]. The indoor environment has an important effect on people's health, quality of life and work efficiency. The harsh thermal environment will affect people's working conditions, and the harsh air environment will cause people to cause respiratory infections, asthma and even cancer. Therefore, designing a real-time monitoring system of the indoor environment is a very meaningful task.

The Internet of Things (IoT) and cloud computing have become hot technologies in this era. Recently, with the rise and maturity of smart home [2] technologies, we have made smart monitoring of the indoor environment possible. Many people like Tsang and Gao has designed an indoor environment system based on ZigBee or Wi-Fi [3, 4]. However, the system they designed was not perfect, not only did it not collect complete environmental parameters, but also did not process the collected environmental parameters by using evaluation models.

The above shows that it is imminent to develop an intelligent, comprehensive, low-cost environmental monitoring system. This article introduces a new indoor intelligent environment monitoring system. This system uses the Internet of Things technology to

K. Deng et al. (Eds.): ICMIR 2018, AISC 856, pp. 305–313, 2019.
https://doi.org/10.1007/978-3-030-00214-5_40

monitor indoor environment parameters such as temperature, humidity, and indoor air quality in real time, and uses Alibaba Cloud Services to transport and process data.

2 System Design

Figure 1 shows our system structure. The system consists of an information collecting layer, a transport layer, and an information processing layer. The information collecting layer uses LinkIt One as the hardware platform. We add sensors in this platform that can collect various environmental parameters such as temperature, humidity, TVOC, and so on. In transport layer we use Alibaba Cloud as a server to transmit the collected information to the server via Wi-Fi. In the information processing layer we process the collected environmental parameters to obtain thermal comfort and air quality evaluation. Then we push the data to the PC and the mobile terminal. People can monitor the indoor environmental conditions in real time.

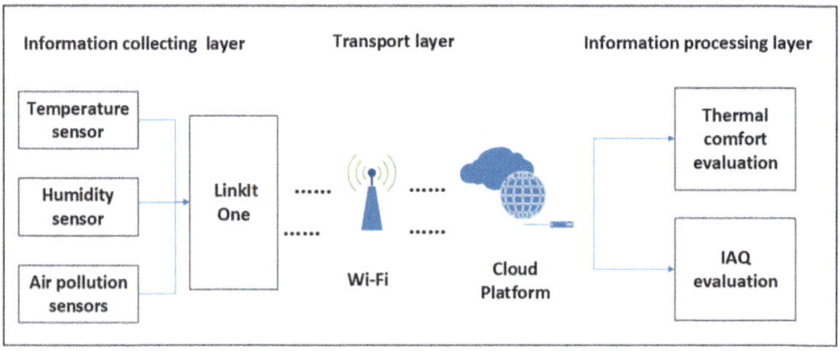

Fig. 1. The system structure

2.1 Information Collecting Layer

The Linkit One platform provides a large number of sensor interfaces. We can place multiple sensors to collect environmental data. In addition, we can programmatically implement data exchange and system control functions. In this system, we use sensors such as SHT31 (Temperature and Humidity Sensor), ZE08-CH20 (Formaldehyde Sensor), CCS811 (TVOC Sensor), iAQ-Core C(TVOC Sensor), MiCS-VZ-89TE (TVOC Sensor), and SMD1013 (NO_x Sensor) to collect various environmental parameters. The unity is controlled by Linkit One.

The Wi-Fi module may enable the Linkit One to connect with the cloud platform through the wireless router so as to achieve communication between the cloud computing platform and the remote terminal. The user can control the running and stopping of the system through the on-off switch of the cloud.

2.2 Transport Layer

There are mainly three steps in the work of the transport layer. First is to add devices in the cloud, the second is to receive data from the devices in the cloud, and the third is to receive cloud instructions from devices.

In the first step, in order to connect the Linkit One platform to the cloud platform, we need to add Linkit One and its attached sensor nodes to the cloud and obtain the Topic of the devices.

In the second step, we monitor the MNS queue in the cloud and obtain the data in the Topic based on the MQTT protocol. In Alibaba Cloud IoT Suite we can subscribe directly to the device and create the queue like 'aliyun-iot-${productKey}'. After the configuration is complete, the IoT suite encapsulates the device data and writes it to the queue in JSON format. The message format is as follows (Table 1):

Table 1. The message format

Item	Content
messageid	Message ID generate by IoT system, 64-bit size
messagetype	Message type, status or upload
topic	Devices' topic which send or receive the message
payload	When messageType = status, the data is the notification data of the device; When messageType = upload, data is the raw data published by the device to the Topic
timestamp	Show the time, expressed as Epoch time

The corresponding device status data refer to the following Table 2:

Table 2. The device status data format

Item	Content
status	'Online' of 'Offline'; the device status
productKey	xxxxxxxxxxx
deviceName	xxxxxxxxxx
time	Notification sending time
lastTime	Last communication time when status changes
clientIp	Public IP address of the device

We access the MNS SDK on the server. We can use the queue mode to subscribe to the message queue of the device created by the IoT to complete the monitoring of the MNS queue, and then complete the steps for the cloud to receive device data.

In the third step, we need to subscribe the device to Topic. We use the SDK to write a device subscription program, and then complete the steps for the device to receive cloud instructions.

2.3 Information Processing Layer

After the transport layer uploads the collected environmental data to the cloud platform, the cloud platform will process the data by use the evaluation model of indoor thermal comfort and air quality to obtain the evaluation results.

3 Evaluation of Indoor Thermal Comfort and Indoor Air Quality

The two main indoor environment factors are the indoor thermal environment and air environment. Therefore, it is necessary to evaluate the indoor thermal environment and air environment. We need to build a model of indoor thermal comfort and indoor air quality evaluation in the information processing layer to complete the processing of environmental data.

3.1 Indoor Thermal Comfort Evaluation Model

The indoor thermal environment refers to the environment of various factors that affect the body's cold and hot sensation. The purpose of studying the thermal environment is to provide the best thermal comfort conditions for human life.

Professor Fanger, a Danish scholar, made important contributions to the study of thermal environment and proposed the internationally recognized thermal comfort theory and thermal comfort equation [5]. Prof. Fanger's Predicted Mean Vote (PMV) thermal comfort index combines the effects of four environmental parameters and two individual body parameter on human thermal comfort. It is the most comprehensive theory for evaluating thermal environment. The PMV indicator has a maximum value of 3 and a minimum value of −3. From large to small, it represents hot to cold, while 0 represents moderate ideality. The mathematical model of PMV is as follows:

$$
\begin{aligned}
PMV = {} & [0.303e^{0.036M} + 0.0275]\{M - W - 3.05 \\
& \times [5.733 - 6.99M - W - P_a] - 0.42[M - W - 58.15] \\
& - 0.0173M(5.687 - P_a) - 0.0014M(34 - t_a) \\
& - 3.96 \times 10 f_{cl}[(t_{cl} + 273) - (t_r + 273)] - f_{cl}h_c(t_{cl} - t_a)\}
\end{aligned}
$$

In this formula, M is the body's metabolic rate; W is the body's external work; t_a is the air temperature; P_a is the water vapor pressure around the body; f_{cl} is the clothing surface area coefficient; t_{cl} is the clothing surface temperature; $t_r$is the average radiation temperature; h_c is the convective heat transfer Coefficient; I_{cl} is the clothing thermal resistance; v_a is the air flow rate; φ is the relative air humidity.

Among them, air temperature, average radiation temperature, clothing surface temperature, and humidity are values that can be easily measured by the sensor. Clothing thermal resistance and air flow speed are not easily obtained, but taking into account the indoor environment, these two can be set several default values depending on the season and the characteristics of the building. The remaining parameters we can obtain the corresponding data through the sensor, then we can evaluate the indoor thermal comfort environment.

3.2 Indoor Air Quality Evaluation Model

The indoor air quality is also called IAQ. At present, the mature IAQ evaluation methods at domestic and abroad mainly include objective evaluation methods, subjective evaluation methods, and comprehensive evaluation methods. But these methods are too complex, because many air quality parameters are difficult to measure, and are not easily monitored in real time. In recent years, some scholars have begun to apply neural networks to the IAQ evaluation. In this article, we will use BP neural network model to evaluate the IAQ.

Figure 2 shows the most basic neural network structure, which includes an input layer, a hidden layer, and an output layer. The basic BP algorithm includes two processes: forward propagation of the signal and back propagation of the error. The error output is calculated from the input to the output, while the adjustment weight and threshold are from the output to the input. The BP neural network is mature in both network theory and performance. Its outstanding advantage is that it has a strong nonlinear mapping capability and a flexible network structure. The number of intermediate layers in the network and the number of neurons in each layer can be arbitrarily set according to specific conditions, and its performance varies with the structure. It is feasible to use BP neural network to evaluate air quality.

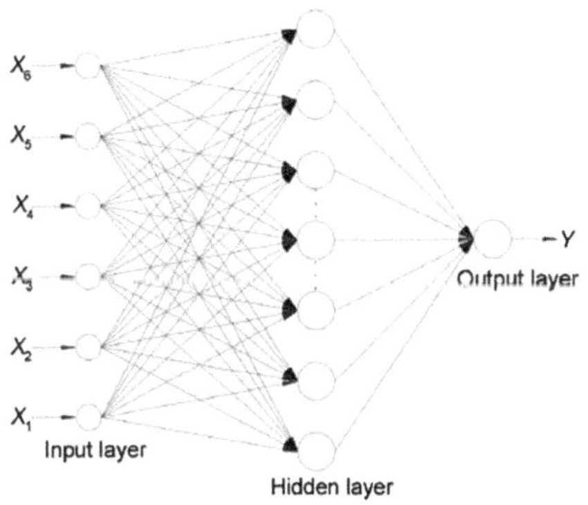

Fig. 2. Basic neural network structure

For the model's input parameters and output results, we take into account the main characteristics of indoor air pollution and made a decision based on the "Indoor Air Quality Standard" GB/T 18883-2002 (Chinese standards) [6].

We select the content of four pollutants, formaldehyde, TVOC, nitrogen oxides (NOx), and carbon dioxide (CO2), as model features. Then, according to the "Indoor

Air Standards", we divide the model output into four levels, which are 0 (clean), 1 (mild pollution), 2 (moderate pollution), and 3 (heavy pollution).

We used air quality testers and sensor components to collect air quality samples of five families over a period of time up to one month, collecting a total of 750 data samples. 350 samples were used to train neural networks and 300 were used for testing. In the training process, we take the learning rate of 0.01, the deviation is 0.03, and the network converges when the number of iterations is 5000. In the actual situation, since the weights and thresholds can be reasonably initialized, the deviation can be appropriately relaxed, so the network can converge faster. We use the test set to test, The evaluation average deviation of the model is 0.21. Figure 3 show model test deviation of the BP neural network model.

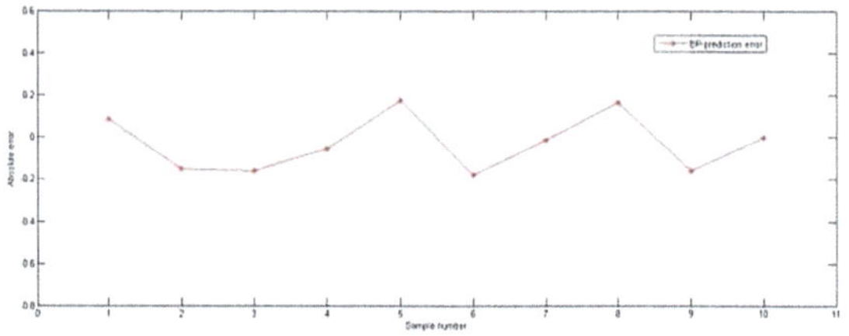

Fig. 3. The model deviation

4 Experiments and Tests

Based on the design of hardware and software above, we designed an easy system as shown in Fig. 4. Put the system in the bedroom, and record the data of temperature, humidity, and air pollution throughout the day. We can monitor the indoor environment in real time on the mobile app as what shown in Fig. 5.

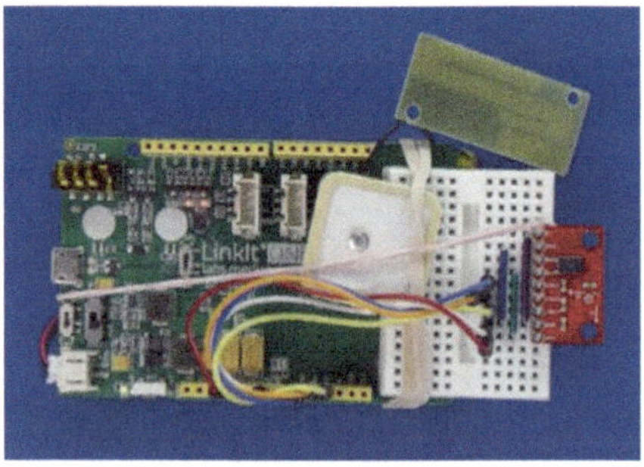

Fig. 4. The indoor environment system

In order to make the measurement results more accurate, we calculated the average of the data recorded within each hour. We chose the five test results of temperature,

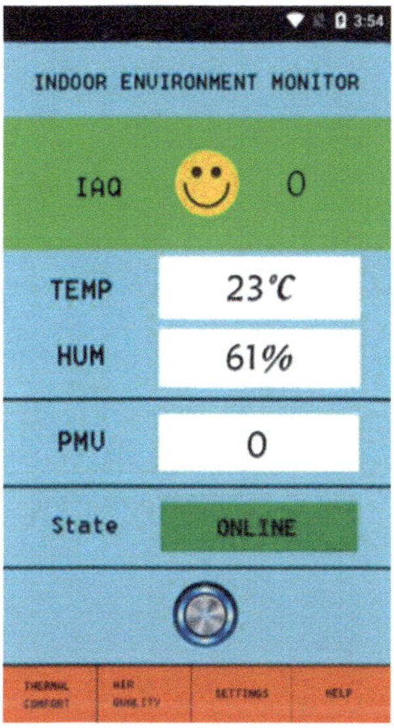

Fig. 5. Indoor environment monitoring APP

humidity, formaldehyde concentration, thermal comfort, and indoor air quality. The test results are compared with the standard thermometer, hygrometer, and formaldehyde detector. The comparing results are shown in Figs. 6, 7, and 8. As we can see, there is an acceptable error between the measured value and the actual value.

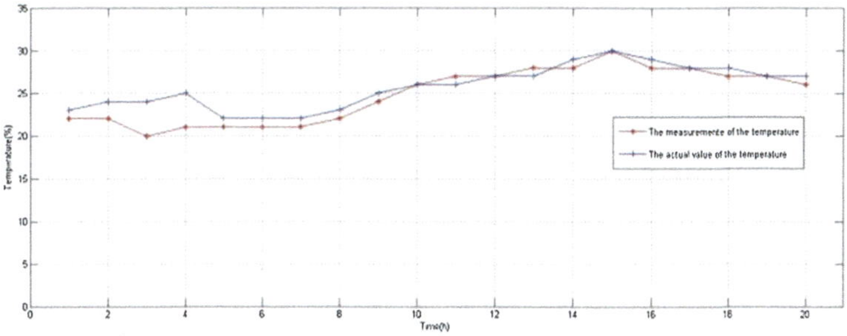

Fig. 6. Temperature contrast

Through the comparison, we can conclude that the accuracy of our indoor environment monitoring system is basically consistent with the single detector and satisfies the actual testing requirements. At the same time, the design not only make up for the lack of a single detection device, but also add two new function of BP neural network based IAQ quality model and remote monitoring. It greatly enhance the practicality.

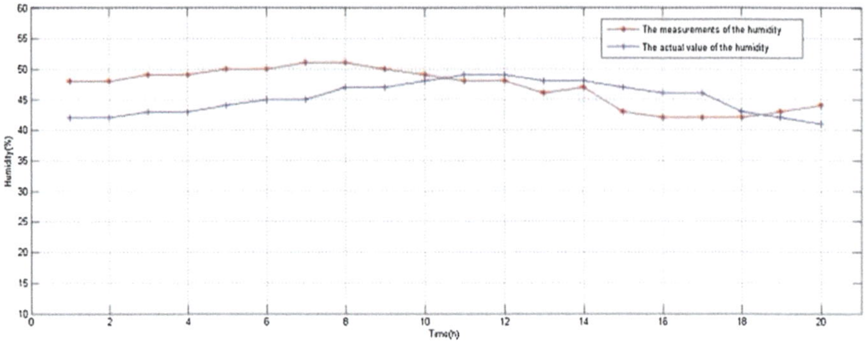

Fig. 7. Humidity contrast

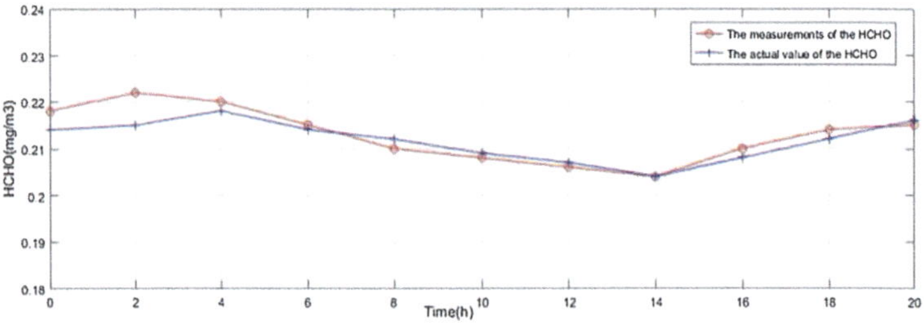

Fig. 8. Formaldehyde concentration contrast

5 Conclusion

In this article, we designed an intelligent indoor environment monitoring system based on Linkit One using the Iof T and cloud platform. The system realizes real-time acquisition and processing of environmental parameters. Users can monitor real-time conditions of indoor thermal comfort and air quality in real time. Experimental tests show that the system has high sensitivity, stable operation, strong real-time performance, low cost, low power consumption, and easy expansion.

References

1. Thakur, R.C., Singh, Y.P., Lamba, S.S.: Liquid level measurement and control using microcomputer. IETE Tech. Rev. **9**, 348–355 (2015)
2. Davidoff, S., Min, K.L., Yiu, C., et al.: Principles of smart home control. In: UBICOMP 2006: Ubiquitous Computing, International Conference, UBICOMP 2006, Orange County, Ca, USA, September 2006, pp. 19–34. DBLP (2006)
3. Tsang, K.F., Chi, H.R., Fu, L., et al. Energy-saving IAQ monitoring ZigBee network using VIKOR decision making method. In: IEEE International Conference on Industrial Technology 2004–2009. IEEE (2016)

4. Gao, Y., Diao, Y., Lin, C., et al.: Monitoring and analysis of thermal environment and harmful gases in mechanically ventilated multistory pig buildings. Trans. Chin. Soc. Agric. Eng. (2018)
5. Fanger, P.O.: Calculation of thermal comfort. ASHRAE Trans. **73**(2), 5–6 (1967)
6. Ministry of Health: GB/T18883-2002 Implementation Guidelines for Indoor Air Quality Standards. China Standard Press (2003)

The Cryptographic Properties of the Autocorrelation Functions for Encryption Algorithm

Yu Zhou[1(✉)], Aili Zhang[2], and Yunfei Cao[1]

[1] Science and Technology on Communication Security Laboratory, Chengdu 610041, China
zhouyu.zhy@tom.com
[2] XinLian College of Henan Normal University (Zhengzhou Campus), Xinxiang 453007, China

Abstract. The security for this Wireless sensor is important, and it is estimated by encryption algorithm. The basic component (named Boolean function) with encryption algorithm is studied, and we obtain the sum of high-order autocorrelation function of Boolean functions in cryptographic algorithms. The definition of the k-th autocorrelation function is given, we give the connections between the sum of the r-th autocorrelation functions and the sum of the $(r-1)$-th autocorrelation functions. We obtain the tightness of the upper or lower bounds on the sum of the r-th autocorrelation function. We give the connection from Walsh transform and the sum of the r-th autocorrelation function.

Keywords: Encryption algorithm · Boolean functions · Autocorrelation GAC · Balanced

1 Introduction

The security of sensors, wireless sensor networks and multi-sensor data fusion is important, but there are some danger and potential problems. This paper focuses on information security, mainly studying the cryptographic properties of the basic components (named Boolean function) in cryptographic algorithms, and providing theoretical support for the analysis and design of cryptographic algorithms.

Boolean functions are widely used in cryptographic algorithms (such as Grain, ZUC, A5/1, etc.), it should satisfy some cryptographical properties, such as balancedness [5], high nonlinearity [1], high algebraic degree, high algebraic immunity [2], good SAC [3] etc. [16] proposed the global avalanche characteristics (GAC) to characterize the all avalanche properties, it includes two indicators (named the sum-of-squares indicator (σ_f) and the absolute indicator (Δ_f)) for a Boolean function $f(x) \in B_n$:

$$\sigma_f = \sum_{\alpha \in \mathbb{F}_2^n} [\Delta_f^{(1)}(\alpha)]^2, \quad \Delta_f = \max_{\alpha \in \mathbb{F}_2^n, wt(\alpha) \neq 0^n} \left|\Delta_f^{(1)}(\alpha)\right|,$$

K. Deng et al. (Eds.): ICMIR 2018, AISC 856, pp. 314–323, 2019.
https://doi.org/10.1007/978-3-030-00214-5_41

where $\Delta_f^{(1)}(\alpha)$ is the 1-th autocorrelation function (see Definition 2 in Sect. 2) with respect to $\alpha \in F_2^n$.

The GAC of one Boolean function have impact connection with other cryptographical properties:

(1) [11] obtained a connection between σ_f and N_f for a balancedness function $f(x) \in B_n$: $N_{\hat{f}} \leq 2^{n-1} - \frac{1}{2}2^{-n/2}\sqrt{\sigma_f}$.
(2) [12] provided a connection between σ_f and N_f for a balancedness function satisfying the propagation criterion [4] for a set.
(3) [13] obtained the lower bound for Δ_f of resilient functions that generalizes the bound of Zheng and Zhang [16].
(4) [8, 9] provided the lower bounds on Δ_f and σ_f of autocorrelation values for small CI.
(5) [18, 19] gave the connection between σ_f and algebraic immunity, and the autocorrelation distributions for a balancedness functions satisfying the minimum σ_f.

Meanwhile, many authors gave lots of constructions of Boolean function with σ_f:

(1) [14] obtained a way to construction balancedness Boolean function with lowGAC and highly nonlinearity.
(2) [15] gave a method for construction Boolean functions $f(x) \in B_n$ with lower the absolute indicator $(\Delta_f < 2^{n/2})$ and high nonlinearity.

Thus, the GAC of one Boolean function are important of designing good cryptographic Boolean functions. At the same time, σ_f can be expressed by the sum of the 2-th autocorrelation functions:

$$\sigma_f = \sum_{\alpha_1 \in \mathbb{F}_2^n} [\Delta_f^{(1)}(\alpha_1)]^2 = \sum_{\alpha_1,\alpha_2 \in \mathbb{F}_2^n} \Delta_f^{(2)}(\alpha_1, \alpha_2),$$

where $\Delta_f^{(2)}(\alpha_1, \alpha_2)$ is the 2-th autocorrelation function (see Definition 2 in Sect. 2). Thus, in Stream cipher cryptographic properties of Boolean functions can be obtained by studying the sum of higher order autocorrelation functions.

In Block cipher, [6, 7] proposed higher order derivatives for differential cryptanalysis, but not study the sum of higher order derivatives functions.

Based on the above all considerations, this paper aims to the sum of the higher order autocorrelation functions (see Definition 2 in Sect. 2), and study the following problems: (1) What is the upper or lower bounds on the sum of the r-th autocorrelation functions? (2) What is the connection between walsh transformation and the r-th autocorrelation function?

The basic definitions and notions are introduced in Sect. 2. The tightness of the bound on the sum of the r-th autocorrelation function is obtained in Sect. 3, and this paper gives the connection between Walsh transformation and the sum of the r-th autocorrelation function. Section 4 concludes this paper.

2 Preliminaries

Let B_n be the set of Boolean functions for n variables, and $\oplus$ denote by the additions. f (x) $\in B_n$ admits a unique algebraic normal form (ANF) as a polynomial over F_2. deg(f) is the number of variables in the highest order term with non-zero coefficient in ANF. The support set Supp(f) = $\{(x_1,\cdots, x_n) \mid f(x_1,\cdots, x_n) = 1\}$. f(x) $\in B_n$ is balanced if hamming weight wt(f) = 2^{n-1}. f(x) $\in B_n$ is affine if def(f) <= 1 in ANF and this set is denoted by A_n.

Definition 1. The Walsh transformation of f(x) $\in B_n$ is defined by

$$\mathcal{F}(f \oplus \varphi_\alpha) = \sum_{x \in \mathbb{F}_2^n} (-1)^{f(x) \oplus \alpha x},$$

where $\varphi_\alpha = \alpha x = \alpha_1 x_1 \oplus \alpha_2 x_2 \oplus \cdots \oplus \alpha_n x_n$. Then the nonlinearity of f is defined by

$$N_f = 2^{n-1} - \frac{1}{2} \max_{\alpha \in \mathbb{F}_2^n} |\mathcal{F}(f \oplus \varphi_\alpha)|.$$

For any f(x) $\in B_n$, the k-th derivatives $D_f^{(k)}(x, (a_1, a_2, \cdots, a_k))$ on $a_1, a_2, \cdots, a_k \in F_2^n$ is defined as:

$$D_f^{(k)}(x, (a_1, a_2, \cdots, a_k)) = D_f^{(k-1)}(x \oplus a_k, (a_1, a_2, \cdots, a_{k-1})) \\ \oplus D_f^{(k-1)}(x, (a_1, a_2, \cdots, a)).$$

For a convenience, we also denote $D_f^{(k)}(x, (a_1, a_2, \cdots, a_k))$ by $f^{(k)}_{\alpha 1, \alpha 2, \cdots, \alpha_k}(x)$.

Definition 2. The k-th (k $\geq$ 1) autocorrelation function f(x) $\in B_n$ is defined by:

$$\Delta_f^{(k)}(\alpha_1, \alpha_2, \cdots, \alpha_k) = \sum_{x \in \mathbb{F}_2^n} (-1)^{D_f^{(k)}(x, (\alpha_1, \alpha_2, \cdots, \alpha_k))}, \alpha_i \in \mathbb{F}_2^n (1 \leq i \leq k \leq n).$$

3 The Tightness of the Bound on the Sum of the r-th Autocorrelation Function

At first, we give the connection between the sum of the r-th autocorrelation function and the sum of the (r − 1)-th autocorrelation function.

3.1 The Connection Between the Sum of the r-th and (r − 1)-th Autocorrelation Function

Theorem 1. For f(x) $\in B_n$.

$$\sum_{\alpha_1, \alpha_2, \cdots, \alpha_r \in \mathbb{F}_2^n} \Delta_f^{(r)}(\alpha_1, \alpha_2, \cdots, \alpha_r) = \sum_{\alpha_1, \alpha_2, \cdots, \alpha_{r-1} \in \mathbb{F}_2^n} \left[\Delta_f^{(r-1)}(\alpha_1, \alpha_2, \cdots, \alpha_{r-1})\right]^2.$$

Proof. From the definition, we have

$$\sum_{\alpha_1,\alpha_2,\cdots,\alpha_r\in\mathbb{F}_2^n}\Delta_f^{(r)}(\alpha_1,\alpha_2,\cdots,\alpha_r)=$$
$$=\sum_{\alpha_1,\alpha_2,\cdots,\alpha_r\in\mathbb{F}_2^n}[\sum_{x\in\mathbb{F}_2^n}(-1)^{D_f^{(r)}(x,(\alpha_1,\alpha_2,\cdots,\alpha_r))}]$$
$$=\sum_{\alpha_1,\alpha_2,\cdots,\alpha_{r-1}\in\mathbb{F}_2^n}\{\sum_{\alpha_r\in\mathbb{F}_2^n}[\sum_{x\in\mathbb{F}_2^n}(-1)^{D_f^{(r)}(x,(\alpha_1,\alpha_2,\cdots,\alpha_r))}]\}$$
$$=\sum_{\alpha_1,\alpha_2,\cdots,\alpha_{r-1}\in\mathbb{F}_2^n}\{[\sum_{x\in\mathbb{F}_2^n}(-1)^{D_f^{(r)}(x,(\alpha_1,\alpha_2,\cdots,\alpha_r^{(0)}))}]+[\sum_{x\in\mathbb{F}_2^n}(-1)^{D_f^{(r)}(x,(\alpha_1,\alpha_2,\cdots,\alpha_r^{(1)}))}]$$
$$+\cdots+[\sum_{x\in\mathbb{F}_2^n}(-1)^{D_f^{(r)}(x,(\alpha_1,\alpha_2,\cdots,\alpha_r^{(2^n-1)}))}]\}$$
$$=\sum_{\alpha_1,\alpha_2,\cdots,\alpha_{r-1}\in\mathbb{F}_2^n}\{[\sum_{x\in\mathbb{F}_2^n}(-1)^{D_f^{(r-1)}(x\oplus\alpha_r^{(0)},(\alpha_1,\alpha_2,\cdots,\alpha_{r-1}))+D_f^{(r-1)}(x,(\alpha_1,\alpha_2,\cdots,\alpha_{r-1}))}]$$
$$+[\sum_{x\in\mathbb{F}_2^n}(-1)^{D_f^{(r-1)}(x\oplus\alpha_r^{(1)},(\alpha_1,\alpha_2,\cdots,\alpha_{r-1}))+D_f^{(r-1)}(x,(\alpha_1,\alpha_2,\cdots,\alpha_{r-1}))}]+\cdots$$
$$+[\sum_{x\in\mathbb{F}_2^n}(-1)^{D_f^{(r-1)}(x\oplus\alpha_r^{(2^n-1)},(\alpha_1,\alpha_2,\cdots,\alpha_{r-1}))+D_f^{(r-1)}(x,(\alpha_1,\alpha_2,\cdots,\alpha_{r-1}))}]\}$$
$$=\sum_{\alpha_1,\alpha_2,\cdots,\alpha_{r-1}\in\mathbb{F}_2^n}\{\sum_{x\in\mathbb{F}_2^n}(-1)^{D_f^{(r-1)}(x,(\alpha_1,\alpha_2,\cdots,\alpha_{r-1}))}[(-1)^{D_f^{(r-1)}(x\oplus\alpha_r^{(0)},(\alpha_1,\alpha_2,\cdots,\alpha_{r-1}))}$$
$$+(-1)^{D_f^{(r-1)}(x\oplus\alpha_r^{(1)},(\alpha_1,\alpha_2,\cdots,\alpha_{r-1}))}+\cdots+(-1)^{D_f^{(r-1)}(x\oplus\alpha_r^{(2^n-1)},(\alpha_1,\alpha_2,\cdots,\alpha_{r-1}))}]\}$$
$$=\sum_{\alpha_1,\alpha_2,\cdots,\alpha_{r-1}\in\mathbb{F}_2^n}\{\sum_{x\in\mathbb{F}_2^n}(-1)^{D_f^{(r-1)}(x,(\alpha_1,\alpha_2,\cdots,\alpha_{r-1}))}[\sum_{\alpha_r\in\mathbb{F}_2^n}(-1)^{D_f^{(r-1)}(x\oplus\alpha_r,(\alpha_1,\alpha_2,\cdots,\alpha_{r-1}))}]\}$$
$$=\sum_{\alpha_1,\alpha_2,\cdots,\alpha_{r-1}\in\mathbb{F}_2^n}\{\sum_{x\in\mathbb{F}_2^n}(-1)^{D_f^{(r-1)}(x,(\alpha_1,\alpha_2,\cdots,\alpha_{r-1}))}[\sum_{\substack{y\in\mathbb{F}_2^n\\ y=x\oplus\alpha_r}}(-1)^{D_f^{(r-1)}(y,(\alpha_1,\alpha_2,\cdots,\alpha_{r-1}))}]\}$$
$$=\sum_{\alpha_1,\alpha_2,\cdots,\alpha_{r-1}\in\mathbb{F}_2^n}[\sum_{x\in\mathbb{F}_2^n}(-1)^{D_f^{(r-1)}(x,(\alpha_1,\alpha_2,\cdots,\alpha_{r-1}))}]^2$$
$$=\sum_{\alpha_1,\alpha_2,\cdots,\alpha_{r-1}\in\mathbb{F}_2^n}[\Delta_f^{(r-1)}(\alpha_1,\alpha_2,\cdots,\alpha_{r-1})]^2.\square$$

From Theorem 1, we have this Corollaries 1 and 2.

Corollary 1 [17]. For $f(x)\in B_n$.

$$\sum_{\alpha\in\mathbb{F}_2^n}\Delta_f^{(1)}(\alpha)=(2^n-2wt(f))^2.$$

When $wt(f)=2^{n-1}$, the sum of 1-th autocorrelation function is equals to zero.

Corollary 2 [16]. For $f(x)\in B_n$.

$$\sigma_f=\sum_{\alpha_1,\alpha_2\in\mathbb{F}_2^n}\Delta_f^{(2)}(\alpha_1,\alpha_2)=\sum_{\alpha_1\subset\mathbb{F}_2^n}[\Delta_f^{(1)}(\alpha_1)]^2.$$

From Corollary 2 and Theorem 6 in [16], we know

$$2^{2n}\leq\sigma_f=\sum_{\alpha_1,\alpha_2\in\mathbb{F}_2^n}\Delta_f^{(2)}(\alpha_1,\alpha_2)=\sum_{\alpha_1\in\mathbb{F}_2^n}[\Delta_f^{(1)}(\alpha_1)]^2\leq 2^{3n}.$$

3.2 The Upper and Lower Bounds

In this subsection, we give the upper or lower bounds on the sum of r-th autocorrelation function.

Lemma 1. For $f(x) \in B_n$. If $\alpha_i = \alpha_j$ for some $(i, j)(1 \le i < j \le k)$, then $D_f^{(k)}(x, (a_1, a_2, \cdots, a_k)) = 0$.

Proof. Note that

$$\begin{aligned} D_f^{(k)}(x,(a_1,a_2,\cdots,a_k)) &= D_f^{(k-1)}(x\oplus a_k,(a_1,a_2,\cdots,a_{k-1}))\oplus D_f^{(k-1)}(x,(a_1,a_2,\cdots,a_{k-1})) \\ &= f(x)\oplus \sum_{0\le i\le k-1} f(x\oplus\alpha_i)\oplus \sum_{0\le i\ne j\le k-1} f(x\oplus\alpha_i\oplus x_j)\oplus\cdots\oplus f(x\oplus\alpha_1\oplus\cdots\oplus\alpha_{k-1} \\ &\oplus f(x\oplus\alpha_k)\oplus \sum_{0\le i\le k-1} f(x\oplus\alpha_k\oplus\alpha_i)\oplus \sum_{0\le i\ne j\le k-1} f(x\oplus\alpha_k\oplus\alpha_i\oplus\alpha_j) \\ &\oplus\cdots\oplus f(x\oplus\alpha_k\oplus\alpha_1\oplus\cdots\oplus\alpha_{k-1}) \end{aligned}$$

For a convenience, suppose $\alpha_k = \alpha_1$, in term of the above equation we have $D_f^{(k)}(x, (a_1, a_2, \cdots, a_k)) = 0$. □

Based on Theorem 1 and Lemma 1, we have the upper or lower bounds on the sum of r-th autocorrelation functions.

Theorem 2. For $f(x) \in B_n$ and $r \ge 2$.

$$2^{rn} \le \sum_{\alpha_1,\alpha_2,\cdots,\alpha_r\in\mathbb{F}_2^n} \Delta_f^{(r)}(\alpha_1,\alpha_2,\cdots,\alpha_r) \le 2^{(r+1)n}.$$

The lower bound holds if and only if $D_f^{(r-2)}(x, (\alpha_1, \alpha_2, \cdots, \alpha_{r-2}))$ are bent functions for any $\alpha_i (1 \le i \le r-2) \in F_2^n$,the upper bound holds if and only if $D_f^{(r-1)}(x, (a_1, a_2, \cdots, a_{r-1})) = 0$ or 1 for any $\alpha_1, \alpha_2, \cdots, \alpha_{r-1} \in F_2^n$.

Proof. We will prove for r using induction method.

(1) For r = 2, from Corollary 2 we have

$$2^{2n} \le \sum_{\alpha_1,\alpha_2} \Delta_f^{(2)}(\alpha_1,\alpha_2) = \sum_{\alpha_1\in\mathbb{F}_2^n} [\Delta_f^{(1)}(\alpha_1)]^2 = \sigma_f \le 2^{3n}.$$

(2) For r = 3, we have

$$\begin{aligned} \sum_{\alpha_1,\alpha_2,\alpha_3\in\mathbb{F}_2^n} \Delta_f(\alpha_1,\alpha_2,\alpha_3) &= \sum_{\alpha_1,\alpha_2\in\mathbb{F}_2^n} [\Delta_f^{(2)}(\alpha_1,\alpha_2)]^2 = \sum_{\alpha_1,\alpha_2\in\mathbb{F}_2^n} [\Delta_{f_{\alpha_1}}^{(1)}(\alpha_2)]^2 \\ &= \sum_{\alpha_1\in\mathbb{F}_2^n} [\sum_{\alpha_2\in\mathbb{F}_2^n} [\Delta_{f_{\alpha_1}}^{(1)}(\alpha_2)]^2 = \sum_{\alpha_1\in\mathbb{F}_2^n} \sigma_{f_{\alpha_1}^{(1)}} \end{aligned}$$

From Corollary 2, we know

$$2^{3n} = 2^n \times 2^{2n} \leq \sum_{\alpha_1,\alpha_2,\alpha_3 \in \mathbb{F}_2^n} \Delta_f(\alpha_1,\alpha_2,\alpha_3) = \sum_{\alpha_1 \in \mathbb{F}_2^n} \alpha_{f_{\alpha_1}^{(1)}} \leq 2^n \times 2^{3n} = 2^{4n}.$$

(3) For r = 4, we have

$$\begin{aligned}\sum_{\alpha_1,\alpha_2,\alpha_3,\alpha_4 \in \mathbb{F}_2^n} \Delta_f(\alpha_1,\alpha_2,\alpha_3,\alpha_4) &= \sum_{\alpha_1,\alpha_2,\alpha_3 \in \mathbb{F}_2^n} [\Delta_f^{(2)}(\alpha_1,\alpha_2,\alpha_3)]^2 \\ &= \sum_{\alpha_1,\alpha_2,\alpha_3 \in \mathbb{F}_2^n} [\Delta_{f_{\alpha_1}}^{(2)}(\alpha_2,\alpha_3)]^2 = \sum_{\alpha_1 \in \mathbb{F}_2^n} [\sum_{\alpha_2,\alpha_3 \in \mathbb{F}_2^n} [\Delta_{f_{\alpha_1}}^{(2)}(\alpha_2,\alpha_3)]^2] \\ &= \sum_{\alpha_1 \in \mathbb{F}_2^n} [\sum_{\alpha_2 \in \mathbb{F}_2^n} [\sum_{\alpha_3 \in \mathbb{F}_2^n} [\Delta_{f_{\alpha_1,\alpha_2}}^{(2)}(\alpha_3)]^2]] = \sum_{\alpha_1 \in \mathbb{F}_2^n} [\sum_{\alpha_2 \in \mathbb{F}_2^n} [\sigma_{f_{\alpha_1,\alpha_2}^{(2)}}]].\end{aligned}$$

From Corollary 2, we know

$$\begin{aligned}2^{4n} = 2^n \times 2^n \times 2^{2n} &\leq \sum_{\alpha_1,\alpha_2,\alpha_3,\alpha_4 \in \mathbb{F}_2^n} \Delta_f(\alpha_1,\alpha_2,\alpha_3,\alpha_4) \\ &= \sum_{\alpha_1 \in \mathbb{F}_2^n} [\sum_{\alpha_2 \in \mathbb{F}_2^n} [\sigma_{f_{\alpha_1,\alpha_2}^{(2)}}]] \leq 2^n \times 2^n \times 2^{3n} = 2^{5n}.\end{aligned}$$

By the same method, we have

$$\begin{aligned}2^{rn} = \underbrace{2^n \times 2^n \times \cdots \times 2^n}_{r-2} \times 2^{2n} &\leq \sum_{\alpha_1,\alpha_2,\cdots,\alpha_r \in \mathbb{F}_2^n} \Delta_f^{(r)}(\alpha_1,\alpha_2,\cdots,\alpha_r) = \sum_{\alpha_1,\alpha_2,\cdots,\alpha_r \in \mathbb{F}_2^n} [\Delta_f^{(r-1)}(\alpha_1,\alpha_2,\cdots,\alpha_{r-1})]^2 \\ &= \sum_{\alpha_1 \in \mathbb{F}_2^n} [\sum_{\alpha_2 \in \mathbb{F}_2^n} [\cdots [\sum_{\alpha_{r-2} \in \mathbb{F}_2^n} \sigma_{f_{\alpha_1,\alpha_2,\cdots,\alpha_{r-2}}^{(r-2)}}]] \cdots]] \leq \underbrace{2^n \times 2^n \times \cdots \times 2^n}_{r-2} \times 2^{3n} = 2^{(r+1)n}.\end{aligned}$$

According to the above proof, we find that:

(i) The right of the above equation holds if and only if $\Delta_f^{(r-1)}(\alpha_1,\alpha_2,\cdots,\alpha_{r-1}) = 2^n$ or -2^n for any $\alpha_1,\alpha_2,\cdots,\alpha_{r-1} \in F_2^n$.

That is, $D_f^{(r-1)}(x,(a_1,a_2,\cdots,a_{r-1})) = 0$ or 1 for any $\alpha_1,\alpha_2,\cdots,\alpha_{r-1} \in F_2^n$.

(ii) The left of the above equation holds if and only if

$$2^{rn} = \sum_{\alpha_1 \in \mathbb{F}_2^n} [\sum_{\alpha_2 \in \mathbb{F}_2^n} [\cdots [\sum_{\alpha_{r-2} \in \mathbb{F}_2^n} [\sigma_{f_{\alpha_1,\alpha_2,\cdots,\alpha_{r-2}}^{(r-2)}}]] \cdots]]$$

if and only if

$$\sigma_{f^{(r-2)}_{\alpha_1,\alpha_2,\cdots,\alpha_{r-2})}} = 2^{2n}$$

if and only if $D_f^{(r-2)}(x,(\alpha_1,\alpha_2,\cdots,\alpha_{r-2}))$ are Bent functions for any $\alpha_i(1 \le i \le r-2) \in F_2^n$. □

Remark 1. From Theorem 2, if $D_f^{(r-1)}(x,(a_1,a_2,\cdots,a_{r-1})) = 0$ or 1 for some $\alpha_1,\alpha_2,\cdots,\alpha_{r-1} \in F_2^n$, then $\{\alpha_1,\alpha_2,\cdots,\alpha_{r-1}\}$ is a linear structure [10].

Corollary 3. For $f(x) \in B_n$ and $r \ge 2$. If $\deg(f) \le r-1$, then

$$\sum_{\alpha_1,\alpha_2,\cdots,\alpha_r \in \mathbb{F}_2^n} \Delta_f^{(r)}(\alpha_1,\alpha_2,\cdots,\alpha_r) = 2^{(r+1)n}.$$

Proof. These are two cases:

(1) When $\alpha_i = \alpha_j$ for $1 \le i < j \le r$.

From Lemma 1, we know that $D_f^{(r-1)}(x,(a_1,a_2,\cdots,a_{r-1})) = 0$, this means $\Delta_f^{(r)}(\alpha_1,\alpha_2,\cdots,\alpha_r) = 2_n$.

(2) When all $\alpha_i \ne \alpha_j$ for all $1 \le i < j \le r$.

Since deg $\left(D_f^{(1)}(x,(\alpha_1))\right) = {}^\circ(f) - 1, {}^\circ\left(D_f^{(r-1)}(x,(\alpha_1,\alpha_2,\cdots,\alpha_{r-1}))\right) = {}^\circ(f) - (r-1) = 0$, that is $D_f^{(r)}(x,(a_1,a_2,\cdots,a_r)) = 0$, we know $\Delta_f^{(r)}(\alpha_1,\alpha_2,\cdots,\alpha_r) = 2^n$.

Combining with the above two cases, we have this result. □

3.3 The Connection Between Walsh Transformation and the r-th Autocorrelation Function

By the same way, Theorem 1 is proved.

Theorem 3. For $f(x) \in B_n$. For any $\omega \in F_2^n$, then

$$\sum_{\alpha_i(1 \le i \le r) \in \mathbb{F}_2^n} \Delta_f^{(r)}(\alpha_1,\alpha_2,\ldots,\alpha_r)(-1)^{\omega(\alpha_1 \oplus \alpha_2 \oplus \ldots \oplus \alpha_r)} = \sum_{\alpha_i(1 \le i \le r-1) \in \mathbb{F}_2^n} \mathcal{F}^2(f^{(r-1)}_{\alpha_1,\alpha_2,\cdots,\alpha_{r-1} \oplus \varphi_\omega}).$$

Proof. From the definition of r-th correlation function,

$$\sum_{\alpha_i(1\le i\le r)\in \mathbb{F}_2^n} \Delta_f^{(r)}(\alpha_1,\alpha_2,\cdots,\alpha_r)(-1)^{\omega(\alpha_1\oplus\alpha_2\oplus\cdots\oplus\alpha_r)} = \sum_{\alpha_i(1\le i\le r-1)\in \mathbb{F}_2^n} \mathcal{F}^2(f^{(r-1)}_{\alpha_1,\alpha_2,\cdots,\alpha_{r-1}}\oplus\varphi_\omega).$$

Proof. From the definition of r-th correlation function,

$$\begin{aligned}
&= \sum_{\alpha_i(1\le i\le r)\in \mathbb{F}_2^n} \Delta_f^{(r)}(\alpha_1,\alpha_2,\cdots,\alpha_r)(-1)^{\omega(\alpha_1\oplus\alpha_2\oplus\cdots\oplus\alpha_r)}\\
&= \sum_{\alpha_i(1\le i\le r)\in \mathbb{F}_2^n} [\sum_{x\in \mathbb{F}_2^n}(-1)^{D_f^{(r-1)}(x,(\alpha_1,\alpha_2,\cdots,\alpha_{r-1}))\oplus D_f^{(r-1)}(x\oplus\alpha_r,(\alpha_1,\alpha_2,\cdots,\alpha_{r-1}))}](-1)^{\omega(\alpha_1\oplus\alpha_2\oplus\cdots\oplus\alpha_r)}\\
&= \sum_{\alpha_i(1\le i\le r-1)\in \mathbb{F}_2^n} \{\sum_{x\in \mathbb{F}_2^n}(-1)^{D_f^{(r-1)}(x,(\alpha_1,\alpha_2,\cdots,\alpha_{r-1}))}[\sum_{\alpha_r\in \mathbb{F}_2^n}(-1)^{D_f^{(r-1)}(x\oplus\alpha_r,(\alpha_1,\alpha_2,\cdots,\alpha_{r-1}))\oplus\alpha_r\omega}]\\
&\quad (-1)^{\omega(\alpha_1\oplus\alpha_2\oplus\cdots\oplus\alpha_{r-1})}\}\\
&= \sum_{\alpha_i(1\le i\le r-1)\in \mathbb{F}_2^n} \{\sum_{x\in \mathbb{F}_2^n}(-1)^{D_f^{(r-1)}(x,(\alpha_1,\alpha_2,\cdots,\alpha_{r-1}))}[\sum_{\substack{\alpha_r\in \mathbb{F}_2^n\\ y=x\oplus\alpha_r}}(-1)^{D_f^{(r-1)}(y,(\alpha_1,\alpha_2,\cdots,\alpha_{r-1}))\oplus y\omega}]\\
&\quad (-1)^{\omega(\alpha_1\oplus\alpha_2\oplus\cdots\oplus\alpha_{r-1})}(-1)^{x\omega}\}\\
&= \sum_{\alpha_i(1\le i\le r-1)\in \mathbb{F}_2^n} \{\sum_{x\in \mathbb{F}_2^n}(-1)^{D_f^{(r-1)}(x,(\alpha_1,\alpha_2,\cdots,\alpha_{r-1}))}[\mathcal{F}(f^{(r-1)}_{\alpha_1,\alpha_2,\cdots,\alpha_{r-1}}\oplus\varphi_\omega)](-1)^{\omega(\alpha_1\oplus\alpha_2\oplus\cdots\oplus\alpha_{r-1})}(-1)^{x\omega}\}\\
&= \sum_{\alpha_i(1\le i\le r-1)\in \mathbb{F}_2^n} [\mathcal{F}(f^{(r-1)}_{\alpha_1,\alpha_2,\cdots,\alpha_{r-1}}\oplus\varphi_\omega)]\{\sum_{x\in \mathbb{F}_2^n}(-1)^{D_f^{(r-1)}(x,(\alpha_1,\alpha_2,\cdots,\alpha_{r-1}))}(-1)^{\omega(x\oplus\alpha_1\oplus\alpha_2\oplus\cdots\oplus\alpha_{r-1})}\}\\
&= \sum_{\alpha_i(1\le i\le r-1)\in \mathbb{F}_2^n} [\mathcal{F}(f^{(r-1)}_{\alpha_1,\alpha_2,\cdots,\alpha_{r-1}}\oplus\varphi_\omega)]\\
&\quad \{\sum_{\substack{x\in \mathbb{F}_2^n\\ y=x\oplus\alpha_1\oplus\alpha_2\oplus\cdots\oplus\alpha_{r-1}}}(-1)^{D_f^{(r-1)}(y\oplus\alpha_1\oplus\alpha_2\oplus\cdots\oplus\alpha_{r-1},(\alpha_1,\alpha_2,\cdots,\alpha_{r-1}))}(-1)^{\omega(x\oplus\alpha_1\oplus\alpha_2\oplus\cdots\oplus\alpha_{r-1})}\}\\
&= \sum_{\alpha_i(1\le i\le r-1)\in \mathbb{F}_2^n} [\mathcal{F}(f^{(r-1)}_{\alpha_1,\alpha_2,\cdots,\alpha_{r-1}}\oplus\varphi_\omega)][\mathcal{F}(f^{(r-1)}_{\alpha_1,\alpha_2,\cdots,\alpha_{r-1}}\oplus\varphi_\omega)]\\
&= \sum_{\alpha_i(1\le i\le r-1)\in \mathbb{F}_2^n} [\mathcal{F}(f^{(r-1)}_{\alpha_1,\alpha_2,\cdots,\alpha_{r-1}}\oplus\varphi_\omega)]^2.\square
\end{aligned}$$

Corollary 4. For $f(x)\in B_n$. For any $\omega\in F_2^n$, and $\mathcal{L}(f^{(r-1)}_{\alpha_1,\alpha_2,\cdots,\alpha_{r-1}}) = \max_{\alpha_1,\alpha_2,\cdots,\alpha_{r-1}\in F\ominus_2^n}\left|\mathcal{F}(f^{(r-1)}_{\alpha_1,\alpha_2,\cdots,\alpha_{r-1}}\oplus\varphi_\omega)\right|$,

we have

$$-\sum\nolimits_{\alpha_1,\alpha_2\in \mathbb{F}_2^n}\Delta_f^{(2)}(\alpha_1,\alpha_2)(-1)^{\omega(\alpha_1\oplus\alpha_2)} = \sum\nolimits_{\alpha_1\in \mathbb{F}_2^n}\mathcal{F}^2(f^{(1)}_{\alpha_1}\oplus\varphi_\omega);$$
$$-\sum\nolimits_{\alpha_i(1\le i\le r)\in \mathbb{F}_2^n}\Delta_f^{(r)}(\alpha_1,\alpha_2,\cdots,\alpha_r)(-1)^{\omega(\alpha_1\oplus\alpha_2\oplus\cdots\oplus\alpha_r)} \le 2^{nr}[2^n - 2N^{(r-1)}_{f_{\alpha_1,\alpha_2,\cdots,\alpha_{r-1}}}]^2.$$

From Theorem 5, we know that the lower nonlinearity $N^{(r-1)}_{f_{\alpha_1,\alpha_2,\cdots,\alpha_{r-1}}}$, the smaller the sum of the r-th autocorrelation functions.

4 Conclusions

We study the sum of high-order autocorrelation function of Boolean functions. We give the definition of the k-th autocorrelation function, and obtained the connections between the r-th autocorrelation function and the (r−1)-th autocorrelation function. Finally, we gave the tightness of the upper or lower bounds on the r-th autocorrelation function. Meanwhile we deduced the connection between Walsh transformation and the sum of the r-th autocorrelation function.

In the next step, we study cryptographical properties of the sum of high-order cross-correlation function of Boolean functions in cryptographic algorithms.

Acknowledgments. Some work was supported by the National Key R&D Program of China (No. 2017YFB0802000). We thank anonymous referees and the editor for comments.

References

1. Rothaus, O.S.: On "bent" functions. J. Comb. Theory Ser. A **20**, 300–305 (1976)
2. Courtois, N., Meier, W.: Algebraic attacks on stream ciphers with linear feedback. In: Advances in Cryptology-CRYPTO 2003. Lecture Notes in Computer Science, vol. 2656, pp. 346–359. Springer (2003)
3. Webster, A.F., Tavares, S.E.: On the design of S-boxes. In: Advances in Cryptology-CRYPTO 1985. Lecture Notes in Computer Science, vol. 219, pp. 523–534. Springer, Heidelberg (1986)
4. Canteaut, A., Carlet, C., Charpin, P., Fontaine, C.: Propagation characteristics and correlation immunity of highly nonlinear Boolean functions. In: EUROCRYPT 2000. LNCS, vol. 1807, pp. 507–522 (2000)
5. Carlet, C.: Boolean functions for cryptography and error correcting codes. In: Crama, Y., Hammer, P. (eds.) Monography Boolean Models and Methods in Mathematics, Computer Science, and Engineering, pp. 257–397. Cambridge University Press (2006). http://www-roc.inria.fr/secret/Claude.Carlet/chap-fcts-Bool-corr.pdf
6. Knudsen, L.R.: Truncated and high order differentials. In: Proceedings of Fast Software Encryption(2). LNCS, vol. 1008, pp. 196–211. Springer (1995)
7. Lai, X.: High order derivatives and differential cryptanalysis. In: Symposium on Communication, Coding, and Cryptography, 10–13 February 1994, Mote-Verita, Ascona, Switzerland (1994)
8. Maitra, S.: Highly nonlinear balanced Boolean functions with very good autocorrelation property. In: Proceedings of the Workshop on Coding and Cryptography 2001. Electronic Notes in Discrete Mathematics, vol. 6, pp. 355–364. Elsevier (2001)
9. Maitra, S.: Autocorrelation properties of correlation immune Boolean functions. In: INDOCRYPT 2001. LNCS, vol. 2247, pp. 242–253. Springer, Heidelberg (2001)
10. Preneel, B., Govaerts, R., Bandewalle, J.: Boolean functions satisfying higher order propagation criteria. In: Advances in Cryptology-EUROCRYPT'91. Lecture Notes in Computer Science, vol. 547, pp. 141–152. Springer, Heidelberg (1991)
11. Son, J.J., Lim, J.I., Chee, S., Sung, S.H.: Global avalanche characteristics and nonlinearity of balanced Boolean functions. Inf. Process. Lett. **65**, 139–144 (1998)
12. Sung, S.H., Chee, S., Park, C.: Global avalanche characteristics and propagation criterion of balanced Boolean functions. Inf. Process. Lett. **69**, 21–24 (1999)

13. Tarannikov, Y., Korolev, P., Botev, A.: Autocorrelation coefficients and correlation immunity of Boolean functions. In: ASIACRYPT 2001. LNCS, vol. 2248, pp. 460–479. Springer, Heidelberg (2001)
14. Tang, D., Zhang, W.G., Carlet, C., Tang, X.H.: Construction of balanced boolean functions with high nonlinearity and good autocorrelation properties. Des. Codes Crypt. **67**(1), 77–91 (2013)
15. Tang, D., Maitra, S.: Construction of n-variable (n = 2mod4) balanced Boolean functions with maximum absolute value in autocorrelation spectra < $2^{n/2}$. IEEE Trans. Inf. Theory (2017). https://doi.org/10.1109/TIT.2017.2769092
16. Zhang, X.M., Zheng, Y.L.: GAC- the criterion for global avalanche characteristics of cryptographic functions. J. Univ. Comput. Sci. **1**(5), 316–333 (1995)
17. Zhou, Y., Xie, M., Xiao, G.: On the global avalanche characteristics of two Boolean functions and the higher order nonlinearity. Inf. Sci. **180**, 256–265 (2010)
18. Zhou, Y., Zhang, W., Li, J., Dong, X., Xiao, G.: The autocorrelation distribution of balanced Boolean function. Front. Comput. Sci. **7**(2), 272–278 (2013)
19. Zhou, Y., Zhang, W., Zhu, S., Xiao, G.: The global avalanche characteristics of two Boolean functions and algebraic immunity. Int. J. Comput. Math. **89**(16), 165–179 (2012)

Research on Behavior of Micro-Particle in Turbid Phantom Based on Interferometry System

Zhang Mei[1], Lin Lin[2](✉), and Yang Lei[1]

[1] College of Electronic Engineering, Dongguan University of Technology, Dongguan 523808, China

[2] School of Information Engineering, Guangdong Medical University, Dongguan 523808, China

992592690@QQ.com

Abstract. The optical properties of micro-particle in turbid phantom is studied and measured based on interferometry system. The transmission characteristics is also measured for comparison with refractive signal. Absorption and scattering effects are distinguished by diameter of particles and the concentration of the dye. Gaussian beam is modeled by hyperboloid to simulate initial status of incident light. Monte Carlo method and Gauss function are used to fit the results. As a conclusion, interferometer system and transmission measurement are a pare of ideal combination to study behavior and optical properties of micro-particles in turbid material.

Keywords: Micro-particle · Interferometry system · Turbid phantom Gauss

1 Introduction

It is a complicated process of micro-particles' diffusion in solution or turbid materials. It could be seen as a stochastic process in which particles interact with the other particles in the medium in a random way. In biomedical tissue, cells, fibers and fat particles also present characteristics of absorption and scattering just like micro-particles, which makes it difficult to trace the movement and optical properties of micro-particles in tissue in vitro or in vivo. the transmission of light too complicated to be predict by analytical method. And several numerical methods have been applied to simulate light transmission, such as Monte Carlo method, finite element method and finite difference method. Optical coherence tomography system has been applied to kinds of clinical diagnosis for its high resolution, imaging speed and non-contact depth detection ability [1–4]. The shortcoming of OCT is that it can not separate out single parameter from depth-dependence signal. Transmission optical system can perform distribution measurement of light beam and obtain scattering property of phantom.

In this paper, we have researched on the behavior of micro-particle in turbid phantoms by spectroscopic optical coherence tomography (SOCT) which works in near infrared waveband and beam-split-analyzer which perform transmission measurement

K. Deng et al. (Eds.): ICMIR 2018, AISC 856, pp. 324–329, 2019.
https://doi.org/10.1007/978-3-030-00214-5_42

[5–8]. The experimental study on the characteristics of the depth dependent coherent light signal is carried out and compared with the distribution of transmissive light. We also have optimized the Monte Carlo method according to the characteristics of focused beam and study the scattering characteristics and the distribution of energy density of the light beam on the focus plane [9–11]. The corresponding mathematical models for Gauss focused beam are established, and the transmission and distribution characteristics of light in turbid sample are numerically simulated and analyzed. The research results have some guiding significance for tracing micro-particle in bio-tissue in vivo.

2 Methods

The structure of SOCT system is shown in Fig. 1 which includes a super luminescent diode with center wavelength of 850 nm, a 50/50 fiber coupler. The reference arm consists of double lens and one reflecting mirror fixed on a stage. In the sample arm, incident light is collimated to a galvanometer and focused on the turbid sample by an objective lens. The galvanometer can perform X-Y scanning.

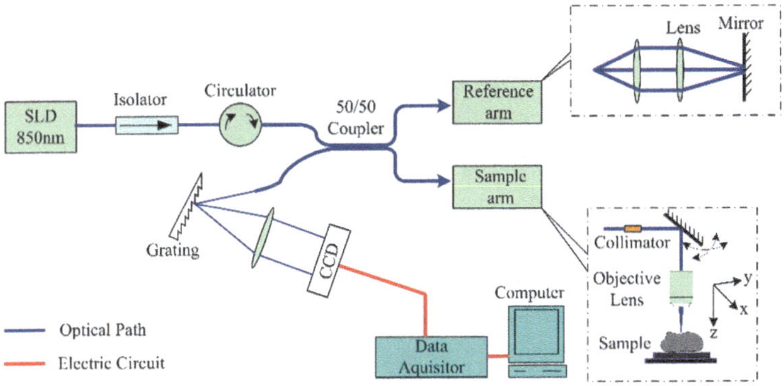

Fig. 1. Schematic presentation of the SOCT at 850 nm wavelength

The transmission measurement system includes a DFB laser of 808 nm, an objective lens, beam analyzer and single-mode fibers. The microsphere polystyrene is used as micro-particle with different diameter and solute in water to make suspension. For a Gauss beam, it spread along the transmission direction which is described by

$$I(p) = \frac{\omega_0^2}{\omega^2(z)} \exp\left(-\frac{p^2}{\omega^2(z)}\right) \tag{1}$$

where ω0 is the waist of Gauss beam, p is distance from the main axis and z is distance from the plane of waist.

3 Results

At first, we obtained spot images of the light beam of the OCT system at the focus plane and several defocus planes as shown in Fig. 2. It can be determined that the waist of incident light is 11.3 μm which is equal to the spatial resolution. Four kinds of PS micro-particle, dissolved in deionized water, used for test. Figure 3 presents the B-scan results by SOCT system at the rate of 20 frames per second. As the scattering coefficient increases with the particle size, one can get deeper depth of detection in suspension with smaller particle.

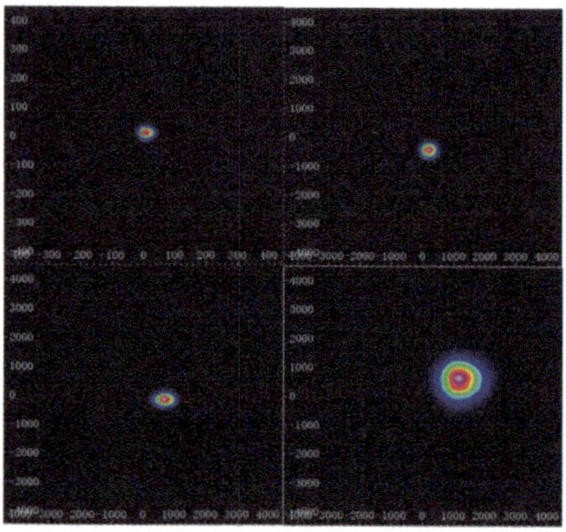

Fig. 2. Spot image of SOCT source measured by beam analyzer. The distance from lens plane is: 34.4 mm (focal plane), 39.5 mm, 29.4 mm and 19.5 mm.

For transmission measurement, we prepare sample in sample pool with 1 mm inner diameter and 1 mm wall thickness also as shown in Fig. 4. The light source is a DFB laser of 808 nm center wavelength with optical fiber output. An objective lens is used to focus light at the sample, behind which there is a beam analyzer to obtain energy distribution of transmissive light. In Fig. 4, f0, f1, f2 and f3 are the focus points in air, behind the first glass, the sample and the second glass respectively. In Fig. 5, it shows that the results of Monte Carlo simulation and Gauss function have good agreement with experimental data but not far away from the center. It suggested that a Gaussian light beam can not maintain initial shape after transmission from turbid sample.

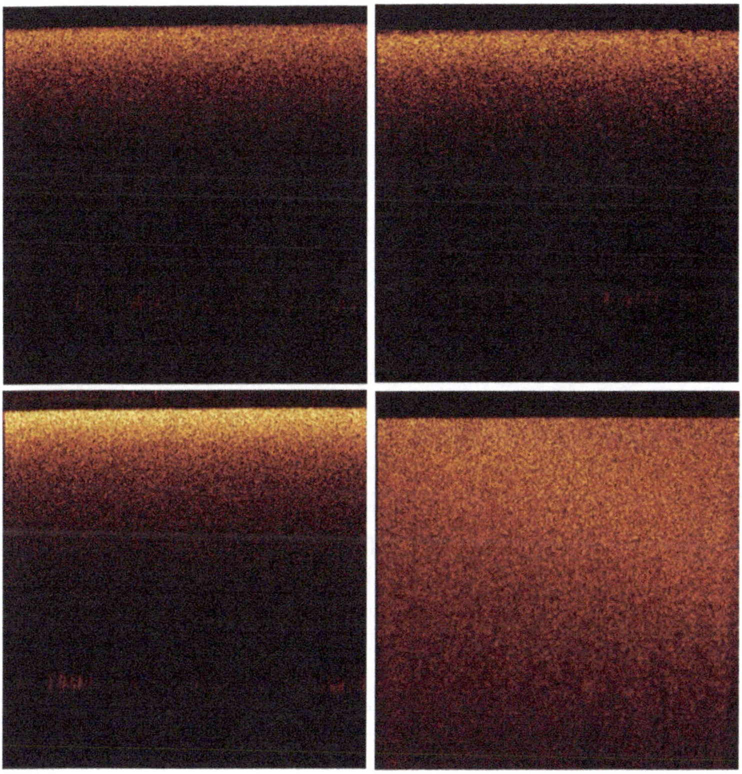

Fig. 3. B-scan result of PS micro-particle by SOCT. The diameters is: 2 μm, 1 μm, 0.5 μm and 0.05 μm.

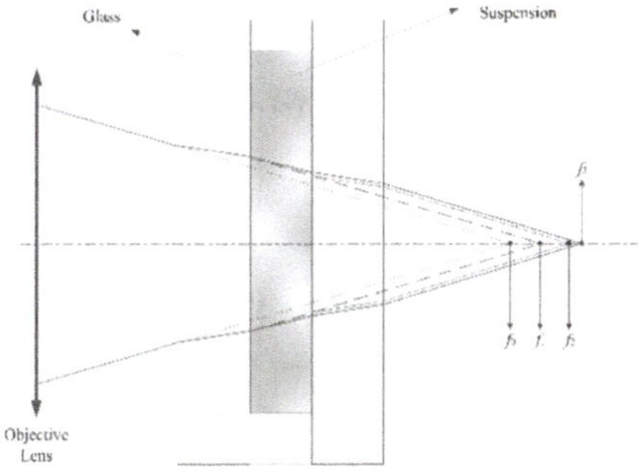

Fig. 4. Diagram of sample pool

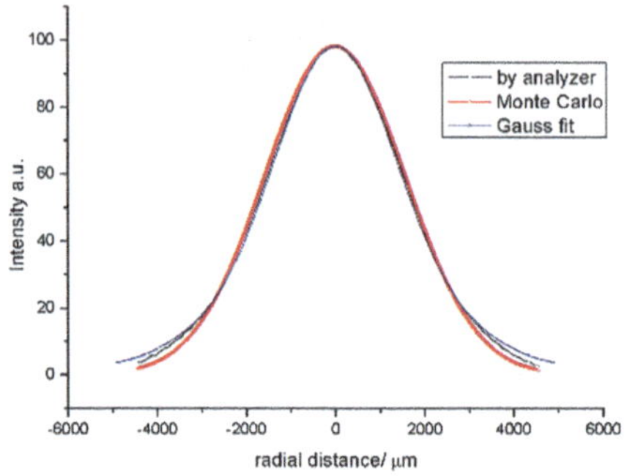

Fig. 5. Experimental result comparison with Monte Carlo simulation and Gauss fit.

4 Conclusion

In conclusion, we have tried to evaluate the signal characteristics of micro-particle by 850 nm SOCT system and 808 nm DFB laser transmission system. The experimental results have shown that the light beam will not maintain its initial distribution shape because of scattering effects. Depth dependence signal give the characteristics of light attenuation along the transmission direction. Transmission distribution reflects the scattering characteristics more than absorption of sample which reshape the beam edge. It should be possible to reconstruct both absorption and scattering coefficients based on two sets of data as mentioned above. We need to study more deeply into correlation analysis of the reflective and transmissive data. Future study will also focus on turbid samples with complex internal structure, such as film structure, and tissue with multiple layers.

Acknowledgements. This work was supported by Guangdong science and technology plan project (2015A020214024) and Educational Commission of Guangdong Province (YQ201402).

References

1. McLaughlin, R.A., Scolaro, L., Robbins, P., Saunders, C., Jacques, S.L., Sampson, D.D.: Parametric imaging of cancer with optical coherence tomography. J. Biomed. Opt. **15**, 046029 (2010)
2. Barwari, K., de Bruin, D.M., Faber, D.J., van Leeuwen, T.G., de la Rosette, J.J., Laguna, M. P.: Differentiation between normal renal tissue and renal tumours using functional optical coherence tomography: a phase I in vivo human study. BJU Int. **110**, E415–E420 (2012)
3. Tomlins, P.H., Adegun, O., Hagi-Pavli, E., Piper, K., Bader, D., Fortune, F.: Scattering attenuation microscopy of oral epithelial dysplasia. J. Biomed. Opt. **15**, 066003 (2010)
4. Zhang, Q.Q., Wu, X.J., Tang, T., Zhu, S.W., Yao, Q., Gao, B.Z., Yuan, X.C.: Quantitative analysis of rectal cancer by spectral domain optical coherence tomography. Phys. Med. Biol. **57**, 10 (2012)

5. Sassaroli, A., Blumetti, C., Martelli, F., Alianelli, L., Contini, D., Ismaelli, A., Zaccanti, G.: Monte Carlo procedure for investigating light propagation and imaging of highly scattering media. Appl. Opt. **37**, 7392–7400 (1998)
6. Schmitt, J.M., Knuttel, A., Bonner, R.F.: Measurement of optical properties of biological tissues by low-coherence reflectometry. Appl. Opt. **32**, 6032–6042 (1993)
7. Wang, L.H., Jacques, S.J., Zheng, L.Q.: MCML-Monte Carlo modeling of light transport in multiple layered tissues. Comput. Methods Programs Biomed. **47**, 131–146 (1995)
8. Hayakawa, C.K., Spanier, J., Bevilacqua, F., Dunn, A.K., You, J.S., Tromberg, B.J., Venugopalan, V.: Perturbation Monte Carlo methods to solve inverse photon migration problems in heterogeneous tissues. Opt. Lett. **26**, 1335–1337 (2001)
9. Chen, C., Lu, J.Q., Ding, H., Jacobs, K.M., Du, Y., Hu, X.: A primary method for determination of optical parameters of turbid samples and application to intralipid between 550 and 1630 nm. Opt. Exp. **14**, 7420–7435 (2006)
10. Thrane, L., Yura, H.T., Andersen, P.E.: Analysis of optical coherence tomography systems based on the extended Huygens-Fresnel principle. J. Opt. Soc. Am. A **17**, 484–490 (2000)
11. Lin, L., Zhang, M., Zhang, H.L.: Numerical simulation of focused optical beam in tissue samples. Chin. J. Med. Imaging Technol. **26**, 2375–2378 (2010)

Optimal Design of Magnetic Shielding Device Based on Seeker Optimization Algorithm

Zhi-feng Lv(✉), Jin-sheng Zhang, Shi-cheng Wang, and Ting Li

High-Tech Institute of Xi'an, Xi'an 710025, China
lvzhifeng1213@163.com

Abstract. The magnetic shielding device has important application value in the fields of geomagnetic navigation, space engineering, space science, national defense industry, microparticle experiment and so on. The optimization of its structure parameters has important theoretical value, economic value and engineering value. In order to optimize the structure parameters, a parameter optimization method of magnetic shielding device based on seeker optimization algorithm is proposed. First, the analytical expression of the calculation of magnetic shielding performance is summarized. Then, because of more optimization variables leading to optimization falling into local optimum, the seeker optimization algorithm (SOA) is used to achieve global optimization. Finally, taking the optimization of structure parameters of the three layer magnetic shielding device as an example, a comparative simulation experiment is carried out. The simulation results show that the seeker optimization algorithm has faster convergence speed, higher precision, and is more easy to find the global optimal solution than the traditional optimization method (genetic algorithm and particle swarm optimization). The method can provide a reference for the structure parameters optimization of the magnetic shielding devices.

Keywords: Magnetic shielding device · Seeker optimization algorithm (SOA) Parameter optimization · Magnetic shielding performance

1 Introduction

For high precision geomagnetic measurement, in order to ensure the normal work of high precision magnetic measuring instruments, it is necessary to eliminate the interference magnetic field produced by the nearby ferromagnetic objects and the electronic control equipment, which requires the establishment of near zero magnetic environment [1–3]. At present, most of the near zero magnetic environment is mainly passive magnetic shielding at home and abroad [4–6], which is to use high permeability material to build a closed magnetic shielding device to ensure that the magnetic field in the internal space of the device is close to "zero magnetic". The magnetic shielding device has important application value in the fields of geomagnetic navigation, space engineering, space science, national defense industry, microparticle experiment and so on. In literature [7], a small geomagnetic environment simulation system is established inside the magnetic shielding device, and the semi physical simulation experiment of geomagnetic navigation is carried out and the performance of the geomagnetic

K. Deng et al. (Eds.): ICMIR 2018, AISC 856, pp. 330–338, 2019.
https://doi.org/10.1007/978-3-030-00214-5_43

navigation matching algorithm is evaluated [8]. In literature [9], a magnetic shielding device is designed based on 1J85 to solve the problem that high precision fiber inertial measurement unit (IMU) is sensitive to magnetic field. It can effectively reduce the magnetic field sensitivity of fiber-optic gyroscope (FOG) and improve the precision of IMU. In literature [10], a large magnetic shielding device was established to study the inherent electric dipole moment of microparticles. It uses the latest design and technology and achieves a good uniformity of the internal magnetic field.

In theory, the more layer number and the larger thickness of the magnetic shielding device, the better the shielding performance. However, it will make the structure of the device complex, and the weight and cost will increase. Therefore, the optimization of structure parameters is one of the key technologies for the design of the magnetic shielding device. Li Pan analyzed and studied the influence of the structure parameters on the magnetic shielding coefficient of multilayer magnetic shield, and optimized the design of the multilayer magnetic shield of nuclear magnetic resonance gyroscope [11]. Zhang Hong obtained the parameters optimization model of the magnetic shielding tube by improving the axial coefficient formula of multi-layer magnetic shielding tube. Through this model, the magnetic shielding tube with small volume, light mass, low cost and large shielding performance can be obtained [12]. Although these studies have achieved good shielding effectiveness, they set the same values of material thickness t, radial interval ΔR and axial interval ΔL in different layers in the process of optimization, which can reduce the optimization variables and simplify the optimization problem. The reduction of variables will reduce the diversity of structure size, which limits the further improvement of magnetic shielding performance.

In order to solve this problem, this paper takes the material thickness t, radial interval ΔR and axial interval ΔL of the different layers as the optimized variables, which increases the number of variables to be optimized. Because of more optimization variables making the optimization more difficult and more easily fall into local optimum, this paper proposes a structure parameter optimization based on the seeker optimization algorithm (SOA) for magnetic shielding device to achieve global optimization, so as to realize the magnetic shielding effectiveness maximization.

2 Magnetic Shielding

2.1 Principle of Magnetic Shielding

The principle of magnetic shielding is to make use of the characteristics of high permeability and low magnetic resistance of ferromagnetic materials. It provides a low reluctance path for magnetic flux. By constructing a closed cavity device, the magnetic field is shunted to the outside, so that the magnetic field entering the inner cavity is greatly reduced, so as to reach the purpose of shielding. The principle is shown in Fig. 1.

The magnetic shielding performance is usually evaluated by magnetic shielding coefficient. For the magnetic shielding device shown in Fig. 1, the magnetic shielding coefficient can be expressed as:

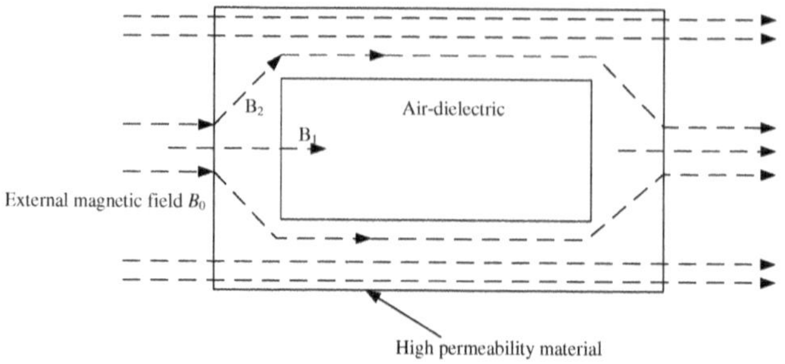

Fig. 1. Magnetic shielding principle diagram

$$S = \frac{B_0}{B_1} \tag{1}$$

In formula (1): B_0 is the magnetic induction intensity before shielding. B_1 is the magnetic induction intensity at the center of shielding device after shielding. The larger the magnetic shielding coefficient is, the better the shielding effectiveness is.

As with the principle of the circuit, two resistors are connected in parallel, even if one of them has a large resistance, there will be current passing through. So, even if the permeability of the shielding layer is high, its internal space can not be absolutely "zero magnetic". Theoretically, the thicker the shielding shell and the higher the permeability is, the better the shielding effect is. In order to achieve the best magnetic shielding performance, multilayer shielding is usually used to shield the residual magnetic flux by layer.

2.2 Calculation of Magnetic Shielding Performance

For a single-layer cylindrical magnetic shielding device, its transverse magnetic shielding coefficient is [11]

$$S_T \approx 1 + \frac{\mu_r t}{2R} \tag{2}$$

In formula (2): μ_r indicates the relative permeability of materials. t indicates the thickness of shielding layer. R indicates the radius of shielding layer.

For the cylindrical shielding device with n layers, its axial section is shown in Fig. 2.

In Fig. 2, t_i, R_i and L_i represent the thickness, radius and axial length of the shielding layer respectively. $\Delta R_{i,i+1}$ and $\Delta L_{i,i+1}$ represent the radial interval and axial interval between the ith layer and the $i+1$th layer respectively, so $R_{i+1} = R_i + t_i + \Delta R_{i,i+1}$, $L_{i+1} = L_i + 2t_i + 2\Delta L_{i,i+1}$.

For shielding device with n layers, the transverse magnetic shielding coefficient is

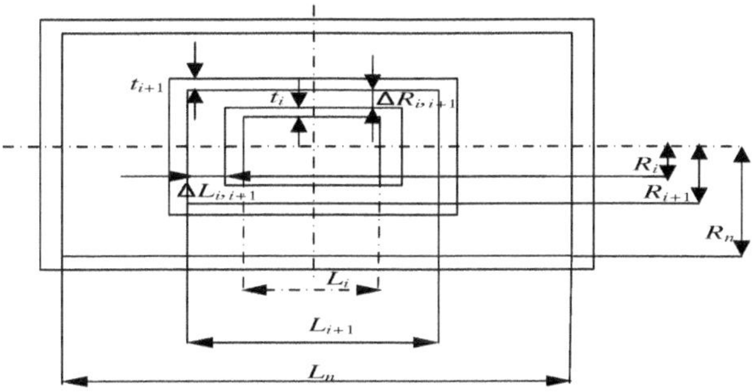

Fig. 2. Axial section of n layers magnetic shielding device

$$S_T \approx 1 + \sum_{i=1}^{n} S_T^i + \sum_{i=1}^{n-1}\sum_{j>i}^{n} S_T^i S_T^j \left[1 - (R_i/R_j)^2\right] + \sum_{i=1}^{n-2}\sum_{j>i}^{n-1}\sum_{k>j}^{n} S_T^i S_T^j S_T^k \left[1 - (R_j/R_k)^2\right] \left[1 - (R_i/R_j)^2\right] + \ldots + S_T^1 S_T^2 \ldots S_T^n \left[1 - (R_1/R_2)^2\right] \ldots \left[1 - (R_{n-1}/R_n)^2\right] \tag{3}$$

In formula (3): i represents the ith shielding layer from the inner to outer.

Since the direction of magnetic field is different, besides the transverse magnetic shielding coefficient, the longitudinal magnetic shielding coefficient should also be considered. For a single-layer cylindrical magnetic shielding device, its longitudinal magnetic shielding coefficient is

$$S_A \approx 1 + S_T \frac{2K}{1 + a + \alpha a^2/3} \tag{4}$$

In formula (4): S_T is the transverse magnetic shielding coefficient. K is

$$K = \left(1 + \frac{1}{4a^3}\right)\beta - \frac{1}{a} + 2a\left[\ln\left(a + \sqrt{1 + a^2}\right) - 2\left(\sqrt{1 + \frac{1}{a^2}} - \frac{1}{a}\right)\right] \tag{5}$$

In formula (5): $a = L/R$. L indicates the axial length of the magnetic shielding device. α and β can be determined by measuring the standard magnetic flux distribution of the cylinder ends and side walls respectively, usually taking $\alpha = 0.85 \pm 0.03$, $\beta = 1.83 \pm 0.06$.

For the cylindrical shielding device with n shielding layers, its longitudinal magnetic shielding coefficient is

$$S_A \approx 1 + \sum_{i=1}^{n} S_A^i + \sum_{i=1}^{n-1}\sum_{j>i}^{n} S_A^i S_A^j \left[1 - \left(L_i/L_j\right)^2\right] + \sum_{i=1}^{n-2}\sum_{j>i}^{n-1}\sum_{k>j}^{n} S_A^i S_A^j S_A^k \left[1 - \left(L_j/L_k\right)^2\right]$$
$$\left[1 - \left(L_i/L_j\right)^2\right] + \ldots + S_A^1 S_A^2 \ldots S_A^n \left[1 - \left(L_1/L_2\right)^2\right] \ldots \left[1 - \left(L_{n-1}/L_n\right)^2\right] \tag{6}$$

From the formula (2) and formula (4), we can find that the magnetic shielding coefficient increases with the increase of the material thickness. And we can also see that the magnetic shielding coefficient increases with the increase of the layers from the formula (3) and formula (6). High permeability material is very expensive and its density is large. If we increase the magnetic shielding coefficient only by increasing the thickness of the material and the number of layers, it will inevitably lead to the increase of the cost and the increase of the weight. Therefore, the structure parameters of magnetic shielding devices need to be optimized to solve the contradiction between shielding cost and shielding performance.

3 Parameters Optimization Based on SOA

The seeker optimization algorithm (SOA) directly simulates the behavior of human intelligence search. Based on the traditional direct search algorithm, SOA takes the search team as population and search location as candidate solution. By simulating human search "experience gradient" and "uncertain reasoning", we can get the best solution to the problem. Its concept is clear and easy to understand. And it's a new swarm intelligence algorithm in the field of evolutionary algorithm research [13].

3.1 The Determination of the Search Step Length

The uncertain reasoning behavior of SOA is to use the approximation ability of fuzzy system to simulate human intelligence search behavior, which is used to establish the connection between perception (that is, the value of the objective function) and behavior (the step length). In many functions, Gauss membership function is generally used to represent search step fuzzy variables:

$$u_A(x) = e^{-(x-u)^2/2\delta^2} \tag{7}$$

In formula (7): u_A is Gauss membership. x is input variable. u and δ are parameters of membership function. When the output variable exceeds $[u - 3\delta,\ u + 3\delta]$, if the membership degree is less than 0.0111, it can be ignored, so set $u_{\min} = 0.0111$.

By using the linear membership function, the membership degree is directly proportional to the order of the function value. That is, the maximum membership value is $u_{\max} = 1.0$ at the best position, the worst position has the minimum membership degree $u_{\min} = 0.0111$ and membership value is $0.0111 < u < 1.0$ in other positions, for example (8):

$$u_{ij} = rand(u_i, 1)\,(j = 1, 2, \ldots, D) \tag{8}$$

In formula (8): u_i is the membership degree of the objective function value i. u_{ij} is the membership degree of the objective function value i at j dimension search space. D is the dimension of search space. The formula (8) is used to simulate the randomness of human search behavior. Function rand $(u_i, 1)$ is the real number distributed uniformly on the interval $[u_i, 1]$. The uncertainty reasoning obtains the membership degree u_{ij} by the formula (8), and the step length can be obtained according to the uncertain reasoning:

$$\alpha_{ij} = \delta_{ij}\sqrt{-\ln(u_{ij})} \tag{9}$$

In formula (9): α_{ij} is the search step of j dimension search space. δ_{ij} is the parameter of Gauss membership function, and its value is determined by formula (10) and (11):

$$\vec{\delta_{ij}} = \omega|\vec{x}_{\min} - \vec{x}_{\max}| \tag{10}$$

$$\omega = (iter_{\max} - iter)/iter_{\max} \tag{11}$$

In formula (10) and (11): $x_{\min}$ and $x_{\max}$ are the minimum and maximum function values in the same population respectively. ω is the inertia weight, decreasing from 0.9 linear to 0.1 with the increase of evolutional generation. *iter* and $iter_{\max}$ are the number of current iterations and the maximum iterations respectively.

3.2 Search Direction Determination

Through the analysis and modeling of human self-interested actions, altruistic behavior and premotion, the egoism direction $\vec{d}_{i,ego}$, altruistic direction $\vec{d}_{i.alt}$ and premotion direction $\vec{d}_{i,pro}$ of the ith search individual are obtained respectively, such as formula (12)~(14):

$$\vec{d}_{i,ego}(t) = \vec{p}_{i,best} - \vec{x}_i(t) \tag{12}$$

$$\vec{d}_{i,alt}(t) = \vec{g}_{i,best} - \vec{x}_i(t) \tag{13}$$

$$\vec{d}_{i,pro}(t) = \vec{x}_i(t_1) - \vec{x}_i(t_2) \tag{14}$$

Taking the above three factors into account, we use random weighted geometric mean of the three directions to determine the search direction:

$$\vec{d}_{ij}(t) - sign(\omega\vec{d}_{ij,pro} + \phi_1\vec{d}_{ij,ego} + \phi_2\vec{d}_{ij,alt}) \tag{15}$$

In formula (12)~(14), $\vec{x}_i(t_1)$ and $\vec{x}_i(t_2)$ are the best positions in $\{\vec{x}_i(t-2), \vec{x}_i(t-1), \vec{x}_i(t)\}$ respectively. $g_{i,best}$ is the best location of the ith search individual's neighborhood in the collective history. $p_{i,best}$ is the best location that the ith search individual has gone through so far. Sign() is a sign function. φ_1 and φ_2 are constants in [0,1]. ω is inertia weight.

3.3 Update of Individual Position

After determining the search direction and step length, update the position. The updating formula is (16) and (17):

$$\Delta x_{ij}(t+1) = \alpha_{ij}(t)d_{ij}(t) \quad (16)$$

$$x_{ij}(t+1) = x_{ij}(t) + \Delta x_{ij}(t+1) \quad (17)$$

3.4 Realization of the Algorithm

SOA optimizes the problem according to the following analysis process.

Step 0: Initialization. Generate s initial locations in the feasible solution domain randomly: $\{\vec{x}_i(t)|\vec{x}_i(t) = (x_{i1}, x_{i2}, \ldots, x_{im})\}$, i = 1, 2, 3, …, s; t = 0.

Step 1: Evaluate and calculate the objective function value of each location.

Step 2: Determine the search strategy and calculate the search direction $d_{ij}(t)$ and step length $\alpha_{ij}(t)$ of each body i in j dimension.

Step 3: Update the location of each searcher according to the formula (17).

Step 4: $t \rightarrow t$+1.

Step 5: Stop search if stop condition is satisfied. Otherwise, go to **Step 1**.

4 Simulation Experiment and Analysis

In general, three shielding layers can achieve better magnetic shielding effectiveness. Therefore, this paper takes the structure parameters optimization of the three-layer magnetic shielding device as an example to carry out the simulation experiment. In order to demonstrate the superiority of SOA compared with other traditional swarm intelligence optimization algorithms, we conduct comparative simulation experiments. The parameters are optimized by genetic algorithm (GA), particle swarm optimization (PSO) algorithm and SOA based on the conditions of the simulation example in literature [11]. By comparing the final magnetic shielding coefficient, we can evaluate the method is good or bad.

4.1 The Determination of Initial Condition, Variables and Objective Function

For optimize the magnetic shielding device, our ultimate goal is to achieve the best magnetic shielding performance under certain constraints (such as cost, weight, volume, etc.). Here, we adopt the same limiting conditions as literature [11]. In the material aspect, the relative permeability is set to $\mu_r = 10000$. In the weight aspect, the total weight $M_{\max}$ is the final optimization result of the three-layer magnetic shielding device in literature [11]. We set $M_{\max}$ as weight constraint and make the total weight of three layers to be optimized satisfied by $M_1 + M_2 + M_3 \leqslant M_{max}$. In the volume aspect, the most inner cavity volume is satisfied by $V_1 = \pi R_1^2 L_1 = 100\,\text{cm}^3$. Besides, the outermost volume of literature [11] is V_{max}. We also set V_{max} as volume constraint and make the outermost volume of the device to be optimized satisfied by $V_3 \leqslant V_{max}$. That is to say, the structure parameters of the device are optimized under the same material, the same weight and the same volume restriction conditions.

Because of the limitation $V = \pi R_1^2 L_1 = 100\,\text{cm}^3$, only one of R_1 and L_1 is independent variable. We set R_1 as the independent variable. In literature [11], the material thickness t, radial interval ΔR and axial interval ΔL of each layer are taken as known quantities and satisfied by $t_1 = t_2 = t_3 = \Delta R_{12} = \Delta R_{23} = \Delta L_{12} = \Delta L_{23} = 1$ mm. So, the variable to be optimized is only R_1 in [11]. Here, in order to improve the magnetic shielding performance, we set $R_1, t_1, t_2, t_3, \Delta R_{12}, \Delta R_{23}, \Delta L_{12}, \Delta L_{23}$ as variables to be optimized.

In reality, the interference of magnetic shielding device is uncertain due to the direction of magnetic field. Therefore, it should follow the "short board effect" to evaluate the shielding effectiveness. That is to say, the overall shielding coefficient is the smaller one of the transverse magnetic shielding coefficient and the longitudinal magnetic shielding coefficient. So, we use $f = \min(S_T, S_A)$ as fitness function. When f reaches the maximum value, the corresponding parameters are the optimal parameters.

4.2 Analysis of Simulation Results

Taking the condition in Sect. 4.1 as the initial condition, the structure parameters of the device are optimized by PSO, GA and SOA algorithm. The optimal fitness function curve is shown in Fig. 3.

From Fig. 3, we can find that the accuracy of SOA and GA is higher than that of PSO algorithm. This is because that the PSO algorithm is easy to fall into the local optimal, so the early maturing phenomenon appears. Besides, the SOA is faster in convergence speed and more accurate than the GA, and the SOA can find the global optimal solution more easily. It shows that the SOA is effective and superior.

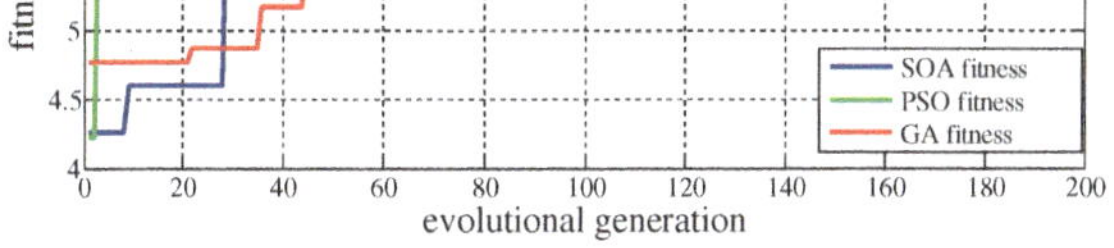

Fig. 3. Fitness function change curve

5 Conclusion

The structure parameters of the magnetic shielding device have an important influence on the magnetic shielding performance. Aiming at the optimization of structure parameters, the SOA is applied to optimize the structure parameters. By compared with genetic algorithm and particle swarm optimization algorithm, the SOA has faster convergence speed and higher optimization precision. The method can provide a reference for the optimization of the structure parameters of magnetic shielding devices.

Acknowledgments. This research is funded by the National Natural Science Foundation (Project Nos. 11602296, 61503393). The authors are greatly thanks to these grants.

References

1. Feng, G., Sun, Y., Zheng, X., et al.: Design and measurement of a magnetic field for precision spectroscopy of helium. Acta Phys. Sin. **63**(12), 123201 (2014)
2. Li, L., Sun, Z., Pan, D., et al.: Status reviews for nearly zero magnetic field environment facility. Trans. China Electrotech. Soc. **30**(15), 136–140 (2015)
3. Guo, J., Meng, J., Li, X., et al.: Dimension design of magnetic shielding cylinder for optical pumped magnetometer. J. Electron. Meas. Instrum. **32**(2), 80–85 (2018)
4. Bork, J., Hahlbohm, H.D., Klein, R., et al.: The 8-layered magnetically shielded room of the PTB: design and construction. Proceedings of 12th international conference on biomagnetism, Espoo, Finland, 13–17 August 2000
5. Dickerson, S., Hogan, J.M., Johnson, D.M.S., et al.: A high-performance magnetic shield with large length-to-diameter ratio. Rev. Sci. Instrum. **83**(6), 065108 (2012)
6. Kajiwara, G., Harakawa, K., Ogata, H., et al.: High-performance magnetically shielded room. IEEE Trans. Magn. **32**(4), 2582–2585 (1996)
7. Lv, Z.-f., Zhang, J.-s., Wang, S.-c., et al.: Design and research of high-accuracy geomagnetic field intensity simulation system. J. Astron. **35**(11), 1284–1290 (2014)
8. Wang, S.-c., Lv, Z.-f., Zhang, J.-s, et al.: Evaluation of ICCP algorithm for geomagnetic navigation based on hardware-in-the-loop simulation. J. Beijing Univ. Aeron. Astron. **41**(2), 187–192 (2015)
9. Li, J., Fang, J.: Magnetic shielding method and experiment study of inertial measurement unit based on high percision fiber-optic gyroscope. Acta Aeron. Astron. Sin. **32**(11), 2106–2116 (2011)
10. Altarev, I., Babcock, E., Beck, D., et al.: A magnetically shielded room with ultra low residual field and gradient. Rev. Sci. Instrum. **85**(7), 075106 (2014)
11. Li, P., Liu, Z., Wang, J.: Optimization design of multilayer magnetic shield for nuclear magnetic resonance gyroscopes. J. Chinese Inert. Technol. **24**(3), 383–389 (2016)
12. Zhang, H., Chen, X., Zou, S.: Parameter optimization design of magnetic shielding cylinders for atomic spin magnetometer. J. Southeast Univ. (Nat. Sci. Ed.) **44**(6), 1177–1180 (2014)
13. Yu, S., Cao, Z.: Optimization parameters of PID controller parameters based on seeker optimization algorithm. Comput. Simul. **31**(9), 347–350 (2014)

Research on Chinese and Vietnamese Bilingual Event Graph Extraction Method

Wanjin Che[1,2], Zhengtao Yu[1,2(✉)], Xian Li[1,2], and Shengxiang Gao[1,2]

[1] School of Information Engineering and Automation, Kunming University of Science and Technology, Kunming 650500, China
ztyu@hotmail.com

[2] Key Laboratory of Intelligent Information Processing, Kunming University of Science and Technology, Kunming 650500, China

Abstract. The Event extraction is the most important research direction in information extraction field. In order to understand the difficulty of bilingual events, the method of extracting bilingual event graphs is proposed. We use trigger words to identify sub-events in bilingual news events. Define an event association graph with a sub-event as an event graph node to construct a single event association. With the help of the Chinese and Vietnamese dictionaries and entities, the relations between the two events is constructed and the event graph is used to calculate the intensity of the event graph. Experiments on the relations between Chinese and Vietnamese events are carried out. The results show that the event extraction graph is used to analyze the correlation between Chinese and Vietnamese events. Finally, it has achieved good results in the identification of event relations.

Keywords: Chinese and vietnamese bilingual · Event relation extraction Event graph

1 Introduction

The relation between Vietnam and China has become more and more important. Cross language research has always been a difficult point in the study of natural language processing. Bilingual event understanding is the key part and the premise of cross language event analysis. It is very important to analyze news events clues in Chinese and Vietnamese news information. At present, it is difficult to analyze bilingual events, because it is difficult to establish the corresponding relations between bilingual languages since they are not in the same space. In terms of monolingual event analysis, Zhao [1] proposed the stage of event type recognition trigger expansion and classification based on the combination of binary classifier and the event element recognition stage based on the maximum entropy method of multiple classifications. Wang [2] proposed to use the news unique characteristic and the heading prompt information for training the classifier to obtain the topic sentence, thus obtains the news essential event and the event element. Xu [3] proposed acquisition method by means of word semantic and trigger element template conversion rule template to improve the template extraction of event element effect. Huang and Riloff [4], trained a multi-level event extraction template, for each seed trigger

K. Deng et al. (Eds.): ICMIR 2018, AISC 856, pp. 339–346, 2019.
https://doi.org/10.1007/978-3-030-00214-5_44

word and then extract the elements through the template. The most important sentence Kastner [5] used semantic and grammatical features and statistical classifier design extraction automatically extracting news points for CNN. In terms of bilingual event analysis, Hong [6] and other named entity types were used to predict event sentences. Bilingual research methods have been applied in many fields, such as sentiment analysis [7, 8]. In the field of event extraction, there are only a few related studies. Based on the analysis of events in Chinese and Vietnamese bilingual news events, Gao Shengxiang [9] proposed a method for generating event clues between Chinese and Vietnamese based on global/local co-occurrence words. Zhu [10] proposed an event extraction method based on Bilingual feature space superposition, a cross language event extraction method based on integer linear programming and a bilingual event extraction method based on active learning. This paper focuses on the relation between all events and events contained in news from news. In bilingual news, there will be the same paragraph describing the news. Therefore, we extract the events as nodes and construct the event graph model based on theco-occurrence relation of events in news text and the event relations of sub-events. Through the event graph model, we can see that the same news in Chinese and Vietnamese reports is different and same.

2 Chinese and Vietnamese Bilingual Event Graphs Construction

2.1 Event Extraction and Event Graph Definition

The candidate events are identified according to the defined set of trigger words. A diagram is a good way to express the relation between things. In this paper, things are independent events and in news texts they are reports in news. The node of the graph represent the events extracted from the news. The edge of the graph represent the coreference relations and temporal relations between events. Among them, some relations are equal, such as coreference relations and time equal, and some relations are unequal. Suppose that the event set is represented by M and R represents the monolingual event set. C represents a bilingual event set. Then, a bilingual event graph can be represented as:

$$\mathrm{G} = (V, E, A, F, m, r, c) \tag{1}$$

V represents node set, *E* represents monolingual undirected edge set, *A* represents monolingual directed edge set, and *F* represents bilingual undirected edge set. *m* represents from the *V* node to the set of instructions in events in the relations between *M*: $m \rightarrow v$, *r* represents an instruction relations to the monolingual side of the $\mathrm{E} \cup \mathrm{A}$ set to the event relation of the *R*: $r \rightarrow \mathrm{e} \vee a$, *c* represents from *F* to bilingual side events in *c* indicate the relations set: $c \rightarrow f$.

2.2 Generation of Nodes of Event Graphs

Each event in *M* may have different meanings in different research backgrounds. It needs to be decided according to the actual situation. In this article, the *M* event refers to the actual events that occur to the real world. Each event m contains event elements. Each event element has its own categories. Such as the incident "Guo Chuan fell into the water due to sudden accidents at 15:00 to 15:30." can be represented as:

$$m \rightarrow v = \left\{ \begin{array}{l} \text{main body} = \text{"Guo Chuan"}, \\ \text{trigger words} = \text{"fell into the water"}, \\ \text{time} = \text{and("15 : 00", "15 : 30")} \end{array} \right\} \quad (2)$$

Graph nodes are linked to events by graph. In monolingual single text environments we do not merge events and similar events are preserved that each node uniquely points to the only event.

2.3 Generation of Edges of Event Graphs

In this paper, we focus on the two relations: temporal relations and coreference relations.

Temporal relations extraction of events. Most of the relevant work defines the range of time extraction into two kinds: one is the relations between the time only selected important events that the time of semantic relation extraction or only a sentence or event; the other is that there are events pointing to the relations between all the events of last event in accordance with the time of relation tree or acyclic directed graph. We will judge whether there is a temporal relations between events.

Coreference relations extraction of events. By extracting the relation of the same event described in the same news to construct an undirected edge of the event graph. We assume that the two events have coreference relations. We categorize events by using SVM. The following features are selected by event sentence similarity based on word bag model and named entity comparative features.

The event elements and trigger words in the event are taken as nodes, the similarity between the events of the two events and the word similarity of the trigger words are taken as the edges and the similarity of words is used as the weight of the edges. We use HowNet based word similarity to compute the similarity between a pair of words (w_i, w_j). By classifying the vertices according to the two events, the similarity computation between the two word bags can be transformed into a weighted two graph (b_1, b_2) maximum matching problem. So the figure $G = (V, E)$ and $V = (V_1, V_2)V_1 \rightarrow w_i \in b_1$, $V_2 \rightarrow w_j \in b_2$.

The goal construct a graph with the largest edge weight and the edge of the graph is the similarity of word meaning. If it is a collection of all possible pairs of words, there are:

$$Sim(b_1, b_2) = \frac{\sum_{(w_i, w_j) \in P} Sim(w_i, w_j)}{Max(|b_1|, |b_2|)} \quad (3)$$

Among them, $|b_1|$ and $|b_2|$ represent the cardinal number of the two word bags respectively.

3 Chinese and Vietnamese Bilingual Event Graphs Computation

The event graph is constructed using the extracted events that the temporal relations between events and the common and cross-language alignment relations. The event graph can be constructed by using the event as the node of the event graph, the association with the events for the cross-language alignment. We introduce edge weight and node weight computation for event graphs.

As for edge weight computation for event graphs, the weight of the edge of event graph reflects the degree of correlation between nodes and the correlation between events. In order to represent the correlation between event elements, each event is represented as a word bag by means of the word bag model. Compute similarity between events.

$$Sim(v_1, v_2) = \frac{\sum_{(e_i, e_j)\in P} Sim(e_i, e_j)}{Max(|v_1|, |v_2|)} \tag{4}$$

The similarity between events is converted to the maximum two-point graph problem. P is the set of all possible event pairs, and. The words consist of two events in bags, $v_i = e_1, e_2, \ldots, e_p$ and $v_j = e_1, e_2, \ldots, e_q$. Hungarian algorithm is used to compute the maximum similarity between two events.

When the similarity between the two events is obtained, then the similarity in a news event set $K_c = \{v_1, v_2, \ldots, v_p\}$ can be expressed as the matrix M:

$$M = \begin{bmatrix} Sim_{1,1} & Sim_{1,2} & \cdots & Sim_{1,i} & \cdots & Sim_{1,p} \\ Sim_{2,1} & Sim_{2,2} & \cdots & Sim_{2,i} & \cdots & Sim_{2,p} \\ \cdots & \cdots & \ddots & \vdots & \ddots & \vdots \\ Sim_{3,1} & Sim_{3,2} & \cdots & Sim_{i,i} & \cdots & Sim_{i,p} \\ \vdots & \vdots & \ddots & \vdots & \ddots & \vdots \\ Sim_{p,1} & Sim_{p,2} & \cdots & Sim_{p,p} & \cdots & Sim_{p,p} \end{bmatrix} \tag{5}$$

As for node weight computation for event graphs, in the constructed event graph, the sorting of events can be transformed into the weight computation of graph nodes then the events are sorted by weight. Using the PageRank algorithm to construct the weight calculation model with event as node. Therefore, the undirected graph node weight model of the news sub-event element can be represented as:

$$SW(i) = (1-d) + d * \sum_{j=1, j\neq i}^{n} M_{i,j} * \frac{SW(j)}{\sum\limits_{k=1, k\neq i}^{n} M_{i,k}} \tag{6}$$

In the formula, the $SW(i)$ represents the weight of the node i, and the d represents the damping coefficient, which is generally set to 0.85. $M_{i,j}$ representing the value in the similarity matrix M, which represents the similarity between the i and the j events in the news A.

4 Experimental Results and Analysis

The experimental data consisted of 150 pairs of news sets as training sets, and 358 pairs as testing sets. The number of events, the number of relations and the total number of events include the training set and test set of the number of temporal relations are 326 and 934, the training set and test set of the number of coreference are 638 and 1974.

The experimental results are evaluated mainly from three aspects of accuracy, recall rate and F value. Precision is the ratio of the number of correctly extracted event temporal relations to the number of relations taken at all times. Recall is the ratio of the number of time events that correctly identify the number of events to the number of events in each event. The number of A assumptions that all events in the corpus, B represents the number of selected events, C represents the number of correct extraction of events, the evaluation precision refers to the accuracy and recall and F measure were defined as follows:

$$\Pr ecision = \frac{C}{B} \times 100\% \tag{7}$$

$$\text{Recall} = \frac{C}{A} \times 100\% \tag{8}$$

$$F = \frac{2 \times \Pr ecision \times \text{Recall}}{\Pr ecision + \text{Recall}} \tag{9}$$

Table 1 shows the accuracy, recall and F value of the temporal relation extraction. Through the experiment, the temporal relation extraction in Chinese on the F value reached 73.51%. In the Vietnamese F value reached 69.59%. This shows that this method is effective and feasible.

Table 1. Experimental results of temporal relation extraction

Method name/evaluation index	Chinese			Vietnamese		
	P (%)	R (%)	F (%)	R (%)	P (%)	F (%)
Method of this paper	75.30	71.81	73.51	73.69	65.92	69.59

Table 2 shows the coreference relation extraction accuracy, recall, F can be seen from the experimental results. The use of lexical and contextual information, event coreference Chinese and Vietnamese relations to identify the F reached 64.56% and 61.78 respectively. The dependency syntax relation is added to the feature, the F value of the event coreference relation recognition is raised to 70.82% and 65.82%, respectively. Purely using morphological and contextual information as features, the accuracy of event coreference relational recognition is very low. After syntactic analysis, the recognition accuracy and recall of events were improved by 6.26% and 4.04% respectively. This is because syntactic analysis is rich in semantic information and provides a good supporting role for the identification of event coreference.

Table 2. Recognition accuracy of event co relation under different features

Method name/evaluation index	Chinese			Vietnamese		
	Precision	Recall	F	Precision	Recall	F
Lexical, contextual, etc. features	68.15%	61.14%	64.56%	65.79%	57.49%	61.78%
Syntactic parsing	73.23%	67.41%	70.82%	70.23%	61.41%	65.82%

In October 2016, Guo Chuan lost contact events as through extraction event relation of coreference relation and temporal relation extraction results and event graph construction results are shown as follows (Table 3).

Table 3. Results of event extraction and node weight calculation

Event serial number	Event	Weight
101	1. Guo Chuan navigation shore team confirmed that Guo Chuan in 25 to 15 between 15:30 due to sudden accidents into the water	0.856
102	2. Over the next three days, the parties were actively involved in air and sea search	0.623
103	3. Guo Chuan's family, relatives and friends have initiated efforts to appeal to civil forces	0.732
104	4. A helicopter travel company in Hawaii is willing to offer five or ten helicopters to boost maritime search efforts	0.598
105	5. Guo Chuan falls from Hawaii at least 900 km.	0.910
201	Ông Guo Chuan mất tích hôm 25.10 khi được cho là ở vào vị trí cách Hawaii khoảng 900 km(In October 25th, the location of Guo Chuan missing was considered at the distance of about 900 km in Hawaii.)	0.786
202	nhà thám hiểm 51 tuổi này xuất phát một mình hồi ngày 18.10 từ San Francisco để đến Thượng Hải với mong muốn lập kỷ lục trong 20 ngày hoặc ít hơn bằng thuyền buồm(On October 18th, the 51 year old Explorer traveled from San Francisco to Shanghai, he wanted to arrive in 20 days and requiring few sailing boats)	0.613
203	Nhóm hỗ trợ của Guo và gia đình ông gần như tuyệt vọng khi lực lượng Tuần duyên Mỹ bỏ cuộc hôm 27.10 sau nhiều nỗ lực tìm kiếm nhưng không thành. (A few days after October 27th, the coast guard was struggling for many days, and the support group's Guo and his family were almost desperate.)	0.875
204	Tân Hoa xã cho biết một công ty du lịch ở Hawaii đề nghị cung cấp gần chục chiếc trực thăng để tìm kiếm ông Guo nhưng họ cần tàu có sân đỗ trực thăng lớn hơn. (A travel company offered nearly half the helicopter search in Hawaii, but it still hasn't been found, Xinhua said.)	0.684
205	Trực thăng chỉ có thể bay khoảng 500 km trong khi nơi xảy ra vụ tai nạn là khoảng 900 km ngoài khơi Hawaii. (Helicopters can fly about 500 km, but the accident is about 900 km from the coast of Hawaii.)	0.617

The example of the event diagram is shown in Fig. 1. In the event diagram, the box represents the extracted event. The solid line indicate an event association between the two events and the dotted line represents an alignment between the cross language events. By using the relation between events as the importance of events, the similarity between Chinese and Vietnamese news events is obtained and the international news events are analyzed for good results are obtained.

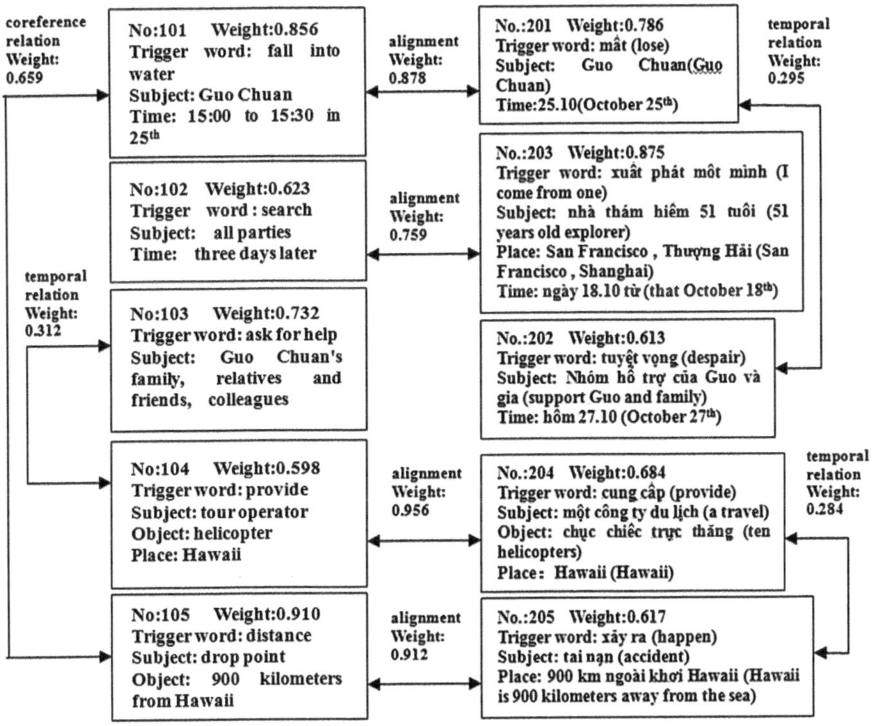

Fig. 1. Bilingual event diagram

5 Conclusion

In this paper, we use SVM model to extract temporal relations by using part of speech, location feature and event element similarity. And the word bag models similarity, entity similarity and semantic feature are used to extract event coreference relations. The event graph model is constructed by taking the event representation node as the side and the temporal relation as the edge. And the weights of the nodes in the graph are solved to realize the importance ordering of events. Experimental results show that the proposed method can extract event co referential relations and temporal relation effectively. It realizes the bilingual event graph extraction of Chinese and Vietnamese, which facilitates the acquisition of bilingual news event information and highlights the focus on news.

Acknowledgments. This work was supported by National Nature Science Foundation (Grant Nos. 61472168, 61672271, 61732005, 61761026), and Science and Technology Innovation Talents Fund Projects of Ministry of Science and Technology (Grant No. 2014HE001), Innovation Team Projects of Yunnan Province (Grant No. 2014HC012).

References

1. Zhao, Y., Chi, W.X., Liu, T.: Chinese event extraction technology research. Chin. Inf. J. **22** (1), 3–8 (2008)
2. Wang, W., Zhao, D., Zhao, W.: Topic identification of key events in Chinese News. J. Peking Univ.: Nat. Sci. Ed. **47**(5), 789–796 (2011)
3. Xu, X., Li, P., Zhu, Q.: Template filtering and conversion methods in semi supervised Chinese event extraction. Comput. Sci. **42**(2), 253–255 (2015)
4. Huang, R., Riloff, E.: Bootstrapped training of event extraction classifiers. In: Proceedings of the 13th Conference of the European Chapter of the Association for Computational Linguistics. Association for Computational Linguistics, pp. 286–295 (2012)
5. Kastner, I., Monz, C.: Automatic single-document key fact extraction from Newswire articles. In: Proceedings of the 12th Conference of the European Chapter of the Association for Computational Linguistics (EACL-09), Athens, Greece (2009)
6. Hong, Y., Zhang, J., Ma, B., et al.: Using cross-entity inference to improve event extraction. In: Meeting of the Association for Computational Linguistics: Human Language Technologies. Association for Computational Linguistics, pp. 1127–1136 (2011)
7. Wan, X.: Using Bilingual Knowledge Ensemble Techniques for Unsupervised Chinese Sentiment and Analysis. In: Proceedings of EMNLP-2008, pp. 553–561 (2008)
8. Li, S., Wang, R., Liu, H., Huang, C.R.: Active Learning for Cross-Lingual Sentiment Classification. In: Proceedings of Natural Language Processing and Chinese Computing, pp. 236–246 (2013)
9. Goldman, X., Yu, Z., Long, W., Ding, W., Yan, C.: Global/local co-occurrence of the distribution of Chinese and Vietnamese bilingual news clues Chinese analysis. J. Inf. Based on 2015, 6 (2015)
10. Zhu, Z.: Research on event extraction method based on Bilingual. Soochow University (2016)

Simulation Analysis of the Affect of Running Speed on the Piggyback-Transportation Safety

He Yinchuan[1], Zhang Qingping[1(✉)], Lin Wu[2], Li Fajia[1], and Li Chen[3]

[1] School of Mechanical Engineering, University of Jinan, Jinan 250022, China
me_zhangqp@ujn.edu.cn
[2] Meishan CRRC Logistics Equipment CO., LTD, Meishan City, China
[3] Research Institute of Highway Ministry of Communication, Beijing 100088, China

Abstract. Safe running speed is an important factor for safety of piggyback transportation. The safety speed of piggyback transport is analyzed by dynamic simulation, which provides a basis for safe running speed of piggyback transport. In this paper, we build a model of piggyback transport vehicle system by Simpack software, and simulate the actual working conditions to get the safe operation speed. In this paper, we build a model of piggyback transport vehicle system by Simpack software, and get the safe operation speed by simulating the actual working conditions.

Keywords: Running speed · Simpack · Piggyback transport Simulation analysis

1 Introduction

Piggyback transportation is one of the multimodal transport modes. It combines the advantages of highway transportation and railway transportation. The way of long-distance transportation of original goods is changed into a way of combined transportation of highways and railways. After loading goods by freight cars or semi trailers, they will be self propelled into or hoisted to special railway transport vehicles, and then long-distance transportation is completed through railway transportation. After reaching the target location. The car automatically drives out or hangs out from the special railway vehicle and goes to the final destination. Railway piggyback transport vehicle is a kind of rail vehicle.

The above studies aim at ordinary rail passenger and freight vehicles. This paper applies Simpack software and the theory related to rail vehicle simulation. The system of railway piggyback transportation vehicle is simulated, and the nonlinear critical speed of the railway piggyback transportation vehicle is obtained under the set of track unevenness, and the safety speed of the railway piggyback transportation vehicle is analyzed.

K. Deng et al. (Eds.): ICMIR 2018, AISC 856, pp. 347–354, 2019.
https://doi.org/10.1007/978-3-030-00214-5_45

2 Simulation Analysis of Piggyback Transportation Vehicle

2.1 Selection of Simulation Model Vehicle

This simulation is selected for a certain type of railway piggyback transport vehicle. The kinetic parameters are shown in Chart 1. The length of the vehicle is about 19.32 m, the length of the loading surface of the vehicle is about 15 m, and the maximum gross weight does not exceed 46 tons, as shown in Fig. 1 (Table 1).

Fig. 1. Simulated piggyback transportation vehicles

2.2 Model Establishment and Simplification

In the rail_track module of Simpack software, the wheelset model, the side frame model, the saddle model, the virtual car body model and the bolster model are set up in turn. The ZK6 bogie model is composed of the above several sub models. Next, the car body model is established under this module. The car body model and the front and rear bogie model assembly constitute the model of railway piggyback transportation vehicle. Because of the influence of running speed of railway piggyback vehicle on the safety of vehicle system is studied in this paper, the road vehicle can be simplified to a mass block model, and the overall model is shown in Fig. 2.

2.3 Impose Constraints

Constraint is applied through the specific articulation of the ZK6 bogie shown in Table 2, and the two system forces existing in the model are defined.

Define the first system force acting between the bearer saddle and side frame of the railway piggyback transportation vehicle. The first system force marking point is built between the bearer saddle and side frame of the railway piggyback transportation vehicle. Define the second system forces acting between the bolster and side frame of railway piggyback transportation vehicle. The second system force marking point is built between the bolster and side frame of the railway carrying vehicle.

Table 1. Parameters table for railway piggyback transportation vehicle

Date of complete vehicle			
Bogie name	ZK6 Bogie	Center distance of axis diameter (m)	1.981
Gauge (m)	1.435	Wheelset medial distance (m)	1.353
Vehicle fixed distance (m)	17	Wheel rolling circle radius (m)	0.42
Load (t)	Single vehicle: 46	Wheelbase (m)	ZK6: 1.83
	Crew: 92		Bogie: 3.2
Axle load (t)	ZK6: 25	Side bearing gap (m)	ZK6: 11
	Frame: 19.1		Bogie: 14

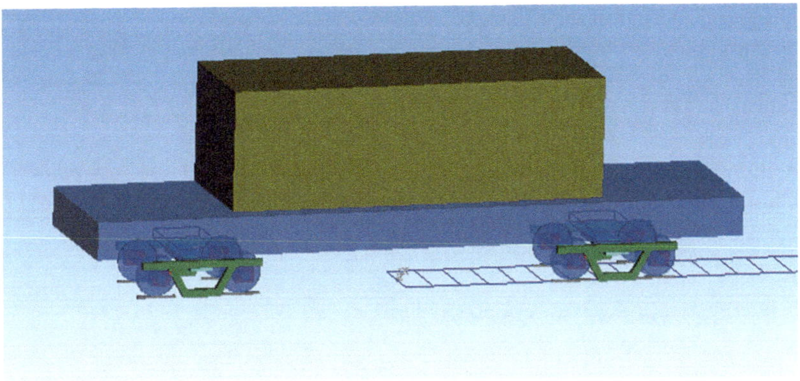

Fig. 2. The model of the railway piggyback transportation vehicle in simpack

2.4 Simulation Condition Setting

In the Simpack software, a new lateral irregularity condition is created, named “Lat”. Select “109: From Input Function (Adv)” in Distance Domain (spatial domain) in Type. Because the track irregularities are generally spatial domain, the excitation in time domain or frequency domain is not applied to the line. In the excitation function, select the function listed in the previous step, “I_Lateral”, and complete the lateral excitation of the track, as shown in Fig. 3. The other three conditions are defined in the same way. The defined incentives are applied to track irregularities. The excitation start position, the excitation smoothing length and the incentive termination location are 50 m, 5 m, and 100 m, as shown in Fig. 4. Then the speed analysis is carried out.

The initial simulation speed is set to 100 km/h, the simulation time is set to 15 s, the sampling frequency is set to 100 Hz, and the off-line integration is carried out, and the change curve of the wheelsets transversal volume with time is obtained.

Table 2. Specific hinged way of ZK6 bogie

Body	Articulated number	Articulated position	Stress elements
Imaginary body	0	Ground	PC, XP
Bolster	25 (α, β, z)	Imaginary body	PC, XP, SS, Fri, Stop, Force
Left and right side frame	07 (6DOF)	Ground	SS, Fri, Stop, JC, PS, Force
Bearer saddle	02 (β)	Wheel	PC Force
Front and back wheelset	07 (6DOF)	Ground	Rail-wheel contacts Force

Note: PC- side bearing force; XP- disc force; Stop- stop force; SS- two system elastic force; PS- system spring force; Fri- friction force; JC- cross pull force; Rail-wheel contacts- wheel rail contact force; α, β, γ-rotational degrees of freedom around x, y, z axis; x, y, z-movement degree of freedom of around x, y, z axis.

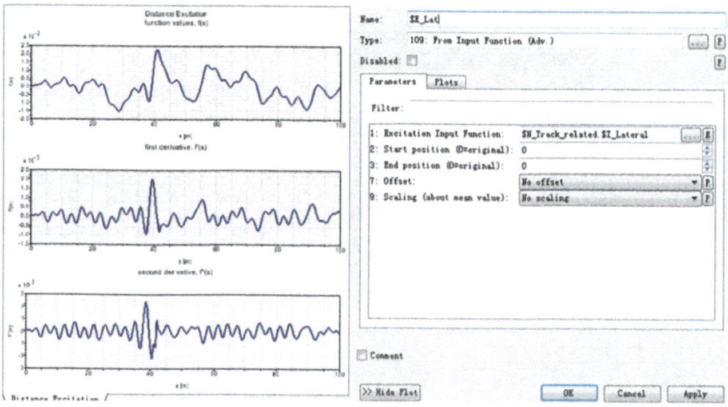

Fig. 3. Definition of lateral excitation

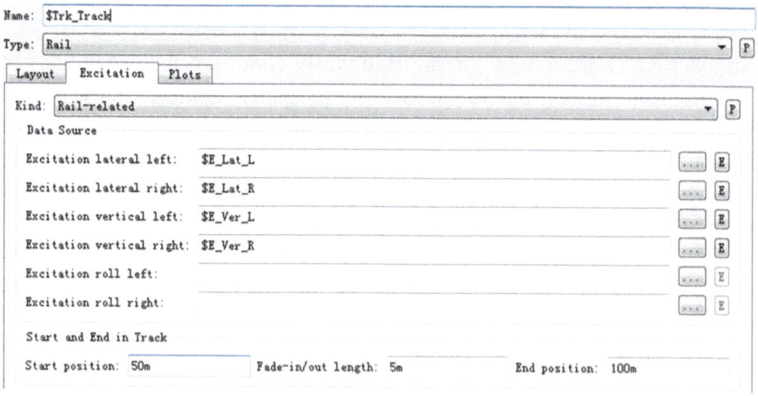

Fig. 4. Imposition of track irregularity conditions

3 Analysis of Simulation Results

When the simulation speed is set to 100 km/h, the change curve of the wheelsets transversal volume with time is shown in Fig. 5.

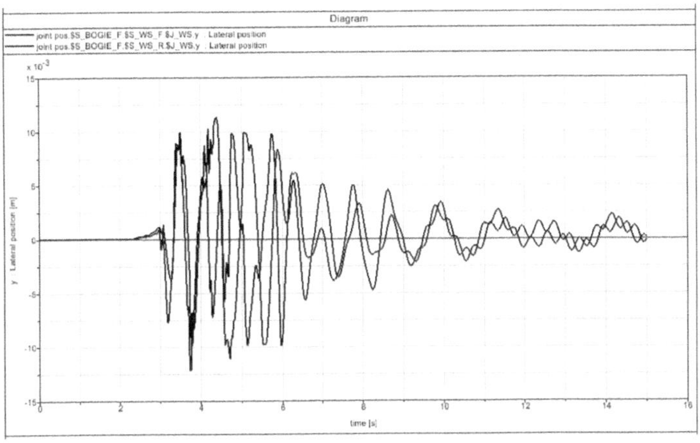

Fig. 5. The change curve of the wheelsets transversal volume with time at 100 km/h

As we can see from Fig. 5: After carrying out 4S simulation, the speed of the railway piggyback transportation vehicle reached 100 km/h. At this time, the wheelsets transversal volume increases with the increase of speed. It can be seen from the figure that the maximum transversal volume is about 12 mm, and then the vehicle is still running safely, the vehicle is at a constant speed in the time of 6 s to 15 s, and the transversal volume is gradually reduced. At this time the vehicle enters the stable state, indicating that the speed 100 km/h is the safe running speed.

As we can see from Figs. 6 and 7: we can see that the number of wheelsets transversal volume is larger and the maximum transversal volume is about 17 mm when the vehicle enters the time of 4 s to 6 s before the steady speed with the increase of the simulation speed. The vehicle is still in a safe range. After 6 s, the speed of the vehicle is stable and the transversal volume is gradually reduced, indicating that the vehicle is still in a stable state under safe running speed.

As we can see from Fig. 8: Before the velocity is at a constant state. The maximum transversal time is around 6 s, the maximum transversal volume is about 43 mm. When the speed is at a constant state, the wheelsets transversal is also in a constant state. According to the schematic diagram shown in Fig. 6, when the speed is 120 km/h, the curve of wheelsets transversal volume after 6 s does not converge or diverge with time, indicating that the speed at this time is critical speed.

As shown in Fig. 9: The simulation speed is set to 121 km/h, when the speed is in steady state, the curve of the wheelsets transversal volume is divergent with time. The maximum transverse time is about 15 s of the simulation time, and the maximum transversal volume is about 43 mm. Then, according to the rule of the curve, the

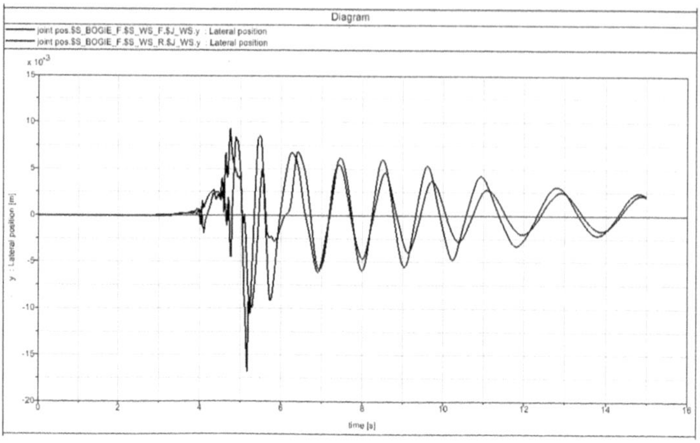

Fig. 6. The change curve of the wheelsets transversal volume with time at 110 km/h

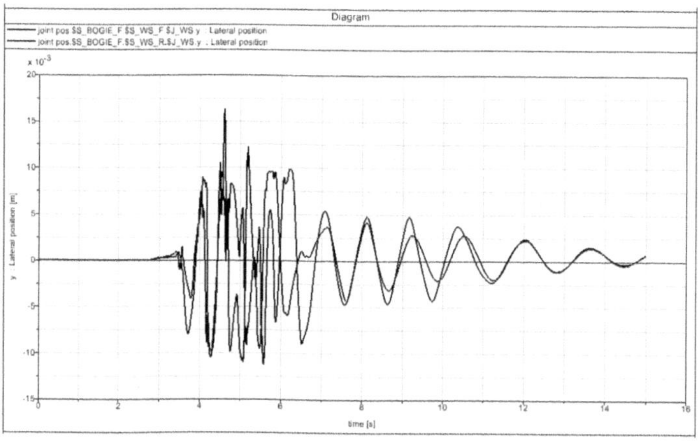

Fig. 7. The change curve of the wheelsets transversal volume with time at 119 km/h

wheelsets transversal volume will increase gradually with time until the transversal volume exceeds the safe transversal volume, and the vehicle will lose its stability.

4 Conclusion

The running speed is calculated by the single integration method. The critical speed of the vehicle system is determined by the result diagram of the wheelsets transversal volume with time. The simulation results show that the maximum speed of the vehicle system is 120 km/h, which has a guiding role for the safe transportation of the vehicle system.

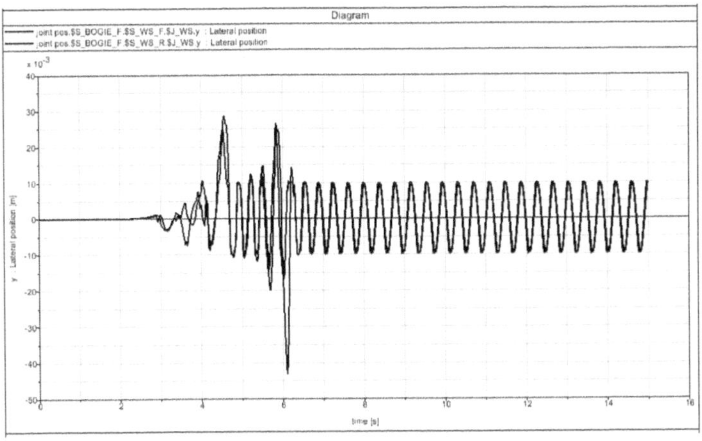

Fig. 8. The change curve of the wheelsets transversal volume with time at 120 km/h

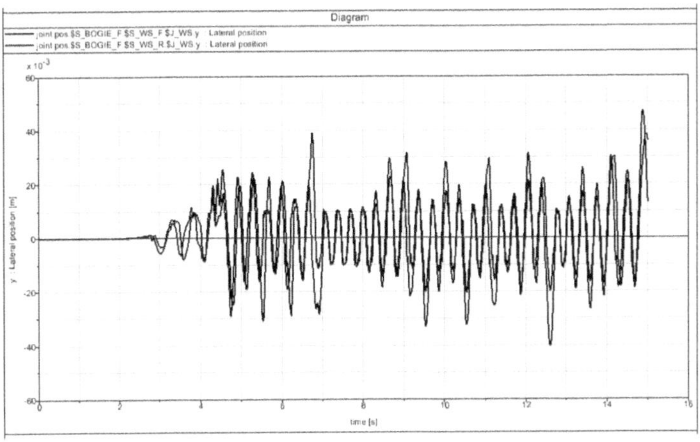

Fig. 9. The change curve of the wheelsets transversal volume with time at 121 km/h

With the gradual upgrading of simulation software and the continuous development of simulation technology, the accuracy of vehicle simulation is more and more demanding. The consistency between simulation results and actual experimental results determines a series of applications for simulation results. Therefore, the actual dynamic test of the piggyback transportation vehicle system is the next task.

References

1. Li, C., He, Y., Zhou, W., Zhang, D., Liu, J.: The development of piggyback transportation technology in Europe and America and Its enlightenment to China. China Transp. Rev. **11**, 80–83 (2017)
2. Wu, X., Bao, P., Yang, Y., Li, Y.: Research on simulation of military wheel-equipment railway transportation impact test by Simpack. J. Syst. Simul. **21**(16), 5025–5029 (2009)
3. Xiaohong, L.: Factors analysis of affecting loading and reinforcing scheme of large goods in railway. World J. Eng. Technol. **04**, 60–68 (2016)
4. Beskou, N.D., Theodorakopoulos, D.D.: Dynamic effect of moving loads on road pavements: a review. Soil Dyn. Earthq. Eng. **31**(4), 547–567 (2011)
5. Ino, T., Ueno, S., Saimoto, A.: A study on the structural fracture of body structure in railroad car. Key Eng. Mater. **577**, 301–304 (2014)

Design of Pipeline Leak Data Acquisition and Processing System

Li Zhonghu(✉), Ma Bo, Wang Jinming, Yan Junhong, and Wang Luling

School of Information Engineering,
Inner Mongolia University of Science and Technology, Baotou 014010, China
lizhonghu@imust.edu.cn

Abstract. A pipeline leakage signal acquisition and processing system based on STM32 and LabVIEW have been designed. In this system, the STM32F103RBT6 is used as the lower computer and processing chip. The signal data was transferred to host computer by ZigBee wireless transmission module. In the host computer, a set of signal acquisition and processing system is designed based on the LabVIEW development platform. The software system can be used in conjunction with MATLAB software. In addition, after every static time the master computer will send the command to make slave computer refresh machine time by GPS modules.

Keywords: Pipeline leakage · Leak location · LabVIEW · Signal acquisition Signal processing

1 Introduction

Pipeline transportation is a new and economical way of transportation. It plays an increasingly important role in the national economics. However, because of the pipeline corrosion, aging, natural disaster and man-made destruction, the pipeline leakage accident has occurred occasionally. The accident will not only cause a lot of resources waste and serious environmental pollution, but also more It will pose a great threat to the life and safety of the related personnel [1]. Therefore, monitoring the condition of pipeline leakage timely and determining the leakage point location accurately, is very important to ensure the normal operation of pipeline and reduce losses.

2 Structure and Principle

The structure of pipeline leakage signal acquiring and processing system is shown in Fig. 1. Firstly, the pipe leakage acoustic emission signal propagates along the pipe. Then it is collected by sensor. The collected signal is stored to the memory of STM32F103RCT6 MCU via signal processing and A/D converting. Second, these data are sent to the master computer wirelessly by ZigBee. ZigBee hosts and master computer are connected by the RS232 serial port [2]. Master computer software system is developed by LabVIEW platform. The software filter and de-noise the data. Then

K. Deng et al. (Eds.): ICMIR 2018, AISC 856, pp. 355–361, 2019.
https://doi.org/10.1007/978-3-030-00214-5_46

time domain analysis, frequency domain analysis and cross-correlation calculation are carried out. Then the time-domain characteristics, frequency-domain characteristics can be acquired and the leak point location can be determined.

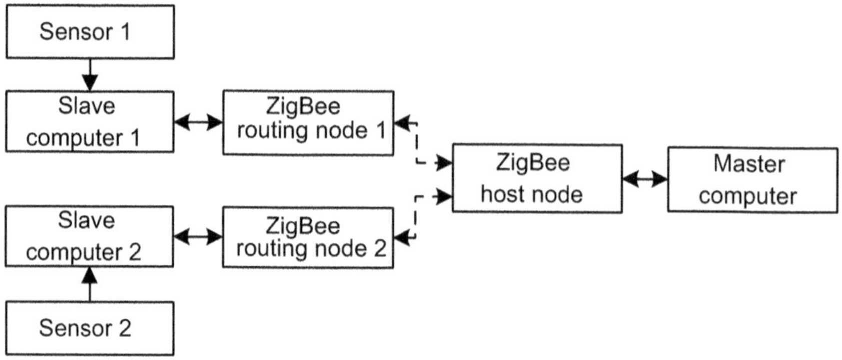

Fig. 1. The principle diagram of the system overall structure

3 The Slave Computer System Design

STM32F103RCT6 is the core of the slave computer. The microcontroller has 12 bit A/D converter, and direct memory access system. It makes the sampling interval more accurate. The principle diagram of the slave computer system is shown in Fig. 2.

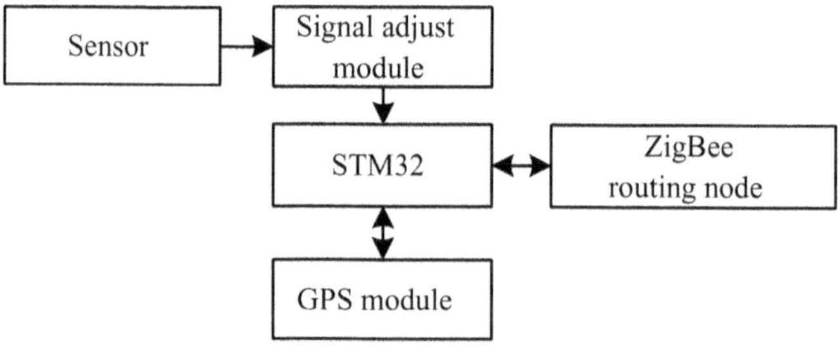

Fig. 2. The principle diagram of the slave computer system

In order to improve the leak point positioning accuracy, it must be ensured that the two clocks of slave computers are the same. This problem can be resolved by using GPS clock timing. The serial port 2 receive and process the message information of GPS module by interrupted mode. Then use the message to calibrate the real time clock of STM32. The minimum error of GPS calibration could be less than 30 ns. The data transport between master computer and slave computer through the ZigBee module. ZigBee module already has a ZigBee protocol. Routing nodes and the host node can work in the same net automatically. The ZA2530 is a long distance transmission chip which is transmission by the mode of serial port. The transmission protocol is the Modbus RTU. When the setting time arrive, the slave computer start sampling. The sampling will stop when sampling points reached 4096. Sampling data will be transmitted through the serial port to ZigBee routing nodes. Then it will transmit to the host node. And eventually it is sent to the master computer. The software of slave computer system was wrote by Keil uVision5. The slave computer can finish the work of signals collection, data storage, data transmission, GPS clock synchronization etc. Software system includes initialization module, A/D conversion module, storage module, wireless transmission module and the timing synchronization module.

4 Master Computer Software System Design

Master computer software system is based on LabVIEW development environment. Master computer software system mainly complete the data reception, filtering, denoising, cross-correlation computation, data storage and other functions. It includes data reception, filtering, wavelet denoising, cross-correlation operation program and database storage procedure.

VISA library is used for the function of data receiving. It can complete functions, such as configuration of serial port, read, write and close port. It can also realize the parse of serial port. LabVIEW filter VI is used for signal filtering. Signal denoising function need to invoke MATLAB script node in LabVIEW. And then using MATLAB wavelet analysis function to complete the wavelet denoising function. Cross-correlation is realised by calling the cross-correlation arithmetic subroutine. Then calculate the pipeline leak point location to realise the leak point location function. The function of database storage is based on Microsoft ActiveX data objects technique which is used for linking database files. data storage and data query functions are realised by using LabVIEW database toolkit.

4.1 The Interface of Master Computer

The interface of master computer is shown in Fig. 3. The left part is for parameter setting. It can set the relevant parameters, such as the distance between the sensors, sampling points, sampling frequency, sound speed on pipeline, the database path, administrator information and filter button, wavelet denoise button, etc. If the filter function is open, it can also set filter type, the cut-off frequency, stop-band attenuation values. Right part is the waveform display area. It can display the original waveform, FFT waveform, filtering and denoising waveform, cross-correlation waveform and the result of locating.

4.2 Filter Subroutine

Digital filter can be divided into finite impulse response (FIR) filter and infinite impulse response (IIR) filter according to the time domain characteristics of discrete system [3]. Under the same bandwidth, high order FIR filter can lead to a large phase shift and bring interference to the waveform time calculation. Pipeline leak signal frequency

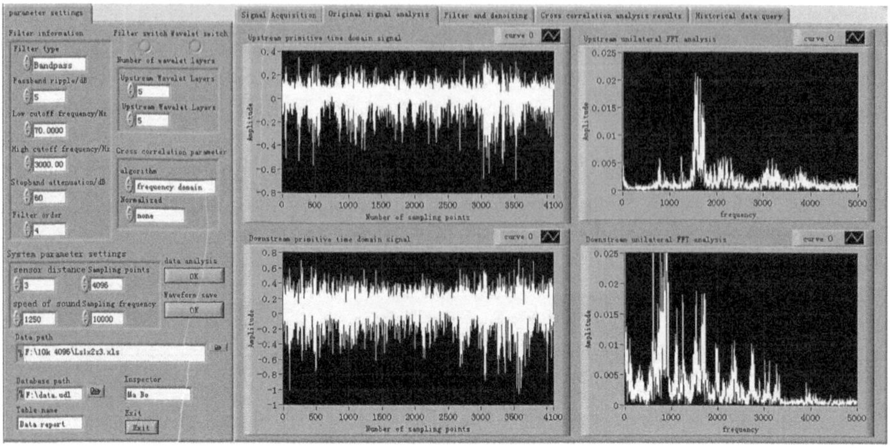

Fig. 3. The interface of master computer

band is mainly concentrated in the 1.5 kHz to 1.75 kHz. There are two small peak near the 0.8 kHz and 1.25 kHz frequency. The attenuation of signal above 3 kHz is large and it can be seen as a high frequency interference. Therefore, lower cut-off frequency is set to 0.07 kHz and high cut-off frequency is set to 3 kHz. Then it can filter the low frequency and high frequency part. Elliptic filter is used in this system, which is one of IIR filters. Compared with butterworth filter and chebyshev filter of the same order, the transition zone of elliptic filter between the passband and stopband is sharp. The ripples of elliptic filter within the passband and stopband have the same amplitude [4, 5].

4.3 Wavelet Denoising Subroutine

Band-pass filter can filter out low frequency and high frequency noise. But in the band which is mixed noise and leak signal, wavelet denoising algorithm is needed. Wavelet is a good way to save the peak and mutation of the useful signal. The key of denoise is quantitative processing of wavelet coefficient. Energy of useful signal is related to the larger amplitude wavelet coefficients and the noise energy is related to small amplitude wavelet coefficient. Noise signal distributing in all coefficient after the wavelet transform. By studying the effect of several wavelet denoising, this system choose Daubechies (dbN) as the wavelet denoising algorithm. In addition the wavelet threshold shrinkage algorithm is also used. To realize Wavelet denoising algorithm, the system calls MATLAB script node in LabVIEW. MATLAB node invokes the Wavelet Toolbox.

4.4 Cross-Correlation Algorithm Program

The purpose of cross-correlation algorithm is to calculate the time difference τ of upstream and downstream signal. The cross-correlation operation and the leak location subroutine are shown in Fig. 4. It is filtered and denoised of the upstream and downstream leak signal. Then call "cross-correlation VI" in LabVIEW to complete cross-correlation algorithm. The max amplitude of the cross-correlation result is the time cross-correlation. The leak position can be calculated by using the result and leak point locating formula.

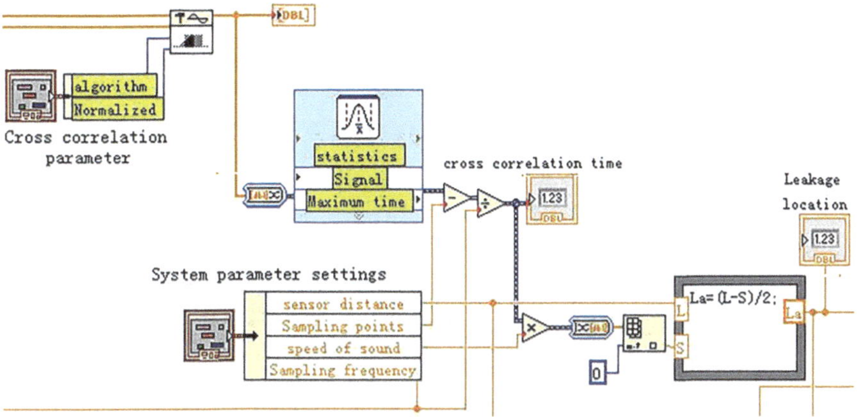

Fig. 4. Cross-correlation operation and the leak location subroutine

Cross-correlation calculation result is shown in Fig. 5.

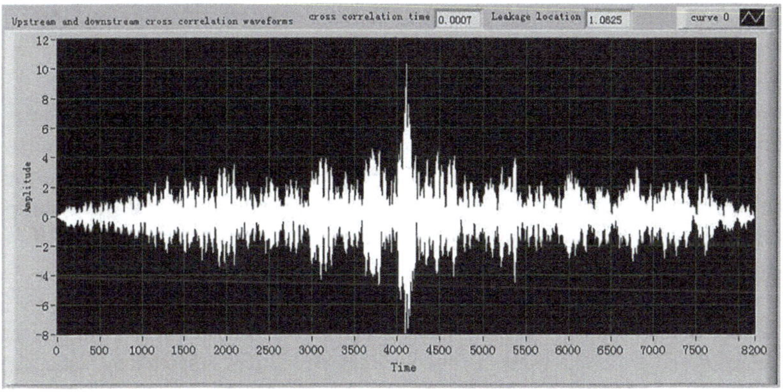

Fig. 5. Cross-correlation calculation result

4.5 Database Storage and Query Subroutine

LabVIEW provides users with specializing toolkit Database Connective to connect Database [6]. The system use the Microsoft ActiveX Data Objects (ADO) technology and Universal Data Link (UDL) to obtain information from database. Database connection is realized by this method. The system uses Microsoft Access as database.

Using VIs of Open Database Connection, Create Table, Insert Data and Select Data to complete operations of database [7]. Database written procedure is shown in Fig. 6.

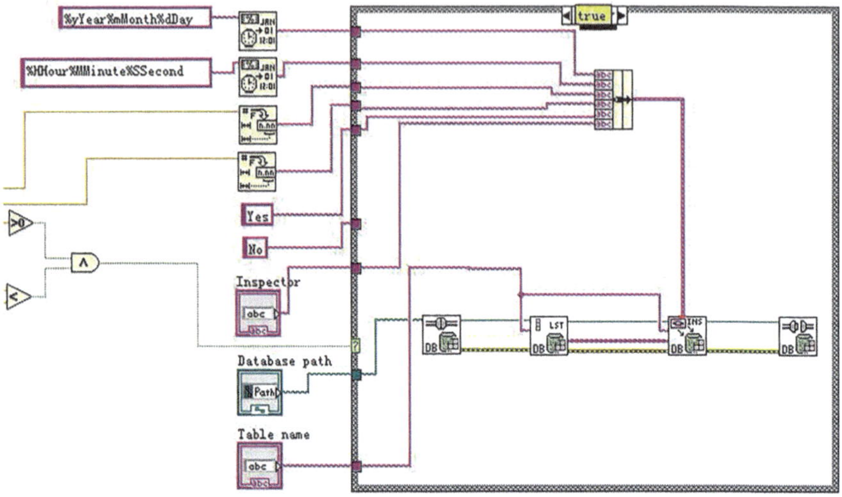

Fig. 6. Database written procedure

5 Conclusion

Based on the actual demand of signal detection and leak point locating, this paper design a signal acquisition and processing system. The system uses STM32F103RCT6 single-chip microcomputer as the core of slave computer and uses LabVIEW to develop master computer software. By combine LabVIEW and MATLAB software in master computer software, it will enrich the system data processing algorithms. The calculation function is powerful. Communication between master and slave computer is based on ZigBee wireless way. It is shown that the system has realized the signal acquisition and processing functions.

Acknowledgements. Project researches on leakage signal mechanism and method of weak signal detection and dispose in water-supply pipelines (61362023) supported by National Natural Science Foundation of China.

References

1. Long, X., Mingxu, Z., Hao, L., et al.: Oil and gas pipeline transportation safety problems and reasonable coping ways. Chem. Enterpr. Manag. **1**, 101–102 (2018)
2. Fan, D., Yongming, Z.: Design of multi-channel wireless data acquisition system based on Zigbee. Instrum. Tech. Sensor **8**, 72–73, 78 (2013)
3. Tao, W.: Application of digital filter in real speech noise reduction. Inf. Commun. **3**, 30–31 (2018)

4. Xiaojie, Y., Huimei, Y.: The design and implementation of IIR digital filter based on FPGA applied in electric energy measurement. Microcomput. Inf. **12**, 201–202 + 240 (2008)
5. Zhou Peng, X., Gang, M.X., et al.: Proficient in Labview Signal Processing. Tsinghua University Press, Beijing (2013)
6. Henan, S.: Rubber extruder state monitoring and control system based on virtual instrument. Tianjin University of Science and Technology, Tianjin (2017)
7. Wentao, L., Cao, Y., Xufang, B., et al.: The realization and application of LabVIEW database access technology. Ind. Mine Autom. **38**(3), 69–72 (2012)

Research on Sentence Semantic Similarity Calculation Method Based on Context

Jun Yu[1,2(✉)], Qingfeng Wei[1,2], and Changshou Luo[1,2]

[1] Institute of Agricultural Information and Economics, Beijing Academy of Agricultural and Forestry Sciences, Beijing 100079, China
Yujun2729@163.com

[2] Beijing Research Center of Engineering Technology on Rural Distance Information Service, Beijing 100079, China

Abstract. By studying how to use domain professional corpus to build application contexts. Through the use of sentence representation layer similarity, sentence word context similarity mining, and calculation of sentence similarity to improve the accuracy of sentence differences, it can provide decision support for the pushing of sentences in the FAQ of Agricultural Science and Technology service. Methods of the sentence similarity calculation were divided into two aspects: surface similarity calculation and semantic level similarity calculation. Cosine similarity algorithm combined with offset similarity calculation method is used to improve the quality of surface similarity calculation. At the same time, the concept of context is introduced, and the similarity calculation of semantic level is realized by the use of graph. Finally, the actual effect of the algorithm is tested by experiments.

Keywords: Context · Sentence similarity · Surface similarity
Semantic level similarity · Graph

1 Introduction

The introduction: During the process of service, Beijing 12396 technology service platform needs to build an automatic answering system to improve the utilization ratio of expert's intelligence. After the study of the iFLYTEK, Turing Robot and various customer service FAQ systems, the research group has found that the answer push is very random in the general question and answer systems; in the business FAQs, the problem is relatively concentrated and universal, such as Taobao and bank's FAQ customer service. However, in the application of Agricultural Science and Technology service, whose problem's coverage are broad, informative and technical, the content of reply needs to be scientific and accurate. At the same time, the currently popular similarity calculation method and framework cannot make good use of the existing structure data in the application, and the data processing and processing are complicated and the workload is large in the practical application. Therefore, the research group proposed the semantic sentence similarity calculation method based on context.

K. Deng et al. (Eds.): ICMIR 2018, AISC 856, pp. 362–370, 2019.
https://doi.org/10.1007/978-3-030-00214-5_47

2 Related Research and Preparation

Experts and scholars have conducted comprehensive researches based on the semantic similarity of words. For example: a statistical method based on the co-occurrence of words. These methods are simple and efficient, but completely ignore the lexical and semantic information of the sentence. The other is based on lexical and semantic information. This method takes related elements of the semantic information into account, but the structure is relatively complex, such as the semantic similarity calculation based on semanteme. Third, corpus training feature extraction method based on neural network has been developed greatly in recent years. Fourthly, comprehensive fusion methods are used for research, such as Sentence Semantic Similarity Calculation Based on Multi-feature Fusion. With the deepening of the research, combining with the application experience, we found that if various methods in practical use are out of the application scene, the algorithm will be either complex or inefficient. With many uncertain factors, it is full of limitations to operate. Therefore, this paper puts forward the research ideas and methods based on the existent FAQ resource in Agricultural Science and Technology service. In this paper, existing FAQ resources are referred to as the domain corpus, and each category formed by them is called an application context. This paper will focus on how to structure an application context by utilizing the professional corpus through the mining of sentence representation layer similarity and sentence word context similarity to calculate sentence similarity and improve difference precision between sentences. Through experimental comparison, algorithm differences before and after improvement are examined, which provides decision support for the pushing of sentences in the FAQ of Agricultural Science and Technology service.

3 Sentence Semantic Similarity Calculation Method Based on Context

3.1 Structure of Context

In the FAQ, questions are usually stored in the groups (classification) which are classified in advance. The group is set to solve a collection of related questions. We call such a similar group as a context of this kind of problems. A series of similar contexts form the context of the entire FAQ. If the user problem can be mapped to such a small context, then the analysis of the semantic similarity of the problem becomes the analysis based on this context. In context analysis, we use the principle of graph to structure context relation undirected graph to reflect the relationship between words. Through the graph, the whole semanteme of the context word is reflected. The context relation undirected graph is a network graph. Words denote the vertex information, while the edge or arc indicates the combination of the two words (or the weight, which is not studied in this paper). Suppose that the context relation undirected graph G has n types of vertexes (which means n types of words), then the adjacent matrix is an n*n square matrix, which is defined as:

$$g[i][j] = \begin{cases} 1, & \mathit{if}\ (Vi, Vj) \in E\ \mathit{or}\ \langle Vi, Vj \rangle \in E \\ 0, & \mathit{otherwise} \end{cases} \tag{1}$$

For example: sentence 1: the method of tomato seeding raising in the greenhouse; sentence 2: the method of tomato seeding raising. Through word segmentation, dicontinuation word removal and key word extraction, four words will be put forword: V1 (tomato), V2 (greenhouse), V3 (seeding raising) and V4 (method); edge relations exist (V1V2), (V1V3), (V2V3), (V3V4), then the context graph (this study does not take the position direction into consideration, so the graph is undirected) is as follows (Fig. 1).

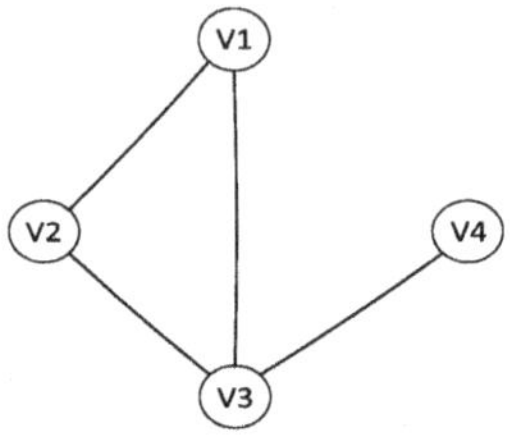

Fig. 1. Graph

The homologous adjacent matrix is as follows (Fig. 2).

	V1	V2	V3	V4
V1	0	1	1	0
V2	1	0	1	0
V3	1	1	0	1
V4	0	0	1	0

Fig. 2. Matrix

After the transformation from the graph to the matrix, any degree of vertex (word) can be obtained (which is equal to the number of related words), which is the sum of the elements of the line i of this vertex Vi in the adjacent matrix. For example: the degree of V1 is 2, the degree of V2 is 2, the degree of V3 is 3 and the degree of V4 is 1. To calculate all adjacent points of the vertex Vi is exactly to scan all elements of the line i in the matrix. If the result of element is 1, the element is the adjacent point. As a result, the word set formed by all the adjacent point refers to the context of the word. For example, the above picture indicates that V2 and V3 is the context word set of V1; V1 and V3 is the context word set of V2; V1, V2 and V4 is the context word set of V3; V3 is the context word set of V4.

The realization of the context relation graph based on corpus is to generate the corresponding context according to the corpus. The steps of the concrete realization:

firstly, the corpus is clustered and the corpus is minimized by clustering, so as to minimize the context. Secondly, segment the corpus and remove the discontinuation words of the corpus. Thirdly, structure a relation graph in sentences, which is based on the relationship between words, so that an adjacent matrix of words' semantic relation is formed.

3.2 Context Mapping

Context mapping is to find out the degree and adjacent point of related words in the context graph by word segmentation of problems inputted by the user. Take Fig. 2 as an example, the user input: tomato seeding raising, then the degree of V1 (tomato) is 2 and the degree of V3 (seeding raising) is 3. The adjacent points and their relationships are as follows. V1 is adjacent to V2 and V3; V3 is adjacent to V1, V2 and V4. The relationship in the context obtained by mapping will be prepared for the next calculation parameter.

3.3 Semantic Similarity Calculation

Sentence semantic similarity calculation involves two levels of calculation: surface similarity calculation and semantic level similarity calculation. Surface similarity refers to the similarity of forms of two sentences, which is measured by the number of the same words or synonyms in two sentences and the relative position in the sentence. Semantic level here refers to the semanteme that is not directly reflected literally and needs to be understood. The common methods used in semantic level processing are semantic dictionaries and semantic contexts. This paper studies the semantic graph based on context.

3.4 Surface Similarity Calculation

There are many methods of surface similarity calculation, such as the method in HowNet, Cosine similarity, general Jaccard similarity and so on. In this paper, cosine similarity is used to calculate the similarity of forms of two sentences, and the semantic offset similarity in How Net is used to calculate the word order similarity.

A. Cosine similarity calculation: Calculate the cosine value of the Angle between two vectors, which is used to indicate the degree of difference between two sentences; Cosine similarity focuses on the similarity and difference in the direction of the vector, which is the similarity and difference of the trend, rather than the absolute distance. The formula is as follows:

$$\mathrm{Cosin}(p_1, p_2) = \frac{x^*y}{\sqrt{x^2 \times y^2}} = \frac{\sum x_i y_i}{\sqrt{\sum x_i^2}\sqrt{\sum y_i^2}} \tag{2}$$

In that, p_1, p_2 are the sentences for comparison of the similarity degree, x_i, y_i are the value of vectors of words in two sentences.

Description: context is a word set graph. After word segmentation, therefore, TF-IDF can be made good use of to calculate the weight of words in the sentence in order to acquire words. After the word acquirement, the similarity measure of the cosine angle of the space vector will not be effected by the index scale. The value of cosine falls into the interval [0, 1]. The larger the value, the smaller the difference. The cosine similarity tends to deliver a better solution.

B. Offset similarity calculation: comprehensively considering the surface similarity of words in a sentence, and the relative position of words in a sentence, it reflects the semantic similarity of words in two sentences in the position relationship.

Suppose p_1, p_2 as two sentences, then the formula of semantic word order similarity of two sentences is as follows:

$$\mathrm{Sim_p}(\mathrm{p_1}, \mathrm{p_2}) = \frac{\sum_{j=1}^{n}\sum_{i=1}^{m}\mathrm{sim}(\mathrm{W_i}, \mathrm{W_j}) \times (1 - |\mathrm{pos}(\mathrm{W_i}) - \mathrm{pos}(\mathrm{W_j})|)}{\mathrm{len}(\mathrm{P_1}) \times \mathrm{len}(\mathrm{P_2})} \tag{3}$$

Description: this formula quotes semantic offset calculation in "A Method of Sentence Similarity Computing Based on Hownet [5]". In that, n and m are the numbers of words in two sentences;

In the formula, $\mathrm{sim}(W_i, W_j)$ is the similarity of two words;

$$\mathrm{Sim}(\mathrm{W_i}, \mathrm{W_j}) = \frac{2 \times \mathrm{T}(\pi(\mathrm{W_i}) \cap \pi(\mathrm{W_j}))}{\mathrm{len}(\pi(\mathrm{W_i})) + \mathrm{len}(\pi(\mathrm{W_j}))} \tag{4}$$

Description: $\pi(\mathrm{W_i})$ refers to the adjacent point of W_i in the context graph; $\mathrm{T}(\pi(\mathrm{W_i}) \cap \pi(\mathrm{W_j}))$ refers to the common adjacent point of W_i and W_j; $\mathrm{len}(\pi(\mathrm{W_i}))$ is the degree of W_i, which is adjacent point data. If W_i, W_j are the same or the synonyms, the result will be 1.

$\mathrm{pos}(W_i)$ means the relative position of the word W_i in the sentence,

$$pos(W_i) = \frac{i}{len(P_i)} \tag{5}$$

i refers to the position of the word in the sentence, $len(P_i)$ means the length of the sentence.

C. Surface similarity calculation formula:

$$\mathrm{Sim_b}(\mathrm{p_1}, \mathrm{p_2}) = \mathrm{Cosin}(\mathrm{p_1}, \mathrm{p_2}) + \alpha_1 \times \mathrm{Sim_p}(\mathrm{p_1}, \mathrm{p_2}) \tag{6}$$

Description: $Cosin(p_1, p_2)$ and $Sim_p(p_1, p_2)$ constitute the surface similarity; α_1 is the influence factor of the offset similarity relative to the whole surface similarity. Through practical analysis, the product of the value of α_1 and offset similarity should be less than the cosine similarity value. At the same time, the product of the value of α_1 and offset similarity will be larger and larger with the cosine similarity value starting from 0, and when the product gets to a certain value, it gets smaller as the cosine similarity value increases. As a result, through practice, we can tell that $\alpha_1 = (1 - Cosin(p_1, p_2)) \times Cosin(p_1, p_2)$.

3.5 The Similarity Calculation of Semantic Level

The similarity calculation of semantic level is mainly used to analyze the similarity of non-inclusion words in the semantic level in two sentences. The formula is as follows:

$$\mathrm{Sim_L}(p_1, p_2) = \alpha \times \frac{T(\pi(P_1) \cap \pi(P_2))}{T(\pi(P_1) \cup \pi(P_2))} \tag{7}$$

Description: When there is no antonym in p_1, p_2, $\alpha = 1$; When there exists an antonym, $\alpha = -1$; $\pi(P_1)$ is the context word set of all words that exists in p_1 that do not exist in p_2(corresponds to a set of adjacent points of a graph). In the same way, $\pi(P_2)$ refers to the context word set of all words that exists in p_2 that do not exist in p_1. $T(\pi(P_1) \cap \pi(P_2))$ refers to the number of common words of the two context word sets (the number of common adjacent points); $T(\pi(P_1) \cup \pi(P_2))$ refers to the number of all words of the two context sets (the number of all adjacent points).

3.6 Sentence Semantic Similarity Calculation Method Based on Context

Based on the calculation of surface similarity calculation and the similarity calculation of semantic level, combined with the actual weight relationship of sentence semantics, the formula is structured as follows:

$$\mathrm{sim}(p_1, p_2) = \mathrm{Sim_b}(p_1, p_2) + \beta_1 \times \mathrm{Sim_L}(p_1, p_2) \tag{8}$$

Description: $Sim_b(p_1, p_2)$ and $Sim_L(p_1, p_2)$ constitute the sentence similarity; β_1 refers to the influence factor of sentence semantic similarity relative to the semantic similarity calculation of the whole sentence. Through practical analysis, the product of β_1 and the semantic similarity should be less than the surface similarity value. At the same time, the product of the value of β_1 and the semantic similarity will be larger and larger with the surface similarity value starting from 0, and when the product gets to a critical point, it gets smaller as the surface similarity value increases. As a result, through practice, we can tell that: $\beta_1 = (1 - Sim_b(p_1, p_2)) \times Sim_b(p_1, p_2)$.

The steps of calculating the similarity of a specific sentence are as follows:

(1) Segment the sentence p_1, p_2 and remove the useless discontinuation words.
(2) Find the antonym in the sentence, according to the antonym library. If the antonym exists, remove it from p_1, p_2, at the same time, $\alpha = -1$ in the formula (6); otherwise, $\alpha = 1$.
(3) Mark the synonyms, according to synonym library. They are considered the same word when calculating.
(4) Calculate the cosine similarity according to formula (1).
(5) Calculate the offset similarity according to formula (2).
(6) Calculate the surface similarity according to formula (5).
(7) Calculate the similarity of semantic level according to formula (6).
(8) Calculate sentence similarity based on context according to formula (7).

The method mentioned in this paper is based on the surface similarity calculation, taking the influence of context on semantics fully into consideration, and do semantic deep mining from the context semantic level.

4 Experiment and Evaluation

The experiment of this paper is based on the network database of Beijing agricultural resources information center. NLPIR Chinese word segmentation system of Chinese academy of sciences is adopted. In this experiment, 3,858 planting problems were collected, and the three types of contexts of tomato, pepper and eggplant were structured. Among them, the number of problems of contexts of tomato planting technology is 325; the number of problems of contexts of pepper planting technology is 244; the number of problems of contexts of eggplant planting technology is 183. C# is used for programming. Compare the test results with the method one [7], method two [8] and the algorithm in this paper. The sample comparison of the experiment is as follows (Table 1).

Table 1. Comparison of the sentences in experiment

Sentence	Method one	Method Two	Method of this paper	Context
Seeding raising and cultivation of tomato in the small arched shed Cutting seeding raising and cultivation of tomato in sunlight green house	0.671	0.713	0.742	Planting technique of tomato
Seeding raising and cultivation of tomato in the small arched shed Soil-free cultivation technique of tomato in green house	0.250	0.334	0.352	Planting technique of tomato
Cutting seeding raising and cultivation of tomato in sunlight green house Soil-free cultivation technique of tomato in green house	0.224	0.284	0.313	Planting technique of tomato
How to prevent navel decay of tomato Key points of fertilizer for pepper	0	0	0	Planting technique of tomato

(*continued*)

Table 1. *(continued)*

Sentence	Method one	Method Two	Method of this paper	Context
Seeding raising and cultivation of tomato in the small arched shed Cutting seeding raising and cultivation of tomato in sunlight green house	0.671	0.713	0.741	Planting technique of tomato, pepper and eggplant
Seeding raising and cultivation of tomato in the small arched shed Soil-free cultivation technique of tomato in green house	0.250	0.334	0.351	Planting technique of tomato, pepper and eggplant
Cutting seeding raising and cultivation of tomato in sunlight green house Soil-free cultivation technique of tomato in green house	0.224	0.284	0.312	Planting technique of tomato, pepper and eggplant
How to prevent navel decay of tomato	0	0	0	Planting technique of tomato, pepper and eggplant

Through the analysis of experiment results, in the context of "tomato planting technology", the value of similarity of different calculation methods differ, and the value of similarity of this method is the highest. From the analysis of the formula, similarity value enhancement is a kind of fine tuning of sentence semantics. This kind of fine tuning is adjusted by the context semantic relation, which is consistent with the actual situation; In the context of "tomato, pepper and eggplant planting technology", the same sentence calculates the similarity, the values of method one and method two is invariable, but the value calculated by method in this paper changes slightly. This kind of change shows the fine tuning of similarity accuracy by the context, which is reasonable compared with the actual situation.

5 Conclusion

In this paper, "sentence semantic similarity calculation method based on context" is combined with corpus in the professional field, and the semantic accuracy is greatly enhanced in the analysis of the semantic similarity of sentences. This method has been used in the agricultural science and technology service FAQ project. The improvement of the similarity accuracy in practice provides a good basis for the pushing of sentences in the FAQ and is more practical. However, the context graph is undirected, and the relevance of the word in the graph, which is the relationship between words and words in the context has been expanded. It would be more appropriate to build semantics if was based on the context of a directed graph. Furthermore, in the analysis of graph relationship, the influence of the connectivity of sentence words on the semantics should be extensionally taken into consideration. So further research is needed in the next step of work.

References

1. Liu, M., Lang, B., Gu, Z., Zeeshan, A.: Measuring similarity of academic articles with semantic profile and joint word embedding. TSingHua Sci. Technol. **22**, 619–632 (2017)
2. Donald, M., Yaniv, B., Bernstein, Y., et al.: Similarity measures for tracking information flow. In: Proceedings of CIKM'05. Bremen, Germany: [s. n.], pp. 517–524 (2005)
3. Li, X., Xie, H., Li, L.: Research on sentence semantic similarity calculation based on Word2vec. Comput. Sci. **44**(9), 256–260 (2017)
4. Zhao, Z., Wu, N., Song, P.: Sentence semantic similarity calculation based on multi-feature fusion. Comput. Eng. **38**(1), 171–173 (2012)
5. Cai, D., Bai, Y., Yu, S., Ye, N., Ren, X.: A context based word similarity computing method. J. Chin. Inf. Process. **24**(3), 24–28 (2010)
6. Cheng, C.-P., Wu, Z.: A method of sentence similarity computing based on Hownet. Comput. Eng. Sci. **34**(2), 172–175 (2012)
7. Xia, X.: Research on C4.5 algorithm based on cosine similarity and weighted pruning strategy. Qingdao University of Science and Technology. Master's thesis (2017)
8. Wang, R., Chi, Z., Chang, B., Bai, X.: Chinese sentence similarity measure based on word sequence length and word weight. J. Chin. Inf. Process. **31**(13), 142–144 (2005)
9. He, W., Wang, Y.: Research on sentence based text representation and Chinese text Classification. J. China Soc. Sci. Tech. Inf. **28**(6), 839–843 (2009)

A Fatigue Accelerated Model of Correlating Proving Ground with Customers Usage for Bus

Lu Zhang(✉), Wei Zhou, and Wen-Liang Li

Key Laboratory of Operation Safety Technology on Transport Vehicles, Ministry of Transport, PRC, Beijing 100088, China
luzh_cau@126.com

Abstract. A new regression model of correlated proving ground with customers usage was proposed based on the mismatch in the current test of customer correlative acceleration fatigue in proving ground, combined with collection and treatment of the bus load spectrum on proving ground and customer road, and establishment of a pseudo damage model based on the acceleration of shaft head. The correlated fatigue characteristics of steering tie rod, stabilizer bar, front axle propons, longitudinal beam, three acceleration of left front wheel shaft and right back wheel shaft on proving ground and customer road was studied in the whole vehicle. The result shows that the model was better than the other two models on the relative error between proving ground and customer goal and the range of relative fatigue. This model could provide a new point of view for reliability test of a new bus.

Keywords: Bus · Customer road · Correlated · Proving ground
Fatigue acceleration

1 Introduction

The reliability enhancement road test of automobile in test field is an important technical measure and means to rapid assessment and evaluation of automobile reliability. Test fatigue acceleration test is the essence of test reliability test, the core of which is to build load spectrum in the users road and proving ground strengthening road, and the connection between the in order to get a different road cycling times.

In the literature [1–5], the mean square root error of the number of loading times is the minimum of the final matching target, and the sensitivity of different size loads to the total fatigue damage is not considered. Rely on users for road and proving ground strengthening road load spectrum by rain flow matrix regress testing ground of various road, which is feasible in theory, but the rain flow collection and processing it is often difficult to achieve high correlation coefficient matrix matching [6, 7]. Based on the measured pseudo damage and damage of rear axle strain signal analysis and comparison the regional road and EVP road load spectrum of equivalent relations, according to the principle of combination of road, have to place the bad roads and road users and EVP mix proportion of different sections. Literature [8] proposed a variable speed reinforcement test associated with the user's actual working condition, and calculated the cumulative fatigue damage corresponding to the torque at all levels of

K. Deng et al. (Eds.): ICMIR 2018, AISC 856, pp. 371–378, 2019.
https://doi.org/10.1007/978-3-030-00214-5_48

the transmission shaft. Literature [9] proposed a method to simplify the application load spectrum encoding method for accelerating the fatigue test of body frame. According to the literature [5], the load spectra of each section of a type of off-road vehicle in the test field are compressed, and the target load spectrum can be used for indoor road simulation test. Literature [10–13] studied the reliability enhancement factor of the automobile test field based on the principle of equal damage, and obtained the strengthening coefficient of the road surface compared with the ordinary road in the test field.

In this paper, through a region of large operating passenger car user information acquisition, as a large gathering user road operating bus load collection basis, large operating reliability of the test site are roads bus load are collected; Pseudo fatigue damage calculation model based on the shaft head acceleration, puts forward a kind of based on the relative damage such as associated user fatigue regression model, the reliability testing ground road and user path tie rod, front suspension stabilizer bar and front axle, longitudinal beam, the first three to the left front wheel acceleration and right rear wheel and axle load spectrum for the first three to the acceleration of, calculate correlation matching the proving ground of fatigue damage and fatigue damage user goals contrast, maximum relative error, the average relative error and testing ground road circulation ratio and reliability test, and with the load spectrum, such as model and damage model such as contrast, compare the accuracy of the three link model.

2 User Information

This passenger transport company is located in the central and eastern part of China. The operation route covers all kinds of roads in China. The starting station of the operation route is the passenger terminal of a city. The terminal station is shanghai, zhengzhou, yantai, hengshui, zaozhuang, suzhou and dalian, and more than 200 operation routes. Based on various operational route of GPS data analysis shows that the vast majority of large operating buses operating line of road type from urban road to the highway and urban road, and large-scale operation of the passenger car running highway and city road mileage rate most about those days. The average mileage of this type of vehicle is 241 km, and about 90% of the operating buses are less than 500 km each way, as shown in Fig. 1.

Large operating passenger on the bus each operation route of GPS data, statistics and analysis for large operating buses on the city roads and highways of velocity distribution as shown in Figs. 2 and 3, known, large-scale operation of the buses on the city roads most running at low speed area, most of the running on the highway speed range is 90 km/h–90 km/h.

In terms of carrying capacity, the probability of the average passenger carrying capacity of large operating passenger vehicles is 90%.

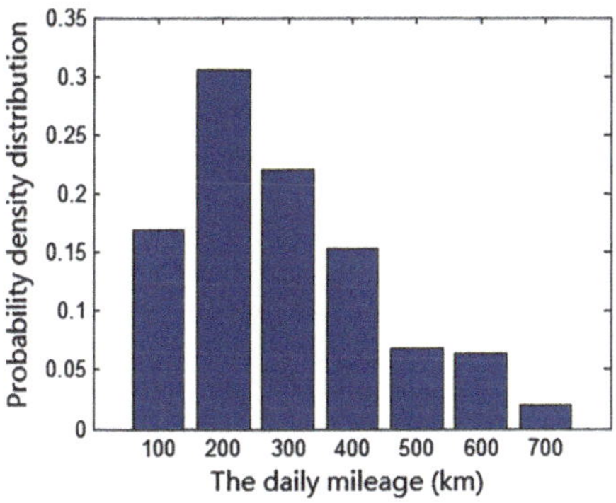

Fig. 1. Distribution of one single operating mileage for large operating passenger

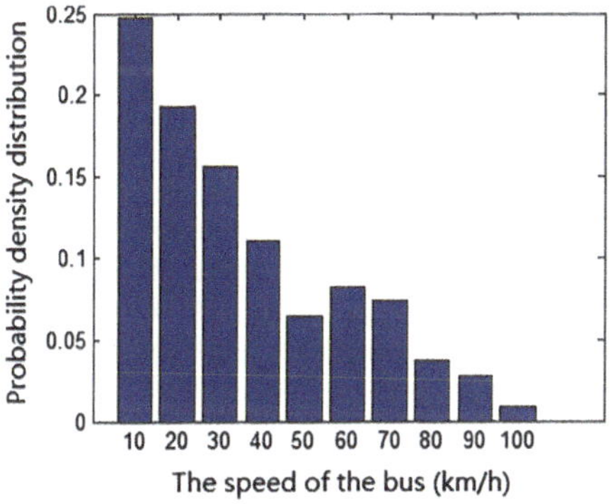

Fig. 2. Velocity distribution of large operating passenger on urban road

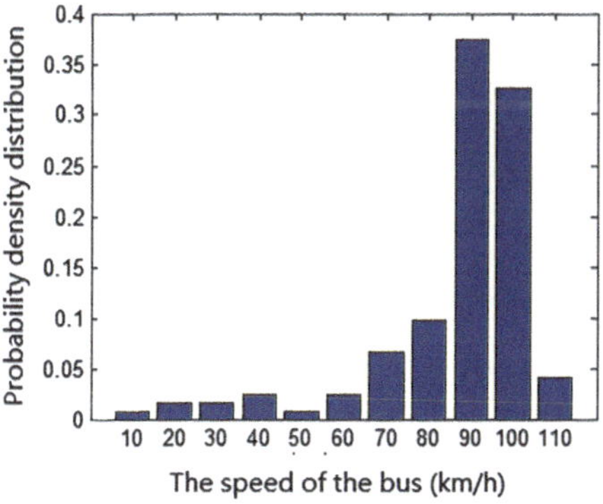

Fig. 3. Velocity distribution of large operating passenger on freeway

3 Collection of Bus Load

In order to test the sample vehicle with a large operating passenger train of 11.4 m, the weight is fully loaded, and the three-direction acceleration sensor is installed at the head of each axle. Three-dimensional modeling and finite element analysis of key parts were carried out, and the strain concentration point was determined as the paste position of strain gauge. Collected representative users road large operating passenger car road load and reliability testing ground road load, among them, the road user acquisition speed requirements mainly refer to the statistical results of user information; Failure mode at the same time, in order to ensure the consistency, should as far as possible to ensure that the user maximum load and road test to strengthen road are equal, so set different speed in different strengthening road, at the same time considering the acceptable speed range test driver, concrete are shown in Table 1.

Table 1. The velocity and distance of each strength roads in proving ground

Sequence	1	2	3	4	5	6	7	8	9
Road type	Twisted way b	Stone road armour	Washboard road armour	Stone road c	Stone road b	Pebble road b	Washboard road b	Long road c	Link load
Speed range (km/h)	5–10	10–20	70–80	50–80	20–30	50–80	60–80	30–40	–
Mileage/m	85	500	300	800	1189	335	700	90	2878

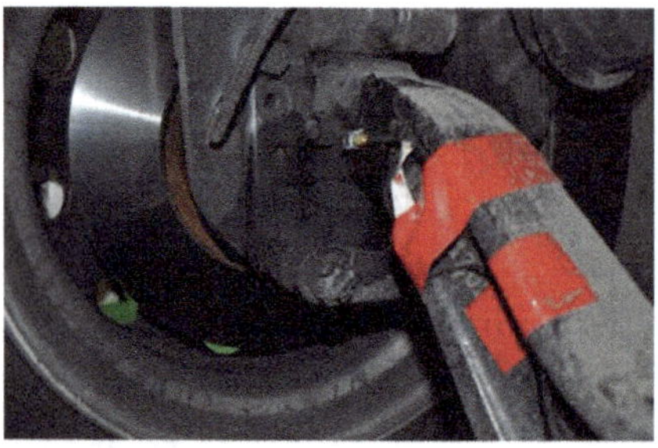

Fig. 4. The test point for vertical acceleration of the left front wheel spindle

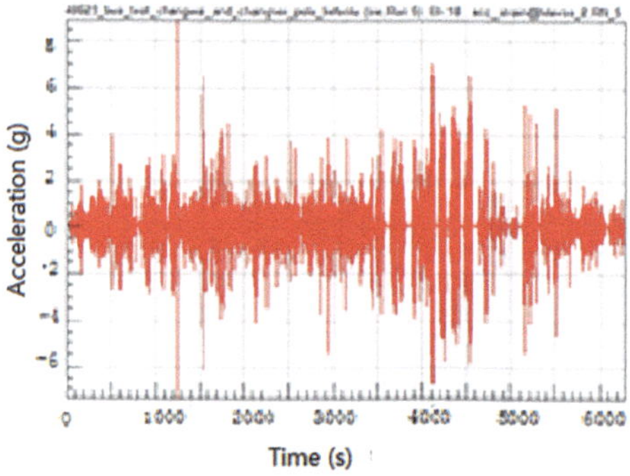

Fig. 5. The vertical acceleration of the left front wheel spindle

A total of 38 groups of signals were measured. Among them: 20 acceleration signals, 13 strain signals, GPS signal, Mark signal, etc. For example, the sensor layout is shown in Fig. 4, and the time history signal collected is shown in Fig. 5.

4 Establishment of Associated User Model

Current user goals on the basis of reliability research road and reliability testing ground road load spectrum matching method is mainly model and damage model, this paper puts forward a kind of relative damage such as correlation model, the model in the user associated proving ground load spectrum matching both considered the key parts in the damage and shaft head acceleration of pseudo strain injury.

4.1 Correlation Model of Load Spectrum

Load spectrum correlation models such as believe in operation if a large passenger input load is the same, largely because of its various components on the load and its signal is proportional to the input load, so the fatigue damage should be caused by the same. Therefore, the method is to ensure that the reliability of the user's road and test field can be well matched to ensure that the damage is equal.

The correlation regression model of road load spectrum of user road and test field is:

$$\begin{pmatrix} P_{11} & \cdots & P_{1k} \\ \vdots & \ddots & \vdots \\ P_{n1} & \cdots & P_{nk} \end{pmatrix} \begin{bmatrix} \alpha_1 \\ \alpha_2 \\ \vdots \\ \alpha_k \end{bmatrix} = \begin{bmatrix} M_1 \\ M_2 \\ \vdots \\ M_n \end{bmatrix} \tag{1}$$

In the formula, P_{ji} is the loop number of the first reliable road rain flow matrix falling in the j interval. α_i is the number of cycles in the reliability path of i species in the test field. M_j is the number of cycles in the j interval for the user's target rain flow matrix. k is the total interval of the number of pavement types equal to the n load amplitude.

From the above equation, it can be concluded that the mean square error of the number of loads is the minimum of the final optimization target, and the sensitivity of different load sizes to the total damage is the same.

4.2 Associated Model of Damage

The damage model is to ensure that the fatigue cumulative damage of each channel is equal to the reliability of the user's road and the test field, namely, the correlation regression model of the damage of the user's road and test field strengthening road is as follows:

$$\begin{pmatrix} D_{11} & \cdots & D_{1k} \\ \vdots & \ddots & \vdots \\ D_{n1} & \cdots & D_{nk} \end{pmatrix} \begin{bmatrix} \alpha_1 \\ \alpha_2 \\ \vdots \\ \alpha_k \end{bmatrix} = \begin{bmatrix} D_1 \\ D_2 \\ \vdots \\ D_n \end{bmatrix} \tag{2}$$

D_{ji} is the damage or pseudo damage of the j-th component or part on the i reliability road of the test field. D_j is the damage/pseudo damage of the j-th component or part of the user's target.

The model is feasible in theory, but because of the need to introduce pseudo parallel regression, and pseudo damage values may not be within 0 to 1, so there are different channels lead to the phenomenon of different objective attend dimensional order of magnitude.

4.3 Relative Damage Correlation Model

For load spectrum related model and problems such as damage associated model, put forward a kind of relative damage such as correlation model, to ensure that each channel user roads and proving ground reliability of relative fatigue cumulative damage has high consistency, such as relative damage correlation regression model as follows:

$$\begin{pmatrix} D_{a(1,1)} & \cdots & D_{a(1,k)} \\ \vdots & \ddots & \vdots \\ D_{a(1,k)} & \cdots & D_{a(n,k)} \end{pmatrix} \begin{bmatrix} \alpha_1 \\ \alpha_2 \\ \vdots \\ \alpha_k \end{bmatrix} = \begin{bmatrix} 1 \\ 1 \\ \vdots \\ 1 \end{bmatrix} \tag{3}$$

Define $D_{a(j,i)}$ as relative damage.

5 Analysis of Reliability of User's Road and Test Field

For the above three models at present basically adopts the multiple linear regression method to solve, but as a result of the actual project in the road test, the above model can exist certain constraints, such as test cycles as a nonzero number, test distance and time requirements, reliability testing ground road cycle sequence and cycle times, test speed, etc. In this assumption, the enhancement coefficient is not less than 20 for the user's road target $L = 600000$ km.

$$\text{The constraint:} \begin{array}{l} (1)\ \sum \alpha_i L_i \leq L/k \\ (2)\ \alpha_i = Z^+ \cup 0 \end{array}$$

In the formula, L is the length of each reliability path of the test field.

Therefore, the multiple regression equation cannot be solved at this time, and the fmincon function of Matlab is used to solve the ratio of the reliability of all sections of the test field under three related models. It can be seen that the cyclic frequency of the isoload spectrum model is greater than that of the equal damage model and other relative damage model. The maximum number of cycles is 12790, which is quite different from the actual experiment. Such as damage model to get in a washboard road cycles are 0, rocks on b and pebble road b lower and higher than the relative damage model respectively, such as relative damage model on distorted b and washboard road armour cycles are 0; The test range of reliability path of the test field under the three models is 59478, 11568, 13267; The comprehensive strengthening coefficient is 30, 155, 135 respectively.

According to matching results, get the range of the components/parts damage/pseudo damage compared with user path as shown in Figs. 6, 7 and 8.

The Fig. 7 shows that adopting such as load matching model to calculate the fatigue damage of the channel in the proving ground road and road user goals of the range of the ratio of the fatigue damage is 12 ~ 781, and the reliability test requirement gap is larger; It can be seen from Fig. 8 that the results obtained by using the equal damage model are much better than the equivalent load spectrum model, and the range of relative fatigue damage is 0.39 ~ 3.43. Such as relative damage model with relatively fatigue damage of the root mean square error as the optimization goal, the result of the match by a combination of the components/parts of a relative damage, the

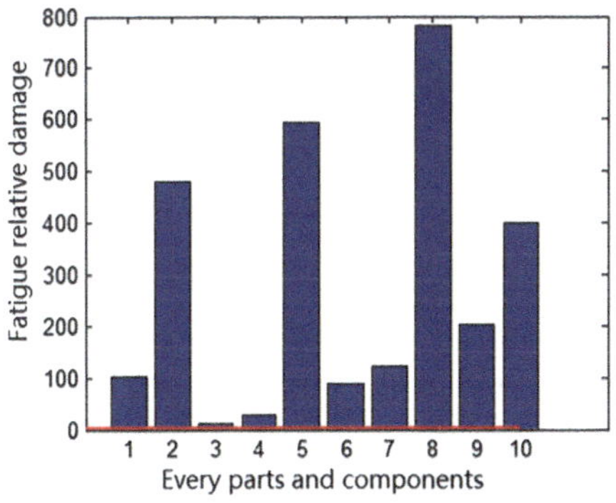

Fig. 6. The comparison of fatigue damage in proving ground and user goals on equal load spectrum model

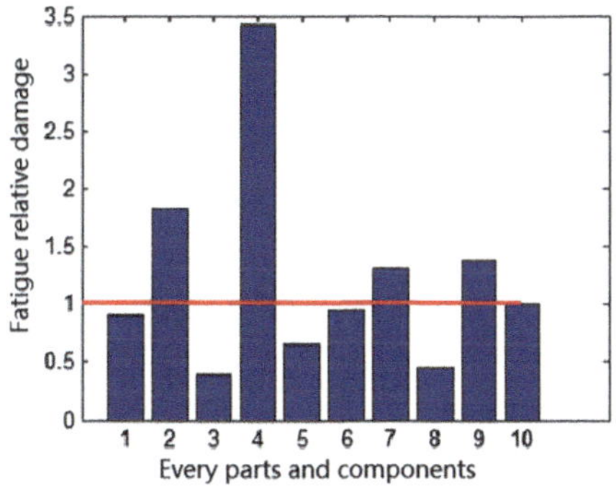

Fig. 7. The comparison of fatigue damage in proving ground and user goals on equal damage model

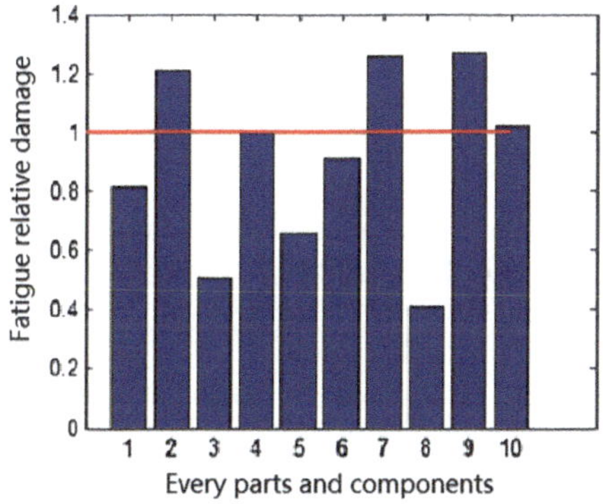

Fig. 8. The comparison of fatigue damage in proving ground and user goals on equal relative damage model

results as shown in Fig. 8, presents the better matching effect, were superior to damage model and load model, such as relative fatigue damage model for relative fatigue damage range of 0.41 ~ 0.41.

Load spectrum model does not take into account different size such as the sensitivity of the load effect on the total fatigue damage is different, lead to calculate the maximum relative error and the average relative error is large (due to too much, this article does not give in the list, in order to "-"); Damage model and relative damage model associated user goals such as roads and proving ground reliability fatigue damage, known, such as damage model and calculate the relative damage models such as the maximum relative error is 242.61% and 59.08%, the average relative error of 54.84% and 22.27% respectively, such as relative damage model in the vast majority of channel damage model such as relative error is less than.

In summary, the relative damage correlation model proposed in this paper is better than the equivalent load spectrum correlation model and other damage correlation model in the reliability test of the user correlation test field.

6 Summary

A method is put forward based on the relative damage of associated user fatigue such as regression model, combined with the reliability testing ground road and user path tie rod, front suspension stabilizer bar and front axle, longitudinal beam, the first three to the left front wheel acceleration and right rear wheel and axle load spectrum for the first three to the acceleration of, calculate correlation matching the proving ground of fatigue damage and fatigue damage user goals contrast, maximum relative error, the average relative error and testing ground road circulation ratio and reliability test range.

The calculation results and the load spectrum, such as the method and comparison between the model and damage model for the model is better than the other two models when correlation matching, the method could be used to provide basis for the establishment of bus reliability test specification.

References

1. Jixiong, Li.: LuoZhaohui. Development of vehicle reliability test specifications base on the principle of vehicle tests equivalence. Autom. Eng. **33**(11), 990–993 (2011)
2. Ensor, D.F.: Automotive test technique correlating with customer usages. China Mech. Eng. **9**(11), 24–26 (1998)
3. Wang, W., Wu, S., Tang, B., et al.: Study on equivalent relation of proving ground and real road. J. Chongqing Univ. Technol. (Nat. Sci.) **24**(12), 15–19 (2010)
4. Yu, H.: Research on Reliability Test Method Correlated Proving Ground with Customers Usage for Automotive Load Carrying System. Jilin University, Changchun (2008)
5. Zhao, X., Zhang, Q., et al.: The compression and extrapolation of load spectrum for a heavy off-road vehicle obtained from proving ground testing. Autom. Eng. **30**(9), 871–875 (2009)
6. Ledesma, R., Jenaway, L.: Development of accelerated durability tests for commercial vehicle suspension components. SAE paper, 2005-01-3565
7. Wei, P.: Study on Loading Spectra Correlation of EVP Proving Ground to Typical Roads and Fatigue Life Digital Drediction. Tongji University, Shanghai (2004)
8. Li, W., Shi, X., Ke, J.: Weave method study of customer correlated load spectrum for driveline durability indoor test. J. Mech. Eng. **50**(20), 143–150 (2014)
9. Men, Y., Yu, H., Guo, P.: Varying-speed enhancement test method on proving ground for passenger car based on correlated customers data. Autom. Eng. **36**(10), 1221–1224 (2014)
10. Guo, H., Chen, W., Fan, X., et al.: Research of enhancement coefficient of automobile reliability enhancement test on proving ground. Chin. J. Mech. Eng. **40**(10), 73–76 (2004)
11. Guo, H.: Research of Enhancement Coefficient of Automobile Reliability Enhancement Test on Proving Ground. Zhejiang University, Zhejiang (2003)
12. Peng, L.: Research on enhancement test coefficient for automobile proving ground. Jilin University, Jilin (2007)
13. Bao, T.-c., Cheng, W.-p., Du, Q., et al.: Enhancement coefficient of motorcycle reliability test on proving ground. J. Tianjin Univ. **42**(7), 591–596 (2009)

Ant Lion Optimizer with Adaptive Boundary and Optimal Guidance

Ruo-an Wang(✉), Yue-wen Zhou, and Yao-yu Zheng

College of Aeronautics and Astronautics Engineering,
Air Force Engineering University, Xi'an, China
550861874@qq.com

Abstract. Aiming at the shortcoming that the ant-lion algorithm has unbalanced exploration and development capability, an improved algorithm with adaptive boundary and optimal guidance is proposed. First, the ant lion adjust the scope of the border in order to balance the exploration and development capabilities. Second, through the adaptive best-guided equation, to improve the convergence speed and global search ability. The simulation results of six standard test functions show that the improved algorithm improves the accuracy and convergence speed of the optimal solution compared with other algorithms.

Keywords: Ant lion optimizer · Adaptive method · Optimal guidance

1 Introduction

Ant Lion Optimizer (ALO) is a new type of cluster intelligent algorithm [1]. It has been applied to the optimal positioning of distributed generation, the establishment of linear discrete filters, the Uav flight path planning and other fields. However, the standard ALO also has a slow convergence speed, easy to get into local optimal and low search precision, thus affecting its actual application.

To solve these problems, literature [2] proposed swirling convergence to improve the ant's random walk and the convergence speed. Literature [3] proposed a chaotic search strategy to obtain a new ant lion and reduce the probability of falling into a local optimum. Literature [4] proposed boundary adjustment and cross-border processing method to improve the search efficiency. Although the above methods make the performance of the ALO algorithm optimized in terms of its emphasis, it is hard to balance the search accuracy, convergence speed and trapping into local optimum of the algorithm [5].

For this problem, this paper proposes ant lion optimizer with adaptive boundary and optimal guidance (ABLALO), based on the interaction between ant and ant lion in ALO.

K. Deng et al. (Eds.): ICMIR 2018, AISC 856, pp. 379–386, 2019.
https://doi.org/10.1007/978-3-030-00214-5_49

2 Standard ALO Algorithm

The ant lion repositions its captured prey and builds a pit to increase the chance of catching another prey. The specific process is as follows:

The initial solution of N ants and M ant lions in D dimensional space was randomly generated using formula (1). And save them in position matrix M_{Ant} and $M_{Antlion}$, where L_d and U_d are the lower and upper bounds of d_th dimensional search space.

$$x_{i,d} = L_d + rand(0, 1)(U_d - L_d) \tag{1}$$

Calculate the fitness value of each ant and ant lion as $F_i = f(x_{i,1}, x_{i,2}, \cdots, x_{i,D})$, and separately store in the fitness matrix M_{OA} and M_{OAL}, where f is the objective function. At the same time record the global optimal position as the elite ant lion.

Ants' random walk array is (where cumsum is cumulative sum, t_k is the number of iterations, rand $\in [0, 1]$ is random number):

$$X_t = [0, \cdots, cumsum(2r(t_k) - 1)] \tag{2}$$

$$r(t_k) = \begin{cases} 1 & \text{if rand} > 0.5 \\ 0 & \text{if rand} \leq 0.5 \end{cases}. \tag{3}$$

Ants search randomly rand the ant lion within a certain boundary and their position x_i^t is converted as follows:

$$x_i^t = \frac{(x_i^t - a_i)(U_i^t - L_i^t)}{b_i - a_i} + L_i^t \tag{4}$$

where b_i and a_i are the maximum and minimum values in the array X_t. U_i^t and L_i^t are the upper and lower boundary of i_th dimension in the t_th iteration, its calculation equation is:

$$\begin{cases} U_i^t = Antlion_i^t + U_i/I \\ L_i^t = Antlion_i^t + L_i/I \end{cases} \tag{5}$$

Where $Antlion_i^t$ is the location of the ant lion, U_i and L_i are the upper and lower bounds of i_th dimensional solution space. The equation for the contraction ratio I of the boundary is:

$$I = 10^w \cdot \frac{t}{T} \tag{6}$$

where t is current iteration, T is maximum iteration. When $t > 0.1T$, $w = 2$; $t > 0.5T$, $w = 3$; $t > 0.75T$, $w = 4$; $t > 0.9T$, $w = 5$; $t > 0.95T$, $w = 6$.

Using the elite strategy, the globally optimal ant lion will affect the random walk of all ants. The position update formula for j_th ant in t_th generation is:

$$\mathrm{Ant}_j^t = \frac{R_A^t + R_E^t}{2} \tag{7}$$

where R_A^t is the random walk around the ant lion which selected by the fitness roulette wheel in t_th iteration using Eq. (4), R_E^t is the random walk around the elite ant lion in t_th iteration.

If the updated ant becomes fitter than the selected ant lion, then the ant lion update its position to the latest ant. Expressed as:

$$\mathrm{Antlion}_i^t = \mathrm{Ant}_j^t \text{ if } f(A) > f(AL) \tag{8}$$

3 ABLALO Algorithm

The standard ALO guarantees its exploration ability due to the randomness of ants walking, facilitating its development ability by ant lions approaching elite ant lion. However, all ant lions in one iteration have the same boundary size, and the ant updated position is always the middle of the selected ant lion and elite ant lion. This is not conducive to balance the exploration and development abilities of the algorithm. So, it is easy to fall into local optimum and has slow convergence. For this problem, an adaptive walking boundary strategy and an adaptive optimal guidance strategy are proposed. The following content explains these strategies.

3.1 Ant Adaptive Walking Boundary Strategy

From Eq. (6), it can be seen that the boundaries of the ants walk decrease as the number of iterations increases. However, in the same iteration, all ants have the same walking boundary. It can't reflect the difference of the ant lions capacity and it is not conducive to balance the exploration and development capabilities of the algorithm [6]. For this purpose, this paper proposes an adaptive walking boundary strategy and its equations are as follows:

$$I = \mu(m) \cdot 10^{w} \cdot \frac{t}{T} \tag{9}$$

$$\mu(m) = 0.5 + \frac{1}{1 + \exp\left(\frac{2m - M}{M}\alpha\right)} \tag{10}$$

where m is the element index of the ant lion ranked using its fitness, M is the total number of ant lions, α adjust the rate of change. From Eq. (10), the worst fitness ant lion is $m = 1$, $\mu(m) \approx 1.5$, the contraction ratio I of the boundary increases with the ant walking margin decreases. This is in line with the principle that worse ant lions should reduce traps to improve local development capabilities and quickly find local optimums [7]. The best fitness ant lion is $m = M$, $\mu(m) \approx 0.5$, I decreases with the ant walking margin increases. This is in line with the principle that better ant lions should expand traps to improve global exploration capabilities and jump out of local optimization.

3.2 Ant Adaptive Optimal Guidance Strategy

In order to balance the exploration and development capabilities, the standard ALO algorithm uses Eq. (7) to update the ant position, bringing the ant lion closer to the elite ant lion. Although the equation has improved the development ability of the algorithm, it does not fully consider that the ALO algorithm needs stronger exploration ability in the initial stage of iteration and requires stronger development capability in the later stage of iteration. In this paper, on the basis of the self-adaptive idea proposed in the literature [8], an adaptive elite ant lion guiding equation is proposed as follows:

$$Ant_j^t = \sigma(t)R_A^t + (1 - \sigma(t))R_E^t \tag{11}$$

$$\sigma(t) = 1 - rand^{(1-t/T)^2} \tag{12}$$

where Ant_j^t, R_A^t, R_E^t are the same as formula (7), $\sigma(t)$ is adaptive coefficient, t is current iterations, T is maximum number of iterations. From Eq. (12), we can see that at the beginning of the iteration, $\sigma(t) \approx 1$, then formula (11) is based on the selected ant lion, which can guarantee a stronger exploration ability in the early period, improve the global search ability of the algorithm, and avoid falling into a local optimum. As the number of iterations increases, $\sigma(t) \rightarrow 0$, then formula (11) is based on the elite ant lion, it ensures that the algorithm refines the optimal area in late period, improving the convergence speed and accuracy of the algorithm. Through the above improvements, ants can adaptively select two different walking patterns, effectively balancing the exploration and development capabilities of the algorithm [9].

3.3 The Description of ABLALO Algorithm

```
Initialize the positons of N ants and M ant lions in D dimensional space using
formula (1)
Calculate its fitness and save the position of the best ant lion
while t≤T
   for each ant
Select an ant lion based on roulette wheel
Calculate the walking boundary using formula (5) and (9)
Create and transform random walks using formula (4)
Update the position of the ant using formula (11)
   end for
   Calculate the fitness of all ants after updating
   Update the ant lion position using formula (8)
   Update the position of the best ant lion
   t=t+1
end while
return the position and fitness of the best ant lion
```

4 Simulation Experiment

The simulation software is MATLAB R2012a, for the three algorithms, the population size N = 40, the number of iteration Maxcycle = 500, the algorithm accuracy value $\varepsilon = 1e-11$, and the experiment times K = 50. By comparing the parameters such as average value, optimal value, worst value and variance in 50 independent experiments, the optimization performance of each algorithm is evaluated [10] (Table 1).

Table 1. Benchmark functions

Function expression	Dimension	Range	Optimum
(Sphere) $f_1(x) = \sum_{i=1}^{D} x_i^2$	50	$[-100, 100]$	0
(Schwefel) $f_2(x) = \sum_{i=1}^{D} \lvert x_i \rvert + \prod_{i=1}^{D} \lvert x_i \rvert$	50	$[-10, 10]$	0
(Sodp) $f_3(x) = \sum_{i-1}^{D} \lvert x_i \rvert^{(i+1)}$	50	$[-10, 10]$	0

(continued)

Table 1. *(continued)*

Function expression	Dimension	Range	Optimum
(Rastrigin) $f_4(x) = \sum_{i=1}^{D}\left[x_i^2 - 10\cos(2\pi x_i) + 10\right]$	50	$[-5.12, 5.12]$	0
(Griewank) $f_5(x) = \frac{1}{4000}\sum_{i=1}^{D} x_i^2 - \prod_{i=1}^{D}\cos\left(\frac{x_i}{\sqrt{i}}\right) + 1$	50	$[-600, 600]$	0
(Rosenbrock) $f_6(x) = \sum_{i=1}^{D}\left[100\left(x_{i+1} - x_i^2\right)^2 + (x_i - 1)^2\right]$	50	$[-2.048, 2.048]$	0

It can be seen from Fig. 1, for the function f_1, f_2, f_3, f_4, f_6, ALO accuracy is higher than ABC. For the function f_5, ALO accuracy is lower than ABC. Overall, the convergence speed and search accuracy of these two algorithms are not high, while the ABLALO algorithm has obvious advantages in the convergence of the six benchmark functions. For the functions f_1, f_2, f_3, f_4, f_5, the improved algorithm can converge to the theoretical accuracy value with less iterations. In function f_6, although ABLALO does not find the optimal value at the 500 iterations, it still has high search accuracy and convergence speed compared with other algorithms.

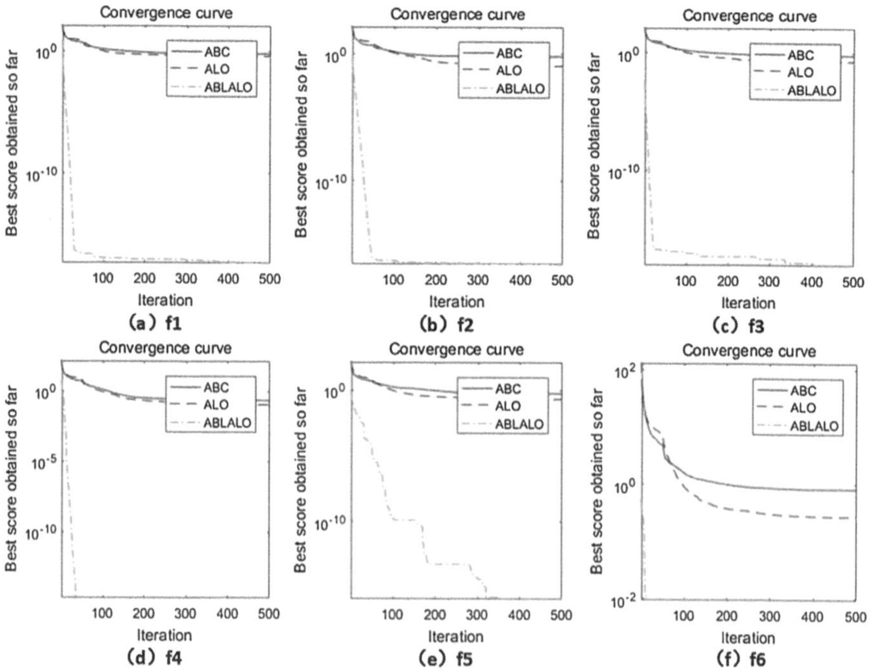

Fig. 1. Test convergence of six reference functions

Table 2. Test results of ABC, ALO, ABLALO algorithms on the $f1 \sim f6$ functions

F	Method	Best	Worst	Mean	Std
$f_1(x)$	ABC	0.000168058	4.36471	0.609207	1.40611
	ALO	5.6799e−06	4.20492	0.337896	0.557556
	ABLALO	1.24554e−18	8.5646e−18	4.2556e − 18	2.7353e−18
$f_2(x)$	ABC	0.000255661	4.64937	0.624723	1.45534
	ALO	0.000506514	3.95975	0.0988384	0.557336
	ABLALO	1.37682e−17	4.0852e−17	2.7737e−17	1.0440e−17
$f_3(x)$	ABC	3.60928e−05	4.04669	0.660491	1.37237
	ALO	0.00316892	3.96689	0.175747	0.778428
	ABLALO	2.88533e−19	1.2435e−18	7.0050e−19	3.9725e−19
$f_4(x)$	ABC	0.0044846	4.64325	0.218889	0.850219
	ALO	9.9695e−07	3.95445	0.0961253	0.557216
	ABLALO	0	0	0	0
$f_5(x)$	ABC	0.000309722	4.5924	0.549343	1.32483
	ALO	0.000459842	4.19349	0.181612	0.801642
	ABLALO	0	3.3306e−16	1.1102e−16	1.9229e−16
$f_6(x)$	ABC	0.00288624	4.49438	0.835884	1.60687
	ALO	0.00184148	4.59364	0.268658	0.997648
	ABLALO	0.00372957	0.0132368	0.00941222	0.00501858

From Table 2, on the one hand, we can see that this algorithm uses self-adaptive boundary strategy, which significantly improves the balance of the exploration and development capabilities. On the other hand, it adopts an adaptive optimal guidance strategy, which increases convergence speed. In summary, with slightly increased complexity, this algorithm effectively overcomes the shortcomings of the ALO algorithm, which easy to fall into local optimum and slow convergence speed.

5 Conclusion

It can be seen that the ABLALO algorithm has greatly improved convergence speed, search accuracy and robustness through the simulation results of the six benchmark functions optimization and the comparison with the ABC algorithm. In addition, as a new algorithm, ALO has many improvement directions. Which improvement strategies can significantly improve the performance of the algorithm with lower complexity is the next step to be studied.

References

1. Mirjalili, S.: The ant lion optimizer. Adv. Eng. Softw. **83**(C), 80–98 (2011)
2. Su, R., Zhang, F., Yan, B., et al.: Optimal power flow calculation in power system based on improved antlion algorithm. Electr. Power Sci. Eng. (9) (2017)
3. Zhao, S., Gao, L., Yu, D., et al.: Ant-lion optimization algorithm with chaos detection mechanism optimizes SVM parameters. Comput. Sci. Explor. **10**(5), 722–731 (2012)
4. Li, Z., Wu, W., Lin, Z.: An image enhancement method based on improved antlion optimization algorithm. Comput. Appl. Res. **3**, 1–2 (2018)
5. Gergel, V., Grishagin, V., Gergel, A.: Adaptive nested optimization scheme for multidimensional global search. J. Glob. Optim. **66**(1), 1–17 (2011)
6. Guan, Z., Liu, Y., Liu, Y., Xu, Y.: Hole cleaning optimization of horizontal wells with the multi-dimensional ant colony algorithm. J. Nat. Gas Sci. Eng. **28**, 347–355 (2016)
7. Wang, S., Yang, J., Chai, S.: Artificial bee colony algorithm based on chaotic catfish effect and its application. Acta Electron. Sin. **42**(09), 1731–1737 (2014)
8. Wu, W., Zhang, J., Lin, Z., et al.: Anthem algorithm with double feedback mechanism. Comput. Eng. Appl. **53**(12), 31–35 (2017)
9. Zhao, H., Li, M.-D., Weng, X.-W.: Artificial bee colony algorithm with self-adaptive globally optimal guided fast search strategy. Control Decis. **11**, 2041–2047 (2010)
10. Zhang, Z., Wang, K., Zhu, L., Wang, Y.: A Pareto improved artificial fish swarm algorithm for solving a multi-objective fuzzy disassembly line balancing problem. Expert Syst. Appl. **86**, 165–176 (2017)

Generating and Drawing of Extended Sherbinski Carpet

Xiao Hu(✉), JiXun Chu, and ZhiGang Zhang

School of Mathematics and Physics, USTB, Beijing 100083, China
445342098@qq.com

Abstract. Fractal is a new discipline in the twentieth century, through the development of the times; it has spread to all areas of natural science. Sherbinski carpet is a fractal graph proposed by the mathematician Sherbinski. In the daily teaching, because of the complexity of its graphics, it is often impossible to intuitively tell the concept. Therefore, it is particularly important to use computer simulation. There are many ways to draw a carpet image, but it is usually drawn in a regular quadrilateral and the method is relatively simple not intuitive. This paper is based on MATLAB, derivation algorithm according to the principle, and to plot any quadrilateral shape Sherbinski carpet point at an arbitrary ratio, which have a role in the popularity of fractal geometry.

Keywords: Fractal · Sherbinski carpet · MATLAB · Drawing

1 Preface

Fractal was first proposed by the American mathematician B. Mandelbrot in 1975 [1]. It is a generalization of the European set, specifically describing irregular and unsmooth shapes that naturally occur in nonlinear systems. With the development of computer technology, the drawing of fractal graphics has received the favor of researchers. With the aid of computers, a variety of fractal graphics can be drawn, which are widely used in pattern modeling and design [2], graphic coding [3] and other directions. At present, the method of drawing fractal images is Mathematica [4, 5], Visual Basic [6], and there are few methods. In 1916, Sherbinski proposed the Sherbinski carpet to promote the development of fractal geometry and expand the popularity of fractal collections.

In this paper, based on the basic modeling, the calculation method is gradually deduced, and the Sherbinski carpet model based on arbitrary quadrilateral and arbitrary proportional points is no longer confined to the regular quadrilateral.

2 Modeling

2.1 Proportion of Point Coordinates

From Fig. 1, If $\frac{AO}{AB} = \lambda$, λ is known.

We have: $\frac{x - x_1}{\lambda} = \frac{x_2 - x_1}{1}$

K. Deng et al. (Eds.): ICMIR 2018, AISC 856, pp. 387–396, 2019.
https://doi.org/10.1007/978-3-030-00214-5_50

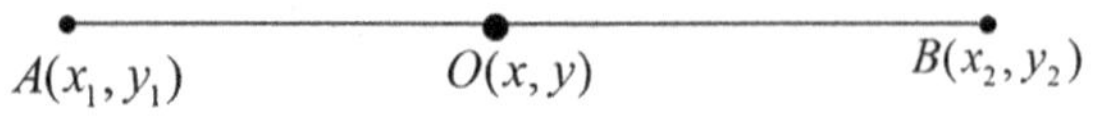

Fig. 1. Example of proportional point coordinates

$$\begin{aligned} & \therefore x - x_1 = \lambda(x_2 - x_1) = \lambda x_2 - \lambda x_1 \\ & \therefore x = (1 - \lambda)x_1 + \lambda x_2 \end{aligned} \tag{1}$$

Similarly can be obtained:

$$y = (1 - \lambda)y_1 + \lambda y_2 \tag{2}$$

By the formula (1) and (2), this is

$$\begin{cases} x = (1 - \lambda)x_1 + \lambda x_2 \\ y = (1 - \lambda)y_1 + \lambda y_2 \end{cases} \tag{3}$$

However, if λ is unknown, it can be obtained from the intersection point.

2.2 Intersection Point Coordinates

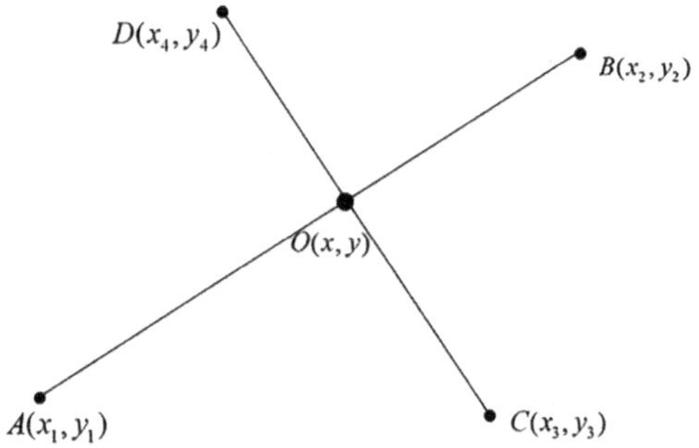

Fig. 2. Example of intersection point coordinates

From Fig. 2, It is easy to know from the nature of similar triangles:

$$\begin{aligned} & \frac{(x_4 - x_3)}{(y_4 - y_3)} = \frac{(x - x_3)}{(y - y_3)} \\ & \therefore (x_4 - x_3)(y - y_3) = (x - x_3)(y_4 - y_3) \end{aligned} \tag{4}$$

From formula (3), put the value of y into formula (4):

$$(x_4 - x_3)(((1 - \lambda)y_1 + \lambda y_2) - y_3) = (((1 - \lambda)x_1 + \lambda x_2) - x_3)(y_4 - y_3)$$

$$\lambda[(x_2 - x_1) \times (y_4 - y_3) - (x_4 - x_3) \times (y_2 - y_1)] = [(x_4 - x_3) \times (y_1 - y_3) - (x_1 - x_3) \times (y_4 - y_3)]$$

$$\therefore \lambda = \frac{[(x_4 - x_3) \times (y_1 - y_3) - (x_1 - x_3) \times (y_4 - y_3)]}{[(x_2 - x_1) \times (y_4 - y_3) - (x_4 - x_3) \times (y_2 - y_1)]} \quad (5)$$

Therefore, knowing the coordinates of the endpoints of the intersecting line segment can calculate the proportional number, and the O point coordinate can be calculated (Fig. 3).

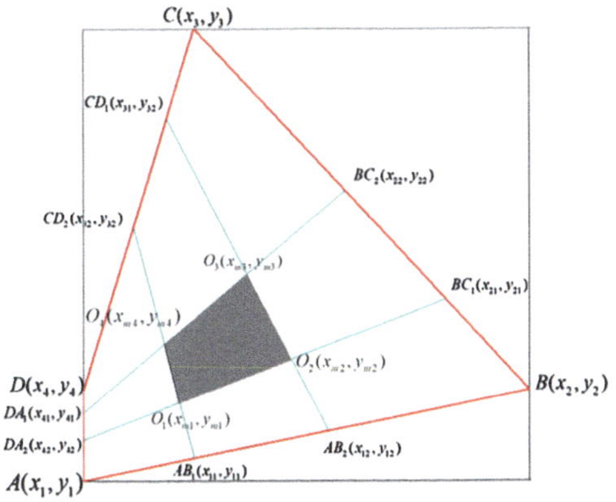

Fig. 3. Model of Sherbinski carpet

2.3 Modeling Coordinates

Now, the existing quadrilateral ABCD with vertices of $A(x_1, y_1)$, $B(x_2, y_2)$, $C(x_3, y_3)$, $D(x_4, y_4)$ is known. Each line segment selects two points in a certain proportion, such as the AB line segment, with a certain ratio to calculate the value of $AB_1(x_{11}, y_{11})$ and $AB_2(x_{12}, y_{12})$, then we should get the coordinates of O_1, O_2, O_3, O_4 and these points are easily known from the formula (3) and (5):

$$O_1 \begin{cases} x_{m1} = (1 - \lambda_1)x_{11} + \lambda_1 x_{32} \\ y_{m1} = (1 - \lambda_1)y_{11} + \lambda_1 y_{32} \end{cases} \quad (6)$$

$$O_2 \begin{cases} x_{m2} = (1 - \lambda_2)x_{12} + \lambda_2 x_{31} \\ y_{m2} = (1 - \lambda_2)y_{12} + \lambda_2 y_{31} \end{cases} \quad (7)$$

$$O_3\begin{cases} x_{m3} = (1-\lambda_3)x_{12} + \lambda_3 x_{31} \\ y_{m3} = (1-\lambda_3)y_{12} + \lambda_3 y_{31} \end{cases} \tag{8}$$

$$O_4\begin{cases} x_{m4} = (1-\lambda_4)x_{11} + \lambda_4 x_{32} \\ y_{m4} = (1-\lambda_4)y_{11} + \lambda_4 y_{32} \end{cases} \tag{9}$$

$$\lambda_1 = \frac{[(x_{42}-x_{21})\times(y_{11}-y_{21})-(x_{11}-x_{21})\times(y_{42}-y_{21})]}{[(x_{32}-x_{11})\times(y_{42}-y_{21})-(x_{42}-x_{21})\times(y_{32}-y_{11})]} \tag{10}$$

$$\lambda_2 = \frac{[(x_{42}-x_{21})\times(y_{12}-y_{21})-(x_{12}-x_{21})\times(y_{42}-y_{21})]}{[(x_{31}-x_{12})\times(y_{42}-y_{21})-(x_{42}-x_{21})\times(y_{31}-y_{12})]} \tag{11}$$

$$\lambda_3 = \frac{[(x_{41}-x_{22})\times(y_{12}-y_{22})-(x_{12}-x_{22})\times(y_{41}-y_{22})]}{[(x_{31}-x_{12})\times(y_{41}-y_{22})-(x_{41}-x_{22})\times(y_{31}-y_{12})]} \tag{12}$$

$$\lambda_4 = \frac{[(x_{41}-x_{22})\times(y_{11}-y_{22})-(x_{11}-x_{22})\times(y_{41}-y_{22})]}{[(x_{32}-x_{11})\times(y_{41}-y_{22})-(x_{41}-x_{22})\times(y_{32}-y_{11})]} \tag{13}$$

3 Simulation

3.1 Queue Thinking

A queue is a special kind of linear table, which only allows delete operation at the front of the table and insert at the rear of the table. From Fig. 4, like the stack, the queue is a linear table whose operation is limited. The end of the insertion operation is called the end of the queue, and the end of the deletion operation is called the leader.

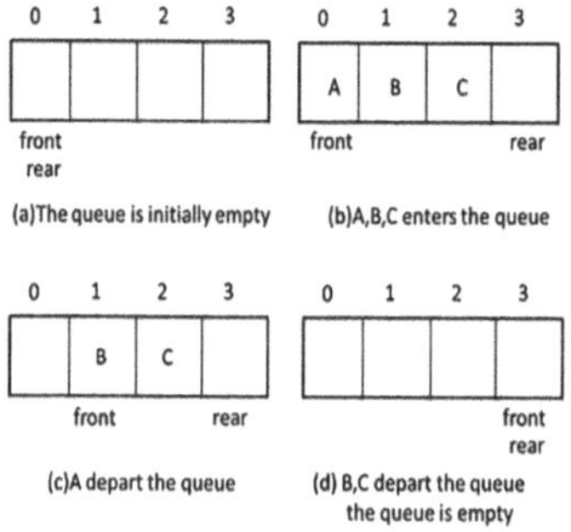

Fig. 4. The principle of the queue

This paper uses a large number of cyclic data calculations, and requires an algorithm with simple thinking, high repeatability, and high utilization. In view of this situation, the nature of the queue can be used intuitively to perform coordinate calculations. First, understand the queue method, as shown in the Fig. 4.

3.2 Sherbinski Triangle

The Sherbinski triangle is also an important part of fractal graphics. Here, the triangle drawing is also described briefly. The drawing method has the same ideas as the carpet. The specific flow is shown in the following figure (Fig. 5):

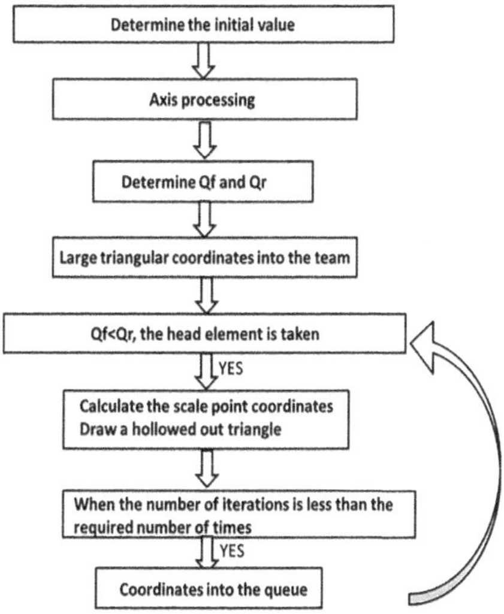

Fig. 5. Sherbinski triangle flow chart

In the first step, put the big triangle coordinates into the queue, code show as below:

```
Qr = Qr + 1;
QSier(Qr, 1).x = x1;
QSier(Qr, 1).y = y1;
QSier(Qr, 2).x = x2;
QSier(Qr, 2).y = y2;
```

In the second step, when there are elements in the queue, that is, when the tail pointer of the queue is smaller than the head pointer, the proportional point of the extracted element is calculated. Code show as below:

```
xm1 = (1 − la) * x1 + la * x2;
ym1 = (1 − la) * y1 + la * y2;
xm2 = (1 − la) * x2 + la * x3;
ym2 = (1 − la) * y2 + la * y3;
```

In the third step, if the number of cycles is less than the required number of times, the three-point coordinate of each triangle is entered into the queue until the required number of times is reached. Code show as below:

```
Qr = Qr + 1;
QSier(Qr, 1).x = x1;
QSier(Qr, 1).y = y1;
QSier(Qr, 2).x = xm1;
QSier(Qr, 2).y = ym1;
```

The fourth step is to draw each triangle.

The triangle drawing process can be summarized as Fig. 6.

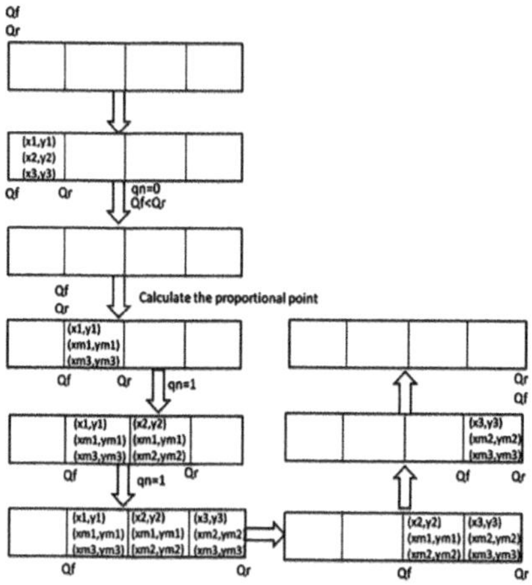

Fig. 6. Sherbinski triangle drawing process

3.3 Sherbinski Carpet

Sherbinski carpet drawing and Sherbinski triangle drawing all use the thinking of the queue. The difference is that the calculation of the point coordinate of the carpet is more complicated. The specific program flow chart is as follows (Fig. 7):

In the first step, the four vertices of the quadrilateral enter the queue, code show as below:

```
Qr = Qr + 1;
QSier(Qr, 1).x = x1;
QSier(Qr, 1).y = y1;
QSier(Qr, 2).x = x2;
QSier(Qr, 2).y = y2;
QSier(Qr, 3).x = x3;
QSier(Qr, 3).y = y3;
```

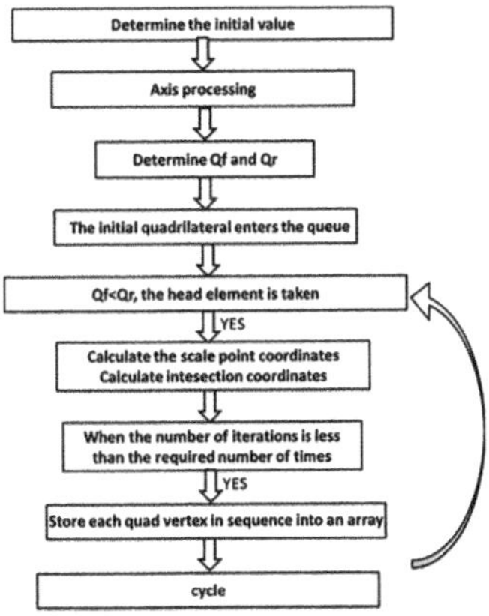

Fig. 7. Sherbinski carpet flow chart

The second step is to take out the elements in the queue, calculate the proportional point of the line segment, for example: $AB_1(x_{11}, y_{11})$ and $AB_2(x_{12}, y_{12})$. The specific code is as follows (for space reasons, only extract part of the code):

```
Qf = Qf + 1;
x1 = QSier(Qf, 1).x;
y1 = QSier(Qf, 1).y;
x11 = (1 − la1) * x1 + la1 * x2;
y11 = (1 − la1) * y1 + la1 * y2;
```

In the third step, the coordinates of the hollowed out quadrilateral must be required. The specific calculation has already been mentioned in (6), (7), (8), (9), (10), (11), (12), and (13). For the sake of calculation, the concept of an array is needed. Code show as below:

```
xt1 = [x11, x12, x12, x11]; yt1 = [y11, y12, y12, y11];
xt2 = [x32, x31, x31, x32]; yt2 = [y32, y31, y31, y32];
xt3 = [x21, x21, x22, x22]; yt3 = [y21, y21, y22, y22];
xt4 = [x42, x42, x41, x41]; yt4 = [y42, y42, y41, y41];
lat = ((xt4 − xt3).*(yt1 − yt3) − (xt1 − xt3).*(yt4 − yt3))./((xt2 − xt1).*(yt4 −
yt3) − (xt4 − xt3).*(yt2 − yt1));
xm = (1 − lat).*xt1 + lat.*xt2;
ym = (1 − lat).*yt1 + lat.*yt2;
```

With the continuous fractal of the Sherbinski carpet, the number of trapezoids hollowed out in the middle gradually increases. To make the data calculation more convenient, the vertices of the hollowed quadrilateral can be divided into 1#, 2#, 3#, and 4# points. as the picture shows (Fig. 8):

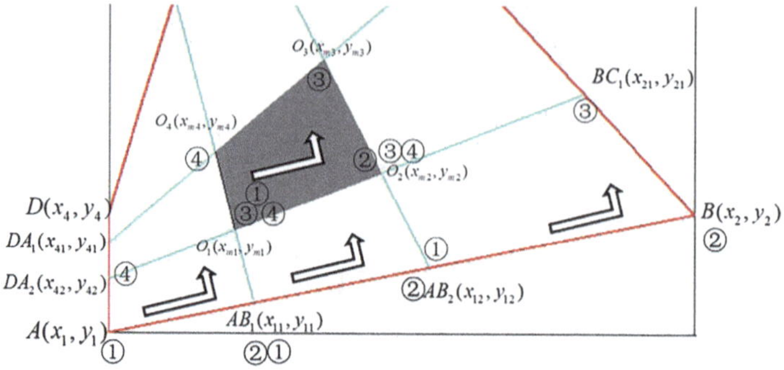

Fig. 8. Sherbinski carpet coordinates

In the fourth step, if the number of loops is less than the specified number of times, then the vertices of each quadrilateral are continuously queued to facilitate subsequent extraction calculations. Code show as below:

```
%1#
xs1 = [x1, x11, x12, x21, x22, x31, x32, x41];
%2#
xs2 = [x11, x12, x2, x22, x3, x32, x4, x42];
%3#
xs3 = [xm(1), xm(2), x21, xm(3), x31, xm(4), x41, xm(1)];
% 4#
xs4 = [x42, xm(1), xm(2), xm(2), xm(3), xm(3), xm(4), xm(4)];
%enter the queue
For j = 1:8
Qr = Qr + 1;
QSier(Qr, 1).x = xs1(j);
QSier(Qr, 1).y = ys1(j);
```

Next, Cycle from here to the second step until the number of loops is equal to the specified number of times.

The fifth step is to draw each triangle.

3.4 Program Results

The first figure has an initial coordinate of (0, 0), (1, 0), (1/2, 0) and a proportional point of 1/3. The second figure has an initial coordinate of (0, 0), (1, 0), (1/2, 0) and a proportional point of 1/5. The third figure has an initial coordinate of (0, 0), (1, 1/3), (1/3, 0) and a proportional point of 1/3. The forth figure has an initial coordinate of (0, 0), (1, 1/3), (1/3, 1) and a proportional point of 1/5 (Figs. 9 and 10).

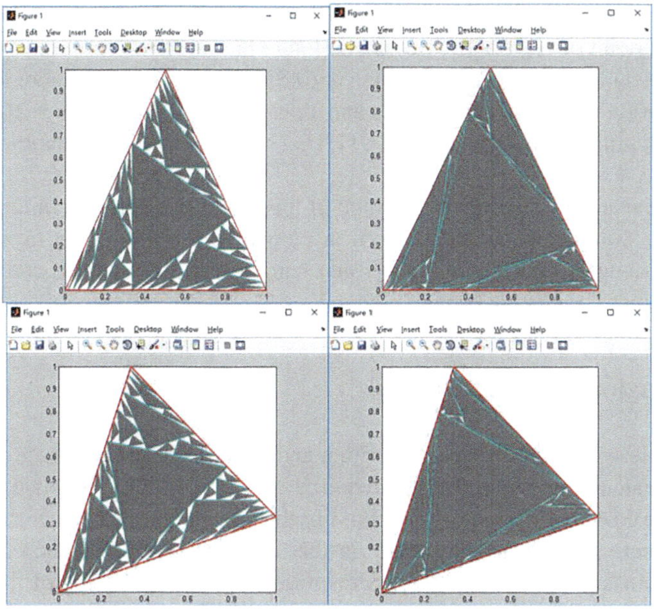

Fig. 9. Sherbinski triangles with different parameters

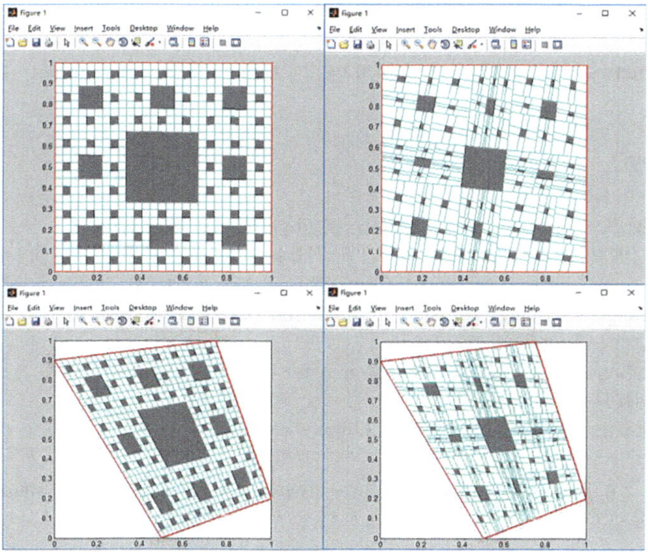

Fig. 10. Sherbinski carpets with different parameters

The first figure has an initial coordinate of (0, 0), (1, 0), (1, 1), (0, 1) and a proportional point of (1/3, 2/3). The second figure has an initial coordinate of (0, 0), (1, 0), (1, 1), (0, 1) and a proportional point of (0.35, 0.55). The third figure has an initial coordinate of (0, 0), (1, 0), (1, 1), (0, 1) and a proportional point of (1/3, 2/3). The forth figure has an initial coordinate of (0, 0), (1, 0), (1, 1), (0, 1) and a proportional point of (0.35, 0.55).

It can be seen that the drawing method of this paper can generate different graphics according to different parameters. From the graph, this method can intuitively draw Sherbinski triangles and quadrilaterals, and can successfully and accurately plot different coordinate points and scale values.

4 Conclusion

This article derives from the basic algorithm and from the base coordinate point, finally obtains the calculation method of the required variables for the Sherbinski carpet, and then uses the MATLAB tool to fully simplify the calculation process of the coordinate points by using the idea of the queue in the data structure. With the computational space, Sherbinski carpets were finally drawn out. From the drawing of the figure, the carpet can be visually observed, which is conducive to the study of its properties. At the same time, this article focuses on arbitrary quadrilaterals and arbitrary proportional points, overcomes specificity, and can modify different parameters to observe carpets of different shapes and proportions, which is of great help to everyday teaching and nature research.

Acknowledgments. This paper was funded by the Beijing Education Union.

References

1. Barnsley, M.F.: Fractals everywhere. Am. J. Phys. **97**(3), 1053 (1988)
2. Zhuang, Y., Xiong, Y., Tan, W., Pan, X.: Fractal-based imagery modeling and deformation of real-world scenes such as fireworks. Comput. Eng. Sci. **6** (2013)
3. Li, D., Zhang, L., Liang, S.: Matlab simulation based on Jacquin fractal image coding. J. Zhejiang Wanli Univ. **5** (2011)
4. Xu, J.: Mathematica's IFS and LS grammar for drawing fractal graphs. J. Neijiang Teach. Coll. **10** (2017)
5. Li, M.: Drawing two-dimensional fractal figures with mathematica. J. Huzhou Teach. Coll. **29** (2) (2007)
6. Xu, S.-Y., Xh, J., Lq, D.: Drawing fractal figures with visual basic. J. Huaibei Coal Ind. Teach. Coll. (2008)

System Architecture and Construction Approach for Intelligent Space Launch Site

Litian Xiao[1,2](✉), Mengyuan Li[1], Fei Wang[1], and Yuliang Li[1]

[1] Beijing Special Engineering Design and Research Institute, Beijing 10028, China
xiao_litian@sina.com
[2] School of Software, TNList, KLISS, Tsinghua University, Beijing 10084, China

Abstract. With the development demand of entering space, frequent space launch requires the support of intelligent technology. We study and put forward the architecture of intelligent launch site with the Internet of Things (IoT), big data and cloud computing, intelligent control, intelligent information application, intelligent transport, intelligent measurement and intelligent launching control. The architecture consists of intelligent technical area, intelligent launching region, intelligent flight range and landing area (or drop zone), intelligent command and control system, and intelligent analysis assessment system. In this paper, we propose the technical approaches of construction technology and goal realization for intelligent space launch site.

Keywords: Space launch site · Intelligent architecture
Construction technology · Internet of Things · Development research

1 Introduction

With the development of launch vehicle technology, China is improving the ability to enter space. China Launch Sites have been developed from Jiuquan, Taiyuan, and Xichang to Wenchang, and continuously improved with the technology development of launch vehicle. However, under the background of the deep integrating new generation of information technology with the traditional industry, e.g. the representative application on Industry 4.0, China Manufacture 2025, IoT, cloud computing, big data, new materials and new power, the launch site systems are still lack of the new intelligent application in China.

According to the requirements of intelligent test and launch, we propose the constructing elements, the technological approaches and the schemes of intelligent launch site system with intelligent technology as the leading factor.

K. Deng et al. (Eds.): ICMIR 2018, AISC 856, pp. 397–404, 2019.
https://doi.org/10.1007/978-3-030-00214-5_51

2 Key Elements of Launch Site System

The key elements of the launch site system are shown in Fig. 1.

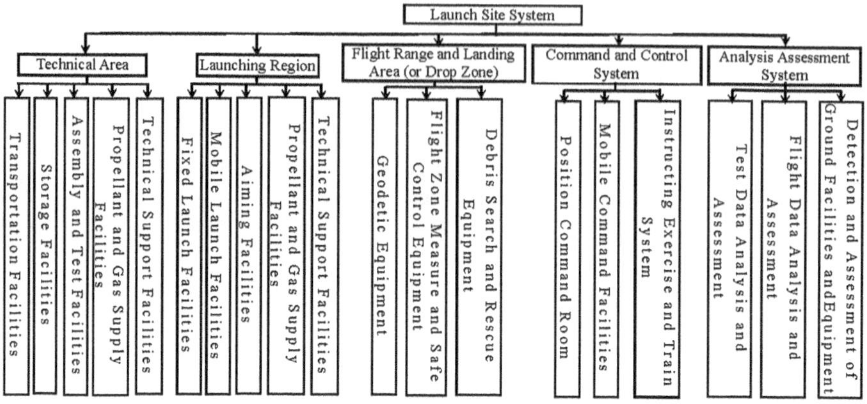

Fig. 1. Typical construction of launch site

Technical Area supplies launch vehicle and payload unloading, handover, storage, assembly and test of instrument and equipment, regular check. Its main works are assembly and test of the launch vehicle, connection and fixation of related cables and air conditioning pipelines, installing satellite fairing, etc.

Launching Region provides facilities for launch vehicles transport, hoisting and docking, docking of satellites and launch vehicle, explosive testing installation, propellant injecting, azimuth aiming, launching; inverse process operation when the launch is delayed or abnormal.

Technical Support System (Logistics System) serves for production and transportation of special gases and propellant, water supply, extinguishing and protection, airport, road, and harbor, etc.

For multi-types of flight mission, Command and Control System ensures the organization, planning, scheduling and command of the flight mission from flight articles entering launch site to orbit insertion.

Analysis Assessment System processes flight data and ground test data, and analyzes differences and failures for comparison of technical indexes. It provides mission satisfaction by testing ground support facilities and equipment and assesses facilities and equipment that need to be repaired after a mission, which is check basis for small, medium and heavy repair.

3 Architecture Construction of Intelligent Launch Site

Internationally, the application of advanced intelligent technology has greatly reduced the test time and crew size in launch site system. The occupation duration of work area has been greatly shortened. Thus it reduces the launching cost and improves the launching efficiency.

3.1 System Architecture of Intelligent Launch Site

For realizing the leapfrogging development of launch sites, the construction needs to be based on IoT, big data and cloud computing, as well as leaded by intelligent autonomous control, intelligent information applications, intelligent communication, intelligent testing and launch control, and intelligent measurement.

The architecture of intelligent launch site consists of intelligent technical area, intelligent launching region, intelligent flight range and landing area (or drop zone), intelligent command and control system, and intelligent analysis assessment system.

3.2 Intelligent Technical Area

Main technological approaches include five aspects.

- Flight articles are auto-loaded and unloaded, auto-assembled and integrated into operational facilities.
- Integrated Intelligent Test and Launch Control System (ITLCS)—It provide a multi-mode interface for test and launch control to meet the needs of the different launch vehicle. It autonomously tests and diagnoses launch vehicle. On the basis of auto-failure diagnosis, it diagnoses the states of its electric systems with data envelopment and threshold, assesses and predicts dynamic data, and has the ability to one-key testing and launching control on the ground.
- Intelligent Propellant Injection and Gases Supply System (IPIGS)—In accordance with the unified electromechanical interface standard, the upper stage or payload is injected or gas-supplied by auto-equipment at the operational station. No manual intervention is required in the process.
- Combined operational and support facilities—According to assembly and test requirements of different missions, facilities are moved and combined into one or more flexible and elastic spaces for use. The operating space can be expanded with the functional compound. It is same as to pipelining in an intelligent plant with facilities.
- Modular locale facility—Setting work areas for different functions in uniform, the modular combination is based on different mission requirements. On-demand combination layout and applications serve auto-equipment, intelligent systems, support facilities, and command and monitoring system.

3.3 Intelligent Launching Region

Main technological approaches include five aspects.

- Intelligent launch workplace—There are two types: the launch tower type and no-tower type. The tower provides auto-lifting vehicle and payloads for auto-docking, assembly, and calibration. It configures multi-mode flame diversion facilities.

With intelligent communication, the vehicle is connected to the ITLCS through adaptors. It can also achieve cable-free testing and launch control by use of wireless sensor network technology. It doesn't need aiming facilities because of autonomous star searching and aiming. Multi-mode adaptive flame diversion and reconfigurable launch pad make that some different vehicles can be launched at the same workplace.

- Unsupported launch workplace—Solid propellant launch vehicles have achieved technically unsupported launching by means of launcher truck.
- IPIGS—Ground pipelines and cables are autonomously joined or disjoined with the vehicle by electromechanical equipment combined with auto-control and intelligent decision. It has the functions of auto-injection and drains back, and auto-emergency disposal.

The whole mechanical structure of the intelligent system can have multiple degrees of freedom and movable location. Its decoupling control combines smart sensor technology with intelligent control strategies.

- Intelligent ground transportation systems and movement launch platform—the overall launch vehicle is automatically transported by using the intelligent vehicle and intelligent mechanical technology. The intelligent transportation system is designed and integrated to the launch platform. It makes movement launch platform intelligently adapt and automatically transform for multi-type launch vehicles, different flame diversion facilities and multiple launch workplaces. The occupying duration of the launch workplace is reduced as much as possible.
- Intelligent support facilities—Intelligent management objects include propellant storage and transportation, gases supply, water supply, air conditioning and so on with intelligent building technology. Intelligent control problems of complex support facilities are solved by the technologies of self-adaptive control, self-organizing and self-learning. Smart sensors collect the health status of the launch site and provide intelligent information for the assessment of ground facilities.

3.4 Intelligent Flight Range and Landing Area (or Drop Zone)

Main technological approaches include two aspects.

- Intelligent autonomous control flight—Autonomous failure detection, failure location and isolation of the vehicle during flight. Ground tracking and telemetry assist external flight guidance with dynamic ground resource allocation and redundant handover during flight.
- Intelligent landing area—Autonomous return is precisely controlled for each sub-stage of the vehicle. Health status of equipment in the vehicle is autonomously diagnosed and assessed. Ground measurement and safety control equipment standby guarantee when a launch vehicle loses its autonomous control.

3.5 Intelligent Command and Control System

Around mission ability, intelligent platforms are unmanned and intelligent command and control system is manned [1, 2]. Command crews implement mission control via intelligent network. This control is command control, not specific operation. Optimal autonomous control is based on the planning instructions or the man-in-loop instructions. According to situational awareness information, the autonomous decision makes the optimal mission scheme for flight mission plan, schedule, flight range layout, choosing sub-stage drop zone, and safety control. The system can simultaneously support multi-type missions at the same launch site or different launch sites.

3.6 Intelligent Analysis Assessment System

By data from smart sensors, situation awareness systems and processing systems, the ground equipment and launch vehicle, as well as system status of the mission facilities, are assessed and given relevant conclusions on basis of knowledge bases and expert systems. Early warning and predictable maintenance rely on intelligent failure diagnosis and prediction with failure bases and model bases. It determines equipment maintenance cycle, residual life, periodic replacement or maintenance, and enhances maintenance support efficiency.

Using big data technology quickly summarizes and refine the mission regularities for improvement and optimization of follow-up missions. It can provide developers for improving flight articles and identification departments as product assessment foundation. The system also is an intelligent tool for quality process management and integrated test identification.

4 Technology and Development of Intelligent Launch Site

In the paper, we only study technical developments for building intelligent launch site that need focused attention.

4.1 IoT

Three aspects of basic research include (1) Identifications of services, devices, flight articles, and facilities, i.e. code standard protocols; (2) Resource discovery, sensors, and nodes utilize a discovery mechanism communicating each other in IoT; (3) Communication Scenarios include event-driven and stimulus in real time. A standard protocol needs to be developed for the IoT at launch sites including selection and customization of communication technology in physical and network layer, discovery and application protocols.

IoT architecture design of intelligent launch site should consider the issues such as scalability, interoperability, business models, security, energy consumption, and mobility, etc. [3, 4]. Issues solutions need study service architecture for dynamic topology, application support system (middleware), publish/subscribe architectures, and cloud architecture. Commercial frameworks are adapt to commercial space and international launch service. We propose the architecture of IoT at the launch site in Reference [5].

4.2 Cloud Computing Data Center

The center mainly consists of basic hardware and software, cloud management platform and IoT application common platform [6]. With the support of data sharing in the center, it can provide different users in customized cloud service mode. In addition to big data processing, storage, and high-performance computing technologies, semantic fusion models need to be established. A large amount of data can be converted to the meaning format of the launch site. Data processing and analysis are as much as possible in the cloud edge node, agent, or gateway.

4.3 Smart Sensors

It is the foundation of the intelligent launch site. Cheap smart sensors need to be developed for autonomous health monitoring of the vehicle and related ground support equipment, warning of possible problems and fast prelaunch operations.

Multidisciplinary sensors and embedded sensors need to be studied and integrated with the abilities including advanced data acquisition, self-calibration, healthy self-diagnosis, fault prediction and correction (self-healing) [7, 8].

4.4 Autonomous Controllable on Intellectual Property and Technical Capacity, and Security

Secure IoT, hardware, and software should achieve the goal with management, technical standards, and specifications. Constructing security situational awareness system implements active defense. In the case of limited equipment resources, the security architecture needs to use lightweight encryption and decryption, trust management, and intrusion detection technologies [9]. Equipment applied in launch site must be verified on network electricity safety admission platform.

4.5 Trusted Verification

The trusted goal of the intelligent system is to propose a running entity that can perform a special action surpassed the preset security rules, and how to verify the entity. Therefore, we need first ensure that the software and hardware systems can be quantified and verified by calculation. The critical core of a trusted system is the trusted computing.

We create a guarantee mechanism for trusted operation on the intelligent system and design the trusted stamp and the verification protocol based on intelligent control models. The approach ensures the trustworthiness of the state migration, so as to obtain the credibility of the system. The trusted state transferring ensures the system trusted operation [10, 11].

4.6 Intelligent Control

Intelligent control technologies have more research foundations mainly including fuzzy control, neural network control and learning control [12]. The core is the integrated design of launch site system and launch vehicle.

4.7 Variable Facility Construction

According to Stitching Architect Research (SAR), three variable elements of the support body, variable body and converter are used as technical bases [13, 14]. The principle of variable facility design is function compound, space expansibility, costs reduction, environmental remediation and site restoration, elements cooperativity.

Intelligent launch site can realize the shared ground facilities for multi-mission, the simplest inflexibility infrastructure, minimum interface between facilities and launch vehicle, and variable and combinable launch facilities. Variable facility construction needs to study smart materials that are another hot topic.

4.8 Parallel Launch Site

It consists of the real launch site and artificial launch site (virtual launch site environment). Based on a combination of virtual and real, the general scheme and intelligent applications of flight mission are verified for rationality and feasibility. It supplies a platform for computation testing and checking launch vehicle system, and simulation training for the mission. The computation testing mainly verifies the regularities and assesses launch system ability, which finds out the unknown phenomenon and the occurrence conditions [15].

For constructing digital environment, multi-agent modeling, intelligent behavior modeling, and computer simulation need integrate virtual and real, and reflect the characteristics of intelligent objects in artificial launch site. The method of infinitesimal perturbation analysis [16] deals with the impact of the expansion and reconstruction of the actual launch site on the artificial launch site.

5 Conclusion

Because the construction of intelligent launch site involves a wide range of technical fields, the paper only focuses on priority technology requirements. We have carried out some researches on intelligent launch site system, big data application, IoT, access gateway and so on. Some applications have been arranged in individual mission projects [5, 10, 11, 15], but there is still a long way to go from the requirements of the intelligent launch site.

Although the US NASA has not clearly put forward the intelligent launch site, it has made clear the related technology development requirements in its future technology roadmaps [17].

Next, we shall formulate specific strategies and full roadmaps for the technologies of the intelligent launch site. It makes the intelligent launch site have the main characteristics of integration, digitization, automation, and intelligence.

Acknowledgments. This work was funded by the project of General Technology on Test and Launch partially sponsored by the 973 program. We would like to thank Tsinghua's Prof. Yu Liu who provided us with the helpful suggestions and some research resources in the research.

References

1. Çağatayhan, Ç.: Tactical command and control systems and network centric warfare. J. Mil. Inf. Sci. **2**(3), 70–73 (2014)
2. Nachet, B., Adla, A.: An agent-based distributed collaborative decision support system. Intell. Decis. Technol. **8**(1), 15–34 (2014). IOS Press
3. Huda, S., Wesam, A., Khair, E.S.: Internet of Things: a review to support IoT architecture's design. In: IT-DREPS Conference, Amman, Jordan, 6–8 December 2017
4. Li, S., Xu, L.D., Zhao, S.: The Internet of Things: a survey. Inf. Syst. Front. **17**(2), 243–259 (2015)
5. Xiao, L., Li, M., Xiao, N.: An architecture of IoT application support system in launch site. In: ICAITA2018 (2018)
6. Gazis, V., Görtz, M., Huber, M., et al.: A survey of technologies for the Internet of Things. In: International Wireless Communications & Mobile Computing Conference, USA (2015)
7. Lin, J., Yu, W., Zhang, N., Yang, X., et al.: A survey on internet of things: architecture, enabling technologies, security and privacy, and applications. IEEE Internet Things J. **99**, 1–11 (2017)
8. Jose, M.: P2.42: Latest development in advanced sensors at Kennedy Space Center (KSC). IEEE Sens. **2**(2), 1728–1733 (2002)
9. Musleh, A.S., Debouza, M., Farook, M.: Design and implementation of smart plug: an Internet of Things (IoT) approach. In: ICECTA (2017)
10. Xiao, L., Li, M., Gu, M., Sun, J.: A hierarchy framework on compositional verification for PLC software. In: IEEE ICSESS, pp. 204–207 (2014)
11. Xiao, L., Li, M., Xiao, N.: An approach for the verification of trusted operation on automatic control system. In: ICPCMM (2018)
12. Qing, H., Ji-kai, S.: Analysis and development of intelligent control technology. Coal Mine Mach. **27**(6), 921–923 (2006)
13. Yuan, H.: The research method of applying the concept of "movable" and "changeable" of architecture to architectural design. Archit. Res. **24**(7), 78–81 (2006)
14. Stephen, K.: Open building: an approach to sustainable architecture. J. Urban Technol. **6**(3), 1–16 (1999)
15. Xiao, L., Li, M., Xiao, N.: A test process model and optimization framework based on network schedule. In: ICAMAM (2018)
16. Wardi, Y., Seatzu, C., Chen, X., et al.: Performance regulation of event-driven dynamical systems using infinitesimal perturbation analysis. Nonlinear Anal. Hybrid Syst. **22**, 116–136 (2016)
17. NASA Technology Roadmaps. http://www.nasa.gov/offices/oct/home/roadmaps

Exploration and Application of Catenary Problems

Lin-tong Zhang(✉), Na Li, and Zhi-gang Zhang

University of Science and Technology Beijing, Beijing, China
17888803421@163.com

Abstract. This article introduces a special curve in mathematics—the catenary line, which clarifies the definition of the catenary line, and deduced the expression of the parameters in the catenary curve equation through Matlab programming, further, it compares the catenary line with the parabola, and concludes that the catenary line is not a parabola; At the same time, the nature of the catenary was explored, and it can be found that the regular polygon wheels can also move. The in-depth analysis of the problem shows that the catenary line happens to constitute the road surface with regular polygon wheels. Through Matlab demonstration, it can achieve a smooth movement of the regular polygon wheels on the catenary. This application can be applied to machinery and engineering, construction and other fields, it has a good application prospects.

Keywords: Catenary · Regular polygon wheels · Matlab

1 Introduction

In mathematics, the catenary is a curve of special significance. Its shape is similar to that of two fixed ropes that sag under gravity. From an innovative point of view, it is programmed with Matlab, and derived the expression of each parameter in the curve equation, thereby comparing the difference between the parabola and the catenary, and applying the special nature of the catenary to the movement of a square wheel.

2 Catenary and Parabola

2.1 Catenary Definition

The definition of a catenary is a curved shape with two ends fixed and freely suspended [1]. The function of a catenary is given by:

$$y = \cosh[a(x + C_1)] + C_2, \text{ where } a,\ C_1, C_2 \text{ is a constant.} \tag{1}$$

Boundary conditions: $\begin{matrix} y(x_1) = y_1 \\ y(x_2) = y_2 \end{matrix}$ is given, Catenary length L is given.

When the coordinates of the two fixed points $(x_1, y_1), (x_2, y_2)$ and the length of the catenary L are known, the value of the constant $a,\ C_1, C_2$ can be derived.

K. Deng et al. (Eds.): ICMIR 2018, AISC 856, pp. 405–414, 2019.
https://doi.org/10.1007/978-3-030-00214-5_52

Naturally requires: $L > \sqrt{(x_2 - x_1)^2 + (y_2 - y_1)^2}$

Using the curve's arc length integration formula, there are:

$$\begin{aligned} L &= \int_{x_1}^{x_2} \sqrt{1 + (y')^2} dx = \frac{1}{a} \int_{x_1}^{x_2} y''(x) dx \\ &= \frac{1}{a} [\sinh a(x_2 + C_1) - \sinh a(x_1 + C_1)] \\ &= \frac{2}{a} \sinh \frac{a(x_2 - x_1)}{2} \cosh \frac{a(x_1 + x_2) + 2C_1}{2} \end{aligned} \tag{2}$$

By formula (1) we can know:

$$\begin{aligned} y_2 - y_1 &= \frac{1}{a} [\cosh a(x_2 + C_1) - \cosh a(x_1 + C_1)] \\ &= \frac{2}{a} \sinh \frac{a(x_2 - x_1)}{2} \sinh \frac{a(x_1 + x_2) + 2C_1}{2} \end{aligned} \tag{3}$$

From this:

$$\sqrt{L^2 - (y_2 - y_1)^2} = \frac{1}{a} \sinh \frac{a(x_2 - x_1)}{2} \tag{4}$$

In the process of solving the equation, it is necessary to indirectly solve the value of a by using a dichotomy or a method of plotting intersection points to indirectly solve the value of a [2].

After the constant a is obtained, the boundary value condition is brought in, and the value of C_1, C_2 can be obtained.

$$y_1 = \frac{1}{a} \cosh a(x_1 + C_1) + C_2 \tag{5}$$

$$y_2 = \frac{1}{a} \cosh a(x_2 + C_1) + C_2 \tag{6}$$

By solving the above equation, the constant C_1, C_2 can be found, so the catenary equation is obtained [3].

2.2 Parabola Definition

A parabola is a trajectory within a plane to a point at which a fixed point F (focal point) and a straight line l (principal line) have equal distances. It has a number of representations, and the function represented by a parameter has the following general formula:

$$y = ax^2 + bx + c, \text{ where } a, b, c \text{ is a constant.} \tag{7}$$

Boundary conditions: $\begin{matrix} y(x_1) = y_1 \\ y(x_2) = y_2 \end{matrix}$ is given, Parabolic length L is given.

When the coordinates of two fixed points $(x_1, y_1), (x_2, y_2)$, and the length of the parabola L are known, the parabola parameters a, b, c can be determined.

By the Pythagorean Theorem:

$$L = \int_{x_1}^{x_2} \sqrt{1 + y'^2} dx = \int_{x_1}^{x_2} \sqrt{1 + (2ax + b)^2} dx \tag{8}$$

Then by the boundary conditions, it can be concluded:

$$y_1 = ax_1^2 + bx_1 + c \tag{9}$$

$$y_2 = ax_2^2 + bx_2 + c \tag{10}$$

With the formulas (8), (9) and (10), the value of the parameter can be obtained, so that the parabolic equation can be found.

2.3 Comparison of Catenary and Parabola

Through Matlab software, the comparison of catenary and parabola is realized. The specific flow is as follows (Fig. 1):

The relevant Matlab commands are as follows:

By formula (4), the parameters of catenary a can be obtained:

y0 = x0/2 * sqrt (L^2 − (y2 − y1)^2) − sinh(x0 * (x2 − x1)/2);

By formula (5) and (6), the parameters of catenary C_1, C_2 can be obtained:

Equation 1 = 1/a * cosh (a * (x1 + c1)) + c2 − y1;

Equation 2 = 1/a * cosh (a * (x2 + c1)) + c2 − y2;

[c1, c2] = solve (Eq 1, Eq 2);

By formula (8), (9) and (10), the parameters of parabola a, b, c can be obtained:

Equation 1 = a * x1^2 + b * x1 + c − y1;

Equation 2 = a * x2^2 + b * x2 + c − y2;

LT1 = 1/(2 * a) * ((2 * a * x1 + b)/2 * sqrt((2 * a * x1 + b)^2 + 1) + ½ * log ((2 * a * x1 + b) + sqrt((2 * a * x1 + b)^2 + 1)));

LT2 = 1/(2 * a) * ((2 * a * x2 + b)/2 * sqrt((2 * a * x2 + b)^2 + 1) + ½ * log ((2 * a * x2 + b) + sqrt((2 * a * x2 + b)^2 + 1)));

Equation 3 = LT2 − LT1 − L;

[a, b, c] = solve (Eq 1, Eq 2, Eq 3);

Through the above process, the catenary and parabola curve equations can be obtained. The following process is implemented through Matlab, and then we can compare the results (Fig. 2).

By comparison, the catenary line is not a special type of parabola, and the parabola does not coincide with the catenary under certain condition of a certain curve length.

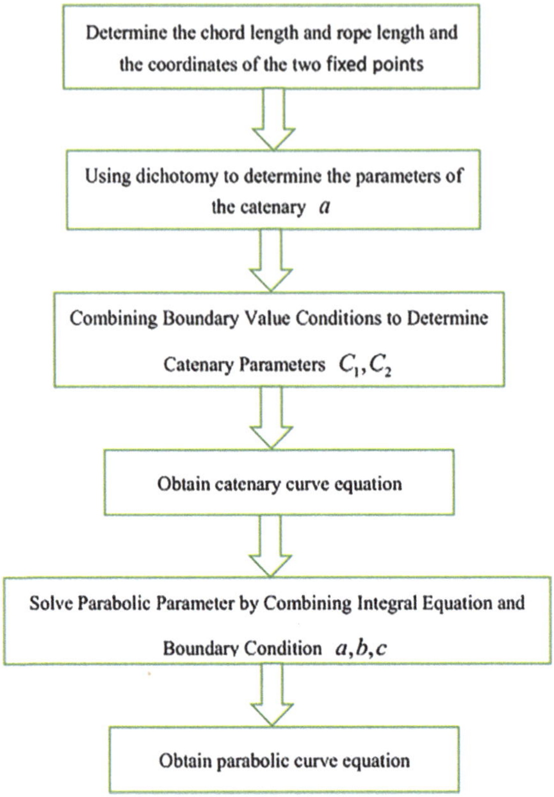

Fig. 1. Matlab flow chart

Under the condition of two fixed point coordinates and certain chord length, increase the rope length. It can be concluded that the value of the abscissa of the lowest point of the catenary changes from small to large, and the ordinate changes from big to small, that is, the lowest point of the catenary moves to the lower right (Fig. 3).

In the condition of certain rope length and a certain fixed point, increase the chord length and change the position of another point. It can be concluded that the value of the abscissa of the lowest point of the catenary changes from small to large, and the ordinate changes from small to large. That is, the lowest point of the chain moves to the upper right.

3 Application of Catenary

3.1 Statement of Problems

It is generally believed that only the round wheels can make our car move forward steadily, but this is only for straight roads. If there are curved roads under ideal

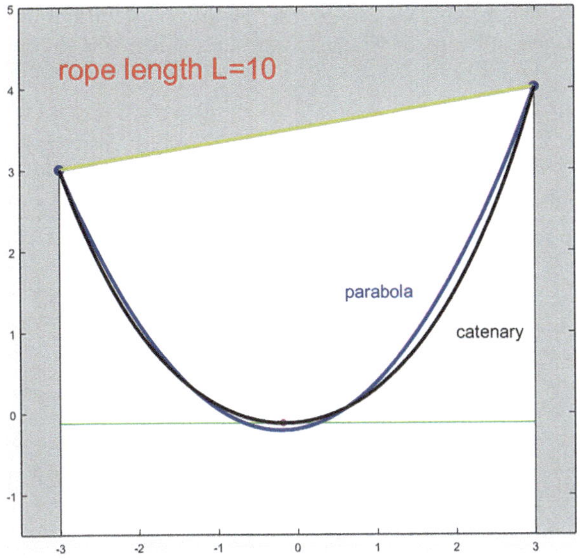

Fig. 2. When the Chord length and the fixed points are known, change the rope length, and plot the parabola and the catenary

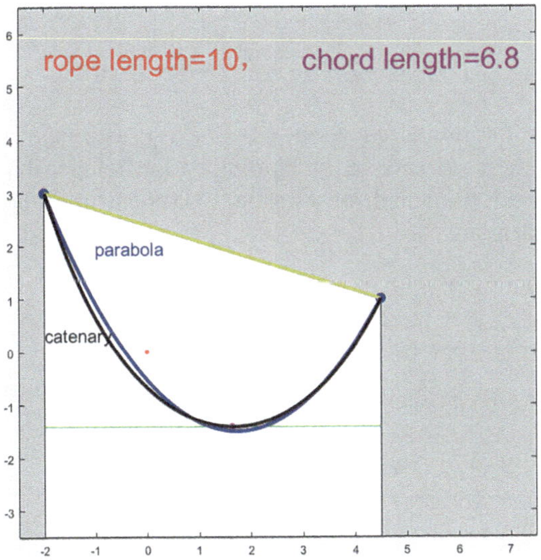

Fig. 3. When the rope length and one fixed point are known, change the chord length and the other fixed point, and plot the parabola and the catenary

conditions, the square wheels can also rotate smoothly. It turns out that as long as a proper road is paved, the square-wheel bicycle can still move smoothly [4, 5].

3.2 Analysis and Exploration

This is a special position of a square wheel. Let the length of the square side be 2. If the x-axis is the ground, the distance between the axle (R-axis) and the ground is $\sqrt{2}$, and the distance between the middle vertex and the ground is $\sqrt{2}-1$ (Fig. 4).

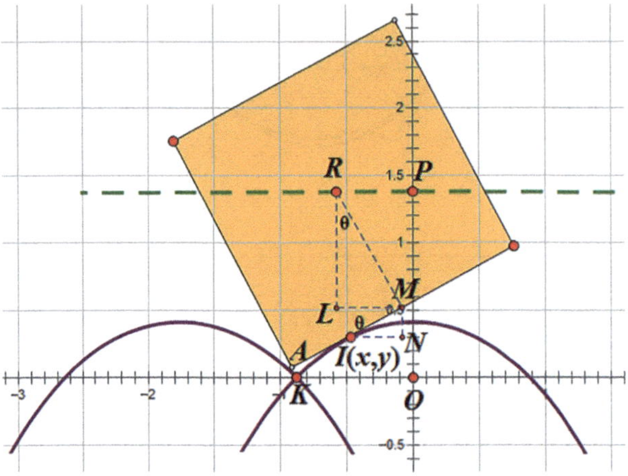

Fig. 4. Square wheel rotation diagram

Let $y = y(x)$ be the orbital equation, where $I(x, y)$ is a cut-off point between the square wheel and the road surface. According to the relationship between the road surface and the wheel, the length of AI should be equal to the length of the arc KI, marked as s, and there are:

$$s = \int_{x_0}^{x} \sqrt{dx^2 + dy^2} = \int_{x_0}^{x} \sqrt{1 + y'^2}dx \tag{11}$$

From the figure we can see:

$$\begin{gathered} AI = s, MI = 1 - s, RM = 1 \\ RM \cos\theta + MI \sin\theta + y = OP \end{gathered} \tag{12}$$

That is, $\cos\theta + (1 - s)\sin\theta + y = \sqrt{2}$.

Solve the s, and bring in $\tan\theta = y^{'}$, you can get:

$$s = 1 + \frac{1}{y'} + y\sqrt{1 + \frac{1}{y'^2} - \sqrt{2} \cdot \sqrt{1 + \frac{1}{y'^2}}} \tag{13}$$

Combine the two equations to get the integral equation:

$$\int_{x_0}^{x} \sqrt{1+y'^2}dx = 1+\frac{1}{y'}+y\sqrt{1+\frac{1}{y'^2}-\sqrt{2}\cdot\sqrt{1+\frac{1}{y'^2}}} \tag{14}$$

After organization:

$$y = -\cosh(x+C)+\sqrt{2} \tag{15}$$

Further reasoning, it can be proved that for any regular polygon, if the length of one side of the regular polygon is $2a$, and the height of the side from the center is h, then the function expression of the path of the wheel is:

$$y = -h\cosh\left(\frac{x}{h}+C\right)+\sqrt{a^2+h^2} \tag{16}$$

3.3 Implementation and Results

Through Matlab programming, the process of moving the regular polygon wheels along the catenary is realized. The specific flow is as follows (Fig. 5):

The relevant Matlab commands are as follows:

Calculate the top Angle of the regular polygon:

```
sit0=(n-2)*pi/n;
```

By formula (16), find the left and right endpoints of the catenary:

```
a = −asinh (1);
b = asinh (1);
x = linspace (a, b, nx);
y = −cosh(x);
```

Calculate the arc length of the catenary:

```
Sall = sinh(b) − sinh(a);
```

Write each of tangent point coordinates:

```
x0 = x(i);
y0 = y(i);
```

Find the arc length on both sides:

```
sx0 = sinh(x0) − sinh(a);
L1 = sx0;
L2 = sall − L1;
```

Determine the coordinates of n vertices of the regular polygon:

```
x1 = x0 − L1 * (1/sqrt(kx0^2 + 1));
y1 = y0 − L1 * (kx0/sqrt (kx0^2 + 1));
x2 = x0 + L2 * (1/sqrt(kx0^2 + 1));
y2 = y0 + L2 * (kx0/sqrt (kx0^2 + 1));
sit = atan(kx0) + pi/2;
x3 = x2 + sall * cos(sit);
y3 = y2 + sall * sin (sit);
x4 = x1 + sall * cos(sit);
y4 = y1 + sall * sin(sit);
```

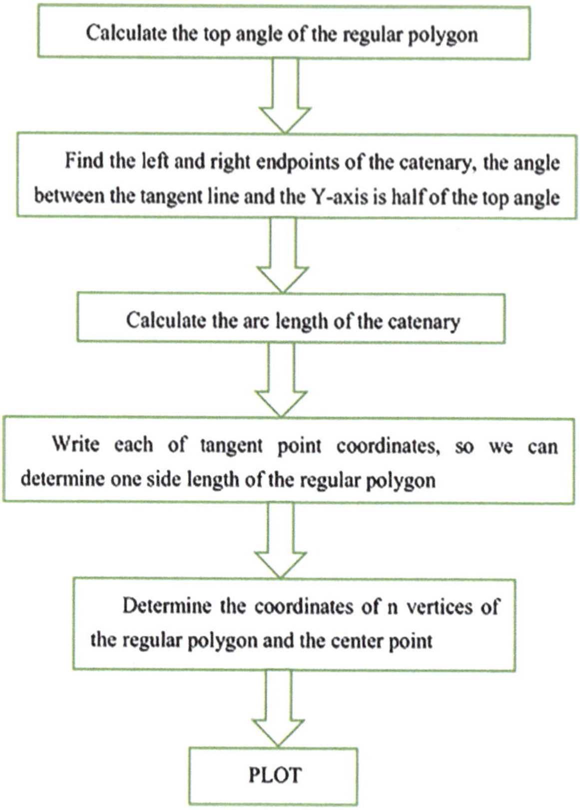

Fig. 5. Matlab flow chart

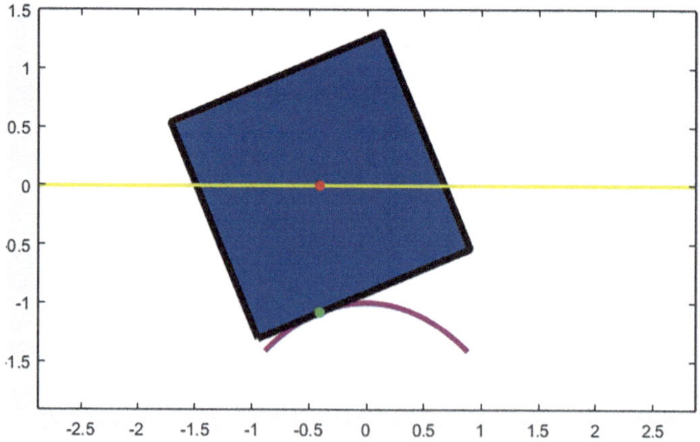

Fig. 6. Square wheel rotation

Through the above process, you can get a regular polygon wheel along the catenary movement process. The following is to achieve the process, and get the results by Matlab (Fig. 6, 7, 8 and 9).

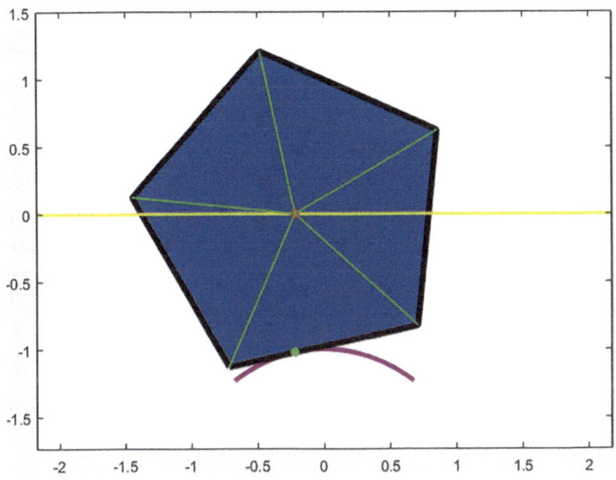

Fig. 7. Pentagon wheel rotation

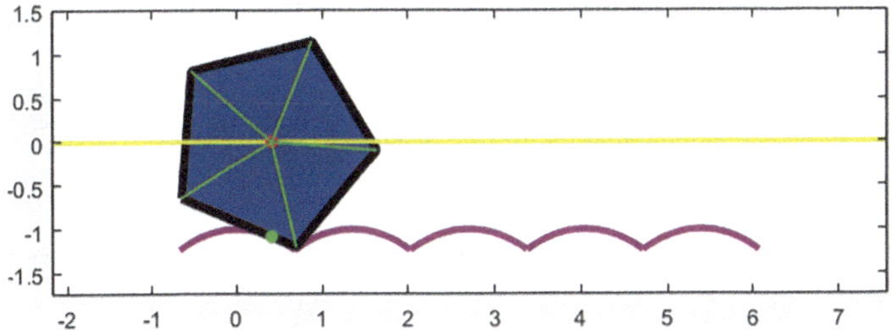

Fig. 8. Square wheel rotation on multistage catenary

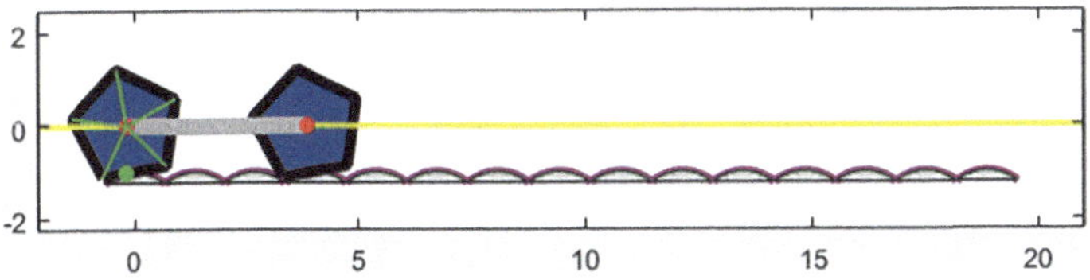

Fig. 9. Pentagon wheel rotation on multistage catenary

4 Summary and Conclusion

Through the study of the catenary, this article deduced the expression of each parameter of the catenary, and confirmed that the catenary is not a parabola by programming with Matlab. Then it extends the catenary to the problem of wheel rotation. Through analysis and exploration, it can be seen that the regular polygon wheels can also be rotated by changing the shape of the road surface. Through mathematical derivation and Matlab programming demonstration, the regular polygon wheels can be rotated on the trajectory of the catenary. This idea breaks through the existing one, that is, the wheel must be circular to be able to turn on a straight road surface. When going forward in the path formed by the catenary, the regular polygon wheels will remain tangent to the arc of the catenary, so that the wheels can move smoothly on the catenary. This result makes me once again marvel at the mystery of nature. That is, the shape of a catenary hanging naturally in the gravitational field has paved the way for a bicycle with the regular polygon wheels.

Acknowledgments. Thank the teachers for giving me patient guidance and practical advice during my writing of this thesis. At the same time, this paper is supported by the Beijing Education Trade Union.

References

1. Liu, Y.: From Da Vinci's question to a famous curve—catenary line. Sci. Technol. Inf. **04**, 136 (2012)
2. Ma, X., Yi, V.: The binary search method for solving catenary problems. Comput. Appl. Softw. **34**(09), 50–56 (2017)
3. Ghafari, H., Dardel, M.: Parametric study of catenary mooring system on the dynamic response of the semi-submersible platform. Ocean Eng. **153**, 319–332 (2018)
4. Xie, F., Sun, M.: Study on the geometric properties of catenary. J. Changchun Inst. Educ. **30** (04), 33–34 (2014)
5. Bankstain, J.G., Huang, X.: A note on catenary lines (in English). J. Yangzhou Vocat. Univ. **13**(01), 1–5 (2009)

Application Research of Process Capability Analysis in Manufacturing Quality Control

Yanming Yang(✉)

Naval Aviation University, Qingdao Campus, Qingdao, China
yymqd@126.com

Abstract. Process capability analysis are widely used in industry to achieve and maintain a high-quality level of manufactured items. Various indices have been proposed, but the most widely used are C_p, C_{pk}, C_{pm} and C_{pmk}. This paper gives the calculation method of the process capability index, discusses the process capability analysis and evaluation method, and analyzes in detail the process countermeasures of the process capability index is too large or too small. Through examples, the normality test method and controlled state test method for sample data are given. That is, the Q–Q probability plot and the Anderson–Daling test method are used to determine whether the sample sequence is normal distribution. The Xbar-R control chart is used to determine whether the sample sequence is in a process-controlled state. Using Minitab software, control charts and process capability diagrams were plotted, and process capability indices C_p and C_{pk} are calculated and analyzed. Finally, according to the quality data of customer feedback, the process capability level and state of the processing process are determined, so that the process capability can meet the customer's requirements.

Keywords: Process capability analysis · Process capability indices
Process monitoring · Quality control

1 Introduction

Process capability analysis is widely used as a measure of process potential and process performance [1]. That is to say, process capability analysis is an important measuring means used by most manufacturing industries to monitor whether the manufactured products conform to their quality standard [2]. Process capability refers to the ability of the production process to maintain product quality under statistical control. In general, the higher the process capability, the smaller the product quality fluctuation; on the contrary, the greater the quality fluctuation. For any quality improvement activity, understanding the process and quantifying process capability are essential. Process capability analysis has become an important part of applying statistical process control to achieve continuous improvement of product quality. The relationship between the actual process capability and the product specification range can be quantified using the process capability index. The role of the process capability index is to measure the ability of the production process to meet the pre-determined product specification level, providing a basis for estimating and predicting process capability. The traditional

K. Deng et al. (Eds.): ICMIR 2018, AISC 856, pp. 415–422, 2019.
https://doi.org/10.1007/978-3-030-00214-5_53

process capability index calculation requires that the production process must be controlled and the process output data fellow normal distribution and independent of each other [3].

2 Process Capability Indices

Process capability indices (PCIs) are one of the monitoring tools widely accepted in the industry [4] to describe and to evaluate the capability of production processes. The most commonly used process capability indices are C_p, C_{pk}, C_{pm} and C_{pmk}, defined as follows [5]:

$$C_p = \frac{USL - LSL}{6\sigma} \tag{1}$$

$$C_{pk} = \min\left\{\frac{USL - \mu}{3\sigma}, \frac{\mu - LSL}{3\sigma}\right\} \tag{2}$$

$$C_{pm} = \frac{USL - LSL}{6\sqrt{\sigma^2 + (\mu - T)^2}} \tag{3}$$

$$C_{pmk} = \min\left\{\frac{USL - \mu}{3\sqrt{\sigma^2 + (\mu - T)^2}}, \frac{\mu - LSL}{3\sqrt{\sigma^2 + (\mu - T)^2}}\right\} \tag{4}$$

where μ is the process average, r is the process standard deviation, USL is upper specification limit, LSL is lower specification limit and T is the target value.

Regarding the research on the process capability indices, Juran [6] first proposed the process capability index C_p in the case of double-sided specification in 1974. Kane [7] has an offset between the distribution center and the tolerance center. Hsiang et al. [8] introduced the ideal value of customer expectations and proposed the Taguchi index C_{pm} in 1985. In 1986, C_p was improved and a correction process capability index C_{pk} was proposed. Pearn et al. [9] combined the advantages of C_{pk} and C_{pm} in 1992, combined the elements of C_{pk} with the denominator of C_{pm}, and proposed the mixed index C_{pmk}. The above studies show that the process capability indices are of great significance to the control of production process capability and has received extensive attention from various industries.

In practice, the process mean and process standard deviation are usually unknown, but they can be estimated from a sample of n measurements $\{x_1, x_2, \ldots, x_n\}$. The most common estimates of μ and σ are [10]:

$$\hat{\mu} = \bar{X} = \frac{1}{n}\sum_{i=1}^{n} X_i \tag{5}$$

$$\hat{\sigma} = S = \left[\frac{1}{n-1}\sum_{i=1}^{n}(X_i - \bar{X})^2\right]^{1/2} \quad (6)$$

3 Process Capability Analysis and Evaluation

3.1 Disposal of Excessive Process Capability Index

When C_p> 1.67, it can be considered that the process capability is too full. If the index of process capability is too large, it means to do rough work with care and carefulness, which will inevitably affect the production efficiency and increase the product cost. At this time, the following measures should be taken to reduce the C_p according to the actual situation.

1. Reduce process capability. For example, if equipment and raw materials with low precision but high efficiency and low cost are used, the process capability index will be reasonably reduced to an appropriate level.
2. Change the design to improve the technical requirements of the product.
3. Adopt methods such as merger or reduction of processes.

3.2 Disposal of Too Small Process Capability Index

When C_p< 1, it means that the level of product quality is low. This is the case for tertiary or quadruple processing. At this time, we must suspend processing, immediately trace the causes and take measures.

1. Strive to improve the accuracy of equipment, make the process more reasonable and effective, further improve the operational skills and quality awareness, improve the quality of raw materials and improve the cutting performance, so that the process capability is properly improved.
2. Revise the standards, if the design allows, can reduce the technical requirements, that is, to deal with the method of relaxing tolerances.
3. In order to ensure the quality of the manufactured products, when the process capability is insufficient, generally reject the unqualified products after the full inspection or implement the grading screening to improve the product quality.

3.3 Appropriate Process Capability Index

When 1 < C_p> 1.67, it indicates that the process capability is adequate, usually one-level or two-stage processing. At this time, process control should be carried out so that the production process is in a controlled or stable state so as to maintain a significant change in process capability, thereby ensuring quality.

4 Case Analysis

4.1 Problem Description

A cable wire manufacturer wants to assess whether the cable diameter accords with the specifications. The cable diameter must be 5.5 ± 0.5 mm to conform to the engineering specifications. Analysts evaluate the capabilities of the process to ensure that it meets the customer's requirements, that is, the C_{pk} is 1.33. Analysts took five cables from the production line every hour to record their diameters. The statistical data of the cable diameter is shown in Table 1. The time series plot of the cable diameter is shown in Fig. 1.

Table 1. The statistical data of the cable diameter (mm)

5.29	5.50	5.55	5.41	5.59	5.43	5.57	5.59	5.81	5.51
4.93	5.34	5.27	5.11	5.65	5.59	5.19	5.62	5.51	5.30
5.45	5.88	5.44	5.61	5.73	6.07	5.32	5.62	5.42	5.49
5.77	5.26	5.46	5.57	5.48	5.46	5.60	5.30	5.64	5.14
5.27	5.45	5.13	5.57	5.25	5.57	5.59	5.29	5.39	5.91
5.38	5.57	5.17	5.21	5.68	5.44	5.50	5.62	5.40	5.37
5.58	5.48	5.32	5.70	5.67	5.60	5.33	5.38	5.67	5.57
5.41	5.34	5.44	5.37	5.74	5.72	5.56	5.60	5.20	5.78
5.43	5.44	5.41	5.26	5.18	5.21	5.32	5.24	5.44	5.23
5.50	5.44	5.45	5.71	5.27	5.36	5.54	5.69	5.31	5.34

4.2 Normality Test

In order to determine whether the data is normally distributed, compare the p-value to the level of significance. In general, the significance level (denoted by α) is 0.05. A significance level of 0.05 means that when the data actually obeys the distribution, the risk of discriminating against data distribution is 5%.

P-value $\leq \alpha$: data disobedience distribution.

If the p-value is not larger than the significance level, the decision is to reject the null hypothesis and conclude that the data does not conform to the distribution.

P-value $> \alpha$: The conclusion that the data disobedience cannot be obtained.

If the *P*-value is greater than the significance level, the decision is that the null hypothesis cannot be denied because there is insufficient evidence to conclude that the data does not conform to the distribution.

As can be seen from Fig. 2, since the *P*-value is 0.794, which is greater than the significance level of 0.05, the decision cannot be negated. That is, the original hypothesis declares that the data is normally distributed.

Examine the Q–Q probability plot and estimate how well the data points fit the fitted distribution line. If the specified theoretical distribution fits the data well, the points will be closely spaced along the line. As can be seen from Fig. 2, the points on the normal Q–Q probability plot are distributed along the fitted line. The normal distribution seems to fit the data well. It should be noted that the fitted distribution line is the straight line in the middle of the figure. The outer solid line on the graph is the confidence interval for each percentile (rather than the entire distribution) and should not be used to assess the fit of the distribution.

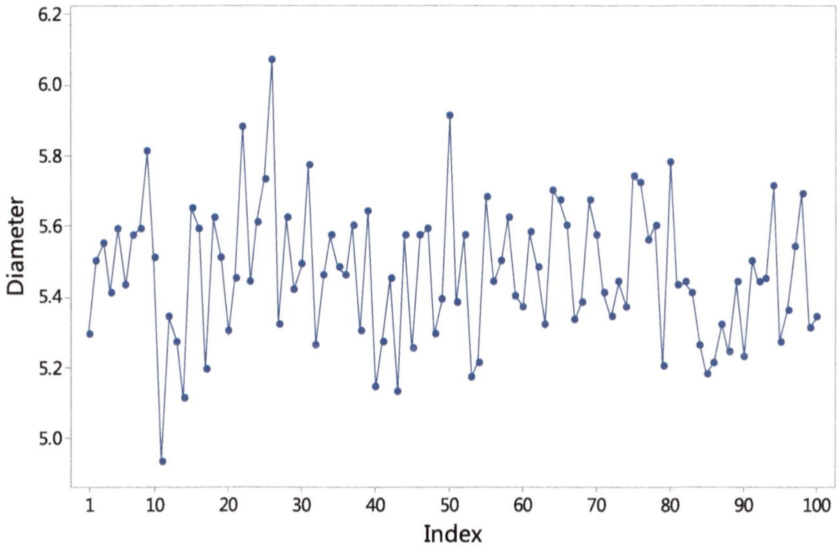

Fig. 1. Time series plot of diameter

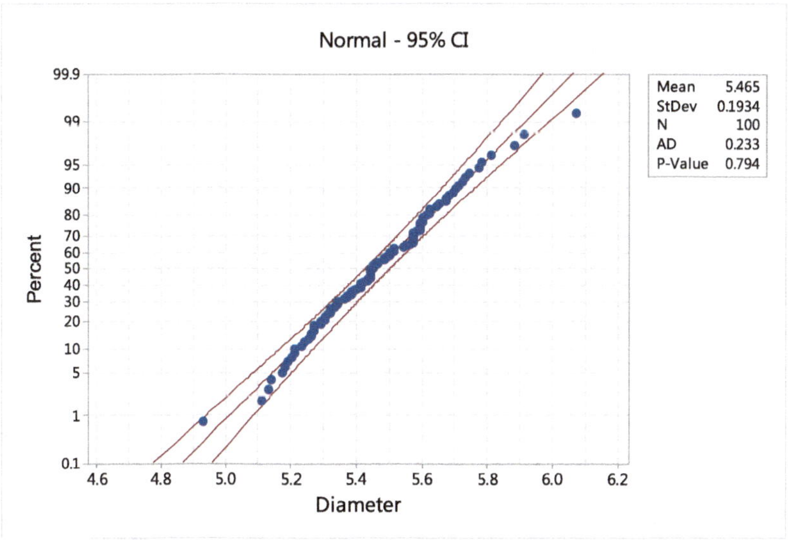

Fig. 2. Q–Q probability plot of diameter

4.3 Process Capability Analysis and Decision

If the process capability statistics are to be interpreted, the sample data should approximately obey the normal distribution. In this case, this requirement seems to have been satisfied, as shown by the histogram with the normal curves.

As can be seen from Fig. 3, the process mean (5.465) is slightly smaller than the target value (5.5). And both tails of the distribution fall outside the specification limits. Which means that it is sometimes found that some cable diameters are less than the 5.0 mm specification limit or greater than the 6.0 mm specification limit.

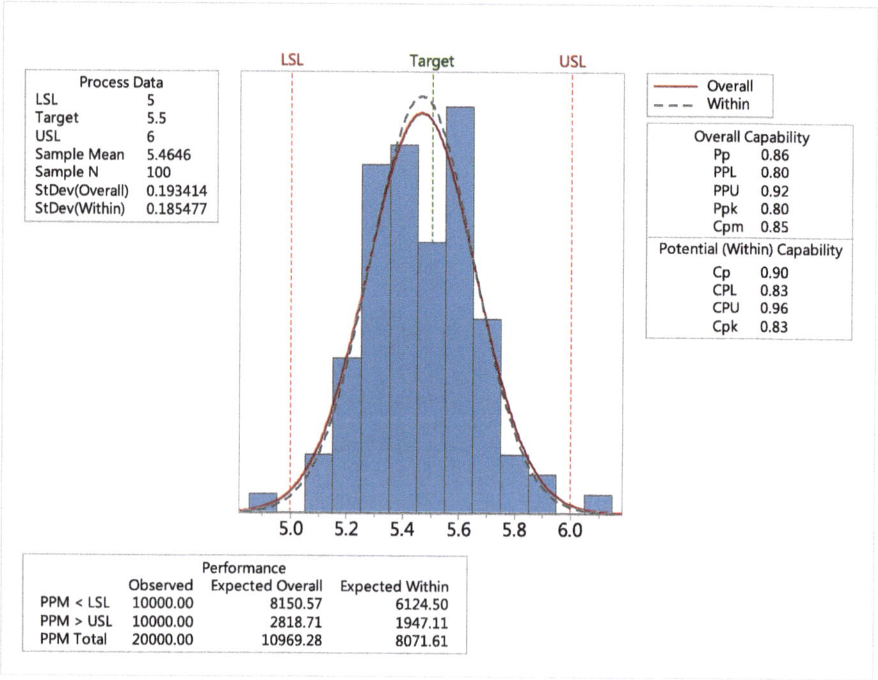

Fig. 3. Process capability report for diameter

The C_{pk} index indicates whether the unit of process production is within tolerance limits. In this case, the C_{pk} index is 0.83, indicating that the manufacturer must improve the process by cutting back the variation and centering the process on the goal. Obviously, the large variation in the process is a far more serious problem for this production line than the process is not centered on the goal.

Similarly, the PPM total is the number of parts per million (10969.28) that are outside the tolerance limits for the features that are of interest. This means that about 10,969 out of every million cables are out of specification.

Therefore, manufacturers do not meet customer requirements and should improve their processes by cutting back process variability.

4.4 Results Further Analysis

We use Minitab software to generate a sixpack report (as shown in Fig. 4), and further analyze the process capability. On the Xbar (Mean value, also denoted by $\bar{X}$) control chart and the R (Range) control chart, the points are randomly distributed between the control limits, indicating that this is a stable process. However, you should also compare the points on the R chart and the $\bar{X}$ chart to see if they are similar to each other. The points in this example are not like this, indicating once again that this is a stable process.

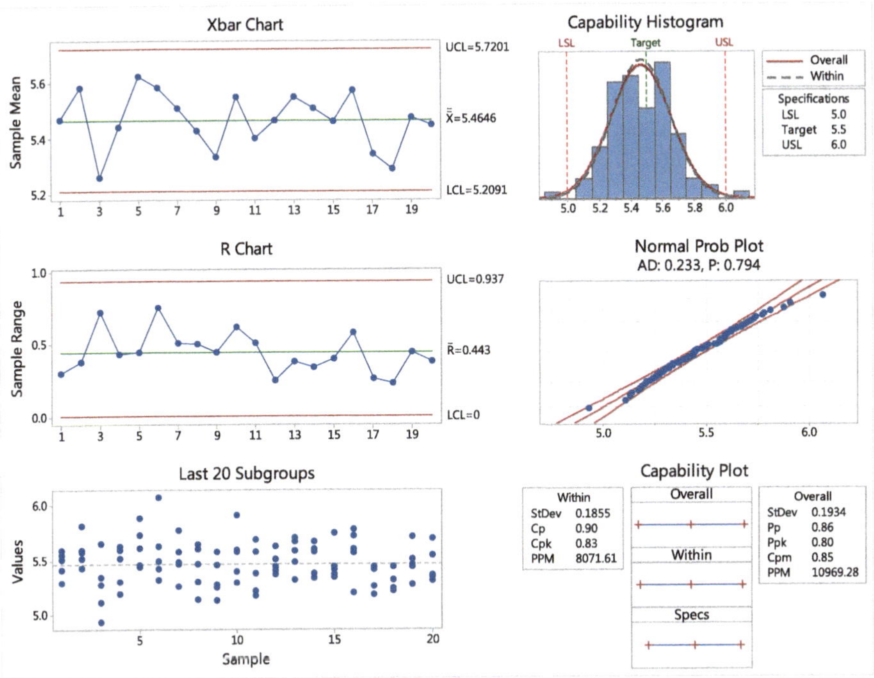

Fig. 4. Process capability sixpack report for diameter

The points on the control charts of the last 20 subgroups randomly spread horizontally without any trend or offset, which also indicates that the process is stable.

If the process capability statistics are to be interpreted, the sample data should approximately obey the normal distribution. On the ability histogram, the data approximately obeys the normal curve. On the normal probability plot, the point is approximately a straight line and falls within the 95% confidence interval. These patterns indicate that the data is normally distributed.

However, it can be seen from the process capability analysis graph that the overall process variation range is wider than the specification limit range (specification). This explains that the cable diameter sometimes exceeds the tolerance limit [5.0, 6.0].

In addition, the value of C_{pk} (0.83) falls below the required target of 1.33, which means that manufacturers need to improve the process.

5 Conclusion

The process capability index has been proposed as a numerical measure for providing process capability in the manufacturing industry, which is an effective and efficient tool for quality assurance [10]. Through the analysis of the process capability of the process, the original process capability of the process was found to be insufficient. After the corresponding rectification measures were taken, the process capability was improved and the customer's requirements were met. The conclusion is: in the case of single-valued data, considering the reliability of the index, in addition to requiring that the sample size is large enough, the process capability index C_{pk} should be the first choice for measuring process capability in the steady state.

References

1. Chen, K.S., Wang, K.J., Chang, T.C.: A novel approach to deriving the lower confidence limit of indices Cpu, Cpl, and Cpk in assessing process capability. Int. J. Prod. Res. **55**(17), 1–19 (2017)
2. Balamurali, S., Usha, M.: Determination of an efficient variables sampling system based on the Taguchi process capability index. J. Oper. Res. Soc. **2**, 1–13 (2018)
3. Koppel, S., Chang, S.I.: a process capability analysis method using adjusted modified sample entropy. Proc. Manuf. **5**, 122–131 (2016)
4. De-Felipe, D., Benedito, E.: Monitoring high complex production processes using process capability indices. Int. J. Adv. Manuf. Technol. **93**(1–4), 1257–1267 (2017)
5. Pan, J.N., Li, C.I.: New capability indices for measuring the performance of a multidimensional machining process. Expert Syst. Appl. **41**(5), 2409–2414 (2014)
6. Juran, J.M.: Juran's Quality Control Handbook, pp. 11–13. McGraw-Hill, New York (1974)
7. Hsiang, T.C., Taguchi, G.: Tutorial on quality control and surface, the Taguchi methods. Joint Meetings of the American Statistical Association, Las Vegas, Nevada (1985)
8. Kane, V.E.: Process capability indices. J. Qual. Technol. **18**(1), 41–52 (1986)
9. Pearn, W.L., Kotz, S., Johnson, N.L., et al.: Distributional and inferential properties of PCIs. J. Qual. Technol. **24**(4), 216–223 (1992)
10. Wu, C.W., Pearn, W.L.: Bayesian approach for measuring EEPROM process capability based on the one-sided indices CPU, and CPL. Int. J. Adv. Manuf. Technol. **31**(1–2), 135–144 (2006)

Research Context of Artificial Intelligence in Recent Three Years

Changshou Luo[1,2(✉)], Liying Zhou[3], Qingfeng Wei[1,2], Jun Yu[1,2], and Sufen Sun[1,2]

[1] Institute of Agricultural Information and Economics, Beijing Academy of Agriculture and Forestry Sciences, Beijing, China
luocs@agri.ac.cn

[2] Beijing Research Center of Engineering Technology on Rural Distance Information Service, Beijing, China

[3] China Agricultural University Library, Beijing, China

Abstract. This paper is based on InCites of Web of Science, using Citing Half-Life, Cited Half-Life and Aggregate Impact Factor, such indices from JCR about the Web of Science Category of "COMPUTER SCIENCE, ARTIFICIAL INTELLIGENCE" (AI subject) in 2003–2016, to analyze the change of research trend and subject impact of AI subject. Based on the annual average cited times of each paper, a small amount of superior papers are selected from all papers of AI subject indexed by Web of Science in 2015–2017, and used as data source. Based on the number of publishing superior papers and the cited times by superior papers, the core journals of AI subject are identified. Based on the number and their cited times of superior papers, the dominant nations and dominant institutions are identified. Based on the superior papers of artificial intelligence in 2015–2017, combing co-word clustering analysis with citation coupling enhancement, key research topics of AI subject are detected. And scale, impact, internal correlation and Half-Life citation, such indices are used to evaluate these research topics.

Keywords: Artificial intelligence · Research profile · Superior papers
Subject topics

1 Introduction

Artificial intelligence (AI, also machine intelligence, MI) is intelligence demonstrated by machines, in contrast to the natural intelligence (NI) displayed by humans and other animals. The research field of AI is a branch of computer science, aims to understand the essence of intelligence and design intelligent machines that can act as human behavior. It is an extensive scientific discipline which enables computer systems to solve problems by emulating complex biological processes such as learning, reasoning and self-correction. It studies theories, methods, and applications with respect to simulation, extension, and expansion of human intelligence for problem-solving. Application domains of AI include robotics, voice recognition, image recognition, natural language processing, and expert systems [1].

K. Deng et al. (Eds.): ICMIR 2018, AISC 856, pp. 423–433, 2019.
https://doi.org/10.1007/978-3-030-00214-5_54

2 Overall Research Context

2.1 Data Source

InCites is a citation-based evaluation tool to analyze and benchmark the scientific productions in specific context. The Journal Citation Reports (JCR) module within InCites provides evaluation and comparison about journals based on their subject categories using citation data drawn from numerous scholarly and technical journals from more than 3300 publishers in over 60 countries/regions. Except for traditional indicators for journals, indicators for categories are also calculated from 2003 [2].

This study uses the InCites as data source, and retrieves related indices about the Web of Science Category of "COMPUTER SCIENCE, ARTIFICIAL INTELLIGENCE" (AI subject) in 2003–2016 from JCR to analyze the change of research trend and subject impact of AI subject.

2.2 The Change of Research Trend

Citing half-life refers to the period of time of the newer half of all references cited by the research object making public. The Aggregate Citing Half-Life is the median age of articles which cited by journals in the subject in the JCR Year based on aggregated journal data. The lower Aggregate Citing Half-Life means that, the articles in the subject category prefer to quote newer citations. Cited half-life refers to the period of time of 50% of the total number of the research object's cited times. The Aggregate Cited Half-Life is the median age, in years, of items in any journal in the category that were cited during the JCR year. The smaller the Aggregate Cited Half-Life, the faster the publications of the subject category were cited [3].

Figure 1 shows the annual change of the Aggregate Citing Half-Life and the Aggregate Cited Half-Life of artificial intelligence. It can be seen that:

(1) From 2003 to 2005, the Aggregate Citing Half-Life of the subject was not only the shortest in the whole analysis period, but also decreased year by year, indicating that this subject developed at the fastest pace in this period, and the knowledge replacement speed was accelerating gradually. In 2005–2008, the biggest increase of the Aggregate Citing Half-Life of the subject was founded in the whole analysis period, which indicated that the subject developed at the slowest pace in this period and the trend of knowledge replacement slowed down. From 2008 to 2014, the Aggregate Citing Half-Life of the subject was basically stable, indicating that the development of the subject was relatively stable during this period. The Aggregate Citing Half-Life of the subject decreased annually from 2014 to 2016, indicating that this subject has regenerated in recent years and was once again entering into the stage of rapid development.

(2) In 2003–2009, the Aggregate Cited Half-Life of the subject was at a relative low level in the whole analysis period, but it increased year by year and the increase was the biggest, indicating that the quality of the research result was relatively higher and the duration of impact on the academia was increasing during this period. In 2009–2013, the Aggregate Cited Half-Life of the subject was basically

stable, indicating that the development was relatively stable during this period. In 2013–2016, the Aggregate Cited Half-Life of the subject decreased year by year, indicating that the knowledge replacement speed of the subject has accelerated significantly in recent years.

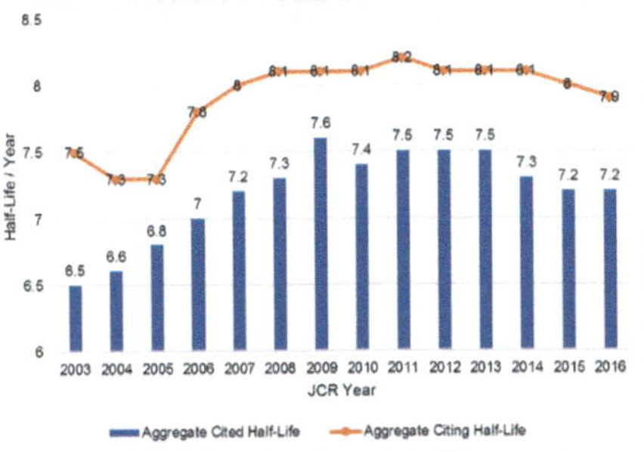

Fig. 1. The annual change of Aggregate Citing Half-Life and Aggregate Cited Half-Life in AI

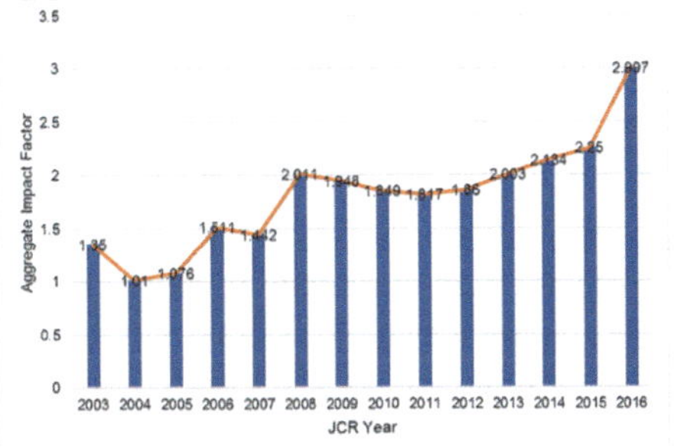

Fig. 2. The annual change of the Aggregate Impact Factor in AI subject

2.3 The Change of Subject Impact

The Aggregate Impact Factor are used to profile the trends of research impact in AI subject. The Aggregate Impact Factor for a subject takes into account the number of articles from all journals in the category and the number of citations to all journals in the subject. An Aggregate Impact Factor of 1.0 means that, on average, the articles in the subject published 1 or 2 years ago have been cited one time in the JCR year [2].

Figure 2 shows the annual change situation of the Aggregate Impact Factor in AI subject. We can find that in 2003–2008, Aggregate Impact Factor of the subject was relatively at a lower level, but its volatility was relatively severe, and had a relatively strong rise. Especially in 2005–2006 and 2007–2008, substantial increase is found twice. From 2008 to 2011, the Aggregate Impact Factor performance of the subject was relatively stable, with a slight decline. From 2011 to 2015, the Aggregate Impact Factor of the subject increased year by year, but the upward trend was relatively flat, and there was no significant volatility. In 2015–2016, the Aggregate Impact Factor of the subject has experienced the most significant increase in the whole analysis period, indicating that the academia have paid more and more attention to AI subject in the past 2 years.

3 Research Context Based on Superior Papers

3.1 Data Source

Above analysis shows that AI subject has developed rapidly in nearly 3 years. Therefore, this paper focus on the development trend of artificial intelligence in the nearly 3 years, based on the academic papers of AI subject indexed by Web of Science in 2015–2017.

3.2 Superior Papers

Compared with the study based on all the paper, the study based on superior papers indicates more accurate description of the capacity of analyzing object academic research, such as the ESI series indices of Web of Science [4], which are widely used in the field of information science. This study learns highly-cited papers indices from ESI, and defines the superior papers as the papers that is cited more than the annual average cited times of AI subject. The thresholds of superior papers in 2015–2017 are shown in Table 1. It can be seen from the table that the total number of papers in AI subject is more than 50,000, but only over 10,000 superior papers.

Table 1. Superior papers overview in 2015–2017

	Papers	Times cited per paper	Superior papers
2015	12896	6.83	3995
2016	14073	3.06	3552
2017	14325	0.81	4429

3.3 Core Journals

According to the Bradford law in information science, 80% of the documents of a subject is concentrated in 20% of the journals. For the reason that Web of Science database cannot obtain papers but also obtain quotations, so this paper does not only consider the journal distribution of papers, but also takes the journal distribution of quotations into account when selecting the core journals of the subject. Table 2 lists the TOP20 core journals based on superior papers and the TOP20 core journals based on superior paper citation in AI subject. We can see from the table that *Neurocomputing*, *Expert Systems with Applications*, *Applied Soft Computing*, *IEEE Transactions on Image Processing*, *IEEE Transactions on Cybernetics*, *Pattern Recognition*, *Knowledge-Based Systems*, *IEEE Transactions on Neural Networks and Learning Systems*, *IEEE Transactions on Pattern Analysis and Machine Intelligence*, *IEEE Transactions on Fuzzy Systems*, *Neural Networks*, etc. Such 11 journals are based on both papers and papers' citation, which are the most important core journals in this subject.

Table 2. TOP20 core journals based on superior papers and their citation in AI subject

Rank	Journal	Papers	Cited journal	Times cited
1	**Neurocomputing**	1742	**IEEE T Pattern Anal**	12415
2	**Expert Systems with Applications**	964	**Expert Syst Appl**	10476
3	**Applied Soft Computing**	779	Proc Cvpr Ieee	9763
4	**IEEE Transactions on Image Processing**	594	**Neurocomputing**	9349
5	**IEEE Transactions on Cybernetics**	535	Lect Notes Comput Sci	8426
6	**Pattern Recognition**	505	Inform Sciences	8405
7	**Knowledge-Based Systems**	486	**IEEE T Image Process**	7746
8	**IEEE Transactions on Neural Networks and Learning Systems**	405	**Pattern Recogn**	6413
9	Soft Computing	292	**IEEE T Fuzzy Syst**	6364
10	**IEEE Transactions on Pattern Analysis and Machine Intelligence**	272	**Appl Soft Comput**	5220
11	Journal of Intelligent & Fuzzy Systems	258	**IEEE T Neural Networ**	5135
12	Neural Computing & Applications	251	Fuzzy Set Syst	4885
13	**IEEE Transactions on Fuzzy Systems**	227	**Knowl-Based Syst**	4239
14	Engineering Applications of Artificial Intelligence	188	IEEE T Syst Man Cy B	4185
15	**Neural Networks**	179	**IEEE T Cybernetics**	4017
16	IEEE Transactions on Knowledge and Data Engineering	170	IEEE T Neur Net Lear	3955
17	Medical Image Analysis	165	Int J Comput Vision	3857
18	Robotics and Autonomous Systems	160	**Neural Networks**	3360
19	Information Fusion	154	J Mach Learn Res	3115
20	Journal of Intelligent Manufacturing	153	Pattern Recogn Lett	2757

3.4 Dominant Nations

For a paper, its major contributors are the lead author and the corresponding author. This paper identifies TOP10 dominant nations of this subject based on the superior papers in AI subject, considering the nation that the lead author or corresponding author belongs to as the evaluation criterion. The results are shown in Table 3. It can be seen that China has the largest number of superior papers, over three times more than the United States, which is in the second place. India also has many superior papers, ranking third. The impact of superior papers in Singapore is the largest, with the average cited times of each paper is 13.50, far more than the second-ranked China. British superior papers are also very influential, ranking third.

Table 3. TOP10 dominant nations in AI subject

Rank	Nation	Superior papers	Times cited per paper
1	Peoples R China	4967	10.89
2	USA	978	9.30
3	India	663	8.78
4	Spain	514	8.71
5	England	411	10.21
6	Iran	403	6.89
7	Italy	375	7.79
8	Australia	372	9.92
9	France	307	6.88
10	South Korea	298	8.81

3.5 Dominant Institutions

This paper identifies TOP10 dominant institutions of this subject based on the superior papers in AI subject, considering the institution that the lead author or corresponding author belongs to as the evaluation criterion. The results are shown in Table 4. It can be seen that Chinese Academy of Science (243) has the highest amount of superior papers. Harbin Institute of Technology (162) ranked second, and Xidian University (152) ranked third. Superior papers of Nanyang Technological University are the most influential, and the average cited times of each paper is 14.91. Sichuan University ranked second, and the the average cited times of each paper is 14.91. Superior papers of Harbin Institute of Technology and Sun Yat Sen University also have a great impact, which are both tied for the third place (Tables 5 and 6).

Table 4. TOP10 dominant institutions in AI subject

Rank	Institution	Superior papers	Times cited per paper
1	Chinese Acad Sci, Peoples R China	243	10.36
2	Harbin Inst Technol, Peoples R China	162	13.10
3	Xidian Univ, Peoples R China	152	8.18
4	Huazhong Univ Sci & Technol, Peoples R China	150	10.53
5	Southeast Univ, Peoples R China	126	12.40
6	Tsinghua Univ, Peoples R China	122	11.25
7	Northeastern Univ, Peoples R China	117	10.09
8	Nanyang Technol Univ, Singapore	111	14.91
9	Sichuan Univ, Peoples R China	95	13.72
10	Tianjin Univ, Peoples R China	95	8.98

Table 5. Key research topics overview in AI subject during 2015–2017

	Key phrases	Papers	Times cited per paper
TP1	Decision making; Fuzzy sets; Multi-criteria decision making; group decision making; Multi-attribute decision making; Technique for order preference by similarity to ideal solution; Intuitionistic fuzzy set; Fuzzy numbers; hesitant fuzzy set; Analytic hierarchy process	778	11.20
TP2	Machine learning; Support vector machine; Feature selection; Ensemble learning; Extreme learning machine	1089	9.86
TP3	Takagi–Sugeno fuzzy model; nonlinear systems; time-varying delay; discrete-time; Fuzzy control; Linear matrix inequality; time delay; Lyapunov functional; fuzzy modeling; Markovian jump; Fuzzy systems; Adaptive control; state estimation; Exponential stability	1047	14.19
TP4	Evolutionary algorithms; Meta-heuristic; Swarm intelligence; particle swarm optimization; Global optimization; Multi-objective optimization; Evolutionary computation; Multi-objective; Artificial bee colony; Differential evolution; Genetic algorithm; local search	1243	9.74
TP5	Deep learning; Convolutional neural networks; Feature learning; image classification; Object detection; Auto-encoder; Object recognition; Computer vision	623	14.54
TP6	Sparse representation; Supervised learning; Semi-supervised learning; Dimensionality reduction; dictionary learning; face recognition; low rank; multi-view; sparse coding; Discriminant analysis; Feature extraction	903	10.33

Table 6. Features of key research topics in AI subject during 2015–2017

	Internal correlation	Citing Half-Life
TP1	0.4372	7.74
TP2	0.3071	8.36
TP3	0.3309	7.09
TP4	0.3508	8.53
TP5	0.2717	7.70
TP6	0.2689	7.87

4 Research Topic Analysis

Based on the 11,976 superior papers in 2015–2017, key research topics in AI subject in recent 3 years are studied.

4.1 Analysis Method

Co-word clustering and citation clustering are major content analysis methods to reveal the research topic of information science. Co-word clustering is based on the frequency of two key words in the same paper, and measures its closeness relationship, then a set of relatively close key words make a research topic. Xu et al. [5] have once used Co-word clustering to detect the research frontier in different fields. Citation coupling is a kind of citation clustering. It is based on the number of references that the two papers both contain, and measures its closeness relationship, then a set of relatively close key words make a research topic. Schiebel [6] have once conducted analysis and study on the research frontier based on citation clustering. Xie [7] have also combined co-word clustering and citation clustering analysis to study the research frontier. The method of this paper in the identification of the key research topic is co-word clustering analysis with citation coupling enhancement [8], combing with the relatively of co-word and citation clustering. Thus, some key words that are less closely related in concurrence dimension but closely related in citation clustering dimension are also discovered and incorporated into a research topic.

The core of identifying the key research topic of the field artificial intelligence using co-word clustering analysis with citation coupling enhancement is based on analysis of the key word relativity based on co-word relationship with citation coupling enhancement. The concrete steps are as follows:

(1) Count the concurrence frequency of every two words and the citation distribution of key words. The matrix of the concurrence frequency and the matrix of the citation distribution frequency of key words are generated.
(2) Analyze the concurrence relativity of key words by the cosine coefficient based on the concurrence frequency of key words. And the matrix of the concurrence of key words is generated. Analyze the citation coupling relativity of key words by the cosine coefficient based on the citation distribution of key words. And the matrix of the citation coupling of key words is generated.
(3) If there are two key words i and j, whose co-word relativity is $CW_{i,j}$, and the citation coupling relativity of which is $CR_{i,j}$, then the relativity of these two key words is $R_{i,j} = \text{Max}\ (CW_{i,j}, CR_{i,j})$.
(4) Based on the matrix of key words with the elements $R_{i,j}$, the system clustering analysis is conducted in order to identify the key research topic of the field of artificial intelligence.

4.2 Identification of Key Research Topics

Based on terms with phrase frequency of more than 10 and professional significance in superior papers, combing the co-word clustering analysis with citation coupling enhancement, six key research topics are identified. The results are shown in Table 3. It can be seen that:

(1) There have been six outstanding research topics in AI subject. They are TP1 (with Decision making; Fuzzy sets; Multi-criteria decision making as key phrases), TP2 (with Machine learning; Support vector machine; Feature selection as key phrases), TP3 (with Takagi–Sugeno fuzzy model; nonlinear systems; time-varying delay as key phrases), TP4 (with Evolutionary algorithms; Meta-heuristic; Swarm intelligence as key phrases), TP5 (with Deep learning; Convolutional neural networks; Feature learning as key phrases) and TP6 (with Sparse representation; Supervised learning; Semi-supervised learning as key phrases).
(2) Among the six research topics, TP4 (with Evolutionary algorithms; Meta-heuristic; Swarm intelligence as key phrases) has the largest scale, with 1,243 related papers. TP2 (with Machine learning; Support vector machine; Feature selection as key phrases) ranks the second, with 1,089 related papers. TP3 (with Takagi–Sugeno fuzzy model; nonlinear systems; time-varying delay as key phrases) ranks the third, with 1,047 related papers.
(3) Among the six research topics, the most influential one is TP5 (with Deep learning; Convolutional neural networks; Feature learning as key phrases), with the citation impact is 14.54 times per paper. TP3 (with Takagi–Sugeno fuzzy model; nonlinear systems; time-varying delay as key phrases) ranks the second, with the citation impact is 14.19 times per paper. TP1 (with Decision making; Fuzzy sets; Multi-criteria decision making as key phrases) ranks the third, with the citation impact is 11.20 times per paper.

4.3 Evaluation of Key Research Topics

In this study, the key phrases relativity based on co-word and citation clustering is the basis of the evaluation of key research topics. The internal correlation of the research topic can be calculated based on the key phrases relativity [9], which can be used as the evaluation index of the maturity of the key research topic. The Citing Half-Life based on references can be used as the evaluation index of the knowledge replacement speed of the key research topic. The internal correlation and Citing Half-Life of six key research topics in AI subject are shown in Table 4. It can be seen that:

(1) TP1 (with Decision making; Fuzzy sets; Multi-criteria decision making as key phrases) has the largest internal correlation, indicating that it is relatively the most mature of key research topics in the field artificial intelligence, with the most close relation with the related research. While TP6 (with Sparse representation; Supervised learning; Semi-supervised learning as key phrases) has the lowest internal correlation, indicating that it is relatively the least mature of key research topics in the field artificial intelligence, with the least close relation with the related research.
(2) TP3 (with Takagi–Sugeno fuzzy model; nonlinear systems; time-varying delay as key phrases) has the shortest Citing Half-Life, indicating that the citation documents of its papers are the latest, and has the relatively highest knowledge replacement speed of key research topics in the field artificial intelligence.

While TP4 (with Evolutionary algorithms; Meta-heuristic; Swarm intelligence as key phrases) has the longest Citing Half-Life, indicating that the citation documents of its papers are the oldest, and has the relatively slowest knowledge replacement speed of key research topics in the field artificial intelligence.

5 Conclusion

This paper is based on InCites of Web of Science, combing with Citing Half-Life, Cited Half-Life and Aggregate Impact Factor, such JCR related indices about the Web of Science Category of "COMPUTER SCIENCE, ARTIFICIAL INTELLIGENCE" in 2003–2016, to analyze the change of research trend and subject impact of AI subject. The analysis results show that this subject has experienced a rapid development in recent years, with higher knowledge replacement speed and increasing attention from the academia.

Based on the annual average cited times of each paper, a small amount of superior papers are selected from academic papers in AI subject during 2015–2017, and used as the data source. Based on the number of publishing superior papers and the cited times by superior papers, the core journals of AI subject are identified; and based on the number and their cited times of superior papers, the dominant nations and dominant institutions are identified. In that, the most significant 11 core journals are *Neurocomputing*, *Expert Systems with Applications*, *Applied Soft Computing*, *IEEE Transactions on Image Processing*, *IEEE Transactions on Cybernetics*, *Pattern Recognition*, *Knowledge-Based Systems*, *IEEE Transactions on Neural Networks and Learning Systems*, *IEEE Transactions on Pattern Analysis and Machine Intelligence*, *IEEE Transactions on Fuzzy Systems*, *Neural Networks*. The most competitive nations are China, the United States, Singapore and the UK. The most significant institutions include Chinese Academy of Science, Harbin Institute of Technology, Xidian University, Nanyang Technological University, Sichuan University and Sun Yat Sen University.

This paper is based on the superior papers in AI subject during 2015–2017 indexed by Web of Science, combing co-word clustering analysis with citation coupling enhancement, to detect the key research topic of this subject. And use scale, impact, internal correlation and Half-Life citation, such indices to evaluate these research topics. The result shows that there have been six outstanding research topics of artificial intelligence, among which TP4 (with Evolutionary algorithms; Meta-heuristic; Swarm intelligence as key phrases) has the largest scale, TP5 (with Deep learning; Convolutional neural networks; Feature learning as key phrases) is the most influential, TP1 (with Decision making; Fuzzy sets; Multi-criteria decision making as key phrases) has the largest internal correlation, indicating that it is relatively the most mature of key research topics in the field artificial intelligence, and TP3 (with Takagi–Sugeno fuzzy model; nonlinear systems; time-varying delay as key phrases) has the shortest Citing Half-Life, with the relatively highest knowledge replacement speed of key research topics in the field artificial intelligence.

References

1. Pavel, H., Johanne, T.: Artificial intelligence in medicine. Metab. Clin. Exp. **69**(S), S36–S40 (2017)
2. Analytics clarivate. InCites J. Cit. Rep. [EB/OL]. http://ipscience-help.thomsonreuters.com/inCites2Live/inCites20.html. 16 Feb 2018
3. Tsay, M.Y., Chen, Y.L.: Journals of general & internal medicine and surgery: an analysis and comparison of citation. Scientometrics **64**(1), 17–30 (2005)
4. Wang, Y., Huang, D., Liu, D.: Statistical indicators' principle and data process of ESI. Libr. Inf. Serv. **35**(9), 73–75 (2006)
5. Xu, Z., Zhao, Y., Zhou, W.: Knowledge mapping analysis of research fronts of international ecology urban: based on view of the analysis of co-word network. Urban Dev. Stud. **2**, 115–124 (2013)
6. Schiebel, E.: Visualization of research fronts and knowledge bases by three-dimensional areal densities of bibliographically coupled publications and co-citations. Scientometrics **91**(2), 557–566 (2012)
7. Xie, P.: Study of international anticancer research trends via co-word and document co-citation visualization analysis. Scientometrics **105**(1), 611–622 (2015)
8. Zhou, L., Leng, F., Zuo, W.: An improved co-word analysis method by bibliographic coupling—research fields divided in ESI agricultural science as a case. Inf. Stud. Theor. Appl. **11**, 120–125 (2015)
9. Zhong, W., Li, J.: The research of co-word analysis (2). J. Inf. **6**, 141–143 (2008)

The FPGA Implementation of the High Speed Batchage Network

Pu Wang, Yuming Zhang, and Jun Yang(✉)

School of Information Science and Engineering, Yunnan University, Kunming 650091, China
junyang@ynu.edu.cn

Abstract. The bazochi-sorting network is one of the most widely used hardware sequencing algorithms, heavily used in the Internet. It is necessary to carry out a large number of complicated internal cabling to realize the batch-odd sorting network. The design method of traditional separation element is large, low reliability, slow and not easy to modify and upgrade. In this paper, FPGA is used to realize the batch-parity network, solve the problems of traditional design, and study the extensibility of the batch-parity network. Experimental results show that the design is correct.

Keywords: FPGA · QuartusII · VHDL · Bazochi-sorting network

1 Introduction

With the various fields of social life of the wide application of computer, and the development of high-speed Internet, a doubling of people need to deal with information, makes the speed of processing information put forward higher requirements [1]. It is that hot spot of compute science research in order to sort as the most basic method of information treatment. But, in the development of hardware system, there are some disadvantages of traditional sorting algorithms; traditional algorithms need to divide different types of input and realize very difficult hardware.

In order to solve the above problems, the batcher sorting network algorithm was born [2]. The parallelism and non-adaptability of the network are very convenient for hardware implementation, and the advantages of hardware parallel processing are also given. After several decades of continuous development, the batcher sort network algorithm has become the most widely used hardware sorting algorithm. Especially in network communication devices, such as high-speed switches, routers and so on, the data processing of packet switching is done using the batch-sorting network.

In this paper, FPGA devices have the characteristics of high speed and online reconfiguration, and a model of the batch-sorting network is designed and constructed [3]. The experimental results show that the performance of the network is improved, and the speed and reliability of the network are improved. Using FPGA to explore the extended application of batcher sequencing network.

K. Deng et al. (Eds.): ICMIR 2018, AISC 856, pp. 434–440, 2019.
https://doi.org/10.1007/978-3-030-00214-5_55

2 The Key Technology

2.1 FPGA

FPGA is a product that is further developed on the basis of programmable devices such as PLD, GAL, EPLD, etc. It is present as a semi-custom circuit in that field of Application Specific Integrated circuit, which overcomes both the disadvantages of the customization circuitry and the limitations of the limit of the original programmable logic device gate Circuits [4].

Modern FPGA chip can be integrated inside millions of logic gate, the user can through the principle diagram input method and VHDL hardware description language is convenient to design a digital system, through software simulation, verify its correctness after downloaded to the chip, realize the logic functions according to user's requirements. FPGA has the static repeatable programming and dynamic characteristics of the online system reconstruction [5], the user can not change the external hardware circuit under the premise of modifying the design of the chip at any time, thus makes the development of hardware as convenient as the software development. Using FPGA to develop the digital system can significantly shorten the design time, reduce the number of peripheral circuits, reduce the development cost, and improve the reliability of the system.

2.2 VHDL

VHDL hardware description language has strong ability of system description and specification of the program design of structure [6], flexible expressions style and multi-level simulation test method, VHDL support top-down and bottom-up or mixed input of design patterns, these features greatly simplifies the task of hardware design, improves the design efficiency and reliability, has become the electronic system design select hardware description language. Based on the QuartusII software development platform, VHDL hardware description language is used to design the batch-odd network.

3 The Design and Implementation of the Batch-Parity Sorting Network Model

3.1 The Bacchichi-Sorting Network

Figure 1 shows a typical sorting process for a four - dollar barchage - sort network. Will first be collating sequence CBDA pairwise comparison, the two elements in the larger one on one end of the arrow, because C > A > B, D is C, B D exchange, exchange of A intermediate results BCAD; We continue to make A comparison between the two, because B BBB 0 A and C > D are still in exchange to get the intermediate result ACBD; The last group of the sequence ACBD is continuously sorted, and the sequence ABCD is sorted.

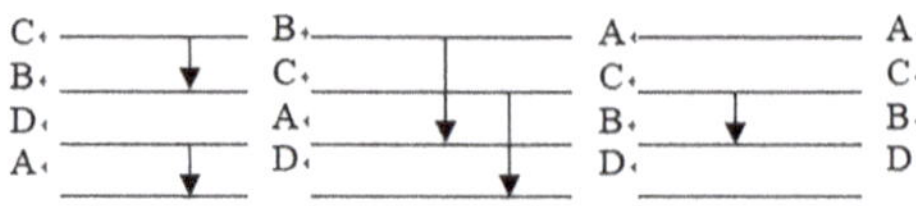

Fig. 1. The four - dollar bacchichi even sort network sorting process

Figure 2 shows is one of the eight yuan batchelor, odd-even sort sorting process of network, it will first eight elements into two groups, respectively, with two to four dollars into parity network and then done to these two sorted sequence merge finally got a network of eight yuan parity sort.

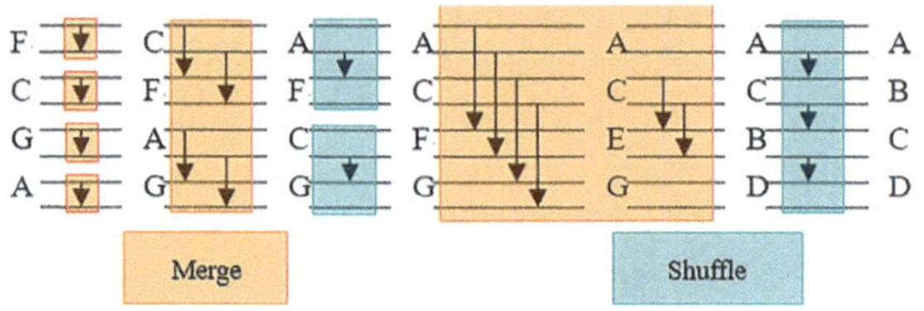

Fig. 2. The eight - dollar bacchichi even sort network sorting process

Thus, the steps of constructing the batch-parity network are as follows:

First to inverse shuffling platoon series are decomposed into two series: one is composed of elements of A and B in the odd number sequence, as OM, sequences, the other is composed of the even number elements in A and B series, EM as I sequence. Then the OM is sent into the odd-merge device, and the EM is sent into the homing device, and a set of ordered odd sequences and a sequence of ordered pairs are obtained. Then, in addition to the first element of the odd sequence, the sequence of the two sequences can be obtained by comparing the two sequences. Among them, an n-yuan merge device is composed of two n/2 yuan merges and a shuffle comparison network. The shuffle comparison is based on the following method: suppose there are 8 nodes, from 0 to 7 No. 01234567. Then divide it in half and place one card in one group on top of the other. Becomes 04152637, so the order of the original two comparison: 0 compared with 4 node, the node 1 and 5 nodes, nodes 2 and No. 6, No. 3 and 7 nodes.

3.2 Based on VHDL Implementation of the Batcher Comparator

In this way, that baysch couple sequence network is made up of a series of bar comparator. It is meant that that maximum value max (x, y) and min (x, y) are output at both outputs if the input x, y is given at both input end.

The batcher comparator is the most basic component of the batch-parity network, which is the basis for building a complete network of barkis.

Figure 3 is a diagram of the batcher comparator module. The a and b ports are 5 bit binary input ports, and the c and d ports are 5 bit binary output ports, which are used to output the comparison results.

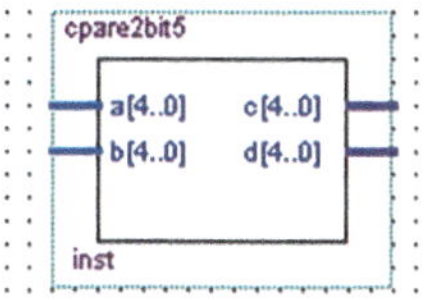

Fig. 3. Batcher comparator module diagram

3.3 Based on FPGA-Based Batcher Network Construction

This section is the core content of this article, that is, based on the FPGA implementation of the batch-odd sorting network. Its design idea is: by calling in the previous section using VHDL language to build the comparator based components, cutting in Altera corporation provide QuartusII development platform, using the top-down design method, using the graphical editor, principle diagram input method to establish the top-level file, define the input and output interface and connect network. In the whole sorting network, the width of the data is defined as 5 bits.

Figure 4 shows the implementation of a complete 8-element batch-parity network model. You can see that it is consistent with the abstract network diagram shown in Fig. 2. A total of 19 barcher comparators are made up. The first 10 comparators constitute two quaternions and the remaining 9 comparators constitute a network of shuffle comparisons. Each of the columns in the diagram works in parallel.

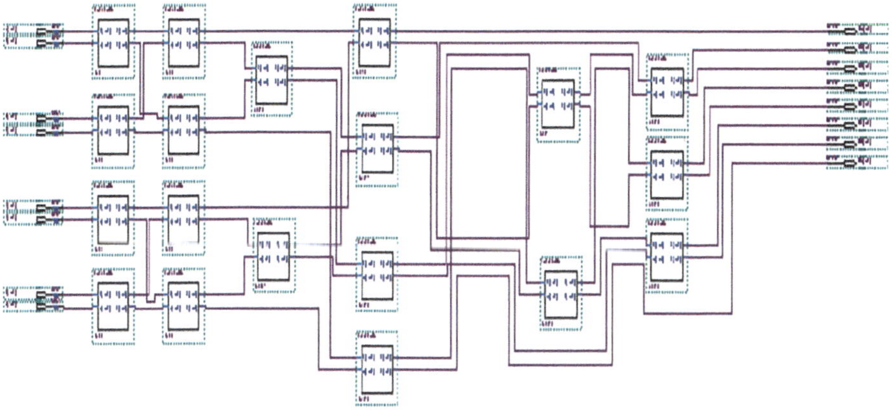

Fig. 4. Schematic diagram of the batch-parity network

Figure 5 shows the module generated after compilation of the program. In Figs. 3 and 4, the pins OF A, B, C, D, E, F, G and H correspond to the input pins in Fig. 4, and the output pins in Fig. 4 OF tube feet OA, OB, OC, OD, OE, OF, OG and OH.

At this point, we have completed the construction of the whole batch-parity network model. Design method based on FPGA, the flexibility to choose schematic descriptions or programming language design each module, and module using the

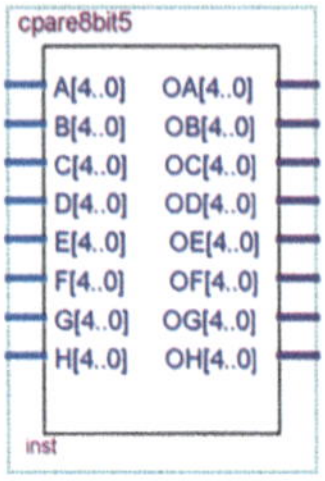

Fig. 5. Batchach network module

reconstruction and online design programmability, can be very convenient to carry on the design changes and upgrades. This design method can greatly reduce the design workload, shorten the development cycle and save the design cost.

4 System Testing and Extension Applications

4.1 The Simulation Test

In order to verify the correctness of the design of the eight - element batcher network, it is necessary to conduct a simulation test of its function and timing property to see whether the design result meets the original design requirements. In this experiment, QuartusII was used to test the sorting network.

The test is divided into two steps:

The first step is to establish the sequence simulation waveform. In the QuartusII platform, the waveform editor is used to create a simulation waveform for eight input pins. Available to test the network effect, the input data according to the cycle is divided into three groups, all of the input sequence is random, due to the sort of data usage in a network width is 5, so all data extraction between 0 to 31. Table 3–6 indicates the data sequence of specific input (Table 1).

Table 1. The test data

Group 1 data	Group 2 data	Group 3 data
8	18	30
5	19	27
3	21	24
2	23	9
4	16	7
1	12	11
7	11	16
6	15	79

The second step is to simulate the input waveform. We used the simulation test tool in QuartusII to simulate the sorting network based on the input waveform established in the previous step. The results are shown in Fig. 6.

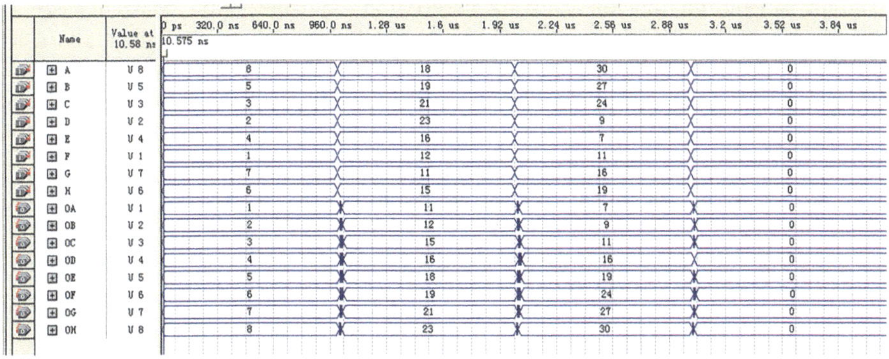

Fig. 6. The simulation results

The results of the simulation results show that the design of the batcher coupling network is correct. Download the ep1k30qc208-2 of the ACEX1K series provided by Altera company to verify the total consumption of 270 logical units. The experimental results are correct and achieve the desired purpose.

4.2 The Extended Application of the Batchage - Sort Network

It's a very easy way to do it on the hardware, and it's a very easy way to do it on the hardware, and it's very easy to do it in parallel, and it's very good for the hardware, the efficiency of the work, the fast and fast, and the most extensive sorting algorithm on the hardware. However, the use of the batchage network is not only limited to sorting data, but also has the function of classifying data. In some high-speed network switches, the use of the batcher sort network and Banyan network are combined to realize the application of packet high-speed switching and forwarding on multiple ports, as shown in Fig. 7.

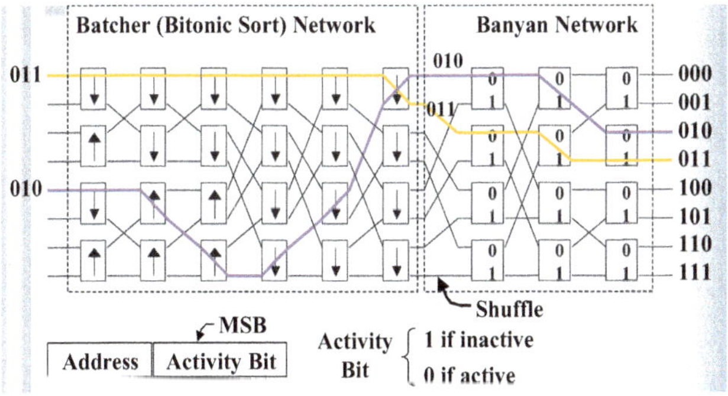

Fig. 7. The application of batcher network in switches

5 Summary

Based on FPGA is introduced in this paper to build batchelor, odd-even sorting network model and extend the application thought and method, through the principle diagram and the mixture of hardware description language input, realized based on FPGA device sort parity network experiment model. Studies have shown that, based on the FPGA device design and cut the odd-even sorting network can well solve the reconfigurable system design, stability, improve the efficiency of the system design, has a certain application value.

Acknowledgements. The author, Pu Wang, thanks the Application of basic research project in yunnan province, "key technology research of automatic real-time detection of solar radio explosion based on video treatment" (No. 2015FB115).

References

1. Xu, L., Yang, G.: Research and implementation of encoding and decoding of LDPC codes. Electron. World (2017)
2. Bhanwala, A., Kumar, M., Kumar, Y.: FPGA based design of low power reconfigurable router for network on chip (NoC). In: International Conference on Computing, Communication and Automation, pp. 1320–1326. IEEE (2015)
3. Bansal, A.: Aviva sorting network. In: International Conference on Power and Energy in Nerist, pp. 1–4. IEEE (2013)
4. Bahoura, M., Park, C.W.: FPGA-implementation of high-speed MLP neural network. In: IEEE International Conference on Electronics, Circuits and Systems, pp. 426–429. IEEE (2012)
5. Bahoura, M.: FPGA Implementation of High-Speed Neural Network for Power Amplifier Behavioral Modeling. Kluwer Academic Publishers, Dortrecht (2014)
6. Graziano, M., Zahir, A., Mehdy, M.A., et al.: VHDL-AMS simulation framework for molecular-FET device-to-circuit modeling and design. Active Passive Electron. Compon. **2018**, 1–18 (2018)

Segmentation and Alignment of Chinese and Khmer Bilingual Names Based on Hierarchical Dirichlet Process

Bingbing Yu[1], Yu Nuo[1], Xin Yan[1](✉), Qingling Lei[1], Guangyi Xu[2], and Feng Zhou[1]

[1] School of Information Engineering and Automation, Kunming University of Science and Technology, Kunming 650504, China
kg_yanxin@sina.com
[2] Yunnan Nantian Electronics Information Co., Ltd., Kunming, Yunnan 650050, China

Abstract. Transliteration is an important foundation of cross-language Natural Language Processing technology. In order to solve the problem of Khmer-Chinese name transliteration, the hierarchical Dirichlet process model allows multi-to-multi alignment, and also solves overfitting problems, and it can involve the influence of the previous syllable alignment on the alignment of the next syllable and effectively mine the latent information in the natural language, Therefore, Chinese and khmer name transliteration based on hierarchical Dirichlet process is proposed in this paper. This paper firstly builds a hierarchical Dirichlet process based on the Chinese-Khmer name alignment model, then, makes the aligned Chinese and Khmer syllables as the training corpus and uses the Mose to train the Chinese-Khmer transliteration model, finally, tests the performance of the Chinese-Khmer name alignment model by the effect of the transliteration model. The results show that the way of aligning Chinese and Khmer names by the alignment model based on hierarchical Dirichlet process firstly and transliterating the names next can get a better performance.

Keywords: Hierarchical dirichlet processes · Co-segment Chinese and Khmer bilingual names · Transliteration

1 Introduction

Chinese and khmer name transliteration is a process of translating khmer person's name into corresponding Chinese characters in terms of words or words. Because mining names from the Internet cannot find translations of names that have not yet appeared, Chinese and khmer name segmentation method based on the hierarchical Dirichlet process overcomes the disadvantages of web-based transliteration and translates them in accordance with the pronunciation of the khmer language into the corresponding Chinese. Unknown words refer to words that have not appeared in existing training corpora and dictionaries, and are also a fundamental and important research in natural language processing. Names are the main components of unknown words, and they are the main carriers of information in natural language processing.

K. Deng et al. (Eds.): ICMIR 2018, AISC 856, pp. 441–450, 2019.
https://doi.org/10.1007/978-3-030-00214-5_56

Transliteration is often adopted in the translation of names in many languages. Transliteration has become an important research hotspot, and it has become an important research hotspot for many kinds of NLPs. Transliteration has been widely used in machine translation, named entity extraction, cross-language information retrieval, and parallel corpus alignment. It is of great significance.

2 Related Work

As early as 2002, researchers began to study transliteration work. The main research methods mainly include the following aspects:

(1) Rule-based approach: This approach requires humans to establish translation rules for specific language pairs and translation directions [1]. Abdul et al. [2] studied the Uyghur Chinese transliteration rules, and on this basis proposed an improved rule for the translation of Uighur-Chinese machines.
(2) Statistical-based methods: This method requires a large number of syllable-segmented training corpora. The scale and quality of corpus directly affect the parameters and performance of the model. The method is similar to the process of machine translation. It needs to train the alignment model and the decoding model, and then obtain the transliteration model according to the result of the alignment and the decoding strategy. Jiampojamarn et al. [3] proposed many-to-many transliteration alignment methods. This method achieves automatic segmentation through multi-to-multi-alignment. Francisco et al. [4] proposed a two-element-tree method to achieve automatic bilingual segmentation; T Fukunishi et al. [5] considered that nonparametric Bayesian model can solve the overfitting problem better, and used nonparametric Bayesian method to achieve many-to-many alignment; Finch et al. [6] proposed the use of non-parametric Bayesian methods to achieve unsupervised bilingual character sequence segmentation.

At present, there are no available bilingual resources that can be used for the study of Chinese-khmer transliteration. This article draws on the above research methods and ideas. Hierarchical Dirichlet processes can solve overfitting problems better, and allow unlimited to multi - to - multiple alignment and can take into account the effect of previous syllable alignment on the alignment of the next syllable. The hierarchical Dirichlet process is an unsupervised bilingual segmentation method that can make full use of undivided Khmer-Chinese bilingual names; If the name of the Khmer is first divided into syllables and then aligned with the corresponding Chinese names, there will be an error amplification - khmer person's name syllable segmentation error, khmer person's name syllables and Chinese characters alignment error. In view of the above characteristics, this paper considers the simultaneous processing of the segmentation and alignment of khmer person's name syllables, which can reduce the accumulation of intermediate errors; Consider the effect of the previous syllable on the alignment of the latter syllable, for example, "ផាន់" in "(ចាន់ សុ ផាន់)" translates to "pan" and in "(សុជិន)", "ផាន" translates as "ban". From this we can see that the alignment of the current syllable pair is related to its previous syllable pair; therefore we put forward a method based on hierarchical Dirichlet process is proposed to realize the multi to multi alignment of the names of khmer to Chinese names, and the Khmer Chinese transliteration is carried out using the MOSES decoder after the bilingual names are aligned.

The main steps are as follows: Step 1, the preprocessing of corpus. Firstly, the Khmer- Chinese pairs were crawled from the Khmer-Chinese parallel webpage, and the name was processed to denoise, eliminate duplicate and filter, obtaining 746 pairs of Khmer-Han names; the second step, Chinese and Khmer names Segmentation process. Khmer names were cut into KCC (Khmer Character Cluster, KCC) [7] using our laboratory's KCC cutting too KCC is a Khmer string and is the smallest unit of Khmer text. Chinese names are segmented by Chinese characters sequence; The third step is to construct HDP (Hierarchical Dirichlet Processes) alignment model. First of all, it randomly aligns the Khmer-Chinese name pair, iterates through the Gibbs sampling loop until convergence, and finally gets the alignment result; the fourth step, the decoding process. The MOSES decoder is used to decode the HDP process.

3 Hierarchical Dirichlet Processes

Hierarchical Dirichlet Processes (HDPs) are proposed by Teh [8] and nonparametric Bayesian models of hierarchical Dirichlet processes [8, 9, 10] provide a solution to achieve multiple documents sharing an infinite number of clusters.

One viewpoint associated with the HDP can be expressed by the Chinese Restaurant Franchise (CRF) which is an extension of the Chinese Restaurant Process (CRP) [8]. The CRF construction process of HDP is the process of assigning tables and dishes to customers. First of all, allocate a dining table for each customer, the customer chooses which table to sit directly proportional to the number of customers of the table, the customer can also choose a new table; second, allocate dishes for each table, the probability of being ordered to a certain dish is proportional to the number of tables that order this dish, and may also choose to new dishes.

4 Construct an Alignment Model

4.1 Parameter Description

Assume that the bilingual person name Corpus consists of a syllable $s_1^m = \langle s_1, s_2, \ldots, s_m \rangle$ in the source language and a syllable $t_1^n = \langle t_1, t_2, \ldots, t_n \rangle$ in the target language. $TUs\,(TransliterationUnits)$ represents a bilingual-aligned unit in a person's name, a bilingual name pair $(s,t) = (\langle s_1, s_2, \ldots, s_m \rangle, \langle t_1, t_2, \ldots, t_n \rangle)$, which can be abbreviated as $x_1 = \left(s_1^m, t_1^n\right)$, and a corpus represented by D, $D = \{x_i\}_{i=1}^{I}$, x_i represents a bilingual name pair, I represents the size of the corpus, and a person's name is divided into $TU(TransliterationUnit)$ collections $\gamma = \{(s_1, t_1), \ldots, (s_h, t_h)\}$ of bilingual alignment pairs.

4.2 HDP Alignment Model

The bilingual alignment model in this paper is defined as follows:

$$\begin{aligned} &(s_i, t_i)|(s_{i-1}, t_{i-1}), G_1 \sim G_1 \\ &G_1|\alpha, G_0 \sim DP(\alpha, G_0) \\ &G_0|\gamma, H \sim DP(\gamma, H) \end{aligned} \tag{1}$$

In formula 1, according to the understanding of the Chinese restaurant problem in the hierarchy, the base distribution G_0 determines the probability of the new *TU* generation, that is, the probability of choosing a new table, and G_1 determines the probability of occurrence of the two combined *TU*, to determine the probability of a new dish, α is used to control the variance, the larger the value, the more similar G_1 and G_0 are. Because G_1 and G_0 cannot be expressed explicitly, after the integration of G_1 and G_0, the bigram distribution can be obtained. Mapping to the khmer-Chinese name alignment model, the Chinese restaurant on the first level has a base distribution of, G_0 and the Chinese restaurant on the second level has a base of G_1.

When constructing an HDP using a Chinese restaurant process, if (s_i, t_i) and

$$(1)\, P(\langle(s_{i-1},t_{i-1}),(s_i,t_i)\rangle \text{ is a novel bigram} \,|\, (s_{i-1},t_{i-1})) = \frac{\alpha}{N(s_{i-1},t_{i-1})+\alpha}$$

$$P(\langle(s_{i-1},t_{i-1}),(s_i,t_i)\rangle \text{ is not a novel bigram} \,|\, (s_{i-1},t_{i-1})) = \frac{N(s_{i-1},t_{i-1})}{N(s_{i-1},t_{i-1})+\alpha}$$

$$(2)\text{ ① a. } P((s_i,t_i) \text{ is a novel syllable pair } |\langle(s_{i-1},t_{i-1}),(s_i,t_i)\rangle \text{ is a novel bigram}) = \frac{\gamma}{N+\gamma}$$

$$P((s_i,t_i) \text{ is not a novel syllable pair } |\langle(s_{i-1},t_{i-1}),(s_i,t_i)\rangle \text{ is a novel bigram}) = \frac{N}{N+\gamma}$$

$$\text{b. i. } P((s_i,t_i) = (\langle s_1,s_2,\ldots,s_m\rangle,\langle t_1,t_2,\ldots,t_n\rangle) | (s_i,t_i) \text{ is a novel syllable pair}) = P_0(s_i,t_i)$$

$$\text{ii. } P((s_i,t_i) \,|\, (s_i,t_i) \text{ is not a novel syllable pair}) = \frac{N(s_{i-1},t_{i-1})}{N}$$

$$\text{②. } P((s_i,t_i)|\langle(s_{i-1},t_{i-1}),(s_i,t_i)\rangle \text{ is not a novel bigram and } (s_{i-1},t_{i-1})) = \frac{N((s_{i-1},t_{i-1}),(s_i,t_i))}{N(s_{i-1},t_{i-1})}$$

(s_{i-1}, t_{i-1}) are occured in the previously generated segment simultaneously, the probabilities are related to the number of simultaneous occurrences of them. The probability of a certain dish is related to the number of food tables. If (s_i, t_i) and (s_{i-1}, t_{i-1}) are not occured simultaneously in previously generated segment, the probability is related to α, that is, the probability of a new dish.

Therefore, it is easy to deduce that the probability distribution of the *i*-th *TU*, as shown in (2):

$$\begin{aligned} p((s_i,t_i)\,|\,(s_{-i},t_{-i}),Z_{-i}) &= \int p((s_i,t_i)\,|\,G_{(s_{i-1},t_{i-1})})p(G_{(s_{i-1},t_{i-1})}\,|\,(s_{-i},t_{-i}),Z_{-i})dG_{(s_{i-1},t_{i-1})} \\ &= \frac{N((s_i,t_i),(s_{i-1},t_{i-1})) + \alpha P_1((s_i,t_i)\,|\,(s_{-i},t_{-i}),Z_{-i})}{N(s_{i-1},t_{i-1}) + \alpha} \end{aligned} \tag{2}$$

$$P_1((s_i,t_i)\mid(s_{-i},t_{-i}),Z_{-i}) = \int p((s_i,t_i)\mid G_0)p(G_0\mid(s_{-i},t_{-i}),Z_{-i})dG_0 \\ = \frac{N(s_{i-1},t_{i-1})+\gamma P_0(s_i,t_i)}{N+\gamma} \quad (3)$$

$$P_0(s_i,t_i) = p(|s|)p(s\,||\,s|) \times p(|t|)p(t\,||\,t|) \\ = \frac{\lambda_s^{|s|}}{|s|!}e^{-\lambda_s}v_s^{-|s|} \times \frac{\lambda_t^{|t|}}{|t|!}e^{-\lambda_t}v_t^{-|t|} \quad (4)$$

In the above equations, α and γ are the parameters of the model, Z_{-i} is the table numbers of all the pair of syllables exclude the i-th syllable pair, and $P_0(s_i,t_i)$ represents the base distribution of generating a new (s_i,t_i) syllable pair, $N((s_i,t_i),(s_{i-1},t_{i-1}))$ represents the number of simultaneous occurrences of (s_{i-1},t_{i-1}) and (s_i,t_i) in the khmer-Chinese syllable, and N represents the number of any two consecutive khmer-Chinese syllable pairs, (any two consecutive khmer-Chinese syllable pairs. hereinafter referred to as a combination), $N(s_{i-1},t_{i-1})$ represents the number of the first syllable pair (s_{i-1},t_{i-1}).

Among them, through multiple tests, α and γ are set to the values as follows: $\alpha = 1000$ $\gamma = 3000$; $|s|$ and $|t|$ are the character lengths of the khmer-Chinese syllable pair to the source language end and the target language, v_s and v_t are the dictionary sizes of the source language end and the target language, after statistics in khmer the number of syllables and syllables in Chinese, v_s and v_t are set to 367 and 474, respectively. λ_s and λ_t are the expected lengths of the source language and the target language, and statistical bilingual corpora are found. Therefore, λ_s and λ_t are set as $\lambda_s = 2, \lambda_t = 1$.

4.3 Gibbs Sampling

Because there may be multiple alignment possibilities for syllable pairs, we assume that there are two alignment possibilities for a pair of syllables in a person's name. They are expressed as follows:

$$S_1 = \beta c_l(s_1,t_1)c_r\gamma \\ S_2 = \beta c_l(s_2,t_2)(s_3,t_3)c_r\gamma \quad (5)$$

Among them, S_1 and S_2 respectively indicate the two alignments of the current syllable pair, C_l and C_r indicate the former and next syllable pairs of the current syllable pair, and β and γ indicate the remaining syllable pairs in one bilingual person's name. Let h denotes the hypothesis of the bilingual syllable pair alignment, h_l denotes the hypothesis that is consistent with S_l, h_2 denotes the hypothesis that is consistent with S_2 and h^- denotes the assumption that h_1 and h_2 are common, that is βs_l and γs_r in formula (5) for these two assumptions, the calculation formulas are as follows:

$$p(h\mid h^-)=\begin{cases} p(\langle c_l,(s_1,t_1)\rangle\mid h^-)\bullet p(\langle(s_1,t_1),c_r\rangle\mid\langle c_l,(s_1,t_1)\rangle,h^-) & h\in h_1 \\ \bullet\, p(z_1\mid\langle c_l,(s_1,t_1)\rangle,h^-) & \\ p(\langle c_l,(s_2,t_2)\rangle\mid h^-)\bullet p(\langle(s_2,t_2),(s_3,t_3)\rangle\mid\langle c_l,(s_2,t_2)\rangle,h^-) & \\ \bullet\, p(\langle(s_3,t_3),c_r\rangle\mid\langle c_l,(s_2,t_2)\rangle,\langle(s_2,t_2),(s_3,t_3)\rangle,h^-) & h\in h_2 \\ \bullet\, p(z_2\mid\langle c_l,(s_2,t_2)\rangle,h^-)\bullet p(z_3\mid z_2,\langle(s_2,t_2),(s_3,t_3)\rangle,h^-) & \end{cases} \quad (6)$$

Approximating Eq. (6) gives Eq. (7),

$$p(h \mid h^-) = \begin{cases} p(\langle c_l, (s_1, t_1)\rangle \mid h^-) \bullet p(\langle (s_1, t_1), c_r \rangle) \bullet p(z_1 \mid \langle c_l, (s_1, t_1)\rangle, h^-) & h \in h_1 \\ p(\langle c_l, (s_2, t_2)\rangle \mid h^-) \bullet p(\langle (s_2, t_2), (s_3, t_3)\rangle \mid h^-) \bullet p(\langle (s_3, t_3), c_r \rangle \mid h^-) & h \in h_2 \\ \bullet\, p(z_2 \mid \langle c_l, (s_2, t_2)\rangle, h^-) \bullet p(z_3 \mid \langle (s_2, t_2), (s_3, t_3)\rangle, h^-) & \end{cases} \tag{7}$$

Among them, the random variables $z_1 z_2 z_3$ are the table numbers of $(s_1, t_1)(s_2, t_2)$, and (s_3, t_3).

Then, the aligned syllable pairs are further obtained from the probability distribution obtained in (2) and the sampling is divided into two steps. First, if the hypothesis is h_1 or h_2, use the formula (8) to sample S:

$$\begin{aligned} & p(S = S_1 \mid h^-) \\ & = \frac{N^{(h^-)}_{(c_l,(s_1,t_1))} + \alpha P_1((s_1,t_1)|h^-)}{N^{(h^-)}_{s_l} + \alpha} \bullet \frac{N^{(h^-)}_{((s_1,t_1),c_r)} + \alpha P_1(c_r|h^-)}{N^{(h^-)}_{(s_1,t_1)} + \alpha} \\ & p(S = S_2|h^-) \\ & = \frac{N^{(h^-)}_{(c_l,(s_2,t_2))} + \alpha P_1((s_2,t_2)|h^-)}{N^{(h^-)}_{s_l} + \alpha} \bullet \frac{N^{(h^-)}_{((s_2,t_2),(s_3,t_3))} + \alpha P_1((s_3,t_3)|h^-)}{N^{(h^-)}_{(s_2,t_2)} + \alpha} \\ & \bullet \frac{N^{(h^-)}_{((s_3,t_3),c_r)} + \alpha P_1(c_r|h^-)}{N^{(h^-)}_{(s_3,t_3)} + \alpha} \end{aligned} \tag{8}$$

Next, sample the tables at (s_1, t_1), (s_2, t_2), and (s_3, t_3). After sampling S, use Eq. (9) to sample z_i, Among them, $K\left((s_i, t_i)^{(h^-)}\right)$ represents the count of all tables assigned to (s_i, t_i) under hypothesis h^-. Formula (9) is as follows:

$$P(z_i = k|(s_i, t_i), h^-) \propto \begin{cases} N_k^{(h^-)} & 1 \le k \le K\left((s_i, t_i)^{(h^-)}\right) \\ \alpha P_1(s_i, t_i) & k = K\left((s_i, t_i)^{(h^-)}\right) + 1 \end{cases} \tag{9}$$

Among them, $N_k^{(h^-)}$ represents the number of customers who choose the Kth table under h^-, and $P_1(s_i, t_i)$ is the probability of selecting a new table, which is expressed as formula (3).

We treats a Khmer-Chinese bilingual name pair as a sequence of multiple *TU*. The key of the alignment model is to find the optimal segmentation; according to the noise channel model and the principle of maximum posterior probability, it is considered that the segment γ is the best segmentation for maximum $p(x_i)$. For a bilingual name in the corpus, $p(x_i)$ is calculated as follows:

$$p(x_i) = \prod_{\gamma \in D} p(\gamma) \tag{10}$$

A bilingual name pair $(s,t) = (\langle s_1, s_2, \ldots, s_m \rangle, \langle t_1, t_2, \ldots, t_n \rangle)$, abbreviated as $x = (s_1^m, t_1^n)$, x_i represents a pair of bilingual names, a person's name is divided into bilingual aligned *TU* set $\gamma = \{(s_1, t_1), \ldots, (s_h, t_h)\}$, *D* is used to represent the corpus, $D = \{x_i\}_{i=1}^{I}$, x_i is a bilingual name pair, *I* is the corpus scale. Calculate the probability of each possible alignment of a bilingual name pair according to Eq. (10) and the forward algorithm, and record the probability of each possible alignment. Then use the backward algorithm to find the path and find each segmentation. The corresponding path uses the Gibbs algorithm to sample the *TU*, determine the optimal path, and decode the path according to the Viterbi algorithm to obtain the best segmentation. The black bold line represents the final alignment result, as shown in Fig. 1 below:

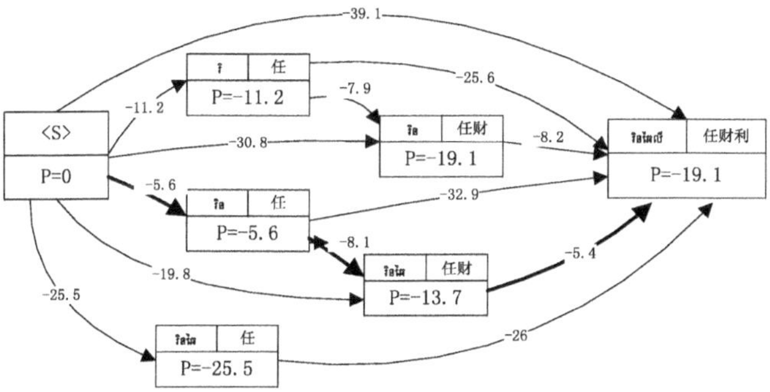

Fig. 1. Chinese and Khmer bilingual alignment process

5 Experiment and Analysis

5.1 Experimental Data

The experimental data was mainly derived from the Khmer-Chinese bilingual parallel web page extraction. A total of 2,746 Khmer-Chinese bilingual names were drawn, of which 80% were for training and the rest were used as test corpora (Fig. 2).

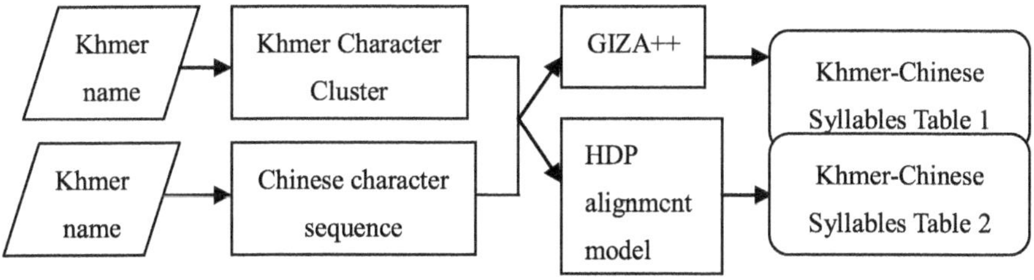

Fig. 2. The process of acquiring the Chinese and Khmer syllable list

5.2 Experimental Evaluation Metrics

In order to evaluate the effectiveness of this method, this paper uses the named entity transliteration evaluation standard in the work sharing white paper of the 2010 Annual Meeting of the Association for Computational Linguistics (ACL) task: ACC (Word Accuracy in Top-1), Mean F- Score (Fuzziness in Top-1), as described below:

N: Total number of test in khmer person's name N
n_i: The number of translations of the ith personal name in the test set
$r_{i,j}$: The jth reference translation of the i th personal name in the test set
$c_{i,k}$: The kth candidate transliteration result of the i-th person name in the test set
Ki: number of transliteration candidates for the i-th personal name in the test set

The ACC is used to evaluate the correct rate of transliteration results ranked first in the list of candidate results obtained by the transliteration model.

$$ACC = \frac{1}{N}\sum_{i=1}^{N}\left\{\frac{1, if\ \exists\, r_{i,j} = c_{i,j}}{0, otherwise}\right. \tag{11}$$

Mean F-score is used to uniformly evaluate the correct rate Pi of the transliteration between the first transliteration candidate and the reference translation and the recall rate Ri. LCS is a function for finding the longest common sequence, argmin represents the value of the argument when the function is minimal, ED represents the edit distance, and the length is calculated in terms of the smallest basic symbol in the language. The formula is as follows:

$$\begin{aligned} LCS(c,r) &= \tfrac{1}{2}(|c| + |r| - ED(c,r)) \\ r_{i,m} &= \arg\min_{j}\left(ED\left(c_{i,1}, r_{i,j}\right)\right) \\ R_i &= \frac{LCS\left(c_{i,1}, r_{i,m}\right)}{|r_{i,m}|} \\ P_i &= \frac{LCS\left(c_{i,1}, r_{i,m}\right)}{|c_{i,m}|} \\ F_i &= 2\frac{R_i \times P_i}{R_i + P_i} \end{aligned} \tag{12}$$

5.3 Contrast Experiment

In order to fully reflect the performance of the hierarchical Dirichlet process alignment model, GIZA++ is chosen as the comparison model because GIZA++ is one of the most widely used, stable, and mature models in machine translation today.

5.4 Experimental Results and Analysis

Experiment 1: To compare the performance of the alignment model and GIZA++, compare the number of syllable pairs extracted from a total of 550 Khmer-Chinese bilingual name pairs in the test data. The experimental results are shown in Table 1 below:

Table 1. Result of experiment 1

Method	Finally get the number of syllables
GIZA++	1663
HDP	904

In the analysis of the results of Experiment 1, the number of syllables pair extracted by GIZA++ is much large than that extracted by HDP. This may be because GIZA++ produces a lot of alignment results in order to fit the data better. HDP overcomes over-fitting problems and therefore produces a smaller syllable table and a shorter syllable.

Experiment 2: In order to compare the effect of the alignment model on the khmer-Chinese transliteration, the accuracy of the transliteration of the Khmer-Chinese name was compared through the manual verification of Khmer students. The experimental results are shown in Table 2 below:

Table 2. Result of experiment 2

Method	ACC	Mean F
(GIZA++) + MOSES	**29.28%**	**67.88%**
Method of this article	32.6%	69.51%

For the analysis of the results of experiment 2, the ACC and Mean F of this method are both higher than those of (GIZA++) + MOSES. The experimental results are not significant, but we can see that this model can overcome the problem of overfitting and can make the transliteration of names align to get better results.

6 Conclusions

This paper proposes a bilingual alignment model based on the hierarchical Dirichlet process and applies it to Khmer-Chinese transliteration of person's name. The multi-to-multi-alignment not only avoids the accumulation of errors caused by syllabic segmentation, but also takes into account the effect of the alignment of the previous aligned syllable on the alignment of the following syllables and eases overfitting problem. The experimental results show that compared with the baseline method, this method can form a smaller syllable table with shorter syllable length, and can obtain a better transliteration effect of the Khmer-Chinese name. The collection of various bilingual names and the construction of a bilingual name transliteration system are the focus of the next step.

Acknowledgements. This work is supported by National Science Foundation of China (NSFC) via grant 61462055, 61562049

References

1. Wan, S., Verspoor, C.M.: Automatic English-Chinese name transliteration for development of multilingual resources. In: Processing of the 17th International Conference on Computational Linguistics, pp. 1352–1356 (1998)
2. Abdulikim, Imamhasan, Tohti, Turdi, Hamdulla, Askar: On rules-based chinese machine translation algorithm for uyghur personal name. Comput. Appl. Softw. **27**(8), 86–87 (2010)
3. Jiampojamarn, S., Kondrak, G., Sherif, T.: Applying many-to-many alignments and hidden markov models to letter-to-phoneme conversion. In: Proceedings of NAACL, pp. 372–379 (2007)
4. Nevado, F., Casacuberta, F., Landa, J.: Translation memories enrichment by statistical bilingual segmentation. In: Proceedings of LREC-04, vol. 2004, no. 15, pp. 335–338 (2009)
5. Fukunishi, T., Finch, A.: A bayesian alignment approach to transliteration mining. ACM Trans. Asian Lang. Inf. Process. **12**(3), 1–22 (2013)
6. Finch, A., Yasuda, K., Okuma, H., et al.: A bayesian model of transliteration and its human evaluation when integrated into a machine translation system. IEICE Trans. Inf. Syst. **94**(10), 1889–1900 (2011)
7. Chea, S., Huor, T., Rithy, R.P., et al.: Word bigram vs orthographic syllable bigram in KHMER word segmentation, vol. 54, no. 251, pp. 156–245 (2008)
8. Teh, Y.W., Jordan, M.I., Beal, M.J., et al.: Hierarchical dirichlet processes. J. Am. Stat. Assoc. **101**(476), 1566–1581 (2006)
9. Teh, Y.W., Jordan, M.I.: Hierarchical bayesian nonparametric models with applications. Bayesian Nonparametrics **2010**(537), 158–207 (2008)
10. Teh, Y.W., Jordan, M.I., Beal, M.J., et al.: Sharing clusters among related groups: hierarchical dirichlet processes. Adv. Neural. Inf. Process. Syst. **37**(2), 1385–1392 (2010)

Product Information Verification Based on Big Data Technology

Shuhao Yu[1], Kangjun Chen[1], Shenglong Zhu[2], and Chengqi Zhu[3](✉)

[1] School of Electronic and Information Engineering, West Anhui University, Lu'an 237012, China
[2] Anhui Electric Power Science and Research, Hefei 230022, China
[3] School of Foreign Languages, West Anhui University, Lu'an 237012, China
1956345025@qq.com

Abstract. The change of production pattern in modern society results in an explosive data. Large amounts of product information data make it hard for people to avoid a dizzying message. Therefore, it is extremely important for us to judge the authenticity of the information. This paper made a series of deep explorations on Python based data crawler technology and "One Material One Barcode" tracing method. The proposed method crawls data from different websites and judges the authenticity of link by regular expression matching algorithm. Then it adds these links to WeChat platform and collects serial number rules of product information. Finally, by activating the serial number of database information it will improve our ability to judge and verify the product information.

Keywords: Big data · Information verification · Python · Web crawler

1 Introduction

In recent years, big data involves a wide range of fields and has a profound impact on social development, specifically reflected in three aspects: big data decision-making has become a new way, the application of big data promotes information technology and deep integration of various industries and big data development promotes the emergence of new technologies and new applications [1].

In the era of big data, real information is in urgent need and big data decisions have become a new way of making decisions. It makes up for the limitations of previous data types and data volumes, and makes decision analysis for a large variety of unstructured massive data. Excavation and utilization of big data is also an important technology for a company or even a person. It reasonably uses existing data resources to complete the collection, storage, processing, and analysis of product information so as to achieve product quality assurance and brand credibility.

K. Deng et al. (Eds.): ICMIR 2018, AISC 856, pp. 451–458, 2019.
https://doi.org/10.1007/978-3-030-00214-5_57

2 Technical Support

2.1 Official Website Link Acquisition and Implementation Methods

(1) Python language data crawling

Today the python language is an easy-to-learn, easy-to-use, object-oriented and interpretable language. There are more and more packages based on the Python language that help us solve problems. The real official website information is the most direct and the most effective channel for verifying products for the majority of users. How to get a real and effective official website link is also one of the main contents in this article. This paper tries to use the Python language-based web crawler technology to get the official website's link address, then filter it according to the Python-based regular expression package, at last remove some dirty data and obtain valuable data.

(2) Web crawler process and basic framework

The web crawler mainly includes the dispatcher, value data, and the main program. The dispatcher plays a role in manipulating part of the operation of the program. The main program includes the URL manager, web page downloader, and web parser. The process of the web crawler is firstly the dispatcher's access to the URL manager and then pass the URL address obtained to the web page downloader at this time. The web page downloader will download the corresponding web page, and then pass the web page to the web parser (DNS) for parsing operation to generate valuable data or a new URL list, after which these data will be returned to the dispatcher again to be saved, and the new URL list will be added to the URL manager again to perform the above operation until there is no URL address to be acquired in the URL manager. Finally we will obtain the value data and the required data format that we need (Fig. 1).

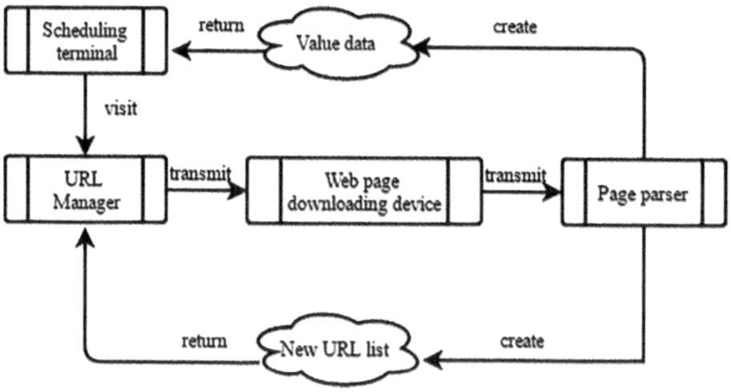

Fig. 1. Web crawler flow figure

(3) The process of crawling the official website link network

First of all, we need to configure the required official website search resources, the scope of the searched content, the search strategy and the level of the content to crawl the site. The field needs to represent the URL address and the URL address whether has been crawled [2]. In this way we can record the data that has been crawled in a table and served as a data source for the web crawler process. The collected URL data will be downloaded to the local one by one through a web page downloader to form a local file or a string format. Then we analyze the value of the data sent from the web page downloader and use third-party plug-in for web parser, because it is commonly used in large parts of china. It can be got more accurate node query by the conversion of HTML and DOM trees. Thus we can intercept more effective information, reduce the redundancy of the data greatly and improve the efficiency of our value information acquisition.

(4) Regular expressions to get the match pattern of the real official website

In the Internet market of china, a domain name is a method used to label information addresses. The registration of a domain name can solve the problem of the address corresponding to the current network, and the user can access the information in the corresponding website through the domain name [3].

In order to increase the authenticity, integrity and accuracy of the domain name and strengthen the domain name registration audit, CNNIC issued the corresponding document to regulate the registration and use of domain names on December 11, 2009. Therefore, the domain name system on the Internet is still very safe and reliably. For the domain name of the official website, it has more strict procedures to apply and comply with the rules of the domain name. In the following, we obtain the basic official website matching model by comparing and searching the official website that Chinese Baidu certified and non-authorized, and we find the official website that Baidu has certified conforms to a basic standard of: http://www.brandname.com/, and non-authorized official website will add a large number of strings after the basic standard or other strings that do not conform to the brand name. It will help us to crawl the real official website information and get regular expression matching based on Python language packages.

At present, the matching tools of regular expressions mainly include DFA and NFA. In this article, we use a matching algorithm based on guessing-verification. The principle of the algorithm is to search for partial characteristics of regular expressions by DFA. And it will complete speculation about the existence of features. When it is guessed that it is possible to match a feature, the NFA is used for verification [4]. This method is exactly corresponding to the basic official website model of the article: Search first, then match to obtain the website link of the real official website.

3 Implementation Procedure

In order to target more groups of users and also fit the security psychology of more people, we have transplanted the official website links to the WeChat platform. The place where the WeChat platform can be transplanted is a small -program, and this interface does not need to complete code compilation by itself. It is only possible to complete the porting of official website data through pay-to-development, one-click generation, and it can also facilitate the use of more people.

3.1 Application of Serial Number

(1) One thing one barcode

The technology of "one thing one barcode" uses cloud platform to connect consumers, enterprises and circulation so as to provide a comprehensive solution of O2O marketing, anti-counterfeiting tracking and warehousing logistics. The logistics anti-counterfeiting sleeve label is printed in advance. According to the labeling operation of the product, the enterprise staff will complete the labeling and pasting of the book and the box one by one. And the product code that corresponds with the package-book can be completed at the same time, then the product is withdrawn from the delivery unit. We can use this way to record each piece of goods outbound and record each piece of goods shipped to the store or area. Therefore, an item corresponds to an anti-counterfeiting code and a logistics code, and a PAD data collector can obtain the product sent to the dealer and the area during the delivery process to prevent cross region sale. In this article, we improve on the application principle of "One Material One Barcode". This is not just the bulk of the book, but the anti-counterfeiting function of the product.

(2) IPhone Serial Number and IMEI Query

At present, many people are using Apple's series of electronic products, and now there are many ways to buy, different channels may lead to price differences, so sometimes we may worry about whether the product is genuine, whether this machine is refurbished machines and so on. As for this, we can use serial number or IMEI to check the status of the product and verify the authenticity of the product. IMEI is an identity code of an international mobile device at present. It is like a human ID card. The IMEI code of a mobile phone is also the ID card of a mobile phone. Each IMEI corresponds to a mobile phone, which can be said to be a "touchstone" for distinguishing the genuineness of the mobile phone. At present, Apple has done a better job in this area. The users who bought the IPhone all know that if you want to use a new phone, you must use the serial number or IMEI code to activate, and then through the activation of the warranty information to determine whether this phone is a real IPhone. This authentication method is also an anti-counterfeiting measure accepted by most users, so the serial number anti-counterfeiting has some basis and is developable.

(3) Creating an enterprise serial number rule

Through the technologies adopted by the above two companies, this article adopts the integration of two technologies to apply to a wider range of areas to achieve product information verification, which is called anti-counterfeiting. Therefore, we first need to create a serial number of enterprise products similar to the IMEI code to ensure that each product has its own unique identity code.

The code-building rules draw on the rules of the book's ISBN code and use 13 digits to divide it into five segments. Each segment represents different information. Different from the group number information of the book, this article sets the first paragraph of the product to represent the production of the product, the second paragraph represents the product type, the third paragraph represents the distribution, the fourth paragraph represents the dealer, the fifth paragraph represents product verification code. The inter-code value is specified by the product and sales division of the company. The combination of letters and numbers can be used to achieve the uniqueness of product information.

The advantage of building code in this way is that you can get the product's specific flow when the product is released, ensure the product's information feedback path, get the product's sales situation, and help the company get the optimal decision after analyzing big data. At the same time, it can also guarantee the authenticity of the product. Using anti-counterfeiting can establish a good corporate reputation.

3.2 Product Information Collection Process

For outbound products, the product information that we need to collect includes product name, product type, production date, time of delivery, product price, area to be distributed, and store. These data will be stored in the server and then passed. Flume performs overall data collection on every server, and the flume will collect the data from each of the sever that has data. The Flume acquisition tool used here is a system with good performance, high reliability, and distributed log collection and transportation provided by Cloudrel. It can implement data collection for each server's data source, and then remove unnecessary dirty data through MapReduce, store the data in Hbase. Then the data warehouse will process and analyze this data, and load it into the data warehouse table. At last the data warehouse table is exported by Sqoop and the data in the table is stored in the MySql database (Fig. 2).

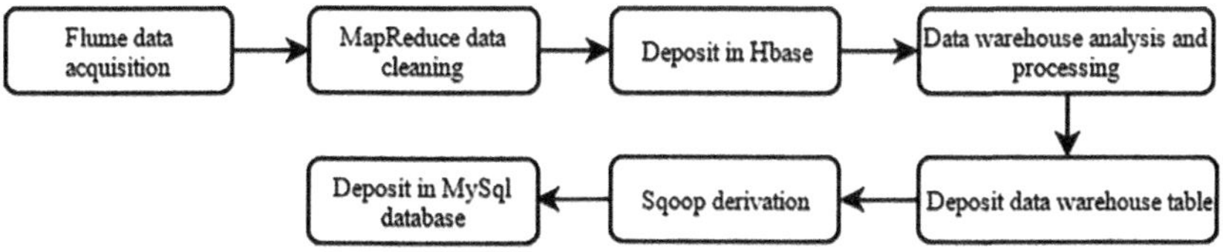

Fig. 2. Product data collection flow

(1) Association of Product Information Data and Serial Numbers

We have previously created a set of serial number rules that belong to the company, and stored the serial number built in hive. The hive can map a structured file into a table, and the metadata will be stored in MySql. Finally, we associate the stored product data with the serial number through the join in the database to complete the one-to-one correspondence between the serial number and the product.

(2) Real-time Query Portal Implementation

At present, most of the current official website is based on web page front-end development and back-end design management. We use IntelliJ IDEA for development tools because it has strong code support functions and can support multiple languages. It also supports the replacement and lookup of regular expressions and is the best Java development tool available today. This article designs the front-end query entry interface by the code, and then using JDBC technology to complete the connection with the database. JDBC is the standard API for Java program database communication. It can easily complete the interaction between Java programs and database content. The specific connection method is to register the database driver provided by the database vendor to the JDBC driver manager.

Next, you need to build a database connection URL. Finally, we will get the "Connection" object, which needs to use the DriverManager object, and then create a Connection object through the getConnection method in the DriverManager [5]. Due to the above technical support, the connection between the front-end query entry and the database can be completed and the query can be completed.

(3) Database Information Feedback

The database information feedback is mainly through the activation of the serial number to complete the product information query. There is a feedback issue here. Our product information is stored in the server through hive, and then in the relational database through the structured table derived from the hive mapping. Because the data warehouse has the function of real-time detection, propagation and data loading, we can complete the data of real-time detection and reload the activated serial number. And the product information data can be modified. At this point, the activated product will generate its own basic information and extension information. The basic information includes the product name, the official price of the product, the product type, the product warranty period, the production date, the location after the distribution, and the store where it is located. These data have been previously collected by flume and stored in the database. After the activation event occurs, the extension information will be generated. The extension information includes the time for the product to be sold, the name of the store to be activated, the activation of the staff code, and sales conditions, so that the user can judge whether the product is a real product or not conveniently.

3.3 Product Information Verification Mode

This paper mainly implements product information verification through the following methods. Firstly, we complete the acquisition of the official website link through the web crawler and regular expression matching algorithm, and integrate the link in the official website. Then we use the small-program interface of the WeChat platform to face a large user community so that more users can quickly and effectively get the most real official website links, and enter the official website for inquiries. Secondly, we research for new anti-counterfeiting methods that are integrated and innovated in order to get a set of serial number rules belonging to the company. Then, with the help of big data technology, we can complete the collection and storage of commodity information. And also the connection between the information query portal and the database will be completed. Thus the seller can activate the product by selling the product, and generate new expansion information after activation.

Information changes are done through the Sql language, and the seller's activation is limited by the database's secure access control. After the user activates the product, he enters the official website through the official website of the WeChat platform, and then enters the serial number to query the basic information of the product and the extended information generated after the activation. The specific anti-counterfeit identification information is in the extension information. The user can judge whether the product is the original manufacturer genuine product by whether the product selling time is consistent with its own purchase time and whether the store where the product is located and the activated store clerk code are consistent with the official website query information (Fig. 3).

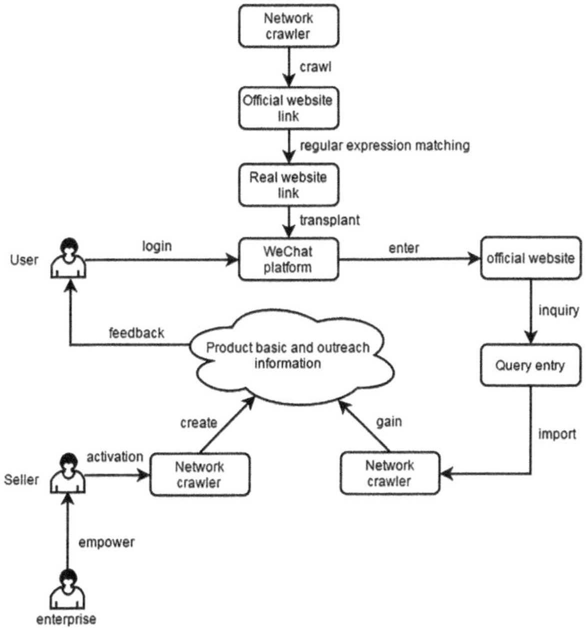

Fig. 3. Product information verification mode flow Figure

The need for the store to activate the product is to ensure that the company can obtain information on the store where the product is sold, which helps the company to analyze the sales situation of the product in the distribution area. And it is conducive to the decision-making of the enterprise and also helps to protect the user's rights and interests.

4 Conclusion

Based on the current anti-counterfeiting technology, this article has made some integrates and innovations to get real official website big data through the matching of web crawler technology and regular expression, and transplants it to the relatively trusted WeChat platform of the masses at the same time. Then it analyzes the principle of sequence number and IMEI code and "One Material One Barcode". On this basic it can form a set of rules for the establishment of the enterprise. In the end, it collects and cleans filters through flume information in big data technology and finally maps into tables and stores them in the database. What's more, it completes the query entry through serial number and database connection so that the product information verification is implemented. The techniques used in the article such as regular expression matching algorithms are still not that perfect, for they cannot fully bring the expected results. There are no more scientific techniques to distribute rules and user groups are not very broad. Whereupon, when it comes to technology, we are ought to select more optimized algorithms and code for building rules and use better techniques to expend the anti-counterfeiting methods and strategies. For example, the scanning of code anti-counterfeiting and anti-counterfeiting of photo image processing, which are more convenient for users. Still, there will be a particular shortage of anti-counterfeiting technology for our society in the future. Thereby, we must have a future understanding of it and invent new anti-counterfeiting technologies.

Acknowledgements. This research was supported by the National Natural Science Foundation of China (NSFC) (No. 61702375) and the Universities Excellent Young Talents Foundation of Anhui Province (No. gxyqZD2016249).

References

1. Lin, Z.: Principle and Application of Big Data Technology. The People's Posts and Telecommunications Press, Beijing (2015)
2. Liu, S., Yue, S.: Crawling technology of network information based on python in big data's time. Electron. Technol. Softw. Eng. **21**, 160–160 (2017)
3. Zhao, J., Shi, Z.: Real name system in internet. Comput. Develop. Appl. **12**, 37–39 (2004)
4. Zhang, S., Luo, H., Fang, B., et al.: A High performance regular expression matching algorithm for Network Security Detection (2010)
5. Eckel, B.: Thinking in Java. China Machine Press, Beijing (2013)

Chinese-Lao Cross-Language Test Similarity Computing Based on WordNet

Sizhuo Li[1], Lanjiang Zhou[2(✉)], Jianan Zhang[3], Feng Zhou[2], Jianyi Guo[2], and Wenjie Huo[1]

[1] School of Information Engineering and Automation, Kunming University of Science and Technology, Kunming 650500, China
[2] The Key Laboratory of Intelligent Information Processing, Kunming University of Science and Technology, Kunming 650500, Yunnan, China
915090822@qq.com
[3] Information Engineering University, Kunming Team of the Three Schools, Kunming 650500, China

Abstract. Text similarity calculation is widely used by information retrieval, question answering system, plagiarism detection and so on. At present, most research just aim at text similarity of the same language, and research on cross-language text similarity calculation is rarely, differences between languages make cross-language text similarity calculation very difficult, in view of this situation, this paper propose a WordNet-based method of Chinese-Lao cross-language text similarity calculation. First, preprocessing and feature selection for Chinese text and Lao text which in medicine, then use the semantic dictionary WordNet to convert the Chinese text and Lao text into a middle layer language, finally, compute the text similarity between Chinese and Lao in the middle layer.

Keywords: WordNet · Middle layer language · Cross-language text similarity

1 Introduction

Text similarity computing has been widely discussed in the fields of linguistics, psychology, information theory and so on. Text similarity calculation aims to compare the correlation between the two texts. In recent years, the method of text similarity computation [1–3] based on the same language is more and more perfect, the algorithm model represented by the Boolean model, vector space model, probability model and so on. However, the research on cross-language text similarity is very rare. Due to the differences in grammar between Chinese and Lao, we can not use the existing method which calculate the similarity of text in the same language to calculate the similarity between Chinese and Lao text.

WordNet is a semantic dictionary using synonym set represents a concept and has multi language version. The Chinese WordNet used in this paper is developed by Southeast University and Lao semantic dictionary is constructed by our laboratory. The synonym set synset_id of WordNet between different language versions are corresponding to each other. Therefore, this paper uses this characteristic and proposes the

K. Deng et al. (Eds.): ICMIR 2018, AISC 856, pp. 459–464, 2019.
https://doi.org/10.1007/978-3-030-00214-5_58

method of Chinese-Lao Cross-Language Test Similarity Computing Based on WordNet in Medicine. This method uses the WordNet to convert the Chinese text and Lao text into a middle layer language, then, compute the text similarity between Chinese and Lao in the middle layer.

2 The Process of Chinese-Lao Text Similarity Computing

2.1 Text Preprocessing

Because the method of this paper needs to analyze the semantic of the word, so it is necessary to deal with some special words, such as names, place names and so on. Then convert these special words into a specific string. In feature selection, these special words are ignored to avoid noise interference.

2.2 Text Feature Selection

The purpose of feature selection is to select the characteristic items, and the selected feature item should be able to express the theme of the original text. In this paper, the word is extracted as the feature of the text, and each document is treated as a word bag. Through the word segmentation and remove the stop words, the Chinese document and the Lao documents can form a feature word set. Then, by using the method of document frequency selection to remove the useless words that interfere with the original text. When DF is greater than a certain threshold value, then remove the t. Because the higher the DF, the more t appears in text. When DF is less than a certain threshold value then removes the t, because t is either a rare word or noise.

2.3 Conversion of Language Space Between Chinese and Lao

This paper uses the WordNet to convert the Chinese text and Lao text into a middle layer language, then, compute the text similarity between Chinese and Lao in the middle layer.

2.4 Semantic Disambiguation

In WordNet, any word may belong to some synonym set at the same time. If we want to calculate the similarity, we should determine the semantic of feature words in context firstly. Semantic relativity is an important basis of the disambiguation algorithm in this paper. We define semantic relativity: If one sense of a word is more related to the senses of other words in context, it is more likely to be correct sense which accord with the semantic context. The disambiguation algorithm defines a semantic hash function $\varphi(x) = h$, h is an integer. Mapping a synonym set x of WordNet to a numeric space. $\varphi(x)$:

(1) if $x_1 \neq x_2$, then $\varphi(x_1) \neq \varphi(x_2)$
(2) if x_1, x_2, x_3 *and* $d(x_1, x_2) < d(x_1, x_3)$, *then* $p(|\varphi(x_1) - \varphi(x_2)| < |\varphi(x_1) - \varphi(x_3)|) > 1 - \theta,\ \theta \to 0$

$d(x_i, x_j)$ is the semantic distance of sense x_i and x_j in WordNet. It represents the shortest weighted path length of the node where x_i and x_j in. The greater the significance of the two senses in the WordNet, the greater the semantic distance. By $\varphi(x)$, you can directly convert $d(x_i, x_j)$ to the difference of $\varphi(x)$.

With the semantic distance and semantic hash, we can define the semantic density to quantify the semantic correlation between a group of words. For a set of synonym sets $w_1, w_2, w_3..., w_n$ their semantic density $(w_1, w_2, w_3..., w_n)$ can be represented by n^3 divided by the volume V_{min} $(w_1, w_2, w_3,..., w_n)$ of the minimal subtree containing all $w_1, w_2, w_3,..., w_n$. As shown in the formula (1).

$$\text{density}(w_1, w_2, w_3, \ldots, w_n) = \frac{n^3}{v_{min}} = \frac{n^3}{1.2\max(b - 72.0)} \quad (1)$$

In formula 1, b is the lowest position in the same position of $\varphi(w_1)$, $\varphi(w_2)$, ..., $\varphi(w_n)$. We can understand the semantic density like this, the greater the semantic density, the more relevant the group of words.

With the semantic density, then we carry out semantic disambiguation. Select any possible senses of all words from disambiguation context window, computing their semantic density. The specific steps are as follows: First, combine synonym set of every word in W into a big candidate set $C = \{s_{i1}, S_{i2}, \ldots\}$. Then sort every synonym set in C into semantic hash value $\varphi(x)$. We get $C = \{s_1, s_2, ..., s_n\}$ and $\varphi(s_1) < \varphi(s_2) < \cdots < \varphi(s_n)$. Calculate the semantic density of any two synonym sets in C, the disambiguation results are the senses which contained by maximum semantic density of the subtree.

2.5 Chinese-Lao Test Similarity Computing

Word semantic similarity computation based on WordNet, there are two major types of algorithms: Based on path, Based on Information Content (IC). In this paper, we use the similarity algorithm based on IC.

The similarity algorithm based on information content is proposed by Resnik for the first time by using the IC value of each concept in WordNet as a parameter. IC is expressed as $-lgp(c)$, it is called self-information in the information theory. According to Resnik, the similarity of two concepts is determined by the deepest public parent node which contains the two concepts. It can get the similarity value of the two concepts only by calculate the eigenvalue of the public parent node. As shown in the formula (2).

$$sim_{RES}(c_1, c_2) = -\lg p(lso(c_1, c_2)) = \mathrm{IC}(lso(c_1, c_2)) \quad (2)$$

$lso(c_1, c_2)$ represents the deepest public parent node of the concept c_1, c_2 in is_a tree.

$p(c)$ represents the probability of an instance of the concept c. This kind of representative algorithm is algorithm of Lin. The semantic similarity algorithm of Lin considers to define a common method of computing similarity. Algorithm model is shown in the formula (3).

$$sim_{LIN} = \frac{2 \times \lg p(lso(c_1, c_2))}{\lg p(c_1) + \log p(c_2)} = \frac{2 \times IC(lso(c_1, c_2))}{IC(c_1) + IC(c_2)} \tag{3}$$

The performance superiority of the similarity algorithm based on IC is mainly determined by the accuracy of the concept IC value and the rationality of introducing the IC parameter into the algorithm. Therefore, improving the IC parameter model can improve the performance of the algorithm. The improved algorithm of Nuno to the IC model is shown in the formula 4.

$$\mathrm{IC}(c) = -\lg p(c) = 1 - \frac{\lg(hypo(c) + 1)}{\lg max_{wn}} \tag{4}$$

hypo(c) represents all child nodes of the concept *c* and max_{wn} represents the number of concepts in the classification tree. The model of Nuno only considers the number of child nodes of the concept, which is limited. In this paper, an improved IC model is proposed, which considers the depth of the concept in the classification tree. Algorithm model is shown in the formula 5.

$$IC_{NEW}(c) = k\left(1 - \frac{\lg(hypo(c) + 1)}{\lg max_{wn}}\right) + (1 - k)\frac{\log(deep(c))}{\log(deep_{max})} \tag{5}$$

K between 0 and 1, in this paper k = 0.5.

Considering the generality of Lin algorithm, we bring formula 5 into formula 3, and get a new similarity model is shown in the formula 6.

$$sim_{NEW}(c_1, c_2) = \frac{2 \times IC_{NEW}(lso(c_1, c_2))}{IC_{NEW}(c_1) + IC_{NEW}(c_2)} \tag{6}$$

The above formula is to solve the similarity of the two concepts in WordNet, the word similarity algorithm is shown in the formula 7.

$$sim_{NEW}(w_1, w_2) = \max\left[sim_{NEW}(c_{1i}, c_{2j})\right] \tag{7}$$

c_{1i}, c_{2j} are some concepts of w_1, w_2.

we convert the feature words $\{CW_1, CW_2..., CW_n\}$ in Chinese text CH into middle layer language, and then get corresponding English senses $\{CE\_W_1, CE\_W_2..., CE\_W_n\}$ after semantic disambiguation. The feature words of Lao text L are $\{LW_1, LW_2..., LW_k\}$. The same operation gets English senses $\{LE\_W_1, LE\_W_2..., LE\_W_k\}$. Based on the formula 7, we obtain the similarity formula between A and B. As shown in the formula (8).

$$\mathrm{sim}(CH, L) = \frac{\sum_{i=1}^{n} \sum_{j=1}^{k} sim_{NEW}(CE\_W_i, LE\_W_j)}{n \times k} \tag{8}$$

The calculated results ranged from 0 to 1, 0 indicates that the two texts are not similar and 1 indicates that the two texts are completely similar. The larger the value, the more similar the two texts.

3 Experiment and Analysis

In this paper, we first test the semantic disambiguation algorithm, providing more accurate feature words for text similarity computation. The experiment uses an open semantic annotation corpus SemCor which tag words based on WordNet. Using part-of-speech tagging (POS tagging) by TreeTagger, then use the annotation results as the input of the disambiguation algorithm. Finally, compare the results and the manual annotation results, and the accuracy of the algorithm is obtained. Table 1 lists the disambiguation accuracy of the first 9 documents in SemCor. As a control, the column of benchmark in the table indicates the accuracy of the random guess. For example, a word has 5 sets of synonyms, and the accuracy of random guesses is 20%.

Table 1. Experimental results of disambiguation

Text	Accuracy rate	Benchmark
1	62.55%	22.14%
2	55.45%	19.28%
3	40.23%	21.56%
4	48.75%	18.19%
5	52.39%	20.09%
6	50.13%	20.81%
7	57.81%	19.98%
8	43.91%	18.31%
9	46.59%	17.45%

The average accuracy of disambiguation algorithm in SemCor is 50.9%.

Next, experiment on Chinese-Lao Cross-Language Test Similarity Computing. The text of this paper is 1000 texts, including 900 Chinese texts and 100 Lao Texts. Among them, 600 Chinese texts are noise texts which constituting the noise set. 300 Chinese texts and 100 Lao texts constitute the standard set, and according to the similarity between the Chinese text and the Lao text can be divided into 20 categories, each category has a range of 13 to 17 Chinese text. It can be understood that there are 13 to 17 Chinese texts which people think similar to Lao text in the standard text. The noise set and the standard set are mixed to the test set. Specific practices are as follows.

Extract a text in a sequence of 100 Lao texts from the standard set, and then calculate the similarity between this Lao text and the Chinese text in the test set. According to the similarity ranking, output result which similarity in the top 17, then observe the output, if the text belongs to Laos and Chinese text of the same class are output, is that the similarity calculation of success. If the Chinese texts that belong to the same category as the Lao text are output, it is considered that the similarity is successful. In this paper we compared the similarity algorithm of spatial cosine with the algorithm in this paper.

Experimental results of the formula: $R = \frac{t}{T} \times 100\%$

R: Correct rate t: Correct text of the test results T: Test which is tested

The experimental data are shown in Table 2.

Table 2. Comparison of experimental results

Computing method	Test Lao	Correct Lao text	Correct rate
Algorithm based on space vector cosine	100	49	49%
Algorithm based on WordNet	100	71	71%

4 Conclusions

This paper presents a similarity calculation algorithm between Chinese and Lao based on WordNet. Use the semantic dictionary WordNet to convert the Chinese text and Lao text into a middle layer language and compute the text similarity between Chinese and Lao in the middle layer. The experimental results show that the proposed method achieves good results.

Acknowledgments. This paper is supported by National Nature Science Foundation Nos. 61662040, 61562049.

References

1. Hong-Lian, L.I., Wei, H.E., Yuan, B. Z.: An kind of chinese text strings' similarity and its application in speech recognition. J. Chin. Inf. Process. (2003)
2. Song, L., Ma, T.: The study on the comprehensive computation of the documents similarity. Comput. Eng. Appl. **42**(30), 160–163 (2006)
3. Yu, J., Zhou, C., Liu, R.: Experimental research on premixed gases explosion in overpressure. J. Dalian Univ. Technol (2005)
4. Lin, D.: An information-theoretic definition of similarity. In: Fifteenth International Conference on Machine Learning, pp. 296–304. Morgan Kaufmann Publishers Inc. (1998)
5. Resnik P.: Using information content to evaluate semantic similarity in a taxonomy. In: IJCAI 1995 Proceedings of the 14th International Joint Conference on Artificial Intelligence, pp. 448–453 (1995)

Optimization of Ship Speed Governor Based on Active Disturbance Rejection Control

Zeng Hong, Li Guanglei(✉), Yang Kenan, Jiang Dingyu, and Wang Hao

College of Marine Engineering, Dalian Maritime University, Dalian, Liaoning, China
ericli1066@dlmu.edu.cn.com

Abstract. In order to handle the characteristics of non-linearity, time-varying and uncertainty, due to the necessary starting and stopping and uncertain external environment interference, in the speed control system of the main diesel engine, the paper established propulsion system model of medium and high speed four stroke diesel engine. We designed a optimization model base on the active disturbance rejection control theory. We also established the control model base on the traditional classical PID control and carried out contract simulation experiment. The results show that, the effect of the optimization model in the aspect of speed control system was good, the dynamic quality and speediness of ship speed control system improved. So this work has a certain reference value for the development of related application system.

Keywords: Marine main engine · Medium and high speed four stroke Speed control · Active disturbance rejection control

1 Introduction

The speed control system is crucial to the performance of the main engine. The classical PID control is widely used in the digital control system of marine engine, which is simple in structure and easy to implement [1]. However, the speed control system has the characteristics of non-linearity, time-varying, and some uncertainty factors such as parking, misoperation and outside environmental interference must be considered, in these cases, the classical PID control system is difficult to adjust the optimal control parameters of the speed control system, even lead to be overshoot of system output, long transition time and oscillation [2]. In order to get the better results, many scholars had tried to use a variety of mathematical simulations and intelligent control algorithms. For example, the fuzzy PID control of precise mathematical model can improve the real-time and openness of the system, but it was difficult to determine and optimize the fuzzy rules [3]; The PID neural network control has nonlinear proximity performance, it improved the stability of the system and the ability to resist interference, but the algorithm was too complicated and the speed of convergence was slowly [4]. Optimal control, robust control, model reference adaptive control and other modern control theory were also proposed to deal with the non-linearity, uncertainty and disturbance in speed control system, but they rely too much on the exact mathematical

K. Deng et al. (Eds.): ICMIR 2018, AISC 856, pp. 465–473, 2019.
https://doi.org/10.1007/978-3-030-00214-5_59

model of the controlled system, there are still a long way to the application of real ship [5, 6].

ADRC (Active disturbance rejection control) technology is a kind of innovation technology, it does not depend on the system model and practical control method, even can solve the nonlinear uncertain model bias so as a range of system interference factors disturbance control quality and stability caused by falling [7]. This paper applied the ADRC in marine engine control system, build high speed four stroke diesel engine nonlinear mathematical model, deal with the problem of uncertainty in the control process, had good performance in restraining the load disturbance and improved the quality and accuracy of dynamic control system.

2 Nonlinear Mathematical Model of Main Engine

The overall structure of the main diesel engine speed control system is shown in Fig. 1.

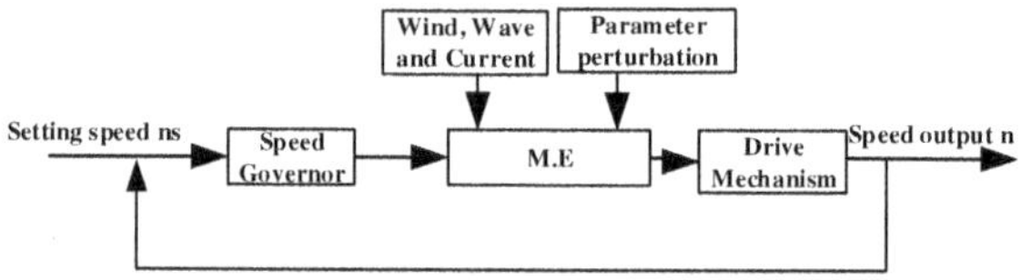

Fig. 1. The main diesel engine speed control system

2.1 Diesel Engine Mathematical Model

This paper took the high speed four stroke diesel engine of the ship named marine engineering ship 618 as the model ship, took the speed control system as research object, and the equation of motion is shown below:

$$J\frac{d\omega}{dt} = Md - Mc \tag{1}$$

J is the moment of inertia that the diesel engines drive to the crankshaft. ω is the angular velocity of the crankshaft; Md is the output torque for diesel engines; Mc for the drag torque. Using the Taylor series to derive the torque increment, and the second and higher order terms are simplified according to the principle of small deviation. Further simplified as:

$$T_a\frac{d\varphi(t)}{dt} + T_g\varphi(t) = K_\eta\eta(t-\tau) - K_1\lambda(t) \tag{2}$$

T_a is the time constant of the diesel engine; $\varphi(t)$ is the speed of diesel engine; h(t) is rack position for controlling fuel supply; λ(t) is external disturbance; τ is ignition delay time; T_g is stable time constant for the diesel engine; K_η is the rack correction constant; K_1 is the disturbance correction factor. Using Laplasse transform:

$$\mathrm{T}_a\Phi(s)s + \mathrm{T}_g\Phi(s) = K_\eta e^{-\tau s}H(s) - K_1\Lambda(s) \tag{3}$$

In the equation, $e^{-\tau s} = \frac{1}{e^{\tau s}} \approx \frac{1}{1+\tau s}$, according to the technical data of marine diesel engine, set $T_a = 0.82$ s, $T_g = 1$, $\tau = 0.13$ s, $K_\eta = 1$, $K_1 = 1$. $\Phi(s)$ is expressed as fellow:

$$\Phi(s) = \left[\frac{1}{1+0.13s}H(s) - \Lambda(s)\right]\frac{1}{0.82s+1} \tag{4}$$

where $\Phi(s)$, $H(s)$, $\Lambda(s)$ is the Laplace transform of $\varphi(t)$, $\eta(t)$, $\lambda(t)$.

2.2 Actuator Mathematical Model

The speed regulator actuator used a reversible DC servo motor, which was adjusted to the rack position by mechanical transmission mechanism to adjust the amount of oil, and then to control the speed. And the motion of motor is shown below:

$$\begin{cases} L_d\frac{di_d}{dt} + R_d i_d + K_e w = u_d \\ J\frac{dw}{dt} + fw = K_m i_d \end{cases} \tag{5}$$

To simplify our modeling, neglecting the influence of inductance and equivalent viscous friction coefficient, then transformed as:

$$\frac{\Omega(s)}{U_d(s)} = \frac{K_m}{R_d J_s + K_m K_e} = \frac{1/K_e}{T_s s + 1} \tag{6}$$

The transfer function of the actuator can be obtained by correcting the series integral as:

$$\frac{H(s)}{H_g(s)} = -K'/K_e T_m \Big/ \left(s^2 + \frac{1}{T_m}s + \frac{K'}{K_e T_m}\right) = w_n^2/s^2 + 2\varepsilon w_n s + w_n^2 \tag{7}$$

where u_d/i_d is voltage/current of motor Armature; L_d/R_d is inductance/resistance of motor armature winding; w is the motor shaft angular velocity; K_e is the motor EMF constant; J is the equivalent rotating inertia of motor shaft; f is the Equivalent viscous friction coefficient; K_m/T_m is torque constant/time constant of motor; w_n/ε is undamped natural frequency, $w_n^2 = K'/K_e$; ε is damping coefficient, $2\varepsilon w_n = 1/T$.

By inquiring the technical documents of the ship and relevant technical data of ship, we set $\varepsilon = 0.707$, $T_m = 0.02$, it can make the governor achieve better dynamic response characteristics, and meet the requirements of high maneuverability and rapidity of marine engineering ship, the transfer function of the oil quantity control mechanism is as below:

$$G(s) = \frac{H(s)}{H_g(s)} = \frac{1250}{s^2 + 50s + 1250} \tag{8}$$

2.3 Interference Factors in Governor Control

The disturbance factor is very complex in the process of diesel engine governor control. How to consider the arrangement reasonably is the key point of controller design. Marine engineering ships often work in terrible sea conditions, the wind, wave, flow and other disturbances caused the ship to sway, swing, and even the propeller out of the water in rough weather. In the case of the engine load changed sharply, the engine speed would maladjusted if oil injection pump could not be adjusted in time. In view of the need for analysis and simulation of these factors, external disturbances are considered as an additional resistance to ship resistance [8].

The model deviation is the main source of uncertainty in the governing system, to facilitate modeling and analysis, it is inevitable to simplify or ignore dynamic segments that less important [9]. We focused on the propulsion system control performance and ignored the internal components of diesel engines. The governor was only composed of the governor model, the diesel engine mathematical model, the transmission device and the interference signal model.

3 Design of Active Disturbance Rejection Controller for Main Engine

Active disturbance rejection control (ADRC) is a control technique first proposed by Han Jingqing [8]. Based on the analysis of the control characteristics and ADRC of the governor of the high speed four stroke diesel engines, we designed the main engine governor based on ADRC.

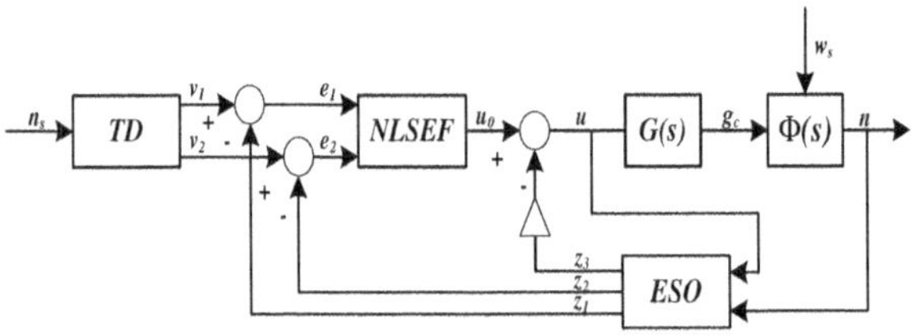

Fig. 2. The structure of ADRC governor

3.1 Principle of Active Disturbance Rejection Control

The following is a brief explanation, ADRC governor consists of Tracking Differentiator (TD), Extended state observer (ESO), Nonlinear state error feedback (NLSEF) (Fig. 2).

ADRC governor suppose that TD tracks expected speed n_s in system, V_1 is the trace of TD's transition process, v_2 is the derivative of v_1. The error signals and differential signals of the system are extracted in real time. ESO can reasonably estimate the state of the speed governing system, the model and the external disturbances, Z_1, Z_1 are ESO estimated value of state variables of speed governing system, Z_3 is the estimation of the unknown disturbance W_S, the problem of precise control of nonlinear uncertain objects with unknown external disturbances has been well dealt with, and making the ESO arrangement has a better effect on suppressing external interference [8]. NLSEF is used to compensate for the estimated total interference, and NLSEF is defined as follows:

$$u_0 = \beta_1 fal(e_1, \alpha_1, \delta) + \beta_2 fal(e_2, \alpha_2, \delta), \quad 0 < \alpha_1 < 1 < \alpha_2 \tag{9}$$

According to the system state error e_1 and e_2, the control rule u_0 of pure integrator series type object is determined, to improve the dynamic characteristics of the main diesel engine speed control system, the final control variable u is determined by the compensation control output disturbance compensation of the error feedback control variable u_0 and the disturbance estimation value Z_3, and $u = u_0 - \frac{z_3}{b_0}$, b0 is the proportional parameter that determines the effect of control. The oil actuator G(s) adjusts the throttle displacement L_s to change the diesel engine cycle fuel supply, and achieve that the main diesel engine Φ(s) is able to adjust the actual speed n.

3.2 Design of Active Disturbance Rejection Controller for Main Engine

In order to prevent the parameters of the ADRC from affecting the final stability, we took the governor and the main diesel engine mathematical model into consideration, the single loop ADRC was used to design the ADRC governor of the main diesel engine [9], the structure is as follows (Fig. 3):

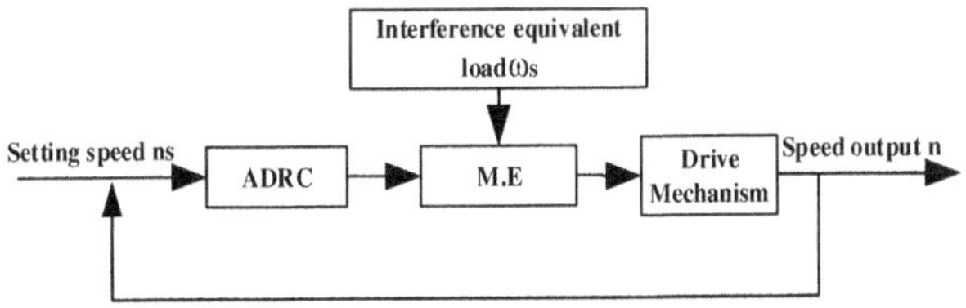

Fig. 3. Design of single-ring ADRC ship governor

Parameter tuning is the difficulty of the governor design, deciding whether the stability and dynamic performance of the governor reach the optimal effect. The adjustment followed the train of thought that the parameter tuning must be considered as an organic system, and the ADRC parts can not be taken apart. We needed to identify the control amount of the object, by observing the changes of some important parameters for the ADRC performance, in order to achieve the optimal control effect of the system [8, 9], we adjusted some performance parameters as dynamic response speed, tracking accuracy and static error. The specific adjustment method is as follows:

Step 1. Seted parameter δ_1 and α_1, α_2 in NLSEF to a smaller number, then adjusted δ roughly;

Step 2. The gain parameters which determine the effect of the ESO control, β_{01}, β_{02}, β_{03}, and β_1, β_2 in NLSEF are adjusted in detail, the adjustment process followed $\beta_{01} \cdot \beta_{02} > \beta_{03}$. According to the output tracking effect of Z_1, we gradually adjusted β_{01}, then by observing the differential effects of both tracking output of Z_1 and the tracking output of Z_2, the β_{02} was adjusted, the adjustment of the β_{03} was implemented to realize the real-time and accurate tracking of the state variables and disturbances of the speed regulation system by ESO. The adjustments of the proportion of parameters β_1, β_2 which determine the effect of control, were based on β_1 and coordinated by β_2.

Step 3. The other parameters of ESO and NLSEF were adjusted simultaneously, and the effect of tracking and output was observed until the satisfactory results appeared.

4 Simulation Experiment

The main technical parameters of the ship used in this paper are as follows (Table 1).

Table 1. Main technical parameters of ship

Rated speed	Speed	Displacement	Lpp	B	D	d	Continuous output	CB
750 rpm	18.28 km	3336t	82 m	22 m	9.5 m	7.7 m	8000 kw	0.55

According to the above parameters tuning method and strategy, and follow the "separation principle" to carry out a large number of simulation tests, ADRC governor of each part of the adjustment parameters are as follows:

TD: $r = 100$, $h_0 = h = 0.01$;

ESO: $\beta_{01} = 20$, $\beta_{02} = 25$, $\beta_{03} = 20$, $\delta = 0.02$;

NLSEF: $\beta_1 = 1.5$, $\beta_2 = 30$, $\alpha_1 = 0.75$, $\alpha_2 = 1.25$, $\delta_1 = 0.05$.

The Classic PID controller parameters: $K_p = 4.5$, $K_i = 3.5$, $K_d = 0.2$.

In order to verify the control effect of ADRC governor on the speed of the main diesel engine, the following simulation experiment was carried out in Simulink environment.

4.1 Step Response of a Governor

Compared with the classic PID controller, the main diesel engine ADRC governor to set the speed of 500 r/min low speed step response test, the test curve shown in below (Fig. 4):

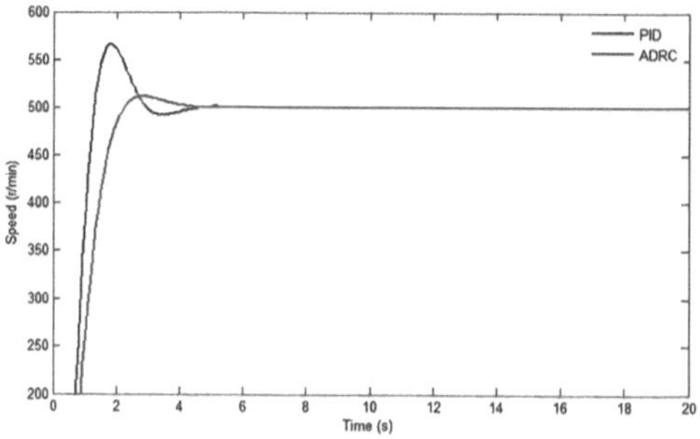

Fig. 4. Step response curve of the governor

By observing the simulation curve that the engine speed in the low speed region, the ADRC governor overshoot is only 1.6%, better than the classic PID controller generated by 13% overshoot, the response speed faster than the classic PID controller and it has a smooth completion of the speed adjustment process. Thus, in the low speed region of ADRC governor by the arrangement of TD allows the system to draw reasonable error signal and differential signal, which is a good deal in the speed control system overshoot and adjust the speed between the contradictions.

4.2 Dynamic Test of Governor

For the analysis of dynamic characteristics of the diesel engine speed control system, the simulation of normal speed governor experiment, we set the system start-up speed of 750 r/min, the speed dropped to 650 r/min at 2 s, when the system is stable at 10 s, the speed reaches 700 r/min, the control effect of comparative simulation curve as shown below (Fig. 5):

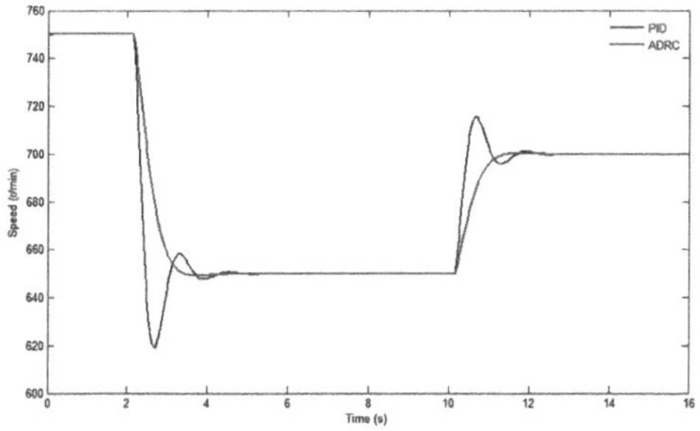

Fig. 5. Simulation curve of engine speed when the set speed changes

When the speed setting value changes firstly, the stability time of the classical PID control system is 3.6 s and ADRC is 2.6 s, the overshoot is 4.62% and the ADRC is only 0.21%; the second change, the PID control system stable time is 2.7 s and ADRC is 1.6 s, overshoot is 2.45% and ADRC is 0.43%. The speed control system with two changes of ADRC is more responsive, and the stability and dynamic performance of the speed control system are much better than those of the classical PID. Load test of governor

Considering the abrupt change of load in bad sea conditions, the load test of the governor is carried out to further verify the dynamic performance. The simulation curve is shown below (Fig. 6):

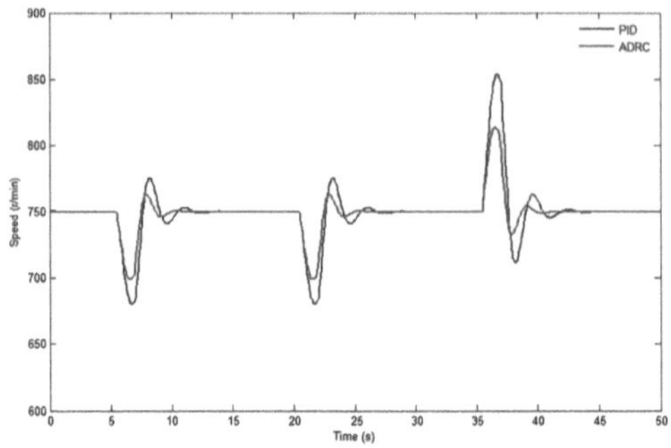

Fig. 6. Simulation curve of engine speed when load changes

The simulation curves, which are reflected by the dynamic testing process, have their dynamic performance indicators, as shown in below (Table 2):

Table 2. Dynamic performance indexes of loading and unloading

Controllers	Load 50% loads at the 1st time		Load 50% loads at the 2nd time		Unload 100% load	
	Overshoot	Stable time	Overshoot	Stable time	Overshoot	Stable time
PID	10%	6 s	10%	6 s	14.77%	9 s
ADRC	6.71%	4 s	6.71%	4 s	8%	5 s

Through the above analysis shows that the ship in the bad sea condition met the sudden increase of load two times, The speed of the main engine was better controlled by ADRC, and the stability of the ship governor was improved.

5 Conclusion

In order to handle the time variation, non-linearity and uncertainty of the speed control system of the main diesel engine, we designed an ADRC governor based on four stroke middle high speed diesel engine in this paper, and have comparison simulation test with classical PID control, the result showed that the effect of ADRC control is faster, smoother, higher precision, and improved the dynamic and steady characteristics of the speed control system of the main diesel engine. It solved the problem that the nonlinear uncertain objects with unknown external disturbances is difficult to be accurately controlled. So this model can better handle of overloading and speed of the main engine. To summarize, this work has some reference value for the design of the governor of four stroke middle high speed diesel engine.

Acknowledgements. The study was supported by "the Fundamental Research Funds for the Central Universities", (No. 3132016316). The authors also gratefully acknowledge the helpful comments and suggestions of the reviewers, which have improved the presentation.

References

1. Fang, M.C., Zhuo, Y.Z., Lee, Z.Y.: The application of the self-tuning neural network PID controller on the ship roll reduction in random waves. Ocean Eng. **37**(7), 529–538 (2010)
2. Han, H.C., Hong, M.Y., Yong, K.: Implementation of evolutionary fuzzy PID speed controller for PM synchronous motor. IEEE Trans. Ind. Inf. **11**(2), 540–547 (2013)
3. Gu, L.L., Zhang, J.D.: Simulation of diesel engine electronic governor based on fuzzy PID controller. J. Dalian Marit. Univ. **40**(1), 57–60 (2014)
4. Shi, Y., Zhang, L.Y., Sun, J., Zhang, H.G.: Research on the speed of diesel engine based on improved BP neural network controller. Appl. Mech. Mater. **281**(1), 105–111 (2013)
5. Huan, T.U., Hui, C.: Modeling and simulation of a large marine diesel propulsion system. Ship Build. China **55**(2), 158–167 (2014)
6. Reza, C.M.F.S., Islam, M.D., Mekhilef, S.: A review of reliable and energy efficient direct torque controlled induction motor drives. Renew. Sustain. Energy Rev. **37**(3), 919–932 (2014)
7. Han, J.: Auto disturbances rejection control technique. Front. Sci. **01**, 24–31 (2007)
8. Ran, M., Wang, Q., Dong, C.: Stabilization of a class of nonlinear systems with actuator saturation via active disturbance rejection control. Automatica **63**, 302–310 (2016)
9. Han, J.: From PID to active disturbance rejection control. IEEE Trans. Ind. Electron. **56**(3), 900–906 (2009)

Modular Management of Safety and Civilization Construction Based on Power Grid Management

Wu Moxuan, Fu Xinyong, Yang Zhenwei, Hu Shihao, and Kang Mei(✉)

State Grid Jiu Jiang Electric Power Company, Jiujiang, China
120162144@qq.com

Abstract. The safety civilization construction is an important indicator for measuring network, this paper mainly in view of the current power grid safety civilization construction of the status quo, put forward the thinking of modularization management, and application verification in XX power grid construction, achieved good effect.

Keywords: Modular management · Safety and civilization construction
Power grid management

1 Introduction

With the rapid development of power grid construction, new higher requirements are put forward for the construction safety of power grid. At present the main problems is still a construction unit standardization construction of safety civilization seriously insufficient, less investment, production and procurement on a standardized safety civilization facilities do not have a unified standard, large difference between two different construction unit management level, the lack of communication each other good experience and methods can't get the promotion in time, safety disease cannot be effectively suppressed. Therefore, it is necessary to standardize the construction of safety civilized construction site according to the modular design, unify the safety and civilization standardization facilities, decorate the technological level, and improve the construction safety level.

2 Modular Management Method

2.1 Overall Planning

The new construction of 110 kV substation is a kind of single project in the construction of infrastructure projects. It is of great importance to carry out the standardization management of safety and civilization construction. XX power supply company around 110 kV substation comprehensive implementation of the newly-built project office, living area, construction area of standardized construction goal, to sum

K. Deng et al. (Eds.): ICMIR 2018, AISC 856, pp. 474–478, 2019.
https://doi.org/10.1007/978-3-030-00214-5_60

up the experiences of site construction management and compiled the 110 kV substation project office area and construction area modular design manual, follows a modular concept created configuration table, contains the main construction method of different type of operation area layout principle, covering the whole process of substation construction.

XX power supply company "clear responsibilities of each unit joining into the management flow chart, according to the system requirements, the prophase planning, implementation, management and inspection appraisal three phases, is divided into 18 process, standardize the ground.

2.2 Zoning Plan, Clear and Modular Configuration

(1) The construction management unit shall, in the contract, list the construction costs of the project safety and civilization, and provide the basic conditions for the construction of the safety and civilization of the project;
(2) the owner project department prepare the project safety management of the overall planning, clear objectives and safety civilization construction management requirements, to modular design of office area, construction area, and to the contractors clarification, site supervise and urge the contractors to carry out the modular design.
(3) the supervision project department shall prepare the work plan of safety supervision, and clarify the construction control measures and key points of the safety civilization;
(4) construction projects in accordance with the owner project department prepare the safety management of the overall planning of the project, the preparation of substation construction safety management and risk control scheme, the construction enterprise audit related functional departments, in charge of the leadership approval, reported to the supervision of the project department review, the owner project department approval and organize their implementation;
(5) to prepare the report plan of safety civilization construction standardization facilities in part of the construction project;
(6) the supervision project department shall examine and approve the safety and civilized construction standardization facilities report plan in part of the construction project;
(7) the project department of the owner shall examine and approve the approval of the construction project in part of the construction project approved by the supervisor;
(8) the construction management unit shall pay the construction fee of the security civilization according to the contract.

2.3 Step-by-Step Implementation, Standardized Application Process

(1) The construction project department shall submit safety facilities, safety protective equipment and civilized construction facilities according to the approval plan of the safety and civilization construction standardization facility;

(2) the supervision project department shall check and accept the entry of the standardization facility for safety and civilization construction;
(3) the owner's project department shall review and accept the entry of the standardization facility for safety and civilization construction;
(4) the construction project department shall arrange special personnel to register and manage the safety and civilized construction standardization facilities, and set up the standard facilities for the construction of safety and civilization construction, and register the accounts of the issuance and inspection tests; Before the construction work, the construction team shall install the standardized facilities for the safety and civilization construction according to the requirements, collect the safety protective equipment and distribute it to the construction personnel.

2.4 Classification Management and Strict Evaluation and Assessment

(1) According to the actual demand of current construction procedure and safety civilization construction, construction management units and the contractors monthly inspection on site safety civilization construction facilities configuration, control configuration table, type and quantity, diagnostic facilities to verify whether can meet the requirements, if you need to add or remove, by the construction projects in accordance with the procedures apply to the supervision, construction project department. Process inspection, construction management unit to the organization supervision, construction units carefully verify the usage of all kinds of facilities cost, purchasing new facilities shall be provided by the construction projects for the purchase invoice and corresponding facilities list. Application of digital photos to confirm the pilot results, work site project department in accordance with the general part of the overlay, project department, production area, the isolation facilities on-site, and civil engineering construction area, the construction bracket construction area, cable channel construction area, electric equipment installation construction area different assignments Sect. 2 photo, respectively, such as named according to the work content, classification and archiving.
(2) check whether there is any need for rectification according to the inspection or inspection results; For the problems that need to be rectified, the closed-loop rectification is completed within the specified time;
(3) the responsible units shall make rectification and closed-loop for the failure to meet the requirements of modular configuration;
(4) the construction management unit shall set up an evaluation team, mainly with the members of the project department of the owner, and appropriately draw the supervision and other project department personnel, and carry out the evaluation once a month. To organize the evaluation of project safety management; The evaluation results are combined with supervision evaluation, regular safety quality inspection and random inspection, etc., and the construction cost of construction safety and civilization is paid according to the score of each project.

3 The Effect of Modularization Management

3.1 Standardizing the Site Safety and Civilized Construction Facilities Configuration

Through the office area, construction area and living area of modular design, general layout, site layout and standardize the construction area, the office space area, function room set, room internal fixation, card and other style and standard, and matches by the visual image design to achieve the visual image to the overall effect of the uniform, neat, beautiful. We have unified the standard facilities for safety and civilization and the level of layout, and achieved standardized management of safe and civilized construction. The modular layout of the office area of 110 kV substation is shown in Fig. 1.

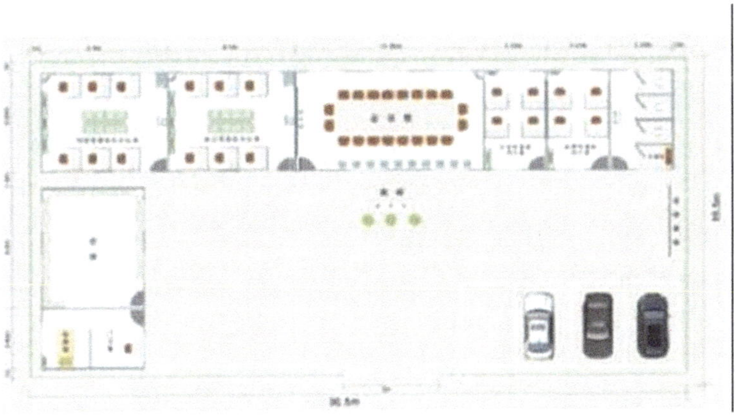

Fig. 1. Modular layout of the office area of 110 kV substation

3.2 The Cost of Project Investment is Saved

Since the modular design is adopted, the standardized configuration of the safety civilization construction of each type of engineering (transformer engineering or line engineering) is uniform, so it can be copied. A after the completion of the project can take the engineering project of whole configuration is transferred to another, to implement the standardization district civilized construction configuration of reuse, can save a large amount of engineering investment cost.

4 Conclusion

Power grid enterprises through safety civilization construction modular management, the standard of the scene safety civilization construction facilities configuration, save engineering investment cost, for enterprise's safety civilization construction has made great contribution.

References

1. Kim, S.R., Wang, Z.J.: On the accuracy and efficiency of calculation with respect to the grid construction methods for unstructured meshes. Soc. Sci. Med. **58**(8), 1523–1530 (2004)
2. Solano, A., Solano, J.: Method of floor construction with a grid system, US (1998)

Research on the Evaluation of Middle-Level Cadres' Job Performance in Universities

Wenxiong Chen[1,2(✉)] and Ge Zhou[3]

[1] Business School, Central South University, Changsha, People's Republic of China
chenwx@csu.edu.cn

[2] Business School, Hunan Normal University, Changsha, People's Republic of China

[3] College of Computer Science, National University of Defense Technology, Changsha, People's Republic of China

Abstract. With the rapid advancement of universities, middle-level cadres became the key force in promoting the construction and reform of universities in China. Middle-level cadres were formally appointed and assessment by the Organization Department. The popular method to evaluate the middle-level cadres is to checking the middle-level cadres' ideological-political moral quality and their job performance. In this paper, based on the considerable quantities of documents and the results from previous studies, combined with the current situation of university administrators, we aims to set up a framework for middle-level cadres' performance evaluation with fuzzy AHP assessment method, and to design an evaluation system with the corresponding index from five aspects of administrative personnel "ethics, ability, diligence, achievement, honesty". It is foreseeable that this work has important theoretical and practical significance for the improvement of the management level of universities.

Keywords: Fuzzy AHP · Middle-level cadres · Comprehensive evaluation Performance

1 Introduction

Most recently, to enhance the comprehensive strength in the university system, most universities and colleges have devoted great efforts to construction and reform, therefore, high quality management is crucial to the development of universities. In China, middle-level cadres became the key force in promoting the development of universities and colleges, and it is necessary to assess job performance of middle-level cadres. In consequence, how to optimization job performance assessment approach, then develop middle-level cadres' managerial ability and efficiency, which is important to promote the development of universities. The traditional assessments are difficult to quantify, and the weight of indexes is usually decided randomly. In evaluative methods, it is necessary to improve the rules by means of optimizing the assessment valuing rules, adopting information-based assessment platform and etc. Consequently, it is necessary to construct a new means to evaluate the performance of middle-level cadres in universities.

K. Deng et al. (Eds.): ICMIR 2018, AISC 856, pp. 479–486, 2019.
https://doi.org/10.1007/978-3-030-00214-5_61

In the past period of time, evaluating job performance has received great attention in academia, and fuzzy techniques and AHP approach have been widely applied. He et al. [1] put forward a method for job performance assessment with the AHP (analytic hierarchy process) theory, which be appropriated for the work level of middle-level cadres. Considering the peculiarities of evaluation levels of human resources management in universities and the shortage of those methods and with the use of multi-layer fuzzy theory, to ally with AHP analysis, a job performance evaluation model of the multivariate goal of middle-level cadres' job performance is established.

Fuzzy set theory is an effective method, especially to solve uncertain and imprecise data, and this method has been widely adopted by researchers in various fields to settle multifarious decision-making matters. In this paper, we strive to build a more effective and accurate performance appraisal index system for middle-level cadres. With respect to five indexes such as "ethics, ability, diligence, achievement, honesty", we qualitatively and quantitatively design sub-indexes, and ascertain the weight of a factor in connection with the multiple-objective decision-making issue. Satty present a novel decision-making approach, which relate to AHP [2]. As a special application of AHP process, fuzzy AHP method can effectively solve multi-level fuzzy decision-making matters. Gungor et al. [3] have created an employee selection platform to assess the most suitable employee, which uses fuzzy AHP to deal with qualitative and quantitative indicators. In this kind of platform, by using fuzzy analytic hierarchy process (AHP), they provide an appropriate index weight system for constructing an effective assessment tool. Therefore, in order to evaluate middle-level cadres' performance, we built a framework with the use of fuzzy AHP and fuzzy synthetical assess approach, to assure the precise of novel assessment system, construct the ability level assessment index system of middle-level cadres in universities.

2 Preliminaries

2.1 On the Basic Definition of Fuzzy Sets

A fuzzy set is a set of elements, and these elements have membership grade. Zadeh [4] proposed fuzzy sets theory as an expand of standard concept of set, subordination function, fuzzy sets theory have been applied recently in diverse domain, for instance, making decision or cluster analysis. In standard knowledge system of set, the role of factors belongs to a set is assessed by analyzing the approach of binary detection in an accordance with a bivalent state, in other words, whether a factor belongs to sets. Different from that, the theory of fuzzy set enables the progressively measurement of the body of factors within sets. In classical theory about fuzzy set, bivalent sets are always named crisp sets. A fuzzy set $\tilde{A} = \{(x, \mu_{\tilde{A}}(x)|x \in X\}$ is a general set of consecutive elements and X as the sub-aggregate of the real numbers R, with $\mu_{\tilde{A}}(\mathrm{x})$ is named the subordinate function and it appoints every element x the rank with body extent from 0 to 1.

The classical theory of fuzzy be put to use in different fields in which information is uncompleted or unprecise. The fuzzy number is a peculiar kind of a fuzzy set of convex normalized. Triangular and trapezoidal fuzzy numbers (TFNs) have been generally used as dispose of the unclearness for finding a solution. On account of its intuitionistic emerge and high execution speed, the trapezoidal fuzzy number has been far-flung applied special function within heaps of fields. TFNs have been generally applied to get the uncertain of the indicators to be bound up with the decision proposing process. On purpose of showing the vagueness that circles the conductors while decision makers manage a pair right comparison matrix, trapezoidal fuzzy number has been appeared by boundaries to substitute for crisp numbers. TFNs denoted as $A = (l, m, u)$, with the membership function as shown below:

$$\mu_{\tilde{A}}(x) = \begin{cases} \frac{x-l}{m-l} & x \in [l, m] \\ \frac{u-x}{u-m} & x \in [m, u] \\ 0 & otherwise \end{cases} \tag{1}$$

The parameter "m" should been the most likely date, the parameter "l" should been the most unlikely value, and the parameter "u" should been the largest likely date. When $l = m = u$, the TFNs becomes a non-fuzzy number.

Consider two trapezoidal fuzzy numbers $\tilde{A}_1$ and $\tilde{A}_2$, $\tilde{A}_1 = (l_1, m_1, u_1)$ and $\tilde{A}_2 = (l_2, m_2, u_2)$. The primary executive rules [5] with two TFNs $\tilde{A}_1$ and $\tilde{A}_2$ are as follows:

$$\tilde{A}_1 \oplus \tilde{A}_2 = (l_1 + l_2, m_1 + m_2, u_1 + u_2) \tag{2}$$

$$\tilde{A}_1 \oplus \tilde{A}_2 = (l_1 l_2, m_1 m_2, u_1 u_2), l_i > 0, m_i > 0, u_i > 0, i = 1, 2 \tag{3}$$

$$\frac{\tilde{A}_1}{\tilde{A}_2} = \left(\frac{l_1}{u_2}, \frac{m_1}{m_2}, \frac{u_1}{u_2}\right), l_i > 0, m_i > 0, u_i > 0, i = 1, 2 \tag{4}$$

2.2 Designing Index System for Performance Evaluation

Through exploration and practice, we design the evaluation index system comply with decision-making process as follows.

2.2.1 Determining the Hierarchical Model

In our study, we develop the hierarchical structure for evaluation in consideration of the clearly defined factors, and then, a team of decision-makers usually established by organization department in universities in China. Those who participate in the decision discussion must be experienced experts, such as administrative in university, and they must make sure the fuzzy relative weights of all the factors.

2.2.2 Discussion on the Scale of Linguistic Variables and Fuzzy Transformation

Determining the importance ratings of characteristics is a group decision-making process with outstanding characteristics. The application of the language method can be availably command the imprecise data in quality function preparation and help decision proposing in deriving the importance ratings. The conductors set up pair factors for the weight among every couple of indexes. In the pair comparison of each level, it can be separated but $\frac{n(n-1)}{2}$ mensuration (n factors in total). It is further applied to constructing positive fuzzy reciprocal comparison matrix. In our work, TFNs are applied to describe subjective comparison of pairs. In order words, "just equal", "equally important", "weakly more important", "strongly more important", "very strongly more important" and "absolutely more important". C. Kahraman et al. [6] proposed the theory of trigonometric fuzzy transformation rules and language rules, which has been applied to convert fuzzy from linguistic date, just like shown in Table 1.

Table 1. Linguistic scales and fuzzy scales for importance

Linguistic scale for importance	Triangular fuzzy scale	Triangular fuzzy reciprocal scale
Just equal	(1, 1, 1)	(1, 1, 1)
Equally important (EI)	(1/2, 1, 3/2)	(2/3, 1,2)
Weakly more important (WMI)	(1, 3/2, 2)	(1/2, 2/3, 1)
Strongly more important (SMI)	(3/2, 2, 5/2)	(2/5, 1/2, 2/3)
Very strongly more important (VSMI)	(2, 5/2, 3)	(1/3, 2/5, 1/2)
Absolutely more important (AMI)	(5/2, 3, 7/2)	(2/7, 1/3, 2/5)

A problem with n factors is involved, in which the comparative significant of index i with j is expressed by TFNs $\tilde{a}_{ij} = (l_{ij}, m_{ij}, u_{ij})$. Just as the conventional AHP approach, with the TFNs comparison matrix $\tilde{A} = \{\tilde{a}_{ij}\}$ could been presented as:

$$\tilde{A} = \begin{bmatrix} 1 & \tilde{a}_{12} & \dots & \tilde{a}_{1n} \\ \tilde{a}_{21} & 1 & \dots & \\ \dots & \dots & \dots & \dots \\ \tilde{a}_{n1} & \tilde{a}_{n2} & \dots & 1 \end{bmatrix} \tag{5}$$

2.2.3 On the Determination of Consistency Index and Consistency Ratio

Saaty [5] constructs an index that can effectively measure consistency, which could be to put in use with express the coherence within pair comparison matrix. In order to make an in-depth study of their consistency, the fuzzy comparison matrix is transformed into a novel matrix [7]. In this paper, we apply the theory presented by Chang et al. [9] in order to get a crisp number out of the triangular fuzzy number. We can calculate the coherence factors of the comparison matrix through the following equations.

$$CI = \frac{\lambda_{\max} - n}{n - 1} \tag{6}$$

The ultimate eigenvalue of the comparison matrix λ_{max} is consistent with the dimension of the matrix n. At the same time, we clearly define a ratio (CR). This ratio is the value among the consistency of the determined metrical matrix and the stochastic matrix.

$$CR = \frac{CI}{RI(n)} \tag{7}$$

where $RI(n)$ is a stochastic factor [10] that be limited to the scale of matrix n, has been presented in Table 2. In case the CR is equal or less than 0.1, it can be acceptable.

Table 2. Random index (*RI*) of random matrices

N	3	4	5	6	7	8	9
RI(n)	0.58	0.9	1.12	1.24	1.32	1.41	1.45

3 On the Establishment of Evaluation System of Index

As administrative staff, the middle-level cadre is a special group in universities, for the reason of the difficulty of the project and not easy to measure, make the work level evaluation is difficult. In the design of the performance evaluation index system, it should be judged from different perspectives, dimensions and levels. Only in this way can the performance of middle-level cadres be revealed effectively. A wide, effective source of raw data helps to improve the objectivity of the evaluation conclusions. In our work, the main sources of the data are school leaders, managers, teachers, students and middle-level cadres. In our study, the index and sub-index been adopted that deliberated and determination by experts and implemented by Organization Department in university. At length, based on the established index system, we construct the hierarchical structure of the evaluation system, which presented in Table 3.

The index system contains five aspects: "ethics, ability, diligence, achievement, honesty", Every last factor can be broken down into three or more than three sub factors. To set up a job performance evaluation method, the weights of all aspects should be taken seriously, denoted by weight coefficient, as shown in the table.

The Research scenario of this thesis is about the job performance evaluation of middle-level cadres in Chinese university. In order to determine the weight of factors (include subfactors), a team of decision-making conductor was established.

We obtain experts' opinions by making and distributing questionnaires. After that, in order to gain the typical comparison matrix from the decision team, we apply the geometric mean method, as they are presented in Table 4, and then calculated the value of one entry of the comparison matrix on the priority stabilization and consistency ratio. With the Eq. (6), the CR value is less than 0.1 within the comparison matrix, consequently, it can be acceptable.

Table 3. The index system of middle-level cadres' job performance

Main Factors	Sub-factors
Ethics (U_1)	Political stand (U_{11})
	Theoretical study (U_{12})
	Professional ethics (U_{13})
	Spirit of devotion (U_{14})
	Work style construction (U_{15})
Ability (U_2)	Learning ability and policy level (U_{21})
	Find, analyze and solve problems ability (U_{22})
	Organization & coordinating ability and execution ability (U_{23})
	Innovation ability (U_{24})
	Ability of expression (U_{25})
	Adaptation ability (U_{26})
Diligence (U_3)	Dedication and sense of responsibility (U_{31})
	Working attitude (U_{32})
	Work efficiency (U_{33})
Achievement (U_4)	Workload (U_{41})
	Quality of work (U_{42})
	Creative work situation (U_{43})
	Great achievements or progress (U_{44})
	Job satisfaction (U_{45})
Honesty (U_5)	Observe the relevant code of conduct and the law (U_{51})
	Put into effect the responsibility system (U_{52})
	Personal integrity image (U_{53})

Table 4. Comparison matrix of the factors

Index	U_1	U_2	U_3	U_4	U_5	Weight coefficient
U_1	1.000	2.000	3.000	3.000	3.000	0.350
U_2	0.500	1.000	1.500	1.500	1.500	0.250
U_3	0.333	0.666	1.000	1.000	1.000	0.133
U_4	0.333	0.666	1.000	1.000	1.000	0.133
U_5	0.333	0.666	1.000	1.000	1.000	0.133

(CI = −5.20005E−06; CR = −4.45987E−06)

The next step we make comparisons all sub-factors related to the factors which belongs to the up stage, to get the outcome and the concrete display of the comparison matrix in Tables 5, 6, 7, 8 and 9. After that, The CR determination values of the comparison matrix below are ascertained by the similar method as Table 4. According the calculation data we found that "ethics" is the most important. As a result, the "ethics" has one of the most important effect on the job performance of middle-level cadres, then the "ability".

Table 5. Comparison matrix of the sub-factors within "ethics"

Index	U_{11}	U_{12}	U_{13}	U_{14}	U_{15}	Weight coefficient
U_{11}	1.000	3.000	1.000	2.000	3.000	0.325
U_{12}	0.333	1.000	0.333	1.500	1.000	0.125
U_{13}	1.000	3.000	1.000	2.000	3.000	0.325
U_{14}	0.500	1.500	0.500	1.000	1.500	0.155

Table 6. Comparison matrix of the sub-factors within "ability"

Index	U_{21}	U_{22}	U_{23}	U_{24}	U_{25}	U_{26}	Weight coefficient
U_{21}	1.000	0.500	1.000	1.000	1.500	1.500	0.162
U_{22}	2.000	1.000	2.000	2.000	3.000	3.000	0.268
U_{23}	1.000	0.500	1.000	1.500	1.500	1.500	0.162
U_{24}	1.000	0.500	1.000	1.500	1.500	1.500	0.162
U_{25}	0.666	0.333	0.666	1.000	1.000	1.000	0.101
U_{26}	0.666	0.333	0.666	1.000	1.000	1.000	0.101

(CI = 3.33332E−06; CR = 2.68816E−06)

Table 7. Comparison matrix of the sub-factors within "diligence"

Index	U_{31}	U_{32}	U_{33}	Weight coefficient
U_{31}	1.000	3.000	1.000	0.560
U_{32}	0.333	1.000	1.500	0.284
U_{33}	1.000	0.666	1.000	0.173

(CI = 0.062525; CR = 0.098546)

Table 8. Comparison matrix of the sub-factors within "achievement"

Index	U_{41}	U_{42}	U_{43}	U_{44}	U_{45}	Weight coefficient
U_{41}	1.000	0.333	0.666	1.000	1.000	0.125
U_{42}	3.000	1.000	2.000	3.000	3.000	0.405
U_{43}	1.500	0.500	1.000	1.500	1.500	0.225
U_{44}	1.000	0.333	0.666	1.000	1.000	0.105
U_{45}	1.000	0.333	0.666	1.000	1.000	0.105

(Cl = −1.60011E−05; CR = −1.33987E−05)

Table 9. Comparison matrix of the sub-factors within "honesty"

Index	U_{51}	U_{52}	U_{53}	Weight coefficient
U_{51}	1.000	1.000	2.000	0.411
U_{52}	1.000	1.000	2.000	0.411
U_{53}	0.500	0.500	1.000	0.178

(CI = −1.20005; CR = −0.455617)

4 Conclusion

The middle-level cadre assessment has been one of the concentrated problems in universities for a long time. Although the traditional assessment is easily operated, the official of the Organization Department's subjective bias and error is inevitable, and actually tends to be not objective. In our work, with the useful of fuzzy AHP and comprehensive evaluation approach, we try to put forward a performance assess factor for middle-level cadres, and establish a performance evaluation system, and carry out case analysis. From the study and practice process, it is indisputable that fuzzy analytic hierarchy process can effectively judge the fuzziness of plan formulation. The purpose is to advance the objectivity and rationality of the weight. It could be anticipated that this kind of method can make the assessment of middle-level cadres' job performance in universities and colleges far more satisfactory. With the advancement of high education and the deepening of the reform, evaluation index system for middle-level cadres' performance and the framework in this paper will meet the requirements for the management of personnel assessment work standardization, scientific to some extent.

References

1. He, X.N., Zhu, Z.Q., Zhou, Y., Lu, G.Y., Liu, Q.Y.: University teaching quality evaluation using fuzzy comprehensive evaluation approach. In: Proceedings of the Second International Workshop on Education Technology and Computer Science, pp. 616–619 (2010)
2. Saaty, T.L.: The Analytic Hierarchy Process: Planning, Priority Setting, Resource Allocation. McGraw-Hill International Book Co., New York (1980)
3. Gungor, Z., Serhadlioglu, G., Kesen, S.E.: A fuzzy AHP approach to personnel problem. Appl. Soft Comput. **9**, 641–646 (2009)
4. Zadeh, L.A.: Fuzzy set. Inf. Control **8**(3), 338–353 (1965)
5. Kaufmann, A., Gupta, M.M.: Introduction to Fuzzy Arithmetic: Theory and Applications. Van Nostrand Reinhold Co., New York (1991)
6. Kahraman, C., Ertay, T., Buyukozkan, G.: A fuzzy optimization model for QFD planning process using analytic network approach. Eur. J. Oper. Res. **171**, 390–411 (2006)
7. Chen, C.F.: Determining the attribute weights of professional conference organizer selection: an application of the fuzzy AHP approach. Tour. Econ. **17**(5), 1129–1139 (2011)
8. Bok, D.C.: Higher Learning. Harvard University Press, Cambridge (1986)
9. Chang, C.W., Wu, C.R., Lin, H.H.: Applying fuzzy hierarchy multiple attributes to construct an expert decision-making process. Expert Syst. Appl. **36**, 7363–7368 (2009)
10. Golden, B.L.: The Analytic Hierarchy Process: Applications and Studies. Springer, New York (1989)

The Design of On-Line Monitoring System of Ventilators

Yueyue Tuo(✉), Gaili Yue, and Sai Du

Xi'an University of Science and Technology, Xi'an, China
1021252491@qq.com

Abstract. The coal mine main fan with the name "mine mine lung" is one of the key equipments in coal mine production. Once the main fan's operating conditions have occurred, it will cause huge losses in the lives of underground personnel and state assets. Therefore, it is necessary to carry out effective automatic monitoring of the coal mine main fan. In this thesis, the CC2530 chip is used as the core processor to build the hardware platform, Zigbee wireless technology and bus data transmission technology as the signal transmission method, and Labview as the monitoring software of the upper computer to realize the real-time monitorings of mechanical vibration, temperature, voltage and current of the main fan, as well as the storage of historical data and query printing. The system has theoretical significance and practical application value. It improves the automation degree of coal mine main fan safety monitoring, and has practical significance and certain application prospects for reducing the occurrence of mine safety accidents.

Keywords: Ventilator · CC2530 · ZigBee · Labview

1 Introduction

The coal mine safety production monitoring system is a coal mine safety production environment monitoring system centered on microcomputers. It is mainly used for real-time monitoring and data preprocessing of safety production information in the coal mine underground environment, fire, ventilation facilities, transportation safety and other aspects. This article uses sensor technology, signal processing technology, computer technology, data communication technology, etc., to monitor the parameters of bearing temperature, mechanical vibration, voltage and current of coal mine main ventilators in real time. At the same time, it also ensures that the equipment can be highly efficient and economical. Steady and continuous operation, and improve the automation of the control of the stable operation of large-scale equipment.

This design utilizes Zigbee wireless technology based on CC2530 and bus data transmission technology to design a low-cost, high-efficiency online monitoring system for coal mine main fan operation status, which solves the real-time monitoring and display problems of coal mine main fan operation status and completes the main The basic technology and debugging of the short-distance wired and wireless transmission and reception, acquisition, transmission and processing of the fan operation status signal will perform system networking between the terminal node and the coordination

K. Deng et al. (Eds.): ICMIR 2018, AISC 856, pp. 487–494, 2019.
https://doi.org/10.1007/978-3-030-00214-5_62

node. The basic technology and debugging of the short-distance wired and wireless transmission and reception, acquisition, transmission and processing of the ventilator operation status signal, and the system networking between the terminal node and the coordination node. The coordinating node communicates with the host computer through RS-232 and uploads the received terminal node data to the host computer. Through the debugging simulation of the system, the feasibility of the system and the correctness of the research results are verified.

2 The Design of On-Line Monitoring System of Ventilator

2.1 Overall Design of Ventilator On-Line Monitoring System

The overall architecture of the system is shown in Fig. 1. It mainly consists of data nodes consisting of data acquisition and sensing modules, power supply modules, and data processing and wireless transmission modules. It consists of a data wireless receiver module, sound and light alarm, and a liquid crystal display module. The coordinating node will upload the data received by the coordinating node to the upper computer in a wired manner.

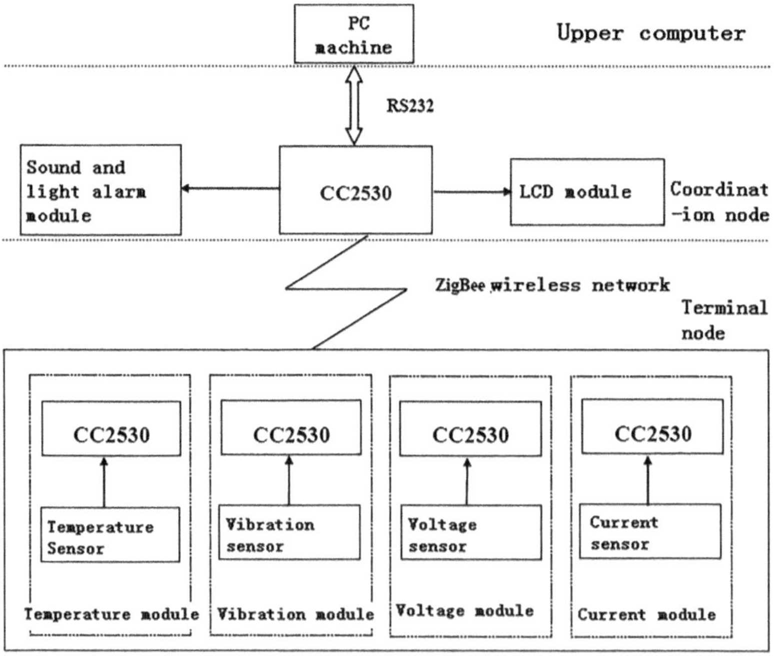

Fig. 1. Overall system design

2.2 Terminal Node Composition

The terminal node consists of a sensor module, a CC2530 RF module, and a power module. The block diagram of the terminal node hardware design is shown in Fig. 2. The sensor module is responsible for monitoring the collection of information in the area. This design uses a single channel to collect analog signals and uses three CC2530 core boards to process and store the analog signals of temperature, vibration, voltage, and current, respectively. The CC2530 RF module is responsible for wireless data communication with the primary coordinator CC2530 for wireless data communication and packaging and sending the collected data. The power module is divided into 5 V power supply and 3.3 V power supply. The 5 V power supply is for the analog sensor module, while the 3.3 V power supply is the temperature module and CC2530 RF module provide the energy required for operation.

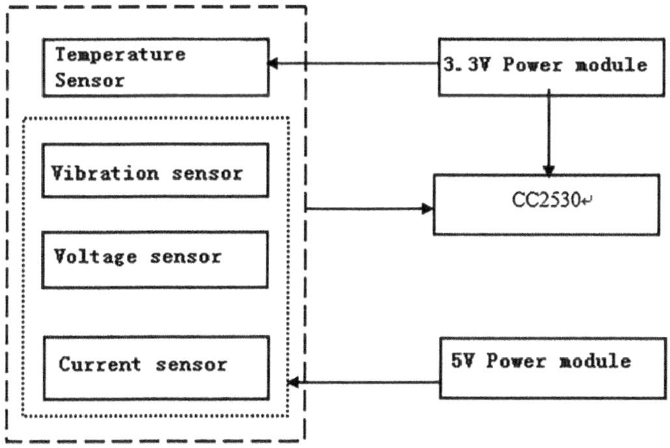

Fig. 2. Block diagram of a terminal node

2.2.1 Analog Sensor Module

The analog sensor circuit diagram for voltage, current and vibration is mainly used to replace the voltage, current value and bearing vibration speed value of the fan with the analog voltage of 0–3 V DC voltage. The output of the sensor signal is simulated by changing the output voltage amplitude. The analog circuit is shown in Fig. 3. The corresponding range of each parameter of the analog sensor module is shown in Table 1.

2.2.2 Temperature Sensor Module and Terminal CC2530 Module

A DS18B20 temperature sensor was used to collect bearing temperature. Use three CC2530 to receive the data collected by the sensor, perform signal processing, and send data to the coordinator. Temperature and vibration use the same CC2530 for data transmission. Each voltage and current uses a CC2530 for data transmission.

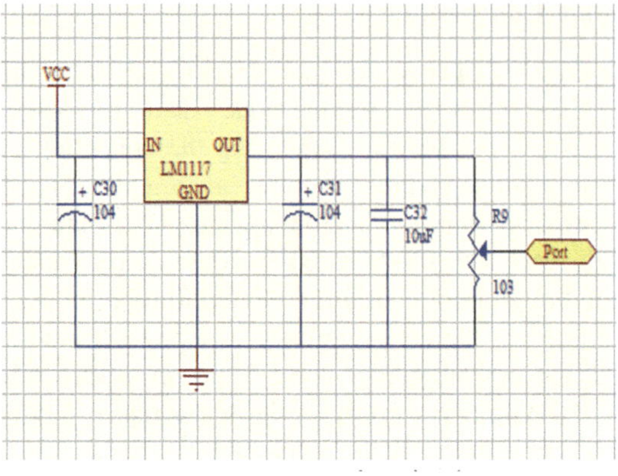

Fig. 3. Sensor analog circuit

Table 1. Sensor parameter table

Sensor	Measurement range	Output voltage signal
Vibration	0–5 mm/s^2	0–3 V
Voltage	0–220 V	0–3 V
Current	0–6 A	0–3 V

2.3 The Design of CC2530-Coordination Node

The hardware design of the coordinating node consists of CC2530 RF module, USB to serial communication module, audible and visual alarm module, and liquid crystal display module. The design block diagram is shown in Fig. 4.

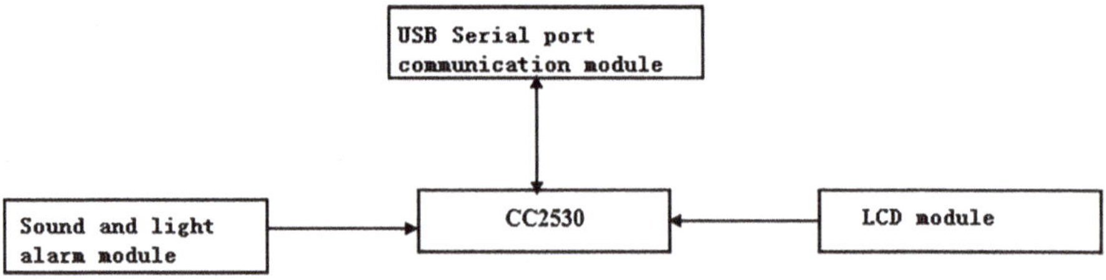

Fig. 4. Coordinated node design block diagram

2.3.1 Serial Communication Module

The system adopts PL2303 chip, realizes the conversion from UART serial port to USB, so as to achieve the function of coordinating the data transmission between node and host computer.

2.3.2 Sound and Light Alarm Module

The system adopts 8550PNP type triodes and light-emitting diodes in parallel to achieve sound and light alarms.

2.3.3 LCD Module

The liquid crystal display module adopts 12864 display screen, adopts the serial port driving method, and is connected with the coordinator CC2530. It mainly realizes the real-time display of signals received by CC2530.

2.3.4 Coordinating the CC2530 Module

Coordinator CC2530 is used to receive the signal sent by the terminal, as a data transfer station, and then send the signal to the host computer.

2.4 Upper Computer Interface Preparation

The preparation of the upper computer program includes two parts: the design of the front panel and the design of the block diagram. The upper computer preparation work block diagram is shown in Fig. 5.

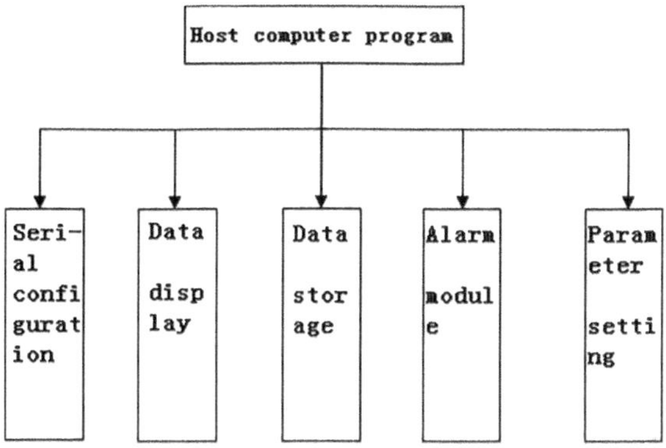

Fig. 5. Upper computer work block diagram

Click to open the Labview software. The front panel is divided into three parts by system configuration, measurement parameter monitoring and historical data records. The system configuration is the configuration of the serial port parameters, including COM port selection, baud rate, data bits, parity bits, and stop bit settings. Measurement parameter monitoring includes graph display of temperature module, air pressure module, vibration module, voltage module and current module, setting of upper and lower limits of parameters and alarm module. Historical data records include historical data records, date and time.

3 System Debugging and Results

First, the four terminal acquisition modules are sequentially powered, and the multimeter is used to monitor the output voltage value transmitted to the module. Then download the entire system program to CC2530 separately. After the download is successful, debug each module to achieve a successful networking. Finally, it is the data reception and display test of the upper computer.

The upper computer interface is written by Labview software. First configure the serial port address parameters. After the correct configuration, the Labview software can run smoothly, as shown in Fig. 6.

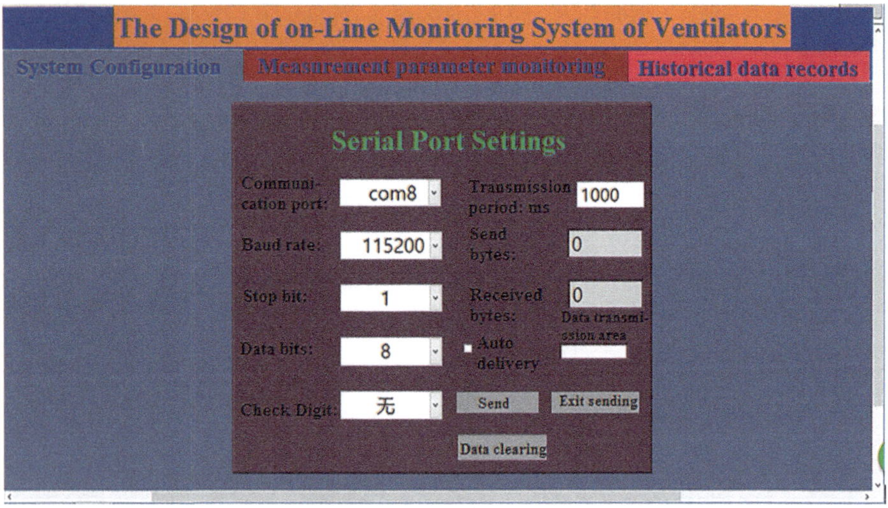

Fig. 6. Serial port configuration

After the serial port is configured, each measurement parameter can be displayed correctly through the display on the measurement parameter monitoring interface. Set the upper and lower limits of each parameter. When the parameter is out of range, the alarm light will be on. If it is not out of range, the alarm light will be gray, as shown in Fig. 7.

In order to facilitate future data analysis, the collected history is saved in the PC Excel table by writing the hard disk, as shown in Fig. 8.

After the overall debugging of the system's terminal nodes, coordinating nodes and host computer, the parameters of the main ventilator of coal mines are collected, sent, displayed and stored to realize on-line monitoring of the main ventilators of coal mines.

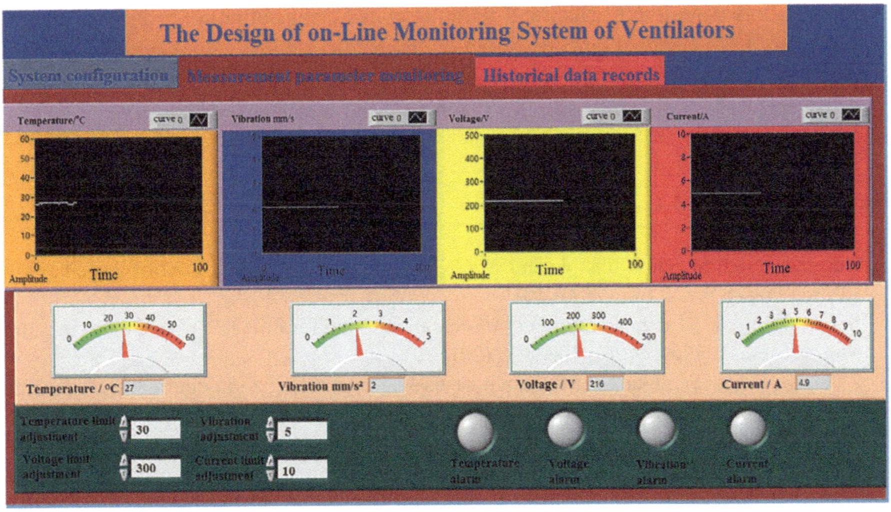

Fig. 7. Display of each parameter

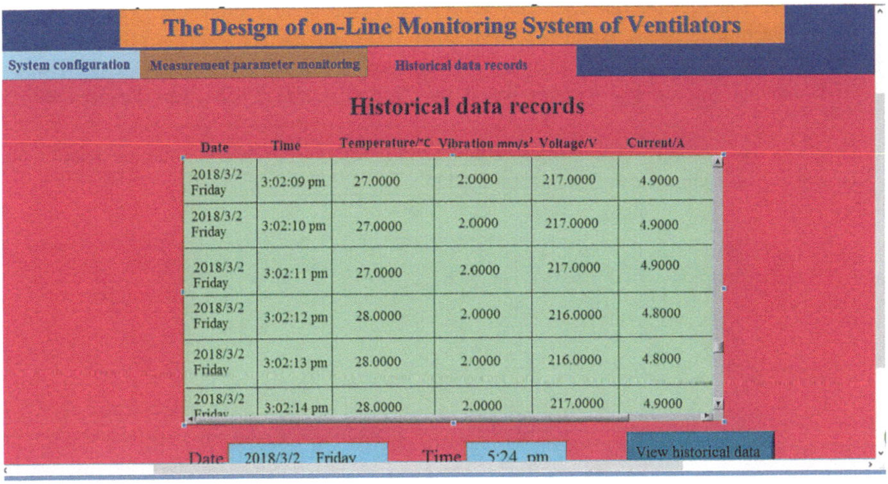

Fig. 8. Historical data storage interface

4 Conclusion

Based on the research of wind turbine engineering, this study designed a kind of on-line monitoring system for coal mine main fan based on CC2530 core Zigbee technology. The system has a strong electromagnetic anti-jamming capability and high reliability. Its functional design is comprehensive and practical, the human-machine interface is simple and clear, and the operation flow is simple and convenient. Its comprehensive monitoring information, monitoring parameters of the motor running state, vibration parameters, electrical parameters, bearing temperature and so on. The system also has a self-diagnostic function that can identify abnormalities such as sensors and databases, facilitate user maintenance, a comprehensive online help system and user manuals, and guide users to better use and operating systems.

References

1. Guo, X., Yang, S.: The Principle and Application of Monitoring and Control System. China Electric Power Press, Beijing (2010)
2. Shi, L., Wang, L., Tang, H.: Design of on-line monitoring for flow rate of main fans in mines. Min. Process. Equip. **39**(06), 103–106 (2011)
3. Zhao, S., Dong, W., Guo, G.: Research of on-line monitoring and control system for mine main fan. Coal Mine Mach. **35**(03), 195–197 (2014)
4. Tu, Y.: Reconstruction and realization of control system for fan in mine. Coal Min. Mach. **35**(03), 147–148 (2014)
5. Wang, Q.: Application of fieldbus and wireless transmission technology. In: Proceedings of the 17th China Electro-Dust Research Conference, vol. 17, pp. 581–591 (2017)
6. Liu, Y.: Automation control system based on ZigBee wireless network technology. Electron. Technol. Softw. Eng. **10**(03), 21–22 (2017)
7. Wang, S., Wu, C.: Design and realization of environmental monitoring system based on CC2530. Comput. Meas. Control **23**(08), 2650–2653 (2015)
8. Wang, X., Yang, R., Guo, C., Zhang, P.: Design and application of wireless sensor network environmental monitoring system based on ZigBee. Chin. J. Electron Devices **40**(03), 760–765 (2017)
9. Long, Y.: Temperature monitoring system design based on Labview simulation. South China University of Technology, Guangzhou (2011)

The Research and Practice of Online Learning Intervention Based on Big Data

Fenyan Guo(✉)

Hunan Electrical College of Technology, Hengyang, China
642102199@qq.com

Abstract. Based on the analysis of research related to learning intervention at home and abroad, this paper combines the characteristics of online learning to explore online learning interventions under big data, and proposes suggestions for online learning interventions and online learning intervention processes. On the basis of this theoretical basis, "E-commerce website construction" online learning was token as an example to practice online learning intervention. From all of this, the aim of this paper is to improve learners' online learning performance, and to provide reference for the study and practice of online learning intervention.

Keywords: Big data · Online learning · Intervention

1 Introduction

Big data has provided unprecedented space and potential for enhancing in-depth and comprehensive insights, affecting people's knowledge structure and behavior. In recent years, the national policies to promote online education have been introduced one after another. Online learning has become an important learning method in the information age. MOOCs, SPOCs, and flipped classrooms have become popular online learning models. Although online learning has developed rapidly, there are also problems such as the limitation of individual efficiency due to individual differences in learning, the need to improve the ability of individualized dynamic teaching adjustments, and the lack of timely management and evaluation of online learning. To solve the above problems and further improve the effectiveness of online learning and improve the quality of online learning, this article starts with education data and studies online learning interventions.

2 Research Status at Home and Abroad

2.1 Current Status of Domestic Research

From the research of China Knowledge Network literature, there are many studies on learning intervention in education and education, and a lot of theories and experiences have been accumulated.

K. Deng et al. (Eds.): ICMIR 2018, AISC 856, pp. 495–501, 2019.
https://doi.org/10.1007/978-3-030-00214-5_63

In the traditional teaching field, learning intervention has always existed as a conventional concept, which includes all interventions that have an impact on learners' learning. However, with the development of learning and analysis techniques, learning interventions based on education big data need to be different from the intervention methods in the traditional teaching environment. Therefore, the concept of learning intervention once again aroused the attention of educational technology researchers. Based on the analysis of the characteristics of the interventions in education and teaching, Zhang Chao defined the learning interventions in the distance learning environment as follows [1]: "Learning intervention is a learning service provider that aims to improve learner learning performance and solve learning problems. The ultimate goal of integrating various indirect access strategies and behaviors is to help learners develop specific knowledge, skills, and attitudes." Chen Shan pointed out [2] that learning intervention is based on the appearance of learners These kinds of difficulties and problems are more targeted for the various types of support they provide, including resources and activities. In the study of learning intervention model construction based on big education data [3], Li also defined learning intervention as "in order to help learners overcome learning difficulties and successfully complete learning, based on the analysis of educational big data based on the learning process. The combination of various supportive strategies and instructional activities implemented for each learner's specific learning status."

Chen Shan applied the intervention design to the third grade primary school science class, and confirmed that the intervention design has a positive effect on the cultivation of students' problem solving skills. Chen Li from the School of Foreign Languages at Xi'an Jiao tong University built an IRT based adaptive reading intervention system for English reading. Tang Li designed the whole process of intervention mechanism [4] in the study of the intervention mechanism based on learning analysis in the smart learning environment.

2.2 Status of Foreign Research

Compared with domestic countries, foreign countries have relatively early intervention studies on learning and analysis, and there are many empirical researches, and there are successful cases. For example, Purdue University's "Signal Project" is dedicated to learning management systems, curriculum management systems, and course grade books. Information is collected to classify dangerous students, and the hazard levels are indicated in green, yellow, and red, and individual guidance is given to students in a "dangerous" state [5]. Northern Arizona University's evaluation performance status system can collect the performance rating of learners in the classroom and send it to learners via e-mail. E-mails mainly include attendance status, academic performance and academic problems [6]. The U.S. SHERPA system can provide students with advice on time management and course selection through analysis of student information; there is evidence that student performance has improved in some cases [7]. But overall, most of the interventions are teachers' manual interventions, and there are few interventions for system automation. Even if there is automatic intervention, the relationship between the content provided and the learning content is not particularly close, and more time and courses are needed. The proposal does not provide individualized interventions for the learner's learning content.

In summary, most studies of domestic learning interventions are focused on theory. There are few studies on the practical application of interventions, and there is a relative lack of relevant research data, and there is less practice of online learning intervention based on big data. Although there are many empirical researches on foreign learning interventions, most of them have not used big data yet, and the interventions have not been personalized enough. Under the educational big data environment, it will be a meaningful attempt to conduct online learning intervention research and practice.

3 Online Learning Intervention Based on Big Data

3.1 Research on Online Learning Intervention

For the range of online learning interventions, the interventions were divided into three types: single intervention, classification intervention, and overall intervention. In different learning situations, combined with system intervention and human intervention, different intervention methods were selected for combination. Before the initial learning, due to the lack of learner's personal learning data, the overall intervention was selected. In the online learning, through the data analysis and the recognition of the learner's state type, the classification intervention and human intervention were selected; after the online learning was completed, the learner was increased on the previous basis. Individual intervention; After collecting the learner-related data for the first time learning, add a single intervention before each learning, and implement different intervention strategies according to the characteristics of the students; use the system intervention at the same time throughout the learning process to perform real-time learning for learners. Learning status feedback. As shown in Fig. 1.

3.2 Design of Online Learning Intervention Process

The online learning intervention process is divided into six important steps. First, online learner basic data, behavioral data, interaction data, performance data, and sentimental data are collected based on certain platforms and tools, and then learners are "learned through behavior analysis techniques". Learning status in terms of "style type", "level of learning progress", "level of learning interaction", and "level of academic achievement" is used to determine whether to intervene and return to the first step of data collection if no intervention is required, otherwise In order to determine the mode of intervention, choose the appropriate intervention method and then implement an intervention strategy for the online learner. Finally, summarize and analyze the effect of the intervention, and feedback the learning situation to the learner, and then begin to enter the next round of intervention process. As shown in Fig. 2.

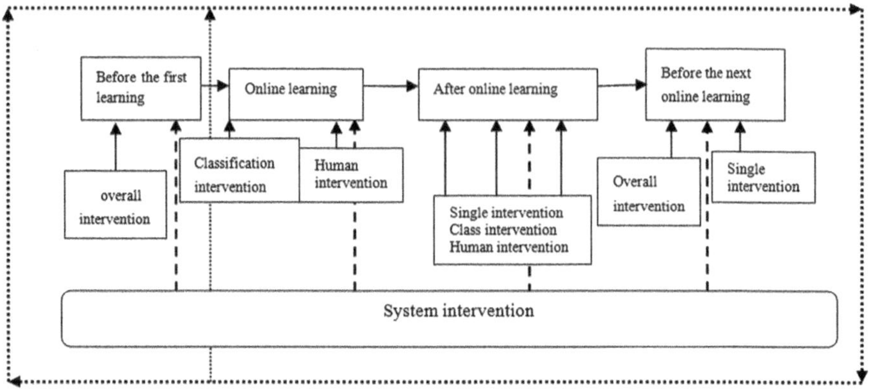

Fig. 1. Combination of interventions in the online learning process

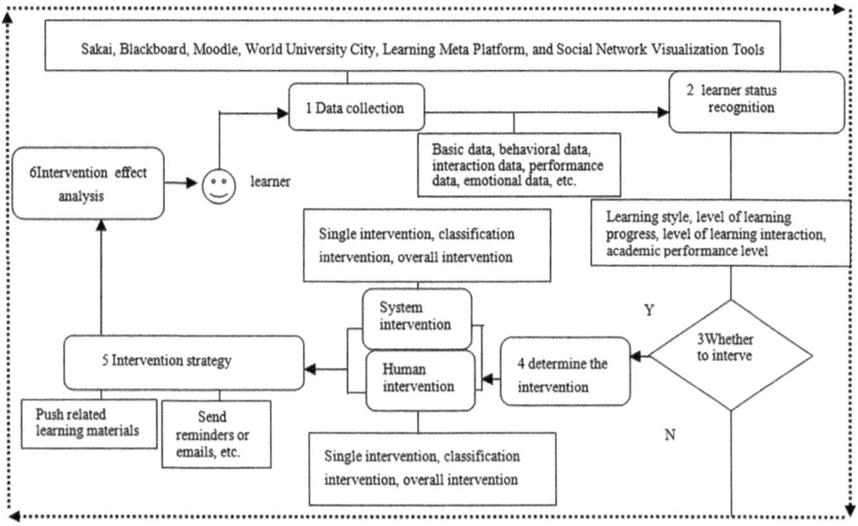

Fig. 2. Online learning intervention process based on big data

3.3 Online Learning Intervention Based on Big Data

Taking the "World University City" space course "e-commerce website construction" as an example, the practice of online learning interventions.

The first step: collecting data. In cooperation with the World University City, collect online learner-related learning data by relying on visual tools such as university city basic application tools, teaching and learning tools, homework test tools, social management tools, and educational entertainment tools.

The second step: recognizing learner status. Statistical analysis of e-learner-related learning data in "e-commerce website construction", and recognition of learner status through knowledge and behavior analysis techniques. As shown in Table 1.

Table 1. Level of learner status recognition

Data Collection		Learner status level recognition									
Learning style scale measurement data, learner browsing record data	The Analyze Technique of Knowledge and Behavior	Learning style type	Visual type			Hearing type			Athletic	Comprehensive	
Learning path, learning content		Learning level	Fast			Medium			Slow		
Interaction frequency, interactive content, social network relationship data		Interaction level	Frequency of Interaction			Depth of interaction			Range of Interaction		
			high	Medium	low	deep	general	shallow	wide	general	narrow
Academic data, incorrect data, and learning duration data		Academic achievement level	The overall level			Learning weaknesses					
			high	Medium	low	Knowledge point 1, 2, 3, …,n					

Intervention strategy design and implementation: Based on recognition of learner learning status, based on analysis engine technology, learners are effectively intervened to help learners adjust learning status to a more optimized level, as shown in Table 2.

Table 2. Intervention strategy design and implementation

Learner status	Intervention strategy	Intervention timing	Intervention method
Learning style type	Push learning materials presented by different organization methods or different media according to learning style characteristics	Learners learn the new knowledge module	Teachers send advice emails or push learning materials through University City social management tools

(*continued*)

Table 2. (*continued*)

Learner status	Intervention strategy	Intervention timing	Intervention method
Learning level	Automatically pop up visual page window, emails or messages to remind learners of slow progress through the system	Learners throughout the learning process	System alerts through University City
Learning interaction level	Encourage learners by visual email or message reminders, homosexual partner referrals, teacher conversations, etc., reminding learners who have low levels of interaction	Learner's learning process, especially the learning modules with high requirements for interaction	Through the University City system messaging or teacher interviews
Academic achievement level	Recommend learning methods to learners with a low level of learning achievement, and push learning resources and instructional suggestions for weak knowledge based on students' weak points of learning	After the learner completes a stage of learning	Teachers send reminder emails or suggestions

4 Comment

There are many researches on learning intervention in education and teaching, but there is not much research on learning intervention based on big data. There is no application of big data in online learning intervention. Through online learner-related learning data analysis, targeted interventions are a new application of big data in learning interventions.

Acknowledgements. This research is founded by Hunan Provincial Education Science Planning subject in 2017, No: XJK17BZY007.

References

1. Yang, X., Wang, L., Tang, S.: Application models and policy suggestions of education big data. Electr. Educ. Res. **09**, 54–61+69 (2015)
2. Zhang, Y.: Research on Application of Big Data in Education Field. East China Normal University, Shanghai (2016)
3. Jin, L.: Big data and information technology teaching reform. China Electr. Educ. **10**, 8–13 (2013)
4. Jiang, Q., Zhao, W., Wang, P., Wang, L.: Personalized adaptive online learning analysis model based on big data and its implementation. China Electr. Educ. **01**, 85–92 (2015)
5. Wu, F., Yan, Z.: Construction and realization of student personalized analysis model based on big data in electronic schoolbag. China Electro Tech. Educ. **03**, 63–69 (2014)
6. Zhao, H., Jiang, Q., Zhao, W., Li, Y., Zhao, Y.: An empirical study of online learning performance early warning factors and intervention strategies based on big data learning analysis. Audio Vis. Educ. Res. **01**, 62–69 (2017)
7. Macfadyen, L.P., Dawson, S.: Mining LMS data to develop an "early warning system" for educators A proof of concept. Comput. Educ. **54**, 588–599 (2010)

Image Segmentation of CV Model Based on Curve Area Constraint

Jianfeng He, Guanhua Guan, and Sanli Yi(✉)

School of Information Engineering and Automation, Kunming University of Science and Technology, Kunming, China
120112624@qq.com

Abstract. Traditional CV models do not take into account the area energy terms inside and outside the evolution curve for active contour segmentation, This slows down the convergence rate of the evolution curve. In this paper, the energy function in the curve is constructed for the CV model problems. An edge function is constructed by using Gauss convolution kernel function, And the edge energy is used to construct the area energy in the evolution curve, the convergence rate of evolution curve is improved by the method proposed in this paper. The experimental results show that, the model proposed in this paper can automatically detect the inner contour of perforated targets, it is robust to the shape and position of the initial contour, and it can quickly calculate the global minimum value of the image. For single target and multi target images, The method of this paper can quickly and accurately segment the image. Compared with the traditional Chan-Vese model, the LBF model and the LIF model, Using the method of this paper to segment images satisfies real-time requirements.

Keywords: Chan-Vese model · Area energy term · Gaussian kernel function Central difference method · Image segmentation

1 Introduction

Image segmentation plays an important role in computer vision, medicine, aerospace and other fields. Image segmentation technology refers to the extraction of regions of interest in images, in which snake model is widely applied to extract image object contour, the snake model [1] is the deformation evolution curve under the action of image force and external constraint, and transforms the image segmentation problem into the mathematical problem of finding the minimum value of the energy function. This transformation makes the automatic segmentation of images possible.

The active contour model is sensitive to the location of the initial contour. After years of development, the researchers have proposed many improvements to the active contour model. According to the way of fitting image information, it can be roughly divided into two categories: One is a model based on the boundary information of the curve [2–4]; Liu et al's CV model based on KL distance weighting and local neighborhood information [5], and a method of satellite remote sensing cloud image segmentation based on edge corrected CV model proposed by Yu et al. [6] are belonged to active contour model based on boundary information; active contour model based on

K. Deng et al. (Eds.): ICMIR 2018, AISC 856, pp. 502–509, 2019.
https://doi.org/10.1007/978-3-030-00214-5_64

curve boundary information uses the boundary information of evolution curve to guide the evolution of curves, it has a good segmentation effect for low noise and strong boundary images. The other is a model based on the image region information [], such as Chan, Vese et al's active contours without edges [8] and Cai et al's [9] proposed adaptive level set image segmentation based on local and global information are belonged to active contour model based on image region; they use the global information of the image to separate the contour of weak boundary and uneven gray image. The CV model not only overcomes the problems that exist in the Snake model,but also the introduction of level set method in CV model enables it to automatically handle topological changes in curve evolution. However, the CV model is only applicable to image segmentation with only two different mean areas including target and background, CV model is not ideal for segmentation of uneven gray images. Also due to the introduction of level set method, CV model needs continuous re initialization in the segmentation process, this greatly reduces the efficiency of the image segmentation algorithm.

In this paper, the area energy function of the fitting curve is proposed to speed up the evolution of the curve. In the process of curve evolution, the closed evolution curve divides the image into two parts, the inside and outside of the curve, with the iteration of the algorithm, the energy of the area inside and outside the closed curve will change accordingly, When the evolution curve reaches the edge of the target contour of the image, the area energy inside the closed curve tends to be stable. Experimental results prove that when the CV model of area energy entry is added to separate single target and multi-target images, it has a higher evolutionary efficiency, and it is robust to the position and shape of the initial contour.

2 CV Model

For the image u_0, it includes two regions of the target R_o and the background R_B, it is assumed that the gray mean values of target R_o and background R_B regions are c_1 and c_2 respectively. The energy in the corresponding region is E_o and E_B, respectively. Now Giving a closed contour curve $C(t) = \{(x, y)\} \emptyset(t, x, y) = 0\}$, the closed curve C divides the image u_0 into the internal region R_o' and the outer region R_B', The energy of internal area R_o' and external area R_B' is E_o' and E_B' respectively. According to the principle of CV model, $E_o < E_o'$, $E_B < E_B'$, the energy of target area and background area always tends to be minimum. the energy function is established to evolve the closed curve according to this. Therefore, the energy functional of the following CV model is defined:

$$F(C, c_1, c_2) = \mu * L(C) + v * S(C) + \lambda_1 * \int_{inside(c)} |u_o - c_1|^2 dxdy + \lambda_2 * \int_{outside(c)} |u_o - c_2|^2 dxdy \tag{1}$$

The first item L(C) in the formula (1) is the length constraint of the curve C to smooth the evolution curve C, the second item is the area constraint of the curve C, its function is to accelerate the evolution of the curve C; the third and fourth items are called fidelity term, they are responsible for attracting the evolution curve C to the outline of the target, $\mu \geq 0$, $v \geq 0$; $\lambda 1, \lambda 2 > 0$ are the weight coefficients of energy terms inside and outside the curve. Only when the evolution curve C is located at the boundary of the two homogeneous region, the energy functional F(C, c1, c2) gets the minimum. Therefore, by minimizing the formula (1), the boundary of the target object can be obtained.

3 CV Model Based on Edge Detection

According to the description above, the essence of the CV model is to establish a continuous mathematical model based on the gray information of the image, using partial differential equations (PDE) related to image gray information to drive the evolution of curves, when the solution of the equation tends to be stable, the curve divides the target area R_o and background area R_B of the image. The main reason for curvilinear motion is that the area $R_o^{'}$ in the curve is incomplete the image target area R_o, As long as there are differences between them, the curve will move; When the closed curve does not reach the edge of the target, The gray variance $E_o^{'}$ and $E_B^{'}$ inside and outside the curve will drive the curve to the edge of the target; In the process of moving the curve to the target edge, the area in the curve and the area of the target area are also different, this is also an important factor in promoting curvilinear motion. In this paper, we propose to combine the local energy of the point on the closed curve to construct the area energy S_{inside} inside the curve, in order to speed up the evolution of the curve, and enhance the segmentation effect of the CV model to the uneven gray image.

It is assumed that the neighborhood of point i of the image is Ω_i, $\Omega_{insidei}$ and $\Omega_{outsidei}$ are The internal and external regions of the neighbourhood Ω_i, respectively. Then, the internal area term $S_{\Omega_{insidei}}$ of the neighborhood Ω_i can be expressed as following:

$$S_{\Omega_{insidei}}(\varnothing) = \int\limits_{\Omega_i} g * H(\varnothing(x, y))dxdy \tag{2}$$

g in formula (2) is an edge detection function, $H(\varnothing(x, y))$ is a Heaviside function, $\varnothing(x, y)$ is a level set function. Due to the calculation of the area of the interior area $S_{\Omega_{insidei}}(\varnothing)$ of point i, $S_{\Omega_{insidei}}(\varnothing)$ only takes into account the local energy of the point near the curve, Therefore, in the formula (2), we need to add Dirac function $\delta(\varnothing)$ to get the area energy inside the whole curve. The energy of the inner area of the curve is as follows:

$$S_{inside} = \int\limits_{\Omega} \delta(\varnothing) * S_{\Omega_{insidei}}(\varnothing)dxdy = \int\limits_{\Omega}\int\limits_{\Omega_i} \delta(\varnothing) * g * H(\varnothing(x,y))dxdy \tag{3}$$

In summary, the final energy equation is:

$$\begin{aligned} F(C, c_1, c_2) = \mu * \int\limits_{\Omega} g * \delta(\varnothing)|\nabla\varnothing|dxdy + v * \int\limits_{\Omega} \delta(\varnothing)S_{\Omega_{insidei}}(\varnothing)dxdy \\ + \lambda_1 * \int\limits_{inside(c)} |u_o - c_1|^2 dxdy + \lambda_2 * \int\limits_{outside(c)} |u_o - c_2|^2 dxdy \end{aligned} \tag{4}$$

According to Euler Lagrange formula, the curve evolution equation is deduced as follows:

$$\begin{aligned} \frac{\partial\varnothing}{\partial t} = \delta(\varnothing)\left[\mu * div\left(g * \frac{\nabla\varnothing}{|\nabla\varnothing|}\right)\right] + v * \int\limits_{\Omega} \delta(\varnothing) * g * H(\varnothing(x,y))dxdy \\ - \lambda_1 * (u_o - c_1)^2 + \lambda_2 * (u_o - c_2)^2 \end{aligned} \tag{5}$$

u_o in formula (5) is the source image, $div\left(\frac{\nabla\varnothing}{|\nabla\varnothing|}\right)$ represents the divergence of the level set function, $\delta_\varepsilon(\varnothing)$ is a Dirac function. To the divergence calculation for the level set function, Dirac function is derivative of Heaviside function, and $H(\varnothing(x,y))$ *and* $\delta(\varnothing(x,y))$ are calculated sa follows:

$$H(\varnothing(x,y)) = \frac{1}{2}\left(1 + \frac{2}{\pi}\arctan\left(\frac{\varnothing}{\varepsilon}\right)\right) \tag{6}$$

$$\delta(\varnothing(x,y)) = \frac{1}{\pi} * \frac{\varepsilon}{\varepsilon^2 + \varnothing^2} \tag{7}$$

In formulas (6) and (7), ε is a constant, and ∅(x, y) is level set function, in the course of curve evolution, the calculation formulas for the gray mean values a and b inside and outside the curve are as follows:

$$c_1(\varnothing(x,y)) = \frac{\int\limits_{\Omega} u_o(x,y)H(\varnothing(x,y))dxdy}{\int\limits_{\Omega} H(\varnothing(x,y))dxdy} \tag{8}$$

$$c_1(\varnothing(x,y)) = \frac{\int\limits_{\Omega} u_o(x,y)(1 - H(\varnothing(x,y)))dxdy}{\int\limits_{\Omega} 1 - H(\varnothing(x,y))dxdy} \tag{9}$$

In Eqs. (8) and (9), u_o is an image function and Ω is the entire image area. During the curve evolution, the length of the evolution curve is calculated by:

$$L(C) = \int_{\Omega} g * \delta(\varnothing) * |\nabla \varnothing| dxdy \tag{10}$$

In formulas (10), $|\nabla \phi|$ is the gradient of the level set function.

4 Experiment and Result Analysis

The experimental environment in this paper is done on Matlab R2014a, the CPU is used as Intel(R) Core (TM) i7-4790 CPU @3.60 GHz RAM 8.00 GB, the operating system is Windows7. In order to fully verify the feasibility of this model in improving image segmentation efficiency, the steps of the implementation of the model level set algorithm in this paper are showing follow:

(a) Initialize level set function $\varnothing_0(x)$ according to formula (11);

$$\varnothing_0(x,y) = \left\{ \begin{array}{ll} -\rho, & \varnothing(x,y) < 0 \\ 0, & \varnothing(x,y) = 0 \\ \rho, & \varnothing(x,y) > 0 \end{array} \right\} \tag{11}$$

(b) Initialize the relevant parameters, time step and weight coefficient of the level set algorithm.
(c) if $|F_{n+1} - F_n| < \eta(\eta \ is \ a \ very \ small \ number)$, then the algorithm ends; otherwise, the following steps (d), (f), (e) are executed iteratively;
(d) According to the Eqs. (7) and (8), we calculate the gray mean value inside and outside the curve respectively;
(e) calculating the area energy in closed curve according to Eq. (3);
(f) Updating the level set function according to the Eq. (5);
(g) End of the algorithm, output image segmentation result.

4.1 First, Verify the Robustness of the Shape of the Initial Active Contour

As shown in the figure below, the source image gives a lung image, the initial contour of different shapes are given in turn, but no matter how the initial outline changes, it can finally converge the outer edge of the image and the edge of the inner contour. It proves that this method is robust to the shape and position of the initial contour. The segmentation results of lung images converge to the edge of target contour after 5 iterations (Fig. 1).

Secondly, in order to verify the advantages of the proposed model compared with the CV model, the advantages of LBF and LIF models in segmentation speed, Aiming at the segmentation experiment of natural and synthetic images, the number of iterations of the model and the running time of CPU are used as evaluation criteria (Fig. 2).

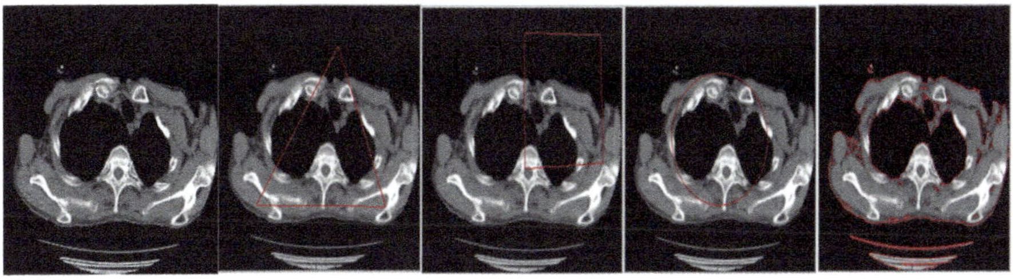

Fig. 1. The robustness of the initial contour shape, from left to right, the above pictures are Source image, Triangular contour, Rectangular contour, Circular contour, Segmentation result.

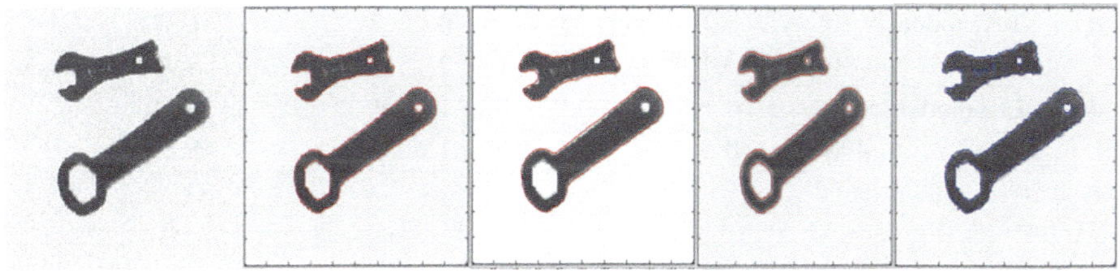

Fig. 2. From left to right, the above pictures are source image, CV model segmentation results, LBF model segmentation results, Text model segmentation results and LIF model segmentation results, respectively

The image given above is a multitarget image with a hole, Compare text model with CV model, text model and CV model not only divide the external contour edges apart, but also segment the edge of inner contour; however, the efficiency of the proposed model is more obvious than that of the CV model; yet, the LBF model can only segment the outer rim of the target, the edge of the inner contour of the target is not separated. For the LIF model, it divides the inner contour edge and the outer contour edge, but it falls into the local minimum value (Fig. 3).

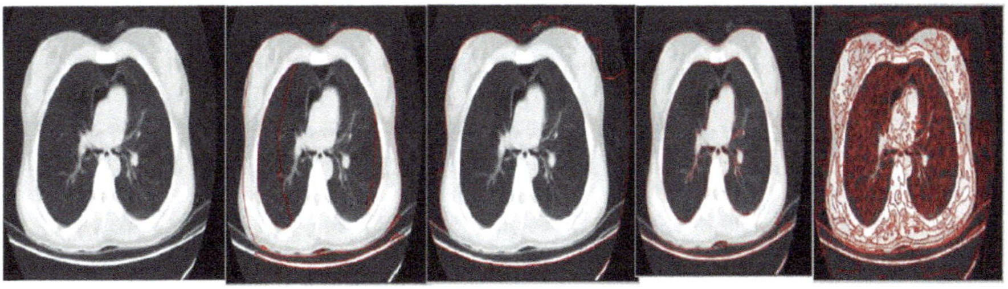

Fig. 3. From left to right, the above pictures are source image, CV model segmentation results, LBF model segmentation results, Text model segmentation results LIF model segmentation results, respectively

The above results give a more complicated lung image, for a complex image, the CV model fails to split the edge of the target. And the segmentation results for the LBF model, it also separates the background. The LIF model failed to divide the contour of the lung image, for the model proposed in this article, it split the outline of the lungs. the following Table 1 gives the parameters setting of each model.

Table 1. Comparison of four models

Models	Iterations	μ	ν	Time-step (Δt/s)	CPU time(s)
CV model	20	0.01 * 255 * 255	0	0.1	1.600
	200	0.01 * 255 * 255	0.2	0.1	51.844
LBF model	300	0.04	1.5	5	6.407
	500	0.04	1.5	4	105.767
Text model	10	0.01 * 255 * 255	1	0.1	1.371
	30	0.01 * 255 * 255	1	2	5.912
LIF model	80	0	0	1	3.010
	400	0	0	1	45.779

5 Summary

The main work of this paper is to use the edge point information of the evolution curve to construct the area energy in the curve. The traditional CV model relies on the grayscale variance inside and outside the curve to drive curvilinear motion, this based on the CV model, an area item S_{inside} is added, the function of the area term S_{inside} is to speed up the speed of evolution of the curve, this can be seen from the experimental comparison of Table 1. Using the model of this paper to segment images, the running time of the algorithm and the number of iterations of the curve convergence have been significantly reduced.

References

1. Kass, M., Witkin, A., Terzopoulos, D.: Snakes: active contour models. Int. J. Comput. Vis. **1** (4), 321–331 (1988)
2. Han, B., Wu, Y., Song, Y.: Segmentation of early fire image of mine based on improved CV mode. J. China Coal Soc. **42**(6), 1620–1627 (2017)
3. Shahverdi, R., Tavana, M., Ebrahimnejad, A., et al.: An improved method for edge detection and image segmentation using fuzzy cellular automata. Cybernet. Syst. **47**(3), 161–179 (2016)
4. Kaushik, A., Mathpal, C.P., Sharma, V.: Edge detection and level set active contour model for the segmentation of cavity present in dental X-ray images. Int. J. Comput. Appl. **96**(9), 24–29 (2014)
5. Liu, Y.J., Lu, Z.T., Feng, Q.J., et al.: KL distance weighted CV model based on local neighborhood information. Acta Electron. Sin. **39**(6), 1447–1451 (2011)

6. Yu, S., Wu, Y., Bi, S.: Satellite remote sensing cloud image segmentation using edge corrected CV model. Acta Opt. Sin. **34**(9), 22–29 (2014)
7. Xiao, W., Chen, W.: A local-based level set algorithm with shape apriori for image segmentation. Int. J. Appl. Math. Stat. **45**(15), 404–412 (2013)
8. Chan, T.F., Vese, L.A.: Active Contours Without Edges. IEEE Press, London (2001)
9. Cai, Q., Liu, H., Zhou, S., et al.: Adaptive level set model based on local and global intensity information for image segmentation. High Power Laser Part. Beams **29**(2) (2017)
10. Li, C., Gore, J.C., Davatzikos, C.: Multiplicative intrinsic component optimization (MICO) for MRI bias field estimation and tissue segmentation. Magn. Reson. Imaging **32**(7), 913–923 (2014)

Theoretical Analysis of Conflict Decomposition Between Isolated Tri-tree and Improved Tri-tree

Zejun Han(✉), Jundong Yang, Liyong Bao, Min He, and Hongwei Ding

Yunnan University of Information, Kunming, China
1046756791@qq.com

Abstract. The information society is the inevitable trend of today's development. In order to ensure the efficient and stable work of the system, it is necessary to improve the channel utilization to ensure the quality of service. Because of the existence of time delay and many other factors, the system will inevitably collide with information and conflict decomposition arises at the historic moment. This paper focuses on the analysis of the isolated tri-tree and the improved tri-tree conflict decomposition to improve the throughput of the system as much as possible to meet our daily needs.

Keywords: Channel utilization · Conflict decomposition · Isolated tri-tree Improved tri-tree

1 Introduction

In network communication, when there are lots of users working intermittently in the network, random multiple access technology is generally adopted to improve the performance of the system. The random multiple access system that uses carrier detection does not completely avoid conflict, and when the load is larger, the probability of conflict in the channel is greater and the system throughput is lower. Therefore, a reasonable conflict resolution algorithm is needed to decompose the packet conflicts in the channel. The tri-tree conflict resolution algorithm is an easy implementation of effective conflict resolution algorithm.

2 Model Description

2.1 Basic Tri-tree Conflict Decomposition Algorithm

The algorithm consists of 3 time slots (SB), first time slots called left time slot (LSB), and second time slots called medium time slots (MSB), and third time slots called right time slot (RSB). It specifies that the active terminal stations that conflict in the same time slot are compiled into a group of conflict decomposition, and other terminal stations that are unrelated to the conflict are temporarily waiting, only when all the active terminal stations of the group are successfully decomposed, after all the

K. Deng et al. (Eds.): ICMIR 2018, AISC 856, pp. 510–515, 2019.
https://doi.org/10.1007/978-3-030-00214-5_65

information packets of all the conflicts are successfully sent, the waiting terminals can be sent send. The algorithm can analyze the detected channel state, terminate the conflict or idle invalid decomposition in advance, so that the decomposition efficiency can be greatly improved. The information packets of each conflict independently do similar Bernoulli experiments, so the probability of grouping information into each slot is equal (1/3).

In the basic protocol, after each collision is heard by the terminal station of the message, all the conflicting information groups will conduct an independent experiment to determine the time slot in the next message, whether it is sent in the left slot or in the right slot. The information packets in the left time slot are sent first, and then the information packets in the right time slot are sent. The model of the basic Trident tree conflict decomposition algorithm is shown in Fig. 1.

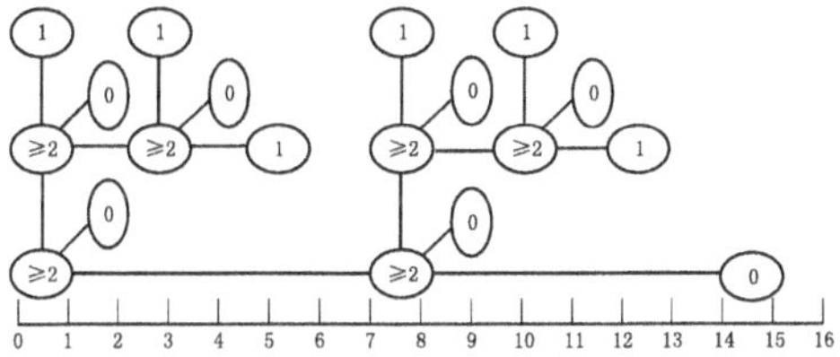

Fig. 1. Diagram of conflict decomposition process of basic trigeminal tree conflict resolution algorithm

In Fig. 1, the time axis is a unit of time slot. (≥2) it indicates that there are 2 or more than 2 terminal stations in this gap to send information packets, that is, conflict occurs; (1) only 1 terminal stations send information packets at this time gap, and it will be sent successfully; (0) there is no terminal station sending information packets to free time slots. Each active terminal station is independent of the result of the Bernoulli experiment. If the result is "0", then the left slots send the letter, if the result is "1", then choose the middle time slot to send the letter, if the result is "2", then choose the right time slot service. When the number of terminal stations in conflict in the marshalling is zero, the decomposition stops.

2.2 Isolated Tri-tree Conflict Decomposition Algorithm

Isolated tri-tree conflict resolution algorithm rules are as follows:

(1) Conflicting information packets are grouped into a set of collision sources in the same slot.
(2) Let each source of conflict belonging to the same group do Bernoulli experiments independently. If the result of the experiment is "0", the collision source selects the left slot to send; if the experimental result is "1", the information packet is sent in the medium time slot. If the result is "2", then the right time slot is selected to send the information packet.

(3) If the collision information sent by the left time slot is used alone, the packet information is transmitted successfully, and a similar Bernoulli experiment is stopped. If the collision source collides again and repeats (1) (2) steps, the colliding source in the left time slot will be successfully transmitted to the next step.

(4) If the collision information sent by the right time slot is occupied independently, the information packet is successfully transmitted, and the private Bernoulli experiment is stopped. If the collision source collides again, repeat (1)–(3) steps until the number of collisions in the decomposition process is 0.

The information packet decomposition process of the algorithm is shown in the graph. Here we take the number of conflicting information packets as 5 (Fig. 2):

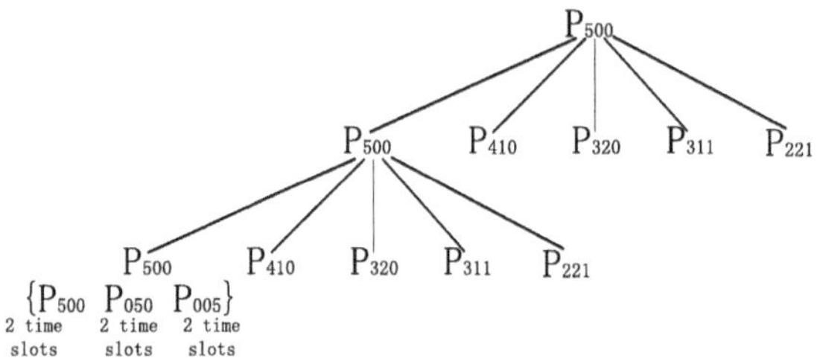

Fig. 2. Schematic diagram of an isolated tree split tree collision resolution at M = 5

The average time slot number required for the decomposition algorithm is as follows:

$$E_5^*(\mathrm{n}) = \sum\nolimits_{n=1}^{\infty} nP_{500}^{n-1}(P_{410}+P_{320}+P_{311}+\ +P_{221}) + \sum\nolimits_{n=1}^{\infty} E_4^*(\mathrm{n})P_{500}^{n-1}P_{410}$$
$$+ \sum\nolimits_{n=1}^{\infty} \left[E_3^*(\mathrm{n}) + E_2^*(\mathrm{n})\right] P_{500}^{n-1}P_{320} + \sum\nolimits_{n=1}^{\infty} E_3^*(\mathrm{n})P_{500}^{n-1}P_{311}$$
$$+ 2\sum\nolimits_{n=1}^{\infty} E_2^*(\mathrm{n})P_{500}^{n-1}P_{211} = \frac{1047}{260}$$

$$\widetilde{L_5^*}\mathrm{n} = 3E_5^*(\mathrm{n}) = \frac{3141}{260}$$

Available decomposition efficiency: $\rho_5^* = \frac{M}{\widetilde{L_5}} = 41.39\%$

2.3 Improved Tri-tree Conflict Decomposition Algorithm

The decomposition process of the improved tri-tree conflict decomposition:

When the number of information packets collided on the transmission channel is, the two conflicting packet data independently do the Bernoulli test. If the test result of the information packet is "0", the information packet is placed in the left time slot; if the experimental value of the information packet is "1", the information packet is put into the middle time slot. If the experimental value of the information packet is "2", the information packet is put into the right slot. The number of conflicting information packets is 5, which will lead to the following situations (Figs. 3 and 4):

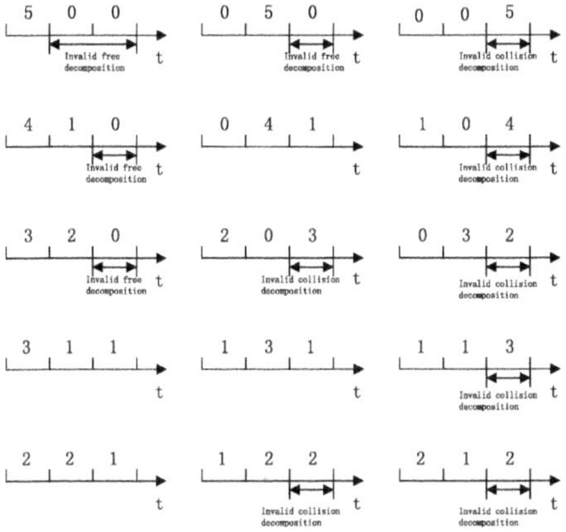

Fig. 3. Schematic diagram of invalid decomposition of "idle" and "collision" in advance at M = 5

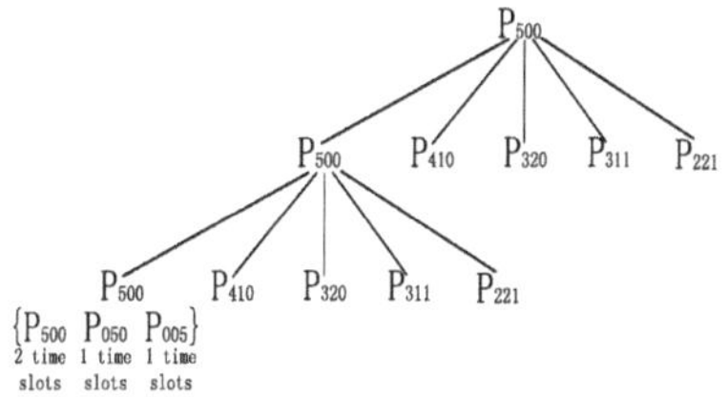

Fig. 4. Schematic diagram of improved tri-tree conflict decomposition tree structure in M = 5

The probability distribution of the decomposition of its conflict:

$$P_{ijk} = C_n^i C_n^j C_n^k p^n (n = i + j + k)$$

The average time slot number required for the decomposition algorithm is as follows:

$$\begin{aligned}\widetilde{L_5} = {} & (1+2+2) * P_{500} + (P_{500} + P_{050} + P_{005}) * \widetilde{L_5} + (2*4+3*2) * P_{410} \\ & + 6 * P_{410} * \widetilde{L_4} + (2*6) * P_{320} + 6 * P_{320} * \left(\widetilde{L_3} + \widetilde{L_2}\right) \\ & + (3*2+2) * P_{311} + 3 * P_{311} * \widetilde{L_3} + (3+2*2) * P_{221} + 2 \\ & * 3 * P_{221} * \widetilde{L_2} = \frac{2945}{312}\end{aligned}$$

Available decomposition efficiency: $\rho_5 = \frac{M}{\widetilde{L_5}} = 52.97\%$

3 Performance Comparison of 2 Algorithms

Here, we make the following explanation:

$E^*_M(\mathrm{n})$: The average frequency of isolated tri-tree conflict decomposition; $\widetilde{L^*_M\mathrm{n}}$: Isolated tri-tree conflict resolution average time slot number; $\widetilde{L_M}$: The average time slot number of the improved tri-tree is decomposed by the conflict; ρ^*_M: Isolated trigeminal tree type conflict decomposition efficiency; ρ_M: The decomposition efficiency of the improved tri-tree conflict decomposition (Table 1).

Table 1. Comparison of the performance of the two algorithms

	$E^*_M(\mathrm{n})$	$\widetilde{L^*_M}\mathrm{n}$	$\widetilde{L_M}$	$\widetilde{L^*_M}\mathrm{n} - \widetilde{L_M}$	ρ^*_M (%)	ρ_M (%)
	The exact value of $E^*_M(\mathrm{n})$	The exact value of $\widetilde{L^*_M}\mathrm{n}$	The exact value of $\widetilde{L_M}$	The exact value of $\widetilde{L^*_M}\mathrm{n} - \widetilde{L_M}$		
M = 2	1.5000	4.5000	3.5000	1	44.44	57.14
	3/2	9/2	7/2	1		
M = 3	2.2500	6.7500	5.3333	1.4167	44.44	56.25
	9/4	27/4	16/3	17/12		
M = 4	3.1154	9.3462	7.3462	2	42.80	54.45
	81/26	243/26	191/26	2		
M = 5	4.0269	12.0808	9.4391	2.6417	41.39	52.97
	1047/260	3141/260	2945/312	317/120		
M = 6	4.9514	14.8543	11.5580	3.2963	40.39	51.91
	38943/7865	116829/7865	72723/6292	7977/2420		
M = 7	5.8748	17.6245	13.6776	3.9470	39.72	51.18
	1293759/220220	3881277/220220	3614495/264264	521587/1321320		
M = 8	6.7927	20.3781	15.7887	4.5894	39.26	50.67
	817503441/120350230	2452510323/120350230	380034559/24070046	276168764/60175115		

4 Conclusion

Through many mathematical analysis, we can draw the following conclusions:

(1) When the number of conflicting information packets becomes larger, the decomposition efficiency is worse than that of ρ^*_M and ρ_M.
(2) The decomposition efficiency of the improved tri-tree conflict decomposition algorithm is better than the isolated tri-tree conflict decomposition algorithm.
(3) The small conflict is easy to decompose, and the big flushing process is difficult to decompose. In other words, the greater the number of conflicting information packets, the greater the cost of decomposition.
(4) When the number of conflicting information packets increasing, the decomposition efficiency of the isolated tri-tree algorithm is slightly larger than $1/3$.

Acknowledgments. Thanks to the National Natural Science Foundation (61461053, 61461054, 61072079), the Yunnan Natural Science Foundation (2010CD023) and the science funding of Yunnan University (XT412004) for support.

References

Zhou, S.: Three Dimensional Probabilistic Random Access for WSN in MAC Layer Protocol. Yunnan University, Kunming (2016)

Nan, J.: Analysis of Random Multiple Access Control Protocol Based on Two Branch Tree Conflict Decomposition Method. Yunnan University, Kunming (2014)

Yang, F.: Two Fork Tree Conflict Decomposition Algorithm and Probability Detection CSMA of Fusion Protocol Analysis. Yunnan University, Kunming (2012)

Ding, H., Guo, Y., Zhao, Y., et al.: Research on the multi-channel probability detection CSMA protocol with sensor monitoring function. Sens. Lett. **13**, 143–146 (2015)

Guo, Y., Nan, J., Ding, H., et al.: Research on the multi-channel P-persistent CSMA protocol with monitoring function. Int. J. Future Gener. Commun. Netw. **8**, 1–29 (2015)

Zhao, Y., Zhou, S., Ding, H., et al.: CSMA/CA MAC protocol with function of monitoring based on binary tree conflict resolution for cognitive radio networks. Int. J. Softw. Sci. Comput. Intell. **8**, 35–51 (2016)

Fusion News Elements of News Text Similarity Calculation

Hongbin Wang[1(✉)], Jingzhen Ye[1], Zhongxi Hou[1], and Liguo Fan[2]

[1] Faculty of Information Engineering and Automation, Kunming University of Science and Technology, Kunming, China
whbin2007@126.com

[2] Branch of Network Security, Bureau of Public Security of Changchun, Changchun, China

Abstract. Text similarity is an effective metric for estimating the text matching degree between two or more texts. Vector Space Model (VSM) is employed for the text similarity calculation in most instances. However, it is insufficient and ill-suited to solve the complex tasks since the high calculation dimension and computational complexity. Therefore, it is crucial to calculate the similarity of two news text, so that whether two reported news is the identical event or the same type of information would be ascertained. According to the analysis of the news reports, five basic factors in terms of "when", "where", "what", "why", and "who" are taken into account for distinguishing a news report. By analyzing these features, in this study, a method to calculate the similarity of news text is proposed. The proposed method fully integrates the influence of the five news feature words into the evaluation of text similarity, which avoids the problem happened in the text interference and computational efficiency to a large extent. There are four steps to execute the proposed method, i.e. extraction of the news elements, classification of these elements, calculation of the similarity, and comparison with available literatures. Experimental results suggest that our proposal outperforms the vector space cosine coefficient method, Jaccard coefficient method and entropy method in terms of the time complexity and computational accuracy.

Keywords: Data fusion · News event · News elements
Text similarity calculation

1 Introduction

Text similarity is a measurement parameter that represents the text matching degree of two or more texts. The greater the similarities, the higher the similarity of the text, whereas the text similarity degree is lower. In the natural language processing field of text clustering, Web intelligent retrieval and question answering system and so on, text similarity calculation is the kernel to solve the problem effectively. The traditional text similarity calculation generally uses the Vector Space Model (VSM) [1, 2] or Latent Semantic Indexing (LSI) model [3, 4], and Attribute Barycenter Subdivision (ABS) model based on attribute theory [5], etc. In VSM, text and being compared text are represented as vectors which take the word as element unit. In addition, each vector

K. Deng et al. (Eds.): ICMIR 2018, AISC 856, pp. 516–525, 2019.
https://doi.org/10.1007/978-3-030-00214-5_66

is weighted according to the word frequency, corresponding to Euclidean space. Finally, the similarity is obtained by utilizing the inner product, Jaccard coefficient, Dice coefficient, or cosine coefficient. In this study, the attribute barycenter model is used to describe the text vector and the contrast text vector. These methods neither consider the dimension of text vector space nor the correlation between words in semantic [6]. Since they are by no means sensitive to the same feature words between texts, the problem of high dimension and time complexity will not affect the accuracy of the calculation results.

In order to optimize the disadvantages of traditional model methods, many researchers are devoted to improving these methods, e.g., Li et al. proposed an improved VSM-based text similarity algorithm by applying the same feature words between texts effect on the text similarity, which largely decrease the low similarity text interference [7]. In addition, Pan et al. took the attribute theory as theoretical basis to establish a text attribute weight analysis model. By presenting the text vector with the query vector in this coordinates, they determined the matching datum between vector and calculate the matching distance so as to establish a matching similarity calculation formula between text and query [5]. Years later, Liu et al. studied on the text similarity calculation based on the event. They focused on the feature that event is the smallest semantic unit of text representation and discussed the text similarity calculation method from two aspects of text type similarity and text content similarity calculation based on the event text representation model. In their study, they took text as concept set in similarity calculation, so the method owns the drawback of lower calculation accuracy [8]. In addition, Xue et al. proposed a method of research on Chinese text similarity based on vector space model. The method used stoplist removal and merger of synonyms to modify file length three strategies and improve the vector space model algorithm. The authors claimed that the work improved the error rate yet the running time is longer [9]. While the work by Piao et al. deemed that the correlation between XMLs structure and content should be built, so they proposed a method of feature dimension reduction and general similarity of XML based on tensor analysis. Such a method can improve the calculation performance for the general similarity of XML [10]. Furthermore, Li et al. proposed a text-similarity method based on entropy to address the traditional text similarity method from the perspective of the similarity measure characters of the text, i.e. regardless of the text similarity factor of the plural common text string within the text [11]. Later, Kenter et al. combined semantic knowledge with word embeddings to calculate the short text similarity. The authors reckoned that the method can improve the calculation performance for short text similarity, but it needs a big number of word vector train corpus [12]. Moreover, Song et al. proposed a method that combines ESA representations and word2vec representations as a technique to generate denser representations which results in a better similarity measure between short texts [13]. Lin et al. proposed a replacement similarity measure to compute the similarity between two documents with relevance a feature [14]. In addition, Christian et al. presented an efficient semantic similarity approach by exploiting explicit hierarchical and transversal relations. The similarity measure provides a significantly higher correlation with human notions of document similarity than comparable measures [15]. Most recently, Venceset al. presented a proposal for searching, retrieving, and recommending documents based on the similarity of their content. By using ontology of the type Friend of a Friend that establishes, the transitivity of similarity between the documents selected by the user and the other documents in the repository would be obtained. Document similarity is calculated as the cosine of the angle between the vectors of terms that represent them. The VSM is used for representing the documents and the weight of the terms is obtained by using the TF-IDF metric [16]. Besides, Lan et al. raised an ontology sparse vector learning model for ontology similarity measuring and ontology mapping in terms of second order cone programming (SOCP) [17].

In this study, we aim at the news text where the news events retrieval needs text similarity to determine the identity of news events. In addition, a news report can be expressed as five elements, i.e. "when", "where", "what", "why", and "who". To determine whether two news reports are the same event or same type, it needs to calculate similarity of the two news text. This study integrates news element feature to calculate news text similarity based on the <<HowNet>> concept model, which was proposed by Pelleg et al. in 2000 [18]. First, the feature of person, places, time, verbs, as well as organization will be extracted, in addition, the five news elements words feature set are established, further, the words similarity in the five news elements words feature set are calculated, and finally the similarity result of two news text will be obtained by utilizing data integration method. The proposal outperforms the traditional text similarity calculation algorithm as suggested in the experiment. In this study, in comparison with the traditional method of text similarity calculation, our method also takes the comparison of semantic similarity into account so that the efficiency can be improved significantly and the vector space dimension and time complexity will be decreased. More importantly, our method can improve the efficiency of the news text similarity calculation, and obtain higher accuracy.

The remainder of this paper is organized as follows. Section 2 introduces the News text similarity calculation method. Section 3 The Experiment and Analysis. The concluding remarks are made in Sect. 4.

2 News Text Similarity Calculation

2.1 News Elements Extraction

A news event is a thing that happens in a specific time and environment, and is represented by the main role of the news event. Formally, the news event can be represented by five elements of the news, set to D_i, defined as a five-tuple array as Eq. (1):

$$D_i = (A_i, O_i, T_i, V_i, P_i) \tag{1}$$

Among them, the five elements in five-tuple array are called news events. The five elements indicate the figure, places, time, action, and organization, respectively [8]. A_i denotes the ith news text figure elements set; Q_i denotes the ith news text place elements set; T_i denotes the ith news text time elements set; V_i denotes the ith news text action elements set; and P_i denotes the ith news text organization elements set. In order to match the characteristics of news, we use ICTCLAS (Institute of Computing Technology, Chinese Lexical Analysis System) [19] Chinese lexical analysis tool to execute word segmentation part-of-speech tagging and named entity recognition for a news text, and make the A_i, O_i, T_i, V_i, P_i five elements constitute a word elements set $\{a_1, a_2, \cdots, a_n\}$, $\{o_1, o_2, \cdots, o_n\}$, $\{t_1, t_2, \cdots, t_n\}$, $\{v_1, v_2, \cdots, v_n\}$, $\{p_1, p_2, \cdots, p_n\}$ after the word segmentation and part-of-speech tagging.

In the same manner, we extract the five-tuple array from compare documents and expressed as $D_j = (A_j, O_j, T_j, V_j, P_j)$, so that we can perform the news texts similarity calculation based on news five elements set.

2.2 News Elements Set Similarity Calculation

The concept of similarity is involved in word morphology, syntactic, semantic, and pragmatics etc. Among them [20], the biggest influence on word similarity is word semantics. The similarity of elements in a set can be divided into three types, where the first is the comparison between two primitives, the second is the comparison between two specific words, and the last is the comparison between primitive and specific word [21].

Primitive similarity is based on the word similarity calculation methods, while the word similarity heavily relies on the subjective concept. There are no explicit objective criteria to measure. Keeping away from the specific application to discuss the word similarity is difficult to obtain a unified definition.

In this study, the words similarity is the degree of syntactic semantic structure of two words, which can replace each other and without changing the others in the context [22]. For two words, if the possibility of the two words replaces each other and don't change the text syntactic semantic structure is greater, the similarity is higher, otherwise the similarity is lower. In this study, the similarity is defined as a real number in the interval from 0 to 1. The word distance and the word similarity have close relation. Word distance and word similarity is a pair of words, which share the same characteristics but express into different forms, that is to say, they can create a simple corresponding relationship between them. For the two words W_1 and W_2, we can denote the similarity as $Sim(W_1, W_2)$ and the word distance is $Dis(W_1, W_2)$, therefore, we can define a simple transformational relation that meets the above-mentioned condition, the formula is defined as Eq. (2):

$$sim(w_1, w_2) = \frac{\alpha}{Dis(w_1, w) + \alpha} \tag{2}$$

Where α is an adjustable parameter to indicate the value of word distance when similarity is 0.5.

However, the transformational relation is able to be defined in other formats. In this study, we just exhibit one possible form of them. In many circumstances, it is difficult to calculate the word similarity. Usually, we calculate the word distance first, and then transform it into words similarity.

The primitive similarity calculation is the basement of the concept of similarity calculation. Due to all of primitives are based on the top and down relation to construct an arborescence primitive hierarchy system, we adopt simple similarity method by semantic distance calculation. Assuming the path distance in the hierarchy system of two primitive is denoted as d, (according to the word similarity Eq. (2)), we can obtain the definition of word distance between two primitives as Eq. (3):

$$sim(p_1, p_2) = \frac{\alpha}{d + \alpha} \tag{3}$$

Where p_1 and p_2 are two primitives. d is a positive integer, which denotes the length of path in primitive hierarchy system, and α is an adjustable parameter. For the degree of similarity of two specific words, if the two words are the same, the similarity is 1, otherwise it is 0. Generally, the similarity of specific word and primitive is a small constant (γ).

For each element set extracted from the essential news elements, we mainly concentrate on the five sets A_i and A_j, O_i and O_j, T_i and T_j, V_i and V_j, P_i and P_j to calculate the corresponding sets similarities because the elements in a set are equal without any order. Similarity calculation model of two sets must meet the following two assumptions:

(1) A set and its self-similarity is 1.
(2) For two sets, suppose there are n elements in a set, respectively, and $m(m < n)$ elements are same. Assume that the similarity of two elements only can be selected as 0 (0 denotes different) and 1 (1 denotes same), so the similarity of the two sets should be m/n.

Before calculating the similarity of two sets, we should first construct the relation for the respective corresponding elements. In this study, we employ the following steps to set up this corresponding relation:

(1) Calculating each similarity of corresponding elements in the two sets.
(2) Selecting the maximum similarity value from all of them, then making the two maximum similarity elements correspond with each other.
(3) Removing those similarity values that have constructed corresponding relation from all similarity values.
(4) Iterating the above steps 2 and 3 until all the similarity values is removed.
(5) Making the do not established corresponding relations elements corresponding empty elements.

According to the one-to-one corresponding relation of two set elements from above steps, the sets similarity is equivalent to the weighted average of the similarities of corresponding elements. Because the elements in a set are equal, so the weighted values of different elements are the same. Consequently, the set similarity is the same with the arithmetic average of all corresponding elements similarities form different sets.

2.3 News Text Similarity Calculation

In order to compare the similarity of text similarity test results in the news text and traditional methods, we need to press the elements of similarity set integration into the similarity of two news of the document, where we use the weighted average method for data fusion.

The set of A_i and A_j, O_i and O_j, T_i and T_j, V_i and V_j, P_i and P_j is the five elements of news text D_i and D_j, The corresponding similarity measure value is X_1, X_2, X_3, X_4, X_5 respectively. Each extracted news element in the proportion of all extracted news elements is the news element weight, and the news element weight denotes as K_n (n = 1, 2, 3, 4, 5), the K_n is shown as Eq. (4):

$$K_n = \frac{A_n}{H_{ij}} \tag{4}$$

Where A_n is the count of figure elements in news text D_i and D_j, H_{ij} is the count of all extracted news elements words in news text D_i and D_j.

Finally, the $Sim(D_i, D_j)$ denotes two text similarity of D_i and D_j, which can be represented as Eq. (5):

$$sim(D_i, D_j) = \sum_{n=1}^{5} X_n K_n \tag{5}$$

To sum up, in this study, the concrete steps of fusion news elements in similarity calculation are as follows:

(1) The corpus of news events articles is converted into text structures.
(2) To convert the text of the focus of the above elements extraction processing each text D_i get A_i, O_i, T_i, V_i, P_i, a collection of five elements.
(3) Calculate the similarity of two documents D_i and D_j, according to the similarity calculation, in this study, we calculate set A_i and A_j, O_i and O_j, T_i and T_j, V_i and V_j, P_i and P_j cooperation group set corresponding to the five parts of speech similarity collection.
(4) Obtain the D_i and D_j two texts five elements similarity X_1, X_2, X_3, X_4, X_5 according to data fusion Eqs. (4) and (5) to calculate the similarity.
(5) Restore the traditional document similarity comparison algorithm, and comparison efficiency and accuracy with our proposed method.

3 The Experiment and Analysis

In order to verify the effectiveness of the proposed method in this study, the experimental setup from the efficiency and similarity is tested respectively, and our method with the traditional Jaccard coefficient and cosine coefficient method to compare the experiment result.

3.1 Efficiency Test Experiment Setup

In terms of efficiency test, we get 300 news text from Tencent network news and select 50, 100, 150, 200, 250, and 300 news text divided into six groups. We use the six groups' test data test efficiency of the proposed method and record the similarity calculation of processing time, and then the traditional Jaccard coefficient and cosine coefficient method to perform the same six groups experiment, and record the time, the efficiency of test results which are shown in Fig. 1.

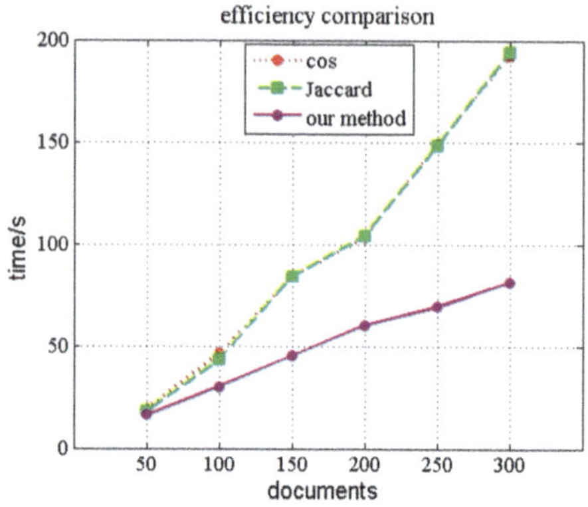

Fig. 1. Computational efficiency contrast

It is trivial to see from Fig. 1 that five groups of news text similarity calculation, according to the time consuming Jaccard coefficient and cosine coefficient method is almost the same in the time consumption, but the consumption of time is more than the consumption of our method time. Along with the number of news text increasing, we can obviously find the efficiency of this method is superior to the traditional Jaccard coefficient and cosine coefficient methods.

3.2 Similarity Comparison Experiment Setup

In terms of the accuracy test, we selected 100 news texts, and use our proposed method to compare similarity between the two. Also restore the Jaccard coefficient based on vector space model and the method of cosine coefficient to calculate the similarity, and the experimental data of these three methods are compared. Below the six groups of typical experimental data using histogram display is shown in Fig. 2.

D_1 and D_2 indicate the Xinhuanet and Tencent news the first time reported to Orient star sinking event, respectively, while D_3 and D_4 are the same time at different aspects of the sinking to Xinhuanet reports. In addition, D_2 and D_3 as Xinhua different time on different aspects of the east star, D_5 and D_6 unrelated events reported from the same place, D_7 and D_8 are two completely different news; D_9 and similar D_{10} is two common text comparison tests.

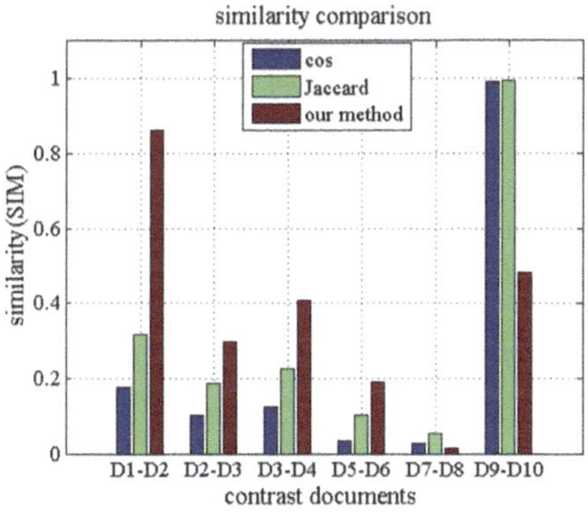

Fig. 2. The news text similarity comparison

From Fig. 2, we can conclude that D_1 and D_2 similarity comparison can be regarded as the same news, this method can perfectly calculate the similarity of two news text. Based on vector space model and calculation method of the Jaccard coefficient and cosine due to the different expression of the same news story and not accurate identification, the D_3 and D_4, D_2 and D_3 similarity comparison indicate this method can also make more accurate judgments. While for D_7 and D_8 two completely different news, Jaccard coefficient and cosine coefficient calculated by the similarity values higher than the method in this study, it also illustrates the method to compare the accuracy of such; D_9 and D_{10} two document similarity comparison with similar sentences and words, Jaccard coefficient and cosine coefficient to calculate the similarity values are higher than many, this method is the cause of this result for the comparison of plain text, plaintext cant extract the five elements of news, do not accord with the characteristics of the news, lead to the method in this calculation is not accurate.

To sum up, through the efficiency contrast experiment results of news text similarity calculation, for judgment the identity and same type of news events, this study uses similarity calculation method which is superior to the traditional Jaccard coefficient, Cosine coefficient and entropy text similarity calculation method. So, through extracting news elements, the fusion of news elements to calculate the news text similarity method is feasible. Our method converts a news text into five elements to perform news text similarity comparison so that the problem of high calculation dimension and complicated calculation are solved. In addition, the efficiency of the news text similarity calculation is improved.

4 Conclusion

In view of the news text, this study combined with the news five elements feature and considered news text five elements influence on text similarity. In similarity calculation, our method extracted news text elements and classified the news elements, then used set similarity computation and data fusion method to calculate the two news text similarity. We restored the traditional text similarity calculation method based on space vector of cosine coefficient or Jaccard coefficient, and then compared the traditional method with our experiment. The experimental results show that compared with the traditional text similarity calculation method, the proposed method can effectively decrease the low text interference and improve the efficiency of text similarity computation. In the future research, we will increase the amount of data to test the news texts, to further improve the correctness of this method, and use the method of news text similarity calculation to study news text classification.

Acknowledgments. This work is supported by the National Natural Science Foundation of China (61462054, 61363044) and the Science and Technology Plan Projects of Yunnan Province (2015FB135).

References

1. Salton, G.: Automatic text processing: the transformation, analysis and retrieval of information by computer. DBLP **41**(3), 753–755 (1989)
2. Wong, S.K.M., Ziarko, W., Wong, P.C.N.: Generalized vector spaces model in information retrieval. In: Fundamenta Informaticae, pp. 18–25 (1985)
3. Deerwester, S., Dumais, S.T., Furnas, G.W.: Indexing by latent semantic analysis. J. Am. Soc. Inference Sci. **1**(6), 391–407 (1990)
4. Dumais, S., Landaner, T., Littman, M.: Automatic cross-linguistic information retrieval using latent semantic indexing. In: SIGIR96 Workshop on Cross-Linguistic Information Retrieval, pp. 1–8 (1996)
5. Pan, Q.H., Wang, J., Shi, Z.Z.: Text similarity computing based on attribute theory. Chin. J. Comput. **22**(1), 651–655 (1999)
6. Zhang, P.Y., Chen, C.M., Huang, B.: Texts similarity algorithm based on subtrees matching. Pattern Recognit. Artif. Intell. **27**, 226–233 (2014)
7. Li, L., Zhu, A., Su, T.: Research and implementation of an improved VSM-based text similarity algorithm. Comput. Appl. Softw. **29**(1), 282–284 (2012)
8. Zhong, Z.M., Liu, Z.T., Zhou, W., Fu, J.F.: Text-based event similarity calculation. J. Guangxi Norm. Univ. Nat. Sci. Ed. **27**(1), 149–152 (2009)
9. Xue, S., Niu, Y.: Research on Chinese text similarity based on vector space model. Electron. Des. Eng. **24**(10), 28–31 (2016)
10. Piao, Y., Jiang, H., Wang, X.: Tensor based approach to XML similarity calculation. Control Decis. **31**(9), 1711–1714 (2016)
11. Li, S., Ling, W., Gong, J., Zhou, C.: Text similarity method based on entropy. Appl. Res. Comput. **33**(3), 665–668 (2016)
12. Kenter, T., De Rijke, M.: Short text similarity with word embeddings. In: ACM International on Conference on Information and Knowledge Management, pp. 1411–1420. ACM, New York (2015)
13. Song, Y., Dan, R.: Unsupervised sparse vector densification for short text similarity. In: Conference of the North American Chapter of the Association for Computational Linguistics: Human Language Technologies, pp. 1275–1280 (2015)
14. Lin, Y.S., Jiang, J.Y., Lee, S.J.: A similarity measure for text classification and clustering. IEEE Trans. Knowl. Data Eng. **26**(7), 1575–1590 (2014)
15. Paul, C., Rettinger, A., Mogadala, A., Knoblock, C.A., Szekely, P.: Efficient graph-based document similarity. In: Proceedings of ESWC16. Springer, Berlin (2016)

16. Vences, R., Gmez, J., Menndez, V.: A document recommendation system using a document-similarity ontology. IEEE Lat. Am. Trans. **14**(7), 3329–3334 (2016)
17. Lan, M., Xu, J., Gao, W.: Ontology feature extraction via vector learning algorithm and applied to similarity measuring and ontology mapping. IAENG Int. J. Comput. Sci. **43**(1), 10–19 (2016)
18. Pelleg, D., Moore, A.: X-means: extending K-means with efficient estimation of the number of clusters. In: The 17th International Conference on Machine Learning (ICML), Palo Alto, Santa Clara, pp. 727–734 (2000)
19. Cao, Y.G., Cao, Y.Z., Jin, M.Z., Liu, C.: Information retrieval oriented adaptive Chinese word segmentation system. J. Softw. **17**(1), 356–363 (2006)
20. Ji, J.Z., Bei, F., Wu, C.S., Chai, Y., Song, C.: Influence of part-of-speeches on the network topic detection of Chinese news and micro-blog. J. Beijing Univ. Technol. **41**(1), 526–533 (2015)
21. Liu, Q., Li, S.J.: Word similarity computing based on how-net. In: The 3rd Chinese Lexical Semantics Workshop Taipei China, pp. 59–76 (2002)
22. Meng, L., Huang, R., Gu, J.: A review of semantic similarity measures in wordnet. Int. J. Hybrid Inf. Technol. **6**(1), 1–12 (2013)

Hybrid Evaluating Algorithm for Service Support Capability of Position Service Unit Based on HMMS

Wen-ming Zhou[1(✉)], Lin Zheng[1], Yao-wu Wu[1], Peng Jia[1], Fei Pei[2], Yun Sun[1], An-shi Zhu[1], Shi-bin Lan[1], and Yun-he Wang[1]

[1] Joint Service Command and Training Centre, Joint Service College, National Defense University of PLA, Taiping Rd, No. 23, Beijing 100858, China
zwmedu@163.com
[2] Military Management College, NDU, PLA, Beijing 100091, China

Abstract. Main influencing factors and key links of the position service support are analyzed. The index system of efficiency evaluation of it is constructed. Several algorithms for calculating all index values of lots of position service units are presented. Based on hidden Markov models (HMMS), propose a kind of hybrid evaluating algorithm for the position service capability. The hybrid evaluating algorithm strongly supported the joint service support information system optimum design and command decision of the position service.

Keywords: Position service · Hidden Markov Models (HMMS) Evaluating · Hybrid algorithm

1 Introduction

The position service is an important support for the operations of the military forces in wartime, especially for missile position service support. It is a very important component part of the logistics service support [1]. The position service support affects directly the weapon use, the process and the result of the battle.

Currently, study reports about position service capability evaluation and simulation is less, the American large-scale simulation systems JWARS [2] and JTLS [3] also didn't simulate the position service support unit and its position service support operation. Due to the merits of the effective non-line method established on statistics, and good classification precision for HMMS, it is widely used in intelligent computing, pattern recognition and classification evaluation. Through analyzing position service process in wartime, a kind of hybrid evaluation algorithm is established for wartime position service capability of position support unit. Evaluation examples of twenty four units are presented to validate its rationality and validity.

K. Deng et al. (Eds.): ICMIR 2018, AISC 856, pp. 526–534, 2019.
https://doi.org/10.1007/978-3-030-00214-5_67

2 HMMS Modeling

HMMS [4, 5] roots in Markov chain, and established on statistical learning theory is an effective method for solving non-lie problems.

2.1 HMMS Terminology

HMMS is a probability model expressed by parameters used to describe random process. The follows account for HMMS:

The states set $ST = \{st_1, st_2, \cdots, st_N\}$, $OO = \{oo_1, oo_2, \cdots, oo_T\}$ is a observation sequence, The hidden state sequence $QQ = \{qq_1, qq_2, \cdots, qq_N\}$, $qq_n \in SS$ and a parameters set $\Theta\Theta = \{\pi\pi, AA, BB\}$:

$\pi\pi = [\pi\pi_1, \pi\pi_2, \cdots, \pi\pi_N]$ is initial probabilities distribution. $\pi\pi_i = P(qq_1 = st_i)$ are initial probabilities of st_i, collected in vector $\pi\pi$.

From state i to state j, transition probabilities: $aa_{i,j} = P(qq_{n+1} = st_j | qq_n = st_i)$. Collected in $AA = \{aa_{i,j} | i,j = 1, 2, \cdots, N\}$.

If the model is in state st_i, for a certain observation oo, its likelihood is characterized by emission probabilities.

For discrete observations, $oo_n \in \{oo_1, oo_2, \cdots, oo_T\}$: $bb_{j,t} = P(oo_n = oo_t | qq_n = st_j)$ is the probabilities when the state in $q_n = st_j$ to observe oo_t, $bb_{j,t}$ collected in $BB = \{bb_{j,k}, j = 1, 2, \cdots, N; k = 1, 2, \cdots, M\}$. Here M is the total number of all possible observation values.

For continuous valued observations, this paper doesn't treat of it.

2.2 HMMS Three Basic Question

For applying DHMMS to intelligent computing, pattern recognition and classification evaluation, three basic questions must be solved.

2.2.1 Calculating Observation Sequence Probabilities

Assume, the observation sequence $OO = oo_1 oo_2 \cdots oo_T$ according to certain probabilities corresponding to $QQ = qq_1, qq_2, \cdots, qq_T$, it's condition probabilities is:

$$\begin{aligned} P(OO|QQ, \Theta\Theta) &= \prod_{t=1}^{T} P(oo_t | qq_t, \Theta\Theta) \\ &= bb_{qq_1}(oo_1) bb_{qq_2}(oo_2) \cdots bb_{qq_T}(oo_T) \end{aligned} \tag{1}$$

Here assumed the observation values were independent respectively, occurrence probability of the state sequence QQ in random process expressed by model $\Theta\Theta$ is:

$$P(QQ|\Theta\Theta) = \pi\pi_{qq_1} aa_{qq_1 qq_2} aa_{qq_2 qq_3} \cdots aa_{qq_{T-1} qq_T} \tag{2}$$

The occurred union probability at the same time of OO with QQ is:

$$P(OO, QQ|\Theta\Theta) = P(OO|QQ, \Theta\Theta) \cdot P(QQ|\Theta\Theta) \tag{3}$$

So, the function of calculating $P(OO|\Theta\Theta)$ is:

$$\begin{aligned} P(OO|\Theta\Theta) &= \sum_{\forall QQ} P(OO, QQ|\Theta\Theta) = \sum_{\forall QQ} P(OO|QQ, \Theta\Theta) \cdot P(QQ|\Theta\Theta) \\ &= \sum_{qq_1 qq_2 \cdots qq_T} \pi\pi_{qq_1} bb_{qq_1}(oo_1) aa_{qq_1 qq_2} bb_{qq_2} \cdots aa_{qq_{T-1} qq_T} bb_{qq_T}(oo_T) \end{aligned} \tag{4}$$

Calculating it directly is very complex, therefore we usually use forward-backward algorithm to calculate it. The detailed process is omitted.

2.2.2 Confirm the Best State Chain

Given a HMMS Θ and observation sequence $OO = oo_1 oo_2 \cdots oo_T$, how to effectively confirm the best corresponding state sequence, here using *Viterbi algorithm* to deal with it.

2.2.3 DHMM Parameters Optimization

That is a training problem, here using parameter revaluation function to solve it. For giving revaluation function, define a forward and backward probability: $\alpha\alpha_t(i)$ and $\beta\beta_t(i)$. $\alpha\alpha_t(i)$ is the probability that the emission sequence symbol reach the state st_i. $\beta\beta_t(i)$ is defined as the probability that the emission symbol sequence $oo_{i+1}, oo_{i+2}, \cdots oo_T$ from state st_i to the end of st_N.

$$\alpha\alpha_t(i) = P(oo_1 oo_2 \cdots oo_t, qq_t = st_i \Theta\Theta) \tag{5}$$

$$\beta\beta_t(i) = P(oo_{t+1} oo_{t+2} \cdots oo_T | qq_t = st_i, \Theta\Theta) \tag{6}$$

Defined $\gamma\gamma_t(i)$ is the probability that system is at the moment t with emission observation vector sequence OO.

$$\begin{aligned} \gamma\gamma_t(i) &= P(qq_t = st_i | OO, \Theta\Theta) \\ &= P(qq_t = st_i | OO, \Theta\Theta) \\ &= \alpha\alpha_t(i)\beta\beta_t(i) / \sum_{i=1}^{N} \alpha\alpha_t(i)\beta\beta_t(i) \end{aligned} \tag{7}$$

Defined $\xi_t(i,j)$ is the transfer probability that from state s_i to state s_j at the moment t.

$$\xi_t(i,j) = P(qq_t = st_i, qq_{t+1} = st_j | OO, \Theta\Theta) \tag{8}$$

Given model $\Theta\Theta$ and observation sequence OO, for making hidden state is in st_i at moment t, and in st_j at moment $t+1$, using forward-backward variety to express as follows:

$$\begin{aligned}\xi_t(i,j) &= \alpha\alpha_t(i)aa_{ij}bb_j(oo_{t+1})\beta\beta_{t+1}(j)/P(OO|\Theta\Theta) \\ &= \alpha\alpha_t(i)aa_{ij}bb_j(oo_{t+1})\beta\beta_{t+1}(j)/\sum_{i=1}^{N}\sum_{j+1}^{N}\alpha\alpha_t(i)aa_{ij}bb_j(oo_{t+1})\beta\beta_{t+1}(j)\end{aligned} \tag{9}$$

$$\gamma\gamma_t(i) = \sum_{j=1}^{N}\xi_t(i,j) \tag{10}$$

And then, reasonable revalued parameters functions are given:

The probability in state st_i at moment $t = 1$

$$\overline{\pi\pi} = \gamma\gamma_t(i) \tag{11}$$

$$\overline{aa_{ij}} = \sum_{t=1}^{T-1}\xi_t(i,j)/\sum_{t=1}^{T-1}\gamma\gamma_t(i) \tag{12}$$

$$\overline{bb_j(k)} = \sum_{o_t=k,t=1}^{T}\lambda\gamma_t(j)/\sum_{t=1}^{T}\gamma\gamma_t(j) \tag{13}$$

The above are the revalued parameters. That is the famous Baum-Welch algorithm.

3 The Index System of Position Service Support Capability Evaluation

The position service support is a multi-factor and multi-index complex system. Its support element and feature index are a little much. Considering the principle of establishing index system, this paper analysis the process of position service support [1, 7], adopt singularity and similarity compressing method of informatics to optimize its index [8]. The position service support capability evaluation index system is as Fig. 1.

4 The Computing Method for Index Value

Based on established optimized index, adopt following methods to comput its value:

Definition 1. *The full rate of position equipment (x_3) is defined as follows:*

$$x_3 = \begin{cases} \frac{n_v}{n_w}, n_v \le n_w \\ 1, n_v > n_w \end{cases} \tag{14}$$

Where n_v is the number of position equipment, n_w is the allocated number.

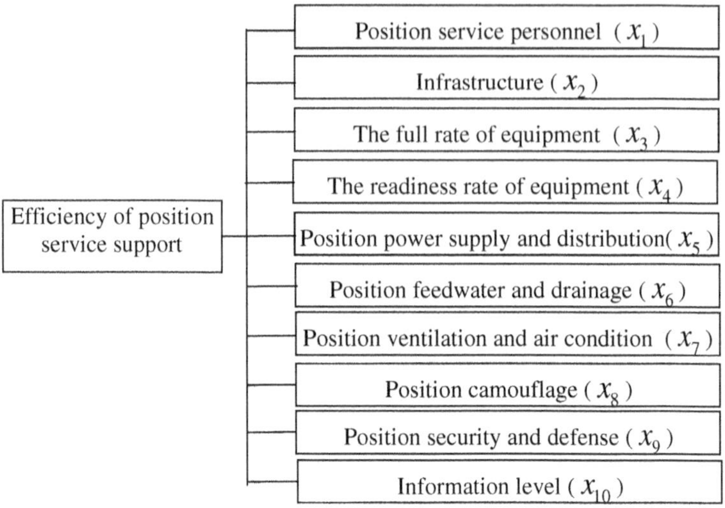

Fig. 1. Efficiency index of the position service support

Definition 2. *The readiness rate of position equipments (x_4) is defined as follows:*

$$x_4 = \begin{cases} \frac{n_1}{n_2}, n_1 \leq n_2 \\ 1, n_1 > n_2 \end{cases} \tag{15}$$

Where n_1 is the number of intact position equipment, n_2 is the allocated number.

4.1 Operations Research Method

Infrastructure of position (x_2) includes civil infrastructure (m_1), power supply and distribution infrastructure (m_2), feedwater and drainage infrastructure (m_3), ventilation and air condition infrastructure (m_4) etc. m_i $(i = 1, 2, \cdots, 4)$ is confirmed by experts.

$$x_2 = \sum_{i=1}^{4} a_i m_i \tag{16}$$

Where a_i is weight coefficient, and $a_1 + a_2 + a_3 + a_4 = 1$.

The factor of position service support personnel always plays the most important role in position service support. Position service staff with adept skill and higher degree will mostly increase the position service efficiency. Personnel is expressed with x_1. Personnel index [9] includes three aspects: knowledge, quality, skill and capacity structure. The respective secondary indexes are omitted here.

Position service information level is specified by x_{10}. It includes command and control information capability, information communication capability, information management capability of position maintenance equipment warehouse, and specialist knowledge database level, etc.

Compute the index value of Position service information level x_1 and the value of position service support personnel x_{10} by the method in literature [10].

4.2 Examination

Through inspection and examination method, the value of power supply and distribution capability (x_5), feedwater and drainage service capability (x_6), ventilation and air condition service capability (x_7), position camouflage service capability (x_8), position security and defense capability (x_9) are all calculated by examination and inspection. Where these indexes all include respective particular second-class indexes.

5 Classification Pattern

Set up classification evaluation pattern of position service support with Delphi method, classify and evaluate and synthesize the ten indexes respectively. Full mark is set according to the one thousand three hundred mark examination system. According to performance, it is sorted as four grades: excellent (A), good (B), moderate (C) and inferior (D), as can be seen in following Table 1.

Table 1. Position service index classifying pattern

Comprehensive index	Four grade pattern			
	A	B	C	D
Position service personnel (x_1)	[400 450]	[350 400]	[300 350]	[0 300]
Infrastructure (x_2)	[90 100]	[80 90]	[70 80]	[0 70]
Full rate of position equipment (x_3)	[95 100]	[85 95]	[75 85]	[0 75]
Readiness rate of position equipment (x_4)	[95 100]	[90 95]	[80 90]	[0 80]
Power supply and distribution (x_5)	[95 100]	[90 95]	[85 90]	[0 85]
Position feedwater and drainage (x_6)	[95 100]	[80 90]	[70 80]	[0 70]
Position ventilation and air condition (x_7)	[90 100]	[80 90]	[70 80]	[0 70]
Position camouflage (x_8)	[90 100]	[80 90]	[70 80]	[0 70]
Position security and defense capability (x_9)	[95 100]	[90 95]	[85 90]	[0 85]
Information level (x_{10})	[45 50]	[40 45]	[35 40]	[0 35]

6 Evaluating Example

6.1 Data Acquisition and Preprocessing

To set up a stable HMMS evaluating model, through mathematics definition, operations research method and examination method, the training data of position support capability of units are collected and processed. Seventy seven samples are acquired. Fifty three of the samples are as training samples to set up the evaluating model, the rest twenty four are as test samples to verify it. Due to the limit to paper length, only part training and test samples can be seen in Tables 2 and 3. Visual training data is given for comparing the difference between respective sample-values as Fig. 2.

Table 2. The value of training samples

Sample \ Index	x_1	x_2	x_3	x_4	x_5	x_6	x_7	x_8	x_9	x_{10}
1	445	92	96	98	96	97	93	91	96	46
2	412	97	97	97	99	95	91	90	98	48
3	428	97	97	100	96	99	95	95	98	46
4	368	87	89	93	95	86	87	84	92	43
5	389	81	86	94	92	85	81	80	91	43
6	363	85	95	94	93	86	84	82	93	41
7	345	71	80	82	89	75	78	72	85	39
8	332	76	76	86	87	77	71	77	90	36
9	335	79	83	83	85	78	75	74	89	40
10	277	69	74	66	78	56	67	68	82	28
…	…	…	…	…	…	…	…	…	…	…
55	285	67	56	75	72	62	56	62	70	28

In order to eliminate data level and reduce error of evaluation, this paper adopts maximum-minimum normalization method to normalize the samples data to [1 20] and round it as follows:

$$f : x \rightarrow y = \frac{x - x_{\min}}{x_{\max} - x_{\min}} \tag{17}$$

Where $x, y \in R^n$, $x_{\min} = \min(x)$, $x_{\max} = \max(x)$.

6.2 Evaluating Position Service Support Capability

Input test samples value to the trained evaluating model, the classification result is as Fig. 3, among twenty four test samples, four is classified error. The accuracy is up to 83.33%.

Table 3. The value of test samples

Sample \ Index	x_1	x_2	x_3	x_4	x_5	x_6	x_7	x_8	x_9	x_{10}
1	416	96	96	96	96	95	97	93	97	45
2	446	95	96	96	95	97	91	90	97	48
3	375	88	88	94	94	89	88	87	93	41
4	390	83	85	93	93	84	80	84	92	40
5	427	94	99	100	97	99	94	93	95	46
6	390	83	85	93	93	84	80	84	92	40
7	354	84	90	95	95	85	87	84	90	41
8	336	75	75	88	90	80	75	75	85	38
9	298	58	70	63	82	60	65	63	76	29
10	269	44	53	61	74	64	59	58	69	31

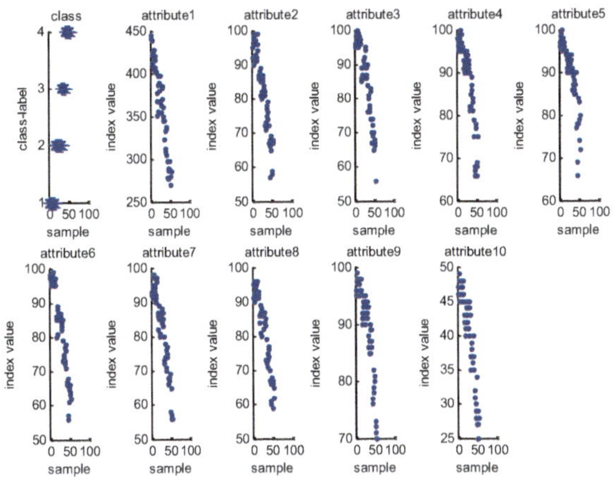

Fig. 2. Visual training data of model

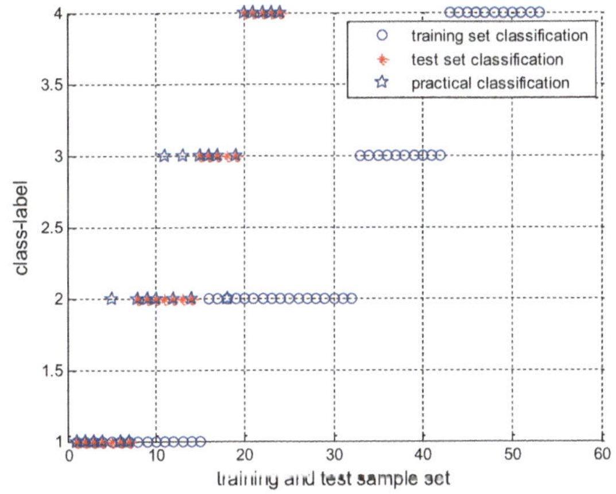

Fig. 3. Training and prediction set classification. O: training set classification; *: test set classification; ☆: practical classification

7 Conclusion

By the evaluating example, the proposed hybrid evaluating position service capability algorithm for position service Unit is applicable and valid. It strongly supports the logistics information system optimum design and command decision making of the position service support.

References

1. Hua, H., Nan, Z.Q., Wang, S.Q.: Logistics Support. Chinese Encyclopedia Publishing House, Beijing (2007)
2. George, F.S., Gregory, A.M.: The Joint Warfare System (JWARS): a modeling and analysis tool for the defense department. In: Proceedings of the Winter Simulation Conference, pp. 78–99 (2001)
3. U.S. Joint Forces Command Joint Warfighting Center. JTLS (3.4.3.0) Version Description Document, 08 June 2010. http://www.rolands.com/jtls/j_vdds./vdd_3200.pdf
4. Schen, X.L.: Speech recognition sys based on HMMS. Master's thesis, Shandong University (2005)
5. Tang, L.: Speech Recognition sys based on HMMS. Master's thesis, National University of Defense Technology (2005)
6. Zhang, J.Y.: Mandarin connected digit speech recognition based on HMMS. Master's thesis, Suzhou University (2005)
7. Sun, X.D., Huang, C.L., Zhang, Q.: Science of Military Logistics. The Publishing House of PLA, Beijing (2002)
8. Chen, Y.S.: Simple in Complex and Certainty in Uncertainty. http://survivor99.com/entropy/chen21.htm
9. Ren, H.Q.: Science of Military Command Decision. National Defense University Publishing House, Beijing (2007)
10. Shi, H.P., Han, T.: Evaluation in support ability of equipment maintenance personnel based on fuzzy comprehensive evaluation. Mod. Electron. Technol. **246**(1), 96–98 (2008)

Robust Tracking Based on Multi-feature Fusion

Zhuma YiZheng, Zhenhong Shang(✉), Hui Liu, and Runxin Li

Faculty of Information Engineering and Automation,
Kunming University of Science and Technology, Kunming, China
shangzhenhong@126.com

Abstract. In target tracking, only the single feature description target and the traditional model updating method are difficult to adapt to the occlusion, deformation and various complex scene changes of the target. To solve this problem, a correlation filter tracking algorithm based on multi-feature fusion and selective update model is proposed. Firstly, the filter model is trained by direction gradient histogram and color feature respectively, and two features are fused according to the peak sidelobe ratio of the different characteristic response maps in the detection stage. The target location of the target is judged to be blocked by the peak sidelobe ratio of the final target position response map of each frame, and the model does not update when the target is blocked. The current model continues to be tracked in the next frame. A comparison experiment is made between 12 challenging video sequences and multiple forward moving target tracking algorithms. The results show that the average center position error is reduced by 25.12 pixels compared with the suboptimal Color Names, CN, and the average tracking accuracy is increased by 29.31%. The experimental results show that the algorithm can track the target steadily and accurately in the case of scale changes, occlusion, and illumination changes.

Keywords: Visual tracking · Correlation filter · Feature fusion
Peak sidelobe ratio · Model update

1 Introduction

Object tracking is a hot topic in the field of computer vision. It is widely used in video surveillance, robot learning, industrial intelligence and so on. Its essence is to find the location and state [1] of a target in a continuous video sequence image. Although great progress has been made in target tracking at present, it is still a challenging problem due to many factors such as occlusion, illumination change and scale change.

In recent years, due to the significant effect of the correlation filter algorithm, many scholars have introduced the correlation filter into the target tracking framework. Thc selection of feature in relevance filter target tracking algorithm has great influence on tracking performance. Among them, the minimum output square error and the (Minimum Output Sum of Square Error, MOSSE) algorithm [2] are proposed by Bolme and so on, and the gray feature is used only for tracking, and Henriques is proposed to extend the gray features of the previous single channel to the multi-channel, and the

K. Deng et al. (Eds.): ICMIR 2018, AISC 856, pp. 535–543, 2019.
https://doi.org/10.1007/978-3-030-00214-5_68

direction gradient feature (Histogram of) is used to track the target. The Kernel Correlation Filter (KCF) algorithm [3] improves the accuracy of tracking. Danelljan et al. added color characteristics in the algorithm and used Principal Component Analysis (PCA) to reduce the dimension of color feature CN [4], and the application effect was good in the color video series. By MOSSE, Danelljan M et al. proposed a DSST algorithm [5] that USES the HOG feature to construct the scale pyramid for the target scale estimation. The above algorithms only use a single feature to describe targets, and cannot express targets in a more comprehensive way. Besides, the above algorithms use the mode by frame by frame to update the filter model, but each frame has different tracking conditions. It is easy to add the wrong information to the target model, which leads to the failure of subsequent frame tracking. Given the above problems, this paper proposes the following strategies based on DSST algorithm that can scale estimation:

(1) The above algorithm uses only one feature (HOG feature or color feature) to describe the target. The feature of HOG is the local feature of the image, which has good adaptability to the fine deformation and illumination change of the target. However, if the target has a large deformation and occlusion, the wrong heel or the heel will occur, and the important perceptual color feature of the human recognition image is based on a global feature of the pixel point and rotates the target. Translation and scale changes are not sensitive, but color characteristics cannot describe the local characteristics of the target very well, and cannot adapt to the change of illumination. In this paper, the two features are fused to describe the target model. In order to obtain the global feature of the target, the local feature of getting can be obtained, and the accuracy of the target detection is improved.
(2) The above algorithm using the target model frame by frame fixed update strategy, if the target block occurs, will continue to update model of incorrect information to add to the model, will lead to the failure of the target tracking. In order to improve the tracking performance, a new strategy is proposed that only meets certain conditions. By judging whether the target is blocked, the model is updated, and the effect of occlusion on target tracking is reduced. Thus the stability of the algorithm is improved.

Based on the DSST algorithm, a scale estimation strategy is proposed. The algorithm first obtains the target location and then estimates the target scale. In order to ensure the integrity of the paper, the DSST algorithm is briefly described below. The detailed description of the algorithm is shown in reference [5].

2 DSST Target Tracking Algorithm

The DSST tracking algorithm proposes to train two filters, position filter, and scale filter respectively to obtain the location and scale of the target, and the two filters are applied in the same way. Firstly, the position of the target response value in the prediction region is found through the position filter. Then, the target scale is estimated at the new position using the trained scale filter.

2.1 Position Estimation

First, the target location is taken as the center of the input image to sample the image block P. Then, extract the HOG characteristic f of P as training sample m. Sample training is to find the optimal filter h, minimizing the squared error between the input sample and the required output. h is composed of filter h^l of each characteristic dimension, and h is obtained by minimizing the mean square deviation.

$$\varepsilon = \left\| \sum_{l=1}^{d} h^l * f^l - g \right\| + \tau \sum_{l=1}^{d} \left\| h^l \right\|^2 \tag{1}$$

In the formula, g is the expected output of training sample m, and τ is a regularization parameter. The training sample f has d dimensions, and f^l is the l dimension, $1 \in \{1, \ldots, d\}$. $*$ represents the loop correlation. The minimum value of the (1) type in the frequency domain is as follows:

$$H^l = \frac{\bar{G}F^l}{\sum_{k=1}^{d} \overline{F^k} F^k + \tau} \tag{2}$$

In the formula, G and F are frequency domain descriptions of g and f, respectively, and $\bar{G}$ and $\bar{F}$ are complex conjugate numbers of G and F, respectively. Since the calculation of Eq. (2) is very time-consuming, to obtain an approximate result, the numerator A_t^l and the denominator B_t of Eq. (2) are separately updated.

$$A_t^l = (1 - \eta) A_{t-1}^l + \eta \overline{G_t} F_t^l \tag{3}$$

$$B_t = (1 - \eta) B_{t-1} + \eta \sum_{k=1}^{d} \overline{F_t^k} F_t^l \tag{4}$$

Where η is the updated learning rate of the filter model.

The above has obtained the optimal filter model and completed the training process. The target will be detected here. For a new frame to get the candidate sample z, calculates the correlation score y between the z and the filter.

$$y = F^{-1} \left\{ \frac{\sum_{l=1}^{d} \overline{A^l} Z^l}{B + \lambda} \right\} \tag{5}$$

The maximum value obtained by y is the new target position.

2.2 Scale Estimates

In order to solve the problem of target scale change, DSST algorithm determines target scale by constructing scale Pyramid training scale filter after target location is determined. After the target location is determined, the 33 image layers of Pyramid are intercepted with the target new position as the center, and the HOG features of these image layers are extracted, and a scale filter H_{scale} is trained to estimate the target scale, and H_{scale} is calculated by the formula (2). In the new frame, in order to get the target scale, we use the formula (5) to get the maximum value of y_s and determine the current scale of the target. Model update (3) and type (4) are used.

3 Multi-feature Fusion Tracking Methods

The DSST algorithm is based on the single HOG feature to describe the target model, and the tracking effect is not ideal when the target has large deformation and occlusion. And the updated model of the algorithm is fixed by frame by frame, but the update strategy cannot satisfy the different changes of each frame in the video sequence, and cannot restrain the model drift. Finally, it may lead to the failure of the tracking.

In this paper, an improvement based on the correlation filter tracking method is proposed. In order to describe the target more fully and reduce the tracking error rate, a PSR (Peak-to-Sidelobe Ration) that trains the filter and the PSR (Peak-to-Sidelobe Ration) which track the results according to the two features can be described in order to describe the target more comprehensively and reduce the tracking error rate. It adapts to two features of fusion. In order to improve the effectiveness of the target model, the tracking results of the algorithm are obscured. If the target is blocked, the target model is not updated, and the model is updated without occlusion, and the performance of the target tracking is improved.

3.1 Peak-to-Sidelobe Ration

In the correlation filter algorithm, the response distribution of the filter can evaluate the confidence of the tracking result. The peak intensity of the ideal response distribution is high, that is to say, the only peak is produced at the center of the target, and the peak is more prominent, and the relative height between the peak and the surrounding area is larger. The peak sidelobe ratio PSR is initially proposed by Bolme and [2]. It is a measure of peak strength, and the PSR method is [2].

$$PSR_t = \frac{y_{t,max} - \mu_t}{\sigma_t} \tag{6}$$

The $y_{t,max}$ in the formula is the peak value of the response graph of the t frame, and μ_t and σ_t are the mean and standard deviation of the surrounding area of the maximum response location. The bigger the PSR_t is, the higher the peak intensity of the response distribution is, the higher the confidence level of the target is.

3.2 Multi-feature Fusion and Position Estimation

Different characteristics have different properties. Different properties have different attributes. When the target is blocked, or the target's light changes rapidly, the HOG or CN features can not be used to track the target well. When the target has obvious illumination changes, the CN feature is used alone, the target is missing, and the CN feature is not enough to deal with the problem of light change. That is, HOG features and

CN features are complementary features. HOG features have good adaptability to illumination changes, and CN features have good adaptability to target deformation and scale changes. In order to describe the target more comprehensively, the above two characteristics will be used to describe the target model and reduce the tracking error rate.

Experiments show that HOG features and CN features are complementary characteristics. HOG features have the better adaptability to illumination changes, and CN features have good adaptability to target deformation and scale changes. In order to describe the target more comprehensively, the above two characteristics will be used to describe the target model and reduce the tracking error rate.

The HOG and CN features of the target are extracted respectively, and two position filters are trained by using (2) formula, and the respective target response diagrams are obtained based on (5) formula. In order to effectively fuse the two features, the tracking confidence of the two features is compared, and the PSR can better measure the tracking confidence. When the PSR value is larger, the target confidence is higher, and the tracking effect is better. This feature weight should be increased in target detection in the next frame.

In the t frame, the CN and HOG features are used to track the target to obtain the filter response $y_{t,cn}$ and $y_{t,hog}$, and feature fusion is performed at the response layer. The fused response y_t is achieved by using the following weighting method.

$$y_t = w_{t,cn} \times y_{t,cn} + w_{t,hog} \times y_{t,hog} \tag{7}$$

The maximum value of y_t in the formula is the last position of the target. The $w_{t,cn}$ and $w_{t,hog}$ are a normalized weight for the CN feature and the HOG feature at frame t, respectively.

$$w_{t,cn} = \frac{PSR^2_{t,cn}}{PSR^2_{t,cn} + PSR^2_{t,hog}} \tag{8}$$

$$w_{t,hog} = \frac{PSR^2_{t,hog}}{PSR^2_{t,cn} + PSR^2_{t,hog}} \tag{9}$$

3.3 Filter Model Update

In the traditional correlation filter algorithm, a fixed frame updating model is adopted. When the target is blocked, the error information introduced in the update process will lead to the drift or even failure of the target tracking. Document [6–9] uses peak sidelobe ratio PSR to determine whether the target is occluded. in the process of target detection, this paper first uses the PSR value of the current frame final location response map to determine whether the target is blocked, and the update rate of the model is redefined by using the PSR.

$$\eta' = \begin{cases} 0.025, & if\ PSR_t > N \\ 0, & if\ PSR_t < N \end{cases} \tag{10}$$

The set of N determines whether the PSR threshold of the filter model should be updated. Through the experimental test of different video sequences, it is found that the PSR value of the target is more than 7 when the target is moving. When the PSR value is less than 7, the target occlude. In order to update the filter model effectively, the N is 7. The η of the last frame in the formula (3) and (4) is replaced by the new model update rate η' and is used in the next frame.

4 Experiments

4.1 Experimental Environment and Parameters

The experimental platform is MATLAB R2014b, all the test videos are consistent with the parameters: the regularization parameter set $\tau = 0.01$, the model update learning rate $\eta = 0.025$, the scale Pyramid 33 layers, the sidelobe area predefined update threshold N to take 7.

4.2 Data and Evaluation Methods

The Experiment select open standard test data set Benchmark [10] 12 periods of video sequences in the experiment, the test video sequence in order for Jogging1, Blurface, Tiger2, Skating2, Coke, Dog, Man, Bird2, Girl, Singer2, Carscale and Trellis, which contains the target a variety of changes, such as shelter, quick movement and rotation change, illumination, dimension, and target jamming and so on, to verify the validity of the algorithm in this paper.

The Center Location Error (CLE), Distance Precision (DP) and frame rate (Frames Per Second, FPS) were used to evaluate the algorithm and other comparison algorithms.

4.3 Comparison with State-of-the-Art

To illustrate the effectiveness of this algorithm, we compare our approach with three state-of-the-art trackers: (CSK, CN, MOSSE).

4.3.1 Quantitative Analysis

Table 1 is the experimental result of the 12 sets of video sequences. The optimal result is represented by the underline. It can be seen that the average CLE and average DP of this algorithm are better than those of other comparison algorithms. Compared with the sub optimal CN algorithm, the average CLE is reduced by 16.74 pixels, and the average DP is increased by 15.16%. But the average frame rate of this paper is 11.27 FPS, which is down compared with other algorithms.

Table 1. Comparison of the algorithm and the comparison algorithm

Method / Algorithm	Median CLE (in pixels)	Median DP (%)	Median FPS (fps)
Ours	**11.26**	**95.31**	11.27
CSK	85.80	43.92	178.52
CN	28.34	80.12	71.36
MOSSE	164.67	36.80	**623.34**

Table 2 respectively enumerate the average CLE of the 8 test sequences, and the optimal results are expressed in bold terms. The average CLE of this algorithm is suboptimal in the Bird2 video only. Therefore, this algorithm is obviously superior to other algorithms.

Table 2. Average CLE (pixels) for some test videos

Video / Algorithm	Jogging1	Tiger2	Skating2	Bird2	Singer1	Carscale	Coke	FaceOcc1
Ours	**4.93**	**12.30**	**34.50**	17.80	**2.43**	**19.10**	**11.50**	**4.75**
CSK	135.00	73.00	153.00	18.60	209.00	81.80	13.50	220.00
CN	40.20	12.70	61.50	**5.99**	8.88	25.60	30.90	7.45
MOSSE	123.00	219.00	157.00	84.50	183.00	139.00	166.00	238.00

4.3.2 Qualitative Analysis

(1) The target occluded. Figure 1(a) is a partial tracking result of the Jogging1 video when the target is completely blocked at the seventy-first frame; in the ninety-eighth frame, the other algorithms follow the failure. Only in this paper, the algorithm can still successfully track the target because it can update the target model adaptively.

(a) Jogging1 (b) Singer2 (c) Carscale

Ours, CSK, CN, MOSSE

Fig. 1. Partial video sequence tracking results

(2) The target light changes. Figure 1(b) is a part of the tracking result of the Singer2 video. In the thirty-sixth frame and the fifty-ninth frame,, the target has intense light change, the contrast algorithm all appears drifting or tracked. failure, only the algorithm can track the target correctly; The color feature is invalid when the illumination changes and this algorithm combines Hog features to adapt to illumination changes.
(3) Change of target scale. In Fig. 1(c) Carscale video sequences, the target scale changes obviously. Only this algorithm can update target scale accurately and track targets accurately.

According to the quantitative and qualitative analysis mentioned above, the algorithm has more robust tracking effect, especially when the target is blocked.

5 Conclusions

In this paper, a tracking algorithm based on correlation filter is proposed for multi-feature fusion and selective update template tracking. First, the CN and HOG features are fused on the response graph level of the filter to improve the accuracy of target tracking. Secondly, in each frame, we adopt PSR adaptive control model update rate to solve the template occlusion problem. The experimental results show that the algorithm has strong robustness when the target occludes, fast motion, illumination change, scale change and class target interference, especially in solving the occlusion problem. However, the average speed of this algorithm is 11.27 frame/s, and the speed is slow, so we need further optimization.

Acknowledgments. This work has been supported by the National Natural Science Foundation of China under Grant No. 61462052.

References

1. Marchand, E., Uchiyama, H., Spindler, F.: Pose estimation for augmented reality: a hands-on survey. IEEE Trans. Vis. Comput. Gr. **22**(12), 2633–2651 (2016)
2. Bolme, D.S,, Ross Beveridge, J.R., Draper, B.A., et al.: Visual object tracking using adaptive correlation filters. In: Proceedings of the 2010 IEEE Conference on Computer Vision and Pattern Recognition (CVPR), pp. 2544–2550. IEEE, Piscataway (2010)
3. Henriques, J.F., Rui, C., Martins, P., et al.: High-speed tracking with kernelized correlation filters. IEEE Trans. Pattern Anal. Mach. Intell. **37**(3), 583–596 (2015)
4. Danelljan, M., Khan, F.S., Felsberg, M., et al.: Adaptive color attributes for real-time visual tracking. In: Proceedings of the 2014 IEEE Conference on Computer Vision and Pattern Recognition (CVPR), pp. 1090–1097. IEEE, Piscataway (2014)
5. Danelljan, M., Häger, G., Khan, F.S., et al.: Accurate scale estimation for robust visual tracking. In: Proceedings of the British Machine Vision Conference 2014, BMVA Press (2014)
6. Li, Y.: Research on Target Tracking Method Based on Nuclear Correlation Filter. Southwest University of Science and Technology, Mianyang (2017)

7. Mao, N., Yang, D.D., Yang, F.C.: Adaptive target tracking based on layered convolution features. Laser Optoelectron. Prog. **53**(12), 201–212 (2016)
8. Xia, Y.X.: Research on Correlated Filtering Target Tracking Algorithm. Huaqiao University, Quanzhou (2017)
9. Zhao, G.P., Shen, Y.P., Wang, J.Y.: Self-adaptive feature fusion target tracking based on cyclic core structure. Chin. J. Opt. **37**(08), 208–217 (2017)
10. Wu, Y., Lim, J., Yang, M.H.: Online object tracking: a benchmark. In: Proceedings of 26th IEEE Conference on Computer Vision and Pattern Recognition, pp. 2411–2418. IEEE Press, New York (2013)

Multi-label K-Nearest Neighbor Classification Method Based on Semi-supervised

Jiaman Ding(✉), Qianjun Song, Lianyin Jia, Jinguo You, and Ying Jiang

Faculty of Information Engineering and Automation, Kunming University of Science and Technology, Kunming 650500, China
tjoman@126.com

Abstract. With the rapid expansion of big data in all fields of science and engineering, multi-label classification has become a more acute problem in real-world data sets, it is notably difficult to develop an efficient model by using traditional supervised learning method because of labeled samples being scarce. In this paper, we proposes a multi-label *k*-nearest neighbor classification method based on semi-supervised learning (SSML-*k*NN in short). SSML-*k*NN firstly proposed a semi-supervised self-training model,and multi-label *k*-nearest neighbor classification based on correlation degree is used to classify the unlabeled datasets. After that, the training intermediate results with high confidence level are selected and added to the training data set. Meanwhile the training model is continuously recycled to expand the labeled data set. Finally, the test set is classified by the trained model. Experiments on publicly emotions datasets and yeasts datasets demonstrate that SSML-*k*NN achieves more prominent results than other related approaches across various evaluation metrics.

Keywords: Multi-label classification · Semi-supervised · Correlation degree *K*-nearest neighbor

1 Introduction

Data classification plays an important role in various fields, such as biomedical [1, 2], image classification [3], data mining and other fields. With the continuous development of research, there are a large number of methods for multi-label classification, which are mainly divided into two categories: problem transformation and algorithm adaptation.

The problem transformation method transforms the multi-label classification problem into a series of relatively independent two-category problems. For example, the Binary Relevance (BR) algorithm proposed in literature [4] is a typical transformation method by decomposing data. This method is intuitive, but does not consider the correlation between labels. The LP algorithm proposed in document [5] classified the label by the label membership of the known label set. The RAkEL algorithm proposed in literature [6] is an improvement algorithm on the basis of the LP. This

K. Deng et al. (Eds.): ICMIR 2018, AISC 856, pp. 544–551, 2019.
https://doi.org/10.1007/978-3-030-00214-5_69

method lead to irrelevant labels in the classification results and reduce the classification performance.

The algorithm adaptation methods refer to those methods which used for the application of multi-label classification by improving the existing classification algorithm. For example, the ML-kNN algorithm proposed in literature [7] uses the maximum posterior probability to predict the label by treating the nearest neighbor information of the classified data, but the algorithm ignores the relationship between the labels. The neural network-based algorithm (BP-MLL) [8] is derived from the popular Back propagation algorithm through employing a novel error function capturing the characteristics of multi-label learning, but it greatly increases the computational complexity.

In the process of model training, a certain amount of data with correct labels is often used as a reference, however, the labeled data often get much more difficult than obtaining unlabeled data. Semi-supervised learning can solve this problems [9, 10]. It is particularly important to de-sign an efficient semi- supervised method to use correlative information between labels for multi-label classification.

To overcome the bottleneck and bridge this gap, we introduce a multi-label k-nearest neighbor classification method based on semi-supervised learning (SSML-kNN in short) method. SSML-kNN firstly proposed a semi-supervised self-training model and multi-label k-nearest neighbor classification based on correlation degree is used to classify the unlabeled datasets. After that, the training intermediate results with high confidence level are selected and added to the training data set. Meanwhile the training model is continuously recycled to expand the labeled data set. Finally, the test set is classified by the trained model. Experiments on publicly emotions datasets and yeasts datasets demonstrate that SSML-kNN achieves more prominent results than other related approaches across various evaluation metrics.

2 Problem Definition

2.1 Multi-label Classification Problem Description

Definition 1. Let's set up $X = \{x_1, x_2,\ldots, x_m\}$ is a data set, $L = \{l_1, l_2,\ldots, l_q\}$ is a set of labels, $A = \{a_1, a_2,\ldots, a_n\}$ is a set of attributes. $X_m = (a_1, a_2,\ldots, a_n, y_m)$, where y_m is the set of labels for the example x_m, which is a nonempty subset of L. The training set for the multi-label classification algorithm is $D = \{(x_i, y_i)|1 \le i \le m, x_i \in X, y_i \subseteq L\}$. y and $y \subseteq L$ of unknown sample x and the corresponding set of labels, y_x indicate the sample label set vector, for any label l extracted from Y, $y_x(l) = 1$ is available when sample x has the label, whereas $y_x(l) = 0$.

2.2 Correlation Analysis

Association rules can effectively carry out the correlation between things and things. Usually three metrics are used to measure an association rule: support, confidence and lift. In the process of correlation measurement, rules that are excavated by support and confidence are sometimes invalid. Lift can be related to two things, and the strength of the association can be judged. However, in the actual application process, the lift is affected by the zero thing too much, so the introduction of *Kulc* metric, that is, the degree of correlation. This metric method takes the average of the confidence of the two things as a condition, reasonably avoids the calculation support, and is not affected by the zero thing [11]. The definition of *Kulc* is as follows:

Definition 2. The degree of association between the label l_j and the l_k is

$$\mathrm{Kulc}(\mathrm{l_j}, \mathrm{l_k}) = \frac{1}{2}(P(l_j|l_k)) + (P(l_k|l_j)) \tag{1}$$

2.3 Classifier Problem Description

Definition 3. The probability that the sample belongs to the l_j class is expressed as:

$$P(H_1^{l_j}) = (s + \sum_{i=1}^{m}(x_i(l_j)))\Big/(s \times 2 + m) \tag{2}$$

Definition 4. The probability that the sample does not belong to the l_j class is expressed as:

$$P(H_0^{l_j}) = 1 - P(H_1^{l_j}) \tag{3}$$

Definition 5. In the case where xi belongs to the class l_j, the conditional probability of the *k* nearest neighbors where exactly u examples belong to class l_j is defined as:

$$P(E_\mathrm{u}^{l_j} H_1^{l_j}) = (s + c[u])\Big/(s \times (k+1) + \sum_{z=0}^{k} c[z]) \tag{4}$$

Definition 6. The posterior probability is defined as follows:

$$\vec{r}_{x_t}(l_j) = \frac{P(H_1^{l_j}) \times \sum_{k=1}^{q}(Kulc(l_j, l_k) \times P(E_{\vec{C}_{x_t}}^{l_j} | H_1^{l_j}))}{\sum_{b \in \{0,1\}}((P(H_b^{l_j}) \times \sum_{k=1}^{q}(Kulc(l_j, l_k) \times P(E_{\vec{C}_{x_t}}^{l_j} | H_1^{l_j})))} \tag{5}$$

When the posterior probability is greater than 0.5, the prediction example x_t belongs to the l_j class, and conversely, does not belong to the l_j.

3 Ssml-Knn

3.1 Semi Supervised Learning

Semi-supervised learning uses data sets X = $\{x_1, x_2, \ldots, x_n\}$ (n = l + u), the data set can be divided into two parts, the first part is the labeled data set $X_l = \{x_1, \ldots, x_l\}$, the second part is the unlabeled dataset $X_u = \{x_{l+1}, \ldots, x_{l+u}\}$. The labeled y_i of the sample point xi in the labeled dataset is known, and the label of the sample points in the unlabeled dataset is unknown [12].

3.2 Global Model

When classifying large scale unlabeled data, it will affect the accuracy of classification if only a small number of labeled data are classified. Therefore, this paper uses semi supervised learning strategy to learn by self-training [13]. By continually adding new prediction to label data into labeled data sets, we can learn again. Self-training models use clustering assumptions and are based on supervised classification methods. The self-training method is to constantly build classifiers with existing experience, predict unlabeled examples, and select the next round of training sets that the example with high confidence adds to improve the model. Multi-label classification of test example i yields a probability of $P(p_1, p_2, \ldots, p_n)$ for each type of label, where n is the number of labels. For the calculation of confidence, the following definitions are given:

$$\text{confidence}(i) = \sum_{i=1}^{n} p_i \log p_i \tag{6}$$

3.3 Algorithm Description

The semi-supervised multi-label k nearest neighbor classification algorithm proposed in this paper is divided into three parts: self-training algorithm, classifier training algorithm, and classifier testing algorithm. The following is a detailed description of the three parts.

Input:Labeled data set D, unlabeled data set U, test data set T

Output: Expected result E

1. for c=1 to 5 do
2. Classifier training: F=Classifier_train(D)
3. Classifier test: C=Classifier_test(F,U)
4. High confidence in selection added to D'
5. D=D+D'
6. U=U-D'
7. end for
8. Classifier test: E=Class_test(F,T)
9. Return result E

Alg.1 Self-training algorithm implementation

The first step in algorithm 1 is to set the maximum number of loops. In step 2, the classifier training algorithm is used to train a small number of labeled data sets D, and the prior probability and the degree of correlation of the labels are added to the set F. In step 3, the classifier test algorithm is used to classify the set F and the unlabeled data set U, and the results are put into the data set C. In step 4 select the result with high confidence in data set C to be added to data set D'. Steps 5 and 6 will add the data in D' to labeled data set D and delete the unlabeled data set U. With D' data, repeat the above steps until the maximum number of cycles is reached. Step 8 tests the test data set to get the expected result E.

Input: Labeled Data Set D

Output: Training Result F

1. for $l_j \in L$ do
2. calculate $P(H_1^{l_j})$ and $P(H_0^{l_j})$
3. end for
4. for i=1 to q do
5. for j=1 to q do
6. calculate the degree of correlation between labels $Kulc(l_j, l_k)$
7. end for
8. end for
9. for i=1 to m do
10. calculate the k nearest neighbors of x_i
11. for j=1 to q do
12. for u=0 to k do
13. calculate $P(E_u^{l_j} | H_0^{l_j})$ and $P(E_u^{l_j} | H_1^{l_j})$
14. end for
15. end for
16. end for
17. Return result E

Alg.2 Implementation of classifier training algorithm

Algorithm 2 calculates the prior probability of the label and the correlation between the labels based on the labeled data set D. Steps 1 to 3 Calculate and for each label according to formula 2 and 3. Steps 4 through 8 calculate the correlation between labels by formula 1. Steps 9 to 16 first find the *k* nearest neighbors of xi, then calculate the conditional probability calculations and according to formula 4 and finally return result set E.

Input: Prior probability set E, unlabeled data set U
Output: Test result C

1.for i=1 to m; $x_i \in$ T
2. calculate the k nearest neighbors of x_i
3.end for
4.for $l_j \in$ L do
5. calculate $\vec{y}_{x_i}(l_j)$and $\vec{r}_{x_i}(l_j)$
6.end for
7.Return result C

Alg.3 Implementation of classifier training algorithm

Algorithm 3 uses the input prior probability set E to classify the input unlabeled data set. Steps 1 to 3 calculate the *k* nearest neighbors to find xi, and steps 4 to 6 calculate and using formula 5 and return the result set to complete the classification.

4 Experimental Results and Analysis

In the multi-label classification process, the commonly used datasets are yeast and emotions. The experiment was based on two data sets. Therefore, according to [9], we first discuss the effect of different *k* on the experimental results. Then, a comparison experiment of different algorithms was performed. Among them, HL represents Hamming Loss, OE represents One-Error, CV represents Coverage, RL represents Ranking Loss, and AP represents Average Precision. ↓ indicates that the smaller the value, the better the performance of the labeling algorithm, and ↑ indicates that the higher the value, the better the algorithm performance.

Experiment 1. The effect of different *k* values on the algorithm.

By changing the value of *k*, the change of multiple evaluation indexes of multi-label classification algorithm performance is discussed. Then find the most stable *k* value.

From Fig. 1, it can be seen that when $k = 7$ is the algorithm, the performance indexes are better.

Experiment 2. Performance Comparison of Different Multi-Label Classification Algorithms.

Table 1 is a comparison of the indicators of the algorithm SSML-*k*NN and common algorithms. Five-fold cross-validation was used in experiments, and Table 1 shows the average of the experimental results.

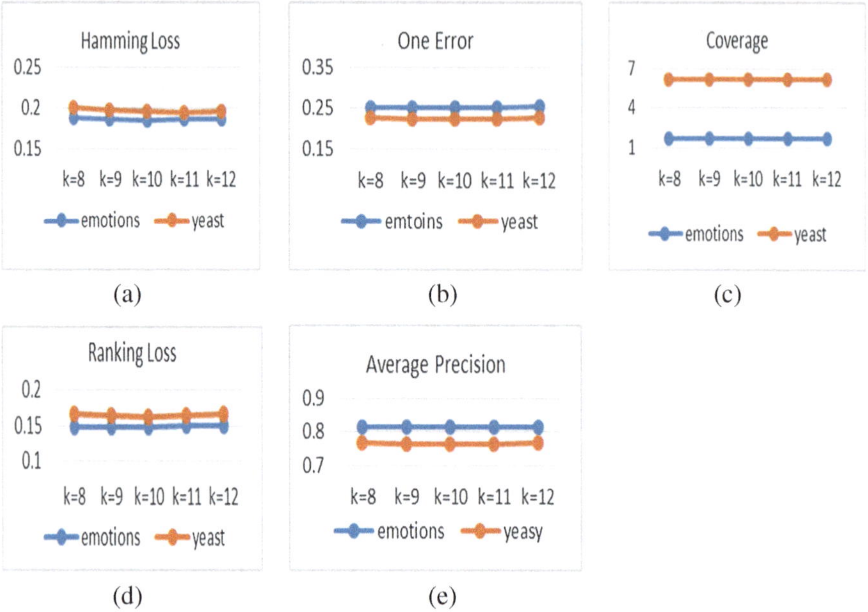

Fig. 1. Different *k* values, the value of each evaluation index

Table 1. Comparison of different multi-label classification algorithms

		HL↓	OE↓	CV↓	RL↓	AP↑
Emotions	SSML-kNN	0.185	0.252	1.701	0.149	0.814
	ML-kNN	0.198	0.265	1.744	0.154	0.803
	BP-MLL	0.192	0.278	1.752	0.155	0.803
Yeast	SSML-kNN	0.196	0.224	6.154	0.163	0.763
	ML-kNN	0.192	0.229	6.181	0.163	0.761
	BP-MLL	0.203	0.234	6.171	0.172	0.762

It can be seen from Table 1. Except for the Hamming loss in the yeast dataset, the ML-*k*NN algorithm does not work well. In all other evaluation indicators, the SSML-*k*NN algorithm performs well. This shows that introducing semi-supervised learning into multi-label classification can improve the classification effect.

5 Conclusion and Future Work

In order to solve the problem of obtaining labeled data much more difficult than unlabeled data, at the same time considering the relationship between labels. In this paper, a semi-supervised multi-label *k*-nearest neighbor classification algorithm is proposed based on semi-supervised learning. Experimental results show that the proposed algorithm improves the utilization of unlabeled data and effectively improves the classification performance. Due to the complexity and diversity of data, and there is a large amount of redundant and noisy data in the era of big data, if the labeled data preprocessing is performed, the data classification performance will be improved again, and further researches can be conducted in the later period for this problem.

Acknowledgements. The paper is supported by Grants from the National Natural Science Foundation of China (Nos. 51467007, 61562054, 61462050).

References

1. Aksoy, S., Shapiro, L.G.: Multi-instance multi-label learning for whole slide breast histopathology. In: Medical Imaging 2016: Digital Pathology. International Society for Optics and Photonics, p. 979108 (2016)
2. Mercan, C., Aksoy, S., Mercan, E., et al.: Multi-instance multi-label learning for multi-class classification of whole slide breast histopathology images. IEEE Trans. Med. Imaging 1 (2017)
3. Zhu, F., Li, H., Ouyang, W., et al.: Learning spatial regularization with image-level supervisions for multi-label image classification. 2027–2036 (2017)
4. Melo, A., Paulheim, H.: Local and global feature selection for multilabel classification with binary relevance. Artif. Intell. Rev. **3**, 1–28 (2017)
5. Ghosh, A., Sekhar, C.: Label correlation propagation for semi-supervised multi-label learning. In: International Conference on Pattern Recognition and Machine Intelligence, pp. 52–60. Springer, Cham (2017)
6. Wu, Y.P., Lin, H.T.: Progressive Random k-Labelsets for Cost-Sensitive Multi-label Classification. Kluwer Academic Publishers, Norwell (2017)
7. Zhang, M.L., Zhou, Z.H.: ML-KNN: a lazy learning approach to multi-label learning. Pattern Recognit. **40**(7), 2038–2048 (2007)
8. Szalkai, B., Grolmusz, V.: Near perfect protein multi-label classification with deep neural networks. Methods **132**, 50–56 (2017)
9. Pham, T.N., Nguyen, V.Q., Dinh, D.T., et al.: MASS: a semi-supervised multi-label classification algorithm with specific features. In: Asian Conference on Intelligent Information and Database Systems, pp. 37–47. Springer, Cham (2017)
10. Benyettou, A., Bennani, Y., Benyettou, A., et al.: Semi-supervised multi-label classification through topological active learning. Int. J. Commun. Antenna Propag. **7**(3), 222 (2017)
11. Han, J., Kamber, M., Pei, J., et al.: Data Mining: Concepts and Techniques. Mechanical Industry Press, Beijing (2012)
12. Board, R., Pitt, L.: Semi-supervised Learning. Springer, New York (2016)
13. Wu, D., Shang, M.S., Luo, X., et al.: Self-training semi-supervised classification based on density peaks of data. Neurocomputing **275**, 180–191 (2017)

Research on Traffic Accidents Classification Oriented to Emergency Call System

Inkyu Lim(✉), Xuefeng Piao, Yue Qin, and Zhenzhou Ji

School of Computer Science and Technology, Harbin Institute of Technology, Weihai, China
noisylik@naver.com

Abstract. This paper proposes a method to classify vehicle traffic accidents according to the classification of road traffic accidents, and the method of judging the accidental vehicle collisions with the target accident vehicles. It designs and implements methods for identifying vehicle traffic accidents and computer information processing system of vehicle traffic incidents loss. And the proposed method of judgment is verified by experiment.

1 Introduction

Vehicle traffic accidents are the most common social hazards that cause accidental casualties in the world today. In order to minimize the accidental accidents and economic losses caused by inaccurate judgments of traffic accidents in road traffic accidents, on-board emergency rescue call system emerged [1, 2], which is abbreviated as e-Call. As shown in Fig. 1, when the vehicle encounters a collision accident or serious failure, it establishes a connection with the emergency call command system manually or automatically through various networks, and transmits the information such as the geographic location, direction of travel, rollover inclination, and energy type of the accident vehicle to the emergency call command system. (Minimum Set of Data, MSD), which can automatically call an emergency rescue phone even if the driver and passenger can not make a call, and report the location of the accident vehicle [3, 4]. As an emergency call command system, after receiving the MSD information sent by the accident vehicle, by analyzing these information, it can accurately and timely determine the accident site situation and the type of the accident, also can locate the accident vehicle according to the received MSD information and notify the rescue agency to start the rescue work immediately [5, 6], which plays a crucial role in the rapid processing of road traffic accidents. It is of great significance to safeguarding the lives and property of personnel.

Currently, for emergency call system, different research results have been proposed from different aspects at home and abroad [1, 7–9], such as judging the types of traffic accidents and estimating the degree of accident damage, etc., Europe applied the system to life early, and it has also achieved better results. In addition, emergency call products that have been marketed abroad include Japan's Toyota's G-BOOK system, U.S. General Motors' OnStar system, Ford's Rescue Car system, Mercedes-Benz's Tele Aid system, and Russia's ERA-GLONASS system [3, 10–12].

K. Deng et al. (Eds.): ICMIR 2018, AISC 856, pp. 552–560, 2019.
https://doi.org/10.1007/978-3-030-00214-5_70

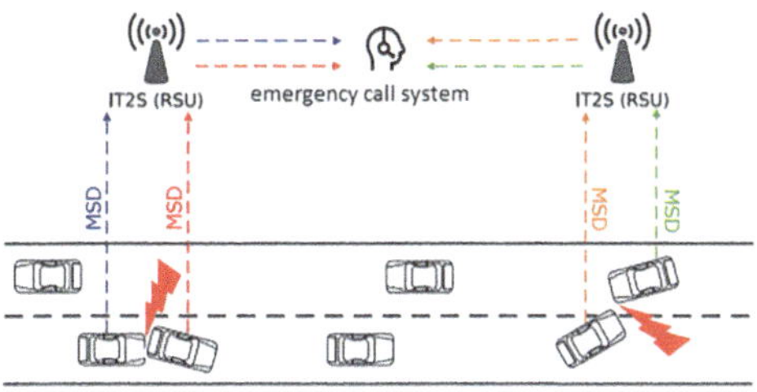

Fig. 1. MSD-based emergency call system

2 Definition Accident Vehicle Type and Damage Location

2.1 Vehicle Traffic Accident Analysis and Accident Type Definition

2.1.1 Analysis of Vehicle Traffic Accidents

Accident vehicles in traffic accidents of vehicles are certain to be damaged regardless of the severity of traffic accidents. They can be classified into different traffic accident types.

1. According to the damaged part of the traffic accident vehicle, the accident vehicle can be divided into a front damaged vehicle, a rear damaged vehicle, a left-side damaged vehicle, and a right-side damaged vehicle. According to whether the traffic accident vehicle is overturned, the accident vehicle can be divided into the rollover accident vehicle and the non-overturned vehicle.
2. According to the number of accident vehicles that collide with each other in a vehicle traffic accident, vehicle traffic accidents can be classified into single-vehicle traffic accidents, double-vehicle (vehicle-to-vehicle) traffic accidents, and multi-vehicle traffic accidents. Single-vehicle traffic accident refers to a traffic accident that occurs when a motor vehicle collides with a person or a object. Double-vehicle (vehicle-to-vehicle) traffic accident refers to a traffic accident that occurs when a motor vehicle collides with another motor vehicle. Multi-vehicle traffic accident refers to traffic accidents caused by the collision of more than three motor vehicles.
3. According to whether there are correlations between adjacent traffic accidents, multiple adjacent traffic accidents can be classified as the same type of traffic incidents. Namely adjacent traffic accidents that are related to each other can form the same type of traffic incident. If there is no relationship or causality between adjacent traffic accidents, it does not belong to the same type of traffic incident.

2.1.2 Definition of Vehicle Traffic Accident Type

Through the above analysis of vehicle traffic accidents, vehicle accidents in road traffic accidents can be classified into the following types of traffic accidents: Single-vehicle traffic accidents, double-vehicle (vehicle-to-vehicle) accidents, and multi-vehicle accidents. The single-vehicle accident can be divided into a damage accident at the front of the bicycle, a damage accident at the rear of the bicycle, a damage accident on the left side of the bicycle, and a damage accident on the right side of the bicycle. For double-vehicle accidents, it can be divided into front accidents, rear accidents, left accidents, right accidents, vehicle scraping accidents. In the case of multi-vehicle accidents, multi-vehicle accidents are characterized by the fact that more than three vehicles are in the same location and always manifests itself in different and complex forms, the paper does not specifically classify the types of accidents. As for whether the accident vehicle is overturned, this paper does not divide it into specific types of accidents but combines the types of accidents defined above as additional options for the types of accidents, such as the accidental damage to the front of Single-vehicle that has fallen. In addition, traffic incidents and non-traffic incidents are classified according to whether there is an associated relationship between adjacent traffic accidents. Table 1 defines types of traffic accidents and specifies the vehicle traffic accident classification criteria.

2.1.3 The Judgment of Accident Vehicle Damaged Part

The principal direction of force (PDOF) of an accident vehicle is commonly used to determine the location of a vehicle in a traffic accident. It refers to the direction of the collision force generated on an accident vehicle. In order to calculate the main direction of the collision force of the accident vehicle, the traditional way is to set the driving direction of the vehicle to 0°, divide the X axis and Y axis as the basis, namely set the driving direction to the X axis and the direction of the positive side of the driving direction to Y axis (as shown in Fig. 2). The main direction of collision force of an accident vehicle is calculated as Eq. (1), where ΔV_x is the maximum speed change in the X-axis direction and ΔV_y is the maximum speed change in the Y-axis direction.

$$PDOF = \begin{cases} arctan\left(\frac{\Delta V_y}{\Delta V_x}\right) & \Delta V_x > 0,\ \Delta V_y > 0 \\ arctan\left(\frac{\Delta V_y}{\Delta V_x}\right) & \Delta V_x > 0,\ \Delta V_y < 0 \\ arctan\left(\frac{\Delta V_y}{\Delta V_x}\right) - 180^\circ & \Delta V_x < 0,\ \Delta V_y < 0 \\ arctan\left(\frac{\Delta V_y}{\Delta V_x}\right) + 180^\circ & \Delta V_x < 0,\ \Delta V_y > 0 \end{cases} \quad (1)$$

Based on the calculated main direction of collision force of the accident vehicle, the damaged site of the accident vehicle is defined according to Table 2 below.

The main direction of the collision force calculated by the traditional method is based on the driving direction of the vehicle, which is convenient for judging the damaged part of the vehicle, but is inconvenient for comparing the main directions of the collision forces. To do this, we need to convert the main direction of the collision force into the azimuth of the main direction of the collision force so that it can be used uniformly. For convenience of explanation, $PDOF_i$ in this specification represents the main direction of the collision force of the vehicle i, and $PDOF_i$* represents the azimuth of the main direction of the collision force of the accident vehicle i.

According to the mechanics theory, $PDOF_i$* and $PDOF_j$* have opposite directions and are on the same line, as shown in Fig. 3.

Table 1. Types of vehicle traffic accident

No.	1	2	3	4	5	6	7	8	9	10
Type of accident	Bicycle front damaged	Bicycle rear damaged	Bicycle right side damaged	Bicycle left side damaged	Double vehicle front damaged	Double vehicle rear damaged	Double vehicle right damaged	Double vehicle left damaged	Double truck damaged	Multi-vehicle damaged
Damaged part	Front	Rear	Right	Left	Front and Front	Rear and Rear; Rear and Front;	Right and Front; Right and Rear	Left and Front; Left and Rear;	Right and Left; Right and Right; Left and Left;	Mixed
Number	1	1	1	1	2	2	2	2	2	More than 3
Driving direction	–	–	–	–	Same or opposite direction	Same direction; Opposite Direction;	–	–	Same direction; Opposite direction; Opposite direction;	–

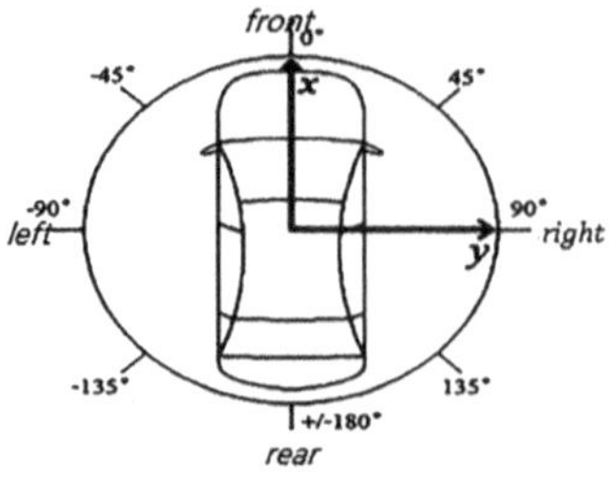

Fig. 2. Main direction of collision force

Table 2. Damaged parts of accident vehicles

PDOF	The damaged parts
–45° ~ 45°	Rear damaged
–45° ~ –135°	Left-side damaged
45° ~ 135°	Right-side damaged
135° ~ –135°	Front damaged

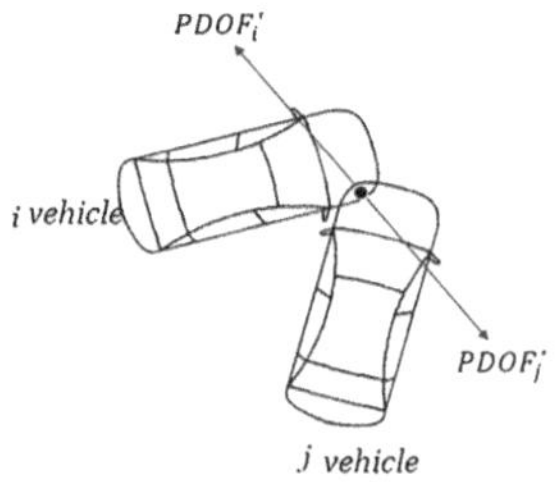

Fig. 3. The colliding force of the vehicle collided in the opposite direction and acts on a straight line

The following formula (2) is a method for calculating the azimuth of the main direction of the collision force from the main direction of collision force of the accident vehicle *i*. Among them, VD_i is the azimuth of the vehicle's driving direction.

$$PDOF_{\mathrm{i}}^{*} = (VD_i + PDOF_i + 360)^{\circ}\%360^{\circ} \tag{2}$$

2.2 Judgment of Collision Vehicle

For convenience of explanation, it is assumed that the vehicle of *i* and the vehicle of *j* collide with each other, and the vehicle of *i* is set as the victim vehicle, and the vehicle of *j* is the injuring vehicle. The geographic location information of the accident vehicle of *i* is composed of latitude and longitude and is represented by $P_{\mathrm{i}}(lon_i, lat_i)$, where loni is longitude information and *lati* is latitude information.

First, because the two vehicles collide, the main direction of their collision force should be opposite, namely, $PDOF_i* = \overline{PDOF_j^*}$, but this study considers the calculation error of the main direction of the collision force. The collision vehicles must satisfy the following formula (3), where θ is Fault tolerance angle range.

$$PDOF_i* = [\overline{PDOF_j^*} - \theta/2,\ \overline{PDOF_j^*} + \theta/2] \tag{3}$$

Second, the two vehicles must be within the effective distance. In the case that the direction azimuth of the collision force of two vehicles meets the formula (3), if the distance between the two vehicles is not within the effective distance, the possibility of colliding the two vehicles can be eliminated. In this study, the effective distance between two vehicles colliding with each other is set to l, namely, the distance $D(i,j)$ between two vehicles must be less than or equal to l, as shown in Eq. (4).

$$D(i,j) \leq l \tag{4}$$

Thirdly, if the accident vehicle of i is set as the target vehicle, then the collision source area of the vehicle may be set as the fan-shaped area marked in Fig. 4, which is the range of areas that the main direction of the collision force may exist when the opponent vehicle j colliding with the vehicle i.

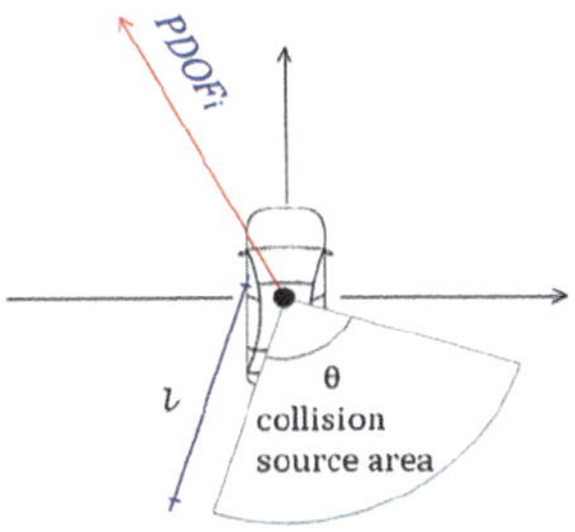

Fig. 4. The collision source area of the accident vehicle i

In other words, it is the range of areas which main collision direction $PDOF_j$ should exist when the opponent vehicle j colliding with the accident vehicle i. The collision source area of the accident vehicle i is represented by $R_i(PDOF_i^*, \theta, l)$, namely the fault tolerance range of the main direction $PDOF_j$ of the collision force to injure the vehicle j, and l is the side length of the sector area, also is the radius of the circle.

If the azimuth of the collision force direction of the accident vehicle j is within the collision source area of the accident vehicle i, that is, $PDOF_j^* \in R_i(PDOF_i^*, \theta, l)$ is satisfied, the accident vehicle and the accident vehicle may collide. In order to judge whether the accident vehicle j exists within the collision source area of the accident vehicle i, it is necessary to determine the azimuth angle between the accident vehicle i and the accident vehicle j, as shown in Fig. 5. $\alpha(i,j)$ indicates the angle between the accident vehicle i and the accident vehicle j. In Fig. 6, the main direction of the collision force of the two accident vehicles is in accordance with Newton's third law, and the distance between the two vehicles is also within the effective distance range, but the two accident vehicles are not collision vehicles.

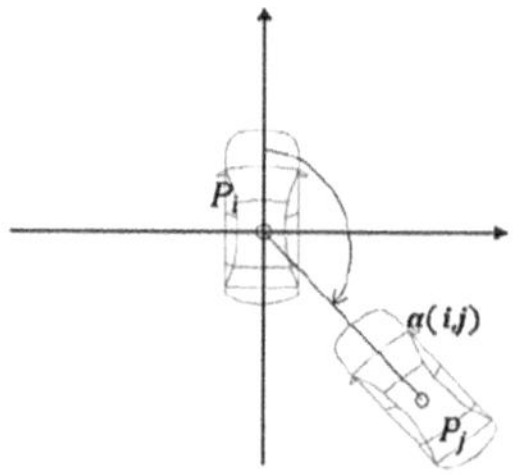

Fig. 5. The angle between the accident vehicle *i* and the accident vehicle *j*

Then, we can use formula (5) or formula (6) to judge whether the collision force direction of the accident vehicle *j* is within the collision source area of the accident vehicle *i*. If neither item is satisfied, the two accident vehicles cannot be collision vehicles.

$$|\mathrm{PDOF}_j^* - \alpha(i,j)| \leq \theta/2 \tag{5}$$

$$360^\circ - |\mathrm{PDOF}_j^* - \alpha(i,j)| \leq \theta/2 \tag{6}$$

2.3 Judgment Procedure for Vehicle Traffic Accident Types

It mainly explains the vehicle traffic accident type judgment procedure we have designed. Based on the above, we can judge the type of accident by determining the location of each accident vehicle damage and the accident vehicle collided. The vehicle emergency rescue call system will receive a certain amount of MSD information sent by the accident vehicle in a certain period of time. The judgment method in the thesis uses these MSD information as input data to determine the damaged part of the accident vehicle and the type of the accident, and store the same type of traffic in a queue. For convenience of description, *n* represents the number of MSD information received in a given time period, and *Q(k)* represents a set queue of the same type *k* accident vehicle. The vehicle emergency rescue call system of the traffic emergency center calculates the main direction of its collision force for each accident vehicle based on the received MSD information and converts it into the azimuth angle of the collision force main direction, and judges the damaged parts of the vehicle according to Table 1. After that, it is determined that each accident vehicle is a target vehicle to determine whether other vehicles collide with each other, and if it is a collided vehicle, they will be classified into the same queue. Finally, determine the type of accident for each traffic accident for each queue according to Table 2. According to this process, we have designed a procedure for determining the types of vehicle accidents.

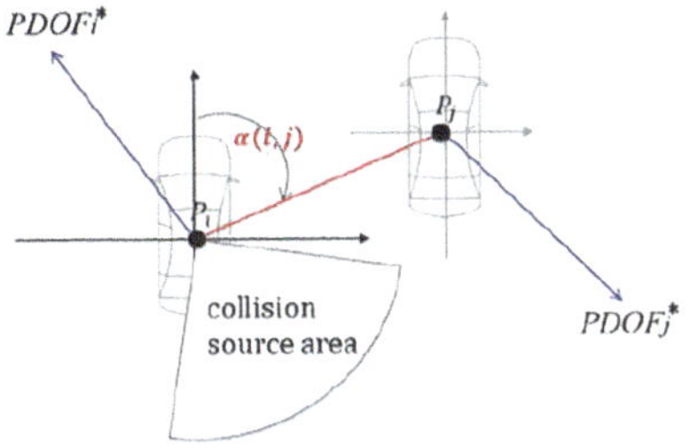

Fig. 6. Example of meeting the force and reaction conditions, but not within range of the collision source area.

2.4 Judgment Procedure for Vehicle Traffic Incidents

Through the above analysis and calculations, we have come to the conclusion that when a vehicle detects a forward traffic accident while driving, it will adopt some emergency avoidance measures. If the emergency avoidance measures taken can not avoid the traffic accident ahead, it will lead to a collision traffic accident ahead, in response to this situation, the vehicle accident judgment method we proposed in 2.3 can be judged and treated it as the same traffic accident. If the emergency avoidance measures taken avoid traffic accidents in front but collide with another vehicle, it is a different traffic accident, but it is classified as the same type of traffic incident because there is a certain correlation between the two traffic accidents.

Firstly, according to the research report, if the distance between two vehicle accidents is within 5 m, there is a possibility of an accident-related relationship. Therefore, we set 5 m as the safety distance D_{safety} and use this as the basic conditions of the relationship to judge whether two vehicle accidents are related. If the distance between any two vehicles in different traffic accidents (different traffic accident queues) is less than the safe distance, we can find out whether there are correlations between these two different traffic accidents.

Secondly, analyze the movement trend of the trajectory of any accident vehicle of different traffic accidents that may be related to each other to determine whether there is an association relationship between different traffic accidents, because the movement trend of a vehicle can reflect the accident emergency avoidance measures before the vehicle accident. In addition, this article judges whether there are correlations between different accidents through the movement trend of accident vehicles. When any one vehicle among the different traffic accidents within the safety distance has a movement tendency of less than or equal to 90°, it is assumed there is an association between the two. Based on these principles, we have designed the relevant identification procedures for vehicle traffic incidents.

3 Conclusions

By studying the research status at home and abroad, we have completed the classification of various vehicle accidents in road traffic and defined 10 types of vehicle accidents based on the analysis results. In this paper, we design and implement the judging procedure for determining the vehicle type of collision traffic accidents, and design and implement the judgment procedures of work vehicle traffic accidents. Experiments show that our program can quickly handle common traffic accidents.

References

1. Žabenský, J., Ščurek, R., Jeremy, T.W.H.: Experimental verification of selected risk factors disrupting ecall system function. Trans. Všb Tech. Univ. Ostrava Saf. Eng. **10**(1), 15–19 (2015)
2. Xiao, W., Li, Z., Chen, W., Lu, Y., Jiang, H.: Trigger algorithm of vehicle automatic crash notification system. Int. J. Automot. Technol. **17**(2), 273–280 (2016)
3. Kononen, D.W., Flannagan, C.A., Wang, S.C.: Identification and validation of a logistic regression model for predicting serious injuries associated with motor vehicle crashes. Accid. Anal. Prev. **43**(1), 112–122 (2011)
4. Stitzel, J.D., Weaver, A.A., Talton, J.W., Barnard, R.T., Schoell, S.L., Doud, A.N., et al.: An injury severity-, time sensitivity-, and predictability-based advanced automatic crash notification algorithm improves motor vehicle crash occupant triage. J. Am. Coll. Surg. **222**(6), 1211 (2016)
5. Aal-Nouman, M., Takruri-Rizk, H., Hope, M.: Efficient message transmission method for in-vehicle emergency service. In: International Conference on Information Communication and Management. IEEE (2016)
6. Izumi, Y., Ogino, H., Feiler, F.: Akkus für e-call-systeme. Atzelektronik **8**(2), 116–121 (2013)
7. Geisler, S., Michelini, M.: A study of crash rates for vehicles with advanced crash avoidance features (2011)
8. Narasimharao, T.L., Venkatrao, G.: Ecall-newly emerging ecal safety, equipment standard and algorithms. Int. J. Eng. Trends Technol. **4**(9), 4064–4073 (2013)
9. Champion, H.R., Augenstein, J., Blatt, A.J., et al.: Automatic crash notification and the urgency algorithm: its history, value, and use. Adv. Emerg. Nurs. J. **26**(2), 143–156 (2004)
10. ÖRni, R., Meilikhov, E., Korhonen, T.O.: Interoperability of ecall and era-glonass in-vehicle emergency call systems. Intell. Transp. Syst. Iet **9**(6), 582–590 (2014)
11. Fernandes, B., Alam, M., Gomes, V., Ferreira, J., Oliveira, A.: Automatic accident detection with multi-modal alert system implementation for its. Veh. Commun. **3**, 1–11 (2016)
12. Blancou, J., Almeida, J., Fernandes, B., Silva, L., Alam, M., Fonseca, J., et al.: eCall ++: an enhanced emergency call system for improved road safety. In: Vehicular NETWORKING Conference, pp. 1–8. IEEE (2017)

The Implementation of Peeping in the Blood Vessels with Tiny Endoscope on Raspberry Pi Platform

Kang Qiu[1], Kuiliang Song[2], and Ling Li[1](✉)

[1] ShenZhen Institutes of Advanced Technology, Chinese Academy of Sciences, Beijing, China
ling.li@sait.ac.cn
[2] ShanDong ZhongBaoKang Medical Implements CO.TLD, Zibo, China

Abstract. In this experiment, the raspberry pi which has carried opencv and mjpg-stream software package is the control center, and make the tiny endoscope work. This experiment has peeped in bionic vascular with tiny endoscope on raspberry pi platform. And in the result of this experiment, we can get the video in bionic vascular and also can get the picture clearly of established pathological tissue. But nowadays, in hospital, all the treatment of vascular disease is based on virtual images synthesized by various medical images through software. In this experiment, we could get more visual pictures in bionic vascular, and it's a way of treating vascular disease in the future.

Keywords: Opencv · Raspberry pi · Tiny endoscope · Bionic vascula

1 Introduction

Vascular disease is one of the most serious diseases in the world today, and its morbidity and mortality have exceeded the tumor disease and the first place. In 2008, global sales of cardiovascular drugs exceeded $100 billion. According to medical statistics report published by the ministry of health in our country, the incidence of cardiovascular disease in the late 1940s in China is only 5.1%, and in the early 1980s to 13.6%, to 2002, people over the age of 18 is 18.8%, and 50–70 - year - old middle-aged people which have vascular diseases, such as hypertension, diabetes, obesity, has more than 70%. According to statistics, in recent years, the population of death from disease has more than 40% which is death from cardiovascular disease. So, vascular disease has become the first cause of death of our population and the first killer of our people.

Nowadays, hospitals often use medical images such as x-rays, CT, MRI and ultrasonic to check and treat various vascular diseases. With the development of science and technology, these high-tech products are becoming more and more precise and intuitive. But at the end of the day, there is a bias in the accuracy and visualization of these things, especially in the context of a less complex image.

But with the miniaturization of semiconductors, the manufacturing process has become more sophisticated, making the miniature cameras shrink in size [2, 5, 6]. The new tiny endoscope can be as small as 1 mm, which is enough to enter the heart of the

K. Deng et al. (Eds.): ICMIR 2018, AISC 856, pp. 561–566, 2019.
https://doi.org/10.1007/978-3-030-00214-5_71

brain. In addition to the small and powerful features of the popular single-board computer raspberry pi, these can perfectly combine to complete certain control experiments.

2 Experimental and Procedure Details

2.1 Introduction of Raspberry Pi

The Raspberry Pi is a computer-generated Raspberry Pi board developed and sold by Eben Upton of University of Cambridge, UK [1]. And It is a microcomputer motherboard based on ARM, with SD/MicroSD card as the memory hard disk, with 1/2/4 USB interface and A 10/100 Ethernet interface (A type without A network port), which can connect the keyboard, mouse and network cable, and the TV output interface of video analog signal and HDMI video output interface. All of the above parts are integrated in a main board that is only slightly larger than the credit card, and the basic functions of all PCS need to be connected to the TV and keyboard, and can perform many functions such as electronic forms, word processing, playing games and playing high-definition video. What's more, Raspberry Pi B provides only computer board with no memory, power, keyboard, chassis or attachment.

One advantage of the Raspberry Pi is that it is a small computer, With relatively small volume, the latest Raspberry Pi 3B version equipped with a 1.2 GHz 64-bit quad-core processor, power consumption reduced to $0.5 \sim 1$ W [1]. Therefore, the Raspberry Pi has a strong computing power, and small size is also easy to become mobile. Simply for the image of the application, Raspberry Pi can be used as the core of data processing, is widely used in video surveillance, the target recognition, and other aspects [2].

Raspberry Pi base on Linux system, has a strong open source feature, users can develop and design according to their needs, its flexibility attracts more and more Raspberry Pi fans [1]. So the tiny endoscopic's control and video's extracted work in this paper is the Linux system (Fig. 1).

Fig. 1. Raspberry Pi B+ microcomputer board

2.2 Instructions of Bionic Blood Vessels and Tiny Endoscopic

The bionic blood vessels in this experiment was based on someone's cerebral arteries. Which was three-dimensional model made with silica gel and other composite materials. And it has most of the properties of real blood vessels (Fig. 2).

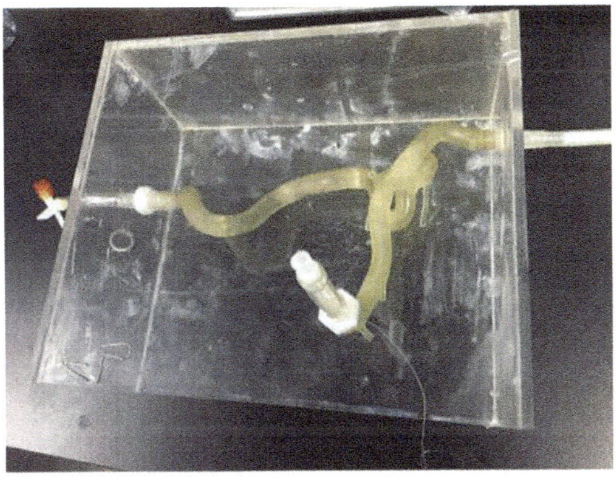

Fig. 2. Bionic blood vessels

Tiny endoscope mainly consists of two parts: tiny camera and micro-controller. What's more, tiny camera are made of the COMS sensor, LED white light source and expandable He–Na laser light source. The specification of tiny camera is 2.0 * 4.5 mm, and it can adapt to −10 to 50 degrees Celsius working environment. And the specific parameters are shown in Table 1:

Table 1. Correlation parameters of tiny endoscope

Camera module	
Item	Description
Power input	DC 3.6 V ± 0.1 V(TV output) ~ 5 V ± 10% (USB output)
Frame rate	30 fps
Power consumption	With TV output Board: 215 mA ± 10 mA (include LED40ma) With USB output Board: 130 mA ± 10 mA (include LED40ma)
Sensor device	Color COMS analog half VGA(400 * 400) Image Sensor
Sensor scan mode	Progressive Scan
Frame rate	30 fps
Light source	LED 0201-Warm color * 4pcs
Operating temperature	−10 °C to + 50 °C
Operating humidity	30% ~ 80%Rh
Cable	OD 1.0 mm * 1 m Cable (Available to 2 m)

2.3 Establishment of Experiment Environment

The establishment of the experimental platform about Raspberry Pi is mainly divided into two parts. One is creating opencv library environment, and another is building video mjpg-stream data stream on Raspberry Pi.

In this experiment, Opencv library's main function is controlling the tiny endoscope, reading and saving the video. And it's main installing include installing compile tools, such as cmake, picture package, video package and management package of python. At last, we also need to pass a series of operation of path and building processing. The main processing as the Table 2:

Table 2. Correlation parameters of tiny endoscope

No.	Installation instructions
1	Sudo apt-get install build-essential pkg-config cmake
2	Sudo apt-get install libjpeg8-dev libtiff5-dev libjasper-dev libpng12-dev..
3	Sudo apt-get install libgtk2.0-dev gfortran
4	wget http://bootstrap.pypa.io/get-pip.py
5	sudo pip install virtualenv virtualenvwrapper
6	Cmake -D CMAKE_BUILD_TYPE = RELEASE…
7	Sudo make install…

However, the mjpg-stream network data stream mainly implements compression of video data from tiny endoscope. And send there data to mount on network port. After all, the realization of this process need to pass steps of collecting video data source, compressing video data and mounting data on network port. The main process as follows:

Step 1: Installing necessary file libraries

sudo apt-get install libjpeg62-dev
sudo apt-get install libjpeg8-dev
sudo apt-get install imagemagick
sudo apt-get install libv4 l-dev
sudo apt-get install cmake
sudo apt-get install git

Step 2: Installing mjpg-streamer

git clone https://github.com/jacksonliam/mjpg-streamer.git
cd mjpg-streamer/mjpg-streamer-experimental

Step 3: Compiling

make clean all
sudo make install
At that, all configuration about Raspberry Pi was finished.

3 Results and Discussion

After configuring all the required environments, we can start relevant command and all the program such as, mjpg-stream data stream and the python program which was based on opencv library. At this step, we should let tiny endoscope enter the bionic blood vessel. And after that, we could see clearly about bionic blood vessels when it move forward slowly. At the same time, all there videos will save in out U disk.

There are two picture as follows: A is the vessel wall while tiny endoscope is moving. B is a specific region of the vessel wall (Fig. 3).

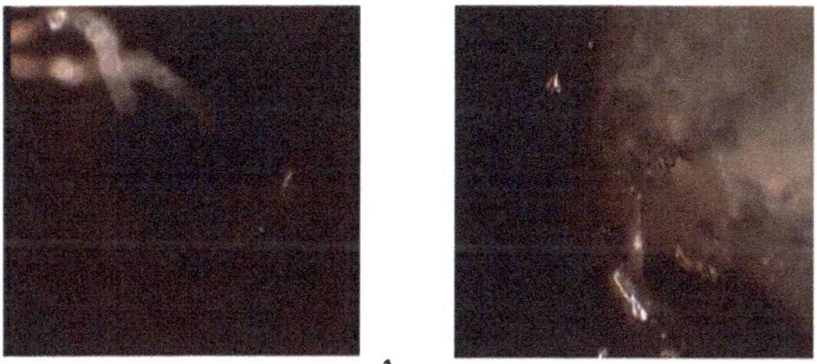

A B

Fig. 3. Picture in bionic blood vessels

From picture A, we can see the bionic blood vessel clearly, and we can easy sure the shape of this vessel. And from picture B, it's a particular area of the bionic blood vessels, we also can get every detail from the picture. So, the video from this experiment give us a more intuitive understand in blood vessels.

4 Conclusions

In this article, the whole experimental control is built on the raspberry pie platform. Raspberry Pi was used to control and drive the tiny endoscope to start and display. At the same time, it's also used to compress and mount video data. In this process, the python program based on opencv will be start to save the video from tiny endoscope in the U disk.

In a word, this experiment has implemented peeping in the blood vessels with tiny endoscope on Raspberry Pi platform. And from the result of experiment, the video from tiny endoscope has a more intuitive understanding of blood vessels, which can provide a more intuitive and accurate solution for some vascular diseases. What's more, there is a clearer understanding and more accurate positioning of vascular lesions.

However, there are still many areas to be improved and inadequate in this experiment:

First, this experiment used bionic blood vessels, and did not add any liquid to simulate the flow of blood.

Second, this experiment was conducted completely in a transparent bionic vascular vessels, which was different from the corresponding animal experiment.

Third, the light source of tiny endoscope is weak, and that made the view of video not clear. And the light source is also very single.

Forth, The low power of the built-in wireless device of the Raspberry Pi and the poor compression efficiency of video image in this experiment resulted in the lack of smooth image of video.

Acknowledgements. This study was financed partially by the National Natural Science Foundation of China (NSFC: 61501444), Shenzhen Technology Development Project Fund under Grant JSGG20160429192140681.

References

1. Feng, Z., Guo, Y., Xing, J., Huo, X.: The implementation of Keren imagie registration system on raspberry Pi platform. In: Proceeding of 36th Chinese Control Conference (2017)
2. Su, X., Chen, W., Yan, S.: Open platform for Internet of things based on raspberry Pi. Electron. Sci. Technol. (2015). https://doi.org/10.16180/j.cnki.issn1007-7820.2015.09.009
3. Su, Y.: Saving strategy based on video data stream from the video of network. Microcomput. Inf. **10** (2005)
4. Li, L., Wang, C., Zhao, F.: A smart home control system integrated the improved streaming media technology and raspberry Pi. In: Proceedings of 2015SSR International Conference on Social Sciences and Information (SSR-SSI 2015 V11) (2015)
5. Chen, R.: Based on open source software and hardware of smart home system design and implementation. Mod. Comput. (2013)
6. Pen, Q., Yang, T., Chen, W.: Strategy of decoding technology and display control in remote video monitoring system. Autom. Power Syst. (2002)

Robotics

Research on Trajectory Planning Technology of Coordinate Measuring Robot

Zhao Jiliang[1(✉)], Ni Bo[2], Su Yu[1], Li Hongbo[1], Shi Hailin[3], and Xin Jindong[3]

[1] Aerospace System Engineering Shanghai, Shanghai, China
ZJL3657253@163.com
[2] Shanghai Precision Measuring and Testing Institute, Shanghai, China
[3] Jiujiang Vocational and Technical College, Jiujiang, Jiangxi Province, China

Abstract. The coordinate measuring robot is a new type of measuring robot which is different from the traditional three coordinate measuring machines. For deeply research on coordinate measuring robot, the kinematics and the trajectory planning technology of the robot should be analyzed and studied. Firstly, according to the theoretical model of the forward and inverse kinematics of the robot which established by D-H method, the simulation model of the forward and reverse solutions of the robot is built by using the SIMMECHANICS and M functions of MATLAB, which verifies the correctness of the kinematics; secondly, random object trajectory is planned, and the object trajectory and the simulation trajectory are compared. Finally, the correctness and feasibility of the principle of trajectory planning are verified in this paper, which lays a foundation for the next work.

Keywords: Coordinate measuring robot · Kinematics · SIMMECHANICS M function · Trajectory planning

1 Introduction

The coordinate measuring machine has the characteristics of versatility, wide measurement range, high accuracy, high efficiency and good performance. It is a new precision measuring instrument developed rapidly in recent years. The current measuring machine is mainly the traditional three coordinate measuring machine (CMM). The traditional three coordinates have the advantages of high measurement precision and simple motion of the head model, nevertheless, both the cost and the measurement requirements are higher [1].

Compared with the traditional measuring machine, the coordinate measuring robot has the advantages of small volume, light weight, low cost and simple structure, and it can achieve high precision measurement at the operation site. Compared with the traditional measurement tools, the measurement robot can measure the physical quantity of the object by remote control, which not only has the advantages of high measuring precision and high degree of automation, but also can realize the measurement of geometric tolerance such as flatness, straightness, etc. Trajectory planning is the basis of robot research, and the kinematics forward and inverse solution of the

K. Deng et al. (Eds.): ICMIR 2018, AISC 856, pp. 569–577, 2019.
https://doi.org/10.1007/978-3-030-00214-5_72

robot is the basis of trajectory planning research. Therefore, in order to further study the measurement robot, we need to analyze and research the kinematics of the measuring robot, and then the robot trajectory planning technology.

2 General Structure of the Robot

The structure diagram of a coordinate measuring robot is shown in Fig. 1. The coordinate measuring robot has three degrees of freedom (3-RRP), including the rotation joint 1, the rotating joint 2 and the straight motion joint.

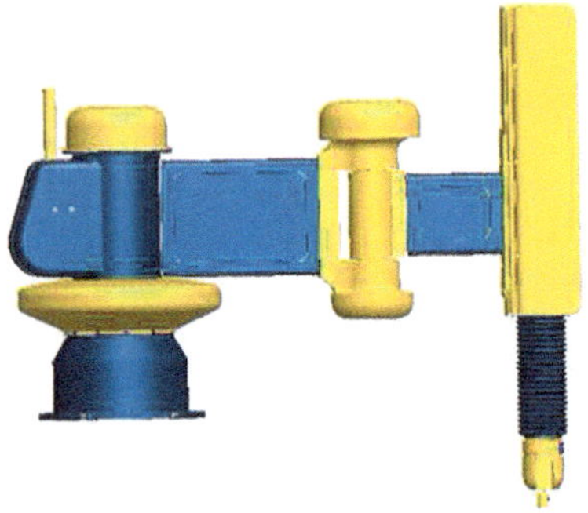

Fig. 1. The structure diagram of a coordinate measuring robot

The measuring robot can measure the target object by the control system in the measuring space. In addition to the measurement of physical quantities such as length and diameter, measuring robots can also measure the tolerance of flatness, straightness and parallelism.

3 Kinematics Forward and Reverse Solution

The kinematic positive solution of a coordinate measuring robot can be solved by means of algebraic method, and the inverse one can be solved by means of geometry or algebra or the combination of the two. Here the kinematic positive solutions are solved by algebraic method, and the kinematics solution is first based on the D-H method to establish the robot kinematics model. The D-H method means that: by establishing the coordinate system in each joint and using a 4×4 homogeneous transformation matrix to describe the spatial relationship of two adjacent joint arms, the equivalent homogeneous transformation matrix of the robot terminal point coordinate system and the base coordinate system are derived, and their kinematic models are established [2–4].

According to the coordinate system established by D-H method, as shown in Fig. 2, the mathematical models in any two adjacent joints of the coordinate measuring robot can be expressed by four parameters (a_{i-1}, α_{i-1}, d_i, θ_i) [5].

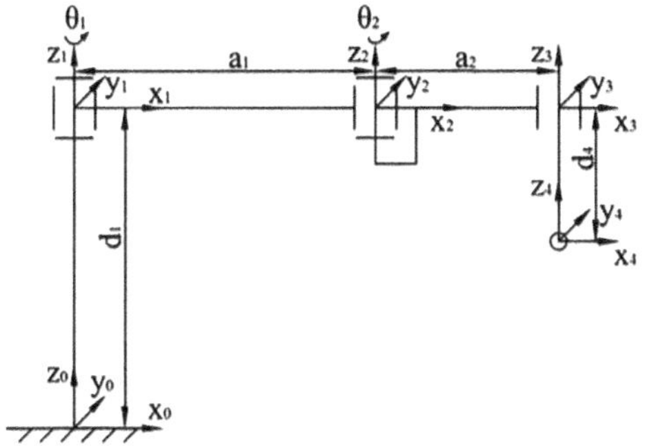

Fig. 2. The establishment of coordinate measuring robot coordinate system

where, a_1` = 400 mm, a_2 = 300 mm, d_1 = 300 mm, d_4 = 50 mm. Therefore, ${}^0_1\mathrm{T}$, ${}^1_2\mathrm{T}$, ${}^2_3\mathrm{T}$ and ${}^3_4\mathrm{T}$ can be determined.

So

$$ {}^0_4\mathrm{T} = {}^0_1\mathrm{T}^1_2\mathrm{T}^2_3\mathrm{T}^3_4\mathrm{T} = \begin{bmatrix} n_x & o_x & a_x & p_x \\ n_y & o_y & a_y & p_y \\ n_z & o_z & a_z & p_z \\ 0 & 0 & 0 & 1 \end{bmatrix} \quad (1) $$

where,

$$ p = \begin{bmatrix} p_x \\ p_y \\ p_z \end{bmatrix} = \begin{bmatrix} \cos(\theta_1 + \theta_2) \times a_2 + \cos(\theta_1) \times a_1 \\ \sin(\theta_1 + \theta_2) \times a_2 + \sin(\theta_1) \times a_1 \\ d_1 - d_3 - d_4 \end{bmatrix} \quad (2) $$

Considering the multiplicity of solutions of the robot, in order to obtain the only analytical solution, the range of the joint motion of the robot must be limited as follows.

$$ \theta_1 \in [0°, 360°), \quad \theta_2 \in [0°, 150°] $$

Based on the range defined by the angle of the joint motion, the inverse kinematics of the robot can be solved by the geometric method:

$$ \begin{cases} \theta_1 = \sin^{-1}\left(\frac{y}{\sqrt{\mathrm{X}^2 + \mathrm{Y}^2}}\right) - \cos^{-1}\left(\frac{\mathrm{X}^2 + \mathrm{Y}^2 + a_1^2 - a_2^2}{2a_1\sqrt{\mathrm{X}^2 + \mathrm{Y}^2}}\right) \\ \theta_2 = \pi - \cos^{-1}\left(\frac{a_2^2 + a_1^2 - (\mathrm{X}^2 + \mathrm{Y}^2)}{2a_1 a_2}\right) \\ d_3 = d_1 - d_4 - \mathrm{z} \end{cases} \quad (3) $$

4 Simulation Verification of Kinematics Forward and Reverse Solution

The kinematics forward and reverse solution is the basis for the study of the robot, and the correctness of the proposed kinematic model can be verified by simulation. The steps of simulation verification are: (1) The input curve is given according to the range of motion of the joint variable; (2) According to the structural characteristics, a mechanism model is built based on SIMMECHANICS [6, 7], and the simulation model of robot inverse solution based on the theoretical model can be derived by MATLAB's M functions; (3) Through the given joint motion curve and the SIMMECHANICS model of the robot, the trajectory of the origin of the robot terminal coordinate system in the basic coordinate system is solved; (4) Using the "MATLAB FUNCTION" module to call the M function according to the trajectories obtained above, the variable values of the joint variables in the joint space coordinate system are obtained; (5) Compared to the variable values of the joint variables obtained by reverse solution and those of joint variable of simulation input, the correctness of the kinematic theory model is verified.

4.1 Simulation Input of Joint Variables

The input curve here is the input value of the joint variable in the joint space. The range of motion of the joint variables, θ1, θ2 and d3 were: [0°, 360°), [0°, 150°] and [0, 150] mm respectively, θ1 and θ2 were expressed with cosine, and d3 was expressed in a straight line. The following Fig. 3 is the joint variable θ1 simulation input curve, the lower Fig. 4 is the joint variable θ2 simulation input curve, and the lower Fig. 5 is the joint variable d3 simulation input curve.

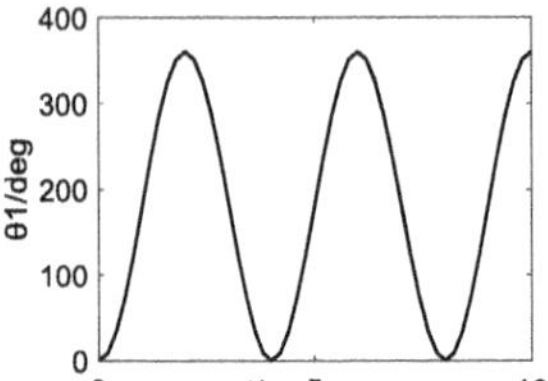

Fig. 3. The input curve of the joint variable θ_1

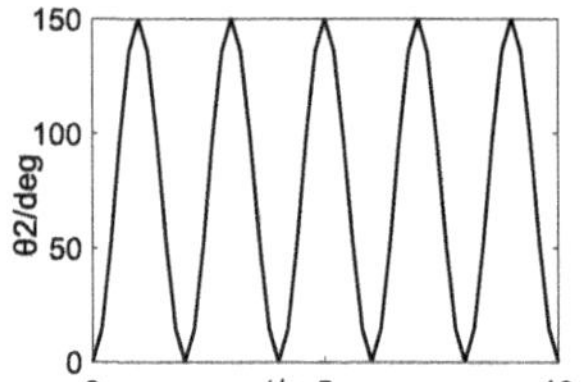

Fig. 4. The input curve of the joint variable θ_2

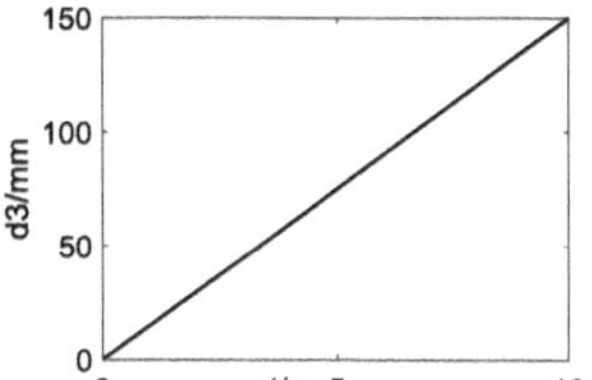

Fig. 5. The input curve of the joint variable d_3

4.2 Simulation Model

The kinematics forward solution simulation model of the mechanism is built by SIMMECHANICS, and the kinematics reverse simulation model of the mechanism is built by M function; then two simulation models are simulated in the same simulation environment. The following Fig. 6 is a simulation model built.

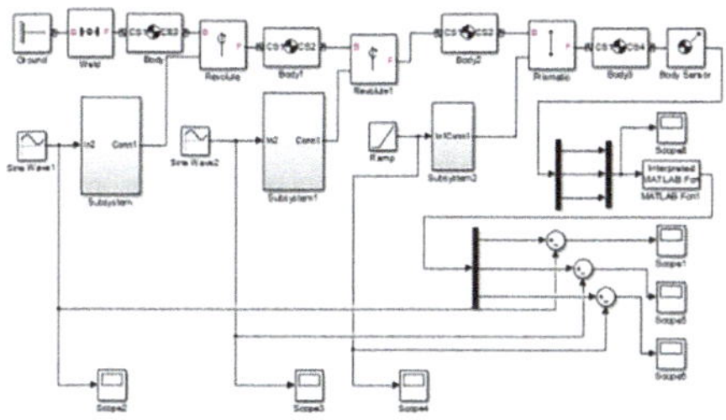

Fig. 6. Kinematics simulation model

4.3 Simulation Output

The simulation model is simulated, and the mechanism in the simulation is shown in the following Fig. 7.

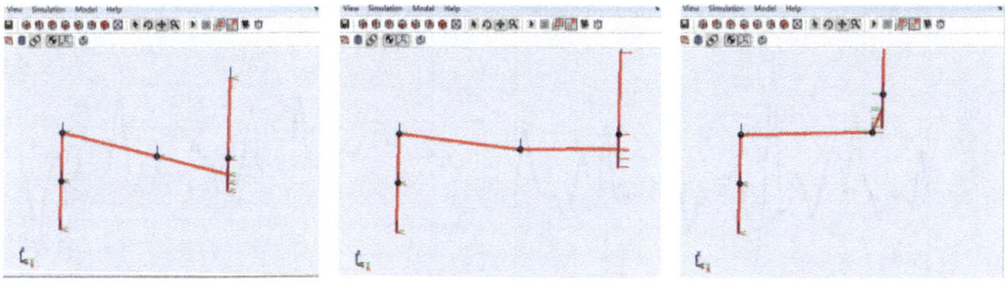

Initial state of mechanism | Simulating process 1 | Simulating process 2

Fig. 7. Robot in the simulation process

In the above model, the "body sensor" module is used to detect the coordinate values of the end of the robot, as shown in the following Fig. 8.

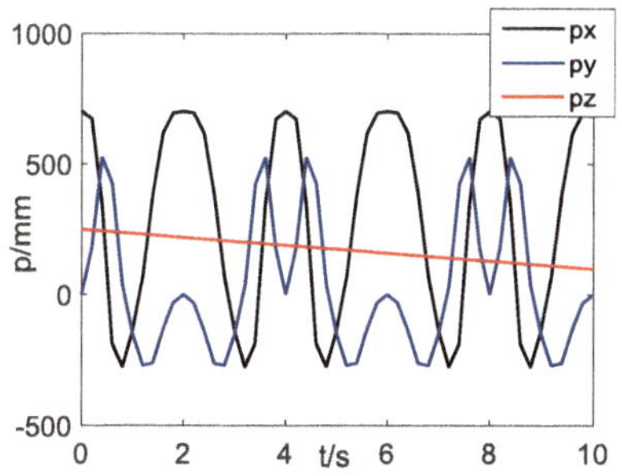

Fig. 8. The coordinate values of the terminal point of the robot

According to the terminal point trajectory of robot, the variable values of joint variables of each joint in the joint space coordinate system are solved by using kinematic inverse solution simulation model built by M function, and the variable values of the joint variables obtained by reverse solution and those of joint variable of simulation input are compared and analyzed. The following Fig. 9 is the error curve of the input value and the inverse solution of the joint variable $\theta_{1,}$ θ_2 and d_3.

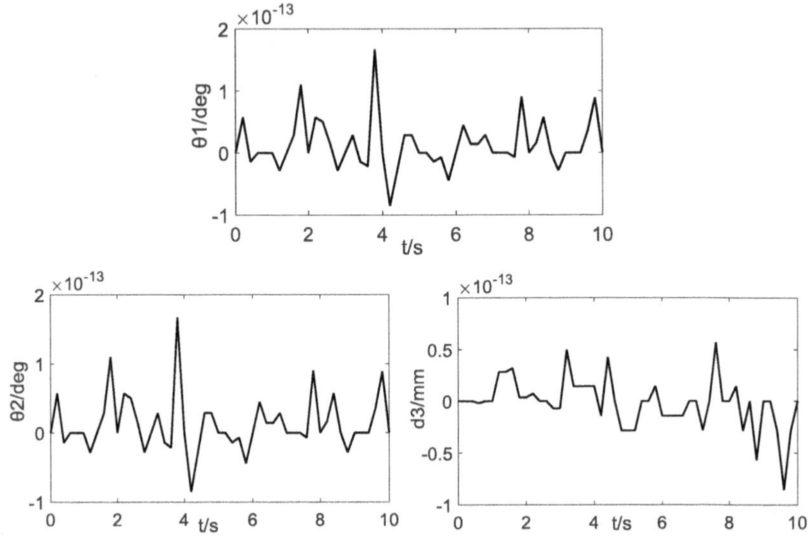

Fig. 9. The error of the input value and the inverse solution of the joint variable

It can be seen from the above Figure that all the error of the input value and the inverse solution of the joint variable are within 10^{-3}, which caused by the calculation error of the computer, and the correctness of the kinematic theory model is verified.

5 Robot Working Space and Trajectory Planning

5.1 Trajectory Planning

The correctness of the theoretical model of the kinematics forward and reverse solution is verified and the workspace of the robot is examined. In the workspace above, planning a trajectory curve arbitrarily, the variable values of joint variables in joint space are solved by using robot kinematics reverse solution model. Inputting the value of this variable into the SIMMECHANICS forward solution simulation model, the error between the simulation trajectory and the target trajectory of the robot terminal point is examined, and the correctness and feasibility of the robot trajectory planning technology principle are investigated. The following Fig. 10 is the simulation model of robot trajectory planning.

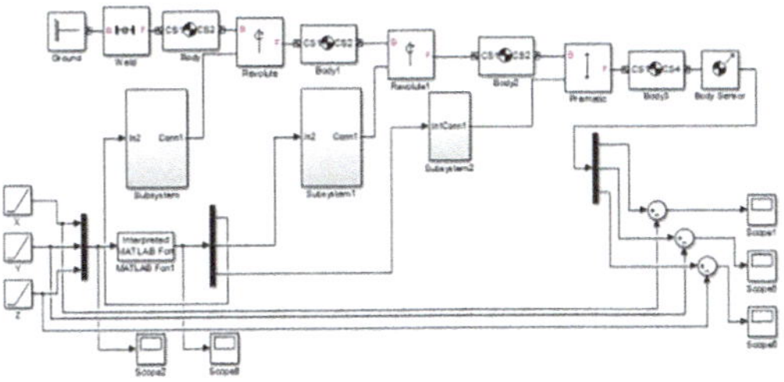

Fig. 10. The simulation model

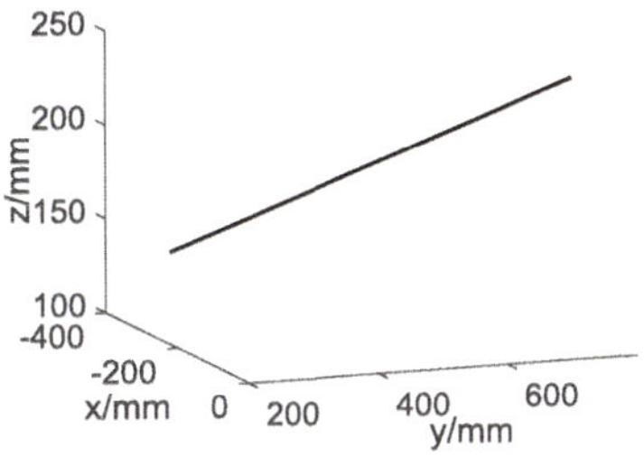

Fig. 11. The target trajectory

Arbitrarily planning a motion trajectory in the workspace of a robot by using MATLAB, it is shown in the following Fig. 11.

According to the target trajectory in the above Figure, the variable value of the joint variable in the joint space is solved by using the kinematics reverse solution model, and below Fig. 12 is the variable simulation curve of every joint variable in joint space.

The joint variable values in Fig. 12 are inputted into the SIMMECHANICS forward solution simulation model to see the consistency of the simulation trajectory at the terminal point of the robot and the target trajectory. The following Fig. 13 is the comparison between the target trajectory and the simulation trajectory.

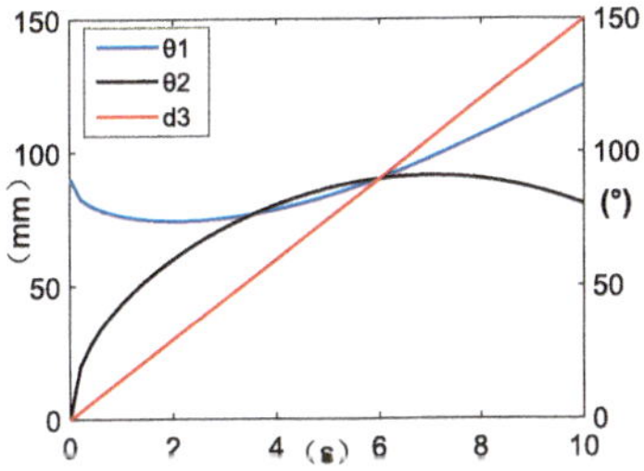

Fig. 12. The variable simulation curve

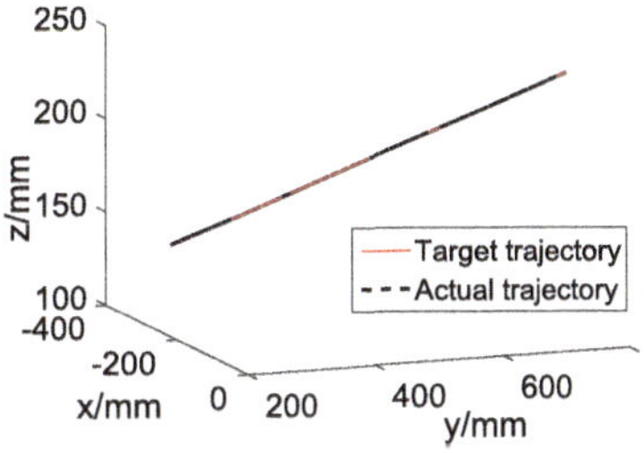

Fig. 13. Comparison between the target trajectory and the simulation trajectory

From the above picture, it can be seen that the target trajectory curve is exactly the same as the simulation track curve. In this paper, the target trajectory and the simulation trajectory are compared and analyzed in the three directions of X, Y and Z, and the error values of the trajectory in the three directions are examined. Figure 14 is the error of the target trajectory and the simulation track in the X, Y and Z direction;

As can be seen from Figs. 13 and 14, the error of the target trajectory curve and the simulation trajectory curve in the three directions of the X, Y, and Z is within 10^{-12},

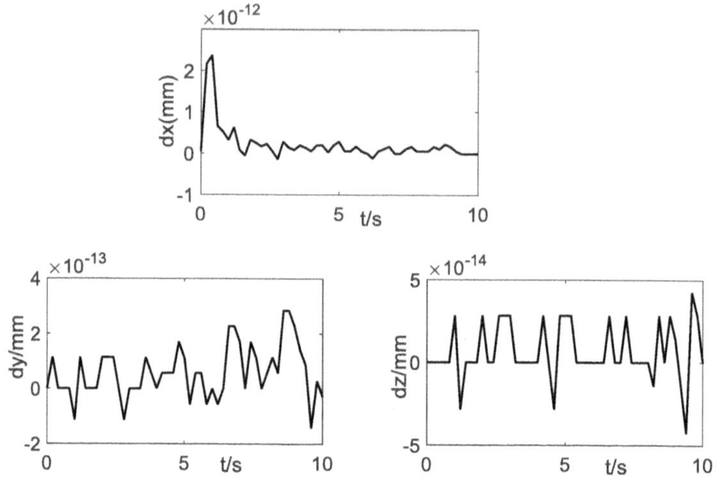

Fig. 14. The error of the target trajectory and the simulation track

which caused by the calculation error of the computer, and the correctness of the kinematic theory model is verified.

6 Conclusion

In this paper, the trajectory planning technology of coordinate measuring robot is analyzed and researched by that: firstly, the theoretical model of the forward and inverse kinematics of the robot is established by D-H method; secondly, the simulation models of the SIMMECHANICS forward solution and M-function reverse solution are established and simulated to verify the correctness of the derivation of the theoretical model; finally, the trajectory of any target is planned. The correctness and feasibility of the principle of trajectory planning are verified in this paper, which lays a foundation for the next work.

Acknowledgments. The authors would like to think the reviewers for their very helpful comments and suggestions which have improved the presentation of the paper.

References

1. Zhang, Z.J., Che, R.S.: New field of bionics metrology vision coordinate measuring machine. Space Meas. Tech. **18**, 13–17 (1998)
2. Denavit, J., Hartenberg, R.: A kinematic rotation for lower pair mechanism based on matrices. ASME J. Appl. Mech. **22**, 215–221 (1995)
3. Wang, X.Y., Liu, S.G., Wang, B., Zhang, G.X.: Mathematical model and error analysis of the articulated arm flexible CMM. Nanotechnol. Precis. Eng. **3**, 262–266 (2005)

4. Xiong, Y.L.: Robotics Foundation, pp. 66–67. Huazhong University of Science and Technology Press, Wuhan (1996)
5. Ye, D., Huang, Q.C., Che, R.S.: A error model for multi joint coordinate measuring machine. Opt. Precis. Eng. **2**(7), 92–95 (1999)
6. Zhang, L.X., Dong, Y.H.: Mechanical and Electrical System Simulation and Design, 1st edn, pp. 5–22. Harbin Engineering University Press, Harbin (2006)
7. Xue, D.Y., Chen, Y.Q.: System Simulation Technology and Application Based on MATLAB/Simulink, 2nd edn, pp. 10–30. Publishing House of Tsinghua University, Beijing (2002)

Optimal Design and Analysis of a Powered Ankle-Foot Prosthesis with Adjustable Actuation Stiffness

Dianbiao Dong, Wenjie Ge(✉), Jianting Wang, Yuanxi Sun, and Donglai Zhao

School of Mechanical Engineering, Northwestern Polytechnical University, Xi'an, China
gwj@nwpu.edu.cn

Abstract. In past decades, the technology of ankle-foot prosthesis is rapidly developing for lower-limb amputees to restore the lost function of the human ankle. Several powered ankle-foot prostheses are proposed with good performance on lever-ground walking at normal speed. However, compared with prosthesis focusing on a specific speed, the adaptability of different walking speeds is still a big challenge for current powered ankle-foot prosthesis. In this paper, the authors present the optimal design and analysis of a novel powered ankle-foot prosthesis by combining the typical SEA actuator with a length-adjustable lever arm. The prosthesis can adjust the system stiffness by changing the length of the lever arm. After the optimization of the length of the lever arm, the novel prosthesis can provide amputee good performance on torque mimicking at low speed, high energy efficiency at normal speed and reduction of peak power at high speed. This paper also gives insight to the benefit of adjustable mechanical system on the performance of the prosthesis to the future design of the powered ankle-foot prosthesis at different walking speeds.

Keywords: Ankle-foot · Actuator · Adjustable stiffness · Prosthesis Optimization

1 Introduction

Due to the increasing requirements of lower-limb amputees in their daily life, the technology of ankle-foot prosthesis is rapidly developing. In recent years, more and more ankle-foot prostheses are designed to restore the ankle function of the lower-limb amputees. Most of the current ankle-foot prosthesis are passive, such as the solid ankle cushion heel (SACH) foot and the energy store and release (ESAR) foot. The SACH foot can only give stable vertical support during locomotion while the ESAR foot can return energy during the control plantarflexion phase of human walking gait, which allows amputee faster walking speed [1]. Due to the development of the actuator technology [2] and the human biomechanics [3, 4], the powered ankle-foot prosthesis is designed to better mimic the human ankle function. Compared with the common commercial passive ankle-foot prosthesis, the powered ankle-foot prosthesis has the ability to provide the amputee with positive ankle work, which is indicated to be

K. Deng et al. (Eds.): ICMIR 2018, AISC 856, pp. 578–587, 2019.
https://doi.org/10.1007/978-3-030-00214-5_73

necessary for human ankle biomechanics. Due to this feature, the powered ankle-foot prosthesis increases the motion range of the ankle and provides amputee more symmetry walking gait [5, 6].

One of the main challenges of the powered ankle-foot prosthesis is the high requirement of actuation systems on peak power and energy consumption [7]. To reduce the requirement, several models combining efficient motor and spring systems are proposed in the rehabilitation community. Hitt [8] proposed a powered ankle-foot prosthesis utilized an elastic part in series with an actuator (SEA). The prosthesis can provide enough positive work by a Maxon RE40 motor at 1.3 m/s walking. Gao [9] presented a novel powered ankle-foot prosthesis combining a spring (elastic element) in parallel with an actuator (PEA). A cam and two linear spring are included in the parallel spring system. With this system, the prosthesis can provide amputee with desired spring-torque profile by using a Maxon DC 35 motor. Au [10] proposed a novel powered ankle-foot prosthesis using a spring in parallel with a controllable actuator with series elasticity (SPEA). It can provide the requirement of normal human biomechanics with an output power of 350 W in acceptable size and weight.

Although several powered ankle-foot prostheses [11–13] have been proven to have a good performance on lever-ground walking in normal speed, the multiple tasks of human motion including variable speeds walking are still giving big challenges to the design of the powered ankle-foot prosthesis. Herr [14] developed a bionic prosthesis using the SPEA technology. The experiment results of seven people showed that it could improve the preferred velocity and give biomechanical patterns across velocities of 0.75–1.75 m/s.

To better mimic the human ankle function in different movement conditions, the human biomechanics should be fully considered. According to the human biomechanics data recorded in [3], the ankle stiffness should be changed based on different walking speeds. Grimmer [15] investigated the adjustment of spring stiffness with an SEA powered ankle-foot prosthesis on peak power and energy consumption. The simulation results show that the variable spring stiffness optimized for specific speed can decrease the demand of peak power and energy consumption. To enhance the adaptability of the powered ankle-foot prosthesis at different walking speeds, the authors introduce the concept design of a novel stiffness adjustable powered ankle-foot prosthesis (SAP) in this paper. By using a length adjustable lever arm, the ankle stiffness can be adjusted to a suitable value, which will improve the performance of ankle-foot prosthesis on different requirements.

2 Biomechanics and System Modeling

2.1 Human Biomechanics

The human biomechanics data used in this paper is from winter's research [16]. As shown in Fig. 1, a common intact human walking gait cycle can be divided into four phases: controlled plantarflexion (CP) phase from initial contact to foot flat, controlled dorsiflexion (CD) phase from foot flat to the max dorsiflexion, powered plantarflexion (PP) phase from the max dorsiflexion to toe off and swing (SW) phase from toe off to the next initial contact. The human ankle has different torque-angle profiles at different speeds, while the torque-angle profile is related to ankle stiffness.

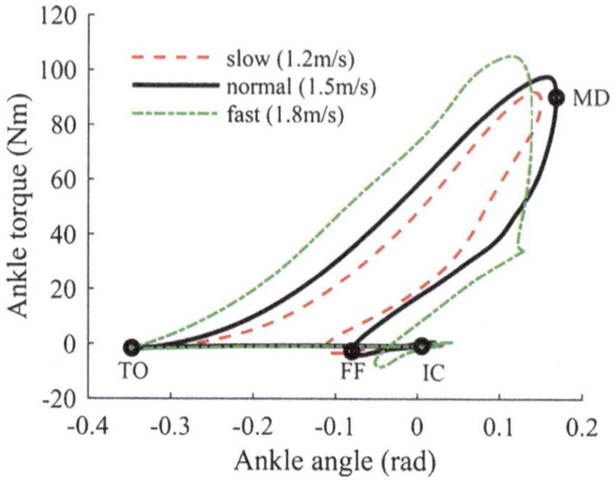

Fig. 1. A human lever-ground walking gait cycle. IC: initial contact, FF: foot flat, MD: max dorsiflexion, TO: toe off.

2.2 Modeling and Calculation of the Novel Mechanism

To reduce the demand of the peak power and energy consumption and improve the adaptability of prosthesis during different speeds, the mechanical model of the novel prosthesis comprises a typical SEA system and an adjustable lever arm (ALA). The schematic of the novel system is shown in Fig. 2.

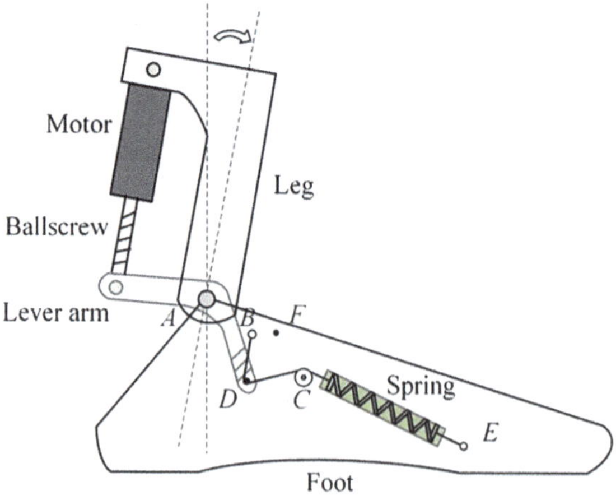

Fig. 2. The schematic of the ankle-foot prosthesis.

The SEA actuator consists of a Maxon EC-4pole 200 W motor, a ball screw, a length adjustable lever arm, a rope and a spring (80 KN/m). The motor and ball screw are hinged between the leg and the lever arm. When the leg rotates forward the ankle joint, the spring will be extended by changing the length of rope between hinge B and pulley C. The hinge B, E and pulley C are fixed on the foot. Besides, a small cylinder is fixed on point F, so that the actuator can rotate the foot and provide torque in counter clock direction. The length of the AD (l_{AD}) can be adjusted by a small motor automatically or manually.

The Force condition of the ALA is depicted in Fig. 3. The rotation angle of the lever arm is determined by the ankle rotation angle θ and the screw driven angle θ^* as:

$$\angle DAD' = \theta + \theta^* \quad (1)$$

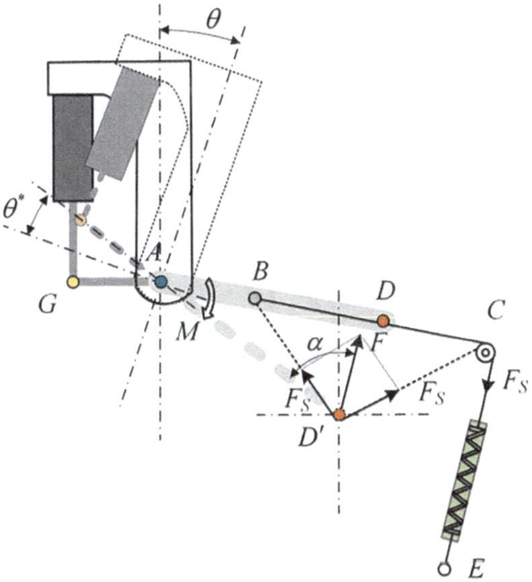

Fig. 3. The force condition of the SAP prosthesis. The solid line depicts the initial position of the prosthesis, while the dot line depicts the current position of the SAP prosthesis after ankle rotates angle θ.

The spring force F_S can be derived from:

$$F_S = k * (l_{BD\prime} + l_{BC}) \quad (2)$$

The impedance torque given by the SEA system is:

$$M = F * l_{AD} * \sin(\alpha) \quad (3)$$

After we match the impedance torque M with human ankle torque T in different tasks, the motor driven angle θ^* and its angular velocity w^* can be determined by above equations. Since the length of AD can be adjusted, we define POS as:

$$\mathrm{POS} = \frac{l_{BD}}{l_{BC}}, \quad 0.1 < \mathrm{POS} < 0.9 \tag{4}$$

After taking the mechanical space into account, POS is set to vary from 0.1 to 0.9. The length of AG (l_{AG}) is set to 0.07 m based on the mechanical size of motor and ball screw. When the initial length of l_{AB} and l_{BC} are set to 0.05 m and 0.05 m respectively, the torque-angle profile of the SEA with different POS can be derived and shown in Fig. 4. The stiffness of the SAP prosthesis system increases with increasing POS.

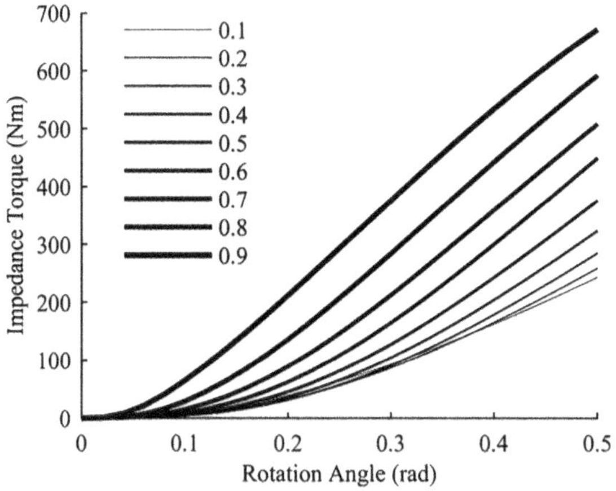

Fig. 4. The torque-angle profile of the SEA system with different POS.

The ball screw has a lead of 4 mm and an efficiency of 90%. The gearbox of the EC motor has a reduction ratio r of 6 and an efficiency of 90%. The ankle torque T is from human data. Hence, the mechanical output of the motor can be derived from:

$$\begin{cases} T_m = \frac{T*0.004}{l_{AG}*r*2\pi*0.9*0.9} \\ w_m = \frac{w*l_{AG}*r*2\pi}{0.004} \\ P_m = T_m * w_m \end{cases} \tag{5}$$

Then, the electrical power P_e and energy E_e can be derived from the FMM model [17] as:

$$\begin{cases} I = (J_m \dot{w}_m + T_m + v_m w_m)/k_t \\ U = L\dot{I} + IR + k_b w_m \\ P_e = UI \\ E_e = \int P_e dt \end{cases} \tag{6}$$

3 Optimization on Multiple Speeds and Discussion

To enhance the adaptability of the prosthesis during different speeds, the ALA is optimized for different proposals based on the specific requirement for different speeds.

3.1 Semi-active Function at Low Speed

When human walks at low speed, the behavior of human ankle is nearly passive [4]. Therefore, to reduce the energy consumption of the EC motor, the prosthesis is designed with a semi-active function at low speed. During CP, the motor follows the rotation of the human ankle. The spring isn't extended in this phase. During CD and PP phase, the motor is locked and the spring is extended. The torque-gait profile of the prosthesis is optimized to match that of the human ankle. During SW phase, the motor is set to the beginning position for the next initial contact.

We use the least squares method (LSM) to evaluate the performance of POS. The length of l_{AB}, l_{BC} and POS are set as optimization variables. After we divided the intact human gait cycle into 1000 point, the optimization objective is defined as:

$$\begin{aligned} &\Phi = \sum_{i=1}^{1000} \left(M_i^{\text{Winter}}(l_{AB}, l_{BC}, \text{POS}) - M_i^{\text{SAP}}(l_{AB}, l_{BC}, \text{POS})\right)^2, \\ &0.02\,\text{m} < l_{AB}, l_{BC} < 0.1\,\text{m}, 0.1 < \text{POS} < 0.9 \end{aligned} \tag{7}$$

The optimal value of l_{AB}, l_{BC} and POS at low speed are 0.055 m, 0.072 m and 0.48. As shown in Fig. 5, POS has low LSM value when it varies from 0.1 to 0.6. Then the LSM value boosts. With an optimal POS value of 0.48, the SAP prosthesis can closely mimic the human ankle torque profile at low speed walking.

3.2 Energy Efficient Active Function at Normal Speed

When human walks at normal speed, the positive ankle work is necessary for providing a symmetry walking gait. Hence, the function of this new prosthesis is designed to be fully active. During CP, the motor is controlled to follow the foot clap until the foot flat. During CD and PP phase, the motor drives the lever arm to further extend the spring to match the human ankle torque-angle curve. The required θ^* and output of the EC motor are given by the optimization. During SW phase, the lever arm is reset to the initial position for the next step.

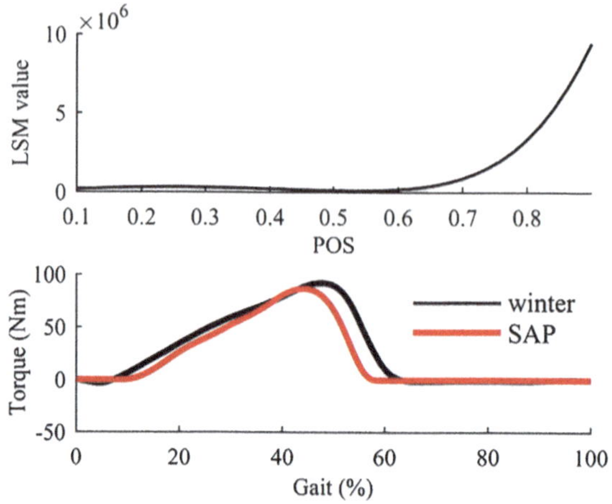

Fig. 5. The optimization results of POS at low speed.

The biggest challenge for prosthesis walking at normal speed is the energy consumption. To extend the walking hours at normal speed, the energy consumption of the SAP prosthesis needs to be optimized. The optimization results of the POS and the electrical power of the EC motor are shown in Fig. 6. The electrical energy consumption is reduced by 34.8% from 77.35 J at POS 0.48 (low speed) to 50.4 J at POS 0.86. The peak power happens at the PP phase with a value of 203 W.

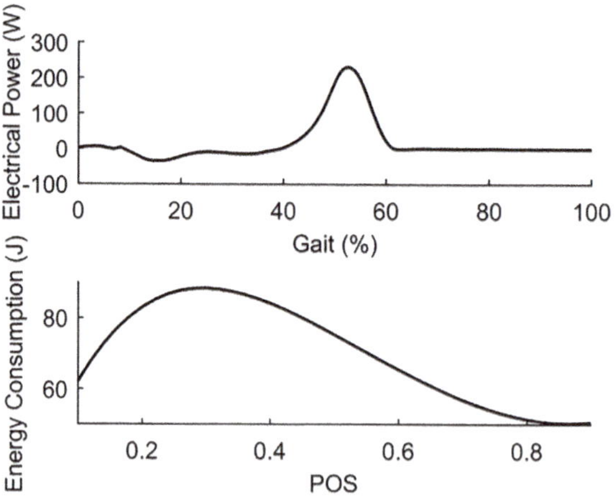

Fig. 6. The optimization results of POS and the electrical power of the EC motor at normal speed.

3.3 Active Function with the Lowest Peak Power at High Speed

The peak power and energy consumption is the main challenge for prosthesis working at high speed walking. In our proposal, we optimized the POS to reduce the electrical peak power of the SAP prosthesis at high speed. To provide amputee with normal gait, the ALA is driven by the EC motor to match the torque profile of human ankle at high speed. During CP, the motor follows the ankle rotation from initial contact to foot flat. During CD and PP, the motor needs to inject energy to provide the high torque. During SW, the ALA is reset to the initial position.

As shown in Fig. 7, the electrical peak power of the motor increases from 510 W at POS 0.1 to 1095 W at POS 0.28, then it decreases to the lowest value of 402 W at POS 0.65. Compared with the electrical power at POS 0.1, the electrical power at optimal POS 0.65 has lower peak power at 9% of the walking gait and higher peak power at 53% of the walking gait. Compared with the mechanical peak power (276 W), the electrical power has higher peak power of 402 W, which is due to the high acceleration velocity in the simulation of electrical power.

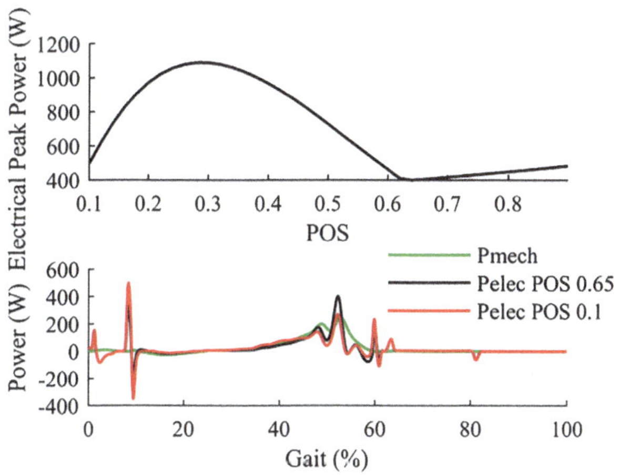

Fig. 7. Optimization results of POS at high speed walking. Pmech and Pelec mean the mechanical and electrical power of the SAP prosthesis respectively.

4 Conclusion

In this paper, the authors propose the concept of a novel stiffness adjustable powered ankle-foot prosthesis by combining a typical SEA actuator and a length adjustable lever arm. Based on the human biomechanics, the optimal length of the adjustable lever arm is given by the optimization of different task among different speeds. The optimization results show that the prosthesis can provide amputee good performance on torque mimicking at low speed walking, energy efficiency at normal speed and reduction of the peak power at high speed. The optimization results also indicate that the adjustable mechanical systems can increase the adaptability of ankle-foot prosthesis at different speeds.

Acknowledgments. Dianbiao Dong is fund by the China Scholarship Council. This work is supported by the National Natural Science Foundation of China (50975230) and Research fund for the Doctoral Program of Higher Education of China (20136102130001).

References

1. Casillas, J.-M., Dulieu, V., Cohen, M., et al.: Bioenergetic comparison of a new energy-storing foot and SACH foot in traumatic below-knee vascular amputations. Arch. Phys. Med. Rehabil. **76**(1), 39–44 (1995)
2. Ham, R., Sugar, T., Vanderborght, B., et al.: Compliant actuator designs. Robot. Autom. Mag. IEEE **16**(3), 81–94 (2009)
3. Bovi, G., Rabuffetti, M., Mazzoleni, P., et al.: A multiple-task gait analysis approach: kinematic, kinetic and EMG reference data for healthy young and adult subjects. Gait Posture **33**(1), 6–13 (2011)
4. Palmer, M., et al.: Sagittal Plane Characterization of Normal Human Ankle Function Across a Range of Walking Gait Speeds. Massachusetts Institute of Technology, Cambridge (2002)
5. Andrysek, J., et al.: Lower-limb prosthetic technologies in the developing world: a review of literature from 1994–2010. Prosthet. Orthot. Int. **34**(4), 378–398 (2010)
6. Versluys, R., Beyl, P.V.M.: Prosthetic feet: State-of-the-art review and the importance of mimicking human ankle–foot biomechanics. Disabil. Rehabil. Assist. Technol. **4**(2), 65–75 (2009)
7. Davis, R., DeLuca, P.: Gait characterization via dynamic joint stiffness. Gait Posture **4**(3), 224–231 (1996)
8. Hitt, J., Bellman, R., Holgate, M., et al.: The sparky (spring ankle with regenerative kinetics) project: design and analysis of a robotic transtibial prosthesis with regenerative kinetics. In: ASME 2007 International Design Engineering Technical Conferences and Computers and Information in Engineering Conference (2007)
9. Gao, F., Liu, Y., Liao, W.: A new powered ankle-foot prosthesis with compact parallel spring mechanism. In: 2016 IEEE International Conference on Robotics and Biomimetics (ROBIO). IEEE (2016)
10. Au, S.K., Weber, J., Herr, H.: Biomechanical design of a powered ankle-foot prosthesis. In: IEEE 10th International Conference on Rehabilitation Robotics IEEE (2007)
11. Cherelle, P., Junius, K., Grosu, V., et al.: The AMP-Foot 2.1: actuator design, control and experiments with an amputee. Robotica **32**(08), 1347–1361 (2014)
12. Shepherd, M., Rouse, E.: The VSPA foot: a quasi-passive ankle-foot prosthesis with continuously variable stiffness. IEEE Trans. Neural Syst. Rehabil. Eng., 1534–4320 (2017)
13. LaPre, A., Wedge, R.: Umberger, B., et al.: Preliminary study of a robotic foot-ankle prosthesis with active alignment. In: IEEE (2017)
14. Herr, H., Grabowski, A.: Bionic ankle–foot prosthesis normalizes walking gait for persons with leg amputation. Proc. R. Soc. Lond. B Biol. Sci. **279**(1728), 457–464 (2012)

15. Grimmer, M., Seyfarth, A.: Stiffness adjustment of a series elastic actuator in an ankle-foot prosthesis for walking and running: the trade-off between energy and peak power optimization. In: 2011 IEEE International Conference on Robotics and Automation (ICRA). IEEE (2011)
16. Winter, D.: Kinematic and kinetic patterns in human gait: variability and compensating effects. Hum. Mov. Sci. **3**(1), 51–76 (1984)
17. Verstraten, T., Furnémont, R., Mathijssen, G., et al.: Energy consumption of geared DC motors in dynamic applications: comparing modeling approaches. IEEE Robot. Autom. Lett. **1**(1), 524–530 (2016)

Suspended Extra Matters Recognition Method Based on Indoor Substation Inspection Robot

Yu Xiaobin[1], Wang Tao[2], Lv Shiliang[1], Mao Kun[1], Dong Xiangning[1], Yu Ningyuan[1], and Yang Guoqing[2(✉)]

[1] State Grid Shandong Electric Power Maintenance Company, Jinan 250118, China

[2] Shandong Luneng Intelligence Technology Co., Ltd., Jinan, China

sdygq2004@163.com

Abstract. Method for identifying suspended extra matters in substations is proposed, which realized real-time monitoring of suspended extra matters in substation. First of all, the digital image processing technology is used to denoise the image. Secondly, the SIFT registration is used to locate the region that analyze and count the number of feature corners in the specified region. Finally, according to the number and the distribution of feature points, the suspended extra matters recognition is carried out. when the feature points statistics exceed specified thresholds, the suspended extra matters is detected and the station staff is warned. Experimental results show that this method has the characteristics of fast calculation, strong adaptability and high accuracy. It can meet the requirement of unattended real-time monitoring and control in substation.

Keywords: Substation · Extra matters recognition · SIFT registration Harris corner

1 Introduction

At present, manual-style inspection method is adopted in the inspection of power substation equipment in our country [1, 2]. That is, substation staffs enter the equipment area and carry out equipment inspection. Current method of manual-style inspection cannot fully meet the requirement of the power industry development, and it brings great difficulties to the security, management and maintenance of power network. The substation is a place with high risk coefficient, and there is a great hidden danger to the staff when the equipment is inspecting, especially in bad weather [3].

Similarly, the inspection of suspended extra matters in substation equipment is mainly carried out by the human eyes. Because of the subjective and fatigue weakness of human eyes, some abnormal situations are sometimes ignored. Moreover, the national electric power dispatching & communication center requires the existing 35 kV, 110 kV and 220 kV to realize unattended operation when the conditions are ripe. Therefore, it is necessary to recognize extra matters based on computer vision. Image matching is one of the basic problem in computer vision [4], and scale invariant feature transform (SIFT) is one of the most active algorithms in the field of image matching. SIFT feature has good invariance for image scaling, translation and rotation,

K. Deng et al. (Eds.): ICMIR 2018, AISC 856, pp. 588–594, 2019.
https://doi.org/10.1007/978-3-030-00214-5_74

and it is also robust to illumination change and affine transformation or 3D projection [5, 6]. So SIFT algorithm has become a popular algorithm in target recognition.

In this paper, the SIFT algorithm is used to match the calibration region of template image and inspection image that collected in the substation. Then, the feature points of the calibration region are detected by the corner detection, and the suspended extra matters is identified by the comparison of the number of feature points and the set thresholds.

2 Image Denoising

In the process of image acquisition and transmission, it is easy to be disturbed by external noise. Therefore, noise removal is a key step in image preprocessing, which directly affects the correct classification of the equipment.

2.1 Median Filter

Median filter is a common nonlinear filtering method, and it is also the most commonly used preprocessing technology in image processing technology. It is very effective in smoothing impulse noise, and it can protect the sharp edges of the image. Median filtering is an optimal filtering method to preserve the boundary and filter the high frequency noise. Especially, it is very effective to deal with the impulsive noise, such as abrupt spikes. The weighted median filter can improve the effect of the edge signal of the median filter [7, 8]. The result of median filter as shown in Fig. 1.

Fig. 1. Median filter

2.2 Gaussian Filter

Gaussian filter is a kind of linear smoothing filter, which is suitable for eliminating the noise of the Gaussian and widely used in image processing [9]. The Gauss filter is a linear filter, which can effectively suppress noise and smooth the image. The function principle is similar to the mean filter, all of which are the mean of the pixels in the filter window as output. The coefficient of the window template is different from that of the mean filter. The template coefficients of the mean filter are all the same as 1, while the Gauss filter template coefficient decreases with the increase of the distance template center. Therefore, the Gauss filter is less blurred than the mean filter [10, 11].

There is a certain relationship between the Gauss filter and the Gauss distribution (normal distribution). A two-dimensional Gauss function is as follows:

$$h(x,y) = e^{-\frac{x^2+y^2}{2\sigma^2}} \tag{1}$$

Where, (x, y) is a point coordinate, which can be considered as an integer in image processing; σ is a standard deviation. The result of Gaussian filter as shown in Fig. 2.

Fig. 2. Gaussian filter

Figures 1 and 2 are the result of median filter and Gauss filter. It can be seen from the figure that the information retained in the median filter is more complete. Therefore, Median filter is selected to de-noise in this paper.

3 Principle of SIFT Algorithm

In 1999, SIFT algorithm [12, 13] was proposed by David G. Lowe in Canada. It is a kind of descriptor based on image local feature, and it was perfected in 2004. SIFT feature points have good invariance for the scale and rotation of images, and can maintain a certain stability for both illumination and 3D viewing angles, so it is recognized as a feature point matching method with good effect by internationally recognized. The main computing steps of SIFT image features are as follows:

Step 1: scale space is structured;
Step 2: precise positioning of key points;
Step 3: the main direction of the feature points is determined;
Step 4: feature vector descriptors are generated.

The two images to be matched are M and N. After the two images are extracted by SIFT feature points, the set of feature points of M is $F_m = \{f_{1m}, f_{2m}, \cdots f_{lm}\}$, and the set of feature points of N is $F_n = \{f_{1n}, f_{2n}, \cdots f_{\mathrm{ln}}\}$. Where, lm and ln are the number of feature points of image M and N respectively. The Euclidean distance of feature vectors is used as a measure of similarity of SIFT feature vectors. The European distance equation is as follows:

$$d(F_{m,}F_n) = \sqrt{\sum_{i=1}^{k} (f_{im} - f_{in})^2} \tag{2}$$

A certain feature point in the matching image is taken, and the nearest distance and the near distance of the Euclidean distance are found in the target image. The pair of matching points are accepted if the distance ratio is greater than a certain threshold. There will be an incorrect matching points after the initial matching is completed in the feature point, and the RANSAC algorithm is used to select the matching points.

4 Simulation Experiment

When there is a suspended extra matters in the calibration area of the substation, the image edge of the calibration area changes obviously. Therefore, the distribution of the feature points can be used for analysis. Therefore, the number and distribution of feature points in the same area of the equipment image and the template image that collected by the inspection robot are compared. If there is a big difference in the number or distribution of feature points, there is a suspended extra matters in this region.

Harris is insensitive to conversion of brightness and contrast, and it has the characteristics of rotation invariance. So the algorithm uses Harris corner to detect the feature point distribution in the region. If a point has a distinct derivative in the X direction and the Y direction, that is, this point has an obvious gradient in the two direction, then the point is considered as a feature point.

The algorithm steps are as follows:

Step 1: region labeling of template image;
Step 2: SIFT registration matching of inspection;
Step 3: preprocessing operations for the calibration area;
Step 4: Harris corner detection;
Step 5: statistical analysis of feature points;
Step 6: compare to threshold;
Step 7: output of test results.

The algorithm flow chart is as follows (Fig. 3):

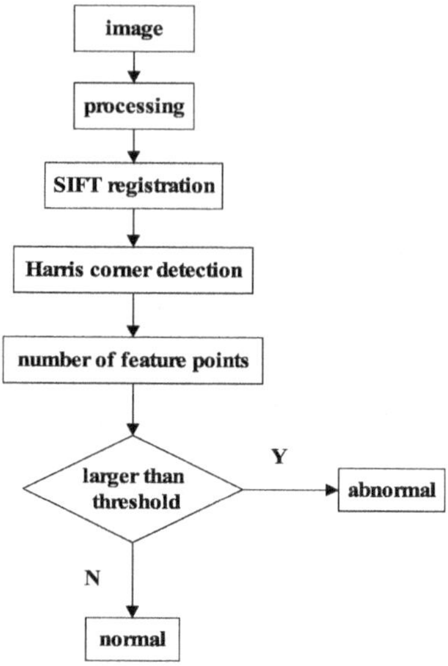

Fig. 3. Flow chart of suspended extra matters detection algorithm

In the experiment, the synthetic image is used to simulate and test the algorithm. As shown in Figs. 4 and 5.

Fig. 4. Template image without suspended extra matters

Fig. 5. Inspection image with suspended extra matters

As shown in Fig. 6, the green rectangle is the device area that needs to be detected. From left to right are template image without suspended extra matters, inspection image without suspended extra matters, and inspection image with suspended extra matters. The experimental results of the algorithm are shown in Fig. 6 ((a) template image without suspended extra matters, (b) inspection image without suspended extra matters, (c) inspection image with suspended extra matters).

(a) (b) (c)

Fig. 6. Test results

As can be seen from the experimental results, the algorithm can output the number of feature points and the number of cluster. Where the number of feature points of the template image is 40 and the cluster is 3 in calibration area. When the inspection image has not suspended extra matters in the calibration area, the number of feature points is 45 and the cluster is 3, which is similar to the template image. When the inspection image has suspended extra matters in the calibration area, the number of feature points is 68 and the cluster is 4, which is obviously different from the number of feature points and the number of cluster in the template image. The number of feature points and cluster detected by the Harris corner is shown in Table 1.

Table 1. Feature points distribution statistics

Test data	Template (without suspended extra matters)	Inspection (without suspended extra matters)	Inspection (exist suspended extra matters)
Feature points	40	45	68
Cluster	3	3	4

From the experimental results, it can be seen that the number of feature points and the number of cluster are changed when there has suspended extra matters, and there is obvious distinguishable. Experimental data show the effectiveness of this algorithm in suspended extra matters recognition.

5 Conclusion

In this paper, Harris feature detection is used to extract feature points in template images and inspection images. The feature points are used to statistical analysis and recognize the foreign objects in the equipment area is achieved. Experimental results show that the algorithm is simple, fast and accurate, and the work efficiency and safety of the staff are improved. The implementation of the algorithm has promoted the unattended operation of substation and realized the automation and intelligence of substation.

Acknowledgements. This work was supported by application research and development of intelligent new technology for substation inspection robot (item serial number: ZY201815).

Thank you for your support of the paper.

References

1. Ding, S.K., Li, J.: Analysis and countermeasures of substation inspection robot in the practical application. Electrotech. Electr. (Jiangsu Electr. Appar.) (2), 57–58 (2016)
2. Niu, Z.F.: Research and Application of the Design Scheme of Substation Equipment Inspection Robot System, pp. 1–5. North China Electric Power University, Beijing (2016)
3. Lu, S.Y., Zhang, Y., Li, J.X., Mu, S.Y.: Application of robot in high voltage substation. High Volt. Eng. **43**(1), 276–284 (2017)
4. Jiang, B., Tang, J., Luo, B.: A survey on graph matching algorithms in computer vision. J. Anhui Univ. (Nat. Sci. Ed.) **41**(1), 29–36 (2017)
5. Li, H.H., Zheng, P., Yang, N., Hu, X.H.: Relative angle distance for image registration based on SIFT feature. J. Northwest. Polytech. Univ. **35**(2), 280–285 (2017)
6. Zhang, X.P., Du, W.Q., Xiao, Z.T., Zhang, F.: Color fundus image mosaic method based on SIFT and Otsu matching. Comput. Eng. Appl. **53**(18), 176–249 (2017)
7. Kleefeld, A., Breuß, M., Welk, M., Burgeth, B.: Adaptive filters for color images: median filtering and its extensions. In: 5th Computational Color Imaging Workshop, pp. 149–158 (2015)
8. Mohan, M., Sulochana, C., Latha, T.: Medical image denoising using multistage directional median filter. In: International Conference on Circuit, Power and Computing Technologies, pp. 1–6 (2015)
9. Yu, H., Zhang, X., Song, S., Wang, S.: An improved Gaussian filter with asynchronously correlated noises. In: 12th IEEE International Conference on Mechatronics and Automation, pp. 1670–1675 (2015)
10. Wang, M., Hu, C.: Extracting oil slick features from VIIRS nighttime imagery using a gaussian filter and morphological constraints. IEEE Geosci. Remote Sens. Lett. **12**(10), 2051–2055 (2015)
11. Han, Z.H., Wang, Y.F.: The difference effect of wavelet transforms and gaussian filter in multi-resolution optical flow. Appl. Mech. Mater. **475–476**, 273–277 (2013)
12. Lowe, D.G.: Distinctive image features from scale-invariant interest point. Int. J. Comput. Vis. **60**(2), 91–110 (2004)
13. Lowe, D.G.: Object recognition from local scale-invariant features. In: International Conference on Computer Vision, Corfu, Greece, pp 1150–1157 (1999)

A Spatial Self-calibration Algorithm for Robot Vision

Huo Yu[2,3], Gong Wang[1,2(✉)], Bingshan Liu[1,2], Zhichao Liu[1,2,3], and Yeifei Liu[1,2]

[1] Technology and Engineering Center for Space Utilization, Chinese Academy of Sciences, Beijing 100094, China
wanggong@csu.ac.cn
[2] Key Laboratory of Space Manufacturing Technology, Chinese Academy of Sciences, Beijing 100094, China
[3] University of Chinese Academy of Sciences, Beijing 100049, China

Abstract. In order to meet the needs of the robot for three-dimensional (3D) reconstruction, an improved binocular vision self-calibration algorithm based on coplanar equation is presented. Binocular stereo vision model is established based on coplanar equation, and through the images of the stereo matching algorithm to find more in the image corresponding to the pixel dot, then according to the method of the polar correction to eliminate the longitudinal error inside the image information. The internal and external parameters of camera are obtained by solving the simultaneous equations. Then these parameters are used to realize the 3D reconstruction of space physics information. Finally, the reconstruction effect of the algorithm on spatial information is verified through the measurement of arbitrary hanging line arcs. The experiment show that this method is high precision, can be in various measure does not depend on any calibration thing better get space coordinate information, and this method basically realizes the real-time 3D reconstruction of robot vision.

Keywords: Robot vision · Coplanar equation
Internal and external parameters · Binocular stereo vision

1 Introduction

Nowadays, the machine vision technology has become an important theoretical basis in the field of robot vision and has been widely used. The function conversion relationship between the two coordinates system depends on the internal and external parameters which are obtained by the camera calibration algorithm.

The traditional camera calibration algorithm [1] mostly depends on the known spatial coordinate information and restricted the space robot system of real-time access to information. Therefore, this paper presents a space self-calibration algorithm that does not depend on any known coordinate information.

K. Deng et al. (Eds.): ICMIR 2018, AISC 856, pp. 595–602, 2019.
https://doi.org/10.1007/978-3-030-00214-5_75

2 Model Building

The three-dimensional information is projected onto the two-dimensional plane through the imaging lens, which is the camera model [2].

2.1 Stereoscopic Vision Model

The traditional imaging model of camera imaging model can be viewed as a hole, and its internal parameters are elements that determine the position of the camera center *S* in the relative position of the photo *P*. In addition, its external parameters are the elements that determine the position and orientation of the beam in the space coordinate system of a given object, including three translation elements and three rotating elements [3].

According to the imaging model, binocular vision model can be obtained as follows (Fig. 1):

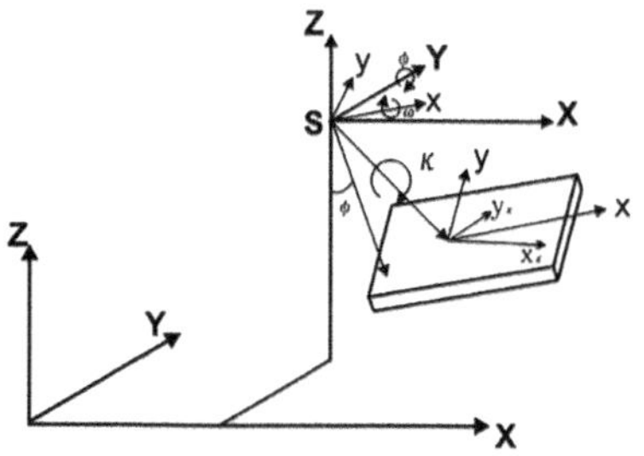

Fig. 1. Schematic diagram of exterior orientation elements

For the general camera model, three rotation matrices can be used to represent:

$$\begin{bmatrix} u \\ v \\ w \end{bmatrix} = R_{\varphi} R_{\omega} R_{\kappa} \begin{bmatrix} x \\ y \\ -f \end{bmatrix} = R \begin{bmatrix} x \\ y \\ -f \end{bmatrix}$$
$$= \begin{bmatrix} \cos\varphi & 0 & -\sin\varphi \\ 0 & 1 & 0 \\ \sin\varphi & 0 & \cos\varphi \end{bmatrix} \begin{bmatrix} 1 & 0 & 0 \\ 0 & \cos\omega & -\sin\omega \\ 0 & \sin\omega & \cos\varphi \end{bmatrix} \begin{bmatrix} \cos\kappa & -\sin\kappa & 0 \\ \sin\kappa & \cos\kappa & 0 \\ 0 & 0 & 1 \end{bmatrix} \begin{bmatrix} x \\ y \\ -f \end{bmatrix}$$
$$\approx \begin{bmatrix} a_1 & a_2 & a_3 \\ b_1 & b_2 & b_3 \\ c_1 & c_2 & c_3 \end{bmatrix} \begin{bmatrix} x \\ y \\ -f \end{bmatrix} \tag{1}$$

Where, R is an orthogonal matrix.

At this point, the volume F of the parallelepiped hexahedron with the three vectors $b, S_1m_1(R_1)$ and $S_2m_2(R_2)$ should be zero:

$$F = b \cdot (R_1 \times R_2) = \begin{vmatrix} b_x & b_y & b_z \\ u_1 & v_1 & w_1 \\ u_2 & v_2 & w_2 \end{vmatrix} = b_x \begin{vmatrix} v_1 & w_1 \\ v_2 & w_2 \end{vmatrix} - b_y \begin{vmatrix} u_1 & w_1 \\ u_2 & w_2 \end{vmatrix} + b_z \begin{vmatrix} u_1 & v_1 \\ u_2 & v_2 \end{vmatrix} = 0 \tag{2}$$

The small value expressions (u_1, v_1, w_1) and (u_2, v_2, w_2) in the above equation are:

$$\begin{bmatrix} u_1 \\ v_1 \\ w_1 \end{bmatrix} = R_{left} \begin{bmatrix} x_1 \\ y_1 \\ -f \end{bmatrix} = \begin{bmatrix} 1 & -\kappa_1 & -\varphi_1 \\ \kappa_1 & 1 & -\omega_1 \\ \varphi_1 & \omega_1 & 1 \end{bmatrix} \begin{bmatrix} x_1 \\ y_1 \\ -f \end{bmatrix} \tag{3}$$

$$\begin{bmatrix} u_2 \\ v_2 \\ w_2 \end{bmatrix} = R_{right} \begin{bmatrix} x_2 \\ y_2 \\ -f \end{bmatrix} = \begin{bmatrix} 1 & -\kappa_2 & -\varphi_2 \\ \kappa_2 & 1 & -\omega_2 \\ \varphi_2 & \omega_2 & 1 \end{bmatrix} \begin{bmatrix} x_2 \\ y_2 \\ -f \end{bmatrix} \tag{4}$$

2.2 Electrode Correction

As shown in Fig. 2, P is a space coordinate in space, and the corresponding image coordinates in the left and right photos are P_l and P_r, and el, er are referred as the poles which are the connection to the projection center respectively with two planes.

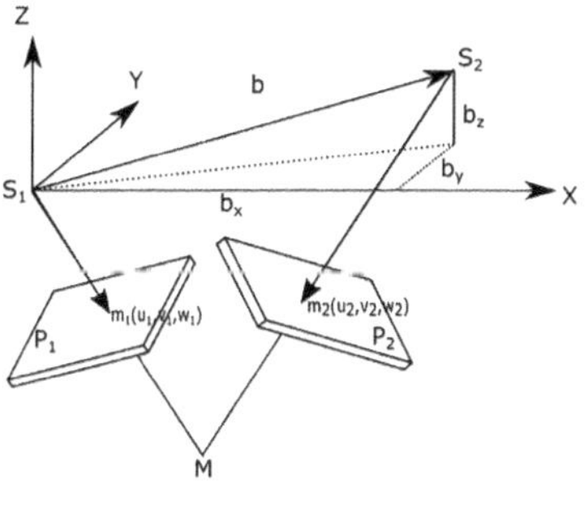

Fig. 2. Coplanar equation model

According to the imaging principle, all points connected on the left image are shown in the right image and vice versa.

So, the constraint equations are obtained:

$$\begin{cases} l_1 = FP_1 \\ l_2 = FP_2 \end{cases} \tag{5}$$

Therefore:

$$\begin{cases} P_2^T F P_1 = 0 \\ F e_l = 0 \\ F^T e_r = 0 \end{cases} \tag{6}$$

F is the basic matrix which be solved by using some corresponding pixel coordinates in the left and the right images (Fig. 3).

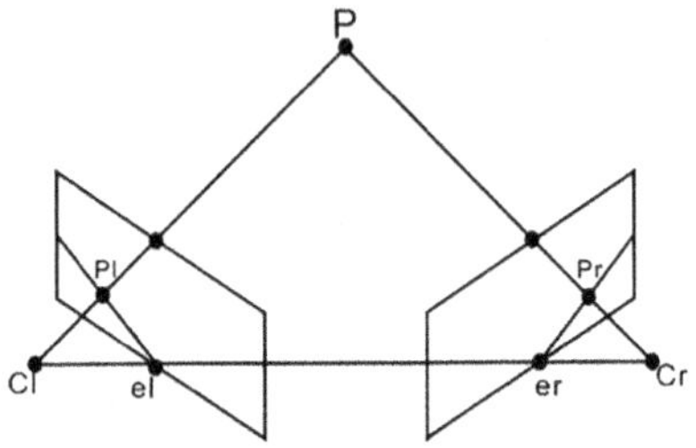

Fig. 3. Polar geometry

Suppose two projection transformation matrices H_1 and H_2:

$$\hat{F} = (H_2^{-1})^T F H_1^{-1} \tag{7}$$

So:

$$(H_2 P_2)^T \hat{F} (H_1 P_1) = 0 \tag{8}$$

The optimal transformation matrix H_1 and H_2 make the corresponding point of the image to eliminate the vertical parallax and correct the image information in different direction. Then the two images will be located in the infinity of pole, and the polar is in a superposition state.

3 Algorithm

In the stereo vision model based on the coplanar equation. The corresponding point image coordinate information of the same space coordinate point in multiple images is obtained through stereo matching [4].

3.1 Solution of Internal Parameters

The internal parameter matrix of the camera is K. In a given two different positions in two images, according to the two images of the absolute conic image to geometric relationships the basic matrix can be expressed as $F = \alpha t'_{\times} K R K^{-1}$, thus:

$$FKKF^T = \alpha^2(t')_\times KRR^T K^T(t')_\times^T = \alpha^2(t')_\times KK^T(t')_\times^T \tag{9}$$

Among them, t' is the pole of the second image, $(t')_\times$ is the anti-symmetric matrix:

$$[t']_\times = \begin{bmatrix} 0 & -t_3 & t_2 \\ t_3 & 0 & -t_1 \\ -t_2 & t_1 & 0 \end{bmatrix} \tag{10}$$

And $\mu = \alpha^2$, $C = KK^T$.
So, Kruppa equation:

$$FCF^T = \mu(t')_\times C(t')_\times^T \tag{11}$$

Where, μ is the scaling factor, C is the positive definite matrix. Only two parameters in the upper form are independent, so a pair of images can only provide two constraints on the C matrix. Since the matrix C has 5 unknowns, it needs to be solved, so at least 3 images are required to provide 6 sets of equations.

For a given set of n images, the following equations can be obtained:

$$\begin{cases} F_1CF_1^T = \mu(t_1')_\times C(t_1')_\times^T \\ F_2CF_2^T = \mu(t_2')_\times C(t_2')_\times^T \\ \quad\cdots \\ F_nCF_n^T = \mu(t_n')_\times C(t_n')_\times^T \end{cases} \tag{12}$$

The traditional solution method based on Kruppa [5] equation is mainly used to calibrate the nonlinear constraints of matrix C elements provided by the above geometric equation.

When the camera makes pure rotation motion, the geometric relations between the two images degenerate into the projective transformation between the two planes [6].

The projective transformation matrix between the two images is H, and the basic matrix of the camera can be expressed as $F = Ht$.

So, if $Det(H) = 1$, then $H = KRK^{-1}$, thus $HKK^TH^T = KK^T$ and $C = KK^T$.

The matrix K is the internal parameter matrix of the camera.

3.2 Relative Orientation of External Parameters

In the binocular vision system based on the coplanar equation, the external parameters of the left image is set as the external parameters of the right image, and the relative orientation is completed.

According the coplanar equation,

$$F = F_0 + \mathrm{dF} = F_0 + b_x x_2 y_2 d\varphi + b_x(f^2 + y_2^2)d\omega + b_x f x_2 d\kappa + b_x f db_y + b_x y_2 db_z = 0 \tag{13}$$

Considering that there is no correlation between the image points (x_1, y_1) and (x_2, y_2) the observed values, the upper left and be replaced by a virtual observed value v_i correction:

$$v_i = C_1 V_{x_1} + C_2 V_{y_1} + C_3 V_{x_2} + C_4 V_{y_2} \tag{14}$$

The rights v_i of P_i':

$$P_i' = \frac{P_i}{C_1^2 + C_2^2 + C_3^2 + C_4^2} \tag{15}$$

According to the forward intersection solution of multiple pictures, the external parameters of camera are obtained based on the principle of minimizing the sum of squares of the photo coordinates of the corresponding points of the photograph.

4 Experimental Results

4.1 Internal Reference

In the camera self-calibration experiment, the Harris corner detection method [7] is used to detect feature points of the image corresponding to the image, and the internal parameters of the camera are obtained.

$$[f/mm, k, s_x/\mu m, s_y/\mu m, u_0/px, v_0/px] = [0.00503,\ -2889.48, 3.75 \times 10^{-6},\ 3.75 \times 10^{-6},\ 664.704,\ 510.015]$$

Where, f is the focal length, k is the radial distortion coefficient, $s_x s_y$ are the pixel sizes, $u_0 v_0$ are the center of the image coordinates.

By using the internal parameters of the calculated camera, the image distortion is eliminated, and the measurement results are qualitatively and reliably. The results are as follows (Fig. 4).

Fig. 4. Reduction result

4.2 Binocular Measurement Experiment

After obtaining the internal parameters of the camera, the principle of the coplanar equation is used to measure the relative orientation of the binocular system. In order to verify the real-time performance of the binocular vision system and the accuracy of the measurement of the space object, the vertical arc sag is taken as the measurement object, and the experimental verification is carried out.

The schematic diagram is shown below. The optical centers of the cameras are O_1, O_2 (Fig. 5).

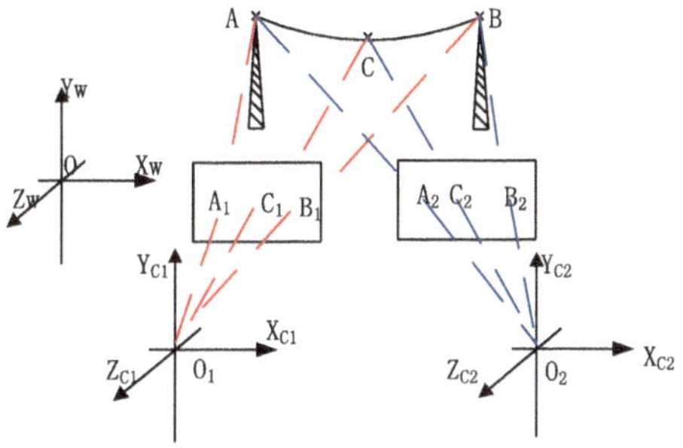

Fig. 5. Sketch of binocular camera model

The measured results are as follows (Fig. 6).

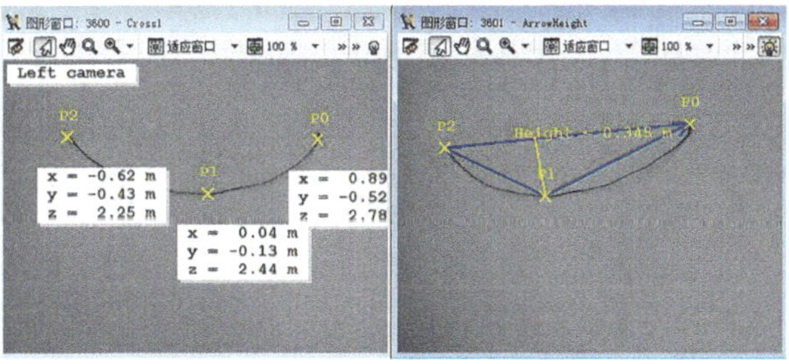

Fig. 6. Binocular measurement results

By comparing the results obtained from many experiments, it is found that the method proposed in this paper can accurately measure the sag height, which basically achieves the purpose of experiment, and the relative error is less than 2% (Table 1).

Table 1. Measurement data

Serial number	Actual data/m	Measured data/m	Relative error
a	0.35	0.345	1.4%
b	0.40	0.401	0.25%
c	0.45	0.443	1.56%
d	0.50	0.505	1.0%
e	0.55	0.561	2.0%

5 Conclusion

According to the robot vision technology in requirements of real-time 3D reconstruction of spatial information, this paper presents a calibration method of non-contact binocular stereo vision based on coplanar equation experiment shows that this method can in calibration object without using any known physical information, quickly and accurately calculate the internal and external parameters of the camera, with high precision and flexibility, satisfied with robot vision measurement and 3D reconstruction needs, all can quickly complete the calibration of binocular stereo vision system is convenient, and has high practicability.

References

1. Zhang, Z.Y.: A flexible new technique for camera calibration. IEEE Trans. Pattern Anal. Mach. Intell. **22**(11), 1330–1334 (2000). (in English)
2. Yang, J., Liu, W., Liu, Y., Wang, F., Jia, Z.: Binocular stereo vision measurement system calibration. Opt. Precis. **24**(2), 300–308 (2016)
3. Feng, W.: Near Shot Photogrammetry. Wuhan University Press, Wuhan (2002)
4. Berg, A.C., Berg, T.L., Malik, J.: Shape matching and object recognition using low distortion correspondences. In: Schmid C., Soatto S., Tomasi, C. (eds.) 2005 IEEE Computer Society Conference on Computer Vision and Pattern Recognition, Proceedings, IEEE Conference on Computer Vision and Pattern Recognition, vol. 1, pp. 26–33. IEEE Computer Soc, Los Alamitos (2005)
5. Hartley, R.I.: Kruppa's equations derived from the fundamental matrix. IEEE Trans. Pattern Anal. Mach. Intell. **19**(2), 133–135 (1997). (in English)
6. Agapito, L., Hayman, E., Reid, I.: Self-calibration of rotating and zooming cameras. Int. J. Comput. Vis. **45**(2), 107–127 (2001). (in English)
7. Yao, G.B., Cui, J., Deng, K.Z., Zhang, L.: Robust Harris corner matching based on the quasi-homography transform and self-adaptive window for wide-baseline stereo images. IEEE Trans. Geosci. Remote Sens. **56**(1), 559–574 (2018). (in English)

A Force Control Robot Screwing System Design for Plastic Workpiece

Xiaohui Xie[1], Ge Li[2], Fengfeng Zhang[1(✉)], and Lining Sun[1]

[1] Robotics and Microsystems Center, Soochow University, Soochow, People's Republic of China
xhxie@suda.edu.cn

[2] School of Continuing Education, Shenzhen Polytechnic, Shenzhen, People's Republic of China

Abstract. Commonly robot screwing system just concentrate on position control and velocity control. Force control in robot usually is not needed to be considered except the torque control in screwdriver. In some cases, the workpiece is weak or the screw is small that the workpiece will be damaged if the force is not controlled in the screwing process, like The SCARA robot is used in screwing a tachograph made of plastic material with small screws. In this case, the force control must be used to avoid the damage in the plastic workpiece. The passive flexible mechanism and positive-reverse rotation method is considered in the robot system. The experiments verify the validity of the force control mechanism and method.

Keywords: Robot screwing · Force control · Plastic workpiece

1 Introduction

Due to the complexity of assemblage like screwing, using large resources and important position in industry production, the labor force on assemblage of American, Japan and west Europe accounted for forty percent in the whole labor. In Germany the proportion is sixty percent. So assemblage automation is important in the industry which can free labor force and the assembling robot like screwing robot must be researched.

1.1 Present Status of Development

American is the birth-land of the robot that its comprehensive research level in robot area leads the world. In 1954, George Waddel developed the first prototype of industry robot which can be electrical programmed and it is also first can be controlled by teaching-playback method. In 1960, Unimation company in American produced the first industry robot by using the patent of Waddel. In twenty-one century, the technology of intelligent robot in American is in the world's leading level, especially in human-intelligence technology area. There are a lot of excellent companies in American like the Boston dynamics company with its big dog, the Google company with its Alpho dog. Japan is another important nation in the robot area with its biggest types

K. Deng et al. (Eds.): ICMIR 2018, AISC 856, pp. 603–609, 2019.
https://doi.org/10.1007/978-3-030-00214-5_76

and quantities in the world. There are a lot of famous robot companies and universities in Japan like Fanuc company and Yaskawa company, Epson and Yamaha company etc.

In China, from 1985, professor Cai Hegao in Harbin institute of technology began to design and develop HT-100A spot welding robot [1]. He solved the key technology like trajectory control precision and locus predictive control of welding robot. In Shanghai jiaotong university, professor Jiang Houzong fulfilled an assemblage robot named precision-1 [2]. The robot can finish robot welding task which using sensor to detect the position of soldering seam. By research on this robot, key technology of two-dimensional vision and six-dimensional force sensor with their processing method has been breached.

1.2 Key Technology of Assemblage Robot

Industry robot is one type of automatic multiple DOF robot faced industry domain. In technology application area, using industry robot can improve quality of product, efficiency of labor force and decrease production cost etc. In assemblage especially electrical production automatic assemblage, SCARA is an important type among robot population. In designing a SCARA assemblage like a robot screwing system, five key points should be considered first.

1. Structure optimization and manufacturing technology: because the robot need have the merits of precision drive and precision positioning, mechanical structure of robot should be light in weight, good rigidity and small inertia [3].
2. Drive and control technology: to deal with robot coupling and non-linearity, control method and strategy of multi-variable control system should be researched deeply.
3. Error evaluation technology: in the four steps of production, assemblage, positioning and control, error of robot can not be avoided. If the high-precision is needed in the application, error modeling and estimation technology should be adopted to decrease the position & posture error and positioning distance error in the system [4].
4. Detection and sensing technology: environment change will feedback to controller by detecting and sensing. By model or parameter optimization according to the sensing information, the robot can work in the best condition.
5. Vision processing and force processing technology: vision and force information can be aggregated to sensor information, but the processing technology are so complex than others which should be studied more seriously.

2 Robot Screwing System Design

2.1 Project Description

The work-piece is a tachograph which is made of plastic. There are fifteen screws need to be screwed on the tachograph. If there are screwed by workers, the screws are small with the diameters 1.7 mm and 2.0 mm respectively that the workers will work with great care and the workload is very big. So the tachograph should be screwed by robot to free labor force. The screwing process is as follows, first screwing three pieces of

1.7 * 5 screws and two pieces of 2.0 * 6 screws to screw the keypad and lens on the front cover of the tachograph like the left in the Fig. 1, that the torque force of 1.7 mm screws is about 1 kg.cm and the torque force of 2.0 mm screws is about 1.2 kg.cm. Then the worker puts the lens on the support and the robot screws four pieces of 1.7 * 3.5 screws and two pieces of 1.7 * 5 screws to screw the lens and support together. In the last, the rear cover of the tachograph is screwed on the front cover by the screwing robot as the right figure of the Fig. 1.

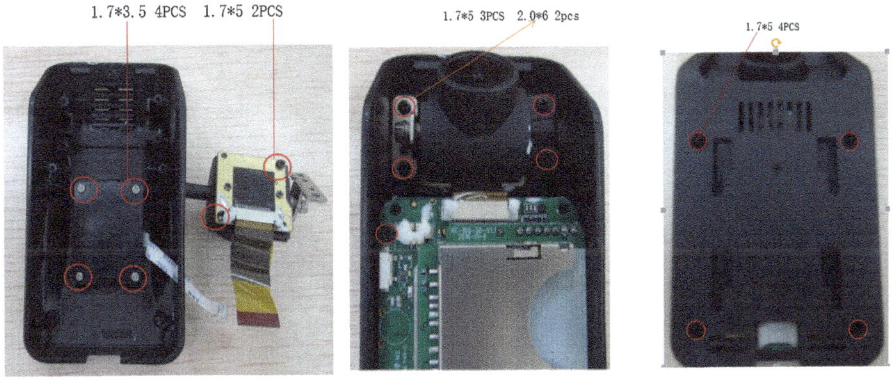

Fig. 1. The tachograph with its screws

2.2 The Screwing Robot System

The robot system uses a SCARA as the main operating tools instead of the worker in traditional hand-work mode. The robot screwing system is made-up of a SCARA, one or two electric screwdrivers and screw feeders, and some load-bearing tools with mobile modules like Fig. 2. The worker puts the workpiece on the screw feeder, then the SCARA will screw the workpiece and after screwed the workpiece will quit out on the screwfeeder.

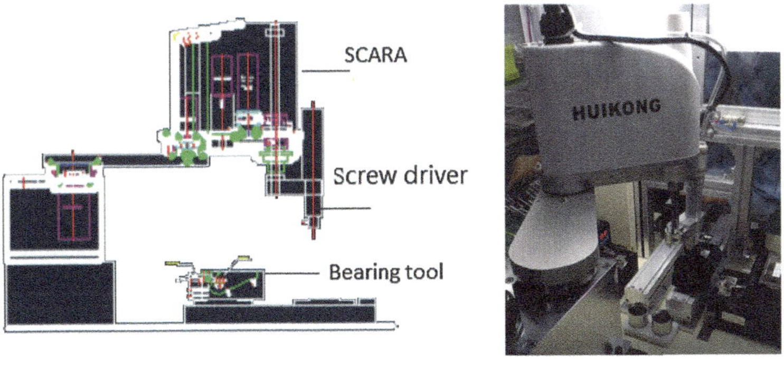

Fig. 2. The robot screwing system

Because the SCARA will screw three different screws, three different robot workstation are designed to satisfy the need. In order to get the good effect of screwing, the fixture is designed according to the profile modeling of the work-piece like Fig. 3.

In the working process, the PLC is the main controller of the system which is in charge of controlling the robot start and stop, controlling the screw feeder, IO information collection and feedback. The SCARA is the main actuator of the screwing process which brings the screwing driver to the feeder and screwing position. The SCARA also controls the screwing process, like entering the tapping hole, tighten the screws and stop moving after getting the torque arrived signal.

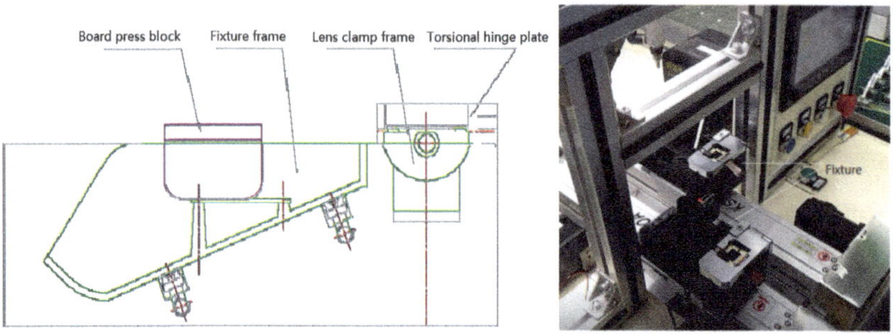

Fig. 3. Fixture design for tachograph

Because the screws in the project are small like 1.7 mm of diameter and the screw hole is the type of self-tapping, in the working process the screw hole will bear large screwing force comparing with the hole's size and strength. If the screw hole is designed not very carefully, the screw hole will be damaged in the working process or the hole will be destroyed slowly after screwed with agglomerating stress in the screwing process.

In the experiments, the robot system is used to screw the work-piece like the fifteen screw holes in Fig. 1. The fifteen screw holes can be screwed successfully but almost 70% of the screw holes in the work-piece will have cracks after screwed in two or three days. In order to solve this problem, the flexible mechanism is designed in the fixture of screw driver to absorb the impact force in the screwing process (Fig. 4).

By using this flexible structure in the fixture between robot and screw driver, the screwing effects get greatly improved. Three machines have been used to test the effects respectively. The first machine is used to screw 290 pieces of tachograph, and the second machine is used to screw 990 pieces of tachograph, and the third machine is used to screw 450 pieces of tachograph. The ratio is 5.17%, 32%, 9.6% respectively which have cracks in the tachograph after screwed in 2–3 days by the three experiments. Though the cracks have been decreased compared with before, the ratio of cracks in the work-piece can not be accepted by manufacturer. In order to solve the problem of cracks, the event-based control method is proposed in the robot system.

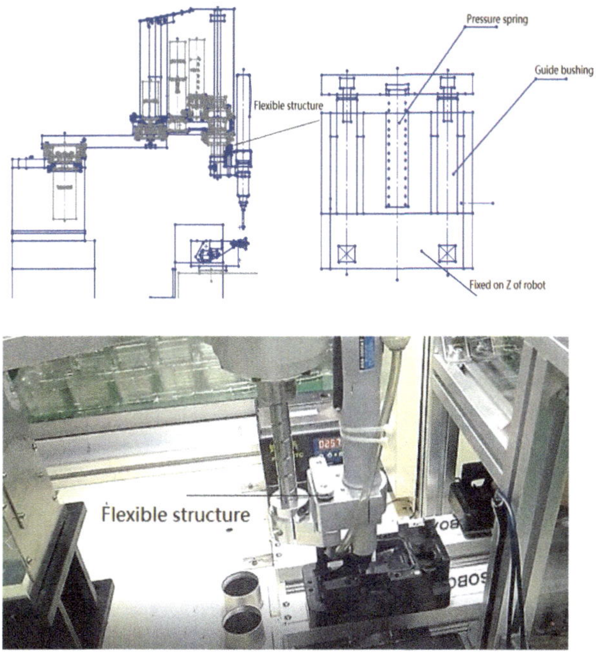

Fig. 4. Flexible structure to solve the plastic stress

2.3 Event-Based Robot Control Method

In traditional control system, all control command and feedback signal are based on time. Because large and variable time delay exists over some system, using time as action reference, the system cannot keep stability. If a parameter which not relating to time is chosen, the system will not based on time but based on the parameter not relating to time, then time delay will not affect the system and the system will keep stability under any time delay. Event-driven method thus is proposed and the event is chosen which does not relate to time [5, 6]. The whole system is driven by event. In event-driven method, the choice of event is important in order that the event is irrelevant to time. In practice, the command number can be picked as the event commonly. Every command issued by the operator is one event. All the operation is driven by it. Except command number, distance, velocity and force etc. can also be chosen as the event. For example, in controlling mobile robot, when the robot moves two centimeter, it triggers one event.

In the robot screwing process, the flexible mechanism is added to the screw driver which can not eliminate the cracks absolutely. The residual plastic stress will make the plastic screw post having some cracks in two or three days after screwed. So the reverse rotation motion in the screwing process is added in the screwing process to eliminate the plastic stress. In order to make the reverse rotation motion moving correctly, the event-based method is used in the process.

In the screwing process like Fig. 5, the event based method is used in order to ensure the system stable. For example, the 1.7 mm with length of 5 mm screw thread is screwed in the schedule. The screw driver are rotating forward at the beginning when getting to the upper edge of the screwing hole. When the screw driver begin going down to screw the hole about 3 mm length, the screw driver will rotate backward about 1 mm to release the plastic stress. When the screw driver rotate backward at 1 mm length, the screw driver will rotate forward to the end in order to screw home the hole. When the torque reached signal is produced, the screw driver will stop and the screw driver will loosening, returning. In this case, the time is not used to justify the process when the rotating forward begin or the rotating backward begin. The length 0 mm, 3 mm, 1 mm and the torque reached signal is used as the event to trigger the beginning of rotating forward or rotating backward.

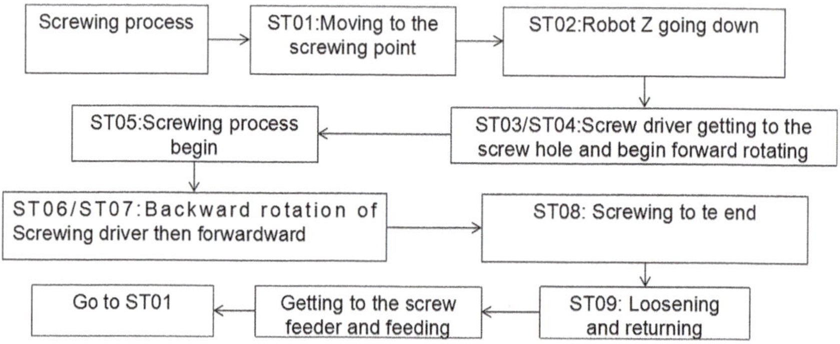

Fig. 5. Event based method of the screwing process with reverse rotation

Table 1. Experiment results of the robot screwing system

Number	Quatity (pcs)	Defective	Percent Defectives	Defective list			
				Floating	Patial	Screwloose	Leakage
1	765	18	97.6%	5		13	
2	908	19	97.9%	14	2	2	1
3	1067	10	99.0%	4	4	1	1
4	423	20	95.3%		2	13	5
	305	12	96.1%	1			11
5	1213	10	99.2%	4		6	
	425	17	96.0%				17
6	1285	20	98.4%	10		10	
	170	0	100.0%				
7	1453	22	98.5%	1	3	16	2
	576	5	99.1%				5

3 Experiments and Conclusion

Screwing experiments with flexible structure and event-based method are done by using a lot of tachographs. The results of experiments are shown in Table 1. The availability of the robot screwing system with the flexible structure and event-based method is proved.

Acknowledgements. Thanks for Supported by National Natural Science Foundation of China (No. 61473200). And thanks for our research team.

References

1. Cai, H.: Design and application of distributed ultrasonic ranging system for autonomous mobile robots. WSEAS Trans. Circuits Syst. **6**(2), 228–234 (2007)
2. Li, Y., Jiang, H.: Dynamics analysis of a 5-DOF robot. Robot **11**(4), 12–16 (1989)
3. Alici, G., Bijan, S.: Enhanced stiffness modeling, identification and characterization for robot manipulators. IEEE Trans. Robot. **21**(4), 554–564 (2005)
4. Abderrahim, M., Khams, A., Garrido, S., et al.: Accuracy and calibration issues of industrial manipulators. Ind. Robot. Program. Simul. Appl. **25**(7), 817–829 (2006)
5. Xie, X., Ma, C., Sun, Q., et al.: Event-based cooperative control of a high-speed bar-tacking sewing machine. Int. J. Cloth. Sci. Technol **25**(4), 317–330 (2013)
6. Imad, E., Xi, N., Fung, W.K., et al.: Supermedia-enhanced Internet-based telerobotics. Proc. IEEE **91**(3), 396–421 (2003)

Development and Design of High-Speed Polishing Robot/Structural Analysis

Jenny Sama Kevin, Shudong Zhuang(✉), and Daoliang Xie

Hohai University, Changzhou, China
zhsd1970@126.com

Abstract. The objective of this paper is to develop and improve the mechanism and spindle of a six degree of freedom high speed polishing robot arm with good precision, mimicking human arm with the ability to solve the difficulty of polish complex thin parts. That is robot that can run with very high speed and get the best finish products, that is to improve the surface quality, improve time efficiency, reduce hazard, reduce cost, and finally optimize the design. The structural analysis simulations test results of the static structural equivalent (von mises) stress, equivalent elastic strain and Total deformations of this robot is been done with Ansys software, and finally the material selection and design is carried out with Solidworks.

Keywords: Polishing technology · Structure design · Mechanism

1 Introduction

Polishing is the finishing process that is used widely in many manufacturing companies including the automobile, aircraft, iron and steel industries. It's a process of using abrasives to smooth or do the final touches of the working surface without affecting its geometrical dimensions [1, 2]. In brief, the purpose of polishing is to reduce roughness average Ra and get the uniform surface evenly distributed throughout the working surface in the previous year's polishing has largely been a manual operation (traditional polishing). The workload was too much, long process time, inefficient, hazardous, highly skill dependent, error prone and high cost. High speed polishing robot is a solution to solved these problems.

Grey cast iron (ASTM A48) is the material used at the robot base and is characterized by its graphitic microstructure. And AISI 1020 Steel is the material selected for lower arm, upper arm, forearm, wrist 1, wrist 2 and end effector.

2 Limitations

Though high speed polishing robot has many advantages as mentioned above, it also has its limitations, and one of the limitations is work envelop(space the robot can reach), only J4 axes and J6 axes motion range can reach 360° but the rest J1, J2, J3, and

Supported by "the Fundamental Research Funds for the Central Universities (BZX/18B101-35)"

K. Deng et al. (Eds.): ICMIR 2018, AISC 856, pp. 610–617, 2019.
https://doi.org/10.1007/978-3-030-00214-5_77

J4 axes can't reach, specific speed as shown below, the payload can't exceed 125 kg, also the end effector is design to handle specific tasks like polishing and grinding, but it can't handle other tasks like welding, pick and drop and so on.so this are the difficulties I find when designing this robot.

Fig. 1. Polishing robot

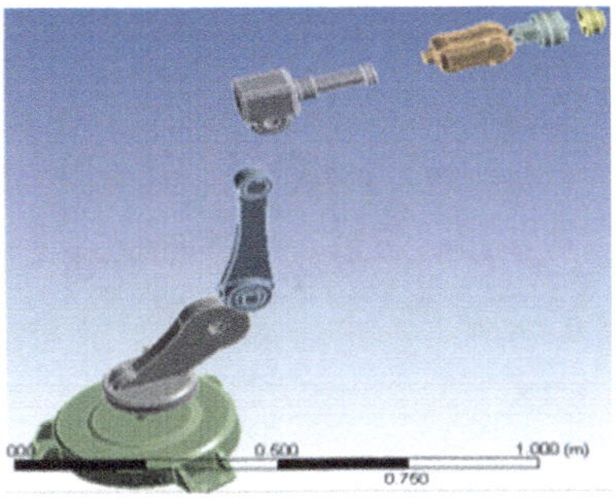

Fig. 2. Robot design

2.1 Structural Design

One of the important aspect of this high speed polishing robot arm is the mechanical design. See Figs. 1 and 2 above. Before designing a robotic arm certain parameters and specifications has to be taken into the consideration like the functionability and reliability along the robot parameter, specifications, motion range, that is the robot work envelop and its applications. Also kinematics and dynamic model, that is direct kinematics and inverse kinematics and their equations, Velocity analysis, degree of freedom and their various equations.

3 Robot Parameter

Robot motion speed is one of the important factors of any robot arm, the J1 has a motion speed of 110°/s (1.92 rad/s), J2 has a motion speed of 110°/s (1.92 rad/s), J3 has a motion speed of 110°/s (1.92 rad/s), J4 has a motion speed of 170°/s (2.97 rad/s), J5 has a motion speed of 170°/s (2.97 rad/s), and J6 has a motion speed of 260°/s (4.54 rad/s).

The robot motion ranges as follows, J1 motion ranges between ± 180°, J2 motion ranges between +75°–60°, J3 motion ranges between +230°–132°, J4 motion ranges between ±360°, J5 motion ranges between ±125°, and J6 motion ranges between ±360°.

4 Robot Work Envelop

Robot Work Envelop see Figs. 3 and 4 below.

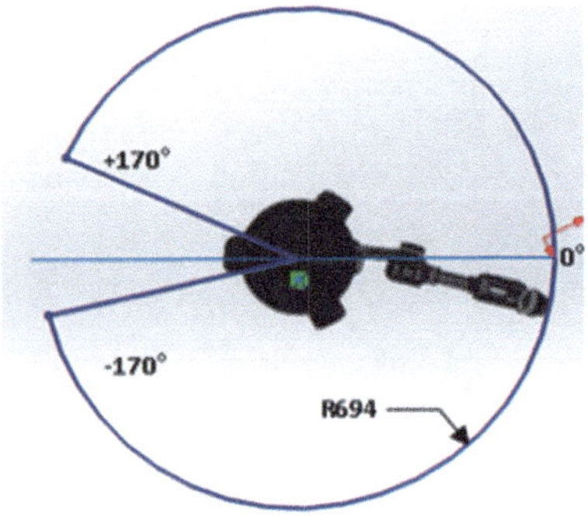

Fig. 3. Robot work envelop diagram A

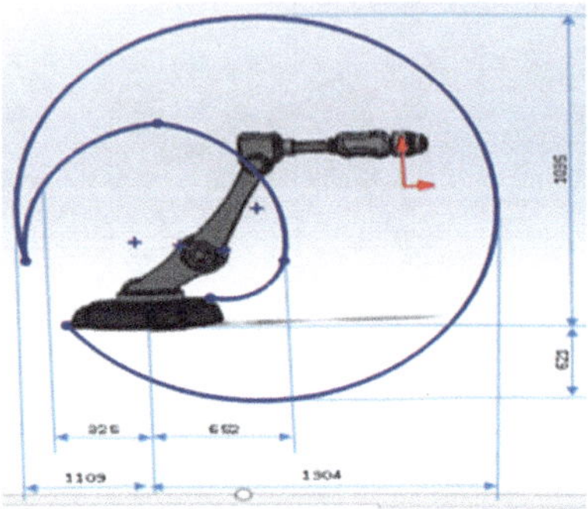

Fig. 4. Robot work envelop diagram B

5 Kinematic and Dynamic Model

Kinematic and Dynamic Model is been shown in Fig. 5 and 6 respectively below.

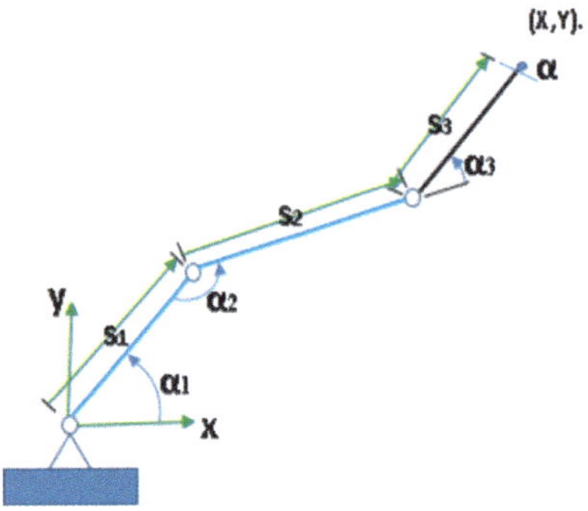

Fig. 5. Direct kinematics

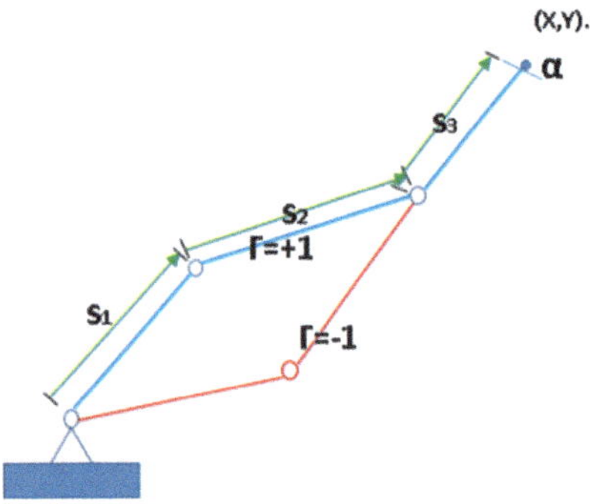

Fig. 6. Inverse kinematics

6 Six Axis Robot/Degree of Freedom

This high speed polishing robot have various axis configurations that feature six axes, also called six degrees of freedom. Six axis robots allow for greater flexibility and can perform a wider variety of applications than robots with fewer axes [3]. See Figs. 7 and 8 respectively below.

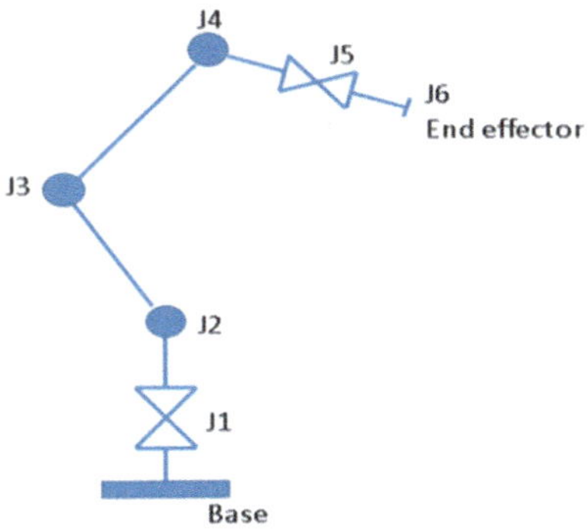

Fig. 7. Schematic mobility diagram

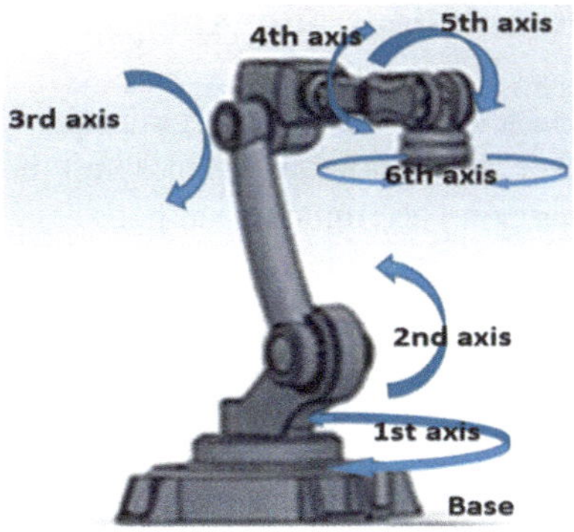

Fig. 8. 3D mobility diagram

7 Materials

Grey cast iron (ASTM A48) is the material used at the robot base and is characterized by its graphitic microstructure, It is the most commonly used cast iron and the most widely used cast material based on weight. Most cast irons have a chemical composition of 2.5–4.0% carbon, 1–3% silicon, and the remainder iron. Grey cast iron has less tensile strength and shock resistance than steel. Details of the materials is shown in Fig. 9 below [4].

AISI 1020 Steel is the material selected for lower arm, upper arm, forearm, whrist 1, whrist 2 and end effector. I selected this material because of is low hardenability and low tensile carbon steel with Brinell hardness of 119–235 and tensile strength of 420.51 MPa. It has high machinability, high strength, high ductility and good weldability. It is normally used in turned and polished or cold drawn condition. AISI 1020 steel can be largely utilized in robot arms and all industrial sectors detail of the materials is shown in Fig. 10 below.

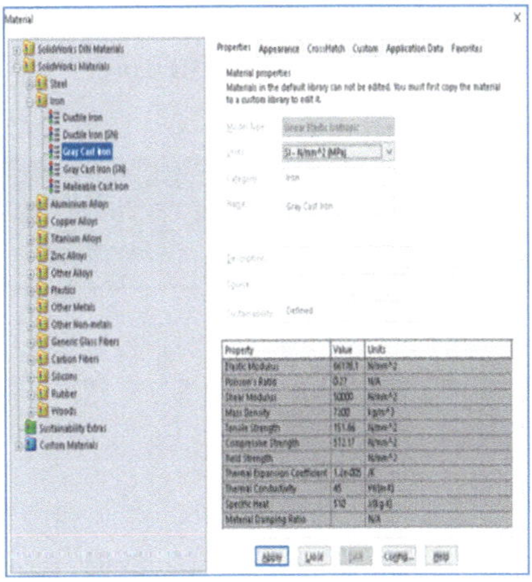

Fig. 9. Grey cast iron (ASTM A48) characteristics

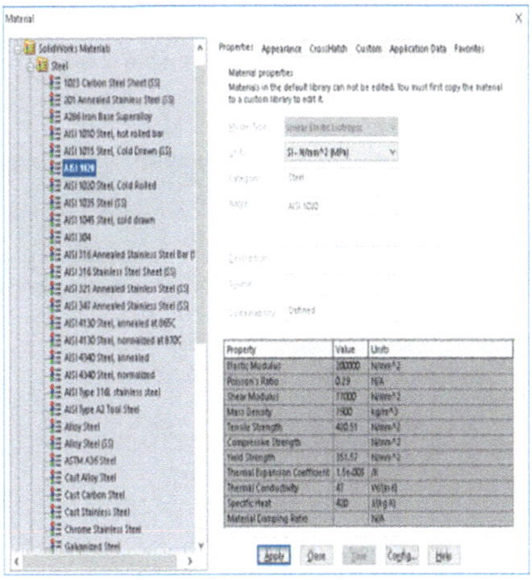

Fig. 10. AISI 1020 steel characteristics

8 Simulations and Analysis

Finite Element Analysis: It is imperative to evaluate mechanical design by simulating the working condition, enables test before manufacturing prototypes of products. Furthermore, determining and improving weak points, so that the model can withstand the load applied to it [5].

ANSYS Workbench is what I used to analyzed this high speed polishing robot, it provides a platform that integrate simulation technologies and parametric CAD systems with unique automation and performance. The simulation process is been shown in Fig. 11 below.

Meshing: one of the most relevant steps in the Finite Element Analysis is the meshing. The speed and the accuracy of the results have a direct connection in how these parts are done. The higher the numbers of nodes are the higher the accuracy of the result, details of these are shown in Fig. 12 below.

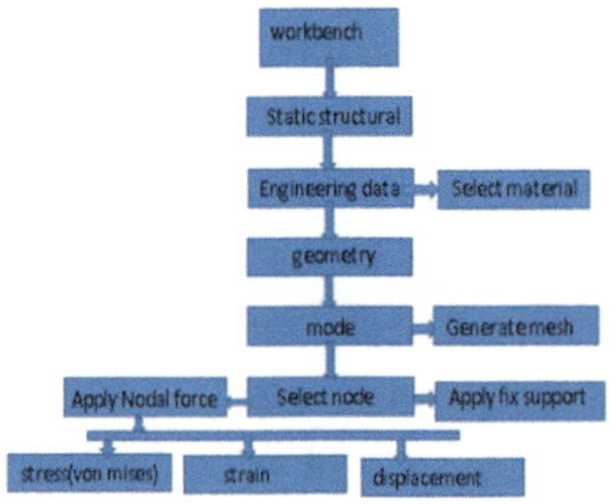

Fig. 11. Flowchart stress, strain & displacement

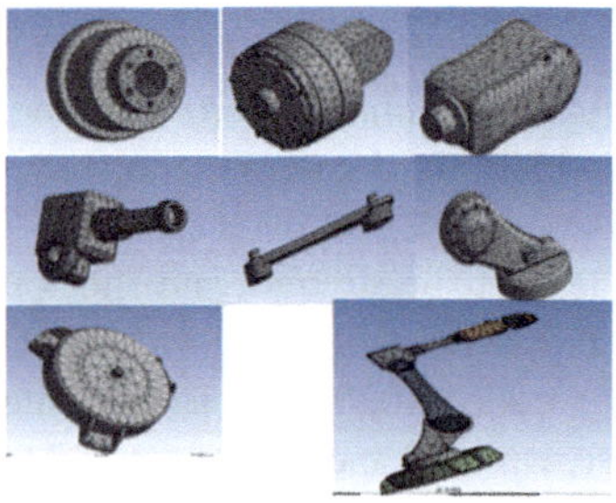

Fig. 12. Meshing diagrams

End effector: The static structural equivalent (von mises) stress, equivalent elastic strain and Total deformations of the end effector is shown below in Fig. 13. The red key indicates maximum and the blue keys indicate minimum, therefore the maximum equivalent (von mises) stress is 4.1694e6 Pa's (pa) and the minimum equivalent (von

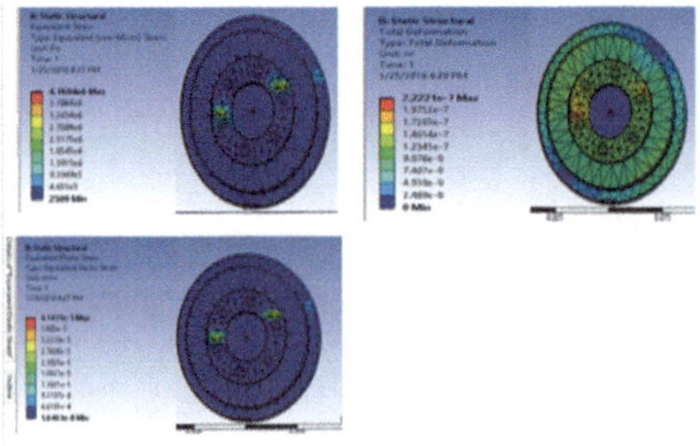

Fig. 13. End effect. Stress, strain & displacement

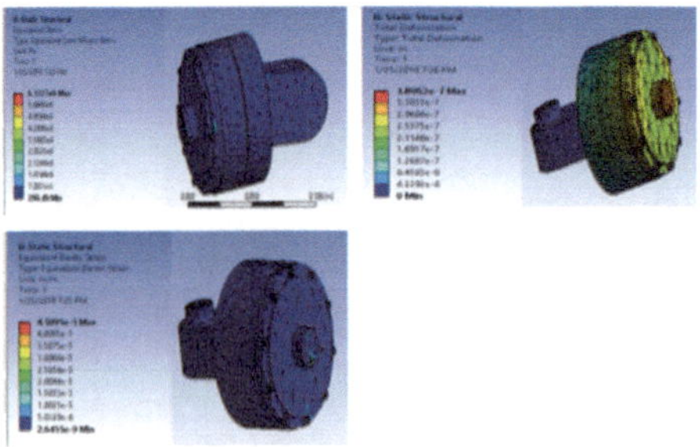

Fig. 14. Wrist 1 stress, strain & displacement

mises) stress is 2509 Pa's(pa), the maximum equivalent elastic strain is 4.1431e−5 m/m and the minimum equivalent elastic strain is 1.6403e−8 m/m, the maximum total deformations is 2.2221e−7 m and the minimum total deformations is 0 m.

Wrist 1: Result, the maximum equivalent (von mises) stress is 6.3727e6 Pa's (pa) and the minimum equivalent (von mises) stress is 290.49 Pa's(pa), the maximum equivalent elastic strain is 4.5095e−5 m/m and the minimum equivalent elastic strain is 2.6455e−9 m/m, the maximum total deformations is 3.8062e−7 m and the minimum total deformations is 0. see Fig. 14 above.

Wrist 2: Result, the maximum equivalent (von mises) stress is 2.9914e6 Pa's (pa) and the minimum equivalent (von mises) stress is 19290.6 Pa's(pa), the maximum equivalent elastic strain is 1.8351e−5 m/m and the minimum equivalent elastic strain is 1.8344e−8 m/m, the maximum total deformations is 2.1662e−6 m and the minimum total deformations is 0 m. see Fig. 15 below.

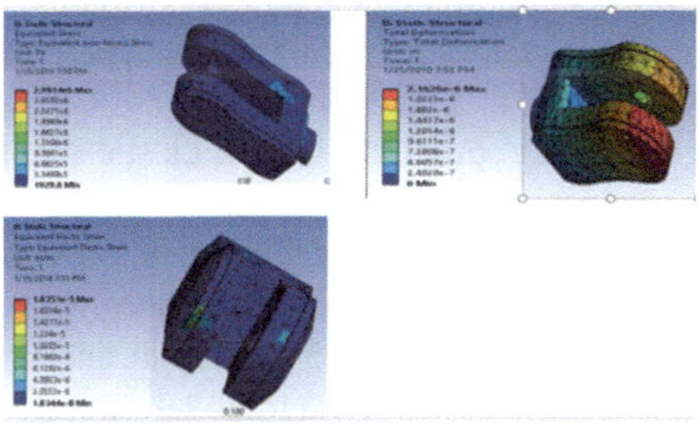

Fig. 15. Wrist 2 stress, strain & displacement

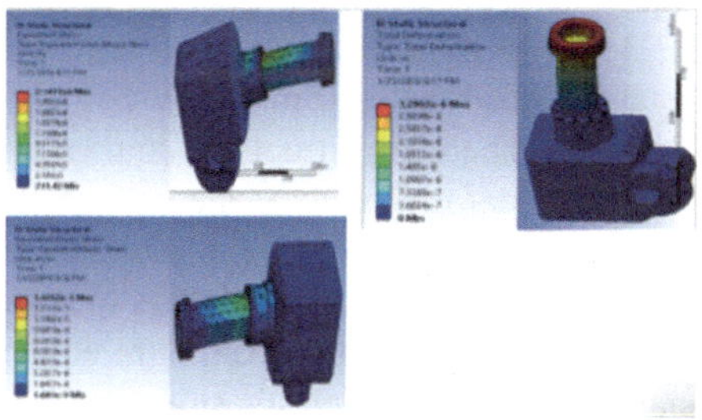

Fig. 16. Forearm stress, strain & displacement

Fore Arm: result the maximum equivalent (von mises) stress is 2.1411e6 Pa's (pa) and the minimum equivalent (von mises) stress is 231.42 Pa's(pa), the maximum equivalent elastic strain is 1.4762e−5 m/m and the minimum equivalent elastic strain is 1.681e−9 m/m, the maximum total deformations is 3.2962e−6 m and the minimum total deformations is 0 m. See Fig. 16 above.

Upper Arm: result, the maximum equivalent (von mises) stress is 5.8155e6 Pa's (pa) and the minimum equivalent (von mises) stress is 2374 Pa's (pa), the maximum equivalent elastic strain is 3.7732e−5 m/m and the minimum equivalent elastic strain is 1.9426e−8 m/m, the maximum total deformations is 0.00013006 m and the minimum total deformations is 0 m. See Fig. 17 below.

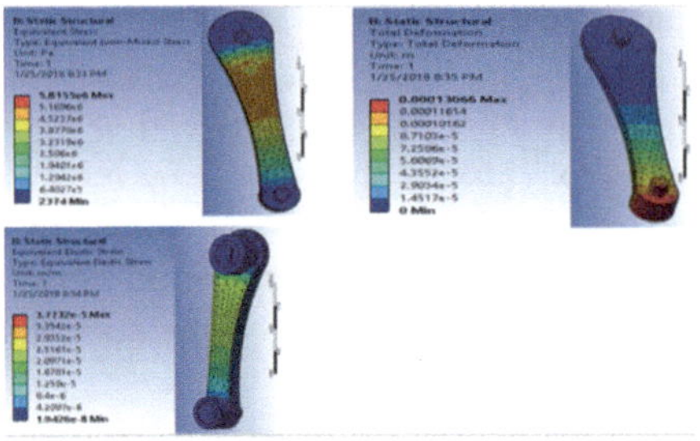

Fig. 17. Upper arm stress, strain & displacement

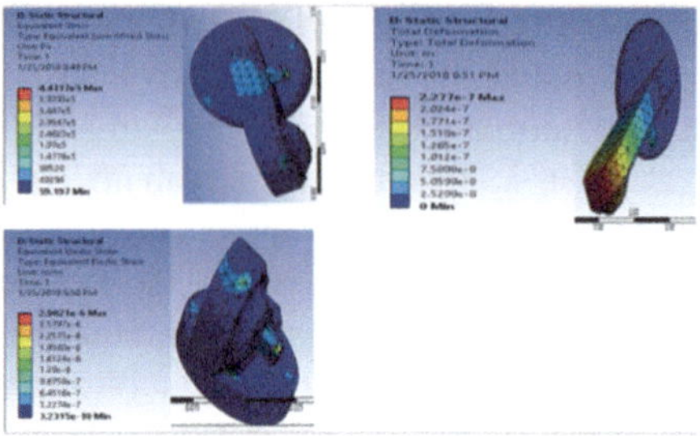

Fig. 18. Lower arm stress, strain & displacement

Lower Arm: Result the maximum equivalent (von mises) stress is 4.4317e5 Pa's (pa) and the minimum equivalent (von mises) stress is 59.197 Pa's(pa), the maximum equivalent elastic strain is 2.9021e−6 m/m and the minimum equivalent elastic strain is 3.2315e−10 m/m, the maximum total deformations is 2.277e−7 m and the minimum total deformations is 0 m. See Fig. 18 above.

Base: The static structural equivalent (von mises) stress, equivalent elastic strain and Total deformations of the base is shown below in Fig. 19, the red key indicates maximum and the blue keys indicate minimum, therefore the maximum equivalent (von mises) stress is 1.623e5 Pa's (pa) and the minimum equivalent (von mises) stress is 1036.6 Pa's (pa), the maximum equivalent elastic strain is 1.462e−6 m/m and the minimum equivalent elastic strain is 8.1926e−9 m/m, the maximum total deformations is 2.7742e−7 m and the minimum total deformations is 0 m.

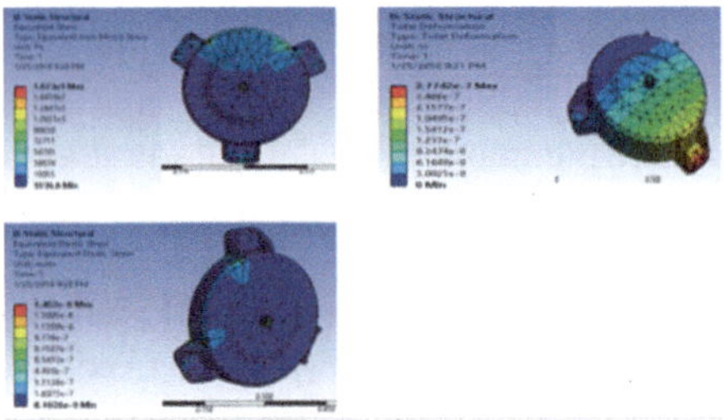

Fig. 19. Base stress, strain & displacement

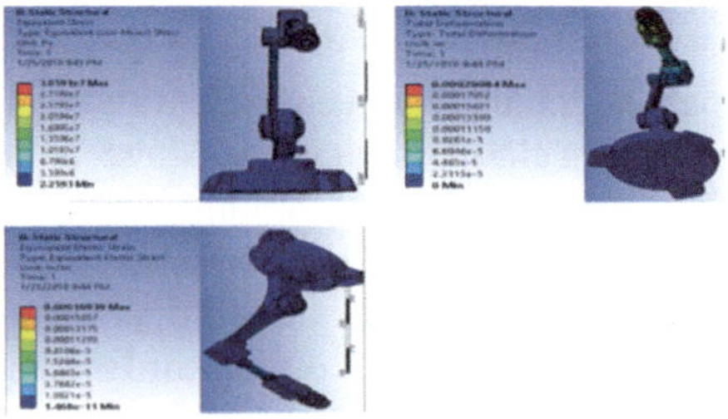

Fig. 20. Robot ass. stress, strain & displacement

Robot Assembly: The complete assemble of the robot stastic structural equivalent (von mises) stress, equivalent elastic strain and Total deformations is shown above in Fig. 20, the red key indicates maximum and the blue keys indicate minimum, therefore the maximum equivalent (von mises) stress is 3.0591e7 Pa's (pa) and the minimum equivalent (von mises) stress is 2.2593 Pa's (pa), the maximum equivalent elastic strain is 0.00016939 m/m and the minimum equivalent elastic strain is 1.468e−11 m/m, the maximum total deformations is 0.00020084 m and the minimum total deformations is 0 m.

9 Conclusion

Great improvement has been done to design this robot with solidworks, details of kinematics and dynamics models is shown above, the degree of freedom of this robot is six, material selected are cast iron and steel, the payload has been optimized to 125 kg, simulations and analysis characteristics and experimental tests was done with ANSYS.

References

1. Marcial, J.A.D., Cuello, A.F., Fontan, D.R.: Rapid finishing of free form surfaces using milling machine and spherical robot. Adv. Sci. Lett. **19**(3), 849–852 (2013)
2. Akbari, A.A., Higuchi, S.: Optimum machining conditions for mirror polishing of aluminum alloy using a robot. JSME Int. J. Ser. C **44**(3), 842–847 (2001)
3. Han, J.: Industrial Robot, 2nd edn. Huazhong University of Science and Technology Press, Wuhan (2012)
4. Zhou, J., Gao, J., Tang, G.: Mechanics of Materials. Machinery Industry Press, Beijing (2012)
5. Tarkian, M., Persson, J., Ölvander, J., et al: Multidisciplinary design optimization of modular industrial robots. In: Paper Presented to ASME-IDETC/CIE, vol. 8 (2011)

Video Stabilization with a Dual System Based on an IMU Sensor for the Mobile Robot

Jutamanee Auysakul(✉), He Xu, and Vishwanath Pooneeth

College of Mechanical and Electrical Engineering,
Harbin Engineering University, Harbin, People's Republic of China
ajutamenee@eng.psu.ac.th

Abstract. Most of the applications in mobile robots use the image data from the monitoring system to compile algorithms. The undesired movement in the video is an obstacle to compute commands, and it requires to reduce this motion for stabilizing the videos. Thus, we presented a dual system to remove the undesired movement from global and local motion by the mechanical system and digital system, respectively. In the mechanical system, we used a PID controller to control the rotation in the roll and the pitch axes in the servo motors to stabilize video in the line of sight, which held the camera' support base by feedbacking the sensed data from an IMU sensor. Additionally, we applied the KLT feature tracking by integrating an IMU data to estimate the motion for removing the local motion. The results of the dual system were demonstrated as this algorithm reduced the undesired movement when compared with a single combined mechanical and digital system.

Keywords: Video stabilization · IMU · Mobile robot · PID controller KLT tracker

1 Introduction

In the surveillance application, the monitoring system is often required to display the line of sight (LOS) with proper stability for using the data without undesired motion (jitter) in applications such as tracking the object. The jitter effect can be reduced by using video stabilization method, that is subdivided into the mechanical and the digital stabilization system [1]. The mechanical stabilization system concenters to remove undesired movement by adjusting the support base of the camera such as an unmanned aerial vehicle (UAVs) using a 2- or 3-axis gimbal mechanism [2]. Also, the digital stabilization system uses reference frames of the original video to estimate the motion in a stabilized video, by using the optical flow of two consecutive frames.

The motion in the video can be separated to global motion and local motion, respectively [3]. Global motion is the influence of the movement of the camera which generates the desired motion from panning and zooming, and due to the undesired motion (jitter). It is a character the motion of the fixed object in the scene such as buildings, roads, and trees. On the other hand, the local motion describes the motion of the moving object in the view such as cars, people, and animals.

K. Deng et al. (Eds.): ICMIR 2018, AISC 856, pp. 618–626, 2019.
https://doi.org/10.1007/978-3-030-00214-5_78

To measure the physical motion, an inertial measurement unit (IMU) is frequently applied: it senses the acceleration and the angular velocity of the object from an embedded accelerometer and gyroscope, respectively. In the video stabilization research, an IMU sensor is used to activate the mechanical stabilization system, such as Antonello et al. [4] who used an IMU to stabilize a pneumatic' support base of a camera. Consequently, the optical motion was estimated by block matching and feature tracking such as KLT [5], SURF [6], SIFT [7] and so on.

In this paper, we used the advantage of the separate motion to remove undesired movement, by proposing a dual adjusting system of the mechanism of the support base combined with a proportional-integral-derivative controller (PID controller) in the global motion and through estimation of the local motion using the KLT feature tracking and an IMU data. The PID controller was used to control the motion of the camera' support base for moving in the LOS, that held the camera with the servo motors moving in the roll and pitch axes, respectively. The PID closed loop used the orientation data from an IMU sensor to send feedback to the system. Moreover, the KLT feature tracking estimated the local motion with the integral data from the IMU sensor to the initial position of the transition matrix.

This paper is provided as follows. Section 2 discusses our proposed mechanical stabilization system with a described of the PID controller by using an IMU data, and Sect. 3 presents the digital stabilization system by using the KLT tracker and an IMU data. Next, the experimental results are presented in Sect. 4. Finally, Sect. 5 concludes our proposed method.

2 Mechanical Stabilization System

The design of the mechanism used in the video stabilization combined two servo-motors and an IMU sensor to control the rotation in roll (ϕ) and pitch (θ) along with an embedded system. The closed loop of the mechanical system was controlled using an onboard microcontroller which sensed the signal from an IMU sensor thereby feed-backing it into the PID controller to smoothen the captured video. Figure 1 shows the block diagram of our design.

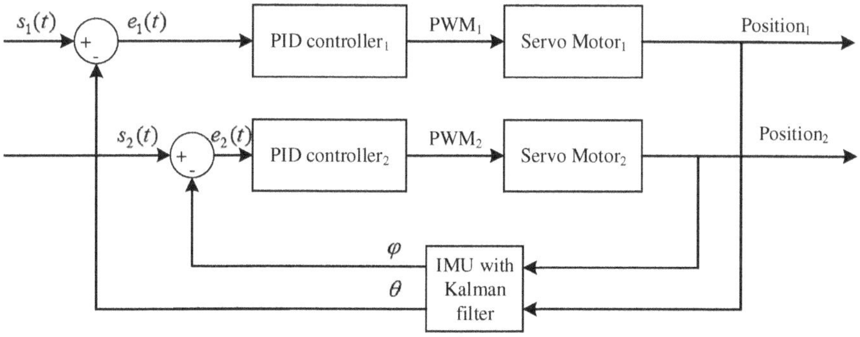

Fig. 1. Block diagram of the mechanical stabilization system.

The values of ϕ and θ depended on the measurable values of the IMU sensor, which was installed on top of the fisheye camera. These values were used to control the motor through the PID controller for monitoring both rotations; we used the servo motor to rotate the camera in each rotational direction.

In this work, the PID controller is used to govern the position of the camera through two servo motors in the roll and pitch, respectively. This controller is a control loop feedback mechanism, that is used in various applications in the continuous system control. The PID controller continuously calculates the error $e(t)$ which is the difference between the process variables (sensing feedback) and the setpoint. The ideal function of PID controller is calculated by the following formula

$$s(t) = k_p e(t) + k_i \int_0^t e(\tau)d\tau + k_d \frac{de}{dt} \tag{1}$$

where $s(t)$ is the control variable. The control variable is summarized the three values in different terms of the error: a proportional term is proportional to the error for correcting the instances of the error, an integral term is proportional to the integral of the error to correct accumulation of the error, and a derivative term is proportional to the derivative of the error to correct the present error versus the previously checked error. The controller variables include a proportional gain (k_p), an integral gain (k_i), and a derivative gain (k_d), respectively. Equation (1) can be parameterized as follows:

$$s(t) = k_p \left(e(t) + \frac{1}{T_i} \int_0^t e(\tau)d\tau + T_d \frac{de(t)}{dt} \right) \tag{2}$$

where T_i and T_d are the integral time constant and the derivative time constant, respectively. To determine the controller parameters, the Ziegler-Nichols method is used to tune for achieving fast response with minimum overshoot responses. Ziegler-Nichols developed techniques to optimize the controller parameters based on a simple experiment which extracts the feature of the dynamic process: the proportional gain is fed to the closed loop process until the system oscillates. The critical proportional gain (k_c) and the period of oscillation (T_c) are used to estimate the controller parameters as follows:

$$\begin{aligned} k_p &= 0.6k_c \\ k_i &= \frac{k_p}{T_i} = \frac{k_p}{0.5T_c} \\ k_d &= k_p T_d = 0.125 k_p T_c \end{aligned} \tag{3}$$

In this work, we used the MPU 6050 model IMU sensor to measure the orientation of the camera and the feedback sent to the PID controller which transfers data through ϕ and θ calculated by the Euler angle, that uses the data from the acceleration meter [8] as shown by the following equations:

$$\tan \varphi = \frac{A_y}{A_z} \tag{4}$$

$$\tan \theta = \frac{-A_x}{\sqrt{A_y^2 + A_z^2}} \tag{5}$$

where A_x, A_y, and A_z are the accelerometer data in x, y, z-axes, respectively. However, to reduce noise and bias in the IMU sensor before feeding it into the system, we applied the Kalman filter to smoothen the data.

3 Digital Stabilization System

The process of digital stabilization includes three main steps: motion estimation, motion filtering and motion compensation [9]. In this work, we proposed the motion estimation by using a KLT feature tracking, which has an initialized position with an IMU data. To estimate motion between two consecutive frames in the first step, the motion vector is selected by the KLT feature tracking because it approaches to find the feature points in the reproducible image [10]. Next, the motion vectors are used to compute the corresponding set with optical flow integrated into the IMU data into the KLT tracker as an initial translation motion and which can be defined as follows:

$$\Delta T = H^{-1} \sum_{w} \left[\nabla I \frac{\partial W}{\partial T} \right]^T [J(x) - I(W(x;T))] \tag{6}$$

$$T_0 = \begin{bmatrix} x_0 - \frac{\varphi W_{image}}{\alpha_\varphi} \\ y_0 - \frac{\theta H_{image}}{\alpha_\theta} \end{bmatrix} \tag{7}$$

where ΔT is a motion vector of a KLT tracker, H is the Hessian matrix, ∇I is the gradient vector, W is coordinate pixels of the current frame, J is the previous frame, $(x_0, y_0)^T$ is the feature position in the previous frames, W_{image} and H_{image} are the width and height of the video. α_φ and α_θ are the maximum rotational degrees of the IMU data in the roll and pitch axes, respectively. Thus, the integrating of the motion vector can be defined as:

$$T_{i+1} = T_i + \Delta T \tag{8}$$

Then, the optical flow between two consecutive frames is used with the affine transform model to estimate the motion. Here, we used the RANSAC algorithm to determine motion in terms of scaling, rotation, and translation in the affine matrix between the current frame and the reference frame.

Moreover, the motion estimation can remove undesired translation motion, that may be from the last step and due to Gaussian noise in the unstable video. We used the Kalman filter in the motion filtering step to predict the next state based on the previous state in real-time because it used only the data from last frames [11]. The dynamic linear model of state X_{k+1} at time k + 1 is represented as,

$$X_k = AX_{k+1} + B_k \tag{9}$$

where X_k is the state system at time k, A is the state of the transition between time k and k + 1, and B_k is the noise of the system. The observation state Z_k of state X_k is given by:

$$Z_k = CX_k + D_k \tag{10}$$

where C and D_k are the observation matrix and observation noise at time k.

Finally, the motion compensation is combined the motion estimation and the motion filtering in the previous step into the affine matrix. Given I (x, y) as the current frame and K (x_1, y_1) as a stable frame to compensate in the affine model, it is illustrated as follows:

$$\begin{bmatrix} x_1 \\ y_1 \end{bmatrix} = \begin{bmatrix} S\cos\phi & -S\sin\phi \\ S\sin\phi & S\cos\phi \end{bmatrix} \begin{bmatrix} x \\ y \end{bmatrix} + \begin{bmatrix} T_x \\ T_y \end{bmatrix} \tag{11}$$

where S and ϕ are the scale and the rotation in the plane, respectively. T_x and T_y are the translation in x and y-axes by integration of a KLT tracker with the initial position in (8) and with a filtered translation.

4 Experimental Results

In this work, we implemented our mechanical stabilization design by controlling the rotational position of the camera through servo motors as shown in Fig. 2. This system is controlled via Arduino microcontroller, which communicates an IMU sensor through the I2C serial protocol and a control servo motor with pulse-width modulation (PWM). Firstly, we tune the PID parameters by increasing k_c until the system oscillates as shown in Fig. 2 which is a separated controller of each motor. Those systems oscillate at k_c as shown in Fig. 3, which can use to calculate T_c. Thus, we computed the PID parameters by using Eq. (3).

To evaluate the efficiency of the video stabilization, we have compared the mean square error (MES) and the peak signal to noise ratio (PSNR) by using the data from two consecutive frames which is given as follows:

$$\text{MSE}(I,J) = \frac{1}{NM}\sum_{i=1}^{N}\sum_{j=1}^{M}(I(i,j) - J(i,j))^2 \tag{12}$$

$$\text{PSNR}(I,J) = 10\log\frac{I_{\max}^2}{\text{MSE}(I,J)} \tag{13}$$

where I and J are the consecutive current and previous frame, respectively. I_{max} is the maximum pixel density in the video. It is calculated in every pixel (i, j) to determine the

Fig. 2. Illustrated the hardware in this experiment.

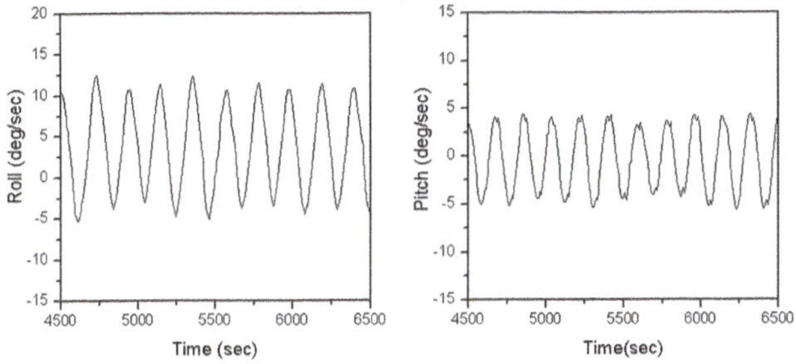

Fig. 3. Oscillation response from the output of $Motor_1$ and $Motor_2$ in the pitch and yaw-axis, respectively.

effectiveness of the stylized video. Moreover, PNSR can be extended to calculate the smoothness of the video by the inter-frame transformation fidelity (ITF) as follows:

$$\mathrm{ITF} = \frac{1}{N_{frame} - 1} \sum_{k=1}^{N_{frame}-1} \mathrm{PSNR}(k) \tag{14}$$

where N_{frame} is the number of frames in the video. IFT represents the overall quality of the stabilized video. If the value of IFT is high, then the stabilized video is high also.

We implemented a dual-system on rough terrain and had random sample frames to show the results in Fig. 4. The result of the mechanical system in Fig. 4(a) is inclined in the original video and is incomplete to stabilize the video in LOS because the servo motor was moving slowly. However, the video in LOS can be stabilized as shown in Fig. 4(b) by integrating our digital stabilization system by using an IMU data.

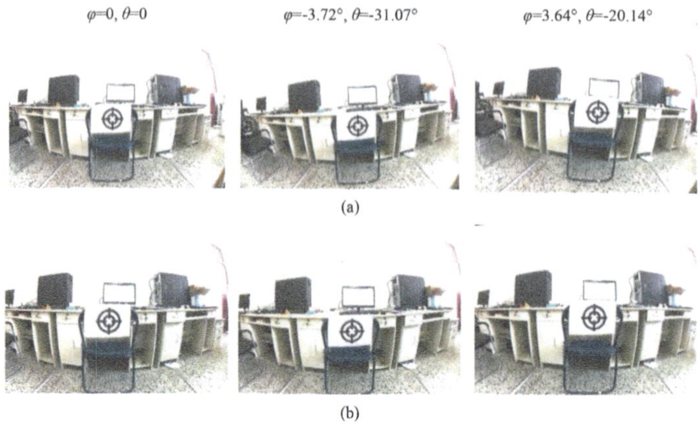

Fig. 4. Comparison the results of (a) original video (left: 1st frame, center: 205th frame, right: 402nd frame) and (b) stabilized video

To compare the effectiveness of our algorithm, we recorded the original video before feeding into the digital stabilization system to test it with the KLT feature tracking with Kalman filter. From Table 1, our proposed approach reduces the average error by 21.77% when compared with original video and by 4.89% when compared with Kalman filter. Furthermore, we compared PNSR of our algorithm and the Kalman filter as shown in Fig. 5 and found our proposed is as high as then an original video and a Kalman filter. It never dropped from the original video and high ITF when compared to other methods as shown in Table 1, that means the proposed method gave the best quality of the stabilized video.

Table 1. Comparison effectiveness of the video stabilization with different methods.

Effectiveness variables	Original video	Kalman filter	Our proposed
MSE maximum	2505.12	2197.09	2169.18
MSE average	899.66	239.88	195.85
ITF	21.11	26.96	28.36

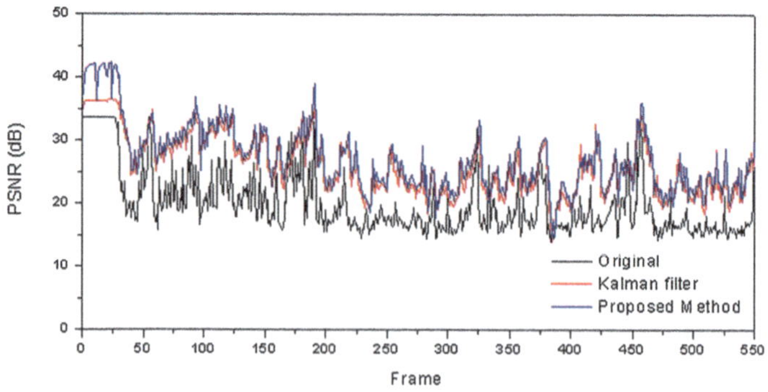

Fig. 5. PSNR of the stabilized video.

5 Conclusions

The video stabilization process applied a dual system entailing the mechanical and the digital video stabilization systems in order to display videos in LOS. The mechanical system uses a PID controller to control the rotational position of the camera in roll and pitch. Nonetheless, only the mechanical system incompletely stabilizes the video in LOS. Thus, the digital system increases the accuracy of LOS into the stabilization video by demonstrating the high ITF and MSE when it was compared to original video and Kalman filter. Hence, this consideration can be applied in the feature applications of video processing.

Acknowledgments. This work was supported by the Natural Science Foundation of the Heilongjiang Province of China (F2016003), Specialized Research Fund for Doctoral Program of Higher Education (20122304110014), the Fundamental Research Funds for the Central Universities (HEUCFX1508), Manipulator Data Processing Development (KY10700150065), and the International Exchange Program of Harbin Engineering University for Innovation-oriented Talents Cultivation.

References

1. García, G.B., Suarez, O.D., Aranda, J.L.E., Tercero, J.S., Gracia, I.S., Enano, N.V.: Learning Image Processing with OpenCV. Packt Publishing Ltd., Birmingham (2015)
2. Odelga, M., Kochanek, N., Bülthoff, H.H.: Efficient real-time video stabilization for UAVs using only IMU data. In: 2017 Workshop on Research, Education and Development of Unmanned Aerial Systems (RED-UAS), pp. 210–215 (2017)
3. Veon, K.L., Mahoor, M.H., Voyles, R.M.: Video stabilization using SIFT-ME features and fuzzy clustering. In: 2011 IEEE/RSJ International Conference on Intelligent Robots and Systems (IROS), pp. 2377–2382. IEEE (2011)
4. Antonello, R., Oboe, R., Ramello, A., Ito, K., Felicini, N., Cenedese, A.: IMU-aided image stabilization and tracking in a HSM-driven camera positioning unit. In: 2013 IEEE International Symposium on Industrial Electronics, pp. 1–7 (2013)
5. Ryu, Y.G., Roh, H.C., Chung, M.J.: Video stabilization for robot eye using IMU-aided feature tracker. In: ICCAS 2010, pp. 1875–1878 (2010)
6. Yu, J., Luo, C., Jiang, C., Li, R., Li, L., Wang, Z.: A digital video stabilization system based on reliable SIFT feature matching and adaptive low-pass filtering. In: CCF Chinese Conference on Computer Vision, pp. 180–189. Springer (2015)
. Jianguo, J., Jiejie, N., Meibin, Q.: Rotational video stabilization algorithm based on SURF and trajectory smoothing. Chin. J. Sci. Instrum. **35**, 550–557 (2014)
8. Alam, F., ZhaiHe, Z., Jia, H.: A comparative analysis of orientation estimation filters using MEMS based IMU. In: Proceedings of the International Conference on Research in Science, Engineering and Technology, Dubai, UAE, pp. 21–22 (2014)
9. Ryu, Y.G., Roh, H.C., Kim, S.J., An, K.H., Chung, M.J.: Digital image stabilization for humanoid eyes inspired by human VOR system. In: 2009 IEEE International Conference on Robotics and Biomimetics (ROBIO), pp. 2301–2306 (2009)

10. Wang, H., Oneata, D., Verbeek, J., Schmid, C.: A robust and efficient video representation for action recognition. Int. J. Comput. Vis. **119**, 219–238 (2016)
11. Aguilar, W.G., Angulo, C.: Real-time model-based video stabilization for microaerial vehicles. Neural Process. Lett. **43**, 459–477 (2016). https://doi.org/10.1007/s11063-015-9439-0

Design and Implementation of OA Management System Based on Rule Engine

Ting Gong(✉), Kun Qu, and Jianfei Shao

School of Information Engineering and Automation, Kunming University of Science and Technology, Kunming 650500, China
ting9210120@163.com

Abstract. In recent years, the application of information technology in business offices has got more and more widespread. Comprehensive use of computer in modern management technology, workflow and knowledge engineering build a practical and efficient office automation system. It is an important means for all enterprises and institutions to implement resource allocation. This thesis puts forward an OA system based on the rule engine,which is on the analysis of the present situation of the small and medium-sized enterprise office system and problems, and combined with the common demand of actual investigation and the mainstream business office. The improved RETE algorithm is applied to the analysis, the design and the implementation of the system, so as to achieve the automation of corporate office management.

Keywords: Rule engine · RETE · OA system

1 Introduction

With the rapid development of computer technology and Internet communication technology, most domestic enterprises adopt the collaborative way when handling work office, which has great limitations. They cannot achieve the sharing of data information and affect the work efficiency of enterprises. Traditional office mode cannot meet the demands of the company's daily office and efficient communication of information. So because of that situation, the Office Automation System came into being. OA system uses fast and inexpensive means of network transmission to realize information sharing function and coordinating the work of various departments, reducing the complex links in the work. At the same time, the system has the characteristics of timely display of internal corporate information, rapid processing of internal affairs of the company, and efficient collaboration among employees; can realize trans-regional and trans-national collaborative work. So that the company's management can make decisions and analysis the company's affairs at any time and place, respond to market changes in a rapid way, and seize the opportunity effectively to enhance the efficiency of the company. Building the enterprise information can enhance the company's core competitiveness, promote the speed and sensitivity of the company response, and promote internal communication capabilities.

So far, most OA management systems still use the traditional "hard programming" development model, embed a large number of business logic in the system, poor

K. Deng et al. (Eds.): ICMIR 2018, AISC 856, pp. 627–634, 2019.
https://doi.org/10.1007/978-3-030-00214-5_79

reusability, this development method leads to the low development efficiency problem, software maintenance upgrades uneasily. However, the demand for OA systems is often changing rapidly. Therefore, some domestic and foreign companies have begun to introduce rules engines in the process of system construction to improve the system's adaptability. The rule engine originated from a rule-based expert system. It peels business rules from the application code out of the program code; describes it in a script language, a decision tree, or other forms; and uses an inference engine to judge and execute business rules [1]. In the 1970s, foreign researchers first proposed rules application to the construction of application systems. After half a century of its research and development, mature theoretical foundations and applications have been formed in the design, implementation, and implementation of rule engine algorithms, rules, and rule engines. In software development, the adoption of rules in rule engines can reduce the difficulty of implementing complex business logic components, the costs of maintenance and scalability of the applications; not only can it separate business logic and application developers' technical decisions, but also can meet the special needs of users and frequent changes in demand.

2 Overall Framework of the System

The B/S architecture is currently the most popular development model. Its advantages are: simple operation, only need to install a common browser; simple maintenance and upgrade methods, zero maintenance on the client, as long as the management server; with distributed features, you can always handle business anywhere. The system is formally based on this development model. In this mode, the user enters the system through the browser page. The system actively accepts the user's request sent to the server, and then after reviewing the user's rights and requests process, and the server can get the data in real time by the result of the process that the server is passing through the model.

The three-tier architecture is commonly used in the application system layered architecture, most of the WEB system is based on this architecture. The traditional three-tier architecture includes the user interface layer also called presentation layer, the business logic layer, and the data access layer, as shown in Fig. 1. The data access layer is mainly the operation layer for non-original data (database or text file, etc.), which provides the specific data service for the business logic layer or the user interface layer. The business logic layer is the "middleware" of the system, connecting the data access layer and the user interface layer. It mainly deals with the data business logic and the response of the result returned by the operation data access layer. The function of the user interface layer is used to receive user input data and display the data required by the user after being processed by the system. The traditional three-layer structure has high cohesion and low coupling; each layer is independent from each other; it is easy to transplant and maintain [2]. Under this architecture, cross-layer calls can-not be made between the system layers and sub-hierarchies, and the user interface layer cannot contain any processing of business logic.

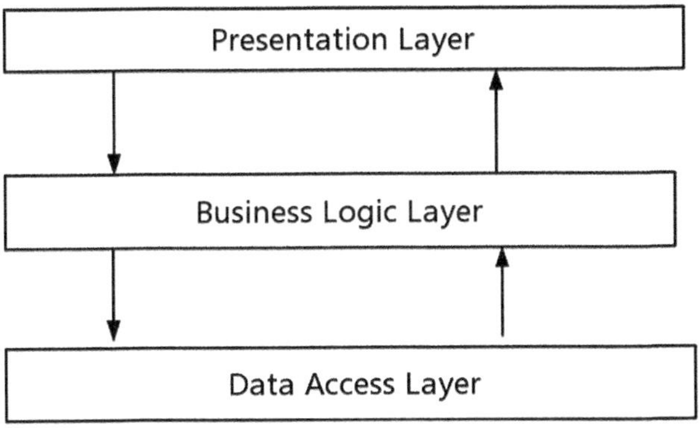

Fig. 1. Three-layer structure of general web system

The system has been improved on the basis of the traditional three tier architecture framework, adding the general rules engine framework and the presentation services layer control layer, as shown in Fig. 2. The presentation layer invokes two interfaces: the business logic layer and the presentation services layer as different forms of foreground presentation. The business logic layer is used to process the system's permission framework; the rules engine is responsible for handling workflow-related business operations, such as the information auditing process; and the user interface control layer is used to generate a corresponding JSP interface according to the result of the rule engine processing.

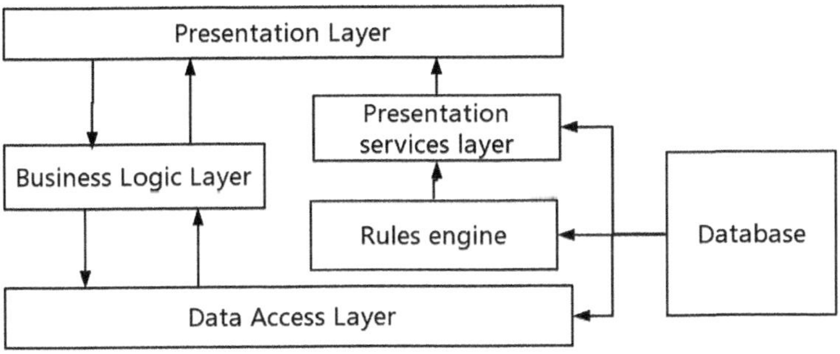

Fig. 2. System three-layer structure

3 Rules Description Files

In this system, the structure of rule's description files is organized in XML format. XML is the most commonly used self-describing text format. It can have complex data structures and good extensibility. It is also normalized and verifiable in format. Table 1 is a description of some of the objects defined in the DTD.

A rule consists of LHS and RHS. In this article, the Workflow and Action objects form the structure of if (LHS) then (RHS), and the left one (LHS) consists of one or more DTDs. The composition of the object attributes, the right side (RHS) consists of one or more actions (action). If a rule contains all the necessary conditions, the rule is satisfied. In the DTD object, all the information of the entity classes involved in inference is described in detail. In general case, the rules need to be modified frequently, but the entity class information is relatively stable, so you can easily modify the rules without having to recompile the source code.

In fact, a rule file can contain multiple rules. After the rule engine is initialized, it loads the rule file, generates a rule set object, and pushes it to the matcher to construct a reasoning network (RETE network).

Table 1. Part of the object defined in the DTD and its description.

Object	Attrition	Necessary	Description
Workflow	W_ID	Yes	Rule number
	W_Name	Yes	Rule name
	W_Priority	Yes	Rule priority
	W_Parent	No	Parent rule object
	W_Level	Yes	Level object
	W_Modern	Yes	Audit method
	W_Type	Yes	Operation type
	Message type	No	Notification type
Operator	Organize	Yes	Authority organization
	Role		Permission role
	Person		Authority collection
Action	A_Classname		Action operation class name
	A_Method		Action processing method
	A_Argument		Front page generation parameters

4 Design and Implementation of Rule Engine

The main task of the rule engine is to match the data objects submitted to the engine with the business rules in the rule set, and activate the business rules in accordance with the current data state, and trigger the operation in the application program according to the execution logic of the rules centrally stated. The rule engine is mainly composed of rule sets, fact memories, matchers, executors, and confliction resolver. The specific structure is shown in Fig. 3.

The fact file is also expressed in XML format, and its DTD structure is basically the same as the rule file, which specifically describes the object characteristics of the workflow. After loading the fact file, the system generates a fact set, each of which is a factual configuration, and the element in fact is pushed in order to the matcher.

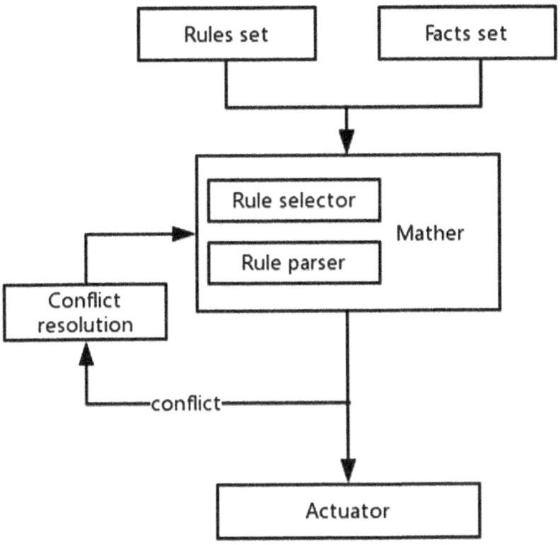

Fig. 3. Rules engine structure diagrams.

4.1 Matcher

The efficiency of the rules engine mainly depends on the matcher, which is used to compare rules and data in the rule base. The matcher is composed of a rule selector and a rule parser. After receiving the rule set object, the rule selector generates an inference network (RETE network) according to the improved RETE algorithm [3], and matches the fact data to the corresponding rules in the inference network. The rule parser is used to store the rules in the working memory and send them to the executor.

The RETE algorithm is a forward fast matching algorithm, however its matching speed has nothing to do with the number of rules. Its main advantages include: exchanging space for time; full sharing of matching results. However, the traditional RETE algorithm is designed for artificial intelligence system, and it is not suitable for the general sense of enterprise application system. In this paper, a class oriented RETE algorithm is adopted, which is an improved algorithm based on RETE.

In RETE network, there are five kinds of nodes, as shown in Table 2.

Table 2. RETE network node description.

Node type	Description
Root node	The entry of all objects into the network and the first node that the token enters
Class node	It is responsible for testing whether the class involved in the tag sent by the parent node is consistent with the class represented by this node
Alpha node	It is used to perform attribute matching (separate into required alpha nodes and unnecessary alpha nodes)
Connected node	Used to connect two or more input nodes in a RETE network
Terminal node	It is used to store action object information and is the end node of the token, which is considered as a rule corresponding to a terminal node

The creation process of RETE network is as follows. The part of RETE network structure generated by this system is shown in Fig. 4.

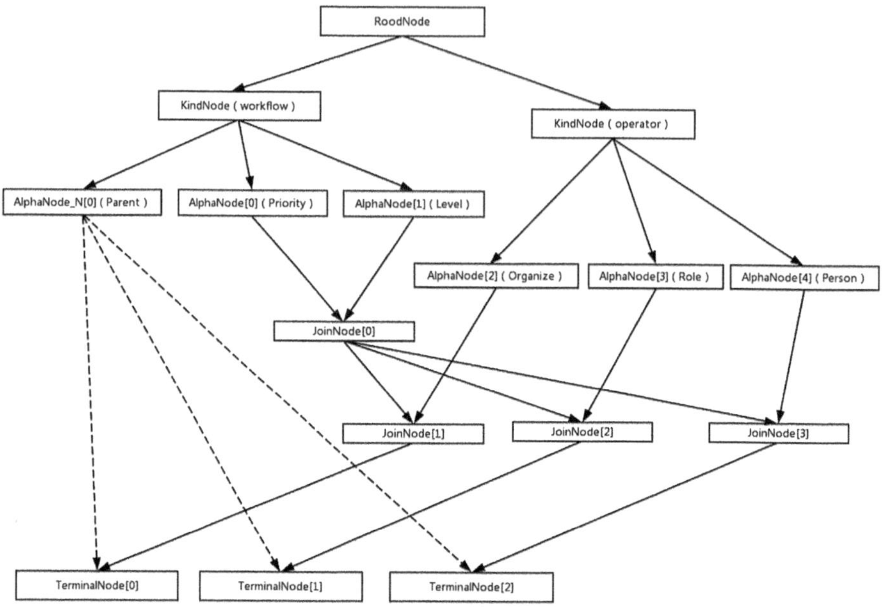

Fig. 4. Part of the RETE network structure generated by this system

(1) Create the root;
(2) Add a rule, take out the DTD object, and check the type of the object in the network. If it is a new object, add a class node. If the object already exists, record the location of the class node. Check the attribute type in the object if the attribute does not exist, add it as an Alpha node to the network. If there is a record Alpha node location, establish the Alpha memory table according to the mode of the Alpha node.
(3) Repeat step 2 until all DTD objects are processed.
(4) Combine connected nodes as follows
   (a) JoinNode[0] left input node is the necessary AlphaNode[0], and the right input node is AlphaNode[1].
   (b) JoinNode[i] left input node is JoinNode[i − 1], and the right input node AlphaNode[i + 1] (i > 1).
   Inline the memory tables of two parent nodes into the memory table of this node.
(5) Repeat step 4 until all JoinNode nodes are processed.
(6) The action object is encapsulated into a termination node as the output node of JoinNode[i];
(7) Repeat step 2–6 until all rules are processed.

For each fact, filtering through the select operation causes the facts to reach the appropriate alpha node along the RETE network. For each alpha node received, separate those appropriate variables. Bring each new set of variables along the RETE network to the appropriate JoinNode. For the Join-Node that receives the new set of variables, the above steps are repeated until the termination node is reached and the conclusion is instantiated and sent to the executor.

4.2 Conflict Resolution

If multiple incompatible rules are activated at the same time in the matcher, it is considered that there is a conflict which needs to be resolved before forwarding to the executor. In this system, what may happen is that of the current rule conflicts with the rules of its parent object. The definition of conflict is as follows:

The confliction of rules means that there are inconsistencies in the conditions of each rule, leading to the existence of reasoning ambiguity.

Assume that A1 and A2 are the matching rules of two configuration instances and can be expressed as

$$A_1 = \bigcap C_{1i} + \bigcap R_{1j}, \quad A_2 = \bigcap C_{2i} + \bigcap R_{2j}. \tag{1}$$

Among them, $\bigcap C_{\bullet i}$ is the conditional part of the rule, and $C_{\bullet i}$ is the ith condition item; $\bigcap R_{\bullet j}$ is the conclusion part, $R_{\bullet j}$ is the jth conclusion item.

If there is a $C_{\bullet i}$,that due to $\bigcap C_{1i}$ and $\bigcap C_{2i}$ to be mutually exclusive, it means that there is a conflict between rules A_1 and A_2.

In the conflict resolver, the conflict rules are sorted according to the priority, and the rule with the higher priority is executed first. If with the same priority, the rules of parent object corresponded. If there is a confliction in the part of action and the action method is in the same class, the parent class is dynamically loaded. Otherwise, the conflict is considered unsolvable and the next fact is executed.

The basic working process of the rule engine is: After the rule engine is initialized, the system reads in the rule file first, automatically generates a rule set object, and pushes it to the matcher; at the same time reads in the fact file, that is, the user configures according to the required format. The data describes the specific characteristics of the workflow. The matcher generates an inference network (RETE network) according to the received rule set object. After determining that the Rete network is correctly generated and the read fact file is not empty, the matcher searches for matching in the inference network until a certain termination is found. The node then considers that the fact has found the corresponding rule. The rule is parsed and instantiated according to the action part of the termination node. The operation class and method are dynamically invoked, and a corresponding foreground display interface is generated. If no termination node is found, the fact is considered as incomplete, and the configuration modification is prompted. If there is a conflict, the conflict resolver processes it and retransmits it to the executor.

5 Summary

This paper designs and implements an OA system based on the rules engine, which revises the traditional RETE algorithm adaptively and proposes an improved method for class objects. XML files are used as a storage method for rules and facts. As well as, this system resolves issues that is difficult to cope with in the traditional "hard programming" development model, so it has fast respond to customer needs. Furthermore, what is required is just modifying the configuration data, without re-encoding and modifying the existing system structure. It greatly reduces the development cycle and reduces redundant work, and have low requirements on the professional qualities of developers.

References

1. Wu, D.: Design and implementation of rule engine for enterprise application. Shandong University (2014)
2. Mitra, S., Using, U.M.L.: Modeling to facilitate three-tier architecture projects in software engineering courses. ACM Trans. Comput. Educ. **14**(3), 1–31 (2014)
3. Zhu Siyuan, Z.: A method of implementing distributed rule engine. Comput. Eng. **36**(12), 158–161 (2015)

Optimization Design and Analytical Analysis of Elliptical Machine

Zhanliang He(✉)

Xinxiang Vocational and Technical College, Xinxiang 453000, Henan, China
zhanliang_he@126.com

Abstract. Elliptical machine is one important of fitness equipments, and becomes more and more popular resulting from people increased emphasis on health problems. Based on ergonomics, the optimization design and analytical analysis of elliptical machine are conducted. The trajectory of the human ankle during running is considered to optimize the machine structure. By analyzing the influence of each member length on the trajectory of the human ankle, the optimal length of each member is determined.

Keywords: Ergonomics · Optimization design · Analytical analysis
Elliptical machine

1 Introduction

In recent years, as the improving of material living standards, human beings pay more attention to the fitness. Elliptical machine, as one of the most important parts of fitness equipment, has attracted wide attention. The ankle trajectory is similar to an ellipse when people are jogging in life. And the pedal trajectory of elliptical machine is also similar to an ellipse by a particular mechanism. During this process, the foot does not leave the pedal completely and there is no impact on the knee joint. Besides, it can exercise the upper and lower limb muscle groups, which is a healthy sport for all the people.

Ergonomics takes the human factor as the primary principle of product design, and studies human-machine interaction to achieve the optimization of the human-machine-environment system [1]. Based on human factors engineering, Wang [2] proposed the general guidelines of elliptical machines designing and improved the design theory of elliptical machine, but no in-depth study was conducted. According to ergonomics, Yang [3] made a preliminary design and motion simulation of the elliptical machine, but machine size was larger.

However, there is no much research on the ergonomics designing of elliptical machine. The existing elliptical machine is mostly designed for commercial gymnasiums, but is not suitable for ordinary home resulting from the size, quality and expensive. Therefore, based on ergonomics, this paper proposes to design a mini-elliptical machine and confirms to the physiological characteristics of people, with low cost and small size.

K. Deng et al. (Eds.): ICMIR 2018, AISC 856, pp. 635–641, 2019.
https://doi.org/10.1007/978-3-030-00214-5_80

2 The Optimization Design of Elliptical Machine

2.1 Optimization Design Requirement

The optimization design requirements of mini-elliptical machine are (a) small size and easy to be moved, (b) pedal trajectory should be consistent with the user's ankle gait movement to ensure the continuity of movement, (c) machine size should meet the needs of different users, and (d) structural parameters should be reasonable to ensure the comfortable of users [4].

2.2 Optimization Design Based on Ergonomics

Based on the ergonomics, the optimization design of elliptical machine, which is first necessary to meet the need of elliptical machine normal operation, and then the pedal trajectory of elliptical machine should conform to the human ankle trajectory to the utmost extent, to improve the comfortable and safety of elliptical machine. The actual ankle trajectory can be fitted with an ellipse when users run. The method is shown below, the distance between every two points in the actual ankle trajectory is calculated, the longest connecting line between two points is major axis, similarly the shortest connecting line between two points is minor axis, and the angle between the major axis and the horizontal direction is ellipse dip angle. The ankle trajectory is different under different conditions. When walking slowly, the ratio between major and minor axis of ellipse is smaller than that of brisk walking. When jogging, the ankle trajectory is closer to the ideal elliptic curve, and the pace becomes larger, thus major and minor axis become longer. Besides, the physical condition of users also affects the ankle trajectory. So selecting different experimenters to jog on the treadmill and recording the ankle trajectory. Table 1 shows the basic situation of the experimenter, and Fig. 1 shows the ankle trajectory.

Table 1. Basic situation of the experimenters

Experimenter	1	2	3	4	5
Sex	Male	Female	Male	Female	Male
Height (cm)	171	162	183	164	166

By analyzing the ankle trajectories of different experimenters in Fig. 1, an ideal elliptical trajectory is fitted. In the fitting process, first to ensure the length of major axis meets the demand of common users, because the major axis length of the ideal ellipse represents stride length. On this basis, the minor axis is also fitted, the major axis length of around 230 mm, the ratio between major and minor axis of 1.86–1.96, and the major axis tilt angle of around 20° are finally determined.

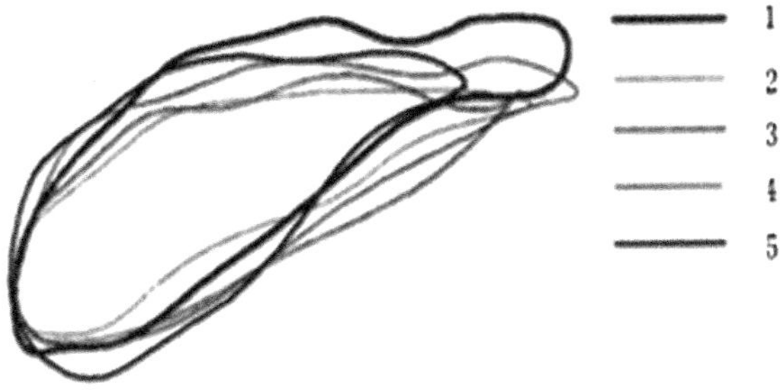

Fig. 1. Participants' ankle trajectories

2.3 Optimization Design of Elliptical Machine Structure

Compared with the conventional elliptical machine, the mini-elliptical machine has the advantage of small dimension and remains the basic functions. The main improvements are removing the handrail structure that occupies space and adding wheels under pedals, which greatly reduce machine dimensions. Finally, as shown in Fig. 2, the mini-elliptical machine is mainly composed of frame, crank, pedal, wheel and raceway. When users stomp on the pedal, the pedal drives the crank to make a circular motion and then rolls back and forth on the raceway. And the pedal moves in an elliptical path, which guides the movement of the users' ankle, so that the fitness movement coincides with the natural stepping of users. Besides, different load can be obtained by adjusting the damping adjuster to achieve different exercise effects.

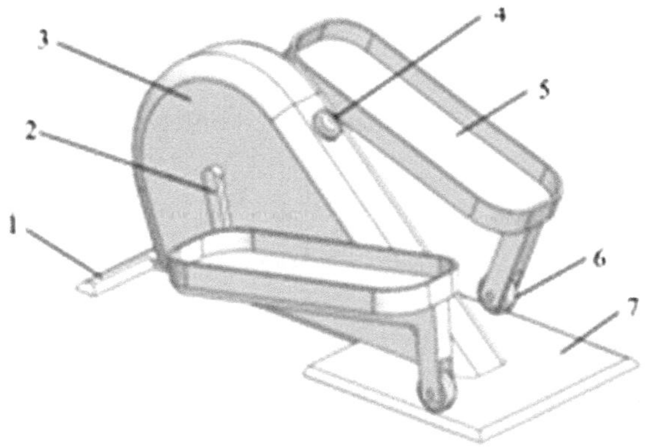

Fig. 2. Structure of elliptical machine. 1. Frame, 2. Crank, 3. Shell, 4. Damping adjustor, 5. Pedal, 6. Wheel, 7. Raceway

3 Analytical Analysis of Elliptical Machine Trajectory

As shown in Fig. 3, the structure of mini-elliptical machine can be simplified into deflection crank slider mechanism. Where *AD* is the pedal, *M* is the simulation ankle point, and *MN* is perpendicular to *AD*. Assuming that *O* is the coordinate origin, the X-axis is horizontal to right, and the Y-axis is vertical to top.

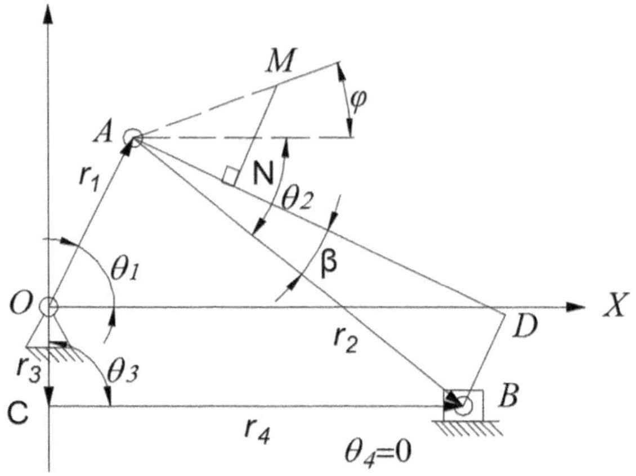

Fig. 3. Kinematic diagram of mini elliptical machine

The vector equation of mechanism:

$$\vec{r}_1 + \vec{r}_2 = \vec{r}_3 + \vec{r}_4. \tag{1}$$

Crank length r_1, connecting rod length r_2 and offset distance r_3 are known. Pedal length is $2b$, $\theta_3 = 90°$, $\theta_4 = 0°$, the initial variable $\theta_1 = \omega t$, separating variables x and y, we get:

$$r_1 \cos\theta_1 + r_2 \cos\theta_2 = r_4. \tag{2}$$

$$r_1 \sin\theta_1 + r_2 \sin\theta_2 = -r_3. \tag{3}$$

From Eqs. (1), (2) and (3), we get:

$$\theta_2 = \arcsin\left(\frac{-r_3 - r_1 \sin\theta_1}{r_2}\right). \tag{4}$$

$$r_4 = r_1 \cos\theta_1 + r_2 \cos\theta_2. \tag{5}$$

Then taking the derivative of t in Eq. (5), we obtain:

$$r_1\omega_1 \sin\theta_1 + r_2\omega_2 \sin\theta_2 = v_3. \tag{6}$$

$$r_1\omega_1 COS\theta_1 + r_2\omega_2 \cos\theta_2 = 0. \tag{7}$$

According to Eq. (7), we obtain:

$$\omega_2 = -\frac{r_1 \cos\theta_1}{r_2 \cos\theta_2}\omega_1. \tag{8}$$

$$\alpha_2 = \frac{r_1\omega_1^2 \sin\theta_1 + r_2\omega_2^2 \sin\theta_2}{r_2 \cos\theta_2} \tag{9}$$

According to Eqs. (4), (5) and (6), the trajectory, velocity and acceleration of simulation ankle point can be deduced.

According to geometrical relationship, the motion equations of simulation ankle point can be obtained:

$$x_M = r_1 \cos\theta_1 + c\cos\varphi. \tag{10}$$

$$y_M = r_1 \sin\theta_1 + c\sin\varphi. \tag{11}$$

Where $c_2 = AN - 2 + MN - 2$, $\varphi = \angle MAN - (\theta_2 - \beta)$.

Taking the derivative of t in Eqs. (10) and (11), it can be obtained:

$$v_{xM} = -r_1\omega_1 \sin\theta_1 - c\omega_2 \sin\varphi. \tag{12}$$

$$v_{yM} = -r_1\omega_1 \cos\theta_1 + c\omega_2 \cos\varphi. \tag{13}$$

$$v_M = \sqrt{v_{xM}^2 + v_{yM}^2}. \tag{14}$$

Taking the derivative of t in Eqs. (12), (13) and (11), it can be obtained:

$$a_{xM} = -r_1\omega_1^2 \cos\theta_1 - c\alpha_2 \sin\varphi - c\omega_2^2 \cos\varphi. \tag{15}$$

$$a_{yM} = -r_1\omega_1^2 \sin\theta_1 + c\alpha_2 \cos\varphi - c\omega_2^2 \sin\varphi. \tag{16}$$

$$a_M = \sqrt{a_{xm}^2 + a_{ym}^2}. \tag{17}$$

From the analysis results, it can be known that the trajectory of *M* is related to each member length, and is also affected by the angle β between pedals. To determine the influence of member length on elliptical trajectory, it is assumed that r_1 = 110 mm, r_2 = 386.85 mm and r_3 = 140 mm, and then each member length is changed in order while other member length is unchanged. Substituting MN = 100 mm and ND = 250 mm into above equations, and using Excel to plot the displacement curve of M.

As shown in Fig. 4, it is found that both major and minor axis increase with an increase in the r_1 while the elliptical dip remains unchanged. As shown in Fig. 5, with an increase in the r_2, the elliptical dip increases, but both major and minor axis increase slowly. As shown in Fig. 6, r_3 has a great influence on the ratio between major and minor axis, the ellipse becomes flat with an increase in the r_3. The optimal length of each member is determined, r_1 = 110 mm, r_2 = 412 mm and r_3 = 140 mm, through comprehensive analysis and calculation. The elliptical dip of 20°, the major axis of 232 mm, the minor axis of 123 mm, and the ratio between major and minor axis of 1.89 are determined accordingly. On this basis, the pedal trajectory conforms to the previous fitted ellipse in Sect. 2.2, and the periodic fluctuation of machine velocity is stable.

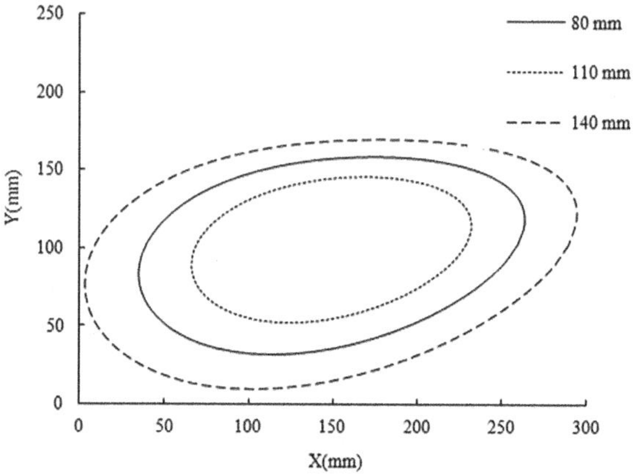

Fig. 4. Analytic curve of elliptical trajectory by changing r_1

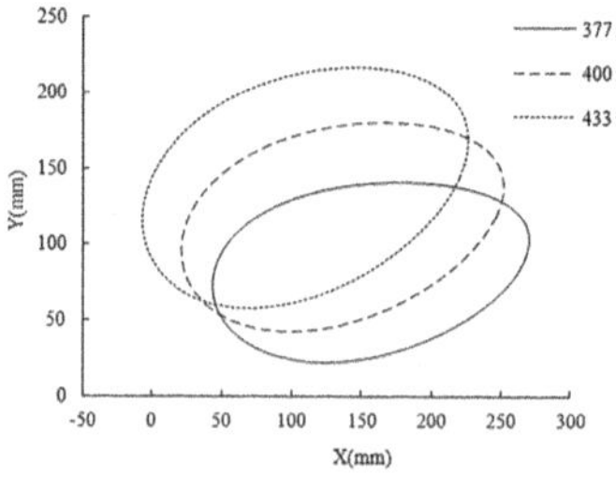

Fig. 5. Analytic curve of elliptical trajectory by changing r_2

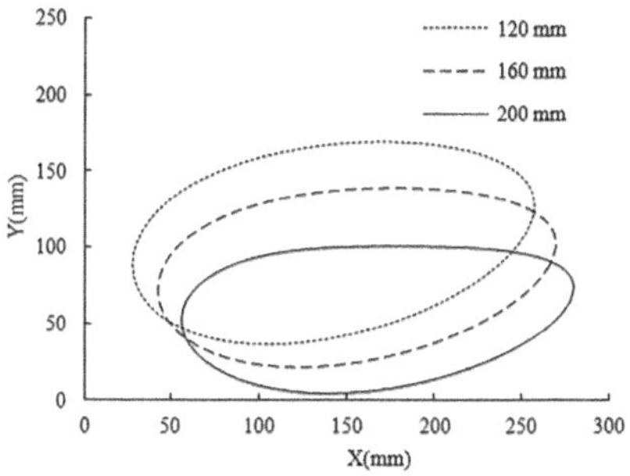

Fig. 6. Analytic curve of elliptical trajectory by changing r_3

4 Conclusion

According to the design theory of elliptical machine and ergonomics, this paper presents an elliptical machine, which has advantages of small size, user-friendly and confirming to ergonomics. And then the optimal elliptical trajectory is determined for r_1 = 110 mm, r_2 = 412 mm and r_3 = 140 mm through analytical analysis.

References

1. Ding, Y.: Ergonomics. Shanghai Jiao Tong University Press, pp. 1–2, 24–31 (2004)
2. Wang, X., Wu, Y.: The design of elliptical cross trainers with ergonomical guidance. J. Gr. **26** (1), 123–128 (2007)
3. Yang, X.: The design and simulation of striding machine with ergonomics guidance. Mach. Des. Res. **33**(06), 104–108 (2017)
4. Tian, H., Li, A., Shang, W.: Optimization design of cycling equipment parameters based on ergonomics. Mach. Des. Manuf. **3**, 192–195 (2015)

Remote Control and Monitoring System of Robotic Arm Using Raspberry Pi

Longkun Xu(✉)

College of Automation and Electronic Engineering,
Qingdao University of Science and Technology, Qingdao, China
longkun_xu@outlook.com

Abstract. In order to meet the requirement of controlling and monitoring the remote robotic arm, a remote control system based on Raspberry Pi was proposed. The system takes the Raspberry Pi 3B as the core and controls the six-axis robot through the network in the upper computer. Meanwhile, the video will be captured by the USB camera and displayed in the master computer. The practice shows that the system can transmit control commands and monitor images reliably, and control the robotic arm to accomplish complex tasks.

Keywords: Robotic arm · Raspberry Pi · Remote control · PWM
Mjpeg

1 Introduction

With the constant development of science and technology, robotic arm has replaced many manual operations with its flexible movements, which is an indispensable part of industrial production. For repetitive motion, we can write the corresponding program for the manipulator to complete the specified action. For those who are in dangerous or narrow work environment and complex movements, inconvenient human operation directly, also hard to let manipulator automatically complete the work task [1]. The use of remote manipulator can ensure the safety of personnel and improve work efficiency.

The Raspberry Pi is a kind of single-chip computer with a credit card size. It is developed by the British Raspberry Pi Foundation, which is cheap, portable, and rich in software resources. It provides 40 GPIO pins for developers to use, and can be connected to an external motor drive module to control the servo action of the robot. There are also Ethernet interfaces and USB interfaces. USB cameras can transmit real-time video streams to the host computer through the network. This article uses the Raspberry Pi as a control core to reduce the use of peripheral devices. The robot connects the PC and the Raspberry Pi over the network, enabling people to remotely operate without distance restrictions, as well as acquiring real-time status through the camera, thus completing complex operations [2].

K. Deng et al. (Eds.): ICMIR 2018, AISC 856, pp. 642–648, 2019.
https://doi.org/10.1007/978-3-030-00214-5_81

2 Hardware Architecture

The entire system mainly includes computer host computer, Raspberry Pi, robot arm, and a USB camera. The Raspberry Pi as a core component is connected to a robot, a camera, and a computer. The camera transmits the collected monitoring image to the PC through the Raspberry Pi. The command to control the manipulator movement is transmitted from the host computer to the Raspberry Pi, which generates PWM to control servo action. The structure of its hardware is illustrated in Fig. 1.

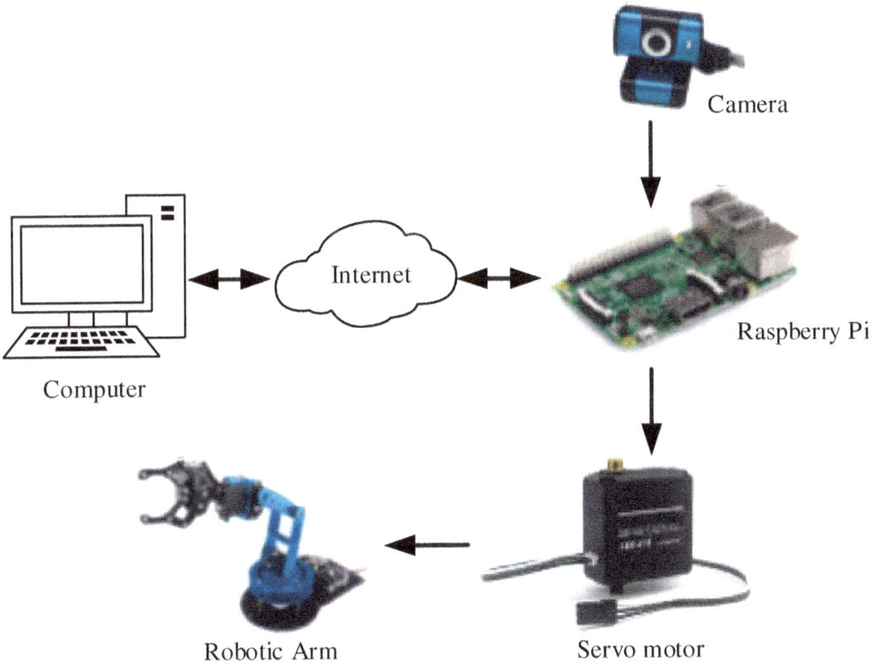

Fig. 1. Hardware architecture of robotic arm

2.1 Raspberry Pi

The Raspberry Pi is a small Linux-based computer equipped with Broadcom's ARM-based processors. For the 3B Raspberry Pi, the processor was upgraded to a 64-bit BCM2837 with a 1.2 GHz ARM Cortex-A53 CPU and 1 GB (LPDDR2) RAM, adding Wi-Fi wireless networking and Bluetooth capabilities. It uses an SD card to store operating systems and user programs and has an Ethernet, four USB ports, an HDMI port, and a 3.5 mm analog audio and video jack. A 40-pin GPIO connector is also provided for user to use in other jobs (Fig. 2).

Fig. 2. Raspberry Pi 3B

The Foundation provides Raspbian, a Debian-based Linux distribution for download, as well as third-party Ubuntu, Windows 10 IoT Core, RISC OS, and specialized media center distributions. It promotes Python and Scratch as the main programming language, with support for many other languages.

2.2 Robotic Arm

The system is controlled by the LeArm robotic manipulator manufactured by the LeHuanSuoEr Corporation, as shown in Fig. 3. The robotic arm is composed of an aluminum bracket and a steering gear, which can mimic the skeletal structure of the human arm [3]. In addition, the robot arm has six degrees of freedom and is directly driven by the steering gear. The servos are located at the base, shoulders, elbows, wrists and claws. There are two servos on the wrist parts, which control the rotation and vertical of the claws. The servo of the pedestal can only rotate around the z-axis of the vertical ground. The servos at the shoulders and elbows rotate around a straight line perpendicular to the z axis. The claw servo controls the closing of the claw.

Fig. 3. Robotic arm

The robot uses a 180° digital servo and can send a single PWM signal to move to the set position [4]. The control signal is a time-base pulse with a period of 20 ms. The high-level time of the pulse is 0.5 ms–2.5 ms corresponding to the rotation of 0°–180° respectively. As shown in Fig. 4.

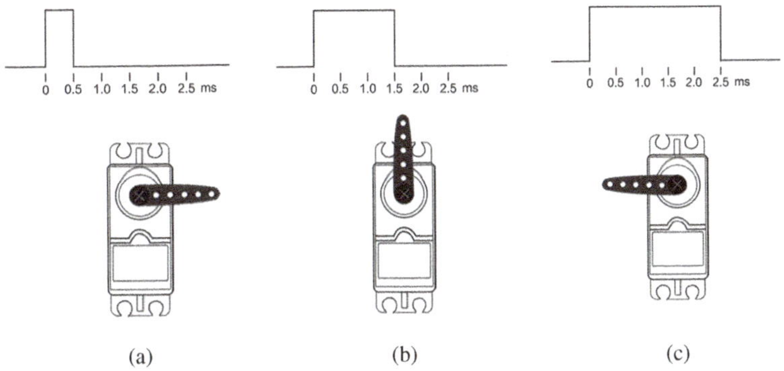

Fig. 4. (a) Rotate 0°, (b) rotate 90°, (c) rotate 180°

2.3 USB Camera

Raspberry Pi provides CSI interface and can connect 5 million pixel OV5647 camera [5]. Because the use of cables is easily affected by external forces, the USB camera is resistant to external forces, easy to move, and plug and play. And the USB camera supports JPEG format output, providing high-definition and low-delay images of 480P, 30FPS without occupying Raspberry Pi computation resources. Therefore, the system uses a USB camera as shown in the Fig. 5.

Fig. 5. USB camera

3 Software Design

The flow chart of the software design is shown in the Fig. 6 [6]. After the Raspberry Pi is powered on, the socket server and the camera is first started. Then start the upper computer to receive monitoring video and send control instructions. Finally, the Raspberry Pi receives instructions to control the manipulator actions.

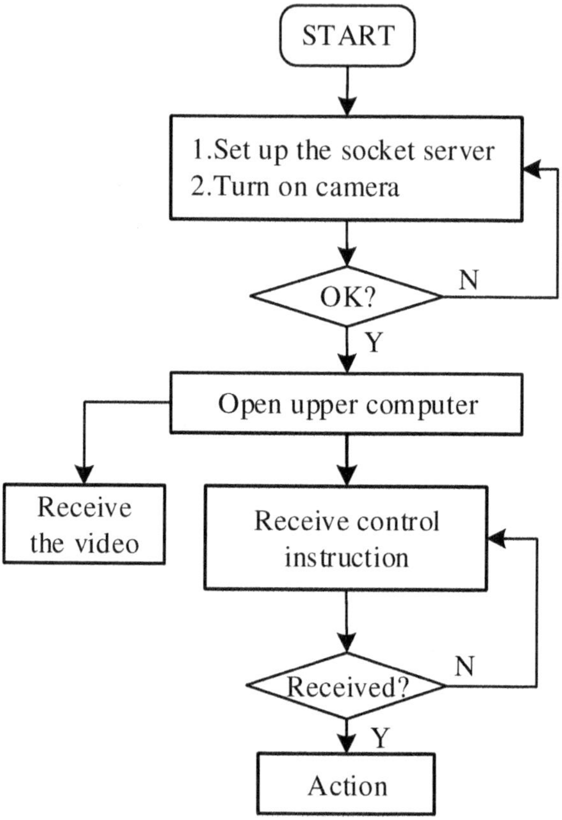

Fig. 6. Flow chart of the software design

3.1 Robotic Arm Control Section

The Raspberry Pi has 26 general-purpose GPIO pins, some of which can be multiplexed as I2C, SPI, and UART serial ports and can be used as normal ports when not multiplexed. WiringPi is a GPIO software library function applied to Raspberry Pi. It is written in C language and can be called by other languages. The WiringPi software library function can be used to control the pins of the raspberry pie. Users can develop more functions by calling the library function interface. WiringPi contains a software-driven

PWM library function that can output PWM signals on any Raspberry Pi GPIO pin. The specific function is:

1. int softPwmCreate(int pin, int initialValue, int pwmRange);
2. void softPwmWrite(int pin, int value);

Where pin is the specific number to generate PWM signal, pwmRange is the PWM period, value is the time of high level in PWM signal.

In order to ensure low CPU utilization, the minimum pulse width of PWM signal generated by GPIO is 100 us, and the PWM period required by the servo is 20 ms, so PWM cycle pwmRange = 20 ms/100 us = 200. The servo specifies that the high level is between 0.5 and 2.5 ms, so the range of value is 5–25.

3.2 Camera Section

Mjpg-streamer is an excellent open source project that can access compatible webcams on Linux via file or HTTP, so as to achieve remote video transmission [7]. After installation and start the MJPG-Streamer in the browser to access the "http://192.168.0. 1:8080/stream_simple.html" can view camera acquisition surveillance images. Among them, "192.168.0.1" is the IP address of the raspberry pie, and 8080 is the access port. The collected images are shown in Fig. 7.

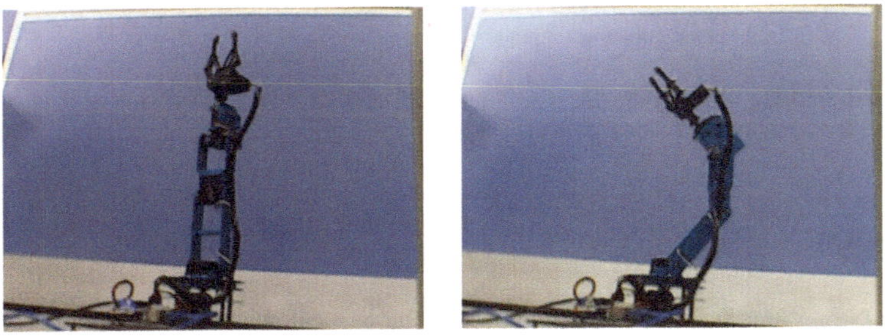

Fig. 7. Camera capture screen

3.3 Human-Machine Interface

The Raspberry Pi communicates with the computer through a network cable and writes a socket server program on the Raspberry Pi to monitor the instructions sent by the computer. After the Raspberry Pi receives an action command, it controls the robot movement.

MFC (Microsoft Foundation Classes) is a class library provided by Microsoft Corporation. It encapsulates many API functions in the form of C++ classes and contains an application framework to quickly develop user interface applications. The human-machine interface of this system is manufactured by MFC, as shown in Fig. 8. The left side of the operation panel displays the monitoring screen collected by the camera. There are 12 buttons on the right side, which control the rotation of the 6 servos respectively. Each click of the current degree will increase/decrease by 1°.

The text box in the middle of the interface shows the current angle of the servo. When you press Enter after entering a new angle, the servo will immediately rotate to the set angle.

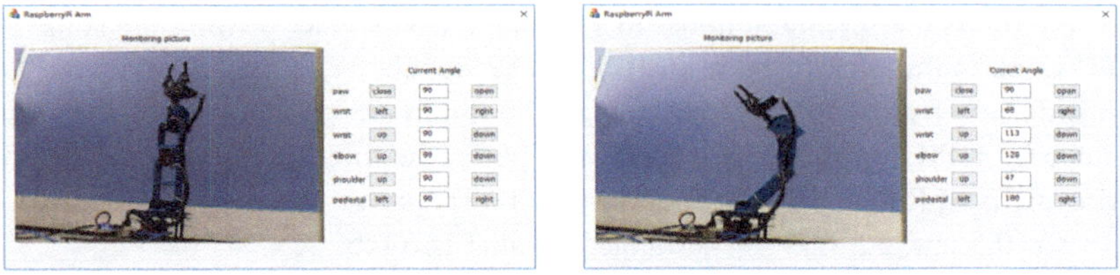

Fig. 8. Human–machine interface

4 Conclusion

This article uses Raspberry Pi, robots, cameras, and upper computers connected through a network to successfully develop a manipulator remote control system. The experimental results show that the system is in good condition and control instructions and video streams can be transmitted stably. At the same time, it has the advantages of low cost and remote operation, which greatly improves people's efficiency in handling complex tasks.

Acknowledgments. We would like to extend our acknowledgement those who have directly and indirectly contributed to our project.

References

1. Zaini, N.A., Zaini, N., Latip, M.F.A., Hamzah, N.: Remote monitoring system based on a Wi-Fi controlled car using Raspberry Pi. In: Systems, Process and Control, pp. 224–229. IEEE (2017)
2. Vanitha, M., Selvalakshmi, M., Selvarasu, R.: Monitoring and controlling of mobile robot via internet through raspberry Pi board. In: International Conference on Science Technology Engineering and Management, pp. 462–466. IEEE (2016)
3. Wan, M.H.W.K., Samin, R.E., Ibrahim, B.S.K.: Internet controlled robotic arm. Proc. Eng. **41**, 1065–1071 (2012)
4. Premkumar, K., Nigel, K.G.J.: Smart phone based robotic arm control using raspberry pi, android and Wi-Fi. In: International Conference on Innovations in Information, Embedded and Communication Systems, pp. 1–3. IEEE (2015)
5. Pereira, V., Fernandes, V.A., Sequeira, J.: Low cost object sorting robotic arm using Raspberry Pi. In: Global Humanitarian Technology Conference - South Asia Satellite, pp. 1–6. IEEE (2014)
6. Siagian, P., Shinoda, K.: Web based monitoring and control of robotic arm using Raspberry Pi. In: International Conference on Science in Information Technology, pp. 192–196 (2015)
7. Ersahin, G., Sedef, H.: Wireless mobile robot control with tablet computer. Proc. Soc. Behav. Sci. **195**(227), 2874–2882 (2015)

Satellite Assembly Technology by Robot Under Visual Guidance and Force Feedback Control

Rui-qin Hu[1,2(✉)], Chang-yu Long[1,2], and Li-jian Zhang[1,2]

[1] Beijing Institute of Spacecraft Environment Engineering, Beijing 100094, China
hrqcast@163.com

[2] Beijing Engineering Research Center of Intelligent Assembly Technology and Equipment for Aerospace Product, Beijing 100094, China

Abstract. Satellites are usually produced in one piece or in small batches, the assembly operation is not repetitive. In order to apply industrial robots to satellite assembly, robot technology under visual guidance and force feedback control was studied, so that the robot can adapt to different working conditions in satellite assembly. In this paper, a satellite assembly scheme based on robot is presented, auxiliary pins are installed on the mounting holes, the robot conveys the component to the tapered guide area of the pins under visual guidance, then the force feedback control is applied to the robot, the component is accurately installed under the pin guidance. Infrared cameras and "rigid body" consisting of cooperative targets were used in the visual guidance system to measure the target pose. A probe tool was designed and a measurement method was given to achieve convenient measurement of the target point. The force/position hybrid control method is used to achieve compliant pin guidance of robot. The experiment results show that under the visual guidance, the robot can transport the component to the guide range of the pin, and can accurately assemble the component in place under pin guidance and force feedback control. The technology can meet the engineering implementation requirements for satellite component assembly.

Keywords: Visual guide · Force feedback control · Satellite assembly Industrial robot

1 Introduction

Satellite assembly operations are usually dominated by manual operations. For heavy components outside the satellite cabin, tools such as overhead cranes can be used to assist the operator in completing the installation. For heavy parts inside the cabins or at special locations, space for observation and operation is limited, it is difficult to use lifting tools, and there is risk of collision.

Industrial robots have been maturely used in industrial assembly, which have large carrying capacity, good working stability, and can realize stable and reliable assembly of heavy-weight components. However, it is generally applied to the production line of batch product and adopts a teaching-repetition mode. Satellite assembly is often not repetitive, and the teaching-repetition mode is not adaptive. In this paper, according to

K. Deng et al. (Eds.): ICMIR 2018, AISC 856, pp. 649–659, 2019.
https://doi.org/10.1007/978-3-030-00214-5_82

the assembly requirements of satellite components, the robot assembly technology under visual guidance and force feedback control is studied, which makes the robot flexible with different working conditions and improves application efficiency of industrial robot in satellite assembly.

For the vision servo of robot, different camera configurations and vision technologies can be used according to the application requirements [1–9]. For the visual guided assembly of satellites, it is necessary to rely on the cooperative mark to make the visual positioning reliable and efficient. On this basis, the position-based visual servo can be used to complete the assembly. Due to the limited space in the satellite cabin, Eye-to-Hand mode is required to fix the camera on the ground.

Due to the visual guidance error, it is difficult to ensure that the component can be accurately assembled in position after the completion of the guidance, and there is a gap or stress at the interface. Therefore, after the contact occurs at the assembly interface, force feedback control is required to ensure that the component is accurately positioned while ensuring that the force at assembly interface is within a safe range.

For force feedback control of robot, there have been mature research results. The methods include impedance control [10–12], force/position hybrid control [13–15] and so on. In force/position hybrid control, the degree of freedom is decomposed into force control and position control in a rectangular coordinate system, and the servo systems independently track the respective targets [16]. For unknown environments, position control is used when the end of the robot is not in contact with the environment, and when it contacts with the environment, robot is switched to force/position hybrid control, that is suitable for satellite assembly applications where operating conditions are not repeated.

In this paper, considering the variable assembly conditions of satellite, the visual guidance and force feedback control are combined on robot to realize automated assembly. The auxiliary pins is installed at the mounting holes, and the visual guide only needs to guide the component into the taper guide area of the pin, then force feedback control is used to achieve the fitting of the mounting surface and alignment of the mounting hole position. Finally, remove the auxiliary pins and install the fasteners.

The adopted scheme has the following advantages:

(1) Using the combination of vision and force information, the vision only needs to guide the component to the guide range of the auxiliary pin, which reduces the precision requirement for visual guidance and makes the system easy to build;
(2) The probe tool can achieve convenient point measurement without the need to stick a target on the product, thereby avoiding additional damage, contamination, and unwanted material to the satellite.

2 System Composition and Working Principle

2.1 System Composition

The robot assembly system is shown in Fig. 1. The infrared cameras are set around the work area, which can realize the positioning of a single target ball or the position and attitude measurement of multiple target ball assemblies (hereinafter referred to as "rigid body"). Several target balls is installed at the end of the robot, with this set of target balls, the coordinate system of the robot can be linked with the coordinate system of the vision system by hand-eye calibration [17]. In addition, the rigid body can be used for visual tracking during assembly. The probe tool is used to the position of the holes on the parts the corresponding holes on the mounting surface, so as to determine the target pose of robot. 6-axis force/torque sensor and pins are used for force feedback control guidance.

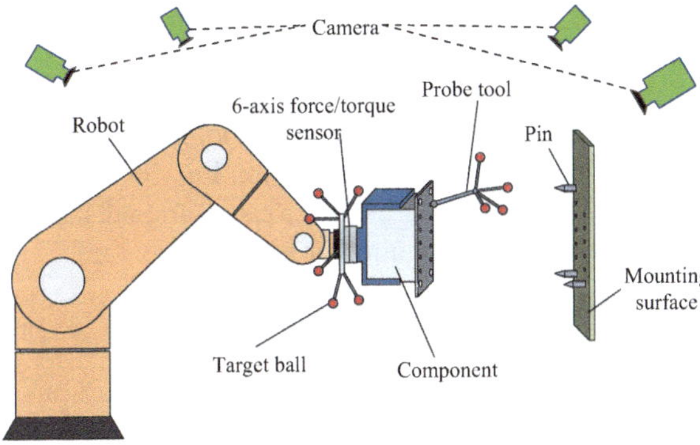

Fig. 1. Block diagram of robot assembly system

2.2 Working Principle

The main steps for applying this system for component assembly are as follows:

(1) Set up the cameras and calibrate to establish a vision measurement field;
(2) Do robot hand-eye calibration, get the conversion relationship between the robot coordinate system and the visual coordinate system;
(3) Use the probe tool to measure the corresponding hole position on the component and the mounting surface to determine the target pose of the robot;
(4) Plan the path of the robot from the initial pose to the target pose;
(5) Perform the visual guidance and force feedback control process of the robot and complete the assembly.

For step (1), (2) and (4), there are already many mature application technologies, which are not described in this article. This article focuses on the techniques used in step (3) and (5), ie the calibration and use of the probe tool, the robot force feedback control under pin guidance.

3 Key Algorithms

3.1 Calibration and Use of the Probe Tool

The schematic diagram of the probe tool is shown in Fig. 2. The assembly of the target balls on the tool is defined as a "rigid body", and coordinate system $O_m\text{–}X_mY_mZ_m$ of the rigid body is established. The pose conversion matrix of $O_m\text{–}X_mY_mZ_m$ relative to the measurement system $O_w\text{–}X_wY_wZ_w$ can be directly obtained by the measurement system. If the coordinate of the measurement point P_m in $O_m\text{–}X_mY_mZ_m$ is known, the position of P_m in $O_w\text{–}X_wY_wZ_w$ can be converted. The measurement point is placed at the position to be measured, and the coordinates of the position in $O_w\text{–}X_wY_wZ_w$ can be obtained, thus the target position is achieved.

In the use of the probe tool, it is necessary to know the coordinates of P_m in $O_m\text{–}X_mY_mZ_m$, which requires the calibration of the tool in advance. This paper presents a calibration method for the probe tool. As shown in Fig. 3, the measurement point P_m is placed in a fixed position, and the tool is rotated around the point, and position information of the "rigid body" is acquired at intervals.

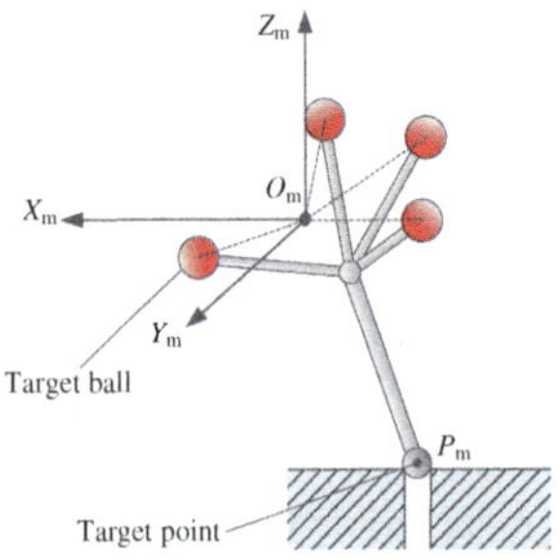

Fig. 2. Framework of probe tool

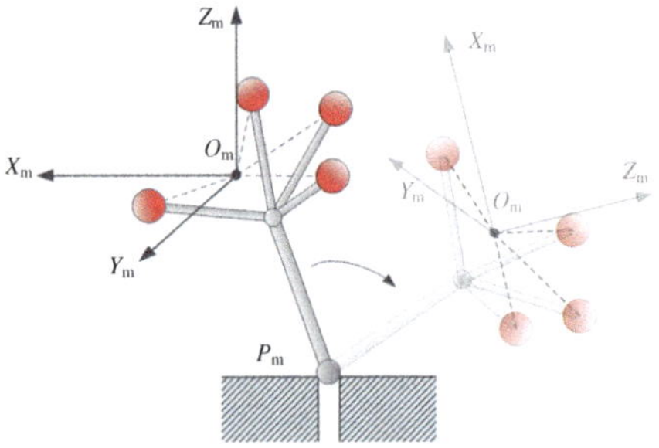

Fig. 3. Framework of calibration for measure probe

The position of P_m in $O_m\text{–}X_mY_mZ_m$ is expressed as:

$$\overrightarrow{{}^{m}P_m} = \begin{bmatrix} X_m & Y_m & Z_m \end{bmatrix}^T \tag{1}$$

And the position of P_m in $O_w\text{–}X_wY_wZ_w$ is expressed as:

$$\overrightarrow{{}^{w}P_m} = \begin{bmatrix} X_w & Y_w & Z_w \end{bmatrix}^T \tag{2}$$

The attitude transformation matrix of $O_m\text{–}X_mY_mZ_m$ with respect to $O_w\text{–}X_wY_wZ_w$ is ${}^{w}_{m}R$, and the translation vector is ${}^{w}_{m}T$, the following equation can be get:

$${}^{w}_{m}R \cdot \overrightarrow{{}^{m}P_m} + {}^{w}_{m}T = \overrightarrow{{}^{w}P_m} \tag{3}$$

As shown in Fig. 3, when the probe tool is at two different poses, the following equation can be get:

$$\begin{cases} {}^{w}_{m}R_1 \cdot \overrightarrow{{}^{m}P_m} + {}^{w}_{m}T_1 = \overrightarrow{{}^{w}P_m} \\ {}^{w}_{m}R_2 \cdot \overrightarrow{{}^{m}P_m} + {}^{w}_{m}T_2 = \overrightarrow{{}^{w}P_m} \end{cases} \tag{4}$$

Since the tool rotates around the point P_m, the position of P_m itself does not change, so the following is obtained:

$${}^{w}_{m}R_1 \cdot \overrightarrow{{}^{m}P_m} + {}^{w}_{m}T_1 = {}^{w}_{m}R_2 \cdot \overrightarrow{{}^{m}P_m} + {}^{w}_{m}T_2 \tag{5}$$

Which is:

$$\left({}^{w}_{m}R_1 - {}^{w}_{m}R_2\right) \cdot \overrightarrow{{}^{m}P_m} = {}^{w}_{m}T_2 - {}^{w}_{m}T_1 \tag{6}$$

Define:

$$\begin{cases} D_R = {}^{w}_{m}R_1 - {}^{w}_{m}R_2 \\ D_T = {}^{w}_{m}T_2 - {}^{w}_{m}T_1 \end{cases} \tag{7}$$

Substitute (7) into Eq. (6):

$$D_R \cdot \overrightarrow{{}^{m}P_m} = D_T \tag{8}$$

If the pose of the tool is changed N times, N different D_R and D_T can be obtained from Eq. (7):

$$\begin{bmatrix} D_{R1} \\ D_{R2} \\ \vdots \\ D_{RN} \end{bmatrix} \cdot \overrightarrow{{}^{m}P_m} = \begin{bmatrix} D_{T1} \\ D_{T2} \\ \vdots \\ D_{TN} \end{bmatrix} \tag{9}$$

Which is:

$$S_R \cdot \overrightarrow{{}^{m}P_m} = S_T \tag{10}$$

and

$$\overrightarrow{{}^{m}P_m} = \left(S_R^T \cdot S_R\right)^{-1}\left(S_R^T \cdot S_T\right) \tag{11}$$

Now the coordinates of P_m in the "rigid body" coordinate system O_m–$X_mY_mZ_m$ is obtained, and the calibration of the probe tool is completed. When the tool has been calibrated, the coordinates of the measured point can be obtained by Eq. (11).

In the study, the hole position on the component and the corresponding hole position on the mounting surface are used as measuring points (as shown in Fig. 4). The probe tool is used for the measurement, under the coordinate system of the rigid body at robot end, a set of hole position on the component $[\vec{P}_1\ \vec{P}_2 \ldots \vec{P}_n]$, and corresponding hole position on the mounting surface $[\vec{P}'_1\ \vec{P}'_2\ \ldots\ \vec{P}'_n]$ can be obtained, and the target position of robot can be calculated, and used as the target of the visual guidance.

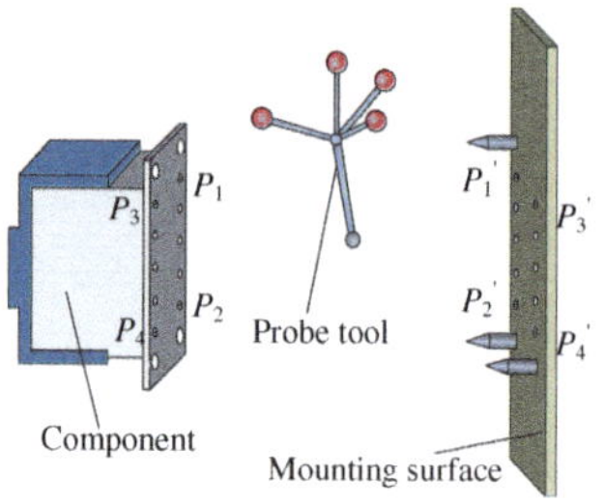

Fig. 4. Framework of coupling for the corresponding holes

3.2 Force Feedback Control

The force feedback control of the robot has been described in previous research results [18]. The pin-guided compliance control is applied in this study.

The robot conveys the component to the pin guide range under visual guidance. When the force between the component and the pin exceeds a predetermined value, the system is switched to force feedback control, and the assembly is completed under the guidance of the auxiliary pin.

Pin compliant schematic diagram shown in Fig. 5. When the component is approaching the mounting surface, it first contacts the pin taper surface. At this time, while maintaining the contact force in the safe range, the position and posture of the component are adjusted according to the contact force and torque information obtained so that the component can follow the pin and move close to the mounting surface until it is unable to continue close under the premise of ensuring the safety contact force, finally the component reaches the assembly position.

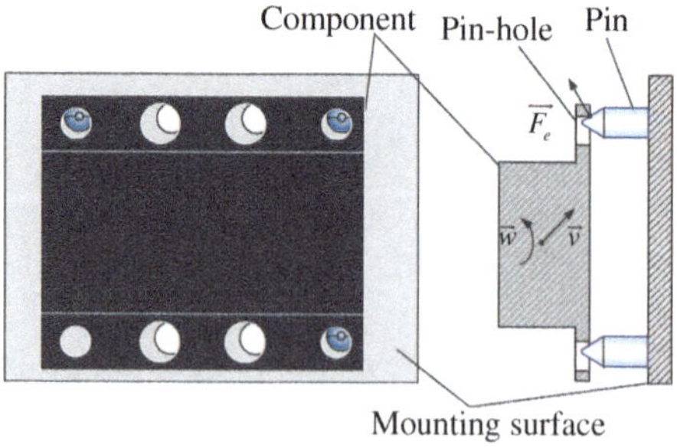

Fig. 5. Compliant assembly under the guidance of pin

Specific control strategies and algorithms have been described in article [18] and are no longer described in this paper.

4 Assembly Experiment

4.1 Experiment Conditions

The experiment system is shown in Fig. 6. The robot uses the KUKA KR210 R2700 robot and the robot end is equipped with ATI's Omega 190 6-axis force/torque sensor. Four OptiTrack Prime 17 W cameras, and cooperative target balls are used. A plurality of target balls installed at the end of the robot form a rigid body for the movement tracking.

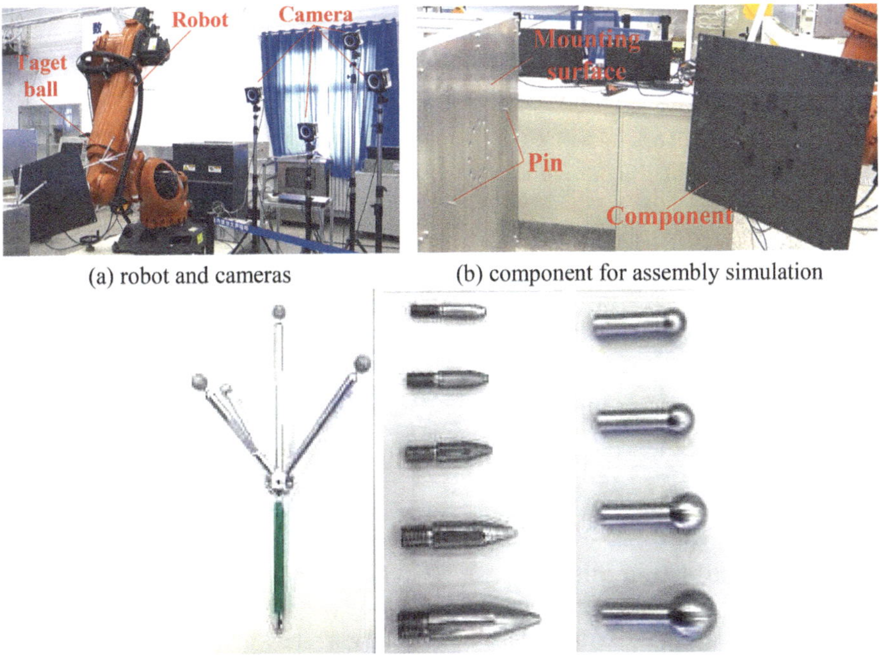

(a) robot and cameras (b) component for assembly simulation

(c) measure tool (d) pin serious (e) ball serious

Fig. 6. Picture of the system for experiment

The experiment uses a simulated component with weight of 30 kg and mounting surface dimension of 440 × 460 mm, 12 mounting holes with diameter of 7 mm distributed on the edge of the component. 12 M6 corresponding tapped holes are arranged on the mounting surface, and 2 of them are selected to install the auxiliary pins for compliant guidance.

As shown in Fig. 6(c), the probe tool has target balls at the end of tree structure and a replaceable metal measuring ball at the bottom for measuring the center position of hole. For the measurement of the hole inside the satellite cabin, special-shaped measuring tools can be designed, the measuring ball is placed at the hole position while the target balls are outside the satellite cabin, so that convenient measurement can be realized even if the measured point is not visible directly by camera. In order to apply the probe tool to different sizes of holes, serialized pins and measuring balls with different diameters are designed, as shown in Fig. 6(d), (e).

Select 8 holes in the left and right sides of the component, and corresponding 8 holes on the mounting surface, use the probe tool to measure and obtain the coordinate values of each point, the results are shown in Table 1. the robot target pose are obtained as in Table 2.

Table 1. Position of holes

No.	Holes on component (mm)	Holes on mounting surface (mm)
1	(−525.3, 86.9, 271.8)	(−987.7, 96.4, 460.0)
2	(−395.9, 85.9, 285.7)	(−951.1, 101.0, 584.9)
3	(−266.6, 84.8, 300.0)	(−913.8, 105.6, 709.2)
4	(−137.1, 83.6, 314.4)	(−875.8, 109.6, 833.9)
5	(−530.6, −332.8, 288.2)	(−996.4, −323.3, 477.6)
6	(−400.6, −333.5, 302.2)	(−959.0, −318.9, 602.2)
7	(−271.9, −335.2, 317.0)	(−921.5, −314.2, 726.5)
8	(−142.7, −336.5, 331.4)	(−884.0, −310.1, 851.0)

Table 2. Target position of robot

Axis	X (mm)	Y (mm)	Z (mm)	A (°)	B (°)	C (°)
Value	1955.7	287.2	511.8	−98.9°	0.9°	−89.8°

According to the obtained robot target pose parameters, the robot's motion is planned, and the robot is driven to perform the assembly movement. Finally, the component is transported to the installation position, as shown in Fig. 7. Before contact with the pin, the robot moves under visual guidance. After contact with the pin, the system feedbacks the force value of the component in real time, when the value exceeds the set threshold, the robot switches to force feedback control.

Before the pin contacts the component, the alignment of the hole position and the pin is shown in Fig. 8. The pin tip is within the boundary of the hole and the centering condition is better, which makes the visual guiding process and the force feedback control process have a smooth connection.

The component force information and robot velocity information of the guide process of pin cone are shown in Fig. 9. The data in the figure is defined in the robot tool coordinate system. The *Z* axis of the robot tool coordinate system is perpendicular to the mounting surface of component.

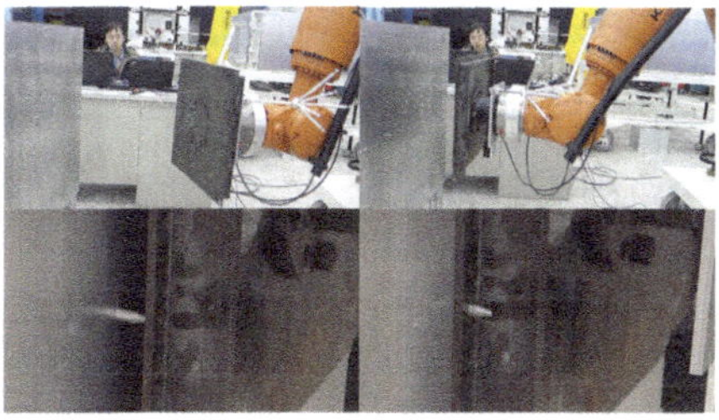

Fig. 7. Video Snapshots of the assembly process

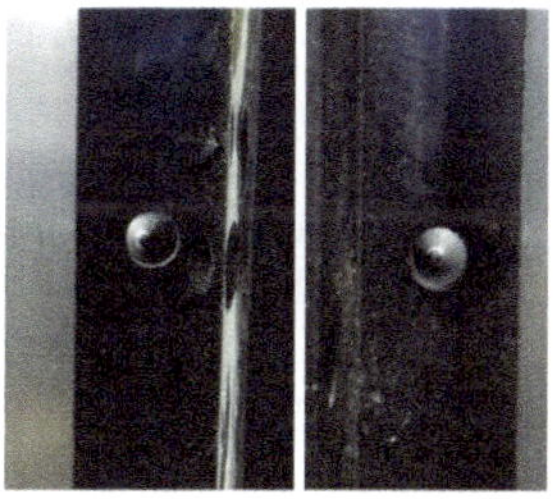

Fig. 8. Pin alignment after vision guidance

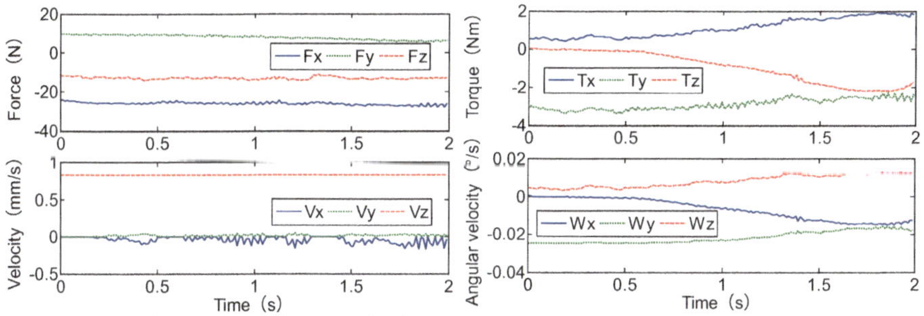

(a)Curve of force and linear velocity (b) Curve of torque and angular velocity

Fig. 9. Data curve of the pin guide process

5 Conclusion

In order to meet the demand for high-reliability assembly of various satellite components, this paper presents a robot assembly method combining vision and force sense. Auxiliary pins are installed at the mounting hole position, and the robot conveys the component to the tapered guide range of the pin by vision guide, then the force-feedback control of the robot is performed under the guidance of the pin to achieve accurate assembly of the component. The composition and working principle of the robot assembly system are introduced. The key contents and algorithms of the robot hand-eye calibration, the calibration and use of the probe tool, the determination of the robot target pose, and the robot force feedback control under pin guidance are given. The experimental results show that the robot can convey the component to the pin's guidance range under visual guidance, and the component can be accurately assembled in place under pin guidance and force feedback control. The technology used can meet the engineering implementation requirements for satellite component assembly.

Acknowledgments. This work is supported by National High-tech R&D Program of China (863 Program) (Grant No. 2015AA043101) and National Natural Science Foundation of China (Grant No. 51705023).

References

1. Fang, S., Huang, X., Chen, H., et al.: Dual-arm robot assembly system for 3C product based on vision guidance. In: IEEE International Conference on Robotics and Biomimetics, pp. 807–812. IEEE (2017)
2. Gu, J., Wang, H., Chen, W., et al.: Monocular visual object-localization using natural corners for assembly tasks. In: IEEE International Conference on Robotics and Biomimetics, pp. 1383–1388. IEEE (2017)
3. Chang, W.C.: Robotic assembly of smartphone back shells with eye-in-hand visual servoing. Robot. Comput. Integr. Manuf. **50**, 102–113 (2018)
4. Turan, M., Almalioglu, Y., Konukoglu, E., et al.: A deep learning based 6 degree-of-freedom localization method for endoscopic capsule robots (2017). arXiv preprint arXiv:1705.05435
5. Lin, C.Y., Son, L.T., Chang, Y.L., et al.: Image-sensor-based fast industrial-robot positioning system for assembly implementation. Sens. Mater. **29**(7), 935–945 (2017)
6. Xu, D., Lu, J., Wang, P., et al.: Partially decoupled image-based visual servoing using different sensitive features. IEEE Trans. Syst. Man and Cybern. Syst. **47**(8), 2233–2243 (2017)
7. Lippiello, V., Cacace, J., Santamaria-Navarro, A., et al.: Hybrid visual servoing with hierarchical task composition for aerial manipulation. IEEE Robot. Autom. Lett. **1**(1), 259–266 (2017)
8. Feng, K., Zhang, X., Li, H., et al.: A dual-camera assisted method of the SCARA robot for online assembly of cellphone batteries. In: International Conference on Intelligent Robotics and Applications, pp. 576–587. Springer, Cham (2017)
9. Wan, W., Lu, F., Wu, Z., et al.: Teaching robots to do object assembly using multi-modal 3d vision. Neurocomputing **259**, 85–93 (2017)
10. Hongan, N.: Impedance control an approach to manipulation: 'part I-theory, part II-implementation, part III-application. J. Dyn. Syst. Meas. Control **107**, 1–24 (1985)
11. Kazerooni, H., Houpt, P.K., Sheridan, T.B.: Robust compliant motion for manipulators. Part I: the fundamental concepts of compliant motion. Part II: design methods. IEEE J. Robot. Autom. **2**(2), 83–105 (1986)
12. Focchi, M., Medrano-Cerda, G.A., Boaventura, T., et al.: Robot impedance control and passivity analysis with inner torque and velocity feedback loops. Control Theory Technol. **14**(2), 97–112 (2016)

13. Raibert, M.H., Craig, J.J.: Hybrid position/force control of manipulators. J. Dyn. Syst. Meas. Control **103**(2), 126–133 (1980)
14. Cheng, G., Wang, Y., Jia, Q., et al.: Hybrid force and position control strategy of robonaut performing peg-in-hole assembly task. Yuhang Xuebao J. Astronaut. **38**(4), 410–419 (2017)
15. Chaudhary, H., Panwar, V., Prasad, R., et al.: Adaptive neuro fuzzy based hybrid force/position control for an industrial robot manipulator. J. Intell. Manuf. **27**(6), 1299–1308 (2016)
16. Chang, J., Wang, Y., Li, B.: Accurate operation control method based on hybrid force/position algorithm for 7-DOF manipulator. Robot **38**(5), 531–539 (2016)
17. Condurache, D., Burlacu, A.: Orthogonal dual tensor method for solving the AX = XB sensor calibration problem. Mech. Mach. Theory **104**, 382–404 (2016)
18. Hu, R., Zhang, L., Meng, S., et al.: Robotic assembly technology for heavy component of spacecraft based on compliance control. J. Mech. Eng. http://kns.cnki.net/kcms/detail/11.2187.TH.20180330.1335.030.html

Path Planning for Large-Scale Spacecraft Components Assembly in Confined Space Using Industrial Robot

Shaohua Meng[1,2(✉)], Lijian Zhang[1,2], Ruiqin Hu[1,2], Peng Zheng[1,2], and Feng Gao[1,2]

[1] Beijing Institute of Spacecraft Environment Engineering, Beijing, China
msh1206@126.com
[2] Beijing Engineering Research Center of Intelligent Assembly Technology and Equipment for Aerospace Product, Beijing, China

Abstract. With the technical targets continuously improving, spacecraft components become more large and heavy, such as optical camera, antenna. It is difficult for operators to hold these large-scale components manually through the entire assembly process. This paper aims to propose a path planning method for industrial robot to accomplish automatic assembly of large-scale components in confined space. Base on the CAD model of spacecraft, robot and components etc., a virtual assembly environment is constructed in accordance with the position and pose of the real scene. In the virtual environment, geometric constraint in the assembly process is recognized and collision detection is implemented using the axis-aligned bounding box (AABB) method. After that, probabilistic roadmap method (PRM) is performed to plan an assembly path in 3D space with obstacles. The PRM planner is divided into two phases: the learning phase and the query phase. In the learning phase, a collision-free roadmap is built. In the query phase, the nodes of the roadmap are connected. The A-star algorithm is accomplished to find a feasible path between the initial node and the goal node and then assembly sequences are produced by robot off-line programmer. Experiment results show that this method can rapidly generate a collision-free assembly path, and control robot to safely and efficiently complete large-scale components installation in the confined space.

Keywords: Robot · Spacecraft components · Path planning · Assembly

1 Introduction

The overall assembly is the final stage and a key link of the spacecraft production. It has a key effect on the quality and performance of the spacecraft [1]. At present, the assembly of spacecraft mainly depends on manual operation and is supported by the simple process equipment such as hoisting tools and adjustable vehicles. It is very difficult to precisely control the pose of spacecraft components during installation. The assembly quality seriously depends on the experience of workers. With the progress of space technology, spacecraft components become more large and heavy, which is difficult for operators to install manually. Moreover, the number of payloads in a cabin

K. Deng et al. (Eds.): ICMIR 2018, AISC 856, pp. 660–668, 2019.
https://doi.org/10.1007/978-3-030-00214-5_83

is increasing and the operation space is extremely confined. It is necessary to develop new assembly technologies and equipment.

Industrial robot has the advantages of large weight, high assembly precision, good assembly consistency, fast response speed and so on. It has been widely used in various fields, such as automobile assembly, aviation manufacturing, electronic product production and so on. However, in the spacecraft production, there are many characteristics such as single pieces production, unfixed assembly position, complex and changeable working condition and high-accuracy requirement. It is not feasible to determine the assembly path through teaching mode for the robot, which is different to the flow process line.

Path planning, as the basis of autonomous execution of robots, has been a hot topic in robotics research field [2]. Scholars have proposed artificial potential field method [3], grid method [4], fast search random tree method [5], intelligent algorithm [6–8] and other methods, effectively solving the problem of robot collision free path planning in a specific environment. However, most of these algorithms take mobile robots as the research object. The path planning methods for the position transformation of 2-dimensional plane or 3-dimensional space don't involve the attitude parameters of the robot, so they cannot solve the path planning problem under the complex constraints of the high dimension system. Industrial robot is a 6-degree-freedom motion mechanism, which has high motion dimensions. Furthermore, both industrial robots and spacecrafts have very complex three-dimensional geometric shapes and it is impossible to establish an accurate constraint model for obstacles. Most of the path planning method mentioned above cannot be implemented in practice for robots with more than three degrees of freedom due to their high combinatorial complexity. Probabilistic roadmap method (PRM) [9] is a sampling-based algorithm, which can effectively accurate modeling of obstacles and is more suitable for solving path planning problems in high dimensional space and complex constraints.

Although industrial robot has made great progress in the field of assembly and manufacturing, it still has a gap to solve the high precision assembly problem of large-scale and heavy components under complex conditions. Therefore, in order to deal with the uncertainties of manual operation, this paper proposes a path planning method for industrial robots to automatically install large-scale spacecraft components in a confined space.

2 Robot Kinematics

Robot kinematics is the foundation of robotics. Its task is to describe the changes of joints and end effector during the movement of robots. It involves two aspects: forward kinematics and inverse kinematics. The robot forward kinematics is the connecting rod parameters and the joint variables of the robot to solve the position and attitude of the end-effector, and the inverse kinematics of the robot is exactly the opposite. The position and attitude of the end-effector are known, and the joint variables of the robot are solved. The inverse kinematics of the robot is much more complicated than the forward kinematics problem, and with the increase of the freedom of the robot, the solution of the inverse kinematics will become more and more complex. In this paper, the Denavit-Harenberg method is implemented for solving the positive solution problem.

3 Path Planning for Robot

The difficulty of industrial robot path planning is that it has high motion dimensions and complex three-dimensional geometric shape. In the process of large-scale components installation, it is not only to avoid the collision between the components and the cabin, but also to ensure that there is no interference between the robot body and environmental objects. This paper implements the PRM based on sampling to carry out the robot assembly path planning.

3.1 Collision Detection

Fast and accurate collision detection is a prerequisite for path planning based on sampling method. The virtual simulation environment is constructed based on CAD models of elements in the physical scene. In the virtual environment, the complex objects are composed of many triangular patches which are surrounded by simple geometries, such as spheres and rectangles. After that, a tree structure data base is constructed to represent the real object, and it can be used to efficiently execute the overlapping test of the ideal geometry.

This paper employs axis-aligned bounding boxes (AABB) technology [10] for collision detection. The AABB is the smallest hexahedron which surrounds complex objects and the sides are parallel to the coordinate axes.

$$\boldsymbol{R} = \{(x, y, z) | x_{\min} \le x \le x_{\max}, y_{\min} \le y \le y_{\max}, z_{\min} \le z \le z_{\max}\} \tag{1}$$

Where $x_{\min}, x_{\max}, y_{\min}, y_{\max}, z_{\min}, z_{\max}$ are the minimum and maximum projection coordinates on X, Y, Z axis.

The intersection test between different AABB bounding boxes is more intuitive. The sufficient and necessary condition for the intersection of two AABB bounding boxes is that the projections on their three axes are overlapping. In the testing process, six comparison operations are required at most. Compared with other algorithms, the most important advantage of the AABB approach is fast intersection test and high computation efficiency.

3.2 Path Planning

The PRM is a sampling-based planner which can achieve some form of completeness. Since its ease of implementation and good performance even for rather high-dimensional problems (5–12 degrees of freedom) [11], the PRM algorithm is a good choice for industrial robot to plan a collision-free path of assembly operations in a complex environment.

For a given robot workspace, the PRM algorithm could be completed in the following two stages: the learning phase and the query phase. Firstly, an undirected graph is constructed as a roadmap, which can be denoted by $G = (V, E)$. The nodes in V are a group of the robot configurations, which are chosen randomly from a uniform distribution on the free space Q_{free}. The edges in E are connecting lines between configuration nodes, which correspond to collision-free paths computed by a local planner.

Secondly, a searching algorithm such as A-star, is employed to find a feasible path for the robot to connect the starting node $\boldsymbol{q}_{\text{init}}$ and the target node $\boldsymbol{q}_{\text{goal}}$.

Learning Phase

In this phase, we need to build a roadmap of robot assembly movement. In the beginning, the roadmap G is empty. Then, robot configurations are sampled randomly with a uniform random distribution over the workspace Q. Each sampling configuration is examined on collision detection, and the collision-free one will be added to the roadmap. The process is repeated until the number of collision-free configurations reaches the specified value n. For every node $\boldsymbol{q}$, a number of closest neighbor nodes are selected to be connected using the local planner Θ according to some metric *dist*. Whenever a feasible path between $\boldsymbol{q}$ and $\boldsymbol{q}'$ is computed by Θ, the edge $(\boldsymbol{q}, \boldsymbol{q}')$ is added to the roadmap. In this paper, Θ is a straight-line planner. The execution process is shown in Algorithm 1 [11].

Algorithm 1: Roadmap Construction Algorithm

Input:

n: number of nodes to put in the roadmap

k: number of closest neighbours to examine for each configuration

Output:

a roadmap $G = (V, E)$

```
V ← ∅
E ← ∅
while |V| ≤ n do
  repeat
    q ← a random configuration in Q
  until q is collision-free
  V ← V ∪ {q}
end while
for all q ∈ V do
  N_q ← the k closest neighbours of q chosen from V according to dist
  for all q' ∈ N_q do
    if (q,q') ∉ E and Θ(q,q') ≠ null then
      E ← E ∪ {(q,q')}
    end if
  end for
end for
```

Query Phase

In this phase, the main task is to search a feasible path between specified configurations $\boldsymbol{q}_{\text{init}}$ and $\boldsymbol{q}_{\text{goal}}$ using the roadmap constructed in the learning phase. Firstly, $\boldsymbol{q}_{\text{init}}$ and $\boldsymbol{q}_{\text{goal}}$ are connected to the roadmap with the local planner Θ according to the metric *dist*.

After that the shortest path is found on the roadmap between $\boldsymbol{q}_{\text{init}}$ and $\boldsymbol{q}_{\text{goal}}$ using A-star algorithm. The execution process is shown in Algorithm 2 [11].

Algorithm 2: Solve Query Algorithm

Input:

q_{init} : the initial configuration

q_{goal} : the goal configuration

k : the number of closest neighbors to examine for each configuration

$G=(V,E)$: the roadmap computed by Algorithm 1

Output:

A path from $\boldsymbol{q}_{\text{init}}$ to $\boldsymbol{q}_{\text{goal}}$ or failure

---

$N_{q_{\text{init}}} \leftarrow$ the k closest neighbours of $\boldsymbol{q}_{\text{init}}$ from V according to *dist*
$N_{q_{\text{goal}}} \leftarrow$ the k closest neighbours of $\boldsymbol{q}_{\text{goal}}$ from V according to *dist*
$V \leftarrow \{\boldsymbol{q}_{\text{init}}\} \cup \{\boldsymbol{q}_{\text{goal}}\} \cup V$
set $\boldsymbol{q}'$ to be the closest neighbour of $\boldsymbol{q}_{\text{init}}$ in $N_{q_{\text{init}}}$
repeat
  if $\Theta(\boldsymbol{q}_{\text{init}}, \boldsymbol{q}') \neq \text{null}$ **then**
    $E \leftarrow \{(\boldsymbol{q}_{\text{init}}, \boldsymbol{q}')\} \cup E$
  else
    set $\boldsymbol{q}'$ to be the closest neighbour of $\boldsymbol{q}_{\text{init}}$ in $N_{q_{\text{init}}}$
  end if
until a connection was successful or the set $N_{q_{\text{init}}}$ is empty
set $\boldsymbol{q}'$ to be the closest neighbour of $\boldsymbol{q}_{\text{goal}}$ in $N_{q_{\text{goal}}}$
repeat
  if $\Theta(\boldsymbol{q}_{\text{goal}}, \boldsymbol{q}') \neq \text{null}$ **then**
    $E \leftarrow \{(\boldsymbol{q}_{\text{goal}}, \boldsymbol{q}')\} \cup E$
  else
    set $\boldsymbol{q}'$ to be the closest neighbour of $\boldsymbol{q}_{\text{goal}}$ in $N_{q_{\text{goal}}}$
  end if
until a connection was successful or the set $N_{q_{\text{goal}}}$ is empty
$P \leftarrow$ shortest path $(\boldsymbol{q}_{\text{init}}, \boldsymbol{q}_{\text{goal}}, G)$
if P is not empty **then**
  return P
else
  return failure
end if

3.3 Implementation for Industrial Robot

As the PRM algorithm is sampled at random in the workspace and is limited by the current computing power, it can only take a finite number of discrete sampling points, thus the generated paths are often jumps. For the complex assembly environment, although the robot has no interference with the stars at the sampling point, the robot cannot guarantee the robot at two adjacent sampling points. There is no interference between stars when moving, so we need to encrypt the generated path to verify that the whole path is feasible. In this paper, the path of planning is encrypted by linear interpolation. The basic idea is to use the path points produced by the PRM algorithm as control point, piecewise linear interpolation, then stitching, and finally realize the entire path encryption.

Considering sampling of the robot workspace, the approach is applied in the discretized C-space of the industrial robot. Then the workspace is converted from the Cartesian coordinate to joint space. Since each revolute joint $J_i(i = 1, \ldots, 6)$ has a limited range of motion, we can suppose that low_i and up_i are the minimum and maximum values, where $low_i < up_i$. Since the robot base is fixed, the C-space of the industrial robot can be represented by $C : \{[low_1, up_1] \times \cdots \times [low_6, up_6]\}$.

Suppose that $(\boldsymbol{q}_m, \boldsymbol{q}_n) \in C \times C$, when the industrial robot is at the configuration $\boldsymbol{q}_m$, we can use $J_i(\boldsymbol{q}_m)$ to represent the position of point J_i in the workspace. The distance between configuration $\boldsymbol{q}_m$ and $\boldsymbol{q}_n$ can be defined as follows.

$$dist(\boldsymbol{q}_m, \boldsymbol{q}_n) = \sum_{i=1}^{6} \|J_i(\boldsymbol{q}_m) - J_i(\boldsymbol{q}_n)\|^2 \tag{2}$$

Where $\|J_i(\boldsymbol{q}_m) - J_i(\boldsymbol{q}_n)\|$ is the Euclidean distance between $J_i(\boldsymbol{q}_m)$ and $J_i(\boldsymbol{q}_n)$.

In the discretized configuration space, a reasonable resolution for the six joints is required and a heuristic method is adopted instead of having uniform resolution. For the J_i,

$$N_i = (\text{int})\left|\frac{J_i^{\max} - J_i^{\min}}{\Delta\theta_i}\right| \tag{3}$$

Where $J_i^{\max}$ and $J_i^{\min}$ are the motion limits of the joint i, and $\Delta\theta_i$ is a pre-set distance that robot moves along the coordinate at one step. Where,

$$(\Delta\theta_1, \Delta\theta_2, \Delta\theta_3, \Delta\theta_4, \Delta\theta_5, \Delta\theta_6) = (1.56, 1.56, 2.72, 5.33, 5.33, 8.56) \tag{4}$$

After setting the resolution of the six joints, we try to connect each node to the k closest nodes in the configure space. On the other hand, too small will affect the success rates to find a collision-free path. In our experiments, $k = 20$ is fair enough.

4 Assembly Experiment

In order to verify the effectiveness of the presented approach, a satellite component was selected to carry out the assembly experiment. The size of the component is more than 2 m, and its weight is 80 kg. It is installed inside a cylindrical cabin which is a confined space. It is a typical scene of the assembly of spacecraft products. It is of practical significance to verify the application of the PRM planner in the assembly stage of spacecraft.

Since the path planning algorithm used in this paper is carried out in the robot joint space, it is necessary to calculate the inverse kinematics of the initial position and pose of the robot and get the corresponding joint positions. Where,

$$\begin{aligned} \boldsymbol{q}_{\text{init}} &= (68, -56, 119, 0, -65, 0) \\ \boldsymbol{q}_{\text{goal}} &= (1, -25, 41, 0, -21, 0) \end{aligned} \tag{5}$$

Considering inevitable errors, the safety distance of 10 mm is set up in the path planning process to prevent the collision in the actual assembly. Figure 1 shows the change curve of the robot joints obtained by path planning. Figure 2 gives the simulation of robot assembly process.

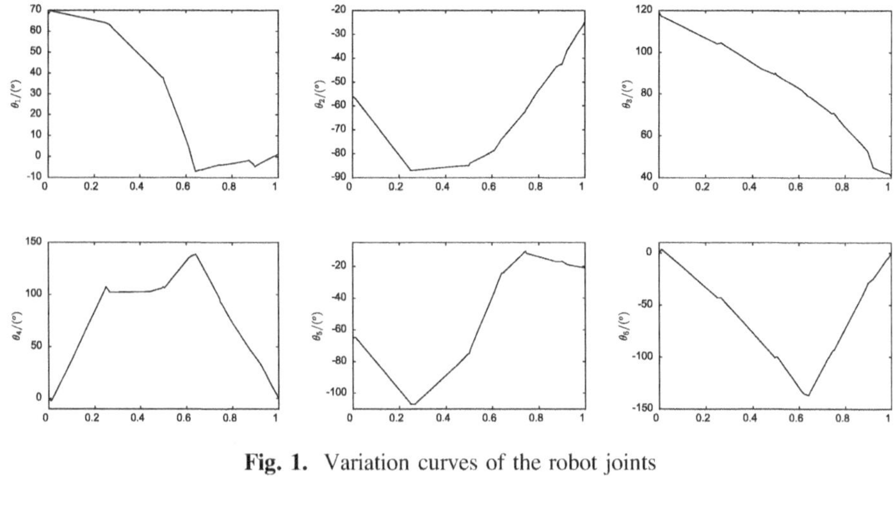

Fig. 1. Variation curves of the robot joints

The simulation result shows that the industrial robot can skillfully change the position and pose of the component, avoid collision with the side wall of the satellite and send it into the confined space, until the target location is reached. During the whole assembly process, the joint angle of the robot is changed more smoothly. The planned path was translated into motion sequences by robot off-line programmer. The real assembly process is shown in Fig. 3.

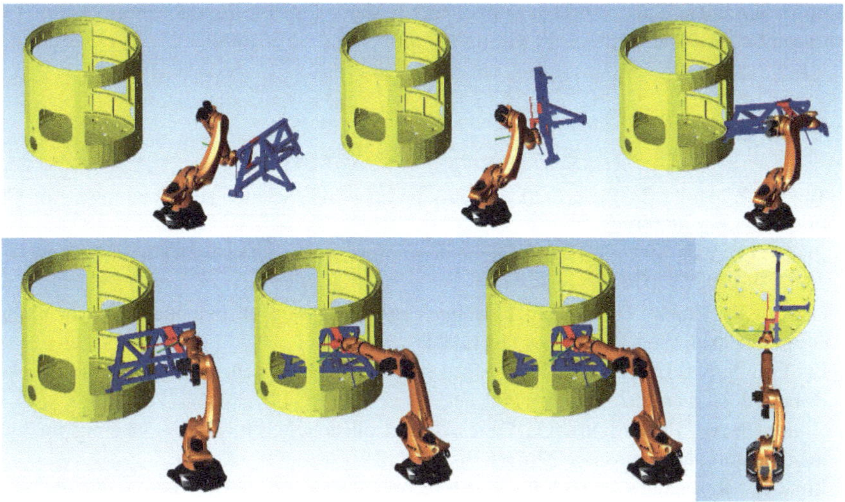

Fig. 2. Simulation results of path planning for robot assisted assembly

Fig. 3. The real assembly scene

5 Conclusion

This paper presented a path planning method for industrial robots installing a large-scale component of spacecraft in a confined space. The CAD models of robot, spacecraft and components were employed in building a virtual environment to simulate the real assembly scene. Then collision detection based on the AABB method was executed in the virtual environment. After that the PRM planner was performed to generate a collision-free path for the industrial robot to move the component from the initial position to the target location. The experiment result shows that the proposed method could quickly plan a collision-free assembly path in a confined space, and then the industrial robot would automatically transport the large-scale component to the assembly location.

Acknowledgments. This project is supported by Major Process Revitalization Plan of China Aerospace Science and Technology Group (Grant No. ZDGY201609).

References

1. Tan, W.Z., Hu, J.G.: Spacecraft Systems Engineering. Science and Technology of China Press, Beijing (2009)
2. Zhu, D.Q., Yan, M.Z.: Survey on technology of mobile robot path planning. Control Decis. **25**(7), 961–967 (2010)
3. Gaol, Y.F., Huang, H.: A path planning algorithm based on potential field for complex environment. Robot **26**(2), 114–118 (2004)
4. Li, T.C., Sun, S.D., Gao, Y.: Fan-shaped grid based global path planning for mobile robot. Robot **32**(4), 547–552 (2010)
5. Gomez-Bravo, F., Carbone, G., Fortes, J.C.: Collision free trajectory planning for hybrid manipulators. Mechatronics **22**(6), 836–851 (2012)
6. Liu, C.L.: Researches on Technologies for Robot Path Planning Based on Artificial Potential Field and Genetic Algorithm. Nanjing University of Science & Technology, Nanjing (2012)
7. Qi, R.L., Zhou, W.J., Wang, T.J.: An obstacle avoidance trajectory planning scheme for space manipulators based on genetic algorithm. Robot **36**(3), 263–270 (2014)
8. Kavrkil, E., Latombe, J.C., Overmars, M.: Probabilistic roadmap for path planning in high-dimensional configuration spaces. IEEE Trans. Robots Autom. **12**(4), 566–580 (1996)
9. Tan, M.: Advanced Robot Control, 1st edn. High Education Press, Beijing (2007)
10. Wang, X.R.: Research on Collision Detection Algorithm Based on AABB Bounding Volume. Hua Zhong Normal University, Wuhan (2007)
11. Howie, C., Kevin, L., Seth, H., George, K., et al.: Principles of Robot Motion: Theory, Algorithms, and Implementation. The MIT Press, Cambridge (2005)

Steady Walking on the Slope of Hexapod Robot Considering Posture

Zhe Wang(✉), Lei Zhang, and Fucai Wang

College of Engineering, Ocean University of China, Qingdao 266000, China
110210222@alu.hit.edu.cn

Abstract. It is one of the important goals of the kinematics research of the foot robot to realize the stable walking of the robot under complex terrain. In order to solve the walking problem of hexapod robots in slope environment, the inverse kinematics equation of leg is deduced from the kinematics analysis of single leg of robot. Considering the motion posture of the body, the parametric motion curve of the foot of the swinging leg is programmed, and the different requirements are met by adjusting the parameters. According to the inclination of the slope, the center motion of the organism is planned. The movement of the center motion of the body is combined with the motion of the foot of the swinging leg to achieve the steady walking of the robot slope. The validity is verified by simulation and physical prototype experiment.

Keywords: Hexapod robot · Foot curve · Body movement planning
Walking on a slope

1 Introduction

It is an important feature of foot robot to have excellent adaptability to complex terrain. Slope is a common complex terrain in nature, in addition to the flat slope terrain, there are some rugged and complex terrain can be equivalent to slope terrain [1]. Many scholars at home and abroad have studied the walking method of legged foot robot in complex terrain. Li Shuanzhu et al. designed the walking gait on the slope of hexapod robot according to the requirement of walking stability [2]. Li Yibin et al. proposed a four-legged robot walking method using diagonal jogging on the slope in combination with posture adaptive adjustment [3, 4]; Han Baoling et al. A method of posture adjustment for quadruped robot is proposed, which uses the gait of foot trot to move on the slope surface [5]. We have planned the motion posture of the quadruped robot in the optimal inclined plane environment based on the omni-directional gait [6]. By establishing the support leg inverse kinematics model and the attitude PID closed-loop control algorithm, the real-time control of the pitch attitude component when the robot walks on the slope is realized [7]. These researches on the walking process of six-legged robots on the slope are mainly based on attitude control and stability requirements, and do not consider the coupling of gravity center position and posture and foot movement during walking.

In this paper, the research of steady walking on the slope of Hexapod robot is carried out, and the concrete realization process is designed. The steady moving target

K. Deng et al. (Eds.): ICMIR 2018, AISC 856, pp. 669–676, 2019.
https://doi.org/10.1007/978-3-030-00214-5_84

of the robot on the slope is transformed into the movement of the joint of the leg. Firstly, the mapping relationship between the robot foot position and the joint angle is established by the transformation relationship between the kinematics and the coordinate system. Then, the motion trajectory of the foot is programmed with the posture. Considering the body posture, slope slope, body center movement and foot Movement, this paper designs the robot's walking method, and completes the steady walking on the slope of the robot with certain additional posture. Finally, the effectiveness of the proposed method is verified by simulation and physical experiments.

2 Hexapod Robot Model

In this paper, the insect-like bipedal hexapod robot shown in Fig. 1 is taken as the research object. Among them, the body is a symmetrical hexagon, each leg has three degrees of freedom, the weight of the robot is concentrated in the center of the body, and the center of gravity of the robot is approximately set to coincide with the center of the body.

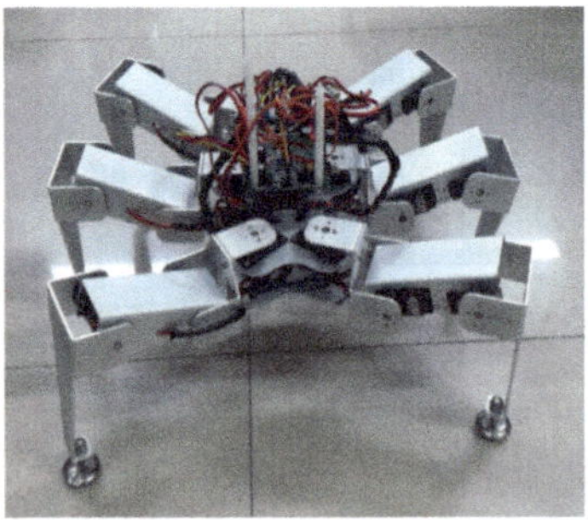

Fig. 1. Platform of hexapod robot

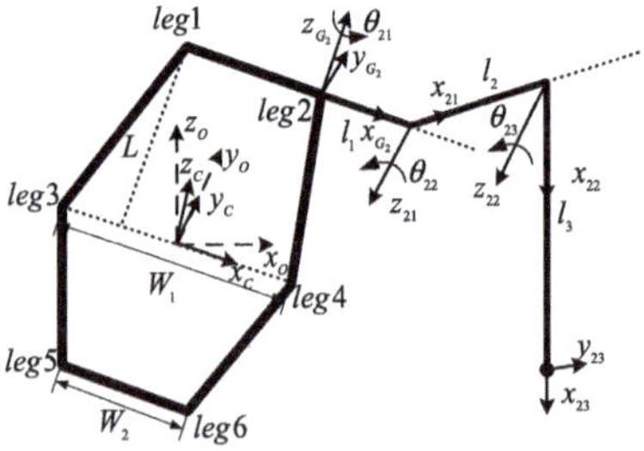

Fig. 2. Structure diagram of hexapod robot

The kinematics model is established using the simplified structure of the robot shown in Fig. 2. The structural parameters are shown in Table 1. The coordinate system is set as follows:

Table 1. Robot mechanism parameter

Structure parameter name	Symbolic representation	Structure parameter name	Symbolic representation
Base section length	l_1	Root joint angle	θ_{i1}
Femur length	l_2	Hip joint angle	θ_{i2}
Tibia length	l_3	Knee joint angle	θ_{i3}
Half body length	L	Body width	W_1, W_2
The i-th leg	$legi(i = 1, 2, \ldots 6)$		

Reference coordinate system $\sum O$: The origin is located at the center of mass of the robot, and the positive direction of the z_O-axis is opposite to the direction of gravity; the y_O-axis points to the front of the robot when the movement is in the initial state and the horizontal direction is parallel; the x_O-axis is determined according to the right-hand rule. This coordinate system is used to simplify the description of the body's posture.

Body coordinate system $\sum C$: The center of mass of the robot is the origin of the coordinates, the z_C-axis is perpendicular to the body's body; the y_C-axis is pointing to the front of the body; the x_C-axis is pointing to the right of the body of the robot.

Root joint coordinate system $\sum G_i$: The origin is the root joint of the i-th leg, and each coordinate axis direction is the same as the coordinate axis of the body coordinate system.

The D-H method is used to obtain the description of the foot position of the i-th leg in the root joint coordinate system:

$$
\begin{aligned}
{}^{G_i}\mathbf{P}_i &= \left[{}^{G_i}x_i, {}^{G_i}y_i, {}^{G_i}z_i\right]^T \\
&= \begin{bmatrix} (-1)^i c_{i1}(l_1 + l_2 c_{i2} + l_3 c_{i23}) \\ s_{i1}(l_1 + l_2 c_{i2} + l_3 c_{i23}) \\ l_2 s_{i2} + l_3 s_{i23} \end{bmatrix}
\end{aligned} \tag{1}
$$

Among them, $c_{ij} = \cos\theta_{ij}$, $c_{i23} = \cos(\theta_{i2} + \theta_{i3})$, $s_{ij} = \sin\theta_{ij}$ $s_{i23} = \sin(\theta_{i2} + \theta_{i3})$, $j = 1, 2$, the same below.

Based on the inverse kinematics equation of one leg, the joint angle matrix $\boldsymbol{\theta}_i = [\theta_{i1}, \theta_{i2}, \theta_{i3}]^T$ can be solved from six foot positions ${}^{G_i}\mathbf{P}_i$, among

$$
\begin{aligned}
\theta_{i1} &= \operatorname{sgn}\left({}^{G_i}y_i\right)\arctan\left(\frac{\left|{}^{G_i}y_i\right|}{\left|{}^{G_i}x_i\right|}\right) \\
\theta_{i2} &= \left|\arccos\frac{l_2^2 + x_i'^2 + {}^{G_i}z_i^2 - l_3^2}{2l_2\sqrt{x_i'^2 + {}^{G_i}z_i^2}}\right| - \arctan\frac{\left|{}^{G_i}z_i\right|}{\left|x_i'\right|} \\
\theta_{i3} &= \left|\arccos\frac{l_2^2 + l_3^2 - \left(x_i'^2 + {}^{G_i}z_i^2\right)}{2l_2 l_3}\right| - \pi
\end{aligned} \tag{2}
$$

where $x_i' = \frac{{}^{G_i}x_i}{cos\theta_{i1}} - \operatorname{sgn}({}^{G_i}x_i)l_1$.

3 Swing Leg Foot Trajectory Planning

During the movement of the hexapod robot, the path points of the swing leg from the current footing point to the foot point between the next footing point are connected to form a robot foot trajectory curve, and the performance of the hexapod robot directly affects the robot sports ability. The sinusoidal curve is used to plan the foot-end trajectory. In the reference coordinate system, the coordinate position of the x_o-axis corresponding to the foot end is unchanged, and the change amount of the coordinate position of the corresponding y_o-axis is equal to half a step $S/2$, and the highest point velocity of the contact point and the foot-end trajectory of the foot-to-ground is zero. The distance from the highest point of the foot trajectory to the ground is h. The trajectory of the swing leg in the reference coordinate system can be expressed as formula (3)

$$\begin{aligned} x(t) &= x_0(0) \\ y(t) &= y_0(0) + \frac{S}{4}\sin\left(\frac{\pi}{T_m}t - \frac{\pi}{2}\right) + \frac{S}{4} \\ z(t) &= z_0(0) + \frac{h}{2}\sin\left(\frac{2\pi}{T_m}t - \frac{\pi}{2}\right) + \frac{h}{2} \end{aligned} \tag{3}$$

In the formula, $\mathbf{P}_0 = [x_0(0), y_0(0), z_0(0)]^T$ represents the position of the initial footfall point of the swing leg in the reference coordinate system during one motion cycle of the robot when walking, and T_m is the time when the endpoint of the swing of the robot swings through this segment of the track.

Defining the hexapod robot's posture matrix $\mathbf{E} = [\alpha, \beta, \gamma]^T$, α, β and γ are the yaw, pitch and roll angles of the robot's posture, respectively. Relative to the reference coordinate system, the rotation relationship between the axes of the body coordinate system is described by the rotation matrix ${}^{O}R_C$:

$$\begin{aligned} {}^{O}\mathbf{R}_C &= Rot(z_o, \alpha)Rot(x_o, \beta)Rot(y_o, \gamma) \\ &= \begin{bmatrix} c_{11} & c_{12} & c_{13} \\ c_{21} & c_{22} & c_{23} \\ c_{31} & c_{32} & c_{33} \end{bmatrix} \end{aligned} \tag{4}$$

Among, $c_{11} = c\alpha c\beta$, $c_{12} = -s\alpha c\gamma + c\alpha s\beta s\gamma$, $c_{13} = s\alpha s\gamma + c\alpha s\beta c\gamma$, $c_{21} = s\alpha c\beta$, $c_{22} = c\alpha c\gamma + s\alpha s\beta s\gamma$, $c_{23} = -c\alpha s\gamma + s\alpha s\beta c\gamma$, $c_{31} = -s\beta$, $c_{32} = c\beta s\gamma$, $c_{33} = c\beta c\gamma$.

The relationship between the various coordinate systems available:

$${}^{O}\mathbf{P}_i = {}^{C}\mathbf{P}_{G_i} + {}^{G_i}\mathbf{P}_i \tag{5}$$

$${}^{O}\mathbf{P}_i = {}^{O}\mathbf{R}_C {}^{C}\mathbf{P}_i \tag{6}$$

Among them, ${}^{C}\mathbf{P}_i$ and ${}^{O}\mathbf{P}_i$ are the position of the foot of the support leg in the body coordinate system and the reference coordinate system, and ${}^{C}\mathbf{P}_{G_i}$ is the position of the heel joint in the body coordinate system, which is determined by the body structure.

Because ${}^{O}\mathbf{R}_C^{-1} = {}^{O}\mathbf{R}_C^T$, from Eq. (6), the position of the foot end in the body coordinate system can be expressed as

$${}^{C}\mathbf{P}_i = {}^{O}\mathbf{R}_C^{T}\,{}^{O}\mathbf{P}_i \tag{7}$$

At this point, the swing foot leg foot planning curve becomes the following formula in the body coordinate system

$$\begin{aligned} x(t) &= c_{11}x_0(0) + c_{21}\left(y_0(0) + \frac{S}{4}\sin\left(\frac{\pi}{T_m}t - \frac{\pi}{2}\right) + \frac{S}{4}\right) + c_{31}\left(z_0(0) + \frac{h}{2}\sin\left(\frac{2\pi}{T_m}t - \frac{\pi}{2}\right) + \frac{h}{2}\right) \\ y(t) &= c_{12}x_0(0) + c_{22}\left(y_0(0) + \frac{S}{4}\sin\left(\frac{\pi}{T_m}t - \frac{\pi}{2}\right) + \frac{S}{4}\right) + c_{32}\left(z_0(0) + \frac{h}{2}\sin\left(\frac{2\pi}{T_m}t - \frac{\pi}{2}\right) + \frac{h}{2}\right) \\ z(t) &= c_{13}x_0(0) + c_{23}\left(y_0(0) + \frac{S}{4}\sin\left(\frac{\pi}{T_m}t - \frac{\pi}{2}\right) + \frac{S}{4}\right) + c_{33}\left(z_0(0) + \frac{h}{2}\sin\left(\frac{2\pi}{T_m}t - \frac{\pi}{2}\right) + \frac{h}{2}\right) \end{aligned} \quad (8)$$

4 Body Motion Planning in Slope Environment

Because the movement of the foot of the swing leg in the ground is actually produced by the movement of the body and the movement of the foot, the key to the stable walking of the robot in the slope environment is how to coordinate the movements of the support legs and the swing legs. Let slope pitch angle be σ. When planning the foot curve in the plane, the step length is S. The swing leg and the support leg foot end move in the reference coordinate system by $S/2$. When the support leg and swing leg move at the same time, the actual movement distance of the foot of the swing leg in the earth is S, so when traveling on the slope, just ensure that the body center changes $\Delta z = S\tan\sigma$ in the Z direction, as shown in Fig. 3.

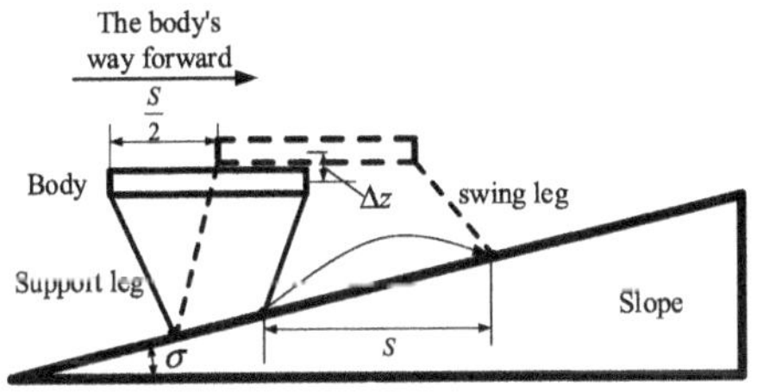

Fig. 3. Schematic diagram of the robot walking on the slope

At this point, the planning curve of the swing leg in the reference coordinate system remains unchanged, and the planned trajectory of the body movement becomes formula (9). Where T_s is the time when the robot supports the foot through this trajectory.

$$\begin{aligned} x &= x(0) \\ y &= \frac{S}{2T_s}t \\ z &= z(0) + S\tan\sigma \end{aligned} \quad (9)$$

Similarly, when the robot walks in a certain posture, the foot curve of the support leg becomes the following formula in the body coordinate system

$$\begin{aligned} x(t) &= c_{11}x_1(0) + c_{21}\left(y_1(0) - \frac{S}{2T_s}t\right) + c_{31}(z_1(0) - S\tan\sigma) \\ y(t) &= c_{12}x_1(0) + c_{22}\left(y_1(0) - \frac{S}{2T_s}t\right) + c_{32}(z_1(0) - S\tan\sigma) \\ z(t) &= c_{13}x_1(0) + c_{23}\left(y_1(0) - \frac{S}{2T_s}t\right) + c_{33}(z_1(0) - S\tan\sigma) \end{aligned} \quad (10)$$

When Eqs. (8) and (10) are combined with the inverse kinematics formula of the leg, the joint angle changes corresponding to the swing leg and the support leg can be obtained. When $T_m = T_s$, the swing phase and the support phase coordinate the motion, and realize the fast uphill of the robot with stable posture.

5 Simulation and Experiment

Taking the three-legged gait walking as an example, the experimental parameters are shown in Table 2. The step length is 6 cm, the maximum lifting height of the swing leg foot is 4 cm, the swing phase movement time is 2 s, the support phase movement time is 2 s, and the slope is taken with an inclination of 6°, the robot is required to walk in a straight line. Select *leg*2, *leg*3 and *leg*6 as the swing leg at the beginning, *leg*1, *leg*4 and *leg*5 as the support leg at the beginning, and Eqs. (3) and (10), the motion curve of the foot in the reference coordinate system during a gait cycle. As shown in Fig. 4. In order to ensure the stability during the movement of the robot, the body plane is used in a posture parallel to the horizontal plane. The movement curve of the swinging leg foot on the ground relative to the starting position is shown in Fig. 5. Figure 6 shows the joint angle change curve of two gait cycles and Fig. 7 shows the angular velocity and angular acceleration of the three joints of *leg*2.

Observing the simulation results, we can see that joint rotation changes continuously and all within the range of joint angle constraints. The robot can achieve planned movement. The angular velocity curve and the angular acceleration curve are continuous and smooth during a gait cycle, which ensures the smoothness of the trunk movement and leg movement of the robot and effectively reduces the impact of force or moment during walking.

Table 2. Experimental parameters

Structural parameter	Value	Structural parameter	Value
l_1/cm	6	Initial $\boldsymbol{\theta}_1$, $\boldsymbol{\theta}_2$/(°)	$[30, 10, -100]^T$
l_2/cm	7	Initial $\boldsymbol{\theta}_3$, $\boldsymbol{\theta}_4$/(°)	$[0, 10, -100]^T$
l_3/cm	15	Initial $\boldsymbol{\theta}_5$, $\boldsymbol{\theta}_6$/(°)	$[-30, 10, -100]^T$
L/cm	10	W_1/cm	4
W_2/cm	3	θ_{i1}/(°)	−30 ~ 30
θ_{i2}/(°)	−90 ~ 90	θ_{i3}/(°)	−150 ~ 0

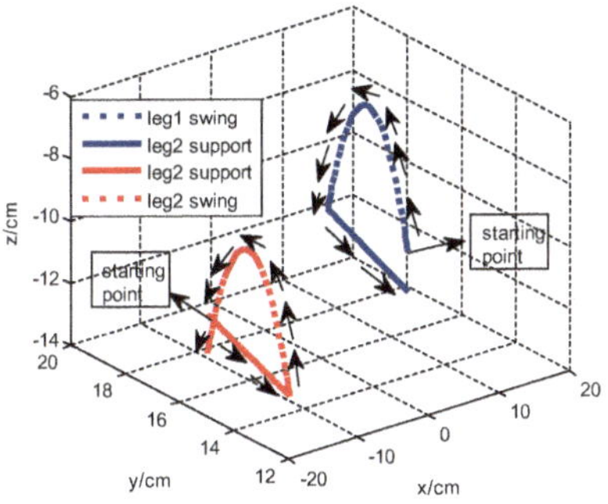

Fig. 4. The motion of the foot of *leg*2 and *leg*1 in the reference coordinate system

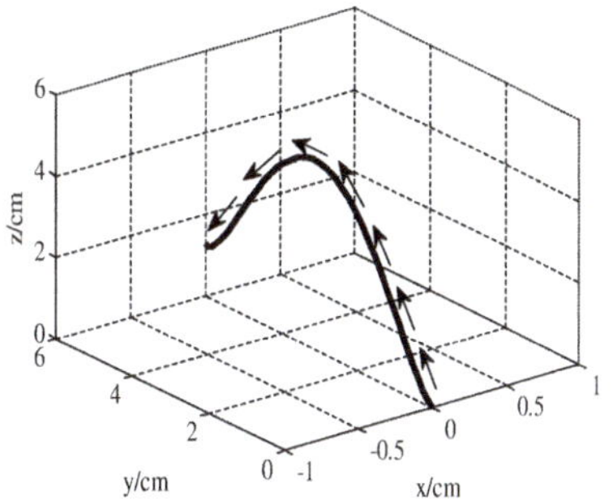

Fig. 5. Swing leg motion relative to starting position

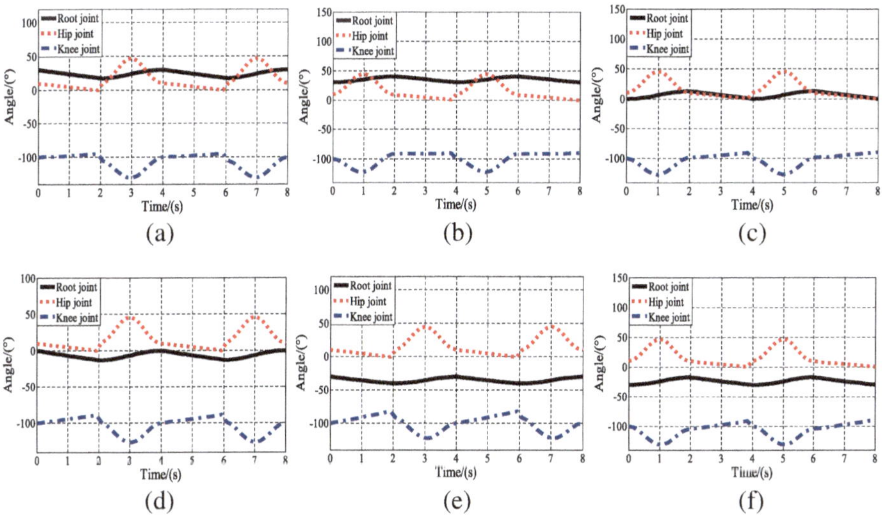

Fig. 6. Joint rotation angles when walking straight on a slope in a horizontal posture. (a) Each joint angle of *leg*1. (b) Each joint angle of *leg*2. (c) Each joint angle of *leg*3. (d) Each joint angle of *leg*4. (e) Each joint angle of *leg*5. (f) Each joint angle of *leg*6

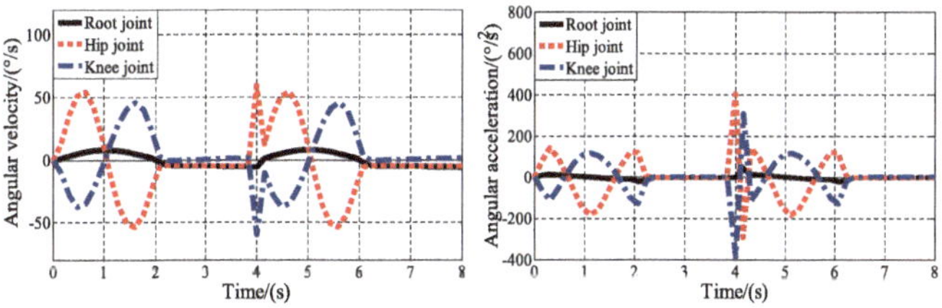

Fig. 7. The angular velocity and angular acceleration of the three joints of *leg*2

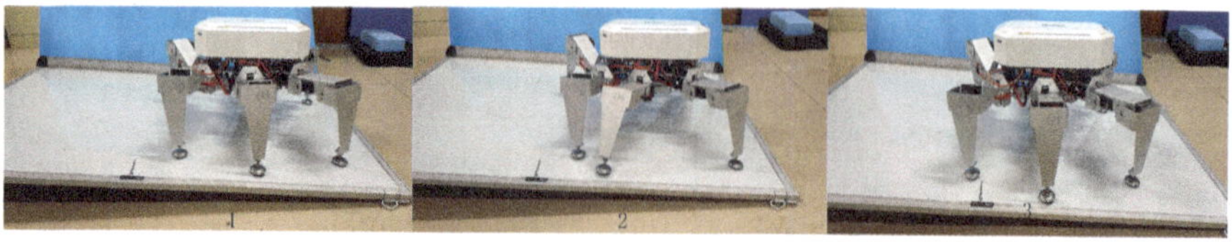

Fig. 8. A walk experiment with a horizontal posture on a slope

The hexapod robot experimental platform used in this article is a hexapod robot experimental prototype developed by the laboratory based on the principle of bionics. The specific parameters are shown in Table 2. According to the previous simulation analysis, each joint angle is input into the experimental prototype. A gait side view video capture taken during the experiment is shown in Fig. 8. Through experiments, it was found that the robot six-legged robot completed the straight motion according to the planned footprint in a horizontal posture. The verification of the method of considering attitude stabilization on the slope is completed.

6 Conclusion

This paper focuses on the stable walking on a slope of a hexapod robot. Considering the combination of body posture, slope gradient, body movement and foot movement, the walking method of the robot is designed to achieve a stable walking of the robot on the slope when considering posture. Finally, the effectiveness of the proposed method is verified by simulation and physical prototyping experiments.

References

1. Wang, G.B., Chen, D.S., Chen, K.W., et al.: The current research status and development strategy on biomimetic robot. J. Mech. Eng. **51**(13), 27–44 (2015)
2. Li, S.Z., Zhong, J.F., Cheng, P., et al.: Design of slope gait for six-legged robot. Mach. Electron. **3**, 53–57 (2013)
3. Meng, J., Li, Y.B., Li, B.: Control method and its implementation of quadruped robot in omni-directional trotting gait. Robot **37**(1), 74–84 (2015)
4. Zhang, G.T., Rong, X.W., Li, Y.B., Li, B.: Control of the quadrupedal trotting based on virtual model. Robot **38**(1), 64–74 (2016)
5. Han, B.L., Jia, Y., Li, H.S., et al.: Posture adjustment for quadruped robot trotting on a slope. Trans. Beijing Inst. Technol. **36**(3), 242–246 (2016)
6. Zhang, L., Ma, S.G., Inoue, K., et al.: Omni-directional walking of a quadruped robot with optimal body postures on a slope. In: IEEE International Conference on Robotics and Automation, pp. 2976–2981. IEEE, Piscataway (2005)
7. Zhang, L., Li, D.D., Yang, F., et al.: Development and attitude control of a hexapod bionic-robot. In: IEEE International Conference on Robotics and Biomimetics, Qingdao, pp. 77–82 (2016)

The Design of Suspension Mechanism and Analysis of Obstacle Ability to Rescue Robots

Shiyue Zhang[2], Xinhua Zhao[2], Weihua Su[1](✉), Hang Wu[1], Zhaokun Dai[1], and Zhuo Chen[2]

[1] National Innovation Institute of Defense Technology, Academy of Military Sciences, Beijing 100000, China
directorsu@126.com
[2] Department of Mechanical Engineering, Tianjin University of Technology, Tianjin 300384, China

Abstract. To improve the ability of rescue robots to move in the complex terrain, in this paper we design the novel suspension structure and propose the corresponding method of enhancing obstacle crossing capability. Firstly, the single longitudinal arm triangular suspension (SLATS) structure for rescue robot is proposed. Secondly, we perform kinematic calculations to determine the basic parameters of the rescue robot. Thirdly, based on the structural parameters and driving conditions, we use the automatic dynamic analysis of mechanical system (ADAMS) to establish the robot model and conduct two obstacle crossing experiments. The experimental results indicate the effectiveness of the proposed method.

Keywords: Rescue robots · Obstacle crossing · Triangular suspension Kinematics · The distribution of gravity center · ADAMS simulation

1 Introduction

The terrains are complex and various in natural disaster environments nowadays. It is low efficient to rely solely on manpower to complete rescue missions. Researching on high topographic adaptability robots has become an important direction for the future development of rescue equipment. The rescue robot we studied is a kind of intelligent robot that is driven by six-wheel motor-in-hub system [1] and equipped with single longitudinal arm triangular suspension (SLATS). As we all know, suspension system plays an important role in improving the terrain adaptability of small mobile robots. But general small mobile robots are rarely equipped with suspension systems, it is attributed to the suspension occupying a lot of space and taking disadvantage of the robot load space layout. By contrasting with independent suspension and dependent suspension and analyzing cross arm, longitudinal arms, multi-link, McPherson, leaf spring, coil spring and air spring suspension type [2, 3], we designed a kind of SLATS structure referring to Fig. 1.

K. Deng et al. (Eds.): ICMIR 2018, AISC 856, pp. 677–685, 2019.
https://doi.org/10.1007/978-3-030-00214-5_85

There are many researchers who have studied the model of obstacle crossing. Bekker [4, 5] did a lot of vertical obstacle crossing experiments. Huang [6] proposed a mining vehicle with four independent driven wheels and found an optimal obstacle performance with the gravity position in the middle of suspension. Lv [7] designed a new type of self-balance obstacle car and used sensors to control the two swing arms for going back and forth. The crusher unmanned vehicle [8] studied by the U.S. military has high maneuverability, dynamic performance and passing performance. We can come to such conclusions: (1) All-wheel-drive vehicles have better capability than non-full-drive vehicles in obstacle crossing aspect. (2) The structured parameters of the ground friction coefficient, the position of gravity center, tire radius and wheelbase have a great influence on obstacle crossing height. (3) The gravity center moves back, the obstacle crossing performance of front wheel improves; the gravity center moves forward, the obstacle crossing performance of rear wheel enhances.

By analysis and research, it is easy to control the position of gravity center when the rescue robot is moving and surmounting obstacles. So, we take advantage of its influence discipline of gravity center and use ADAMS software to establish the virtual model to explore whether the change of gravity center can further improve the obstacle crossing capability. The obstacle crossing analysis can provide theoretical and technical supports for the structural design of rescue robot in the future.

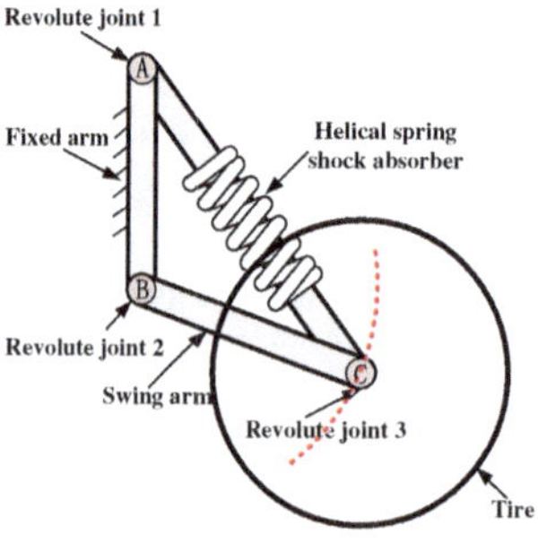

Fig. 1. SLATS schematic diagram

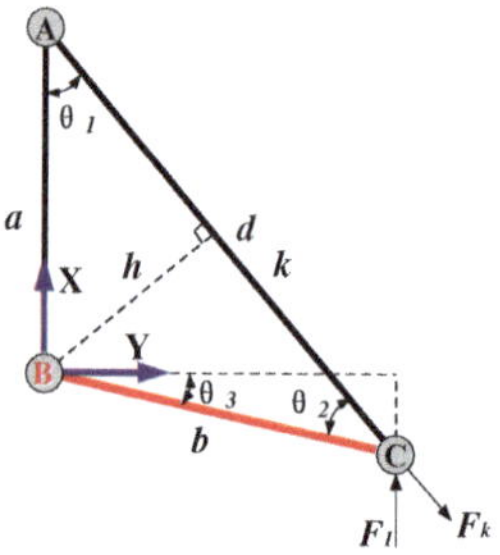

Fig. 2. SLATS force analysis

2 The SLATS Model and Kinematic Analysis

As is shown in Fig. 1, SLATS is mainly composed of revolute joint, fixed arm, swing arm and helical spring shock absorber. The fixed arm is connected to the frame of the rescue robot. The swing arm is mainly used to limit the trajectory of the tire. This suspension type always maintains a stable triangular shape during movement and increases the tire travel space and probability of tire contacting with the ground [9]. So the rescue robot equipped with SLATS has a high driving stability and terrain adaptability.

Figure 2 shows SLATS force status in a special condition. We set up a coordinate system *XBY*, the length of fixed arm is a, the length of swing arm is b, the length of helical spring shock absorber is d and its original length is L. The angles are defined such as θ_1, θ_2 and θ_3, h is set to the vertical distance from point B to the connected line between point A and point C, F_1 is set to the vertical force between the tire and suspension, k is set to the stiffness coefficient of the spring, F_k is set to the spring force. We establish force equilibrium equations to explore the relationship of initial suspension posture.

We set the swing arm as the research object and point B as torque center. The torque balance, triangular relation and Hooke law equation are as follows:

$$F_k h = F_1 b\cos(\theta_3) \tag{1}$$

$$b/\sin(\theta_1) = a/\sin(\theta_2) = d/\sin(\pi - \theta_1 - \theta_2) \tag{2}$$

$$h = a\sin(\theta_1) \tag{3}$$

$$\theta_1 + \theta_2 + \theta_3 = \pi/2 \tag{4}$$

$$F_k = k(L - d) \tag{5}$$

Thus

$$k = \frac{F_1 b\cos[\pi/2 - \theta_1 - \arcsin(a\sin(\theta_1)/b)]}{a[L - b\sin(\pi - \theta_1 - \arcsin(a\sin(\theta_1)/b))/\sin(\theta_1)]\sin(\theta_1)} \tag{6}$$

3 The Basic Structural Parameters of the Rescue Robot

According to the integral structure parameters of the rescue robot, we derive the partial parameters of the suspension. The calculation and simulation are essential in the solving process. Please refer to Fig. 5, we define a stable driving condition: (1) each suspension is evenly loaded. (2) Swing arms parallel to horizontal ground. We can get restraint equations which are

$$F_1 = (Mg + F_L - 6mg)/6 \qquad \tan(\theta_1) = b/a$$

Besides, we bring all the structure parameters into the Eq. (6), and get the value of k. Finally, we get the basic structure parameters shown in Table 1.

Table 1. The basic structure parameters

Robot	Total mass (M)	Dimensions		Load (F_L)	Tire radius	Each tire mass (m)
	420 kg	1650 * 1400 * 800 mm^3		600 N	225 mm	10 kg
Suspension	a	b	L	F_1	θ_1	k
	270 mm	308 mm	515 mm	700 N	48.8°	10 N/mm

4 The Rescue Robot Obstacle Crossing Experiment

4.1 The Necessity of Researching on Obstacle Crossing

Rescue robots should reach the target quickly when performing tasks. On the one hand, robots need to have an optimal path planning algorithm to detour the obstacle [10, 11]. On the other hand, robots can directly surmount obstacles. The combination of two aspects can play a crucial role in improving task efficiency. The integral control rules are illustrated in Fig. 3. It includes two functional modules which are the path planning control module and the obstacle crossing control module. We conduct a deeper study on the obstacle crossing control module. Based on our research on the effected parameters of obstacle crossing height, we mainly control the position of gravity center to enhance the obstacle crossing capability.

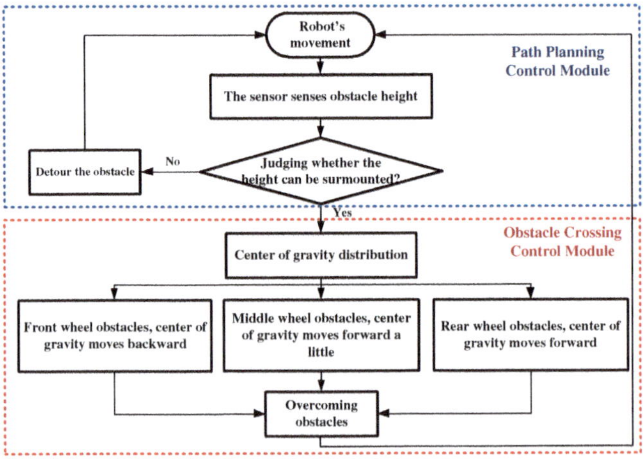

Fig. 3. Integral control rules

4.2 The Establishment of Virtual Model

Based on the basic structure parameters, the virtual rescue robot model equipped with SLATS is established using ADAMS and the imposed conditions and kinematic pairs are shown in Fig. 4. The function of imposed conditions is to simulate real movement situation, such as the contacted type between tires and ground, drive motor, tire mass, robot's body mass and spiral spring. The function of kinematic pairs is to constrain the relationship between the components, such as fixed joint, translational joint and revolute joint.

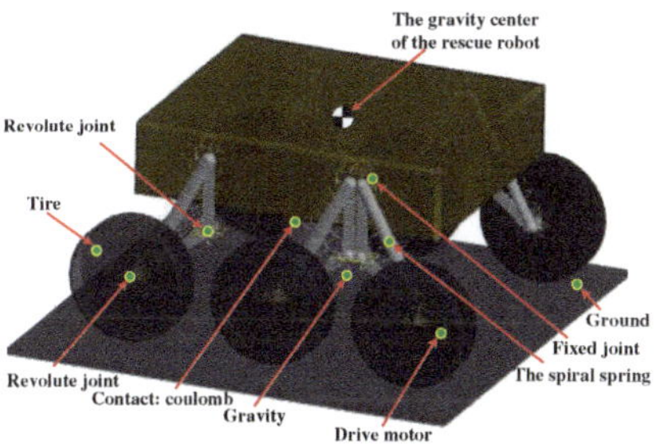

Fig. 4. The virtual model of rescue robot

4.3 The Explored Experiment of Obstacle Crossing Capability

We set the experimental conditions: (1) The vertical obstacle height (H_m) is set to 270 mm; (2) The rescue robot's body mass is set to 420 kg (including the load), the tire mass is set to 10 kg and its radius is set to 225 mm; (3) The speed of robot is set to 196 mm/s. (4) The suspension system mass is set to 0, all spring stiffness coefficients are 10 N/mm; (5) The ground friction coefficient is set to 0.55; (6) The position of gravity center is controlled by sensors. Figure 5 indicates the sensor location and control flow. Sensors include two functional parts these are inductor and actuator. The inductor is to sense the distance from every wheel axis to the ground base and transmit signal to actuator. The function of the actuator is to receive the signal from inductor and change the position of the gravity center. Through an experiment simulation, all wheels of the rescue robot cannot surmount the obstacle high of 270 mm. We use the method of changing the gravity center position to explore whether the rescue robot can surmount the obstacle height of 270 mm. Please refer to Fig. 5, the control commands of obstacle crossing are as follows:

1. Sensor1-inductor is mainly to sense whether the robot in current driving state can surmount the obstacle. If the robot can surmount the obstacle, it won't adjust own structure. If not, the robot needs to adjust its gravity center. According to our experiment setup, the rescue robot in current driving state can't surmount the obstacle and needs to adjust its gravity center.
2. The function of sensor1-actuator is to receive the signal from sensor1-inductor and move the gravity center of robot body to point A.
3. Sensor2-inductor is to sense the distance from front wheel axle to the ground base (H_1). When H_1 is greater than 450 mm (ensuring that the wheel can go over the obstacle), it will make sensor2-actuator reaction.
4. The function of sensor2-actuator is to receive the signal from sensor2-inductor and move the gravity center of the robot body to point B.
5. Sensor3-inductor is to sense the distance from middle wheel axle to the ground base (H_2). When H_2 is greater than 450 mm, it will make sensor3-actuator reaction.
6. The function of sensor3-actuator is to receive the signal from sensor3-inductor and move the gravity center of the robot body to point C.

7. Sensor4-inductor is to sense the distance from rear wheel axle to the ground base (H_3). When H_3 is greater than 450 mm, it will make sensor4-actuator reaction.
8. The function of sensor4-actuator is to receive the signal from sensor4-inductor and move the gravity center of the robot body to point R, the rescue robot returns normal state to drive.

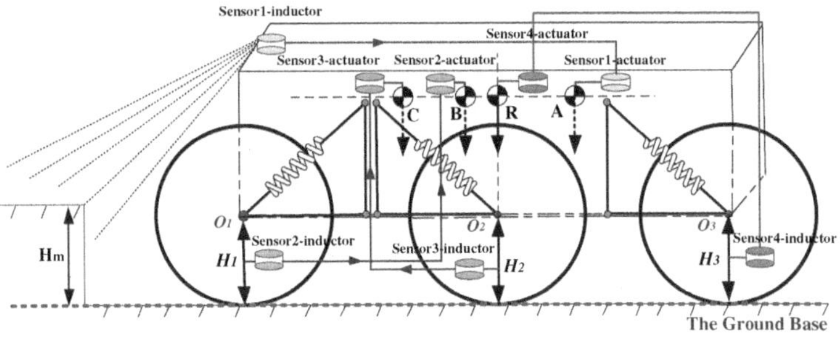

Fig. 5. The sensor location and control flow

Running the step 1–8, the rescue robot can perfectly surmount the obstacle by controlling the position of the gravity center actively. We can analyze the obstacle crossing state of the rescue robot referring to Fig. 6. State_1 and state_2 stand for front wheel obstacle crossing and its gravity center is located at point A. State_3 and state_4 stand for middle wheel obstacle crossing and its gravity center is located at point B. State_5 and state_6 stand for rear wheel obstacle crossing and its gravity center is located at point C. State_6 stands for the stable driving state and its gravity center is located at point R.

However, when the robot can surmount the obstacle in current driving state, is it necessary to control the gravity center actively? We change the obstacle height to 250 mm. Through experiment simulations, the rescue robot equipped with or without sensors can all surmount the obstacle. Under such conditions, we conduct a comparative experiment to research the posture and capability of obstacle crossing. Figure 6 describes the obstacle crossing status comparison diagram. Point O is set to the position of robot body center. Point R is set to the position of gravity center. CH marked in state_1 stands for the vertical distance from the point O to the ground base. θ marked in state_2 stand for the pitch angle of the robot. CH and θ play an important role in the process of obstacle crossing [12], and its changeable curve is measured respectively. Figure 7 is the comparison diagram of CH. Figure 8 is the comparison diagram of θ.

As can be seen from Figs. 6 and 7, the front wheel surmounts the obstacle when $t \in (19.5, 23.5)$. Middle wheel surmounts the obstacle when $t \in (23.5, 27)$. Rear wheel surmounts the obstacle when $t \in (27, 30)$. Through the analysis of obstacle crossing time, the rescue robot without sensors costs 10.5 s for surmounting the obstacle, yet the rescue robot with sensors costs 9.5 s. This demonstrates that controlling the position of the gravity center actively can improve the efficiency of obstacle crossing. As is shown in Fig. 8, due to the change of the gravity center, the rescue robot with sensors has a

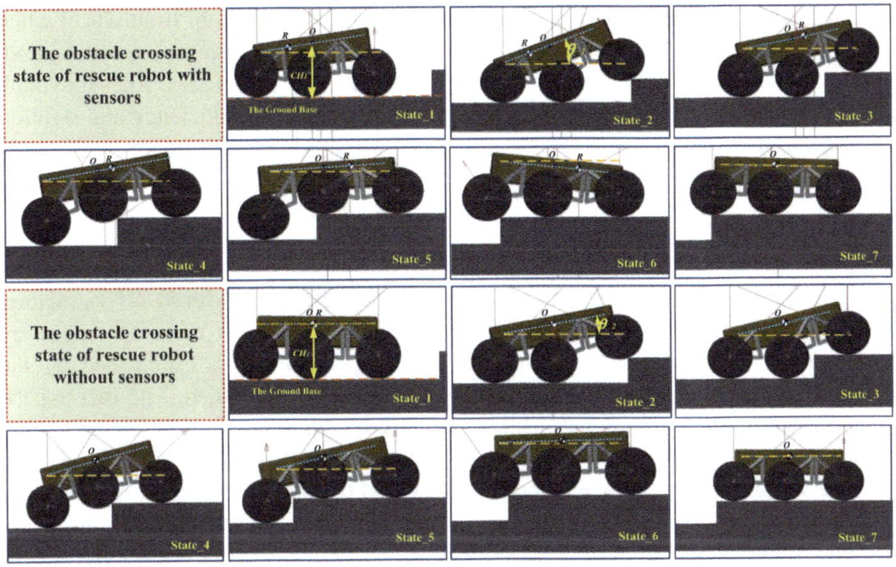

Fig. 6. The obstacle crossing status comparison diagram

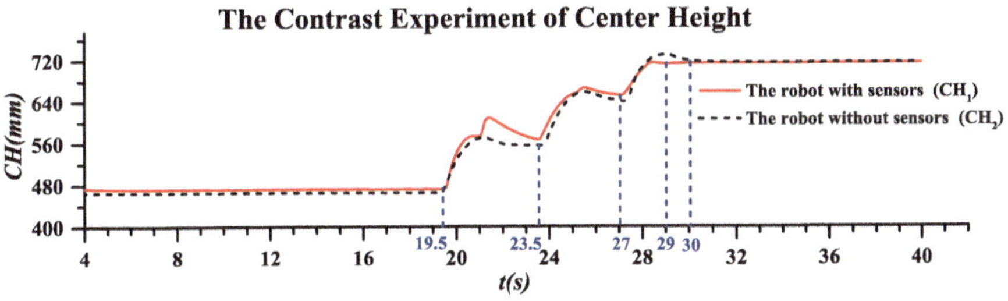

Fig. 7. The *CH* comparison diagram

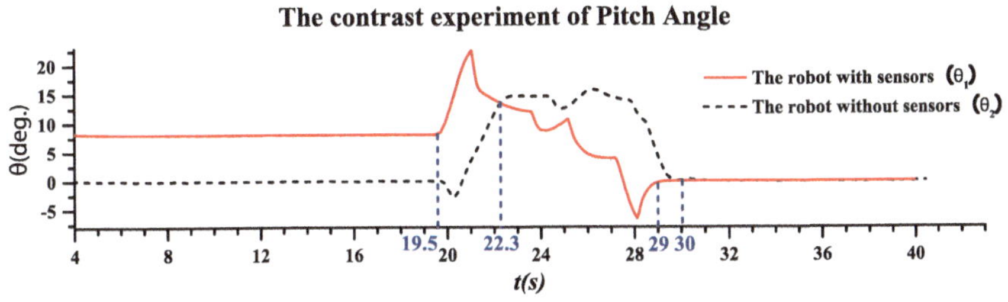

Fig. 8. The pitch angle comparison diagram

bigger pitch angle for improving the obstacle crossing capability of front wheel when $t \in (19.5, 22.3)$, and its pitch angle becomes smaller than the rescue robot without sensors for keeping a stable obstacle crossing posture when $t \in (22.3, 30)$. Above all, the distribution of gravity center plays an important role to the efficiency and stability of obstacle crossing.

5 Conclusions

This paper is based on the research of the rescue robot equipped with SLATS structure. Through kinematic calculations, the basic structural parameters of rescue robot are determined. What's more, the virtual model of rescue robot is established by the ADAMS, and the imposed condition and kinematic pairs are added. The first experimental purpose is to explore the capability of obstacle crossing by changing the gravity center via some sensors. Its results indicate the effectiveness of the proposed methods.

At the end, we design a comparative experiment that the robot can all surmount the obstacle in the states of controllable or uncontrollable gravity center. The results show that controlling the position of the gravity center actively can improve the efficiency of obstacle crossing and make the rescue robot keep a stable posture. I assure the distribution method of gravity center to improve the obstacle crossing capability will have a wide range of application. This research can play an important role in promoting the development of high topographic adaptability of robots.

Acknowledgements. This work is supported by the Science and Technology Pillar Program, Tianjin, china, under Project 16YFZCSF00590.

References

1. Velardocchia, M., Rondinelli, E.: Design and development of an in-hub motors hybrid vehicle for military applications. In: SAE 2010 World Congress and Exhibition (2010)
2. Bastow, D.: Car Suspension and Handling. Society of Automotive Engineers Inc., Warrendale (1993)
3. Hong, K.S.: Modified skyhook control of semi-active suspensions: a new model, gain scheduling, and hardware-in-the-loop tuning. ASME J. Dyn. Syst. Meas. Control **124**(1), 158–167 (2002)
4. Bekker, M.G.: Theory of Land Locomotion. The University of Michigan Press, Ann Arbor (1956)
5. Bekker, M.G.: Introduction to Terrain-Vehicle Systems. The University of Michigan Press, Ann Arbor (1960)
6. Huang, Z.H., Liu, S.J., Xie, Y.: Obstacle performance of cobalt-enriching crust wheeled mining vehicle. J. Cent. South Univ. **13**(2), 180–183 (2006)
7. Lv, Q.Y., Yang, L.: Modeling and Design of a new type of self-balancing obstacle vehicle. In: International Conference on Chemical, Material and Food Engineering (2015)
8. Stentz, A., Bares, J., Pilarski, T.: The crusher system for autonomous navigation. Auvsis Unmanned Syst. N. Am. **7**(2), 31–36 (2007)

9. Mihon, L., Lontiş, N., Deac, S.: The behaviour of a vehicle's suspension system on dynamic testing conditions. **294**(1), 012083 (2018)
10. Mohamed, A., Ren, J., Sharaf, A.M., et al.: Optimal path planning for unmanned ground vehicles using potential field method and optimal control method. Int. J. Veh. Perform. **4**(1), 1–14 (2018)
11. Yen, C.T., Cheng, M.F.: A study of fuzzy control with ant colony algorithm used in mobile robot for shortest path planning and obstacle avoidance. Microsyst. Technol. **24**(8), 1–11 (2016)
12. Tseng, H.E., Xu, L., Hrovat, D.: Estimation of land vehicle roll and pitch angles. Veh. Syst. Dyn. **45**(5), 433–443 (2007)

Analysis on Obstacle Performance of Three-Axle Radially Adjustable Pipeline Robot

Yue Li(✉), Lei Zhang, and Hao Gao

Department of Automation and Control, Ocean University of China, Qingdao, China
729774953@qq.com

Abstract. A pipeline robot design based on a parallel four-bar linkage reduction mechanism, screw nut and flexible device structure is proposed to establish a model for the mechanical characteristics of the robot during obstacle movement in the pipeline, and a simulation experiment is conducted. The experimental results show that the structure design has good barrier performance.

Keywords: Pipeline robot · Linkage reduction mechanism · Screw nut Obstacle movement · Flexible device

1 Introduction

With the popularization of pipelines in daily life and industrial production, regular overhaul of aging pipelines can effectively prevent pipeline leakage and explosion accidents. However, there are many difficulties in the maintenance of underground pipelines and narrow pipelines erected at high altitudes by manpower. Therefore, there is an increasing demand for controllable mobile platforms that can achieve stable and efficient movement within pipelines [1–3].

In recent years, domestic and foreign scholars and engineering and technical teams have made a lot of progress in the development of pipeline robots. Choi [4] developed the MRINSPECT series of pipeline robots by adopting a passive adaptive mechanism and a multi-motor independent driving method. The speed and differential movement of the robot can be achieved by independently controlling the speed of the motor. Li et al. [5, 6], etc. designed spiral-driven pipeline robots. Based on a variable constraining mechanism using a single motor as a drive source, a camera carrier platform mechanism is proposed to enable the screw-driven robot to steadily detect the pipeline environment at the front in the axial and circumferential directions; Deng et al. [7, 8] Axially staggered symmetry mechanical structure, developed six independent wheel drive in-tube inspection robot, its traction force generated to enhance the self-centering function when walking; Zhang [9], etc. developed a three-axis differential type pipeline robot, using six synchronized reduction, Front and rear wheel speed synchronization design; Zhang [10–12] and other developed based on the parallelogram wheel leg structure, using screw nut adjustment method of the robot diameter adjustment mechanism.

K. Deng et al. (Eds.): ICMIR 2018, AISC 856, pp. 686–691, 2019.
https://doi.org/10.1007/978-3-030-00214-5_86

The three-axis radially adjustable tracked pipe robot has the characteristics of stable movement in the pipeline and strong adaptability to different pipe diameters [13, 14]. At the same time, differential motion can be achieved by means of independent control of multiple motors, and good passage is achieved in the bent pipe section [15, 16]. Based on the above advantages, in order to solve the problem of robot interference in the pipeline and the compensation of the full-rigid structure to the pipe wall pressure, a pipe based on a parallel four-bar linkage reduction mechanism, screw nut and elastic device adjustment method is proposed. The robot design analyzes its working principle and mechanics characteristics when crossing obstacles. It analyzes its working principle and mechanics characteristics when crossing obstacles, and carries out relevant experimental research.

2 Pipeline Robot Structure

The robot structure design is shown in Fig. 1. The three-axis symmetrical structure is adopted to improve the stability of the robot's tube movement, and the movement of the screw nut is driven by the stepping motor which connecting the screw to rotate, so as to control the parallel four-bar linkage mechanism to realize the simultaneous zoom of the robot structure. When the robot is walking in the pipe, the radial dimension of the robot is changed through the direct control of the stepper motor, so that the robot has a good adaptability to the pipe diameter in the design range.

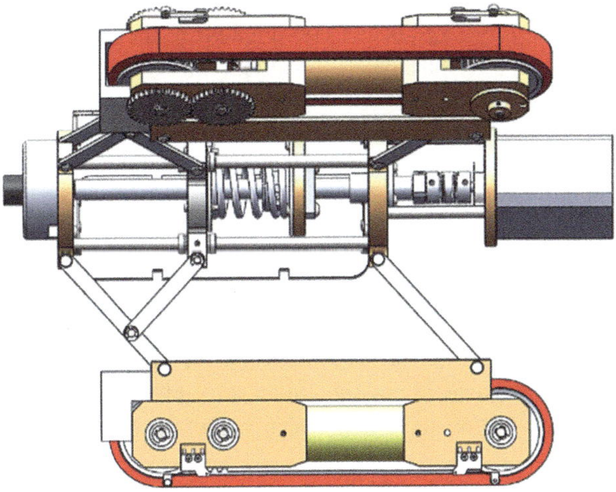

Fig. 1. Structure design schematic of robot

The pressure on the wall caused by the robot's own gravity is an important factor that affects the traction and ensures the robot's stable movement in the pipeline. As the angle of inclination of the pipe increases, the self gravity provides a decrease in the adhesion that results in motion instability. Therefore, it is necessary to provide additional pressure compensation in the radial direction of the pipe to ensure that the robot can move stably under different inclination angle conditions.

Aiming at the problem that the robot compensates the pressure in the pipe wall, the spring adjustment mechanism is designed. The design takes into account the traction force adjustment and the active adaptation of the pipe diameter, and has a certain flexibility and redundant space. When the three axes of the robot are completely attached to the pipe wall, the stepping motor continues to drive the screw nut to work to compress the spring, to provide additional driving force to the pipe wall through the connecting rod structure, so as to achieve pressure compensation of the pipe wall. Compared to the full-rigid structure design, the robot has a certain amount of obstacle-free redundant space. When a small obstacle is encountered, the robot can complete the movement in the form of a passive compression spring, which enhances the robot's environmental adaptability.

The value of the wall pressure of the robot can be measured by the pressure sensor detecting the pressure feedback value of the spring adjusting mechanism. It is possible to design the dynamic adjustment of the flexible link and realize the precise control of the robot motion state.

3 Obstacle Analysis of Robot

Due to the uncertainty of the physical environment of the pipeline, the obstruction performance of the pipeline robot has an important influence on the passage of the robot in the pipeline. The design of the adjustable-adjustment mechanism makes this pipeline robot have a very large active coordination in the movement space. And the design of the elastic support unit makes the three axes of the robot still have redundant motion space for obstacle-obstructing movement when completely fitting on the pipe wall. Compared with the traditional three-axis pipe robot rigid link structure design, this design has greater advantages in barrier performance.

There are mainly two kinds of obstacles encountered by robots in pipelines: raised step-type obstacle and sulcus-type obstacle. In the case where the step-type obstacle is larger in size, the robot can be decomposed into two steps: stepping up and down.

Because the pipeline robot moves slowly in obstacles, the movement state will not produce abrupt changes. Therefore, the static state of the force in the process is analyzed. As shown in Fig. 2, the obstacle between the front wheel and the ground is overcome when the obstacle is crossed. Fulcrum replaced. According to the actual pipeline environment, the obstacle height can be set no higher than the drive wheel radius.

The parameter settings in Fig. 2 are shown in Table 1.

It can be seen from the force balance:

$$\begin{cases} (F_{N1} - f_1)\cos\omega + N_1 \sin\omega + N_2 - F_N = 0 \\ (F_{N1} - f_1)\sin\omega - N_1\cos\omega + (F_{N2} - f_2) = 0 \end{cases} \tag{1}$$

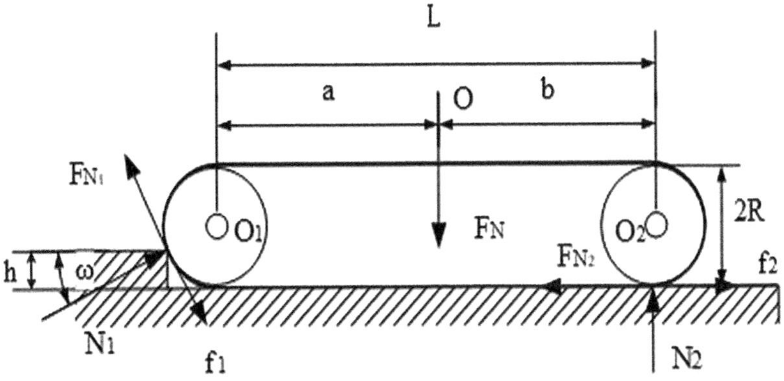

Fig. 2. Robot obstacle analysis

Table 1. Obstacle force parameter setting

Parameter	Parameter setting
O	The barycenter of the robot
L	Horizontal distance between front and rear round center
F_N	Robot force on the wall
N_1, N_2	Responsiveness of the front and rear rounds
F_{N1}, F_{N2}	Front and rear wheel static friction
f_1, f_2	Front and rear wheel rolling friction
a, b	The horizontal distance between the barycenter and the center of the front and rear wheels
R	Drive wheel radius
h	Obstacle height
ω	Obstacles to the wheel's reaction force and horizontal angle

In the process of obstacle crossing, the output torque of the motor should not be less than the resistance torque of the robot:

$$T \geq F_N a + (F_{N2} - f_2)h - N_2 L \tag{2}$$

Among the formula:

$$\begin{cases} F_{N1} = \mu_1 N_1 \\ F_{N2} = \mu_1 N_2 \end{cases} \quad \begin{cases} f_1 = \mu_2 F_1 \\ f_2 = \mu_2 F_2 \end{cases}$$

μ_1 and μ_2 are the static friction coefficient and the rolling friction coefficient between the track and the pipe wall.

Since the track of the robot is made of hard wear-resistant rubber material, the rolling friction coefficient $\mu_2 \approx 0$ can be set. Then we can get the relationship between the maximum obstacle height of the robot, the output torque of the motor and the pressure of the robot on the wall of the pipe:

$$h \leq \sqrt{L + \left(1 - \frac{\mu_1 R}{2Z}\right)^2} + 1 - \frac{\mu_1 R}{2Z} \tag{3}$$

Among the formula:

$$Z = \left(\frac{T}{F_N} - a\right)\left(1 + \mu_1^2\right)$$

Substituting the robot design parameters for its obstacle-obstacle ability simulation experiment results are shown in Fig. 3. The simulation results show that the pipeline robot with the structure proposed in this paper has good obstacle-obstacle performance; meanwhile, the obstacle-obstacle effect is directly related to the robot's movement parameters. In order to ensure that the robot has good obstacle avoidance performance in the pipeline and does not cause additional energy consumption and motor work load, the robot's sports performance parameters can be controlled according to actual conditions.

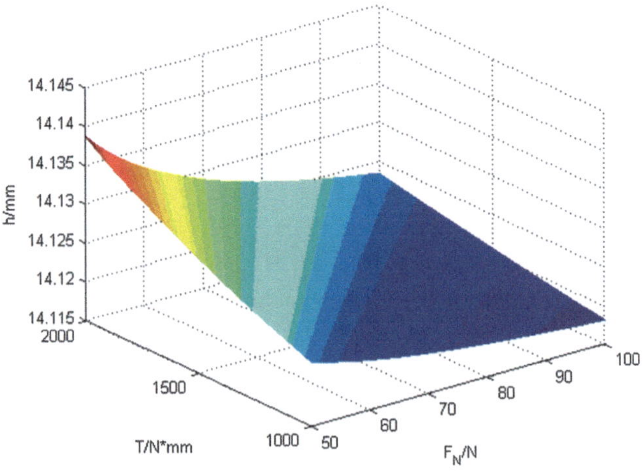

Fig. 3. Robot obstacle simulation

4 Conclusion

In this study, a self-adaptive 3-axis tracked pipe robot design based on a parallel four-bar linkage reduction mechanism, screw nut and elastic device adjustment method is proposed. The analysis of the mechanical model of the obstacle crossing and the results of simulation experiments show that this design can improve the obstruction ability of the pipeline robot. This study enriches the theoretical research of three-axis differential pipe robots.

References

1. Hirose, S., Ohno, H., Mitsui, T., Suyama, K.: Design of in-pipe inspection vehicles for Φ25, Φ50, Φ150 pipes. In: Proceedings of IEEE International Conference on Robotics and Automation, 1999, Detroit, USA, pp. 2309–2314. IEEE (1999)
2. Park, J., Hyun, D., Cho, W.-H., Kim, T.-H., Yang, H.-S.: Normal-force control for an in-pipe robot according to the inclination of pipelines. IEEE Trans. Ind. Electron. **58**, 5304–5310 (2011)
3. Roh, S.G., Choi, H.R.: Differential-drive in-pipe robot for moving inside urban gas pipelines. IEEE Trans. Rob. **21**(1), 1–17 (2005)
4. Choi, H.R., Ryew, S.M.: Robotic system with active steering capability for internal inspection of urban gas pipelines. Mechatronics **16**(12), 713–736 (2002)
5. Li, P., Ma, S., Li, B., Wang, Y.: Screw drive in-pipe robot with axial and circum-directional inspection ability. J. Mech. Eng. **46**(21), 19–28 (2010)
6. Li, P., Ma, S., Li, B., Wang, Y.: Design and motion analysis of an in-pipe robot with adaptability to pipe diameters. J. Mech. Eng. **45**(1), 154–161 (2009)
7. Deng, Z., Chen, J., Jiang, S., Sun, D.: Traction robot driven by six independent wheels for inspection inside pipeline. J. Mech. Eng. **41**(9), 68–72 (2005)
8. Tang, D., Li, Q., Jiang, S., Deng, Z., Liu, H.: Design and analysis of a pipeline robot with the function of differential movement. J. Mech. Eng. **47**(13), 1–8 (2011)
9. Zhang, X., Deng, Z., Jia, Y., Jiang, S., Jiang, X.: Design and research of a triaxial differential drive unit for in-pipe robot. Robot **30**(1), 22–28 (2008)
10. Zhang, Y., Yan, G., Ding, G., Xu, X.: Pipe-diameter adapting and adjusting mechanism of gas pipeline inspection robot. J. Shanghai Jiaotong Univ. **39**(6), 822 (2005)
11. Zhang, Y.: Research on Gas Pipeline Inspection Robot and Its Motion Control. School of Electronic Information and Electrical Engineering, Shanghai Jiaotong University, Shanghai (2007)
12. Zhang, Y., Yan, G.: In-pipe inspection robot with active pipe-diameter adaptability and automatic tractive force adjusting. Mech. Mach. Theory **42**(12), 1618–1631 (2007)
13. Yu, J., Lu, Y.: Kinematics and statics analysis of an novel pipeline robot. Mech. Sci. Technol. Aerosp. Eng. **36**(10), 1530–1535 (2017)
14. Zhang, L., Wang, X.: Stable motion analysis and verification of a radial adjustable pipeline robot. In: Proceeding of the 2016 IEEE International Conference on Robotics and Biomimetics, Qingdao, China, 3–7 December 2016
15. Zhang, L., Meng, S.: Analysis of traveling-capability and obstacle-climbing capability for radially adjustable tracked pipeline robot. In: Proceeding of the 2016 IEEE International Conference on Robotics and Biomimetics, Qingdao, China, 3–7 Dec 2016
16. Fei, Z., Zhang, J., Zhang, L., Wang, X., Lu, Q.: Research on design and simulation of tracked pipeline robot. J. Qingdao Univ. (E&T) **31**(2), 37–42 (2016)

GPSR Compliance

The European Union's (EU) General Product Safety Regulation (GPSR) is a set of rules that requires consumer products to be safe and our obligations to ensure this.

If you have any concerns about our products, you can contact us on ProductSafety@springernature.com

In case Publisher is established outside the EU, the EU authorized representative is:

Springer Nature Customer Service Center GmbH
Europaplatz 3
69115 Heidelberg, Germany

Batch number: 10761593

Printed by Printforce, the Netherlands

Advances in Intelligent Systems and Computing

Volume 856

The series "Advances in Intelligent Systems and Computing" contains publications on theory, applications, and design methods of Intelligent Systems and Intelligent Computing. Virtually all disciplines such as engineering, natural sciences, computer and information science, ICT, economics, business, e-commerce, environment, healthcare, life science are covered. The list of topics spans all the areas of modern intelligent systems and computing such as: computational intelligence, soft computing including neural networks, fuzzy systems, evolutionary computing and the fusion of these paradigms, social intelligence, ambient intelligence, computational neuroscience, artificial life, virtual worlds and society, cognitive science and systems, Perception and Vision, DNA and immune based systems, self-organizing and adaptive systems, e-Learning and teaching, human-centered and human-centric computing, recommender systems, intelligent control, robotics and mechatronics including human-machine teaming, knowledge-based paradigms, learning paradigms, machine ethics, intelligent data analysis, knowledge management, intelligent agents, intelligent decision making and support, intelligent network security, trust management, interactive entertainment, Web intelligence and multimedia.

The publications within "Advances in Intelligent Systems and Computing" are primarily proceedings of important conferences, symposia and congresses. They cover significant recent developments in the field, both of a foundational and applicable character. An important characteristic feature of the series is the short publication time and world-wide distribution. This permits a rapid and broad dissemination of research results.

More information about this series at http://www.springer.com/series/11156

Kevin Deng · Zhengtao Yu
Srikanta Patnaik · John Wang
Editors

Recent Developments in Mechatronics and Intelligent Robotics

Proceedings of International Conference on Mechatronics and Intelligent Robotics (ICMIR2018)

Set 2

Springer

Editors
Kevin Deng
School of Transportation Science
and Engineering
Beihang University
Beijing, China

Srikanta Patnaik
Department of Computer Science
and Engineering
Soa University
Bhubaneswar, Odisha, India

Zhengtao Yu
School of Information Engineering
and Automation
Kunming University of Science
and Technology
Kunming, China

John Wang
Department of Information Management
and Business Analytics
Montclair State University
Montclair, NJ, USA

ISSN 2194-5357 ISSN 2194-5365 (electronic)
Advances in Intelligent Systems and Computing
ISBN 978-3-030-00213-8 ISBN 978-3-030-00214-5 (eBook)
https://doi.org/10.1007/978-3-030-00214-5

Library of Congress Control Number: 2018953052

This Springer imprint is published by the registered company Springer Nature Switzerland AG
The registered company address is: Gewerbestrasse 11, 6330 Cham, Switzerland

Preface

I, Prof. Srikanta Patnaik, Programme Chair of the Conference, welcome all the participants to the International Conference on Mechatronics and Intelligent Robotics (ICMIR2018) held at Kunming, China, during 19–20 May 2018. This annual conference is being organized each year by Interscience Research Network, an international professional body, in association with International Journal of Computational Vision and Robotics, and International Journal of Information and Communication Technology, published by Inderscience Publishing House. I must welcome this year's General Chairs, Prof. Dr. Zhengtao Yu, Dean, School of Information Engineering and Automation, Kunming University of Science and Technology, China, and Prof. Feng Qiao, Shenyang Jianzhu University, Shenyang, China, for their generous contribution to make ICMIR2018 a success.

Like every edition, ICMIR2018 was academically very rich, and we had two eminent professors as keynote speakers, namely Prof. John Wang, Department of Information Management and Business Analytics, School of Business, Montclair State University, USA, and Dr. Nilanjan Dey, Department of Information Technology, Techno India College of Technology, Kolkata, India.

There has been a paradigm shift in this domain after the introduction of IoT and Web-enabled systems. This area covers a wide range of applications by now, such as IoT-enabled manufacturing, Industry 4.0, robot-assisted manufacturing; advanced mechanisms and robotics; systems modelling and analysis; instrumentation and device control; automation systems; intelligent sensing and control; medical robotics; and autonomous and complex systems. New technologies are constantly emerging, which are enabling applications in various domains and services.

Mechatronics and Intelligent Robotics is no longer a functional area within the department of mechanical or electronics, but is an integral part of the manufacturing function of any organization. In recent times, Mechatronics and Intelligent Robotics is probably the single most important facilitator of the manufacturing process. The strong influence of Mechatronics and Intelligent Robotics is now the deciding factor in shaping the process of globalization, particularly in the productive, manufacturing and commercial spheres, and is also widely recognized. Creating economic opportunities and contributing to monotony reduction is another thrust area for the emerging epoch of Mechatronics and Intelligent Robotics.

This edition of ICMIR covers the following areas but is not limited to: Intelligent mechatronics, robotics and biomimetics; Novel and unconventional mechatronics systems; Modelling and control of mechatronics systems; Elements, structures, mechanisms of micro- and nano-systems; Sensors, wireless sensor networks and multi-sensor data fusion; Biomedical and rehabilitation engineering, prosthetics and artificial organs, AI, neural networks and fuzzy logic in mechatronics and robotics; Industrial automation, process control and networked control systems; Telerobotics, human–computer interaction and human–robot interaction; Artificial intelligence; Bio-inspired robotics; Control algorithms and control systems; Design theories and principles; Evolutional robotics; Field robotics; Force sensors, accelerometers and other measuring devices; Healthcare robotics; Human–robot interaction; Kinematics and dynamics analysis; Manufacturing robotics; Mathematical and computational methodologies in robotics; Medical robotics; Parallel robots and manipulators; Robotic cognition and emotion; Robotic perception and decision; Sensor integration, fusion and perception; Social robotics. This volume covers various articles covering the recent developments in the area of Mechatronics and Intelligent Robotics categorized into six tracks, such as follows:

1. Intelligent Systems
2. Robotics
3. Intelligent Sensors and Actuators
4. Mechatronics
5. Computational Vision
6. Machine Learning and Soft Computing

Srikanta Patnaik
Programme Chair: ICMIR2018

Acknowledgement

Like every year, this edition of ICMIR2018 was also attended by more than 130 participants, and 155 papers were shortlisted and published in this proceedings. The papers covered in this proceeding are the result of the efforts of the researchers working in broad domains of Mechatronics and Intelligent Robotics. We are thankful to the authors and paper contributors of this volume.

We are thankful to the editor in chief and the anonymous review committee members of the Springer series on *Advances in Intelligent Systems and Computing*, for their support to bring out the proceedings of ICMIR2018.

We are thankful to our friends, namely Prof. John Wang, School of Business, Montclair State University, USA, and Dr. Nilanjan Dey, Techno India College of Technology, Kolkata, India, for their keynote addresses. We are also thankful to the experts and reviewers who have worked for this volume despite the veil of their anonymity.

We should not forget to inform you that the next edition of the conference, i.e. 3rd International Conference on Mechatronics and Intelligent Robotics (ICMIR-2019), shall be held at Kunming, China, in association with Kunming University of Science and Technology, Kunming, China, during the last week of April 2019.

It was really a nice experience to interact with the scholars and researchers who came from various parts of China and outside China to participate in ICMIR2018. In addition to the academic part, the participants must have enjoyed their stay, participation and presentation during the conference and sightseeing trip at Kunming.

I am sure that the readers will be benefited from this volume of AISC series volume on "Recent Developments in Mechatronics and Intelligent Robotics: Proceedings of International Conference on Mechatronics and Intelligent Robotics (ICMIR2018)".

Contents

Robotics

Intelligent Sensor and Actuator

Mechatronics

Machine Learning Soft Computing

Intelligent Sensor and Actuator

Temperature Control of Medium Frequency Induction Heating Furnace Based on Neural Network

Wang Yang(✉) and Wang Qi

School of Information Engineering, Nanchang Hangkong University, Nanchang 330063, China
973491538@qq.com

Abstract. In the field of smelting, compared with the traditional heating furnace with high pollution and low energy efficiency, the intermediate frequency induction furnace has the advantages of environmental protection, easy operation and high energy utilization, which is playing an increasingly important role over time. Taking the intermediate frequency induction furnace controlled by solid relay SSR-220D03 as the object of study, we design the controller based on BP neural network which is derived from to the Voltage- Temperature transfer function. The simulation results manifest that the system has the merits of fast temperature control, quick response, high accuracy and strong robustness.

Keywords: Induction heating furnace · Neural network · Temperature control Micro-controller

1 Introduction

Intermediate frequency (IF) furnace is widely used in metal quenching, melting and other heat treatment aspects [1, 2]. Its working principle is using electromagnetic induction phenomenon, transferring the power with frequency AC into intermediate frequency (300–1000 Hz) and then heating the metal. Compared with other foundry equipment, the IF induction furnace has the advantages of high thermal efficiency, short melting time, less burning loss of alloying elements, wide melting material, less environmental pollution and precise control of the temperature. Changchun Institute of Optics and Fine Mechanics, Chinese Academy of Sciences and Xiangtan Power Plant developed a series of widely used IF furnace-related products. However, in the design and manufacture of dual melting process for induction heating power supply of heat treatment system, China is still far behind the developed countries such as the United States and Germany [3]. Development of more energy-efficient, more reliable, more stable IF induction furnace is the core task in the field of metal smelting.

Furnace temperature control is an important part of IF furnace heating. Feeding speed, the temperature of the environment, the specific heat of the heating material, voltage and other factors all have impacts on the furnace temperature [4]. IF furnace has the characteristics of large lag, large inertia and time-variant transfer function parameters, which are complex and nonlinear. Although some IF models have been

K. Deng et al. (Eds.): ICMIR 2018, AISC 856, pp. 695–701, 2019.
https://doi.org/10.1007/978-3-030-00214-5_87

proposed, due to the nonlinear and time-varying nature of these models, the traditional PID control can not meet the exact requirements of furnace temperature control. With strong self-organization and adaptability, neural networks are suitable for dealing with systems that are difficult to describe by rules or processes, especially for nonlinear and time-varying systems [3]. Therefore, the neural network is suitable for the furnace temperature control of the IF furnace, and can guarantee the robustness and stability of the control process. In this paper, we take STC12 series microcontroller as the MCU, and use SSR-220D03 single-phase AC solid state relay on the intermediate frequency furnace to realize temperature control.

2 System Hardware Design

The overall system block diagram is shown in Fig. 1. The control core STC12C5A60S2 micro-controller and its peripheral circuit form an IF furnace temperature measurement and control system.

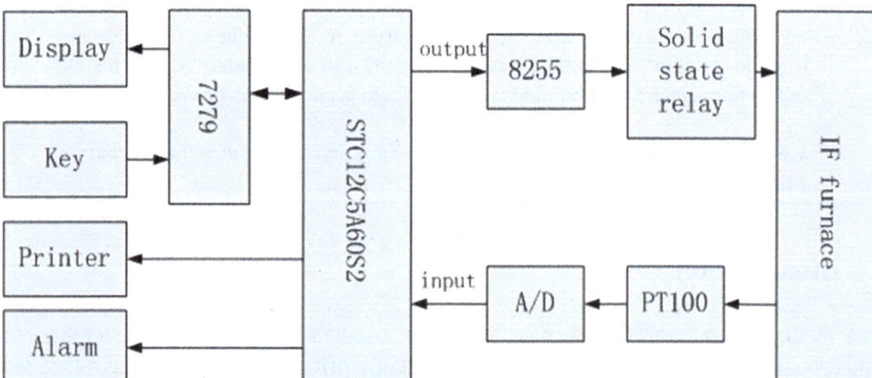

Fig. 1. Temperature controller block diagram

Human-computer interaction system includes micro-controller, display, keyboard, printer and alarm circuits. Its function is to detect the working condition of the temperature controller, giving feedback to the operator and offering the operator a input/testing platform. The display and keyboard part use the programmable keyboard/display interface device 8279, so that it can save I/O ports and improve the MCU efficiency. Display section consists of 16 8-segment common cathode LED digital tube. In order to reduce energy consumption and display more digits, we take the use of dynamic scanning. Keyboard is the 4 × 4 matrix keyboard, with the functions of 0–9 numbers, check, enter, start and so on. It takes the keyboard to enter temperature control curves, such as oven temperature size and temperature holding time. It can also enter different commands through the keyboard to display the A/D, D/A chip output value, as well as the relay switch state, for debugging and testing.

In addition the keyboard can also control the printer to print the temperature control curve. If abnormal error occurs, the buzzer will automatically send an alarm signal.

3 Temperature Automatic Control

3.1 Mathematical Model of Temperature Control System

IF furnace temperature control system block diagram is shown in Fig. 2. Literature [4] derived all parts of IF furnace transfer function, including the voltage to output power, output power to coil magnetic induction, induction heating session and other aspects of the transfer function. The final open-loop Voltage-Temperature transfer function is:

$$G(s) = \frac{Ke^{-\tau s}}{T_m s + 1} \tag{1}$$

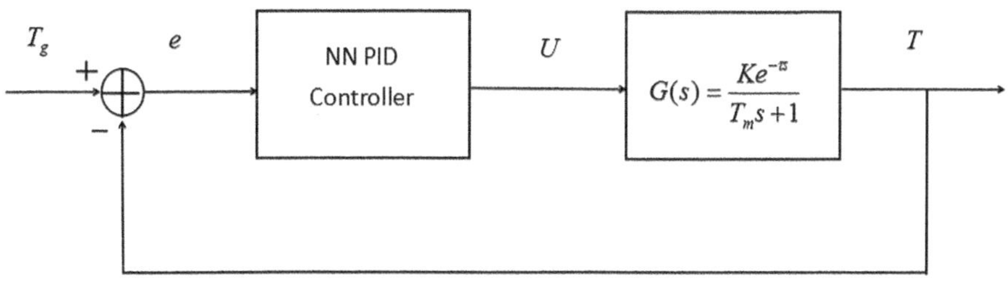

Fig. 2. Temperature control system block diagram

The $\frac{K}{T_m s + 1}$ is the first-order inertia part, the pure hysteresis part is $e^{-\tau s}$.

Formula (1) shows that the transfer function is not easy to design a conventional PID controller (usually need to add the Smith estimated compensation). Moreover, IF furnace is a time-varying system, the parameters K, T_m, τ will change over time, so here we use the nonlinear Neural Network (NN) PID controller.

3.2 BP Neural Network PID Control

The classic incremental PID control algorithm is as follows:

$$\begin{aligned} u(k) =& u(k-1) + K_p(e(k) - e(k-1)) + K_I e(k) \\ &+ K_D(e(k) - 2e(k-1) + e(k-2)) \end{aligned} \tag{2}$$

The error $e(k)$ is the difference between the actual temperature and the ideal temperature at the time k, that is

$$e(k) = T(k) - T_g(k) \tag{3}$$

K_P, K_I, K_D are the proportion, integral and differential coefficient respectively.

The classical PID control parameters K_P, K_I, K_D are fixed, so it can not adapt to the nonlinear time-varying IF furnace transfer function. Therefore, this paper uses BP neural network model. Via the neural network algorithm, the PID parameters adaptively change with the input value. So that the actual temperature will quickly get close to the ideal temperature and reaches a steady state.

As is shown in Fig. 3, the BP neural network input is

$$x_i = e(k - i + 1), i = 1, 2, 3 \tag{4}$$

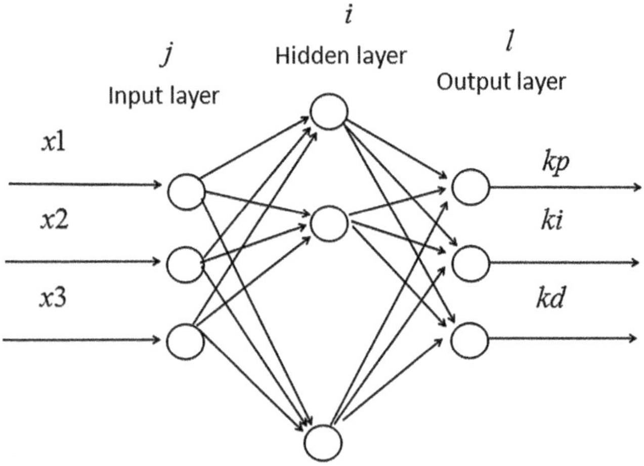

Fig. 3. BP neural network model

The final output are the PID ratio, integral, differential coefficient. Neural networks include the input layer, the hidden layer and the output layer. Calculation process are shown in Fig. 3 according to the arrow from left to right, layer by layer. Literature [1] introduced the details of neural network computing process.

4 Experimental Results

This article designed two experiments. The first experiment is used to test the system startup characteristics and steady-state characteristics, the second experiment is used to test the system robustness and anti-interference ability. Experiments use the software Matlab 7.10.0 (R2010a) to do simulation. In order to simulate the real IF furnace work scenarios, the system transfer function parameter $T = 1.2 * (1 - e^{-0.1k})$, slowly changes with time (slow time-varying parameters). Hysteresis parameter $\tau = 1$. This article take classic PID control as a comparative experiment, the coefficients K_P, K_I, K_D were 0.3,0.5,0.1 respectively.

As shown in Fig. 4, the ideal temperature set in Experiment 1 is constant with size 1, and the total temperature control time is 500 s. The first 100 s is the rapid temperature rise of the heating process. In this process, the temperature rise time of BP neural network is about 70 s, while the heating time of classical PID algorithm is about 100 s, so with the BP neural network method, the temperature can be heated to the ideal value more quickly when the system starts. So it has a better start characteristic. The temperature control curve of BP neural network is basically consistent with the ideal temperature after 100 s and the overshoots of classical PID and BP neural network method are almost 0. So both of the two methods have good steady-state characteristics.

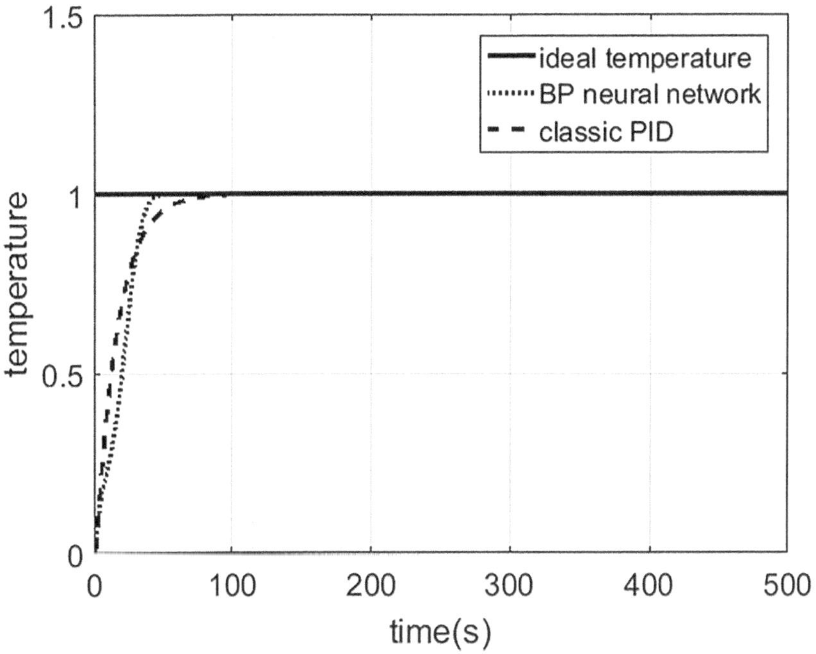

Fig. 4. Temperature control curve with constant ideal temperature

The ideal temperature set in Experiment 2 was 2.5 initially, changed to 1.5 after 250 s and then stabilized at 1.5. The experimental results are shown in Fig. 5. It can be seen from the figure that the BP neural network method has a convergence time of about 280 s − 250 s = 30 s and a conventional PID algorithm of about 330 s − 250 s = 80 s when temperature suddenly change. Compared to the conventional PID control algorithm, the BP neural network method makes the furnace temperature to converge to the ideal temperature more quickly. Therefore, the BP neural network method has strong robustness and anti-interference ability against mutation and interference (Fig. 6).

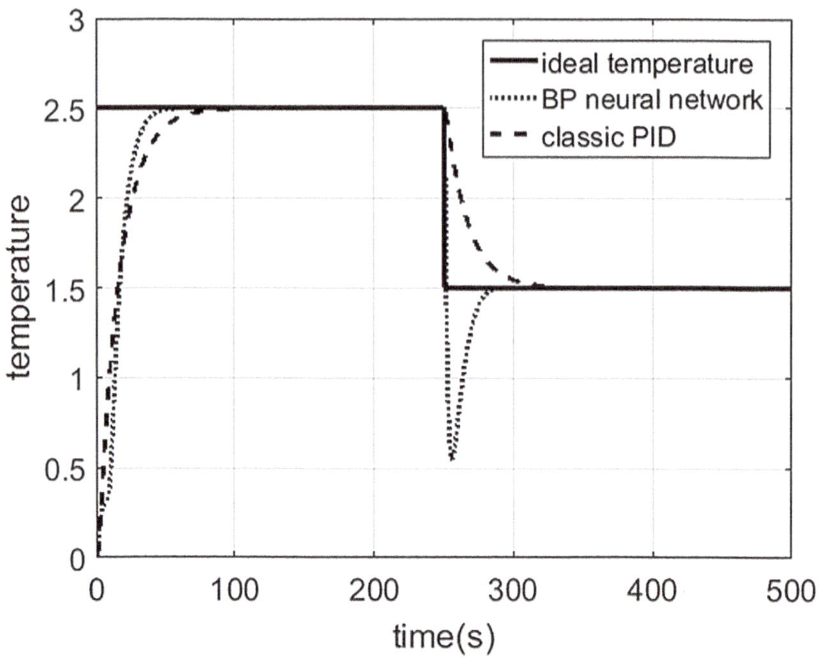

Fig. 5. Temperature control curve when ideal temperature suddenly change

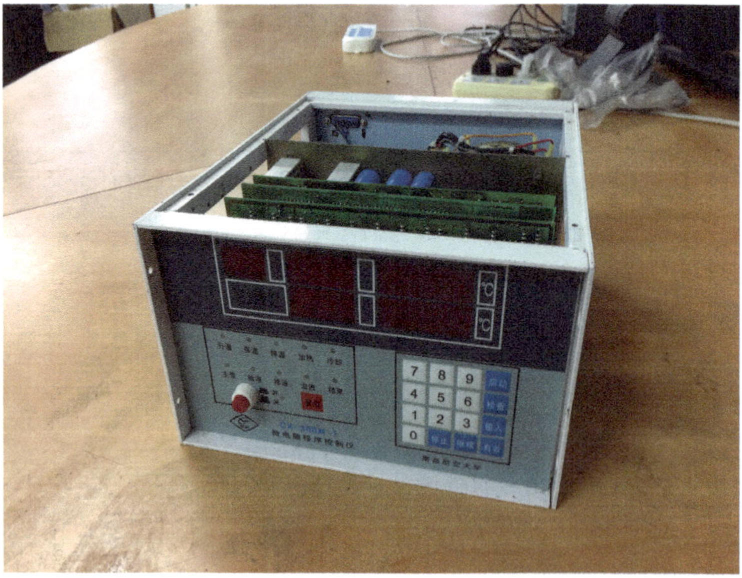

Fig. 6. Temperature controller product

In the end, we have produced a temperature controller product according to the method mentioned above, which is sold on the market with some reputation because of its fast and stable temperature control.

5 Conclusion

The core chip of this system is STC12 series microcontroller, which improves the operation speed when it compared with the traditional 8051 microcontroller. PT100 bridge circuit temperature measurement achieves accurate and reliable temperature measurement. In addition, the design of the DC power supply for the stable temperature controller guarantee temperature controller to a long stable working time. For temperature control, this paper use the BP neural network PID control algorithm. This algorithm can adaptively calculate the proportional, differential and integral coefficients of PID control. Taking BP neural network algorithm as a control method, temperature control has the advantages of accuracy, fast response and with strong robustness and anti-interference ability.

References

1. Yu, Y., Yin, D.: Application of the BP neural network PID algorithm in heat transfer station control. In: Xie, A., Huang, X. (eds.) Advances in Computer Science and Education, pp. 223–228. Springer, Berlin Heidelberg (2012)
2. Kang, J.J., Cui, Y.J., Wang, Z.Q., et al.: Medium frequency induction furnace temperature control based on BP neural network PID. In: IEEE International Conference on Information and Automation, pp. 1110–1115. IEEE (2017)
3. Zhang, X.: Application research of BP neural network algorithm on induction heating based on mid-frequency power supply source. Xiangtan University (2012)
4. Liu, Z.: Research on temperature control technology for medium frequency heating furnace. Xi'an Shiyou University (2013)
5. Zheng, Z., Wan, G., Liu, X.: An intelligent monitoring system of medium-frequency induction furnace based on fuzzy control. In: International Conference on Intelligent System Design and Engineering Application, pp. 261–264. IEEE (2011)

Transient Voltage Suppressor's Electromagnetic Pulse Effect Prediction Based on Elman Neural Network

Jinjin Wang(✉), Gang Wu, Xin Nie, and Wei Wang

State Key Laboratory of Intense Pulsed Radiation Simulation and Effect, Northwest Institute of Nuclear Technology, Xi'an 710024, China
wangjinjin@nint.ac.cn

Abstract. Elman neural network is a recurrent neural network with local memory units and local feedback connections. In order to prevent electromagnetic pulse from entering the electronic equipment, we usually add transient voltage suppressors as surge protection devices. However, we can only know its response to electromagnetic pulse through experiments. In order to reduce experiment times and guide experiment process, we proposed using neural network to predict electromagnetic pulse effect of transient voltage suppressor for the first time. Through the analysis and experiments, we use Elman neural network to build the model and predict the effect. Experiments show that our Elman neural network model on transient voltage suppressor which is an effective interference suppression device has better prediction results.

Keywords: Elman neural network · Transient voltage suppressor
Electromagnetic pulse · Predict

1 Introduction

TVS (transient voltage suppressor) is an effective protective device which is widely used in circuit protection [1–3]. It prevents precision components from being destroyed by electromagnetic pulse [4–6]. The main method of detecting its protective performance is conduct experiments. We try to use neural network algorithm [7–9] to predict TVS's electromagnetic pulse effect. This method can provide guidance for the experiment, reducing the number of experiments and save the cost. Now there are few people have researched it before.

Elman feedback neural network [10–12] is a kind of neural network which has a stronger computing power. It has a dynamic memory function and is very suitable for the time series prediction problem. As some of our experimental waveforms have a thousand magnitude sampling points, so we need stronger computing power. And the prediction is also time related. Compared with BP (Back propagation) neural network [13, 14], Elman network has a connection layer which is used to form local feedback. The connection layer can remember the state of the past.

In our research, we have first proposed our Elman neural network and used it in TVS's electromagnetic pulse effect prediction. It has three steps. First, we carried out

K. Deng et al. (Eds.): ICMIR 2018, AISC 856, pp. 702–711, 2019.
https://doi.org/10.1007/978-3-030-00214-5_88

TVS's electromagnetic pulse experiments [15–17] obtained the waveforms data. Second, we preprocessed the waveforms and extracted parameters of the experiments. After processing, we build our Elman neural network model based on the training waveforms. Third, we predict the unknown testing waveforms with our neural network. Experiments show that our Elman neural network model on TVS in different applications has better prediction results. We can use the model to guide our electromagnetic experiments. The rest of the paper is organized as follows. Section 2 gives the introduction of our build Elman neural network in detail. In Sect. 3, we demonstrate our method using two kinds of TVS in different applications. Section 4 gives the conclusion.

2 Elman Neural Network Modeling

2.1 Elman Neural Network

Elman algorithm is a feedback neural network which was proposed in 1990 by Elman [10]. Elman network has a special structural unit except input layer, hidden layer and output layer which is connection layer. The connection layer keeps an account of the previous output value of hidden layer. The hidden layer not only receives the data from the input layer, but also receives the data from the connection layer. The input layer reflects the spatial information of the signal, while the connection layer reflects the temporal information of the signal. Elman feedback neural network has memory function as a result of connection layer's feedback. It widely used in sequence analysis, system identification, control and other fields.

$$y(k) = g(W^3 x(k)) \tag{1}$$

$$x(k) = f(W^2 x_c(k) + W^1 u(k-1)) \tag{2}$$

$$x_c(k) = x(k-1) \tag{3}$$

u is the input vector of r dimension, y is the output vector of m dimension, x is n dimension vector for intermediate layer, x_c is n dimension feedback vector, W^1 is the connection weight from input layer to hidden layer, W^2 is the connection weight from connection layer to hidden layer, W^3 is the connection weight from hidden layer to output layer. $g(x)$ is the transfer function of output neuron which often uses linear combination function. $f(x)$ is the transfer function of intermediate layer which often uses Sigmoid function. They are showed in Eqs. 1–3.

$$E = \frac{1}{2}\sum_{j=1}^{m} (d(k) - y(k))^2 \tag{4}$$

Elman algorithm use error function to correct weights. The error function is showed in Eq. 4. $y(k)$ is the predicted output of the network, $d(k)$ is the expected output of the network. The process is as follows. Take the partial derivative to W^3 which is showed below.

$$\frac{\partial E}{\partial w_{ij}^3} = -(d_i(k) - y_i(k))\frac{\partial y_i(k)}{\partial w_{ij}^3} = -(d_i(k) - y_i(k))g_i'(\cdot)x_j(k) \tag{5}$$

$$\delta_i^0 = (d_i(k) - y_i(k))g_i'(\cdot) \tag{6}$$

$$\frac{\partial E}{\partial W_{ij}^3} = -\delta_i^0 x_j(k) \quad (i = 1, 2, \ldots, m;\ j = 1, 2, \ldots, n) \tag{7}$$

Similarly, take the partial derivative to W^2

$$\frac{\partial E}{\partial W_{jq}^2} = -\delta_j^h u_q(k-1) \quad (j = 1, 2, \ldots, n;\ q = 1, 2, \ldots, r) \tag{8}$$

$$\delta_j^h = \sum_{i=1}^{m} \left(\delta_i^0 W_{ij}^3\right) f_j'(\cdot) \tag{9}$$

Taking the partial derivative to W^1

$$\frac{\partial E}{\partial W_{jl}^1} = -\sum_{i=1}^{m} \left(\delta_i^0 W_{ij}^3\right) \frac{\partial x_j(k)}{\partial W_{jl}^1} \quad (j = 1, 2, \ldots, n;\ l = 1, 2, \ldots, n) \tag{10}$$

$$\frac{\partial x_j(k)}{\partial W_{jl}^1} = f_j'(\cdot)x_l(k-1) + \alpha \frac{\partial x_j(k-1)}{\partial W_{jl}^1} \quad (j = 1, 2, \ldots, n;\ l = 1, 2, \ldots, n) \tag{11}$$

Define η_1, η_2, η_3 as W^1, W^2, W^3's learning step. Then:

$$\Delta W_{ij}^3 = \eta \delta_i^0 x_j(k) \quad (i = 1, 2, \ldots, m;\ j = 1, 2, \ldots, n) \tag{12}$$

$$\Delta W_{jq}^2 = \eta_2 \delta_j^h u_q(k-1) \quad (j = 1, 2, \ldots, n;\ q = 1, 2, \ldots, r) \tag{13}$$

$$\Delta W_{jl}^1 = \eta_1 \sum_{i=1}^{m} \left(\delta_i^0 W_{ij}^3\right) \frac{\partial x_j(k-1)}{\partial W_{jl}^1} \quad (j = 1, 2, \ldots, n;\ l = 1, 2, \ldots, n) \tag{14}$$

2.2 Waveforms Preprocessing

Our TVS's effect data are waveforms. The max sampling point of the data is 3500. We have carried out the electromagnetic pulse experiments on two kinds of TVS in different applications. We conducted experiments on different voltage levels. The waveform's abscissa is time and ordinate is the current intensity. The unit of time is nanosecond and current is ampere. We used filter algorithm to smooth the waveform and filtered the interference components first. The voltage level is an important experimental condition. We use it in the input layer as a layer node. Each time corresponds to current amplitude. What we care about is the response value of the current under different voltage amplitudes. Time is also an important input item. So the input layer is combined with voltage level and time value. The output layer is the value of current in a waveform.

2.3 Modeling

We divided the data into training samples and testing samples. We trained Elman neural network on the training samples. Then we predict the response value of current on the testing samples. Then we will not conduct the experiment and know the value of the current. It can guide our experiment plan and implementation. Now process of modeling on each TVS experiment is below.

3 The Elman Feedback Neural Network's Modeling Results Analysis

3.1 TVS's Model is 1.5KE100CA

We find Elman feedback neural network is more suitable for TVS's predicting modeling by experiment. Our experimental TVS's model is 1.5KE100CA of some manufacturer. Its nominal VBR is between 95 V and 105 V. We use double exponential wave generator to test TVS's effect of electromagnetic pulse in large dynamic voltage range. Its partial experiment results are shown in Fig. 1.

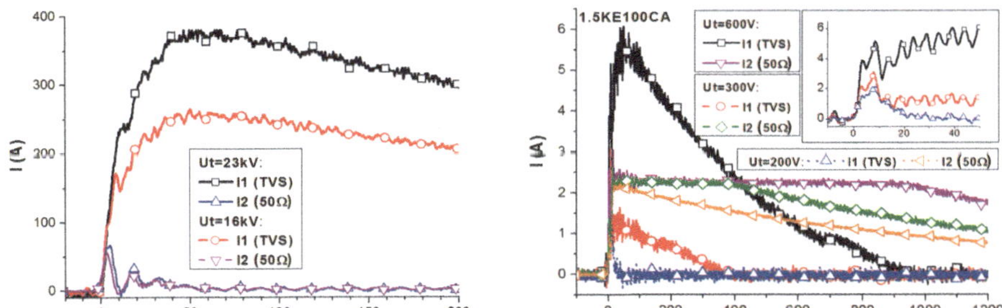

(a) Operation of the pulse generator voltage from 200 to 500V

(b) Operation of the pulse generator voltage from 16 to 23kV

Fig. 1. Nominal VBR of 100 V's TVS's experiment waveforms of double exponential wave generator

We treat the experiment results data at 2, 4, 8, 16, 23 kV as our algorithm's modeling data. The data set is divided into training data and testing data. We use 5-fold cross validation method to modeling and testing. In order to improve the accuracy of modeling, the data were filtered before and after the experiment. Figure 2 is the waveforms of all data.

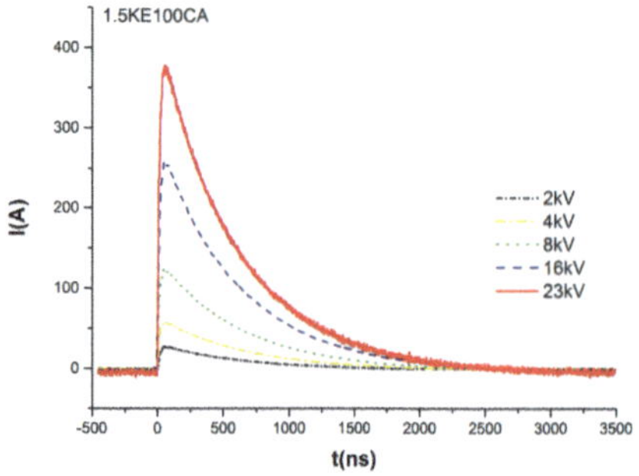

Fig. 2. TVS 1.5KE100CA's effect data of current of double exponential wave generator at different voltage levels.

When we predict the effect data of the TVS at 8 kV voltage, we use the other data to train our Elman neural network model. The result is shown in Fig. 3. It is the comparison of the original and predicted waveforms. Figure 4 is the predicted result of the TVS at 16 kV voltage. Our predicted results of 8 kV and 16 kV show that the predicted waveforms are consistent with the original experimental waveforms. The mean square error is small. Figures 5 and 6 are the results of 4 kV and 23 kV. We can get from that our Elman algorithm's predicted results are basically consistent with the original waveforms. But we also found that the 4 kV's result is derived from the original waveform after 600 ns. For analysis, it caused by the large voltage span. In the experiment, we may experiment on TVS on large voltage span. We'll further improve the prediction accuracy in our research.

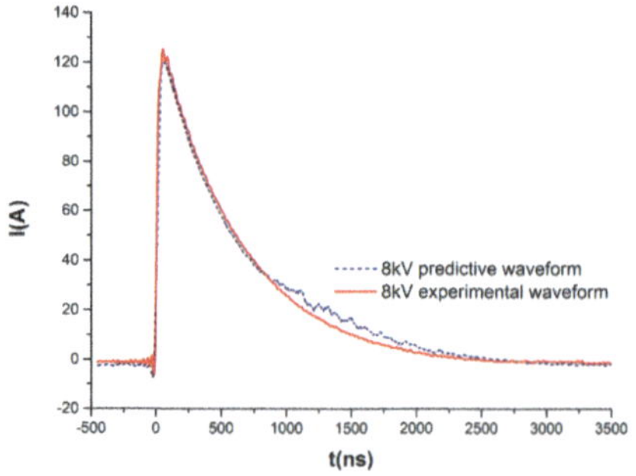

Fig. 3. The comparison of the predicted and original waveforms at 8 kV voltage

The prediction of 23 kV is an attempt. It is beyond the data scope of the training sample. The lower prediction accuracy of this kind of problem is the disadvantage of neural network, including Elman network. But the wave front and half width is roughly same, it also can guide the next experiment.

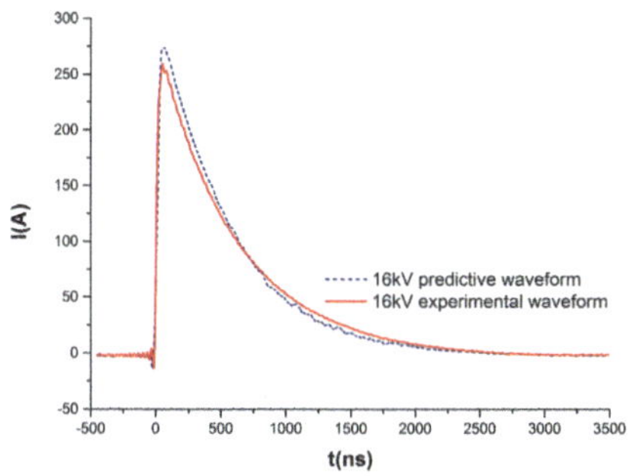

Fig. 4. The comparison of the predicted and original waveforms at 16 kV voltage

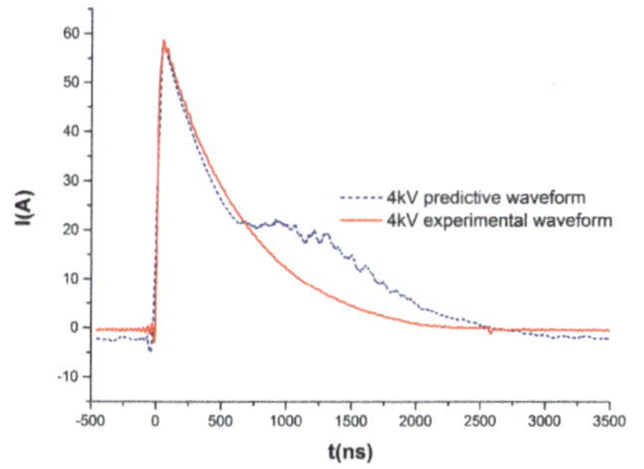

Fig. 5. The comparison of the predicted and original waveforms at 4 kV voltage

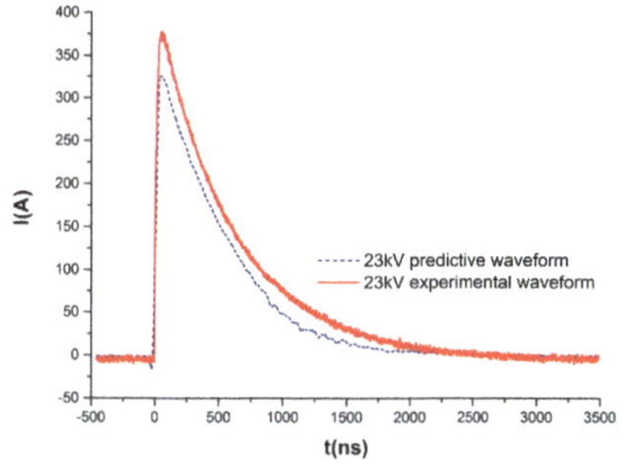

Fig. 6. The comparison of the predicted and original waveforms at 23 kV voltage

3.2 Two Applications of TVS's Model is P6K12CA

(1) Paralleling 50 Ω resistor
In the experiment of electromagnetic pulse current injection, it paralleled 50 Ω resistor with TVS at back-end circuit. Figure 7 is the waveforms of all back-end current data we get from the pulse current injection experiments. Data preprocessing is similar to the previous TVS. We can see that it sustained decay oscillation process in 70 ns at 1.5 kV voltage. Then its current maintained at 0.22A after oscillation. Figures 8 and 9 are the current predicting results of 12 kV and 6 kV voltage in pulse current injection. We can see that predicted waves are similar to the original experimental waves.

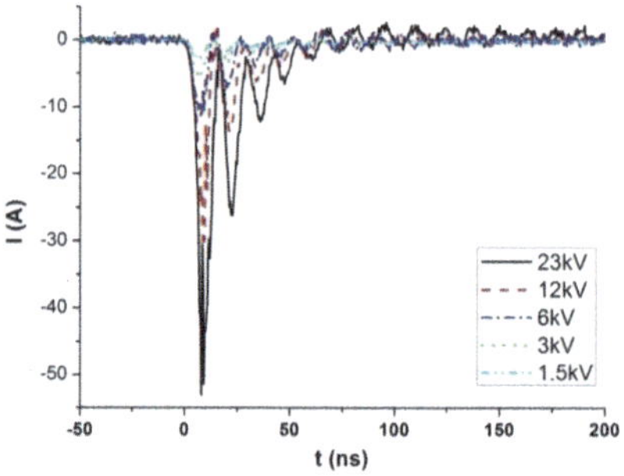

Fig. 7. TVS's response current in pulse current injection experiments paralleling 50 Ω resistor

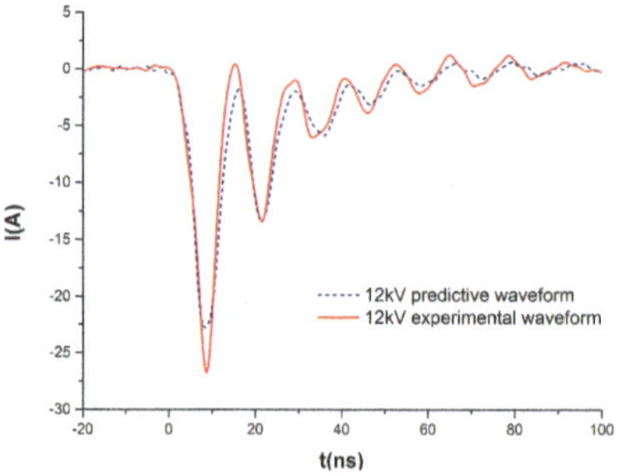

Fig. 8. The comparison of the predicted and original waveforms at 12 kV voltage

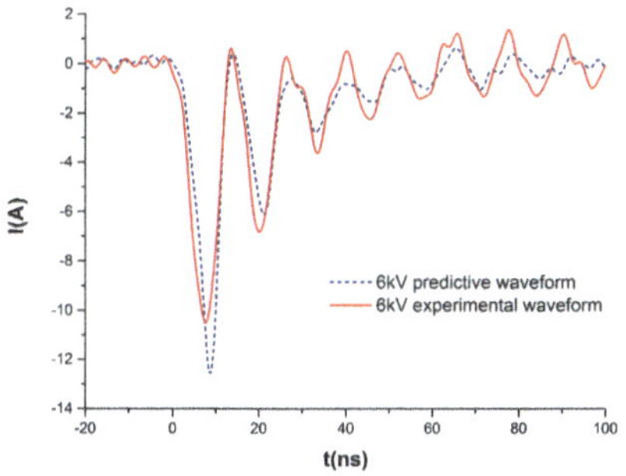

Fig. 9. The comparison of the predicted and original waveforms at 6 kV voltage

(2) Paralleling 200 Ω resistor
In the experiment of electromagnetic pulse current injection, it paralleled 200 Ω resistor with TVS at back-end circuit. Figure 10 is the waveforms of all back-end current data we get from the pulse current injection experiments. We can see that it sustained decay oscillation process larger than 100 ns. Then its current maintained at 60 mA after oscillation. Our predicted results of 12 kV and 6 kV show that the predicted waveforms are consistent with the original experimental waveforms in Figs. 11 and 12.

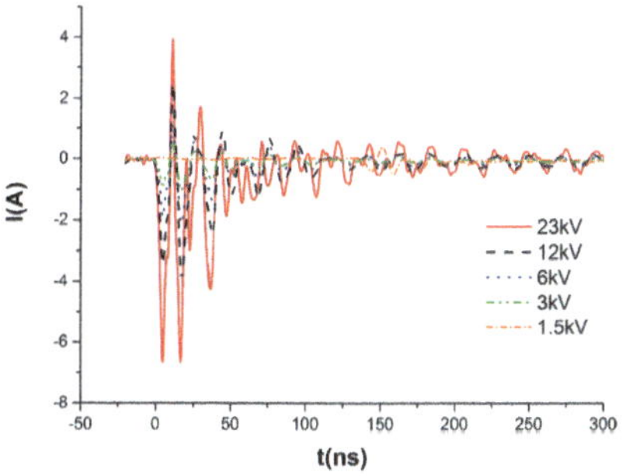

Fig. 10. TVS's response current in pulse current injectionexperiments paralleling 200 Ω resistor

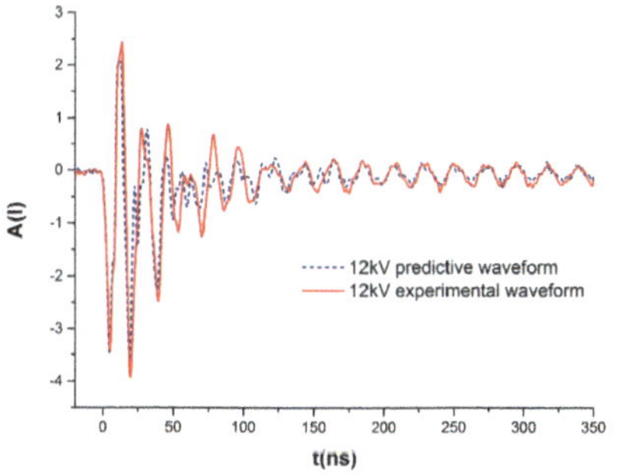

Fig. 11. The comparison of the predicted and original waveforms at 12 kV voltage

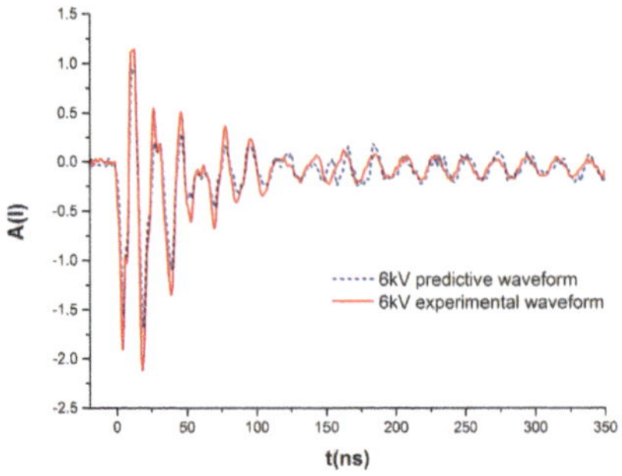

Fig. 12. The comparison of the predicted and original waveforms at 6 kV voltage

4 Summary

In our paper, we have first proposed our Elman neural network and using in TVS's electromagnetic pulse effect prediction. We preprocessed the waveforms of the experiments and divided them into training data and testing data. We trained our Elman neural network model using the training data. Then we had set up our Elman neural network model. Third, we used our model to predict electromagnetic pulse effect in different voltage levels. Experiments showed that our Elman neural network model on TVS in different applications has better prediction results. Our Elman neural network model on TVS has better prediction results.

References

1. Laurent, S., Tropeano, A.I., Boutouyrie, P.: Pulse pressure reduction and cardiovascular protection. J. Hypertens. Suppl. **24**(3), S13–S18 (2006)
2. Chundru, R., Li, Z., Pommerenke, D.: An evaluation of TVS devices for ESD protection. In: IEEE International Symposium on Electromagnetic Compatibility, vol. 1353, no. 1, pp. 62–67 (2011)
3. Pharkphoumy, S., Khurebastar, Z., Bouangeun, D.: ESD protection properties of TVS. **62** (1), 11–28 (2015)
4. Baum, C.E.: Proc. IEEE **80**(6), 789–817 (1992)
5. Lanoz, M., Nicoara, B.I.C., Radasky, W.A.: IEEE Trans. Electromagn. Compat. **38**(3), 400–413 (2002)
6. Lanoz, M., Nicoara, B.I.C., Radasky, W.A.: Modeling of an EMP conducted environment. IEEE Trans. Electromagn. Compat. **38**(3), 400–413 (2002)
7. Bishop, C.M.: Neural Networks for Pattern Recognition. Oxford University Press, Oxford (1995)
8. Zhang, X.Q., Wang, H.B., Hong-Zhen, Y.U.: Neural network based algorithm and simulation of information fusion in the coal mine. Int. J. Min. Sci. Technol. **17**(4), 595–598 (2007)

9. Graves, A., Wayne, G., Reynolds, M.: Hybrid computing using a neural network with dynamic external memory. Nature **538**(7626), 471–476 (2016)
10. Elman, J.L.: Finding structure intime. Cogn. Sci. **14**, 179–211 (1990)
11. Song, C., Zhao, Y.: Suspension test system based on modified Elman neural network. China Mech. Eng. **27**(1), 1–6 (2016)
12. Zhang, B., Zhou, J.: Video-based face recognition by auto-associative Elman neural network. In: International Congress on Image and Signal Processing, pp. 89–93 (2015)
13. Xiao, Z., Ye, S.J., Zhong, B., Sun, C.X.: BP neural network with rough set for short term load forecasting. Expert Syst. Appl. Int. J. **36**(1), 273–279 (2009)
14. Yi, J., Wang, Q., Zhao, D.: BP neural network prediction-based variable-period sampling approach for networked control systems. Appl. Math. Comput. **185**(2), 976–988 (2007)
15. Wu, G.: Shape properties of pulses described by double exponential function and its modified forms. IEEE Trans. Electromagn. Compat. **56**(4), 923–931 (2014)
16. Lu, X., Wei, G., Pan, X.: A pulsed differential-mode current injection method for electromagnetic pulse field susceptibility assessment of antenna systems. IEEE Trans. Electromagn. Compat. **57**(6), 1435–1446 (2015)
17. Lin, J.Y., Dong, W.J., Zhang, C.X.: Investigation of injury effects to strong electromagnetic pulse on typical flying control system. **7363**(1), 1532–1535 (2012)

Design and Simulation of PLC Control Timer

Jing Zhao[1(✉)], Lan-Qing Wang[2], Yun-da Xu[1], and Ye Cao[1]

[1] School of Physics and Electronic Technology, Liaoning Normal University, Dalian 116029, China
zjmilk@126.com
[2] School of Mathematics and Computer Science, Shanxi Datong University, Datong 037009, China

Abstract. In view of the need of timing of small-car competition, a timer is designed and produced, which is accurate and reliable operation. Using the photoelectric switch as the sensor and using the programmable controller PLC to complete the design and manufacture skillfully. This timer is used for timing competition and does work very well.

Keywords: Timer · Photoelectric switch · Programmable logic controller

1 Introduction

Timers are widely used, such as game timing, speech timing, performance timing, competition timing, and so on. The timer designed in this paper is used for competition timing. The main purpose is to make the competition results fair, just and open. Taking the racing small-car as an example, a precise and reliable timer is designed and made.

The small-car includes a timer, photoelectric sensor, PLC and digital display circuit, as shown in Fig. 1, the photoelectric sensor is usually disconnected and arranged on the circuit timing, when the small-car through the timing starting point, photoelectric sensor to the small-car through the signal, the signal after PLC start timing and display time in seconds. When the small-car passes the end point of the timing, the photoelectric sensor collects the end signal of the small-car. The signal is processed by PLC and ends the time and outputs the timing result.

The output signal of photoelectric sensor is connected to PLC, which is the start time signal and the stop time signal, plus a reset button to re prepare the timing.

2 The Specific Requirements of the Timer

The X1 is connected with the photoelectric switch. It is used to detect whether or not there is a moving object. When the photoelectric switch is blocked for the first time, that is X1 is turned on, PLC starts to time. When the photoelectric switch is occluded second times, PLC will stop immediately and keep and display the time at that time.

X2 is a reset switch to ensure that PLC starts from 0 at the next time of time.

X0 is a master switch. The timing function can be realized only when the X0 is on.

K. Deng et al. (Eds.): ICMIR 2018, AISC 856, pp. 712–717, 2019.
https://doi.org/10.1007/978-3-030-00214-5_89

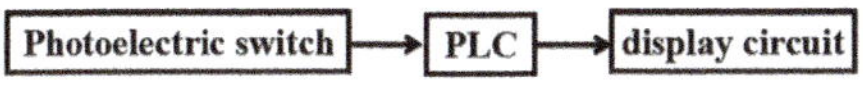

Fig. 1. PLC control timer structure diagram

The program has been used for small-car racing timing. The transmitter and receiver of the photoelectric switch are placed both sides of the small-car runway. The runway is an ellipse. The timer is accurate and can be modified online at any time.

The A point is both the starting point and the ending point. The B point is the passing point. The small-car is driving around the designated line, the ellipse line in the picture. When the A point starts, the timer switches, and the photoelectric switch turns on the signal to PLC and PLC starts timing.

This control is used for the small-car to run around a designated track in a circle, and it is also suitable for the two points competition time (Fig. 2).

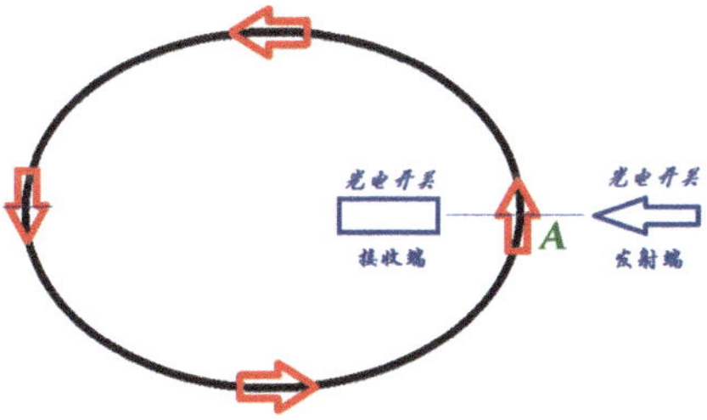

Fig. 2. PLC control timer diagrammatic sketch

3 PLC-I/O Table

PLC-I/O distribution is shown in Table 1, X0–X2 is input and Y0–Y6, Y10–Y16 is output.

Table 1. PLC-I/O distribution

INPUT		OUTPUT	
X0	The total switch	Y0	High 7-segment of a
X1	The photoelectric switch	Y1	High 7-segment of b
X2	The reset switch	Y2	High 7-segment of c
		Y3	High 7-segment of d
		Y4	High 7-segment of e
		Y5	High 7-segment of f
		Y6	High 7-segment of g
		Y10	Low 7-segment of a
		Y11	Low 7-segment of b
		Y12	Low 7-segment of c
		Y13	Low 7-segment of d
		Y14	Low 7-segment of e
		Y15	Low 7-segment of f
		Y16	Low 7-segment of g

Y0–Y6 is controlling of high 7-segment and Y10–Y16 is controlling of low 7-segment.

4 PLC Control Timer Flow Chart

According to the specific requirements of the timer and I/O table, PLC control timer flow chart is in Fig. 3.

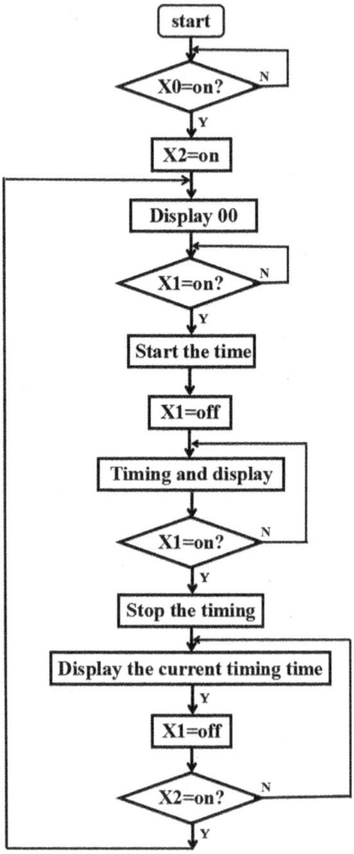

Fig. 3. PLC control timer flow chart

5 PLC Control Timer Ladder Diagram

According to the specific requirements of the timer and I/O table, the ladder diagram be written (Fig. 4).

R0 is an intermediate relay, and when the R0 is on, PLC stops the timing.

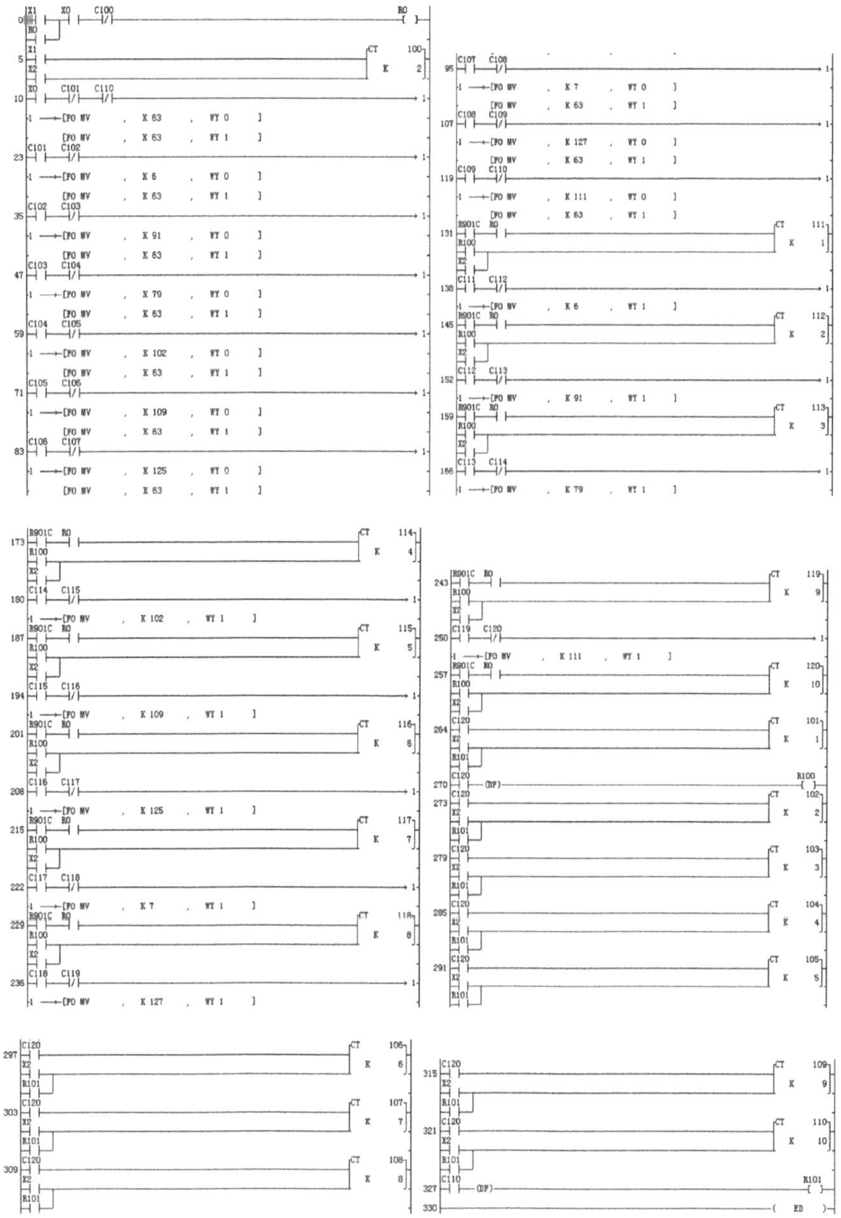

Fig. 4. PLC control timer ladder diagram

CT100 is an internal counter of PLC. When PLC receives second signals from the photoelectric switch, the CT100 count is full 2 times, and PLC stops the timing.

F0 is high-level instruction. The K63 is 7-segment code. The K63 is copied to WY0 and WY1. K63 is moved 0 to 7-segment display (Fig. 5).

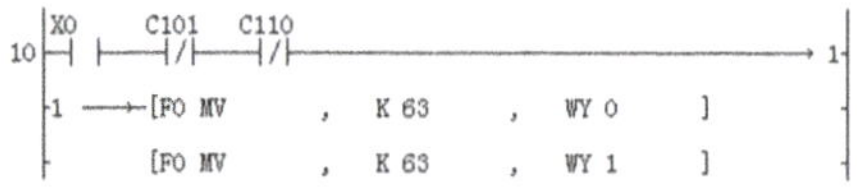

Fig. 5. K63 is moved 0 to 7-segment display

Others is shown in Table 2.

Table 2. 7-segment code and display character

Display character	7-segment code
0	K63
1	K6
2	K91
3	K79
4	K102
5	K109
6	K125
7	K7
8	K127
9	K111

The timing unit of the timer is 0.1 s, and the 7-segment shows the number of time in 1 s.

If the photoelectric switch is placed outside the small-car track, when the small-car passes the A for the first time, PLC starts to timing. CT100 minus 1. When the small-car passes the B, CT100 minus 1 again. When the small-car goes round the track to the

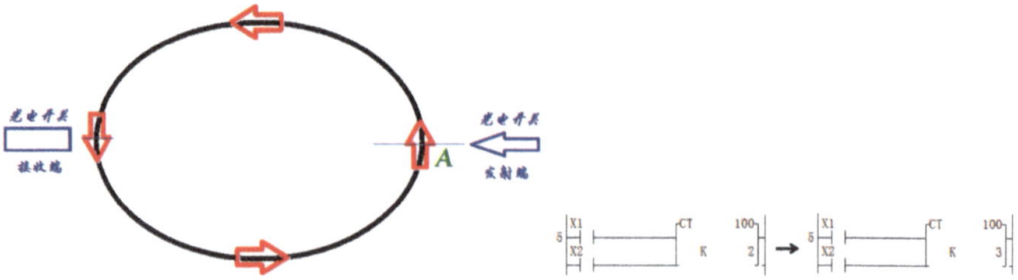

Fig. 6. PLC control timer diagrammatic sketch

A, that is, the second time passes through the A, the CT100 minus 1 again, at this time CT100 value is reduced to 0, and PLC stops the time and displays the time. So the CT100 constant setting 2 can be changed to 3, and the others are not changed (Fig. 6).

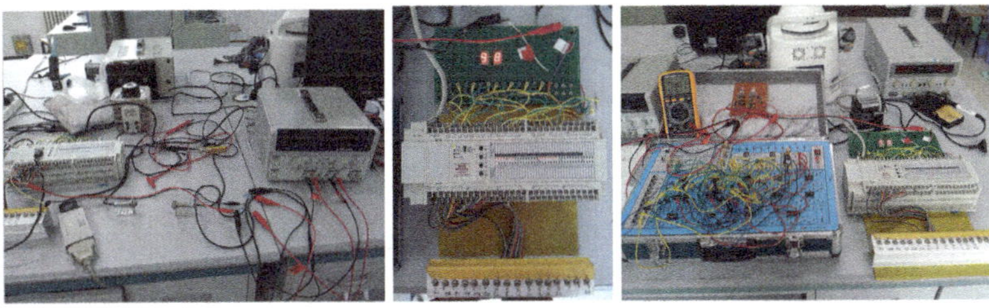

Fig. 7. PLC control timer experiment

6 Conclusion

Figure 7 is the PLC control timer experiment. This timer in this paper has been tested and passed in Panasonic PLC, which can be used directly.

References

1. Zhao, J.: Design of PLC control system for traffic lights. In: Mechatronics and Manufacturing Technologies, August 2016
2. Zhao, J.: Research on the measurement system of liquid weighing based on PLC. In: International Conference Energy and Mechanical Engineering, October 2015
3. Zhao, J.: PLC counter applications technique. In: Progress in Mechatronics and Information Technology, October 2013

Classification Algorithm for Human Walking Gait Based on Multi-sensor Feature Fusion

Hongfang Wu, Xiaoling Zhang, Jian Wu, and Lei Yan(✉)

School of Technology, Beijing Forestry University, Beijing, China
mark_yanlei@bjfu.edu.cn

Abstract. The accurate gait mode classification is the prerequisite and basis for judging movement intention and making controlling strategies of lower limb exoskeleton robot. In our research, we developed a wearable multi-sensor human gait information acquisition system. A kind of the human walking gait classification algorithm was established to collect human lower limb movement information. The algorithm could classify four walking gait modes, including standing, horizontal walking, up the stairs and down the stairs, which combined time domain and frequency domain characteristics. The acceleration and the angular acceleration signal was extracted by windowing, and the time domain feature (mean, variance, rms, max, minimum) and the frequency domain feature (the peaks of the fast Fourier transform (FFT)) were extracted from a single time window. And the SVM classifier with different kernel functions was used for classification and identification. The results showed that the four different gait modes could be distinguished by the feature extraction method combining the time domain with the frequency domain. When the window size was 300 ms, the average recognition rate of the Polynomial kernel function classifier reached to 96.956%.

Keywords: Gait classification · SVM · FFT · Information fusion

1 Introduction

Lower limb power-assisted exoskeleton is a kind of electromechanical device which is worn by a human operator. It provides force and assistance to the human while walking. In the past few decades, the technology of lower extremity exoskeleton robot has been developing rapidly, and it has been widely applied in military, medical rehabilitation and industry [1, 2]. However, the researchers face a great challenge in the control of exoskeleton robots. With the wearer's gait mode changing, it is difficult for the robot to make accurate judgments about the wearer's behavior and intension, not to mention automatically adjusting the control parameters to make the wearer safer and more comfortable. So, it is vitally important to accurately identify the wearer's gait mode in the control of the exoskeleton.

The classification of different motion signals is a category of gait recognition. Gait recognition, according to the way to obtaining human physiological information, can be divided into image-based and sensor-based gait recognition. The image-based recognition, with the processing of a series of video/image sequence, realizes the action

K. Deng et al. (Eds.): ICMIR 2018, AISC 856, pp. 718–725, 2019.
https://doi.org/10.1007/978-3-030-00214-5_90

identification [3]. Sensor-based recognition is based on the motion sensor (speed, acceleration, displacement, pressure). The sensors are placed on some parts of the body, such as the waist, thighs, arms, neck, wrist, etc., to extract the body's motion signals and classify different modes of motion by processing the signals [4]. Compared with image-based gait recognition, sensor-based recognition has more advantages such as small volume, light weight, easy to carry and cheap. Among them, the representative includes the electroencephalography (EEG), electromyography (EMG) and mechanical sensing signals. The mechanical sensing signal is measured by the mechanical sensor, such as inertial measurement unit and pressure sensor [5]. In the field of rehabilitation robots, EEG allows paralyzed patients to control the robot to follow his own brain, but can not tell the unconscious movement of the human brain. EMG mainly measures the electrical signals generated by the muscle cells 100 ms before the muscle's movement [6], which can predict the human gait in advance. HAL, a kind of walking-assistant robot, identified the human body's motion intention mainly by collecting the EMG signals [7]. Although it had the predictability, the differences among different individual's EMG were relatively large, which could cover up the real physiological differences of the measured object [8].

Mechanical sensors are mainly placed on the surface of the human body or the exoskeleton. The method is simple and easy to achieve interference. Varol et al. presented an activity-mode intent recognition framework for a powered lower limb prosthesis, which collected signals of joint angles and angular velocities of the knee and ankle, socket sagittal plane moment, foot forces of heel and ball, and classified the patterns as standing, sitting, or walking [9]. But the accuracy of the overall recognition was not mentioned.

In this paper, we propose a wearable multi-sensor body motion information acquisition system based on inertial measurement unit (IMU). The information of human's four different gait modes was collected by the IMU installed on the leg. The four gait modes included standing, horizontal walking, up the stairs and down the stairs. The signals were pre-processed and then the acceleration and angular acceleration signal were extracted using the sliding window. The time domain parameters, including mean, variance, root mean square, maximum, minimum, and the peaks of the fast Fourier transform were extracted from the single time window. The SVM classifier was used for training. The results indicate that this method has a high recognition accuracy and can provide a reference for the control strategy of the lower extremity exoskeleton.

2 Methods

2.1 Gait Information Collection System

In order to collect gait information in different modes, we developed a wearable gait information collection system, as is shown in Fig. 1. The hardware of the system included four IMUs modules, a control module, a wireless receiving module and a computer. To ensure the consistency of data timing, CANbus was used to connect the control module with four IMUs modules. The wireless receiver module received data from the control module wirelessly. The wireless receiver module was directly connected to the computer with USB. The sampling frequency of the system is 50 Hz.

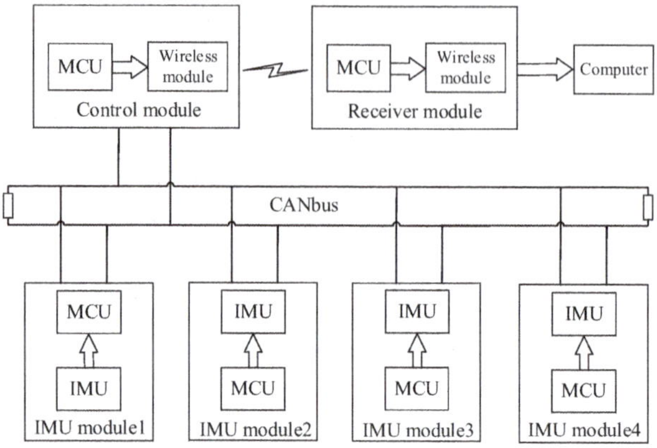

Fig. 1. The structure of data transmission in the designed system using CANbus networks.

In our research, four IMUs were used to collect data. Each IMUs unit had a triaxial accelerometer and a triaxial gyroscope, so the sensor unit obtained 3-axis acceleration and 3-axis angular acceleration data. Four IMUs modules were placed on the left and right leg above the knee and above the ankle joints respectively. The specific placement of the sensor is shown in Fig. 2.

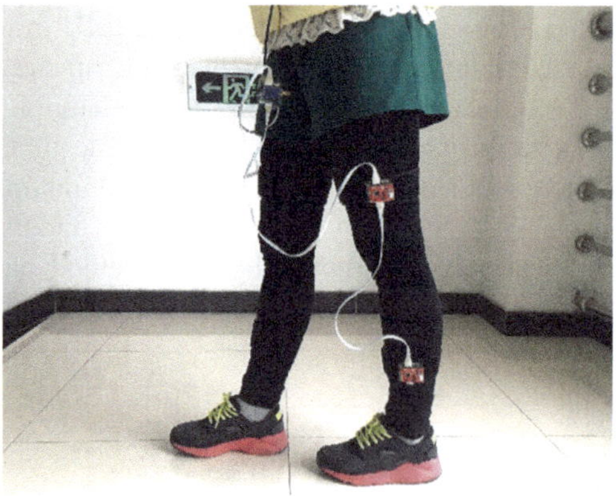

Fig. 2. The distribution of sensors on human body

2.2 Data Collection

Five healthy young people volunteered to participate in the experiment, and they all signed informed consent. They were asked to walk on different road conditions (standing, horizontal walking, up the stairs, down the stairs) wearing a data collection system, as is shown in Fig. 3. The gait speed was about 1 m/s. 5 experiments per person were recorded. The order of the experiment was standing—up the stairs—horizontal walking—down the stairs. The duration of each experiment is 35 s to 40 s.

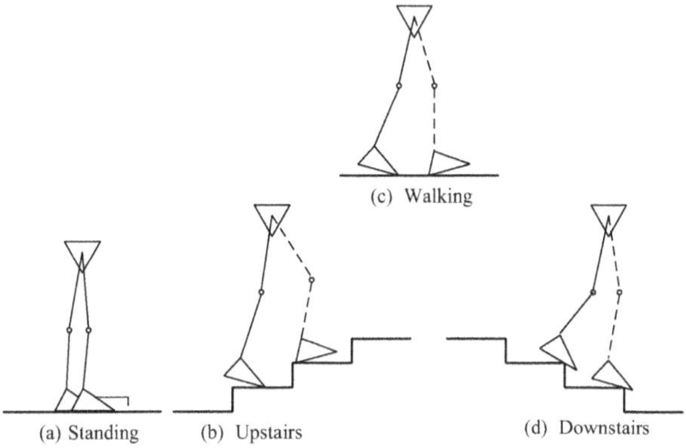

Fig. 3. Common walking patterns

2.3 Preprocessing and Feature Extraction

After obtaining the sensor data for each gait mode, a 1 × 24 element discrete time series sample was obtained using the windowing method. The window time was 300 ms and the sampling frequency was 50 Hz. To get more samples, we moved the window at 20 ms intervals. Thus, 250 samples were obtained for each gait, and a total of 1000 samples of four gait. 10,000 samples could be obtained in ten experiments. 5 volunteers could contribute to 50,000 samples. We marked the standing as the category 1, up the stairs as the category 2, horizontal walking as the category 3, down the stairs as the category 4. Each category includes 12,500 samples.

Time domain characteristics refer to some of the characteristics associated with practice in the process of sequence changing. Commonly used time domain features include mean, variance, root mean square, maximum and minimum. The frequency domain feature is usually used to obtain the periodic information in the signal. Because the gait modes, such as walking, up the stairs, are typical cyclical movement. And different gait modes in the spectrum also have great difference. The equators are listed in Table 1. FFT is the fast algorithm for discrete Fourier transforms that transfer a signal into the frequency domain. Some signals in the time domain is difficult to extract the eigenvalues. However, it would be easier to extract features when the signal was transferred into the frequency domain. That is why a lot of signal analysis use FFT transform. The gait features presented in this paper include: mean, variance, root mean square, maximum, minimum and the peaks of the fast Fourier transform.

6 features were extracted from each window, including mean, variance, root mean square, maximum, minimum and the peaks of the fast Fourier transform. Each window had 24-dimensional data and could extract 144 features (24 × 6). Thus the dimension

Table 1. Feature for walking pattern recognition

Feature	Definition
Mean	$\bar{x} = \frac{1}{n}\sum_{i=1}^{n} x_i$
Variance	$\sigma^2 = \frac{1}{n}\sum_{i=1}^{n}(x_i - \bar{x})^2$
Root mean square	$x_{rms} = \sqrt{\frac{x_1^2 + x_2^2 + \cdots + x_n^2}{n}}$
Maximum	$x_{max} = max\{x_1, x_2, \cdots, x_n\}$
Minimum	$x_{min} = min\{x_1, x_2, \cdots, x_n\}$
FFT	$s_{FFT}(k) = \sum_{i=0}^{n-1} s_i e^{-2j\pi ki/n}, k = 0, 1, \cdots, n-1$

of the eigenvector was 144 × 1. All the features were normalized to [0, 1] to be classified.

Since the initial feature set was quite large (144), not all features were equally useful for classification. We investigated different ways of feature selection and reduction methods. In this research, the number of features was reduced from 144 to 20 by principal component analysis (PCA). Principal component analysis (PCA) was a common data dimension reduction method with the aim of reducing the amount of calculation by converting high-dimensional data to low-dimensional data under the premise of little loss of information [10].

2.4 SVM Classifier

SVM [11] is a kind of pattern recognition method derived from statistical learning theory. The linear non-separable samples in low-dimensional space are transformed into high-dimensional feature spaces by kernel function to make them linearly separable and construct the optimal hyperplane using structural risk minimization theory in feature space. The classification of the samples was realized through the decision plane [12]. By training the training samples, the decision function was obtained as follows.

$$f(x) = sgn\left(\sum_{i=1}^{ns} a_i y_i K(x_i, x) + b\right) \tag{1}$$

In this function, ns is the number of support vectors and b is the threshold selected from the training sample, a constant. As for a test vector, if

$$\sum_{i=1}^{ns} a_i y_i K(x_i, x) + b > 0 \tag{2}$$

classify x as 1, otherwise classify x as −1. $k = K(x_i, x)$ is kernel function. Generally speaking, the kernel function includes linear, polynomial and Radial Basis Function

(RBF). The use of different kernel functions can achieve different types of machine learning. The results of classification are also varied.

In our research, the SVM classifier based on C-support Vector classification (C-SVC) was established. The kernel functions, linear, polynomial and Radial Basis Function (RBF), was adapted.

3 Results and Discussion

The classification accuracy of 96.956% was achieved by using the SVM classifier based on multi-sensor fusion in 4 different gait modes. The confounding matrix of the classification recognition results in different gait patterns are shown in Fig. 4. The window size was 300 ms and the kernel function was polynomial.

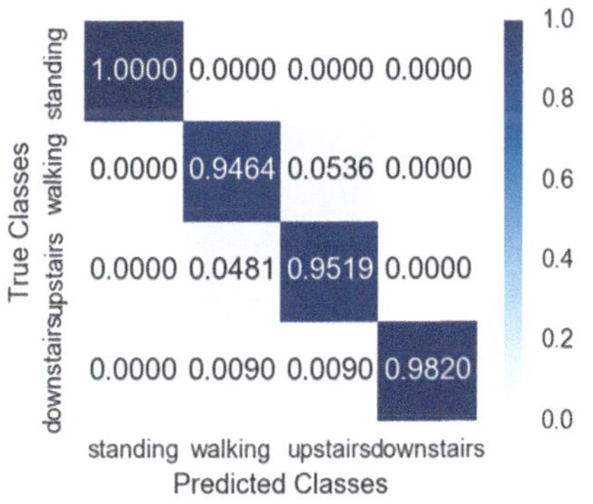

Fig. 4. The confounding matrix of the classification recognition results

The SVM classification performance was closely related to the choice of kernel function. However, there was no good way to guide the selection of the kernel function in specific issue. In this paper, three SVM classifiers were established for the commonly used kernel functions. They were SVM1 based on linear, SVM2 based on polynomial and SVM3 based on Radial Basis Function (RBF). By comparing the accuracy of three models' classification, we found the suitable kernel function for this study. After simulation, the accuracy of the classification based on different kernel functions is shown in Table 2.

Table 2. The accuracy of the classification based on different kernel functions

Gait mode	SVM1 (Linear)	SVM2 (Polynomial)	SVM3 (RBF)
Standing	100.00%	100.00%	100.00%
Horizontal walking	86.79%	94.64%	94.44%
Up the stairs	72.41%	95.19%	89.19%
Down the stairs	85.71%	98.20%	99.07%

According to Table 1, the accuracy of the classification based on the RBF and the polynomial are both more than 95%. The Polynomial is much better in the comprehensive index. And the accuracy of the classification based on the linear is below the average. Above all, Polynomial was used in SVM algorithm as the kernel function.

To achieve real-time gait recognition, the smaller the window, the better. But the smaller window means less data, and less data means that we will get less accurate features, which have an bad impact on the classification accuracy. As the size of the window increases, the performance tends to be better (as is shown in Fig. 5). To find the balance between the real-time performance and the classification accuracy, the effect of different window sizes on classification accuracy was measured. The window size was from 100 ms to 600 ms with the interval of 20 ms. Above all, 150 ms was chosen as the best window size.

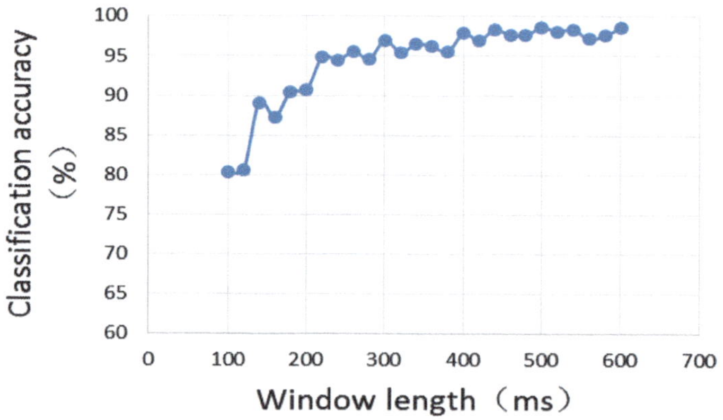

Fig. 5. The effect of different window sizes on classification accuracy

4 Conclusion

In this paper, we propose a feature extraction method for human gait mode based on multi-sensor fusion, and use SVM to classify with the accuracy up to 95%. Experiments show that the use of Polynomial kernel function and 300 ms window can lead to the best classification with the accuracy of 96%. The study can provide a good reference for the research about the control of the lower limb exoskeleton.

There is still much to be done. For example, the gait data are relatively small. Thus, more data need to be collected to improve the analysis; The speed of four walking modes is fixed, not taking the effects of different speed into account. In addition, when the amount of data is relatively large, more advanced depth learning algorithm should be considered.

Acknowledgments. This research was financially supported by the Fundamental Research Funds for the Central Universities (Grant No. 2015ZCQ- GX-03).

References

1. Li, N., Yan, L., Qian, H., Wu, J., Men, S., Li, Y.: A design and simulation analysis of an improved lower limb exoskeleton. J. VibroEng. **16**, 3655–3664 (2014)
2. Li, Y., Yan, L., Qian, H., Wu, J., Men, S., Li, N.: Dynamics and kinematics analysis and simulation of lower extremity power-assisted exoskeleton. J. VibroEng. **16**, 781–791 (2014)
3. Kong, W., Saad, M.H., Hannan, M.A., Hussain, A.: Human gait state classification using artificial neural network, pp. 1–5. IEEE (2015)
4. Young, A.J., Simon, A.M., Hargrove, L.J.: A training method for locomotion mode prediction using powered lower limb prostheses. IEEE Trans. Neural Syst. Rehabil. Eng. **22**, 671–677 (2014)
5. Novak, D., Riener, R.: A survey of sensor fusion methods in wearable robotics. Robot. Auton. Syst. **73**, 155–170 (2015)
6. Collinger, J.L., Wodlinger, B., Downey, J.E., et al.: High-performance neuroprosthetic control by an individual with tetraplegia. Lancet **381**, 557–564 (2013)
7. Kawainoto, H., Lee, S., Kanbe, S., et al.: Power assist method for HAL-3 using EMG-based feedback controller. In: IEEE International Conference on Systems, Man and Cybernetics, Washington, vol. 2, pp. 1648–1653 (2003)
8. Sensinger, J.W., Lock, B.A., Kuiken, T.A.: Adaptive pattern recognition of myoelectric signals: exploration of conceptual framework and practical algorithms. IEEE Trans. Neural Syst. Rehabil. Eng. **17**, 270–278 (2009)
9. Varol, H.A., Sup, F., Goldfarb, M.: Multiclass real-time intent recognition of a powered lower limb prosthesis. IEEE Trans. Biomed. Eng. **57**, 542–551 (2010)
10. Ahsan, M., Elahi, E., Farooqi, W.A.: Superscalar power efficient fast Fourier transform FFT architecture. In: International Conference on Computer, Control and Communication, pp. 1–4 (2009)
11. Kamruzzaman, J., Begg, R.K.: Support vector machines and other pattern recognition approaches to the diagnosis of cerebral palsy gait. IEEE Trans. Biomed. Eng. **53**, 2479–2490 (2006)

End-to-End Mandarin Speech Recognition Using Bidirectional Long Short-Term Memory Network

Yu Yao(✉) and Ryad Chellali

College of Electrical Engineering and Control Science, Nanjing Tech University, Nanjing 211816, China
yangguangel@163.com

Abstract. In view of the unreasonable conditional hypotheses of hidden Markov model in speech recognition, an acoustic model based on bidirectional long short-term network was proposed after further studying the ability of recurrent neural network. We have successfully applied the training criterion based on connectionist temporal classification to the training of this acoustic model, and built our end-to-end Chinese speech recognition system without hidden Markov model. Meanwhile, a speech decoding method base on weighted finite-state transducer was designed to effectively solve the problem that lexicon and language model are difficult to integrate into the decoding process. Compared with the traditional GMM-HMM system and the hybrid DNN-HMM system, the experimental results show our end-to-end system significantly reduces the recognition error rate, while at the same time speeding up decoding dramatically. It is shown that the acoustic model proposed in this paper can effectively enhance the model discrimination and optimize the structure of speech recognition system.

Keywords: Speech recognition · Bidirectional long short-term memory network · Connectionist temporal classification · Weighted finite-state transducers · End-to-end system

1 Introduction

In the past 30 years, the automatic speech recognition system has been extensively studied by many universities and research institutions, and has basically met the practical requirements under the quiet environment of standard pronunciation. During this period, the acoustic model construction paradigm based on Gaussian mixture model-hidden Markov model (GMM-HMM) became the mainstream framework of speech recognition, where HMM act to deal with the temporal variability of speech and GMM complete the mapping between the input signals and the HMM states [1]. In paper [2], multilayer perceptron was used to calculate the emitting probability of the HMM states instead of GMM, which has also slightly upgraded this recognition framework. However, because of the high error rate by the mediocrity model discrimination, speech recognition had not been able for larger-scale commercial.

K. Deng et al. (Eds.): ICMIR 2018, AISC 856, pp. 726–735, 2019.
https://doi.org/10.1007/978-3-030-00214-5_91

In 2006, the deep belief network was proposed by Geoffrey Hinton from Toronto University, which effectively solved the problem of deep neural network (DNN) training and promoted the rise of deep learning. With the powerful learning and presentation capabilities of DNN, the hybrid DNN-HMM acoustic model has once greatly improved the accuracy of automatic speech recognition [3–5].

However, in the training process of this hybrid system, GMM is still required to force the training data to be aligned to further train the DNN, which is obviously not conducive for global optimization of whole utterance and also increase the complexity and threshold of speech recognition system. In addition, the HMM-based recognition framework is not very perfect in theory, because HMM belongs to the generative model existing irrational conditional independence hypothesis [6]. For sequence labelling tasks, Graves et al. [7] proposed to use the connectionist temporal classification (CTC) object function to train recurrent neural network (RNN), so that RNN can automatically complete sequence alignment.

On the basis of Graves' work, we further study deep recurrent neural network and propose a new acoustic model based on bidirectional long short-term memory (BLSTM), and successfully applies the CTC function to this acoustic model training. In addition, we design a Chinese speech decoding approach based on the weighted finite-state transducer (WFST) for the difficulty of integrating linguistic knowledge such as lexicon and language model into the process of speech decoding [8]. This decoding method encodes the CTC labels, lexicon and language model into a single comprehensive WFST network, and then finds out the path with highest score as the recognition result by beam search. On the THCHS-30 data set, our experimental results show that the BLSTM-based end-to-end recognition system in the paper achieve a PER of 11.16% and a WER of 24.92%, and surpasses the GMM-HMM-based system and hybrid DNN-HMM system in recognition accuracy, while the decoding speed is speech up more than twice. Compared with RNN-based and LSTM-based acoustic model, the word error rate is reduced by 11.61% and 3.84% respectively with the advantage of utterance context information.

2 The Acoustic Model Based on Bidirectional Long Short-Term Memory

Compared with feedforward neural network, RNN is a special type network that allows the existence of self-feedback connections in hidden layer. Cyclical connections enable the neural nodes to have the ability to remember stimulus from the hidden layer at the last time, which will finally affect the output layer thought forward pass. Therefore, it also makes RNN particularly suitable for dealing with sequence input data [9]. However, for some tasks such as handwriting recognition and speech recognition, we not only need the past information, but also hope to have access to future context information to predict the network state at the present time. Bidirectional RNN provide an elegant solution about taking advantage of the past and future context [10]. As shown in Fig. 1, The structure of BRNN contains forward and backward two separate hidden layers, which act together on the same output layer.

Given an input sequence $\boldsymbol{X} = (\boldsymbol{x}_1, \ldots, \boldsymbol{x}_T)$, a bidirectional RNN computes the forward hidden layer states $\boldsymbol{H}^f = \theta(\boldsymbol{h}_1^f, \ldots, \boldsymbol{h}_T^f)$ by iterating from $t = 1$ to T:

$$\boldsymbol{h}_t^f = \theta(\boldsymbol{W}_{ih}^f \boldsymbol{x}_t + \boldsymbol{W}_{hh}^f \boldsymbol{h}_{t-1}^f + \boldsymbol{b}_h^f) \tag{1}$$

where $\boldsymbol{W}_{ih}^f$, $\boldsymbol{W}_{hh}^f$ represent the input-to-hidden weight matrix and the hidden-to-hidden weight matrix respectively, $\boldsymbol{h}_{t-1}^f$, $\boldsymbol{b}_h^f$ are hidden output vector at time $t - 1$ and bias vector, and $\theta(.)$ is the activation function of RNN. Meanwhile the backward hidden layer states can be computed by iteration from $t = T$ to 1:

$$\boldsymbol{h}_t^b = \theta(\boldsymbol{W}_{ih}^b \boldsymbol{x}_t + \boldsymbol{W}_{hh}^b \boldsymbol{h}_{t+1}^b + \boldsymbol{b}_h^b) \tag{2}$$

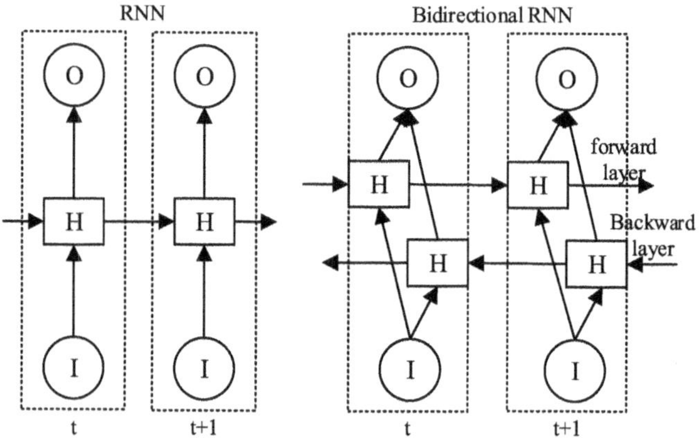

Fig. 1. Comparison of standard and bidirectional RNN.

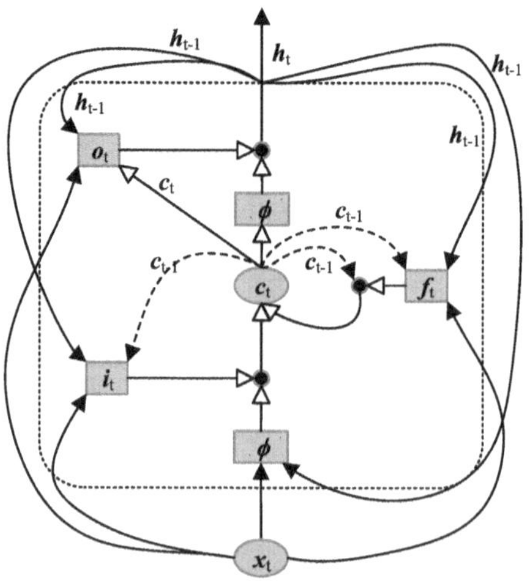

Fig. 2. A memory block of LSTM.

At each frame t, we calculate the current hidden state, and then concatenate the forward and backward hidden outputs $[\boldsymbol{h}^f, \boldsymbol{h}^b]$ as the inputs for the next recurrent layer. Finally, we can calculate the states of output layer units by iteration.

In this paper, Back-propagation through time (BPTT) algorithm is applied to learn the connection weights between each layer in recurrent neural network. Unfortunately, vanish gradient problem for standard RNN makes networks difficult to learn long-term temporal dependency of context [11]. In this case, the LSTM units proposed by Hochreiter *et al.* [12] were used as the building blocks of RNN. An LSTM memory unit mainly contains memory cells storing networks state, and three control gates—the input, output and forget gates, which provide write, read and reset function for the cells. As shown in Fig. 2, the dotted arrows are used for connecting the memory unit, the control gates are able to time output precisely, and the solid circles represent the multiplication unit. At time t, the output of LSTM unit is as follows,

$$\boldsymbol{i}_t = \sigma(\boldsymbol{W}_{ix}\boldsymbol{x}_t + \boldsymbol{W}_{ih}\boldsymbol{h}_{t-1} + \boldsymbol{W}_{ic}\boldsymbol{c}_{t-1} + \boldsymbol{b}_j) \tag{3}$$

$$\boldsymbol{f}_t = \sigma(\boldsymbol{W}_{fx}\boldsymbol{x}_t + \boldsymbol{W}_{fh}\boldsymbol{h}_{t-1} + \boldsymbol{W}_{fc}\boldsymbol{c}_{t-1} + \boldsymbol{b}_f) \tag{4}$$

$$\boldsymbol{c}_t = \boldsymbol{f}_t \odot \boldsymbol{c}_{t-1} + \boldsymbol{i}_t \odot \phi(\boldsymbol{W}_{cx}\boldsymbol{x}_t + \boldsymbol{W}_{ch}\boldsymbol{h}_{t-1} + \boldsymbol{b}_c) \tag{5}$$

$$\boldsymbol{o}_t = \sigma(\boldsymbol{W}_{ox}\boldsymbol{x}_t + \boldsymbol{W}_{oh}\boldsymbol{h}_{t-1} + \boldsymbol{W}_{oc}\boldsymbol{c}_t + \boldsymbol{b}_o) \tag{6}$$

$$\boldsymbol{h}_t = \boldsymbol{o}_t \odot \phi(\boldsymbol{c}_t) \tag{7}$$

where $\boldsymbol{i}, \boldsymbol{o}, \boldsymbol{f}, \boldsymbol{c}$ represent the input gates, output gates, forget gates and memory cells. $\boldsymbol{W}_{\cdot x}, \boldsymbol{W}_{\cdot h}, \boldsymbol{W}_{\cdot c}$ are the weight matrices connecting to the input layer, the hidden layer before, and the memory cell respectively. $\sigma(.)$ represents sigmoid activation function, and $\phi(.)$ represents tanh activation function. At the same time, the LSTM output of the backward hidden layer can be calculated in the same way.

3 End-to-End System Training and Decoding

3.1 End-to-End Training with Connectionist Temporal Classification

In large vocabulary continuous speech recognition, the embedded training method is widely applied in the acoustic model training process, i.e. we need to use the HMM automatically align the speech segmentations and phonemes, and then train the acoustic feature mapping model, such as GMM, DNN, which undoubtedly increase the complexity of the recognition system. Moreover, the limitation of HMM is not conducive to the further development of speech recognition technology. In this paper, CTC technology is applied to end-to-end train our BLSTM-based acoustic model without aforehand forced alignments.

CTC training is based on the CTC objection function in the output layer of RNN, and completes the alignment between input sequence and output labels automatically. For a sequence labeling task, the label set L contains K unique label elements, e.g., about 47 syllables for Chinese. Assume that input sequences $\boldsymbol{X} = (\boldsymbol{x}_1, \ldots, \boldsymbol{x}_T)$ and corresponding label sequences $z = (z_1, \ldots, z_U)$, the target of CTC training is to maximize the logarithmic probability of label sequences z given the input sequences $\boldsymbol{X}$ by adjusting the internal network parameters, i.e., $\max(\ln P(z|\boldsymbol{X}))$. The length of $\boldsymbol{X}$ is greater than the length of z, i.e., $T \geq U$. the output label $z_u \in L \cup \{<blank>\}$, where $<blank>$ corresponds to an blank label used to represent mappings of those labels out of the L set, such as silence, pause and so on. The output layer of CTC networks is a Softmax layer with $K+1$ output units corresponding to the extended set $L' = L \cup \{<blank>\}$. At time t, the output vector of Softmax layer is y_t, where y_t^k is the posterior probability of label k. Since the label sequences and the input sequences are not aligned, it is difficult to calculate the likelihood of the label sequences directly from the RNN outputs. To solve the alignment between input and output, a frame-level CTC path $\boldsymbol{p} = (p_1, \ldots, p_T)$ was introduced into contacting z and $\boldsymbol{X}$, which allows repeated occurrence of the blank label and non-blank labels. The probability of the entire CTC path can be formed by the probability of each frame corresponding to the label.

$$P(\boldsymbol{p}|\boldsymbol{X}) = \prod_{t=1}^{T} y_t^{p_t} \tag{8}$$

By defining mapping $\Phi : L^{\leq T} \mapsto L'^{T}$, the label sequences z can be mapped to CTC path $\boldsymbol{p}$. This is a many-to-one mapping because many CTC paths can respond to the same label sequence. For example,$\Phi(\text{A A - - } BC-) = \Phi(\text{ - A A } B-CC) = \text{A } BC$. In the way, the conditional probability of z can be calculated by summing the probability of all possible CTC paths:

$$P(z|\boldsymbol{X}) = \sum_{p \in \Phi(z)} P(\boldsymbol{p}|\boldsymbol{X}) \tag{9}$$

However, all the possible conditions of CTC paths will increase exponentially with the increasing input scale. Fortunately, after reviewing traditional method, we can apply Forward-backward Algorithm to efficiently calculate the likelihood of path in the trellis networks. First, the blank label indexed by zero need to be inserted into the beginning and the end of z, while it is also inserted between every output label z_u. After, we get an augmented label sequence $\boldsymbol{l}=(l_1, \ldots, l_{2U+1})$, which is very similar to z, just a lot of blank inserted. In the trellis of Fig. 3, using Forward-backward Algorithm, the forward variable α_t^u and the backward variable β_t^u can be evaluated, which represents the probability of all forward paths ending with label l_u at frame t and the probability of all backward paths beginning with l_u at frame t.

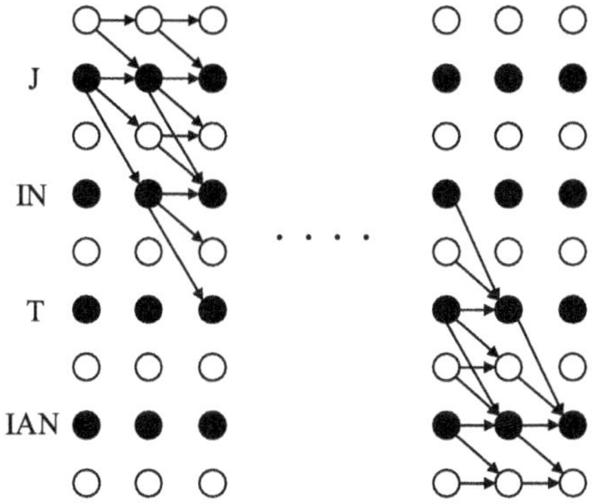

Fig. 3. A trellis network representing Chinese "J IN T IAN" meaning TODAY.

The posterior probability of z can be calculated as:

$$P(z|\boldsymbol{X}) = \sum_{u=1}^{2U+1} \alpha_t^u \beta_t^u \quad (10)$$

where t can be any frame $1 \leq t \leq T$. According to the CTC objective function $\ln P(z|\boldsymbol{X})$, the derivative of the objective function with respect to y^t can be computed as:

$$\frac{\partial \ln P(z|\boldsymbol{X})}{\partial y_t^k} = \frac{1}{P(z|\boldsymbol{X})} \frac{1}{y_t^k} \sum_{u \in \gamma(\boldsymbol{l},\mathrm{k})} \alpha_t^u \beta_t^u \quad (11)$$

Where $\gamma(\boldsymbol{l},k) = \{u|l_u = k\}$ returns the index set of positions where label k occurs in $\boldsymbol{l}$. With the differentiability of CTC objective function, every layer gradient can be calculated through the Softmax layer by BPTT algorithm.

3.2 Decoding with Weighted Finite-State Transducer

The speech decoding process is very dependent on the knowledge of linguistic, but [13, 14] doesn't integrate grammar knowledge very well in decoding stage. In this paper, the characteristic of Chinese pronunciation and linguistic knowledge are studied. We build a comprehensive decoding graph containing lexicon and word-level language model using WFST, which ensures the integrity of the knowledge of phonetics and linguistics and improves the decoding efficiency greatly. The WFST networks consist of the CTC token transducer, the lexicon transducer and the grammar transducer, where complete the corresponding CTC labels to lexicon units, encodes the mapping from lexicon units to words, and encodes the permissible word sequences in a language level. Figure 4 shows the representation of linguistic knowledge in three different levels.

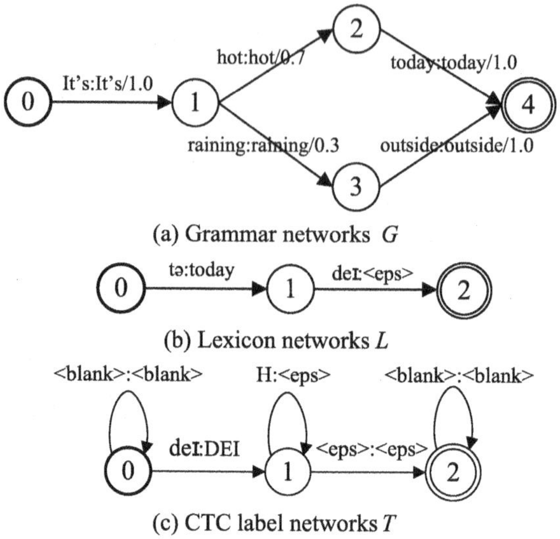

Fig. 4. Three types of WFST network examples.

In practical application, after generating token networks, lexicon networks and grammar networks respectively, we compile these three individual transducers to a decoding graph with composition, determinization and minimization algorithm [15, 16]:

$$S = T \circ min(det(L \circ G)) \tag{12}$$

where $\circ$, *det* and *min* represent the operations of composition, determinization and minimization. The search graph S can map the CTC labels out of the acoustic model into the corresponding word sequence and get the probability score of each possible path, and finally select the word sequence with the highest overall score as the recognition result.

4 Experiments

4.1 Experimental Dataset

We trained and evaluated our end-to-end recognition system on the THCHS-30 corpus (available http://www.openslr.org/18/). This dataset contains 35 h of recorded speech signals from 50 participants, where the sampling rate is 16,000 Hz and the sample size are 16 bits. 25 h (10,000 sentences) of recorded data were used as a training set, about 2 h (893 sentences) as the development set, and the remaining about 6 h (2495 sentences) as the test set. The dataset also includes a 3-gram language model based on words and a 3-gram language model based on phonemes, and corresponding dictionaries and phonemes dictionaries. Since the vocal units of Chinese are syllables, consisting of 23 consonants and 24 vowels, each vocal unit contains five different kind of tones. Therefore, a total of 218 acoustic vowels are used as the basic modelling unit to test the performance of different acoustic models.

4.2 Acoustic Model Setup

(1) **GMM-HMM**. The input feature parameters are 39 dimensional Mel frequency cepstral coefficients (MFCC) with deltas and double deltas, while the modeling units are monophone and triphone. After tree-based state tying, the number of context-dependent state based on triphone is 1658.
(2) **DNN-HMM**. The input feature parameters are 40 dimensional Mel-scale filter banks (FBanks), and the modeling units are monophone and triphone. DNN consists of 4 hidden layers, and 1024 units at each layer. The input layer are 11 neighboring frames of filterbank features about 440 units. the output layer has 656 output nodes which the modeling units are syllables, and 1658 output nodes which the modeling units are context dependent state. In addition, the training data were forced to be aligned by GMM-HMM acoustic model.
(3) **BLSTM-CTC**. The input feature parameters are 120 dimensional FBanks with deltas and double deltas, and the modeling units is monophone. Our acoustic model based on 4 bidirectional LSTM layers. Each hidden layer contains a total of 640 LSTM units of both the forward and the backward layers, and the output layer has 220 output units. The CTC criterion is used as the objective function of bidirectional LSTM trained by BPTT algorithm.

4.3 Recognition Results and Performance Evaluation

All experiments in this paper are based on the same training set, development set, test set and language models, and are tested on GTX1070 graphics card and i7 CPU. As we can see from Table 1, our end-to-end system based on BLSTM-CTC acoustic model achieves 11.16% on phoneme error rate (PER) and 24.92% on word error rate (WER). Compared with GMM-HMM model based on monosyllable and context dependent unit, the PER is reduced by 21.04% and 9.13% respectively, and the WER is reduced by 25.91% and 11.02% respectively. Meanwhile, compared with the hybrid DNN-HMM system, it reduces 4.7% on PER and 4.43% on WER, which is close to the recognition performance of hybrid DNN-HMM system based on context dependent unit. The experimental results reveal that our CTC-trained BLSTM-based model

Table 1. Comparisons of PER and WER between different systems.

Acoustic model	Modeling unit	PER%	WER%
GMM-HMM	Monosyllable	32.20	50.83
GMM-HMM	Context dependent unit	20.29	35.94
DNN-HMM	Monosyllable	15.85	29.35
DNN-HMM	Context dependent unit	10.27	23.69
BLSTM-CTC	Monosyllable	11.16	24.92

outperform the traditional HMM-based models in the case of the same modeling unit, and the recognition accuracy is very closed to the state-of-the-art hybrid DNN-HMM system based on context dependent unit.

In addition, another important performance evaluation of speech recognition system—decoding speed is compared and analyzed, as shown in Table 2. The decoding time of our end-to-end system is 1.3 times less than that of the hybrid DNN-HMM system, and the search graph (TLG) size is only 0.61 times of the hybrid system's search graph. It is confirmed that the end-to-end technology proposed in this paper simplifies the structure of speech recognition system, compresses the size of decoding graph and speeds up the speech recognition process dramatically.

Table 2. Comparisons of decoding speed between different systems.

Model	RTF	Graph size
DNN-HMM	0.56	718
BLSTM-CTC	0.24	439

From Table 3, compared with RNN-CTC and LSTM-CTC model, the WER of BLSTM-CTC-based model is reduced by 11.61% and 3.84% respectively. Because our BLSTM model can have access to future context information, it shows better sequence labeling ability.

Table 3. Comparisons of WER between different end-to-end systems.

Acoustic model	Modeling unit	PER%
RNN-CTC	Monosyllable	36.53
LSTM-CTC	Monosyllable	28.76
BLSTM-CTC	Monosyllable	24.92

5 Conclusions

In this work, we have further explored deep bidirectional long short-term memory network, and built an BLSTM-CTC-based end-to-end Chinese speech recognition system. CTC training criterion was successfully applied in our BLSTM-based acoustic model training, which makes our system get rid of the dependency on HMM. Meanwhile, the WFST-based decoding recipe was designed to make the knowledge of lexicon and language model integrate into searching process efficiently and effectively. On the THCHS-30 data set, the experimental results show that our end-to-end system not only enhance recognition accuracy greatly, but also improve the decoding speech dramatically. The superiority of the BLSTM-based acoustic model proposed in this paper has been verified. On the other hand, the accuracy of speech recognition system is also influenced by external environment and the speaker factor. How to reduce the influence of these factors and how to further improve the differentiation and robustness of the model will be the focus of the next stage of our research.

References

1. Hinton, G., Deng, L., Yu, D., et al.: Deep neural networks for acoustic modeling in speech recognition: the shared views of four research groups. IEEE Signal Process. Mag. **29**(6), 82–97 (2012)
2. Morgan, N., Bourlard, H.A.: Continuous speech recognition: an introduction to the hybrid HMM/connectionist approach. IEEE Signal Process. Mag. **12**(3), 24–42 (1995)
3. Hinton, G.E., Osindero, S., Teh, Y.W.: A fast learning algorithm for deep belief nets. Neural Comput. **18**(7), 1527–1554 (2006)
4. Mohamed, A.R., Hinton, G., Penn, G.: Understanding how deep belief networks perform acoustic modelling. In: IEEE International Conference on Acoustics, Speech and Signal Processing, pp. 4273–4276 (2012)
5. Veselý, K., Ghoshal, A., Burget, L., et al.: Sequence-discriminative training of deep neural networks. In: Interspeech, pp. 2345–2349 (2013)
6. Blasiak S, Rangwala H, A hidden Markov model variant for sequence classification. In: Proceedings of the 7th International Joint Conference on Artificial Intelligence, vol. 22, no. 1, p. 1192 (2011)
7. Graves, A., Gomez, F.: Connectionist temporal classification: labelling unsegmented sequence data with recurrent neural networks. In: International Conference on Machine Learning, pp. 369–376 (2006)
8. Mohri, M., Pereira, F., Riley, M.: Speech Recognition with Weighted Finite-State Transducers, pp. 559–584. Springer, Berlin (2008)
9. Graves, A., Mohamed, A., Hinton, G.: Speech recognition with deep recurrent neural networks. In: 2013 IEEE International Conference on Acoustics, Speech and Signal Processing, ICASSP, pp. 6645–6649 (2013)
10. Graves, A., Jaitly, N., Mohamed, A.R.: Hybrid speech recognition with deep bidirectional LSTM. In: IEEE Conference of Automatic Speech Recognition and Understanding, pp. 273–278 (2014)
11. Wollmer, M., Schuller, B., Eyben, F., et al.: Combining long short-term memory and dynamic bayesian networks for incremental emotion-sensitive artificial listening. IEEE J. Sel. Top. Signal Process. **4**(5), 867–881 (2010)
12. Hochreiter, S., Schmidhuber, J.: Long short-term memory. Neural Comput. **9**(8), 1735–1780 (1997)
13. Maas, A., Xie, Z., Dan, J., et al.: Lexicon-free conversational speech recognition with neural networks. In: Conference of the North American Chapter of the Association for Computational Linguistics: Human Language Technologies, pp. 345–354 (2015)
14. Hannun, A., Case, C., Casper, J., et al.: Deep speech: scaling up end-to-end speech recognition. arXiv preprint arXiv:1412.5567 (2014)
15. Povey, D., Ghoshal, A., Boulianne, G., et al.: The Kaldi speech recognition toolkit. In: Conference of the Automatic Speech Recognition and Understanding, Idiap, EPFL-CONF-192584 (2011)
16. Droste, M., Kuich, W., Vogler, H.: Handbook of Weighted Automata. Monographs in Theoretical Computer Science An Eatcs, vol. 380, no. 1–2, pp. 69–86 (2009)

Design and Analysis of the Excavation and Picking Equipment of Radix Pseudostellariae

Shungen Xiao[1,2], Mengmeng Song[1(✉)], Enbo Liu[1], and Wei Yang[1]

[1] School of information, Mechanical and Electrical Engineering, Ningde Normal University, Ningde, China
544824964@qq.com

[2] School of Mechatronic Engineering and Automation, Shanghai University, Shanghai, China

Abstract. Radix pseudostellaria is widely cultivated because of its good pharmacology of traditional Chinese medicine. However, due to the lack of excavation device that suitable for digging operation in rolling foothills country, which leads to the low efficiency of harvesting of radix pseudostellaria in harvest season. Based on the characteristics, in this paper the specific design scheme of automatic rotary excavator is carried on and the design idea of the structure of disengaging device and excavation device is analyzed. In the end of article, the performance of excavation device is analyzed by using Finite Element Software. The experimental results show that the design of the excavation equipment of radix pseudostellariae meets the actual requirements.

Keywords: Radix pseudostellaria · Structural design · Finite element analysis Excavation

1 Introduction

Radix pseudostellariae is an excellent Chinese medicine, which tasted slightly bitter and with invigorating the spleen, moistening lung, tonifying Qi and other characteristics [1]. Radix pseudostellariae is mainly cultivated in Fujian, Jiangsu, Shandong and Anhui, among which is the largest planting base of radix pseudostellariae located in Zherong County in Fujian Province of China. Radix pseudostellariae has very high economic status in Zherong County, the Planting area of radix pseudostellariae is accounts for 20% of total planting area of crops [2]. Nowadays, it has quickly become the third largest crop in the region and the pillar industry of local agriculture. However, the way farmers harvest radix pseudostellariae still follow the traditional production mode, that is, manual excavating and manual separation. Such a production model often requires a large number of labor, and always brings the disadvantage of high labor intensity, mass labor time consuming and low production efficiency. Due to the limited labor force, it often leads to harvest extension, affecting its economic benefits and medicinal value [3, 4]. Therefore, it is of great significance to design a series of

K. Deng et al. (Eds.): ICMIR 2018, AISC 856, pp. 736–742, 2019.
https://doi.org/10.1007/978-3-030-00214-5_92

excavating, transporting, separating, screening and collecting device for the harvesting of radix pseudostellariae.

2 The Excavation and Picking Mechanism of Radix Pseudostellariae

The scheme design of the excavation and picking mechanism of radix pseudostellariae is shown in Fig. 1. It is mainly composed of propulsive excavator, conveying device, drum screen separation device, dithering screen separation device, collecting devices, diesel engines and frame, in which diesel engine is used as the whole prime motor. Sprockets transmission and gear transmission are selected in the whole power transmission [5, 6]. First of all, the mixture of radix pseudostellariea soil is excavated by propulsive excavator. The excavating shovel and the conveying parts are connected by the screw conveyor to make the soil reach the screw conveyor structure by extrusion action of screw conveyor, which not only improves the transport efficiency, but also spreads the excess soil directly fell to the ground. Then the mixture in the screw conveyor will reach the conveying parts, and then to the drum screen separating device. After the secondary soil separation treatment of the drum screen separating device and the dithering screen separation device, the mixture of radix pseudostellariea and soil will eventually be separated. Processed radix pseudostellariae eventually arrived at and stored in the collection device. The mechanism greatly improves the harvesting efficiency, reduces the working time, and realizes the whole series process of excavating, separating and collecting of radix pseudostellariea, which solves the shortage of the harvesting of radix pseudostellariea. Obviously, the structure improves the working efficiency, reduces the power consumption, and achieves the goal of high efficiency and low intensity.

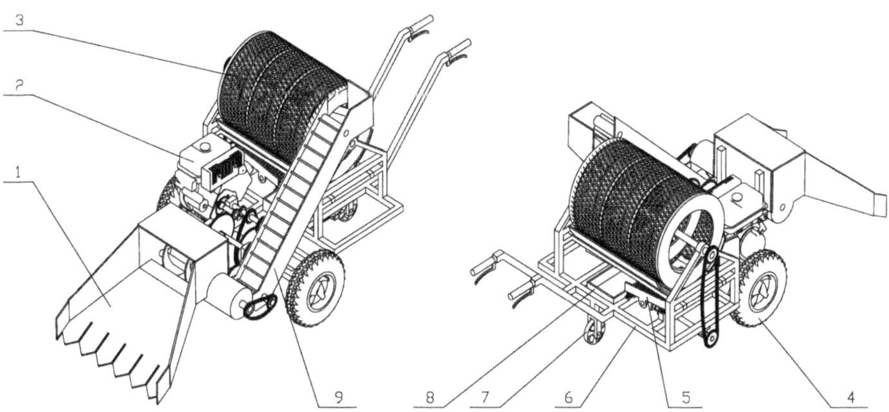

Fig. 1. The integral structure of excavation and picking mechanism of radix pseudostellariae. 1-propulsive excavator; 2-diesel engines; 3-drum screen separation device; 4-wheel; 5- dithering screen separation device; 6- frame; 7-auxiliary balance wheel; 8-collecting devices; 9-conveying device

3 Design of Key Parts

3.1 Design of the Separating Mechanism

The structure of the separating mechanism is shown in Fig. 2, which consists of two parts: the drum screen and the vibrating screen. The mixture of radix pseudostellariea and soil is transported to the drum screen through the conveyor belt. The drum screen separates radix pseudostellariea from soil by peripheral force of Circular motion, and then through vibrating screen to make sure that radix pseudostellariae and soil can be separated fully (secondary separation).

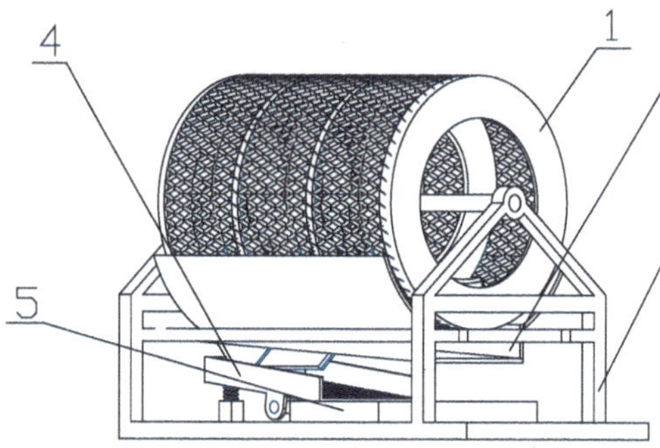

Fig. 2. Separation device diagram. 1-drum screen; 2-soil exportation; 3-frame; 4-vibrating screen; 5-collecting cartoon

3.2 The Design of Excavation Device

(1) Selection of excavating shovel

The purpose of excavating shovel is to dig out the mixture of radix pseudostellariea and soil. The most ideal condition of excavation is that the radix pseudostellariea cannot be damaged and the soil should be excavated as little as possible. In order to separate the soil more conveniently, the ability of excavating shovel to break soil should be considered in the design of mechanical structure. At the same time, the excavated device is required to have small traction resistance and good wear resistance. So, 65Mn structural steel is used to make excavating shovel which is composed of several triangular shovel blades.

(2) Main working parameters of excavating shovel

From Fig. 3, the main parameters of the shovel plate are as follows: θ is the edge angle, α represent the angle of inclusion, L is length, B denote width.

In order to ensure that the angle of cutting edge does not damage radix pseudostellariae, the following inequalities must be satisfied:

$$R_{1t} \geq R_{1n}\mu 1 = R_{1n} \tan \phi_1 \tag{1}$$

In this formula, R_1 denote the resistance acting on the shovel, μ_1 is friction coefficient of stem to steel ($\mu_1 \approx 0.85 \sim 0.9$).

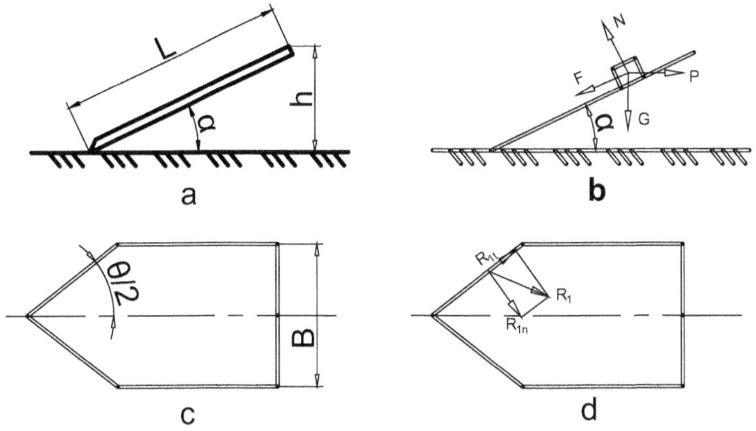

Fig. 3. The working parameters of excavating shovel and force analysis charts

As can be seen from the figure:

$$R_1 \cos\frac{\theta}{2} \geq R_1 \sin\frac{\theta}{2} \tan\phi_1 \tag{2}$$

$$\frac{\theta}{2} \leq 90^\circ - \phi_1 \tag{3}$$

Normally, $\theta/2 = 45^\circ \sim 55^\circ$, $\theta = 100^\circ$.

Therefore, according to the design requirements, the shovel width is 900 mm, the thickness is 5 mm, the single shovel width B is 140 mm, and the gap between the products is 12 mm.

From the force analysis of Fig. 3, we can get:

$$P\cos\alpha - T - G\sin\alpha = 0 \tag{4}$$

$$N - G\cos\alpha - P\sin\alpha = 0 \tag{5}$$

$$T = \mu N \tag{6}$$

Where P represents the force required to dig a shovel; N *is* the force of the shovel to the soil; G denotes the gravity of the soil on the shovel surface; T is the friction force of the soil to the shovel; μ_1 represents the friction coefficient of the soil to the digging shovel.

Simultaneous Eqs. (4)–(6):

$$\alpha = \arctan\left(\frac{P - \mu G}{\mu P + G}\right) \tag{7}$$

$$P = G\tan(\alpha + \varphi) \tag{8}$$

The gravity G of the soil is mainly the weight of the soil supported by the excavating shovel. Gravity calculations are as follows:

$$G = \rho gbdl \tag{9}$$

In formula (9), l is the length of shovel surface (m), ρ represents the density of soil (kg/m^3), d denotes excavating depth (m).

The working resistance of the excavating shovel is mainly composed of two parts, moving resistance of excavated soil by shovel, cutting resistance. The total resistance of shovel can be expressed in follow relation:

$$R = P + P_1 = G\tan(\alpha + \varphi) + KA \tag{10}$$

In formula (10), φ, K represents Friction angle of soil to steel and specific resistance of excavating soil respectively. Light soil: K = 16000–20000 (N/m^2), Mid light soil: K = 20000–24000 (N/m^2), Mid solid soil: K = 24000–30000 (N/m^2) cross section area of excavating soil (m^2).

Figure 3 shows the length of the shovel:

$$L = h/\sin\alpha \tag{11}$$

Radix pseudostellariae can be fully dug out. Finally, according to the calculation, the inclination angle α = 25°, friction angle φ = 30° and the length of the shove L = 47 cm, short-cut process L = 50 cm, and the maximum excavating resistance is calculated to be 4310 N. A schematic diagram of the excavating shovel structure is shown in Fig. 4.

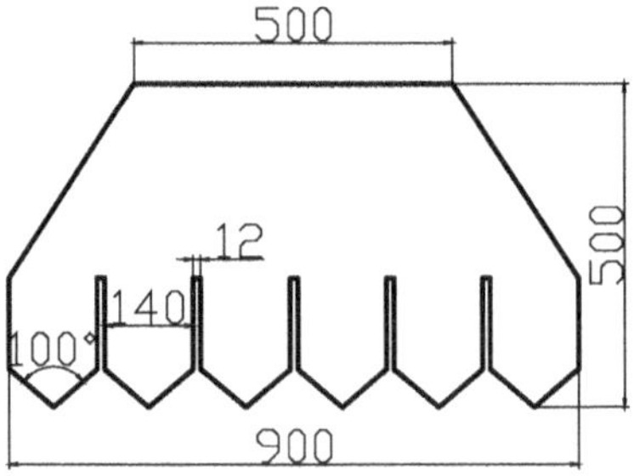

Fig. 4. Schematic diagram of the excavating shovel structure

4 The Finite Element Analysis of Excavating Shovel

The excavating shovel model is determined according to the parameters of the excavating shovel structure, application of load and boundary constraints. The material of the excavating shovel is defined as 65Mn structural steel, the density is 7810 kg/m^3, the elastic modulus E is 1.9 × 10^{11}, the Poisson's ratio is 0.29, and the angle of the excavating shovel is 25°. By solving the model of excavating shovel by ANSYS [7], the deformed shape and stress contour plot of excavating shovel in this design are obtained, which are shown in Figs. 5 and 6, respectively.

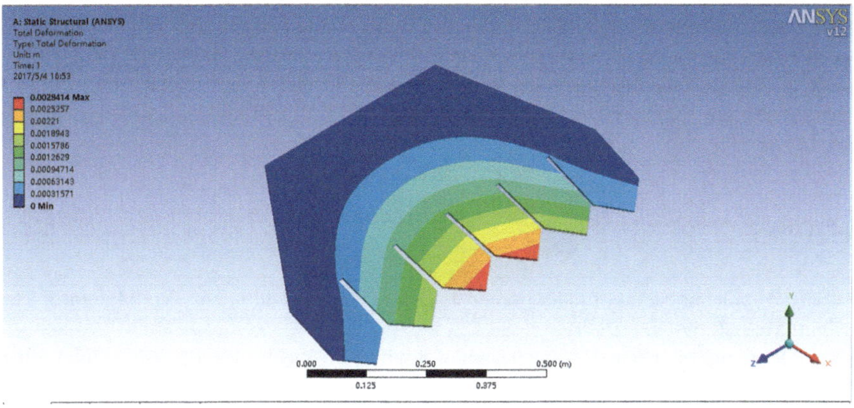

Fig. 5. The deformed shape of excavating shovel

Figure 6 shows that the maximum deformation of the excavating shovel is 2.84 mm, which exerts in the blade part of the shovel plate. This value is negligible in the range of plastic deformation of the material relative to the whole excavating shovel. The maximum stress of the excavating shovel is 54.77 MPa, which exerts at the welding position between side of the shovel and the baffle. It shows that the stress mainly appears in the welding position. Therefore, the welding position should be fully welding and heat treatment and strengthened during processing. The minimum principal stress is 8.3×10^{-3} MPa.

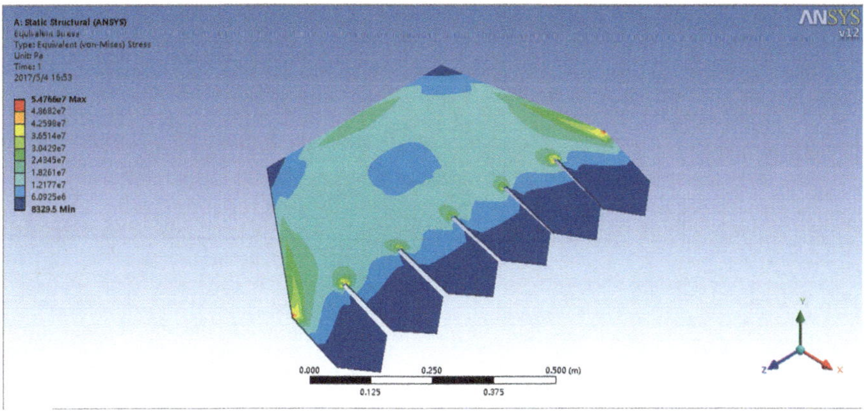

Fig. 6. The stress contour plot of excavating shovel

Therefore, the deformation and stress intensity of the excavating shovel are within the limit range of the material. So, the design of the excavating shovel meets the structural requirements.

Acknowledgments. This work is supported by the Education Science Project of Young and Middle-aged Teachers of Universities in Fujian Province (Grant No. JZ160396), Young Teachers Special Research Project of Ningde Normal University (Grant No. 2017Q104), major training project of Ningde Normal University (Grant No. 2017ZDK18) and Collabortive Innovation Center of Ningde Normal University (Grant No. 2017Z01). These supports are gratefully acknowledged.

References

1. Wang, Y.: Identification of radix pseudostellariae and its pseudo-traits. Nei Mongol J. Tradit. Chin. Med. **2**, 39–40 (2014)
2. Xiao, S., Song, M., Chen, C.: The design of disengaging mechanism of radix pseudostellariae and soil. IOP Conf. Ser. Mater. Sci. Eng. **274**, 1–8 (2017)
3. Li, Z., Cao, C., Lou, S., et al.: Design of radix pseudostellariae combine harvester. J. Agric. Mech. Res. **10** (2017)
4. Wang, T., Liao, Y., Yang, Y., et al.: The design and mechanical analysis of digging shovel of cassava harvest machine. J. Agric. Mech. Res. **10** (2015)
5. Lu, M.Q.: Development of Small-Sized Corydalis Tuber Harvester. Nanjing Agricultural University, Nanjing (2013)
6. Ding, Z.K., Zhou, F.J., Guan, Z.J.: Design of swill sorting machine components. J. Agric. Mech. Res. **9**, 139–142 (2011)
7. Zhang, Z.Q., Chang, Y.: Structure design of vibrating eccentric and its finite element analysis. J. Jiangnan Univ. (Nat. Sci. Ed.) **5**, 573–577 (2011)

Virtual Reality Tracking Registration and Pose Estimation Method

Chao Zhang[1], Yange Wang[1], Cheng Han[1], Jianping Zhao[1], and Ye Bai[2(✉)]

[1] School of Computer Science and Technology, Changchun University of Science and Technology, Changchun 130022, Jilin, China
[2] College of Optical and Electronical Information, Changchun University of Science and Technology, Changchun 130022, Jilin, China
308744267@qq.com

Abstract. There are still problems with high algorithm complexity due to the high precision of the tracking method for special markers, a position and pose estimation method based on rigid body motion analysis is proposed in this paper. This method used a rigid body structure to register and track virtual scenery and observation viewpoints. According to the spatial beam convergence idea, it can accurately track the movement of a rigid body structure, making the pose estimation process more flexible and results more accurate. This method uses a large amount of spatial analytical geometric theory knowledge to make it reliable in theory. The experiment also verified the method's anti-jamming, positioning accuracy, modularity, etc. The experimental results show that the use of rigid body structure to simulate the three-dimensional virtual observation point of view can accurately control the first-person viewing angle change, and the user experience during the roaming process is obvious higher than traditional visual interaction.

Keywords: Rigid body movement · Pose estimation · Rigid body structure Observation point of view · Registration tracking · Beam reconstruction

1 Introduction

Augmented Reality [1, 2] technology has rapidly evolved into a research hotspot for comprehensive applications with the development of image processing, computer vision and other research fields. It is composed of a number of key technologies including registration, tracking, location, interaction and display, among which location tracking exists as the main supporting technology. In augmented reality positioning and tracking technology [3], Koller et al. [4] used a single camera and used an initial calibration method to estimate the camera position and achieved user interaction with the three-dimensional model. Stato et al. proposed a hybrid tracking method combining a visual sensor and an electromagnetic tracker, which makes good use of the stability of the electromagnetic tracker and the accuracy of the visual sensor. Azuma et al. studied the high mixing degree visual tracking system based on local differential optical flow algorithm and applied it to outdoor environment. In the computer vision system, in

K. Deng et al. (Eds.): ICMIR 2018, AISC 856, pp. 743–754, 2019.
https://doi.org/10.1007/978-3-030-00214-5_93

order to restore the real scene 3 d information, need to get through visual unit calibration model parameters, geometric imaging and triangulation and perspective projection method is used to describe the depth of the 3 d real scene, distance and other information. It is a bridge of two-dimensional space transition to three-dimensional space, It is widely used in industrial detection, robot positioning and navigation, visual monitoring and other fields [5].

In the vision sensor tracking technology, the tracking method based on special markers has the advantages of small calculation load, high precision, and good real-time performance. Tracing methods based on natural features have problems with large computational load and poor real-time performance [6]. Just like this, tracking methods based on special markers are widely used in various types of augmented reality systems, and the user share is showing a rising trend. However, the special marker-based tracking method [7, 8] still has the problem of high accuracy and high algorithm complexity. In other words, there is an inverse relationship between tracking accuracy and real-time performance. Therefore, this paper proposes a pose estimation method for rigid-body structures that uses the principle of multi-view beam convergence to achieve tracking algorithms with low algorithm complexity while maintaining the original high-precision calculations.

2 Posture Estimation Method Based on Rigid Body Motion Analysis

2.1 Fixed Point Beam Reconstruction

For the straight-line fitting problem [9, 10] of the measured space discrete point set $U_{i=1}^{n} P_i$, the space straight line [11] that satisfies the best square approximation principle should satisfy:$\sum_{i=1}^{n} (\varepsilon_{xi}^2 + \varepsilon_{yi}^2 + \varepsilon_{zi}^2) = \min$. Among them, $\mathcal{E}_{xi}$, $\mathcal{E}_{yi}$ and $\mathcal{E}_{zi}$ respectively represent errors in the three coordinate axes of the discrete point $p_i(x_i, y_i, z_i)$ in the X, Y, and Z directions.

Assume that the projective form of the linear equation of the space can be represented by Eq. (1), where a, b, c, d are the projective parameters of the space-linear equation. The matrix representation of the linear equation of space can be expressed by Eq. (2).

$$\begin{cases} x = az + b \\ y = cz + d \end{cases} \tag{1}$$

$$\begin{bmatrix} \mathrm{x} \\ y \end{bmatrix} = \begin{bmatrix} z & 1 & 0 & 0 \\ 0 & 0 & 1 & z \end{bmatrix} \begin{bmatrix} a & b & c & d \end{bmatrix}^T \tag{2}$$

It can be known from the space straight line fitting process that the error amounts are implicit in the parameters of the spatial discrete point set and the space line equation. According to the matrix representation of the space line equation, the error equation of the spatial line can be obtained as in Eq. (3) show. Among them, $\widehat{a}, \widehat{b}, \widehat{c}, \widehat{d}$ are the parameters of the space equation containing the error.

$$v = \begin{bmatrix} z & 1 & 0 & 0 \\ 0 & 0 & 1 & z \end{bmatrix} [\widehat{a}\ \ \widehat{b}\ \ \widehat{c}\ \ \widehat{d}]^T - \begin{bmatrix} x \\ y \end{bmatrix} \tag{3}$$

According to the formula (4), the space linear parameter $[\widehat{a}\ \ \widehat{b}\ \ \widehat{c}\ \ \widehat{d}]$ fitted by the spatial discrete point set can be obtained. The total error equation of the space line is expressed by formula (4) as:

$$v = \begin{bmatrix} z_1 & 1 & 0 & 0 \\ 0 & 0 & 1 & z_1 \\ \vdots & \vdots & \vdots & \vdots \\ z_n & 1 & 0 & 0 \\ 0 & 0 & 1 & z_n \end{bmatrix} [\widehat{a}\ \ \widehat{b}\ \ \widehat{c}\ \ \widehat{d}]^T - \begin{bmatrix} x_1 \\ y_1 \\ \vdots \\ x_n \\ y_n \end{bmatrix} \tag{4}$$

TLS is a method of adjustment that takes into account measured data and parameter matrix errors. Calculate the statistically optimal solution of the parameters of the space equation by using the iterative algorithm of total least squares TLS.

From the perspective of the omnidirectionality of spatial three-dimensional lines, the three-dimensional imaging model has a bidirectional path mapping feature, namely: the image coordinate mapping of the three-dimensional coordinates and the three-dimensional coordinate mapping of the image coordinates. For image coordinate mapping of three-dimensional coordinates, any three-dimensional coordinates can be uniquely mapped into one image coordinate. That is, an image coordinate on an imaging plane in a three-dimensional imaging model can be regarded as a camera pointing from a certain three-dimensional coordinates. Intersection of a light beam at the optical center with the imaging plane. For three-dimensional coordinate mapping of image coordinates, each image coordinate cannot determine its unique three-dimensional coordinate. In this mapping relationship mode, the image coordinates and the three-dimensional coordinates are absolute mapped into a space beam emitted from the optical center. That is, there is a one-to-many mapping relationship between the image coordinates and the three-dimensional coordinates.

Figure 1 shows the mapping relationship between image coordinates and three-dimensional coordinates. Where: C_i denotes the optical center of the i-th camera. I_o^i denotes the i-th image coordinate of the o-th camera imaging plane. p_o^{ij} represents the three-dimensional coordinates of the i-th image coordinate in the spatial plane Y = j on the imaging plane of the o-th camera.

Assuming that the three-dimensional Y-axis value sequence of the image coordinate $I_o^i(x_o, y_o)$ is $Y_{set} = (j, \ldots, l, \ldots n)$, according to the three-dimensional imaging model formula (5) and the nonlinear distortion formula (6), the mapping relationship between the spatial three-dimensional point set and the image coordinates can be obtained as (7).

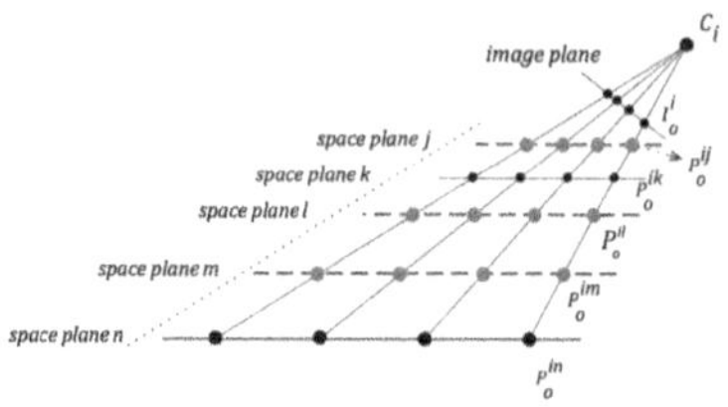

Fig. 1. The mapping relationship between image coordinates and 3D coordinates

$$\lambda \begin{bmatrix} u \\ v \\ 1 \end{bmatrix} = \begin{bmatrix} f_u & 0 & c_u \\ 0 & f_v & c_v \\ 0 & 0 & 1 \end{bmatrix} [R \quad T] \begin{bmatrix} X \\ Y \\ Z \\ 1 \end{bmatrix} = K[R \quad T] \begin{bmatrix} X \\ Y \\ Z \\ 1 \end{bmatrix} \tag{5}$$

$$\begin{cases} \Delta u = u(k_1 r^2 + k_2 r^4 + k_3 r^6) + \rho_1 (r^2 + 2u^2) + 2\rho_2 uv \\ \Delta v = v(k_1 r^2 + k_2 r^4 + k_3 r^6) + \rho_2 (r^2 + 2v^2) + 2\rho_1 uv \\ r^2 = u^2 + v^2 \end{cases} \tag{6}$$

$$f(x_o, y_o, Y_{set}) \Rightarrow (X_{set}, Z_{set}) \tag{7}$$

Among them, in formula (5), (u, v) is the image coordinate. (x, y, z) is world coordinate, f_u and f_v indicate the effective focal lengths of the u-axis and v-axis directions of the imaging plane coordinate system, respectively. (c_u, c_v) is the camera's main point coordinate, K is the camera's internal reference matrix. R and T respectively describe the rotation matrix and translation vector between the camera coordinate system and the world coordinate system. In formula (6), k_1, k_2, and k_3 are radial distortion coefficients, p_1 and p_2 are tangential distortion coefficients.

Calculate the sequence X_{set} and Z_{set} of the X-axis and Z-axis by mapping the relationship (7). The space points can be obtained: $U_{idx=1}^{cnt}(x_{idx}, y_{idx}, z_{idx}) = (X_{set}, Y_{set}, Z_{set})$, According to the overall error equation of the space light beam, the parameters of the spatial beam equation are calculated to be $[\hat{a} \; \hat{b} \; \hat{c} \; \hat{d}]$. Then the general equation of the spatial light beam equation is Eq. (8), and then the evaluation function for introducing the inner point of the spatial beam is shown in Eq. (9): among them, $\vec{n_1} = (1, 0, -\hat{a})$, $\vec{n_2} = (0, 1, -\hat{b})$. With the help of the evaluation function of the spatial light beam inside point, the reliability of the spatial light beam fitting y is defined as shown in Eq. (10):

$$\begin{cases} x - \hat{a}z - \hat{c} = 0 \\ y - \hat{b}z - \hat{d} = 0 \end{cases} \tag{8}$$

$$\Delta_{idx} = \frac{\left| (x_{idx} - \hat{a}z_{idx} - \hat{c})\vec{n_2} - (y_{idx} - \hat{b}z_{idx} - \hat{d})\vec{n_1} \right|}{\left| \vec{n_1} \times \vec{n_2} \right|} \tag{9}$$

$$\gamma = \frac{1}{\max(\Delta_{set}) - \min(\Delta_{set})} \tag{10}$$

To a certain extent, the credibility reflects the contribution of fitting errors caused by the interior points of the spatial light beam, so the credibility of the optimal solution parameter $[\widehat{a}\ \widehat{b}\ \widehat{c}\ \widehat{d}]$ is the most reliable.

2.2 Rigid Body Structure Positioning and Tracking

According to the basic principle of multi-view geometry, the three-dimensional space coordinates cannot be calculated from the two-dimensional image coordinates under a single viewing angle condition, and at least the three-dimensional coordinates must be clearly defined in order to calculate the three-dimensional coordinates. From the perspective of beam convergence, the intersection of two intersecting light beams of space is the spatial three-dimensional point to be solved.

In a three-dimensional space, multiple beams with convergence potential are defined as a cluster of beams. It is therefore that each beam cluster must form all intersections of a cluster of beams in a certain three-dimensional envelope space. As shown in Fig. 2, a cluster diagram of a cluster of beams is shown.

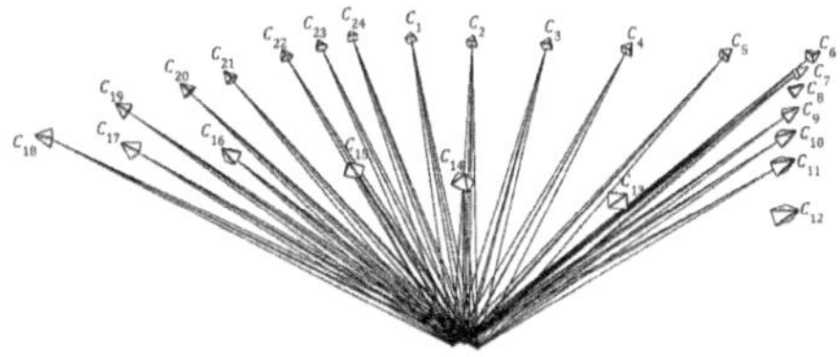

Fig. 2. Light beam clustering

Assume that the general equation of two spatial light beams is expressed as:

$$l_1 : \begin{cases} x - \widehat{a}_1 z - \widehat{c}_1 = 0 \\ y - \widehat{b}_1 z - \widehat{d}_1 = 0 \end{cases} \quad l_i : \begin{cases} x - \widehat{a}_i z - \widehat{c}_i = 0 \\ y - \widehat{b}_i z - \widehat{d}_i = 0 \end{cases} \tag{11}$$

Then, the intersection metric of two spatial light beams l_i and l_l is introduced as Eq. (12). Among, $\vec{m}_1 = (1, 0, -\widehat{a}_1)$, $\vec{m}_2 = (0, 1, -\widehat{b}_1)$, $\vec{m}_3 = (1, 0, -\widehat{a}_i)$, $\vec{m}_4 = (0, 1, -\widehat{b}_i)$, $(\vec{m}_i, \vec{m}_j, \vec{m}_k)$, denotes the mixed product of $\vec{m}_i, \vec{m}_j$ and $\vec{m}_k$.

$$\tau_i = \frac{\left|(\vec{m_1}, \vec{m_3}, \vec{m_4})\vec{m_2} - (\vec{m_2}, \vec{m_3}, \vec{m_4})\vec{m_1}\right|}{|M|}, \quad \text{where} \quad M = \begin{vmatrix} 1 & 0 & -\widehat{a_1} & -\widehat{c_1} \\ 0 & 1 & -\widehat{b_1} & -\widehat{d_1} \\ 1 & 0 & -\widehat{a_i} & -\widehat{c_i} \\ 0 & 1 & -\widehat{b_i} & -\widehat{d_i} \end{vmatrix} \tag{12}$$

According to the qualification condition of intersection factor $\tau_i < 0.15$, we get the beam cluster $L = \{l_1\} \cup \{l_p, \ldots, l_q\}$ belonging to beam l_1, and calculate the approximate intersection point of every two beams in beam cluster L, and form the intersection point set $P_L = \{P_{L1}, \ldots, P_{LN}\}$ of beam cluster. For the point set P_L, the concentration determination function that introduces the best P_L subset is shown in Eq. (13):

$$\max\left(\frac{N}{(X_{\max} - X_{\min})(Y_{\max} - Y_{\min})(Z_{\max} - Z_{\min})}\right) \tag{13}$$

At this time, the three-dimensional point density of the P_L subset P_S is the largest, so that the optimal cluster center coordinates of the point set P_L are calculated as:

$$O(x, y, z) = \frac{\sum_{i=1}^{N} p_{si}}{N} \tag{14}$$

Because the rigid body structure has the relatively unchanged internal structure, the relative physical properties between the nodes in the rigid-body structure are used to constrain the position identification of the rigid body structure. In this way, the matching relationship between the rigid body structure G_C in the current state and each node in the rigid body structure G_I in the initial state can be obtained:

$$U_{i=1}^{n} \rightleftharpoons U_{i=1}^{n} G_{Ii} \tag{15}$$

Since the rigid body structure has geometric shape invariance, the spatial transformation relationship of the rigid body structure can be decomposed into the synthesis of translation and rotation. The calculation relationship between the rotation matrix and the translation vector parameter between the two calculated from the matching relationship of the rigid body node is:

$$(G_{CX}, G_{CY}, G_{CZ})\Theta(G_{IX}, G_{IY}, G_{IZ}) = [R_G \quad T_G] \tag{16}$$

For a translational motion of a rigid body structure, it can be simply represented by a one-dimensional vector (T_{GX}, T_{GY}, T_{GZ}). However, for the rotational motion of a rigid body structure, the rotation angle calculation method using the rotation matrix directly will not only cause the linearization error, but also have the limitations of computational complexity, rotation non-uniformity and balance lock. Since the application of a rotation matrix to the transformation of a real three-dimensional scene is relatively complicated and limited, the expression of the rotation matrix is converted into an abstract quaternion representation of the movement around the axis. It not only avoids the limitation caused by the rotation matrix expression method, but also has the advantages of clear geometric meaning and simple calculation.

Use quaternion to describe the angle of rotation of a rigid body structure around an arbitrary axis in a three-dimensional space. Using the quaternion method to describe the rotation α Angle of rigid body structure in three-dimensional space around any axis A (a_x, a_y, a_z). Its quaternion Q is represented by Eq. (17).

$$Q(q_w, q_x, q_y, q_z) = \left[\cos\frac{\alpha}{2}, Ha_x, Ha_y Ha_z\right], \quad H = \sin\frac{\alpha}{2} \tag{17}$$

Then the expression of the quaternion [12] of the rotation matrix is formula (18). The pose parameters of the rigid body structure can be obtained by using the following calculations.

$$R_G = \begin{bmatrix} r_{11} & r_{12} & r_{13} \\ r_{21} & r_{22} & r_{23} \\ r_{31} & r_{32} & r_{33} \end{bmatrix} = \begin{bmatrix} 1 - 2q_y^2 - 2q_z^2 & 2q_xq_y + 2q_w q_z & 2q_xq_z - 2q_wq_y \\ 2q_xq_y - 2q_w q_z & 1 - 2q_x^2 - 2q_z^2 & 2q_yq_z + 2q_wq_x \\ 2q_xq_z + 2q_wq_y & 2q_yq_z - 2q_wq_x & 1 - 2q_x^2 - 2q_y^2 \end{bmatrix} \tag{18}$$

3 Experimental Results and Mathematical Analysis

In multi-view environment, the calculation of three-dimensional data greatly depends on the accuracy of the internal and external parameters of multiple cameras. Only by obtaining correct and reliable parameter values can the validity of the calculated data be fundamentally guaranteed. As shown in Fig. 3, the relative pose relationship between multiple cameras. From the picture, it can be clearly seen from the camera space position that their external parameters are displayed.

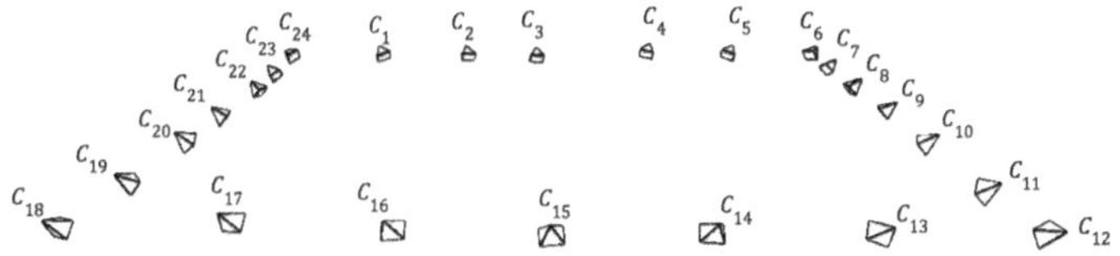

Fig. 3. Camera pose relationship

The process of locating and tracking the rigid body structure [13, 14] is to first shoot infrared images with multiple views through multiple infrared cameras. Then extract the three-dimensional coordinates of the reflective marker ball in the infrared image. Next, locate the rigid body structure and find the transformation relationship of the rigid body structure. Finally, the stereoscopic composite image is viewed in real time through the binding of the stereo virtual viewpoint. As shown in Fig. 4:

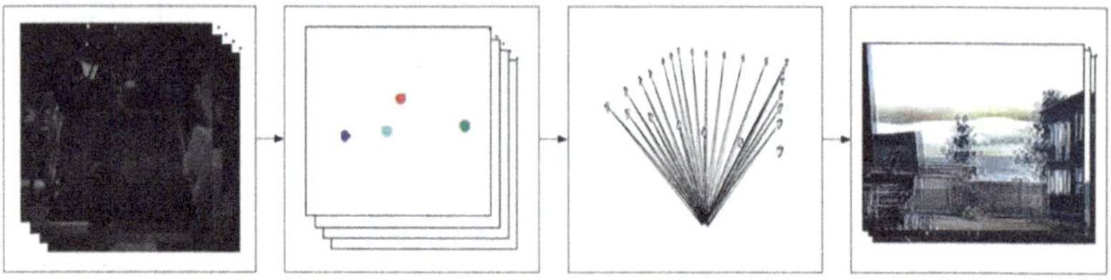

(*a*) Multi-angle image(*b*) Connected area marker(*c*) Rigid body recognition(*d*) Motion tracking

Fig. 4. Rigid body structure positioning and tracking

3.1 Rigid Body Structure Positioning and Tracking

In order to understand the three-dimensional coordinates of the reflective marker ball in space, it is necessary to photograph the reflective marker ball in different attitudes in a three-dimensional space from multiple angles. As shown in Fig. 5, infrared images captured by multiple viewing angle infrared cameras contain rich posture information of each reflective marker sphere, which provides basic guarantees for accurate coordinates of reflective marker spheres.

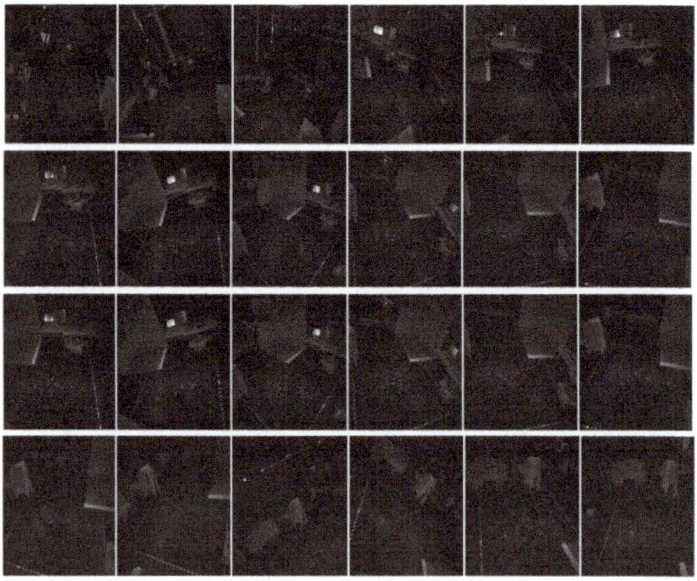

Fig. 5. Infrared image of every angle of view at a certain time

Because spherical reflectors are used, reflective markers should have a consistent two-dimensional imaging appearance from every angle. It is also true that the connecting area of the reflective marker sphere of any angle in the two-dimensional infrared image is almost similar to a circle. When marking the connected area, use the 8-neighborhood Two-Pass method to perform a two-pass scan on the binary image and mark all the connected areas in the binary image. Figure 6 shows the connected area marker diagram. It can be seen from the figure that the positions of different reflective marker balls in the binary image are given different colors.

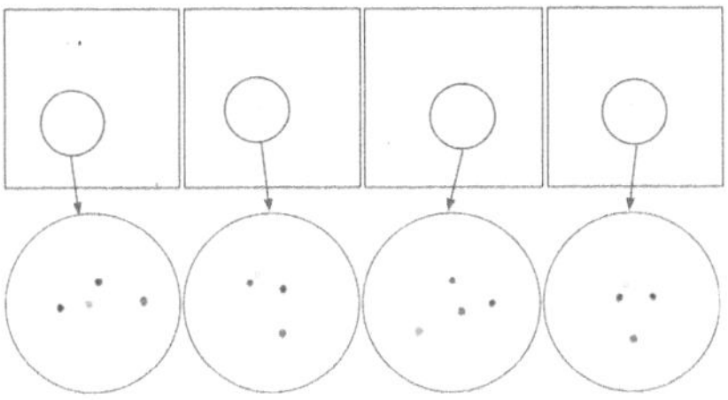

Fig. 6. Connected area marker

In the presence of interference markers, reflective markers marked by connected regions may be affected. Therefore, the coordinate position of the reflective marks of each connected region is limited by scale ratio of approximately 1.0, it can be expressed by Eq. (19):

$$(x, y) = \frac{\sum_{i=1}^{n}(x, y)}{n} \leftarrow \frac{x_{\max} - x_{\min}}{y_{\max} - y_{\min}} \approx 1.0 \tag{19}$$

3.2 Rigid Body Pose Estimation

The physical property constraints of the rigid body structure provide the criteria for identifying it, as shown in Fig. 7 for rigid body recognition.

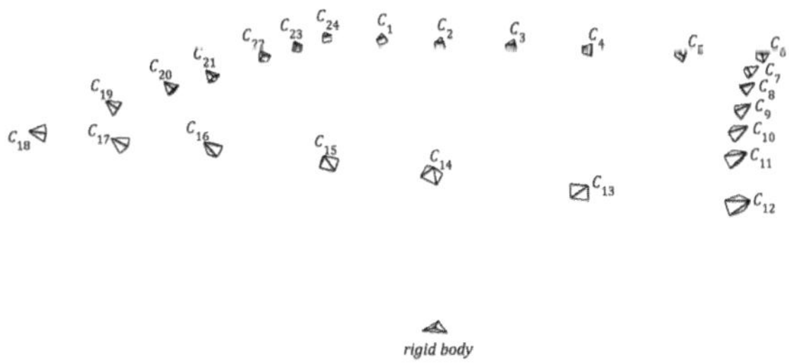

Fig. 7. Rigid body registration

In Table 1, the rigid body physical property calculation error records the rigid body size calculation value that starts at 1200 frames and is separated by 200 frames each time. Calculated by means of the mean of all recorded data: $error_{mean} \approx 0.12$ mm. From the calculated value of the rigid body size in Table 1, it can be seen that the error

Table 1. Rigid body physical property calculation error (mm)

Acquisition frame	Minimum error	Maximum error	Average error
1200	0.11	0.32	0.13
1400	0.08	0.26	0.09
1600	0.09	0.30	0.11
1800	0.13	0.29	0.11
2000	0.11	0.28	0.12
2200	0.09	0.33	0.13

between the true physical size value of the rigid body and the actual calculated value is very small, which indicates that the rigid body structure has a better reduction degree.

3.3 Virtual Observation Viewpoint Simulation

Using a rigid body motion, the first person perspective in a three-dimensional virtual scene simulates the combined motion of two virtual viewpoints, completes the capture and rendering of the stereoscopic composite image, and enhances the flexibility and efficiency of the virtual viewpoint for viewing the three-dimensional scene. The two viewpoints are placed in a parallel structure, and they are bound to a rigid body structure. The two virtual viewpoints will follow the rigid body for free movement. There is only horizontal disparity between the left and right two virtual viewpoints, that is, the relative translation vector between the two virtual viewpoints is (t_x, 0.0, 0.0). t_x determines the control of the stereoscopic effect. As shown in Fig. 8, the effect of

(a) Perspective1

(b) Perspective2

(c) Perspective3

Fig. 8. Three-dimensional virtual camera

stereoscopic picture output in different visual angles of stereoscopic visual points is shown. In a three-dimensional scene, the size of the virtual object is adjusted to be substantially the same as the actual size. The current value of t_x is 65 mm, which simulates the true pupil distance of the human eye.

4 Conclusion

This paper mainly studies the pose estimation method based on rigid body motion analysis. The method consists of two steps: fixed-point beam reconstruction, rigid body structure positioning and tracking. The posture estimation method based on rigid body motion analysis uses a rigid body structure for tracking registration of virtual scenery, making the registration process more flexible and results more accurate. Through the verification experiment of the pose estimation of rigid body structure, it can be seen that using the rigid body structure to simulate the three-dimensional virtual observation point of view can accurately control the first person perspective change, and the user experience in the roaming process is obviously higher than the traditional visual interaction mode.

Acknowledgments. This work is supported by the technology Program of Jilin Province, China (20170203004GX, 20170203003GX). Thanks to the reviewers for their comments and suggestions.

References

1. Carmigniani, J., Furht, B., et al.: Augmented reality technologies, systems and applications. Multimed. Tools Appl. **51**(1), 341–377 (2011)
2. Oskiper, T., Chiu, H.P., Zhu, Z.W., et al.: Stable vision-aided navigation for large-area augmented reality. In: Proceedings of the 2011 IEEE Virtual Reality Conference, pp. 63–70 (2011)
3. Luo, B., Wang, Y, T., et al.: Sensor fusion based head pose tracking for lightweight flight cockpit systems. Multimed. Tools Appl. **52**(1), 235–255 (2011)
4. Koller, D., Klinker, G., Rose, E., et al.: Real-time vision-based camera tracking for augmented reality applications. In: Proceedings of the ACM Symposium on Virtual Reality Software and Technology, pp. 87–94 (1997)
5. Huo, J., Yang, N., Yang, M.: Attitude measurement of spatial moving object based on vectors of light beams. J. Photonics **44**(7), 1–6 (2015)
6. Luo, B., Wang, Y.T., Shen, H., et al.: Overview of hybrid tracking in augmented reality. Acta Autom. Sin. **39**(8), 1185–1201 (2013)
7. Zhao, Y., Jingjiao, L.I., Wang, A., et al.: Tracking and registration algorithm of augmented reality on unknown scene based on IEKF-SLAM. Comput. Eng. (2016)
8. Shen, K., Peng, T.L.: Tracking registration of augmented reality marker based on SURF. Comput. Eng. **36**(13), 254–257 (2010)
9. Lara-Alvarez, C., Romero, L., Gomez, C.: Multiple straight-line fitting using a Bayes factor. Adv. Data Anal. Classif. **11**(1), 1–14 (2016)
10. Tsai, J.R.: Interval estimation for fitting straight line when both variables are subject to error. Comput. Stat. **28**(1), 219–240 (2013)

11. Shen, Y., Li, B., Chen, Y.: An iterative solution of weighted total least-squares adjustment. J. Geodesy **85**(4), 229–238 (2010)
12. Bird, J., Arden, D.: Indoor navigation with foot-mounted strap down inertial navigation and magnetic sensors. IEEE Wirel. Commun. **18**(2), 28–35 (2011)
13. Foytik, J., Asari, V.K.: A two-layer framework for piecewise linear manifold-based head pose estimation. Int. J. Comput. Vis. **101**(2), 270–287 (2013)
14. Ershadi-Nasab, S., Noury, E., Kasaei, S., et al.: Multiple human 3D pose estimation from multiview images. Multimed. Tools Appl. **12**, 1–29 (2017)

An ORB Corner Tracking Method Based on KLT

Qi Naixin[1(✉)], Li Xiaofeng[1], Yang Xiaogang[1], Li Chuanxiang[1], Cao Lijia[2], and Zhang Shengxiu[1]

[1] Xi'an Research Institute of High-Tech Xi'an, Xi'an, China
qinaixin2015@sina.com
[2] College of Automation and Information Engineering, Sichuan University of Science and Engineering, Zigong, Sichuan, China

Abstract. In order to solve the wrong corner tracking problem, a novel corner tracking method is proposed combining the ORB algorithm and the KLT algorithm. In this proposed method, the ORB detector is introduced to extract the corners, and the ORB descriptor is used to describe the tracked corners in KLT tracking process. The descriptors of pre-frame and cur-frame can be obtained which is used to calculate the similarity coefficient by brute force matching method. Then the wrong tracked corners can be removed by judging the similarity coefficient. Several simulations are made to examine the performance of the proposed method. The experimental results show that the proposed method can remove the wrong tracked corners and achieve robust corner tracking performance, which can be better used in the visual odometer.

Keywords: ORB · KLT · FAST · BRIEF · Visual odometer

1 Introduction

Visual odometer [1] (VO) can estimate camera's motion by computing the corners' position in the image sequences. It is widely used in robotics and driverless car. The VO method requires extracting and tracking corners in an image stream. At present, the main corner extracting methods are FAST [2] (Features from Accelerated Segment Test), Harris, SIFT [3], SURF [4], and ORB [5] etc. The ORB corner detector is proposed in [6]. It is based on the well-known FAST corner detector [7] and the developed BRIEF descriptor [8]. As a result, ORB can operates faster than SIFT at two orders of scale and has a good performance.

The kernel of corner matching problem is finding the corresponding of corners in space. The corner matching strategy can be summarized as two categories: corner matching based on Full Search (FS) and corner matching based on optimization. The former calculates good matching of corners by searching the matches between two known corner sets, such as Harris corner matching [9] and SIFT feature matching [10] etc. Those methods are sensitive to corners and search the matches by traversal method; the prior information in the image stream is not utilized. As a result, the computational of those methods are complex, which are not used in a real time system in general. The corner matching methods based on optimization use the corners set in current image to

K. Deng et al. (Eds.): ICMIR 2018, AISC 856, pp. 755–763, 2019.
https://doi.org/10.1007/978-3-030-00214-5_94

estimate the best position of corners of the next image, such as Optical Flow algorithm. KLT is a typical corners tracking method based on Sparse Optical Flow [11]. The KLT algorithm extracts the sparse and distinct corners firstly and then completes the corners matching by optimizing the positions of those extracted corners in the image.

At first, the KLT tracking algorithm is proposed by Lucas Kanade [12]. It is a typical corner tracking algorithm based on the continuous information between image streams, which extracts the corners by the KLT detector and matches those corners by KLT matcher. With the development of research, the KLT tracking algorithm has been more and more widely used. Ajmal S. Mian et al. realize the tracking of the military planes with simple structure based on the KLT tracking algorithm [13]. F. Abdat et al. utilize the KLT algorithm to achieve the facial expression recognition [14]. Meghna-Singh et al. improve the robustness of the KLT algorithm [15]. In the optimal position matching of corners, the mean weight function is substituted by the Gaussian weight function, which is robust to noise.

In real applications, the KLT algorithm has two problems: first, the KLT detector need calculate the second derivative of the image and smoothes it using Gaussian kernel, as a result of that the extracted corners are not accurate; second, the precision of matching will be reduced when the objective moves quickly, which can lead to tracking inaccuracy problem. In this paper, the ORB algorithm and KLT algorithm are combined to solve the above problems, and achieve corners tracking robustly.

2 Robust Corner Tracking

2.1 Corner Detecting

ORB is consisted of the Oriented FAST detector and the BRIEF descriptor, which is invariant to image rotation, scale, affine. Because the normalized image centrality is set as the corner location, the process of FAST feature detecting can be ignored, which improving the computational speed.

In order to add the directions for the original FAST detector, the moments $\boldsymbol{m}$ of a patch is defined as:

$$\boldsymbol{m}_{pq} = \sum_{x,y} x^p y^q I(x,y), \tag{1}$$

where $I(x, y)$ is the gray value of the point (x, y). The intensity centroid can be calculated by the moments $\boldsymbol{m}$:

$$C = \left(\frac{\boldsymbol{m}_{10}}{\boldsymbol{m}_{00}}, \frac{\boldsymbol{m}_{01}}{\boldsymbol{m}_{00}}\right). \tag{2}$$

In this way, the direction of angular point θ is defined as:

$$\theta = \operatorname{atan2}(m_{01}, m_{10}), \tag{3}$$

where atan 2 is a function calculating the arc tangent based on the signs of x and y. When the main direction of angular point is obtained, the BRIEF descriptor can be extracted. Then a binary test ϖ in image patch P, is defined by:

$$\varpi(P;x,y) := \begin{cases} 1 : P(x) < P(y) \\ 0 : P(x) \geq p(y) \end{cases}, \tag{4}$$

In which $P(x)$ is the intensity at point x. The BRIEF descriptor is a vector of n, which can be defined by:

$$f_n(P;x,y) := \sum_{i=1}^{n} 2^{i-1} \varpi(P;x,y). \tag{5}$$

In order to make BRIEF descriptor perform rotation invariant, "Steer BRIEF" is used, which is a binary code string with length n by n point pairs. We define a matrix $\boldsymbol{S}$ based on these $2n$ points (x_i, y_i), $i = 1, 2, \ldots, 2n$ as shown in Eq. (6).

$$\boldsymbol{S} = \begin{pmatrix} x_1, \cdots, x_n \\ y_1, \cdots, y_n \end{pmatrix}. \tag{6}$$

The corresponding rotation matrix $\boldsymbol{R}_\theta$ can be gained by the patch orientation θ:

$$\boldsymbol{R}_\theta = \begin{pmatrix} \cos\theta & -\sin\theta \\ \sin\theta & \cos\theta \end{pmatrix}. \tag{7}$$

A "steered" version $\boldsymbol{S}_\theta$ of $\boldsymbol{S}$ can be adopted as:

$$\boldsymbol{S}_\theta = \boldsymbol{R}_\theta \boldsymbol{S}. \tag{8}$$

Now the ORB descriptor with rotation invariance can be described as:

$$g_n(P,\theta) := f_n(P) | (x_i, y_i) \in T, \tag{9}$$

where T is a set consisted of $\boldsymbol{S}_\theta$.

2.2 Corner Tracking

The KLT algorithm is a classical matching method based on features. It assumes a character window W with feature texture information. The change of pixels is described by a translation model. $\boldsymbol{I}(x, y, t)$ is defined as the image frame at time t, and $\boldsymbol{I}(x, y, t+\tau)$ is defined as the image frame at time $t+\tau$ with their locations satisfy the following property:

$$\boldsymbol{I}(x, y, t+\tau) = \boldsymbol{I}(x - \Delta x, y - \Delta y), \tag{10}$$

where $\boldsymbol{d} = (\Delta x, \Delta y)$ is called the point displacement at $\boldsymbol{X}(x, y)$.

Then we chose $\boldsymbol{d}$ to minimize the residue error ε defined by SSD (Sum of Squared intensity Differences). Given two images $\boldsymbol{I}$ and $\boldsymbol{J}$, the residue error ε is defined by:

$$\varepsilon = \iint_W [\boldsymbol{J}(\boldsymbol{X}+\boldsymbol{d}) - \boldsymbol{I}(\boldsymbol{X})]^2 \omega(\boldsymbol{X}) \mathrm{d}\boldsymbol{X}, \tag{11}$$

where W is the given window, and $\omega(\boldsymbol{X})$ is a weighting function which can be set to 1 in the simple case. In general, the value of vector $\boldsymbol{d}$ is smaller than that of $\boldsymbol{X}$, which can be set to zero. We then linearize the resulting system $\boldsymbol{J}(\boldsymbol{X}+\boldsymbol{d})$ at point $\boldsymbol{a} = [a_x, a_y]^{\mathrm{T}}$ using Taylor expansion:

$$\boldsymbol{J}(\xi) \approx \boldsymbol{J}(\boldsymbol{a}) + (\xi_x - \boldsymbol{a}_x)\frac{\partial \boldsymbol{J}}{\partial x}(\boldsymbol{a}) + (\xi_y - \boldsymbol{a}_y)\frac{\partial \boldsymbol{J}}{\partial y}(\boldsymbol{a}), \tag{12}$$

where $\xi = [\xi_x, \xi_y]^{\mathrm{T}}$.

Assuming $\boldsymbol{X}+\boldsymbol{d} = \xi$, $\boldsymbol{X} = \boldsymbol{a}$ and removing the higher order term based on the definition of derivative, the Eq. (13) can be gained.

$$\boldsymbol{J}(\boldsymbol{X}+\boldsymbol{d}) \approx \boldsymbol{J}(\boldsymbol{X}) + \boldsymbol{d}_x\frac{\partial \boldsymbol{J}}{\partial x}(\boldsymbol{X}) + \boldsymbol{d}_y\frac{\partial \boldsymbol{J}}{\partial y}(\boldsymbol{X}). \tag{13}$$

If the first step Taylor coefficient g is defined as:

$$g = \left[\frac{\partial \boldsymbol{J}}{\partial x}(\boldsymbol{X}) \frac{\partial \boldsymbol{J}}{\partial y}(\boldsymbol{X})\right]^{\mathrm{T}}, \tag{14}$$

then the Eq. (13) can be rewritten as

$$\boldsymbol{J}(\boldsymbol{X}+\boldsymbol{d}) \approx \boldsymbol{J}(\boldsymbol{X}) + \boldsymbol{g}^{\mathrm{T}} \cdot \boldsymbol{d}. \tag{15}$$

Thus, Eq. (11) is rewritten as:

$$\begin{aligned}\boldsymbol{\varepsilon} &= \iint\limits_W [\boldsymbol{J}(\boldsymbol{X}) - \boldsymbol{I}(\boldsymbol{X}) + \boldsymbol{g}^{\mathrm{T}} \cdot \boldsymbol{d}]^2 \omega(\boldsymbol{X}) \mathrm{d}\boldsymbol{X} \\ &= \iint\limits_W (\boldsymbol{h} + \boldsymbol{g}^{\mathrm{T}} \cdot \boldsymbol{d})^2 \omega(\boldsymbol{X}) \mathrm{d}\boldsymbol{X}\end{aligned} \tag{16}$$

where $\boldsymbol{h} = \boldsymbol{J}(\boldsymbol{X}) - \boldsymbol{I}(\boldsymbol{X})$. The derivative of $\boldsymbol{d}$ is calculated by (16):

$$\frac{\partial \boldsymbol{\varepsilon}}{\partial \boldsymbol{d}} = \iint\limits_W [\boldsymbol{J}(\boldsymbol{X}) - \boldsymbol{I}(\boldsymbol{X}) + \boldsymbol{g}^{\mathrm{T}} \cdot \boldsymbol{d}] \boldsymbol{g}(\boldsymbol{X}) \omega(\boldsymbol{X}) \mathrm{d}\boldsymbol{X}. \tag{17}$$

In order to compute the displacement vector $\boldsymbol{d}$, we assume $\frac{\partial \varepsilon}{\partial \boldsymbol{d}} = 0$, in which $\boldsymbol{\varepsilon}$ achieves a minimum. Equation (17) is transferred into:

$$\begin{aligned}\iint\limits_W [\boldsymbol{J}(\boldsymbol{X}) - \boldsymbol{I}(\boldsymbol{X})] \boldsymbol{g}(\boldsymbol{X}) \omega(\boldsymbol{X}) \mathrm{d}\boldsymbol{X} &= - \iint\limits_W \boldsymbol{g}^T(\boldsymbol{X}) \boldsymbol{d} \boldsymbol{g}(\boldsymbol{X}) \omega(\boldsymbol{X}) \mathrm{d}\boldsymbol{X} \\ & \left[\iint\limits_W \boldsymbol{g}^T(\boldsymbol{X}) \boldsymbol{g}(\boldsymbol{X}) \omega(\boldsymbol{X}) \mathrm{d}\boldsymbol{X}\right] \boldsymbol{d}\end{aligned} \tag{18}$$

If we define:

$$\begin{aligned}\boldsymbol{Z} &= \iint\limits_W \boldsymbol{g}^T(\boldsymbol{X}) \boldsymbol{g}(\boldsymbol{X}) \omega(\boldsymbol{X}) \mathrm{d}\boldsymbol{X} \\ &= \iint\limits_W \begin{bmatrix} g_x^2 & g_x g_y \\ g_x g_y & g_y^2 \end{bmatrix} \omega(\boldsymbol{X}) \mathrm{d}\boldsymbol{X}\end{aligned} \tag{19}$$

$$\begin{aligned}e &= \iint\limits_W [\boldsymbol{I}(\boldsymbol{X}) - \boldsymbol{J}(\boldsymbol{X})] \boldsymbol{g}(\boldsymbol{X}) \omega(\boldsymbol{X}) \mathrm{d}\boldsymbol{X} \\ &= \iint\limits_W [\boldsymbol{I}(\boldsymbol{X}) - \boldsymbol{J}(\boldsymbol{X})] \begin{bmatrix} g_x \\ g_y \end{bmatrix} \omega(\boldsymbol{X}) \mathrm{d}\boldsymbol{X}\end{aligned} \tag{20}$$

Equation (18) is rewritten as:

$$\boldsymbol{Zd} = \boldsymbol{e}. \tag{21}$$

The iteration of Newton processes using Eq. (21) until the result precision reaches the threshold. Now the corner tracking is achieved and the good matching is gained. The final solution is obtained by:

$$\boldsymbol{d}_{k+1} = \boldsymbol{d}_k + \begin{bmatrix} g_x^2 & g_x g_y \\ g_x g_y & g_y^2 \end{bmatrix} \bullet \left[\iint_W [\boldsymbol{I}(\boldsymbol{X}) - \boldsymbol{J}(\boldsymbol{X})] \begin{bmatrix} g_x \\ g_y \end{bmatrix} \omega(\boldsymbol{X}) \mathrm{d}\boldsymbol{X} \right], \tag{22}$$

where $\boldsymbol{d}$ is the translation of character window, and $\boldsymbol{d}_k$ is the result of $\boldsymbol{d}$ calculated by Newton iteration at time k. In the iteration process, $\boldsymbol{d}$ is initialized by $\boldsymbol{d}_0$ which is set to 0 in the paper.

If objective moves slowly, the translation between frames will be small. In this case, KLT method not only reduces the searching region, but also has high matching accuracy.

2.3 Procedure of the Proposed Method

The KLT algorithm can speed up the matching process by reducing the searching region, and the ORB algorithm can improve the accuracy of the corners' location. In this paper, we combine their advantages to achieve a robust corners tracking, which is described as follows:

1. Input the initializing image frame, and extract corners using ORB algorithm as the initializing corners to be tracked in next frame;
2. Track the corners in cur-frame (current frame image) using KLT algorithm;
3. Describe every corner tracked in step 2 using ORB descriptor;
4. Calculate the similarity coefficient by brute force matching between descriptor of pre-frame (previous frame image) and descriptor of cur–frame for every corner. The distance is the Hamming distance;
5. If the coefficient calculated in step 4 reaches the value of threshold, the corner is chosen as the good tracked corner, vice versa. Removing the bad tracked corners;
6. Check the number of the good tracked corners. If the number is lower than the value of threshold, extract new corners using ORB detector. The full corners are set as the initializing corners to track in next frame.

3 Experimental Results

In this part, we evaluate the performance of the proposed method using two datasets: image sequence collected in dynamic scene and image sequence collected in static background. For each image sequence, we extract corners using ORB detector in initial and track corners using the KLT algorithm and the proposed algorithm (ORB-KLT). The experimental results can be seen in Figs. 1, 2, 3 and 4.

Figures 1 and 2 shows the results for the image sequence collected in dynamic scene. In this part experience, the image sequence is collected by the moving camera. The sequence includes 240 images. The corners are tracked by KLT algorithm and the proposed algorithm, respectively. In the results, eight frames of the tracked results are given. The first frame shows the initial corners, and others show the tracked corners. In the last six frames, we can find that there are wrong tracked corners in the results obtained by KLT algorithm, which are removed in the results obtained by ORB-KLT. The corners tracked by ORB-KLT are more stable than that tracked by KLT.

Figures 3 and 4 shows the corners tracking results for image sequence collected in static background. In this part experiment, the image sequence is collected by the camera in static. The sequence includes 173 images. The corners are also tracked by KLT algorithm and the proposed algorithm, respectively. In the results, six frames of the tracked results are given. The first frame shows the initial corners, and the others show the tracked corners. From the 61th, 91th, 121th, and 151th frame, we can find that some wrong tracked corners exist. If the tracked corners are used in the visual odometer, estimating errors will be lead-in, which situation can be avoided by using the proposed algorithm.

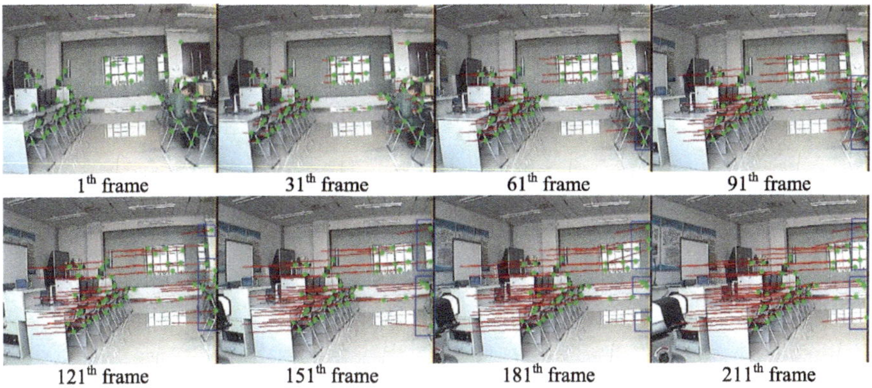

Fig. 1. Results for the image sequence collected in dynamic scene using KLT algorithm

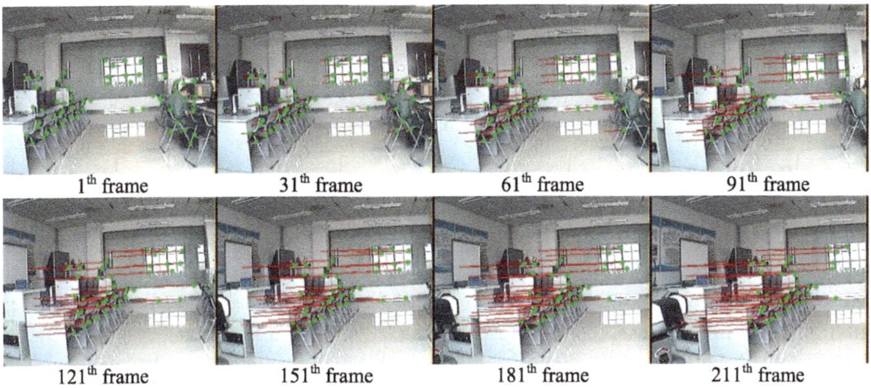

Fig. 2. Results for the image sequence collected in dynamic scene using ORB-KLT algorithm

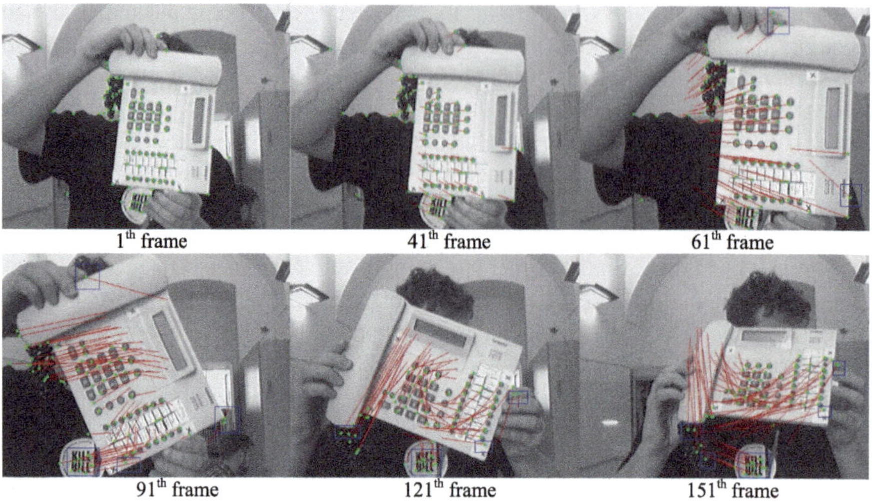

Fig. 3. Results for the image sequence collected in static background using KLT algorithm

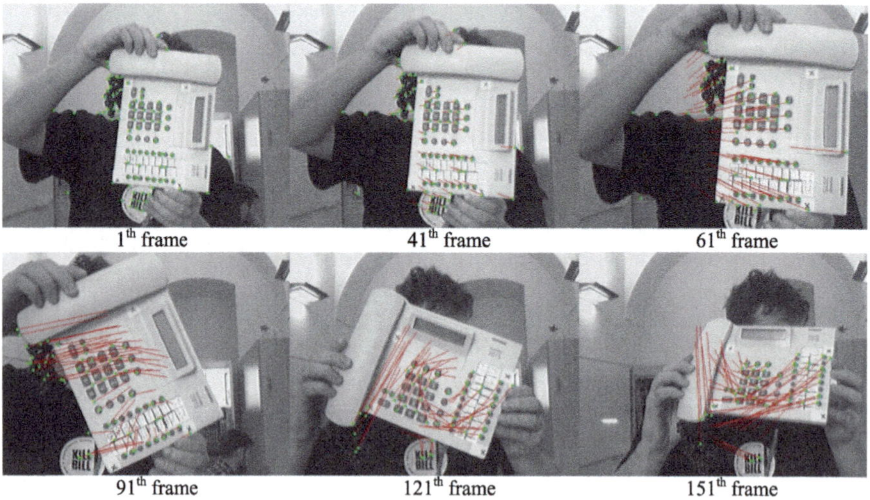

Fig. 4. Results for the image sequence collected in static background using ORB-KLT algorithm

4 Conclusion

KLT algorithm can track the corners based on optimization, but the tracking inaccuracy problem will be lead-in when calculating the second derivative of the image and smoothing using Gaussian kernel. In order to solve the problem, we proposed a novel method combining the ORB algorithm and KLT algorithm in this paper. In the basis of KLT algorithm, the ORB detector is introduced to extract corners and the ORB descriptor is utilized to describe the tracked corners. In the end, the wrong tracked corners are removed by judging the dependency of descriptors between pre-frame and cur-frame. The proposed method can achieve good corner tracking performance highlighted by the experimental results and can be better used in VO.

Acknowledgments. This work is supported by Natural Science Foundation of China under Grant 61773389, and Research Foundation for the Introduction of Talent under Grant 2018RCL18.

References

1. Davide, S., Friedrich, F.: Visual odometry: Part I: the first 30 years and fundamentals. IEEE Robot. Autom. Mag. **18**(4), 80–92 (2011)
2. Luo, Y., Fu, Y., Zhang, Y.: A monocular-vision real-time matching algorithm based on FAST corners and Affine-improved random ferns. ROBOT **36**(3), 271–278 (2014)
3. Chen, Y.H., Lin, H.Y.S., Su, C.W.: Full-frame video stabilization via SIFT feature matching. In: 2014 Tenth International Conference on Intelligent Information Hiding and Multimedia Signal Processing (IIH-MSP), pp. 361–364. IEEE (2014)
4. Jia, C., Panfeng, H.: Research of a real-time feature point tracking method based on the combination of improved SURF and P-KLT algorithm. Acta Aeronautica et Astronatica Sinica **34**(5), 1204–1214 (2013)
5. Zhuo, L., Geng, Z., Zhang, J., et al.: ORB feature based web pornographic image recognition. Neurocomputing **173**, 511–517 (2016)
6. Rublee, E., Rabaud, V., Konolige, K., et al.: ORB: an efficient alternative to SIFT or SURF. In: 2011 IEEE International Conference on Computer Vision (ICCV), pp. 2564–2571. IEEE (2011)
7. Rosten, E., Drummond, T.: Machine learning for high-speed corner detection. In: Computer vision–ECCV 2006, pp. 430–443 (2006)
8. Calonder, M., Lepetit, V., Strecha, C., et al.: Brief: binary robust independent elementary features. In: Computer Vision–ECCV 2010, pp. 778–792 (2010)
9. Huang, K.Y., Tsai, Y.M., Tsai, C.C., et al.: Feature-based video stabilization for vehicular applications. In: 2010 IEEE 14th International Symposium on Consumer Electronics (ISCE), pp. 1–2. IEEE (2010)
10. Lowe, D.G.: Distinctive image features from scale-invariant keypoints. Int. J. Comput. Vis. **60**(2), 91–110 (2004)
11. Nawaz, M.W., Bouzerdoum, A., Phung, S.L.: Optical flow estimation using sparse gradient representation. In: 2011 18th IEEE International Conference on Image Processing (ICIP), pp. 2681–2684. IEEE (2011)
12. Shi, J.: Good features to track. In: 1994 IEEE Computer Society Conference on Computer Vision and Pattern Recognition, Proceedings CVPR 1994, pp. 593–600. IEEE (1994)
13. Mian, A.S.: Realtime visual tracking of aircrafts. In: 2008 Computing: Techniques and Applications, DICTA 2008, Digital Image, pp. 351–356. IEEE (2008)
14. Abdat, F., Maaoui, C., Pruski, A.: Real time facial feature points tracking with Pyramidal Lucas-Kanade algorithm. In: 2008 the 17th IEEE International Symposium on Robot and Human Interactive Communication, RO-MAN 2008, pp. 71–76. IEEE (2008)
15. Singh, M., Mandal, M., Basu, A.: Robust KLT tracking with Gaussian and laplacian of gaussian weighting functions. In: 2004 Proceedings of the 17th International Conference on Pattern Recognition, ICPR 2004, vol. 4, pp. 661–664. IEEE (2004)

Graphite Line on Paper as an Aqueous Chemical Sensor

Jasem Alenezi, Jennifer Czechowski, Justin Niver, Sulaiman Mohammad, and Reza Rashidi(✉)

Mechanical Engineering Technology, State University of New York, Alfred State College, Alfred, NY 14802, USA
rashidr@alfredstate.edu

Abstract. This work investigates the ability of a graphite line drawn on a sheet of paper to be used as a sensor to detect different types of liquid chemicals. To the best of our knowledge, no work has been reported on identifying liquid chemicals based on their electrical resistance using graphite on paper. The sensor is capable of distinguishing between different types of liquids including acids, bases, solvents and water. A graphite pencil was used to sketch a 40 mm long by 6 mm wide line on a piece of white paper. The graphite on paper test strip was placed in a test fixture equipped with electrical contacts. The electrical contacts were attached to a data acquisition device (DAQ) along with LabVIEW software to measure and record the voltage outputs which were later used to calculate the electrical resistance. To make a consistent measurement in different samples, a test fixture was designed and fabricated using a 3D printing machine. The solvent tested in this experiment was acetone, the acid tested was nitric acid, and the base tested was tetramethylammonium hydroxide. Two droplets of each chemical were dropped onto the paper through a hole in the test fixture at room temperature. The voltage was recorded every millisecond for about 10 s after the liquid was dropped on the paper. The plots of electrical resistance values versus time showed a distinct curve for each type of liquid which can be used for liquid identification purposes.

Keywords: Chemical sensor · Graphite on paper · Chemical detection
Acid · Base · Solvent

1 Introduction

Graphite on paper sensors have attracted lots of attention because of their potential to produce disposable, environmentally friendly, low cost, thin, and malleable devices such as tactile sensors [1], piezoresistive force sensor [2], strain sensors for human motion detection [3, 4], thermal flow sensors for noninvasive respiratory monitoring [5], and temperature sensors [6] and other flexible electronics [7]. Unlike usual substrates such as glass or plastic, papers are porous and allow for the graphite to be deposited over a large surface area. When the graphite is drawn on the paper, graphite material enter into the pores of the paper surface. This graphite on paper can then be used as the sensing element in devices.

K. Deng et al. (Eds.): ICMIR 2018, AISC 856, pp. 764–770, 2019.
https://doi.org/10.1007/978-3-030-00214-5_95

Graphite is an allotrope of carbon and has a layered structure of carbon atoms. Its atoms have three covalently bonding sites and one unrestricted electron to move which makes graphite electrically conductive [8]. In a recent study, a highly sensitive graphite on paper sensor with a laminated plastic coating was determined to be a humidity-resistive and durable mechanical sensor that detects tactile pressure. The study determined that the plastic lamination and production technique of the graphite on paper sensor could enable the mass production of a low cost micro-electro-mechanical (MEMS) device [1].

In this paper, it is demonstrated that graphite on paper can also be used as an aqueous chemical sensor capable of detecting different types of liquid chemicals such as acids, bases, and solvents. The advantages of using graphite on paper for chemical detection are low cost, disposability, and easy fabrication, however, as a drawback, graphite in such devices may also be sensitive to the surrounding temperature and vibrations which may affect the sensor reading.

2 Design, Fabrication and Experiments

In order to conduct consistent experiments, a special test fixture was designed to hold the graphite on paper test strip, as shown in Fig. 1. The fixture was designed using SOLIDWORKS and fabricated using an Alunar M505 3D printer. The material used for the fixture was black Pxmalion 1.75 mm printer plastic. The fixture consisted of a 30 mm long × 10 mm width × 12 mm high upper part and included a 6 mm diameter hole located in the center. The upper part also had two 2.5 mm diameter holes located at both ends where conductive copper wires were soldered in place for electrical connection. The lower part of the fixture has the dimensions of 50 mm long × 12 mm wide × 3 mm thick. The graphite on paper test strip was then placed in the lower piece with the upper piece resting on top. This avoided the copper wires to be directly in contact with the chemical.

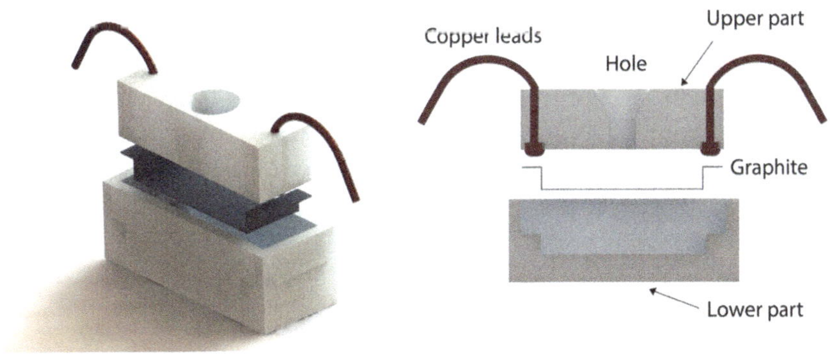

Fig. 1. Test fixture designed to hold the graphite on paper test strip.

A 40 mm long × 6 mm wide line was sketched on a piece of white acid-free 75 g/m^2 MagTec sheet of paper using a Bic #2 graphite pencil. The paper was placed in the test fixture equipped with electrical contacts. The electrical contacts were attached to a data acquisition device (DAQ) along with LabVIEW software to allow for the recording of changes in voltage which was later used to calculate the electrical resistance.

A National Instruments USB 6003 multifunctional data acquisition device (DAQ) was set up to measure the changes in voltage. The circuit was supplied with an input of 10 V and a known 3.84 kΩ resistor in series. The circuit was wired in series with the test fixture contacts as shown in Fig. 2. The voltage drops across the known resistor and the graphite on paper test strip in the circuit were measured by the DAQ card and recorded by the LabVIEW software every millisecond for 10 s. Changes in the resistance of the graphite on paper sensor in the presence of different liquids were computed using the measured voltages based on Kirchhoff's law.

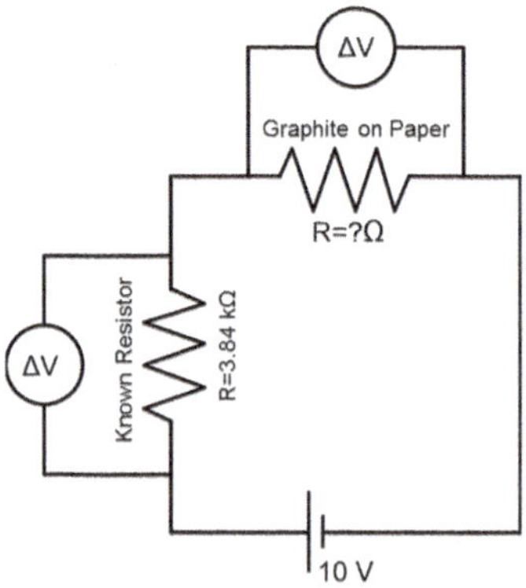

Fig. 2. Measurement setup including the NI DAQ device, 3.84 kΩ resistor and the test fixture used for computing the resistance of the graphite on paper.

Fig. 3. Experimental setup used for testing different types of liquids. The graphite on paper test fixture connected in series to the data acquisition device (DAQ) for measuring the voltage across the strip.

Figure 3 shows the experimental setup used for testing different types of liquids. Two droplets of each liquid were dropped onto the graphite test strip using a disposable precision pipette through the center hole of the test fixture. The voltage was then recorded every millisecond for 10 s after the liquid was dropped on the paper. After each test, the fixture was cleaned and dried. Each liquid was tested using a new test strip. All experiments were conducted at room temperature.

3 Results and Discussion

The experiments performed on the chemicals are categorized to three groups of solvents, acids, and bases in this section. The solvent tested in this experiment was 98% acetone. Furthermore, 70% nitric acid (HNO_3) as the common acid was tested. In addition, 25% tetramethylammonium hydroxide (TMAH) was tested in this experiment. The results showed that the resistance versus time curves obtained from the experiments could be closely duplicated for each liquid. The experimental results were analysed in two different ways. First, the resistance was analysed one to two seconds after a change in the graphite resistance was observed and the change was considered as a criterion for chemical detection. Second, the final resistance about 10 s after observing a change in resistance was considered as the second criterion for chemical detection. Figure 4 shows the results of the experiments on the three aforementioned liquids. The resistance in the curve associated with the acetone initially decreased and then gradually increased when the graphite was exposed to the solvent. The curve associated with the nitric acid showed a significant decrease in the resistance of the graphite on paper when the graphite is exposed to the acid. The resistance drop in the curve associated with the TMAH was gradual.

In order to quantitatively distinguish the chemicals with the second aforementioned criterion, the final resistance of the graphite on paper in 10 s after the change in resistance was calculated. Figure 5 shows the initial resistance of the graphite and its final resistance after 10 s for each chemical. The figure also includes the final resistance values after 1 s for a comparison. Furthermore, Fig. 6 shows the percentage change in resistance after 1 s and 10 s of initial change. The largest resistance drops occurred in the nitric acid and the smallest resistance drops occurred in the TMAH. In addition, the acetone showed an increase in the resistance.

The data shows the acetone had increasing final resistances after 10 s. This may be due to the fact that the acetone evaporate quickly at room temperature. The heat required for the acetone to evaporate is captured from the surrounding area. Therefore, the immediate drops in temperature due to the evaporation of acetone caused a sharp increase in the resistance due to the negative temperature coefficient of resistance of the graphite [5] which is reflected in the graph. The data collected from this experiment shows that the temperature of the fluid has a direct effect on the resistance.

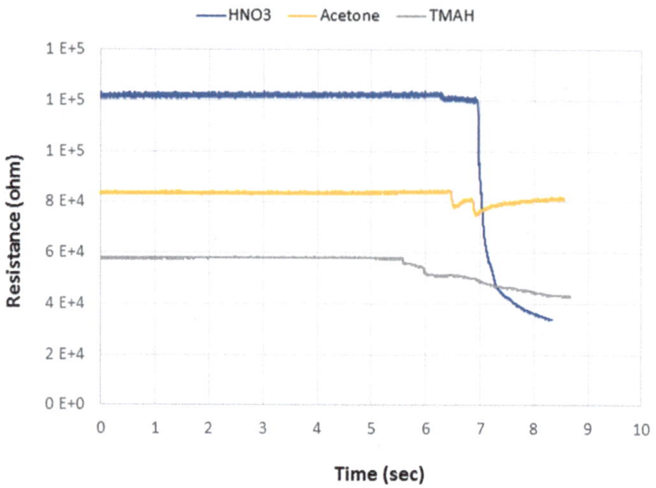

Fig. 4. Resistance of graphite on paper exposed to different solvent, acid and base.

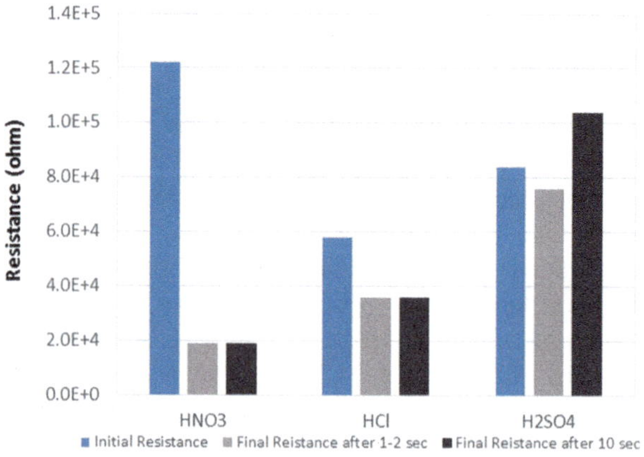

Fig. 5. Initial resistance of graphite on paper and its resistance after 1–2 s and 10 s exposure to different chemicals.

The nitric acid in this experiment showed an initial drop in resistance. Acids are electrolytes and contain a concentration of dissolved hydrogen ions (H^+). The ions act as charge carriers and conduct electricity. Therefore, the resistance of the graphite on paper would decrease when exposed to the acid. Some of the strong acids such as nitric acid dissociates completely and produce more H^+ ions and thus indicate a significant drop in resistance. The TMAH is a basic electrolyte and has a concentration of dissolved hydroxide ions (OH^-). The ions act as charge carriers and conduct electricity. Weak or strong bases also affect their electrical conductivity. The TMAH had dropping its resistance value.

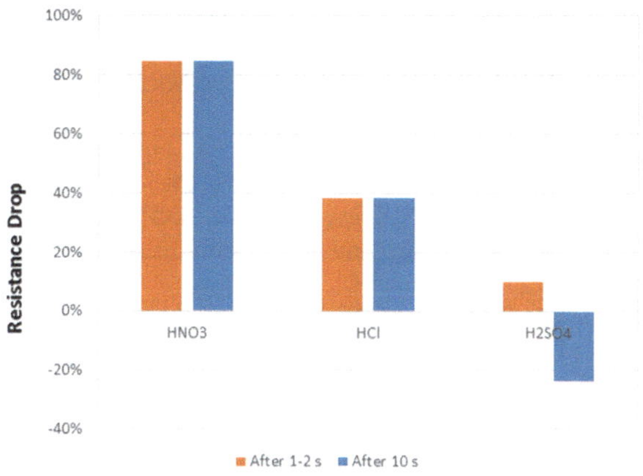

Fig. 6. Change in resistance of graphite on paper exposed to different chemicals after 1–2 s and 10 s.

4 Conclusions

This work investigated the ability of a graphite line drawn on a sheet of paper to be used as a sensor to detect different types of aqueous chemicals. The sensor is capable of distinguishing between solvents, acids, and bases. Resistance of graphite line on paper showed different behavior when exposed to different types of chemicals. The sensor was determined to be highly sensitive to different chemicals. The amount of ionic materials dissolved in the fluid combined with graphite material had a direct impact on the resistivity. The solvent detection principle was based on drops in temperature of thermosensitive graphite due to its evaporation at room temperature. A test fixture was developed to consistently detect chemicals by dropping certain amount of each chemical onto the graphite on paper test strip through the inlet of the device. The developed sensor can be used in several industries where a routine test of chemicals is required.

References

1. Phan, H., Dao, D.V., Dinh, T., Brooke, H., Qamar, A., Nguyen, N., Zhu, Y.: Graphite-on-paper based tactile sensors using plastic laminating technique. In: 28th IEEE International Conference on Micro Electro Mechanical Systems (MEMS), pp. 825–828 (2015)
2. Ren, T.L., Tian, H., Xie, D., Yang, Y.: Flexible graphite-on-paper piezoresistive sensors. Sensors **12**(5), 6685–6694 (2012)
3. Yamada, T., Hayamizu, Y., Yamamoto, Y., Yomogida, Y., Izadi-Najafabadi, A., Futaba, D. N., Hata, K.: A stretchable carbon nanotube strain sensor for human-motion detection. Nat. Nanotechnol. **6**(5), 296–301 (2011)
4. Liao, X., Liao, Q., Yan, X., Liang, Q., Si, H., Li, M., Wu, H., Cao, S., Zhang, Y.: Flexible and highly sensitive strain sensors fabricated by pencil drawn for wearable monitor. Adv. Funct. Mater. **25**(16), 2395–2401 (2015)

5. Dinh, T., Phan, H.P., Qamar, A., Nguyen, T.K., Woodfield, P., Zhu, Y., Nguyen, N.T., Dao, D.V.: Environment-friendly wearable thermal flow sensors for noninvasive respiratory monitoring. In: IEEE 30th International Conference on Micro Electro Mechanical Systems (MEMS), pp. 993–996 (2017)
6. Dinh, T., Phan, H., Dao, D.V., Woodfield, P., Qamar, A., Nguyen, N.: Graphite on paper as material for sensitive thermoresistive sensors. J. Mater. Chem. C **3**(34), 8776–8779 (2015)
7. Dinh, T., Phan, H.P., Qamar, A., Nguyen, N.T., Dao, D.V.: Flexible and multifunctional electronics fabricated by a solvent-free and user-friendly method. RSC Adv. **6**(81), 77267–77274 (2016)
8. Delhaes, P.: Graphite and Precursors. CRC Press, Boca Raton (2014)

Heating Fuel Thermal Reactor Thermal Efficiency Analysis Method

Qin Wang(✉), Wang-zhi Qiu, Jun-wei Tian, and Zhi-yi Jiang

School of Mechatronic Engineering, Xi'an Technological University, Xi'an 710021, China
wq992514@163.com

Abstract. Fuel thermal reactor is a widely used power equipment, aiming current development of small and medium thermal reactor fuel, to accelerate the transformation of technology, in order to save energy, reduce pollution, improve fuel thermal reactor thermal efficiency purposes, fuel thermal reactor thermal efficiency calculation method of analysis to further improve the system provides a theoretical basis. This paper derived fuel melt through the use of anti-tank heat balance method efficiency and heat transfer efficiency is equal to the product of the combustion efficiency of the new expression minus the fixed loss calculation, verification of the results with conventional counterbalance subtraction formula identical results.

Keywords: Thermal reactor · Anti-balance method · Thermal efficiency

1 Introduction

Fuel thermal reactor is mainly used for heating marking paint in high speed road marking paint marking the construction. Thermal efficiency is the synthetic index of the Fuel thermal reactor, which embodies the comprehensive performance of the Fuel thermal reactor as a kind of energy conversion equipment. Fuel thermal reactor are widely used and fuel consumption is large. With the improvement of energy saving and emission reduction requirements, it is particularly important to carry out energy-saving and emission reduction in the Fuel thermal reactor. The main method to judge the energy consumption of Fuel thermal reactor energy consumption is thermal test, The AQSIQ has issued a number of thermal testing standards, and has stipulated the relevant requirements for the application of the test methods. In this paper, the formula for calculating thermal efficiency of Fuel thermal reactor is obtained by analyzing and calculating to make quick judgments on the energy efficiency of Fuel thermal reactor in practice [1–3]. There are two common methods for calculating the thermal efficiency of industrial boiler, one is the calculation of positive balance method, the other is the calculation of anti-balance method. In this paper, the calculation method of industrial boiler thermal efficiency is applied to calculation of thermal efficiency of Fuel thermal reactor [2]. Derive the formula conclusion by assuming the known condition. In practice, the thermal efficiency value of Fuel thermal reactor can be quickly and accurately obtained by measuring the corresponding known conditions in the sensor [4].

K. Deng et al. (Eds.): ICMIR 2018, AISC 856, pp. 771–776, 2019.
https://doi.org/10.1007/978-3-030-00214-5_96

This article mainly introduces the calculation of thermal efficiency of fuel thermal reactor by anti-balance method. Among them, the excess air coefficient of flue gas outlet at flue gas is an important parameter to ensure the economical operation of the fuel thermal reactor, and it is an important measure to maintain the boiler thermal efficiency, The value of excess air coefficient is determined according to the O_2 quantity, on one hand, the value of the O2 quantity can reflect the air leakage in the Fuel thermal reactor, On the other hand, it also can reflect the matching of the blast quantity and fuel quantity of the Fuel thermal reactor [5]. Through the sensor monitoring flue gas containing O_2 quantity, we can understand the ratio of fuel and air, adjust the value of excess air coefficient in time, can make the fuel fully burn to achieve energy saving purposes [6–8].

2 Positive Balance Method and Anti-Balance Method

2.1 Balance Method

The positive balance method is applied to the thermal efficiency calculation of the Fuel thermal reactor by introducing the thermal efficiency calculation method of the traditional coal-fired boiler. After repeated verification of theory and practice, this subject continues to derive an inverse anti-balance calculation method for thermal efficiency of fuel thermal reactor, through mathematical logic deduction, and the result is the same as the positive balance method, which can be applied to calculate the thermal efficiency of fuel thermal reactor in practical work.

The positive balance method to calculate the fuel thermal reactor thermal efficiency as shown in type (1)

$$\eta_1 = \frac{D_{sc}(h_{bq} - h_{gs} - \frac{rw}{100})}{BQ_r} \times 100\% \quad (1)$$

In the formula η_1 stand for the positive balance efficiency (%) of the fuel thermal reactor; D_{sc} stand for the Output steam of the fuel thermal reactor (kg/h); h_{bq} is Saturated steam enthalpy (kJ/kg); h_{gs} is Steam boiler feedwater enthalpy (kJ/kg); r is Latent heat of vaporization (kJ/kg); w stands for Steam humidity (%); B is Fuel consumption (kg/h); Q_r represents Diesel combustion heat value (kJ/kg). Diesel combustion heat value is 42700 kJ/kg.

0 # Diesel Oil is selected according to the actual fuel thermal reactor, according to the diesel unit price 7.49 yuan/liter, its density is 0.84 g/cm^3, so every 40 min need 100–120 yuan Diesel.

The diesel quality is seen in the type (2)

$$m = \rho \frac{a}{b} \quad (2)$$

The actual diesel consumption is seen in the type (3)

$$B = \frac{m}{h} \tag{3}$$

a stands for burning 40 min diesel costs, b stands for diesel unit price and ρ means diesel density. Type of diesel 11.2–13.46 kg, diesel oil consumption $B = 16.79\,\text{kg/h}$. There are seven parameters to be measured in the formula, which can be classified into three cases. (1) D_{sc} and B can be obtained directly from the field sensor. (2) h_{bq}, h_{gs} and r can be based on pressure P or temperature T as a form of vapor or water properties. (3) Q_r and w are based on different grades of diesel, different fuel thermal reactor model and operating mode by the test.

2.2 Anti-Balance Method

The formula for calculating thermal efficiency of fuel thermal reactor with inverse equilibrium method is:

$$\eta_2 = 1 - q_2 - q_3 - q_4 - q_5 - q_6$$

In the formula, q_2 stands for Heat loss of smoke exhaust, q_3 represents Heat loss of gas incomplete combustion, q4 stands for Heat loss of solid incomplete combustion, q_5 means Loss of heat dissipation and q6 means Physical heat loss of ash residue.

combustion efficiency:

$$\eta_r = 100\% - (q_3 + q_4) \tag{4}$$

The heat absorbed by the boiler (the effective use of heat) and the actual heat emitted by the fuel minus the ratio of fixed losses, called heat transfer efficiency η_c that can be described as follows:

$$\eta_c = 1 - \frac{q_2}{\eta_r - (q_5 + q_6)} \tag{5}$$

$(q_5 + q_6)$ means fixed loss, the smaller can be omitted, so $\eta_c = 1 - q_2/\eta_r$, and then $q_2 = \eta_r(1 - \eta_c)$. Surrogate η_r and q_2 into anti-balance thermal efficiency subtraction formula:

$$\eta = 100\% - (q_2 + q_3 + q_4 + q_5 + q_6) = \eta_r \cdot \eta_c - (q_5 + q_6)$$

make $(q_5 + q_6) = q_g$, we can obtain another expression for calculating thermal efficiency:

Thermal efficiency:

$$\eta = \eta_r \cdot \eta_c - q_g \tag{6}$$

Combustion efficiency:

$$\eta = 100 - (q_3 + q_4) \tag{7}$$

Heat transfer efficiency:

$$\eta_c = 1 - \frac{q_2}{\eta_r} \tag{8}$$

Fixed loss:

$$\eta_g = (q_5 + q_6) \tag{9}$$

Fuel actual (incomplete) combustion of the heat emitted from the total combustion can emit the ratio of heat q_2 is called combustion efficiency, we can also call it smoke exhaust heat loss):

$$q_2 = (m + n\alpha_{py})\left(\frac{t_{py} - t_{lk}}{100}\right) \tag{10}$$

In the formula, m and n stand for calculation coefficient, For fuel thermal reactor, the calculation coefficient is generally selected m = 0.5, n = 3.45. α_{py} stands for excess air coefficient in smoke exhaust, Combustion diesel α_{py} = 1.2; t_{py} is the temperature of smoke exhaust, make t_{py} = 600 °C; t_{lk} represents cold air temperature in furnace; make t_{lk} = 25 °C; q_4 stands for heat loss of solid incomplete combustion; q_4 = 0 in fuel oil. Exhaust heat loss is one of the main heat losses of the fuel-heat melting furnace, up to 26.68%. Smoke exhaust heat loss depends mainly on exhaust temperature and excess air coefficient.

q_3 stands for heat loss of gas incomplete combustion, is based on the CO content in flue gas according to Table 1

The CO content in the flue gas directly decides the heat loss of gas incomplete combustion q3 in the fuel thermal reactor, and if the CO content is higher, it can cause larger with q3 which reminds us that the fuel thermal reactor is not mixed or the blast and fuel do not match.

Table 1. (q3) the heat loss of gas incomplete combustion

Subject	Company	Value		
CO	% (ppm)	CO ⩽ 0.05 (CO ⩽ 500)	0.05 < CO ⩽ 0.1 (500 < CO ⩽ 1000)	CO > 0.1 (CO > 1000)
q3	%	0.2	0.5	1

As for the fuel thermal reactor, q4 is 95%.

The loss of heat dissipation q_5 can be selected directly by Table 2 when the actual operating output of the fuel thermal reactor is not less than 75%% of the rated output:

Table 2. Heat dissipation loss of boiler rated output (q_5)

q_5	t/h	≤ 4	6	10	15	20	35	≥ 65
	MW	≤ 2.8	4.2	7.0	10.5	14	29	≥ 46
	%	2.9	2.4	1.7	1.5	1.3	1.1	0.8

When the fuel thermal reactor actually operates less than 75% of the rated output, the following formula is amended:

$$q_5 = q_{5ed} \frac{Q_{ed}}{Q_{sc}} \tag{11}$$

When the actual operation of the boiler is lower than the rated output 30%, the condition of 30% output is corrected; When the fuel thermal reactor output can not able to measure, the actual output of the rated output of 65% calculation.

3 Conclusion

There are eight parameters to be measured in the calculation process, which can be divided into two cases to obtain.

(1) O2, CO, t_{py}, t_{lk}, D_{sc} or Q_{sc} Can be obtained directly from the field sensor.

(2) As for q5ed and q3 You can save table 1, table 2 to your computer.

As with positive balance efficiency calculations, the computer can calculate and display instantaneous anti-balance thermal efficiency after obtaining all the calculated parameters. This paper integrated fuel thermal reactor heating characteristics. In the heat loss of the fuel thermal reactor, the q2, q3 and q5 are controlled, and the heat loss is controlled basically. Therefore, the calculation and analysis of thermal efficiency and fitting out the formula for calculating thermal efficiency of Shaanxi fu el thermal reactor is calculated by analysis and calculation. Few parameters in formulas are easy to carry out the actual test work.

Funds. The department of project in shanxi province (NO. 2016GY-175); Shaanxi provincial education office fund Natural special (NO. 387); Wei Yang district of Xi'an Central Science and Technology Fund (NO. 201713).

References

1. Duan, H., Zhang, H.: Energy efficiency analysis of heating in secondary refining processes. Energy Conserv. Technol. **30**(4), 324 (2012)
2. Wen, Z., Ge, S., Zhang, X.: The testing and evaluation of thermal efficiency of oil-fired or gas-fired boiler in use. Ind. Boil. **130**, 14–16 (2011)
3. Wang, G.: Discussion on construction equipment for highway route signs. North. Commun. **4**, 191–194 (2010)

4. Chen, M., Feng, G., Liu, J.: The application of spraying thermo-melting road mark coatings to Yuantai Expressway. Shanxi Sci. Technol. Commun. **6**, 77–78 (2011)
5. Zhang, G.: Two methods to improve the thermal efficiency of industrial boilers. Energy Res. Util. **1**, 1–2 (2011)
6. Chen, Z., Sun, B.: Thermal efficiency analysis of the industry boiler. Sci. Technol. Innov. Product. **5**, 65–66 (2007)
7. Yingying, L., Wanhua, S., Song, W., Ei, Y.: Investigation on combustion paths and their effects on thermal efficiency and emissions of a high BMEP diesel engine. Trans. CSICE **29** (1), 1–2 (2011)
8. Wenfei, W., Feng, Y., Zhang, X.: Zonal method solution of radiative heat transfer in a one-dimensional long roller-hearth furnace in CSP. J. Univ. Sci. Technol. Beijing **14**(4), 307–311 (2007)

An Azimuth Self Correction of Combined Vector Underwater Acoustic Sensor

Li Qidong(✉), Bi Jiayu, Gong Zhanjiang, Sun Likai, and Zhang Peng

The 49th Research Institute of China Electronics Technology Group Corporation, Harbin 150001, China
liqidong722@163.com

Abstract. This paper introduces an independent orientated magnetism-multiple three-dimensional MEMS vector hydrophone that can be used for target differentiating. The vector hydrophone is designed by a three-dimensional accelerometer and magnetism sensor. The accelerometer is close mounted to the signal processor so that the noise can be reduced and signal attenuation can be restrained. Since the single vector hydrophone cannot achieve a high orientate precision, it is solved by this magnetism-multiple three-dimensional vector hydrophone. The vector hydrophone has been finished the underwater standing wave field test and the non-magnetism rotate platform test. The testing result validates adopting magnetism sensor to compensate the direction of vector hydrophone is feasible. The vector hydrophone which can be used to pinpoint the underwater target and implement acoustic detection, underwater orientation and navigation represents the characters as small size, simple structure, high stability and reliability.

Keywords: Vector hydrophone · Magnetism sensor
Underwater sound detection

1 Introduction

With the development of underwater acoustic sensor technology, the use of vector hydrophones to acquire underwater acoustic information and underwater target detection methods is gradually becoming an effective means for marine environmental monitoring, detection and underwater target recognition [1, 2]. In 1997, the first time through the introduction of technology in China, systematically launched a special study on vector hydrophone technology. In 1998, the first domestic vector underwater acoustic sensor outfield test was conducted, and the first domestic vector underwater acoustic sensor sea trial was conducted in 2000, it made important contributions to the engineering application of vector underwater acoustic sensors [3, 4]. The developments of Vector hydrophone based on MEMS technology are gradually becoming miniaturized and integrated. The single vector hydrophone developed by MEMS technology can achieve the target orientation function, but how to make a single vector underwater acoustic sensor to better target the space target is the focus and difficulty of the current research.

K. Deng et al. (Eds.): ICMIR 2018, AISC 856, pp. 777–783, 2019.
https://doi.org/10.1007/978-3-030-00214-5_97

This paper provides a magnetic composite three-dimensional MEMS vector underwater acoustic sensor with self-orientation and target discrimination. It uses a combination of magnetic sensor and 3D MEMS vector underwater acoustic sensor structure composite design technology and signal processing technology. This vector hydrophone can improve detection of underwater acoustic signals with three-dimensional sound signal detection, low frequency detection capability, and target orientation capability. And it has the characteristics of simple structure, miniaturization, stability and reliability.

2 Theory

2.1 Acoustic Theory [5–7]

Acoustic waves are acoustic waves that propagate in a vibrating manner in elastic media, Acoustic measurement is a method of determining sound field characteristics or sound source characteristics through detection of sound field parameters. For example, the sound pressure distribution in the sound field can be obtained through the sound pressure test, and the sound source position can be indirectly measured. Sound source direction can be judged by measuring particle velocity or acceleration. Therefore, the purpose of the intended measurement can be achieved most effectively by parameter measurement, and optimization must be performed by analyzing and understanding the parameters.

The propagation of sound waves forms the sound field, The sound propagation properties of the free field under ideal conditions can be given by acoustic theory derivation [7, 8]. The propagation characteristics of a point in the sound field can be measured by the following parameters, including sound pressure, vibration velocity, and sound intensity. From the point of view of measurement, sound pressures at different spatial points are known to estimate the position of the sound source. The known velocity can be used to determine the propagation direction. The known sound intensity can be used to determine the strength of the sound source.

The sound pressure in the sound field is

$$p(r\,,t) = p_0\,\mathbf{e}^{j(\omega t - kr)} \tag{1}$$

Particle vibration speed

$$v(r\,,t) = v_0\,\mathbf{e}^{j(\omega t - kr)} \tag{2}$$

Sound intensity

$$I = \frac{1}{T}\int_0^T \mathrm{Re}(p)\mathrm{Re}(v)dt \tag{3}$$

In the formula, p is the sound pressure in the sound field, v is the sound pressure in the sound field.

The motion characteristics of a point in the sound field are measured by other parameters, such as particle velocity, acceleration, and displacement. From the measurement point of view, the force of the space-equivalent particle in the sound field can trigger the movement of the particle. Knowing one of the particle vibration velocity, acceleration, and displacement can push out the other two; Conversely, if the acceleration or displacement of the measured particle motion can indirectly determine the vibration velocity of the particle, the sound source can be indirectly known through the vibration velocity of the particle. After understanding the parameters and relationships between the sound field propagation characteristics and the particle motion characteristics, you can learn more about the design points of the sensor. Therefore, a suitable measurement parameter and measurement sensor can be selected according to the characteristics of the amplitude and frequency range of the acoustic signal. The need for sound source detection can be achieved indirectly through transformation, even if the acoustic measurement effect is maximized.

2.2 Magnetic Signal and Acoustic Vector Signal Orientation Algorithm

To establish a plane rectangular coordinate system, the directivity of a vector underwater acoustic sensor is shown in Fig. 1. The velocity and sound pressure of the particles in the three directions of the vector underwater acoustic sensor output orthogonal to each other are v_x, v_y and p; Azimuth angle φ of magnetic azimuth sensor output with respect to magnetic north pole.

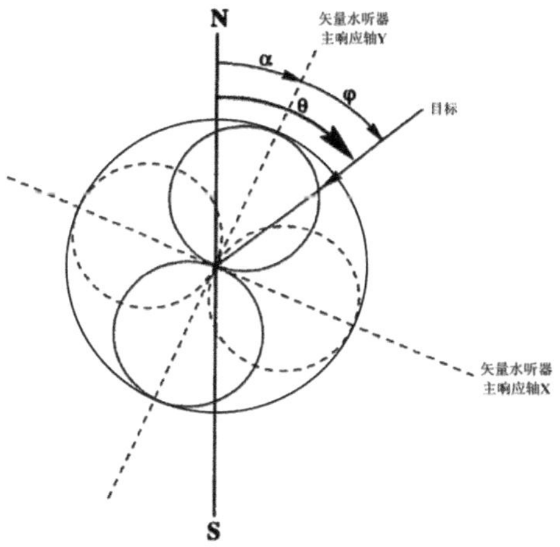

Fig. 1. Vector hydroacoustic sensor directivity map

FFT transformation of sound pressure $p(t)$ and two orthogonal vibration components $v_i(t)$, Get the corresponding spectrum is $P(f)$ and $V(f)$, The cross-spectral function of the sound pressure signal and the vibration velocity component is [8, 9]:

$$S_{pv_i}(f) = P(f)V_i^*(f) \tag{4}$$

Symbol * indicates conjugate operation;

The sliding window function is used to time-average the cross-spectral functions and the cross-spectrum of Eq. (1) is averaged to obtain the average period map output as:

$$\langle S_{pv_i}(f) \rangle = \langle P(f) \cdot V_i^*(f) \rangle \tag{5}$$

$\langle \cdot \rangle$ Means sliding window average period diagram;

The horizontal orientation $\alpha(f)$ of the main response axis of the target sound source relative to the vector hydroacoustic sensor is:

$$\alpha(f) = \arctan \frac{\mathrm{Re}\left\{\left\langle P(f)V_y^*(f)\right\rangle\right\}}{\mathrm{Re}\{\langle P(f)V_x^*(f)\rangle\}} \tag{6}$$

Underwater target θ to be measured is calculated relative to the horizontal azimuth $\alpha(f)$ of the main response axis of the vector hydrophone and the azimuth ϕ of the main response axis of the vector hydrophone relative to the magnetic north. Realize the orientation of underwater targets

$$\theta = \alpha(f) + \varphi. \tag{7}$$

3 Design

Piezoelectric principle and ceramic materials are used to make acoustically sensitive primitives, and acoustically sensitive primitives are made into a vector hydroacoustic transducer using a multi-model cascading method. The magnetic orientation sensor is parallel to the XOY plane, and the north pole is oriented parallel to the y-axis of the vector hydrophone sensor. Azimuth Self-Modified Combined Vector Underwater Acoustic Sensor Fig. 2.

Using acoustic signal processing circuits to convert, amplify, and filter acoustic signals, Convert the underwater acoustic signal measured by the vector underwater acoustic sensor into an acoustic analog signal; The magnetic orientation processing circuit realizes the alternating excitation and signal amplification, filtering, A/D conversion of the magnetoresistive bridge, and the three-dimensional magnetic field components are processed to obtain the true magnetic azimuth of the north magnetic field.; The digital processing circuit is composed of digital devices such as high-speed A/D converters and DSP processors. It uses high-speed A/D converters to convert

Fig. 2. MEMS vector acoustic sensor schematic

analog acoustic signals and analog magnetic signals into digital signals, and the collected digital signals are transmitted to DSP processors.

4 Design Sample Performance Testing

Vector Underwater Sensor Design Operating Frequency from 10 Hz to 300 Hz. After completion of the development, the vector underwater acoustic sensor is sent to the primary metering station for acoustic performance measurement. Received sensitivity measurement in underwater test using standing wave field comparison method.

Mainly for the sound pressure sensitivity measurement of vector underwater acoustic sensors, measurement uncertainty U = 1.5 dB. Vector Acoustic Sensors Receive Sensitivity of −198 dB (@100 Hz).

A Sensitivity and Directivity

Using a vector underwater acoustic sensor calibration device of the defense underwater acoustics leveling station, a preliminary calibration of the MEMS single vector underwater acoustic sensor was performed using the standing wave field comparison method, mainly including the sound pressure sensitivity and the pointing performance of the underwater acoustic sensor. Vector underwater acoustic sensor is fixed on a circular frame for sensitivity testing. The volumetric water acoustic sensor is vertically distributed in the underwater acoustic field of the circular frame, rotates along the axial direction of the frame, and rotates the angle while measuring the output voltage of the sensor. The directivity of the sensor at 100 Hz frequency is shown in Fig. 3.

B Magnetic Orientation Test

Non-magnetic rotary table is used for performance testing and calibration of the direction angle measured by the magnetic sensor. Select 24 points uniformly on a circle including −180° and 180° points. The sensor output is recorded on the RS485 bus using a computer, and a standard magnetic compass is used as a test comparison

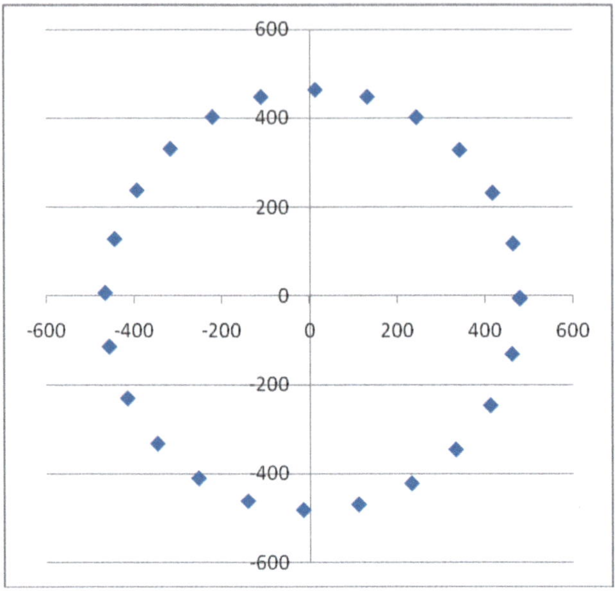

Fig. 3. Magnetic sensor direction angle measurement curve

standard. The measurement data is recorded and the error is corrected. The result is shown in Fig. 3.

5 Conclusion

The test results in this paper demonstrate the feasibility of designing and fabricating an independent orientated magnetism-multiple three-dimensional MEMS vector hydrophone by combining magnetic sensor and vector hydroacoustic sensor structural composite design technology and signal processing technology. The combined vector hydrophone can easily realize the precise orientation of underwater targets. And in the harsh marine environment, the combined vector hydrophone has the better environmental adaptability, because it can compensate the azimuth angle in real time, more accurate orientation and tracking target. This vector hydrophone provides a new development idea for passive orientation of underwater targets.

References

1. Chen, H.: Vector Sensor. Harbin Engineering University Press, Harbin (2006)
2. Chen, L., Zhang, P., Xu, X., Wang, F.: Overview of vector hydrophone. Transducer Microsyst. Technol. **25**(6), 5–8 (2006)
3. Yang, D.C.: Vector Hydrophone Lake Test Report. China National Defense Science and Technology Report (1998)

4. Yang, D.C.: Vector Hydrophone Test Report. China National Defense Science and Technology Report (2000)
5. Chen, L.J., Yang, S.: Vector water acoustic sensor test model and discussion on vibration pickup conditions. Appl. Acoust. **25**(6), 334–339 (2006)
6. He, Z., Zhao, Y.: Theory of Acoustics. National Defense Industry Press, pp. 315–326 (1981)
7. Kim, K., Gabrielson, T.B., Lauchle, G.C.: J. Acoust. Soc. Am. **116**, 3384–3392 (1991)
8. Hu, J., Hu, J.: Vector Sound Signal Processing Basics. National Defense Industry Press, pp. 57–59 (2009)
9. Zhang, L., Tian, T., Meng, C.X.: Single-vector underwater acoustic sensor based on cross-spectrum direction finding multi-target discrimination. Ship Sci. Technol. **31**(10), 18–20 (2009)

Discriminant Analysis of Different Kinds of Medicinal Liquor Based on FT-IR Spectroscopy

Yang Liu[1,2], Fan Wang[2], Chunfu Shao[1,2], Wei You[2], Qi Chen[2], and Yujie Dai[1(✉)]

[1] Key Laboratory of Industrial Fermentation Microbiology (Tianjin University of Science & Technology), Ministry of Education, College of Bioengineering, Tianjin University of Science and Technology, Tianjin 300457, People's Republic of China
yjdai@126.com

[2] Tasly Academy, Tasly Group, Tianjin 300410, People's Republic of China

Abstract. The discriminant analysis model of different medicinal liquor was established based on Fourier transformed infrared spectroscopy (FT-IR) combined with support vector machine (SVM) and principal component analyses (PCA) with the validation accuracy of 99% and training accuracy of 100%. The model was also tested by the external samples with the prediction accuracy of 97%. The accuracy data of the experimental showed that Fourier transform infrared spectroscopy (FT-IR) can be applied well for the classification of medicinal liquor.

Keywords: Fourier Transformed Infrared Spectroscopy (FT-IR) Medicinal liquor · Species · Qualitative

1 Introduction

Medicinal liquor is one of the oldest liquor with long history and culture in China, which belongs to a traditional medicinal preparation. It combines liquor with Chinese medicinal herbs used in the treatment of many human diseases. It is a kind of stable, safe, effective and convenient preparation. Because alcohol is a good semi polar organic solvent, the various effective ingredients of Chinese medicine are soluble in it. The curative effect will be improved with the aid of alcohol. For example, ginseng based medicinal materials are rich in saponins and polysaccharides, and have the effects of anti-proliferation, anti-apoptosis, anti-inflammatory and tumor metastasis [1]. The deer antler, not only contains glycolipids, hormones, fatty acids, amino acids and proteins, but also contains calcium, magnesium, sodium and other ingredients. The deer antler has a good physical effect of velvet warm but not dry, and can excite and improve the function of the human body for body weakness patients with chronic illness [2]. Chinese wolfberry contains a variety of amino acids, polysaccharide, and some special nutrients such as betaine, which has the effect of reducing blood sugar and cholesterol [3]. Chinese angelica, contains phthalides, phenolic acids, flavonoids and

K. Deng et al. (Eds.): ICMIR 2018, AISC 856, pp. 784–791, 2019.
https://doi.org/10.1007/978-3-030-00214-5_98

other compounds with the effect of enriching and harmonizing blood, regulate menstruation pain, moisturizing pathogenic dryness to loosen the bowel, anti-cancer, anti-aging and improving the immune ability [4]. Timely and appropriate drinking of the medicinal liquor made of these Chinese medicinal herbs can achieve better health care for people.

Fourier transform infrared spectroscopy (FT-IR) is a fast developing determination technique in recent years with the characteristics of quick, simple and accurate. It has been well applied and developed in the fields of agriculture, food, petrochemical and other fields. It can be used for qualitative identification [5, 6] and the quantitative analysis [7, 8] of Chinese medicinal liquor.

The infrared spectrum can be divided into near, mid and far infrared spectroscopies according to the difference of wavelength range. The mid infrared region (2.5–25 μm; 4000–400 cm^{-1}), which is the most widely used in the infrared region [9], can well reflect the physical changes within the molecule and the molecular structure [10] and it can effectively distinguish and identify the composition and structure of the material, thus, in this experiment, the mid infrared spectrum coupled with support vector machine (SVM) and principal component analysis (PCA) was employed to construct models for distinguishing different medicinal liquors.

2 Experimental

2.1 The Preparation of Medicinal Liquors

The medicinal liquors were prepared by impregnation method. According to the Chinese Pharmacopoeia (edition 2015) [11], the amount of liquor and raw Chinese herbal medicine materials were calculated and shown in Table 1. The medicinal materials washed clean and placed in glass bottles separately. After the liquor was added, the mouth of the bottle was sealed for impregnation. The preservation temperature of the samples is 25 °C. The soaking time is about three months.

Table 1. Dosage of herbal medicines and base liquor

Medicinal	Raw material weight (g)	Liquor (ml)
White ginseng	100.75	500
	100.62	500
Chinese angelica	150.46	500
	150.52	500
Chinese wolfberry	150.13	500
	150.44	500
Deer antler	50.63	500
	50.41	500
Red ginseng	100.37	500
	100.56	500

2.2 Samples

Five kinds of Chinese herbal medicines such as white ginseng, red ginseng, Chinese angelica, Chinese wolfberry, deer antler, were used as raw materials to make medicinal liquor. The five kinds of raw materials were confirmed by the identification method by Chinese Pharmacopoeia (edition 2015). A total of 576 infrared spectra samples were collected from the five kinds of medicinal liquors and the spectrum acquisition results are shown in Fig. 1, including 116 samples from each of the white ginseng, Chinese wolfberry, angelica liquor and antler liquors respectively, 112 samples from red ginseng liquor (Because the soaking time of red ginseng liquor is two days later than other medicinal liquors). 20 samples, which was taken every 4 days, of each medicinal liquor were taken as testing set (100 total testing samples), and the rest were used for modeling (476 samples).

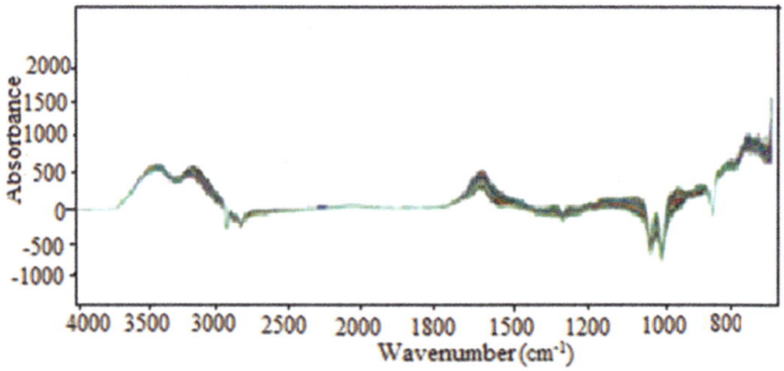

Fig. 1. Infrared spectra of all samples

2.3 Acquisition of the Mid Infrared Spectra of Medicinal Liquors

All infrared spectra were collected by Bruker Tensor 27 Fourier transformed infrared spectrometer (Bruker, Germany) with attenuated total reflection (ATR) accessory and OPUS spectrum acquisition software. The data were treated with *The Unscrambler 10.3* (CAMO, Norway) spectrum processing software. The wave frequency range is 4000–650 cm^{-1}. For the spectrum collection, scanning times is 32, The scanning speed is 10 kHz with the resolution of 8 cm^{-1}. The medicinal liquor was filtered by membrane of 0.22 μm so as to reduce the damage of ATR from insoluble impurity in liquor, then the medicinal liquor was injected into the ATR accessory with a syringe and the infrared spectra of the liquor was collected. After the data is collected, the vacuum pump is opened and the medicinal liquor is sucked, subsequently, the ATR accessory is washed with anhydrous ethanol for the spectra collection of next sample.

2.4 Establishment of Model for Identification of Different Medicinal Liquors

The support vector machine (SVM) and principal component analysis (PCA) algorithms were used to establish the model by using the wave number in the infrared scanning of the sample as the independent variable, the absorbance value as the dependent variable and each kind of liquor as a class.

Main spectral processing methods were listed in Table 2. The accuracy of different spectral processing methods were compared (Accuracy refers to the ratio of the samples which predicted correctly to the total samples used to predicted). The processing results from the higher accuracy spectral processing methods were further treated with primary component analysis. The wave number bands which have greater influence on the results were selected from the X-loading plot in PCA. It can be seen that the wave number bands at the peaks have great contribution to the principal component. Subsequently, the SVM model is redone based on the selected bands of the principal component and the accuracy was revalued. In the accuracy evaluation of the obtained models, the training accuracy (means using one of the samples which have been used for modeling as a prediction sample to evaluate the accuracy of the model) as well as the validation accuracy (means using samples beyond for modeling as prediction samples to evaluate the accuracy of the model) were applied to evaluate the model. The validation accuracy was used as the main index while the training accuracy was utilized in case that the validation accuracies of two processing methods were the equal. The wave number band with the highest validation accuracy is considered as the effective wave number band for a specific processing model method [12]. The optimal spectra processing method combined with the optimal bands were selected to establish the search grid of the SVM model to find the orthogonality between the C value and gamma value. The SVM model with the highest validation accuracy was selected.

Table 2. Spectra processing methods for the modeling to discriminate different kinds of medicinal liquor

Methods	Training accuracy (%)	Validation accuracy (%)
RAW	79	78
Baseline	86	84
Spectroscopic	94	93
SNV	95	94
Normalize	34	36
Detrend	79	77
Detrend + SNV	96	94
SNV + Detrend	96	93
Smooth + Gap.dervatives(1st)	29	27
Smooth + Gap.dervatives(2st)	20	19
Smooth + Gap.dervatives(1st) + SNV	**99**	**97**
Smooth + Gap.dervatives(2st) + SNV	99	94

2.5 Validation of the Model

The model is tested by the samples outside the training set. 20 samples, which was taken every 4 days, of each medicinal liquor were taken as testing set (100 total testing samples). According to the comparison between the predicted and actual data, we can get the prediction accuracy, so as to evaluate the prediction effect of the established model.

3 Results and Discussion

3.1 The Selection of Spectral Processing Methods

12 kinds of infrared spectrum data processing methods were tested respectively for modeling. The training and validation accuracies of each spectrum processing method were shown in Table 2. The highest validation accuracy of 97% was obtained by using the processing method of *Smooth + Gap.dervatives + SNV (1st)*. Because the validation accuracies of other processing methods are much lower than this, thus, the *Smooth + Gap.dervatives(1st) + SNV* method was selected as the processing method and the processing results were used to construct models by SVM analysis for discriminating different kinds of medicinal liquor.

3.2 The Selection of Effective Wave Number Bands

After the models for discriminating different kinds of medicinal liquor was obtained from SVM Analysis, PCA analysis was applied in order to get a better model. The scatter plot of the PCA results were shown in Fig. 2a. It can be seen that the contribution of the first principal component is 41% to the model and that of the second principal component is 27%. In order to reduce the processing time and the complexity of the models, the wave number band, which has greater contribution to the model, was chosen from the first principal component in X-loading plot (Fig. 2b). Because infrared absorbances of the main components of liquor are mainly distributed smaller than 1800 cm^{-1}, which contribute much to the model, thus the effective wave number band was chosen from this range. Considering some areas of impurities and noise should be the effective wave number band, 1168–767 cm^{-1}, was chosen as the effective wave number band finally and the SVM modeling was repeated. The maximum validation accuracy of 98% and training accuracy of 100% were obtained based on the effective wave number band.

3.3 Grid Search for the Optimization of the Model

Grid search was carried out to find the optimal C and gamma values to improve the validation accuracy of the SVM model. Gamma and C value are of very importance to SVM model. The C value, also named penalty parameter [13], was known as the tolerance of the errors [14]. If the C is too high, the algorithm will over fit the training data, if the C is too small, then insufficient training will occur [15]. Gamma is a parameter with the RBF (Radial Basis Function). The higher the gamma value, the less of the number of support vectors. The results are shown in Table 3. It can be found that the highest orthogonal value is 99%, and for the value of C and gamma are 100 and 0.01. Then, the SVM model was rebuilt with the condition of optimal values of C and

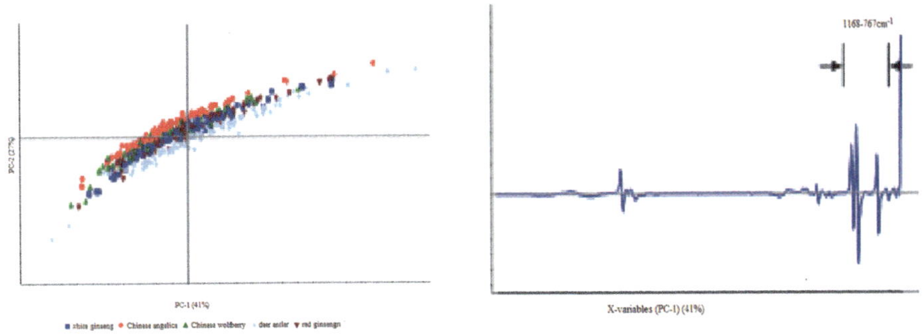

(a) (b)

Fig. 2. The PCA results under the *Smooth + Gap.dervatives(1st) + SNV* method (a) scatter plot (b) X-loading plot

gamma and the model obtained after grid search has the training accuracy of 100% and validation accuracy of 99%.

Table 3. Grid search for spectral processing results (Accuracy, %)

Gamma	C				
	0.01	0.1	1	10	100
0.01	58	74	90	98	**99**
0.1	59	89	96	98	98
1	48	74	93	92	92
10	23	23	36	39	40
100	23	22	20	21	21

The final PCA plot, after the processing method of *Smooth + Gap.dervatives (1st) + SNV* at the effective number band of 1168 767 cm^{-1}, is shown in Fig. 3. In the PCA plot, it can be distinguished well from the clustering of each medicinal liquor, which means the medical liquors could be discriminated well.

3.4 Validation of the Qualitative Classification Model for Discriminating Different Kinds of Medicinal Liquor

20 samples outside the training set were taken from each kind of medicinal liquor for model testing (100 total testing samples). A testing sample was taken every four days approximately. The results were shown in Table 4. The prediction accuracy of the 100 samples was 97%, which indicates that the model can be used to discriminate medicinal liquors successfully.

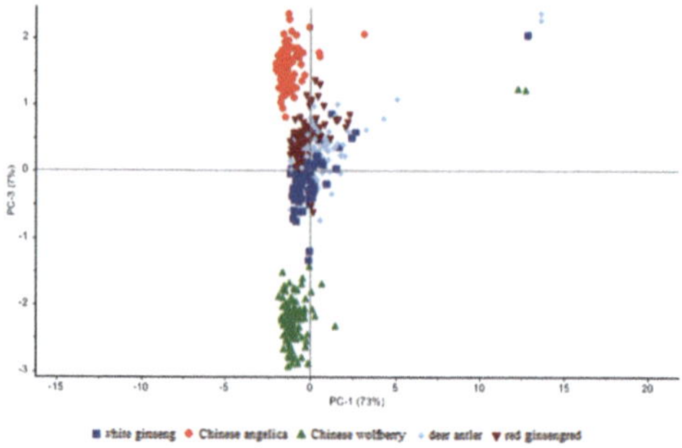

Fig. 3. The PCA analysis results with the processing method of *Smooth + Gap.dervatives (1st) + SNV* and effective wave band of 1168–767 cm^{-1}

Table 4. The results of SVM model for the classification of different kinds of medicinal liquors

Actual	Predictive				
	White ginseng	Chinese angelica	Chinese wolfberry	Deer antler	Red ginseng
White ginseng	19	0	0	0	1
Chinese angelica	0	20	0	0	0
Chinese wolfberry	0	0	20	0	0
Deer antler	0	0	0	20	0
Red ginseng	2	0	0	0	18

4 Conclusion

In this study, the model for discriminant of different medicinal liquors was established based on Fourier transformed infrared spectroscopy (FT-IR) combined with support vector machine (SVM) and principal component analyses (PCA). with the validation accuracy of 99% and training accuracy of 100%. The model was also tested by the external samples and the prediction accuracy of 97% was obtained. The accuracy data of the experimental showed that Fourier transform infrared spectroscopy (FT-IR) can be applied well for the classification of medicinal liquors.

Acknowledgments. This work was supported by the National Natural Science Foundation of China (Grant No. 21272171).

References

1. Quan, K., Liu, Q., Li, P., et al.: Advance of ginsenosides in anticancer activities. J. Med. Postgrad. **28**(4), 427 (2015)

2. Hu, T.C., Liu, Y.M., Tao, R.S., et al.: Advances on the chemical components and pharmacological activities of Velvet Antler. J. Econ. Anim. **19**(3), 156–162 (2015)
3. Yu, G.H., Pei, W.X., Sun, H.J., et al.: Neuroprotective effect and relative mechanisms of lycii fructus polysaccharide. Chin. J. Exp. Tradit. Med. Formulae **24**(9), 225–231 (2018)
4. Gong, W.X., Zhou, Y.Z., Li, X., et al.: Research progress in antidepressive active ingredients and pharmacological effects of Angelicae Sinensis Radix. Chin. Tradit. Herbal Drugs **47**(21), 3905–3911 (2016)
5. Urbano-Cuadrado, M., Luque De Castro, M.D., Perez-Juan, P.M., et al.: Near infrared reflectance spectroscopy and multivariate analysis in enology: determination or screening of fifteen parameters in different types of wines. Anal. Chim. Acta **527**(1), 81–88 (2004)
6. Cozzolino, D., Kwiatkowski, M.J., Parker, M., et al.: Predictionof phenolic compounds in red wine fermentations by visible and nearinfrared spectroscopy. Anal. Chim. Acta **513**(1), 73–80 (2004)
7. Cozzolino, D., Kwiatkowski, M.J., Waters, E.J., et al.: A feasibility study on the use of visible and short wavelengths in the near-infrared region for the non-destructive measurement of wine composition. Anal. Bioanal. Chem. **387**(6), 2289–2295 (2007)
8. Sauvage, L., Frank, D., Stearne, J., et al.: Trace metal studies of selected white wines: an alternative approach. Anal. Chim. Acta **458**(1), 223–230 (2002)
9. Coimbra, M.A., Barros, A.S., Coelho, E., et al.: Quantification of polymeric mannose in wine extracts by FT-IR spectroscopy and OSC-PLS1 regression. Carbohydr. Polym. **61**(4), 434–440 (2005)
10. Qin, O., Chen, Q., Zhao, J.: Intelligent sensing sensory quality of Chinese rice using near infrared spectroscopy and nonlinear tools. Spectrochim. Acta Part A Mol. Biomol. Spectrosc. **154**, 42–46 (2016)
11. Chinese Pharmacopoeia Commission: Chinese Pharmacopoeia (edition 2015), pp. 8–9, 133–134, 249–250, 323. China Medical Science Press, Beijing (2015)
12. Bona, E., Marquetti, I., Link, J.V., et al.: Support vector machines in tandem with infrared spectroscopy for geographical classification of green arabica coffee. LWT-Food Sci. Technol. **76**, 330–336 (2017)
13. Luo, Y., Wang, X.C., Deng, D.W.: Near infrared spectroscopy technology and its application in food sensory analysis. Food Sci. **30**(7), 273–276 (2009)
14. Rinaldo, S.B., Duhan, D.F., King, K., et al.: Tasting and evaluating aroma of wine: frontal lobe measurement using near infrared. Dev. Market. Sci. Proc. Acad. Market. Sci. 653 (2013)
15. Lu, Y.Z., Du, C.W., Yu, C.B., et al.: Classifying rapeseed varieties using Fourier transform infrared photoacoustic spectroscopy (FTIR-PAS). Comput. Electron. Agric. **107**, 58–63 (2014)

Mechatronics

Predictive Analysis of Faults in Tower Crane

Zhe Zhang[1(✉)] and Jianli Yu[2]

[1] College of Science, Zhongyuan University of Technology, Zhengzhou, China
zzxq2010@126.com
[2] Zhengzhou Institute of Aeronautical Industry Management, Zhengzhou, China

Abstract. At present, the research of crane fault diagnosis is focused on the faults prediction. In recent years, artificial neural network is a new interdisciplinary and broad-application-prospect subject, which has been used in more and more practical problem predictions, particularly in prediction of the future behavior referring to the nonlinear system. Therefore, this paper uses the neural network method to study the identification of running tower crane. In addition, this paper proves that the BP network is an effective method to identify the running status of the tower crane. Finally, conclusion could be gained that neural network method has high prediction accuracy.

Keywords: Tower crane · Artificial neural network · Fault diagnosis
Prediction analysis

1 Introduction

Crane is necessary in modern machinery and it is an essential mechanical equipment in production process mechanization, automation to improve material transportation conditions and labor productivity. It is also important for the development of national economy and people's material and cultural life. China's crane production history is not long, "Seven Five" period used technology-trade combination ways to introduce small and medium tonnage crane production technology from Germany and Japan respectively. Compared with the world's advanced countries, domestic crane tonnage is still small, and the degree of serialization as well as the technology is also lower in the design and manufacture. In recent years, the development of domestic cranes and the technical level increase rapidly, the series have been continuous improved.

If the crane did not work, it would cause great harm. Thus, it is necessary to predict the fault of the crane. There are many new methods and ideas about the crane fault analysis research in worldwide, including fault analysis based on fuzzy theory [2], fault tree analysis [3], gray theory and neural network [4–6]. Such many methods could solve many problems of crane fault analysis in different ways.

K. Deng et al. (Eds.): ICMIR 2018, AISC 856, pp. 795–800, 2019.
https://doi.org/10.1007/978-3-030-00214-5_99

2 Basic Theory of Artificial Neural Network

Artificial neurons are highly simplified biological neuron models that simulate the structure and function of biological neurons from the perspective of engineering implementation. Artificial Neural Network (ANN) is a widely interconnected network of simple information processing units (neurons, processing elements, electronic components) called nodes. It is the abstraction, simplification and simulation of the human brain, which could reflect the basic characteristics as like human brain. Neural networks are networks that are widely interconnected by large numbers of neurons. According to the different connection methods, there are two types of neural networks: forward network and feedback network. After several decades development, hundreds of artificial neural networks have been built up.

BP network use error back propagation algorithm or its variant form of network model. The standard BP network is a multi-layer network that uses the gradient descent algorithm to carry out weight training for nonlinear differentiable functions. BP neural network is usually composed of input layer, output layer and several hidden layers. Each layer consists of many nodes. Each node is a neuron, the upper node and the lower nodes connect through weight. A typical BP network is a three-layer network with a hidden layer. After training, the neural network establishes a mapping, which is the approximation of the previous feature space to the decision space map. In the mid-1980s, Rumelhart and McClelland proposed a back-propagation algorithm (BP) for the multi-layer feedforward network, which is a gradient descent method in a multilayer feedforward network in application. And its structure is same as that of multi-layer network sensor except Sigmoid function is used for the excitation function of each hidden layer node. Therefore, BP artificial neural network can be called multi-layer perceptron using Sigmoid function as excitation function.

3 Work State Identification of Tower Crane Based on BP Network

Tower cranes are one of the major construction machinery indispensable in high-rise buildings. The machine has high accident rate and considerable risk of operation. Also, it has disadvantages such as the lack of intelligent monitoring function of the online security system due to crane's own safety system, which also leads to many accidents. Based on this, the BP network analyze seven state parameters of the tower crane to identify the training neural network. After trained, networks diagnose the working status of tower cranes. Here BP neural network is applied on prediction and analysis of tower crane working state parameters.

In this network, the seven parameters of the tower crane in different working states are the input vector p. There are no evaluation criteria for the various working conditions of the tower crane. In this study, the fuzzy theoretical evaluation result was selected as the target output. The main factors of the crane safety status are seven indexes, such as weight, stability coefficient, lifting moment, wind speed, working range, motor temperature and hoisting height. Each of them has different effect on the working status of the crane. Using the fuzzy evaluation results given in [9, 10] as the target output.

Fuzzy evaluation result y is between [0, 1]. In this paper, [0, 1] is divided into five subsets, where [0.85, 1] indicated tower crane was in absolutely safe state and calibration bit value is '1'. [0.5, 0.7) indicated tower crane was in relatively safe state and calibration bit value is '0.5'. [0.3, 0.5) indicated tower crane was in dangerous state transitions to a safe state and calibration bit value is '0.25'. [0, 0.3) indicated tower crane was in fault state and calibration bit value is '0'. the fuzzy state of the tower crane is roughly categorized to five modes, which are digitally calibrated to 1, 0.75, 0.5, 0.25 and 0. The part of the fifty groups of training samples were selected, as shown in Table 1.

Table 1. 50 groups of training samples

Working state	Target input (status parameter)							Fuzzy evaluation result	Expected output
	Stability	Lifting weight	Lifting torque	Lifting height	Lifting amplitude	Wind speed	Motor winding temperature		
1	1.15	6375	686	38.6	10.8	10.9	109	0.63	0.5
2	1.08	6565	666	39	10.1	15	88	0.52	0.5
3	1.046	6214	650	38.2	10.5	14.3	102	0.67	0.5
4	1.023	6483	650	39.4	10	15.7	114	0.43	0.25
5	1.007	6214	678	39.8	10.9	13.1	114	0.42	0.25
6	1.15	6565	666	39.8	10.1	14.3	118	0.51	0.5
7	1.08	5882	615	39.4	10.5	10.9	109	0.78	0.75
8	1.08	6375	678	38.2	10.6	15.7	118	0.54	0.5
9	1.023	5882	686	38.6	11.7	14.3	88	0.5	0.25
10	1.007	6483	615	39	9.5	13.1	102	0.52	0.5
11	1.15	5882	678	39	11.5	15.6	102	0.66	0.5
12	1.046	6375	615	39.8	9.6	15.7	88	0.56	0.5
13	1.007	6375	650	38.6	10.2	10.9	118	0.53	0.5
14	1.007	6565	686	39.4	10.4	15	109	0.25	0
15	1.15	6483	650	38.2	10	15	109	0.67	0.5
16	1.08	6214	686	39.8	11	15	114	0.49	0.25
17	1.046	6483	666	39.8	10.3	10.9	102	0.51	0.5
18	1.023	5882	615	38.6	9.4	14.3	114	0.62	0.5
19	1.023	6565	678	38.2	10.3	10.9	88	0.48	0.25
20	1.15	6214	650	39.4	10.5	13.1	88	0.75	0.75
21	1.08	6565	615	38.6	9.4	14.3	114	0.61	0.5
22	1.046	6483	686	39	10.6	13.1	118	0.46	0.25
23	1.023	6375	678	39.4	10.6	14.3	102	0.45	0.25
24	1.007	6214	666	38.6	10.7	15.7	109	0.47	0.25
25	1.17	6130	640	39.2	10.4	12.4	86	0.8	0.75
26	1.17	6435	640	15	9.9	14.7	106	0.66	0.5
27	1.07	6400	640	15	10	14	100	0.7	0.75
28	2	1500	300	38.2	20	4	80	0.97	1

(continued)

Table 1. (*continued*)

Working state	Target input (status parameter)							Fuzzy evaluation result	Expected output
	Stability	Lifting weight	Lifting torque	Lifting height	Lifting amplitude	Wind speed	Motor winding temperature		
29	1.26	5682	604	20	10.6	9	86	0.96	1
30	1.058	6565	686	38.5	10.4	15.2	106	0.41	0.25
31	1.13	5936	633	18	10.7	9	94	0.89	1
32	1.007	6501	671	39.8	10.3	15.2	115	0.31	0.25
33	1.08	6204	621	38.2	10	10	97	0.82	0.75
34	1.07	6400	640	22	10	14	100	0.7	0.75
35	1.007	6170	680	20	11	14.6	111	0.49	0.25
36	1.2	6320	686	39.5	10.9	13.8	115	0.58	0.5
37	1.066	6327	649	38.2	10.3	14	102	0.69	0.5
38	1.07	6381	637	38.2	10	10	100	0.74	0.75
39	1.09	6423	671	39.8	10.4	14.6	100	0.54	0.5
40	1.05	6362	657	24	10.3	13	103	0.67	0.5
41	1.046	6417	664	38.3	10.3	13	107	0.6	0.5
42	1.037	6444	666	38.4	10.3	14.3	108	0.55	0.5
43	1.037	6454	670	38.6	10.4	14.3	109	0.51	0.5
44	1.017	6492	673	38.8	10.4	14.3	112	0.42	0.25
45	1.009	6554	682	39	10.4	14.3	113	0.31	0.25
46	1.02	6423	686	39.2	10.7	12.8	88	0.43	0.25
47	1.02	6320	661	38.5	10.5	15.7	118	0.49	0.25
48	1.036	6501	661	25	10.2	13.8	88	0.58	0.5
49	1.58	3500	671	38.5	19	12.8	106	0.81	0.75
50	1.058	6170	661	39.8	10.7	10.9	111	0.43	0.5

Select five parameters working states of the tower crane (Table 2) for network verification:

Table 2. Tower crane five working state parameters

Working state	Target input (status parameter)							Fuzzy evaluation result	Expected output
	Stability	Lifting weight	Lifting torque	Lifting height	Lifting amplitude	Wind speed	Motor winding temperature		
1	1.08	6375	678	38.2	10.6	15.7	118	0.54	0.5
2	1.15	6483	650	38.2	10	15	109	0.67	0.5
3	1.007	6214	666	38.6	10.7	15.7	109	0.47	0.25
4	1.07	6381	637	38.2	10	10	100	0.74	0.75
5	1.037	6444	666	38.4	10.3	14.3	108	0.55	0.5

After establishing BP network, it is required to do repeated training. Using MATLAB software, quasi-Newton algorithm and 'trainbfg' network training function were selected to finish the test. Then based on the BP network program, MATLAB simulation experiment and the results are shown as followings:

Select the desired network accuracy goal is 0.001, run test in order from 1 to 50 among hidden layer neurons. After repeated trial, it is concluded that when the number of hidden neurons is larger than 21, the convergence rate is relatively fast, and the evaluation result is more accurate. Here, I choose 22, the network was quickly converged. When training up to 329 times, the standard numerical error is 1e-3, the network training chart was shown in Fig. 1:

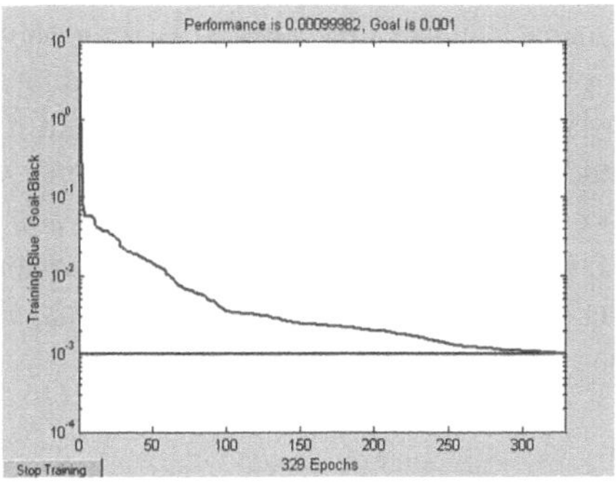

Fig. 1. Network training chart

Next, in the 50 tower cranes training samples, five working states parameters of the tower crane were selected for network verification (Table 3):

Table 3. Tower crane five working state parameters and BP network verification results

Working state	Target input (status parameter)							Fuzzy evaluation result	Expected output	Network evaluation results
	Stability	Lifting weight	Lifting torque	Lifting height	Lifting amplitude	Wind speed	Motor winding temperature			
1	1.08	6375	678	38.2	10.6	15.7	118	0.54	0.5	0.44
2	1.15	6483	650	38.2	10	15	109	0.58	0.5	0.53
3	1.007	6214	666	38.6	10.7	15.7	109	0.47	0.25	0.25
4	1.07	6381	637	38.2	10	10	100	0.74	0.75	0.73
5	1.037	6444	666	38.4	10.3	14.3	108	0.55	0.5	0.42

From Table 3, the evaluation result of BP neural network and the result of fuzzy evaluation could support each other, and the error is small, which indicates that BP neural network is an effective method for fault detection of tower crane working state.

4 Conclusion

Based on BP neural network model, the following conclusions could be gained: BP neural network has a good function approximation for nonlinear systems. BP neural network is feasible with forecasting accuracy and can be used in this field. Having considered the current situation of crane fault diagnosis research, this paper further studies the working status of tower crane based on neural network, and get some important conclusions.

It is proved that BP network is an effective method to predict and analyze the breakdown of tower crane, but the training time of BP network is long. In the use of neural networks for fault diagnosis, the selection of training samples is particularly important. Training samples should be enough and representative in broad area; In the BP network, there are many training functions. It is impossible for one function can adapt to all the training process, so the function need to be replaced constantly during training until you find the most suitable one. BP algorithm has the following shortcomings: the convergence rate is slow, the local extreme value, it is difficult to determine the number of hidden layer and hidden layer nodes. So, there are many improved algorithms. The algorithm used in this paper is quasi-Newton algorithm. Hidden layers of neurons are from 1 to 50 in turn test. A relatively accurate number was found after repeated trial.

References

1. Jiang, Z.: Introduction to Artificial Neural Networks. Higher Education Press, Beijing (2001)
2. Liu, J., et al.: Based on the fuzzy theory of gantry crane safety assessment of the comprehensive situation. Mach. Manuf. Autom. (2014)
3. Wu, R.: Analysis of mechanical failure of crane based on gray relational analysis. Coal Mine Mach. (2008)
4. Gu, Y., Wang, Q., Zhang, L.: Fault diagnosis of ship electric AC crane based on fault tree analysis. J. Jincheng Vocat. Tech. College (2009)
5. Yang, X., et al.: Study on fault diagnosis system of gantry crane based on RBF neural network. J. China Constr. Mach. (2013)
6. Shang, J., et al.: Application of BP neural network based on genetic algorithm in tower crane fault diagnosis. Lift. Transp. Mach. (2012)
7. Patan, K., Witczak, M., Korbicz, J.: Towards robustness in neural network based fault diagnosis. Appl. Math. Comput. Sci **18**(4), 443–448 (2008)
8. Tang, Y., et al.: A bridge crane fault diagnosis method based on bayesian network. Lift. Transp. Mach. (2017)
9. Zheng, X., et al.: Application of RBF neural network and uniform design in safety state pattern recognition of tower crane. Mach. Des. Manuf. 78–79 (2005)
10. Chang, X., et al.: Tower crane safety state pattern recognition based on RBF neural network. Build. Mech. 41–43 (2005)

Numerical Study on Universal Joint Under Thermo-Mechanical Loadings

Jin Li[1(✉)], Haifang Dong[1], Yafeng Li[2], and Wei Yuan[1]

[1] Wuhan Second Ship Design and Research Institute, Wuhan 430064, Hubei, China
lijinhuake@163.com
[2] Wuhan Wuchang District Business Bureau, Wuhan 430000, Hubei, China

Abstract. The paper is about the finite element analysis of the contact pressure between axial hole and sliding bearing of the universal joint in the Tower Soft Yoke Single Point Mooring System. The working temperature of the universal joint is −40 ∼ 80 °C. The contact pressure of Universal joint is related to the working temperature, loading and the sliding bearing thickness. In this paper, the numerical simulation of the contact pressure of Universal joint is carried out under different working temperature, loading and the sliding bearing thickness. In the simulation analysis, Sub-modeling is applied to save computing time, and the frictional force is modeled by the classics Coulomb friction model. The results show that the contact pressure of the Universal joint has nonlinear relationship with the different temperature, loading and the sliding bearing thickness. Under the condition of the same interference fit, the contact pressure of the Universal joint grows with the increase of the working environment temperature, loading and the sliding bearing thickness.

Keywords: Tower soft yoke single point mooring system · Universal joint Contact pressure · Numerical simulation · Sub-modeling

1 Introduction

Recently, the State Development and Reform Commission agreed to set up the national energy major scientific and technological innovation project of marine nuclear power demonstration project platform in the charge of 719th Research Institute of CSIC. Located in the Bohai Gulf, the platform is moored far from the coast and carries out power generation operations on a fixed-point basis. The soft yoke single point mooring system is a key part of the project. The single point mooring system consists of a bow mooring support, a mooring leg, a mooring rigid arm, an upper block and a fixed jacket. The mooring leg is a tubular steel member for connecting the mooring bracket and mooring rigid arm, which has three rotational degrees of freedom to adapt to the platform in motion when the wind flow comes into action. The mooring leg is mainly composed of the upper universal joint, the mooring leg steel pipe and the lower universal joint, at the lower end, connected to the lifting lugs on the mooring rigid-arm ballast tank via a universal joint, at the upper end by a universal joint with axial thrust

K. Deng et al. (Eds.): ICMIR 2018, AISC 856, pp. 801–809, 2019.
https://doi.org/10.1007/978-3-030-00214-5_100

roller bearings and lifting lugs on the mooring brackets. Upper and lower gimbals are of cross-shaped structure in the form, through two perpendiculars to the rotation axis to ensure that it has two degrees of freedom of rotation, the use of lubricating oil lubrication of the bearing; in the same time, the lower end of the connecting arm is provided with a rotary joint. As a result, the mooring leg has three rotational degrees of freedom, and the swivel joint bearing adopts the daily manual oiling method. The universal joint is made up by the U-shaped coupler, and the upper and lower mooring legs fork, pin, sliding bearings and other components. The specific structure is shown in Figs. 1 and 2 is about assembling the connection arm and universal joint.

Fig. 1. Universal joint structure modal

Fig. 2. Arm link universal joint assembly

2 Research Object Description

The working conditions of the universal joint of the mooring leg are low speed, heavy load and variable torque, so the universal joint shaft hole and the outer surface of the bearing adopt interference connection. Through the universal joint shaft hole and the sliding bearing interference with torque transmission and axial load, interference connection is able to achieve the purpose of reliable connection. Compared with the key, the thread and other connections, the utility model has the advantages of simple structure, good centering precision, large carrying capacity, good impact performance, no need to open the keyway to reduce the shaft, etc. Interference connection is widely used in heavy machinery, lifting and conveying machinery, marine engineering, railway locomotives and other general machinery, and generally suitable for medium and large size. Taking CATIA V5R20 as the 3D modeling platform and Abaqus as the simulation platform, this paper analyzes and calculates the interference of the universal joint shaft hole and the outer surface of the sliding bearing, and takes into all these factors account like the influence of the different working environment temperature on the contact stress. U-type coupling, with the upper and lower mooring legs fork head material is 42CrMo, sliding bearing is made of copper alloy CuAl10Fe5Ni5. The material properties of the two materials are shown in Table 1.

Table 1. Universal joint and sliding bearing material properties

Material	42CrMo	CuAl10Fe5Ni5
Elastic modulus (GPa)	210	110
Poisson's ratio	0.3	0.34
Linear expansion coefficient (10^{-6}/K)	12.1	18
Density (g/cm^3)	7.85	7.86
Yield strength (MPa)	930	350

3 Interference Fit Calculation and Analysis

3.1 Interference Fit Calculation Model

When the universal joint shaft hole and the sliding bearing are in interference fit, the positive pressure will be generated on the mating surface to expand the inner diameter of the universal joint shaft hole and compress the inner and outer diameters of the sliding bearing. The magnitude of the interference between the universal shaft and the sliding bearing is not only closely related to the transmission capacity between the shaft bore and the sliding bearing, but also related to the motion state between the universal shaft and the sliding bearing. At low speed, the effect of centrifugal force is almost negligible. However, at high speed, centrifugal force is an important factor for the joint of universal joint shaft hole and plain bearing. Figure 3 shows the interference fit calculation model between universal joint shaft hole and sliding bearing. The contact stress between the contact surfaces of the gimbal is related to the interference fit of the joint shaft hole and the sliding bearing, which can be simplified as the interference fit of two thick-wall cylindrical sleeves. According to the calculation method of thick-walled cylindrical sleeve of material mechanics and elasticity, the pressure at the joint surface due to interference is calculated as follows:

$$p_{\max} = \frac{\delta_{\max}}{d_0\left(\frac{C_1}{E_1} + \frac{C_2}{E_2}\right) \times 10^3} \tag{1}$$

$$C_1 = \frac{d_0^2 + d_1^2}{d_0^2 - d_1^2} - v_1 \tag{2}$$

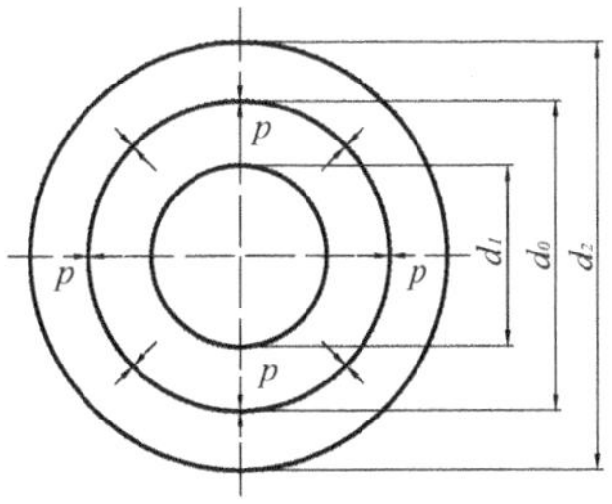

Fig. 3. Calculation model of thick-walled cylinder

$$C_2 = \frac{d_0^2 + d_2^2}{d_0^2 - d_2^2} + v_2 \tag{3}$$

In the type: p_{max} is the radial pressure between the surfaces, MPa; δ_{max} is the interference amount of interference connection, μm; d_0 nominal diameter for mating, mm; d_1 and d_2 are the diameter of the inner diameter of the sliding bearing and the outer diameter of the universal joint, mm; C_1 and C_2 are the rigidity coefficients of the sliding bearing and the universal joint respectively; E_1 and E_2 are the elastic modulus of the sliding bearing and the universal joints, respectively, MPa; v_1 and v_2 are the Poisson's ratio of the sliding bearing and the universal joint.

According to mechanical design manual, the part of the outer diameter of the sliding bearing and universal joint shaft hole with the base axis system, the inner diameter of the sliding bearing with the use of the base hole system. With different parts, different load conditions and the temperature of the work environment, the tolerance level and the amount chosen correspondingly are also different. The conditions of the inner and outer diameters of the sliding bearing are different, and the requirements are not the same, and since the universal joint is used to transmit the larger torque, a very reliable connection needs to be achieved. The size table shows the sliding bearing outer diameter and universal joint shaft with interference fit $U7/h6$, Sliding bearing bore and pin with clearance fit $H8/f8$.

3.2 Equations of Temperature Field

In actual work process, universal joint, changes with the work environment temperature and seasons, and as a result of two different thermal expansion coefficients, the deformation caused by the temperature is also different, so the temperature stress is generated in the parts. When the temperature of the object is kept as a constant distribution, that is, it does not change with time, it then becomes a steady state problem, there are:

$$\frac{\partial T}{\partial t} = 0 \tag{4}$$

For the discrete element in thermal analysis, it is also necessary to express the temperature field of the unit $T(\mathrm{x}, \mathrm{y}, \mathrm{z})$ as the interpolation relation of the temperature of the node.

$$T(\mathrm{x}, \mathrm{y}, \mathrm{z}) = N(\mathrm{x}, \mathrm{y}, \mathrm{z}) \cdot q_T^e \tag{5}$$

In the form, $N(\mathrm{x}, \mathrm{y}, \mathrm{z})$ is a shape function matrix. q_T^e is a node temperature array:

$$q_T^e = [\, T_1 \quad T_2 \quad \ldots \quad T_n \,]^T \tag{6}$$

The T_1 , T_2 , ... , T_n is the node temperature value.

Based on the functional extremum principle of heat transfer, the finite element equation that can get the heat transfer problem is:

$$K_T^e \cdot q_T^e = F_T^e \tag{7}$$

$$\begin{aligned} K_T^e = & \int_{\Omega^e} \left[\kappa_x \left(\frac{\partial N}{\partial x}\right)^T \left(\frac{\partial N}{\partial x}\right) + \kappa_y \left(\frac{\partial N}{\partial y}\right)^T \left(\frac{\partial N}{\partial y}\right) + \kappa_z \left(\frac{\partial N}{\partial z}\right)^T \left(\frac{\partial N}{\partial z}\right) \right] d\Omega \\ & + \int_{S_3^e} \bar{h}_c N^T N dA \end{aligned} \tag{8}$$

$$F_t^e = \int_{\Omega^e} \rho Q N^T d\Omega + \int_{S_3^e} \bar{h}_c T_\infty \cdot N^T d\Omega + \int_{S_3^e} \bar{q}_f \cdot N^T dA \tag{9}$$

The first type of S_1: a given temperature distribution:

$$T(\mathrm{x}, \mathrm{y}, \mathrm{z}, t) = \bar{T}(t) \tag{10}$$

The second type of S_2: a given heat flow condition:

$$\kappa_x \frac{\partial T}{\partial x} n_x + \kappa_y \frac{\partial T}{\partial x} n_y + \kappa_z \frac{\partial T}{\partial z} n_z = \bar{q}_f(t) \tag{11}$$

The third type of S_3: convective heat transfer conditions for a given environment:

$$\kappa_x \frac{\partial T}{\partial x} n_x + \kappa_y \frac{\partial T}{\partial x} n_y + \kappa_z \frac{\partial T}{\partial z} n_z = \bar{h}_c (T_\infty - T) \tag{12}$$

$$\partial\Omega = S_1 + S_2 + S_3 \tag{13}$$

In the formula, K_T^e is called unit heat transfer matrix, ρ is material density (Kg/m^3), F_T^e is unit node equivalent temperature load array, T_∞ is environment temperature, κ_x, κ_y and κ_z are heat conduction coefficient along x, y and z direction respectively $(W/(m \cdot K))$. $\bar{h}_c$ is the convection heat transfer coefficient on the boundary between the object and the surrounding medium $(W/(m^2 \cdot K))$. n_x, n_y and n_z are the cosine of the normal line outside the boundary, $\bar{q}_f$ is the heat flux density given on the boundary (W/m^2), Q is the heat source intensity inside the body (W/kg), and $\bar{T}(t)$ is the given temperature on the boundary (°C).

At the room temperature of 20 °C, the amount of interference of universal joint is determined, but temperature range of its working environment in the Bohai Sea is −40 ~ 80 °C, and the linear expansion coefficients of the two materials are different.

The interference should be fixed according to the working ambient temperature. The formula is as follows:

$$\begin{cases} x_{z\,\max} = x_{G\,\max} + d_0[\alpha_z(t_z - t) - \alpha_k(t_k - t)] \\ x_{z\,\min} = x_{G\,\min} + d_0[\alpha_z(t_z - t) - \alpha_k(t_k - t)] \end{cases} \tag{14}$$

In the form, $x_{z\,\max}$ and $x_{z\,\min}$ are the maximum and minimum assembly interference, mm; $x_{G\,\max}$ and $x_{G\,\min}$ are the maximum and minimum work interference, mm; d_0 is the nominal diameter of the coordination, mm; α_z and α_k are the linear expansion coefficient of the shaft and hole materials, °C^{-1}; t_z and t_k are the working temperature of the shaft and the hole, °C. t is the temperature at the time of assembly, °C (Table 2).

Table 2. Assembly under different temperature working conditions

Temperature (°C)	−40	−20	0	20	40	60	80
Maximum assembly interference (mm)	0.316	0.352	0.389	0.426	0.462	0.499	0.535
Minimum assembly interference (mm)	0.223	0.259	0.296	0.333	0.369	0.406	0.442

4 Finite Element Modeling

Friction is a highly nonlinear physical phenomenon; especially its discontinuous discontinuity is one of the difficult problems in numerical simulation analysis. Taking the single point mooring universal joint as an example, this paper uses the CAE software Abaqus to establish the finite element simulation model of the interference fit of the universal joint. The friction between the shaft joint and the journal bearing is based on the classical Coulomb friction model. The friction coefficient is 0.15, the maximum allowable shear stress is related to the positive pressure generated by the interference fit between the universal joint shaft hole and the sliding bearing, and the relationship between the value and the friction factor is:

$$\tau_{\max,j} = \mu \cdot p_j \tag{15}$$

In the form: j a node between the contact surfaces.

In the finite element calculation, it is important that the universal joint shaft bore and the plain bearing unit establish a good contact behavior. In order to ensure the convergence of the results and reduce the calculation time, the mesh of the contact surface needs to be finely divided and the cell shape is good. Sliding bearings using 20-node hexahedron sweep method, the other components using hexahedral-oriented method of dividing the grid. By setting the size of the contact between the two contact surfaces, the two contact surfaces establish a shared grid unit so as to more accurately simulate the stress gradient of the contact area.

Considering the large size of the model, if the entire model is refined to get the grid-independent solution, the number of grids to be divided is large and more computational resources are consumed. At the same time, the universal joint has a total of eight sliding bearings, U-type coupling and its associated sliding bearings (Fig. 1, the upper bearing) contact stress maximum. Therefore, the paper uses sub-model technology to study it in order to save calculation time. The sub-model technique is based on the Saint-Vacant principle and has the following advantages: (1) Reduces or even eliminates the complex load transfer area required in a finite element solid model; (2) Analyzes of different designs in the contact area of interest; (3) Verify whether the result has been grid-independent solution. A total of 379837 grids and 1202546 nodes are solved using the entire model. The selection of sub-model is only 75785 grids and 274258 nodes (Fig. 4).

Fig. 4. Integral meshing and sub-model meshing

5 Finite Element Simulation

Through the simulation, we can get the result of sliding at the same interference, different working environment temperature (−40 °C, −25 °C, −10 °C, 5 °C, 20 °C, 35 °C, 50 °C, 65 °C, 80 °C) Bearing contact stress curve, shown in Fig. 5. Can be seen from the figure, the same amount of interference, the contact bearing sliding contact with the work of the ambient temperature rises almost straight up. The results of the finite element analysis and the theoretical calculation are in good agreement with each other because the linear expansion coefficients of the materials used for the plain bearing and the universal joint are different. In the work environment temperature changes, the deformations of the two materials are different, but at the same time because of contact constraints, temperature-induced deformation is limited, resulting in temperature stress.

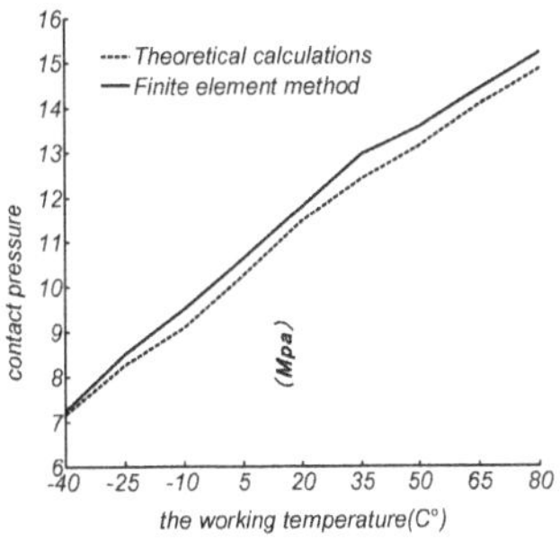

Fig. 5. Variation curve of contact stress at different operating ambient temperatures

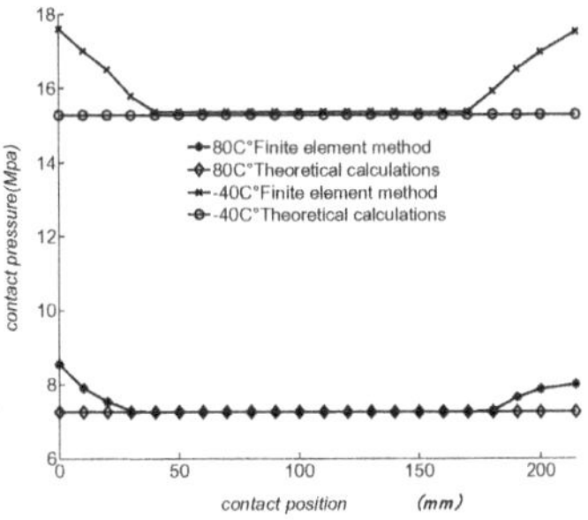

Fig. 6. The relation between contact stress and contact position

As can be seen from Fig. 6, the trend of contact stress is almost the same under the working conditions of −40 °C and 80 °C. At the two edges of sliding plain bearings with the largest contact stress, stress concentration occurred and the contact stress in the middle part changed little. The difference between finite element analysis and theoretical calculation in the middle section is very small.

From the analysis in Sect. 2, we know that when the material is selected, the size of the contact stress mainly depends on the thickness of the plain bearing. Therefore, in the calculation, the thickness of the plain bearing is firstly divided into 2.5 mm, 5 mm, 10 mm, 15 mm and 20 mm. The situation, the amount of interference taken as 0.4 mm. The relationship between the pretensioning contact stress and the thickness of the plain bearing with different thickness of the plain bearing is given. As the wall thickness increases, the contact stress increases substantially linearly (Fig. 7).

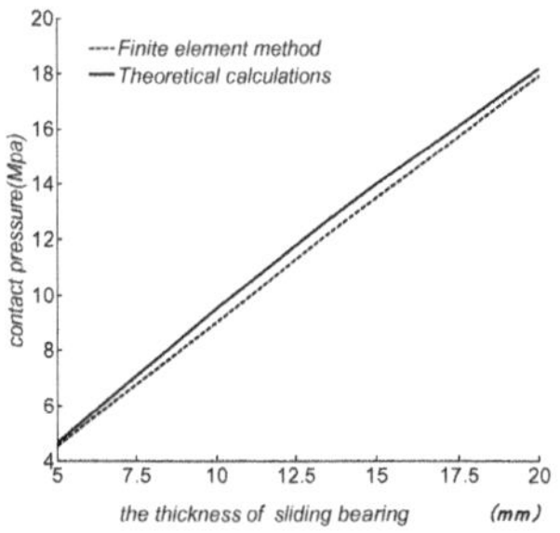

Fig. 7. The contact stress varies with the thickness of the sliding bearing

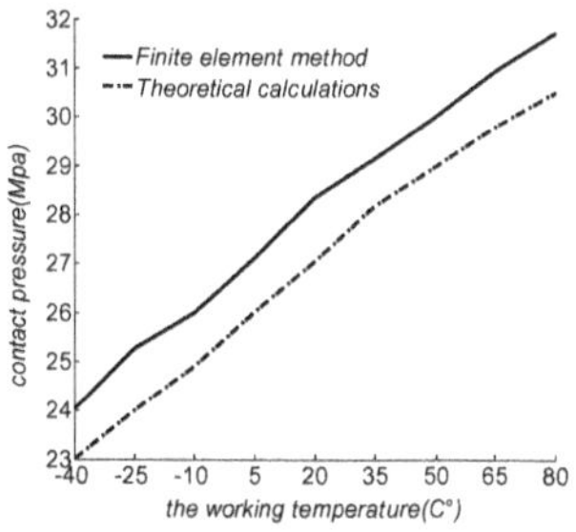

Fig. 8. Apply external load contact pressure with temperature curve

When the external load was applied, it was found that under the same interference, different working environment temperature (−40 °C, −25 °C, −10 °C, 5 °C, 20 °C, 35 °C, 50 °C, 65 °C, 80 °C) Sliding bearing contact stress curve, shown in Fig. 8. It can be seen from the figure that after the external load is applied, the contact stress of the universal joint shaft hole and the sliding bearing changes with the temperature and is similar to that when no external load is applied, and the stress increases as the working ambient temperature increases. The result is roughly the sum of the stress caused by the temperature and the stress caused by the external load.

6 Conclusions

(1) At the same interference, different temperatures, the contact stress of the plain bearing increases approximately linearly with the working ambient temperature. (2) The stress concentration occurs on the two contact edges of the sliding bearing, and the contact stress in the middle part almost has no changes. (3) After determining the interference and material, the size of the contact stress between the contact surfaces depends mainly on the thickness of the plain bearing. When the interference and other conditions are the same, the relationship between the contact stress and the thickness of

the plain bearing under different thickness of the plain bearing is that the contact stress increases with the plain bearing's wall thickness. (4) When the external load is applied, the contact stress of the universal shaft and the sliding bearing changes with the temperature, which is similar to that without the external load, and increases with the working environment temperature. The results are approximately the sum of the stresses induced by the temperature and the external load.

References

1. Yang, G., Xie, J., Li, Q.: Key parameters of fretting damage under shrink fit. J. Mech. Eng. **46**(16), 53–59 (2010)
2. Baldanzini, N.: A general formulation for designing interference-fit joints with elastic-plastic components. J. Mech. Des. **126**(4), 737–743 (2004)
3. Oswald, F.B., Zaretsky, V., Poplawski, J.V.: Interference-fit life factors for roller bearings. Tribol. Trans. **52**(4), 415–426 (2009)
4. Croccol, D., deAgostinis, M., Vincenzi, N.: Static and dynamic strength evaluation of interference fit and adhesively bonded cylindrical joints. Int. J. Adhes. Adhes. **30**(5), 359–366 (2010)
5. Han, C., Zhang, J.: Analyzing fretting contact of hollow shafts with interference fit. Huazhong Univ. Sci. Tech. (Nat. Sci. Edn.) **41**(5), 21–27 (2013)
6. Cao, H., Li, B., Chen, X.: Centrifugal expansion of high-speed spindle and its influences on bearing dynamic characteristics. J. Mech. Eng. **48**(19), 59–64 (2012)
7. Zeng, F., Chen, G.: Fretting analysis of interference fitting of wheel-set based on ANSYS. J. Mech. Eng. **47**(5), 121–125 (2011)
8. Jia, G., Gong, J.: Effect of interference fit on fretting wear of gear shaft shoulder. J. Hunan Univ. Nat. Sci. **40**(5), 31–36 (2013)

Design of Osculum Type Pneumatic Unloading Manipulator Based on PLC Control

Chengxiang Li(✉), Bin Wang, Yaolei Zhang, Haiyan Li, and Juying Wen

China Academy of Launch Vehicle Technology Research and Development Center, Beijing 100076, China
chengxiang_li@126.com

Abstract. Aiming at the low efficiency of traditional U-shape refrigerator shell, a kind of osculum type pneumatic unloading manipulator based on PLC control is designed. Based on the analysis of the functions and working principles to be realized, the design scheme of mechanical system and pneumatic drive system is introduced in detail. Using the osculum type pneumatic unloading manipulator, the smooth unloading of large sheet metal parts was realized. Using Siemens S7-300 PLC, the automatic control principle of the manipulator was designed, the automatic production of production equipment was realized. The practical results show that the equipment produces a product not more than 35 s, and the appearance of the product is beautiful, the dimension precision meets the use requirement. It improves the working conditions, reduces the labor intensity of the workers, and increases the labor productivity. This technology can be popularized and applied in the design of other industrial manipulator.

Keywords: Pneumatic manipulator · Structural design · Sheet metal parts PLC

1 Introduction

The manipulator is the basic equipment used in industrial manufacturing. The program controls the grasping, transporting and releasing of manipulator, take the place of the worker's heavy labor, realize the automation of production process, increase labor productivity. Pneumatic manipulator is a typical integration device of mechanical, electronic and pneumatic. It has many advantages such as low cost, easy operation and maintenance [1]. The U-shaped refrigerator shell is a sheet metal structure,the thickness of sheet material is about 0.5 mm. The refrigerator shell production line traditionally used manpower for unloading, the Production efficiency is low, the human cost is expensive. In this paper, taking the U-shaped refrigerator shell unloading as the task, designing a set of pneumatic unloading manipulator equipment. The Siemens S7-300 programmable controller is adopted for production control, the automation of production tasks has been realized.

K. Deng et al. (Eds.): ICMIR 2018, AISC 856, pp. 810–820, 2019.
https://doi.org/10.1007/978-3-030-00214-5_101

2 The Overall Scheme and Working Principle of Pneumatic Manipulator

The equipment is 4-dof manipulator, can be moved horizontally, up and down, rotated and telescopic, also can hold tight and loosen the shell. Pneumatic manipulator is composed of mechanical system, position detection system, pneumatic transmission system and electrical control system. Figure 1 shows the schematic diagram of unloading manipulator. The manipulator uses two side suction cups to pick up the u-shaped shell, and place it in the specified position according to the posture of assembly requirement [2]. After completing the folding task, the horizontal mobile platform driven by motor moved up to the folding machine. Vacuum adsorption device fetched shell, the up and down mobile device is driven by locking cylinder lifting off the countertop, and then the horizontal mobile platform carried out translation T1, the adsorption device carried out rotation R1 by the driven cylinder, the sheet rotate 90°, changed from vertical to horizontal type, and then it carried out translation T2 driven by locking the cylinder. The adsorption device put the shell on the rotating platform, then it rotated, drives by the cylinder, the lifting platform lift up and down, transmission belt transfer the shell from the rotating platform to the assembly line [3].

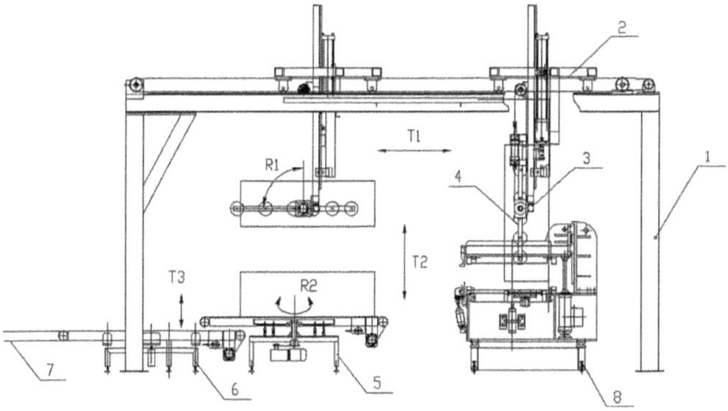

Fig. 1. Schematic diagram of unloading manipulator. 1. Gantry bracket device, 2. Horizontal mobile platform, 3. Up and down mobile device, 4. Mechanical arm system, 5. Rotating platform, 6. Lifting platform, 7. Assembly line, 8. Folding machine

3 Mechanical System Design

The mechanical system of the manipulator mainly includes the gantry bracket device, the horizontal mobile platform, the up and down mobile device and the mechanical arm system.

3.1 Gantry Bracket Device

The gantry bracket device is composed of gantry bracket, chain drive mechanism and reduction machine. The gantry bracket is welded by square steel tube and Q235 steel plate, provided movement guide and structural support for horizontal mobile platform. Chain drive mechanism consists of chain, chain wheel, transmission shaft and bearing seat. The chain drive mechanism drives the horizontal movement of the horizontal mobile platform, which is driven by the SEW deceleration motor.

3.2 Horizontal Mobile Platform

The horizontal mobile platform consists of wheel rack, drive cylinder, four sets of mobile wheels, four sets of limit pulleys and limit bearing. The wheel rack is used to install other components, which is welded with steel tube and Q235 steel plate, used to fix other components. The horizontal mobile platform is located in the guide of gantry bracket, which is composed of four moving wheels. The chain is fastened to the chain plate, and pulls the horizontal mobile platform moving. The driving cylinder selects the CDNSI125-1000DZ73 lock cylinder of SMC Company, which drives up and down mobile device moving up and down. The guide groove of up and down motion is fixed between V shape limit pulley and limit bearing, which to prevent itself horizontal sloshing, making sure it moved up and down steadily (Fig. 2).

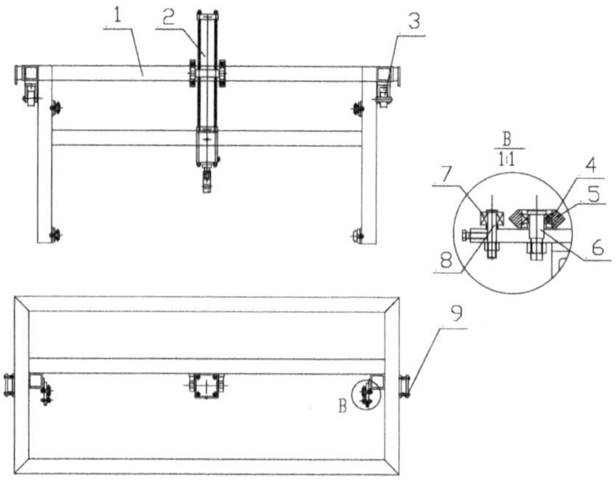

Fig. 2. Schematic diagram of horizontal mobile platform. 1. The wheel rack, 2. Drive cylinder, 3. Mobile wheels, 4. Limit pulleys, 5. Bearing, 6. 8. Axes of bolt, 7. Limit bearing, 9. Chain plate

3.3 Up and Down Mobile Device

Up and down mobile device consists of mobile bracket, cylinder connector, V shaped guide, bearing, transmission shaft, sprocket wheel, chain, the drive cylinder and adjusting the screw. Mobile bracket is welded with steel tube, which is used to install other components. Up and down mobile device is connected with the drive cylinder of horizontal mobile plant through the cylinder connector. Rotating mechanism consists of transmission shaft, sprocket wheel, chain, the drive cylinder, adjusting the joint, adjusting the screw and the system of mechanical arm, which is used for rotating of mechanical arm. Rotating mechanism is driven by the drive cylinder, and is used to chain transmission with the ratio of 1. The left and right arm synchronicity rotation is realized through the shaft transmission. Adjusting the screw is adopted with internal thread of both ends. Adjust the loose and tight of the chain by adjusting the joint of both ends (Fig. 3).

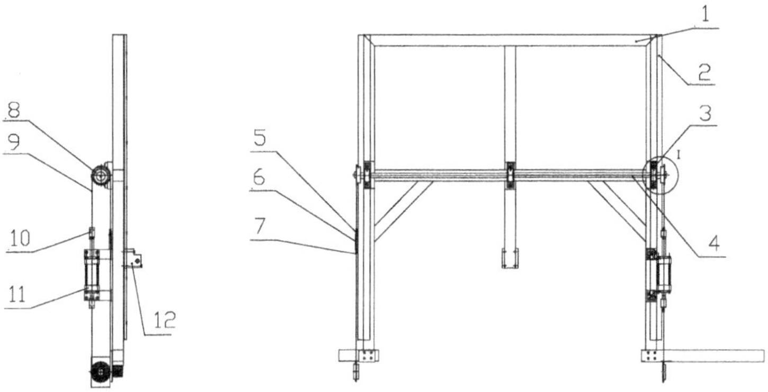

Fig. 3. Schematic diagram of up and down mobile device. 1. Mobile bracket, 2. V shaped guide, 3. Bearing, 4. Transmission shaft, 5. 7. Adjusting the joint, 6. Adjusting the bolt, 8. Sprocket wheel, 9. Drive chain, 10. Cylinder joint, 11. The drive cylinder, 12. Cylinder connector

The diameter of sprocket wheel:

$$d = p/\sin(180^\circ/z) \tag{1}$$

The rotation angle of the sprocket wheel:

$$\theta = 360^\circ/\pi d = \frac{\sin(180^\circ/z)L}{\pi p}\cdot 360^\circ \tag{2}$$

Meanings of formulary:

p—chain pitch,
z—number of teeth of the sprocket wheel,
L—action distance of the drive cylinder.

Chain is the chain of 10B, the chain pitch is 15.875 mm, number of teeth of the sprocket wheel is 25, the pitch diameter is 126.66 mm, the cylinder selects two-way acting cylinder of SMC CDA2WL80-100, the action distance is 100 mm, which can drive left and right arm rotate 90° simultaneity by sprocket wheel [4].

3.4 The System of Mechanical Arm

The system of mechanical arm includes left mechanical arm and right mechanical arm, holding it from left and right sides of the U shape refrigerator shell separately. Structural composition and working principle of the left and right arm is same, as shown in Fig. 4. The mechanical arm is fixed to up and down mobile device by sliding block. The axis vice of ball spline is consist of the ball spline shaft and convex type ball spline, which has the characteristics of compact structure, assembles easily, high stiffness. Under the drive of sprocket wheel, it can realize the rotation of the mechanical arm and a zero clearance to rotation direction. It also work as bush of the linear ball bearing, the linear motion is carried out under the action of thrust cylinder, which realizes stretch out and back of mechanical arm [5]. The left mechanical arm used two-way lock cylinder of MDBL50-575-D-Z73 type, which action distance is 75 mm. The right mechanical arm used two-way lock cylinder of MDNBL50-575-D-Z73 type, which action distance is 575 mm and has the function of stopping at any position. It is suitable for producing the side panel of different width refrigerator.

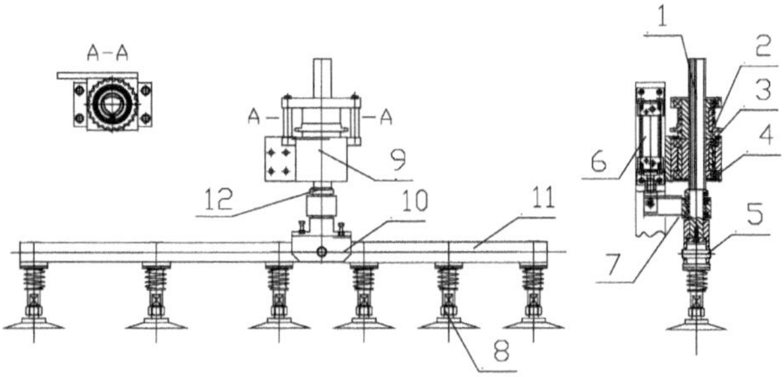

Fig. 4. Schematic diagram of structure of mechanical arm. 1. Ball spline shaft, 2. Sprocket wheel, 3. Bearing, 4. Convex type ball spline, 5. Hinge pin, 6. Thrust cylinders, 7. Thrust arm, 8. Vacuum sucker, 9. Slider block, 10. The rotating arm joint, 11. The rotating arm, 12. Adjustable sleeve

4 Design the System of Pneumatic

The principle of manipulator's pneumatic is shown as Fig. 5, which is consist of air source, filter reducing value, three pieces of pneumatic, control valve of pneumatic and kinds of cylinder. Control devices of pneumatic are selected from the SMA Company of Japan. Vacuum sucker is controlled by the two-position three way of solenoid valve, which is selected the type of VP342-5D-02A. In front of vacuum sucker installed the filter, and each solenoid valve equipped with muffler. Lift lock cylinder vertical mounting. The piston pull is used to balance the product's gravity. The pull of right mechanical arm locking cylinder is installed horizontally. In order to ensure both sides of the piston balance, pressure-reducing valve of refulgence is installed between reversing valve and rod less cylinder, make sure that the pressure of rod less cylinder lower than rod cylinder. When the cylinder is locked, both sides of piston are balance.

Locking cylinder controls action of locking and releasing by single solenoid valve of two position five-way, which controls the action of cylinder by middle location solenoid valve of two position three-way. Lift locking cylinder is driven by solenoid valve of two position five-way, which selects the type of VS7-8-FPG-D-3NA03. The right mechanical arm locking cylinder is driven by solenoid valve of t two position five-way, which selects the type of SY7320-5DZ-02. The other solenoid valve select the type of SY7220-5DZ-02 [6].

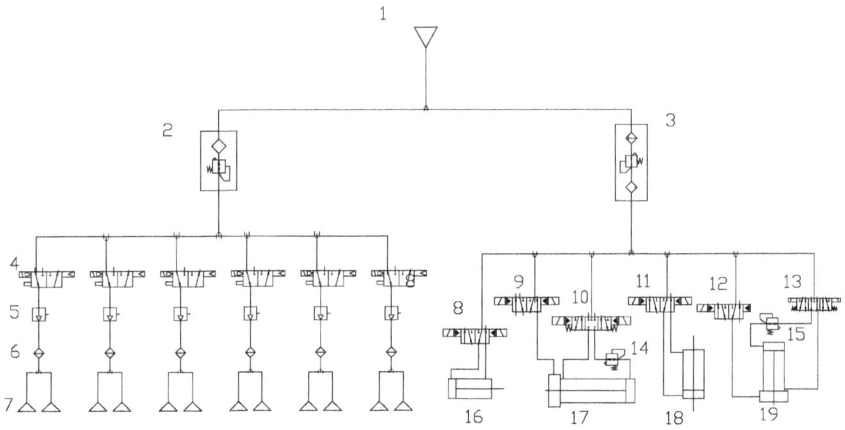

Fig. 5. Diagram of the principle of manipulator's pneumatic. 1. Air source, 2. Filter reducing value, 3. Three pieces of pneumatic, 4. Solenoid valve of two position three-way, 5. Vacuum generator, 6. Filter, 7. Vacuum sucker, 8. 10. 11. 12. Solenoid valve of two position five-way, 9. 13. Solenoid valve of three position five-way, 14. 15. Reducing valve, 16. Locking cylinder of left mechanical arm, 17. Locking cylinder of right mechanical arm, 18. Rotating cylinder, 19. Lift locking cylinder

The adsorption object of vacuum sucker is sheet metal, the material of which is cold-roll steel sheets. The biggest size of sheet is 5200 mm × 1000 mm × 0.6 mm. The gravity of the sheet:

$$W = \rho V g = 7.85 \times 520 \times 100 \times 0.06 \times 9.8 \times 10^{-3} = 240N \quad (3)$$

The calculation formula of sucker's diameter [7]:

$$D \geq \sqrt{\frac{4WK_1K_2K_3}{\pi np}} \quad (4)$$

Meanings of formulary:

W—the gravity of adsorption object,
K_1—safety factor, normal $K_1 = 1.2–2$, take $K_1 = 2$,
K_2—work situation factor, normal $K_2 = 1–3$, take $K_2 = 3$;

K_3—sucker posture factor, vertical position, $K_3 = 1/\mu$, μ: friction coefficient between sucker and adsorption object, normal $\mu = 0.4$–0.8, take $\mu = 0.5$, $K_3 = 2$;
n—number of sucker, in this device: n = 12;
p—vacuum of sucker, take 60–95% of the maximum vacuum p_v of vacuum generator, take 65%, select ZH15DS vacuum generator.

$$p = 0.65\, p_v = 0.65 \times 88\ KP_a = 0.055\ MP_a \tag{5}$$

The diameter of Vacuum sucker:

$$D \geq \sqrt{\frac{4WK_1K_2K_3}{\pi np}} = \sqrt{\frac{4 \times 240 \times 2 \times 3 \times 2}{3.14 \times 12 \times 0.055}} = 74.56\ \text{mm} \tag{6}$$

According to the calculation, selecting universal rotational vacuum sucker of Japan Convum, the type of PUTSB-80-30, the diameter is 80 mm.

5 Design of PLC Control System

Using PLC for control, not only the reliability is high, but also the control is flexible. When the operation sequence changes, the control program can be re-used quickly, only needed to modify and upload. It not only improves productivity, but also lowers the cost [8].

The main task of the unloading manipulator is to complete U-shaped refrigerator shell automatic unloading. It mainly controls the horizontal movement of the platform, the up-down movement and the joint movement of the manipulator, the rotation and transmission of the rotating platform, and the status detection [9].

5.1 PLC Selection and I/O Interface Allocation

In the manipulator, used the trip switch to detect and determine whether the mobile platform is in place, used the proximity switch to determine whether the manipulator is scraping the product, used magnetic switch to detect and judge whether the cylinder is in place. Each cylinder has two positions that need to be detected, therefore, two magnetic limit switches are installed on each cylinder, 17 input nodes are required. The movement of the suction cup, motor and all cylinders requires manual operation, 11 input nodes are required. Added the system hand/auto switch, start, stop and reset button, 28 input nodes are required. On the output side, need cylinder moving up and down, mechanical arm turning and reversing, mechanical arm sticking out and retraction, running indicator light, stop indicator light and failure indicator light signals, and the frequency conversion motor variable frequency output, frequency conversion motor failure and the vacuum sucker on-off signals, 14 output nodes are required.

This manipulator is a station of the U-shaped refrigerator shell production line, each station of the production line has dozens of address signals. It involves a lot of input and output data, and need a PLC that can support powerful operations, Siemens S7-300 meets the requirement [10]. The instrument connection adopts the bus and the dividing line. The control bus is introduced from the main control cabinet or the input and output module of PLC, through the electrical dividing box of each station, connected to the various sensors, air valves and other actuators [11]. The I/O port of the equipment is connected with pneumatic manipulator and control operation box. The I/O allocation table of PLC is shown in Table 1.

Table 1. The I/O address distribution table of the manipulator PLC

Name	Code	Address	Name	Code	Address
Manual/Auto	SA1	I33.0	Left arm front limit detection	SB10	I36.1
Start operation	SA2	I33.1	Right arm front limit detection	SB11	I36.2
Stop operation	SA3	I33.2	Right arm back limit detection	SB12	I36.3
Reset operation	SA4	I33.3	Left arm back limit detection	SB13	I36.4
Left move of the platform	SA5	I33.4	Detection of the right arm suction	SB14	I36.5
Right movement of platform	SA6	I33.5	Detection of the left arm suction	SB15	I36.6
Lifting cylinder drop	SA7	I33.6	Detection of rotating cylinder up	SB16	I36.7
Lifting cylinder rises	SA8	I35.0	Detection of rotating cylinder down	SB17	I37.0
Left arm cylinder out	SA9	I35.1	Frequency conversion output 1	U	Q29.0
Left arm cylinder bake	SA10	I35.2	Frequency conversion output 2	V	Q29.1
Right arm cylinder out	SA11	I35.3	Frequency conversion output 3	W	Q29.2
Right arm cylinder bake	SA12	I35.4	Frequency conversion failure	YA4	Q29.3
Rotating cylinder lift	SA13	I35.5	Lifting cylinder drop	YA5	Q30.0
Rotating cylinder return	SA14	I35.6	Lifting cylinder rises	YA6	Q30.1
Sucker suction	SA15	I35.7	Arm forward	YA7	Q30.2
Loading stop limit	SB1	I34.0	Arm reverse	YA8	Q30.3
Discharge stop limit	SB2	I34.1	Right arm extend	YA9	Q30.4

(*continued*)

Table 1. (*continued*)

Name	Code	Address	Name	Code	Address
Loading limit	SB3	I34.2	Right arm retract	YA10	Q30.5
Discharge speed limit	SB4	I34.3	Sucker sucked	YA11	Q30.6
Waiting speed limit	SB5	I34.4	Lift arm extend	YA12	Q30.7
Loading speed limit	SB6	I34.5	Lift arm retract	YA13	Q31.0
Waiting stop limit	SB7	I34.6	Running indicator	YA14	Q31.1
Loading upper limit	SB8	I34.7	Stop indicator	YA15	Q31.2
Loading lower limit	SB9	I36.0	Failure indicator	YA16	Q31.3

5.2 PLC Programming

Control programs include manual program and automatic program, manual program is a personnel operation executive program. According to the configuration data selection, automatic program to achieve automatic operation of equipment. Unloading manipulator program flow shown in Fig. 6, parts of the control program shown in Fig. 7 [12, 13].

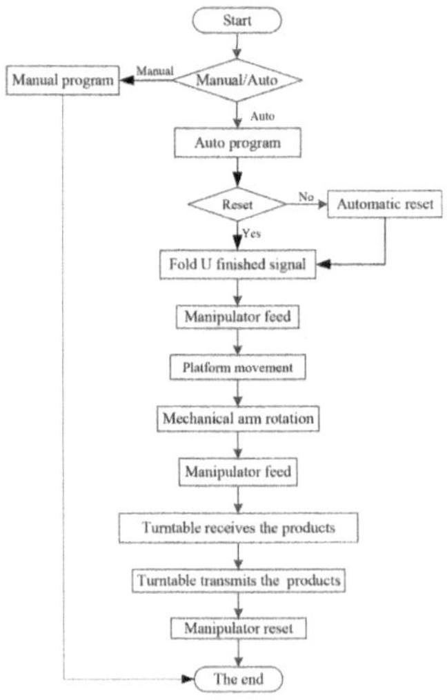

Fig. 6. Program flow diagram for unloading manipulator

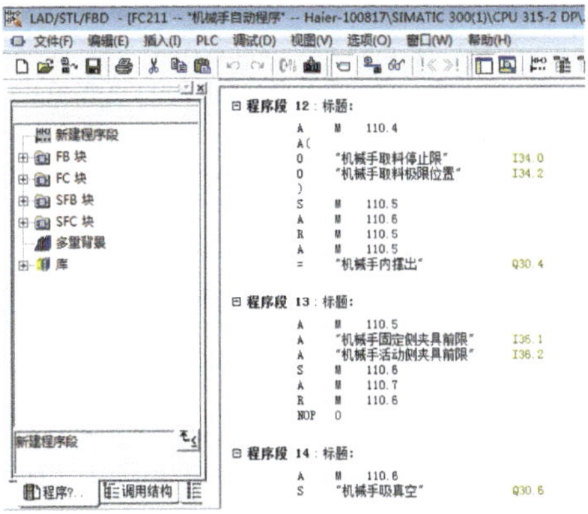

Fig. 7. Program example for unloading manipulator

6 Conclusion

In this paper, according to the characteristics of sheet metal products, the pneumatic manipulator based on PLC control is designed, which used vacuum sucker to replace the traditional grasp. It can easily hold and release products. This paper introduces the design scheme of mechanical system and pneumatic transmission system of manipulator, and the programmable logic controller is adopted to control the system. The osculum type pneumatic unloading manipulator realizes the grasping of the metal piece of the U-shaped refrigerator shell, and solves the problem of holding the thin plate, which is difficult to realize by hand grasping manipulator. The application of production practice shows that the manipulator is stable, well run and easy to maintain, and the product has a beautiful appearance. It realizes automatic production, improves labor conditions, reduces labor intensity of workers, ensures safe production, saves enterprise labor cost, and improves labor productivity. This technology has broad application prospect in many sheet metal production industries.

References

1. Zhang, C.: Assembly, Debugging and Programming of Automatic Production Line. Chemical Industry Press, Beijing (2011)
2. Daxu, Z.: Dynamic analysis on a feeding and unloading manipulator using Kane's method combined with screw theory. In: 2012 International Conference on Industrial Control and Electronics Engineering, pp. 999–1002 (2012)
3. Zhang, Q.: Design of pneumatic manipulator based on PLC. Chin. Hydraul. Pneum. **5**, 121–122 (2012)
4. Sun, H.: Mechanical Principles. Higher Education Press, Beijing (2001)
5. Zhou, P.: Mechanical Design Utility and Installation Atlas. Mechanical Industry Press, Beijing (2007)
6. Ji, H.: Hydraulic Pressure Transmission and Control. Huazhong University of Science and Technology Press, Wuhan (2014)

7. Li, W., Feng, Y., Zhang, J.: Design of suction cup type pneumatic manipulator based on PLC control. Hydraul. Pneum. Seals **12**, 35–38 (2016)
8. Islam, F., Faraz, Syed A.: PLC based intelligent toll road traffic control using. Int. J. Comput. Theory Eng. **6**(4), 353 (2014)
9. Xie, Y.: Study of 5-freedom modular pneumatic manipulator based on PLC. Mach. Des. Manuf. **1**, 180–181 (2009)
10. Li, J., Liu, L.Q., Xu, X.D., Liu, T., Li, Q., Hu, Z.J., Wang, B.M., Xiong, L.Y., Dong, B., Yan, T.: Development of a measurement and control system for a 40 l/h helium liquefier based on Siemens PLC S7-300. Phys. Proced. **67**, 1181–1186 (2015)
11. Zhang, Y.: Design of control system of a production line for side shell of refrigerator based on S7-300 PLC. Autom. Panor. **9**, 88–91 (2016)
12. Liao, C.: S7-300/400 PLC Application Technology, 4th edn. Mechanical Industry Press, Beijing (2016)
13. Singh, G., Agarwal, A., Jarial, R.K.: PLC controlled elevator system. IEEE (2013)

Design and Simulation of PLC Control Elevator

Zhao Jing[1(✉)], Wang Lanqing[2], Cao Ye[1], and Xu Yunda[1]

[1] School of Physics and Electronic Technology, Liaoning Normal University, Dalian 116029, China
zjmilk@126.com
[2] School of Mathematics and Computer Science, Shanxi Datong University, Datong 037009, China

Abstract. In this paper, the elevator control system is designed for 4 floors and 4 station elevators using Panasonic PLC. The functions are realized: (1) at the beginning, the elevator is on any floor. (2) when a call signal comes, it can remember and respond to it. If the call signal is larger than the current floor, the elevator will go up, or go down. (3) when the multiple call signals come, first the elevator will response to the first call.

Keywords: Elevator · Control systems · Programmable logic controller

1 Introduction

As we all know, nowadays, elevators are already indispensable vertical transportation means for high-rise buildings, shops, multi-storey factories and warehouses. That is to say, elevators have become an indispensable part of our life. Imagine a dozens of floors of buildings, such as 30 floors. If there were no elevators, what was the concept of climbing from the first floor to the twenty-five floor.

The elevator is a complex system, which is integrated with mechatronics. The field of elevator not only involves mechanical transmission, electrical control, civil engineering and other engineering fields, but also needs to consider a series of problems such as its reliability and comfort. To be more precise, the most basic should be highly safe for modern elevators.

Elevator electric control technology is a command center to coordinate all systems of elevator. The difference of its functions directly affects the safety and reliability of elevator operation.

PLC is a special microcomputer used for automatic control. It is widely used because of its reliability, strong anti-interference, easy to use and easy to expand.

K. Deng et al. (Eds.): ICMIR 2018, AISC 856, pp. 821–827, 2019.
https://doi.org/10.1007/978-3-030-00214-5_102

2 The Specific Requirements of the Elevator

The elevator can achieve the following specific requirements:

(1) If the elevator is on the 1 floor and 4 floor is called, the elevator will go up to the 4 floor, during which the 2 and 3 floors are kept on.
(2) If the elevator is on the 2 floor and 4 floor is called, the elevator will go up to the 4 floor, during which the 3 floors are kept on.
(3) If the elevator is on the 3 floor and 4 floor is called, the elevator will go up to the 4 floor.
(4) If the elevator is on the 4 floor and 4 floor is called, it is invalid.
(5) If the elevator is on the 1 floor and 3 floor is called, the elevator will go up to the 3 floor, during which the 2 floors are kept on.
(6) If the elevator is on the 2 floor and 3 floor is called, the elevator will go up to the 3 floor.
(7) If the elevator is on the 3 floor and 3 floor is called, it is invalid.
(8) If the elevator is on the 4 floor and 3 floor is called, the elevator will go down to the 3 floor.
(9) If the elevator is on the 1 floor and 2 floor is called, the elevator will go up to the 2 floor.
(10) If the elevator is on the 2 floor and 2 floor is called, it is invalid.
(11) If the elevator is on the 3 floor and 2 floor is called, the elevator will go down to the 2 floor.
(12) If the elevator is on the 4 floor and 2 floor is called, the elevator will go down to the 2 floor, during which the 3 floors are kept on.
(13) If the elevator is on the 1 floor and 1 floor is called, it is invalid.
(14) If the elevator is on the 2 floor and 1 floor is called, the elevator will go down to the 1 floor.
(15) If the elevator is on the 3 floor and 1 floor is called, the elevator will go down to the 1 floor, during which the 2 floors are kept on.
(16) If the elevator is on the 4 floor and 1 floor is called, the elevator will go down to the 1 floor, during which the 3 or 2 floors are kept on.
(17) If the elevator is on the 1 floor, the 2 and 3 floors are called, first the elevator will go up to the 2 floor, pause 5 s, and then will go up to 3 floor.
(18) If the elevator is on the 1 floor, the 3 and 4 floors are called, first the elevator will go up to the 3 floor (during which the 2 floors are kept on), pause 5 s, and then will go up to 4 floor.
(19) If the elevator is on the 1 floor, the 2 and 4 floors are called, first the elevator will go up to the 2 floor, pause 5 s, and then will go up to 4 floor (during which the 3 floors are kept on).
(20) If the elevator is on the 2 floor, the 3 and 4 floors are called, first the elevator will go up to the 3 floor, pause 5 s, and then will go up to 4 floor.
(21) If the elevator is on the 2 floor, the 1 and 3 floors are called, first the elevator will go down to the 1 floor, pause 5 s, and then will go up to 3 floor (during which the 2 floors are kept on).

(22) If the elevator is on the 2 floor, the 1 and 4 floors are called, first the elevator will go down to the 1 floor, pause 5 s, and then will go up to 4 floor (during which the 2 or 3 floors are kept on).
(23) If the elevator is on the 2 floor, the 3 and 1 floors are called, first the elevator will go up to the 3 floor, pause 5 s, and then will go down to 1 floor (during which the 2 floors are kept on).
(24) If the elevator is on the 2 floor, the 4 and 1 floors are called, first the elevator will go up to the 4 floor (during which the 3 floors are kept on), pause 5 s, and then will go down to 1 floor (during which the 3 or 2 floors are kept on).
(25) If the elevator is on the 3 floor, the 2 and 1 floors are called, first the elevator will go down to the 2 floor, pause 5 s, and then will go down to 1 floor.
(26) If the elevator is on the 3 floor, the 2 and 4 floors are called, first the elevator will go down to the 2 floor, pause 5 s, and then will go up to 4 floor (during which the 3 floors are kept on).
(27) If the elevator is on the 3 floor, the 4 and 2 floors are called, first the elevator will go up to the 4 floor, pause 5 s, and then will go down to 2 floor (during which the 3 floors are kept on).
(28) If the elevator is on the 3 floor, the 1 and 4 floors are called, first the elevator will go down to the 1 floor (during which the 2 floors are kept on), pause 5 s, and then will go up to 4 floor (during which the 2 or 3 floors are kept on).
(29) If the elevator is on the 3 floor, the 4 and 1 floors are called, first the elevator will go up to the 4 floor, pause 5 s, and then will go down to 1 floor (during which the 3 or 2 floors are kept on).
(30) If the elevator is on the 1 floor, the 2 and 3 and 4 floors are called, first the elevator will go up to the 2 floor, pause 5 s, and then will go up to 3 floor, pause 5 s, and then will go up to 4 floor.
(31) If the elevator is on the 2 floor, the 1 and 3 and 4 floors are called, first the elevator will go down to the 1 floor, pause 5 s, and then will go up to 3 floor (during which the 2 floors are kept on), pause 5 s, and then will go up to 4 floor.
(32) If the elevator is on the 2 floor, the 3 and 4 and 1 floors are called, first the elevator will go up to the 3 floor, pause 5 s, and then will go up to 4 floor, pause 5 s, and then will go down to 1 floor (during which the 3 or 2 floors are kept on).
(33) If the elevator is on the 2 floor, the 3 and 1 and 4 floors are called, first the elevator will go up to the 3 floor, pause 5 s, and then will go up to 4 floor, pause 5 s, and then will go down to 1 floor (during which the 3 or 2 floors are kept on).
(34) If the elevator is on the 2 floor, the 4 and 1 and 3 floors are called, first the elevator will go up to the 3 floor, pause 5 s, and then will go up to 4 floor, pause 5 s, and then will go down to 1 floor (during which the 3 or 2 floors are kept on).
(35) If the elevator is on the 3 floor, the 4 and 1 and 3 floors are called, 3 floor call invalid. first the elevator will go up to the 4 floor, pause 5 s, and then will go down to 1 floor (during which the 3 or 2 floors are kept on).
(36) If the elevator is on the 3 floor, the 1 and 2 and 4 floors are called, first the elevator will go down to the 2 floor, pause 5 s, and then will go down to 1 floor, pause 5 s, and then will go up to 4 floor (during which the 2 or 3 floors are kept on).

(37) If the elevator is on the 3 floor, the 1 and 4 and 2 floors are called, first the elevator will go down to the 2 floor, pause 5 s, and then will go down to 1 floor, pause 5 s, and then will go up to 4 floor (during which the 2 or 3 floors are kept on).
(38) If the elevator is on the 3 floor, the 2 and 4 and 1 floors are called, first the elevator will go down to the 2 floor, pause 5 s, and then will go down to 1 floor, pause 5 s, and then will go up to 4 floor (during which the 2 or 3 floors are kept on).
(39) If the elevator is on the 4 floor, the 1 and 2 and 3 floors are called, first the elevator will go down to the 3 floor, pause 5 s, and then will go down to 2 floor, pause 5 s, and then will go up to 1 floor.

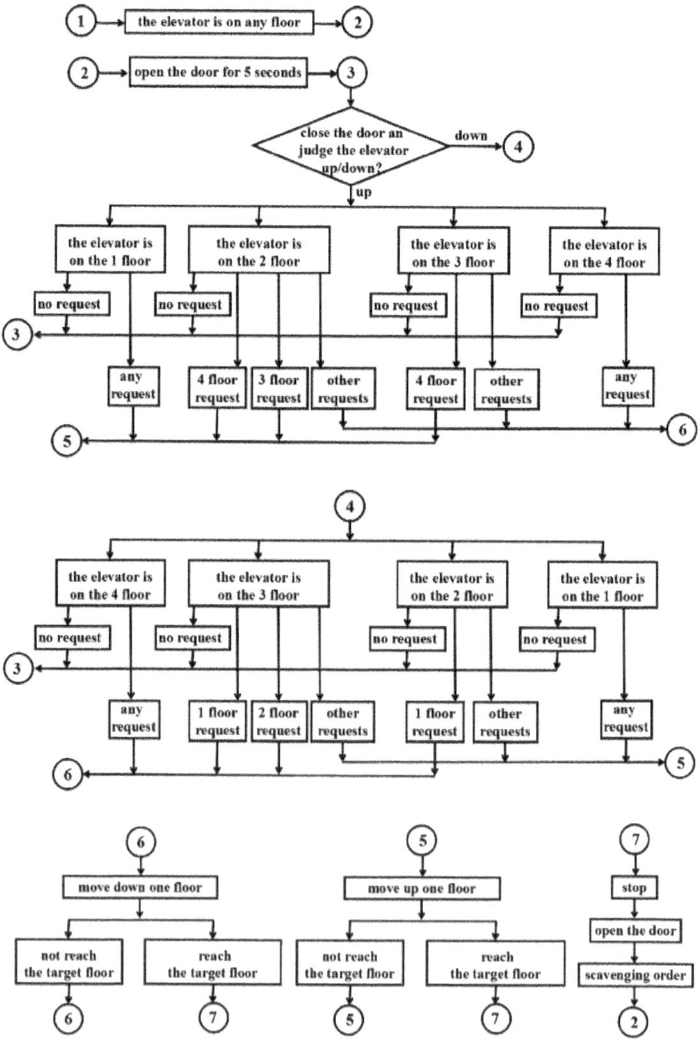

Fig. 1. PLC control elevator flow chart

3 PLC Control Elevator Flow Chart

PLC control elevator flow chart is in Fig. 1. When the elevator stops in the N floor (N = 1, 2, 3, 4), the M floor has a call signal, M > N, the elevator goes up; otherwise goes down.

4 PLC-I/O Table

PLC-I/O distribution is in Table 1, X0–X8 is input and Y0–Y6 output.

Table 1. PLC-I/O distribution

Input		Output	
X0	Starter button	Y0	1 floor signal indicator light
X1	1 floor sensor switch	Y1	2 floor signal indicator light
X2	2 floor sensor switch	Y2	3 floor signal indicator light
X3	3 floor sensor switch	Y3	4 floor signal indicator light
X4	4 floor sensor switch	Y4	open the door indicator light
X5	4 floor call down button	Y5	close the door indicator light
X6	3 floor call down button	Y6	4 floor call down indicator light
X7	3 floor call up button	Y7	3 floor call up indicator light
X8	2 floor call down button	Y8	3 floor call down indicator light
X9	2 floor call up button	Y9	2 floor call up indicator light
XA	1 floor call up button	YA	2 floor call down indicator light
XB	4 floor call button	YB	1 floor call up indicator light
XC	3 floor call button	YC	4 floor call indicator light
XD	2 floor call button	YD	3 floor call indicator light
XE	1 floor call button	Y10	2 floor call indicator light
XF	open the door button	Y11	1 floor call indicator light
X10	close the door button		

5 PLC Control Elevator Ladder Diagram

According to the specific requirements of the elevator and flow chart and I/O table, the ladder diagram be written (Fig. 2).

Fig. 2. PLC control elevator ladder diagram

Fig. 3. PLC control elevator experiment

6 Conclusion

Figure 3 is the PLC control elevator experiment and model. The case in this paper has been tested and passed in Panasonic PLC, which can be used directly.

References

1. Zhao J.: Design of PLC control system for traffic lights. Mech. Manuf. Technol., 2016
2. Zhao J.: Research on the measurement system of liquid weighing based on PLC. In: International Conference Energy and Mechanical engineering, 2015
3. Zhao J.: PLC counter applications technique. In: Progress in Mechatronics and Information Technology, 2013

Optimal Design of a Geared Five-Bar Prosthetic Knee Considering Motion Interference

Yuanxi Sun, Wenjie Ge(✉), Jia Zheng, Dianbiao Dong, and Donglai Zhao

School of Mechanical Engineering, Northwestern Polytechnical University, Xi'an, China
gwj@nwpu.edu.cn

Abstract. The prosthetic knee is the essential rehabilitation device for the above-knee amputees to recover from mobility impairments. This paper presents an optimal design of a geared-five bar (GFB) prosthetic knee considering the motion interference between its mechanism and internal actuating area. To ensure that the GFB prosthetic knee can provide the amputees with the expected knee centrode while there is no internal motion interference during knee flexion, the motion interference is solved by the separating axis theorem and added into the optimization process as the penalty function. The final optimization result shows that the centrode of the GFB prosthetic knee is close to the ideal one and the internal cavity for the actuation is guaranteed, which will be beneficial to the realization of the miniaturization of the powered geared five-bar prosthetic knee.

Keywords: Geared five-bar mechanism · Prosthetic knee · Motion interference
Separating axis theorem

1 Introduction

The prosthetic knees are the essential rehabilitation devices for the above-knee amputees. According to their motion characteristics, the prosthetic knees can be divided into two categories: the uniaxial ones and the multiaxial ones. The uniaxial prosthetic knee uses a single hinge to realize the primary rotation function of the knee, while the multiaxial one uses the multi-bar mechanisms or the other mechanisms to achieve the complex function of the knee. Compared with the uniaxial prosthetic knee, the multiaxial prosthetic knee can provide the amputees with better bionic performance and better motion stability.

The most commonly seen multiaxial prosthetic knees are the four-bar [1] prosthetic knee and the six-bar [2] prosthetic knee. These prosthetic knees can replicate the motion of the human knee (the motion of the human knee is not the simple rotation, but the compound motion formed by rolling and sliding), thus generating the trajectory of instantaneous center of rotation (ICR) between the thigh and the shank, i.e. the centrode [3].

However, the centrode of the four-bar prosthetic knee has absolute deviations from the expected one, and the six-bar knee prosthetic knee has the deficiency of considerable

K. Deng et al. (Eds.): ICMIR 2018, AISC 856, pp. 828–837, 2019.
https://doi.org/10.1007/978-3-030-00214-5_103

size. Hence, a geared five-bar (GFB) [4] prosthetic knee is proposed to improve these shortcomings. Apart from its compact size and better centrode, the most significant advantage of the GFB prosthetic knee is its adjustable bionic performance. Unlike the conventional four-bar or six-bar prosthetic knee, the kinematics and dynamics of the GFB prosthetic knee can be easily adjusted by changing its gear ratio [5] while the dimension of the GFB mechanism remains unchanged. This unique feature of the GFB prosthetic knee will significantly enhance its adaptability to multiple motions of different amputees.

To apply the GFB mechanism in the prosthetic knee, its dimension and gear ratio must be regulated to fit the function of the knee best. In addition, the cavity of the GFB mechanism must provide enough space for the installation of actuation devices such as the motor or the hydraulic/pneumatic actuator. The former objective can be solved by optimizing the parameters of the GFB mechanism to realize the specific requirement of the prosthetic knee, while the latter objective cannot be easily achieved by position planning because the relative position of each bar in the GFB prosthetic knee will change during walking. Therefore, this paper presented a separating-axis-theorem-based penalty function for the GFB mechanism, which can enable its bionic optimization without motion interference between its inner cavity and the actuating device.

The rest of this paper is organized as follows: Sect. 2 presented the optimization of the GFB prosthetic knee considering the motion interference solved by the separating axis theorem; Sect. 3 analyzed the optimization result as well as the walking of the GFB prosthetic knee; Sect. 4 concluded this paper.

2 Optimization of the GFB Prosthetic Knee

2.1 Model of the GFB Prosthetic Knee

The mathematical model of the GFB prosthetic knee is shown in Fig. 1. The bar 5 is fixed on the X-axis, and the hinge O_1 is set as the origin of the coordinates. The GFB mechanism has eight parameters, namely the lengths $[l_1, l_2, l_3, l_4, l_5]$ of five bars, the initial angles $[\alpha_1, \alpha_4]$ of bar 1 and 4, and the gear ratio of gear g_1 to g_4. The bar 5 vertically mounted on the shank is set as the shank bar. The bar 2 is set as the thigh bar, and its angle is α_2 when the thigh is vertical.

$$l_1 \cos\theta_1 + l_2 cos\theta_2 = l_5 + l_4 \cos\theta_4 + l_3 \cos\theta_3 \tag{1}$$

$$l_1 \sin\theta_1 + l_2 \sin\theta_2 = l_4 \sin\theta_4 + l_3 \sin\theta_3 \tag{2}$$

$$\theta_4 - \alpha_4 = n(\theta_1 - \alpha_1) \tag{3}$$

In Eqs. (1)–(3), the unknown variables are $[\theta_1, \theta_3, \theta_4]$, and the known variable is θ_2, which equals α_2 plus the angle of knee flexion. By solving these nonlinear equations, the positions of the GFB mechanism under different knee flexion can be then solved [6]. The general initial guess for the iterative solutions of Eqs. (1)–(3) is $X_{\text{init}} = [108°, 144°, 72°]$.

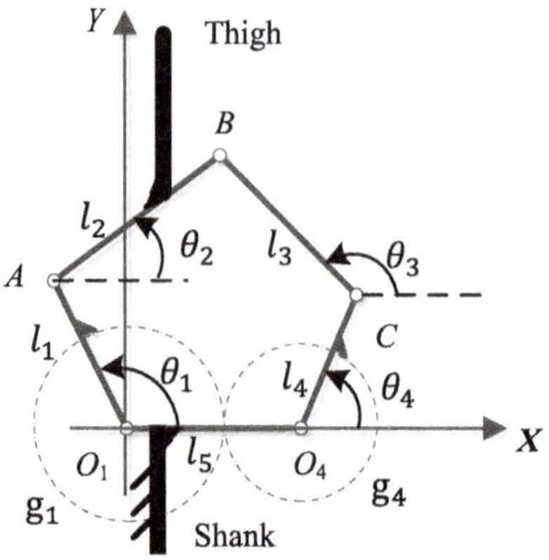

Fig. 1. Model of the GFB prosthetic knee

2.2 ICR Analysis of the GFB Prosthetic Knee

According to the Kennedy-Aronhold Theorem, the ICR of the thigh (bar 2) and the shank (bar 5) can be solved as the flowchart shown in Fig. 2.

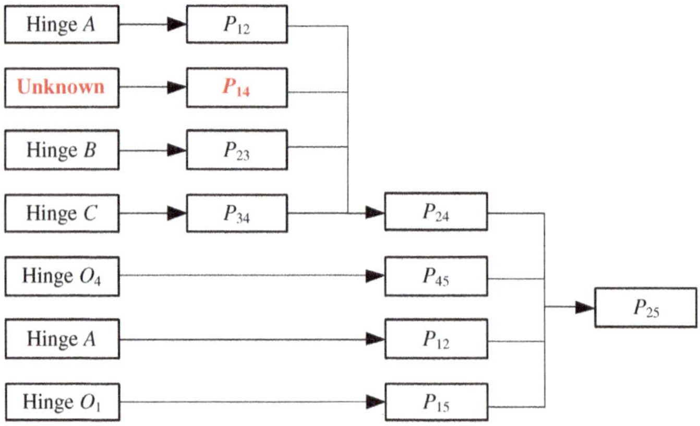

Fig. 2. The flow chart of solving the ICR of the GFB mechanism using the Kennedy-Aronhold Theorem

The ICR P_{25} is the intersection of the line $P_{24} \rightarrow P_{45}$ and the line $P_{12} \rightarrow P_{15}$, while the ICR P_{24} is the intersection of the line $P_{23} \rightarrow P_{34}$ and the line $P_{12} \rightarrow P_{14}$. Among all the ICRs, only the ICR P_{14} of the bar 1 and bar 4 is unknown. Because the bar 1 is fixed with the gear O_1 and the bar 4 is fixed with the gear O_4, the ICR P_{14} is actually located at the position where two gears mesh.

The specific position of the ICR P_{14} along the X-axis is shown in Eq. (4).

$$P_{14x} = \begin{cases} -\omega_4/(\omega_1 - \omega_4)l_5, \omega_1/\omega_4 > 1 \\ \omega_4/(\omega_4 - \omega_1)l_5, 1 > \omega_1/\omega_4 > 0 \\ \omega_4/(\omega_1 + \omega_4)l_5, 0 > \omega_1/\omega_4 \end{cases} \tag{4}$$

Let the rotation speeds of the gear O_1 and the gear O_4 be ω_1 and ω_4. Then the graphical locations of the ICR P_{14} along the X-axis can be drawn as shown in Fig. 3.

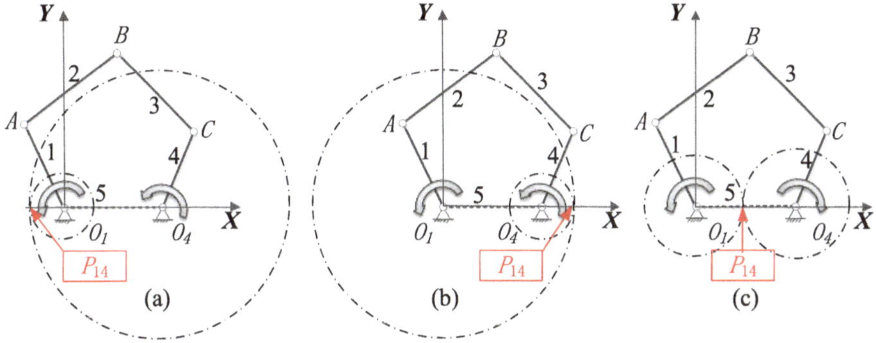

Fig. 3. The location of the ICR P_{14} (a) $\omega_1/\omega_4 > 1$, (b) $1 > \omega_1/\omega_4 > 0$, (c) $0 > \omega_1/\omega_4$

2.3 Establishment of the Optimization Objective of the GFB Prosthetic Knee

The optimization objective of the GFB prosthetic knee is to make its centrode most approximate to the expected one, i.e., the centrode of the human knee. Assuming that the range of the knee flexion angle corresponding to the expected centrode is $[0, \varphi]$, and there are k ICRs in the discrete centrode. Then, when the GFB prosthetic knee generates the ith ICR, its knee flexion angle will reach at $(i-1)\varphi/(k-1)$, and the angle of the bar 2, i.e. θ_2, will equal to $\alpha_2 + (i-1)\varphi/(k-1)$. Where i is the index of the ICR in the discrete expected centrode.

Let the position of the ith ICR generated by the GFB prosthetic knee be $[x_i, y_i]$. Then it can be expressed by the following Eq. (5), where F represents the process of solving the ICR of the GFB prosthetic knee.

$$(x_i, y_i) = F(l_1, l_2, l_3, l_4, l_5, n, \alpha_1, \alpha_4, \theta_2) \tag{5}$$

Let $[X_i, Y_i]$ represent the position of the ith expected ICR. Then the optimization objective can be described as the least square difference between the discrete centrode of the GFB prosthetic knee and the expected centrode, as shown in Eq. (6).

$$\min \sum_{i=1}^{k} \left[(x_i - X_i)^2 + (y_i - Y_i)^2 \right] \tag{6}$$

2.4 Separating-Axis-Theorem-Based Penalty Function of Motion Interference for the Optimization of the GFB Prosthetic Knee

For the GFB prosthetic knee, its internal cavity must provide enough space to install the necessary actuating device. Assuming that the bottom of the actuating device is located at the center of the bar 5, and the height and width of the actuating area are H and W, respectively, as shown in Fig. 4(a). Then, when the GFB prosthetic knee slightly flexed, as shown in Fig. 4(b), none of its bars interferes with the actuating area. However, when the GFB prosthetic knee flexed at no small angle, as shown in Fig. 4 (c), some of its bars will interfere with the actuating area. Therefore, the optimization of the GFB prosthetic knee must consider the motion interference between the GFB mechanism and its internal actuating area.

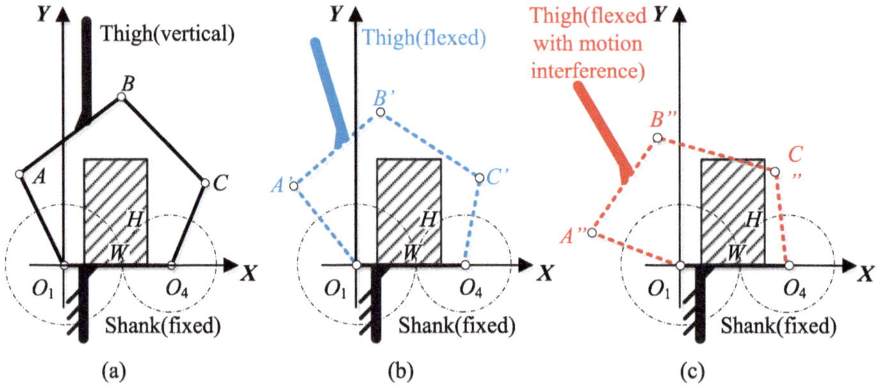

Fig. 4. GFB prosthetic knee interfered with its inner actuating area during motion

Although the bars of the GFB prosthetic knee can be made outward-curving to avoid such motion interference, the increased dimension of the GFB prosthetic knee will reduce its advantage of compact size, and the specially shaped bars will also cause manufacturing difficulties. Therefore, this paper will turn the problem of motion interference between the GFB mechanism and its actuating area into the penalty function during the optimization, thus achieving the best bionic performance of the GFB prosthetic knee while its inner motion interference is entirely eliminated.

To judge whether each bar in the GFB prosthetic knee has collided with the actuating area during knee flexion, this paper used the separating axis theorem (SAT) to realize the judging process, as shown in Fig. 5.

In Fig. 5, the actuating area is projected along the separating axes of each bar (the projection along the separating axis of the bar 5 is excluded because the actuating area is located at the bar 5). Then, if there is a gap between the projection P_i of the ith bar and the projection S_i of actuating area along the ith separating axis, i.e., $P_i \cap S_i = \emptyset$, it means that the ith bar does not interfere with the actuating area. If all the bars excluding the bar 5 do not interfere with the actuating area, the GFB prosthetic knee will not interfere with the actuating area. Hence, the penalty function f of the motion interference during knee flexion can be expressed as the following process shown in Table 1.

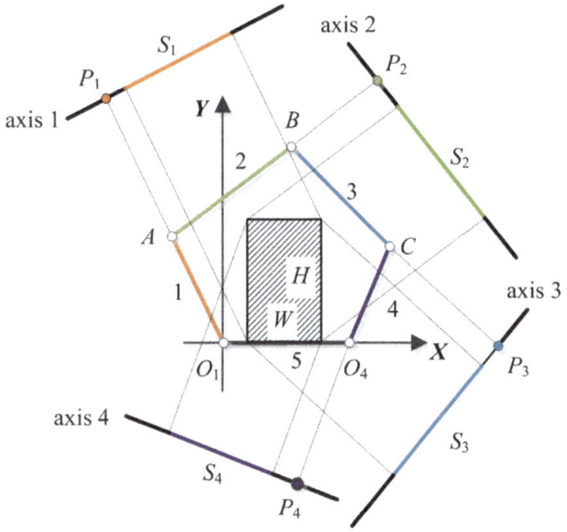

Fig. 5. Projections of the actuating area along separating axes of each bar

Table 1. The process to obtain the penalty function f for the optimization

1.	$f = 0$
2.	For $i = 1$ to k
3.	If $(P_{i1} \cap S_{i1}) \cup (P_{i2} \cap S_{i2}) \cup (P_{i3} \cap S_{i3}) \cup (P_{i4} \cap S_{i4}) \neq \varnothing$
4.	$f = +\infty$; Break
5.	End If
6.	End For

*: k is the number of ICRs in the discrete expected centrode; P_{ij} and S_{ij} $(j = 1,2,3,4)$ represent the projections of the jth bar and the actuating area along the jth separating axis when the GFB prosthetic knee generates the ith ICR.

2.5 Constraints for the Optimization of the GFB Prosthetic Knee

After the penalty function f is established, the final optimization objective of the GFB prosthetic knee can be obtained as

$$\min \sum_{i=1}^{k} \left[(x_i - X_i)^2 + (y_i - Y_i)^2 \right] + f \tag{7}$$

The optimization constraints are given as

$$\text{s.t.}\begin{cases} 16\,\text{mm} < l_i < 125\,\text{mm} (i = 1, 2, 3, 4) \\ -\pi < \alpha_j < \pi \; (j = 1, 2, 4) \\ -10 < n < 10 \\ W < l_5 < 125\,\text{mm} \end{cases} \tag{8}$$

Where W represents the width of the actuating area.

In Eq. (8), each bar is limited to shorter than 125 mm to reduce the overall dimension of the GFB prosthetic knee. The bar 5 must be long enough to enable the actuating area, i.e. $l_5 > W$. This paper intends to install a motor in the GFB prosthetic knee, so the size of the actuating area is set as $H = 45$ mm, $W = 25$ mm. The gear ratio of the GFB prosthetic knee is constrained between $[-10, 10]$ to decrease the manufacturing difficulty of gears. The initial angle of the bar 1, 2 and 4 are limited between $[-\pi, \pi]$ to ensure that they can turn over.

3 Optimization Result and Motion Analysis of the GFB Prosthetic Knee

3.1 Optimization Result of the GFB Prosthetic Knee

To avoid the difficulty of finding the initial guess for the optimization, this paper uses the improved genetic annealing algorithm (IGAA) combined with the region search method to obtain the global optimum [7]. After multiple runs of the optimization, the final optimization result of the GFB prosthetic knee is obtained in Table 2.

Table 2. Optimization result of the GFB prosthetic knee (length unit: mm, angle unit: rad)

l_5	l_1	l_2	l_3	l_4	n	α_1	α_4	α_2
66.3	117.0	21.9	123.8	22.3	−6.5	1.5	1.1	0.3

It is known from Table 2 that the GFB prosthetic knee is of suitable size, and the initial angles of bar 1, 2 and 4 are above zero and below ninety degrees, which will be beneficial to manufacturing and assembling. It is also noticed that the gear ratio of the GFB prosthetic knee is below zero and reaches −6.52, which means that two gears will rotate in different directions, and one gear will be considerably larger than the other one. To realize the large gear ratio within small space, the bigger gear in the GFB prosthetic knee can be practically implemented by the incomplete gear. The graphical optimization result of the GFB prosthetic knee is shown in Fig. 6.

It is known from Fig. 6 that, the GFB prosthetic knee will not interfere with its inner actuating area even when the knee flexion reaches ninety degrees, and the centrode of the GFB prosthetic knee is close to the expected one, which means that the optimization with penalty function is successful. In addition, the shape of the GFB prosthetic knee remains almost rectangular during knee flexion, which will be much beneficial to avoiding external interference with other lower prosthesis.

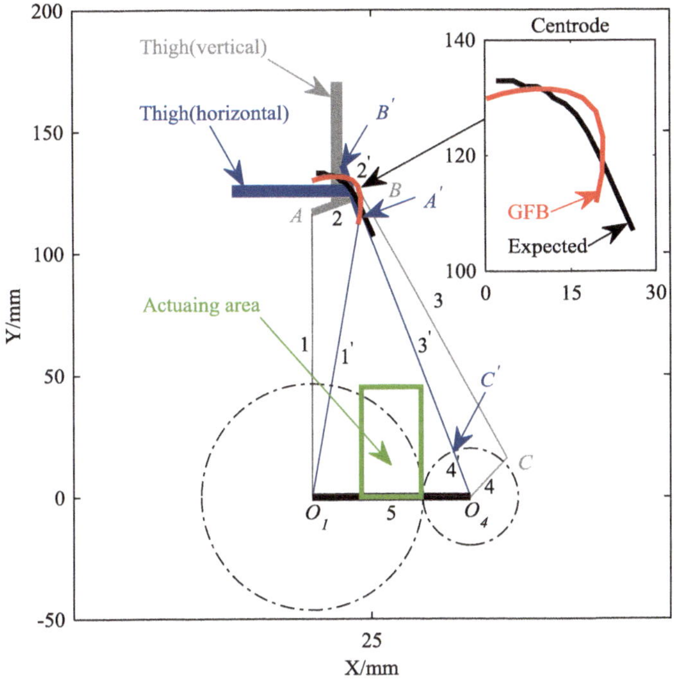

Fig. 6. Graphical result of the optimization result

3.2 Walking Analysis of the GFB Prosthetic Knee

To explore the walking motion of the GFB prosthetic knee, this paper applied the walking gait of the able-bodied volunteer into the GFB prosthetic knee. The walking results in the stance and the swing phase are respectively shown in Figs. 7 and 8.

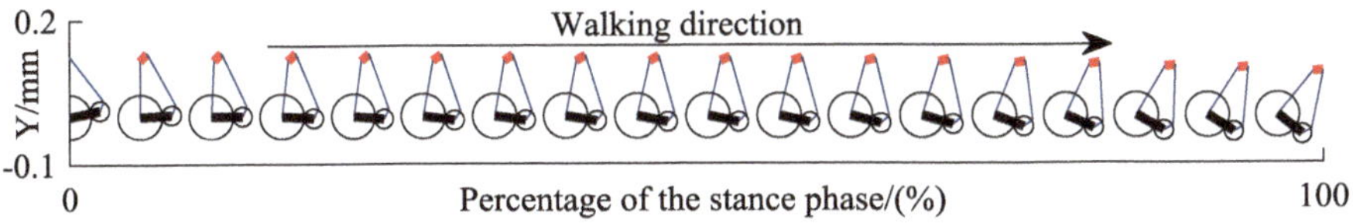

Fig. 7. Motion of the GFB prosthetic knee in the stance phase

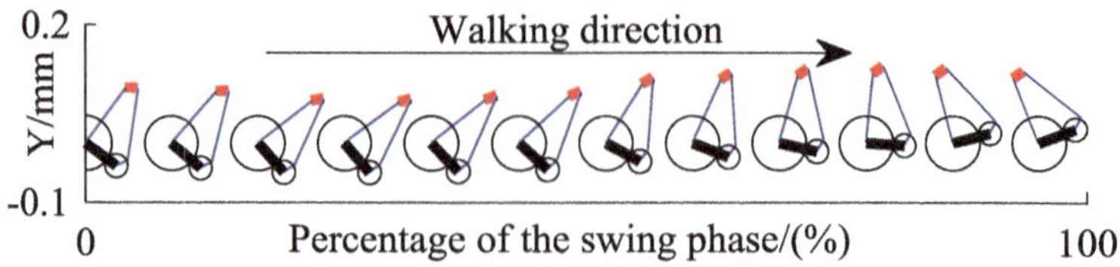

Fig. 8. Motion of the GFB prosthetic knee in the swing phase

It is known from Fig. 7 that, in the early stance phase, the shank bar 5 primarily remains horizontal, while the thigh bar 2 slightly rotate from tilting to the flat state. When the GFB prosthetic knee enters into the late stance phase, both the shank bar 5 and the thigh bar 2 are leaning forward, which means that the lower limb is in the backswing state and will transition into the next swing phase.

In Fig. 8, when the GFB prosthetic knee enters into the swing phase, the GFB mechanism deforms significantly. The reason for this phenomenon is that the shank swings fast from the backward tilting state to the forward tilting state, thus resulting in the rapid changes in knee flexion.

4 Conclusion

This paper presented the optimal design of a geared five-bar (GFB) prosthetic knee considering its motion interference with its inner actuating area. The optimized GFB prosthetic knee has a compact size and can provide the above-knee amputees with expected centrode. In addition, by adding penalty function of the actuating area in the optimization process, the optimized GFB prosthetic knee can freely flex while its inner actuating space is guaranteed.

Acknowledgments. This work was supported by the National Natural Science Foundation of China under Grant number 50975230, and Research Fund for the Doctoral Program of Higher Education of China under Grant number 20136102130001.

References

1. Hobson, D.A., Torfason, L.E.: Optimization of four-bar knee mechanisms–a computerized approach. J. Biomech. **7**(4), 371–376 (1974)
2. Jin, D., Zhang, R., Dimo, H.O., Wang, R., Zhang, J.: Kinematic and dynamic performance of prosthetic knee joint using six-bar mechanism. J. Rehabil. Res. Dev. **40**(1), 39–48 (2003)
3. Sun, B.: Multi-objective optimization design of the institutions of the prosthetic knee joint. Chin. J. Tissue Eng. Res. **16**(22), 4010–4014 (2012)
4. Sun, Y., Ge, W., Zheng, J., Dong, D.: Design and evaluation of a prosthetic knee joint using the geared five-bar mechanism. IEEE Trans. Neural Syst. Rehabil. Eng. **23**(6), 1031–1038 (2015)

5. Dong, D., Ge, W., Liu, S., Xia, F., Sun, Y.: Design and optimization of a powered ankle-foot prosthesis using a geared five-bar spring mechanism. Int. J. Adv. Rob. Syst. **14**(3), 1729881417704545 (2017)
6. Sun, Y., Ge, W., Zheng, J., Dong, D.: Solving the kinematics of the planar mechanism using data structures of Assur groups. J. Mech. Robot. **8**(6), 061002 (2016)
7. Sun, Y., Ge, W., Zheng, J., Xia, F., Dong, D.: Optimization of actuating torques in multi-bar prosthetic joints with springs. Eng. Optim. **49**(7), 1183–1196 (2017)

Research and Design of Computer AGC Control System in Hydropower Plant

Hongliang He(✉)

City College, Kunming University of Science and Technology, Kunming, China
1916657988@qq.com

Abstract. In this paper, the principles and algorithms of automatic generation control system (AGC) in hydropower plants are deeply studied and analyzed. Based on different control methods, the system control model is established. Two methods of economic distribution of generator load are analyzed and demonstrated: Proportional distribution according to unit capacity and distribution according to equal consumed energy ratio. On this basis, use modern computer hardware, software and network technology to design AGC system. Finally, the paper puts forward some suggestions to improve the economic distribution accuracy of load between units and reduce the adjustment error under the equivalent control mode.

Keywords: Hydropower plants · Automatic generation control system AGC · Equal consumed energy ratio · Economic load distribution

1 Introduction

The automatic generation control system (AGC) of hydropower plants is put forward from the actual production with the development and application of the computer-based hydropower plant automation system [1]. Its purpose is to satisfy the power system needs by controlling the active power of the entire power plant quickly and economically while meeting the constraints. The control process must follow two basic principles: First, safe and reliable; Second, rapid economic. Safe and reliable controls include meeting the requirements of the power system for safe power generation and the safe operation of the unit. Rapid economic control is the rapid response to active power changes at the power plant, changes in system frequency, and some non-normal conditions. The computer must make a reasonable start and stop arrangement and load distribution to the generator set so as to save the water consumption of the entire plant. AGC control system is directly related to the frequency quality of the entire power system.

1.1 The Structure and Control Methods of Hydropower Plant Automatic Generation Control System (AGC)

Hydropower station automatic generation control system is mainly composed of hydro-generator, power plant monitoring system, control signal receiving device, power regulation device and data acquisition device. According to the different control

K. Deng et al. (Eds.): ICMIR 2018, AISC 856, pp. 838–846, 2019.
https://doi.org/10.1007/978-3-030-00214-5_104

methods, hydropower station automatic generation control system can be divided into two control modes [2]: stand-alone control mode and plant equivalent control mode. Stand-alone control mode, the dispatch center directly issued a given power to each unit. Under the equivalent control mode, the dispatch center delivers the equivalent power of the whole plant, and the given power of each unit is distributed by the computer monitoring system or the AGC system in the plant. Block diagram of the two control modes are shown in Figs. 1 and 2, respectively.

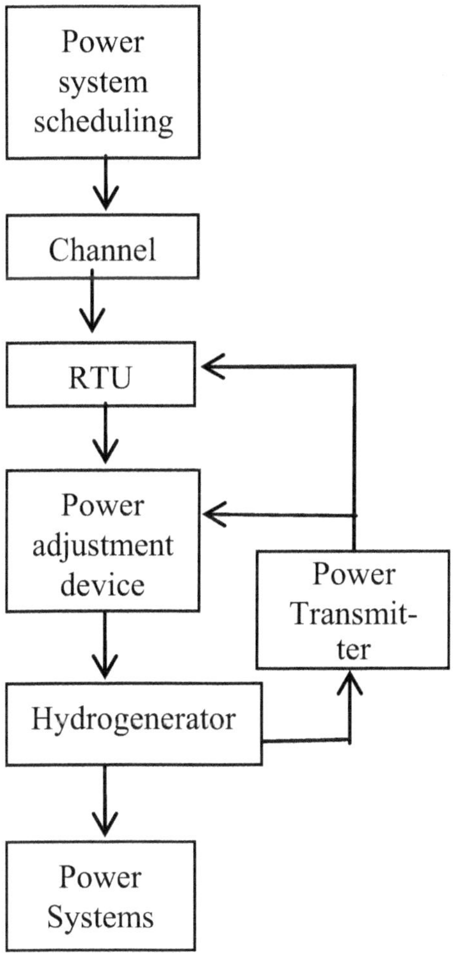

Fig. 1. Stand-alone control mode

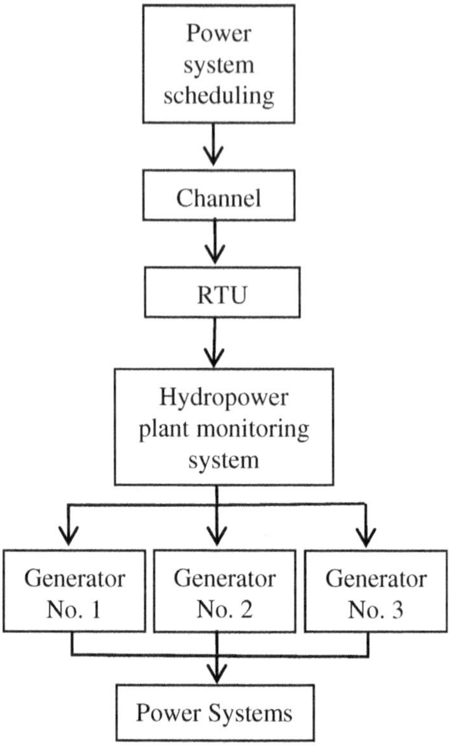

Fig. 2. Equivalent control mode

In the equivalent control mode, according to the different control levels, automatic generation control system is divided into the AGC system in the whole power grid and the AGC system in every power plant. The AGC system in every power plant is designed to prevent the actual operation of the power plant from changing, the communication between the power station and the dispatching center being damaged, and the abnormal AGC of the power system, resulting in the failure of the power station to realize economic operation.

1.2 The Model of Hydropower Plant Automatic Generation Control System (AGC)

The design of AGC control model needs to fully consider the following aspects [3]: (1) Take the principle of not discarding or discarding a small amount of water, make maximum use of reservoir water quantity, and try to keep the power plant running at a higher water head. (2) Completion of daily load curve setting or real-time given power generation task. (3) Calculate the instantaneous deviation of grid frequency or the integral value of deviation, determine the total output of power plant and maintain the grid frequency. (4) Avoid the vibration area and cavitation area of the unit, and meet the output restriction conditions of the unit under the current water head [4].

The AGC regulation mode includes the power regulation mode and the frequency regulation mode. In the power regulation mode, the active power setting value of the whole plant can be set according to the daily load curve or a given load. According to the difference of AGC adjustment mode, the model of AGC active power control is classified and designed.

(1) the first way to adjust the active power.

The daily load curve is set by the mid-range or operator. The AGC uses the active power value of the daily load curve as the whole plant load setting to control the active power.

When the system frequency is normal:

$$P_{AGCSet} = P_{Curvest} - \overline{P_{AGC}} \quad (1)$$

Where: $P_{Curvest}$ is the whole plant active load set by the load curve; $\overline{P_{AGC}}$ is the active power issued by the unit that did not participate in AGC.

When the system frequency exceeds the emergency frequency modulation section:

$$P_{AGCSet} = P_{ACT} + K_{fE}\Delta f - \overline{P_{AGC}} \quad (2)$$

Where: P_{ACT} is the active power that the whole factory actually sends at present moment; K_{fE} is the emergency frequency modulation coefficient; Δf is the difference between the system frequency and the standard frequency; $\overline{P_{AGC}}$ is the active power issued by the unit that did not participate in AGC.

(2) The second way to adjust the active power.

Due to unpredictable load on the power grid, the daily load curve can not be provided at this time, and the load dispatching command provided by the intermediate is always received as the factory load set value at any time. Similarly, setting the AGC load for the entire plant in the two ways described above, the only difference is that the entire plant load setting is provided by the schedule command.

(3) the way to adjust the system frequency.

Sometimes the power plant without the need to ensure active power, only need to monitor the bus frequency at any time to complete the task of FM. When the frequency exceeds the normal rated frequency section, the AGC increases or decreases the load of the AGC unit until the system frequency returns to the normal section or reaches the load limit of the AGC unit under the front head. In this case, the two models are the same. When the frequency is in the normal section, K_{fE} is the rated frequency adjustment coefficient. When the frequency crosses the normal section, K_{fE} is the emergency frequency adjustment coefficient.

$$P_{AGC} = P_{ACT} + K_{fE}\Delta f - \overline{P_{AGC}} \quad (3)$$

Where: P_{ACT}, K_{fE}, $\overline{P_{AGC}}$ have the same meaning as they do in (2).

2 Economic Allocation of Load in the Power Plant

2.1 Proportional Distribution According to Unit Capacity

The active power distributed to each generator is distributed according to the ratio, that the capacity of each generator to the capacity of all participating AGC generators. For example, the i-th generator participating in the AGC adjustment will get the active power Pi as:

$$P_i = P_{AGC} \frac{P_{iN}}{\sum_{i=1}^{n} P_{iN}} \tag{4}$$

Where: P_{AGC} is the control target value of AGC; P_{iN} is the rated power of the i-th unit participating in the AGC regulation; n is the number of units participating in the AGC regulation.

2.2 Equal Consumed Energy Ratio

The goal of reasonable distribution of active power load in hydropower plants is to save water consumption as much as possible while meeting certain constraints [5]. The relationship between energy consumed per unit time of power generation equipment and the active power delivered, that is, the relationship between the input and output of power generation equipment is called the water consumption characteristics of generator. As shown in Fig. 3, the vertical axis represents the amount of water consumed per unit of time and the horizontal axis represents the active power Pg [6].

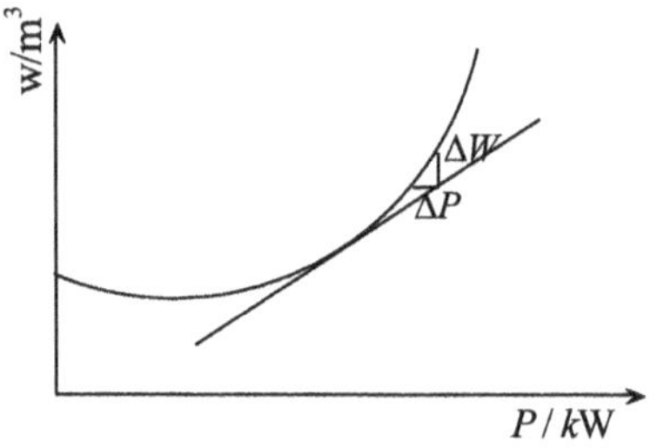

Fig. 3. Water consumption curve

The ratio of the vertical and horizontal coordinates of a point on the water consumption characteristic curve, that is, the ratio of input energy to output power per unit time is called the specific consumption μ, $\mu = W/P$. The slope of a tangent at a point on the consumption characteristic curve is called the rate of increase in consumption. It is the ratio of the increment of input energy to the increment of output power in unit time, that is $\lambda = \Delta W/\Delta P = dW/dP$, and the change rule is shown in Fig. 4.

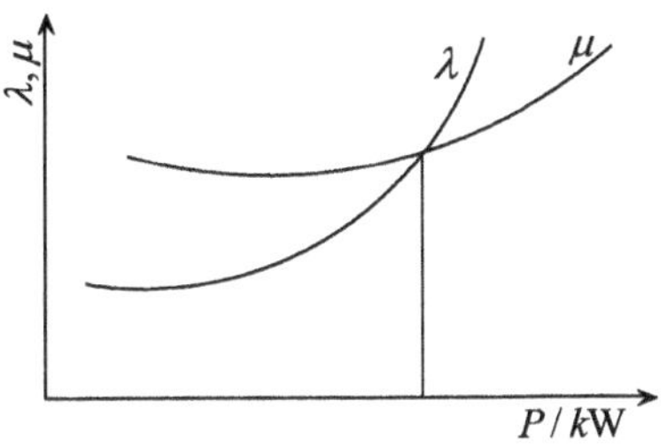

Fig. 4. Specific consumption and Rate of increase in consumption curve

The goal of AGC control is to minimize the overall water consumption of hydropower plants. The total water consumption of hydropower plants is only related to the active power Pi generated by each generator. The objective function is:

$$\begin{aligned} Q_{\sum} &= Q_1 + Q_2 + \ldots Q_n \\ &= \sum_{i=1}^{n} Q_i \end{aligned} \tag{5}$$

Where: n is the number of generators participating in the AGC; Q_i is the water consumption per unit time of the i-th generator. $Q_{\sum}$ is the total water consumption per unit of time for all generators.

Equation (7) should satisfy the following constraints:

(1) The active power P_{gi} and the active power of the load to be sent by the hydropower plant should be equal:

$$P_{g1} + P_{g2} + \ldots P_{gn} = P_L \tag{6}$$

Where: P_{gn} is the actual active power generated by the n-th generator; P_L is the total active power that should be generated by the hydropower plant.

(2) Constraints on active power P_{gi}:

$$P_{gimin} \leq p_{gi} \leq P_{gimax} \tag{7}$$

Where: P_{gimin} is the generator minimum active power; P_{gimax} is the generator maximum active power.

Then: Apply Lagrange multiplier method to solve the above extreme problem:

$$\begin{aligned} C' &= C(P_{g1}, P_{g2}, \ldots, P_{gi}, \ldots, P_{gn}) - \lambda f(P_{g1}, P_{g2}, \ldots, P_{gi}, \ldots, P_{gn}) \\ &= \sum_{i=1}^{n} Q_i(P_{gi}) - \lambda(P_{g1} + P_{g2} + \ldots + P_{gi} + \ldots + P_{gn} - P_L) \end{aligned} \tag{8}$$

Where: $f(P_{g1}, P_{g2}, \ldots, P_{gi}, \ldots, P_{gn})$ is equal constraints; λ is a Lagrange multiplier. Find the minimum value of C', it should be:

$$\frac{\partial C'}{\partial P_{g1}} = 0, \ldots \frac{\partial C'}{\partial P_{gi}} = 0, \ldots \frac{\partial C'}{\partial P_{gn}} = 0, \frac{\partial C'}{\partial \lambda} = 0 \tag{9}$$

After calculation, get the conclusion:

Let $\frac{dQ_i}{dP_{gi}} = q_i$, Then: $q_1 = \lambda, q_i = \lambda, q_n = \lambda, \sum_{i=1}^{n} P_{gi} = P_L$

So: $q_1 = \ldots = q_i = \ldots = q_n = \lambda$. When the active power P_L that the hydropower station needs to issue is certain, in order to minimize the total water consumption, the loads should be distributed among generators according to the equal consumed energy ratio.

3 The Design of Hydropower Plant Automatic Generation Control System (AGC)

The use of advanced computer hardware, software and network technology for the overall design and system configuration [7]. When the monitoring system receives the active power parameter, the monitoring system will allocate this active power to the LCU on each unit according to the actual situation of each unit of the hydropower station. The LCU of the unit will make a reasonable calculation of the given active power and the unit real active power, and then send the action command to the governor [8]. Governor gradually adjust the guide vane opening, change the input of the prime mover, so as to achieve the purpose of regulating active power.

System hardware includes PC, touch screen, PLC (Programmable logic controller), governor. Figure 5 shows the specific architecture.

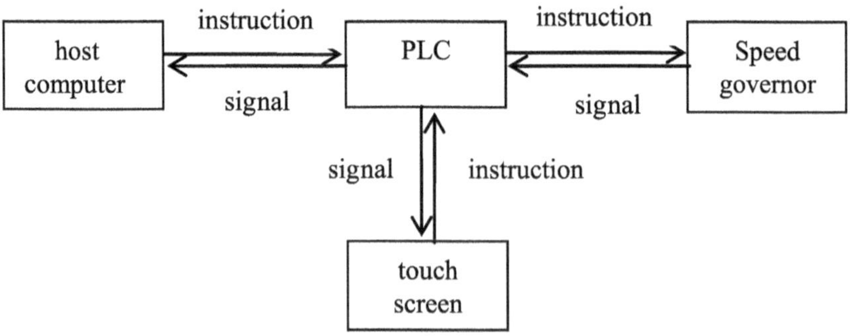

Fig. 5. Hardware block diagram

The digital signal reflecting the system frequency is compared with the unit power digital output signal after analog-digital conversion. After the PID adjustment process, the digital output signal is converted into the unit power analog control value, and sent to the hydraulic follow-up system through power amplification to control the opening of the guide vanes so as to achieve the purpose of power regulation. Block diagram shown in Fig. 6.

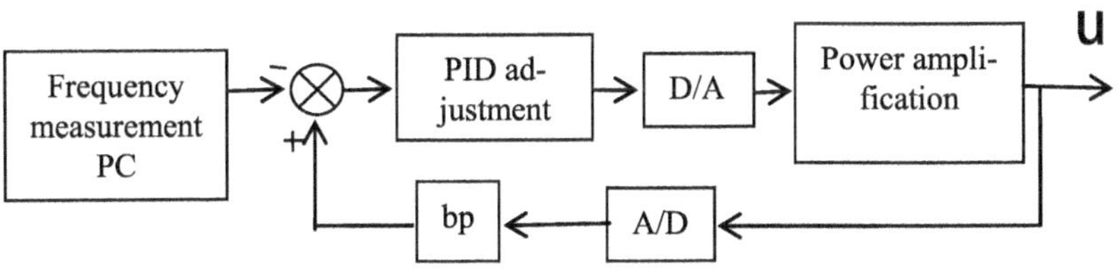

Fig. 6. PLC adjustment principle diagram

To prevent some units abnormal communication, leading to the power plant can not meet the target power. In the equivalent control mode, the general use of small control loop in the power plant to increase the small closed loop to solve this problem [9]. That is, adding PID integral in the control loop and selecting the appropriate PID parameters make the unit finish a large amount of power increment adjustment within a very short time (30 s) after receiving the given power sudden change value. And within 1 min to complete the entire adjustment process. If the overshoot is too large, "brake" pulse can be used to eliminate or reduce the overshoot, which can not only make up for the power deviation caused by communication failures, but also improve the adjustment accuracy of the total plant power. Improve the control performance of the whole plant.

4 Conclusion

Automatic Generation Control System (AGC) of hydropower plants is an important guarantee measure to maintain the power quality of the power grid [10]. In this paper, the existing control system of automatic generation of hydropower plants is analyzed, the control model is optimized, the system is designed by using modern computer hardware, software and network technology, and the control system is optimized under the premise of ensuring the safety and economy of the system. Combined with the requirement of modern power grid for power quality, some suggestions are put forward to improve the power regulation accuracy and reduce the error of power regulation caused by unit response delay. It has a certain reference value to the design of modern power plant automatic generation control system.

References

1. Wang, J., Chen, M., Xiang, T.: Study on automatic generation control of hydropower plant based on genetic algorithm. China Rural Water Hydropower **1**, 115–117 (2007)
2. Shi, C., Yu, X., Wu, Z., Wang, H.: Engineering design and analysis of automatic generation control for hydropower plants. Autom. Electric Power Syst. **25**(4), 57–59 (2001)
3. Cheng, C., Wang, P., Wang, G., Chen, M., Li, G.: Xiaolangdi hydropower station automatic generation control application and management. Hydropower **37**(12), 58–60 (2011)
4. Wang, J.: Briefing on the intra-factory economic operation of Wan'an hydropower plant. Electr. New Energy **10**, 42–47 (2017)
5. Xu, W., He, Y., Wang, C.: Adjustment of vibration area processing strategy in automatic generation control operation of hydropower plant. Power Grid Technol. **36**(11), 122–128 (2012)
6. Sun, G., Ding, N., Lu, M., Zhang, X., Sun, J.: Practical research on energy saving load allocation at factory level AGC. Zhejiang Electr. Power **36**(9), 57–61 (2017)
7. Gong, Y.: Design and application of SCADA automatic generation control function in Guangzhou pumped storage power station. Hydropower Autom. Dam Detect. **37**(6), 9–10 (2008)
8. Wu, E.: Design and implementation of AGC function in Changzhou hydraulic project. Hongshuihe **28**(6), 46–48 (2009)
9. Wang, Y.: Research on automatic generation control technology in small hydropower plants. Heilongjiang Water Conserv. Sci. Technol. **42**(12), 15–17 (2014)
10. Zhang, J.: Operational impact analysis and control strategy of hydropower plant AGC. Low Carbon World **20**, 37–38 (2017)

Research on Early Fault Detecting for Wheelset of High-Speed Train Based on Metal Magnetic Memory Method

Zhenfa Bi[1(✉)], Le Kong[1], and Jun Zhang[2]

[1] School of Railway Transportation, Shanghai Institute of Technology, Shanghai, China
bizhenfa@sit.edu.cn
[2] Electronic Information Engineering College, Jiangsu University, Jiangsu, China

Abstract. The wheelset is the critical component of the walking department of the high-speed train, and ground detection pattern can only be used for the flaw which has been macro existed afterwards and not for the early fault during operation and the trend of tracking of it. Aiming at the deficiency of the testing, the real-time detecting system is proposed to meet safety requirements of train operation. The system consists of three main parts, namely detection mode selection, data acquisition system and large capacity data transmission system, and signal denoising and feature extraction. The research shows that the system can improve the safety of train operation, and provide the reference for the application of metal magnetic memory method in the field of nondestructive testing.

Keywords: High speed train · Safety test · Metal magnetic memory method
Early failure · Real-time detecting system

1 Introduction

With the development of the railway to the direction of high speed and heavy load in China, the wheelset in service bears more and more dynamic load, static load, assembly stress, thermal stress between brake shoe and brake, centrifugal force through curve, so there would be stress concentration and fatigue crack initiation (both called early fault, is the abnormal symptom of early small crack) among the wheel-set tread, rim, spoke and plate hole and this would cause deterioration of the dynamic performance [1, 2]. When the vibration and noise are repeated to the wheelset, it accelerates stress concentration and fatigue initiation to the development of the macro defects. The rapid formation of crack failure even leads to broken wheelset, which results in significant safety accidents. For example, in 1998 due to the tiny fault of wheelset can't be diagnosed as early as possible, which results in ICE catastrophic accident in Germany. Therefore, it is of great practical significance to detect the early fault on the wheelset of the high-speed train.

K. Deng et al. (Eds.): ICMIR 2018, AISC 856, pp. 847–856, 2019.
https://doi.org/10.1007/978-3-030-00214-5_105

At present, there are many nondestructive detecting methods for wheelset such as ultrasonic wave, electromagnetic ultrasonic, eddy current, and light section image [3]. There is no applicable measurement tool for dynamic access to the quality status of the wheelset. Many detections are done in repair base called ground detection [4, 5]. It can't show the actual working state under the load condition, and it can find the flaw which has existed afterward. The low efficiency will lead to waste of man power resources, material resources, and financial resources.

To achieve the goal of real-time dynamic prediction of early fault, the real-time detecting system is proposed based on metal magnetic memory method. The integrated intelligent platform is built for the information acquisition, multiple information processing, quantitative evaluation and comprehensive analysis.

2 Selection of Detection Mode

2.1 Dynamic Detection of Early Fault

Metal magnetic memory method is a new nondestructive testing method in 21 Century, first proposed by the professor Doubov from Russia. It bases on the residual magnetic field (RMF) of a component or equipment which allows the localization of defects (such as cracks, inclusions, blowholes) or stress concentration zones [6–8]. As shown in Fig. 1, the maximum value of magnetic flux leakage field *Hp* exists in the concentrated area of stress and de-formation. It is said that the tangential component of magnetic leakage field *Hp(x)* has the maximum value, and the normal component *Hp(y)* changes sign and has zero point. The irreversible change in the magnetic state will continue after the elimination of the working load. It can determine the stress concentration area (fatigue crack initiation), and realize the early diagnosis of the micro fault by the measurement of the normal component of the scattered magnetic leakage field.

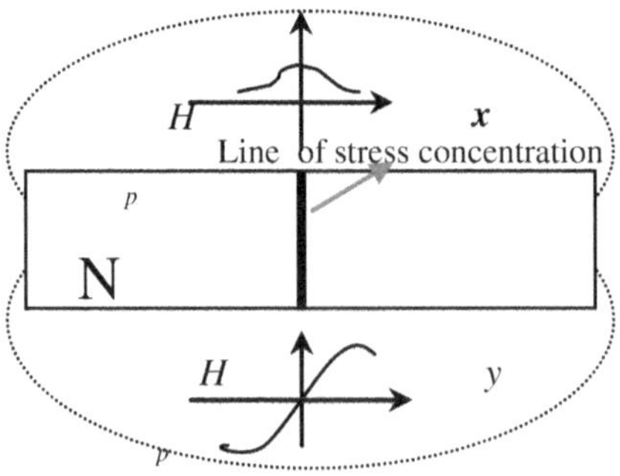

Fig. 1. The principle diagram of MMMM

Compared with the traditional nondestructive testing method, metal magnetic memory method can find the microscopic faults such as stress concentration, which can be applied to the prediction of failure [9–11]. The dynamic detection of the early fault on the wheelset of the high-speed train is shown in Fig. 2.

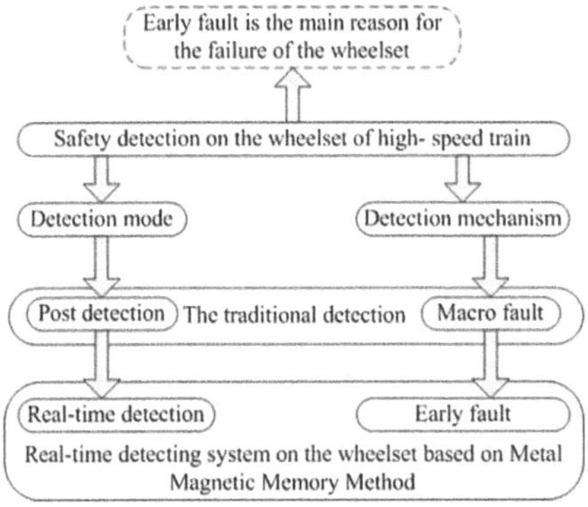

Fig. 2. Analysis of the detections on the wheelset

2.2 Complexity of Detection

Research on theory and engineering of metal magnetic memory is mainly concentrated in Russia, Eastern Europe, and China. This method has been applied in the power station, the pipeline, the pressure vessel, the chemical industry and so on, but there is less research on the wheelset. The environment is very complex, first of all, the wheelset is the high-speed rotating part during operation. Then the load of the wheelset is very complex, such as force, hot and vibration. The superposition of multiple loads aggravates the fatigue damage of the wheelset. If the traditional detection method is used, the state of the wheelset during operation is very hard to be simulated, the stress concentration in the multi-coupling amount will be ignored, and it will bring a lot of risk to the safety of the train.

3 Composition of Detecting System

Because of the complex environment and dynamic load of the high-speed train, the dynamic and quantitative diagnosis with the metal magnetic memory method on the wheelset is difficult, mainly in the following aspects as Fig. 3.

(1) Real-time detection on the wheelset by multi-sensors. The core of the real-time detection of the wheelset is the state in real-time load condition, and the detection of this state requires a variety of types of sensors, such as magneto-resistive sensor, acceleration sensor, force sensor, speed sensor, temperature sensor, location sensor, etc. Through the sensor information, the coupling relationship between the magnetic memory information and various loads is known, and the load condition can be correctly located when the abnormal information occurs.

(2) Large capacity information transmission technology. At present, the high-speed train has 16 marshalling. The use of multiple sensors on the wheelset for online

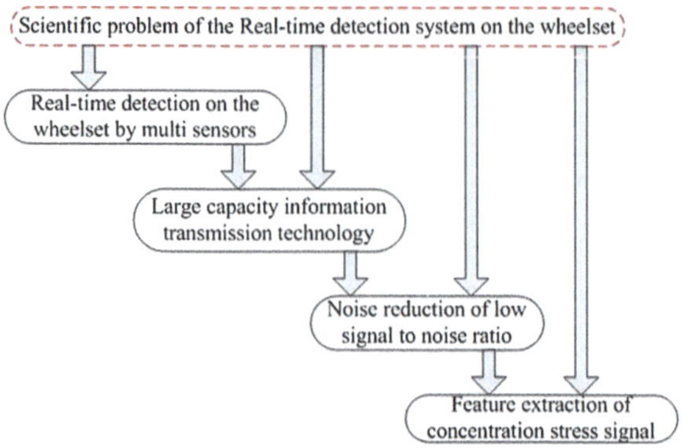

Fig. 3. Scientific problem of the real-time detection system

detection, will produce a number of data information, such as magnetic memory information, location information, temperature information, control information, vibration information, acceleration information, etc. All the information should be sent to the central control station for analysis and processing through the field bus. To know the running state of the wheelset in real-time, the information should be transmitted and processed in time, so the communication load of data transmission network system should be improved. The detecting frame on the wheelset is shown in Fig. 4.

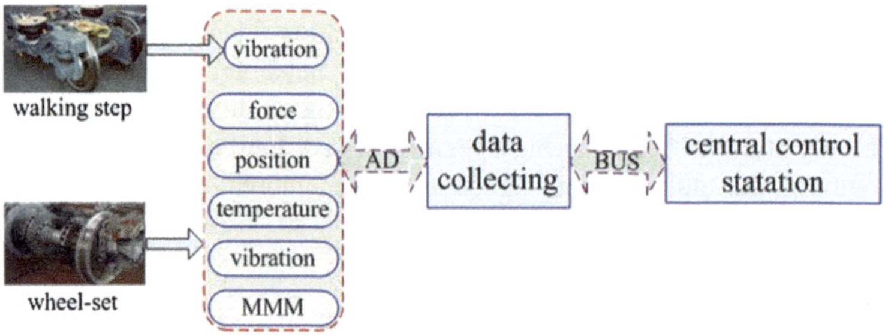

Fig. 4. Detecting frame on the wheelset

(3) Noise reduction of low signal to noise ratio [12–14]. Magnetic memory sensor is mainly used for detecting the state of the wheelset, and a plurality of magnetic memory sensors are installed on the wheelset. Other types of sensors are used to study the coupling relationship between the physical quantity and the magnetic memory information at the same time. Because the magnetic memory signal shows the leakage magnetic field distribution on the ferromagnetic part surface under the influence of the geo-magnetic field, it is sensitive to the external environment. Here is the adaptive de-noising technique proposed for extraction from the magnetic field signal (shown in Fig. 5).

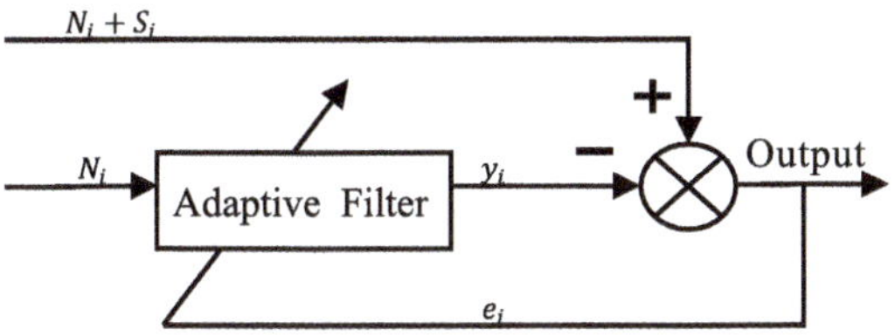

Fig. 5. Schematic of the adaptive noise reduction system

Where y_i is the input of magnetic memory sensor, i is the position number of sensor; S_i is the useful signal and N_i is the noise signal. If the output of filter is equivalent to N_i, by the expression of $y_i = N_i$, then the output of the system is equivalent to the useful signal, by the expression as following.

$$e_i = S_i + N_i - y_i \approx S_i \tag{1}$$

(4) Feature extraction of concentration stress signal. The decision rules on concentration stress area is good in the experimental study. But the application of the feature value easily leads to false detection in the engineering application of the wheelset detection because of the complex circumstance.

4 Results of Research

A amount of work has been done in the research of real-time detecting system of early fault on the wheelset of the high-speed train based on metal magnetic memory method.

(1) An intelligent data collecting controller on based wireless technology has been developed. This controller consists of four main parts, namely, X86 platform, WiFi Module, 12 analog input and 4 G Module. The accuracy of A/D acquisition can be less than 5/10000 by testing the current calibration equipment. WiFi Module is used for communicating between the sensors and the controller. 4G Module is used for communicating between the controller and the internet. The controller is shown in Fig. 6.

Fig. 6. Data collecting system

(2) A large number of studies have been carried out on the force-magnetic effect of the wheelset material 25CrMo4 under static load. The tensile test was carried out on the flat specimen by controlling the tensile testing machine. The leakage magnetic field intensity was collected along the longitudinal direction of specimen after each stretching. The research results of this part is shown as following under different tensile load such as 50 kN, 100 kN and 150 kN in Fig. 7.

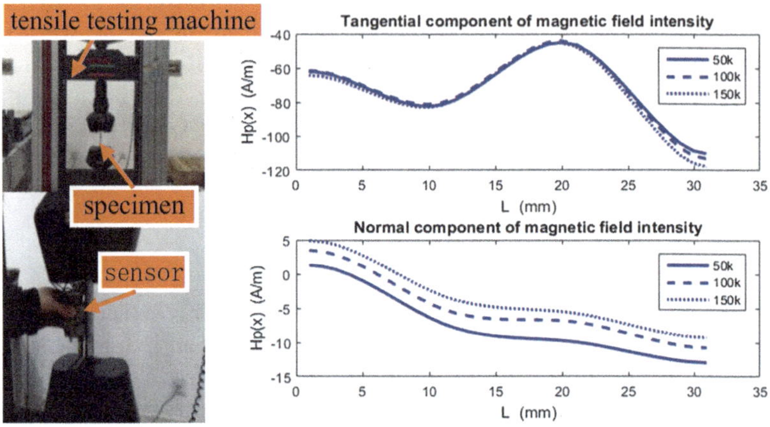

Fig. 7. The testing of leakage magnetic field intensity

The kurtosis is a dimensionless parameter because it is particularly sensitive to shock signals and is particularly useful for surface damage problems, especially for early fault diagnosis. The kurtosis is expressed as following:

$$K = \int_{-\infty}^{+\infty} \left(\frac{x - \bar{x}}{\sigma}\right)^4 P(x)dx \tag{2}$$

Where K is the kurtosis of the signal, $x(t)$ is the instantaneous magnetic field intensity value, the $\bar{x}$ is the amplitude mean, $p(x)$ is the probability density, and σ is the standard deviation.

Table 1. Original value and kurtosis of magnetic field intensity

Load	X-Max	X-Min	X-Mean	Y-Max	Y-Min	Y-Mean	$K(x)$	$K(y)$
50 kN	−45.08	−110.09	−71.73	1.31	−12.91	−7.57	1.97	1.71
100 Kn	−44.13	−113.06	−71.57	3.50	−10.75	−5.18	2.57	1.75
150 kN	−45.24	−117.59	−73.78	4.91	−9.25	−3.62	2.43	1.79

The peak and mean values of the tangential component (X-direction) and normal component (Y-direction) and the kurtosis of the leakage magnetic field intensity are shown in Table 1.

According to data information from Table 1, it shows that the tangential component of the leakage magnetic field intension increases with the increase of the static load, and the tangential component decreases with the increase of the static load.

(3) Acquisition and analysis of magnetic memory signal of the wheelset in different stages of the life cycle.

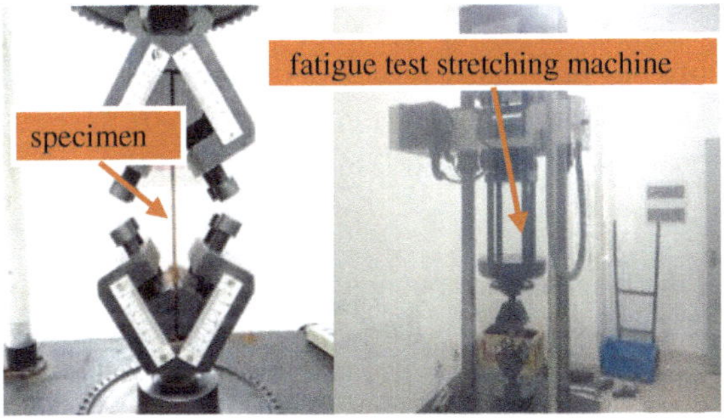

Fig. 8. Fatigue test stretching machine

In order to explore the relationship between the metal magnetic memory detection signal and the wheel life cycle, the fatigue load test of the material was carried out. Three pieces of the specimen in the Fig. 8 shows an experimental apparatus for fatigue stretching.

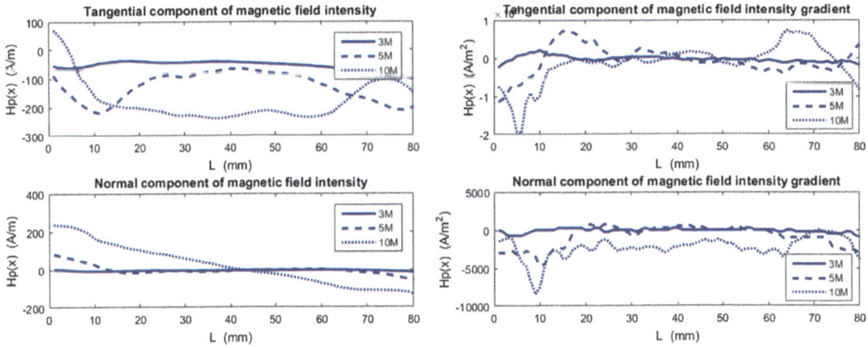

Fig. 9. Original value and gradient of the MMM signal

The specimens were stretched 3 million times, 5 million times and 10 million times under constant force. The collected data and gradient is shown in Fig. 9.

According to Fig. 9, it can be seen that the number of alternating loads has a great influence on the magnetic memory signal of the specimen. With the increase of the number of alternating loads, the tangential component of magnetic memory signal shows a significant increase trend, the normal component of the magnetic memory signal is more discretized and always has zero points, and a certain degree of stress concentration may occur.

Table 2 show the peak and mean values of the magnetic memory signals.

Table 2. Original value and kurtosis of magnetic field intensity

Times	X-Max	X-Min	X-Mean	Y-max	Y-Min	Y-Mean	$K(x)$	$K(y)$
3M	−84	−245	−144.59	−8	−49	−15.89	2.02	5.40
5M	−152	−554	−308.67	126	−105	−16.33	2.37	5.19
10M	−206	−487	−358.60	346	−182	−21.73	3.79	2.29

The kurtosis of the tangential component of the magnetic memory signal shows a significant increase, which shows that the probability density of the signal has a large discrete phenomenon. The kurtosis of magnetic memory signal is shown in Table 2.

5 Conclusion

This paper proposed a real-time detecting system of early fault on the wheelset based on metal magnetic memory method to meet safety requirements of train operation. In this system, detection mode se-lection, data acquisition system and large capacity data transmission system, and signal de-noising and feature extraction have been introduced. Hope that these studies can provide the reference for nondestructive testing.

(1) An intelligent data collecting controller on based wireless technology has been developed. The accuracy of A/D acquisition can be less than 5/10000 to meet data acquisition requirements.
(2) Static tested for the wheelset material shows that as the static load increases, the tangential component of the leakage magnetic field on the surface of the specimen increases, but the magnitude of change is not significant and the linear correlation is not significant. When the variation is between 50 kN and 150 kN, the corresponding mean value of the magnetic memory signal range of −73.778 A/m to −71.568 A/m. The normal component has increased obviously. When the variation is between 50 kN and 150 kN, the corresponding mean value of the magnetic memory signal range of −7.6 A/m to −3.6 A/m. By comparing and analyzing the tangential and normal components of the leakage magnetic field strength, the tangential component and the normal component have basically negative correlations.

(3) Wheelset materials were fatigue tested. The results show that with the increase of the number of alternating loads, the magnetic memory signal shows more obvious dispersion, and the values of the tangential and normal signals of the magnetic memory signal are all increased, and the mean tangential change range is −144.586A. From/m to −358.597 A/m, the normal variation ranged from −15.885 A/m to −21.726 A/m. The gradient and kurtosis of the magnetic memory signals all have an influence. In particular, with the increase of the number of alternating loads, the kurtosis of the magnetic field curve changes more, which has important reference value in the practical application of wheelset dynamic detection.
(4) The service life of the wheelset can be predicted. From the research of detecting system and the test of the influence of static load and alternating load on the material of wheelset, the signal characteristics of the wheelset at different stages of use can be described.

The research work has been financially supported by the National Natural Science Foundation of China (51405303), the City University Youth Teacher Training Fund (ZZyy15110) and the Shanghai University of Applied Technology Talent Fund (YJ2014-17).

References

1. Chen, X., Li, L., Hu, B., et al.: Magnetic evaluation of fatigue damage in train axles without artificial ex-citation. Insight Non Destr. Test. Cond. Monit. **48**(2006), 342–345 (2006)
2. Tao, G.Q., Wang, H., Zhao, X., et al.: Re-search on wheel tread damage mechanism based on interaction of wheel and rail. J. Mech. Eng. **49**(2013), 23–29 (2013)
3. Li, X., Ding, H., Bai, S.: Research on the stress-magnetism effect of ferromagnetic materials based on three-dimensional magnetic flux leakage testing. NDT&E Int. **62**(2014), 50–54 (2014)
4. Maciej, R.: Metal magnetic memory testing of welded joints of ferritic and austenitic steels. NDT&E Int. **44**(2011), 305–310 (2011)
5. Demetrion, M.A., Polycarpou, M.M.: Incipient fault diagnosis of dynamical systems using online approximators. IEEE Trans. Autom. Control **43**(1998), 1612–1617 (1998)
6. Yao, K., Shen, K., Wang, Z., Wang, Y.: Three-dimensional finite element analysis of residual magnetic field for ferromagnets under early damage. J. Magn. Magn. Mater. **544**, 112–118 (2014)
7. Doubov, A.A.: Screening of weld quality using the metal magnetic memory. Weld. World **1998**, 196–199 (1998)
8. Dubov, N., Sergey, K.: The metal magnetic memory method application for online monitoring of damage development in steel pipes and welded joints specimens. Weld. World **57**(2013), 123–136 (2013)
9. Leng, J., Xu, M., Zhang, J.: Magnetic field variation induced by cyclic bending stress. NDT&E Int. **42**, 410–414 (2009)
10. Leng, J., Xu, M., Zhou, G., Wu, Z.: Effect of initial remanent states on the variation of magnetic memory signals. NDT&E Int. **52**(2012), 23–27 (2012)

11. Roskosz, M., Gawrilenko, P.: Analysis of changes in residual magnetic field in loaded notched samples. NDT&E Int. **41**(2008), 570–576 (2008)
12. Sablik, M.J., Wilhelmus, J.G., Smith, K., et al.: Modeling of plastic deformation effects in ferromagnetic thin films. IEEE Trans. Magn. **46**, 491–494 (2010)
13. Shi, C.L., Dong, S.Y., Xu, B.S., He, P.: Metal magnetic memory effect caused by static tension load in a case-hardened steel. J. Magn. Magn. Mater. **2010**, 413–416 (2010)
14. Wang, Z.D., Yao, K., Deng, B., Ding, K.Q.: Theoretical studies of metal magnetic memory technique on magnetic flux leakage signals. NDT&E Int. **43**(2010), 354–359 (2010)

The Intermittent Fault Diagnosis Method Based on WPD-SRC for the Mechanical System

Fan Wang, Jianfeng Qu(✉), Yi Chai, Zeping Wang, and Chunyan Li

College of Automation, Chongqing University, Chongqing 400030, China
qujianfeng@cqu.edu.cn

Abstract. A special intermittent fault was caused by the complex operating environment and the damage of its components in varying degrees in the mechanical equipment system. Thus, it is essential to eliminate the hidden troubles caused by intermittent faults, whose diagnosis method based on sparse representation classification and wavelet packet decomposition (WPD-SRC) is proposed. The energy of every band of multi-channel vibration signals has been extracted by WPD as the feature vector. The sparse learning dictionary was composed of the feature vector of different fault states. The normalized test sample and the sparse dictionary were used as the input of the SRC and for the reconstruction of the error vector with the sparse representation coefficient to diagnose the intermittent fault. The experiment result shows that the intermittent fault can be identified by adopting WPD-SRC, and then what type it belongs to is diagnosed.

Keywords: Intermittent fault diagnosis
Wavelet packet decomposition (WPD) · Learning dictionary
Sparse representation classification (SRC)

1 Introduction

With the development of electronic and computer technology, a special fault—the intermittent fault which is different from permanent faults and transient faults, has gradually aroused public interest. [1] An intermittent fault is a kind of deciduous fault which can only last for a short time; which can be repeated; and which can be eliminated without treatment [2]. In mechanical systems, intermittent faults are mainly caused by wear and cracks in bearings, gears, and so on [3]. The bearings and gears are periodically rotated by the motor, therefore, most of these intermittent faults are periodic [4]. Given the generation and intensification of wear and cracks of bearings and gears take a long time, in a short time, the fault diagnosis of the mechanical system can be considered that the degree of wear and fracture of equipment is changeless [5, 6]. Therefore, the magnitude of such intermittent faults is constant [7].

The literature [8] proposed a Dynamic Bayesian Network (DBN)-based fault diagnosis methodology in the presence of TF and IF for electronic systems. DBN was used to model the dynamic degradation process of electronic products, and Markov chains were employed to model the transition relationships of the states; The literature

K. Deng et al. (Eds.): ICMIR 2018, AISC 856, pp. 857–865, 2019.
https://doi.org/10.1007/978-3-030-00214-5_106

[9] studied the intermittent fault diagnosis problem of digital circuits with the statistical model method firstly, and used the two-state Markoff model to describe the dynamic characteristics of intermittent faults. In the literature, [10] the continuous wavelet transform and Hilbert spectrum analysis was used to collect the variation characteristics of signals under intermittent faults [11].

Wavelet packet decomposition is a signal analysis method which takes into account the advantages of both Fourier transform and wavelet transform. In view of rotating machinery equipment, the vibration signal of intermittent fault conditions features by being periodic and non-stationary [12]. And WPD is significantly good for the periodic non-stationary signal [8]. Thus, the fault feature can be effectively extracted in a more refined band [14].

Sparse representation has been widely used in the field of pattern recognition because of its high recognition rate and robustness [5]. The significance of sparse representation of a particular signal lies in representing the signal with as few atoms as possible in a given over-complete dictionary [13]. On the one hand, the data-driven method does not need to study the complex dynamics of intermittent fault systems [14]; on the other hand, the intermittent faults with random, intermittence and repeatability have sparseness in some domains or spaces.

2 Methodology

2.1 Wavelet Packet Decomposition and Sparse Representation Classification

Figure 1 illustrates an overview of the WPD-SRC intermittent fault diagnosis method. As shown in Fig. 1. The WPD-SRC is used to identify states of intermittent faults produced in a gear or bearing in a mechanical system. Different vibration signals are considered as data sources. First, the feature is extracted by WPD, and the sparse dictionary is constructed by the energy eigenvalues under all the states. Then, the sparse representation coefficients are solved with SRC. Finally, the minimum error vectors are reconstructed to realize the classification of the states.

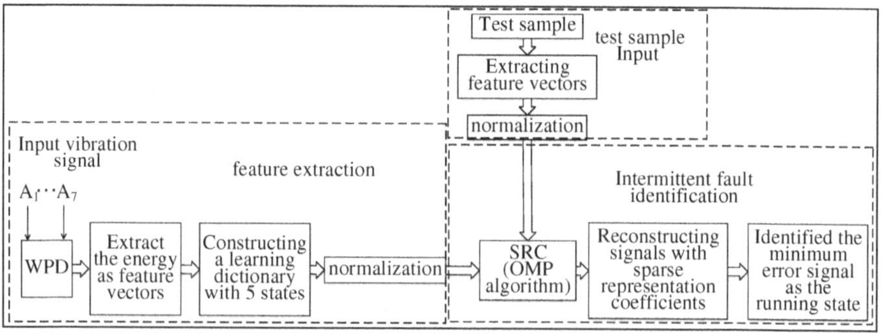

Fig. 1. The framework diagram of WPD-SRC algorithm

2.2 Feature Extraction Based on WPD

WPD adopts orthogonal wavelet, and can decompose the vibration signals into several bands with the equal bandwidth. Take the three wavelet decomposition as an example. The node (0, 0) is the original vibration signal, and the second layer nodes (1, 0) and (1.1) represent the low and high frequency components of the original signal. The third layer $X(3,j), j = 0, 1, 2, \ldots, 7$ represents the discrete amplitude of the reconstructed signal corresponding to the node (3, j). Supposed that the reconstruction signal corresponding to the wavelet coefficients $C_{j,k}$ of each node is $S_{j,k} \quad (j = 0, 1, 2, \ldots, 2^i - 1)$.

The energy $E_{jk}(j = 0, 1, 2, \ldots, 2^i - 1)$ can be represented as:

$$E_{jk} = \int \left|S_{jk}(t)\right|^2 dt = \sum_{k=1}^{n} |X_k|^2 \tag{1}$$

where X_k is the amplitude of discrete points of reconstructed signals S_{jk}.

Suppose that the total energy is: $E = \sum_{j=0}^{2^i-1} E_{ij}$, and then the normalized energy eigenvector is:

$$T = [E_{j0}/E, E_{j1}/E, E_{j2}/E, \ldots, E_{j2^i-1}/E] \tag{2}$$

That is $T = [e_{j0}, e_{j1}, e_{j2}, \ldots, e_{j2^i-1}]$, where i is the number of WPD.

As for the feature extraction for multi channels vibration signals,

$$T = [e_{10}, e_{11}, e_{12}, \ldots, e_{12^i-1}; e_{20}, e_{21}, e_{22}, \ldots, e_{22^i-1}; \ldots; e_{j0}, e_{j1}, e_{j2}, \ldots, e_{j2^i-1}] \tag{3}$$

where j is the channel number of the vibration signals.

2.3 Sparse Representation Classification (SRC)

There are a total of j vibration signals, and samples in class i are recorded as:

$$T = [e_{10}, e_{11}, e_{12}, \ldots, e_{12^i-1}; e_{20}, e_{21}, e_{22}, \ldots, e_{22^i-1}; \ldots; e_{j0}, e_{j1}, e_{j2}, \ldots, e_{j2^i-1}] \in R^{d \times n_i} \tag{4}$$

A sample to be tested is given, $y \in R^d$. Assume that it belongs to class i, y will be represented by linear combinations. Where is

$$y = \alpha_{10}e_{10} + \alpha_{11}e_{11} + \cdots + \alpha_{12^i-1}e_{12^i-1} + \alpha_{j0}e_{j0} + \alpha_{j1}e_{j1} + \cdots + \alpha_{j2^i-1}e_{j2^i-1} \tag{5}$$

That is $y = Ax_0$, A represents $n \times k$ dimension over-complete dictionary;x_0 represents Sparse Vector Representation. From Eq. (5), we can see that x_0 may have the form shown below:

$$x_0 = [0, \ldots, 0, \alpha_{k1}, \ldots, \alpha_{k2^i-1}, 0, \ldots, 0]^T \in R^n \tag{6}$$

That is, the nonzero item of x_0 corresponds to the known i class sample, where k is kind of a class. In order to obtain the sparse solution of Eq. (6), the formula needs to add the regular term to solve the problem as follows:

$$\min_{x_0} \|x_0\|_1 \quad s.t\ (y - Ax_0) < \varepsilon \tag{7}$$

where $\|\cdot\|_1$ is ℓ^1 norm, ε is upper limit of error.

Ideally, SRC was used to identify and classify the intermittent fault signals is mainly based on the position of non-zero items in Eq. (6). In order to make a quantitative diagnosis, the reconstruction error vector of the sample to be measured is required.

$$E_r(i) = \|y - A\delta(x_0)\|_2 \quad i = 1, 2, \ldots, m \tag{8}$$

Each element of the reconstructed error vector E_r corresponds to a fault category, and the minimum corresponding to the category is the category which y belongs to.

3 Experiments and Analysis

3.1 Gear Fault Data

The gear vibration data are obtained through the rotating machinery vibration analysis with a QPZZ-II test platform, as shown in Fig. 2(a). The system can simulate the multi states and vibration of rotating machinery rapidly, and can be used for the comparative analysis and the diagnosis of various states. There are 5 kinds of states are selected, which are Normal, Wear fault, Intermittent wear fault, Broken fault in gearbox, Intermittent broken fault in gearbox. Each state has 40 training samples and 52 testing samples. The mechanical structure of the gear box is displayed intuitively in Fig. 2(b). It can be seen that the gear is a point of occurring intermittent faults, and different degrees of damage can produce different fault states.

There are some other unseen points of fault. The intermittent fault caused by the lubricating oil or the presence of small hard particles in bearings is non-uniform or too little. The tooth surface will generate micro fatigue cracks when the working surface of the gear is subjected to cyclic shearing stress. This operating state will lead the system to demonstrating an intermittent broken fault in the gearbox state, etc.

3.2 The Experiment and Result

In the experiment of this paper, the original vibration signal in all states is shown in Fig. 3. The original signal is input, and the db5 wavelet function is chosen to make 3 layer wavelet decomposition of the original vibration signal. And then, wprcoef function is employed to calculate the frequency band energy of the third layer. Then, the normalized feature vector is used as the sparse representation dictionary.

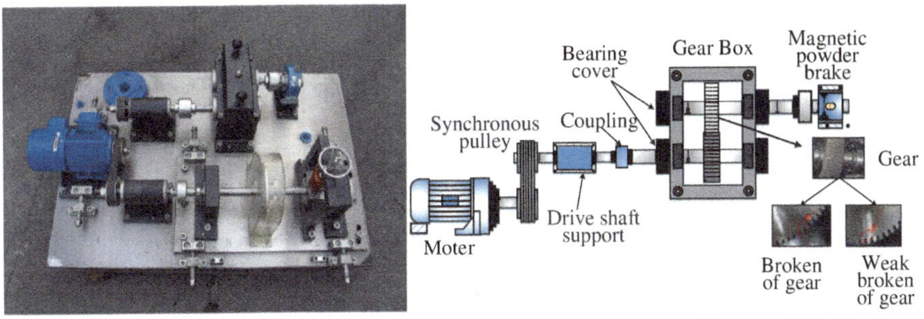

(a) External View of Mechanical Equipment (b) Mechanical Structure Diagram of the Gearbox

Fig. 2. Mechanical drawing of gear

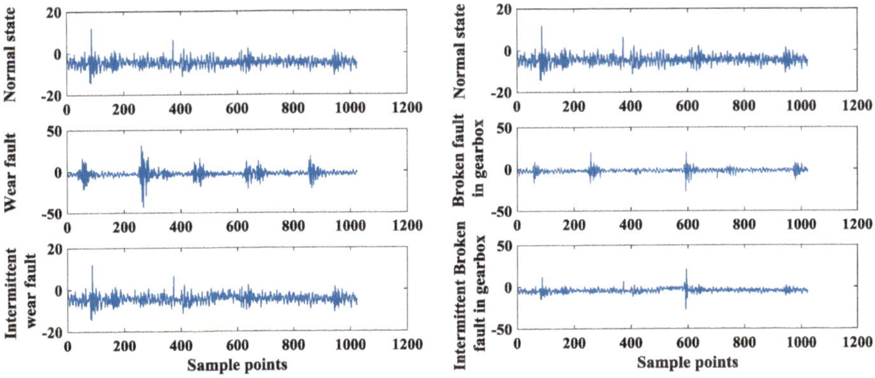

Fig. 3. Time domain waveform of the original vibration signals of various running states

The OMP (Orthogonal Matching Pursuit) algorithm is used to solve sparse representation coefficients. One sample in 5 running states is taken out as the tested sample, and the sample as well as dictionary A is the inputs of SRC. The output is the sparse representation coefficients. As shown in Fig. 4 there are the sparse representation coefficients of five samples to be tested in dictionary A. It is obvious that sparse representation coefficients are sparse. It can be seen from Fig. 4 that the sample is recognized as the area where the coefficients are densely concentrated in the graph. If the identification result is the same as the labeled test sample, the recognition will be considered successful. Otherwise, the error is identified.

The above-mentioned recognition process is only a qualitative description. In order to identify quantitatively, the reconstruction error vector must be introduced. The reconstructed error vector of the test sample is obtained through the application of the Eq. (9) and the sparse representation coefficients.

$$E_r = [E_r(1), E_r(2), E_r(3), E_r(4), E_r(5)] \tag{9}$$

Table 1. Reconstruction error vectors of 5 test samples

	$E_r(1)$	$E_r(2)$	$E_r(3)$	$E_r(4)$	$E_r(5)$	Diagnosis result
Figure 4(a)	**0.587**	1.027	0.996	0.988	0.999	Normal
Figure 4(b)	0.996	**0.467**	1.027	1.033	1.006	Wear
Figure 4(c)	1.227	1.049	**0.048**	1.037	0.791	Intermittent wear
Figure 4(d)	1.009	0.994	1.035	**0.513**	0.998	Broken fault in gearbox
Figure 4(e)	0.743	0.907	1.024	0.893	**0.243**	Intermittent broken fault in gearbox

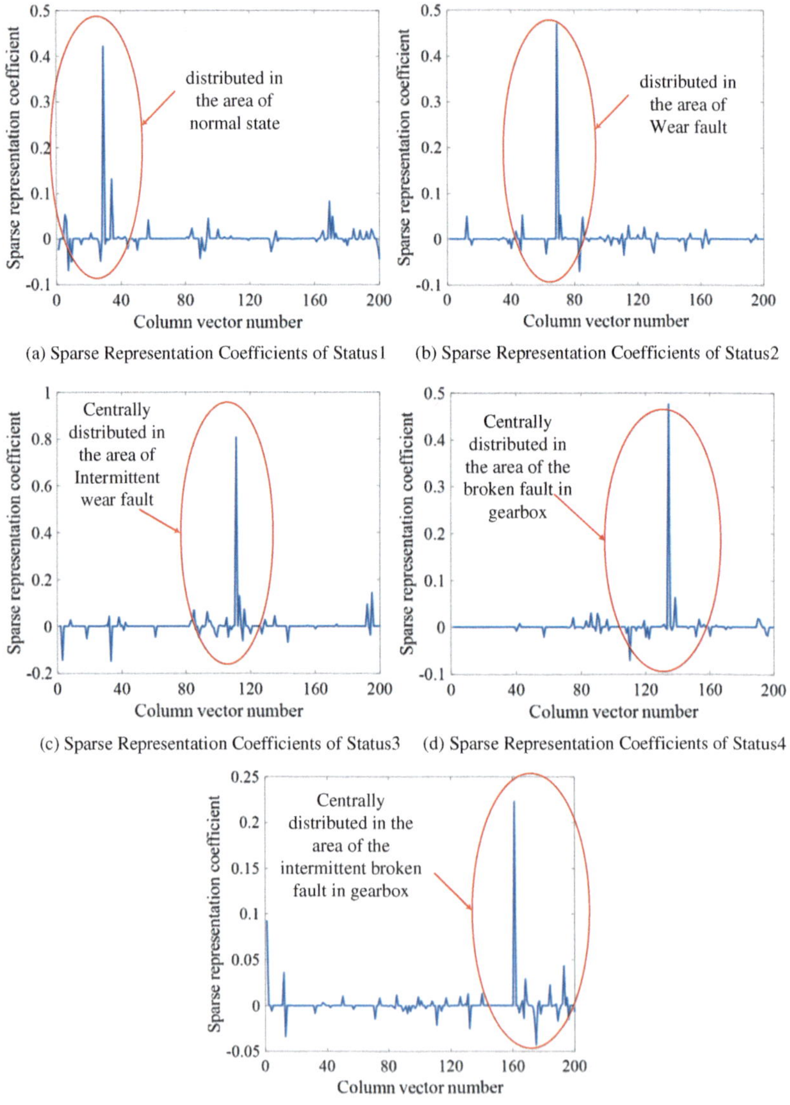

(a) Sparse Representation Coefficients of Status1 (b) Sparse Representation Coefficients of Status2

(c) Sparse Representation Coefficients of Status3 (d) Sparse Representation Coefficients of Status4

(e) Sparse Representation Coefficients of Status5

Fig. 4. Sparse representation coefficients of test samples of all states

The five elements of E_r make the input signals subtract the five groups of sparse reconstructed signals. The location of the smallest element in E_r corresponds to the fault category which is the category of the test sample. The result is correct if it is consistent with the category of the labeled test sample. If not, the error is identified. Table 1 shows the reconstructed error vector calculated by the sparse representation coefficients obtained from the test sample corresponding to Fig. 4. As what can be seen from the Table 1. The results show that the sample recognition is correct. Thus, the algorithm presented in this paper can identify different fault states of vibration signals.

To verify WPD-SRC, the number of the sample in every running state increased to 52. The sample of all states added up to 260. The specific processing method is the same as the experiment of a single test sample. Figure 5(a) shows the calculated sparse representation coefficients. It can be seen that the nonzero elements in the corresponding sparse coefficients for test samples are clustered with respect to a specific fault severity level. This fault category corresponding to the non-zero elements exactly indicates the fault severity level of the test samples.

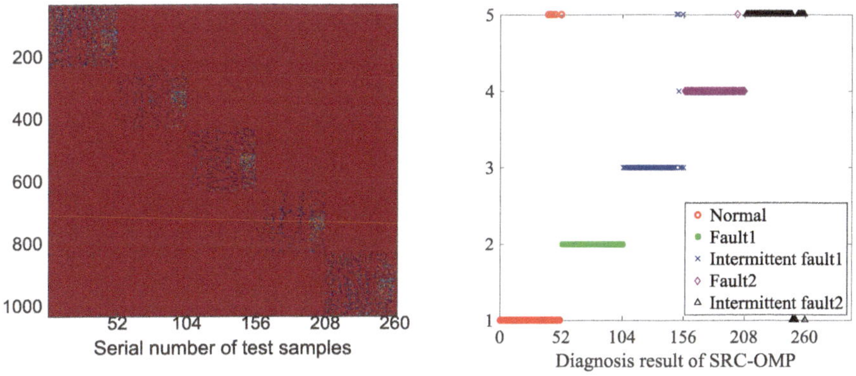

(a) The Sparse Vector of the Test Samples **(b)** Diagnosis Results of SRC (OMP)

Fig. 5. The result of WPD-SRC. **(a)** The sparse vector of the test samples **(b)** Diagnosis results of SRC (OMP)

The recognition rate in each running state is obtained by using the statistical data except the total number of samples. The specific recognition rate of each running state is shown in Table 2. The result is displayed vividly in Fig. 5(b). What can be seen are not only the correct identification of each class of samples, but also the identification of errors. The distribution of identification error samples can be seen clearly. Most of the samples which have wrong recognition are distributed in the fifth type of regions, which shows that the characteristics of the fifth state are higher than those of other classes.

Table 2. The recognition rate of each running state

Status	Status1	Status2	Status3	Status4	Status5
Recognition rate	88.46%	100.00%	92.31%	98.08%	90.38%

4 Conclusions

In this paper, a diagnosis method for intermittent faults is proposed on the basis of the general fault diagnosis, and an intermittent fault diagnosis method of sparse classification is based on wavelet packet decomposition, mainly including: (1) three layers wavelet packet decomposition is for the sample data, and the energy value of the third layers is extracted as the feature; (2) the extracted feature values of different running states are used to organize the sparse learning dictionary, and the intermittent faults are classified by using SRC and the reconstruction error vector.

In the later experiments, the simulation should add compound faults and intermittent compound faults, so as to consummate the recognition and diagnosis of different operating states of equipment.

Acknowledgements. This work was supported by the Natural Science Foundation of Chongqing City, China(cstc2016jcyjA0504).

References

1. Zhou, D.H., Shi, J.T., He, X.: Review of intermittent fault diagnosis techniques for dynamic systems. Acta Autom. Sin. **40**(2), 161–171 (2014)
2. Correcher, A., Garcia, E., Morant, F., et al.: Intermittent failure dynamics characterization. IEEE Trans. Reliab. **61**(3), 649–658 (2012)
3. Guo, M.W., Ni, S.H., Zhu, J.H.: Diagnosing intermittent faults to restrain BIT false alarm based on EMD-MSVM. Appl. Mech. Mater. **105–107**, 729–732 (2012)
4. Zhu, K., Song, X., Xue, D.: A roller bearing fault diagnosis method based on hierarchical entropy and support vector machine with particle swarm optimization algorithm. Measurement **47**(1), 669–675 (2014)
5. Zhang, H., Zhaohui, D.U., Fang, Z., et al.: Sparse decomposition based aero-engine's bearing fault diagnosis. J. Mech. Eng. **51**(1), 98–105 (2015)
6. Chen, H., Chen, P., Chen, W., et al.: Wind turbine gearbox fault diagnosis based on improved EEMD and hilbert square demodulation. Appl. Sci. **7**(2), 128 (2017)
7. Zhang, W., Jia, M., Yan, X., et al.: Weighted sparsity-based denoising for extracting incipient fault in rolling bearing. J. Mech. Sci. Technol. **31**(10), 4557–4567 (2017)
8. Cai, B., Liu, Y., Xie, M.: A Dynamic-bayesian-network-based fault diagnosis methodology considering transient and intermittent faults. IEEE Trans. Autom. Sci. Eng. **PP**(99), 1–10 (2017)
9. Liu, J., Zhang, X.: Detection method of intermittent faults in electronic systems based on Markov Model. In: Fourth International Symposium on Computational Intelligence and Design, pp. 216–219. IEEE Computer Society (2011)
10. Guo, M., Ni, S., Zhu, J.: Intermittent fault diagnosis for built-in test system based on EMD and HMM. J. Vib. Meas. Diagn. **32**(3), 467–470 (2012)

11. Xing, Z., Qu, J., Chai, Y., et al.: Gear fault diagnosis under variable conditions with intrinsic time-scale decomposition-singular value decomposition and support vector machine. J. Mech. Sci. Technol. **31**(2), 545–553 (2017)
12. Rabbi, S.F., Little, M.L., Saleh, S.A., et al.: A novel technique using multi-resolution wavelet packet decomposition for real time diagnosis of hunting in line start IPM motor drives. IEEE Trans. Ind. Appl. **PP**(99), 1 (2016)
13. He, Q., Ding, X.: Sparse representation based on local time–frequency template matching for bearing transient fault feature extraction. J. Sound Vib. **370**, 424–443 (2016)
14. Khilar, P.M., Mahapatra, S. Intermittent fault diagnosis in wireless sensor networks. In: International Conference on Information Technology. IEEE Computer Society, pp. 145–147 (2007)

Design and Contact Nonlinear Analysis on Automotive Electrical Power Steering Torsion Bar

Shujuan Fang(✉)

Hulunbuir University Mining College, Inner Mongolia Hulunbuir 021008, China
32577140@qq.com

Abstract. Torsion bar is the key part for Electrical power steering (EPS). In this paper, mathematical model on the torsion bar without outer spline is built, according to the automotive parameters. The 3D model and finite element model on EPS sub-assembly (torsion bar, input shaft, and output shaft) are developed, and then the contact nonlinear analysis based on interference fit between torsion bar and output shaft is done. The EPS are developed and the function test and durability test are done, and the tests are passed. This kind of structure can lead to cost down and time saving for torsion bar manufacturing. The contact non-linear analysis method can be used for complicated and nonlinear system analysis when doing mechanical design.

Keywords: Automotive · EPS · Torsion bar · Contact nonlinear

1 Introduction

In 1988, the first C type EPS (Electrical Power Steering) was developed and assembled in Japan. The EPS assembly rate increases by 8% ~ 10% annually, more and more EPS are assembled in automotive all over the world till now. During this period, the requirements for key components of EPS are also increasing. Many scholars have studied the finite element method on EPS analysis, eg., Lee et al. [1] study hall effect torques sensor by using 3D finite element method and its characteristics with respect to four different shapes of the stator teeth are compared. Gao et al. [2] use finite element method to optimize the dynamic characteristics on EPS. Yuan et al. [3] create the modeling and analyze the vibration for EPS and get its natural frequency and stress distribution.

In this work, the torsion bar deformation is studied by the principle of material mechanics. The 3D model of EPS sub-assembly is built, and then the finite element is meshed. The torsion bar stress and strain of the EPS sub-assembly is studied by nonlinear analysis method. Then the results are verified by the function and durability tests, and the test results are passed.

K. Deng et al. (Eds.): ICMIR 2018, AISC 856, pp. 866–872, 2019.
https://doi.org/10.1007/978-3-030-00214-5_107

2 Classification of Electrical Power Steering

EPS is classified as following three types, according to the different installation position in the vehicle: column type EPS (C-EPS), pinion type EPS (P-EPS), rack type EPS (R-EPS). That means, C-EPS is mainly installed on the steering column that is under the steering wheel, and the C-EPS is fit for A0 or A-class car; P-EPS is mainly installed on the pinion of the gear and rack type steering gear that is in the car foreside, and the P-EPS is fit for B or C-class car; R-EPS is mainly installed on the rack of the gear and rack type steering gear that is near the engine, and R-EPS is fit for C or above C-class car or truck [4]. See Fig. 1.

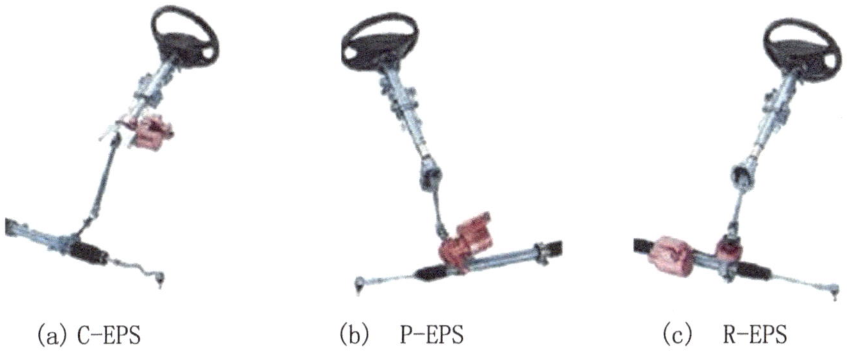

Fig. 1. Classification of EPS

3 Working Principle of Electrical Power Steering

EPS mainly includes in the Electrical Control Unit (ECU), motor, sensor and deceleration mechanism gearbox [5]. The torsion bar is an important part of the deceleration mechanism gearbox, which is used for converting the changing of steering wheel torque to the changing of mechanical angle, and then EPS sensor sends out signal after testing the mechanical angle changing. When the vehicle is started, EPS ECU do self-check firstly, which is included in the checks of engine signal, power supply voltage signal, vehicle speed signal, and torque signal, then ECU begins to control the motor to drive the power steering according to the internal algorithm based on these signals.

4 Torsion Bar Design

To simplify assembly with input shaft and output shaft, the torsion bar of EPS is usually designed as the external spline for one end, and the other end is shaft with pin hole. There is transition arc part between the slender rod structure in the middle of the torsion bar and the rod end. When being assembled, the end of the torsion bar with external spline is pressed into the output shaft by the hydraulic press. Then the torque sensor is installed. After the midpoint of the torque signal is adjusted in the EPS system, the input shaft is installed, and the pin is inserted into the pin hole of the input shaft and the torsion bar. The structure of the torsion bar is shown in Fig. 2.

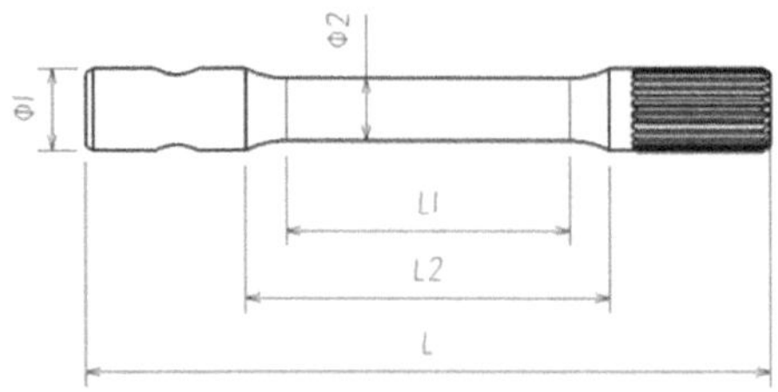

Fig. 2. Torsion bar structure

However, to save the processing cost of torsion bar and simplify the processing steps, design a torsion bar without outer spline. Because there is interference fit between torsion bar and the output shaft, the torque can be transferred from the torsion bar to the output shaft.

To simplify the calculation, the torsion bar is considered as the bar structure with the same cross section diameter, which can be called as rod. The torsion deformation of the rod is studied by the principle of material mechanics, and the stiffness formula of the rod is obtained as follows.

$$\varphi = \frac{TL}{GI_\rho} \times \frac{180^\circ}{\pi} \tag{1}$$

φ is the torsion angle (degrees), T represents the torque of rod section, and L is the length of the rod. G is shear modulus of materials, I_ρ is the polar moment of inertia, and GI_ρ is torsion stiffness.

So, the φ is as following if considered as solid rod,

$$\varphi = \frac{TL}{GI_\rho} = \frac{T \times L \times 32}{G \times \pi \times d^4} \times \frac{180^\circ}{\pi} \tag{2}$$

When the rod is affected by the torque T, the shear stress is calculated as the following formula

$$\tau = \frac{T}{W_p} \tag{3}$$

$$W_p = \frac{\pi d^3}{16} \tag{4}$$

Strength condition requirement is

$$\tau_{max} \leq [\tau] \tag{5}$$

$[\tau]$ is the allowable shear stress [6].

Consider A0-class car as target model, and the torque applied to the torsion bar by steering wheel is generally less than 10 Nm. If consider the safety factor as 1.5, then T_{max} = 10 Nm × 1.5 = 15 Nm. The torsion bar material is $60Si_2MnA$, G = 79000 MPa, $[\tau] = 590\,\text{MPa}$. Then, come to the conclusion d = 5.01 mm. So, the value d_2 = 6 mm.

Generally, the torsion angle of the torsion bar is set to be ±6°. So $\varphi = 6°$ and d_2 = 6 mm, then can come to the conclusion that $L_a = 22.3\,\text{mm}$ based on the formula (3). The approximate value 23 mm is taken as the working length. The recommended diameter of cross section is $d_1 = d_2 \times (1.2 \sim 1.3) = (7.2 \sim 7.8)$ mm, and d_1 = 7.8 mm is taken.

The position between the end and the working length of the torsion bar is called as the transition section. The size of the general transition section is gradually changed in order to minimize the stress concentration.

Transition section length $L_g = \frac{L_2 - L_1}{2} \approx \frac{d_1 - d_2}{2\tan 15^{\circ}} = 3.4\,\text{mm}$

Transition section radius $R = \frac{L_g^2}{d_1 - d_2} + \frac{d_1 - d_2}{4} = 6.87\,\text{mm}$, set as R = 6.9 mm.

In addition, the transition section is generally divided into effective length and non effective length, effective length

$L_e = \frac{L_g}{3}\left[\frac{d_2}{d_1} + \left(\frac{d_2}{d_1}\right)^2 + \left(\frac{d_2}{d_1}\right)^3\right] = 2.06\,\text{mm}$, set as $L_e = 2.1\,\text{mm}$

The length on the slender rod structure in the middle of the torsion bar is

$$L_1 = L_a - 2 \times L_e = 18.8\,\text{mm}$$

Set the end length as $L_d = 20\,\text{mm}$

Then the total length is as following

$$L = L_1 + 2 \times L_g + 2 \times L_d = 65.6\,\text{mm}.$$

5 Torsion Bar Structure Analysis

Generally, the mechanical design is based on experience, but the stress and strain need to be analyzed carefully for the complex structure and strict safety required components. In addition, the interference fit stress and strain of the two contact surfaces between the torsion bar and the output shaft cannot be tested directly, and the mating surfaces are often subjected to cyclic loading. Therefore, the contact nonlinear finite element analysis method is used [7, 8].

5.1 The Establishment of Geometric Model

3D models (input shaft, torsion bar and output shaft) are built by three-dimensional software, and then transferred to the middle format files and imported into the finite element analysis software. The torsion bar and input shaft are fixed by pin. The EPS sub-assembly is shown in Fig. 3.

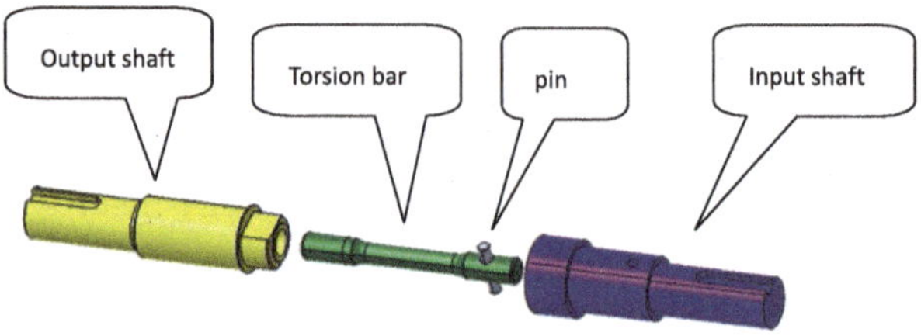

Fig. 3. EPS sub-assembly (input shaft, output shaft, torsion bar, and pin)

5.2 Finite Element Meshing [9, 10]

The traditional linear analysis method is to mesh the contact surface by the same node, and assume that the contact force was transferred by the two rigid bodies' same nodes. We need to know the exact location of the contact, then the boundary conditions for the linear analysis method is defined, but it cannot simulate the real friction condition. For the contact nonlinear analysis, the hexahedral elements are used as much as possible, and the wedge elements are appropriately used in the complex parts of the models, then the finite elements are meshed by free method.

5.3 Material Attributes

The torsion bar material is spring steel $60Si_2MnA$, and the material density is 7850 kg/m^3 and elastic modulus is 213 GPa, Poisson's ratio 0.29; the input shaft and the output shaft material is 40Cr, and the material density is 7850 kg/m^3 and elastic modulus is 206 GPa, Poisson's ratio 0.3.

5.4 Boundary Conditions

When the nonlinear analysis is done, it is necessary to set up the contact surface on important parts, and the rest parts can be fixed by the coupling of the nodes. The boundary conditions are as following

(1) The nodes between the outer diameter of the torsion bar and the inner diameter of the input shaft are fixed, so the two contact surfaces are considered as linear systems. And the outer diameter of the output shaft is fully constrained to simulate the ground resistance moment. These are the linear systems.

(2) When the contact nonlinear analysis is done, the contact surfaces attribute is defined on the contact parts between the torsion bar and the output shaft. The contact part of the torsion bar is defined as the CONTA173 contact surface. The contact part of the

output shaft is defined as the TARGE170 target surface. The contact theory is used to search for the separation or contact between the torsion bar outer surface and the output shaft inner surface.

(3) The contact stiffness ratio factor FKN is 0.1. The friction coefficient at the contact position of the torsion bar and the output shaft takes 0.11. (Refer to the condition of the pressure between the steel and the high quality steel in the material mechanics).

5.5 Loads

The steering torque 15 Nm which is applied on the input shaft cannot be simulated directly by finite element analysis software. So, the 1000 N force is applied to the tangent direction of the outer diameter of the input shaft to simulate the steering torque 15 Nm.

5.6 Analysis Results

Interference fit is a typical contact nonlinear behavior in finite element analysis. Different contact algorithms have different analysis methods. The contact is set up in finite element analysis, in theory, that means a constraint [11] is applied to two components which are contacted. Strengthening Lagrange algorithm is used as the default calculation method.

The analysis results show that torsion bar maximum stress is 5751 N and maximum tangential displacement is 0.22 mm, and the location is at the edge of the end of the torsion bar. The strain analysis cloud is shown in Fig. 4.

The analyses results conclude that the mechanical design meet the application requirements.

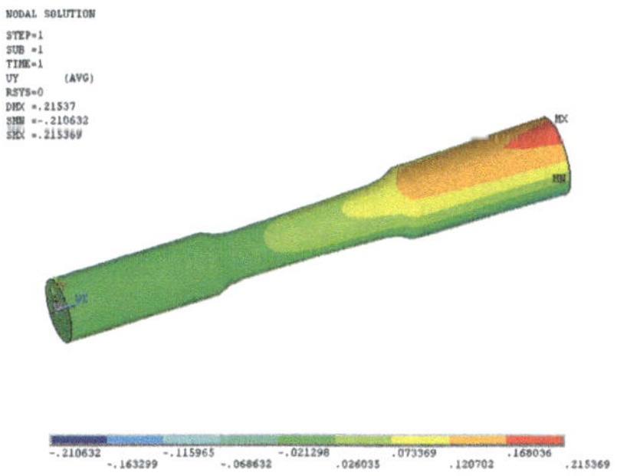

Fig. 4. Strain analysis cloud

6 Test Verification

According to the theoretical calculation results, the torsion bar structure and related parameters are suitable for the target vehicle, and the EPS system is developed. Based on the EPS test specification, the system function and durability tests are done, and the test results meet the requirements.

7 Conclusions

Based on the specific requirements of the vehicle, a new torsion bar structure without external spline is designed, and the geometric model and finite element model on EPS sub-assembly is established, and then the contact nonlinear analysis is done. The analysis results meet the requirements of application.

According to the results of the design, the EPS sub-assembly samples are built and are verified by the function and durability tests, and the test results are passed.

References

1. Lee, B., Kim, Y.S., Hong, S.C., Park, I.H.: 3D FE analysis of hall effect torque sensor and shape design of its stator teeth. In: International Conference on Electrical Engineering (2009)
2. Gao, X., Wang, Y.X.: Modal analysis of electronic power steering based on finite element method. Adv. Mater. Res. **452**, 354–358 (2012)
3. Yuan, C., Meng, G., Li, Y.: Finite element analysis for the advance of power steering system based on ANSYS, Automobile Parts (2001)
4. Kim, J.H., Song, J.B.: Control logic for an electrical power steering system using assist motor. Mechatronics **12**(3), 447–459 (2002)
5. Song, J., Boo, K., Seobkim, H., Lee, J.: Model development and control methodology of a new electrical power steering system. Drive Syst. Tech. **218**(9), 967–975 (2006)
6. Liu, M., Li, G., Li, Y.: Contact analysis between cam and roller of inner-displaced indexing mechanism based on FEM. J. Mach. Des. **32**(4), 14–18 (2015)
7. Lv, Q.T., Xue, S., Zhu, W.Q, et al.: Pull-rod bent focusing mirror subassembly engineering analysis based on contact nonlinear analysis method. SPIE, 7385:73850E_1-73850E_9 (2009)
8. Mahmoud, F.F., El-Shafei, A.G., Attia, M.A., Pahman, A.A.A.: Analysis of quasistatic frictional contact problems in nonlinear viscoelasticity with large deformations. Int. J. Mech. Sci. **66**(4), 109–119 (2013)
9. Ghio, M., Contini, P., Ubezio, G., Ansaldi, F., et al.: Node sampling for nonlinear vibration analysis of structures with intermittent contact. AIAA J. **48**(9), 1903–1915 (2010)
10. Gottlieb, M., Habib, G.: Non-linear model-based estimation of quadratic and cubic damping mechanism governing the dynamics of a chaotic spherical pendulum. J. Vib. Control **18**(4), 536–547 (2012)
11. Lv, Q.T., Xue, S., Peng, Z.Q., Zhu, W.Q., Fu, Y.: Comparison of linear and contact non-linearity method for rod-beam compacting focusing mirror assembly. Opt. Precis. Eng. **18** (8), 1842–1848 (2010)

Pose Error Alignment with On-machine Measurement for Grinding Process of the Aspheric Mirrors

Jianpu Xi(✉), Qing Yang, and Yadong Li

Zhongyuan University of Technology,
No. 41, Zhongyuan Road, Zhengzhou, China
xjpyq2010@163.com

Abstract. In this paper, the evaluation mathematic model of the characteristic parameters of aspheric surface was built based on abstract the data of profile through cross-section's measurement of aspheric surface. The center's coordinates of each measuring section's least square circle is calculated according to the least square method used in the evaluation of roundness for the measuring data of each measuring section, then the least square axis' parameters of aspheric surface are determined based on the center's coordinates of all measuring section least square circles, and the part coordinate system is built. Based on it, through parallel movement and rotary transformation, sampling data coordinates were transformed into the measuring coordinate system, the unknown parameters of aspheric surface are determined by using nonlinear least square iterative method and the linearization processing.

Keywords: Cylindrical coordinate · Aspheric mirror · Profile error
Pose error

1 Introduction

With the development of ultra-precision optical fabrication and measurement technology, the large-diameter aspheric optical components are being used more and more widely in modern high technology regions: astronomical optics, space optics, detection and high power laser fusion devises et al. [1–3]. Because the component can correct aberrations, enhance the image quality, enlarge the field of view, reduce the weight of optical systems. Therefore, an optical system using aspheric elements has the potential to be smaller than an equivalent optical system with spherical lenses. Applying aspheric allows for elimination of these aberrations, combined with many other advantages. So their application requires new techniques for optical aspheric manufacturing and measuring [4]. Brittle and hard materials are generally applied to the large aspheric mirrors, and due to surface shape and manufacturing difficulty, the sub micrometer profile accuracy will be achieved by the ultra-precision NC grinding and polishing process successively [5, 6]. The grinding stage is a key the technology of processing for large-diameter aspheric mirrors, through measuring, re-processing and re-measuring to meet the profile accuracy requirements [7].

K. Deng et al. (Eds.): ICMIR 2018, AISC 856, pp. 873–880, 2019.
https://doi.org/10.1007/978-3-030-00214-5_108

As the key measurement method of aspheric for grinding stage, the accuracy of coordinate measurement is the key factor affecting the manufacturing accuracy and efficiency. To achieve high profile accuracy of aspheric surface during ultra-precision grinding process has always been challenging as many factors can affect the evaluating accuracy of the profile error. The measurement method of profile error include on-machine and off-line measurement, because the on-machine measurement accuracy being limited by the precision of machine tool, so in order to improve the measurement accuracy and reduce manufacturing time for large mirrors, the off-line measurement method was developed by some scholars [8, 9]. The profile accuracy of the machined aspheric surface is seriously affected by the residual form errors. These errors might be caused by sampling path, sampling points and evaluating method [10].

In this paper, a new evaluating measurement method, which can be used the nonlinear least-square fitting algorithm and error compensation method. A new method based on Gauss–Newton algorithm was developed to calculate the bias of estimated aspheric surface parameters, i.e., the radius, conic constant, polynomial coefficients. The eccentricity and tilt deviation was calculated based on the analysis of the error of the measurement coordinate and workpiece coordinate, and the corresponding deviation value was computed by error analysis. The proposed measuring method and mathematical model are effective to improve profile measurement accuracy in the grinding stage.

2 Data Acquisition and Error Analysis

2.1 The Measurement Theory

The measurement process of aspheric surface was performed on a cylindrical coordinate mode with the 300-mm-diameter machine, as shown in Fig. 1. Before the measurement of aspheric workpiece, it is necessary to determine aspheric vertex coordinates, thus in the radial direction aspheric surface can be sampling through the cylindrical coordinate way. The cylindrical coordinates of measurement points including radius direction, circumferential direction and Z direction coordinate values, the radius direction coordinate value need to be determined according to the symmetry relations of aspheric surface. Firstly, the value of rotary center line was confirmed, the radial value is determined according to the movement of the X axis and the Y direction, and the position of the measuring starting point is determined. After determining the starting position of the measurement, the rotation axis is adopted to determine the circumferential angle. According to the principle of roundness measurement, the data of circle were sampled along the radial direction of aspheric surface. After measuring a cross section, the position of the X axis is moved, and the measurement of the second section is carried out. The Z coordinate is the relative movement of the digital length gauge. The radius of the cylindrical coordinate moves in the X axis direction, so the coordinate points is determined by the coordinates of the center point and the amount of movement, the radial coordinate ρ_i can be shown according to the center coordinates x_c and X direction movement x_i, the coordinates of radial, ρ_{ij}, can be computed using the following formulate:

$$\rho_{ij} = x_{ij} - x_c \tag{1}$$

Where i is the ith measurement section, j is jth measurement point of the measurement section.

Fig. 1. On-machine measurement system for large-diameter aspheric grinding machine

2.2 Analysis of Measurement Eccentricity and Tilt Error

According to the measurement path of aspheric mirror, the measurement process is sampled with concentric circle mode. The ideal circle parameters, center coordinates (x_c, y_c), circle radius r_c, were determined for each section. The least square method is used to obtain the center coordinate, (x_c, y_c) and circle radius r_c, the results can be calculated as follow.

$$\begin{cases} x_c = \frac{2}{n}\sum\limits_{j=1}^{n} s_j \cos\theta_j \\ y_c = \frac{2}{n}\sum\limits_{j=1}^{n} s_j \sin\theta_j \\ r_c = \frac{1}{n}\sum\limits_{j=1}^{n} s_j \end{cases} \tag{2}$$

The eccentricity and tilt error is caused by coincide of workpiece coordinates and measurement coordinates, so the corresponding relationship is established between the workpiece coordinates and measurement coordinates. Firstly, the angle is calculated between them in order to translation the data in coordinates. It is assumed that the angle are α, β, γ between the ideal axis line and x, y, z coordinates axis, let $p = \cos\alpha, q = \cos\beta, w = \cos\gamma$, so $p^2 + q^2 + w^2 = 1$. The axis line as shown:

$$\begin{cases} x_a = x_0 + pt \\ y_a = y_0 + qt \\ z = z_0 + wt \end{cases} \tag{3}$$

The parameters x_0, y_0, z_0, p, q, w, can be calculated through the least square method, when the axis line was determined, the least square line was used to instead the aspheric ideal axis line, so the workpiece coordinate system is the ideal axis as the Z axis, the axis direction was determined by the parameter, p, q, w, which can build the workpiece coordinates. The center point of measurement coordinates and workpiece coordinates can be obtained according to the known parameters, x_0 and y_0.

The workpiece coordinates is rotated and translated to the measurement coordinates as the least square axis line, the rotation angle, θ_x, θ_y, can be obtained from the point of the least square axis line. The parameters of Eq. (3) had been determined, with the known value of the spatial axis line point $x_0 = 0$, $y_0 = 0$, $t = 1$, the coordinates point (x_a, y_a, z), which would be rotated and translated to the measurement coordinates as coordinate point, $(0, 0, z')$, then the rotation and translation process is completed. When the z is positive value, the x, y value plus and minus is only considered. According the rotation matrix, θ_x, θ_y, can be acquired for the rotation angle of workpiece coordinate and measurement coordinate system (Fig. 2).

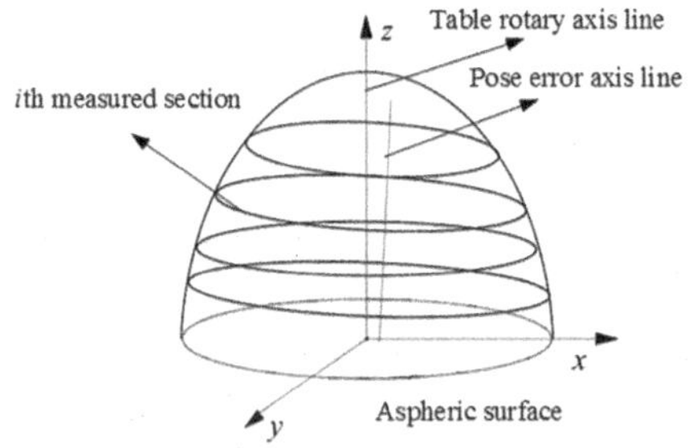

Fig. 2. The chart of pose error

2.3 The Sampling Data Processing

The measurement of aspheric mirror was performed on the off-line device, assuming the measured cross section is m, the measured point of each section is n, the measured coordinate is $(\rho_{ij}, \theta_{ij}, z_{ij})$, which will convert to the rectangular coordinate system, (x_{ij}, y_{ij}, z_{ij}). Because the measurement of each section is an equal angle, so the translation relationship as follows:

$$\begin{cases} x_{ij} = \rho_{ij}\cos((j-1)2\pi/n) \\ y_{ij} = \rho_{ij}\sin((j-1)2\pi/n) \end{cases} \tag{4}$$

Where $z_{ij} = z_i$. The measured data can not be directly fitted, and coordinate need to translated to the measurement coordinate. Owing to the eccentricity, e is existed between the workpiece and measurement coordinates, and the Eccentric coordinates is $e(x_0, y_0)$. Before rotating the coordinate need to translation, the new coordinate points, x'_{ij}, y'_{ij}, can be obtained, the translation relationship as follows:

$$\begin{cases} x'_{ij} = x_{ij} - x_0 \\ y'_{ij} = y_{ij} - y_0 \\ z'_{ij} = z_{ij} \end{cases} \tag{5}$$

Rotation give the new coordinate, $(x''_{ij}, y''_{ij}, z''_{ij})$:

$$\begin{bmatrix} x''_{ij} & y''_{ij} & z''_{ij} \end{bmatrix} = \begin{bmatrix} x'_{ij} & y'_{ij} & z'_{ij} \end{bmatrix} \times R_x R_y \tag{6}$$

where $R_x R_y$ is the rotation matrix product. If the rotation angle is positive value, simplification, coordinate point as follow:

$$\begin{cases} x''_{ij} = x'_{ij}\cos(\theta_y) + y'_{ij}\sin(\theta_x)\sin(\theta_y) + z'_{ij}\cos(\theta_x)\sin(\theta_y) \\ y''_{ij} = y'_{ij}\cos(\theta_x) - z'_{ij}\sin(\theta_x) \\ z''_{ij} = -x'_{ij}\sin(\theta_y) + y'_{ij}\sin(\theta_x)\cos(\theta_y) + z'_{ij}\cos(\theta_x)\cos(\theta_y) \end{cases} \tag{7}$$

After rotation and translation the polar coordinate as follow:

$$\rho''_{ij} = \sqrt{x''^2_{ij} + y''^2_{ij}} \tag{8}$$

3 The Evaluation Model of Aspheric Parameters

3.1 The Parameter of Aspheric Fitted by the Nonlinear Least Square Method

Rotationally symmetric aspheric surface include two shapes: concave and convex surface, the aspheric equation is at the origin of the peak, so the measurement data is the spatial coordinate point. The machining error of aspheric surface will be subject to the Z direction deviation, in order to get the contour deviation, and need to fitting aspheric surface, the deviation can be analysis between the fitting form and the measurement data.

Assuming the measured cross section is m, the measured point of each section is n, so the ith section and jth point coordinate is (s_{ij}, z_{ij}). For all point of each section, the z coordinate is equal, $z_{ij} = z_i$. The parameter of aspheric surface was fitting using the least square method through let the shortest distance for z axis. The conic aspheric surface was analyzed as example, the aim function as follow:

$$F(k, r) = \sum_{i=1}^{m}\sum_{j=1}^{n}(z_{ij} - (-r + \sqrt{r^2 - ks^2_{ij}})/k)^2 \tag{9}$$

Where $k = 1 + k$. Using the least square method and letting the deviation minimum:

$$F(k,r,h) = \sum_{i=1}^{m}\sum_{j=1}^{n}(z_{ij} - (-r + \sqrt{r^2 - ks_{ij}^2})/k)^2 = \min \tag{10}$$

From Eq. (10), the function is a nonlinear least square question, it is not use the linear least square to deal with the function, and the parameter k, r can be calculated using the nonlinear least square method.

The Gauss-Newton iteration approach is used to solve the non-linear equations, the objective function as follows [11]:

$$f_{ij}(k,r) = z_{ij} - ((-r + \sqrt{r^2 - ks_{ij}^2})/k) \tag{11}$$

can be obtained according to Gauss-Newton can be described as follow:

$$\begin{bmatrix} \sum_{i=1}^{m}\sum_{j=1}^{n}(\frac{\partial f_{ij}}{\partial k})^2 & \sum_{i=1}^{m}\sum_{j=1}^{n}\frac{\partial f_{ij}}{\partial k}\frac{\partial f_{ij}}{\partial r} \\ \sum_{i=1}^{m}\sum_{j=1}^{n}\frac{\partial f_{ij}}{\partial k}\frac{\partial f_{ij}}{\partial r} & \sum_{i=1}^{m}\sum_{j=1}^{n}(\frac{\partial f_{ij}}{\partial r})^2 \end{bmatrix}\begin{bmatrix} k \\ r \end{bmatrix} = -\begin{bmatrix} \sum_{i=1}^{m}\sum_{j=1}^{n}\frac{\partial f_{ij}}{\partial k}f_{ij} \\ \sum_{i=1}^{m}\sum_{j=1}^{n}\frac{\partial f_{ij}}{\partial r}f_{ij} \end{bmatrix} \tag{12}$$

The parameters k, r can be determined through solve the equation.

3.2 The Model of Aspheric Residual Error

According to the sampling data and fitting method, the surface data was obtained and fitted through cylindrical coordinate system path for the grinding aspheric mirror, which will acquire the ideal aspheric surface parameter. The residual error is defined as the deviation of aspheric surface along the symmetric axis direction, so that the residual error which can be used to compensation grinding the aspheric between measurement point and ideal aspheric point was calculated. So, it is necessary to optimize the results in the calculation process. The measurement coordinate, $S(s_{ij}, z_{ij})$, the ideal aspheric coordinate, $S1(s1_{ij}, z1_{ij})$. The residual error ΔE, can be computed using the following formulate:

$$\Delta E = z_{ij} - z1_{ij} \tag{13}$$

Where z_{ij} is the X and Y direction coordinate points, x_{ij} and y_{ij}. $Z1_{ij}$ is the X and Y direction coordinate points, $x1_{ij}$ and $y1_{ij}$. In order to accurately calculate the residual error in the Z direction, that need to meet $x_{ij} = x1_{ij}$, $y_{ij} = y1_{ij}$. The residual error can be obtained owe to the ideal aspheric equation, the described as follow:

$$\Delta E = \left| z_{ij} - cs_{ij}^2/(1 + \sqrt{1 - (1+k)c^2 s_{ij}^2}) \right| \tag{14}$$

where $s_{ij} = (x_{ij}^2 + y_{ij}^2)^{1/2}$.

The process have two steps: firstly, the parameter, k, c, can be estimated through the nonlinear least square method, then the residual error can be obtained according the Z direction deviation.

4 Conclusions

In this paper, the pose error alignment process is conducted according the cross-section's measurement data of aspheric surface through the least square method. To improve the machining accuracy and efficiency as well as solve the problem of profile measurement and evaluating in the grinding stage, high precision profile measurement solution for large-diameter aspheric was proposed based on the cylindrical coordinate measurement mode for on-machine measurement with contact probe. Firstly, the path planning is carried out for the coordinate measurement on the sampling method, the eccentricity and tilt deviation was calculated based on the analysis of the error of the measurement coordinate and workpiece coordinate, and the corresponding deviation was obtained, the influence of deviation was eliminated to improve the measurement accuracy according to the coordinate transformation theory. the un-known parameters of aspheric surface are determined by using nonlinear least square iterative method and the linearization processing.

Acknowledgements. This project is supported by National Natural Science Foundation of China (Grant No. 51705546) and the Henan Colleges and Universities Key Research Projects Foud (Grant No. 18A460037).

References

1. Shore, P., Cunningham, C., DeBra, D., Evans, C., Hough, J., Gilmozzi, R., Kunzmann, H., Morantz, P., Tonnellier, X.: Precision engineering for astronomy and gravity science. CIRP Ann. **59**(4), 694–716 (2010)
2. Komanduri, R., Lucca, D.A., Tani, Y.: Technological advances in fine abrasive process. CIRP Ann. **46**(2), 545–596 (1997)
3. Brinksmeier, E., Mutlugünesa, Y., Klocke, F., Aurich, J.C., Shore, P., Ohmori, H.: Ultra-precision grinding. CIRP Ann. **59**(2), 652–671 (2010)
4. Leadbeater, P.B., Clarke, M., Wills-Moren, W.J., Wilsonf, T.J.: A unique machine for grinding large, off-axis optical components: the OAGM 2500. Precis. Eng. **11**(4), 191–196 (1989)
5. Comley, P., Morantz, P., Shore, P., Tonnellier, X.: Grinding meter scale mirror segments for the E-ELT ground based telescope. CIRP Ann. **60**(1), 379–382 (2011)
6. Lin, X.H., Wang, Z.Z., Guo, Y.B., Peng, Y.F., Hu, C.L.: Research on the error analysis and compensation for the precision grinding of large aspheric mirror surface. Int. J. Adv. Manuf. Technol. **71**(1–4), 233–239 (2014)
7. Huang, H., Chen, W.K., Kuriyagawa, T.: Profile error compensation approaches for parallel nanogrinding of aspherical mould inserts. Int. J. Mach. Tools Manuf. **47**(15), 2237–2245 (2007)

8. Chen, F.J., Yin, S.H., Ohmori, H., Yu, J.W.: Form error compensation in single-point inclined axis nanogrinding for small aspheric insert. Int. J. Adv. Manuf. Technol. **65**(1–4), 433–441 (2013)
9. Lee, E.S.: A study of the development of an ultraprecision grinding system for mirror-like grinding. Int. J. Adv. Manuf. Technol. **16**(1), 1–9 (2000)
10. Wegener, K., Hoffmeister, H.-W., Karpuschewski, B., Kuster, F., Hahmann, W.-C., Rabiey, M.: Conditioning and monitoring of grinding wheels. CIRP Ann. **60**(2), 757–777 (2011)
11. Sun, W., Hill, M., Mcbride, J.W.: An investigation of the robustness of the nonlinear least-squares fitting method to small segment angle surfaces. Precis. Eng. **32**(1), 55–62 (2008)

Taxi Carpool Technology and Its Algorithm Implementation

Shao Xinye[1], Wang Ruixiang[2], Li Jianlan[3], and Shao Jianlong[1(✉)]

[1] School of Information Engineering and Automation, Kunming University of Science and Technology, Kunming 650500, People's Republic of China
sj-long@163.com

[2] Xian Uniview Information Technologies Co., Ltd., Xi'an 710100, People's Republic of China

[3] Yunnan Vocational College of Mechanical and Electrical Technology, Kunming 650203, People's Republic of China

Abstract. In response to the current domestic and international traffic conditions and the status of taxi operations, such as the difficulty of taxiing, refusal to load passengers, flooding of illegal taxi, and urban traffic jams, a kind of multi-window taximeter system based on Windows CE embedded operating system is designed and simultaneous metering, timing and billing algorithms for multiple groups of passengers proposed. The ARM9 is used as the control core of the system. The application software is a kind of Windows visualization program, developed in the environment of Microsoft Visual Studio 2005.

The system includes the carpool meter system, the display control system and the unlock tool. The carpool meter system can set different price parameters according to different regions, and it can charge preferentially for multiple passengers at the same time according to the parameters and the mileage pulse received per second, and prints the paper through the printer. The wireless inspection module identifies whether a taxi is a legitimate vehicle, and sends the traffic information to the relevant management. In the display control system interface can be handwritten to input the name of the destination. By pressing the button to select passenger information available in the car, then the LED display strip screen on the top of the taxi will display the passenger information. At the same time, it has the function of robbing the alarm to warn the outsider to help the driver to call the police. Also, unlock tool is designed for the carpool meter system.

The preliminary application results show that the multi-window taximeter that meets the design requirements is very convenient. It can fundamentally solve the difficulty of taking taxi, rejection to passenger, the spread of the illegal car, and it provides a solution to solve the problem of city traffic pressure and taxi reasonable operation etc.

Keywords: Windows CE · ARM · Carpool meter · Display control

K. Deng et al. (Eds.): ICMIR 2018, AISC 856, pp. 881–888, 2019.
https://doi.org/10.1007/978-3-030-00214-5_109

1 Introduction

As the pursuit of quality of life, comfort, convenience and speed increasingly enjoys popular support, more and more people would prefer to take a taxi. Additionally, the price of taxis has been frozen for nearly twenty years in China, so the cost has become relatively cheap, making more ordinary people join in the ranks of the taxi. As more people prefer to take a taxi, and the actual traffic conditions are not optimistic, so that at some time passengers simply cannot hit the car, so the driver could choose who he wants to take. And that leads to pick off, bargaining, rejection and other non-standard business practices. The unruly chaos caused great dissatisfaction with the passengers. There were a lot of complaints, and the management department felt a lot of pressure. Traffic management departments feel there are little they can do about the emergency of unlicensed taxi and their growing trend. And this leads to the driver suffer dual pressure from the passenger and the traffic administration. In this case, it is easier to make passengers dissatisfied. This is a vicious circle that has a bad effect on all parties.

Under such conditions, taxi carpooling came into being. At present, the carpool is only in a small number of cities, and the taxi driver can carry out the carpool when the passengers are allowed, and the price should be discussed in advance. And new problems arise increasingly, such as inconvenient communication between drivers and passengers and between occupants, meanwhile it is unable to print invoices for each passenger, what's worse, it have inconsistent, unreasonable and opaque charging, etc. All of those problems restrict the pace of the taxi-pooling scientific operation. In order to solve the those problems, and facilitating the public travel, reducing the taxi driver working pressure, improving the service efficiency of the taxi, saving energy and reducing the pressure on the urban traffic problems, this paper puts forward a set of design scheme - the multi-window taximeter system based on Windows CE embedded operating system, which can charge preferentially for multiple passengers at the same time and prints invoices for every passenger, according to the requirements of the traffic department, meanwhile regulates the charge standard of the taxi-pooling [1].

2 The Overall Design Scheme of the Multi-window Taximeter System

The multi-window taximeter presented in this paper includes the display control system, the carpool meter system and the unlock tool. Both the display control system and the carpool meter system are independent of each other, and can be used alone, and both of them include the hardware and the software. Unlock tool is designed for the carpool meter, just a software, used by the traffic management. The hardware of the carpool meter includes the LJD-eWin7 touch controller, the printer, the wireless inspection and the signal acquisition circuit. The hardware of the display control system includes the LJD - eWin6 touch controller, the LED display and the alarm signal detection circuit. Both the software parts of the taxi-pooling system and the display system include the Windows visualization program and the single-chip microcomputer program. The Windows visualization program is based on message queue, realized by using MFC. When a function key is pressed, it will execute the corresponding function [2].

The entire program design block diagram shown as Fig. 1.

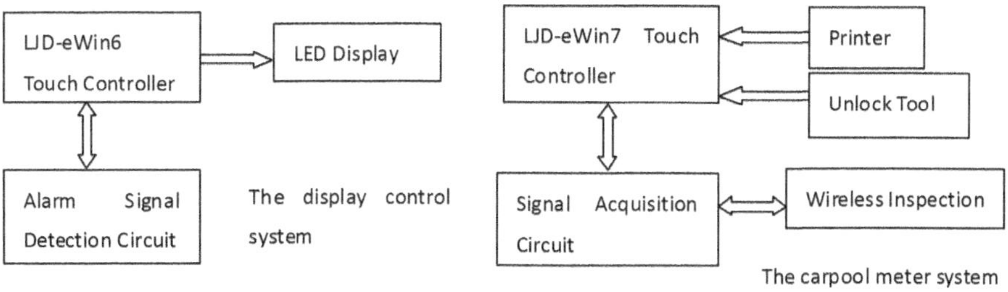

Fig. 1. The entire design scheme

3 The Hardware Design

3.1 The Design of the Carpool Meter System

3.1.1 The Control Core Design of the Carpool Meter System

The system adopts the LJD-eWin7 touch controller which has the Windows CE operating system as the control core. The Windows application program of this system is developed in the Microsoft Visual Studio 2005. And it can charge for multiple passengers at the same time.

The LJD-eWin7 touch controller uses SAMSUNG S3C2440-40 as the main control chip, the main frequency reaches 400 MHz, which can quickly respond to the request of the external equipment and improve the system running speed. Two RS232 serial ports can easily communicate with the printer and signal acquisition circuit to ensure the normal work of all parts of the system. The capacity of the FRAM is 8 KB, which can write data at the bus speed. Not only the writing number is almost infinite, but the data also won't be lost, so it can store the taxi information, the billing information and the passenger's details safely. The 4-wire high-precision resistive touch screen can be clicked 1 million times, makes it convenient to operate the Windows visualization program [3].

3.1.2 The Design of the Signal Acquisition Circuit

Signal acquisition circuit using STC11F04E microcontroller as the control center, is responsible for collecting and processing mileage pulse signal and detecting printer ticket black spot position, and LJD-eWin7 touch controller connected to the RS232 serial port COM1. The LJD-eWin7 touch controller receives the mileage pulse number which is collected by the signal acquisition circuit per second, and counts mileage and waiting-time according to the mileage pulse number.

It needs to detect the black spot's position of the paper when prints the bill. When the black spot is detected, the message will be sent to the LJD-eWin7 touch controller through the serial port, and then the LJD-eWin7 touch controller will send the instruction of printing the passenger's information to the printer through the RS232 serial port COM2. After receiving the instruction, the printer will print the passenger's information correctly, make sure the information and the data are corresponding. Before this, the printer has been receiving the paper instruction, and the printer will print the passenger's information directly if the black spot haven't been detected after 40 s, on this occasion, the information and data of the bill may not correspond, but it increases the fault tolerance of the system.

3.1.3 The Design of Wireless Inspection

Wireless inspection takes NRF24L01 as the core of control, and its SPI interface can be connected to the I/O port of low cost MCU. The transmission rate in the air is up to 2Mbps and the transmission time is very short, which greatly reduces the collision phenomenon in wireless transmission [4].

The wireless inspection will be installed on the top of the taxi equipped with the carpool meter system. When the legal taxi passes by the inspective car, the receiver of the NRF24L01 will receive the instruction of inquiring the license plate number from the inspective car, and then it will send the information to the microcontroller through the SPI interface, after that the microcontroller will read the license plate number information that is stored in the LJD-eWin7 through the RS232 serial port COM1, and sent the information to the NRF24L01 transmitter through the SPI interface. Then the information will be sent to the inspective car, which can judges whether the taxi is legal or not according to received information.

When the taxi passes through the traffic management, the receiver NRF24L01 will receive the instruction of checking traffic information and then send the information to the microcontroller through the SPI interface. After received the information, the microcontroller will read the traffic information that is stored in the LJD-eWin7 through RS232 serial port COM1, and send the traffic information to the NRF24L01 transmitter through the SPI interface. Finally, the traffic information will be sent to the traffic management for storage.

3.2 The Design of the Display Control System

3.2.1 The Control Core Design of the Display Control System

This system uses the LJD-eWin6 touch controller as the control core. Its hardware configuration and the development of the application program are the same as the LJD-eWin7 touch controller expect for no FRAM, the different board size and the distribution of the serial port [5].

According to the design requirements of this part, it doesn't need to store some important information, so the FRAM is removed to save cost. This system can receive the alarm signal coming from the alarm signal detection circuit and control the LED to display the passenger's message and the robbery alarm information.

3.2.2 The Design of the Alarm Signal Detection Circuit

The alarm signal detection circuit uses the STC15F104E single chip computer as the control core, and is connected with the RS232 serial port COM1 of the LJD-eWin6 touch controller. Meanwhile, it can realize the robbery alarm function. When the taxi which has the display control system is robbed, the driver can touch the alarm switch and maintain 10 s or more while the robbers are unwitting. When the microcontroller detects the signal, it will send information to the LJD-eWin6 touch controller through the serial port. When the LJD-eWin6 touch controller received the alarm information, it will send instruction to the LED display through the RS485 serial port COM3, the text of "robbed, please call the police!" will be displayed on the LED screen. At the same time, the LJD-eWin6 touch controller will return the information to the STC15F104E microcontroller through the COM1.By the time the STC15F104E microcontroller receives the returned information, the LED inside the car will be lighted to notice the driver that the alarm is successful. Meanwhile, the alarm at the roof will begin to flash, which can further warning the outsiders to call the police.

4 The Software Design

4.1 The Application Program of Carpool Meter

4.1.1 The Diagram of Application Structure

Carpool meter system application program can set charge parameters, vehicle parameters and other information. It can also display, print and save passenger traffic information, detect taxi driving information, etc. The structure diagram of the program is shown as Fig. 2, and the main operating interface is shown as Fig. 3.

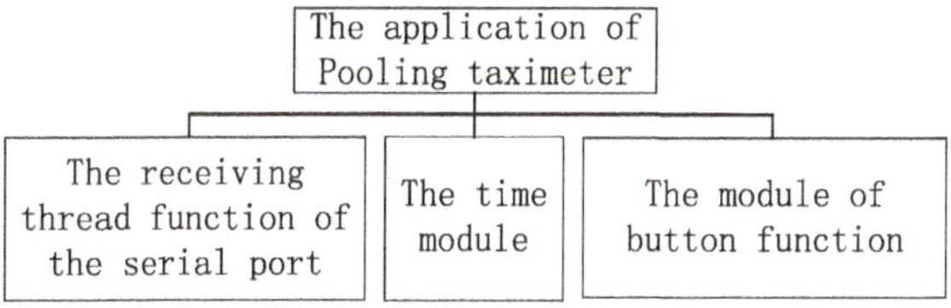

Fig. 2. The application structure diagram of the carpool meter system

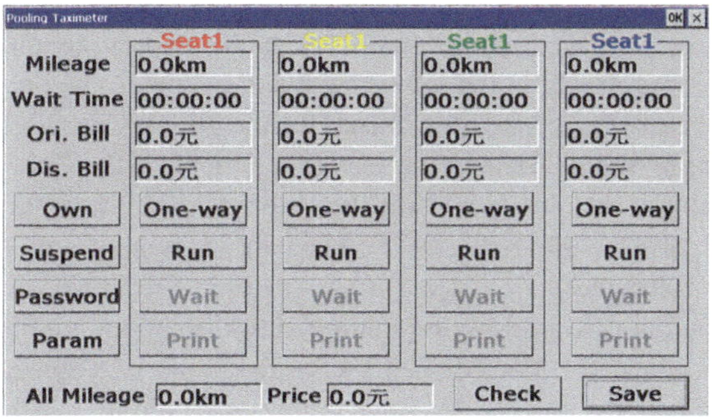

Fig. 3. The main operating interface of the carpool meter system

Every passenger on the seat has four buttons, including "One-way", "Run", "Wait" and "Print". When the passenger gets off the taxi, and if the passenger doesn't need the invoice, then the driver will click the "Wait" button, so the passenger's information will only be saved without printing. It can ensure that a passenger's information can only be printed and stored once.

If the passenger doesn't want to carpool after getting on the taxi, the driver can click the "Own" button. At this time, the charge is more expensive than usual. If the passenger agrees to carpool with others, there will be a big discount, including the starting price, the pooling mileage and the waiting time, and they all have different discount according to the passenger's number. This kind of charge rule reflects the idea that the more passengers, the more discount.

When click the "Param" button, there will appear another interface shown as the Fig. 4. The parameter in the interface can only be set by the management. In this way, the charge is transparent, and there will be no bargaining with the passengers any more.

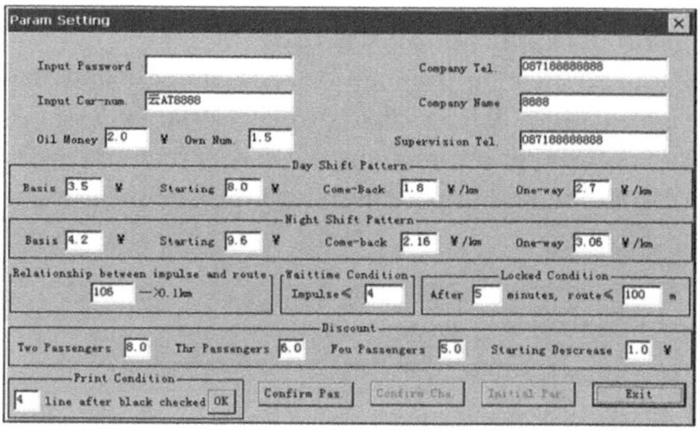

Fig. 4. The charge parameter and the taxi information initialed by the system

4.1.2 The Implementation of the Charging Algorithm

The passenger is mobile while the seat on the taxi is fixed so that the seat can be used instead of the passenger. Charging for the seat begins when the passenger sits on the seat. So it is easily to charge preferentially for multiple passengers at the same time. Because all passengers travel in the same shape and behavior, the passengers sitting on a fixed seat can be abstracted as four objects of the BillingClass, set up separately as Seat1, Seat2, Seat3, Seat4. Meanwhile, set four flag bits, flagup1, flagup2, flagup3, flagup4 as the public variables outside of the BillingClass. When the passenger sits on the seat, set the corresponding flag as 1; when the passenger gets off, set the corresponding flag as 0 [6].

CountWatitimeFun, CountRouteFun and CountMoneyFun are the public members of the BillingClass, and can be accessed by the four objects. Then calculates the each part cost, the original cost and the discounted cost of each object. After calculation, the original cost and the discounted cost will be displayed in the corresponding edit box.

The calculation formula of the discounted cost shown as Eq. (1):

$$DB = DP + \sum_{i=1}^{4}\left(\frac{DT_i}{150} + \frac{DR_i}{0.5}\right) * \frac{DDP1}{2} + \sum_{i=1}^{4}\frac{DOR_i}{0.5} * \frac{DDP2}{2} + \sum_{i=1}^{4}\left(\frac{NT_i}{150} + \frac{NR_i}{0.5}\right) * \frac{NDP1}{2} + \sum_{i=1}^{4}\frac{NOR_i}{0.5} * \frac{NDP2}{2} \quad (1)$$

In the Eq. (1), the *DB* is the discounted cost. The *DP* is the discounted basic price of the day or night according when the passengers get on. The *DDP1* is the discounted come-back price of the day shift. The *DDP2* is the discounted one-way price of the day shift. The *NDP1* is the discounted come-back price of the night shift. The *NDP2* is the discounted one-way price of the night shift. The subscript i can be 1, 2, 3, 4 and respectively represents one, two, three and four person carpooling. The DT_i represents respectively each part of the waiting time in the day shift. The DR_i represents respectively each part of the come-back mileage in the day shift. The DOR_i represents respectively each part of the one-way mileage beyond 10 km in the day shift. The NT_i represents respectively each part of the waiting time in the night shift. The NR_i represents respectively each part of the come-back mileage in the night shift. The *NORi* represents respectively each part of the one-way mileage beyond 10 km in the night shift.

The calculation formula of the original cost shown as Eq. (2):

$$B = P + \left(\frac{DT}{150} + \frac{DR}{0.5}\right) * \frac{DP1}{2} + \frac{DOR}{0.5} * \frac{DP2}{2} + \left(\frac{NT}{150} + \frac{NR}{0.5}\right) * \frac{NP1}{2} + \frac{NOR}{0.5} * \frac{NP2}{2} \quad (2)$$

In the Eq. (2), the *B* is the original cost. The *P* is the basic price of the day shift or nightshift according when the passengers get on. The *DP*1 is the come-back price of the day shift. The *DP*2 is the one-way price of the day shift. The *NP*1 is the come-back price of the night shift. The *NP*2 is the one-way price of the night shift. The *DT* is the waiting time of the day shift. The *DR* is the come-back passenger mileage of the day shift. The *DOR* is the one-way passenger mileage beyond 10 km of the day shift. The *NT* is the waiting time of the night shift. The *NR* is the come-back passenger mileage of the night shift. The *NOR* is the one-way passenger mileage beyond 10 km of the night shift.

4.2 The Application Design of Display Control System

The application of the display control system receives the alarm signal, and in the operation interface the passenger's information can be inputted and selected. It can also control the LED display screen to display. The structure diagram of the application is shown as the Fig. 5, and the main operation interface is shown as the Fig. 6.

In the display control system application interface, enter the name of the destination, handwriting input or pinyin input are both available. Quickly select the rest seat of the taxi by button, then click the "Display" button to have the passenger's information shown on the LED screen, makes it very convenient for passengers to choose the suitable taxi. The names of places where the passenger want to be and passing by can be exchanged by operating the key.

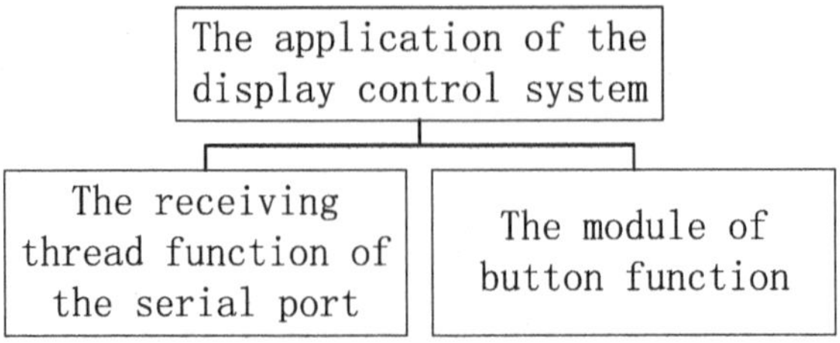

Fig. 5. The application structure diagram of the display control system

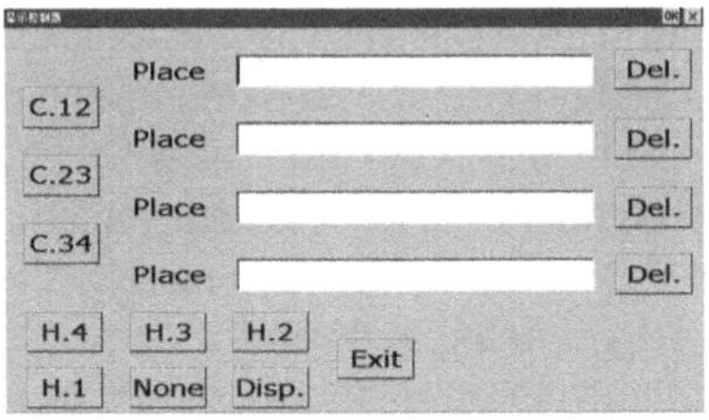

Fig. 6. The main operation interface of the application

5 The Conclusion

The multi-window taximeter system provides a solution to solve the problems, such as the urban traffic congestion, too much illegal taxi, difficulty of taking taxi, rejection to passengers and the unregulated carpooling. It not only mobilizes the enthusiasm of the driver by allowing drivers to get more income through carpooling, but the passenger can also benefit from carpooling. What's more, it is one of the best ways to make the society to be "resource-saving and environment-friendly", and to benefit our generations to come.

Acknowledgments. This work was supported by the Education Technology Research Project of Kunming University of Science and Technology [No. 2506100219].

References

1. Wang, Y.: The design and implementation of the intelligent taxi system based on the ARM. Xi'an University of Electronic and Engineering, Xi'an (2011)
2. Jin, Y: The multi-window taximeter. China. 201120141740.9. 2011.10
3. Beijing Blue Ocean Microchip Co. Ltd.: The LJD-eWin7 user manual (2011)
4. Shenzhen YunJia Technology Co. Ltd.: The NRF24L01 user manual about the embedded wireless transmission module specification (2012)
5. Beijing Blue Ocean Microchip Co. Ltd.: The LJD-eWin6 user manual (2011)
6. Zhang, S.: The C++ Language Program Design and Practice. Tsinghua University Press, Beijing (2010)

Speech Recognition Method Based on Spectrogram

Yingying Li(✉), Siyuan Pi, and Nanfeng Xiao

School of Computer Science and Engineering,
South China University of Technology, Guangzhou 510006, China
Crystaliyy@foxmail.com

Abstract. Deep learning makes a great breakthrough in the field of artificial intelligence. Currently, the robustness of the speech recognition in time domain performs poorly, and the spectrogram complexity of the speech recognition in frequency domain also needs to be reduced greatly. Therefore, this paper presents a faster R-CNN-based target detection method to recognize the spectrogram for the speech recognition in the time and frequency domain. The presented method only focuses on the local interest regions (obvious voiceprint) of the spectrogram, which filters the high frequency noise to improve performance. The experimental results show that the presented method has higher accuracy and robustness than existing methods, and which can perform well evenly in the noisy factory.

Keywords: Speech recognition · Spectrogram · Target detection
Faster R-CNN

1 Introduction

Gaussian Mixture Model-Hidden Markov Model (GMM-HMM) has been dominant in automatic speech recognition (ASR) [1]. Even today, many speech recognition systems still use HMM to process the speech temporary variables and adopt GMM to confirm the states produced by HMM [2]. While in recent years, because of the powerful feature extraction capabilities and modeling capabilities of deep neural networks (DNNs) [2], they begin to replace gradually the traditional GMM to calculate the output probability [3, 4] and combine with HMM to compose DNN-HMM.

However, the speech signals are non-stationary process with the range of time and frequency [1]. HMM models of the speech recognition systems have poor robustness, and they mainly focus on the analysis of the time dimension. Although their performances are very well in the noiseless environments, the performances are still very poor in the noisy environments. Since there are great differences between the human voices and the noises, it is a good choice in the speech recognition that recognizes the spectrogram in the time domain and the frequency domain. Generally, the spectrogram is a short time Fourier transformation (STFT) [5], the STFT must transform each original speech frame, and recognize the time domain and the frequency domain by convolutional neural network (CNN) [6], such as deep convolutional neural network (DCNN) and deep fully convolutional neural network (DFCNN) [7, 8].

K. Deng et al. (Eds.): ICMIR 2018, AISC 856, pp. 889–897, 2019.
https://doi.org/10.1007/978-3-030-00214-5_110

This paper proposes a target detection method based on faster R-CNN [9] to recognize the spectrogram for the speech recognition, the combination of the speech recognition with the target detection is to train the boxes marked by each word within a certain frequency range in the spectrogram as the detection targets. And the experimental results also show that the speech recognition accuracy is the highest in the regions of obvious voiceprint, and different people speak the same words with the same voiceprint. In other words, the system corresponds to non-specific person.

2 Related Works

Standard speech recognition systems based on HMM are actually statistical pattern recognition methods based on statistics, and the systems usually include five parts: acoustic feature extraction, acoustic model, language model, decoder and post-processing fusion technology [10]. The corresponding models of the five parts are calculated according to the probability hypothesis, which are all the focus in the speech recognition, for example, the acoustic model has undergone a series of changes. For a long time, the acoustic models are based on GMM-HMM. However, due to the deep learning development [11], DNN-HMM as the acoustic model of the speech recognition system has gradually become a mainstream because of its lower word rate. Although the deep learning makes the speech recognition system based on HMM achieve better results, its complex structure is not satisfactory. As a result, the speech recognition systems based on end-to-end models [3, 4] emerged. In other words, the deep learning network implements mapping from the audio to the results.

Currently, the image recognition research focused on CNN in recent years. However, it is also found that CNN can also be learned and helpful in the speech recognition, that is to say, the speech recognition can combine with the image recognition. Since the image recognition method is used to identify the speech, firstly, it is necessary to converts the audio signal into a spectrogram by STFT for input, and then identify the spectrogram. Although CNN performs well in the image recognition, and it can also be used to identify the spectrogram, the experiments proved that the deeper CNN's convolution layer numbers, the better CNN's recognition effect. Therefore, DCNN and DFCNN become popular framework in the speech recognition.

Although these frameworks can achieve very good results, the computational complexity increases with increasing CNN's network depth. As far as the spectrogram is concerned, not all the contents of the whole spectrogram are valid information. For example, people are not talking at all times in the time domain. Also in the frequency domain, the range arrived by people's speech is also limited. In other words, in addition to the intersection of information generated by the effective time and the effective frequency is valuable, the other contents of the spectrogram are irrelevant, which is the noises.

Based on the above idea, this paper proposes the target detection method based on faster R-CNN. The proposed method only focuses on the local interest region, which filters the noises that has a great impact on the recognition performance. The effectiveness of speech recognition system will be improved in the noise environment, especially the differences between the sound frequency and the human voice frequency are very huge, such as factory operations sounds.

3 Speech Recognition Principle

The proposed target detection method can identify the spectrogram by the following two steps: (1) change the audio into the spectrogram, (2) identify the spectrogram via faster R-CNN.

3.1 Spectrogram

The speech signal generation is not a smooth process, in which the channel can be seen as a resonant cavity which is always in motion. It is generally known that the acoustic signal is approximate in a short period time (usually 10–30 ms), that is the acoustic signal has a short stability period [1]. Sampling a certain moment of the original acoustic signal and analyzing the short-term speech signal near this moment can obtain a spectrogram. Continuous sampling and analyzing acoustics, a 2D spectrogram can be obtained, which is time for the abscissa and frequency for the ordinate. In the spectrogram, the color depth represents the acoustic signal strength at that time and that frequency. From this point, the spectrogram can be seen as a means or a way to express information.

The acoustic signal analysis method can be processed in the time domain or the frequency domain, but they have their limitations: the time domain analysis is not sensitive to the frequency, and the frequency domain analysis cannot change over the time. Therefore, it is necessary to combine both analysis in the time domain and the frequency domain, which is time-frequency representation [12]. And the STFT is the most commonly used for the time-frequency analysis, it is based on the Fourier transformation (FT) [13] which is a method to analyze signal to be used to move the windows function [12] over the time domain to calculate the power spectral density function at the different moments. The time-frequency analysis method is defined [12]:

$$STFT(t,\omega) = \int_{-\infty}^{+\infty} s(t')w(t'-t)e^{(-j\omega t')}dt' \tag{1}$$

where, $s(t)$ is a raw signal, $w(t)$ is a window function, and $\omega = 2\pi f$ is an angular frequency. The meaning of Eq. (1) is to use the window function in the time domain to intercept the original signal, and the signal in the window is seen as a smooth signal. And then the local information is transformed by FT to get the spectrum at that moment. Continuous operation (moving the window over) of Eq. (1) and changing t, that is STFT. In order to facilitate the computer processing, Eq. (2) is generally discretized as [5, 14]:

$$STFT(\mathrm{n},\omega) = \sum_{m} s(m)w(m-n)e^{(-jm\omega)} \tag{2}$$

where, n is discretized and ω is continuous, that is the discrete time Fourier transformation (DTFT) which is discretized in the time domain and continuous in the frequency domain.

3.2 Faster R-CNN

Different from R-CNN [15] and fast R-CNN [16], faster R-CNN as an improved version uses a deep CNN, which is the only part in this framework. It not only improves the mAP, but also improves the calculation speed a lot with GPU. The end-to-end network combines feature extraction, classification and regression, as shown in Fig. 1.

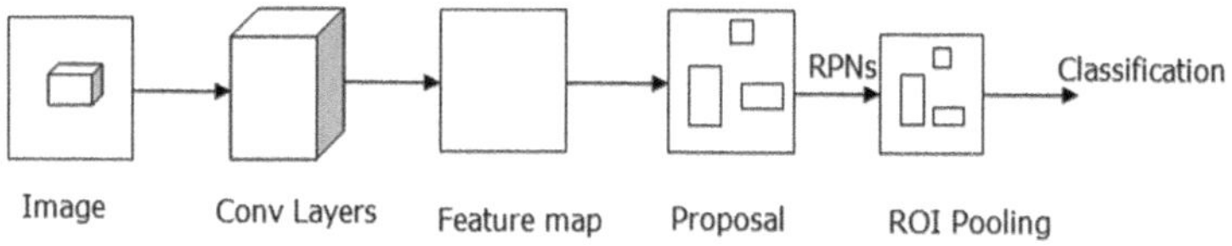

Fig. 1. Faster R-CNN framework

Faster R-CNN can be considered as a combination of region proposal network (RPN) and fast R-CNN. For RPN part, there are some concepts that should be understood.

(1) Region Proposal. It is different from the traditional sliding window algorithm. The region proposal method is to select the possible areas as pre-selected box, then detect the pre-selected areas and judge. It makes the calculation simpler because it focuses on where it should be interested.
(2) IOU. There is an important indicator for the target detection, which is intersection-over- union (IOU) [18], as expressed below:

$$IOU = (A \cap B)/(A \cup B) \tag{3}$$

where, A represents the detection box area and B represents the ground truth area shown in Fig. 2. For example, S_{DB} represents the detection box area, S_{GT} represents the ground truth area, and S_I represents the intersection of S_{DB} and S_{GT}. Equation (3) is changed as:

$$IOU = S_I/(S_{DB} + S_{GT} - S_I) \tag{4}$$

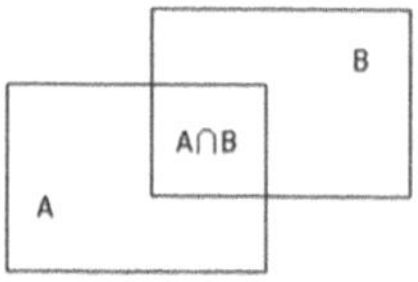

Fig. 2. IOU

(3) RPN. RPN is a convolutional neural network that trains any size input image as the feature map. The convolution results are expanded into a multidimensional vector and are sent to the two fully connected layers at the same time. One of the two fully connected layers is for classification, and the other is for boundary regression. RPN predicts the targets and their boundaries through a series of target scores. And the loss function of RPN can be seen as a rule for classifying labels. This rule is to judge whether the inside of the boundary box is the target or the background, it can be expressed as flow: when the overlap of anchors and ground truth boxes is higher than 0.7, these anchors belong to positive label (target); when it is less than 0.3, these anchors belong to negative label (background). The anchors mentioned above are multi-size and multi-scale rectangular boxes [9]. And the anchor numbers for each center is multiplied by the size numbers and the scale numbers.

For fast R-CNN part, it shares the network layer with RPN, in order to reduce computational complexity.

(1) Output layer. After the RPN, the feature maps are sent to the two fully connected layers at the same time. One of the two layers, called box-classification layer (cls), is to classify the target, and the other, called box-regression layer (reg), is to adjust the location of the region proposals [9]. For the cls layer, it's to classify the targets which are judged by RPN. Therefore the outputs of this layer are the probability array of all targets in the region proposal boxes, which are called region of interesting (ROI) [9]. If there are k classes, the length of the array is k + 1 (including background). The output array can be expressed as (p_0 is always background probability):

$$P = \{p_0, p_1, p_2, \ldots\ldots, p_k\} \tag{5}$$

For the reg layer, it is to refine the location of the region proposals. A region proposal box can be represented by a list: (x, y, w, h), where, x and y indicate the center point coordinates of the anchor, and w and h indicate the length and width of the anchor.

(2) Loss function. For faster R-CNN, the loss function shown in below [9]:

$$L(\{p_i\}, \{t_i\}) = \frac{1}{N_{cls}} \sum_i L_{cls}(p_i, p_i^*) + \lambda \frac{1}{N_{reg}} \sum_i L_{reg}(t_i, t_i^*) \tag{6}$$

where, i indicates the i-th gradient descent of the batch gradient descent, and p_i indicates the it belongs to class i. $p*_i$ is the ground truth label, when the anchor belongs to the positive label, $p*_i = 1$, and to the negative label, $p_i^* = 0$. t^i is the list of k class anchor box, and $t*_i$ is the list of the ground truth box. In addition, the loss function L_{cls} and L_{reg} is modified as follows:

$$L_{cls}(p_i, p_i^*) = -p_i \log p_i^* - (1 - p_i) \log(1 - p_i^*) \tag{7}$$

$$L_{reg}(t_i, t_i^*) = -\mathrm{t_i} \log t_i^* - (1 - t_i) \log(1 - t_i^*) \tag{8}$$

(3) Regression. This is the regression of the bounding box, whose purpose is to make the local of region proposal closer to the ground truth box. As shown in Fig. 3, the box1 refers to the proposal region, the box3 refers to the labeled ground truth, the box2 refers to the objective of the bounding box regression. The purpose is to find a way to narrow the gap between them, and the final prediction result is box2.
(4) Feature sharing. In faster R-CNN framework, RPN and fast R-CNN share the convolutional layer [9]. In other words, the two modules train alternately. This way of sharing convolution features greatly improves the training and testing speed of the network, and accelerates convergence.

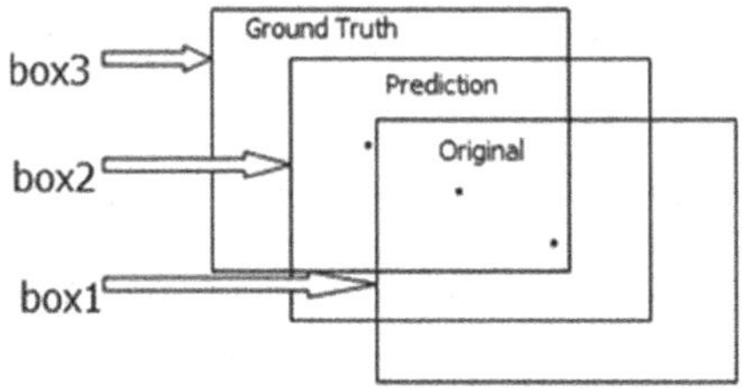

Fig. 3. Bounding Box Regression

4 Experiments

The CNN models commonly used in the deep learning are respectively LetNet, AlexNet, ZF, Google-Net, VGG and DRL (deep residual learning), and so on. Among the CNN models, the big model VGG16 is much better than other models in faster R-CNN, but it occupies too much GPU memory [9]. And ZF as a small network is very suitable for the small data sets. The experiments in this paper mainly test the two networks: the small ZF model and the middle VGG16 model.

4.1 Experiments on Kaldi Yesno

Kaldi is a powerful speech recognition toolkit, it supports a wide range of speech recognition models, including GMM-HMM and DNN-HMM. For all the data sets, yesno is a tiny data set, which is used to verify the recognizing spectrogram method via faster R-CNN primitively and quickly.

The first is to convert all yes/no audio files into the spectrogram. Among them, the super parameter is frame-length = 384, sampling-rate = 44100, Window-size = 384, overlap = 288.

The second is to recognize the spectrogram by faster R-CNN. In this data set, the ground truth box is labeled in the range 0–5000 Hz, because human speech is usually around 5000 Hz. The experiment tests on the small ZF model and the comparison

Table 1. WER: word error rate

Method	GMM-HMM	Spectrogram via fast R-CNN
WER1	6.48%	0.12%
WER2	6.39%	1.04%
WER3	6.50%	0.95%
WER4	6.51%	0.83%
WER5	6.51%	0.88%

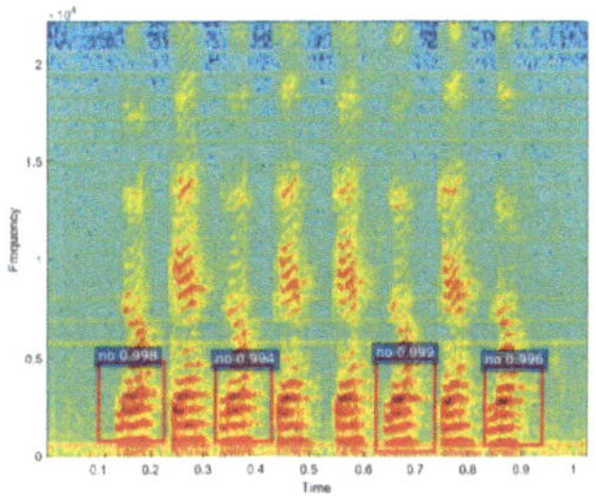

Fig. 4. proposals of "no"

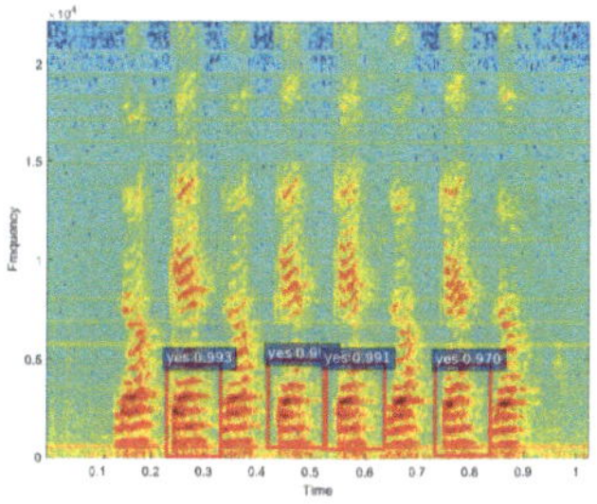

Fig. 5. proposals of "yes"

results are shown in Table 1, where WER means the word error rate. Figures 4 and 5 are the test results, respectively.

4.2 Experiments on Chinese Voice

After the above experiments, the proposed method is tested on the Chinese voice data set, such as robot voice instructions. Factory robots have common actions, such as grab (pronounce "zhua" in Chinese), push (pronounce "tui" in Chinese). The experimental dataset includes 20 different action pronunciations, recorded by 5 different people. And the recording environment is divided into quiet conditions and noise conditions. This experiment tests on both the small ZF model and the middle VGG_CNN_M_1024 model. The results are shown in Tables 2 and 3.

Table 2. Quiet environment

Model	ZF model	VGG16 model
WER1	4.35%	2.34%
WER2	4.52%	2.10%
WER3	4.46%	2.65%
WER4	4.73%	2.48%

Table 3. Nosy environment

Model	ZF model	VGG16 model
WER1	11.85%	8.22%
WER2	11.56%	8.18%
WER3	11.65%	8.13%
WER4	11.75%	8.20%

5 Conclusions

This paper proposes a method to implement speech recognition based on the spectrogram via faster R-CNN, the proposed target detection method is used to identify the spectrogram, and the targets are the boxes marked by each word of obvious voiceprint in the spectrogram. Finally,

Acknowledgments. This research is funded by the National Natural Science Foundation (Project No. 61573145), the Public Research and Capacity Building of Guangdong Province (Project No. 2014B010104001) and the Basic and Applied Basic Research of Guangdong Province (Project No. 2015A030308018), the authors are greatly thanks to these grants.

References

1. Benzeghiba, M., Mori, R.D., Deroo, O., et al.: Automatic speech recognition and speech variability: a review. Speech Commun. **49**(10–11), 763–786 (2007)
2. Hinton, G., Deng, L., Yu, D., et al.: Deep neural networks for acoustic modeling in speech recognition: the shared views of four research groups. IEEE Signal Process. Mag. **29**(6), 82–97 (2012)
3. Miao, Y., Gowayyed, M., Metze, F.: EESEN: end-to-end speech recognition using deep RNN models and WFST-based decoding, 167–174 (2015)
4. Bahdanau, D., Chorowski, J., Serdyuk, D., et al.: End-to-end attention-based large vocabulary speech recognition. Comput. Sci. (2015)
5. Costa, D.C., Lopes, G.A.M., Mello, C.A.B., et al.: Speech and phoneme segmentation under noisy environment through spectrogram image analysis. In: IEEE International Conference on Systems, Man, and Cybernetics, pp. 1017–1022. IEEE (2012)
6. Zhang, S., Liu, C., Jiang, H., et al.: Feedforward sequential memory networks: a new structure to learn long-term dependency. Comput. Sci. (2016)
7. Qian, Y., Bi, M., Tan, T., et al.: Very deep convolutional neural networks for noise robust speech recognition. IEEE/ACM Trans. Audio Speech Lang. Process. **24**(12), 2263–2276 (2016)
8. Zhang, Y., Chan, W., Jaitly, N.: Very deep convolutional networks for end-to-end speech recognition (2016)
9. Ren, S., He, K., Girshick, R., et al.: Faster R-CNN: towards real-time object detection with region proposal networks. In: International Conference on Neural Information Processing Systems, pp. 91–99. MIT Press (2015)
10. Chan, W., Jaitly, N., Le, Q., et al.: Listen, attend and spell: a neural network for large vocabulary conversational speech recognition. In: 2016 IEEE International Conference on Acoustics, Speech and Signal Processing (ICASSP), pp. 4960–4964. IEEE (2016)
11. Lecun, Y., Bengio, Y., Hinton, G.: Deep learning. Nature **521**(7553), 436–444 (2015)
12. Wang, S., Chen, X., Cai, G., et al.: Matching demodulation transform and synchrosqueezing in time-frequency analysis. IEEE Trans. Signal Process. **62**(1), 69–84 (2013)
13. Rekik, S., Guerchi, D., Selouani, S.A., et al.: Speech steganography using wavelet and Fourier transforms. EURASIP J. Audio Speech Music Process. **2012**(1), 20 (2012)
14. Klára, V., Viktor, I., Krisztina, M.: Voice disorder detection on the basis of continuous speech. In: 5th European Conference of the International Federation for Medical and Biological Engineering. Springer, Berlin (2011)

15. Girshick, R., et al.: Rich feature hierarchies for accurate object detection and semantic segmentation. In: Proceedings of the IEEE Conference on Computer Vision and Pattern Recognition (2014)
16. Girshick, R.: Fast R-CNN. In: IEEE International Conference on Computer Vision, pp. 1440–1448. IEEE (2015)

Design and Optimization of a Constant Pressure Hydraulic Accumulator Using a Cam Mechanism

Donglai Zhao, Wenjie Ge(✉), Xiaojuan Mo, Dianbiao Dong, and Yuanxi Sun

School of Mechanical Engineering, Northwestern Polytechnical University, Xi'an, China
gwj@nwpu.edu.cn

Abstract. Hydraulic accumulators are used widely in the industry field since been developed. Traditional hydraulic accumulators have two major deficiencies, one is that the fluid pressure decreases when supplying additional fluid power into the hydraulic system, and the other one is that the energy density is lower than fuels and lithium batteries. In this paper, we design a constant pressure hydraulic accumulator (CPHA) using a cam mechanism which can maintain pressure in a constant value and achieve a higher energy density. The structure design and mathematical models of the CPHA are presented. The eight design parameters of the CPHA are optimized using a genetic algorithm. Based on the optimized results, the energy density of the CPHA is analyzed and compared to a conventional accumulator.

Keywords: Constant pressure · Cam mechanism · Hydraulic accumulator High energy density

1 Introduction

As the hydraulic actuation system obtains the widespread application, accumulators play important roles in hydraulic circuits. A conventional accumulator is a rigid tank which is partitioned into two regions, one is filled with nitrogen gas and the other is full of hydraulic fluid. The nitrogen gas and the hydraulic fluid are separated by a piston or a bladder. Therefore the hydraulic pressure in the accumulator is a function described by the gas state equation. The hydraulic accumulator primarily plays two roles in a hydraulic system, one is to store energy and provide additional fluid power, and the other is to reduce pressure fluctuations and absorb shock.

However, some inherent weaknesses limit a better use of these conventional accumulators. One major flaw is the heat loss in the gas cavity. When the nitrogen gas in the accumulator is compressed rapidly, the gas temperature will rise considerably. The temperature difference between the gas and the accumulator wall results in a thermodynamically irreversible transfer of heat and an energy loss in the accumulator.

The second major insufficience of hydraulic systems is the low energy density for a traditional accumulator [1]. As discussed in the literature review, the initial gas pressure

K. Deng et al. (Eds.): ICMIR 2018, AISC 856, pp. 898–907, 2019.
https://doi.org/10.1007/978-3-030-00214-5_111

of an accumulator must be lower than the system's working pressure decided by a relief valve. When a hydraulic system works, the fluid oil enters into the accumulator and compresses the gas cavity until the pressure balances. When the accumulator supplies extra fluid power, the fluid pressure and the energy density both decrease with the fluid oil returning to the pipeline.

People have made plenty of approaches to decrease the heat loss and improve the energy storage density of a hydraulic accumulator. Pourmovahed [2] analyzed the accumulator efficiency with and without elastomeric foam. The result shows the addition of an elastomeric foam can improve the efficiency and reduce thermal losses. Van de Ven [4, 5] designed a hydraulic flywheel accumulator (HFA) which integrated rotating kinetic and pneumatic energy storage. Pedchenko and Barth [3] proposed a strain energy accumulator which stored energy through the strain energy of the stretched elastomer material. Van de Ven [6–8] also invented a novel constant pressure hydraulic accumulator. By using a variable area piston, the accumulator can improve the energy density and keep the hydraulic pressure constant.

In this paper, a novel constant pressure hydraulic accumulator (CPHA) with a high energy density is designed. Contrary to a traditional accumulator, the CPHA's fluid cavity and gas cavity are separated. Each cavity is sealed by a piston and the piston rods are connected by a translating cam mechanism (TCM). Using the nonlinear characteristic of the TCM, the hydraulic fluid can remain constant pressure while the gas pressure varies as a function of volume. The first two sections of this paper describe the structure and the mathematical model of the CPHA. Then the optimized design parameters of the CPHA are obtained using a genetic algorithm. The energy density of the CPHA is also analyzed compared with a conventional bladder accumulator.

2 Design and Optimization of the CPHA

When compressed, the inner gas's pressure of the accumulator can be calculated by the state equation with a specified initial condition. Although the change of the pressure is nonlinear, we can get a constant pressure through anther nonlinear transformation. The usually nonlinear transformation can be achieved by using link mechanism, crank shift and cam mechanism. Considering the size and the number of the parts, we choose a cam mechanism as a compact structure to achieve this constant pressure accumulator. The CPHA model presented here are considerably simpler than what would be required for a final design validation.

2.1 Structure Design of the CPHA

Contrary to a traditional accumulator, the new accumulator is designed as a split type accumulator with one fluid cavity and two gas cavities. The hydraulic fluid cavity and the gas cavity are separated which means the fluid pressure can be different from the gas pressure. As illustrated in Fig. 1, all three cavities are fixed on the base. The fluid cavity's piston rod is designed as a translating cam and the gas cavity's piston rod is designed as a cam roller. When the hydraulic oil enters the fluid cavity, the translating cam will move along the x-axis pushing the cam roller and the piston to compress the gas chamber correspondingly. To balance the translating cam's radial force, two gas cavities are symmetric distributed along the fluid cavity's axis with a fixed angle. The reaction force between the translating cam and the cam roller can decompose a radial force of the gas cavity's piston rod. To decrease this influence, an extension spring with preload connects two gas cavity's piston rods. To decrease the seal ring's contact pressure and the friction, a bladder is used as an inner container in the cavity. By this way, the friction between the piston and the cylinder can be ignored. The calculation procedure of the parameters of the system including initial gas volume, gas initial pressure, fluid pressure etc. is presented as follows.

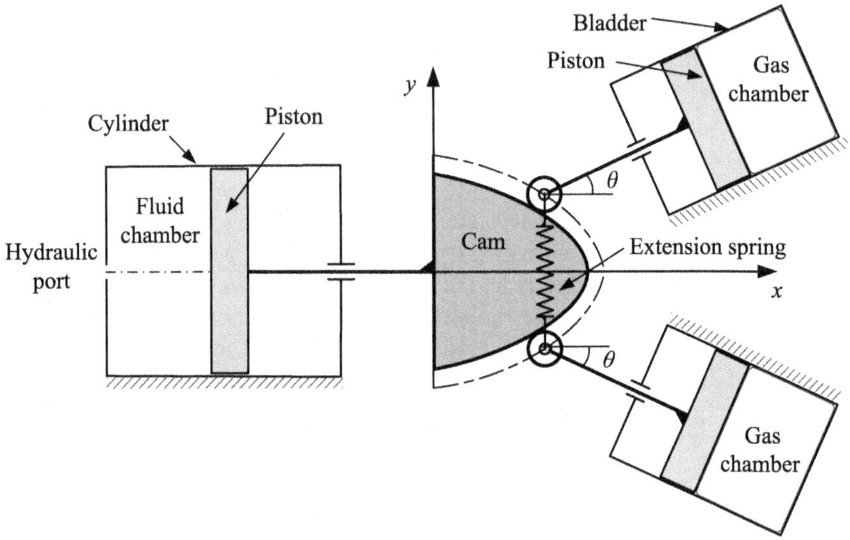

Fig. 1. A schematic diagram of the CPHA

The hydraulic store energy E in the accumulator can be expressed in terms of the fluid cavity's piston displacement S_1 and the gas cavity's piston displacement S_2 as:

$$E = E_f(S_1) = E_g(S_2) + E_k(S_2) \tag{1}$$

Where E_f is the energy stored by the hydraulic fluid, E_g is the energy stored by the gas and E_k is the energy stored by the extension spring. And these energies can be expressed separately as:

$$E_f(S_1) = \int_0^{S_1} P_f A_f ds \tag{2}$$

$$E_g(S_2) = 2 \times \int_0^{S_2} P_g A_g ds \tag{3}$$

$$E_k(S_2) = \int_0^{S_2 \sin\theta} F_k ds \tag{4}$$

Where A_f and A_g are the cross-sectional area of the fluid cavity's piston and the gas cavity's piston, and P_f and P_g are the compressed pressure in the fluid cavity and the gas cavity respectively, F_k is the force of the extension spring, θ is the fixed angle of the gas cavity in the coordinate system.

As discussed in the literature, the isothermal operation is possible by placing elastomeric foam in the gas cavity. The gas pressure P_g as a function of the piston displacement S_2 can be described by:

$$P_g = \frac{P_{g0} V_{g0}}{V_{g0} - A_g S_2} \tag{5}$$

Where P_{g0} and V_{g0} are the initial gas pressure and volume respectively.

Since the translating cam is symmetrical along the x-axis, as seen in Fig. 1, the force condition of the translating cam and the roller can be analyzed only in the upper part of the cam. When the constant pressure accumulator works, the roller moves from initial position P_0 to a certain position P_1 along the cam profile. The plane coordinate system is fixed on the translating cam.

As shown in Fig. 2, the cam profile fixed in a plane coordinate system and a roller with the radius r is connected to cam profile. The theoretical cam profile can be calculated as:

$$\begin{cases} x = x_0 - S_1 + (l_0 + S_2)\cos\theta \\ y = S_2 \sin\theta + y_0 \end{cases} \tag{6}$$

Where x_0 and x_1 are the intersection points of the gas piston rod's axis and the x-axis at the initial position and the compressed position, l_0 is the distance between the point (x_0, 0) and the roller's center at the initial position, y_0 is the distance between the roller's center and the x-axis at the initial position, $S_1 = x_0 - x_1$ is the fluid cavity's piston displacement.

The final cam profile considering the roller's radius r_0 is solved in the plane coordinate as:

$$\begin{cases} x' = x - r_0 \cos\alpha \\ y' = y - r_0 \sin\alpha \end{cases} \tag{7}$$

Where α is the angle between the connected force and the x-axis, the angle can be calculated by $\cot\alpha = dy/dx$.

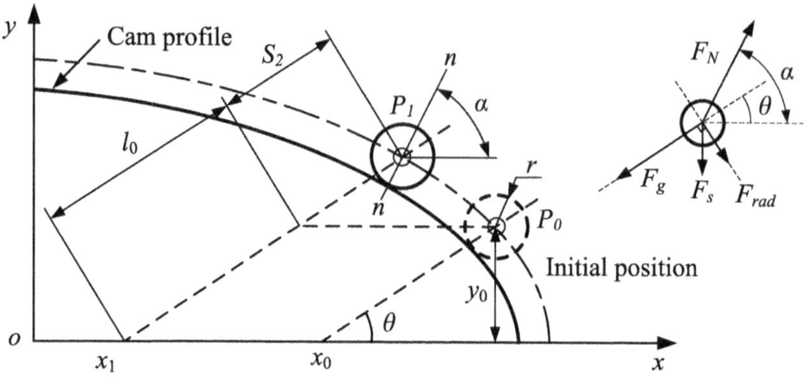

Fig. 2. Coordinate of the constant accumulator

By substituting Eqs. (2)–(5) into (1), the hydraulic store energy in the new accumulator can be expressed as:

$$P_f A_f S_1 = 2P_{g0} V_{g0} \left[\ln V_{g0} - \ln(V_{g0} - A_g S_2) \right] + F_0 S_2 \sin\theta + \frac{k}{2} S_2^2 \sin^2 \theta \tag{8}$$

Where F_0 is initial force of the extension spring and k is the spring stiffness.

The Eq. (8) established the equilibrium of S_1 and S_2, the cam profile can also be calculated using this equation.

Using a bearing in the roller, the friction and gravity can be ignored compared with the fluid force. As shown on the right side of the Fig. 2, the forces acted on the roller are the connected force F_N, the compressed gas force F_g, the extension spring force F_s, and the radial force F_{rad}. Since the fluid pressure keeps constant, the translating cam's force equation can be expressed as:

$$2F_N \cos\alpha = P_f A_f \tag{9}$$

Therefore, the radial force we concerned here can be expressed as:

$$F_{rad} = F_N \sin(\alpha - \theta) - F_s \cos\theta = \frac{P_f A_f \sin(\alpha - \theta)}{2\cos\alpha} - (F_0 + kS_2 \sin\theta)\cos\theta \tag{10}$$

2.2 Optimization of the Design Parameters of the CPHA

As the friction force is a dissipative force, the radial force of the gas cavity's piston rod F_{rad} which increases the seal ring's friction should be minimized. The presented mathematical models of the CPHA are used to optimize the six design parameters. This section presents the approach and results of the optimization.

(a) The optimization objective of the CPHA

When all design parameters are set, the radial force F_{rad} of the CPHA can be obtained using Eq. (10). If the radial force for each discrete position in the whole compressing cycle is expressed as $F_{rad}(i)$, the optimization objective for the CPHA can be expressed as:

$$fval = \min \sum_{i=1}^{n} |F_{rad}(i)| \tag{11}$$

Where n is the number of positions in the compressing cycle (n = 40).

(b) The constraints of the six parameters

Lower bounds of all hydraulic property parameters, the first two parameters, are defined to insure the size of the CPHA and the size of a conventional accumulator in the same magnitude. Lower bounds of θ and y_0 are defined to avoid geometrically infeasible solutions. Lower bounds of the extension spring parameters k and F are defined to insure the spring force F_s and the radial force F_{rad} in the same magnitude. The upper bounds of all six design parameters are defined to insure the reasonable sizes of the CPHA. Compared with the dimensions of the Rexroth 1L bladder-type accumulator HAB1-350-6X, the lower and upper bounds of all eight design parameters is listed in Eq. (12).

$$s.t.\begin{cases} 17.5\ \text{MPa} \le P_{g0} \le 35\ \text{MPa} \\ 200\ \text{cm}^3 \le V_{g0} \le 500\ \text{cm}^3 \\ \frac{\pi}{12}\ \text{rad} \le \theta \le \frac{\pi}{2}\ \text{rad} \\ 8\ \text{cm} \le y_0 \le 20\ \text{cm} \\ 10000\ \text{N/m} \le k \le 300000\ \text{N/m} \\ 500\ \text{N} \le F_0 \le 5000\ \text{N} \end{cases} \tag{12}$$

(c) The optimization method

The optimization function for the radial force is highly nonlinear and implicit which reduces the effectiveness of gradient-based optimization methods. Therefore, a genetic algorithm is used in this paper because of its robustness and convenient implementation.

2.3 The Result of the Optimization

Because the fluid cavity and the gas cavity of a conventional accumulator are in the same container, when the accumulator works, the hydraulic oil can't occupy the entire volume of the container. The volume of the fluid oil in a normal accumulator is decided by the pressure of the hydraulic system and the precharge pressure of the gas. The performance of a conventional accumulator is studied by Quan [9]. The simulation and the experiment show that the best precharge pressure of the gas is the pressure when the accumulator's damping ratio is 0.707. In Quan's experiment, the best performance of the accumulator can be achieved when the precharge pressure of gas is 16.5 MPa at 17.5 MPa system pressure. In this situation, the volume of fluid oil stored in 1 L bladder accumulator is only 57 ml.

Compared with a traditional accumulator discussed by Quan, the fluid cavity's volume of the CPHA is designed as 57 ml and the fluid pressure P_f keeps 17.5 MPa. The available internal diameter r_f of the fluid cavity in engineering field is finite and discrete. Considering the size and weight of the cavity, the internal diameter r_f is chosen as 40 mm and then the cross-sectional area A_f is 12.56 cm^2. Also the internal diameter of the gas cavity is chosen as 63 mm and the cross-sectional area A_g is 31.06 cm^2.

The optimization results are shown in Table 1, as a requested hydraulic pressure 17.5 MPa. As shown in Fig. 3, the hydraulic force on the fluid cavity's rod can keep constant while the axial force of the gas cavity's rod increases. Using extension spring and optimization, the radial force of the gas cavity's rod can decrease markedly comparing with the hydraulic force and the gas force.

Table 1. The optimized results of the CPHA

P_{g0}	V_{g0}	θ	y_0	k	F
18.00 Mpa	224.85 cm^3	1.38 rad	12.39 cm	296240.00 N/m	2674.60 N

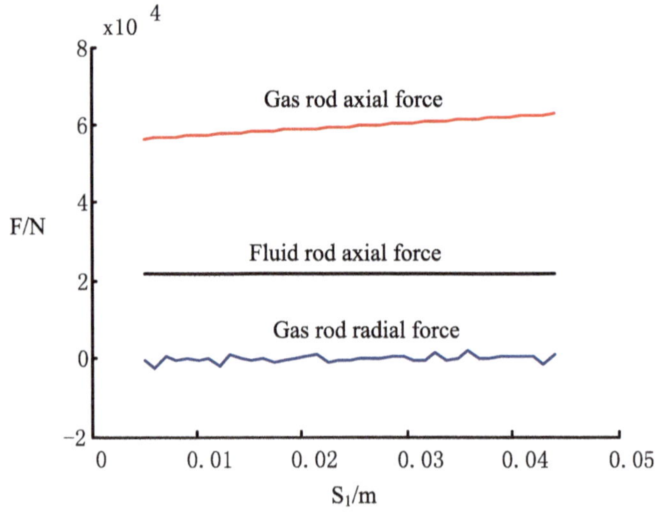

Fig. 3. The forces on the fluid and gas rod

3 Energy Density Analysis of the CPHA

As discussed in the literature review, the hydraulic accumulator's energy storage density is low which limits the application of a hydraulic storage systems. In this section, the energy density of a CPHA using a cam mechanism will be developed and compared to a conventional accumulator.

Define the volumetric energy density of a gas-charged accumulator to be:

$$\omega = \frac{\text{Stored Energy}}{\text{Total Volume}} \tag{13}$$

By assuming an ideal gas with isothermal operation again, the energy stored in a conventional accumulator can be expressed as:

$$E_1 = -\int_{V_0}^{V} P_g dV = P_{g0} V_0 \ln \frac{V_0}{V} = P_{g0} V_0 \ln r \tag{14}$$

Where and P_{g0} and P_g are the initial and compressed gas pressure, V_0 and V are the initial volume and final volume of the gas respectively, and r is the volume ratio defined as $r = V_0/V$.

The total volume of a conventional accumulator is defined as the sum of the initial gas volume and the displaced fluid volume [8]:

$$V_{total} = V_0 + (V_0 - V) = 2V_0 - V \tag{15}$$

Therefore, the energy density of a conventional accumulator can be expressed as:

$$\omega_1 = \frac{p \ln r}{2r - 1} \tag{16}$$

For the CPHA in this article, the energy storing process in the gas cavity is the same as a conventional accumulator. While the fluid cavity's pressure remains constant, the volume of hydraulic fluid displaced for a certain energy is lower than a conventional accumulator. The hydraulic fluid volume can be calculated as:

$$V_f = \frac{P_{g0} V_0 \ln r}{p_f} \tag{16}$$

Therefore, the energy density of the CPHA can be expressed as:

$$\omega_2 = \frac{E_1}{V_0 + V_f} = \frac{P_{g0} P_f \ln r}{P_f + P_{g0} \ln r} \tag{17}$$

Based the optimization result on the previous section, the energy densities of a conventional accumulator and the CPHA can be seen in Fig. 4, as the maximum pressure of fluid is 17.5 MPa. With the volume ratio r increasing, the CPHA has a better energy density property than a conventional accumulator.

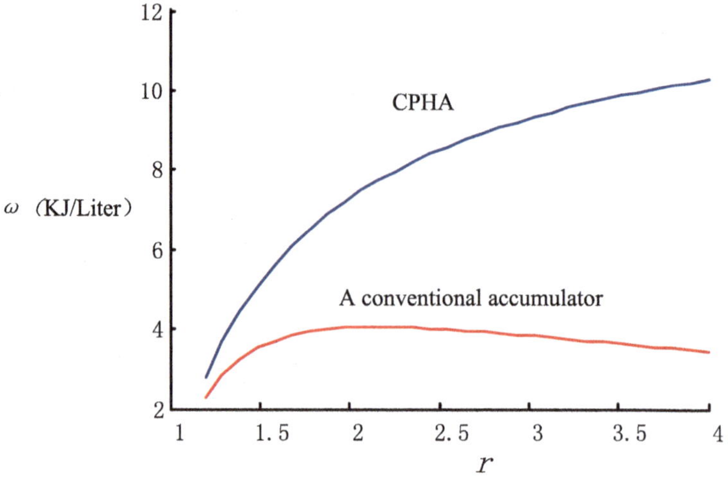

Fig. 4. Energy density of the CPHA and a conventional accumulator

4 Conclusion

This paper presents a novel accumulator which achieves a constant hydraulic pressure using a split structure and a translating cam mechanism. The energy-balance equations of the fluid and the gas are obtained based on the ideal gas state equation. The structure model and the force conditions of the translating cam mechanism are establised and optimised to minimize the radial force of the gas cavity's piston rod. The optimal parameters of the CPHA are obtained using a genetic algorithm. The energy density of the CPHA is compared with a Rexroth 1L bladder-type accumulator. The result shows that the CPHA has a higher energy density than a conventional accumulator and can achieve a better performance in a hydraulic system. The further research will focus on the dynamic property of the CPHA and reduces the system's size and mass.

Acknowledgments. This work was supported by the National Natural Science Foundation of China under Grant number 50975230, and Research Fund for the Doctoral Program of Higher Education of China under Grant number 20136102130001.

References

1. Pourmovahed, A.: Sizing energy storage units for hydraulic hybrid vehicle applications. In: ASME Dynamic Systems and Controls Division, pp. 231–246 (1993)
2. Pourmovahed, A., Baum, S.A., et al.: Experimental evaluation of hydraulic accumulator efficiency with and without elastomeric foam. J. Propuls. Power **4**(2), 185–192 (1988)
3. Pedchenko, A., Barth, E.J.: Design and validation of a high-energy density elastic accumulator using polyurethane. In: ASME Dynamic Systems and Control Conference and Symposium on Fluid Power and Motion Control, pp. 283–290 (2009)

4. Van de Ven, J.D.: Increasing hydraulic energy storage capacity: flywheel-accumulator. Int. J. Fluid Power **10**(3), 41–50 (2009)
5. Strohmaier, K.G., Van de Ven, J.D.: Constrained multi-objective optimization of a hydraulic flywheel accumulator. In: ASME/BATH 2013 Symposium on Fluid Power and Motion Control. V001T01A014 (2013)
6. Li, P.Y., Van de Ven, J.D., Sancken, C.: Open accumulator concept for compact fluid power energy storage. In: ASME 2007 International Mechanical Engineering Congress and Exposition, pp. 127–140 (2007)
7. Cole, G.A., Van de Ven, J.D.: Design of a variable radius piston profile generating algorithm. In: ASME 2009 International Mechanical Engineering Congress and Exposition (2009)
8. Van de Ven, J.D.: Constant pressure hydraulic energy storage through a variable area piston hydraulic accumulator. Appl. Energy **105**(1), 262–270 (2013)
9. Quan, L., et al.: Theory and experiment of accumulator absorbing pressure pulsation without regard to its entrance characteristics. Chin. J. Mech. Eng. **43**(9), 28–32 (2007)

Research on DC Motor Control Based on Predictive PI Algorithm

Jinpeng Zhou[1(✉)], Xu Ma[2], Zhang Xu[3], and Qi Zhou[4]

[1] Tianjin University of Technology, Tianjin 300384, China
173122312@stud.tjut.edu.cn

[2] Tianjin Key Laboratory for Control Theory and Applications in Complicated System, the School of Electrical and Electronics Engineering, Tianjin University of Technology, Tianjin 300384, China

[3] School of Mechanical Engineering, Tianjin University of Technology and Education, Tianjin 300130, China

[4] School of Mechanical Engineering, Hebei University of Technology, Tianjin 300130, China

Abstract. In minimally invasive surgery, manual manipulation of the catheter is inaccurate which requires the worker to manipulate the catheter flexibly and effectively and exposes the surgeon to intense radiation. The author's goal was to develop a robotic catheter operating system that would be hand-accurate on behalf of the surgeon. With the increasing demand for flexibility and rapid response to control systems, predictive PI control can play an important role. We proposed a practical application which is based on predictive PI control to the remote operation of this development system. And compare with the traditional PID control method for simulation. Demonstrate the feasibility and effectiveness of the control method. In this remote operation, in the synchronous control performance, use predictive PI control is much better than control using traditional PID control.

Keywords: Predictive PI control · Manipulation performance
Minimally invasive surgery (MIS) · Remote operation
Robotic catheter manipulation system (RCMS)

1 Background Introduction

It is expected that intravascular intervention will be more and more popular in medical practice because it can be used both for diagnosis and surgery. However, like a new technology, it requires a lot of skillful operations. More, it requires a doctor with extensive skills and experience to do manually insert the catheter. Not only is the machine less flexible than human hands and lacks tactile feedback [1, 2].

The Niobe system developed and produced by Steretaxis is a remote magnetic navigation robot system with a magnetic field inside to control the movement of the catheter tip. The magnetic field is formed by two permanent magnets that are fixed at both ends of the movable table [3, 4].

K. Deng et al. (Eds.): ICMIR 2018, AISC 856, pp. 908–915, 2019.
https://doi.org/10.1007/978-3-030-00214-5_112

Compare to the above-mentioned robotic system, our system has a slave machine consisting of a two-degree-of-freedom catheter manipulation platform and a rotatable platform that can be steered. In addition, there is a newly developed force feedback measurement mechanism at the slave to monitor the proximal thrust [5–7].

2 Introduction of Robot Catheter Manipulation System

The structural design of RCMS is divided into primary and secondary. The surgeon's console is the main side of the system and the catheter's operating is the slave of the system. Therefore, the surgeon can operate the system smoothly and easily. The console sends the handle's axial displacement and rotation angle to the catheter's operating [8].

2.1 Mechanical Structure of the System

The system mechanical structure used in this design adopts a master-slave structure. The movable parts of the surgeon's console and catheter manipulator maintain the same displacement, such a design will make the operation visible and easy to start. On the other hand, this structure can achieve mechanical feedback from the surgeon. Its mechanical structure is shown in Figs. 1 and 2 [9].

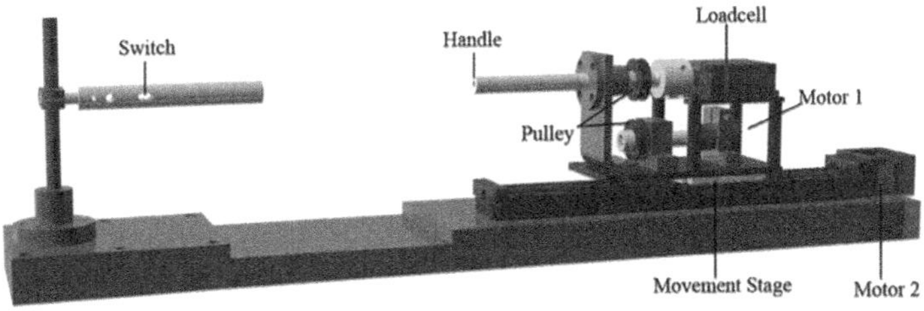

Fig. 1. Structure of the surgeon console

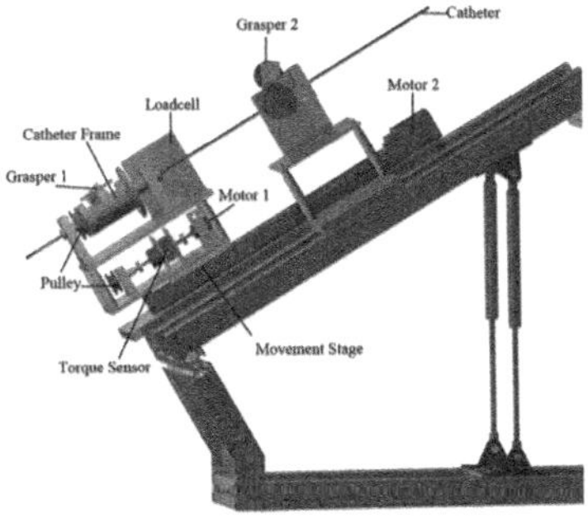

Fig. 2. Catheter manipulator

Figure 2 shows the catheter manipulator. This part is placed next to the patient. The catheter is inserted by using this device.

The surgeon console is shown in Fig. 1 and shows the surgeon's console, RCMS. The surgeon console is the primary and the entire system is operated by an experienced surgeon. The right handle is used to detect the surgeon's behavior [10, 11].

2.2 Control System Design

As mentioned in the background introduction, the system uses a master-slave operating system. A DSP was used as the control unit for each surgeon's console and catheter robot side. The Internet-based communication is established between the surgeon's console and the catheter manipulator, and rotation angle laterally to the catheter manipulator [12, 13].

The design of the interventional robotic system is shown in Fig. 3.

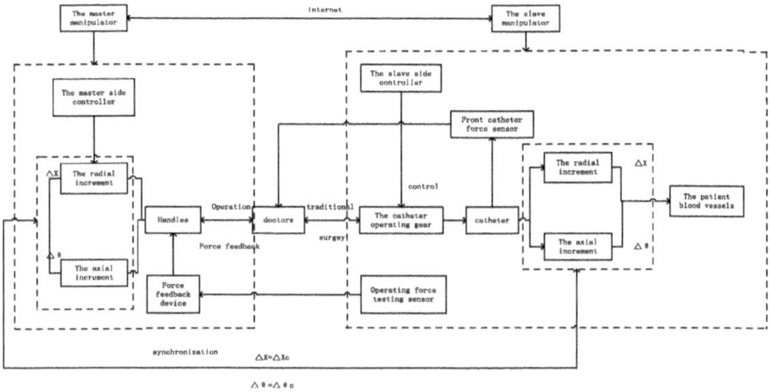

Fig. 3. Schematic diagram of the interventional robot operation control system

3 Methods

3.1 Classic PID Control Algorithm

The principle of the classical PID control algorithm is that the sensor converts the measured parameter into a uniform standard signal and sends it to the regulator. Compared with the set value in the regulator, the difference e(t) of the comparison is sent to the actuator after PID operation, and the output is changed to achieve the purpose of automatic adjustment [14].

Its control law can be described as follows:

$$u(t) = K_p e(t) + K_i \int_0^t e(\tau) d\tau + K_d \frac{de(t)}{dt} \quad (3-1)$$

Or further written in the form of a transfer function:

$$G(s) = \frac{U(s)}{E(s)} = K_p\left(1 + \frac{1}{T_i S} + T_d S\right) \quad (3-2)$$

The classical PID controller can be represented by the following function: $G(s) = K_p\left(1 + \frac{1}{T_i S} + T_d S\right)$. The Ziegler-Nichols empirical method sets the parameters of the PID controller [15].

The Cohen-Coon method is mainly used to set the parameters of the PID controller of the first-order pure time-delay model (FOPDT). The principle of this method is to set the system's dominant pole so that the excessive curve of the controlled object is reduced by 4:1. Decay is performed to obtain the optimal parameter setting value.

FOPDT model: $G_p(s) = \frac{K_p}{Ts+1}e^{-\tau s}$, where Kp, T and τ are the system's gain coefficient, time constant, and pure lag time, respectively [16].

3.2 Smith Predictive Control Algorithm

The main characteristic of Smith's predictive control are model-based and predictive compensation. The adverse effects of coming to improve the control quality of the time-delay control system. The control system diagram is shown in Fig. 4 [17].

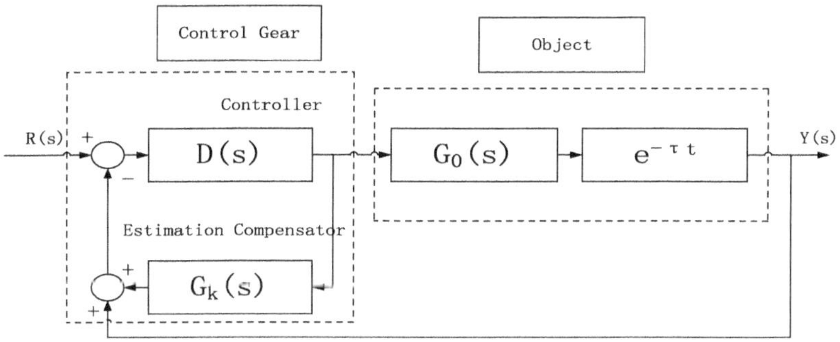

Fig. 4. Block diagram of the control system

In the formula, $\hat{G}_0(S)$ represents the mathematical model of the object with no pure lag process; $\hat{\tau}$ is the estimated pure lag time.

Therefore Fig. 4 can be transformed into Fig. 5 [18, 19].

Available from Fig. 5

$$\begin{aligned}\hat{Y}(s) &= G_k(s)U(S) + G_0(s)e^{-\tau s}U(s) \\ &= \hat{G}_0(s)U(s) + G_0(s)e^{-\tau s}U(s) - \hat{G}_0(s)e^{-\tau s}U(s)\end{aligned} \quad (3-3)$$

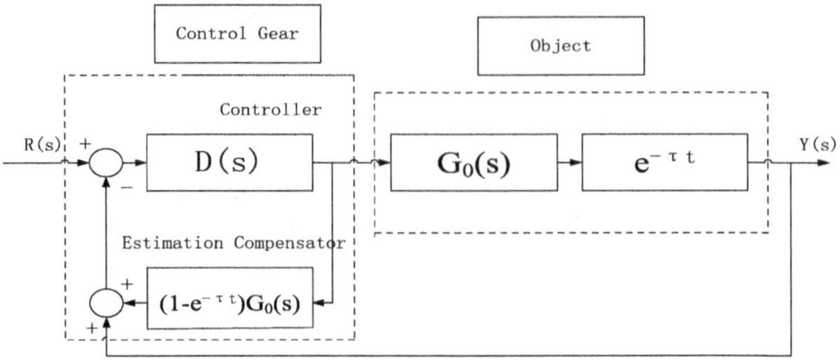

Fig. 5. Structure of compensation control system

If $\hat{G}_0(s) = G_0(s)$ and $\hat{\tau} = \tau$, then

$$\hat{Y}(s) = \hat{G}_0(s)U(s) \tag{3 − 4}$$

Therefore, the working principle of the predictive controller is that the controller will estimate the response output of the control system under the action of the disturbance in advance, and feed back the τ to the controller in advance.

3.3 Prediction PI Control Algorithm

In order to overcome the shortcomings of traditional control methods for brushless DC motors. Predictive PI control block diagram is shown in Fig. 6 shows [20, 21].

In the process of predicting PI control of brushless DC motors, it is necessary to firstly predict the motor speed. This is the most important part of predicting PI control. The predicted speed of the motor is obtained.

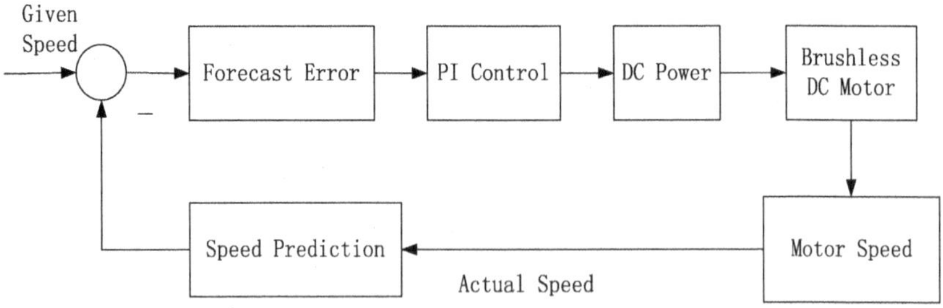

Fig. 6. Brushless DC motor prediction PI control block diagram

4 Experiments

This section will simulate the above three algorithms and compare the simulation results to analyze their advantages and disadvantages [22].

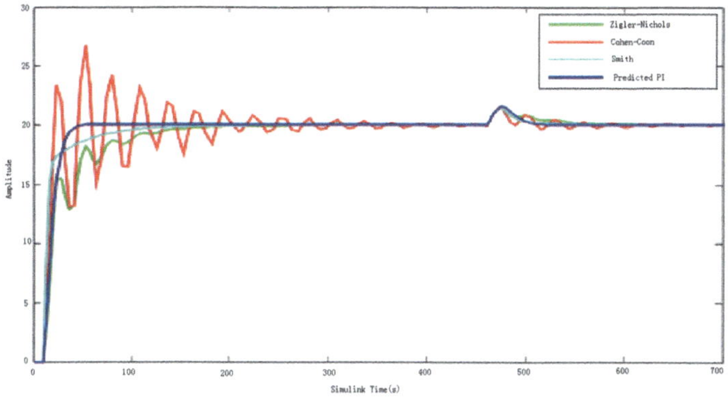

Fig. 7. PID, Smith estimation and prediction PI simulation results

Assume that the transfer function of a simple first-order lag system can be expressed as $G(s) = \frac{1}{7S+1}e^{-10s}$. After the system reaches a steady state, an output disturbance with a step of 1 is added at 450 s. The system simulation results are shown in Fig. 7.

It can be seen from the above figure that the PID controller system set by Cohen-Coon method has a large overshoot and the adjustment time is long, and the PID controller obtained by the Ziegler-Nichols method makes the system rise time longer and the system response lower. There is oscillation and slow in the system.

Simulation results show that the predicted PI control algorithm not only has good dynamic performance but also has good steady-state performance.

5 Conclusions and Future Work

This paper presents a new type of robotic catheter manipulation system (RCMS). We designed a new force-feedback mechanism to measure the proximal force, so the surgeon could feel force feedback to avoid injury during surgery. Secondly, we designed a predictive PI controller for the manipulator in the end of the catheter to improve the accuracy of axial and compared with the traditional PID control method. In the future work, we will reconstruct the rotational motion dynamics model of the robotic catheter operating system to improve the accuracy and obtain the endovascular force of the catheter distal surgery [23, 24].

Acknowledgment. This research is supported by Natural Science Foundation of Tianjin (14JCYBJC42700).

References

1. Fu, Y.L., Gao, A.Z., Liu, H., Guo, S.: The master-slave system for positioning the steerable catheter. Int. J. Mechatron. Autom. **1**(3–4), 143–152 (2011)
2. Carpentier, A., Adams, D.H., Filsoufi, F.: Carpentier's Reconstructive Valve Surgery. Saunders Elsevier, Maryland Heights (2010)
3. Kshirsagar, N., Patil, S., Kshirsagar, R., et al.: Review on application of nanorobots in health care. **3**(5), 472–480 (2014)
4. Guo, J., Guo, S., Xiao, N., Ma, X., Yoshida, S., Tamiya, T., Kawanishi, M.: A novel robotic catheter system with force and visual feedback. Int. J. Mechatron. Autom. **2**(1), 15–24 (2012)
5. Kshirsagar, N., Patil, S., Kshirsaga, R.: Review on application of nanorobots in health care. World J Pharm Pharm Sci **3**(5), 472–480 (2014)
6. Jones, L.A., Hunter, I.: A perceptual analysis of stiffness. Exp. Brain Res. **79**, 150–156 (1990)
7. Ganji, Y., Janabi-Sharifi, F.: Catheter kinematics for intracardiac navigation. IEEE Trans. Biomed. Eng. **56**, 621–632 (2009)
8. Hokayem, P.F., Spong, M.W.: Bilateral teleoperation: an historical survey. Automatica **42**, 2035–2057 (2006)
9. Kesner, S.B., Yuen, S., Howe, R.: Ultrasound servoing of catheters for beating heart valve repair. In: Information Processing in Computer Interventions, pp. 168–178 (2010)
10. Yuen, S.G., Vasilyev, N.V., del Nido, P.J., Howe, R.D.: Robotic tissue tracking for beating heart mitral valve surgery. Med. Image Anal. **17**, 1236–1242 (2010)
11. Yuen, S.G., Perrin, D.P., Vasilyev, N., del Nido P.J., Howe, R.D.: Force tracking with estimation for beating heart surgery. In: IEEE Transactions on Robotics, vol. 26. pp. 888–896 (2010)
12. Kesner, S.B., Howe, R.D.: Position control of motion compensation cardiac catheters. In: IEEE Transactions on Robotics, pp. 1–11 (2011)
13. Somasundaram, D.: Differential Geometry: A First Course. Alpha Science International Ltd, Oxford (2005)
14. Chitragari, G., Mahler, D.B., Sumpio, B.J., et al.: Prosthetic options available for the amputee. Clin. Podiatr. Med. Surg. **31**(1), 173–185 (2014)
15. Webster, J.G.: Tactile Sensors for Robotics and Medicine. Wiley, New York (1988)
16. Xu, M., Jinpeng, Z., Miao, L., Xu, Z., Jinchu, M., Zuxin, P.: A stable control method for minimally invasive surgery system. In: 2017 IEEE International Conference on Mechatronics and Automation, ICMA 2017 (1 of 1)
17. Yu, F.T.S., Yin, S.: Fiber Optic Sensors. Marcel Dekker, New York (2002)
18. Objet Geometries Ltd.: FullCure870 VeroBlack Data Sheet (2010)
19. de Visser, H., Passenger, J., Conlan, D., Russ, C., Hellier, D., Cheng, M.: Developing a next generation colonoscopy simulator. Int. J. Image Graph. **10**(2), 203–217 (2010)

20. Patrad, D., Senguptas, S., Duanw, W., et al.: Intelligent, self-powered, drug delivery systems. Nanoscale **5**(4), 1273–1283 (2013)
21. D'Amico, T.A.: Robotics in thoracic surgery: applications and outcomes. J. Thorac. Cardiovasc. Surg. **131**, 19 (2006)
22. Zhao, D.-P., Liu, D.: Fuzzy fusion of force feedback of vascular interventional robotic system. Robots **35**(1), 60–66 (2013)
23. Gonzalez, M.J., Vadera, S., Mullin, J.: Robot-assisted stereotactic laser ablation in medically intractable epilepsy: operative technique. Neurosurgery **10**, 167–173 (2014)
24. Nakamura, N., Sugano, N., Nishii, T., et al.: A comparison of cementless total hip arthroplasty. Clin. Orthop. Relat. Res. **468**(4), 1072–1081 (2010)

Design of DC Charging Pile Based on Intelligent LED Lamp Rod

Zhenhuan Gu, Kai Yang(✉), Xin Wang, and Qingyang Meng

College of Optical and Electronic Technology, China Jiliang University, Hangzhou, China
yangkai@cjlu.edu.cn

Abstract. In this paper, an intelligent charging pile based on intelligent LED lamp pole is designed, which can occupy small public space, widely distribute, charge the electric vehicle quickly, operate easily, and have a high man-machine interaction experience. What's more, it can through the network communication module to achieve timely maintenance and feedback user views. In the hardware design, both LED lamp pole and intelligent charging pile are supplied by low-voltage DC power supply. The process of power conversion between them is the same. It is necessary to convert the input AC power supply into DC power supply through three stages of voltage reduction, rectifier and steady current. The protection module is adopted to prevent the fluctuation of the power grid from damaging the equipment, so that the two can realize the integrated design scheme.

Keywords: DC charging pile · Network communication module
Human computer interaction system

1 Introduction

1.1 Background

In recent years, investigations by environmental organizations show that one of the main causes of air pollution is excessive emission of vehicle exhaust. With the rapid development of economy and the improvement of people's living conditions, automobile has become a must for most families. But fuel-powered cars emit a lot of CO_2. In addition, a large number of pollutants from cars will pollute water resources, and they also cause noise pollution. Therefore, although fuel-powered cars have brought convenience to people's lives, at the same time, they will bring harm to the environment and human health. To sum up, in the modern society with rapid economic development, environmental pollution is becoming more and more serious, and the global reserves of resources are becoming smaller and smaller. Therefore, the development and use of non-polluting new energy vehicles is an irresistible trend. At present, electric vehicles powered by electric energy have been developed and manufactured in some western countries with developed economy and technology. The emergence of electric vehicles will effectively alleviate air pollution [1]. The research and development of new clean fuels and clean fuel vehicles is not only of great

K. Deng et al. (Eds.): ICMIR 2018, AISC 856, pp. 916–922, 2019.
https://doi.org/10.1007/978-3-030-00214-5_113

significance to the sustainable development of China's energy and automobile industries, but also can promote the development of related industries and technologies.

1.2 Necessity

Although the prospect of charging piles is very considerable, the various kinds of charging piles that are popularized in the market are limited by the required fixed space, with low safety protection, inconvenient maintenance, poor man-machine interaction, and complicated operation. The existence of hidden security problems and the inability to effectively and timely feedback user information and opinions, not only brought a lot of convenience to the general public, but also brought more trouble. The intelligent charging pile based on intelligent LED lamp pole developed by this project can effectively solve these problems. At the same time, relying on the widely distributed intelligent LED lamp pole, the environment adaptability is enhanced, and the system is stable and high. And through the network communication module to each intelligent charging pile to achieve real-time monitoring and maintenance [2]. Compared with the traditional charging pile, the intelligent charging pile based on intelligent LED lamp pole designed in this paper has the advantages of fast charging speed, high charging efficiency, good safety protection, simple operation, convenient maintenance and less occupation. The system has many advantages, such as high stability and wide distribution, so it can give users an intelligent experience [3]. The project can provide an efficient charging experience for the majority of electric vehicle users, also make a certain contribution to energy saving and emission reduction in China, and provide effective support for the realization of sustainable development strategy in China.

2 Main Content

2.1 The Overall Design Scheme

Replacing traditional street lamp with intelligent LED lamp pole can save more than 50% electric energy on the premise of guaranteeing road lighting quality. This project puts forward: in the course of street lamp transformation, the standard DC power supply is obtained by using the high power rectifier to centralize the 220 V AC rectifier to supply DC power to the intelligent LED lamp pole, and at the same time, the intelligent LED lamp pole charging pile integrative equipment is built. The charge and discharge of electric vehicle can be realized. The plan is shown in Fig. 1. This scheme is widely used in intelligent LED lampposts which are widely used in cities, which makes it possible to quickly build electric vehicle charging piles with low cost and small space.

The intelligent LED lamp pole and intelligent charging pile integrated equipment has the function of detecting and communicating, and can transmit the status of street lamp and electric vehicle to the manager system in real time or at regular time through the communication channel. At the same time, integrated equipment can install cameras to monitor the surrounding environment in real time. At the same time, the manager system can not only monitor the switch state of the intelligent LED lamp pole, but also

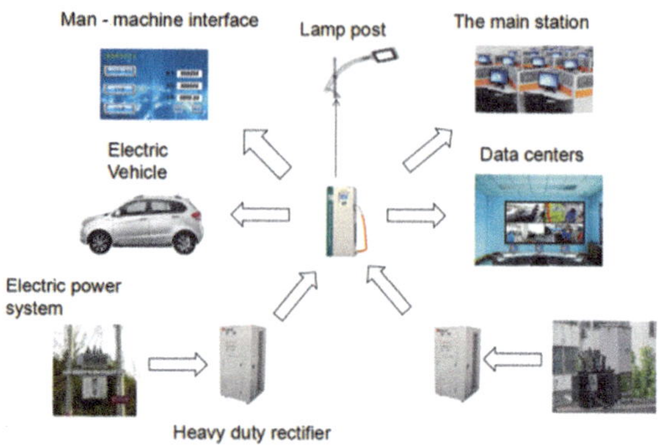

Fig. 1. The overall design scheme

send the switch command to the intelligent LED lamp pole in the monitoring range, so as to realize the collective control of the street lamp. After the revamping of many road sections, it can be interconnected to realize the low voltage DC power supply network.

2.2 Design of DC Charging Pile Module

When charging electric vehicle with AC charging pile, the electric power is input through the charger of electric vehicle, while the DC charging pile is the input voltage of AC380V ±15, the input voltage of frequency 50 Hz is rectified and filtered to supply the DC input voltage to the IGBT bridge. The DC voltage after rectifying and filtering is converted to AC voltage by driving circuit. The AC voltage is pulse width modulation. Then, the AC voltage is isolated by the high frequency transformer, and the DC pulse is obtained by rectifying and filtering again, and then the battery pack is charged. At the same time, the controller implements the control strategy according to the detected terminal voltage and charging current, and changes the charging parameters in real time [4]. The over-current, over-voltage and over-temperature protection circuits are designed to ensure the safety and reliability of the system. The principle diagram of DC charging pile charging module is as follows. DC charging pile charging mode schematic diagram is shown in Fig. 2.

The overall design of the hardware system of the intelligent charging pile is shown in Fig. 3. The hardware structure is divided into three layers. The upper layer is the components related to the interactive function of the human machine or the network. The touch screen, the reading card module, the meter module, the network module are connected to the upper board through RS485, the power supply refers to the indicator light, the running indicator, and the fault indicator through the IO port and the upper board. The middle layer is the control circuit board layer, the upper plate is divided into the upper plate and the lower plate, the upper plate is mainly used to realize the man-machine interaction and the network monitoring function. The lower board is mainly used to realize the charging control and the fault protection function. The upper and lower plates are connected through the RS485 interface; the bottom layer is the power conversion and the related circuit, in which the lightning protection module, the urgent

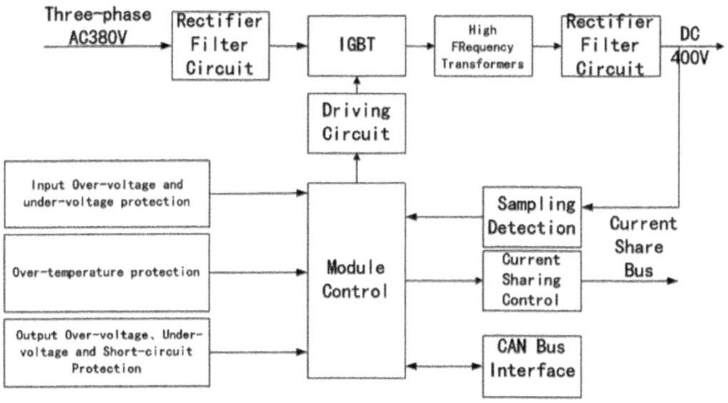

Fig. 2. Schematic diagram of DC charging pile module

stop circuit, the entrance guard are used. Circuit and so on through the IO mouth down board feedback signal, input contactor [5], discharge circuit, auxiliary power contactor, output contactor, output contactor, charge gun electronic lock controlled by lower plate IO port, module power supply group, charging gun connected to lower plate through CAN bus, insulation monitoring module connected to lower plate through RS485 interface, current sampling circuit, voltage sampling circuit, charging circuit, charging The gun detection point is connected to the lower plate through the AD port.

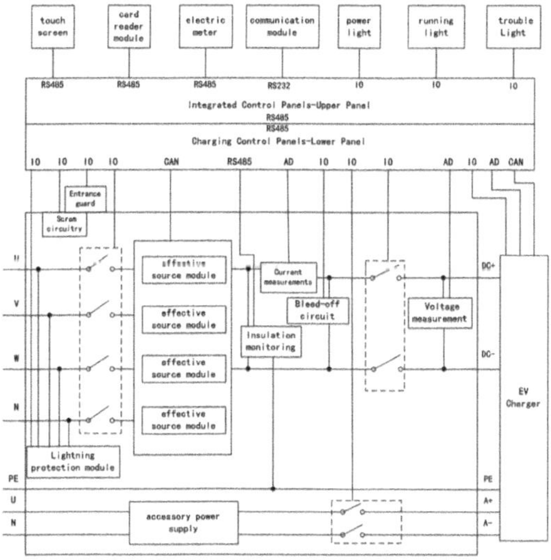

Fig. 3. Structure block diagram of DC charging pile system hardware

2.3 Design of Network Communication Module

The network communication module realizes the transparent data transmission between the control device and the network management platform. This paper adopts the industrial grade Ethernet to serial port module of Jinan people's Internet of things Technology Co., Ltd. The model is UCR-TCP232-T24. The module meets the requirements of industrial use, Ethernet interface has 2 kV electromagnetic isolation, adaptive 10/100 m transmission rate, and full-duplex asynchronous serial interface transmission baud rate can be set in the range of 300–256000 bps. The schematic diagram of the circuit is shown in Fig. 4.

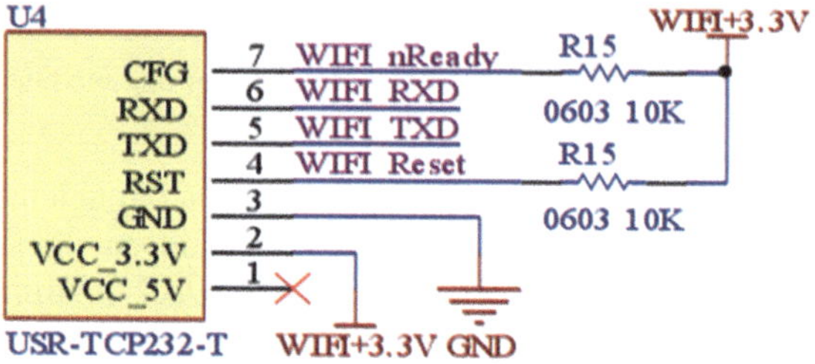

Fig. 4. USR-TCP232-T hardware circuit schematic diagram

2.4 Human-Computer Interaction System Design and Microcontroller Selection

Charging pile is directly oriented to users and should provide friendly and convenient operation experience. Therefore, this project has a graphical interface design of human-machine interaction module. Considering that the charging pile is arranged outdoors, the industrial grade resistive touch display screen of Beijing Diwen Technology Co., Ltd is selected as DMT80480T050-02WT. Its working brightness is 900 nit, supports LED backlight, anti-magnetic, anti-static [6]. Can store BMP pictures and field data combination display; can be easily upgraded through SD card display program. Built-in active buzzer, can be used directly for operation instructions and alarm. Touch screen for +5 V power supply. The rate requirement is 2 W, and the schematic diagram of the hardware circuit is shown in Fig. 5.

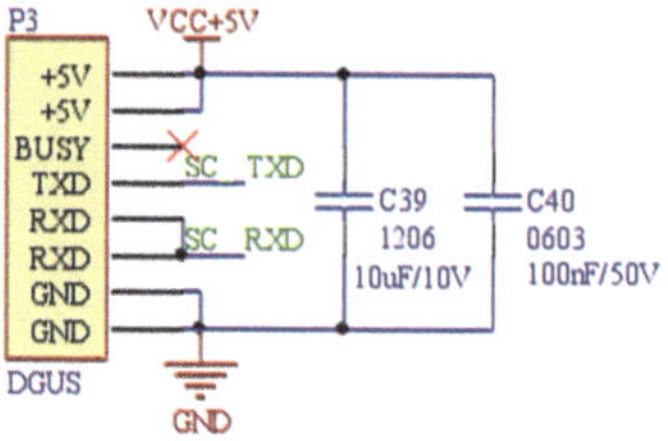

Fig. 5. Hardware circuit schematic of touch display screen

Touch display screen is the medium for user or administrator to operate charging pile device. On the one hand, the user or administrator displays the property information, status information and transaction information of charging pile; on the other hand, the user or administrator performs parameter setting and process control. Parameter setting refers to the input and identification of charging process parameters and charging pile property parameters through the soft keyboard on the display screen [7]. The process control means that the parameters confirmation and interface switching are completed through the logic keys on the display screen. The basic interface diagram of human-computer interaction designed in this paper is shown in Fig. 6.

Fig. 6. The basic operating interface diagram of human-computer interaction

In order to meet the requirements of stability, computing capacity, memory, peripheral resources, price and power consumption, St company's 32-bit industrial microcontroller, STM32F103VET6, is selected. The chip is based on ARM Cortex-M3, the highest main frequency is 72 MHz, the operation speed is 1.25 DMIPS, and the internal fast interrupt controller can speed up the system response speed. 512KB FLASH and 64KB SRAM can meet the requirements of function expansion and application programming. Rich peripherals, including 3 SPIs, Five USARTs, one CAN2.0B, two 12-bit ADCs, 12 DMA channels, one RTC, 112 external I/O. Retail price is not more than 20 yuan, high price performance [8]. The microcontroller control diagram is shown in Fig. 7.

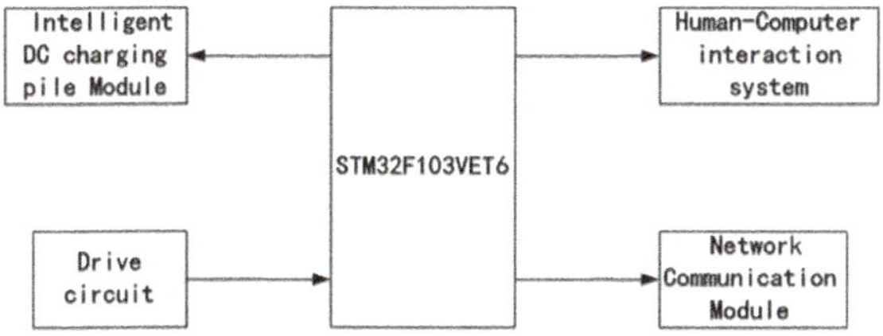

Fig. 7. Microcontroller control chart

3 Future Work

This paper creatively combines the intelligent LED lamp pole with the intelligent charging pile, and realizes the integrated design of the two, which not only reduces the occupation of public space, but also solves the problem that the uneven distribution of charging piles hinders the promotion of electric vehicles. In the future, this design will further contribute to urbanization.

Acknowledgments. First and foremost, I would like to show my deepest gratitude to my supervisor, Yangkai, a respectable, responsible and resourceful scholar, who has provided me with valuable guidance in every stage of the writing of this thesis. Without his enlightening instruction, impressive kindness and patience, I could not have completed my thesis. I would also like to thank all my teachers and my friends especially my three lovely roommates, for their encouragement and support.

References

1. Liu, Z., Chen, J., Lin, K., Zhao, Y., Xu, H.: Dometic and foreign present situation and the tendency of electric vehicles. Electr. Power Constr. **36**, 25–32 (2015)
2. Cui, Y.-f., Zhang, L.-s., Yang, Q., Wang, J.: Development status of electric vehicles and charge technology research at home and abroad. Yunnan Electr. Power **38**, 9–12 (2010)
3. Luo, Y., Tang, X., Huai, J., Chen, J.: Development status and trend of electric vehicle at home and abroad and its impact on vehicle fuel. Int. Petrol. Econ. **22**, 64–70 (2014)
4. Pang, H., Lo, E., Pong, B.: DC electrical distribution systems in buildings. In: International Conference on Power Electronics Systems and Applications, pp. 115–119. IEEE (2007)
5. Lu, M.Z., Luo, Y.S., Song, W.: The design and implementation of control system for electric vehicle direct current charging spot. Electron. Sci. Technol. **4**, 101–107 (2017)
6. Liu, X.S.: Research on the Design and Control Method of Electric Vehicle Charging Pile. Guangxi University of Science and Technology, Guangxi (2013)
7. Wang, Y.: Research on electric Vehicle Intelligent Charging Pile System. Changchun University of Technology, Changchun (2016)
8. Zhong, W.X.: Design and implementation of wireless card swipe terminal based on STM32. Electron. World, 153–154 (2012)

Piezoelectric-Based Monitoring of Restless Legs Syndrome (RLS)

Kevin Bower, Rafael Colon, Chris Karnyski, Jacob Minkel, and Reza Rashidi(✉)

Mechanical Engineering Technology, State University of New York, Alfred State College, Alfred, NY 14802, USA
rashidr@alfredstate.edu

Abstract. Restless Legs Syndrome (RLS) is a very common disease that affects millions of people each year; however, there is not an effective way to monitor it. By creating a device that is strapped to an individual's leg, RLS can be easily monitored. This device measures the amount of vibration a person gives off while uncontrollably shaking their leg. A piezoelectric cantilever beam was selected to monitor these vibrations and a protective housing was designed to comfortably fit around a human leg. The device was fabricated using 3D printing techniques and was tested by machine to ensure its operation. A variable speed Scotch Yoke apparatus with a displacement of 10 mm which oscillates at different rates was used for testing the device. By wiring the developed sensor, and a reference accelerometer to a Data Acquisition Card, comparisons were made between the true movement and the measured voltage of the sensor. The piezoelectric cantilever beam was then strapped on a volunteer to attempt to replicate the results. The experimental results showed a consistent frequency output with the reference one and can be used as a reference for monitoring restless legs syndrome.

Keywords: Piezoelectric · Restless legs syndrome (RLS)
Piezoelectric monitoring · 3D printing · Biomedical engineering
Data Acquisition Card (DAQ Card)

1 Introduction

One of the most common diseases among humans is Restless Legs Syndrome (RLS). RLS is a neurological disorder that can be treated, but not fully cured. This illness causes an individual to have a strong urge to move their legs, due to an unpleasant feeling found within them. The symptoms will either start or worsen when an individual is inactive and weaken when the individual is exercising, or even stretching [1, 2]. Medication, change in diets, and constant exercise can help alleviate the symptoms of RLS. A deficiency of Iron, along with a diet of refined sugars, alcohol consumption, caffeine, or nicotine can lead to symptoms of RLS. Medications that can help these symptoms are: Ropinirole (Requip), Pramipexole (Mirapex), Gabapentin Enacarbil (Horizant), and Rotigotine (Neupro) [2]. The elderly is at greater risk of being diagnosed than younger individuals, but there have been many cases of a child

K. Deng et al. (Eds.): ICMIR 2018, AISC 856, pp. 923–930, 2019.
https://doi.org/10.1007/978-3-030-00214-5_114

being diagnosed [1]. Current diagnostic techniques are inconvenient requiring multiple electrodes to be attached to the patient and a computer for monitoring [2, 3]. A smaller and more versatile monitoring device would better diagnose RLS and allow data to be collected without altering the patient's normal activity.

An ideal device for RLS monitoring would be a piezoelectric cantilever beam. Sensors utilizing this technology are both small and extremely sensitive to vibration [4]. Piezoelectric material can either be vibrated by applying an electrical voltage, and it can also produce a voltage by being vibrated. Most applications of piezoelectric materials deal with collecting voltage from vibrating the piezoelectric material [5, 6]. One application of voltage collecting is monitoring the structural health of building and planes, preventing structural failure [7]. Voltage collection can also be used for energy harvesting to power small devices [8]. The concept of vibration sensing was used for the work presented in this paper. Other applications similar to this include wrist worn gesture input devices [9], respiration sensors [10], and biodegradable force sensors [11]. Through this same principle of operation, the rapid twitching of leg muscles in an RLS event can be converted into electrical voltage. These movements can be recorded by wearing the device on the leg [12]. The changes in output voltage can then be recorded and create a history of movement. This history can be used by doctors to better diagnose and treat RLS in patients.

2 Design

The design for the mounting device for the piezoelectric motion monitor is inspired by a fitness tracker worn on the wrist. The mounting platform is slightly rounded on the inside to fit comfortably around the test subject's calf. The device is secured to the test subject with a strap to provide comfort. There are mounting loops for the straps, depicted in the model in Fig. 1. For testing, a mount is fabricated to attach the device to a scotch yoke apparatus. The piezoelectric plate is mounted on an elevated anchor with a strong epoxy. This allows the beam to deflect freely. Figure 2 portrays a working industrial drawing for the device housing, which is 3D printed.

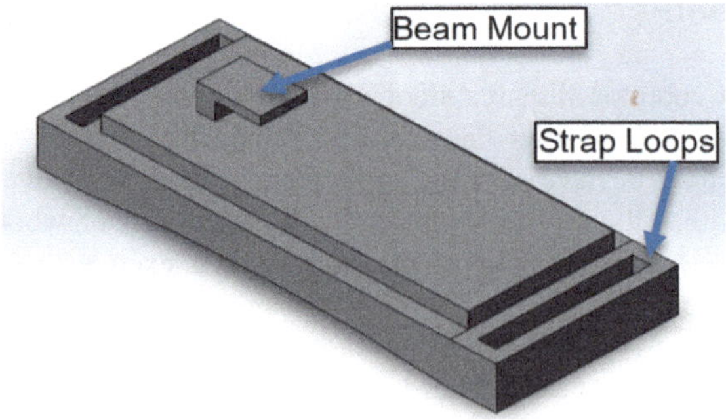

Fig. 1. Isometric view of mount

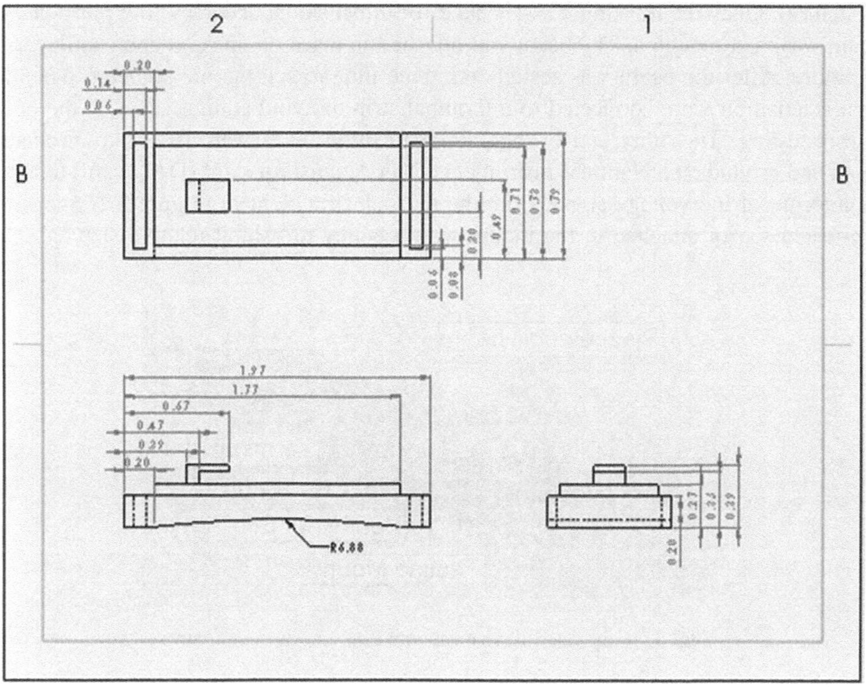

Fig. 2. Industrial drawing for the device. Dimensions are in inch.

Piezoelectric devices operate on the piezoelectric effect. Certain crystals and ceramics, known as piezo materials, can generate a small voltage when mechanically stressed. The stress will be induced in the piezo sensor through the vibrations of a cantilever beam. Through computer monitoring of the sensor's output voltage, graphs can be created and trends observed to help determine if a patient has RLS. The specific crystal being used in this experiment is item number 686 plate made of the synthetic PZT 850 material obtained from APC International Ltd. The 686 plate has dimensions of 32 mm long, 4.5 mm wide, and 0.5 mm thick and an electromechanical conversion factor (K) for PZT is 0.72.

3 Fabrication

In order to fabricate the piezoelectric measuring device, the first step was to acquire the necessary materials. The device requires the piezoelectric element, a mounting area, and a housing for protection. A 3D printer was used to create the housing with a suitable mount designed into it. The printer used was an AnyCubic Plus using 1.75 mm filament being extruded at 0.4 mm. Additionally, solder was used to attach the beam to wires used for monitoring and epoxy was used to adhere the piezo beam to the housing. Tape was used to temporarily secure the device in place on the test subject and to the Scotch Yoke apparatus. The housing was modeled using SOLIDWORKS® (Dassault

Systèmes). Once the housing was checked for dimensional accuracy, the piezoelectric beam was then mounted. The beam was attached in place on one end and cantilever on the other. After the beam was seated and given time to set, the pre-soldered wires for data acquisition were connected to a terminal strip to avoid adding stress to the beam during testing. The other end of the data acquisition wires were routed to an analog input and ground on a National Instruments Data Acquisition card (DAQ card) to allow monitoring of the voltage signal from the piezoelectric element. Figure 3 conveys the fabricated sensor attached to the mount with a penny to exhibit relative size.

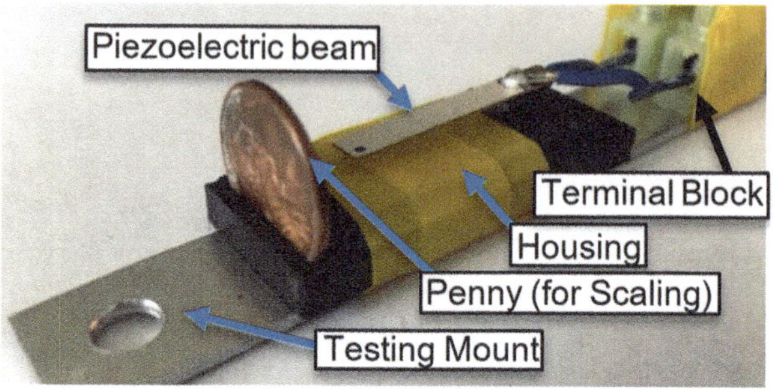

Fig. 3. Fabricated device mounted on an Aluminum strip.

4 Experimental Setup

Testing was performed with the device to determine its performance. A Scotch Yoke testing apparatus was used to produce baseline readings to test against. The RLS sensing device was affixed to the machine as depicted in Fig. 4, and was run at 10 different frequencies. After testing on the machine, human testing was done. A volunteer person was selected to simulate the twitching motion of Restless Legs Syndrome. For the test, the fabricated piezoelectric sensing device was affixed to the front of the test subject's foreleg. The device was held against the tibia with tape as portrayed in Fig. 5. Once secured, the test subject laid his leg horizontal, and samples were collected while the subject exhibited no motion, up and down motion in the leg, and also rotary motion in the leg.

The DAQ card (NI USB-6003) was used to sample the information from the device. While the piezoelectric element vibrated, its electric potential increased relative to the amplitude and frequency of the vibration. LabVIEW, a National Instruments software program, ran on the computer connected to the card in order to monitor the input, plotted the data against time elapsed, and stored the measured data into a text file for reference. This file was used for additional plotting and analysis on the data to find intervals of increased activity and to determine if any trends are associated with Restless Legs Syndrome in the test subject.

Fig. 4. Device mounted to Scotch Yoke.

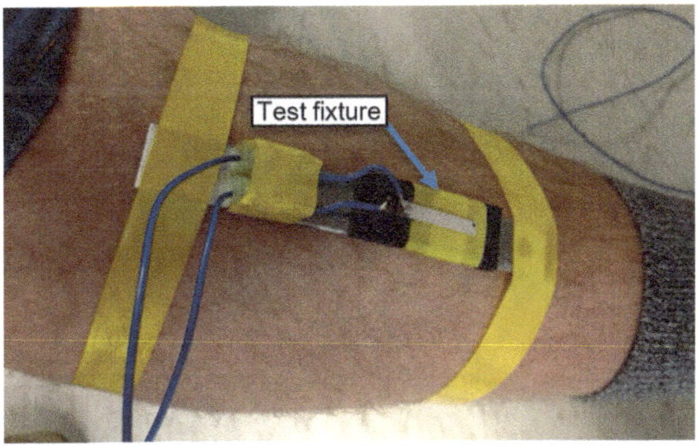

Fig. 5. Device affixed to subject's leg.

5 Results and Discussion

It was found through testing that the piezoelectric beam worked at detecting the small leg twitches of RLS. Figures 6, 7 and 8 display sampled data from the scotch yoke. An accelerometer was used to quantify the cyclic motion of the machine. Figure 6 conveys the accelerometer data, with which the motor speed can be determined as approximately 15 Hz or 900 rpm. Figure 7 depicts the recorded output voltages of the piezoelectric beam under the same conditions. The high frequency component of the signal is noise and is around 60 Hz and most of the signal amplitude. When focusing on the lower frequency component of the signal, the same number of peaks can be found as with the accelerometer in Fig. 6. Figure 8 conveys the signal from the beam with low pass filtering to reduce the effect of noise.

Figures 9, 10 and 11 illustrate the human tests and a control test of no motion. The high frequency noise present in the machine tests was not present in the human tests. It is suspected that the 60 Hz noise signal was caused by some component of the Scotch

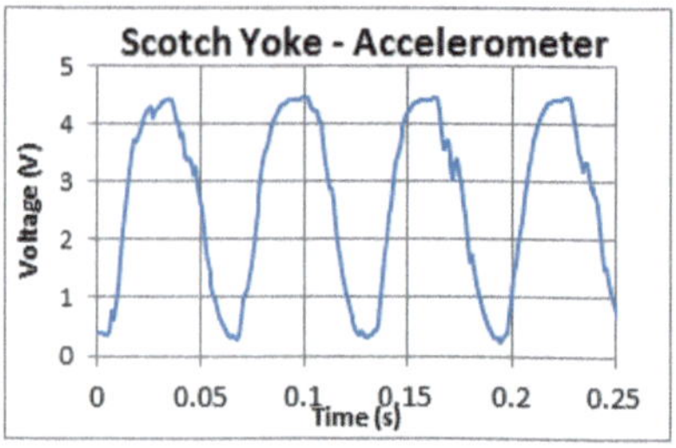

Fig. 6. Accelerometer on scotch yoke.

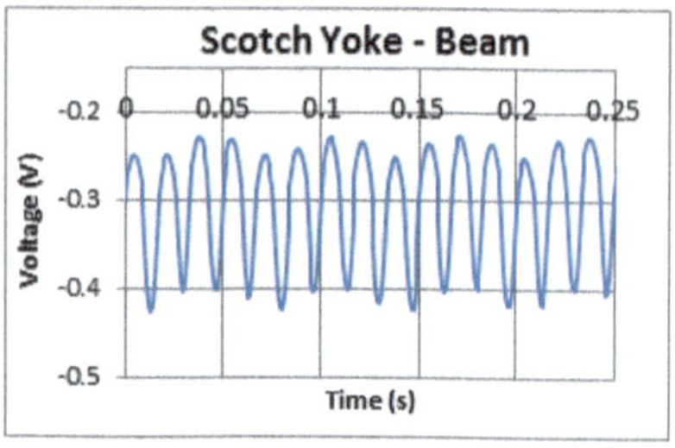

Fig. 7. Beam on scotch yoke, noise signal.

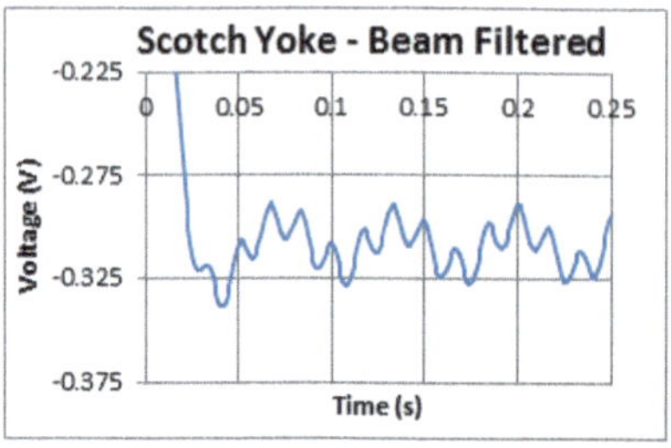

Fig. 8. Beam on scotch yoke, filtered noise.

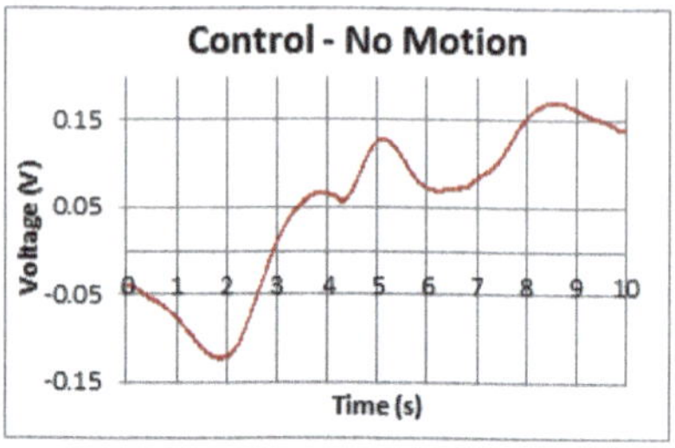

Fig. 9. Beam on flat surface, no motion.

Yoke. Figure 9 displays the beam's signal while horizontal and stationary; there are large, continuous changes in voltage, but note the smoothness of the graph. Figure 10 portrays the beam's signal while the subject had his leg horizontal, resting on a chair, and exhibiting up and down motion by raising and lowering his leg, at a frequency of about 1 Hz, while keeping the leg fully extended. The large-scale motion of the signal is of little significance because there is similar fluctuation with the no motion control signal. What are significant are the local spikes exhibited in smaller time intervals, on the order of several spikes per second, as evident in Fig. 10. These local spikes correspond to the motion of the test subject. Figure 11 conveys the beam's signal while the subject had his leg horizontal like before, but this time exhibiting rotary motion. Similar local spikes are perceivable in Fig. 11 as well.

Although this design worked well, there was a lot of noise in the beam's signal. These larger waves made it difficult to see the vibrations from a twitching leg, but they can be removed with the use of a filter program in LabVIEW. Another improvement would be to use a thinner and longer piezoelectric beam. This would give a larger and easier deflection of the beam, yielding a larger output voltage spike. It would also help to detect smaller and weaker twitches that would be an early warning of RLS. A less intrusive and bulky prototype with built in recorder would aid in more accurate detection of vibrations by being less restrictive to the patient. The subject could go through their normal day without being attached to wires and a computer, giving a truer representation of symptoms.

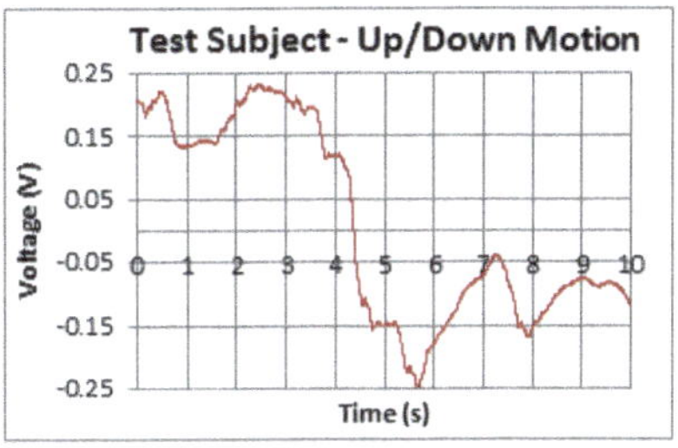

Fig. 10. Beam on subject, up/down motion.

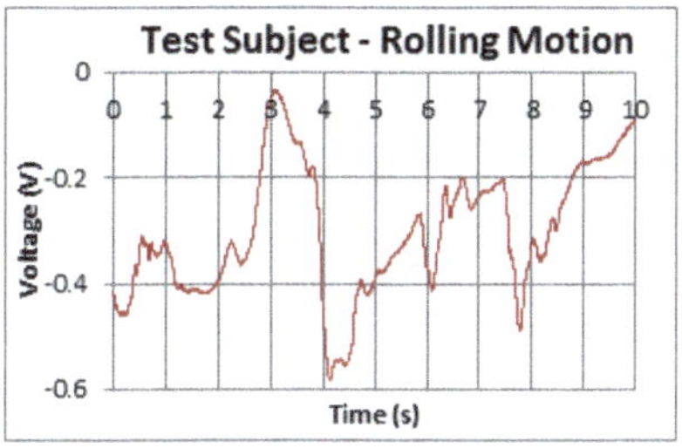

Fig. 11. Beam on subject, rotary motion.

6 Conclusions

RLS is an incurable neurological disease that is difficult to detect with current techniques. The rapid twitching of muscles in RLS patients can generate small vibrations that can be detected by a piezoelectric cantilever beam. Data is collected over a period of time and is plotted to show trends in fluctuating voltage. Testing proved that monitoring of RLS is a viable application of piezoelectric materials, although more refinement of the technique is necessary, specifically with regards to quantifying leg motions from the signal. Using a piezoelectric beam is more compact and less intrusive than the current monitoring machines and through tracking changes in output voltages, the symptoms of RLS can be more easily detected and diagnosed.

Acknowledgements. We thank Thomas Kunkel for assistance in 3D printing of the prototype housing.

References

1. Ekbom, K., Ulfberg, J.: Restless legs syndrome. J. Intern. Med. **266**, 419–431 (2009)
2. Earley, C.J.: Restless legs syndrome. N. Engl. J. Med. **348**(21), 2103–2109 (2003)
3. Hammock, M.L., Chortos, A., Tee, B.C., Tok, J.B., Bao, Z.: 25th anniversary article: the evolution of electronic skin (e-skin): a brief history, design considerations, and recent progress. Adv. Mater. **25**(42), 5997–6038 (2013)
4. Giurgiutiu, V.: Structural Health Monitoring with Piezoelectric Wafer Active Sensors. Elsevier, Amsterdam (2005)
5. Huang, Y., Dong, W., Huang, T., Wang, Y., Xiao, L., Su, Y., Yin, Z.: Self-similar design for stretchable wireless LC strain sensors. Sens. Actuators A **224**, 36–42 (2015)
6. Lipomi, D.J., Vosgueritchian, M., Tee, B.C., Hellstrom, S.L., Lee, J.A., Fox, C.H., Bao, Z.: Skin-like pressure and strain sensors based on transparent elastic films of carbon nanotubes. Nat. Nanotechnol. **6**(12), 788 (2011)
7. Jaffe, B., Cook, J.W.R., Jaffe, H.: Piezoelectric Ceramics. Academic Press Inc., New York (1971)
8. Sodano, H.A., Inman, D.J., Park, G.: A review of power harvesting from vibration using piezoelectric materials. Shock Vib. Dig. **36**(3), 197–206 (2004)

9. Booth, R., Goldsmith, P.: A wrist-worn piezoelectric sensor array for gesture input. J. Med. Biol. Eng. **38**(2), 284–295 (2018)
10. Mahbub, I., Pullano, S.A., Wang, H., Islam, S.K., Fiorillo, A.S., To, G., Mahfouz, M.R.: A low-power wireless piezoelectric sensor-based respiration monitoring system realized in CMOS process. IEEE Sens. J. **17**(6), 1858–1864 (2017)
11. Curry, E.J., Ke, K., Chorsi, M.T., Wrobel, K.S., Miller, A.N., Patel, A., Kim, I., Feng, J., Yue, L., Wu, Q., Kuo, C.L.: Biodegradable Piezoelectric Force Sensor. Proc. Natl. Acad. Sci. **115**, 909 (2018)
12. Dong, W., Xiao, L., Hu, W., Zhu, C., Huang, Y., Yin, Z.: Wearable human–machine interface based on PVDF piezoelectric sensor. Trans. Inst. Meas. Control **39**(4), 398–403 (2017)

The Influence of the Clearance on the Motion of the Reciprocating Compressor Connecting Rod

Shouguo Cheng[1,2](✉)

[1] School of Mechatronics Engineering and Automation, Shanghai University, Shanghai, China
chengshouguo@163.com
[2] Department of Mechanical and Electrical Engineering, Jiangyin Polytechnic College, Jiangyin, China

Abstract. The effect of revolute joint clearance on reciprocating compressor's motion is studied. The dynamic equations of the system are established through the Lagrange equation, and the clearance is equivalent to a rod without mass. The effects of clearance size and crank speed on the acceleration of the connecting rod in horizontal and vertical directions and the angular acceleration are discussed. The results show that the clearance has great influence on the motion parameters of the connecting rod. At last, the Lyapunov exponent and phase portrait are utilized in the analysis of nonlinear.

Keywords: Reciprocating compressor · Clearance · Massless virtual link
Lyapunov exponent · Connecting rod

1 Introduction

Reciprocating compressor is a kind of machine widely used by petrochemical enterprises. Many processes need to be pressurized by compressor. Because of the structural characteristics of reciprocating compressor and the complexity of its operating conditions, the failure rate of reciprocating compressor is very high in the process of operation. There are a lot of reports and reasons for the failure and accident of compressor in the enterprise every year. It is understood that in some petrochemical enterprises, two or three cases of reciprocating compressor occurred in one year, such as the fracture of the piston rod, which caused a great threat to the normal operation of the production. The forced parking repair caused by the accident has brought huge economic losses to the enterprise. Therefore, in order to establish a relatively perfect on-line condition monitoring and protection system for reciprocating compressors, the accident rate will be significantly reduced.

In the actual operation of the equipment, such as the correct early prediction and diagnosis of the hidden faults, the fault location can be accurately judged under the condition of the compressor without decomposition. By means of or relying on advanced sensor technology and dynamic testing technology and computer signal processing technique, the abnormal parts and causes of the equipment are analyzed.

K. Deng et al. (Eds.): ICMIR 2018, AISC 856, pp. 931–939, 2019.
https://doi.org/10.1007/978-3-030-00214-5_115

Reducing and preventing accidents and improving the economic benefits of production play a great role in promoting.

The study of clearance mechanism can provide a research basis for joints wear fault. Many scholars have done a lot of research on dynamics of mechanisms with clearance. Dubowsky [1, 2] discussed the dynamics of joint clearance sizes of four bar linkage decades ago. Flores et al. [3–5] has done a great deal of work on the dynamic research of clearance mechanism, including the clearance of the revolute joints and the clearance of the translational Joints, the two-dimensional plane mechanism and the three-dimensional space mechanism, and the effect of lubrication and friction on the clearance mechanism. Khemili and Romdhane [6] have discussed the influence of joint clearance on the kinematic characteristics of crank slider mechanism through software simulation and experimental verification. Jie and Qi [7] investigated the finite element method for describing the dynamics of multi-body systems, and established the contact force model based on the friction force and penalty function method.

Flores [8] has proposed a method to quantify joint wear in revolute clearance. The contact force model is used to analyze the uneven distribution of wear phenomena on the joint surface. Tian et al. [9] proposed a comprehensive method for studying the coupling dynamics of a multi body gear system supported by ElastoHydroDynamic lubricated cylindrical joints. Zhang et al. [10] proposed an optimized design method based on the experiment to optimize the performance of a mechanical system with a revolute joint clearance within the range of different design parameters. Zhao et al. [11] Used the engineering application software to model and analyze the reciprocating compressor and optimize the system parameters with genetic algorithm. Finally, the effectiveness of the method is verified by experiments.

Most of the researches make use of slider or crosshead analysis to illustrate the effect of clearance on the system, which is also convenient for measurement. In this paper, the effect of clearance on the connecting rod is mainly studied. The massless link is used as the clearance model, and the system equation is established by using the Lagrange equation, and the solution is solved by the matlab/simulink. Finally, the results and the nonlinear characteristics of the system are analyzed.

2 Reciprocating Compressor Model with a Clearance

2.1 Joint Clearance Model

The role of clearance in motion mechanisms is often considered as the following three models.

The classical collision model of clearance joint is a collision model based on the unilateral constraint, which divides the motion of the secondary elements into the 3 states of the collision, collision and separation.

The continuous contact force model of clearance joint is a collision model based on force constraint. The model divides the motion state of the accessory elements with clearance into free motion state and contact deformation state. When the two components contact, the contact deformation and contact force between the components are changed from zero.

The continuous contact model of the clearance pair is proposed by Earles. It is assumed that the gap collision process is very short, and the component is mainly in contact state during the system movement, so the gap is described with the mass rod. The continuous contact model treats the gap as a rigid rod without mass, that is, the clearance bar. The original clearance mechanism is transformed into a multi rod and clearance mechanism [12]. The massless link model is used to show the effect of clearance on the system.

2.2 The Dynamic Equation of the System

The simplified reciprocating compressor model is shown in Fig. 1. The closed chain schematic diagram of the system is shown in Fig. 2. Because of the special clearance model, the mechanism has two degrees of freedom.

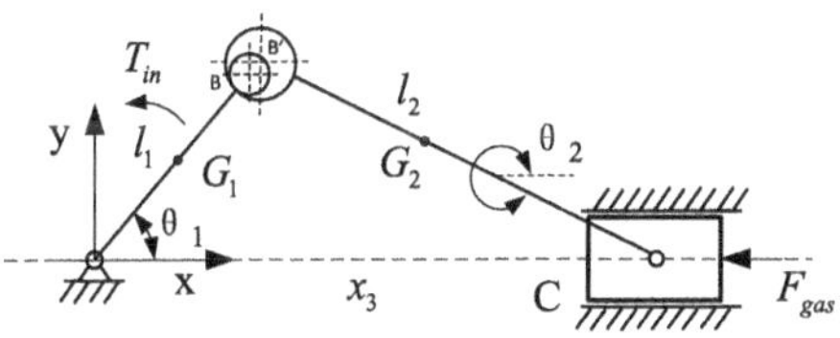

Fig. 1. Simplified reciprocating compressor model with a joint clearance

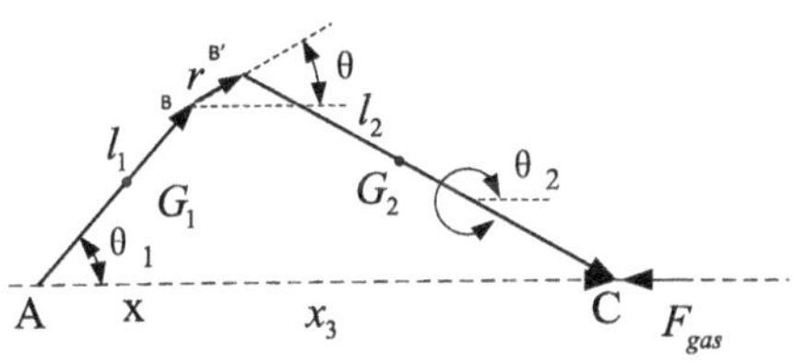

Fig. 2. Closed chain schematic diagram of the system with a joint clearance

In Figs. 1 and 2, θ_i represents the angle between the component and the horizontal. l_i represents the length of the component, and G_i represents the centroids of the component, i = 1, 2. The velocity and acceleration of the two directions of the centroid of the connecting rod and the angular velocity and angular acceleration of the connecting rod rotation are considered.

Assuming that the crank rotates uniformly, the displacement of the crosshead as generalized coordinate, it is only necessary to list a Lagrange equation:

$$\frac{d}{dt}\left(\frac{\partial E}{\partial \dot{q}}\right) - \frac{\partial E}{\partial q} + \frac{\partial U}{\partial q} = Q \tag{1}$$

Where E is the kinetic energy of the system, U is the potential energy of the system, q is the generalized coordinate, and Q is the corresponding generalized force.

The simulation model of the system is built by using Simulink in MATLAB software. After debugging and running, the relationship between the kinematic parameters of the connecting rod and time is calculated.

3 Results and Discussions

The influence of crank speed and clearance size on the motion of connecting rod is discussed in this section. The parameters used for compressor are Table 1 [11].

3.1 Simulation and Validation at Different Driving Speeds

The model is selected as the case study to investigate the dynamic characteristics of the mechanism with the joint clearance at different driving speeds. As shown in Figs. 3, 4 and 5, the rotational speeds are 300 rpm, 480 rpm and 600 rpm, respectively and the clearance size is 0.2 mm. Each figure compares the acceleration of the ideal joint with the clearance joint.

Figure 3 represents Connecting rod X direction acceleration at different driving speeds. Figure 4 represents Connecting rod X direction acceleration at different driving speeds. Figure 5 represents Connecting rod X direction acceleration at different driving speeds.

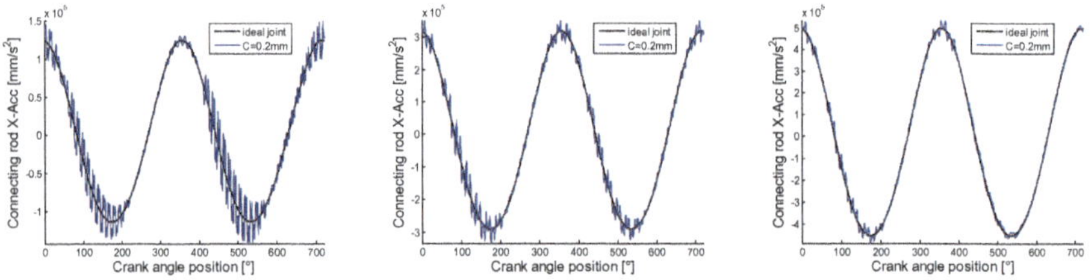

(a) 300 rpm (b) 480 rpm (c) 600 rpm

Fig. 3. Connecting rod X direction acceleration at different driving speeds

It can be seen from Figs. 4, 5 and 6 that in the case of 0.2 mm clearance, the influence of crank speed on the motion of the connecting rod is disproportionate, because the length of the ordinate is not the same. It is easy to draw the conclusion that the influence of clearance to the connecting rod Y direction acceleration is greater than the X direction acceleration and the influence of clearance on the rotation angular velocity of the connecting rod is greater than Y direction acceleration.

3.2 Influence of Clearance Size

As manufacturing and assembly will cause clearances in the joint, with the continuous operation of reciprocating compressors, friction and wear will occur, making the clearance increase. The joint clearance less than 0.3 mm is generally considered acceptable, and the clearance greater than 0.3 mm must be shut down for maintenance. When the clearance is 0.1 mm, 0.3 mm, and 0.5 mm, Figs. 6, 7 and 8 show the connecting rod X direction acceleration, connecting rod Y direction acceleration and angular acceleration of connecting rod in the two cycles.

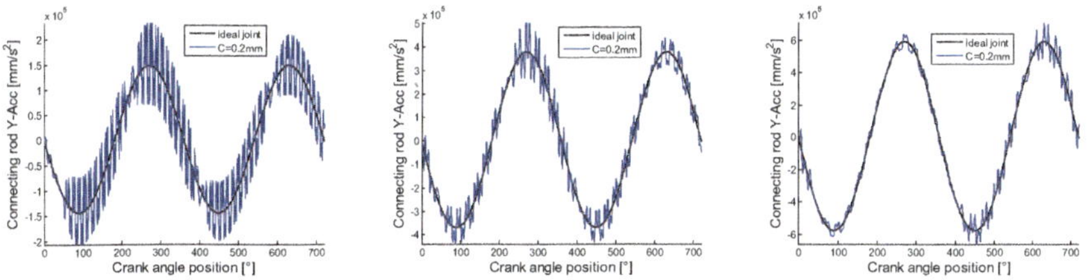

(a) 300 rpm (b) 480 rpm (c) 600 rpm

Fig. 4. Connecting rod Y direction acceleration at different driving speeds

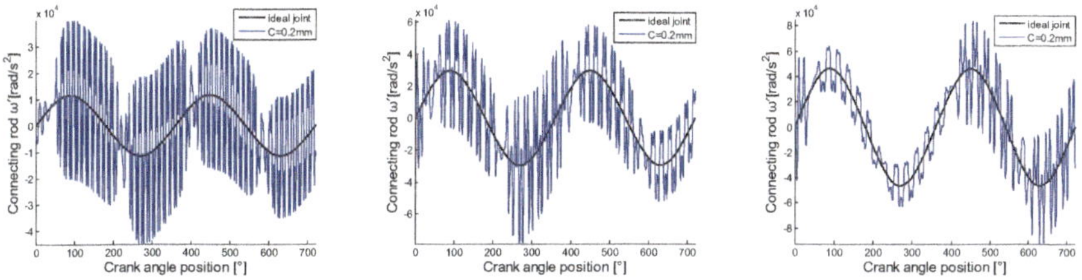

(a) 300 rpm (b) 480 rpm (c) 600 rpm

Fig. 5. Angular acceleration of connecting rod at different driving speeds

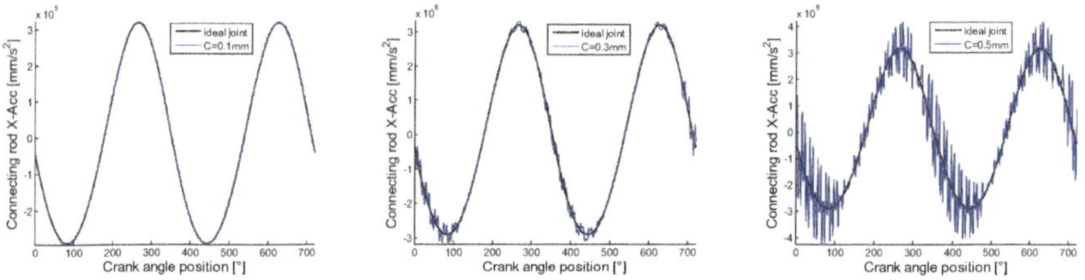

(a) (b) (c)

Fig. 6. Connecting rod X direction acceleration at different clearance size

It can be seen from Fig. 8 that the acceleration in the Y direction of the connecting rod varies with the increase of clearance.

It can be seen from Fig. 7 that the connecting rod X direction acceleration changes more and more with the increase of clearance.

Similarly, with the increase of clearance, the angular acceleration of connecting rod is changing more and more obvious, just like Fig. 9.

Compared with Figs. 6, 7 and 8, the influence of clearance on the centroid acceleration of the Y direction of the connecting rod is greater than that of the X centroid acceleration of the connecting rod. The effect of clearance on the rotational acceleration is greater than that of the accelerometer.

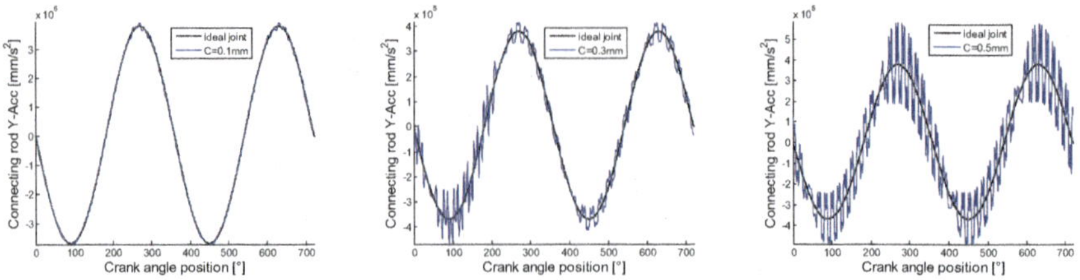

(a) (b) (c)

Fig. 7. Connecting rod Y direction acceleration at different clearance size

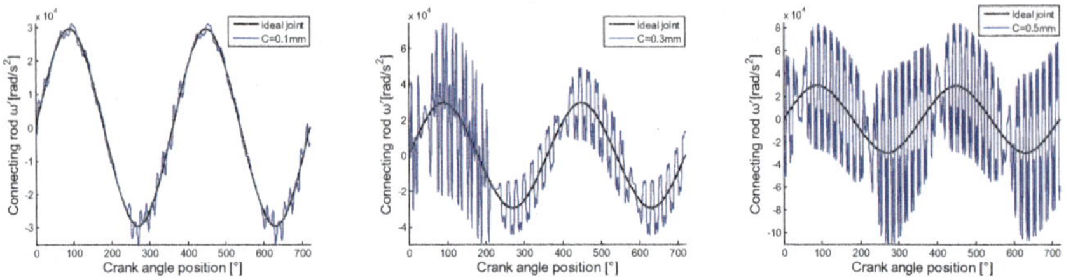

(a) (b) (c)

Fig. 8. Connecting rod angular acceleration at different clearance size

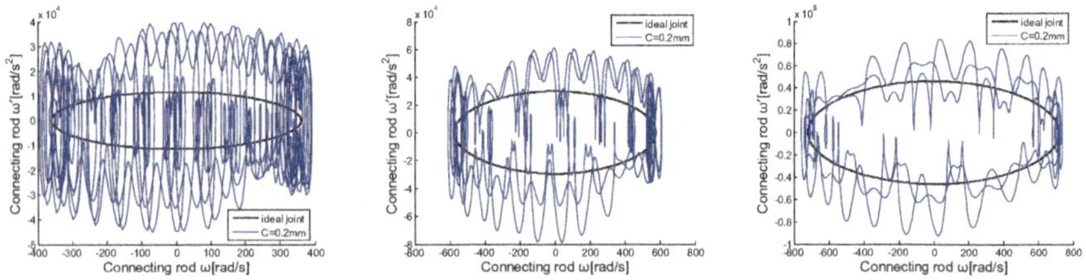

(a) 300 rpm (b) 480 rpm (c) 600 rpm

Fig. 9. Phase diagram of the rotational speed and angular acceleration of the connecting rod at different driving speeds

4 Nonlinear Analysis of Calculation Results

Phase portrait and maximum Lyapunov exponent are often used as judgement basis for system nonlinearity. According to the dynamic analysis of systems with clearance, it is easy to draw phase portrait and calculate maximum Lyapunov exponent. From the previous analysis, the angular acceleration of the connecting rod is strongly influenced by the clearance, and the next phase diagram is calculated by the angular velocity and angular acceleration of the connecting rod, and Lyapunov exponent is used to calculate the angular acceleration of the connecting rod.

Figure 9 is a phase diagram of the rotational speed and angular acceleration of the connecting rod under different crank speeds. Figure 10 is a phase diagram of the rotational speed and angular acceleration of the connecting rod at different clearance size.

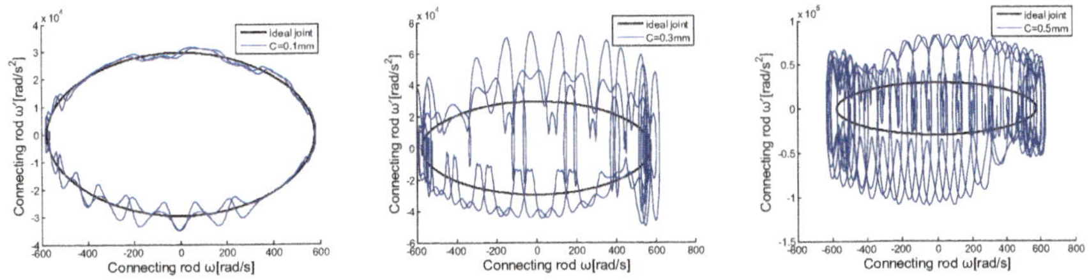

(a) (b) (c)

Fig. 10. Phase diagram of the rotational speed and angular acceleration of the connecting rod at different clearance size

It can be seen from the phase diagram that with the increase of clearance and the increase of crank rotation speed, it may show chaotic behavior. The values of maximum Lyapunov exponent are calculated by Wolf method in different states of system. The value of Fig. 11(a) calculated is 0.0553, and the value of Fig. 11(b) is 0.1203, and the value of Fig. 11(c) is 0.1507. The value of Fig. 12(a) calculated is 0.1090, and the value of Fig. 12(b) is 0.1335, and the value of Fig. 12(c) is 0.1664.

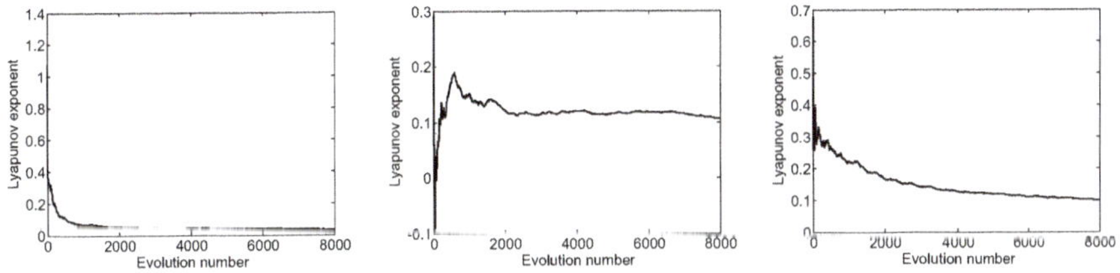

(a) (b) (c)

Fig. 11. Lyapunov exponent at different clearance size

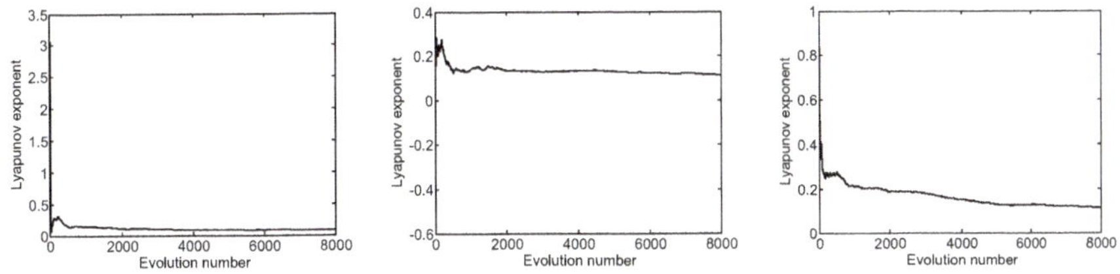

(a) 300 rpm (b) 480 rpm (c) 600 rpm

Fig. 12. Lyapunov exponent at different driving speeds

It can be seen from Figs. 11 and 12 that the maximum Lyapunov exponent of the system is greater than zero, indicating that the system is in chaotic.

5 Conclusions

The objective of this work is to discuss how the joint clearance and the crank velocity affect the performance of the connecting rod on the reciprocating compressor. The system equation was built using the Lagrange methodology. The results show that the influence of these parameters on the connecting rod of the reciprocating compressor exists. Through the phase portrait and maximum Lyapunov exponent, it can be seen that the bigger the clearance is, the more obvious the chaos is. The distance between the connecting rod and the connecting rod is closer than that of the crosshead, and the phenomenon is more obvious. When the crank is rotating, the connecting rod moves linearly in two directions and rotates. The kinematic parameters of the connecting rod are much richer than the parameters on the crosshead. Studying the change of connecting rod parameters can provide effective basis for diagnosing reciprocating compressor faults.

Acknowledgments. This work was supported by National Natural Science Foundation of China (51575331).

References

1. Dubowsky, S.: On predicting the dynamic effects of clearances in planar mechanisms. J. Eng. Ind. **96**(1), 317–323 (1974)
2. Dubowsky, S., Norris, M., Aloni, E.: An analytical and experimental study of the prediction of impacts in planar mechanical systems with clearances. J. Mech. Transm. Autom. Des. **106**, 444–451 (1984)
3. Flores, P., Ambrosio, J., Claro, J.C.P.: A study on dynamics of mechanical systems including joints with clearance and lubrication. Mech. Mach. Theory **41**, 247–261 (2006)
4. Machado, M., Costa, J., Seabra, E., Flores, P.: The effect of the lubricated revolute joint parameters and hydrodynamic force models on the dynamic response of planar multibody systems. Nonlinear Dyn. **69**(1–2), 635–654 (2012)
5. Flores, P., Ambrósio, J., Claro, J.C.P., Lankarani, H.M.: Spatial revolute joints with clearance for dynamic analysis of multibody systems. Proc. Inst. Mech. Eng. Part K J. Multibody Dyn. **220**(4), 257–271 (2006)
6. Khemili, I., Romdhane, L.: Dynamic analysis of a flexible slide-crank mechanism with clearance. Eur. J. Mech. A. Solids **27**, 882–898 (2008)
7. Zhang, J., Wang, Q.: Modeling and simulation of a frictional translational joint with a flexible slider and clearance. Multibody Syst. Dyn. 1–23 (2015)
8. Flores, P.: Modeling and simulation of wear in revolute clearance joints in multibody systems. Mech. Mach. Theory **44**, 1211–1222 (2009)
9. Tian, Q., Xiao, Q., Sun, Y., Hu, H., Liu, H., Flores, P.: Coupling dynamics of a geared multibody system supported by ElastoHydroDynamic lubricated cylindrical joints. Multibody Syst. Dyn. **33**, 259–284 (2015)

10. Zhang, Z., Xu, L., Flores, P., Lankarani, H.M.: A Kriging model for the dynamics of mechanical systems with revolute joint clearances. J. Comput. Nonlinear Dyn. **9**(3), 1–13 (2013)
11. Zhao, H.Y., Xu, M.Q., Wang, J.D., Li, Y.B.: A parameters optimization method for planar joint clearance model and its application for dynamics simulation of reciprocating compressor. J. Sound Vib. **344**, 416–433 (2015)
12. Seneviratne, L., Earles, S.: Chaotic behaviour exhibited during contact loss in a clearance joint of a four-bar mechanism. Mech. Mach. Theory **27**, 307–321 (1992)

Research on the Clamping Manipulator of Drill Pipe Automatic Transmission System

Yongbai Sha[1(✉)], Xiaoying Zhao[2], and Tiancheng Ouyang[1]

[1] School of Mechanical Science and Engineering, Jilin University, Changchun, China
shayb@jlu.edu.cn

[2] Changchun Vocational Institute of Technology, Changchun, China

Abstract. Clamping manipulator is the key component of drill pipe automatic transmission system. On the basis of the analysis of the design requirements of the clamping manipulator, the clamping manipulator was designed, and its mechanism and working principle are introduced, and the force of the gripper of the clamping manipulator was analyzed, and the required driving force was calculated.

Keywords: Drill pipe automatic transmission system · Clamping manipulator Working principle · Driving force · Simulation analysis

1 Introduction

The drill pipe automatic transmission system is the processing system of a high automation level in the petroleum machinery industry and the deep drilling industry, mainly to achieve clamping, lifting and transporting drilling rod and so on in the tripping process [1]. Clamping manipulator is the end actuator of drill pipe automatic transmission system, and it is one of the key parts. Because the drill pipe automatic transmission system is suspended under the racking platform which is higher than the drill floor, and the working environment is bad, once the manipulator has problems, it will directly affect the safety of the whole drilling system. Therefore, the research of clamping manipulator is of great importance [2, 3]. Whether it can work safely or reliably will directly affect the safety of drilling workers and drilling efficiency.

2 The Overall Structure of the Drilling Pipe Automatic Transmission System and the Design Requirements of the Clamping Manipulator

2.1 The Overall Structure of the Drilling Pipe Automatic Transmission System

The clamping manipulator is the end actuator of the automatic transmission system of the drill pipe. Automatic transmission system is arranged under the racking platform, and it will take out the drill-pipe stand from the placing rack and move smoothly to the

K. Deng et al. (Eds.): ICMIR 2018, AISC 856, pp. 940–946, 2019.
https://doi.org/10.1007/978-3-030-00214-5_116

wellhead in the process of drilling, and it will move the drill pipe from the wellhead to the placing rack and discharged in order [4, 5]. The drill pipe automatic transmission system is showed as Fig. 1, which is mainly composed of translational mechanism, slewing mechanism, stretching mechanism, rectifying mechanism, lifting mechanism and clamping manipulator, and the clamping manipulator is the end actuator of the drill pipe automatic transmission system. The clamping manipulator is responsible for grasping the drill pipe, and in the process of transportation, the drill pipe should be run smoothly and cannot fall off or slide. Considering of the long drill-pipe stand, two clamping manipulators are clamped at the same time in order to ensure the reliability of the drilling pipe automatic transmission system. The distance between the two clamping manipulators is constant, and the displacement in vertical direction is constant during the drill pipe migration, thus drill rod's shaking is reduced and the operation process is more stable. In addition, the upper and lower clamping manipulator bear the gravity of drill pipe together. When one clamping manipulator fails, another clamping manipulator can prevent accidents.

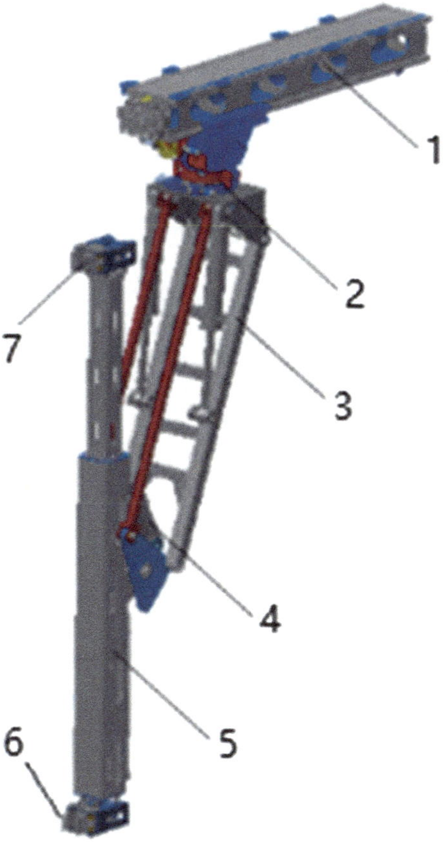

Fig. 1. Structure of drill pipe automatic transmission system. 1-translational mechanism; 2-slewing mechanism; 3-stretching mechanism; 4- rectifying mechanism; 5- lifting mechanism; 6,7-clamping manipulator

2.2 The Design Requirements of the Clamping Manipulator

(1) Parameter requirements: the diameter range of clamping drilling rod is Φ89–Φ 168 mm, and the maximum clamping weight is 2000 kg.
(2) The structure of the clamping manipulator is as compact as possible to avoid collision to the adjacent drill rod in the drilling pipe placing rack during the clamping process.
(3) The clamping manipulator should have sufficient clamping force and ensure the stability and reliability of the clamping action.
(4) Realizing automatic operation and improve work efficiency [6, 7].

3 Structure and Working Principle of the Clamping Manipulator

The structure of the clamping manipulator is shown in Fig. 2, and the front cover plate of the box is removed. It is mainly composed of box, connection block, hydraulic cylinder, block, mechanical gripper, friction block, clamp spring and sensor connection block.

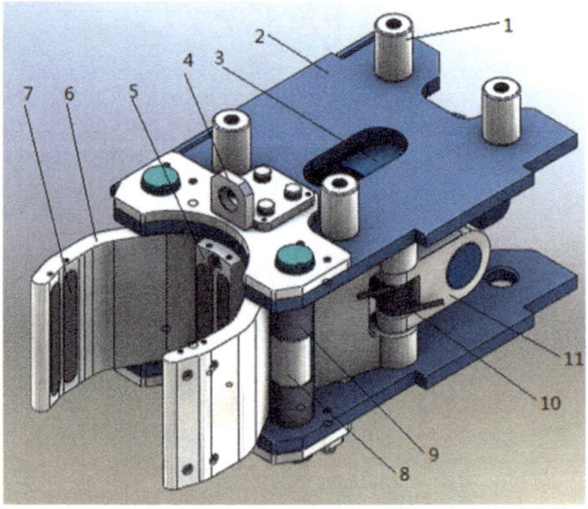

Fig. 2. Structure of clamping manipulator. 1- connection block; 2- box; 3- hydraulic cylinder; 4- sensor connection block; 5- block; 6- mechanical gripper; 7- friction block; 8-spacer sleeve; 9- bearing; 10- coil spring; 11- connecting plate

Clamping manipulator is equipped on both end of the sliding frame of the lifting mechanism, and the two manipulators are connected by bolts and connecting blocks. The piston rod of the hydraulic cylinder is fixed to the box, and the cylinder barrel can do expansion movement relative to the piston rod, and the movement of the mechanical gripper is carried through the connecting plate. Because of the effect of the coil spring torque, the outer surface of the mechanical gripper will be attached to the roller bearing. When the cylinder extends backward, the mechanical gripper will slide backward relative to the bearing, and the two mechanical grippers will contract, and then hold the drill pipe tightly; when the cylinder is retracted forward, the mechanical paw moves forward relative to the bearing, and the distance between the two gripper increases, thereby loosening the drill pipe. The clamping manipulator adopts three point clamping mode, and the friction block installed inside of the gripper can increase the friction force of the clamping drill rod, and ensure the reliability of the clamping. The sensor connection block is used to install the proximity switch. When the manipulator is near the drill pipe, the sensor detects the signal and then controls the action of the clamping manipulator. In the clamping and loosening process, this kind of clamping manipulator can ensure the synchronicity of the movement of the two sides of the mechanical paw, and without changing the gripper, it can clamp different sizes of drill pipe with a diameter of Φ89–Φ168 mm.

4 Calculation of Driving Force of Clamping Manipulator

Because of the load of the clamping manipulator - the weight of the drill pipe stand is larger, to ensure the reliability of the work, the clamping manipulator must provide sufficient driving force. Using the maximum diameter of Φ168 mm drill pipe to conduct stress analysis, when clamping the drill pipe, the drill rod is subjected to the force of three different directions from the clamping manipulator on the horizontal plane. The three force is the joint force F of the each mechanical paw to the drill pipe (its size is equal), and the pressure N of the block to the drill pipe, and the angle between the two clamping forces is 140°, as shown in Fig. 3.

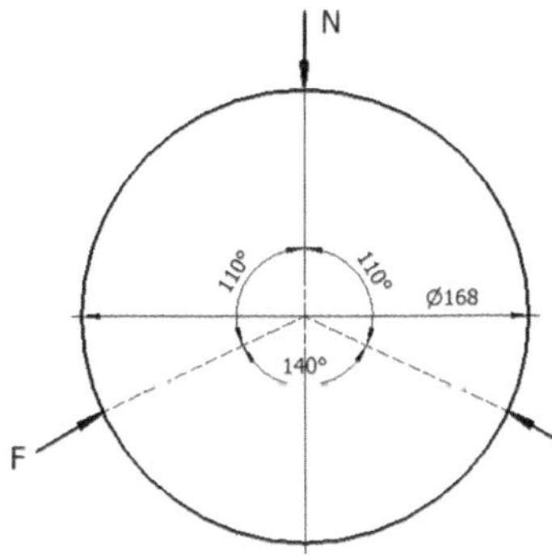

Fig. 3. Force diagram of drill pipe under clamping state

According to the relationship in the figure, it can be obtained:

$$N^2 = 2F^2 - 2F^2 \cos 40^\circ \tag{1}$$

In which N is the positive pressure of the block to the drill pipe, and F is the positive pressure of the mechanical gripper to the drill pipe.

According to the formula (1), it can be obtained:

$$N = 0.68F \tag{2}$$

The friction force of the clamping manipulator to the drill rod is as follow:

$$f_1 = N\mu,$$
$$f_2 = F\mu = 1.47f_1$$

In which f_1 is the friction force of the block to the drill pipe, and f_2 is the friction force of the mechanic gripper to the drill pipe, μ is the friction coefficient between the friction block and the drill pipe where 0.15 is taken.

The gravity of the drill pipe stand is balanced with the friction force, that is:

$$f_1 + 2f_2 = 3.94f_1 = mg \tag{3}$$

In which m is the quality of the drill pipe stand, and m = 2000 kg.

Figure 4 shows the force situation of the mechanical gripper in the horizontal direction in the state of the clamping manipulator clamping the drill pipe.

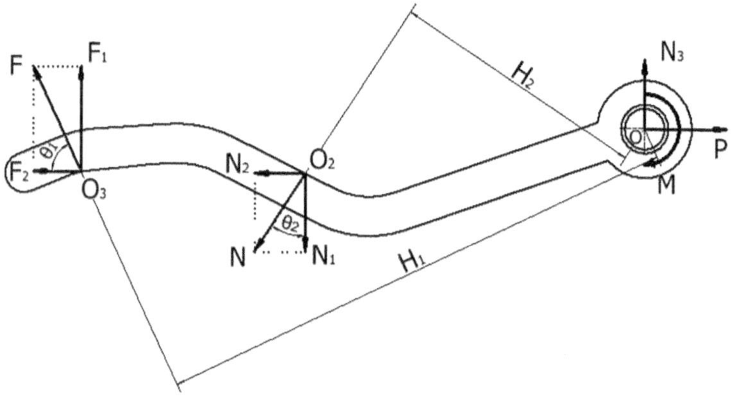

Fig. 4. Force situation of the mechanical gripper in the state of the clamping state

In the clamping state, the force of the mechanical gripper is balanced, and the equations can be listed as follow:

$$\begin{cases} F \cdot H_1 + M = N \cdot H_2 \\ F_2 + N_2 = P \\ F_1 + N_3 = N_1 \end{cases} \tag{4}$$

In the figure and the forum, F is the positive pressure of the drill pipe to the mechanical paw; and H_1 is the force arm of the positive pressure F and H_1 equals 270 mm; and N is the pressing force of the roller bearing to the mechanical paw; and H_2 is the force arm of the pressing pressure N and H_2 equals 130 mm; and M is the bending moment produced by the coil spring at point O and taken as 10 kN mm; and P and N_3 are the forces in the horizontal and vertical direction of the gripper at the hinge point O, and P is half of the total driving force of the hydraulic cylinder; and F_1, F_2, N_1, N_2 are respectively the vertical and horizontal component forces of the positive pressure F and the pressing force N, and θ_1, θ_2 are the angles of the corresponding force and the join force, in which θ_1 equals 70° and θ_2 equals 25°.

It can be seen from the diagram that:

$$F_2 = F\cos\theta_1, N_2 = N\sin\theta_2 \tag{5}$$

Combining the above formula (1)–(5) and substitution the known parameters, it can be obtained that $P = 71$ kN, so the driving force of the hydraulic cylinder is $P_{\text{all}} = 142$ kN.

5 Simulation Analysis

Simulation analysis is carried out to the designed clamping manipulator. Figure 5 is the driving force simulation curve of clamping hydraulic cylinder, and Fig. 6 is the horizontal displacement curve of cylinder of the clamping hydraulic cylinder, and the simulation time is 2.5 s.

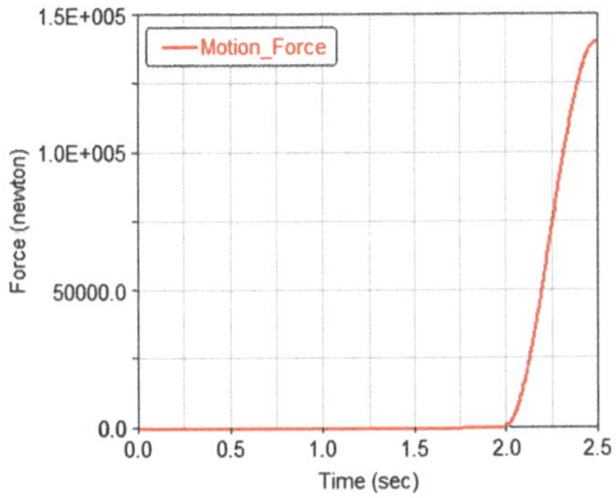

Fig. 5. Driving force F simulation curve

According to the simulation results, when t is from 0 to 2 s, the driving force is very small, and the cylinder starts to move along the horizontal direction, and the motion of the cylinder drives two grippers to move backward relative to the rolling bearing, so that the two paws are closed. At $t = 1$ s, two paws collide with the drill pipe and reach the specified position, and the displacement of cylinder is 53 mm. At $t = 2$ s, the hydraulic system pressure increased, and the driving force of the hydraulic cylinder increased

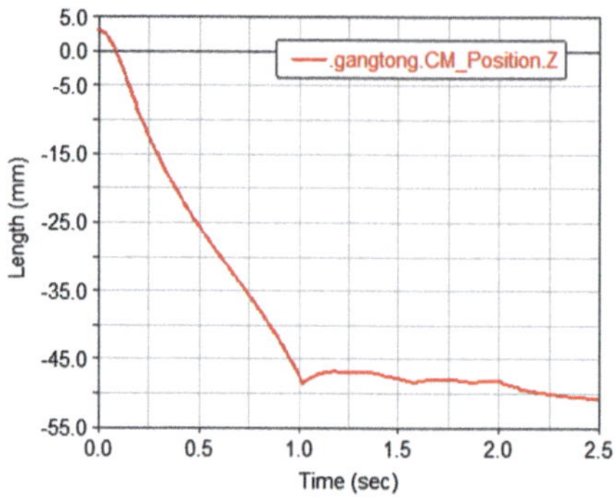

Fig. 6. Cylinder horizontal displacement curve

rapidly. When t = 2–2.5 s, the two paws gradually clamped the drill pipe stand, and the driving force increased to 142 kN.

From the simulation results, it can be seen that in the working process, the hydraulic cylinder moves quickly and runs smoothly, and the driving force changes evenly and the fluctuation is small, which meets the design requirements.

6 Conclusion

According to the requirements of load and action, the clamping manipulator of drill pipe automatic transmission system is designed. The working principle is analyzed, and the driving force of the clamping manipulator is calculated, and the performance of the mechanism is verified through simulation analysis.

Acknowledgments. This paper is funded by the project (SinoProbe-09-05) from the Ministry of Land and Resources and expressing thanks for this.

References

1. Shibang, M., Rongjie, H., Dangqin, X., Linhai, Z.: Design and analysis of drill pipe racking manipulator. China Pet. Mach. **43**(11), 48–53 (2015)
2. Wenfeng, Y., Qian, L., Hu, Y., Xiaoshi, M.: Motion simulation and strength analysis of drill pipe handing manipulator. Mech. Transm. **40**(12), 82–84, 138 (2016)
3. Zohoor, H., Khorsandijou, S.M.: Dynamic model of a flying manipulator with two highly flexible links. Appl. Math. Model. **32**, 2117–2132 (2008)
4. Jiliang, Z.: Structural Design and Simulation Analysis of Pipe Automatic Transferring System. Jilin University, Changchun (2012)
5. Sha, Y., Zhu, J., Zhao, X., et al.: Structural research of pipe automatic handing system of land rig. Mater. Eng. Autom. Control **562**, 958–961 (2012)
6. Jiaan, L., Lin, Y., Siyun, L., Huangyuan, W.: A new type of automatic pipe arranging manipulator. China: CN203515388U, 2014-04-02
7. Johnson, J.: Better tube transport. Motion Syst. Des. **52**(12), 14–15 (2010)

Design and Simulation of a Novel FBG Accelerometer

Qi Jiang, Yongxin Zhu(✉), and Xiuyu Li

School of Control Science and Engineering, Shandong University, Jinan, China
zhuyongxin1206@163.com

Abstract. Vibration signals contain a lot of information that reflects the working status of the mechanical arms, which is of vital significance for vibration suppression monitoring and fault diagnosis. Hence, it is meaningful to develop a new type of sensor that detects vibration signals accurately and conveniently. Fiber Bragg Grating (FBG) sensors are increasingly applied because of their many advantages like high sensitivity, low weight and small size, anti-electromagnetic interference and easy multiplexing, etc. In this work, the model of an FBG accelerometer has been designed which contains a rhombus hollow lattice cylinder. Finite Element Simulations was used to analyze dynamic behavior of the sensor with a simulated resonance frequency of 297 Hz and good frequency response range from 10 to 100 Hz. Theoretical sensitivity of the sensor of was 71.56 pm/g. The sensor is supposed to be used for vibration monitoring of low frequency of mechanical arms.

Keywords: Accelerometer · Fiber Bragg Grating (FBG) · Dynamic monitoring Finite Element Simulations

1 Introduction

In recent years, FBG vibration sensor have gained a lot of applications in structural health monitoring, mechanical fault diagnosis, and earthquake prediction [1, 2]. The principle of FBG acceleration sensor can be divided into intensity modulation type and wavelength modulation type. The intensity modulation methods mainly include edge filtering, matched filtering, etc., with advantages of high sensitivity and high detection frequency [3]. However, it can be easily affected by the intensity of light source and temperature fluctuations, without the advantages of serial multiplexing. Wavelength demodulation has been the main signal detection method for fiber grating sensors, which uses the wavelength information as the output signal and is not affected by the fluctuation of the light intensity. This paper will mainly introduce FBG acceleration sensors by wavelength demodulation.

The theory of FBG accelerometer is generally inertial principle. The inertial mass is excited by the external vibration signal, resulting in the deformation of the sensor's sensitive area. This strain is converted by the sensor structure to axial expansion or bending of the FBG, resulting in a drift in the Bragg wavelength, which indirectly indicates acceleration [4].

K. Deng et al. (Eds.): ICMIR 2018, AISC 856, pp. 947–953, 2019.
https://doi.org/10.1007/978-3-030-00214-5_117

The structure of the sensor can be divided into many types, among which there are mainly cantilever structures, hinge structures, mandrel structures, and embedded structures. In 2014, Khan et al. [5] proposed a temperature self-compensating FBG accelerometer based on a special cantilever beam structure. The structure adopted an all-padding package scheme combined with double-grating temperature compensation technology to realize acceleration-temperature-insensitive measurement. The results showed that in the measurement range of 0–9 g, the flat zone was 45–145 Hz, the sensitivity of acceleration was 306 pm/g, and the temperature-insensitive characteristics of the acceleration sensor were verified in the range of 25–85 °C. In 2017, Dai et al. [6] proposed a FBG accelerometer with two symmetric bending spring plates. The FBG produced axial strains of equal magnitude but opposite directions, resulting in a doubled wavelength shift. Test results showed that the accelerometer's sensitivity was more than 1000 pm/g in the frequency range from 0.7 Hz to 20 Hz, which was suitable for low-frequency vibration detection.

Li et al. [7] proposed a new type of diaphragm-based FBG vibration sensor with temperature compensation based on the lateral characteristics of the fixed optical fiber with both ends fixed, with a resonance frequency of 300 Hz. The sensitivity in the range of 10–150 Hz is 31.25 pm/g, and the linear and relative sensitivity errors were 2.21% and 10%, respectively.

In this paper, the model of a novel acceleration sensor using composite materials was designed and optimized. By finite element analysis software ANSYS, the dynamic behavior of the sensor has been simulated, which provides reference for structure optimization and manufacturing.

2 Sensing Principle and Structural Design

2.1 Sensing Principle

The Bragg formula is given by $\lambda_B = 2n_{eff}\Lambda$, where λ_B is the center wavelength of the back-reflected light (Bragg wavelength), n_{eff} is the effective refractive index, and Λ is the periodicity of the refractive index modulation [8]. These parameters will change if the Bragg grating is subjected to temperature change or mechanical deformation or variation. This influence of temperature and mechanical deformations on the Bragg wavelength value can be expressed as Eq. (1):

$$\frac{\Delta\lambda_B}{\lambda_B} = 2\left(\Lambda\frac{\partial n_{eff}}{\partial l} + n_{eff}\frac{\partial\Lambda}{\partial l}\right)\Delta l + 2\left(\Lambda\frac{\partial n_{eff}}{\partial T} + n_{eff}\frac{\partial\Lambda}{\partial T}\right)\Delta T \tag{1}$$

Equation (2) shows the strain effect on the Bragg wavelength shift. This strain depends on changes in the refractive index modulation period and the photoelastic effect. It can be expressed as follows:

$$\Delta\lambda_B = \lambda_B(1 - p_e)\varepsilon \tag{2}$$

where p_e is the effective elastic-optic coefficient of the fiber.

FBG-based sensors measure physical quantity by a transformation from magnitude (acceleration or displacement, etc.) to strain or temperature of the optic grating. The proposed sensor in this work consist of a body of rhombus hollow lattice cylinder, with an inertial mass on the top [9]. The accelerometer can be modelled as a one degree of freedom (DOF) mass-spring system (Fig. 1).

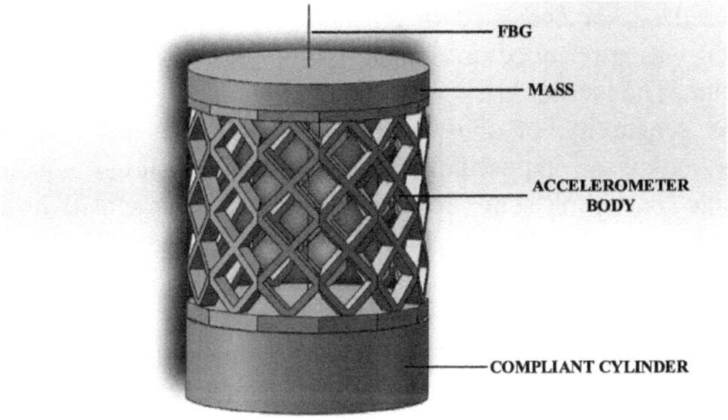

Fig. 1. Proposed accelerometer model

There are two parameters that are natural frequency and sensitivity which determined its performance. In addition, the damping ratio should be also considered. According to the simple 1-DOF harmonic oscillator (without damping) mechanism, the natural frequency (f_n) of the system can be given by

$$f_n = \frac{1}{2\pi} \frac{k_{eff}}{m_{eff}} \tag{3}$$

where k_{eff} and m_{eff} are the effective stiffness and mass.

According to the Newton's second law and the Hook's law, acceleration (a) can be formulated in Eq. (4). Sensor sensitivity (S) is defined as the ratio between the wavelength shift of the FBG and the acceleration value. Then, sensitivity can be expressed as a function with strain as an independent variable (Eq. (5)):

$$a = \frac{F}{m_{eff}} = \frac{k_{eff}\Delta L}{m_{eff}} \tag{4}$$

$$S = \frac{\Delta\lambda_B}{a} = \frac{\lambda_B(1 - p_e)\varepsilon}{\frac{k_{eff}\Delta L}{m_{eff}}} = \frac{\lambda_B(1 - p_e)\varepsilon m_{eff}}{k_{eff}\Delta L} \tag{5}$$

where ΔL is the increment of length of the sensor body.

2.2 Structural Design

In the design of a sensor, resonance frequency is an important parameter that determines the application range of the sensor. The designed FBG accelerometer is supposed to apply in the measurement of low frequency vibration of mechanical arms. The device should be designed to have less rigidity and damping, and minimize its transverse sensitivity to improve its transverse anti-jamming capability.

The proposed sensor consisted of an inertial mass, a sensor body, a compliant cylinder. Fiber was embedded in sensor body, pre-tensioned and boned in both extremes by glue. The inertial mass was 5 mm in length and 45 mm in diameter, which influences the sensitivity of the sensor together with the compliant cylinder. Accelerometer body was composed of 36 rhombus lattice hollows, which was designed using polyamide, providing some elasticity to increase the sensitivity of the sensor. Silica gel was chosen as compliant cylinder because of its good flexibility and low cost [10]. It was considered that the stress–strain behavior of silica gel is linear within a small range.

In general, the sensor proposed in this paper has the advantages of low weight and small size. It is 50 mm in length and 45 mm in diameter. Total weight of the sensor calculated by ANSYS is about 100 g. It will be processed by additive manufacturing in the next step of the study.

3 Finite Element Modeling and Simulations

ANSYS software was used as Finite element analysis tool to develop the numerical simulations of the accelerometer. Suitable materials was arranged for each part of the sensor in the simulation environment. The inertial mass used structural steel material. The body of the sensor designed in SolidWorks was imported into ANSYS. The mesh consisted of 201656 nodes and 105073 elements (Fig. 2). The material properties settings were shown as follows. Polyamide: Young's Modulus 1.59 GPa, Poisson's ratio 0.35 and density 1 g/cm^3; Silica gel: Young's Modulus 4.0 MPa, Poisson's ratio 0.47 and density 0.98 g/cm^3 [11].

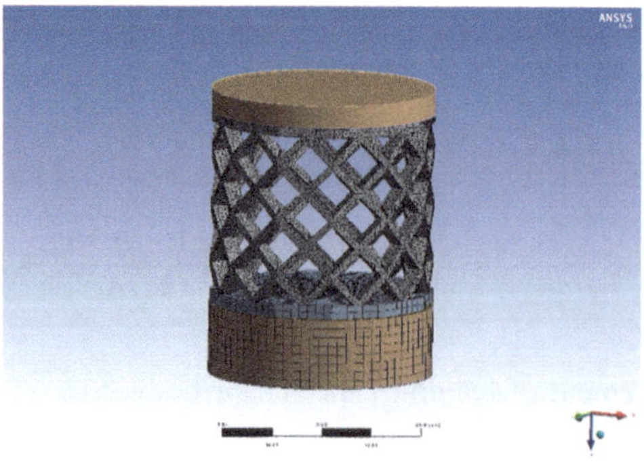

Fig. 2. Mesh of the accelerometer finite element model

3.1 Modal Analysis

Modal simulations was developed to obtain the vibration mode shapes and the resonance frequencies of the first four vibration modes, as shown in Fig. 3. The resonant frequencies obtained from the analysis were 84.20, 84.26, 202.87 and 297.45 Hz. The first and the second frequencies are corresponding to bending mode shapes of vertical directions. The third one is corresponding to a torsional mode around the vertical direction. The fourth frequency correspond to an axial mode which is main vibration mode of the sensor. In order to prevent other modes from disturbing the main mode, the first three frequencies should be avoided in the working frequency band.

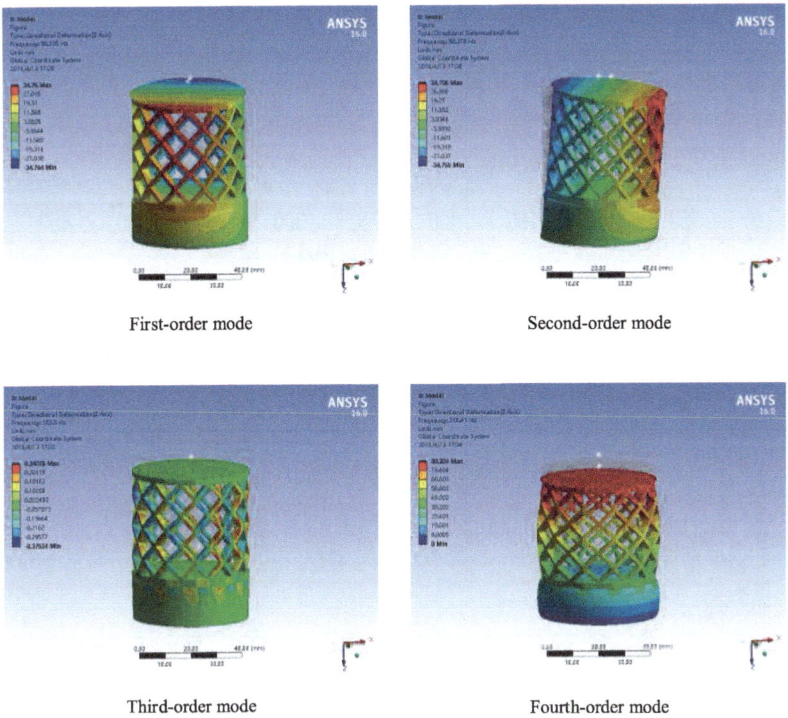

Fig. 3. First four modes of the sensor

3.2 Harmonic Response

In order to obtain the steady-state response of the sensor to different frequency sinusoidal signals, a sinusoidal acceleration signal with an amplitude of 1 g and a frequency range of 10–500 Hz was applied to the entire sensor with a frequency step of 10 Hz. Frequency response is shown in Fig. 4. The ordinate represents the axial deformation of the fiber among different frequencies. It shows clearly a peak at about 300 Hz, which is consistent with modal analysis (297 Hz). According to modal analysis, the interference negative effect of the bending and torsional mode shapes on the frequency response should be considered. Based on that, the working frequency band was selected in the flat area of 10–100 Hz (Fig. 5). The average deformation in the range from 10–100 Hz was 3.09×10^{-3} mm. Therefore, the sensitivity of the accelerometer of was 71.56 pm/g, computed from the fiber strain by Eq. (2) with values of p_e and λ_B equal to 0.19 and 1550 nm, respectively.

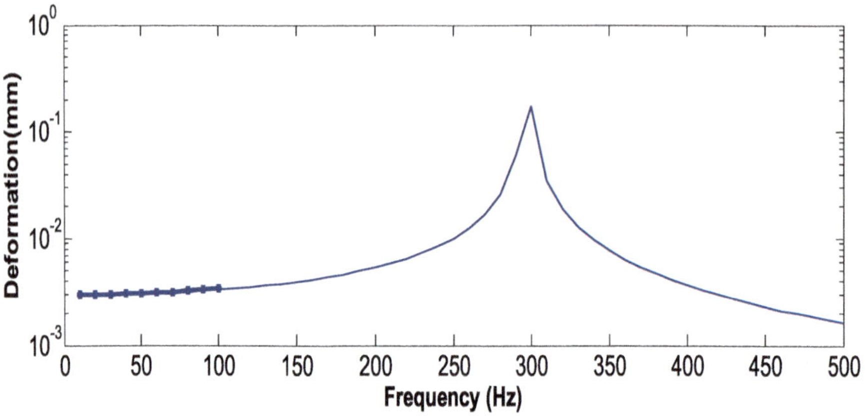

Fig. 4. Frequency response of 10–500 Hz

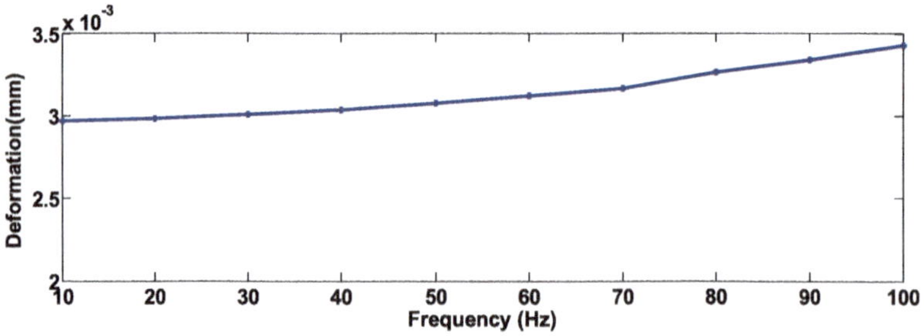

Fig. 5. Frequency response of 10–100 Hz

4 Conclusions

In the control of the mechanical arms, it is critical to suppress the vibration of the robot arm to achieve good control effects of overall system. Vibration signals are generally introduced as feedback to the control system. Therefore, it is meaningful to develop a new type of sensor that accurately and conveniently detects vibration signals. By using FBG sensors which are increasingly mature, acceleration signal can be measured well.

The proposed FBG accelerometer has many advantages like low weight (100 g) and small volume (45 mm diameter and 50 mm long). The theoretical sensitivity of the sensor is 71.56 pm/g and its simulated resonance frequency is 297 Hz. It will be processed by additive manufacturing and applied in vibration suppression monitoring of mechanical arms in further study.

Acknowledgments. This work was financially supported by National key research and development program (No. 2017YFB1302102).

References

1. Guo, Y., Zhang, D., Zhou, Z., et al.: Research progress in fiber Bragg grating accelerometer. Laser Optoelectron. Prog. **50**(6), 1–8 (2013)
2. Zhou, Z., Tan, Y., Liu, M., et al.: Actualities and development on dynamic monitoring and diagnosis with distributed fiber Bragg grating in mechanical systems. J. Mech. Eng. **49**(19), 55–69 (2013)
3. Liu, Q.: Research on Technology of Fiber Bragg Grating Acceleration Geophone in Crosswell Seismic. Northwestern Polytechnical University, Xi'an (2015)
4. Zhang, Y., Qiao, X., Liu, Q., et al.: Study on a fiber Bragg grating accelerometer based on compliant cylinder. Opt. Fiber Technol. **26**, 229–233 (2015)
5. Khan, M.M., Panwar, N., Dhawan, R.: Modified cantilever beam shaped FBG based accelerometer with self temperature compensation. Sens. Actuators A **205**(205), 79–85 (2014)
6. Liu, F., Dai, Y., Karanja, J.M., et al.: A low frequency FBG accelerometer with symmetrical bended spring plates. Sensors **17**(1), 206 (2017)
7. Li, T., Shi, C., Tan, Y., et al.: A diaphragm type fiber Bragg grating vibration sensor based on transverse property of optical fiber with temperature compensation. IEEE Sens. J. **99**, 1 (2017)
8. Zhang, X., Rong, Q., Sun, H., et al.: Low-frequency fiber Bragg grating accelerometer based on a double-semicircle cantilever. Opt. Fiber Technol. **20**(3), 190–193 (2014)
9. Gutiérrez, N., Galvín, P., Lasagni, F.: Low weight additive manufacturing FBG accelerometer: design, characterization and testing. Measurement **117**, 295–303 (2017)
10. Zhang, Y.: Research on Optical Fiber Vibration Sensing Technology in Micro Seismic. Xi'an Shiyou University, Xi'an (2016)
11. Antunes, P., Lima, H., Monteiro, J., Andre, P.S.: Elastic constant measurement for standard and photosensitive single mode optical fibers. Microw. Opt. Technol. Lett. **50**, 2467–2469 (2008)

Three Clusters of Mechatronics and Intelligent Robotics Research: A Scientometrics Exploration Based on the WoS

Xiaojuan Geng[1(✉)] and Wenjie Zhou[2]

[1] College of Finance and Economics, Gansu Agricultural University, Lanzhou, China
gengxiaojuan2003@163.com
[2] Business School, Northwest Normal University, Lanzhou, China

Abstract. Present research shed light on the structure of Mechatronics and Intelligent Robotics research community based on a Scientometrics exploration process. The findings include that: (a) There are 3 clusters existed in the research community of the Mechatronics and Intelligent Robotics. (b) The landscape of research topics in the Mechatronics and Intelligent Robotics research community further show the detail of the structure of this field. (c) The structure of the Mechatronics and Intelligent Robotics research community under a evaluation process.

Keywords: Mechatronics and Intelligent Robotics · Research community Structure

1 Introduction

Scientometrics is a very useful measure to identify the structure, hot topics and research tendency of specific scientific research community. Through a Scientometrics exploration procedure, researchers could understand the whole knowledge landscape of research area in a relatively simple way via visualization of the connection of the concept in this area. Many research communities, Mechatronics and Intelligent Robotics among them, are booming these days for the rapid development of technology. Thus, it is not easy for those researchers who are interested her/himself in a specific research area or try to understand this area on whole because of the too much research papers available and the structure of this area seems too complicated to be understood. For this point, present research aims to conduct an exploration research which will focus on the whole landscape, concept connectedness and hop research topics in the research area of Mechatronics and Intelligent Robotics within recent 5 years (2012–2017) by a visualization analysis based on VOS Viewer and Ucinet, both them are Scientometrics programs, and Web of Science (WoS), an advanced scientific literature database.

K. Deng et al. (Eds.): ICMIR 2018, AISC 856, pp. 954–961, 2019.
https://doi.org/10.1007/978-3-030-00214-5_118

2 Problem Statement

This study is significant for a comprehensive bird view on Mechatronics and Intelligent Robotics research. The presentation helps in exploring the structure of research community of Mechatronics and Intelligent Robotics in one comprehensive set of results over time. The study is drived by the following research questions:

1. What types of Mechatronics and Intelligent Robotics literature are explored in studies of structure in this research community?
2. What kinds of research topics have been explored in Mechatronics and Intelligent Robotics research community?
3. Which hot topics have been covered in Mechatronics and Intelligent Robotics research?

3 Research Design

3.1 Literature Search and Screening

Web of Science (formerly known as Web of knowledge) was originally created by the Institute for Scientific Information(ISI) and is now maintained by Clarivate Analytics (formerly the intellectual property and science department of the Thomson Reuters), which provides access to multiple cross-disciplinary database, enabling in-depth exploration of specialized sub-fields in academic or scientific disciplines thereby. To achieve the goal of the study, we set the target literature as "articles" published in 2016 or 2017. Nearly 30,000 literature references are obtained from the Wos database, in which only 2561 papers meet the requirements of this study (including abstracts rather than book reviews or related non-academic works). Therefore, the natural language processing (NLP) is based on these 2561 academic papers.

3.2 High Frequency Words Revision and Standardization

A NLP is used to extract high-frequency-word from abstract by means of the computer language named Python, considered as a general tool for text data mining. It is essential to standardize words/terms used to express similar concepts since the same concepts can be described by different words/terms (Guan et al. 2018).

3.3 Ucinet, VOS Viewer and Visualization of High Frequency Words Network

Both Ucinet and "VOS viewer" are computer programs used to map co-word structures that are based on corpora from the full-text database (Callon et al. 1986). In recent years, as a result of the development of graph theory and matrix operation, many researchers strive to study the co-word clusters (Courtial 1986; Xavier 2005).

3.4 Co-word Analysis

"Co-words", the co-occurrence of words, has been long regarded as the carrier of meaning in different fields of science and technology research (Callon et al. 1983). Co-word analysis is an important content analysis method by calculating the number of words appearing or the number of co-occurrence in the same article, by selecting words according to the topic and then making of the similar matrix, based on that, clustering the terms and giving interpretation on the structure for the clusters represented finally (Ying 2006).

4 Findings

4.1 Three Clusters of Mechatronics and Intelligent Robotics Research

Table 1 shows that there are 3 clusters existed in the research community of Mechatronics and Intelligent Robotics. Due to the hot research topics among the first bunch of cluster include control, performance, and algorithm and so on, this means that the first cluster of research in this field is focus on the maximum the performance of control system based on the algorithm improvement. And the second cluster, as show here in Table 1, is focusing on the software design and machine learning. Furthermore, the third cluster tends to pay more attention on the task and environment of robot and their intelligence.

In short, based on the scientometrics exploration, 3 clusters are found among the research community of Mechatronics and Intelligent Robotics. Hot research topics for each cluster were showing in Table 1. Thus, we answered the first research issued in the second part of this article.

4.2 Landscape of the Research Community of Mechatronics and Intelligent Robotics

Aiming to understand the landscape of the research community of Mechatronics and Intelligent Robotics on whole, we visualized the concept connections in this field. Figure 1 shows that there also 3 bunches of concept connections in this research community. For Fig. 1, each kind of color represents a kind of concept connections. And the connecting of specific conception means that they are closely related. Compare Fig. 1 to Table 1, we can find that the structure of the Mechatronics and Intelligent Robotics research community is clearly and stably. Thus, we answered the second research question issued in part two of this article.

4.3 The Structure of the Mechatronics and Intelligent Robotics Research Community

The third research question we are interested in is that how the structure of the Mechatronics and Intelligent Robotics research community is formulated. Figure 2 is showing the structure of the Mechatronics and Intelligent Robotics research on different levels.

Table 1. Three cluster and typical research topics in Mechatronics and Intelligent Robotics research

Word	Degree	Weighted centrality	Closeness centrality	Betweenness centrality	Clusters
Control	406	34190	1	1.98E−04	1
Performance	406	18604	1	1.98E−04	1
Controller	406	11066	1	1.98E−04	1
Algorithm	404	12825	0.995073892	1.74E−04	1
Time	406	13904	1	1.98E−04	1
Motion	402	10392	0.990196078	1.54E−04	1
Simulation	406	10737	1	1.98E−04	1
Structure	406	7955	1	1.98E−04	1
Position	404	7604	0.995073892	1.77E−04	1
Actuator	400	5397	0.985365854	1.65E−04	1
Design	406	22590	1	1.98E−04	2
Learning	404	7651	0.995073892	1.77E−04	2
Software	406	8809	1	1.98E−04	2
Machine	404	7465	0.995073892	1.92E−04	2
Platform	406	5581	1	1.98E−04	2
Integration	406	6684	1	1.98E−04	2
Tool	406	5465	1	1.98E−04	2
Manufacturing	396	5034	0.975845411	1.56E−04	2
Hardware	404	4721	0.995073892	1.97E−04	2
Laboratory	394	3244	0.971153846	1.61E−04	2
Robot	406	23517	1	1.98E−04	3
Environment	406	16212	1	1.98E−04	3
Task	406	11331	1	1.98E−04	3
Sensor	406	10730	1	1.98E−04	3
Intelligent	406	10452	1	1.98E−04	3
Vehicle	398	5166	0.980582524	1.65E−04	3
Network	400	6958	0.985365854	1.49E−04	3
Architecture	404	7603	0.995073892	1.78E−04	3
Behavior	404	5816	0.995073892	1.92E−04	3
Vision	400	4624	0.985365854	1.66E−04	3

Figure 2a–f show the connection of concepts when ties at 30, 50, 100, 150, 200, 250 and 300. Generally speaking, Fig. 2 answered the third research question of this article, how the structure of the Mechatronics and Intelligent Robotics research community formulated.

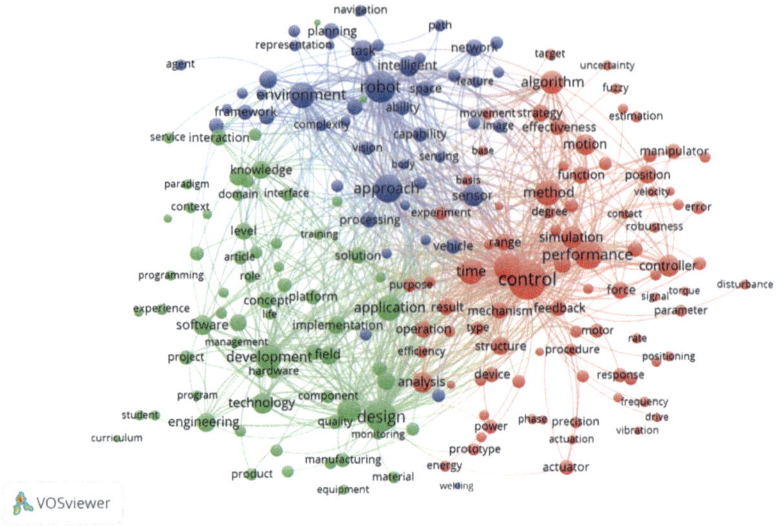

Fig. 1. Landscape of Mechatronics and Intelligent Robotics research community

4.4 The Evaluation of Research Topics in the Field of Mechatronics and Intelligent Robotics

Based on the above analysis, we have already identified the structure and its formulation process in Mechatronics and Intelligent Robotics researches. Furth more, for the purpose of reveal the evaluation of structure in the Mechatronics and Intelligent Robotics research community, we visualized the top level hot research topics and their evaluation process as Fig. 3.

Figure 3 Shows that control performance is the pioneer of the Mechatronics and Intelligent Robotics research and machine learning or engineering is followed. That is to say, among the Mechatronics and Intelligent Robotics research community, some topics are noticed in the very beginning of the research as well as some topics followed. In short, the structure of the Mechatronics and Intelligent Robotics research community under an evaluation process.

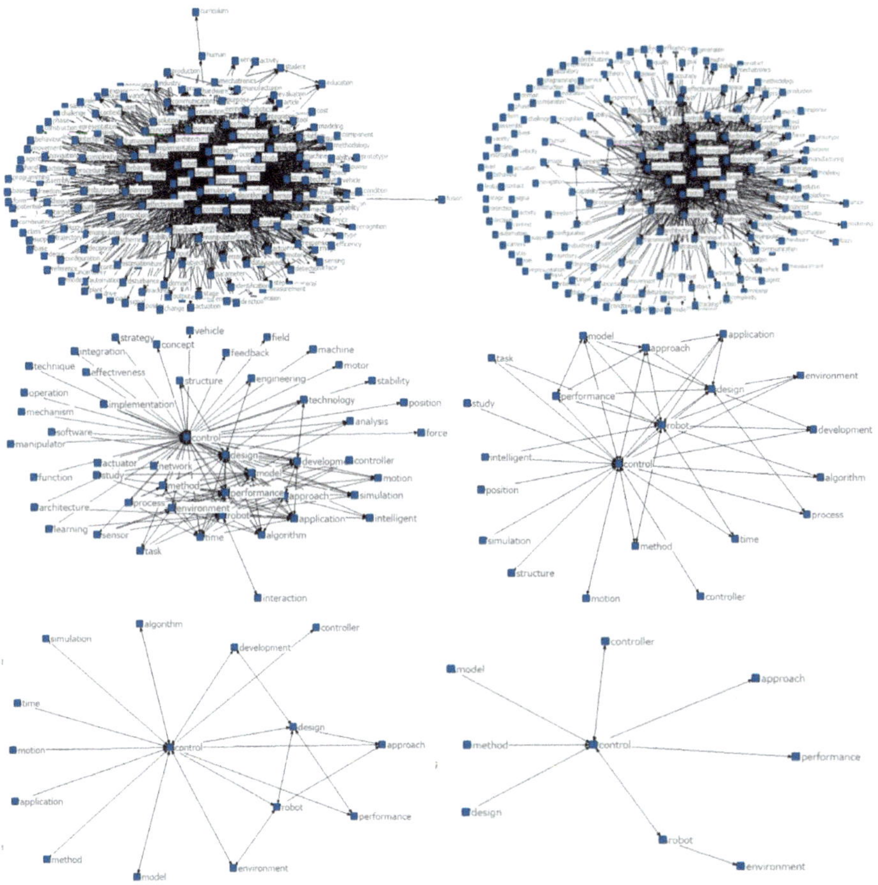

Fig. 2. Concept connections in Mechatronics and Intelligent Robotics research at different ties level

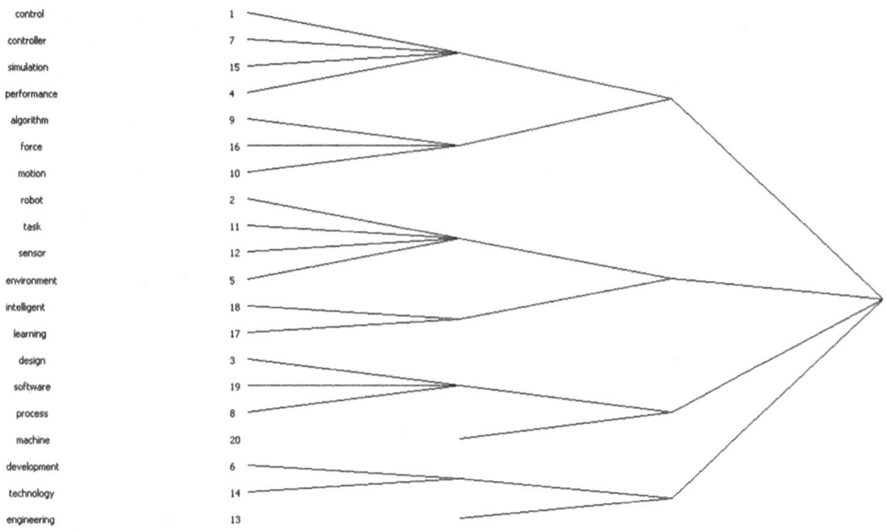

Fig. 3. The evaluation process of hot focused topics in Mechatronics and Intelligent Robotics research

5 Conclusion

This project conducted a Scientometrics exploration on the Mechatronics and Intelligent Robotics research community. The findings include that:

(a) There are 3 clusters existed in the research community of the Mechatronics and Intelligent Robotics.
(b) The landscape of research topics in the Mechatronics and Intelligent Robotics research community further show the detail of the structure of this field.
(c) The structure of the Mechatronics and Intelligent Robotics research community under a evaluation process.

Acknowledgments. This research was supported by the National Natural Science Foundation of China (project number: 71563042) and the Humanities and Social Sciences of Ministry of Education Planning Fund (project number: 14YJC790037). "The reliability and validity of co-words analysis in scientometrics".

References

An, X., Wu, Q.: Co-word analysis of the trends in stem cells field based on subject heading weighting. Scientometrics **88**, 133–144 (2011)

Guan, A., Gao, C., Zhang, T., Zhou, W.: Exploring the metaphor of information under the context of information science research: a co-words analysis. Procedia Comput. Sci. **131**, 739–745 (2018)

Bauin, S., Michelet, B., Schweighoffer, M.G., Vermeulin, P.: Using bibliometrics in strategic analysis: "understanding chemical reactions" at the CNRS. Scientometrics **22**(1), 113–137 (1991)
Bawden, D.: Organised complexity, meaning and understanding: an approach to a unified view of information for information science. Aslib Proc. New Inf. Perspect. **59**(4/5), 307–327 (2007)
Budd, J.M.: Meaning, truth, and information: prolegomena to a theory. J. Doc. **67**(1), 56–74 (2011)
Callon, M., Courtial, J.P., Turner, W.A., Bauin, S.: From translations to problematic networks: an introduction to co-word analysis. Soc. Sci. Inf. **22**(2), 191–235 (1983)

Computational Vision

Efficiently Extracting Dominant Planes from Massive 3D Points Based on Scene Structure Priors

Wei Wang(✉), Guoheng Ren, and Liyong Chen

School of Network Engineering, Zhoukou Normal University, Zhoukou, China
wangwei@zknu.cn

Abstract. Extracting reliable dominant planes from 3D points plays an important role in modeling large-scale urban scenes. However, traditional local and global methods frequently appear powerless when massive noisy 3D points are present. In order to solve the problem, the paper proposes an efficient multi-plane extraction method to explore potential plane relations by detecting 2D line segments based on scene structure priors in a low-dimension projection map derived from 3D points, and then generate planes from the resulting plane relations. Experimental results confirm that our method can efficiently produce sufficient and reliable dominant planes from massive 3D points with high noise levels (only about 6s on 2000K 3D points).

Keywords: Plane fitting · Multi-view stereo · Urban scene

1 Introduction

Piecewise planar assumption is commonly used to approximately give a complete structure interpretation for modeling urban scenes from multiple images. In this case, sufficient and reliable dominant planes need to first be extracted from initial 3D points in order to further infer the optimal plane for each scene patch. However, most existing methods [1–3] tend to independently extract dominant planes from only initial 3D points corresponding to the image under consideration. This frequently results in an unreliable plane inference because these extracted planes could be not complete enough to better depict initial scene structures. In addition, traditional multi-plane fitting methods [4–7] (including local and global methods) usually need too many storage resources and too much computation time (especially for global methods) when massive 3D points are present. In addition, the relations between planes are not fully taken into account in these methods (e.g., global methods simply force two neighboring 3D points to be the same plane), which frequently also leads to larger errors.

Figure 1(a)–(c) show an example which initial 3D points are reconstructed using the PMVS (Patch-based Multi-view Stereo) method [8] from 3270 images captured from an urban scene. However, these massive 3D points with high noise levels seem to be very jumble, from which extracting reliable dominant planes is challenging for traditional methods.

K. Deng et al. (Eds.): ICMIR 2018, AISC 856, pp. 965–973, 2019.
https://doi.org/10.1007/978-3-030-00214-5_119

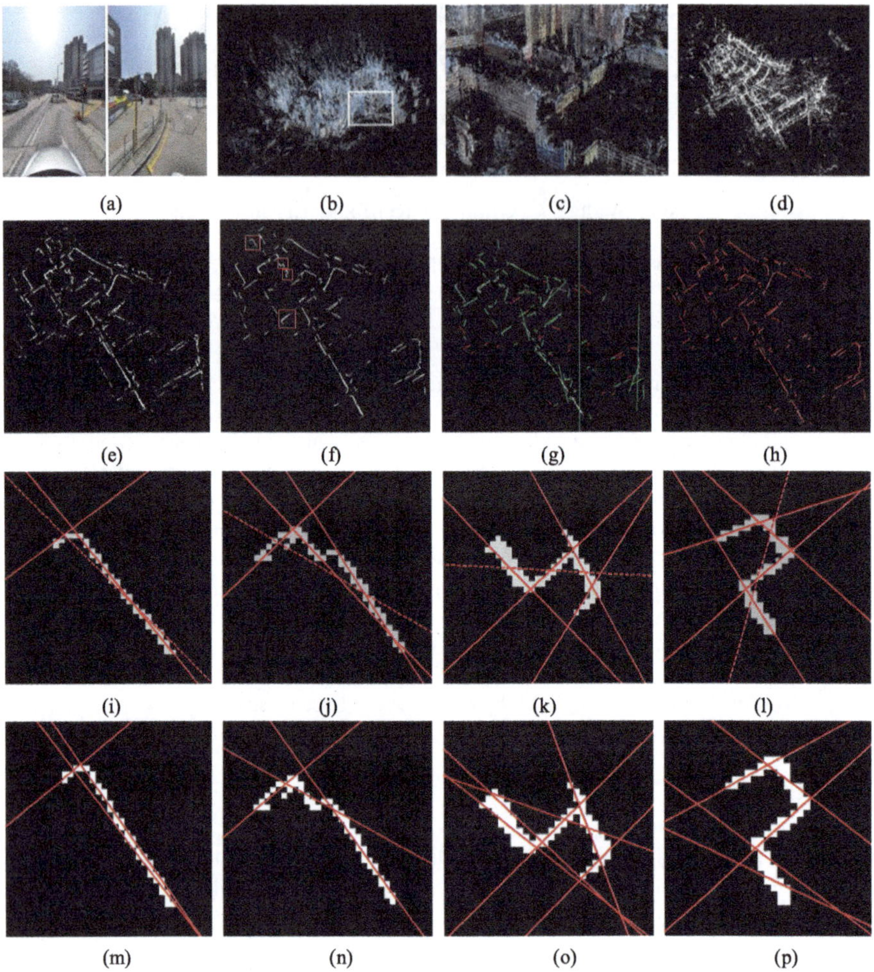

Fig. 1. Multi-line detection: (a) two sample images; (b) initial 3D points; (c) close-up in the rectangle in (b); (d) initial projection map; (e) 2D grids after eliminating the 2D grids with low values; (f) regions in C_1 and C_2; (g) errors caused by the least squares method (red lines are reliable lines); (h) lines produced by our method; (i)–(l) samples in rectangles in (f) (red and dotted lines are respectively produced by our method and the least squares method); (M)–(P) results produced by the PEaRL method [6].

In order to solve the problem, this paper proposes an effective multi-plane extraction method that utilizes scene structure priors to explore potential planes from 3D points. The proposed method can efficiently extract reliable dominant planes from massive noisy 3D points (only about 6s on 2000K 3D points).

2 Our Method

Our proposed method comprises four main components: projection map generation, connected-region detection, multi-line detection, and multi-plane generation. Next, we first introduce scene structure priors used in our method, and then elaborate each component in the subsequent sections.

2.1 Scene Structure Priors

In Urban scenes, the following scene structure priors commonly exist (implicitly or explicitly): (1) Plane prior. For a small enough scene patch, its geometric model frequently can be approximated as a plane. In this paper, the plane prior (denoted as M) is defined as the set of reliable 2D lines. (2) Angle prior. The intersection angles between the planes are frequently fixed values. The angle prior (denoted as A) is set to the pre-specified angles [30°, 45°, 60°, 90°, −60°, −45°, −30°] in this paper. More angles help to detect more lines (or planes), but may also incur heavy computational load.

2.2 Projection Map Generation

For an urban scene, the buildings are always perpendicular to the ground, and the vertical scene direction can be estimated by the vanishing point detection in an image or multiple images [1]. Then, a horizontal plane (denoted as H) can be determined by the vertical direction and some arbitrary initial 3D points. Next, as shown in Fig. 1(d), a discretized projection map (denoted as Φ) is thus generated through two steps: (1) Projecting initial 3D points to H. (2) Rounding off and grouping the resulting 2D points using the 2D grid map where each grid (predefined size $x \times x$) is used to count the number of the 3D projected points and handle together. In our experiment, the grid size is set to 1×1. Note that, for the robustness against the deviation of the vertical direction, a large grid size should be introduced such as 2×2. However, too larger grid size could also reduce the accuracy of the next line segment detection.

As a result, in Φ, the 2D grids with high values frequently derive from the 3D points in the same plane perpendicular to H, and gather to the regions with dominant line structures. Here, in Fig. 1(e), the 2D grids with low values generally correspond to noisy 3D points where the plane prior is violated, and can be eliminated subsequently.

2.3 Connected-Region Detection

In order to reliably detect line segments in Φ, we first obtain the value T by mean of all grid values, and then detect three types of connected-regions through two-level threshold (p_1, p_2) (set to (0.7, 0.5) in this paper) as follows:

(1) Connected-regions C_1 where the value of each 2D grid is larger than or equal to $p_1 * T$. In general, the set C_1 has a small number of regions, most of which have dominant line structures (or contains fewer line segments).

(2) Connected-Regions C_2 where the value of each 2D grid is larger than or equal to $p_2 * T$ but smaller than $p_1 * T$. Each region in the set C_2 frequently contains more line segments (about four in our experiments).
(3) Other regions where the value of each 2D grid is smaller than $p_2 * T$. These regions generally derive from noisy 3D points and are directly filtered out.

Next, we will focus on how to effectively detect 2D line segments in each region in the sets C_1 and C_2 based on scene structure priors.

2.4 Multi-line Detection in Regions

At first, we define the following inlier ratio $l^*(L_s, P)$ to select an optimal line l^* from a set of lines L_s:

$$l(L_s, P) = \arg\max_{l \in L_s} \left(\sum_{p \in P} \delta(d(p, l) < \kappa) / |P| \right) \tag{1}$$

where P denotes the set of 2D grids in the current region; $d(p, l)$ denotes the distance of the point $p \in P$ to the line $l \in L_s$, κ is the distance threshold (set to 0.8); $\delta(x)$ is the indicator function that equals 1 if x is true, and 0 otherwise; L_s denotes the set of all possible lines determined by scene structure priors (i.e., M and A) at the key 2D grid $s \in P$. Note that, s is selected from the set of 2D grids down-sampled from P in order to improve the efficiency of the line detection, and thus called as the key grid.

Then, in order to detect multiple lines in the current region, the above process needs to be repeated several times after removing the inliers of the detected lines from P at each time. Then, we define the following stop condition:

$$T_{prior} = (\upsilon_i < \theta) \wedge \left(\sum_{i=1}^{k} \upsilon_i > \zeta \right) \tag{2}$$

where i denotes the ith repetition, and the constants θ (set to 0.1) and ζ (set to 0.9) are the pre-defined thresholds. υ_i is defined as:

$$\upsilon_i = \arg\max_{s \in P} l_i^*(L_s, P) \tag{3}$$

where $l_i^*(L_s, P)$ denote the $l^*(L_s, P)$ value at the ith repetition, and l_i^* is the corresponding line.

Algorithm. 1: Multi-line detection based on scene structure priors

Input: the sets C_1 and C_2, the set A, temporary sets L and C_0.
Output: the set M.
Initialization: $M = null$, $C_0 = null$.
1. For each region $c \in C_1$
2. Initialize R using the RANSAC method and $L = null$.
3. Generate L_s at each key grid $s \in P$ according to A and R.
4. Compute υ_i and save the corresponding line l_i^* to L.
5. Remove the inlier grids of l_i^* from P.
6. If $\upsilon_i \geq \theta$, Goto 3.
7. ElseIf $T_{prior} = ture$, save L to R.
8. Else, save c to C_0 and Goto 1.
9. Save R to M.
10. End for
11. For each region $c \in C_2$
12. Initialize R according to M and $L = null$.
13. Steps similar with 3-9.
14. End for
15. Set $C_2 = C_0$, $C_0 = null$ and Goto 11 until $C_0 = null$.
16. Output M.

Obviously, when T_{prior} is true, most of 2D grids in the current region have been fitted by several lines, and the set of detected lines is more reliable. Note that, at each repetition, P is updated by removing the inliers of detected lines, and L_s is updated according to each angle in A and the slope of each line in $R \subset M$ at the grid $s \in P$, where R is initialized in different manners for different regions.

(1) For each region in C_1, R is directly initialized by the line produced using the RANSAC method. Because the region has dominant line structure, and C_1 has fewer amounts, this process is generally fast.
(2) For each region in C_1, R is initialized by the optimal line selected from M made up of the lines detected in all regions in C_1. In fact, the region in C_2 is closed with some regions in C_1. Therefore, M is reliable for initializing R.

The proposed method is described in Algorithm 1.

Figure 1(h) shows the detected lines for all the regions in C_1 and C_2 using Algorithm 1. Apparently, Algorithm 1 can produce satisfactory results even for some complex regions shown in Fig. 1(k), l. In fact, as shown in Fig. 1(m)–(p), without the guidance of scene structure priors, the global multi-model fitting method (i.e., PEaRL) also results in larger errors because too many noisy points are present in each region.

2.5 Multi-plane Generation

Lines detected from connected-regions correspond to the planes of the scene, and the angles between lines can be taken as the plane relations. Then, in order to rapidly generate planes from these plane relations, we first select a line with more inliers in an arbitrary region to fit the projected 3D points corresponding to its inliers, and then generate other planes according to the plane relations (i.e., the angles between other lines and the line). As shown in Fig. 2, the extracted planes and inliers are satisfied. Moreover, the process is fast because the time-consuming plane fitting is avoided.

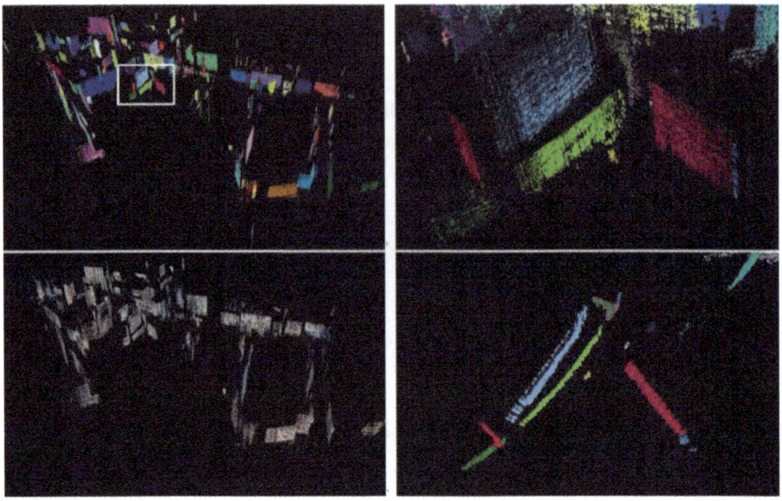

(a) (b)

Fig. 2. Multi-plane generation: (a) planes (different colors denote different planes) and the textured inliers; (b) close-up and top-view of planes in the rectangle in (a). Best viewed in color.

3 Experiment

To evaluate the performance of our method, we conducted experiments on three our own data sets of real-world urban scenes. The three data sets respectively contain 3270 (scene #1), 4976 (scene #2) and 4922 (scene #3) images (1224 × 1848). The initial 3D points are produced by the PMVS method (scene #2 and scene #3 are shown in Fig. 3 and scene #1 is shown in Fig. 1), from which extracting reliable planes is challenging for traditional methods because of too many noise points.

In order to evaluate our method quantitatively, for each scene, we randomly select fifty images and manually divide the building regions of each image into several sub-regions, each of which corresponds to a real plane. Then, the inliers (i.e., 3D points) that are projected into a specific sub-region are assigned to a real plane. Thus, a plane extraction accuracy measure K is defined as

$$\mathrm{K} = \frac{1}{N \cdot m} \sum_{k=1}^{N} \sum_{i=1}^{m} \delta(f_i = f_i^*) \tag{4}$$

where f_i^* is the true plane label, f_i is the label obtained using our method, $\delta(\cdot)$ is defined in (1), N and m respectively denote the number of the selected images and the number of the inliers of all sub-regions (or planes) in the current image.

(a) (b) (c) (d)

Fig. 3. Large-scale urban scenes: (a) two sample images; (b) 3D points produced by the PMVS method; (c) close-up of the 3D points in the rectangles; (d) regions in C_1 and C_2.

Urban scenes exhibit strong structural regularities and initial 3D points frequently distribute along the building facades. Therefore, the 2D points projected from these 3D points always arrange along the specified lines. As shown in Fig. 3(d), line segments become more dominant when 2D grids with low values are filtered out from the projection map. Thus, reliable line segments shown in Fig. 4(a) can be effectively detected in each connected regions based on scene structure priors.

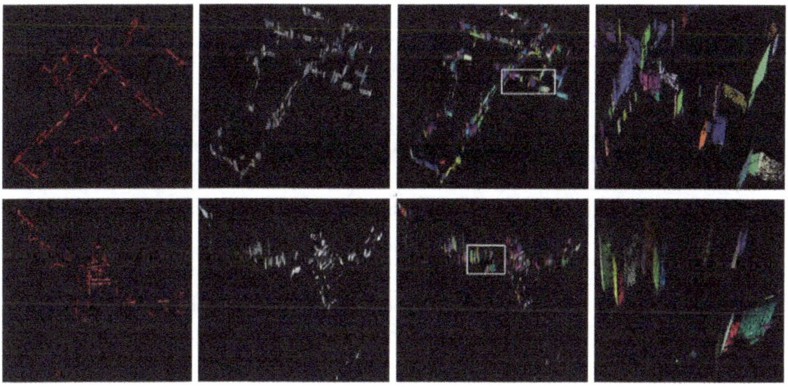

(a) (b) (c) (d)

Fig. 4. Effective multi-plane extraction: (a) multi-line detection; (b) all inliers; (c) all extracted planes (different colors denote different planes); (d) close-up of the planes in the rectangles in (c).

Finally, the planes can be efficiently generated from the resulting line segments, and the corresponding inliers also extracted from initial 3D points. It can be clearly seen that the inliers shown in Fig. 4(b) better represent the skeletons of the scenes, and the extracted planes shown in Fig. 4(c) are sufficient for explaining the initial structures of the scenes. Figure 4(d) shows some details of the extracted planes, especially for some small planes, our method still produces satisfactory results.

Table 1 shows the initialization and the quantitative results for each urban scene, including the number of initial 3D points, the number of connected-regions (CR), the number of extracted planes, and the inlier ratio. We also provide the approximate running time costs in seconds. From Table 1, it can be seen that our method performs well in both accuracy and efficiency. The reasons mainly lie in: (1) the multi-line detection based on scene structure priors can obtain reliable plane relations in advance, which makes the next plane generation very fast; (2) the global optimization commonly used in computer vision is better encoded and simplified by the simpler and more effective multi-line detection. This also helps to produce better results (e.g., all 3D points in the same plane can be globally considered as a single 2D line, instead of time-consuming plane sampling and optimizing in most existing multi-model fitting methods).

Table 1. Quantitative results of different method (Time: second)

Our method							RCMSA	
Scene	Points	CR	Planes	Inlier ratio	K	Total time	K	Mean time
#1	2727503	602	64	0.81	0.88	6.7	0.86	14.5
#2	1431703	809	75	0.79	0.87	3.8	0.85	16.3
#3	1579809	813	62	0.78	0.84	4.1	0.81	12.8

To further validate the performance of the proposed method, we compared it with the state-of-the-art multi-model fitting method, i.e., RCMSA [7]. In our experiments, RCMSA performs poorly on massive 3D points, and is thus deployed on only the inliers (i.e., 3D points) corresponding to each single image for comparison (e.g., about 4K 3D points).

From Table 1, it can be seen that the accuracy of our method is higher than that of RCMSA. Actually, for RCMSA, certain real planes could be omitted in the generation of candidate planes, which then lead to an unreliable plane optimization. Moreover, the noise points could also be an important factor that influences the reliability of the whole method. Note that, in Table 1, for each scene, the total time of our method performing on all 3D points corresponding to all images is lower than the mean time of RCMSA independently performing on the inliers corresponding to each single image. This indicates that our method is significantly faster than RCMSA.

In summary, with the guidance of scene structure priors, our method can efficiently extract sufficient and reliable planes from massive noisy 3D points, and the resulting planes and inliers can be used to further model large-scale urban scenes.

4 Conclusion

In this paper, we proposed an effective multi-plane extraction method suitable for massive 3D points. Instead of the time-consuming plane fitting directly over 3D points, the proposed method performs significantly fast by detecting 2D line segments based on scene structure priors. It overcomes many challenging difficulties (e.g., inefficiency, missing some planes) and can efficiently extract sufficient and reliable planes from massive 3D points.

Acknowledgements. This work is supported by Open Projects Program of National Laboratory of Pattern Recognition (201700004), Natural Science Foundation of Henan (162300410347), College Key Research Project of Henan (17A520018, 17A520019), Key Technology Research and Development Project of Henan (172102310727) and School-based project of Zhoukou Normal University (zknuc2015103).

References

1. Mičušík, B., Košecká, J.: Multi-view superpixel stereo in urban environments. Int. J. Comput. Vis. **89**(1), 106–119 (2010)
2. Verleysen, C., Vleeschouwer, C.D.: Piecewise-planar 3D approximation from wide-baseline stereo. In: Proceedings of Computer Vision and Pattern Recognition, Las Vegas, pp. 3327–3336 (2016)
3. Bodis-Szomoru, A., Riemenschneider H., Van Gool, L.: Fast, approximate piecewise-planar modeling based on sparse structure-from-motion and superpixel. In: Proceedings of Computer Vision and Pattern Recognition, Columbus, pp. 469–476 (2014)
4. Chin, T.J., Yu, J., Suter, D.: Accelerated hypothesis generation for multi-structure data via preference analysis. IEEE Trans. Pattern Anal. Mach. Intell. **34**(4), 625–638 (2012)
5. Toldo, R., Fusiello, A.: Robust multiple structures estimation with J-linkage. In: Proceedings of European Conference on Computer Vision, Marseille, pp. 537–547 (2008)
6. Isack, H., Boykov, Y.: Energy-based geometric multi-model fitting. Int. J. Comput. Vis. **97** (2), 123–147 (2010)
7. Pham, T.T., Chin, T.J., Yu, J., Suter, D.: The random cluster model for robust geometric gitting. IEEE Trans. Pattern Anal. Mach. Intell. **36**(8), 1658–1671 (2014)
8. Furukawa, Y., Ponce, J.: Accurate, dense, and robust multi-view stereopsis. IEEE Trans. Pattern Anal. Mach. Intell. **32**(8), 1362–1376 (2009)

Application Prospect of 3D Printing Technology in the Food Intelligent Manufacturing

Hong Zhang[1,2], Min Fang[2], Yongming Yu[2], Qiannan Liu[1], Xiaojia Hu[1], Liang Zhang[1,2], Honghai Hu[1], Chunjiang Zhang[1], Feng Huang[1], Fang Tian[1], Nana Zhang[1], and Xiaofeng Dai[1,2(✉)]

[1] Institute of Food Science and Technology, Chinese Academy of Agricultural Sciences/Comprehensive Key Laboratory of Agro-Products Processing, Ministry of Agriculture, Beijing 100193, China
daixiaofeng@caas.cn
[2] Academy of Food Nutrition and Health Innovation, CAAS, Hefei 238000, China

Abstract. 3D printing, also known as additive manufacturing, had been used in many fields, such as medicine, construction, automobile industries and so on. In recent years, 3D printing technology in food industry had developed rapidly, as with the diversification of the 3D printing forms. This paper introduced the working principle, application range, advantages and limitations, market application of the extrusion, powder binding and inkjet food 3D printing. Three food materials property related critical factors were suggested here for the rational design of 3D food structures: printability, applicability and post-processing. At present, the factors restricting the development of food 3D printing technology were the limitations and liquidity of printing materials, the printing speed, the stability of the printed model and so on. We suggest that it is urgent to strengthen the combination of food nutrition science with 3D printing and searching breakthroughs on personalized nutrition food manufacturing. We look forward to seeing a new revolution in the food intelligent manufacturing industry launch by 3D printing technology.

Keywords: Food 3D printing · Food intelligent manufacturing
Forming principle · Personalized nutrition · Development status
Application prospects

1 Introduction

3D printing is a new technology expedited by rapid prototyping technology. 3D printing technology, also known as additive manufacturing, can create predefined shapes utilizing metal, plastic, rubber, gypsum and other materials by melting extrusion, powder sintering, photosensitive reaction and other ways. Because that the technology can digitize the complicated processing, and has the advantages of high accuracy, high speed and low cost, so it is considered as an important technological method to promote the revolution of Industry 4.0. Although 3D printing technology has

K. Deng et al. (Eds.): ICMIR 2018, AISC 856, pp. 974–984, 2019.
https://doi.org/10.1007/978-3-030-00214-5_120

firstly made breakthroughs in biological organs, automobile manufacturing, housing construction and other aspects, it has also emerged in the field of food manufacturing.

Currently, consumers are mainly concerned with the nutrition, taste, appearance and convenience of food. The imbalance of dietary nutrition structure is an important factor affecting the health in China, so more and more consumers pay more attention to the nutrition of food (Xu et al. 2015; Carlson and Frazão 2014). Drawing 3D printing technology into the field of nutrition and health food processing, mixing protein, fat, carbohydrates, vitamins, minerals and other functional factors, the nutrient composition can be balanced according to the demand ratio (Sun et al. 2018). On the premise of satisfying the processing suitability of the raw material 3D, new food with balanced nutrition, delicious, beautiful appearance and convenient eating would be printed. It has broad prospects for group catering or personalized precise nutrition (Yang et al. 2017).

2 The Type of Food 3D Printing

2.1 Extrusion Food 3D Printing

Under digital control, the extrusion-based 3D food printer can print complex three-dimensional food layer by layer. It is the most popular method in food printing, which provides an engineering solution for digitalized food design and nutrition control (Sun et al. 2018). The type of printing is easy to operate, but it has a higher demand for the flow characteristics of food materials.

According to the requirements, the first step of extrusion-based food 3D Printing is to build the basic model and three-dimensional model through 3D software. Secondly, input the model into the 3D printer system, and set the operation parameters on the 3D printer. Lastly, establish the processing procedure according to the material processing characteristics. Broadly speaking, a food printer is a machine that can turn digital recipes into edible and possibly delicious morsels (Sher and Tutó 2015).

2.1.1 The Coordinate Type of Extrusion-Based Food 3D Printing

The driving path of extrusion-based food 3D Printing is realized through the coordinate system. At present, there are four forms of coordinate systems that were applied to food 3D printers: cartesian coordinates, delta coordinates, polar coordinates and scara coordinates (Sun et al. 2018).

Cartesian Coordinate

As seen in Fig. 1(1), Cartesian coordinates have three axes of X, Y and Z, the nozzle and model can realize relative motion between front to back, left to right, and up to down. This coordinate is applied in 3D printing, and there are three common implementation forms. The first form is that the bottom plate is fixed, and the above printing nozzle moves along three coordinate axes of X, Y and Z in accordance with the predetermined procedure. For example, food 3D printer Foodini of the Natural Machines in Spain and the food 3D printer developed by institute of food science and technology, CAAS in Fig. 2(2). The second is that the bottom plate moves along Z axis, and the above printing nozzle moves along the axes of X and Y. For instance, the 3D cream printer developed by Tiebo Sun in food and drug career technical college of

Jiangsu province. The third is the bottom plate moves along Y axis, and the above printing nozzle moves along the axes of X and Z (Sun et al. 2018).

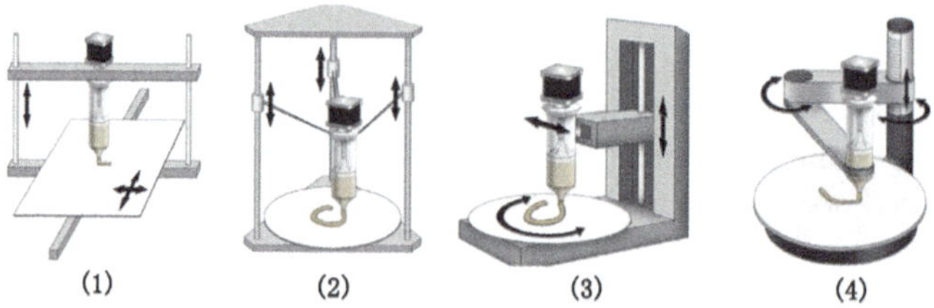

Fig. 1. Extrusion-based food 3D printer coordinate, (1) Cartesian; (2) delta; (3) polar; (4) scara

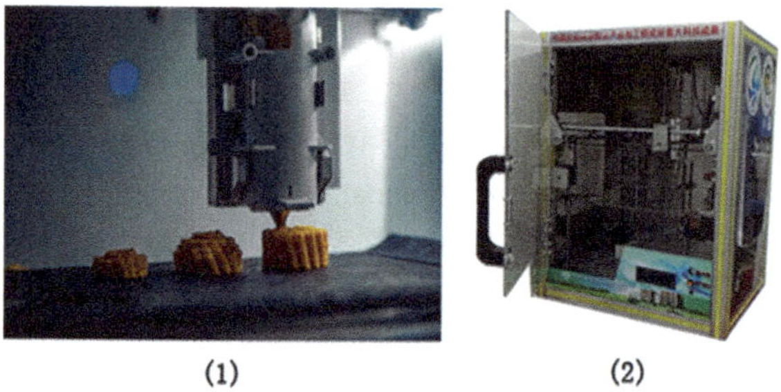

Fig. 2. Food 3D printer of Cartesian coordinate, (1) Foodini 3D printer; (2) Food 3D printer by institute of food science and technology, CAAS

As the design of the cartesian-based 3D food printer is simple, easy to maintain and calibrate. Meanwhile, the control procedure is accurate and uncomplicated. So the initial 3D food printers mostly use the coordinate system to finish printing. But the factors of slow printing speed, complex organization and relatively expensive price restrict the development of the coordinate system printer.

Delta Coordinate

The delta-based food 3D printer nozzle is controlled by three mechanical arms, similar to the manipulator. As seen in Fig. 1(2), the bottom plate is fixed, the nozzle is derived by three mechanical arms to realize the geometric motion of space. Currently, the common delta-based food 3D printers on the market include Pinya3 and Tytan 3D in Poland (Krassenstein 2016).

Compared with the cartesian-based food 3D printer, the delta-based food 3D printer makes the 3D printer cheaper, because of its simple structure. The 3D printer of delta coordinates has faster printing speed, and can produce a larger volume of food in a short time, but the accuracy of printing is relatively low. Meanwhile, the control procedure is more complex. Because the delta-based food 3D printer shave quickly acceleration or deceleration, it is easy to cause vibration and is not conducive to control the stability of printing (Sun et al. 2018).

Polar Coordinate

The food 3D printer of polar coordinate is made of bottom plate, nozzle and strut. The bottom plate can complete rotary motion, the nozzle can realize the movement of up or down and rotation of the strut (Sun et al. 2018). At present, the polar-based food 3D printers are consisted of TNO food 3D printer (Van der Linden 2015; Van Bommel 2014) which is from Netherlands and XOCO chocolate 3D printer (Michiel Cornelissen Ontwerp 2016) on the market.

According to the mechanical characteristics of the polar-based food 3D printer, it has fast printing speed, relatively. But the shape of printing food is limited, which is only based on the circle of space.

Scara Coordinate

The design concepts of the scara-based food 3D printer originated are from the arm of the robot, two revolving connecting rods can realize the motion of the X and Y direction, the connecting rod mechanism is fixed on the main frame to realize the up and down motion on the Z direction. The type of food 3D printer was designed by the Columbia University and has been used to make a lot of products (Lipton et al. 2010, 2015). The structural design of the food 3D printer in this form is relatively simple and the cost of manufacturing is relatively low.

2.1.2 Extrusion Principle of 3D Printing in Extruded Food

There are three main extrusion principles for extrusion-based 3D printers: syringe-based extrusion, air pressure driven extrusion and screw-based extrusion (Sun et al. 2018).

The Principle of Piston-Extrusion

The extrusion system of the food 3D printer is composed of electric machinery, syringe, plunger and extrusion nozzle. The syringe cavity can be used to store the food materials. The plunger moved downward by electric machinery, so that the food material is extruded evenly through the extrusion nozzle. The food 3D printers using this extrusion principle are mainly include CocoJet developed by the Choc Creator in UK, jointly developed by the 3D Systems and Hershey (Shandrow 2015), and the XYZ food 3D printer of Taiwan. The syringe-based food 3D printer is suitable for semisolid and solid materials.

The Principle of Air Pressure Driven Extrusion

The extrusion system of food 3D printer is made up of air pump, material cavity and extrusion nozzle. The material is stored in the material cavity, and the material is extruded through the extrusion nozzle under the push of the compressed air in the air pump. The present food 3D printer, which is used for the extrusion principle, has a paste 3D printer developed jointly by TNO and Barilla (Van der Linden 2015). The principle of pressure extrusion food 3D printer is suitable for printing fluid-like materials, material flow better. Compressed air needs to be connected to the food 3D printer after sterilization and filtration.

Extrusion Principle of Screw Type

The food 3D printer extrusion system of screw extruder is composed of electric machinery, screw, material cavity, feed inlet and extrusion nozzle. The material enters the ink cartridge through the entrance, then the motor drives the screw forward to extrude the material, and finally the material is extruded molding by extrusion nozzle (Sun et al. 2018). This type is seldom used as the control of extrusion space is not easy to realize. There is a food 3D printer that is used for the extrusion principle developed by Kunshan porimy 3D Printing Technology Co. Ltd. in China.

2.1.3 Forming Principle of 3D Printing Extruded Food

For the extrusion food 3D printers, with the change of print status of raw materials, the molding principle is also different. There are mainly three types including the room temperature extrusion, hot-melt extrusion and hydrogel-forming extrusion, as indicated (Sun et al. 2018).

Room Temperature Extrusion Molding

The material used for room temperature 3D printing extruded food must provide certain viscosity which makes them bond with each other. The products printed in this form usually need to be cooked through baking, frying and steaming, as shown in Figs. 3 (Lipton et al. 2010) and Fig. 4. Printed products include pizza, cookies, pancakes, etc. The materials used are mainly butter, cheese, flour, eggs, milk, etc. And then the above raw materials need to be mixed evenly to form a uniform solid, semi-solid or fluid type printing material.

(1) (2)

Fig. 3. 3D printing food of room temperature extrusion, (1) Status after extrusion; (2) Status after cooking

Hot-Melt Extrusion Molding

Chocolate beans and chocolate powder are the most common materials used in hot-melt extrusion 3D printing (Hao et al. 2010; Mantihal et al. 2017). Chocolate bean and chocolate powder could be heated to molten state with excellent liquidity and formability. Chocolate is solidified at room temperature after extrusion, which enable the shape of 3D printed construct to be maintained for a long time. The finished product generally does not need to be cooked, can be eaten directly.

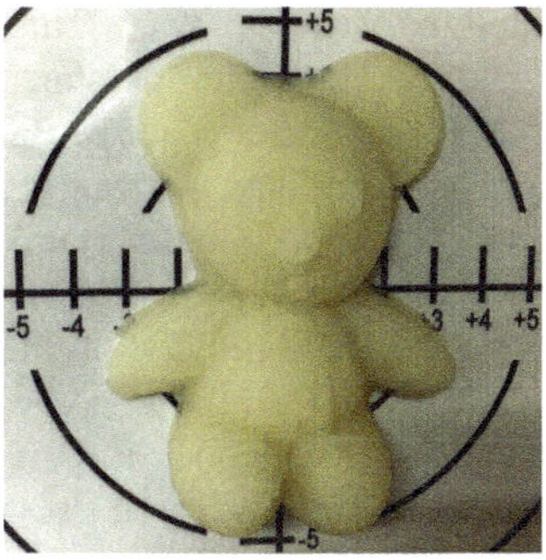

Fig. 4. Mashed potatoes printed by food 3D printer of institute of food science and technology, CAAS

Hydrogel-Forming Extrusion Molding

The hydrogel-forming extrusion 3D printing materials contain constituents with gel forming ability, which show different states at different temperatures. Adding ingredients with gelling properties to 3D print materials, such as AGAR, and prepare solutions with different ratio according to the formulas. Put the prepared solution into 3D printer and it will be produced through an extrusion molding process. Sometimes this process need to finished in ice-water bath (Serizawa et al. 2014; Gong et al. 2014). This kind of hydrogel-forming extrusion 3D printer was invented by Yamagata University in Japan. Another widely used apparatus was candy extrusion molding machine, such as 3D printer and 3D candy of Katjes in Germany.

2.2 Powder Binding Food 3D Printing

Powder binding printing is another kind of commonly used form of food 3D Printing. It forms a complete 3D printing product by binding the powder layer by layer according to assigned models. Compared with the extrusion 3D printing, this type is faster and can print more complex food.

Powder binding printing can be divided into three type: Selective Laser Sintering (SLS), Selective hot air sintering and melting (SHASAM) and Liquid binding (LB) (Godoi et al. 2016; Peltola et al. 2008; Diaz et al. 2016; Wegrzyn et al. 2012; Bredt and Anderson 1999). The first step is to determine the shape of the model and decompose the graph layer by layer. Then, the layers of powder are fused coagulation or bonding upon the application of heat sources (laser or hot air) or liquid binder. After completing a cross-section of the graph, layer of powder goes down one layer and a new layer of powder was added. The process is repeated until the pre-set model is finalized (Godoi et al. 2016).

2.2.1 Selective Laser Sintering

Selective laser sintering applies an infrared laser as a powder source, sintering powder materials. Food 3D printing can also use laser as a heat source. The pre-designed food model is built by laser sintering of food power. This printing process has been described above. Since this kind of 3D printing is done through depositing different powder layer by layer, each layer of powder could be different food material components. This method can be applied to produce food with balanced nutrition (Diaz et al. 2016). The choice of laser has a great impact on the quality of printed food (Kruth et al. 2007). The main reason is that the laser has a great influence on the melting degree of powder materials, for example, the absorptivity of materials is affected by laser wavelength and the powder densification is related to energy density (Gu et al. 2012).

By using selective laser sintering technique, TNO has created many 3D printed food of different shapes, including sugar powder, chocolate powder and curry powder.

2.2.2 Selective Hot Air Sintering and Melting (SHASAM)

Selective hot air sintering and melting 3D printing technology is similar to selective laser sintering, just replace the laser heat source with hot air. The process was described above and the printed food was shown in Fig. 5. Air temperature and velocity is essential for this type of printed food.

Fig. 5. 3D printing food based on SHASAM technology

2.2.3 Liquid Binding (LB)

Liquid binding food 3D printing combine the 2D powder chips into complex 3D model through a liquid with bonding property (Russell et al. 2015). Liquid binder can also be used as regulator of taste and color, enrich the flavor and color of printed food. At present, this technology is widely applied in the field of powdered sugar, making candy products with various colors and shapes.

2.3 Inkjet Printing (IJP)

Inkjet food 3D printing is adding decorations to existing foods and it is an extension of 2D printing in food industry. It dispenses a stream of droplets from a thermal or piezo electric head to certain regions for the surface filling or image decoration on food surfaces, such as cookie, cake, and pizza. The ornament materials need to maintain liquidity during the process of printing and it is pushed onto the surface of food from the nozzle. Unlike the extrusion food 3D printing, inkjet food 3D printing is not printing layer by layer, but using the form of local printing to complete the final shape, the principle is shown in Fig. 6. The advantage of this method is that the printing speed is faster and it is easy for industrialization. Some inkjet food 3D printing is self-supporting form, such as graphic decoration of Foodjet company. Others need to solidify the decorative materials by original food cavity, such as cavity depositing and surface filling.

Fig. 6. 3D printing principle of IJP

3 Material Requirement of 3D Printing

In order to obtain high-quality printed food, we need to select the proper materials besides choosing the right 3D printing method. 3D printed food materials should have three properties: printability, applicability and post-processing (Godoi et al. 2016). For extrusion-based 3D printing, the printability mainly refers to the rheological properties, thermal properties (melting point and glass transition temperature) and gelation properties of the materials. Particle size distribution, wettability, flowability (Shirazi et al. 2015) and bulk density have a great effect on the printability of powder-based 3D printing (Berretta et al. 2013; Schmid et al. 2014). In addition to the processing factors (laser type, laser power, laser spot diameter, etc.), the material properties are critical to the printing precision and accuracy of fabricated parts (Shirazi et al. 2015). Applicability means whether the materials capable of ensuring the mechanical stability of 3D construct and whether we can get the desired shape according to the pre-set procedure. Post-processing is the manufacturing process of extrusion-based 3D printed products, such as baking, steaming and frying. Generally, inkjet printing handles low viscosity materials that do not possess enough mechanical strength to hold 3D structure. Therefore, it is usually used to print two-dimensional images. From the point of view of printing precision and accuracy, the compatibility between ink and substrate surface, viscosity and rheological properties of ink, temperature and printing rate, are important to a successful printing (Liu et al. 2017a). The stability of 3D printed products are closely related to the rheological, thermal and mechanical properties of food materials.

Gelation properties of starch can change the viscosity of materials for food 3D Printing, thereby affecting the printability and applicability of 3D printing. The viscosity of food materials reduce and the mechanical strength and stability increase when starch content increase (Liu et al. 2017a, b). Fiber-rich food materials is easy to clog the nozzle because fiber is difficult to be crushed sufficiently. Adding fat-rich materials to the fiber-rich food materials could improve the rheological property because fat act as plasticizer or lubricant. The viscosity and rheological property of paste material relate to the category of protein it contains. Therefore, it is essential to choose the proper kind and dose of protein according to the desired rheological properties when adding protein to improve the flow characteristic of 3D printing materials. Particle composition, particle size and shape of insoluble materials also have an impact on the rheological and mechanical properties of paste 3D printing materials. The paste extruded materials demands low viscosity and easy to be extruded under high speed shearing. In addition, the material must be able to restore high viscosity quickly after deposition so as to keep the stability of 3D construct. The composition of paste 3D printing materials has a great influence on post-processing, such as heating drying and cooling drying rate, processing time and the state of final products (Lille et al. 2017).

4 Food 3D Printing and Personalized Nutrition

With more and more people begin to pay attention to nutrition, health and customization, it opens a new era for food 3D Printing. As a unique form of processing, food 3D printing can realize digital control of materials composition and flavor change and being increasingly appreciated by people. And it contributes to the development of food processing techniques that have the potential to be used in a wide range of applications such as space missions, aged-care facilities, and hospitals. FLM has a lot to offer for these applications due to the potential to modify flavor, nutrition, and texture through the incorporation of combinations of food additives (Hamilton et al. 2018).

For example, in order to solve the problems of swallowing and chewing in the elderly, researchers in many countries have used food 3D printing technology (Foodjet) to produce several kinds of soft aged-food with special texture and easy to swallow (Kouzani et al. 2017). Specific nutrient supplements were used during the manufacture processing of these food to meet the nutritional needs of the elderly. Thus, on the basis of previous studies, we can combine Dietary Guide lines of different populations with food 3D printing technology. Under the premise of meeting printing conditions and according to the personalized nutrition and energy demand of different people, the scientific collocation could be made by analyzing the nutritional composition and energy of foodstuffs. Therefore, food 3D printing can satisfy both the visual and the personalized nutrition of consumers.

5 Prospect and Future Trend

Food 3D printing technology is becoming more and more mature from emerging to nowadays. After 20 years' development, the profit of leading enterprises increases significantly, in addition, market recognition and industry income rose rapidly. According to the industry life cycle theory, technology industry shows the characteristic of accelerated growth from import period to growing period, thus 3D printing industry is entering the accelerated growth period. Yet, there still exist many factors that restrict the development of it, such as the limitation of materials, flowability, printing speed and the stability of printed food. On the premise of solving the above constrains, we need to strengthen the combination of food nutrition science and 3D printing. Meanwhile, in order to realize the industrialized and personalized production, it is wise to introduce the concept of nutrition and health and the idea of Industry 4.0 to it. In addition, it is necessary to strengthen the publicity of food 3D printing, making the public to get more information about this emerging technology and to recognize its role in improving people's lifestyles and health conditions. By the time 3D printing can be widely applied in food industry and be used by everyone, the precision nutrition food industry can implement the remote personalized services for families. We can predict that just as the printing in 1450, the steam engine in 1750 and the transistor technology in 1950, food 3D printing technology will change the tongue of the world in the future.

Acknowledgements. The authors are grateful to the financial support from Central Public-interest Scientific Institution Basal Research Fund (Grant No. S2016CB02) which enable us to finish this article.

References

Carlson, A., Frazão, E.: Food costs, diet quality and energy balance in the United States. Physiol. Behav. **134**, 20–31 (2014)

Xu, X., Hall, J., Byles, J., Shi, Z.: Assessing dietary quality of older Chinese people using the Chinese Diet Balance Index (DBI). PLoS ONE **10**(3), e0121618 (2015)

Sun, J., Zhou, W., Yan, L., Huang, D., Lin, L.Y.: Extrusion-based food printing for digitalized food design and nutrition control. J. Food Eng. **220**, 1–11 (2018)

Sher, D., Tutó, X.: Review of food 3D printing. Temes de Disseny **31**, 104–117 (2015)

Yang, F., Zhang, M., Bhandari, B.: Recent development in food 3D printing. Crit. Rev. Food Sci. Nutr. **57**(14), 3145–3153 (2017)

Krassenstein, B.: Tytan 3D to Launch a Multi-material 3D Printer Which Prints in Salt, Paper, Food, Adhesives & More [EB/OL] (2016)

Van Bommel, K.J.C.: Food 3D Printing [EB/OL] (2014)

Van der Linden, D.: Food 3D Printing Creating Shapes and Textures (2015)

Michiel Cornelissen Ontwerp. XOCO Chocolate Printer [EB/OL] (2016)

Lipton, J.I., Cutler, M., Nigl, F., Cohen, D., Lipson, H.: Additive manufacturing for the food industry. Trends Food Sci. Technol. **43**(1), 114–123 (2015)

Lipton, J., Arnold, D., Nigl, F., Lopez, N., Cohen, D.L., Noren, N., Lipson, H.: Multi-material food printing with complex internal structure suitable for conventional post-processing. In: Solid Freeform Fabrication Symposium, pp. 809–815 (2010)

Shandrow, K.L.: CocoJet: 3-D Printing and Hershey's Chocolate, Together at Last [EB/OL] (2015)

Hao, L., Mellor, S., Seaman, O., Henderson, J., Sewell, N., Sloan, M.: Material characterisation and process development for chocolate additive layer manufacturing. Virtual Phys. Prototyp. **5** (2), 57–64 (2010)

Mantihal, S., Prakash, S., Godoi, F.C., Bhandari, B.: Optimization of chocolate 3D printing by correlating thermal and flow properties with 3D structure modeling. Innov. Food Sci. Emerg. Technol. **44**, 21–29 (2017)

Serizawa, R., Shitara, M., Gong, J., Makino, M., Kabir, M.H., Furukawa, H.: 3D jet printer of edible gels for food creation. In: Proceedings of SPIE Smart Structures and Materials Nondestructive Evaluation and Health Monitoring, 90580A-90580A (2014)

Gong, J., Shitara, M., Serizawa, R., Makino, M., Kabir, M.H., Furukawa, H.: 3D printing of meso-decorated gels and foods. In: Materials Science Forum, vol. 783, pp. 1250–1254. Trans Tech Publications (2014)

Godoi, F.C., Prakash, S., Bhandari, B.R.: 3D printing technologies applied for food design: status and prospects. J. Food Eng. **179**, 44–54 (2016)

Peltola, S.M., Melchels, F.P., Grijpma, D.W., Kellomäki, M.: A review of rapid prototyping techniques for tissue engineering purposes. Ann. Med. **40**(4), 268–280 (2008)

Diaz, J.V., Van Bommel, K.J.C., Noort, M.W.J., Henket, J., Briër, P: U.S. Patent Application No. 14/894, 231 (2016)

Wegrzyn, T.F., Golding, M., Archer, R.H.: Food layered manufacture: a new process for constructing solid foods. Trends Food Sci. Technol. **27**(2), 66–72 (2012)

Bredt, J.F., Anderson, T.: U.S. Patent No. 5,902,441. U.S. Patent and Trademark Office, Washington (1999)

Kruth, J.P., Levy, G., Klocke, F., Childs, T.H.C.: Consolidation phenomena in laser and powder-bed based layered manufacturing. CIRP Ann. Manuf. Technol. **56**(2), 730–759 (2007)

Gu, D.D., Meiners, W., Wissenbach, K., Poprawe, R.: Laser additive manufacturing of metallic components: materials, processes and mechanisms. Int. Mater. Rev. **57**(3), 133–164 (2012)

Shirazi, S.F.S., Gharehkhani, S., Mehrali, M., Yarmand, H., Metselaar, H.S.C., Kadri, N.A., Osman, N.A.A.: A review on powder-based additive manufacturing for tissue engineering: selective laser sintering and inkjet 3D printing. Sci. Technol. Adv. Mater. **16**(3), 033502 (2015)

Schmid, M., et al.: Flowability of powders for selective laser sintering (SLS) investigated by round robin test. In: High Value Manufacturing: Advanced Research in Virtual and Rapid Prototyping, pp. 95–99 (2014)

Berretta, S., Ghita, O., Evans, K.E., Anderson, A., Newman, C.: Size, shape and flow of powders for use in selective laser sintering (SLS). In: High Value Manufacturing: Advanced Research in Virtual and Rapid Prototyping, p. 49 (2013)

Liu, Z., Zhang, M., Bhandari, B., Wang, Y.: 3D printing: printing precision and application in food sector. Trends Food Sci. Technol. **69**, 83–94 (2017)

Liu, Z., Zhang, M., Bhandari, B., Yang, C.: Impact of rheological properties of mashed potatoes on 3D printing. J. Food Eng. (2017b)

Lille, M., Nurmela, A., Nordlund, E., Metsä-Kortelainen, S., Sozer, N.: Applicability of protein and fiber-rich food materials in extrusion-based 3D printing. J. Food Eng., 1–8 (2017)

Kouzani, A.Z., Adams, S., Whyte, D.J., Oliver, R., Hemsley, B., Palmer, S., Balandin, S.: 3D printing of food for people with swallowing difficulties. KnE Eng. **2**(2), 23–29 (2017)

Russell, D., Hernandez, A., Kinsley, J., Berlin, A.: Methods and apparatus for 3D printing [P]. US: 20050280185 (2005)

Hamilton, A.C., Alici, G., in het Panhuis, M.: 3D printing vegemite and marmite: redefining "breadboards". J. Food Eng. **220**, 83–88 (2018)

An Economical High Precision Machine Vision System and Its Performance Test Using Ceramic Substrate

Zhen Cai[1], Huawei Qin[2], Shiwei Liu[2], and Jiwan Han[3(✉)]

[1] Ocean College, Zhejiang University, Zhoushan 316021, Zhejiang, China
[2] Institute of Mechatronic Engineering, Hangzhou Dianzi University, Hangzhou 310018, Zhejiang, China
[3] Institute of Biological, Environmental and Rural Sciences, Aberystwyth University, Aberystwyth SY23 3DA, Wales, UK
jihl7@aber.ac.uk

Abstract. Machine vision technology has developed rapidly due to the improvement of computer industry, however its applications haven't been widely used in industry because of the high price of the main devices. This paper proposed an economical machine vision dimension measurement system consisting of a low-cost machine vision camera and light source. A typical application of this system is to measure the sizes of ceramic substrates. The influence caused by the hardware were analyzed, and the calibration algorithms were developed to improve the measurement precision. The random noise and dark noise were reduced, and a uniform algorithm was developed to calibrate the pixel gray values of image background. Experimental results showed that the maximum error of measured length was within 2 μm for the measuring range of 70 mm, which could meet the precision requirement of practical applications. The proposed system can achieve high measurement precision using low-cost machine vision camera and light source.

Keywords: Low-cost · Dimension measurement · Ceramic substrate Uniform algorithm

1 Introduction

At present, various applications of machine vision are utilized to obtain the information of objects due to the advantages of fast, high accuracy and stability [1], moreover machine vision systems can achieve measurements of the objects which are hard to metered by manual methods, such as the hot large parts [2–4]. Much research has been conducted on the dimension measurement of different objects with machine vision systems [5, 6]. Churchill et al. [7] compared the dimension results measured by machine system and manual method, and the results showed the consistency of machine vision system was greater than manual measurements. Mohamed et al. [8] described a vision-based inspection system to measure the roundness error of a cylindrical part, in which a new approach of the set-up and image acquisition technique

K. Deng et al. (Eds.): ICMIR 2018, AISC 856, pp. 985–993, 2019.
https://doi.org/10.1007/978-3-030-00214-5_121

was utilized. Sun et al. [9] developed the method based on the processing of a digital image to measure the diameter of a shaft, and its relative errors was about 0.005 mm.

Most studies developed various algorithms or schemes on the basis of the stable, high performance and costly devices. However, equipping machine vision systems with that devices is expensive and inefficient in industry. Many measurements are still achieved by traditional manual or semiautomatic methods in factories, and the main reason is that the cost of equipment is much higher than manual methods. Therefore, a relatively economical but high precision solution of dimension measurement is needed. This paper has proposed a machine vision system to measure the dimensions of thin slices accurately with low-cost equipment by analyzing and eliminating the influence factors caused by the hardware in the measurement process. In order to validate the accuracy and practicability of the proposed system, ceramic substrates which are used as insulating materials in circuits were chosen as research object, the sizes were measured, and its process was discussed.

2 Experimental Set-Up

Economical hardware, including a CMOS camera, lens and LED illumination device were used in the proposed system. The CMOS camera is the acA2500-14gm by Basler Inc., whose resolution is 2594 × 1944, and the pixel size is 2.2 μm. For comparison, another stable CMOS camera, UI-2880CP-M-GL by IDS Inc. whose resolution is 3088 × 2076 was utilized. The lens is M3514-MP by Computar Inc., whose focal length is 35 mm. The LED illumination device is CV-FL-100 × 100R by MORITEX Inc., whose optical surface dimension is 10 cm × 10 cm. To reduce power fluctuation, all devices were powered by the AC power supply, whose voltage output accuracy is within 1%. The length of a ceramic substrate is about 70 mm, the width is about 60 mm, and each pixel in the CMOS sensor represents approximately 0.0314 mm. The set-up of the proposed system are shown in Fig. 1.

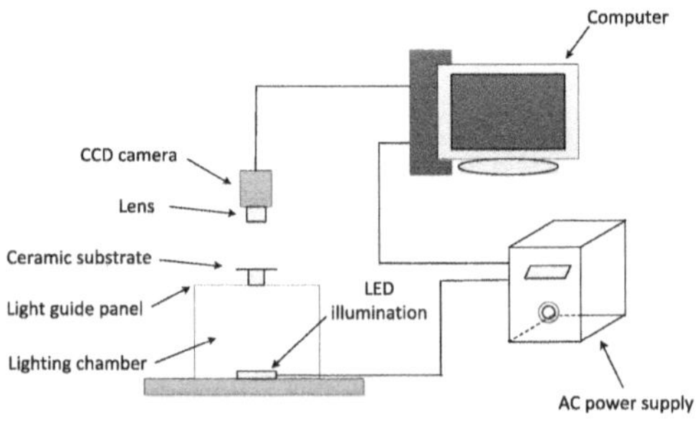

Fig. 1. The set-up of the proposed measurement system.

3 Image Calibration

The gray value is one of the keys to guarantee the precision of output results. A typical workflow of machine vision systems are shown in Fig. 2. The illumination is sensitive to the performance of hardware and environmental influence factors, and the uniformity and stability of illumination would be affected by these factors. Image quality would be reduced by various noises generated by the camera in the photoelectric conversion part. In addition, algorithms applied in image processing would influence the accuracy of final results and the speed of calculations.

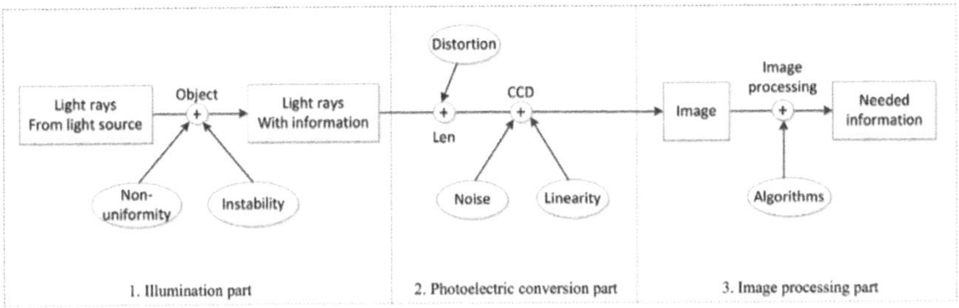

Fig. 2. The main steps of typical machine vision systems.

3.1 Noise Reduction

During image acquisitions, the stability and uniformity of pixel gray values were mainly affected by the random noise and dark noise. The reduction algorithms of random noise can be broadly divided into two categories [10, 11], one is based on a single image, and the other is based on an image sequence. Reducing random noise according to the gray values of surrounding pixels by single image algorithms is unsuited to edge areas. While the algorithms based on image sequence are more suitable as the reliability and consistency of gray values can be improved by multiple images acquired at the same condition. Hence averaging images were applied to reduce random noise in the experiments, and the number of images captured was tested. 25 images were acquired with an interval of 10 s, the x-coordinate was selected randomly, and the corresponding y-coordinates of edges of all images were calculated. The results are shown in Fig. 3. The maximum error of y-coordinate is approximately 1.406 pixels (0.044 mm) when the number of averaged images is 4, which can meet the accuracy requirement, and the accuracy improvement from averaging 4 images to 5 images is limited.

Dark noise is a signal-independent noise mainly related to the working temperature of the CMOS sensors [12]. The image acquirement was performed to obtain the dark noise characteristic when the camera temperature was stable. The aperture of the camera was closed, the exposure time was 50 ms, and an image sequence (50 images) was acquired with an interval of 10 s. There was no light received by the camera, so the

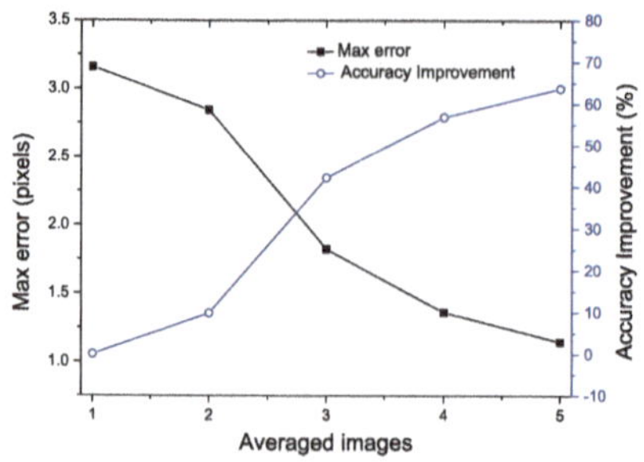

Fig. 3. The results of different averaged images.

gray values of the acquired images were only related to dark noise and random noise. The calibration matrix of dark noise was calculated:

$$\bar{G}_d(i,j) = \sum_{m=1}^{n} (r_m(i,j) + d_m(i,j)) \Big/ n \tag{1}$$

where n is the number of images, $r_m(i,j)$ is the gray value of random noise in (i,j) of the m-th image, and $d_m(i,j)$ is the dark noise gray value in (i,j) of the m-th image.

The random noise was reduced using the averaging method, and the dark noise calibration matrix $\bar{G}_d$ was subtracted from the image to be calibrated. From the result shown in Fig. 4, the number of pixels whose gray value greater than 1 is decreased, and the greater the gray value is, the better reduction of dark noise.

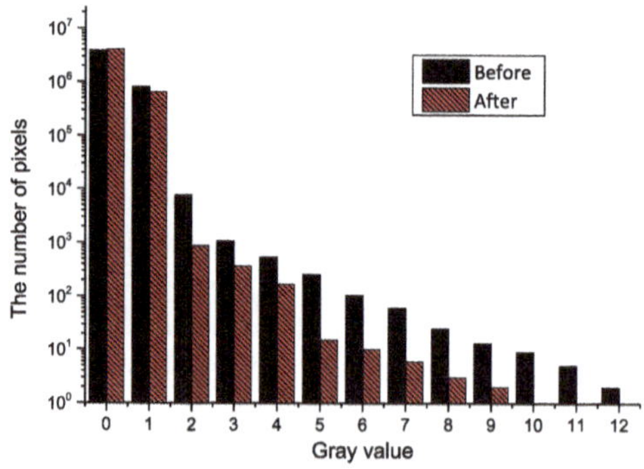

Fig. 4. The result histogram of dark noise reduction.

3.2 Uniformity Calibration

As the light guide panel placed on illumination device couldn't make illumination uniform enough, the intensity of light received in the center area of CMOS sensor was maximum, and it became weaker from the center to edges. To reduce the effects of this phenomenon, an uniform algorithm based on the edge gray values was developed.

An image sequence (50 images) of a ceramic substrate was acquired, all edges were extracted, and the average gray value of all edge pixels was calculated:

$$\bar{G}_{edge} = \sum_{m=1}^{n} \sum_{i=1}^{p_m} E_{[m,i]} \Big/ (p_m \times n) \tag{2}$$

where n is the number of images, p_m is the number of pixels of the m-th image and $E_{[m,i]}$ is the gray value of the i-th pixel of the edge E of the m-th image.

Then 50 images without ceramic substrates were acquired, $\bar{G}(i,j)$ was the average gray value of pixel (i,j) in the image sequence, and the calibration matrix C was calculated:

$$C(i,j) = G_{edge} / \bar{G}(i,j) \tag{3}$$

All pixels of the image to be measured were multiplied by the corresponding pixel in matrix C, the image after calibration was:

$$f_{calibrated}(i,j) = C(i,j) \times f(i,j) \tag{4}$$

The results of uniform calibration are shown in Fig. 5. Pixel gray values of the calibrated image background are more uniform, and the gray value gradient from the center to edges is reduced. According to the (c) and (f) in Fig. 5, the gray value change of one line pixel in the background area is much smaller.

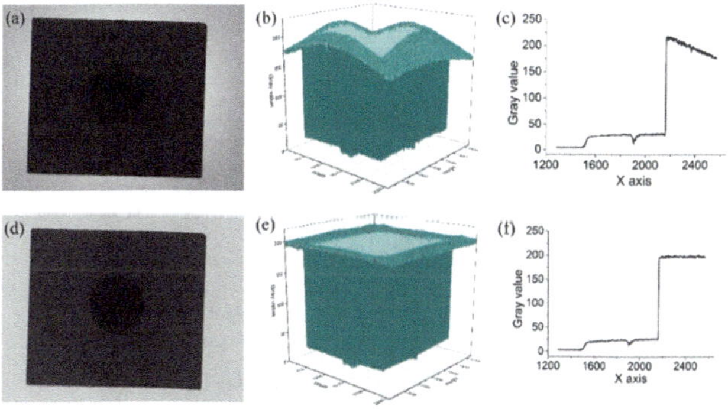

Fig. 5. (a) The original image, (b) 3d plot of the original image, (c) gray values of a line pixels from center to edge of the origin image, (d) the calibrated image, (e) 3d plot of the calibrated image, (f) gray values of a line pixels from center to edge of the calibrated image.

4 Measuring Method

As the contour of ceramic substrates is nearly a rectangle which is a simple shape for measurement, measuring dimensions of ceramic substrates is calculating the pixel distances of two parallel edges, and one of the key to obtain the precise results was positioning the measuring points on edges exactly. The calibrating and measuring procedures are summarized in the following steps.

Step 1: Obtain calibration matrixes

The calibration methods proposed above are used to compute the calibration matrixes of the machine vision system before measurements.

Step 2: Spatial calibration

Put a standard ceramic substrate whose size information is already known on the measuring platform. The pixel lengths of 4 edges are calculated by the method from step 3 to step 6 described below. Due to the distortion of camera, the conversion factor of each edge converting pixel to millimeter are calculated separately:

$$T = D_{pixel} / D_{mm} \tag{5}$$

Step 3: Image acquisition and calibration

All ceramic substrate images to be measured are acquired in the same condition without changing the illumination and positions of all components, and the acquired images are calibrated by the noise reduction algorithms and uniform algorithm.

Step 4: Edge detection

The edges of the ceramic substrate are extracted using edge detection algorithms based on Sobel operator, and binary images including edge information are obtained, the image are shown in Fig. 6.

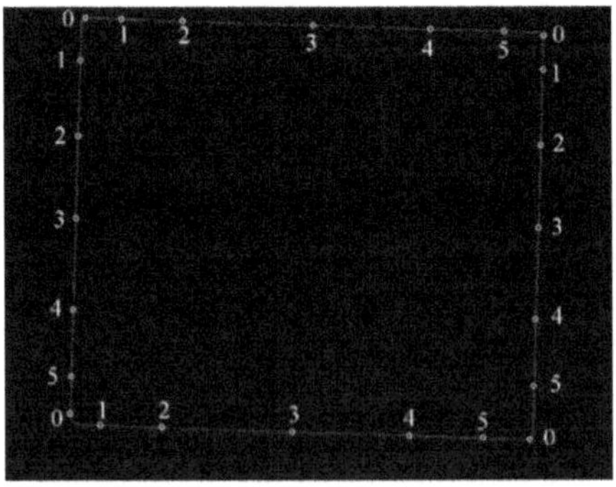

Fig. 6. The binary image generated during measurements.

Step 5: Vertexes detection

In manual measurements, the point measured by caliper is positioned according to the distance from the vertex on the edge. To simulate this measuring method and make sure the measured points in each measurement are the same, the vertex coordinates of the ceramic substrate are calculated.

a. Obtain the coordinates of 4 vertexes roughly by scanning the binary image, and these coordinates are used to define the searching range in step 5(b) which are marked as circle 0 in Fig. 6.
b. Five points on each edge are selected in the range of corresponding vertex coordinates of the binary image computed above. In order to reduce random errors, coordinates of 10 surround pixels of each pixel are recorded by scanning, and these 11 pixels form a pixel sequence.
c. Subpixel coordinates of all 11 points in each pixel sequence are calculated according to the coordinates and original image, and they are averaged to represent the point scanned in step 3(b), which are marked as circle 1 to 5 in Fig. 6.
d. The linear equations of each edge are calculated based on average coordinates of 5 points by least square method.
e. Vertex coordinates are obtained by calculating the intersection point coordinates of every two edges.

Step 6: Coordinate calculation of measured points

The point coordinates to be measured are calculated based on the vertex coordinates and distances from the vertexes. The final point coordinates are computed by using the averaging method proposed in step 5(c).

Step 7: Measuring

Calculate the distance between two points, and convert the pixel to millimeter by using the conversion factors calculated in step 2.

5 Results and Discussion

A large number of experiments were performed in order to validate the proposed system in terms of accuracy and repeatability. The ceramic substrate to be measured was first metered using a micrometer in experiments. The 50 measurement results of one ceramic substrate measured by acA2500-14gm are shown in Fig. 7, and the result

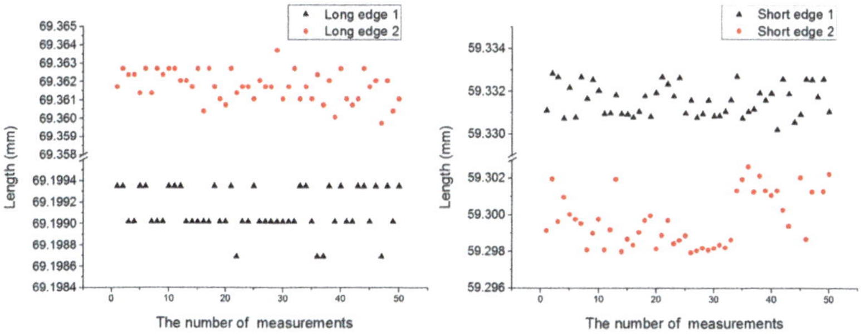

Fig. 7. The measurement results of a ceramic substrate.

Table 1. The statistics of acA2500-14gm and UI-2880CP-M-GL

Edge	acA2500-14gm			UI-2880CP-M-GL		
	Average (mm)	Max error (μm)	Std Dev (mm)	Average (mm)	Max error (μm)	Std Dev (mm)
1	69.1989	0.68	1.63E−04	69.1981	0.5	1.44E−04
2	69.3615	0.83	1.51E−04	69.3602	0.59	1.37E−04
3	59.3315	1.652	3.67E−04	59.3328	0.94	2.16E−04
4	59.3	0.72	2.03E−04	59.3007	0.42	1.26E−04

statistics of acA2500-14gm and UI-2880CP-M-GL are shown in Table 1. The lengths measured by the acA2500-14gm are stable, the maximum error of long edges is within 1 μm, and the maximum error of short edges is within 2 μm. According to Table 1, the differences of measurements between the two cameras are within 0.8 μm, which demonstrate that the precision of the proposed measurement system with developed algorithms is improved and it can meet the requirements of the practical application.

6 Conclusion

This study has proposed a low-cost machine vision system whose precision was improved by software-based methods. The influences factors of result precision, such as illumination performance, CMOS noises were analyzed, the noise reduction algorithms were applied, and the background gray value calibration algorithm and dimension measurement method were developed. A large quantity of measurements of ceramic substrates were performed, and the results showed the maximum error of measured sizes was within 2 μm, which could meet the precision requirement of the practical application. The measuring time of a ceramic substrate was approximately one second, which guaranteed the efficiency of the production line. Apart from measuring ceramic substrates, the sizes of other thin-sheet can also be measured using this system.

References

1. Dickmanns, E.D.: The development of machine vision for road vehicles in the last decade. In: 2002 Intelligent Vehicle Symposium, vol. 1, pp. 268–281. IEEE (2002)
2. Dworkin, S.B., Nye, T.J.: Image processing for machine vision measurement of hot formed parts. J. Mater. Process. Technol. **174**(1), 1–6 (2006)
3. Jia, Z., Wang, B., Liu, W., et al.: An improved image acquiring method for machine vision measurement of hot formed parts. J. Mater. Process. Technol. **210**(2), 267–271 (2010)
4. Rafajłowicz, E., Wnuk, M., Rafajłowicz, W.: Local detection of defects from image sequences. Int. J. Appl. Math. Comput. Sci. **18**(4), 581–592 (2008)
5. Igathinathane, C., Pordesimo, L.O., Batchelor, W.D.: Major orthogonal dimensions measurement of food grains by machine vision using ImageJ. Food Res. Int. **42**(1), 76–84 (2009)

6. Igathinathane, C., Pordesimo, L.O., Columbus, E.P., et al.: Sieveless particle size distribution analysis of particulate materials through computer vision. Comput. Electron. Agric. **66**(2), 147–158 (2009)
7. Churchill, D.B., Bilsland, D.M., Cooper, T.M.: Comparison of machine vision with human measurement of seed dimensions. Trans. ASAE **35**(1), 61–64 (1992)
8. Mohamed, A., Esa, A.H., Ayub, M.A.: Roundness measurement of cylindrical part by machine vision. In: 2011 International Conference on Electrical, Control and Computer Engineering (INECCE), pp. 486–490. IEEE (2011)
9. Sun, Q., Hou, Y., Tan, Q., et al.: Shaft diameter measurement using a digital image. Opt. Lasers Eng. **55**, 183–188 (2014)
10. Sarode, M.V., Deshmukh, P.R.: Reduction of speckle noise and image enhancement of images using filtering technique. Int. J. Adv. Technol. **2**(1), 30–38 (2011)
11. Benesty, J., Chen, J., Huang, Y.A., et al.: Study of the Wiener filter for noise reduction. In: Speech Enhancement, pp. 9–41. Springer, Berlin (2005)
12. Cervera, C., Ribet-Mohamed, I., Taalat, R., et al.: Dark current and noise measurements of an InAs/GaSb superlattice photodiode operating in the midwave infrared domain. J. Electron. Mater. **41**(10), 2714–2718 (2012)

Feature Extraction of Rolling Bearing Incipient Fault Using an Improved SCA-Based UBSS Method

Shihao Cai, Jianfeng Qu(✉), Zeping Wang, and Chunyan Li

College of Automation, Chongqing University, Chongqing 400044, China
qujianfeng@cqu.edu.cn

Abstract. Incipient fault signal of rolling bearing tends to have weak amplitude and low signal-to-noise ratio. Bearing vibration signal collected by sensor is generally a mixture of several unknown source signals. It is difficult to extract incipient fault features from bearing vibration signals directly. The blind source separation (BSS) algorithm is used to restore source signals from the mixed signals. However, ordinary BSS algorithm requires that the number of sensors cannot be less than the number of source signals. To solve these problems, an improved underdetermined blind source separation (UBSS) method is proposed in this paper. The bearing vibration signals are preprocessed by short-time Fourier transform (STFT). Then the fuzzy C-means (FCM) clustering and the weighted minimal L1 norm method are respectively used to evaluate the mixing matrix and recover the separated signals. At last, the features of rolling bearing incipient fault are extracted from the recovered signals by envelope demodulation. Experimental results verify the effectiveness of the proposed method.

Keywords: Incipient fault · Feature extraction · UBSS · Rolling bearing SCA

1 Introduction

Rolling bearing plays an important role in mechanical equipment. Vibration signal is often used to analyze bearing failure. Generally, the vibration signals of bearing are collected by vibration sensors mounted on the bearing pedestal. Weak signals of rolling bearing incipient faults have the characteristics of weak amplitude and low signal-to-noise ratio [1]. Without detecting incipient fault, bearing failure may cause equipment performance degradation, or even result in a catastrophic loss [2]. Therefore, it is very significant to detect the incipient fault of rolling bearing.

Bearing vibration signal acquired by sensor is a mixture of several unknown source signals. Due to this reason and the characteristics of incipient fault, it is difficult to extract incipient fault features directly from bearing sensor data using most existing bearing fault diagnosis methods.

Blind source separation (BSS) is an algorithm that restores the source signals with only the observed signals when the source signals and the transmission channels are unknowable [3]. The BSS, as a signal processing method, has been widely used in

K. Deng et al. (Eds.): ICMIR 2018, AISC 856, pp. 994–1002, 2019.
https://doi.org/10.1007/978-3-030-00214-5_122

condition monitoring and fault diagnosis. Gelle applied BBS algorithm for the vibration acceleration signals of rotating machines [4]. Wu developed a BSS algorithm to reduce the influence of adjacent cylinders on the signals, thus the status of a specific cylinder can be monitored well through an acoustic emission (AE) sensor installed near it [5]. With extensive researches, a lot of algorithms have been applied to the BSS problem.

Most of the BBS algorithms are applied to the case where the quantity of sensors is above the quantity of source signals. In practice, however, the condition is often hard to satisfy. The underdetermined blind source separation (UBSS) is more in line with practical applications without the limitation of the condition. Rickard proposed a DUET algorithm for recovering an arbitrary number of source signals by two observed signals [6]. Tengtrairat put forward a new framework for solving single-channel blind source separation (SCBSS) problem which didn't need random initial inputs or any predefined structure on the sensors [7].

In this paper, the sparse component analysis (SCA) [8] based approach is used for the UBSS. The SCA is a method that uses the sparse decomposition to solve the BSS problem. Bofill put forward a two-stage scheme using SCA for the UBSS, which was widely researched and applied [9]. For the purpose of improving the precision of the two-stage scheme, this paper presents an improved UBSS method. Taking advantage of the UBSS, this paper combines the improved SCA-based UBSS method with envelope demodulation for feature extraction of rolling bearing incipient fault. Experimental results reveal the effectiveness of the proposed method in this paper.

This paper is organized as follows: Sect. 2 introduces the structure of the proposed method in this paper. Section 3 introduces the model of UBSS and the improved SCA-based UBSS method. In Sect. 4, a set of experiments are carried out to validate the method in this paper. The conclusions are given in Sect. 5.

2 The Proposed Incipient Fault Feature Extraction Method

The UBSS algorithm can separate different source signals from several observed signals, even though the quantity of observed signals is less than the quantity of source signals. Motivated by the advantages of the UBSS, a method based on an improved SCA-based UBSS algorithm and envelope demodulation is proposed for feature extraction of rolling bearing incipient fault.

Firstly, the bearing vibration signals collected by sensors are transformed by short-time Fourier transform (STFT). Secondly, the mixing matrix is estimated by the approach based on the fuzzy C-means (FCM) clustering. Thirdly, using the estimated mixing matrix, the source signals are recovered by the weighted minimal L1 norm method. Finally, the recovered signal with the biggest kurtosis is selected for envelope demodulation, and features of rolling bearing incipient fault can be extracted.

3 Improved SCA-Based UBSS Method

3.1 Underdetermined Blind Source Separation

The instant mixing model of the BBS can be defined as the following equation:

$$X(t) = AS(t) + N(t),\ t = 1, 2, \ldots, T \tag{1}$$

where $X(t) = \left[x_1(t)x_2(t), \ldots x_m(t)^T\right]$, represents a column vector representing m observed signals form sensors, and $S(t) = \left[s_1(t), s_2(t), \ldots s_n(t)^T\right]$, represents a column vector representing n source signals. Let A be an $m \times n$ matrix representing the unknown mixing matrix. $N(t) = [n_1(t), n(t), \ldots n_m(t)]^T$ denotes the additive noise. t is the discrete time. The model of the BSS is shown in Fig. 1.

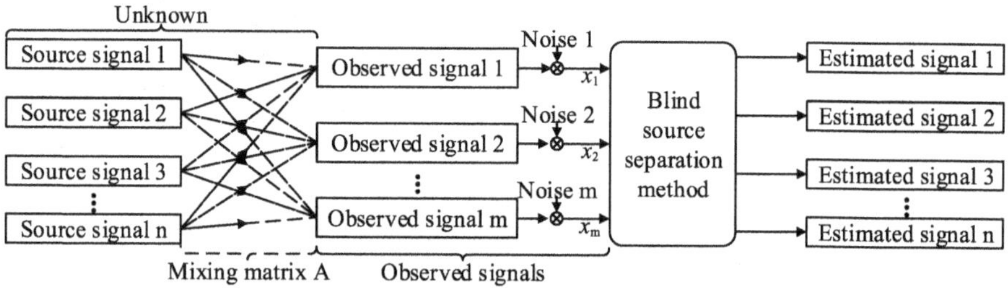

Fig. 1. Model of the BSS

It is so called the UBSS when $m < n$. Even if the mixing matrix is known, source signals cannot be estimated directly by observed signals. Sparse signal refers to that the signal value is zero or close to zero at most sampling points [10]. It is so called the SCA that using signal sparse representation. If the sparsity of time domain signals is not enough, signals can be sparse by proper transformation, and then separated in the transform domain. The UBSS problem is generally decomposed into two stages: estimating the mixing matrix A and restructuring the source signals $S(t)$.

3.2 Fuzzy C-Means Clustering Algorithm

Fuzzy theory is introduced to the clustering process of FCM clustering algorithm. According to the membership degree of the data, FCM clustering algorithm determines which cluster the data belongs to.

Let $D = \{d_1, d_2, \ldots d_l\}$ be a dataset to be classified, and FCM algorithm will generate a membership vector for any sample in the dataset. Let $u_{ij} = u_i(d_j) \in [0, 1]$ be an element in the membership vector, indicating the membership degree of the j-th sample point which belongs to the i-th classification in the dataset D. The sum of the membership degrees of all sample points in the same class is constant to 1. The goal of FCM is to minimize the following objective function [11]:

$$J = \sum_{i=1}^{c} \sum_{j=1}^{l} u_{ij}^{w} \left\| d_j - v_i \right\|^2 \tag{2}$$

where v_i is the cluster center of the i-th class, $w \in [1, \infty)$ is the weighted index and the membership vectors of all sample points can be combined to form a $c \times l$ matrix.

The cluster centers and the memberships of sample points can be updated by (3) and (4).

$$v_i = \frac{\sum_{j=1}^{l} u_{ij}^{w} d_j}{\sum_{j=1}^{l} u_{ij}^{w}} \tag{3}$$

$$u_{ij} = \frac{\left\| d_j - v_i \right\|^{\frac{-2}{w-1}}}{\sum_{i=1}^{c} \left\| d_j - v_i \right\|^{\frac{-2}{w-1}}} \tag{4}$$

3.3 Mixing Matrix Estimation

In this paper, a method using the peak point of frequency energy and FCM clustering is applied to evaluate the mixing matrix. The time domain signals are converted into the time-frequency signals by STFT. STFT can be expressed as (5). In this paper, the window function used by STFT is a rectangular window.

$$X(t,f) = \int x(t) m(\tau - t)\, e^{-j2\pi f \tau}\, d\tau \tag{5}$$

Here, $m(\tau - t)$ represents the window function, $x(t)$ is in the time domain and $X(t,f)$ is the converted time-frequency signal.

Then calculate the sum of the energies of all observed signals at each frequency in the frequency domain [12]. The sum of energies can be expressed as (6)

$$E(f) = \sum_{i=1}^{m} \int_{-\infty}^{+\infty} (R(X_i(t,f)))^2 + (I(X_i(t,f)))^2 dt \tag{6}$$

where $R(X_i(t,f))$ and $I(X_i(t,f))$ denote real part and imaginary part of observed signal $X_i(t,f)$.

Let ε be the energy threshold, and then find the frequency f_k corresponding to the peak value which is greater than ε in $E(f)$. After that, for each f_k, the time-frequency points of all the observed signals at the same time are normalized. Cluster centers of the observed signals at frequency f_k on the scatter plot are determined by FCM clustering algorithm, and one column of the mixing matrix can be achieved by the direction vector constructed from the cluster centers. Finally, the whole mixing matrix can be obtained.

3.4 Weighted Minimal L1 Norm Method

Taking the two dimensional case as an example, the theory of the weighted minimal L1 norm method is illustrated. The mixing matrix A is estimated according to the procedure described above, and a_i represents the column vector of the mixing matrix. Let $\vec{x}(t)$ be the vector of observed signal at any moment t. The concrete steps to restore source signals by the weighted minimal L1 norm method are described below:

Step 1: Calculate the direction angle $\theta(t)$ of the observed signal vector $\vec{x}(t)$ and α_i of the mixing matrix column vector a_i according to (7) and (8)

$$\theta(t) = \arctan(\vec{x_1}(t)/\vec{x_2}(t)) \tag{7}$$

$$\alpha_i = \arctan(a_{1i}/a_{2i}),\ i = 1, \ldots, n \tag{8}$$

where $\vec{x_1}(t)$, $\vec{x_2}(t)$, a_{1i} and a_{2i} are coordinates of $\vec{x}(t)$ and a_i respectively.
Step 2: Calculate the absolute error $\Delta\theta$ between $\theta(t)$ and α_i, and compare $\Delta\theta$ with the threshold γ to get all the potential decompositions of $\vec{x}(t)$. The threshold γ should be slightly less than the maximum deviation between the direction angles of the adjacent column vectors of the mixing matrix A. If $\Delta\theta \leq \gamma$, take the corresponding a_i as a potential decomposition of $\vec{x}(t)$. And only the vector combinations containing vectors above and below $\vec{x}(t)$ can be selected.
Step 3: Let $A_j = \{a_1, \ldots, a_j | 1, \ldots, j \in (1, \ldots, n), j \leq n\}$ be all the decompositions of $\vec{x}(t)$, and two vectors that satisfy the conditions are selected from A_j to compose 2×2 matrix B_k (a total of C_j^2 matrices). Then figure out all possible solutions $\hat{S}^{(k)}(t) = \left[\hat{S}_1^{(k)}(t), \ldots, \hat{S}_n^{(k)}(t)\right]^T, k = 1, \ldots, C_j^2$ by (9).

$$\widehat{S}^{(k)}(t) = B_k^{-1}\vec{x}(t) \tag{9}$$

Step 4: Work out L1 norm J_k and the weighted coefficient q_k for each $\hat{S}^{(k)}(t)$ by (10) and (11).

$$J_k = \sum_{i=1}^{n} \left|\widehat{S}_i^{(k)}(t)\right|,\ k = 1, \ldots, C_j^2 \tag{10}$$

$$q_k = \frac{1/J_k}{\sum_{i=1}^{C_j^2} (1/J_i)} \tag{11}$$

Step 5: An estimation $\hat{S}(t)$ of the source signal at the moment t is obtained by superimposing the products of q_k and $\hat{S}^{(k)}(t)$ by (12).

$$\widehat{S}(t) = \sum_{k=1}^{C_j^2} q_k \cdot \widehat{S}^{(k)}(t) \quad (12)$$

Step 6: For all moments, the estimation of source signals is obtained by step 1–5.

4 Experimental Researches

4.1 Simulation Experiments

A simulation is performed with five simulated source signals which are expressed in (13). The simulated source signals are shown in Fig. 2 with a sampling rate of 1024 Hz.

$$\begin{cases} s_1 = (\cos(2\pi f_0 t) + 1) \times \sin(2\pi f_1 t) \\ s_2 = \sin(2\pi f_2 t) \\ s_3 = (\cos(2\pi f_0 t) + 1) \times \sin(2\pi f_3 t) \\ s_4 = \sin(2\pi f_4 t) \\ s_5 = e^{-\lambda t} \times \sin(2\pi f_5 t) \end{cases} \quad (13)$$

Here $f_0 = 10\text{Hz}, f_1 = 50\text{Hz}, f_2 = 100\text{Hz}, f_3 = 150\text{Hz}, f_4 = 200\text{Hz}, f_5 = 400\text{Hz}$ and t represents the sampling moment, and $\lambda = 50$ is a parameter that controls the attenuation rate of the impulse signal.

As shown in Fig. 3, white Gaussian noise is added to the observed signals. Time domain waveforms of the recovered signals are represented in Fig. 4, and the frequency spectrums are displayed in Fig. 5. The UBSS method in this paper can restore the source signals well in the noise case, and frequency features of the source signals can be clearly distinguished in the spectrum.

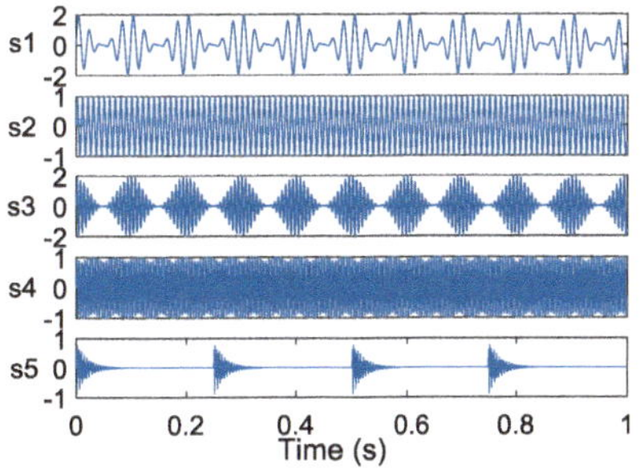

Fig. 2. Waveforms of simulated source signals

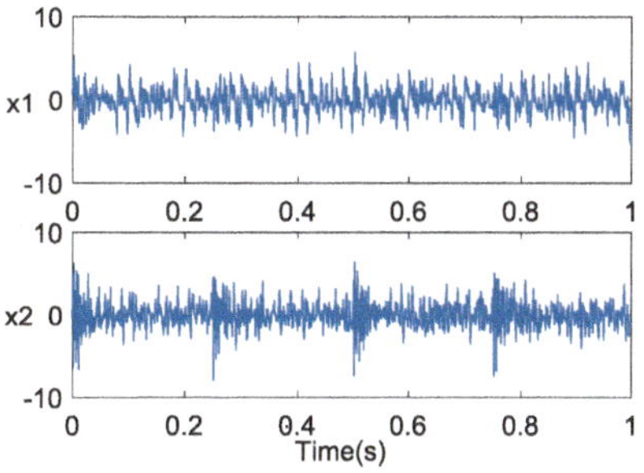

Fig. 3. Waveforms of observed signals with noise

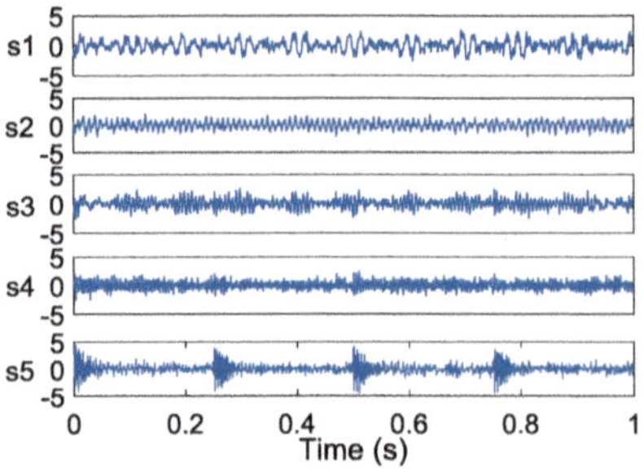

Fig. 4. Waveforms of five recovered signals

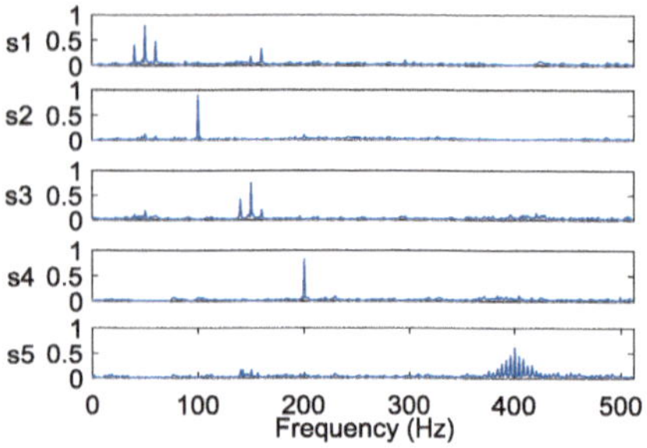

Fig. 5. Spectrums of the recovered signals

4.2 Validation with Bearing Data

The proposed method in this paper is applied to the incipient fault feature extraction of a realistic motor bearing test system consisting of a 2 hp motor, a torque sensor and a dynamometer. The sensor data of bearing vibration is offered by the Bearing Data Centre of Case Western Reserve University [13]. Using electro-discharge machining, single point faults with different fault diameters (7, 14 and 21 mils) are introduced to the test bearings. The motor drives bearings with speed of 1720 to 1797 rpm.

This experiment considers that the inner race fault with fault diameter of 7 mils is an incipient fault. The defect frequency of ball pass frequency of inner race (BPFI) is 5.4152 (multiple of running speed in Hz). The bearing vibration data is measured by accelerometers which are attached to the housing with magnetic bases, at 12,000 samples per second. The vibration signals collected at the drive end are used as the observed signals, and white Gaussian noise is added to the observed signals. The bearing rotation rate is 1772 rpm and the rotation frequency f_r of the rolling bearing is 29.53 Hz. The waveforms of the observed signals in the time domain are shown in Fig. 6a. The envelope demodulation spectrums of the observed signals are shown in Fig. 6b. The inner race fault frequency ($\mathrm{BPFI} = 5.4152 \times 29.53 = 159.92\,\mathrm{Hz}$) cannot be distinguished form Fig. 6b, which indicates that it is infeasible to apply envelope demodulation on the vibration signals of rolling bearing incipient fault directly.

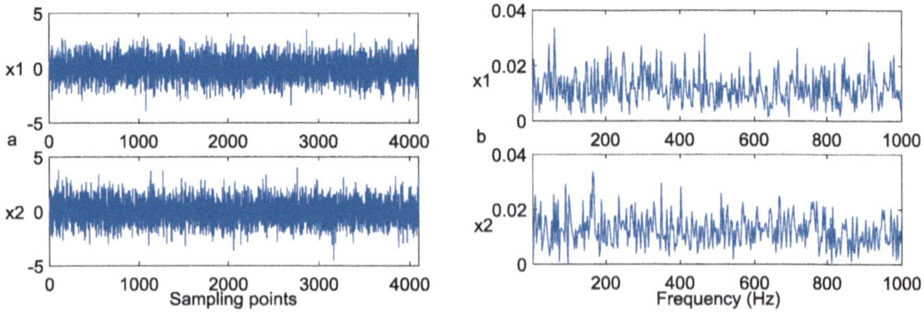

Fig. 6. Waveforms of two observed signals and the corresponding envelope demodulation spectrums: (a) time domain waveforms; (b) corresponding envelope demodulation spectrums.

Apply the improved SCA-based UBSS algorithm in this paper to the observed signals in Fig. 6a. As shown in Fig. 7, five recovered signals are separated by the UBSS method in this paper. By calculation, the signal e_4 has the biggest kurtosis, which indicates that it contains much more transient impulse components than the other recovered signals in Fig. 7. In order to observe the features of signal better, envelope demodulation is applied to signal e_4, and part of the envelope demodulation spectrum is displayed in Fig. 8. In Fig. 8, the inner race fault frequency (BPFI) is extracted clearly. In addition, the side frequencies are also extracted.

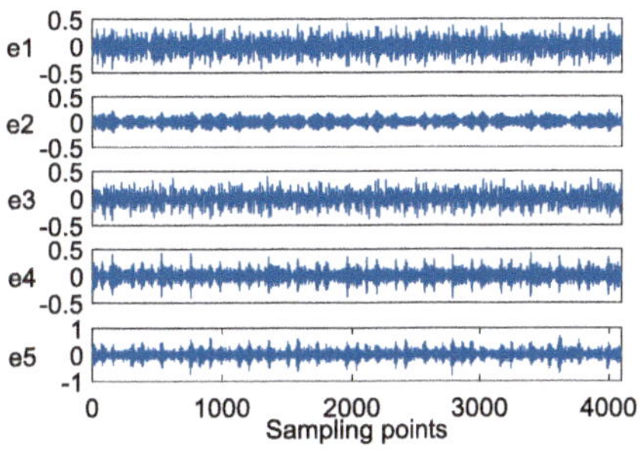

Fig. 7. Waveforms of the recovered signals

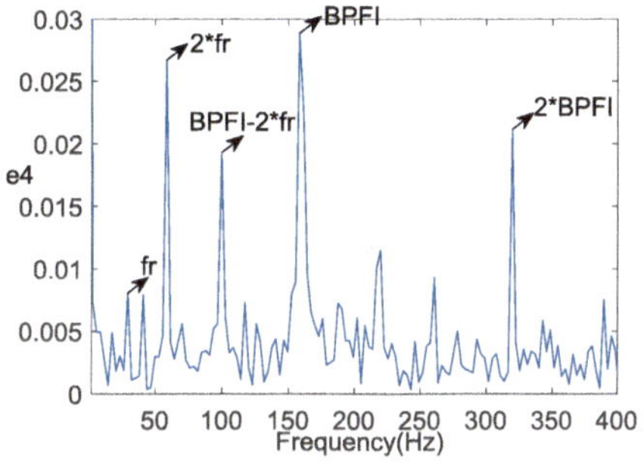

Fig. 8. Envelope demodulation spectrum of e_4

5 Conclusions

The incipient fault detection of rolling bearing is of great importance to the maintenance of mechanical equipment. In this paper, a method that combines the improved SCA-based UBSS algorithm and envelope demodulation is put forward for feature extraction of rolling bearing incipient fault. The simulation result validates the improved SCA-based UBSS method. The experiments with bearing vibration data demonstrate that the proposed method in this paper is effective for feature extraction of rolling bearing incipient fault. With respect to the future work, it is necessary to further research the BSS of nonlinear mixed signals and focus on optimizing the performance of the proposed method.

6 Acknowledgments

This paper was supported by the Natural Science Foundation of Chongqing City, China (cstc2016jcyjA0504).

References

1. Wang, H., Chen, J., Dong, G.: Feature extraction of rolling bearing's early weak fault based on EEMD and tunable Q-factor wavelet transform. Mech. Syst. Signal Process. **48**(1–2), 103–119 (2014)
2. Yang, Y., Yu, D., Cheng, J.: A fault diagnosis approach for roller bearing based on IMF envelope spectrum and SVM. Measurement **40**(9–10), 943–950 (2007)
3. Comon, P., Jutten, C.: Handbook of Blind Source Separation: Independent Component Analysis and Separation, vol. 4, pp. 823–831. Academic Press, London (2010)
4. Gelle, G., Colas, M., Delaunay, G.: Blind sources separation applied to rotating machines monitoring by acoustical and vibrations analysis. Mech. Syst. Signal Process. **14**(3), 427–442 (2000)
5. Wu, W., Lin, T.R., Tan, A.C.C.: Normalization and source separation of acoustic emission signals for condition monitoring and fault detection of multi-cylinder diesel engines. Mech. Syst. Signal Process. **64–65**, 479–497 (2015)
6. Rickard, S.: The DUET blind source separation algorithm. In: Blind Speech Separation, pp. 217–241 (2007)
7. Tengtrairat, N., Woo, W.L.: Single-channel separation using underdetermined blind autoregressive model and least absolute deviation. Neurocomputing **147**(147), 412–425 (2015)
8. Yu, K., Yang, K., Bai, Y.: Estimation of modal parameters using the sparse component analysis based underdetermined blind source separation. Mech. Syst. Signal Process. **45**(2), 302–316 (2014)
9. Bofill, P., Zibulevsky, M.: Underdetermined blind source separation using sparse representations. Signal Process. **81**(11), 2353–2362 (2001)
10. Cai, T.T., Wang, L.: Orthogonal matching pursuit for sparse signal recovery with noise. IEEE Trans. Inf. Theory **57**(7), 4680–4688 (2011)
11. Havens, T.C., Bezdek, J.C., Leckie, C., et al.: Fuzzy c-means algorithms for very large data. IEEE Trans. Fuzzy Syst. **20**(6), 1130–1146 (2012)
12. Yu, G., Zhou, Y., Liu, L., et al.: Identification of modal parameters using under-determined blind source separation. J. Vib. Meas. Diagn. **36**(4), 722–728 (2016)
13. Bearings Vibration Data Set, Case Western Reserve University. http://csegroups.case.edu/bearingdatacenter/home

Application of Gesture Recognition Interaction to Digital Virtual Reconstruction of Yunnan Bronze Drum

Xu Wu[1], Guo Wei[1(✉)], Jin Chunjie[1], Wu Ping[1], and Shao Xinye[2]

[1] School of Electrical and Information Engineering, Yunnan Minzu University, Kunming, China
18487089465@163.com

[2] Faculty of Information Engineering and Automation, Kunming University of Science and Technology, Kunming, China

Abstract. At present, our country attaches great importance to the protection and development of intangible culture. However, it would be destroyed and difficult to reserve in traditional way. This paper, starting with the analysis of the above problems, makes use of the virtual reality technology and the somatosensory interactive equipment to analyze the specific object of Yunnan bronze drum, so as to achieve the purpose of giving full play to its humanistic value and economic value. First, the bronze drum pictures and audio data are collected, and the three-dimensional model of the bronze drum is made with 3dsMax. Second, the gesture recognition of the bronze drum knocking is completed by Kinect, and the virtual bronze drum is performed. Finally, the classification recognition method based on the convolution neural network is used to identify the finger tip, to control the rotation of bronze drum. The better protection and inheritance can be achieved with the digital virtual reconstruction of Yunnan bronze drum.

Keywords: Audio acquisition · Gesture recognition
Convolution neural network

1 Introduction

There is colorful cultural heritage of ethnic minorities in Yunnan province, especially the bronze drum culture. The earliest bronze drum in China was found in the Chunqiu period at Wanjia dam in Yunnan. Bronze drum is a popular percussion instrument in minority areas, which were often used for military command in ancient times, and in the banquets, music and dance. Bronze drum is one of splendid culture in ancient China, which is a symbol of the ancestors' wisdom of Chinese minorities, and of the characteristics of Oriental Art. It is definitely a treasure of world culture and art.

At present, most of the studies on Yunnan bronze drum are from the perspective of ethnic religion and culture. However, there is little research on the digital protection of Yunnan bronze drum, which results that the artistic and economic values of bronze drum culture are not fully explored and utilized. Traditional relic restoration and protection methods are usually carried out in physical and chemical way. But there are

K. Deng et al. (Eds.): ICMIR 2018, AISC 856, pp. 1003–1012, 2019.
https://doi.org/10.1007/978-3-030-00214-5_123

still problems of secondary damage and difficulties in maintenance and reservation [1]. In view of this, the digital protection based on virtual reality technology emerges at the historic moment.

Digital protection is a new relic protection, which has a great advantage over traditional protection and transmission. Transforming cultural relics into digital cultural heritage through modern computer technology can protect cultural relics more efficiently and continuously [2]. Taking the Internet as the medium to break the limitation of time and space, we can better carry forward and inherit bronze drum culture. At the same time, it will transform traditional culture into the tourism attraction and promote the integrated development of tourism and culture in Yunnan province. Finally, it will become a new engine of Yunnan economic development.

2 The Data Acquisition and 3D Modeling of Yunnan Bronze Drum

It can use a 3D scanner for physical scanning of Bronze drum to learn the shape, color, texture, decoration and other key information. Unlike the data collection of other cultural relics, it is crucial to record the knocking sound on the bronze drum in the original data collection of Bronze drum.

2.1 The Collection of Bronze Drum Percussion Sound

The collection of bronze drum sound is based on the combination of microphone recording, audio processing software capture interception and noise reduction optimization. The GL100-FX professional instrument is used to collect the sound, and the percussion sound is transmitted to the computer via the Mic port. Considering the different knocking sounds at different positions, the surface of bronze drum can be separated by polar coordinate system, thus, 37 sampling points can be selected for each drum. Then, knocking the bronze drum lightly or heavily, 74 audio files are obtained. The sound acquisition process is shown in Fig. 1.

Fig. 1. Sound acquisition process

2.2 Spectrum Analysis

As for the collected audio files, we need to analyze the spectrum accurately to classify and save them by Cool Edit Pro audio processing software. The software is a multi-track recording and audio processing software. It can carry out audio analysis, fast output spectrum and clearly reveal the difference of each voice [3]. We choose three points on different positions of bronze drum to strike with the same intensity. At the same time, we choose a point to knock with two kind of strength. The collected audio is analyzed by Cool Edit Pro audio processing software. As shown in Fig. 2.

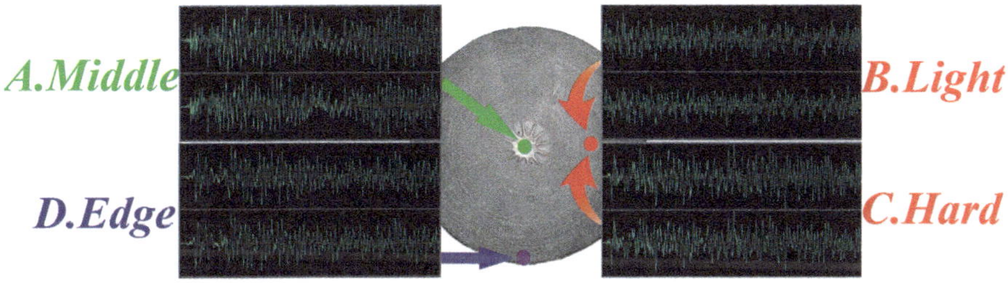

Fig. 2. Locations and sounds

It can be clearly seen from the diagram that there is a certain frequency difference of the sound produced by knocking at different positions of the bronze drum, especially at the edge of the bronze drum and the center of the bronze drum. When different intensity knocks on a same point of the bronze drum, the amplitude of the sound spectrum is higher under the heavier percussion.

2.3 3D Modeling and Optimization

The construction of 3D virtual model is the foundation of digital protection of bronze drum. We can use the professional modeling software 3dsMax, which is a powerful, well expanded and with vivid images software that can cooperate with other image processing software smoothly and stably. The three-dimensional model of Yunnan bronze drum is made by 3dsMax. We should pay attention to the shape, color and decoration of bronze drum. Through the combination of texture and illumination, we can render the bronze drum model to achieve vivid and realistic results.

In order to improve the fluency of the environment loading and the relationship between the browser rendering speed and the modeling authenticity, the LOD can be used to optimize the model [4]. When observing bronze drum at a long distance, we can use rough models and texture maps to describe the bronze drum, so that we can quickly draw the complex scenes. When observing bronze drum at a close distance, we use a detailed model to accurately reflect the details of the drum.

3 Recognition of Percussion Gestures

After completing the 3D modeling and collecting the percussion sounds, users' actions and sounds should be matched. Traditional input devices, such as mouse, can satisfy the needs of interaction to some extent. However, this interaction is poor. Human gestures can be identified directly by Kinect based on deep image recognition, replacing the mouse to achieve the interaction with the virtual scene.

3.1 Brief Introduction of Kinect

Kinect depth sensor can be regarded as pinhole camera model. Coordinate system of camera takes optical center as the origin. The X axis and the Y axis represent the diffusion boundary of the detected object. The Z axis indicates the distance from the project to origin [5]. Figure 3 is the Kinect camera coordinate system.

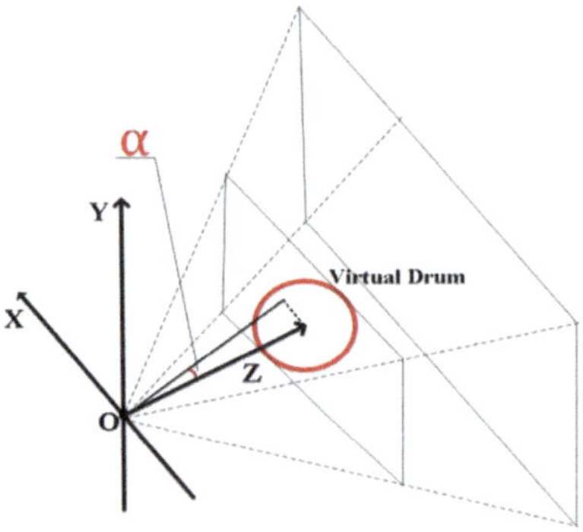

Fig. 3. Kinect camera coordinate system

3.2 Gesture Detection

Depth information is got via collecting color image, depth image and skeleton data. Then, filtering can be used to remove voids and noises. Hands Generator provided by OpenNI is also used to detect and track gestures which can get the position of the palm [6].

We define the position coordinates of gestures as Hand (x, y, z). Considering that the surface of the bronze drum is roundness, the cone characteristic can be found through the origin of coordinate and the surface imaging of bronze drum. By analysis the elevation angle α between the gesture position from the origin of the coordinate system and defining the furthest and nearest distance from the origin point, we can learn whether the gesture is captured or not. The schematic figures are shown in Figs. 3 and 4.

First, determine the location and center coordinates of the virtual bronze drum. The virtual plane with an R radius, is located at the origin of the distance Z_0. When the user extends the arm forward until the center of the palm is overlapped with the center of the drum, the initializing process of the system is completed, and then the gesture detection of the stroke of the drum starts.

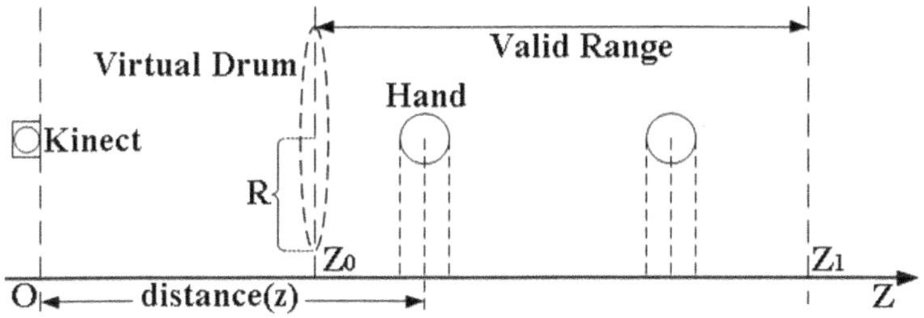

Fig. 4. Distance detection

When hand moves, $\alpha = \arctan(\sqrt{x^2+y^2}/z)$. If $\alpha \leq \alpha_0 = \arctan(R/Z_0)$, and the hand moves between the valid range, from Z_0 to Z_1, thus the gestures are captured.

$$Get\_Hand(\alpha) = \begin{cases} 0 & otherwise \\ 1 & \alpha \leq \alpha_0 \cap Z_0 \leq z \leq Z_1 \end{cases} \tag{1}$$

When Get_hand = 1, gestures are valid. And map it to the depth map, which is regarded as the gray value of the pixel [7]. When Get_hand = 0, gestures are invalid.

3.3 Judgment of Position and Strength

Knocking at different positions of the drum will produce different sounds. When z = distance(Z_0), polar coordinates distance between the percussion point and the center of the drum can be calculated through $G(r,\theta) = f(x, y)$. Then match the location of percussion with the collection point of bronze drum, which can determine the location.

Knocking drum with different strength will also produce different sounds. Through detecting the waving range of the hand, that is, the change of the linear distance between the waving hand and sensor, recording the actuation time to judge the speed, we can simulate the magnitude of force when the drum is struck. Define the motion of the gestures on the Z axis as distance (z, t).

Hands may shake while knocking [8], as process from point Start to point Real Start shown in Fig. 5. However, a normal complete process of knocking is the process from point Real Start to point Knock. In this process, $\partial distance(z, t)/\partial z \leq 0$. If there are multiple points to satisfy the formula, set $T = T_{max}$. Thus, we can define the percussion strength.

$$P = (Z_{max} - Z_0)/(T_0 - T_{max}) \tag{2}$$

The maximum percussion sound and the maximum percussion strength (P_{max}) can be obtained through practical tests. Using P/P_{max} to define the intensity of each knock, the effect of different volume on different intensity can be simulated.

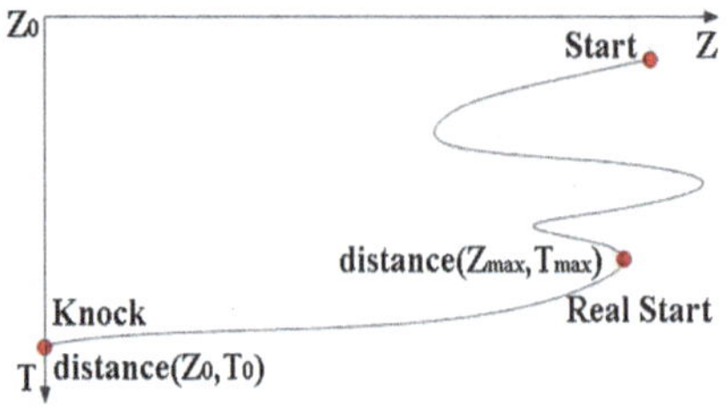

Fig. 5. Motion trajectory

3.4 Sound Call

When the specific knock information, the intensity and the location, is obtained, the audio file can be called to play. The core codes are as follows:

```
(1) Set variables to record the knocking position and time
<field name="gu_position" type="SFInt32" accessType="initializeOnly"/>
<field name="gytime" type="SFTime" accessType="outputOnly"/>
(2) Judge the percussion situation
<ROUTE fromNode="touchg1y1" fromField="isActive" toNode="Mover"
toField="qiao"/>
More percussion
<ROUTE fromNode="touchg1y74" fromField="isActive" toNode="Mover"
toField="qiao"/>
(3) Display knocking actions
<ROUTE fromNode="Mover" fromField="gytime" toNode="sound"
toField="startTime"/>
<ROUTE fromNode="Mover" fromField="gytime" toNode="chuizi-TIMER"
toField="startTime">
<ROUTE fromNode="chuizi-TIMER" fromField="isActive" toNode="Mover"
toField="chuiziting">
(4) Display percussion sounds
function qiao(value,time)
{gytime=time;
if(gu_position==101) {string[0]="sound/g1y01.mp3"; sound.url=string;}
More percussion
else if(gu_position==174) {string[0]="sound/g1y74.mp3"; sound.url=string;}}
```

4 Fingertip Identification of Rotary Bronze Drum

As mentioned above, we have achieved the simulation of knocking bronze drum by the detection of hand movement via Kinect. In order to have an all-around appreciation of the bronze drum, we need to identify the action of the fingertips so that we can rotate and scale the bronze drum.

4.1 The Principle of Convolution Neural Network

There are a lot of common gesture recognition methods. Commonly used methods are neural network recognition, geometric feature recognition and hidden Markov model recognition. The advantages and disadvantages of the three methods are shown in Table 1.

Table 1. Advantages and disadvantages analysis of gesture recognition method

Recognition method	Advantages	Disadvantages
Neural network	Strong ability to recognize and classify	Easy to produce fitting phenomenon
Geometric characteristics	A good stability	The number of samples increase and the recognition rate decreases sharply
Hidden Markov model	Ability to describe the space-time variation of gesture	Slow recognition speed

With the rapid development of machine learning and deep learning in computer vision, the recognition method based on neural network is getting more and more attention. In order to solve the problem of fitting phenomenon, we can use the deep convolution neural network (CNN) model. CNN does not need artificial design features, which can directly input image data into network models and give classification results at output ends. CNN has the characteristics of local connection, weight sharing, deep hierarchical results and automatic feature extraction, which can bring new ideas for gesture recognition [9].

4.2 Image Preprocessing

CNN can directly input the original image, whereas it can improve the recognition effect by preprocessing. Graying and binarizing the image, we get a picture of 48 * 48 size as the final input data of the experiment. As shown in Fig. 6.

Fig. 6. Image preprocessing

4.3 Algorithm Structure Design Based on CNN

In the algorithm structure design of CNN, 8 tier or 6 tier structure is usually adopted. The 6 tier structure design is usually applied to tasks with less classification. In this paper, we take use of 8-tier structure convolution neural network, including input tier, 2 level convolution tier (C), 2 level sampling tier (S), 2 level full link tier (F) and output tier. The output of this project is more accurate [10]. As shown in Fig. 7.

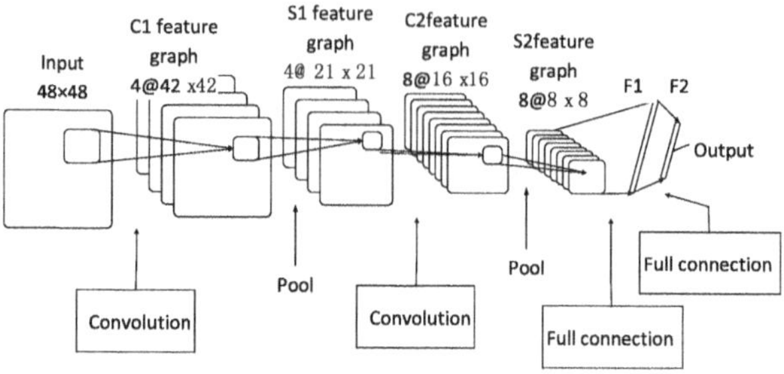

Fig. 7. Structure of gesture recognition network

In the input layer, the fingertip action image with the size of 48*48 can be directly read and transmitted to the neural network. We can get the characteristic diagram of fingertip action image by computing with convolution layer. In this paper, we use 7 * 7 convolution kernel to extract feature from image. There are four characteristic graphs of C1 layer, and the size is 36 * 36. It is eight characteristic graphs of C2 layer, and the size is 16 * 16. The output of convolutional layer is subsampled in the pooling layer to reduce the number of feature mapping and to reduce the dimension of the feature, recognizing the direction and bending degree of the fingertip. Thus we can get an enhanced and effective recognition feature of the finger movement image. The size of the pooling window is 3 * 3. It is four characteristic graphs of the S1 layer that the size is 21 * 21. It is eight characteristic graphs of the S2 layer that the size is 8 * 8. Each neuron in the fully connected layer is connected to all the output neurons in the upper layer. Finally, the F2 layer quantifies the feature data and connects it to the output layer. The output layer is a classifier, and it designs 14 different fingertips recognition action. Thus, there are 14 neurons in the output layer, of which 8 movements represent the rotation of the bronze drum on the up, down, left, right, upper left, lower left, upper right and lower right, and another 6 gestures magnify or narrow the bronze drum to varying degrees.

The operation and appreciation process of the different position of the bronze drum is realized by matching the corresponding action of each fingertip operation image of the classified output. Take the upward rotation of the bronze drum as an example. We stretch our palms up horizontally and bend four fingers up to an angle, which can be defined as the upturn of bronze drum. As long as the output of CNN is classified as this category, it can achieve the upward rotation. All kinds of fingertip gestures and the corresponding meanings are shown in Table 2.

Table 2. Fingertip actions and corresponding operations

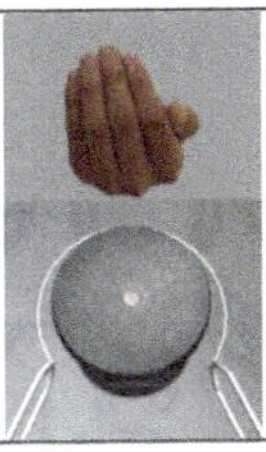	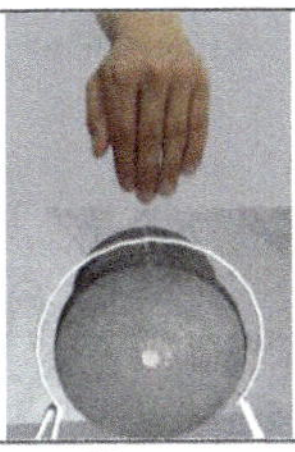	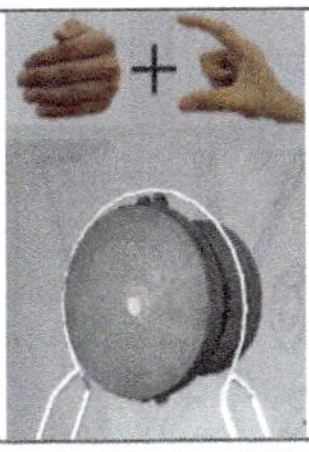	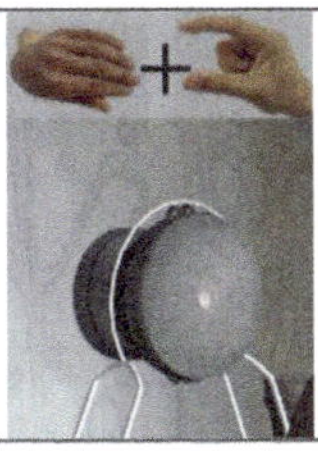
Up	Down	Left+Shrink90%	Right+Shrink60%
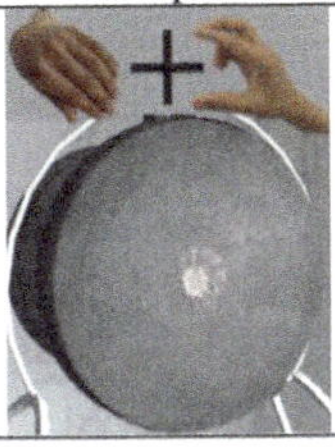	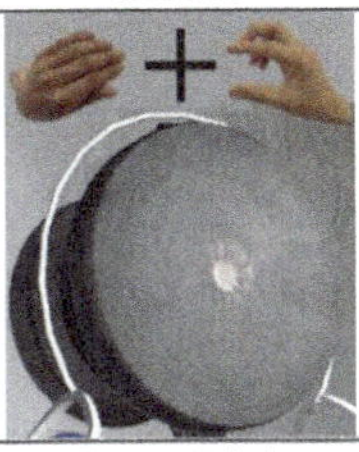	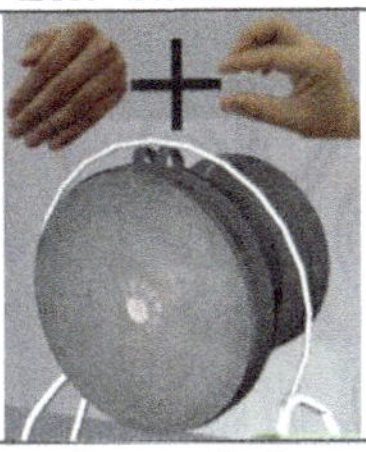	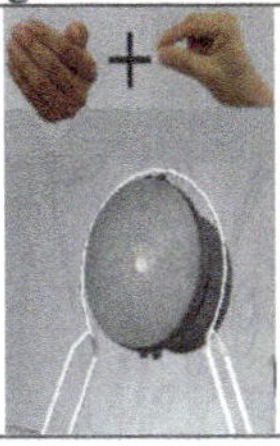
Lower Right +Enlarge200%	Upper Right +Enlarge175%	Lower Left +Enlarge175%	Upper Left +Shrink30%

5 Conclusions

In this paper, the digital protection of Yunnan bronze drum is studied by VR technology. Using scanning and sound acquisition, we can get valuable information and model. Then, we make use of Kinect somatosensory recognition technology to achieve the function of multi-dimensional interaction of visitors to knock the drum, so as to complete the virtual reconstruction of the drum's virtual performance. At the same time, we use CNN to identify finger movements, and complete the transfer of turning and zooming of bronze drum. It achieves the way that completely detaches from the mouse, and finally completes the all-round display of the virtual bronze drum, and enhances the users' experience effect. Then, the digital research results of bronze drum can be used in the tourist attractions and museums, which can promote the heritage and protection of bronze drum culture, and further facilitate the integration development of economy and culture in Yunnan province.

Acknowledgments. This research is supported by National Natural Science Foundation of China ("contact + non-contact" hybrid interface design strategy and its key human factors research. Grant: 61741206), and also supported by "The Key Laboratory of Ethnic Culture Resources Informatization".

References

1. Xia, W.: Digital research on the art of ancient Yunnan bronze. Yunnan Normal University (2016)
2. Xu, W., Jin, C., Su, Y.: Reaearsh on the digital communication and development of Yunnan Bai embroidery. In: IOP Conference Series: Earth and Environmental Science, vol. 100 (2017)
3. Jianke, W.: Processing technology of digital audio based on Cool Edit Pro Software. Exp. Sci. Technol. **11**(6), 38–40 (2013)
4. Peiyao, L.: Transformation method for automatic generation of buildings in CityGML LOD1-LOD4 using IFC models. Southwest Jiaotong University (2017)
5. Chen, L.: Research on complex gesture recognition technology based on Kinect. Nanjing University of Science & Technology (2017)
6. Zhili, F., Weiliang, W., Xinyu, G., et al.: Performance optimization of Kinect somatosensory control. Comput. Eng. Des. **35**(12), 4350–4355 (2014)
7. Fangjian, Y., Yixuan, W., Yigang, W., et al.: Fingertip detection and gesture identification based on Kinect depth image. Electron. Sci. Technol. **27**(8), 6–10 (2014)
8. Yi, Z., Shuo, Z., Yuan, L., et al.: Gesture track recognition based on Kinect depth image information and its applications. Appl. Res. Comput. **29**(9), 3547–3550 (2012)
9. Yang, H., Xuan, S., Mo, Y.: Hand gesture recognition based on the convolution neural network. Computer Technology and Development (2018)
10. Bin, Z., Xufei, S., Yipeng, W.: Design and implementation of gesture recognition algorithm based on convolution neural network. Microcomput. Appl. **36**(20), 51–53 (2017)

Development of Multi-process for Video Stitching in the AVM Applications Based on OpenCV

Jutamanee Auysakul(✉), He Xu, and Vishwanath Pooneeth

College of Mechanical and Electrical Engineering, Harbin Engineering University, Harbin, People's Republic of China
ajutamanee@eng.psu.ac.th

Abstract. Around-view monitor applications have increased the value of monitoring technology for the automotive market. Most previous applications have used a bird's eye viewpoint that offers a limited field of view for the driver. In this study, we propose an algorithm of multi-process to stitch the full panorama view that was captured from multiple cameras for implementation in real-time. Our development algorithm was created using the OpenCV's libraries to improve the live stream. The process includes online and offline steps aiming to reduce the recalculation of parameters for stitching the images. Furthermore, we use a multi-threaded approach to increase the efficiency of multi-camera input. The experimental results demonstrated the effectiveness of the stitching process to reduce the computation load and can provide full around view and efficient live stream information.

Keywords: Around view monitor (AVM) · OpenCV · Video processing
Video stitching

1 Introduction

The around view monitor (AVM) system is a feature of Advanced Driving Assistance Systems (ADAS) that is designed to enhance driving safety and improve the view of the driver. Recently, ABI Research [1] has forecasted that the ADAS market, between the period of 2014 and 2020, has grown 22.8% each year due to increasing consumer demand for safety and comfort on the road while driving or parking a vehicle. This stimulation is also related to an increase in government regulations ensuring safety for drivers. The AVM system can reveal blind spot areas with image information data from a large field of view (FOV) by taking advantage of the capabilities of a fisheye camera [2–4]. One important reason for the popularity of AVM systems is that previous methods were embedded for surround view [3, 5], and the inexpensive nature of this product allowed installation in low-cost cars. Furthermore, the AVM system is applied in the surveillance robot to use the image data in the typical tasks such as mapping and tracking object.

Most image processing in AVM applications occurs using a pipeline of registration, calibration, and composition. The overlapping area between images is stitched together

K. Deng et al. (Eds.): ICMIR 2018, AISC 856, pp. 1013–1020, 2019.
https://doi.org/10.1007/978-3-030-00214-5_124

by using pairs of the keypoints that estimate the homography matrix with RANSAC (random sample consensus) technique. Each key point is indexed by descriptors such as SIFT [6], SURF [7], or ORB [8] feature detectors. Next, calibration is performed through a registration process that aligns images into wrapping planes. Finally, image composition is used to combine images into a large canvas. The blending of the color in a stitched image in a region of overlap is achieved by multiband-blending, and the image is then converted to a polar view to monitor the around view. All of the stitching processes can be performed using OpenCV (open source computer vision) library with a stitching module and detailed sub-module. This library serves as an aid in programming the stitching procedure and implementing the AVM.

The live streams from multiple cameras are needed to report a real-time observation. However, the images streamed from different devices may have a lag. The threaded approach is an execution of the smallest sequence of programming instructions that are divided into a single threaded and a multi-threaded approach. The single thread approach operates one command at a time. This is different than a multi-threaded approach that can be processed into an array of commands in a single process that is executed independently and shares processing resources. The multi-threaded approach can be applied for multi-camera applications to speed-up the live stream as demonstrated in [9].

In this research, we focused on the real-time video feature of AVM, which includes online and offline processing to reduce the processing time when stitching videos from the individual cameras based on OpenCV libraries. In the following section, we describe the multi-process method of the real-time stitching and then present the algorithm design that is based on OpenCV. We then discuss the experimental results by comparing a frame per second (fps) with different threaded approaches in multi-process and a single process, and the final monitor of polar view in the Results section. Finally, the findings and potential for future studies are summarized in the last section.

2 Multi-process Approach to AVM Systems

Our AVM system design, mounted on a vehicle is focused on a single monitor, that displays the surrounding environment through a combination of four live streams from four fisheye cameras. The algorithms are formulated with the C++ language based on the OpenCV's libraries. To improve processing time, we proposed a multi-process approach consisting of offline and online processes, as shown in Fig. 1. The offline process is used to estimate the stitching parameters by capturing images in real-time. Then, all of the parameters are utilized in the online process for warping and blending the four live streams. Finally, the panorama image is created and mapped onto the lookup table (LUT) of polar view.

2.1 Offline Process

The offline process is used to estimate all of the parameters which convert a single image from each camera into a panorama image. The stitching detail module in the OpenCV's libraries has been used to implement this phenomenon in our research.

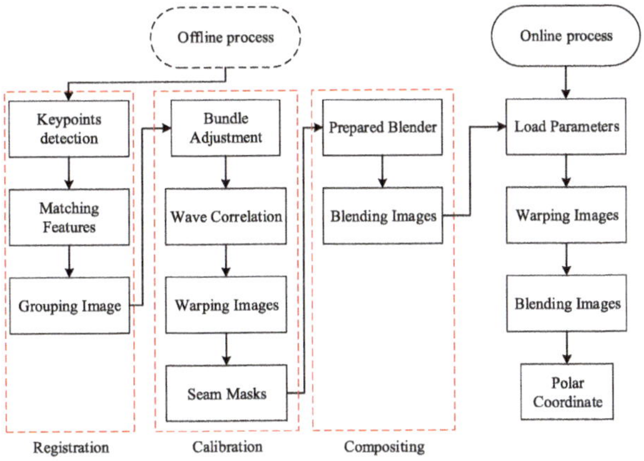

Fig. 1. Block diagram of the process for AVM applications.

The stitching process consists of three main steps: registration, calibration, and composition. Registration is used to estimate the camera parameter by using the matching feature to detect the corresponding point in overlapping area between two images. The SURF is applied to search the feature point on input images. In order to eliminate the distorted effect in fisheye images, we needed to increase the confidence of matching step from default for eliminating irrelevant matches. Then a subset of the panorama image is attained by grouping the best match, and the homography matrix on each camera is created.

Next, in the calibration step, we focused on minimizing the difference between the optical and ideal cameras. The bundle adjustment is used to calculate the camera parameters by minimizing a summary between the camera center and a feature through which the rays pass. It is used to define focal length. However, the confidence of threshold is reduced to a half of default value because the position of cameras on the mobile robot is very different in viewpoint because the overlapping region from the distorted image is limited. Thus, camera parameters: intrinsic matrix (K), radial distortion coefficients (R) and tangential distortion coefficients (T) are estimated, and a camera setting is improvised by wave correlation in horizontal. Then, warped images are transferred to the cylinder projection plane that shows around view in the horizontal plane. The corners, sizes, and masks of warped images are fed to a prepared blender step. The best area of the overlapping region can be optimized by seaming masks with the graph cut technique [10].

In composition process, the blender is created to combine all warped images to make a panorama image. This blender had prepared using only the first stitching and the multi-band blending. Multi-band blending is a very useful tool to stitch images without blurring and ghosting [11] on overlapping areas. Finally, the panorama images are built with the composition of the warped images on the created blender. Hence, all estimated parameters are contained for the online process to preserve time as the position and angle of cameras is fixed.

2.2 Online Process

After creating the first stitched panorama view, the estimated parameters of cameras are fed to the online process. The live stream from each camera is warped and blended and results in the first stitching. The single threaded or the multi-threaded approach can be used to process the input of live streams. The single threaded approach means that the four cameras all run at the same time, but this wastes CPU time. However, modern programming allows us to process multiple tasks by using the multi-threaded approach. This avoids halting the process if some threads encounter a problem [12]. Thus, the multi-camera application efficiency of the program is increased due by using a multi-threaded approach.

For AVM applications, we also needed to represent the final display in a circular view to monitor the around view. To do this, the blended panorama view stream in Cartesian coordinates was converted to polar coordinates. The transformed pixels use bilinear interpolation to calculate the weighted color in the input of the four nearest pixels. The mapping of the polar view is kept in the LUT to reduce the processing time.

3 Algorithm Design

In this section, we discussed the algorithm as shown in Table 1 designed to process the live stream for AVM. The OpenCV is an open source of a library for programming functions that are proposed to develop computer vision in real-time. We applied only the functions in this library to make the AVM.

The problem of stitching for an overlapping area is complex because of the distorted images obtained from fisheye cameras and the inconsistent viewpoints. We removed the fisheye lens distortion correction (LDC) by using OCamCalib's toolbox [13]. Then, we can stitch it by adjusting the confidence of feature matching and the threshold for stitching two images with a detailed sub-module. In addition, the type of warping plane needs to be appropriate to the application as the cylinder plane is used to project the image. This warping can be observed in the surround view, and it reduces the distorted image by maintaining the horizontal FOV. The real-time monitoring of AVM applications can be done with OpenCV's library functions as discussed in the previous section. The stitching module and the stitching detail sub-module were used as components of our algorithm. The live streams are needed for warping and to check the seam mask in every frame before feeding to the blending step, before converting to a polar view. The repetition of the live stream feed is used to determine the storage parameter that is calculated in the offline process. The algorithm design is based on the module stitching detail in OpenCV.

Table 1. Algorithm of Multi-process based on OpenCV's libraries

Input: live streams from four fisheye cameras
Output: live stream of AVM
Offline process: 1. Capturing live streams from four cameras 2. Registration 2.1 Detecting feature points by SURF feature 2.2 Matching and grouping images with the nearest matching neighbor 2.3 Estimate homography matrix with RANSAC technique 3. Calibration 3.1 Bundle adjustment to estimate camera parameter 3.2 Warp correlation with wave correction 3.3 Warping image with a cylindrical projection 3.4 Seam estimation with graph cuts 4. Compositing 4.1 Prepare a blender with multi-band blending: 4.2 Blend the current images and extract the blended result with the feed and blend function Online process: 5. Repeat step 3.3, 3.4 and 4.2 by keeping K and R parameters 6. Convert to polar coordinate $r = \sqrt{(x^2 + y^2)}$ *x, y*: pixel values of the panorama image $\theta = \tan^{-1}(x / y)$

4 Experimental Results

In this study, we developed an algorithm on C++ platform using the OpenCV's libraries to improve surround view monitoring. Four fisheye cameras mounted on a mobile robot provided a FOV of 175 degrees, as shown in Fig. 2. The position and the angle of those cameras were pre-fixed. For these experiments, we employed Intel NUC Core i5, 1.6 GHz CPU, and 4 GB RAM, based on the Microsoft Visual C++. NUC is a mini computer that is installed on the mobile robot and can be monitored via the wireless network through Team Viewer software [14]. The fisheye cameras were connected with the NUC by USB port and supplied with voltage via an external battery. It is easy to control a mobile robot for the surveillance of an environment by an individual mobile robot.

The images were captured with a resolution of 480 × 640 pixels as shown in Fig. 2. Calibration with OCamCalib's toolbox was applied to estimate the LDC parameters that remove the distortion due to the fisheye effect. The overlapping region of undistorted images needs to be stitched to make a panorama image. The parameters of the stitching process are referenced with OpenCV's stitching detail library. In the offline process, the first step of image stitching is to detect the keypoints on undistorted images with detectors SURF, and then to match features in the overlapping area between two images with OpenCV's FLANN (Fast Library for Approximate Nearest Neighbor).

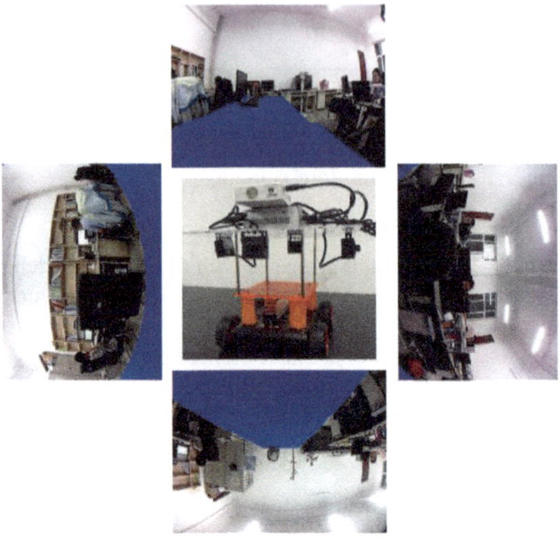

Fig. 2. Illustration of a surround view captured by four fisheye cameras with FOV175 degrees mounted on a mobile robot.

Finally, we projected all images onto the cylindrical plane. Projection planes allow the target to be observed in around view. Next, we blended the images together with a multi-band blender. The total time for stitching the image was 5.760 s then we compared the processing time in each process. The registration process includes the operation time a 76% more than the warping and blending process due to it needs to divide the stitching process of the multiple cameras into multi-process. The offline process is necessary for the first stitching process and is needed to recalculate any change in camera position.

The corrected parameters were loaded from the offline process into the online process to warp all the live streams onto the cylindrical plane. The warping process used the camera parameters that were refined by the homography matrix in each camera to determine the warp images, corners, and masks of the overlapping area. Then, the warped images prepared in the offline process were fed into the blender to create a live stream panorama view. This was done as shown in Fig. 3 to enhance AVM application. The polar view used a bilinear interpolation to recalculate the pixel values to find appropriate color for color smoothing in the converted pixels.

Furthermore, we had to implement the frame rate of polar stream view and a computation load by comparing between the single threaded and the multi-threaded approach in the different process method as shown in Table 2. The multi-process with the multi-threaded approach has effectiveness better than other methods in term of frame rate and computation load. However, the single process that was implemented the only online process with the multi-threaded approach is as effective as the multi-process with the single threaded approach. Thus, use of the multi-threaded approach influences the performance of the real-time stream in AVM applications.

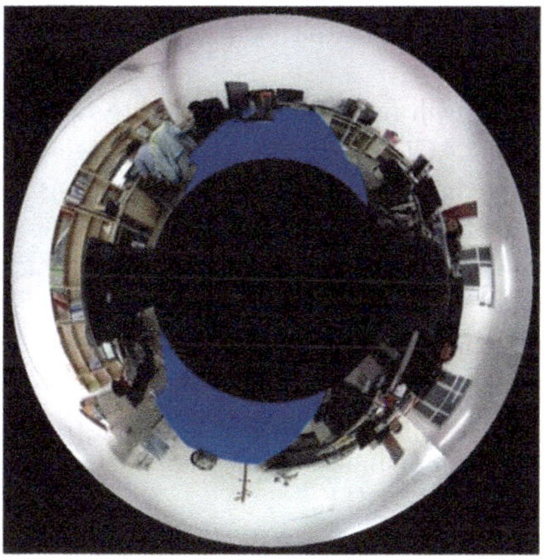

Fig. 3. The polar view of the AVM applications.

Table 2. Comparison the effectiveness of the different process and threaded approach method.

Parameters	Single process + Single threaded	Single process + Multi-threaded	Multi-process + Single threaded	Multi-process + Multi-threaded
fps	12.7	15.3	14.5	18.4
%CPU	67.1	55.6	59.7	47.5
%Memory	56.4	42.3	45.1	35.8

5 Conclusions

Based on the functions in the OpenCV libraries, we developed an AVM algorithm. An offline process is used to estimate parameters to create the first panorama. The estimated parameters are forwarded to the online process, followed by warping and blending of the multi-stream data from cameras installed on the mobile robot. The panorama view converts to a polar view to allow presentation of an around view. This approach increases the performance of frame rate in real-time and can reduce the processing time and computation load. Our algorithm was successful in our experiment on a mobile robot. This approach can be extended to working on a large platform such as a vehicle or surveillance robot, or other applications.

Acknowledgments. This work was supported by the Natural Science Foundation of the Heilongjiang Province of China (F2016003), Specialized Research Fund for Doctoral Program of Higher Education (20122304110014), the Fundamental Research Funds for the Central Universities (HEUCFX1508), Manipulator Data Processing Development (KY10700150065), and the International Exchange Program of Harbin Engineering University for Innovation-oriented Talents Cultivation.

References

1. ABI Research Predicts Growth for Advanced Driver Assistance Systems Market in China after 2017, Driving 36% CAGR Revenue Growth through 2020. https://www.abiresearch.com/press/abi-research-predicts-growth-advanced-driver-assis/
2. Su, C.L., Lee, C.J., Li, M.S., Chen, K.P.: 3D AVM system for automotive applications. In: 2015 10th International Conference on Information, Communications and Signal Processing (ICICS), pp. 1–5 (2015)
3. Zhang, B., et al.: A surround view camera solution for embedded systems. In: 2014 IEEE Conference on Computer Vision and Pattern Recognition Workshops, pp. 676–681 (2014)
4. Cheng, R., Zhao, Y., Li, Z., Jiang, W., Wang, X., Xu, Y.: Panorama parking assistant system with improved particle swarm optimization method. J. Electron. Imaging **22**, 041123 (2013). https://doi.org/10.1117/1.jei.22.4.041123
5. Chen, Y.Y., Tu, Y.Y., Chiu, C.H., Chen, Y.S.: An embedded system for vehicle surrounding monitoring. In: 2009 2nd International Conference on Power Electronics and Intelligent Transportation System (PEITS), pp. 92–95. IEEE (2009)
6. Lowe, D.G.: Distinctive image features from scale-invariant keypoints. Int. J. Comput. Vis. **60**, 91–110 (2004)
7. Bay, H., Ess, A., Tuytelaars, T., Van Gool, L.: Speeded-up robust features (SURF). Comput. Vis. Image Underst. **110**, 346–359 (2008). https://doi.org/10.1016/j.cviu.2007.09.014
8. Rublee, E., Rabaud, V., Konolige, K., Bradski, G.: ORB: an efficient alternative to SIFT or SURF. In: Proceedings of the 2011 International Conference on Computer Vision, pp. 2564–2571. IEEE Computer Society, Washington, DC, USA (2011)
9. Shieh, W.Y., Huang, J.C.: Speedup the multi-camera video-surveillance system for elder falling detection. In: 2009 International Conference on Embedded Software and Systems, pp. 350–355 (2009)
10. Kwatra, V., Schödl, A., Essa, I., Turk, G., Bobick, A.: Graphcut textures: image and video synthesis using graph cuts. In: ACM SIGGRAPH 2003 Papers, pp. 277–286. ACM, New York, NY, USA (2003)
11. Juan, L., Oubong, G.: SURF applied in panorama image stitching. In: 2010 2nd International Conference on Image Processing Theory, Tools and Applications, pp. 495–499 (2010)
12. Silberschatz, A., Galvin, P.B., Gagne, G., Silberschatz, A.: Operating System Concepts. Addison-Wesley, Reading (1998)
13. Scaramuzza, D., Martinelli, A., Siegwart, R.: A flexible technique for accurate omnidirectional camera calibration and structure from motion. In: Fourth IEEE International Conference on Computer Vision Systems (ICVS'06), pp. 45–45 (2006)
14. John, C.: Teamviewer new technology. Bus. Wire **33**, 25–28 (2013)

The Application Research of Object Detection and Image Segmentation Based on Improved Canny Algorithm

Yanxia Wei[1,2(✉)] and Zhen Jiang[1]

[1] Department of Precision Mechanical Engineering, Shanghai University, Shanghai, China
sditwyx_1999@163.com
[2] School of Mechanical and Automotive Engineering, Liaocheng University, Liaocheng, China

Abstract. Edge detection and recognition is one of the most critical procedures in image processing and pattern recognition. Traditional Canny algorithm has some advantages, such as simple and fast execution times. It is not good enough when traditional Canny algorithm is applied in complex background images for objects edge detection. It is sensitive to noise and low location accuracy. To solve the shortcomings of traditional Canny algorithm, we propose an improved edge detection and recognition algorithm. The proposed edge detection algorithm can remove the influence of complex background images in the process of edge detection and characters segmentation. The effectiveness of the proposed algorithm is demonstrated by the experimental results. The proposed algorithm can eliminate the complex background of credit card and detect the meaningful edges of its numbers. The experimental results show the improved algorithm can obtain accurate edge position and good segmentation results.

Keywords: Edge detection · Image segmentation · Canny algorithm
Mathematical morphology algorithm

1 Introduction

Credit card numbers recognition is a new and hot topic of computer vision and pattern recognition. There are no corresponding national standards in credit card numbers about the numbers bits, font and printing style. Different banks usually adopt different ways, such as numbers of 16 or 19 bits, printed numbers or raised numbers etc. Because the credit cards with raised numbers are not easy to removed, and these are frequently used in our daily life. The credit cards with raised numbers have a severe problem that the color of numbers is similar or same to the background of credit cards, which brings heavy difficulty in detecting edge and segment numbers. To solve the problem, we present the improved edge detection and recognition based on Canny algorithm and mathematical morphology to detect edges of credit card numbers and segment them.

With the high speed development of network technology, it is very popular for consumers to use computer or phone payment. Especially using phone payment is

K. Deng et al. (Eds.): ICMIR 2018, AISC 856, pp. 1021–1029, 2019.
https://doi.org/10.1007/978-3-030-00214-5_125

much more convenient than computer payment or traditional payment. In some cases, phone payment even more convenient than cash. In the course of payment, it is critical step to input the credit card numbers for complete payment, especially to the application of needing binding the credit card. In the course of inputting and binding, there are some shortcomings, such as wasting time, slow speed, error prone, poor user experience, and so on. If the situation is on, payment application service providers may eventually lose their customers. So for those application providers, it becomes particularly urgent to recognize the credit card numbers in their applications in the era of mobile internet. By the driving of the Alipay and WeChat payment, more and more commercial users want to integrate credit card recognition into their payment application system for simplifying the user operations and enhancing the user's convenience.

The contents of the paper are organized as follows. Section 2 introduces the edge detection of Canny algorithm and the mathematical morphology algorithm, Sect. 3 presents our proposed algorithm based on Canny algorithm and mathematical morphology algorithm. Section 4 shows the experiment results, the image segmentation of credit card numbers with proposed algorithm. In last, we conclude our work in Sect. 5.

2 Related Work

As a comprehensive review on edge detection algorithms beyond the scope of this paper, we mainly review the methods related to our work in this section.

2.1 Canny Edge Detection Algorithm

Generally, the purpose of edge detection is to find a set of pixel points where the pixel brightness either varies abruptly or has the discontinuous point. Canny algorithm is an important method in aspect of edge detection. Canny algorithm can reach an expected tradeoff between the algorithm performance and the computational burden. The Canny algorithm is widely regarded as the edge detection standard, which is used extensively in computer vision system and robot vision system to seek abrupt brightness changes and to find object contours in an image [1–5].

In Canny algorithm, we adopt two threshold values to detect image edge: a high threshold value (HTV) and a low threshold value (LTV). The criterion for determining a single pixel as an edge pixel is that the gradient magnitude of this pixel is either greater than the HTV, or between the HTV and the LTV and connected to other decided edge pixels. [6] The strategy of the HTV and the LTV selects more effective edge pixels comparing with other edge detection algorithm using only a single threshold, such as Robert operator, Sobel operator etc.

The Canny algorithm based edge detection mainly aims at achieving:

(1) Low error rate

If we want to detect real image edge, we must lower the error detection rate as possible. When the value of SNR is larger, the quality of the detection edge is higher.

(2) Good positioning accuracy

The value of Loc is higher, the positioning accuracy is better.

(3) Single edge response

To improve the edge accuracy, the pixels has singly one corresponding response to edge detection.

2.2 Mathematical Morphology Based Algorithm

Mathematical morphology is an effective and robust strategy applied in image processing and analysis [7]. Mathematical morphology is similar to the focus of attention of human visual characteristics. The mathematical morphology operation involves two types of images, original images as to be processed images and kernel images as structure elements. When the structure elements move over the original images, we can use the structure elements to collect image information [8, 9]. By these operations, we can review the relationship between image parts and understand the structure of the image. To detect and study the structure characteristics of the image, structural elements can directly carry some prior knowledge, such as shape, size, and even add gray-scale information. The combination of two basic operations, erosion operation and dilation operation, constitutes some other mathematical morphological processing, such as open operation, close operation, hit/miss operation, thin operation and so on. Different morphological operations have different effects in image [4, 10–12].

The algorithm of the mathematical morphology is represented with the help of the set theory. Two basic morphological operations, erosion and dilation, are showed as below.

(1) Gray scale erosion

The gray scale erosion operation of two images is written as following

$$(F \ominus S)(x,y) = min\left\{ F(x+x', y+y') - S(x',y') \middle| \begin{matrix} (x+x', y+y') \in D_F \\ (x',y') \in D_S \end{matrix} \right\} \tag{1}$$

where F is the gray scale image. S denotes the structural element, that is, the scale filter. D_F and D_S are the image regions of F and S, respectively.

(2) Gray scale dilation

The gray scale dilation operation of two images is computed by

$$(F \oplus S)(x,y) = max\left\{ F(x-x', y-y') + S(x',y') \middle| \begin{matrix} (x-x', y-y') \in D_F \\ (x',y') \in D_S \end{matrix} \right\} \tag{2}$$

Though the Canny algorithm has been significantly applied in image processing and pattern recognition, it is still inevitable in terms of edge losses, localization errors, and noise sensitivity in detecting complex background images [6, 13]. To some extent, mathematical morphology exits the same problems too. In other words, small scale filters can detect edges well but introduce the influence of noise, on the contrary large scale filters can suppress noise effectively but could not find image contour details perfectly [10]. To overcome or at least to alleviate these problems, the improved algorithm is naturally proposed. The method is mainly applied for edge detection and characters segmentation of credit card numbers.

3 Proposed Methods

Gray-scale transformation, filtering image, detecting image edge, mathematical morphology, removing noisy edge segments, checking endpoints, linking discontinuity edges, pruning and segmentation constitute a complete improved edge detection and segmentation algorithm. In summary, our objectives were as follows:

Step 1: Gray-Scale Transformation and Binarization
To make the image more easily processed and analyzed, we firstly adopt gray-scale transformation or binarization. By above technical methods, we can obviously reduce the processing computation.

Threshold selection plays an important part in image processing. The selection methods of the threshold value can be separated into two types, local method and global method. Local method divides an image into a set of sub-images and then selects each threshold values for each of the sub-images. Global method is to divide an entire image with a single threshold using the Otsu algorithm or other algorithms, which are easy to implement. So we adopt the latter in our paper.

Step 2: Filtering Image for Sharpen the Edge of Objects
There usually have some noise in the image without filtering, which contaminate the edge feature information of the image. Image filtering is mainly used to enhance the boundary of gray level jump of image by special templates. If the original image is fuzzy, we can employ filtering method to make image clearer.

Step 3: Mathematical Morphology to Eliminate Background
Mathematical morphology is the mathematical tool of image analysis and preprocessing based on morphological structure elements. Its basic ideas are to employ specific forms of structural elements to detect and recognize the edge information or the appearance contour of image for accomplishing the objective of image processing and image analysis. The application of mathematical morphology can simplify the image data and obviously reduce the computational load, eliminate the influence of the irrelevant structure and preserve the appearance shape of objects [12].

We adopt edge detection method based on morphology gradient shown as follows:

$$Edge = (((F \circ S) \bullet S) \oplus S) - ((((F \circ S) \bullet S) \bullet S) \ominus S) \quad (3)$$

where, "o", "•" and "-" respectively represent open operation, close operation, and element-wise subtraction.

Step 4: Detecting Image Edge with Canny Algorithm
Image edge is the most basic feature of the image. The purpose of the edge detection and recognition is to extract the numbers of pixels with special meaning and connect them to appearance contour. A pixel is considered as an edge point if its gradient value is greater than those of pixels at both its sides in the direction of maximum intensity change.

Step 5: Image Location with Horizontal Projection Histogram
The horizontal and vertical directions projection histogram contains the integral information of the image pixel distributions [14]. To locate the upper and lower borders position, we employ the horizontal projection histogram.

$$S^i = \sum_{j=1}^{n} Edge(i,j), \quad i = 1,2,3\ldots m \quad (4)$$

where, S^i is the sum of the *i-th* row of *Edge*(*x*, *y*), which represents the gray-scale image to be located. The symbols of *m* denotes rows of image.

Step 6: Image Segmentation with Vertical Projection Histogram
To segment the credit card numbers, we must employ the vertical projection histogram.

$$S^j = \sum_{i=1}^{m} Edge(i,j), \quad j = 1,2,3\ldots n \quad (5)$$

where, S^j is the sum of the *j-th* column of *Edge*(*x*, *y*), the symbols of *n* denotes the columns of image.

After above steps, we locate and segment the credit card numbers into separate characters to be recognized. Vertical projection histogram of the binary image [15] are employed for segmenting the credit card numbers. As can be seen from Fig. 6, by means of the deep valley between each two successive numbers in the vertical projection histogram, we can easily separate the two numbers. We use the threshold of deep valleys, combined with prior knowledge, to segment the characters of credit card numbers [15]. After all of number to be separate into small parts, the rectangular box of each number is determined.

4 Experiment Results

To evaluate the robustness of the algorithm proposed in this paper, matching experiments are made with MATLAB. The visible image of credit card comes from network. It can be seen that the edge detection results with traditional Canny algorithm in credit card are not good enough because there are great fatal disadvantages about foreground-background separation. Compared to the traditional method, our proposed method performance is superior.

From Fig. 1, we can obtain the image by image preprocessing, such as gray scale conversation and binarization. Figure 1(a) exhibits an original image of credit card, Fig. 1(b) displays the gray-scale image, Fig. 1(c) shows its binary image with Otsu algorithm.

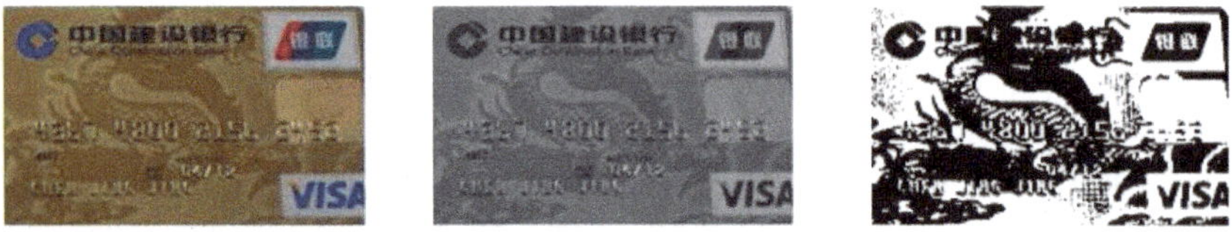

(a) Original Image (b) Gray Scale Image (c) Binary Image

Fig. 1. The example of image preprocessing to credit card

We adopt various traditional edge detection methods to search for the foreground objects contours. It can be seen from Fig. 2(a–e), the results of edge detection are not enough good so as to extract the credit card numbers contours. We can find that the foreground of numbers to be segmented is similar to the background of the credit card. So we can't get the features of credit card numbers by the traditional edge detection methods. Figure 2(f) shows that the edge detection image with proposed algorithm. By improved algorithm, we can easily remove the influence of background. Figure 3 shows the horizontal projection histogram of image from Fig. 2(f).

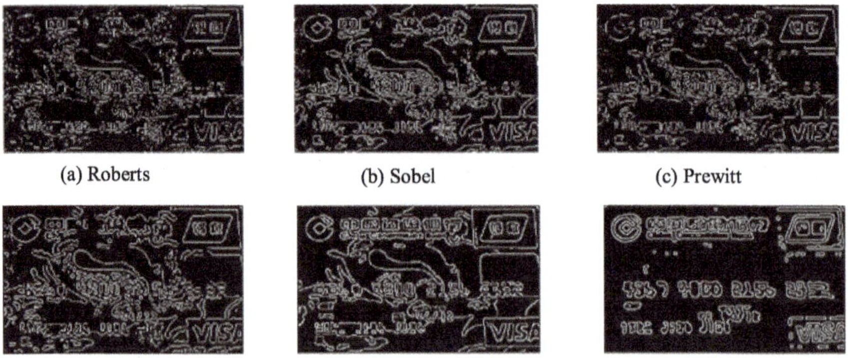

(a) Roberts (b) Sobel (c) Prewitt

(d) Log (e) Canny (f) the proposed algorithm

Fig. 2. The example of edge detection with various methods

Combined with prior knowledge, we can coarsely locate the position of credit card numbers in its bottom. Figure 4 shows that the position of credit card numbers by vertical projection histogram to the above image.

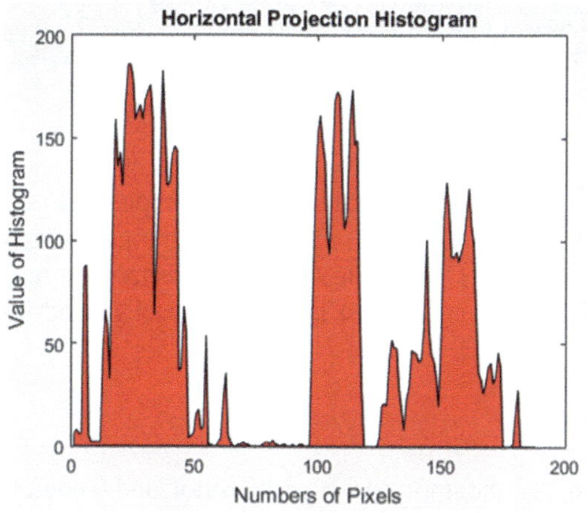

Fig. 3. Horizontal projection histogram of above edge image

Fig. 4. Accuracy position of credit card numbers

In last, combined with the prior knowledge that the width of character is equal to the half height of character, we separate the credit card numbers into small images, and restrict these characters into small rectangle to be recognized. Figures 5 and 6 shows that the vertical projection histogram to the image from Fig. 4.

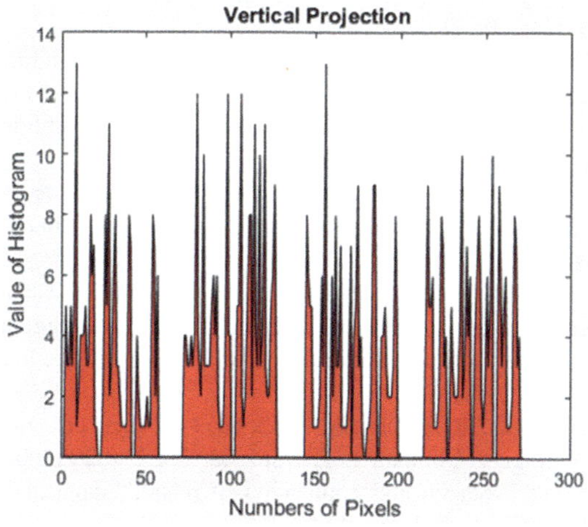

Fig. 5. Vertical projection histogram of above edge image

Fig. 6. Experiment results of segmentation

The experiment results demonstrate that our proposed method can detect and recognize the edge information and segment the character of credit card numbers. Compared to the traditional Canny edge detector, the method has a good result. The proposed method is reliable and robust in complex background images.

5 Conclusions

In this paper, we propose the improved edge detection and recognition based on Canny and morphology algorithms. Our method can effectively detect the objects edge in complex background image. From the experiment results, we can draw a conclusion the tradition Canny algorithm can't undertake the tasks of credit card with raised numbers because credit card background is so complex that conventional algorithms can't obtain the features of numbers. Experiment results demonstrate that the improved algorithm can correctly detect the numbers edge and segment the numbers of credit card with high accuracy, robustness and applicability.

Acknowledgments. Our research work is supported by Natural Science Foundation of China (61433016).

References

1. Canny, J.: A computational approach to edge detection. IEEE Trans. Pattern Anal. Mach. Intell. PAMI-8 **6**, 679–698 (1986)
2. Mokrzycki, W., Samko, M.: Canny edge detection algorithm modification. In: Computer Vision and Graphics, pp. 533–540 (2012)
3. Zhe, L., Wang, F.L., Chang, Y.Q.: An improved canny algorithm for edge detection. J. Northeast. Univ. **75**(5), 1516–1523 (2007)
4. Yu, H.: An improved Canny edge detection algorithm. Comput. Eng. Appl. 577–582 (2004)
5. Vianna, C.J., De Souza, E.A.: Combining Marr and Canny operators to detect edges, vol. 2, pp. 695–699 (2003)
6. Chang, S.H., et al.: Small retinal vessel Eztraction using modified Canny edge detection. In: International Conference on Audio, Language and Image Processing, pp. 1255–1259 (2008)
7. Shih, F.Y., Cheng, S.: Adaptive mathematical morphology for edge linking. Inf. Sci. **167**(1–4), 9–21 (2004)
8. Wang, H.F., Zhan, G.L.: Research and application of edge detection operator based on mathematical morphology. Comput. Eng. Appl. **45**(9), 223–226 (2009)
9. Jiang, M., Chen, G.: Research and application of metallographical image edge detection based on mathematical morphology. IEEE (2009)

10. Chen, W.B., Zhang, X.: Improved edge detecting algorithm based on mathematical morphology. In: International Conference on Machine Vision and Human-Machine Interface, pp. 199–202. IEEE Computer Society (2010)
11. Li, S., Pu, F., Li, D.: An improved edge detection algorithm based on area morphology and maximum entropy. In: International Conference on Innovative Computing, Information and Control, pp. 536–536 (2007)
12. Zhang, W., Shi, D., Yang, X.: An improved edge detection algorithm based on mathematical morphology and directional wavelet transform. In: International Congress on Image and Signal Processing, pp. 335–339. IEEE (2016)
13. Agaian, S.S., Nercessian, S.C.: Shape-dependent canny edge detector. Opt. Eng. **50**(8), 087008 (2011)
14. Ryu, H., et al.: Use of projection histograms for tracking faces under out-of-plane rotation. Opt. Eng. **47**(9), 957–966 (2008)
15. Nejati, M., Majidi, A., Jalalat, M.: License plate recognition based on edge histogram analysis and classifier ensemble. In: Signal Processing and Intelligent Systems Conference, pp. 48–52. IEEE (2015)

Clinical Assistant Diagnosis System Based on Dynamic Uncertain Causality Graph

Xusong Bu[1(✉)], Zhan Zhang[2], and Qin Zhang[1]

[1] School of Computer Science and Engineering, Beihang University, Beijing, China
buxusong@buaa.edu.cn
[2] Beijing Tsingrui Intelligence Technology Co., Ltd., Beijing, China

Abstract. Artificial intelligence clinical assistant diagnosis system has become one of the mainstream in the field of medical research at present. The main challenge of this system is to have both high diagnostic accuracy and good interpretability for diagnostic results. Dynamic Uncertainty Causality diagram is a probability graph model, which can explain the calculation results in a graphical form. This paper introduces the clinical assistant diagnosis system based on DUCG, and shows the system diagnosis process through a case study. Experiments show that: the system with high accuracy and good interpretability.

Keywords: Assistant diagnosis · Causal inference · Probability model

1 Introduction

In the past decade, many artificial intelligence systems have been developed for clinical assistant diagnosis, the models used for clinical assistant diagnosis system [1, 2] can be divided into two categories: non-probability model and probabilistic model. Neural network [3] model is the typical representative of non-probabilistic model. This model is derived from the training data through learning, and the performance is affected by the quality of the training set. The inability to explain the inference process and the conclusion of the "black box" feature also limit the model's interpretation for diagnostic results. The typical representative of probability model is Bayesian Network [4]. This model based on CPTs for reasoning calculations and its reasoning accuracy depends heavily on CPTs, and the computation amount of inference with CPTs is an NP hard problem with complex model, which means that the computation amount is exponential to the scale of problem. However, the medical knowledge base is usually complicated and has few case data. These characteristics affect the application of these two models in clinical assistant diagnosis.

Dynamic Uncertainty Causality Graph [5] is a new probabilistic model. It can graphically express the causal relationship among events and allows incomplete representation of knowledge. The algorithm has low requirements on the accuracy of the parameters, so the parameters can be given by the experts based on either experience or machine learning.

K. Deng et al. (Eds.): ICMIR 2018, AISC 856, pp. 1030–1038, 2019.
https://doi.org/10.1007/978-3-030-00214-5_126

In this paper, a case study of jaundice related disease diagnoses illustrates how DUCG performs as a clinical assistant diagnosis system, and how the DUCG knowledge base is built. In this case, the correctness of diagnoses reaches 99%.

2 Brief Review of DUCG

DUCG can: (1) represent complex causalities explicitly and easily with graphical symbols including logic gates representing any logic relationship among variables; (2) combine separately constructed sub-DUCGs into a whole DUCG, so that the maintenance of DUCG knowledge base is easy and each sub-DUCG is simple and well understood; (3) reduce the scale of problem for observed evidence significantly, so as to find all possible hypotheses and achieve high inference efficiency. DUCG contains S-DUCG and M-DUCG; S-DUCG is used for single-valued variables and M-DUCG is used for multi-valued variables. The model used in the clinical assistant diagnosis system is M-DUCG. The following paper mentioned that DUCG all refer to M-DUCG. In DUCG, B_i (square) indexed by i indicates the root cause event and the hypothesis in the inference calculation, it can only be used as root cause variable. X_i (circle) indicates the consequence or child variable, it also can be used as the cause variable. G_i (logical gate) indicates a logical gate variable, used to represent the logical relationship among its parent variables. The logical relationship among parent variables is described by the logic gate specification table LGS_i. D_i (pentagonal) represents the default event of X_i. The causal relationships among variables are represented by the weight function event F, which is drawn as the red directional arrow. the reasoning process of DUCG is as follows:

(1) **Simplification based on General DUCG**

Based on the evidence E received and Rules 1–10 [5], Rule 16 [6], the DUCG is first simplified. By initial simplification, irrelevant variables and relations are removed, reducing the hypothetical space and the computational complexity of reasoning.

(2) **Further simplification based on B_i**

According to the unified assumption, the simplified DUCG is decomposed into sub-DUCG, each sub-DUCG contains only one B-type variable, and the simplification rules 1–9 are used to simplify the subgraph. All $B_{i,j}$ in the remaining sub-DUCG will be considered as hypotheses, and j represents the state of Bi. If there is only one hypothesis to be determined, the reasoning will finish without calculations.

(3) **Expanding of multiple events**

When the hypothetical space contains more than one hypotheses, it need to expand E and $H_{kj}E$ according to the sub-DUCG for further calculating $P\{H_{kj}|E\}$. By repeatedly using the Eq. (1) and rule 11–15 [5],

$$X_{nk} = \sum_{i} F_{nk;i} V_i = \sum_{i} (r_{n;i}/r_n) A_{nk;i} V_i = \sum_{i,j} (r_{n;i}/r_n) A_{nk;ij} V_{ij} \tag{1}$$

E will be ultimately expressed as a form of sum-of-products only contains {B-, D-, A-, r-}-type variables, so do $H_{kj}E$.

(4) **Probability calculation**

Solving for *P{E}* and *P{H*$_{jk}$*E}* by calculating the expressions E and $H_{kj}E$, and then using the Eq. (2):

$$h_{kj}^{s} = \xi_k \Pr\{H_{kj}|E\} = \xi_k \left(\Pr\{H_{kj}E\}/\Pr\{E\}\right) \quad (2)$$

To calculating the conditional probability of each hypothesis h_{kj}. $\xi_k = \Pr\{E_k\} \Big/ \sum_k \Pr\{E_k\}$ is the weight of the sub-DUCG.

3 System Architecture Design

According to the different functions the system is divided into four main modules: knowledge edition module, assistant diagnosis module, inference engine module and data storage module. The structural design is shown in Fig. 1.

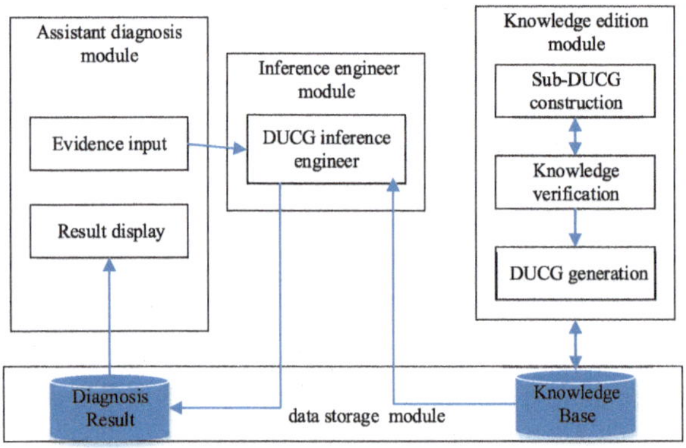

Fig. 1. System architecture

3.1 Knowledge Edition Module

Knowledge edition module provides users with DUCG building tools, this module contains three sub-functions:

(1) **Sub-DUCG construction**

Sub-DUCG construction module allows users to divide the DUCG into many sub-DUCGs for construction, and the sub- DUCGs can be compiled into a complete DUCG by DUCG generation module. This method of construction reduces the difficulty of constructing complex DUCG and facilitates the cooperation of multi-user.

As shown in Fig. 2, the user can express the complete causal logic of the disease in a single DUCG, when a large number of variables are contained in one DUCG, this DUCG construction method is not clear enough. Figure 3 shows another way to construct a DUCG: (a) represents the causal relationship between Malaria and laboratory tests, symptoms, physical signs; (b) represents all physical signs contained in Malaria; (c) represents all laboratory tests of Malaria; (d) represents all symptoms of Malaria. These four sub-DUCG can generate a complete Malaria DUCG through the DUCG generation function.

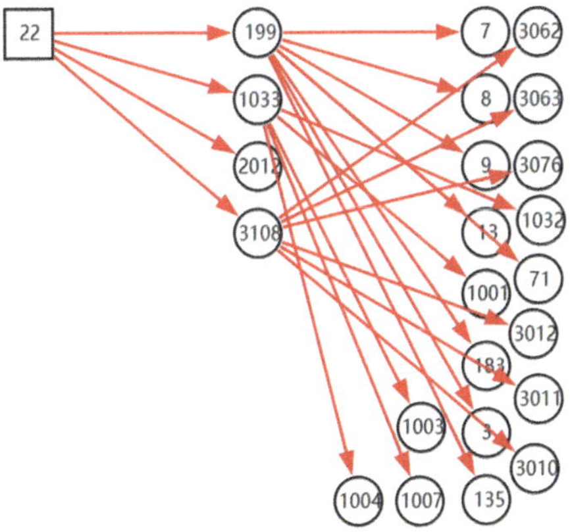

Fig. 2. A DUCG of Intrahepatic bile duct stone

(2) **Knowledge verification**

Knowledge verification module is used to verify whether the user-constructed DUCG meets the DUCG rules during the sub-DUCG construction and DUCG generation. Such as, if the parameters are normalization, and if the relationships among father and son are correct.

(3) **DUCG generation**

The function of the DUCG generation module is to remove duplicate variables, relationships in sub-DUCG and compiles the sub-DUCGs into complete DUCG.

Figure 4 is a jaundice DUCG containing Intrahepatic bile duct stone and Malaria, and it consists of a total of five sub-DUCG in Figs. 2 and 3. during the generation the duplicate variables X_7, X_{3063}, etc. are removed.

3.2 Inference Engine Module

Inference engine's Function is causal reasoning calculation according the received evidences and returns the result of inference calculation.

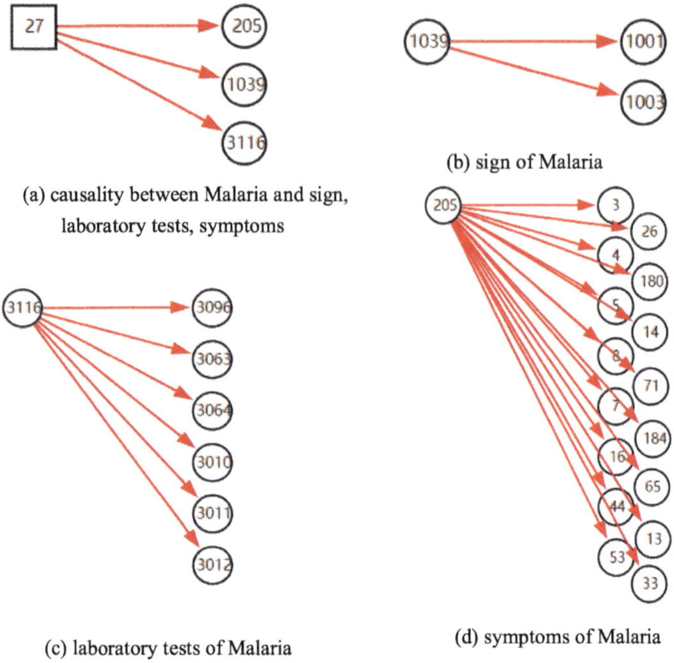

Fig. 3. Construct a Malaria DUCG in a modular way.

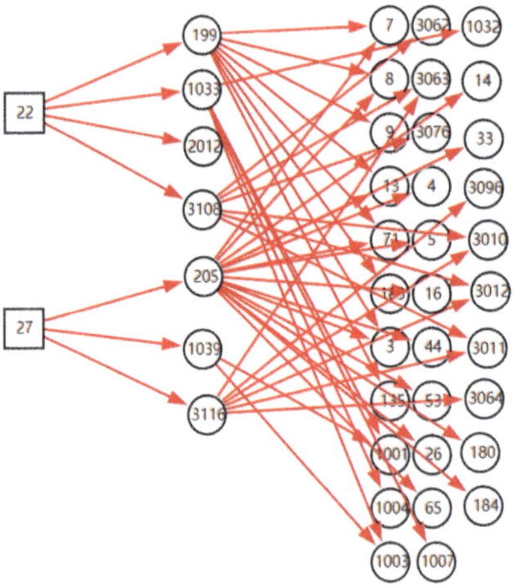

Fig. 4. A DUCG of Intrahepatic bile duct stone and Malaria

3.3 Assistant Diagnosis Module

The assistant diagnosis module mainly contains two sub-modules:

(1) **Evidence input module**

This module is used to collect user information. In order to allow users to easily enter the evidence, we divided all evidences into four categories: symptoms, physical signs, imaging diagnosis, laboratory tests. According Fig. 4, If the patient's consultation information is anorexia, fervescence, hepatomegaly, ascites, both leucocyte and leucocyte are on the high side. Then the system gets the reasoning evidence $E = X_{3,1}X_{7,1}X_{1004,1}X_{1007,1}X_{3010,1}X_{3076,1}$.

(2) **Result display module**

The functions of Result display module are to show the ordering probability of assistant diagnostic results and graphical explanations for the results. According the evidence received from Evidence input module, the diagnostic results are shown in Table 1, and Fig. 5 is the graphical explanation for the results.

Table 1. Diagnostic results based on DUCG in Fig. 4

Disease	Probability
Intrahepatic bile duct stone	83.876%
Malaria	16.128%

As we can see from Table 1, when $E = X_{3,1}X_{7,1}X_{1004,1}X_{1007,1}\ X_{3010,1}X_{3076,1}$, the patient is more likely to suffering from Intrahepatic bile duct stone. The graphical explanation shows a good explanation for this diagnosis. In Fig. 5, Intrahepatic bile duct stone can explain all the evidence, but Malaria can only explain part of the evidence. This is the greatest advantage of the system, it not only tells the doctor the results, but also tells the doctor why the results.

3.4 Data Storage Module

The data storage module is used to store knowledge bases and user symptoms and diagnoses, patient information and assistant diagnostic results.

4 A Real Case of System Application

We worked with medical experts of the First Affiliated Hospital of Zhejiang University to establish a jaundice knowledge base to test this system. Jaundice is a common and complex clinical symptom potentially occurring in hepatology, general surgery, pediatrics, infectious diseases, gynecology, and obstetrics, and it is fairly difficult to distinguish the cause of jaundice in clinical practice, especially for general practitioners in less developed region [7].

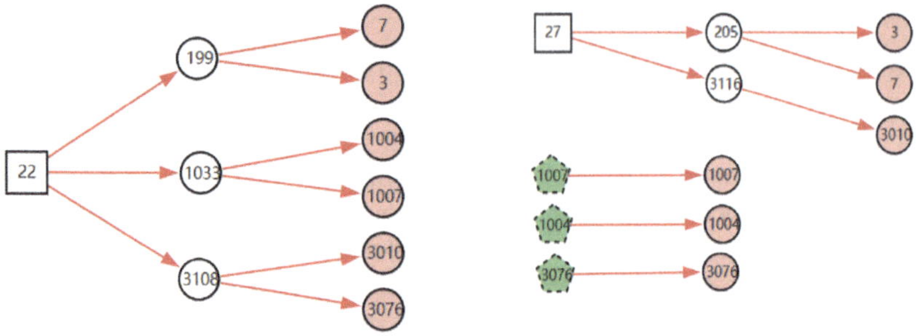

(a)graphical explanation of Intrahepatic bile duct stone

(b) graphical explanation of Malaria

Fig. 5. Graphical explanation for diagnostic results

4.1 DUCG of Jaundice

As shown in Fig. 6, this knowledge base contains 27 jaundice-related diseases: Hepatitis A, Hepatitis B, Hepatitis C, Hepatitis D, Hepatitis E, Undefined viral hepatitis, Alcoholic liver disease, Autoimmune hepatitis, NAFLD, Cirrhosis, Primary liver cancer, Exanthematous typhus, Severe systemic bacterial infection, Infectious mononucleosis, Dubin-Johnson syndrome, Gilbert syndrome, PNH, AHE, Common bile duct stone, IBDS, Pancreatic cancer, Primary biliary cirrhosis, Hyperthyroidism, DILI, Malaria, Jaundice of pregnancy, Hepatolenticular degeneration, and they are represented by different B-type variables. 348 X-type variable are used to express the disease-induced symptoms.

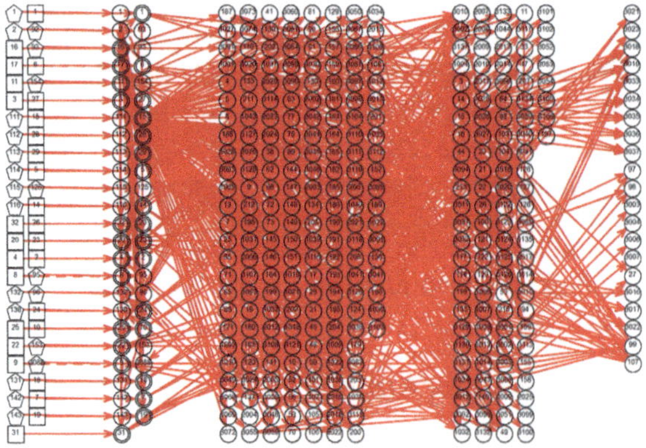

Fig. 6. The DUCG of jaundice

4.2 Result

To test the accuracy of this system, we randomly selected 203 clinical cases from 4,000 clinical cases. Each case includes the patient's symptoms, physical signs, laboratory tests and imaging data. Test results was shown in Table 2. Results show that the system diagnostic accuracy rate of 99%, the diagnostic accuracy has reached clinical standards.

Table 2. Overall accuracies of the DUCG diagnostic system

Disease	Accuracy rate
Hepatitis A	100%
Hepatitis B	90%
Hepatitis C	100%
Hepatitis D	90%
Hepatitis E	100%
Undefined viral hepatitis	100%
Alcoholic liver disease	100%
Autoimmune hepatitis	100%
Nonalcoholic fatty liver disease	100%
Cirrhosis	100%
Primary liver cancer	100%
Exanthematous typhus	100%
Hepatolenticular degeneration	100%
Drug-induced liver injury	100%
Malaria	100%
Severe systemic bacterial infection	100%
Dubin-Johnson syndrome	100%
Gilbert syndrome	100%
Paroxysmal nocturnal hemoglobinuria	100%
Autoimmune hemolytic anemia	100%
Common bile duct stone	100%
Intrahepatic bile duct stone	100%
Pancreatic cancer	100%
Primary biliary cirrhosis	100%
Jaundice of pregnancy	100%
Hyperthyroidism	100%
Infectious mononucleosis	100%
Total	**99.01%**

5 Conclusion

This paper briefly introduces the theory of DUCG and presents a clinical assistant diagnostic system based on DUCG theory, and establish jaundice knowledge base to test the system. The experiment proved that the system not only has high diagnostic accuracy but also has good interpretability for the diagnosis results. In the future, we will build more disease knowledge base to test the system.

References

1. Shortliffe, E.H.: Knowledge engineering for medical decision making: a review of computer-based clinical decision aids. Proc. IEEE **67**(9), 1207–1224 (1979)
2. Suzuki, K.: Massive-training artificial neural network coupled With Laplacian-eigenfunction-based dimensionality reduction for computer-aided detection of polyps in ct colonography. IEEE Trans. Med. Imaging **29**(11), 1907–1917 (2010)
3. Miller, A.S.: Review of neural network applications in medical imaging and signal processing. Med. Biol. Eng. Comput. **30**(5), 449–464 (1992)
4. Costa, C.A.: From complex questionnaire and interviewing data to intelligent Bayesian network models for medical decision support. Artif. Intell. Med. **67**, 75–93 (2016)
5. Zhang, Q.: Dynamic uncertain causality graph for knowledge representation and reasoning: discrete DAG cases. J. Comput. Sci. Technol. **27**(1), 1–23 (2012)
6. Zhang, Q., Geng, S.: Dynamic uncertain causality graph applied to dynamic fault diagnosis of large and complex systems. IEEE Trans. Reliab. **64**, 910–927 (2015)
7. Shao-rui, H.A.O.: Intelligent diagnosis of jaundice with dynamic uncertain causality graph model. Zhejiang Univ. Sci. B (Biomed. Biotechnol.) **18**(5), 393–401 (2017)

Exploring Illumination Removal of Face Image Via Logarithmic Gradient Penalty for Accurate Facial Expression Recognition

Xiaohe Li(✉) and Xingming Zhang

Department of Computer Science and Engineering, South China University of Technology, Guangzhou, China
li.xiaohe@mail.scut.edu.cn

Abstract. Facial expression recognition system requires sufficient detailed image textures to distinguish emotions. However, the traditional Face Illumination Removal approach seem to break this condition providing illumination invariant inputs with heavy corrupted textures. To overcome this problem, we employed l_0 gradient minimization approach to separate illumination variance and invariant components in the logarithmic face image. The experiment shows that, with carefully designed logarithmic gradient pruning mechanism, our method could boost the distinguish ability of any classifiers in any facial expression levels, especially the lower ones.

Keywords: Illumination removal · Weber's law · Lambert reflection Logarithmic gradient penalty

1 Introduction

Illumination is an inevitable natural phenomenon for face imaging. However, the distribution of illumination in face image reduce classification performance dramatically, i.e., the dark part of face image hide local features which should be treated equally to normal part by classifiers. On the contrary, too bright region of face image also compress local image gradients producing unwanted sensing result. Therefore, a lot of methods trying to alleviate the impact of light for face image sensing were proposed [1–10]. These methods can be classified according to whether they have a prior model, saying the model constrained and model free approaches. The model constrained approaches [1–3] tried to establish illumination removal methods taking advantage of linear subspace where the face images lied in, under normal or different illumination condition, i.e., the 9 point light source approach [1]. The model free approaches [4–10] also can be divided in two categories: local pattern based [1–7] and reflection surface based [8–10] approaches. At present, Deep Learning have successfully learn local features together with low dimensional sensing vectors and continuous outperforms traditional approaches in uncountable vision problems. Thus, illumination removal research should focus on reflection surface based approaches which produce raw pixel illumination invariant image, since the deep learning sensing framework only requires raw pixel picture inputs and performs better without illumination turbulence.

K. Deng et al. (Eds.): ICMIR 2018, AISC 856, pp. 1039–1046, 2019.
https://doi.org/10.1007/978-3-030-00214-5_127

Nowadays, three state of the art reflection surface based illumination removal approaches have been brought out: the WeberFace [8], Cos [9] and LTV [10]. All of them assume that human face are smoothly composed by Lambert reflection surface, thus illumination component in human face image can be removed or subtracted by smoothness constraints together with Lambert reflection theory. By directly using the Weber's Formula, the Weberface provided a light independent, perceptual intensity image; Based on DCT (Discrete Cosine Transform) of logarithmic face image, Cos have proposed a low frequency component selection method to eliminate the affection of illumination. Depends on the Meyer's G-norm theory, LTV successfully applied TVL1 (Total Variance model with first order normalized loss function) to separate illumination components from logarithmic face image. Among these approaches, only the LTV model considered that the brightness of human face are regional differently, i.e., the brightness of the eyeball is distinctly different from the face skin. Thus, smoothness should be applied inside each region with different reflective ratio, and they proved that an adjustment of parameters in TVL1 model can produce more reasonable light separate result. One things they may ignore is that, the brightness region of face image becomes very differently between the regions of face with different reflection ratio under some illumination conditions, especially the dim environment with single light source. Under these kind of illumination environment, we will observe that the brightness difference regions become very fewer containing different reflections ratios parts together with some kind of shadows inside them. On the other hand, the reflection ratio inside each detailed region in human face image could be considered illumination invariant and should be left to the illumination invariant detailed face image.

From the discussion above, it should be much better if we could capture the large proportional brightness regions in human face image, ignoring the smaller regions with different reflection ratio. Based upon this opinion, we will introduce a more suitable modifier of the famous TV (Total Variance) [11] model trying the solve the illumination problem in human face image: the G_0-TV [12] (Gradient Minimization Based Total Variance Model), because the boundaries of different brightness regions of face image is obvious spares enough to apply this model.

2 Methodology

2.1 Face Image Analysis Upon Lambert Reflection

The Lambert reflection theory is always used to explain the brightness intensity I of a give surface observed from the frontal view, proportional to the environment light intensity L and inversely proportional to the cosine of the angle between light orientation and the surface's normal vector $R : I = RL$. Taking into the view of a given human face image, we can always write:

$$I(x, y) = s(x, y)L(x, y) \tag{1}$$

$s(x, y)$ is the albedo of the part inside human face surface where pixel (x, y) was generated, it could be regarded as illumination invariant without any problem. But the light changing component $L(x, y)$ could be more difficult to be explained. When taking into account the surface's norm located at the quotient point of (x, y), it can be divided into surrounding light intensity of (x, y)'s quotient point, denoted as $B(x, y)$, and the cosine of the angle between the surrounding light orientation and the surface's norm located at (x, y)'s quotient point, denoted as $A(x, y)$. Thus, we should better rewrite Eq. (1) with the following form:

$$I(x, y) = s(x, y)A(x, y)B(x, y) \tag{2}$$

Both $A(x, y)$ and $B(x, y)$ represent the changing of light. However, they provided different properties to $L(x, y)$. Since we always assume that the human face are smoothly continuous composed by small Lambert reflection surface, $A(x, y)$ should be always smooth enough. However, considering the surrounding light intensity around each (x, y)'s positions, saying the $B(x, y)$, the shadow and exposure affects should be obvious observed.

Summarizing the analysis above, we could say that the light changing component $L(x, y) = A(x, y)B(x, y)$ could be regarded as a regional smooth component with obvious boundaries between each regions. Consequently, we should think about how to separate the light dependent $L(x, y)$ from single face image $I(x, y)$. By applying log function on both side of Eq. 1, we get:

$$\ln(I(x, y)) = \ln(s(x, y)) + \ln(L(x, y)) \tag{3}$$

This simple mathematically transformation tells us that, in the logarithmic face image, the light dependent part of face image could be separated linearly. In the next part of this section, we will describe about how the G_0-TV model works upon the regional smoothness assumption and the logarithmic trick we discussed in this subsection.

2.2 The G_0-TV Model Revisited

The G_0-TV model is the most successfully image smoothing model as far as we concerned until now. It was originally designed to create cartoon image from real word photo applying simple algorithm. We found that the objective of this model is quite reasonable to express the regional smoothness assumptions of light dependent components in logarithmic face image:

$$u = \arg\min_u \ \|f - u\|_2^2 + \lambda\|\Delta u\|_0 \tag{4}$$

f is the original signal, i.e., in the case of face image illumination removal we are discussing in this paper, it represent the logarithmic face image. Δu is the first-order differential of signal u. In the scope of one dimensional discrete signals, this means the convolution operating result applying the convolution kernel $(-1, 1)$. In the scope of two dimensional discrete signals, i.e., the face image, both horizontal and vertical derivation should be applied: $\Delta u = \partial_x u + i\partial_y u$. The image coefficients of $\partial_y u$ ensures horizontal and vertical derivations should be treated in separate numeral space.

In practice, it is more convenient to compute Δu using the convolution theorem, which states that dot product of two signal in the Fourier domain is equivalent to convolution of the two signals in real domain. The 0-order norm $\|\Delta u\|_0$ is proportional to the number of nonzero gradients of u, denoted as $C_{\#}(\Delta u)$, yet it is always close to infinity by the definition of p-order norm. Although the 0-order norm or the alternative $C_{\#}()$ is non-differentiable at any place, an solvable approximation of the objective was given:

$$u = \arg\min_u \|f - u\| + \beta(\|\partial_x u - h\| + \|\partial_y u - v\|) + \alpha C_{\#}(v^2 + h^2) \tag{5}$$

h and v are the dummy variables, representing the horizontal and vertical derivations u should have. The last term $C_{\#}(v^2 + h^2)$ controls the number of nonzero values both in h and v, hence the number of nonzero gradients of u is indirectly controlled.

The approximation objective above could be solved with an iterative optimization strategy, dividing the main goal with two sub problems: Given h and v, the first problem was written as:

$$\arg\min_u = \|f - u\| + \beta(\|\partial_x u - h\| + \|\partial_y u - v\|) \tag{6}$$

It has a close form solution:

$$u = \mathcal{F}^{-1}\left(\frac{\mathcal{F}(f) + \beta(\mathcal{F}(\partial_x).\mathcal{F}(h) + \mathcal{F}(\partial_y).\mathcal{F}(v))}{1 + \beta(\mathcal{F}(\partial_x)^2 + \mathcal{F}(\partial_y)^2)}\right) \tag{7}$$

$\mathcal{F}$, $\mathcal{F}^{-1}$ are the Fourier and inverse Fourier transformation. $\mathcal{F}(\partial_x)$ is the optical transfer of the point spread function $\partial_x(\cdot)$, just the Fourier transforms of centralized horizontal convolutional kernels $(-1, 1)$ with the same size as f . $\mathcal{F}(\partial_y)$ was computed with kernel $(-1, 1)^T$ in the similar way.

Applying the Lagrange multiplier, the solution could be proved with two ways: 1, using the convolution theorem; 2, or using the Eigen decomposition of circulant matrix in the more precise way inside the scope of discrete values. Given u, the second problem was written as:

$$h, v = \arg\min_{h,v} \|\partial_x u - h\| + \|\partial_y u - v\| + \frac{\alpha}{\beta} C_{\#}(v^2 + h^2) \tag{8}$$

It have been proved that when:

$$h(p)|v(p) = \begin{cases} 0 & \partial_x u(p)^2 + \partial_y u(p)^2 > \alpha/\beta \\ \partial_x u(p)|\partial_y u(p) & otherwise. \end{cases} \tag{9}$$

we will arrive at the global optimal solution of the second problem expressed by Eq. (8).

2.3 The Parameter Analysis of G_0-TV

For a theoretically analysis of the first sub problem's solution, we define:

$$g(x,y) = f(x,y) + \beta(\partial_x^{-1}h + \partial_y^{-1}v) \tag{10}$$

From the definition above, we can reformulate Eq. (6) with Rudin-Oscher-Fatemi model's objective function:

$$u = \arg\min_u \|g - u\| + \beta(\|\partial_x u\| + \|\partial_y u\|) \tag{11}$$

It is obvious when observing Eq. (6). Where the meaning of parameter β is well studied by Meyer's G-norm [13]. With Meyer's theory, a Banath space was constructed by a general function v, such that the divergence of each function $\bar{g}$ in G_v is v. Thus, the G-norm $\|v\|_G$ of v was defined with:

$$\vec{g}^{\dagger} = \arg\inf\{\left\||\vec{g}|_{l^2}\right\|_{L^\infty} : v = div.(\vec{g})\} \tag{12}$$

Using the G-norm, we could determine β with the smallest scale of illumination defined by it. For example, if we define a non-regular shape on image f. We can get:

$$\begin{aligned} \|f(S)\|_G = &\quad \max\left\{\sqrt{g_1(x,y)^2 + g_2(x,y)^2}\right\}, s.t. \\ g_1(x,y) = &\quad \int_{-\infty}^{x} fs(u,y)du \\ g_2(x,y) = &\quad \int_{-\infty}^{y} fs(u,y)dv \end{aligned} \quad . \tag{13}$$

Where the solution of Eq. (11) could be directly written by:

$$u = g - P_{G\beta/2}(g) \tag{14}$$

$P_{G\beta/2}(g)$ is the orthogonal projection to the space that $\{v \in G : \|v\|_G \leq \beta/2\}$, with respect to the L_2 scalar product. And the scale of the non-regular shape S could be founded by:

$$scale(S) = \frac{\beta}{2\|g(S) - u(S)\|_G} \tag{15}$$

At first, the scale of S is estimated given a random guessed small β. Since the scale of S is fixed, the algorithmic behavior of TV is controlled by changing β. That is, if we want $u(S)$ to wipe all the features inside it and replace it by an desired smoothed one: $\tilde{u}(S)$. β is directly estimated: $\beta \geq 2 * scale(S) * \|g(S) - \tilde{u}(S)\|$. Thus, given a small scaled shape S_m which we want to ignore its textures and a large scaled shape S_l where we want to retain it's textures, we may guarantee:

$$scale(S_m) * \|g(S_m) - \bar{g}(S_m)\| \leq \frac{\beta}{2} \leq scale(S_l) * \|g(S_l) - \bar{g}(S_l)\| \tag{14}$$

3 Experiments

The image result of our approach is shown in Fig. 1. The original 17 face images of the first person in Extended-YaleB dataset were arranged from dim to the bright in the first row. The corresponding processed results of each images were in the second row. As we can see, although the original dark images have exposure regions, our approach is able to recovery the image details both in the dim and bright parts, preserving the intrinsic surface textures thorough the whole image.

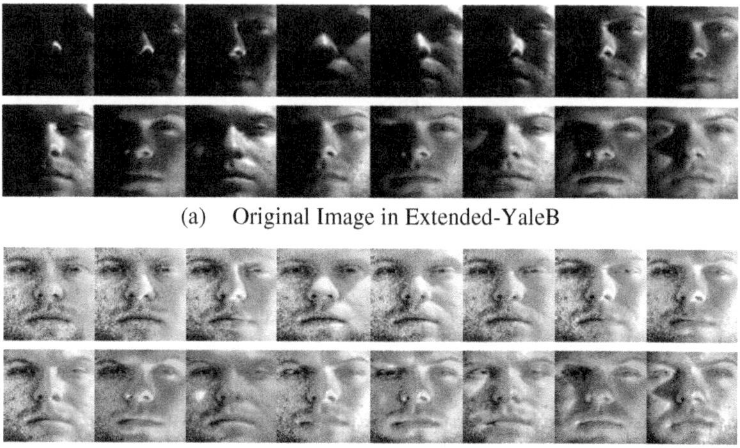

(a) Original Image in Extended-YaleB

(b) Processing Results

Fig. 1. The image processing results of our approach

In order to verify the effectiveness of our approach, we perform 8-fold cross verification using the simplest deep learning model: MLP with CK+ [14] Dataset. As shown in Table 1, our approach outperforms the other state of art illumination removal algorithms in expression recognition with a good margin. Gives us strong evidence that stronger illumination processing approaches are need in the scenario of facial expression recognition.

Table 1. Comparison with expression recognition

	G1	G2	G3	G4	G5	G6	G7	G8	Avg.
Raw	0.25	0.23	0.26	0.31	0.21	0.16	0.23	0.15	0.23
Ours	0.85	0.84	0.91	0.91	0.85	0.89	0.86	0.58	0.84
WeberFace	0.83	0.85	0.88	0.84	0.85	0.89	0.79	0.54	0.80
LTV	0.73	0.82	0.83	0.83	0.85	0.84	0.81	0.53	0.77
Cos	0.75	0.84	0.81	0.83	0.87	0.87	0.79	0.54	0.78

4 Conclusion

In this paper, we proposed large regional smoothness assumption of illumination component in face image. Taking advantage of the powerful G_0TV model, our assumption was implemented.

Our experiments showed that, the G_0TV model was able to automatically distinguish the large gradient caused by incidentally changing of local surrounding illumination environment between the gradients caused by different albedo in face image, with a proper parameter adjustment. Compared to other lambert reflection based illumination removal algorithms, our approach significantly improved the accuracy of facial expression recognition task, due to the reasonable assumption of illumination pattern in face image.

The shortcoming of our approach is that, the illumination removed face image still have certain amount of noisy points in the dark part of the original face image. We can further smooth them out with image denoising technologies. However, since our goal is to recognize facial expression on the face image, balancing the local image features between dim and bright park of face image should be more important than smoothing the noisy points out. Therefore, a further research is eliminated. For the usage of G_0-TV model, if we can add more reasonable differential convolution kernel along different directions, a better result may be accomplished.

Acknowledgments. The author declares that there is no interest conflicts in this paper.

References

1. Lee, K.-C., Ho, J., Kriegman, D.: Nine points of light: acquiring subspaces for face recognition under variable lighting. In: Proceedings of the 2001 IEEE Computer Society Conference on Computer Vision and Pattern Recognition, CVPR 2001, vol. 1, p. I IEEE (2001)
2. Lu, Y., Lu, C., Qi, M., Wang, S.: A supervised locality preserving projections based local matching algorithm for face recognition. In: Lecture Notes in Computer Science (Including Subseries Lecture Notes in Artificial Intelligence and Lecture Notes in Bioinformatics). LNCS, vol. 6059, pp. 28–37 (2010)
3. Cao, F., Hu, H., Lu, J., Zhao, J., Zhou, Z., Wu, J.: Pose and illumination variable face recognition via sparse representation and illumination dictionary. Knowl. Based Syst. **107**, 117–128 (2016)
4. Lowe, D.G.: Object recognition from local scale-invariant features. In: Proceedings of the Seventh IEEE International Conference on Computer Vision, vol. 2, pp. 1150–1157. IEEE (1999)
5. Hong, P., Si-Yu, X.: Illumination invariant face recognition based on improved local binary pattern. In: 2011 30th Chinese Control Conference (CCC), pp. 3268–3272 (2011)
6. Dalal, N., Triggs, B.: Histograms of oriented gradients for human detection. In: 2005 IEEE Computer Society Conference on Computer Vision and Pattern Recognition (CVPR 2005), vol. 1, pp. 886–893. IEEE (2005)

7. van de Weijer, J., Khan, F.S.: An overview of color name applications in computer vision. In: Lecture Notes in Computer Science (Including Subseries Lecture Notes in Artificial Intelligence and Lecture Notes in Bioinformatics), vol. 9016, pp. 16–22 (2015)
8. Wu, Y., Jiang, Y., Zhou, Y., Li, W., Lu, Z., Liao, Q.: Generalized Weber-face for illumination-robust face recognition. Neurocomputing **136**, 262–267 (2014)
9. Chen, T., Yin, W., Zhou, X.S., Comaniciu, D., Huang, T.S.: Total variation models for variable lighting face recognition. IEEE Trans. Pattern Anal. Mach. Intell. **28**(9), 1519–1524 (2006)
10. Chen, W., Er, M.J., Wu, S.: Illumination compensation and normalization for robust face recognition using discrete cosine transform in logarithm domain. IEEE Trans. Syst. Man Cybern. B Cybern. **36**(2), 458–466 (2006)
11. Rudin, L.I., Osher, S., Fatemi, E.: Nonlinear total variation based noise removal algorithms. Physica D **60**(1–4), 259–268 (1992)
12. Xu, L., Lu, C., Xu, Y., et al.: Image smoothing via L0 gradient minimization. ACM **30**(1), 1–12 (2011)
13. Gilles, J., Osher, S.: Bregman implementation of Meyer's G–norm for cartoon + textures decomposition. UCLA Cam Report (2011)
14. Lucey, P., Cohn, J.F., Kanade, T., et al.: The extended Cohn-Kanade Dataset (CK+): a complete dataset for action unit and emotion-specified expression. In: Computer Vision and Pattern Recognition Workshops, pp. 94–101. IEEE (2010)

Research and Implementation of Distributed Image Storage Based on Mongo

Xiao Yunhua(✉), Xue Juan, and Li You

School of Information Technology and Management,
Hunan University of Finance and Economics, Changsha 410205, Hunan, China
13181172@qq.com

Abstract. With the development of the Internet, accessing the website has become an essential Internet application in people's lives. In order to attract more users, it is the goal of the pursuit of the website that to ensure the stability of the site and improve the site's access speed. Picture is the main reason of the website loading speed in the condition of good network condition. This paper proposes to use the Mongo database to store pictures, can greatly improve the reading speed of the picture, speed up the site's access speed, and reduce the waiting time of users.

Keywords: Picture storage · Mongo · Slicing cluster · Copy set

1 Introduction

With the development of the Internet, people's lives can not be separated from the Internet. The ratio of accessing websites is growing in a variety of Internet applications. It is the goal of the pursuit of the website that to ensure the stability of the site and improve the site's access speed. Picture is the main reason of the website loading speed in the condition of good network condition, because the picture file is relatively large and time consuming is relatively long from the back-end image server to the front. At present, many web servers use distributed storage. It mainly uses MFS (distributed file system) technology and mysql+ disk storage technology (For example: Douban and to8to use the MFS Technology). These two technologies are stored image on the server disk. The bottleneck of the current server is IO because reading and writing of the picture speed is relatively low on the disk. Mongo database is the NoSQL database that appeared in the last few years. It is very suitable that storing pictures in mongo database. The image distributed storage in the Mongo database can speed up the reading speed of the image and the site's access speed.

This paper is supported by Hunan educational reform project: Research on the CDIO talent training mode of computer science and technology major with the dual features of the local and the industry (Hunan Education [2017] File452, NO. 561). Hunan University of Finance projects (2017xjjq015).

K. Deng et al. (Eds.): ICMIR 2018, AISC 856, pp. 1047–1053, 2019.
https://doi.org/10.1007/978-3-030-00214-5_128

2 Mongo Storage Model

Mongo database supports JSON (Object Notation JavaScript) BSON (Binary). It can store more complex data types. Mongo storage is a form of key-value. This paper is prepared to store binary image in the Mongo database directly.

In order to ensure the security, reliability, scalability and rapid access of data, this paper store the image by the characteristics of Mongo cluster. Mongo provides two cluster modes. They are copy set and slicing cluster. Copy set replicate data to other nodes asynchronously. Each time only one node copies data from the master node asynchronously. Slicing cluster divided data into parts and stored on each node. The storage node is working at the same time. These will speed up the storage of data.

Each node in the Mongo cluster is an instance of a Mongo data block. In the Mongo database storage, each database contains a file of databasename.ns and data files. The file of databasename.ns contains a namespace that holds the relevant information about the database. Databasename.{0..} stores the actual file data. This paper will store the binary image in Databasename.{0..}. There is a virtual memory behind DISK. Virtual memory and physical memory mapping, which is the Mongo storage engine model. Mongo storage engine is MMAP memory mapping. The Mongo data maps the disk storage location of data corresponding into the memory. It caches the file to memory after finds the file path from the memory. It reads the data from the memory directly in the future, which can speed up the reading image.

3 The Design of Mongo Based on Distributed Image Storage

3.1 The Analysis of Mongo Copy Set

Mongo copy set only operates the main node when the data is stored. Its storage speed is the same as the storage speed of non-clustered Mongo. It reads data from all nodes in cluster. So it will speed up the reading of the data. Mongo copy set has data backup In terms of data security and reliability. The data will not be lost when a node is failure. Mongo copy set depends on disk storage size and it cannot be extended in terms of extensibility. The storage structure of mongo copy set as shown in Fig. 1.

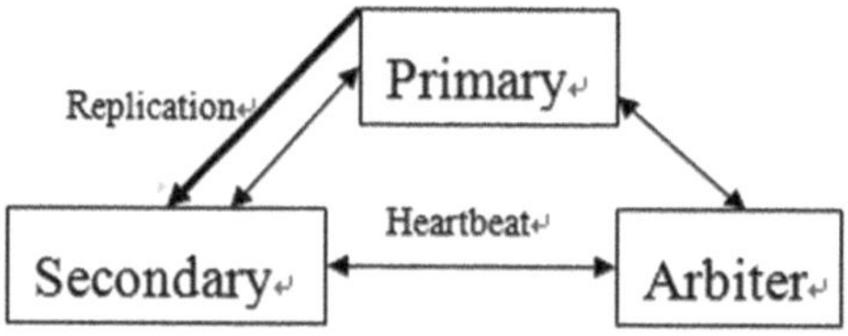

Fig. 1. Copy set architecture

Mongo copy set command is as follows:
Execute as the following code 3.1, 3.2, 3.3 on the server that the node is located:

```
[root@to8tosvn ~] # mongod --shardsvr --replset shard1 --prot 27017 --dbpath
/data/shard1_1 --logpath /data/shard1_1/shard1_1.log --logappend --fork
```

Code 3.1 The copy set node Configure

```
[root@to8tosvn ~] # mongod --configsvr --dbpath /data/config --port 27101 --
logpath /data/config/config.log --logappend --fork
```

Code 3.2 The mongo config server configure

```
[root@to8tosvn ~] # mongod --configdb 192.168.1.83:27101,192.168.1.84:27101,
192.168.1.85:27101 --port 27001 --chunksize 2 --logpath /data/mongos.log --
logappend --fork
```

Code 3.3 The mongo route configure

3.2 The Analysis of Mongo Slicing Cluster

Data will be distributed to the rest of the nodes if there is a failure node in the Mongo slicing cluster in terms of stability; Mongo slicing cluster will distribute the data distributed to each node in order to accelerate data storage in terms of storage; It reads the data from all nodes as well as the replica set in terms of reading data; Mongo slicing cluster will average distribute the data to the other node when a node is failure in terms of safety and reliability; It can achieve the goal of increasing the storage capacity by increasing the node in terms of scalability. The storage architecture of mongo slicing cluster is shown in Fig. 2.

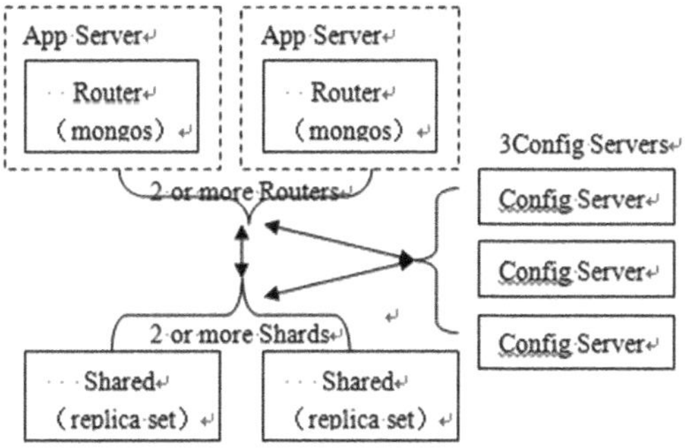

Fig. 2. The storage architecture of mongo slicing cluster

Mongo slicing cluster divides the data into a small piece. Each data block is called a chunk. The specific data distribution of mongo slicing cluster is shown in Fig. 3.

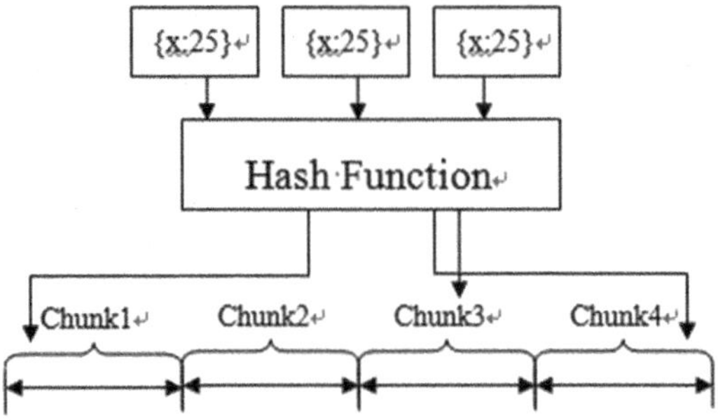

Fig. 3. The specific data distribution of mongo slicing cluster

The build code of mongo slicing cluster is shown in code 3.4:

The build code of mongo slicing cluster is shown in code 3.4:

```
[root@to8tosvn ~] # mongod 192.168.1.83:27018
MongoDB shell version:2.6.4
Connecting to:192.168.1.83:27018/test
Shard1:PRIMARY>db.runCommand({addshard:"shard1/192.168.1.83:27017,192.168.1.84:27017, 192.168.1.853:27017"});
```

Code3.4 The build code of mongo slicing cluster

We find that Mongo copy set is not optimized for fast storage and cannot expand storage space, But it can achieve the needs of stable and fast data reading. Mongo slicing cluster can achieve the needs of rapid storage 、 reading data and scalable, but it is not absolutely safe. The data can not be recovered after it is damaged. The two kinds of clusters have different characteristics, but each has the insufficiency. The two clusters can achieve the desired goal when they are combined. The storage method that the slicing cluster and copy set are combined is shown in Fig. 4.

Fig. 4. The storage of mongo slicing cluster and copy set

The data copy to the second slice in Fig. 4. These data are distributed to other nodes by hash. It is not copy from ShardA to the corresponding ShardA node.

After the deployment of a good copy of the set with the split, Mongo copy set combined with the key commands to build a cluster of clusters such as code 3.5.

The key build code of the slicing cluster and copy set are combined is shown in code 3.5 after the slicing closet and copy set are deployed.

```
[root@to8tosvn ~] # mongod 192.168.1.83:27018
MongoDB shell version:2.6.4
Connecting to:192.168.1.83:27018/test
Shard1:PRIMARY>db.runCommand({addshard:"shard1/192.168.1.83:27017,192.
168.1.84:27017, 192.168.1.85:27017"});
Shard1:PRIMARY>db.runCommand({addshard:"shard2/192.168.1.83:27018,192.
168.1.84:27018, 192.168.1.85:27018"});
```

Code3.5 The build code of mongo slicing cluster and copy set

The configuration of the Mongo in the Java is shown in code 3.6.

```
The configuration of the Mongo in the Java is shown in code 3.6
Mongodb.host=192.168.2.191:27001,192.168.2.192:27001
Mongo_user=
Mongo_passwd=
Mongo_database=image
```

Code 3.6 the mongo configuration

Java connected to the Mongo by reading the above configuration. The initialization code is shown as code 3.7.

```
DB mdb=MongoClient.getDB("image_config");
BasicDBObject obj= new BasicDBObject();
Obj.append("_id",md5("to8to3d"));
Obj.append("category", "to8to3d");
Obj.append("ns", "to8to3d");
Mdb.getCollection("image_config").insert(obj)
```

Code3.7 Java connect to the mongo code

The core code that Java store image is shown as code 3.8.

```
DB mdb=MongoClient.getDB(db);
Mdb.getCollection(ns).createIndex(new BasicDBObject("meta._org_id",1));
Retrun mdb.getCollection(ns).save(image);
```

Code 3.8 The core code that Java store image

The preliminary function is achieve by the code, but there is also a single point of failure in the framework. It is a single point of the drive that the Java connection Mongo cluster.

4 The Comparison of Experimental Data

Experimental environment: Operating system: Centos 6.5, 64bit; Kernel version: 2.6.18; CPU: 8core, 2.4GH; Memory: 16G; Mongo version: 2.6.4; Mysql version: 5.5; Mysql engine: Innodb.

We test the Mongo and MySQL performance. We operates 10000 times the insertion, deletion and change separately. Test results are shown in Tables 1 and 2.

Table 1. The test results of mysql

	First	Second	Third	Average
Insert	1.430	1.433	1.437	1.435
Delete	0.0088	0.0091	0.0091	0.009
change	0.766	0.762	0.767	0.765

Table 2. The test results of mongo

	First	Second	Third	Average
Insert	0.315	0.319	0.316	0.317
Delete	0.00074	0.00076	0.00072	0.00074
change	0.078	0.079	0.077	0.079

The histogram contrast chart of the average results is shown in Fig. 5.

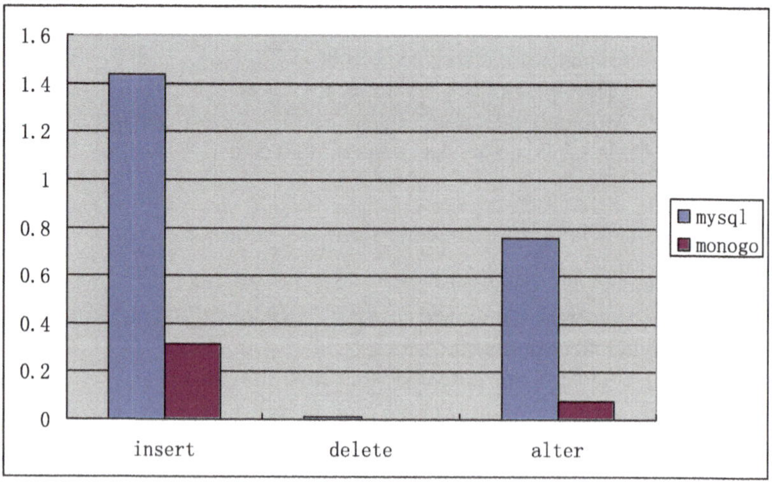

Fig. 5. The histogram contrast chart of the average results

From the above we can know that the storage speed of Mongo is 4 ~ 5 times faster than MySQL. The delete speed of Mongo is faster a order of magnitude than the MySQL. The update speed of Mongo is about 10 times faster than MySQL. The Mongo database stored the binary image file. The data is got directly from the mongo when it is queried. The MySQL stored the location of the picture. The data is got from the corresponding server node.

5 Conclusion

We can see that the performance of Mongo is much higher than MySQL by the above experimental data analysis and comparison. There is a small part of the picture is Mongo in to8to website. Its performance is several times faster than the MFS. The data of this experiment is in a small amount and this experiment don't operate for the growth of the data. So It can not guarantee the performance of mongo is much higher than mysql after the data is grew.

Someone on the Internet directly test a single instance of Mongo and find the performance of mongo is decreased a lot when the data is exceed the allocated memory. But they did not test the Mongo split cluster and copy set. These deficiencies will continue to research next.

References

1. Chuanhui, Y.: Large Scale Distributed Storage System: Principle Analysis and Architecture. Machinery Industry Press (2013)
2. Coulouris, G., Dollimore, J., Kindberg, T., Blair, G.: Distributed System: Concept and Design, 5th edn. Machinery Industry Press (2013)
3. Yuanwei, G.: Big Data Storage DB Mongo Actual Combat Guide. The Posts and Telecommunications Press (2015)
4. Chodorow, K., Dirolf, M., Qiang, D., Minghui, W.: MongoDB Definitive Guide, 2nd edn. The Posts and Telecommunications Press (2014)
5. Wensheng, Z., Hanhua, Z.: Research on constructing high performance web technology based on MongoDB. J. Jilin Norm. Univ. (Nat. Sci. Edn) (2013)

Influence of the Novel Rapid Preparation Method on Sample Imaging Evaluation for Pathogenic Microorganisms Scanning Electron Microscopy

Mingzhu Song[1,2](✉), Xirui Wang[3], Wei Yao[1], and Yang Su[4]

[1] School of Medicine, Chengdu University, Chengdu 610106, People's Republic of China
songmingzhu@cdu.edu.cn

[2] School of Materials Science and Engineering, Southwest Jiaotong University, Chengdu 610031, People's Republic of China

[3] School of Tourism, Economics and Management, Chengdu University, Chengdu 610106, People's Republic of China

[4] Sichuan Academy of Medical Sciences and Sichuan Provincial People's Hospital, Chengdu 610212, People's Republic of China

Abstract. This study provides a novel rapid and effective sample processing method for scanning electron microscopy observation of bacterium and fungi. We investigated the relationships between image quality and instrument operating parameters. The procedure consists of seven stages, including fixing, cleaning, dehydration, replacement, drying, spraying and observation. The incorporation of tertiary butyl alcohol (TBA), which played important roles in both dehydration and cleaning, was optimized using six distinct tertiary butyl alcohol concentrations (30%, 50%, 70%, 85%, 95% and 100%, v/v). Electrically conductive adhesive transfer tape was chosen as the carrier on which samples were mounted for SEM observation. Saved the time of preparation and cut down the use of expensive toxic reagents. The methodology established in this study helps to improve the efficiency for identifying the ultra microstructures of pathogenic microorganisms. The method describes the complete method development and validation process and is well suitable for pathogenic microorganisms scanning electron microscopy sample imaging evaluation.

Keywords: Biofilm · Pathogenic microorganisms · Sample preparation SEM

M. Song and X. Wang—Authors contributed equally to this work, as co-author of the first author.

K. Deng et al. (Eds.): ICMIR 2018, AISC 856, pp. 1054–1060, 2019.
https://doi.org/10.1007/978-3-030-00214-5_129

1 Introduction

The biological membrane of pathogenic microorganisms is known as biofilm existing studies show that the existence of the biofilm is closely related to the severity of the disease [1]. Dissection the structure of biofilm is important to understand its biological functions and how it would contribute the pathogenic of microorganisms.

Electron microscope is one of the important tools to observe the microbial biofilm. SEM electron beam is not able to pass through the sample, only scanned the sample surface and excited secondary electrons thereby imaging. The image has a strong three-dimensional effect, which provides direct visual information in the early stages of biofilm studies [2].

Currently, SEM has become a powerful tool for the study of the microscopic world, many important discoveries and research results are obtained through observation. However, when we take the actual observation and operation, considering the SEM is a highly sensitive instrument with complicated structure, there will be many factors that affect the imaging quality. For instance, contrast grade will be reduced due to the surface structure restricted by various factors. Therefore, only by choosing the appropriate parameters and sample preparation method, a fine structure of the outer periphery of the cell can be clearly observed [3].

2 Materials and Methods

2.1 Preparation of Pathogenic Microorganisms Cell Suspension

Candida albicans (CCTCC AY 206001) and *Staphylococcus aureus* (CCTCC AB 205220) were obtained from the China Center for Type Culture Collection (CCTCC) (Wuhan, Hubei Province, China) [4]. The specific steps performed in our pathogenic microorganism cell suspension processing method are shown in Table 1. The conventional steps performed in our pathogenic microorganisms samples processing method are shown in Table 2.

Table 1. Comparison of preparation procedures and conditions for distinct samples

Specimen	*Staphylococcus aureus*	*Candida albicans*
Main steps	Streak inoculation, incubation, centrifugation, dilution, resuspended sample	
Temperature	37 °C	30 °C
Speed	180 rpm	200 rpm
Time	12 ~ 18 h	12 ~ 24 h
Medium	Broth medium	PDA medium

Table 2. Comparison of preparation steps and times for SEM

Main steps	*Candida albicans*		*Staphylococcus aureus*	
	Conventional method	Improved method	Conventional method	Improved method
2.5% Glutaraldehyde fixed	4 °C,2 h	4 °C,4 h	4 °C,2 h	4 °C,4 h
Cleaning	PBS	Deionized water	PBS	Deionized water
1% osmium tetroxide	4 °C,1 h	NO	4 °C,1 h	No
Washing	PBS	NO	PBS	No
Distinct alcohols dehydration	Ethanol	TBA	Ethanol	TBA
Dehydration time	10 min/time, once 20 min/time, twice	10 min/time, once 20 min/time, twice	10 min/time, once 20 min/time, twice	10 min/time, once 20 min/time, twice
Replacement	20 min/time, twice	No	20 min/time, twice	No
Drying	55 °C, baking 6 h	Vacuum drying 55 min	55 °C, baking 6 h	Vacuum drying 55 min
Spraying	10.0 Pa, 15 mA, 130 s		10.0 Pa, 15 mA, 130 s	
Observation	15 kV/5 kV, 11.5 ~ 13.4 mm		5 kV, 12.2 ~ 12.4 mm	

2.2 SEM Observation of the Samples

The prepared samples were placed into the sample chamber. Determined the vacuum degree of the sample chamber become qualified it can be chosen the accelerate voltage and adjust the working distance. We then optimized the conditions for ion sputtering and metal spraying. Setting the working current and working distance to be constant, the other parameters were optimized as follows: vacuum degree 10 Pa, currentintensity15 mA, coating time 130 s and deposition thickness 10 nm. The appropriate distance between specimen holder and gold target is 5 cm.

3 Results

3.1 Comparison of Distinct Preparation Methods

For the sake of comparison, pathogenic microorganisms *S. aureus* and *C. albicans* were prepared by using of a conventional method and an improved method. The samples were observed and taken pictures by SEM. We can see clearly in the process of fixation. Due to fixed time of conventional method relatively short lead to the samples were surrounded by pathogenic microorganism secretions (Fig. 1 A1 and B1). However, the improved method of preparing sample not only found no pathogenic microorganism secretions but also simplifies routine double fixed steps (Fig. 1 a1 and b1).

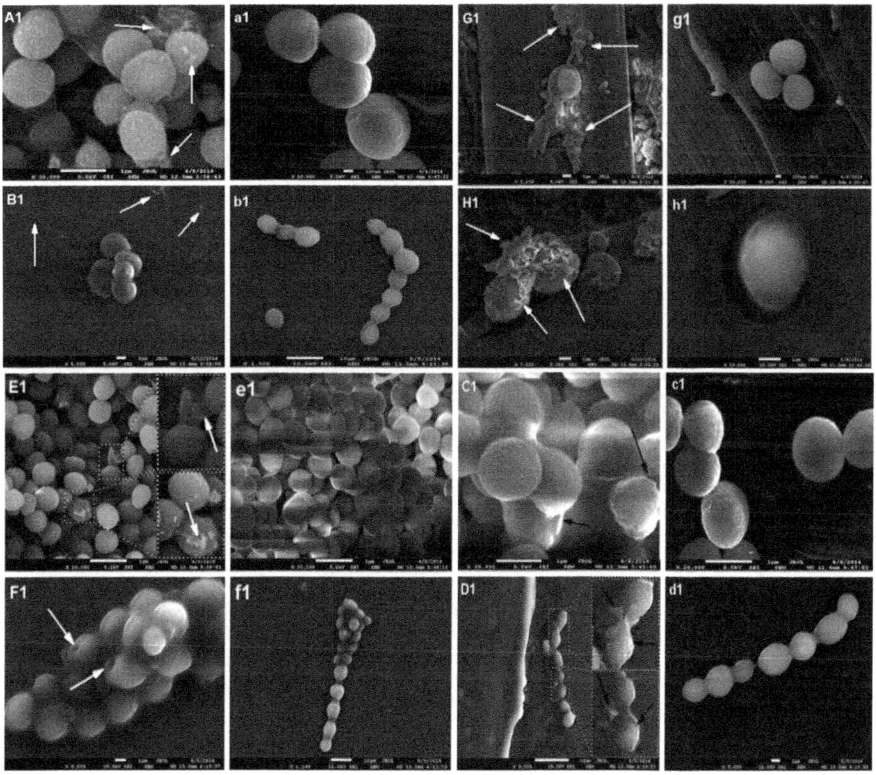

Fig. 1. A1, B1, C1, D1, E1, F1, G1 and H1were adopted conventional preparation methods. a1, b1, c1, d1, e1, f1, g1 and h1 were adopted improved preparation methods. A1, a1, G1, g1, E1, e1, C1 and c1 were SEM for *S. aureus*. B1, b1, H1, h1, F1, f1, D1 and d1 were SEM for *C. albicans*. This figure shows comparison of distinct fixed methods, distinct washing methods, distinct dehydration methods, distinct drying methods.

The sample was covered by hydrochloric acid particle residue when used PBS (pH7.2) in the process of cleaning (Fig. 1 G1 and H1). However, improved methods used in sterile deionized water when cleaning the sample. After drying, not only did not appear residual impurities but also did not cover the sample (Fig. 1 g1 and h1).

The conventional method used different concentration gradient ethanol to dehydration. However, the surface structure of sample was influenced by dehydration time interval. Observation indicates that pathogenic microorganism biofilm surface was cracking or different degrees of shrinkage sunken or even fracture and become serious damage, when samples were dehydrated 10 min/time (Fig. 1 E1 and F1). This dehydrant of improved method was tertiary butyl alcohol instead of ethanol. But the dehydration time was 5 min/time. This method could be maintained the smoothness and plumpness against the surface structure of pathogenic microorganism biofilm. Besides, the cells did not appear shrinkage and crack. We got a distinct gradation sample with well-preserved tissue micro structure. The method meets the needs of observing by using of SEM, because the original features of sample was maintained (Fig. 1 e1 and f1).

These are samples of processed by different drying methods. The results showed that the samples of drying at 55 °C for 6 h. The conventional drying method prepared tiny deformation of samples and the shedding of biofilm (Fig. 1 C1 and D1). However, the samples used tertiary butyl alcohol dehydration and vacuum drying for 55 min (Fig. 1 c1 and d1).

3.2 The Settings of SEM Parameter

When the accelerating voltage value was15 kV, we obtained the best contrast to give a high quality image in the fine structure of the pathogenic microorganism biofilm surface. When the accelerating voltage is 5 kV, pathogenic microorganism biofilm surface look more real and get more surface structure. When the lens aperture was set to be 50 μm and the working distance was 12.4 mm, we can get maximise the high-resolution information in three-dimensional images.

4 Discussion

The sample preparation method has direct influence on the quality and authenticity of scanning electron microscopy. In the process of dehydration of the sample of microorganisms, the bacteria or fungi cells might form secretions when stimulated by the dehydration. However, the fixing agent can inhibit these changes. The complexity of the convetional method and the high price as well as toxicity of osmic acid was not enough to be applied [5]. By comparison, we should use the method of glutaraldehyde instead of double-fixed, which could avoid the mutual reaction between the two reagents. The glutaraldehyde-fixed way eliminates the multiple buffer washing steps. In addition, it can obviously shorten the time of the sample preparation. This method not only improves efficiency of sample preparation, but reduces the cost [6].

Accordingly, once the sample is fixed, it should use distilled water to clean because it is economical, substantial and convenient. Moreover, the sample cleaning also reduces the pollution on the lumen of SEM and delays the fusing of filament.

The time control of dehydration is very important. The sample might have had certain shrinkage or deformation before drying, which is unfavorable to preserving the structure of the sample. On the contrary, it is more likely to cause insufficient dehydration and unclear image. The shrinkage rate is apparently related to the time of dehydration [7].

By comparison, it was not hard to find that the TBA used as dehydrating agent in place of ethanol can reduce tissue contraction and lead to more thorough dehydration. In the improved method for preparation of sample, we have adopted TBA of different concentration as the dehydrating agent [8].

The high temperature will hasten water evaporation so that the sample can dehydrate by the oven-drying method. Under low vacuum environment, by the sublimation of medium, sample can achieve the goal of drying, which rules out the damages against the cells through the liquid-gas interface [9].

The sample chamber must meet certain degree of vacuum and the current strength, when the ion sputtering and coating. The vacuum degree should be set to10.0 Pa [10], the input current should be less than 15 mA [11].

The sample preparation of pathogenic microorganism is related to the quality of the images. The proper setting of parameters will be critical factors in whether or not we can get an ideal image. Eventually, we will select a comparatively best value to observing. The size of 50 μm or 30 μm was used at high magnification. The working distance refers to the distance between the sample and lower end of objective [12]. Another important factor that affects the image quality, and the variable range is 0.1 ~ 3 mm [13, 14]. We generally choose the distance between 10 ~ 15 mm to achieve small beam diameter and reduce the spherical aberration effect. A diaphragm opening of 50 μm and a working distance of 12.5 mm are selected in this experiment. We can gain clear depth of focus and three-dimensional images.

Acknowledgments. We are grateful to Canquan Mao Professor from Laboratory of Molecular Evolution and Applied Biology, School of Life Science and Engineering, Southwest Jiaotong University, Chengdu, P. R. China for helping us with language editing and research platform. This work was financially supported by Sichuan Provincial Department of Education 2017 Natural Science Key Project (No. 17ZA0095); Sichuan Provincial Key Laboratory of Medicinal and Edible Plant Resources Development Fund Project (No. 10Y201705).

Conflict of Interest. Mingzhu Song and Xirui Wang contributed equally to this work, as co-author of the first author. All authors have no conflicts of interest.

References

1. Rajesh, P.S., Ravishankar Rai, V.: Quorum quenching activity in cell-free lysate of endophytic bacteria isolated from Pterocarpus santalinus Linn., and its effect on quorum sensing regulated biofilm in Pseudomonas aeruginosa PAO1. Microbiol. Res. **169**, 561–569 (2014)
2. Ali Ahmed, A.B., Taha, R.M.: Chapter 31—The perspectives of the application of biofilm in the prevention of chronic infections. In: Se-Kwon, K. (ed.) Advances in Food and Nutrition Research, pp. 403–416. Academic Press, Cambridge (2011)
3. Tollkamp-Schierjott, C.: The influence of different SEM parameters on CD measurement results. Microelectron. Eng. **6**, 673–678 (1987)
4. Song, M.Z., Wang, X.R., Mao, C.Q., Yao, W.: The discovery of a potential antimicrobial agent: the novel compound natural medicinal plant fermentation extracts against *Candida albicans*. IOP Conf. Ser. Mater. Sci. Eng. **301**, 012026 (2018)
5. Prakash, D., Nawani, N.N.: A rapid and improved technique for scanning electron microscopy of actinomycetes. J. Microbiol. Methods **99**, 54–57 (2014)
6. Jiang, W., Pan, H., Wang, F., Jiang, M., Deng, X., Li, J.: A rapid sample processing method to observe diatoms via scanning electron microscopy. J. Appl. Phycol. **27**, 243–248 (2014)

7. Cheng, C.F., Wu, K.M., Chen, Y.T., Hung, S.L.: Bacterial adhesion to antibiotic-loaded guided tissue regeneration membranes - a scanning electron microscopy study. J. Formos. Med. Assoc. Taiwan yi zhi **114**, 35–45 (2015)
8. Yamada, H., Chikamatsu, K., Aono, A., Mitarai, S.: Pre-fixation of virulent *Mycobacterium tuberculosis* with glutaraldehyde preserves exquisite ultrastructure on transmission electron microscopy through cryofixation and freeze-substitution with osmium-acetone at ultralow temperature. J. Microbiol. Methods **96**, 50–55 (2014)
9. Munnecke, A., Servais, T.: Scanning electron microscopy of polished, slightly etched rock surfaces: A method to observe palynomorphsin situ. Palynology **20**, 163–176 (1996)
10. Faganello, J., Arruda, W., Schrank, A., Vainstein, M.H.: An alternative method to prepare samples of the pathogenic yeast *Cryptococcus neoformans* for scanning electron microscopy analysis. J. Microbiol. Methods **64**, 416–419 (2006)
11. Sarmiento, P.L., Ciarmela, M.L., Sanchez Thevenet, P., Minvielle, M.C., Basualdo, J.A.: Comparison of preparation techniques of mixed samples (fungi-helminth eggs) for scanning electron microscopy by critical point drying. Parasitol. Res. **99**, 455–458 (2006)
12. Priester, J.H., Horst, A.M., Van de Werfhorst, L.C., Saleta, J.L., Mertes, L.A., Holden, P.A.: Enhanced visualization of microbial biofilms by staining and environmental scanning electron microscopy. J. Microbiol. Methods **68**, 577–587 (2007)
13. Jiang, W.K., Pan, H.Q., Wang, F.X., Jiang, M.L., Deng, X.Y., Li, J.B.: A rapid sample processing method to observe diatoms via scanning electron microscopy. J. Appl. Phycol. **27**, 243–248 (2015)
14. Thai, T.Q., Nguyen, H.B., Saitoh, S., Wu, B., Saitoh, Y., Shimo, S., Elewa, Y.H., Ichii, O., Kon, Y., Takaki, T., Joh, K., Ohno, N.: Rapid specimen preparation to improve the throughput of electron microscopic volume imaging for three-dimensional analyses of subcellular ultrastructures with serial block-face scanning electron microscopy. Med. Mol. Morphol. **49**, 154–162 (2016)

Visual Recognition of Traffic Accident Scene's Image Research

Bi-xiang Li(✉) and Lan Fang

Department of Information Engineering,
City College of Wuhan University of Science and Technology,
Wuhan 430083, China
lbxl53@163.com

Abstract. The optimization of the visual image of traffic accident scene can effectively improve the efficiency of traffic accident treatment. To identify the visual image, we need to obtain the morphological characteristics of the scene image, and establish the sector feature vector of the scene of the traffic accident image to optimize the image recognition. The traditional method extracts a reasonable threshold by extracting the neighborhood image features of the key points of the image, but ignores the establishment of the image sector region feature vectors, resulting in low recognition accuracy. Intelligent vision based visual image recognition method for traffic accident scene. The scene of the accident to extract the background image using mean background modeling theory, the background image is normalized, and the accident scene contour image into two value image, the traffic accident scene image filtering, to obtain morphological characteristics of scene images of different sizes, the establishment of traffic accident scene image regional feature vector, combined with local Discrimination Mapping Theory dimensionality reduction of image feature vectors of traffic accident scene, complete the traffic accident scene visual image recognition. Experimental results show that the proposed method can effectively improve the visual image recognition rate of traffic accident scene, and the recognition accuracy is high.

Keywords: Traffic accident scene · Visual recognition · Optimal identification

1 Introduction

Recently, with the continuous improvement of national economic income, and with the rapid development of vehicle industry in China, the number of traffic accidents is also increasing year by year, the traffic accident scene investigation and analysis level also put forward higher requirements [1]. With the development of computer vision technology and digital image processing technology, it is possible to make quantitative analysis of the scene of traffic accident by means of camera equipment [2]. However, due to the interference of the external environment such as camera vibration and ambient light for camera collecting traffic accident scene image, the image usually blurry [3, 4]. In this case, how to use visual image identification for traffic accident scene effectively becomes the primary task of image information processing field,

K. Deng et al. (Eds.): ICMIR 2018, AISC 856, pp. 1061–1068, 2019.
https://doi.org/10.1007/978-3-030-00214-5_130

which applying in traffic accident scene identification, so more and more experts and scholars are cantering in this field, which has been become the focus of intelligent transportation, many researches has acquired great achievement [5].

Tavernier [6] proposed a method of visual image recognition of traffic scene based on coordinate extreme theory. The method recognized the concavity and convexity of a single pixel wide closed contour of the visual image of traffic accident scene firstly, secondly, the local and regional contour of the concave region was determined by ray method, the visual image contour of traffic accident scene without concave point was identified by coordinate extreme theory. This method could improve the level of automatic recognition effectively, but the method existed poor quality of image recognition. Yamada [7] proposed abased on control point theory to identify the visual image of traffic accident scene. This method was based on the camera calibration firstly, then the camera lens distortion was corrected by the visual image of the traffic accident scene, the position and attitude of the camera in the world coordinate system were determined by the control point theory, then according to the imaging theory of pinhole camera model, the accident scene was projected onto the image, and the forming visual image's feature was extracted, analyzing the visual image features based on image color components, the based on this method, the visual image of traffic accident scene was identified. The method could avoid the interference of traffic accident scene environment effectively, but the method was a low recognition accuracy problem. Wang Dong [8] proposed a method which based on feature matching for traffic accident scene image recognition. this method selected the average value of a certain number of correctly identified traffic accident scene visual as the standard image firstly, then obtaining the key points of visual images of standard traffic accident scene, combining the histogram of gradient direction histogram to extract the feature of the key points in the scene of the traffic accident, by calculating the similarity between the feature vectors of the image to be identified and the corresponding key point of the standard traffic accident scene image, then setting reasonable threshold to identify the visual image of traffic accident scene which was to be detect. This method could remove the scene image of traffic accident effectively, but the method existed poor identification reliability [9, 10].

Upon above problems, this paper proposes a method which is based on intelligent vision to recognize visual image of traffic accident scene. The experiment result shows that this method could improve the visual image recognition rate of traffic accident scene effectively, and has high recognition accuracy.

2 The Visual Image Recognition Method Basing on Intelligent Vision of Traffic Accident Scene

2.1 Establishing the Multi-scale Feature Vector of Visual Image in Traffic Accident Scene

In the process of visual image recognition, firstly, extracting the background image in mean background modeling theory, then normalizing the background image in traffic accident scene, at the same time, the contour image of accident scene is converted into binary image, the image filtering is used by the size of the circle to conduct traffic accident scene, then obtaining the morphological characteristics of different sizes of the scene images, building the feature vector of the sector in traffic accident scene, the detailed process is as follows:

When acquiring the image of traffic accident scene, then extracting the background image which is based on mean background modeling theory, normalizing the center of the traffic accident scene image, computing the centroid of traffic accident image, setting the image as the center coordinate:

$$\begin{cases} x = \sum_{i=1}^{N}\sum_{j=1}^{N} i * f(i,j) / \sum_{i=1}^{N}\sum_{j=1}^{N} f(i,j) \\ y = \sum_{i=1}^{N}\sum_{j=1}^{N} j * f(i,j) / \sum_{i=1}^{N}\sum_{j=1}^{N} f(i,j) \end{cases} \tag{1}$$

As is shown in the formula (1), i, j represents the coordinate of pixel in traffic accident image, $f(i,j)$ represents the pixel value of pixel point (i,j), (x, y) represents the center point coordinate of image in traffic accident scene, N represents the size of visual image in traffic accident scene.

The multi-scale feature of traffic accident scene is extracted by image filtering operation, using the fixed circle for dynamic moving in the scene of traffic accident, the area which isn't traverse is the residual area of image in traffic accident scene, the formula (2) is as follows:

$$S = A - B \tag{2}$$

As is shown the formula (2), A represents the area of identification of image in traffic accident scene, formula (3) represents the area traversed by a fixed size circle in the scene of a traffic accident:

$$B = \cup \underset{r=1,2,\ldots,N}{\arg} \; R_r(x, y)\, and\, R_r(x, y) \in A \tag{3}$$

As is shown the formula (3), $R_r(x, y)$ represents the circle area of the center is (x, y), and r is the radius, the residual area of the scene of the traffic accident describes the characteristic of the different scale circle, which is the image by identified in traffic accident scene, setting $\Delta\theta$ as step, on the sector area of image center of traffic accident scene, the image is divided as $N = 360/\Delta\theta$ in traffic accident scene, the pixel moment inertia of every residue image area is computed by every partition, and it is regarded as the characteristic value of the sector area, which is represented by formula (4):

$$J = \sum m \times w^2 \tag{4}$$

As is shown the formula (4), m is the pixel value of pixel point, w represents the distance between pixel point and the center of sector area, setting the center point of sector is (x, y), which is divided by traffic accident scene, and the point represents the pixel point coordinate of residue area in sector area, the feature vector is built by this feature value, which is represented by formula (5):

$$f = \{J_1, J_2, \ldots, J_N\} \tag{5}$$

It is built feature vector of visual image in traffic accident scene, computing the Euclidean distance between the feature vector and template vector, which is represented as formula (6):

$$d = \sqrt{\sum_{k=1}^{N} (f(k) - m(k))^2} \tag{6}$$

As is shown in formula (6), f and m represents the feature vector and template vector of image in traffic accident scene, $f(k)$ represents the k vector value of image feature vector in traffic accident scene, $m(k)$ represents the k vector value of template vector, through computing the minimum Euclidian distance between the vector and template vector of image feature vector in traffic accident scene, it is based on the number of times n of moving the feature vector, it is acquired direction difference between the identified image of traffic accident and template image, Which is represents by formula (7):

$$\hat{l} = \mathop{\arg\min}_{l=1,2,\ldots,N} d(l) \tag{7}$$

As is shown in formula (7), $d(l)$ represents the direction difference between feature vector and template vector which is built by traffic accident scene. It is direction identification and corrected by camera, and it is used by image multi-scale analysis identification of image in traffic accident scene, multi-scale identification process is the filtering operation of size changing circle of image in traffic accident scene, and building multi-scale feature vector of sector of image in traffic accident scene.

2.2 Visual Image Identification in Traffic Accident Scene

In the process of visual image identification in traffic accident scene, it is based on by 3.1 which is acquired in multi-scale feature vector in accident scene, combing the local identification mapping theory, it is reduced in dimension of multi-scale feature vector of image in traffic accident scene, and it is based on above process, the visual image identification is combined by k-nearest neighbor classifier in traffic accident scene, the special process is as follows:

$$\vartheta = \mathop{\arg\min}_{A^T XX^T A = T} tr\{A^T XLX^T A\} \tag{8}$$

As is shown the formula (8), $A^T XX^T A$ represents the generalized inverse transformation of matrix of image area in traffic accident scene, computing the generalized eigenvalue decomposition problem of visual image in traffic accident scene is as follows formula (9):

$$XLX^T a = \lambda XX^T a \tag{9}$$

As is shown the formula (9), $\lambda XX^T a$ represents the eigenvalues of orthogonal mapping matrix of image in traffic accident scene, λ represents characteristic coefficient, a represents the eigenvectors corresponding to the eigenvalues of the orthogonal mapping matrix, ordering $a_1, a_2, \ldots, a_d$ represents the forward d eigenvectors corresponding to the smallest eigenvalue in traffic accident scene in formula (9), for the object function of visual image in traffic accident scene in formula (8) reaching mapping matrix minimization A, the formula (10) is as follows:

$$A = [a_1, a_2, \ldots, a_d] \tag{10}$$

So one sample x_{new} is mapped by low dimension data y_{new} of image in traffic accident scene, the formula (11) is as follows:

$$y_{new} = A^T x_{new} \tag{11}$$

As is shown the above formula (11), $x_{new} \in R^D$, $y_{new} \in R^d$, $d \ll D$, d represent the low dimension data which is mapped by visual image sample in traffic accident scene, on the basis of the above method, completing the visual image recognition in traffic accident scene.

3 Empirical Simulation and Validation

For validating the identification's effective which is based on the visual image in traffic accident scene, it is simulated with experiment. The experiment data is selected 100 pictures from traffic management bureau of public security department of Guangdong Province, which was recorded in May to October, 2015, 50 pictures is test set, and 50 pictures is train set.

The intelligent visual method and the statistical feature clustering method are used to identify the visual image of traffic accident scene. In the process of obtaining the visual image of traffic accident scene, the observed image sample data will be noisy or image data missing, the representation of noise environment is given: (x_t, y_t) represents the contour coordinates of image in t time in traffic accident scene, $(\tilde{x}_t, \tilde{y}_t)$ represents the value of after noise, ξ_t, η_t represents satisfying the gauss noise of standard normal distribution, reflecting image pollution in traffic accident scene:

$$\begin{cases} \tilde{x}_t = x_t + w\xi_t, \ \xi_t \sim N(0, 1) \\ \tilde{y}_t = y_t + w\eta_t, \ \eta_t \sim N(0, 1) \end{cases} \tag{12}$$

On the basis of formula (12), in the case of noise, 2 methods are used to identify the robustness of the visual image in traffic accident scene (%), the result as follows in Fig. 1.

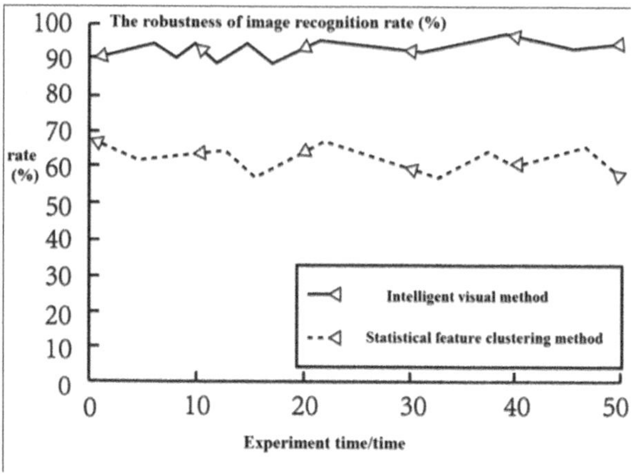

Fig. 1. Comparison of robustness of different methods for image recognition

Through the analysis of Fig. 1, under noisy environment, the robustness of image recognition by using intelligent vision method is better than the statistical feature clustering method, this is mainly because of the intelligent vision method to transform the scene image of the accident scene into binary image, it is obtained the morphological characteristics of different sizes of the scene images, the dimension reduction of image feature vector in traffic accident scene based on local discrimination mapping theory, the visual image recognition of traffic accident scene, the robustness of intelligent visual method for traffic accident scene visual image recognition is better.

The comparison experiment is identified of image recognition between intelligent visual method and statistical feature clustering method of different methods for image recognition. The two methods are used to identify the visual image recognition (s) in traffic accident scene, the comparison result is as Table 1.

Table 1. Time consuming comparison of accident scene visual image recognition between two methods

Method	Morphological characteristics/number	Feature eigenvector dimension/dimension	Time/s
Intelligent visual method	20	150	0.025
Statistical feature clustering method	20	150	0.04

As is analyzed as the Table 1, the use of intelligent visual method for image recognition is shorter than the statistical feature clustering method, this is mainly because of the use of the intelligent visual method to reduce the dimensionality of the image feature vector of the traffic accident scene by using the theory of local differential mapping, this process greatly reduces the time-consuming of the subsequent K-nearest neighbor classifier for visual image recognition, the intelligent visual method to identify the scene of the traffic accident is shorter.

The comparison experiment is identified of image recognition between intelligent visual method and statistical feature clustering method of different methods for image recognition. The two methods are used to identify the degree of similarity SD (%) between the visual image of traffic accident scene and the actual image of traffic accident scene, it is represented by the formula (13):

$$SD = SR - SM \tag{13}$$

As is shown the above formula (13), represents the similarity between the visual image of the acceptable traffic accident scene and the visual image of the actual traffic accident scene, the greater the degree of indemnification, the easier it is to identify the false visual image, the identification and recognition rate (%) of the visual image of the accident scene by the two methods, the comparison result is shown as Fig. 2.

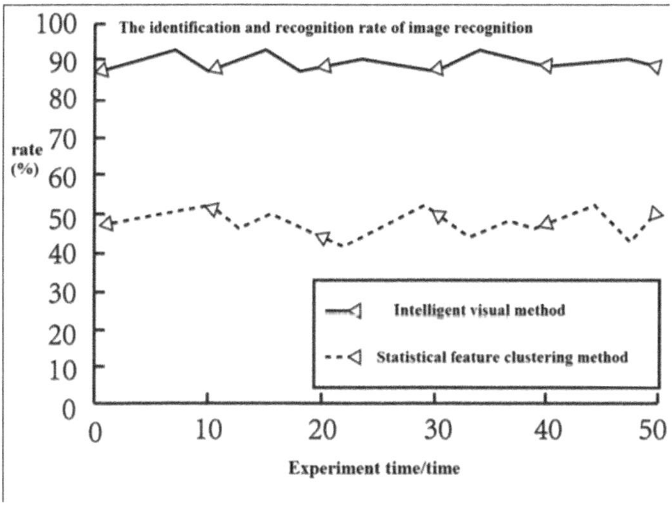

Fig. 2. Comparison of identification and recognition rate of different methods for image recognition

Through the analysis of Fig. 2, the recognition of intelligent vision method is higher than the statistical feature clustering method, this is mainly because the intelligent visual method first uses the mean background modeling theory to extract the scene background image, it is normalized for the traffic accident scene background image, meanwhile, the accident scene contour image is converted in to binary image, building the feature vector of the sector area in traffic accident scene, combining the local discrimination mapping theory, it is dimensionality reduction of visual image feature vector in traffic accident scene, this process greatly reduces the time-consuming of the subsequent K-nearest neighbor classifier for visual image recognition, the visual image recognition is higher of intelligent visual method to identify the traffic accident scene.

4 Conclusions

In summary, in view of the current method of visual image recognition in traffic accident scene, it is difficult to obtain the morphological characteristic of the different sizes of the scene images. This paper presents a method of visual image recognition for traffic accident scene based on intelligent vision, the experimental results show the proposed method can effectively improve the visual image recognition rate in traffic accident scene, and the recognition accuracy is higher than the other methods.

Acknowledgment. This research was supported by Innovation and Entrepreneurship Project in City College of Wuhan University of Science and Technology, Hubei Province, China (Promoting student employment plan through reform of practical teaching system and innovation of science and technology competition (No: 13)).

References

1. Bagué, A.V.: Traffic Accident Data Recorder and Traffic Accident Reproduction System and Method: U.S. Patent 6,246,933, pp. 6–12 (2001)
2. Franke, E.A., Kinkler, E.S., Magee, M.J., et al.: Method and Apparatus for Traffic Incident Detection: U.S. Patent 6,411,328, pp. 6–25 (2002)
3. Hänninen, M.: Bayesian networks for maritime traffic accident prevention: benefits and challenges. Accid. Anal. Prev. **73**, 305–312 (2014)
4. Yao, S.: Vehicle Abnormal Behavior Detection in Traffic Surveillance Video. Nanjing University of Posts and Telecommunications, Nanjing (2015)
5. Wei, Y., Jianqiang, G., Lang, W.: Recognition of vehicle image with multi feature information. J. Chang'an Univ. Nat. Sci. Ed. **36**(4), 79–85 (2016)
6. Tavernier, B.C., Paire-Ficout, L., Charnay, P., et al.: Emotional experiences one year after a traffic accident: an exploratory study of verbatim accounts of the ESPARR cohort. Inj. int. J. Care Injured **48**(3), (2017)
7. Yamada, A., Unuma, K., Kojima, H., et al.: Subarachnoid hemorrhage caused by a traffic accident: de novo aneurysm ruptured 30 years after surgical neck clipping. J. Forensic Leg. Med. **22**, 90–92 (2014)
8. Dong, W., Zhongqi, Z.: Identification of traffic accidents on the internet of things based on visual features. Comput. Simul. **31**(10), 196–199 (2014)
9. Sharma, B., Katiyar, V.K., Kumar, K.: Traffic accident prediction model using support vector machines with Gaussian kernel. In: Proceedings of Fifth International Conference on Soft Computing for Problem Solving, pp. 1–10. Springer, Singapore (2016)
10. Abellán, J., López, G., De OñA, J.: Analysis of traffic accident severity using decision rules via decision trees. Expert Syst. Appl. **40**(15), 6047–6054 (2013)

Research on Effectiveness Evaluation of Integrated Space-Ground Information System Oriented Aircraft Carrier Strike

Ren Wei[1], Zhang An[2(✉)], Zhou Ding[2], and Bi Wenhao[2]

[1] School of Electronics and Information, Northwestern Polytechnical University, Xi'an 710129, China
[2] School of Aeronautics, Northwestern Polytechnical University, Xi'an 710072, China
zhangan@nwpu.edu.cn

Abstract. The integration of Space-Ground Information Systems is centered on information center, and links a variety of the ground-based, sea-based, air-based and space-based communication and information platform, and processes and transmits information efficiently with a high speed which makes a great change to battlefield pattern. On the basis of the constructing principles of the evaluation index system and the framework of the Space-Ground Information System, this paper constructs the effectiveness evaluation index system of Aircraft carrier strike task oriented Space-Ground Information System, models the effectiveness evaluation index, and makes a optimization of Integrated Space-Ground Information System. The index weight is analyzed by the fuzzy analytic hierarchy process and the entropy method to get the effectiveness value.

Keywords: Integrated Space-Ground Information System · System integration
Combat effectiveness · Integrated weighting method

1 Introduction

With the change of war form, the world military powers have built various military information systems of different uses. It becomes crucial for our army to build the Integrated Space-Ground Information System that can realize the full use of information. In combat, a complex mission is decomposed into multiple task been executed. Even if the effectiveness index is the same, the performance index will be different when the task is not at the same time; Even if the performance index is the same, the performance index may also be chosen in different ranges when we deal with different tasks. The index system should be discussed according to different operational missions and different operational tasks. In this paper, the research on effectiveness evaluation method of Integrated Space-Ground System Oriented Aircraft Carrier Strike has not only plays an important role in the study of Integrated Space-Ground System application model, but also instructs the construction and realization of the Integrated Space-Ground Information System. Using the integrated Air-Space-Ground information network, we can realize a integrated intelligent information system which has the

K. Deng et al. (Eds.): ICMIR 2018, AISC 856, pp. 1069–1078, 2019.
https://doi.org/10.1007/978-3-030-00214-5_131

ability of manipulating the information in intelligent processing, high-speed transmission and fast routing [1, 2]. The results of this paper can provide reference for the application mode and optimization of the Integrated Space-Ground Information System.

2 Research on Effectiveness Evaluation of the Integrated Space-Ground Information System Oriented Aircraft Carrier Strike

The efficient transmission and utilization of information makes the combat system present a flat network structure. Index system on effectiveness evaluation of integrated Space-Ground system oriented aircraft carrier strike can still be described in vertical tree structure. Operational effectiveness index includes two parts: system effectiveness index and application effectiveness index. System effectiveness, operational effectiveness and application effectiveness have the different effectiveness index.

2.1 Effectiveness Evaluation Index System Model

(1) Information Reconnaissance Capability
(1) Global Coverage Ratio

When we calculate single satellite coverage ratio, we can assume that a satellite is illuminated by a beam pointing to the center of earth, and the covering area is circular [3]. The single satellite coverage area diagram is shown in Fig. 1. Among them, α is visual elevation angle and β is the coverage area angle. Coverage area geocentric angle β is

$$\beta = 90^\circ - \alpha - \arcsin\frac{R_e \cos\alpha}{R_e + H} \tag{1}$$

Among them, H represents the satellite height and R_e represents the earth radius respectively. Therefore, the distance between the edge of the coverage area and the satellite $D = \sqrt{R_e^2 + R^2 - 2R_eR\cos\beta}$, the coverage area of earth surface is

$$S_i = 2\pi R_e^2(1 - \cos\beta) = 4\pi R_e^2 \sin^2\frac{\beta}{2} \tag{2}$$

Global coverage ratio C is

$$C = \frac{\sum_{i-1}^{n} S_i}{S} \tag{3}$$

Among them, S represents the total earth area and n represents the total satellite number respectively. For the key areas, the degree of coverage should be guaranteed to ensure the comprehensiveness of satellite detection. If a region is covered by multiple layers, the region only calculates the coverage area of one of the satellites.

(2) Detection Probability

The Integrated Space-Ground Information System detects targets mainly through a variety of Space-Based reconnaissance satellite constellations. The imaging reconnaissance satellite can calculate the confidence level of the computational identification results through the combination of satellite imaging characteristics and target geometric features, as well as target recognition and imaging resolution. The detection probability is affected by satellite orbit height, solar altitude angle, cloud and other factors. In the Integrated Space-Ground Information system, the detection probability of the i th satellite in m_1 imaging reconnaissance satellites is

$$P_{cai} = \rho_E \rho_\theta \rho_c \left[1 - (1-g)^{\frac{T_{target}}{G_{\max}}}\right] \tag{4}$$

Among them, ρ_c represents meteorological factors, ρ_θ represents solar altitude angle factors, ρ_E represents other comprehensive factors, T_{target} represents activity time of the target in the detection area, $G_{\max}$ represents the maximum revisit time of satellite and g represents the detection probability of each transit respectively.

The sensitivity of the electronic reconnaissance satellite and the frequency coverage of the satellite have a significant influence on its recognition ability. Here we assume that the working frequency range of the target radar can all be received by the electronic reconnaissance satellite. Through the i th frequency range, the detected probability of the target P_{ti} is as follows:

$$\begin{aligned} P_{ti} = & \frac{1}{2}\left(1 - erf\left(\sqrt{\beta} - \sqrt{SN}\right)\right) + \frac{\exp(-(\sqrt{\beta} - \sqrt{SN})^2)}{4\pi\sqrt{SN}} \\ & \cdot \left|\frac{3}{4} - \frac{\sqrt{\beta}}{4\sqrt{SN}} + \frac{1 + (\sqrt{2\beta} - \sqrt{2SN})^2}{16\sqrt{SN}}\right| \end{aligned} \tag{5}$$

$$erf(x) = \frac{2}{\sqrt{\pi}} \int_0^\pi \exp(-t^2)dt \approx \sqrt{1 - \exp\left(-\frac{4x^2}{\pi}\right)} \tag{6}$$

Among them, SN represents the signal to noise ratio of radiation source $\beta = -\ln(P_{fa})$. P_{fa} represents the false alarm probability of satellite payload. Formula (6) is the error function of the formula (5). Through the full frequency range of Space-Borne Radar of the i th satellite in m_2 electronic reconnaissance satellites, (n radar frequency range), the target detection probability is

$$P_{dai} = 1 - \prod_{i=1}^{n} (1 - P_{ti}) \tag{7}$$

The detection probability of the Integrated Space-Ground Information System is

$$P_t = 1 - \left[\prod_{i=1}^{m_1}(1 - P_{cai})\right]\left[\prod_{i=1}^{m_2}(1 - P_{dai})\right] \tag{8}$$

Among them, m_1 and m_2 represents the number of imaging the reconnaissance satellite and electronic reconnaissance satellite respectively.

(2) Navigation and Location Capability

The configuration of the navigation and location constellation determines its coverage characteristics and other performance. In the service area, the usability of satellite/constellation is guaranteed by ensuring the satellite coverage and certain Position Dilution of Precision (PDOP).

The coverage range of space signal refers to the coverage of a single satellite that ensure the receivers can receive the minimum power under a given cut-off height angle

Table 1. Coverage rates of different satellites at high cut-off angles

Satellite type	Orbit height	Coverage rate of different height angles (maximum angle and area ratio)				
		0°	5°	10°	15°	20°
GEO/IGSO	35786 km	17.38°	17.31°	17.11°	16.79°	16.33°
		42.44%	38.19%	34.08%	30.15%	26.40%
MEO	21528 km	26.40°	26.30°	25.99°	25.49°	24.78°
		38.58%	34.43%	30.47%	26.72%	23.22%

[4]. According to the relevant literature, coverage rates of different satellites at different heights are illustrated in Table 1.

(3) Information Security and Protection Capability
(1) Anti-Destruction Capability

Anti-destruction capability refers to the ability to maintain the network structure or restore its function when the system is attacked. The transmission efficiency of information make system as a measure of the anti-destruction capability. Global efficiency is defined as

$$R = \frac{1}{N(N-1)}\sum_{i \neq j}\frac{1}{d_{ij}} \tag{9}$$

Among them, N represents the total nodes number in topology network of the Integrated Space-Ground Information System and d_{ij} represents the length of the shortest path between any two nodes in the network, respectively. When the topological structure of the space and earth integrated information system is been destroyed, the degree of each node represents the number of other nodes connected with it.

Random attack and selective attack can be used to attack the network nodes of the Integrated Space-Ground Information System [5]. Global efficiency R is the reciprocal of the average path length between the nodes of the topology network of the Integrated Space-Ground Information System.

(2) Anti-Network attack capability

The anti-network attack of the Integrated Space-Ground Information System includes four sub-indexes, anti-information acquisition attack capability, anti-privilege attack capability, anti-integrity destruction attack capability and anti-availability destroy attack capability.

The indexes of anti-network attack include the probability of acquiring the accurate IP address with its amount of information and the required time, the represent the network performance of the Integrated Space-Ground Information System. So we use the global efficiency of the Integrated Space-Ground Information probability of acquiring the accurate service with its amount of information and the required time and the probability of acquiring the accurate account password with its amount of information and the required time. The indexes of anti-privilege attack capability include the required time for the operating system to enhance authority, the required time for the data base to enhance authority, hidden level, firewall penetration rate, and intrusion detection evasion rate. The indexes of anti-integrity destruction attack capability include the required to steal files, delete files and change files, change the amount of software, hardware, other important information as well as the time required to change the software. The indexes of anti-availability destroy attack capability include network availability, network latency, network throughput, average network response time, network recovery time, host CPU occupancy rate, host service response latency and host recovery time.

For qualitative index, it is difficult to measure with accurate mathematical models, so the gray method is used to measure these qualitative indexes. For quantitative index, include both positive and negative indexes. When the index is dimensionless and normalized, make sure the index value are in the range of $[0, 1]$. And in the normalization process, the negative index is transformed into the positive index to be processed.

(4) Grid Service Quality

The stability of grid service is not only related to the stability of the two infrastructures of grid node software and hardware, but also related to many external factors, such the attack of the exotic weapons in a combat. Therefore, the grid service stability index is a random variable and random variables can be evaluated by means of exponential probability density distribution function. So the stability of grid service S_G can be expressed as

$$S_G = 1 - e^{-\lambda x} \tag{10}$$

Among them, λ can be obtained from statistical data and x represents the number of grid nodes. The effectiveness evaluation process is as Fig. 2.

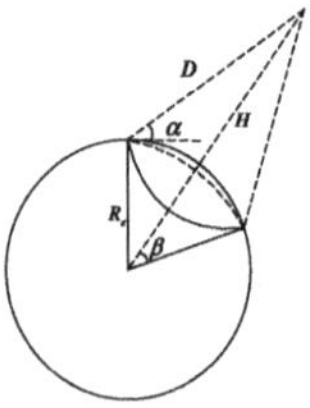

Fig. 1. Sketch map of single satellite coverage area

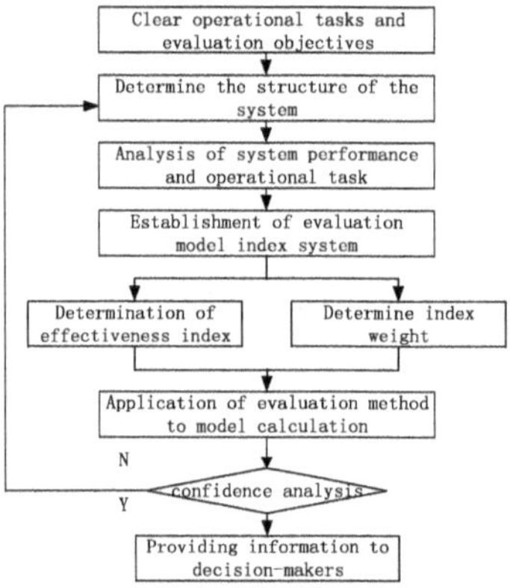

Fig. 2. Effectiveness evaluation process

2.2 Application Effectiveness Index Evaluation Model

(1) Target Discovery Probability

Assume that, in the Integrated Space-Ground Information System, multiple reconnaissance satellites arrive at the detection area at the time point $t_1, t_2, \ldots, t_n, t_{n+1} \ldots$. The location accuracy of the satellites are $\varphi_1, \varphi_2, \ldots, \varphi_n, \varphi_{n+1} \cdots$ respectively. Suppose the aircraft carrier is the center of the circle, ε is the radius. The satellite group in this particular area can detect the target completely. In the range of $(\varepsilon, \lambda]$, the activities of aircraft carriers obey the normal distribution with λ as a one-sided confidence limit. But in the range of $(\lambda, +\infty]$, the aircraft cannot be detected. Therefore exist the threshold value $T_1 = \frac{\varepsilon}{v_{\max}}, T_2 = \frac{\lambda}{v_{\max}}$. According to the actual situation of the target been detected by the satellites [6].

$$\Delta t \cdot v_{\max} > |\varphi_i, \varphi_j| \tag{11}$$

Among them, Δt is the time interval between i and j satellites that have passed a certain-target $\Delta t = t_{n+1} - tWinEnd_n$, and $v_{\max}$ represents the maximum speed of the aircraft carrier.

(2) Penetration Probability

Data transmission efficiency includes data transmission rate V_d and data format conversion efficiency E_t. According to the relevant data [7], when the transmission format is voice, low speed data, formatted message, the transmission speed is $(2.4\ \text{kB/s}, 9.6\ \text{kB/s})$; when the transmission format is high speed data, the transmission speed is $[9.6\ \text{kB/s}, 8\ \text{mB/s})$; when the transmission format is dynamic image, the transmission speed is [8 mB/s, 30 mB/s); and when the transmission format is virtual reality, the transmission speed is [30 mB/s, + ∞), V_d has the membership function.

$$R_V = \begin{pmatrix} 0.3 & V_d \in (2.4\ \mathrm{kB/s}, 9.6\ \mathrm{kB/s}) \\ 0.5 & V_d \in [9.6\ \mathrm{kB/s}, 8\ \mathrm{mB/s}) \\ 0.8 & V_d \in [8\ \mathrm{mB/s}, 30\ \mathrm{mB/s}) \\ 1 & V_d \in [30\ \mathrm{mB/s}, +\infty) \end{pmatrix} \tag{12}$$

$$R_E = \begin{pmatrix} 1 & x \in (0, 3) \\ 0.8 & x \in [3, 4) \\ 0.7 & x \in [5, 8) \\ 0.3 & x \in [8, +\infty) \end{pmatrix} \tag{13}$$

Because the maximum transmission unit of the link is different, assume x as the conversion number of the maximum transmission unit. Thus E_t has the membership function (13).

(3) Target Attack probability

Use the capture probability of the target to measure the probability of hitting the target. Assume that the missile motion obeys the Gauss distribution $N(0, \sigma^2)$ and the capture probability after the missile enters the radar detection area of the Integrated Space-Ground Information System is 1. Then one hit target capture probability P_{B1} is

$$P_{B1} = \frac{1}{\sqrt{2\pi}\sigma} \int_{-\varpi}^{\varpi} \exp\left(-\frac{x^2}{2\sigma^2}\right) dx \tag{14}$$

Among them, σ is the longitudinal distribution of missile and ϖ is longitudinal capture width of radar. Under normal circumstances, in the three scans, if target is captured twice, it recognized the target is captured. Therefore, the capture probability of the targeted aircraft carrier can be calculated by the formula

$$P_B = 1 - C_3^3(1 - P_{B1})^3 - C_3^2(1 - P_{B1})^2 P_{B1} \tag{15}$$

2.3 Evaluation Method

Fuzzy AHP method is too subjective and the entropy method relies too much on the limitations of sample. Comprehensive weighting method can not only weaken the dependence of fuzzy analytic hierarchy process on subjective factors, but also weaken the dependence of entropy weighting method on the number of sample. The specific step of Fuzzy AHP and entropy weighting method refer to the [8, 9].

3 Simulation Analysis

The battle scenario is as follows: The blue side using the missile group under the Integrated Space-Ground Information System as strike weapon to strike the American aircraft carriers which is cruising on the Pacific on the red side. The Integrated Space-Ground Information System includes space-based satellite network and positioning radar network. In the space-based satellite network the imaging reconnaissance satellite constellation and the electronic reconnaissance satellite constellation are all 4 networks, and the positioning radar network contains 5 radars. There are N = 12 important nodes. The topological map is shown in Fig. 3. Select some mobile communication network nodes as the research object. The topological map of each mobile communication network node is shown in Fig. 4.

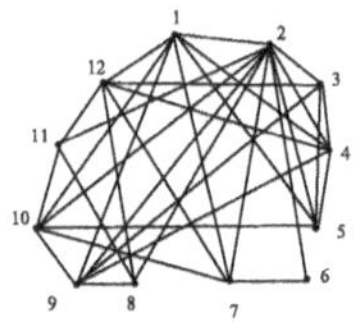

Fig. 3. Topological map of important nodes in the Integrated Space-Ground Information System

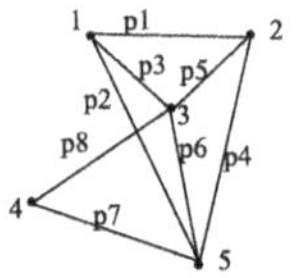

Fig. 4. Topology map of node in mobile communication network

Do the simulation under the Matlab environment and input 4 sets of parameters to the model. Due to the fact that the Integrated Space-Ground Information System has not been deployed, its global coverage varies with the constellation's regression cycle. Therefore, according to the global coverage rate of the relevant satellite constellation abroad, it is assumed that the deployment of the U. S. aircraft carrier deployed on the Pacific Ocean is in the range of $65^{\circ}44'N \sim 85^{\circ}33'S,\ 78^{\circ}08'W \sim 99^{\circ}10'E$. Assume that the Integrated Space-Ground Information System can detect the carrier's target by intelligence reconnaissance from the space-based satellite network. Although the country has already have the Integrated Space-Ground Information System to detect the related deployment of satellite constellation, however, some satellite parameters such as satellite deployment orbit height, deployment number and satellite coverage capacity parameters are still confidential information. Therefore, this paper use the American system *SBIRS*, the global coverage of middle and high orbit constellations as the global coverage of intelligence reconnaissance in the Integrated Space-Ground Information System. In the Integrated Space-Ground Information System, assuming reconnaissance satellites and electronic reconnaissance satellites are all 4 networks. The radar net accuracy of locating the target aircraft carrier is (0.01 km, 2°, 2°). The real location of the aircraft carrier is (40, 30, 30). The positioning radar network consists of 5 radars, and the radar deployment position is (100, 0, 10000), (300, 10, 30000), (−50, 0, 0)(20, 0, 0) and (−100, 0, 0). Assume radius range that the satellite group can completely detect the aircraft carrier target is $\varepsilon = 2$ km. One side confidence limits of Gauss distribution is $\lambda = 20$ km. The maximum velocity of the aircraft carrier is $v_{\max} = 50$ km/h. Suppose the time window of the target aircraft is observed by the satellite group is Win = $\begin{bmatrix} 09:23:33.043 \\ 09:30:37.756 \end{bmatrix} \begin{bmatrix} 08:34:19.798 \\ 08:42:32.868 \end{bmatrix}$ and $\begin{bmatrix} 07:45:05.762 \\ 07:53:47.093 \end{bmatrix}$. Missile is the preferred weapon for aircraft carrier attack. The cruise missile launched from land base and ballistic missile launched from land-based or naval vessels consist of the missile group to attack the aircraft carrier. When the missile is engaged in the combat mission, the related index of the Integrated Space-Ground Information System that determines its operational capability is the data transmission capacity. Assume that the data rate in the data transmission capability is $V_d = 20$ mB/s; the conversion number of the maximum transmission unit is 5 times; the longitudinal distribution of missile is 10 m, and the longitudinal capture width of radar is 10 m. There exist many network nodes in the Integrated Space-Ground Information System, so here $N = 12$ important nodes are selected to study the anti-destruction capability of the network composed of them. Because there are many nodes in mobile communication network in the Integrated Space-Ground Information System, this paper selects some mobile communication network nodes as the research object. When no nodes are damaged, the ability of nodes to be secured by mobile communication networks can be represented by a value $C_{NA} = \beta_1 A_C \beta_1^T$. Assume the damage probability of nodes and links are all 0.01. The signal to noise ratio *SN* of the radiation source of the electronic reconnaissance satellite is 15 dB, 20 dB, 18 dB, 12 dB, respectively. Load false alarm probability P_{fa} is taken 10^{-6}. Each index value of B layer is obtained as shown in Table 2.

Table 2. Index effectiveness evaluation value of layer B

Index	Index value			
B1	0.9759	0.6493	0.8519	0.6519
B2	0.8269	0.7652	0.6753	0.6823
B3	0.8637	0.8953	0.7970	0.9005
B4	0.8637	0.8253	0.4721	0.6852
B5	0.6999	0.3562	0.6875	0.7658
B6	0.1576	0.1575	0.0535	0.1735
B7	0.7734	0.8965	0.5453	0.8777
B8	0.6895	0.9578	0.9871	0.9732
B9	0.8503	0.8665	0.8775	0.8956

According to the fuzzy AHP method, each index weigh in B layer can be obtained. $W_B = (0.1715,\ 0.0846,\ 0.1416,\ 0.0535,\ 0.08, 21,\ 0.1643,\ 0.1357,\ 0.0461,\ 0.1206)$. Each index weight in B layer can be obtained by entropy weight method is $W_{B'} = (0.1072,\ 0.1073,\ 0.1073,\ 0.1109,\ 0.1103, 0.1141,\ 0.1195,\ 0.1072,\ 0.1072)$. The comprehensive weight of performance index layer index $W = 0.6541 \times W_B + 0.3459 \times W_{B'} = (0.1492,\ 0.0924,\ 0.1297,\ 0.0765,\ 0.0919,\ 0.1469,\ 0.1301,\ 0.0672, 0.1159)$, and the final operational effectiveness was 0.7332, 0.6841, 0.639 and 0.7088 respectively.

The results show that the model is basically reliable, which can provide a quantitative analysis method of operational effectiveness evaluation of the Integrated Space-Ground Information System for aircraft carrier combat mission.

4 Concluding Remarks

Based on the introduction of the framework of the Integrated Space-Ground Information System, the paper constructs the evaluation index system of the operational effectiveness of the Integrated Space-Ground Information System for the aircraft carrier attack task, and puts forward the model of each effectiveness evaluation index. Finally, the fuzzy AHP and the entropy weight method are used to analyze the weights, and the operational effectiveness values are obtained. The results of this paper can provide reference for the application mode and optimization of the Integrated Space-Ground Information System. This work is supported by National Nature Science Foundation under Grant 61573283.

References

1. Jiang, H., Liu, X., Hu, Y., Ding, Y., Fu, Q., Zhu, Y., Wang, L., Zhao, Y., Wang, T.: Thoughts on several key issues of the integrated space-ground information network. Acta Armamentarii **S1**, 96–100 (2014)
2. Min, S.: Conception of the integrated space-ground information network in China. Satell. Appl. **01**, 27–37 (2016)
3. Hu, L., Li, H., Yan, S., Zhang, L.: Construction of evaluation index system for missile early warning satellite detection effectiveness. J. Air Force Early Warn Col.l **02**, 97–99+107 (2014)
4. Hu, Z.: Performance Evaluation Theory and Experimental Verification of Beidou Satellite Navigation System. Wuhan University, Wuhan (2013)
5. Li, X.: Research on comprehensive effectiveness and invulnerability evaluation of command information system based on SOA. Ship Research Institute of China (2015)
6. Wang, Y., Wang, X., Yue, L., Mo, Q.: Research on joint detection capability of multiple satellites to ship targets. Comput. Simul. **07**, 48–51 (2010)
7. Youngjie, Q., Liang, S., Xingbo, C., Wei, H.: Effectiveness evaluation of operational activity of missile early warning system. Fire Control and Command Control **06**, 120–123 (2011)
8. Guo, W., Shufang, Z., Wang, Z.: A method to evaluate radar effectiveness based on fuzzy analytic hierarchy process. In: Control and Decision Conference, pp. 1920–1924 (2008)
9. Ma, H., Wei, S., Qin, G.: Research on Determining the Index Weight of DS/FH TT&C Anti-jamming Effectiveness Evaluation Based on Entropy Method. Atlantis Press, Amsterdam (2015)

View-Based Collaborative Pattern for Spacecraft Assembly Process

Changyu Chen[1(✉)], Qian He[2], Bin Zhang[1], and Wenxing He[1]

[1] Beijing Institute of Satellite Environment and Engineering, Beijing, China
chenchangyu_cast@163.com
[2] China Tianchen Engineering Corporation, Tianjin, China

Abstract. Spacecraft assembly consists of a series of collaborative tasks of high concentration and overlap of resources and constantly updated information. In order to accomplish tasks according to quality and quantity, personnel need to rely on a large amount of awareness information in order to fully and continuously grasp the assembly process situation. For different requirements of spacecraft designer, assembly planner, assembly worker, three configuration information views are defined: view of spacecraft design configuration (ADC), view of assembly plan configuration (APC), and view of assembly implementation configuration (AIC). Through the comparisons of different configuration view, a collaborative pattern of spacecraft assembly is proposed, which makes changes and situation of spacecraft assembly configuration easier to perceive.

Keywords: Spacecraft assembly · Configuration management
Collaborative awareness · Computer supported cooperative work

1 Introduction

Spacecraft assembly accounts for 30%–50% of the total development time for research and development of a spacecraft. It plays a decisive role in the success or failure of the product [1, 2]. Assembly process consists of a series of collaborative tasks of high concentration and overlap of resources and information, involving roles of assembly process planning and simulation, equipment design, assembly testing, etc. Typical process includes instrument assembly, cable assembly, pipe assembly, thermal control component assembly, product lifting, product transport and transportation operations. These tasks have the characteristics of large number, suddenness [3] and urgency, multi-professional coordination, complex task assignment interface, high degree of parallelism, and strict quality requirements. In order to accomplish tasks according to quality and quantity, personnel need to rely on a large amount of awareness information in order to fully and continuously grasp the assembly process situation.

In spacecraft assembly process, the most important awareness for all the roles is configuration information, which defined as the product's functional characteristics and physical characteristics. Configuration management is a management activity for specifying and maintaining the functional characteristics and physical characteristics of products during the product life cycle. Technology state management has been widely applied in large-scale engineering management and has formed a series of theories and

K. Deng et al. (Eds.): ICMIR 2018, AISC 856, pp. 1079–1084, 2019.
https://doi.org/10.1007/978-3-030-00214-5_132

methods [4]. Configuration management is an important means to ensure the quality of final assembly of spacecraft [5, 6].

The current state of configuration management in assembly process relies mainly on the baseline management method. This method divides the assembly process of spacecraft into several phases and sets the state of the products that should be possessed during the arrival phase; when implementation of the assembly begins, coordinated operation of the chain member units and the assembly and implementation units ensures that the spacecraft can reach the state goals defined in the corresponding phases at the scheduled time.

In every phase (Fig. 1), spacecraft configuration control is completed by the collaboration of design unit, assembly process plan unit and assembly implementation unit. The design unit is responsible for editing technical route and requirements of assembly for each phase. Most documents are distributing in type of unstructured; the assembly process planner unit reads and fully understands this type of documents and combines the specific technical requirements of each phase. The technical route is refined to form an assembly process including three-layer structure: process volumes, processes, operation steps. The assembly process planner should promised, when the assembly process is applied to the process corresponding to the stage baseline, the configuration of the spacecraft can fully comply with the technology requirements released by the design unit [5]. Due to the strong randomness of the spacecraft development, the actual configuration parameters often deviate from the expectations of spacecraft designers and assembly process planners, and the revision of the configuration requirements cannot be avoided essentially. Therefore, during the entire assembly process, spacecraft designers, assembly process planners, and assembly workers need to know and follow the control situation of the technical state of each other, so as to achieve the effective control of the state of the assembly.

In order to support the collaboration of different roles, this paper propose a view-based collaboration method. Following chapter expresses the concept of views corresponding to spacecraft designer, assembly planner and assembly worker. Then, a new collaborative pattern based on view comparisons is introduced.

2 Views of Spacecraft Assembly Configuration

By deriving the basic concept of configuration, spacecraft assembly configuration can be defined as the function and physical characteristics of all the parts and related assembly process equipment in the assembly stage of a single spacecraft product. The parts include structures, instruments, cables, pipes and so on. The equipment includes the types of parking, hoisting, transporting, testing and so on. The configuration management of spacecraft assembly is to control all the variable functions and physical attribute parameters of the above objects.

Three types of spacecraft assembly configuration view are defined: ADC (view of Assembly Design Configuration), APC (view of Assembly Plan Configuration), AIC (view of Assembly Implementation Configuration).

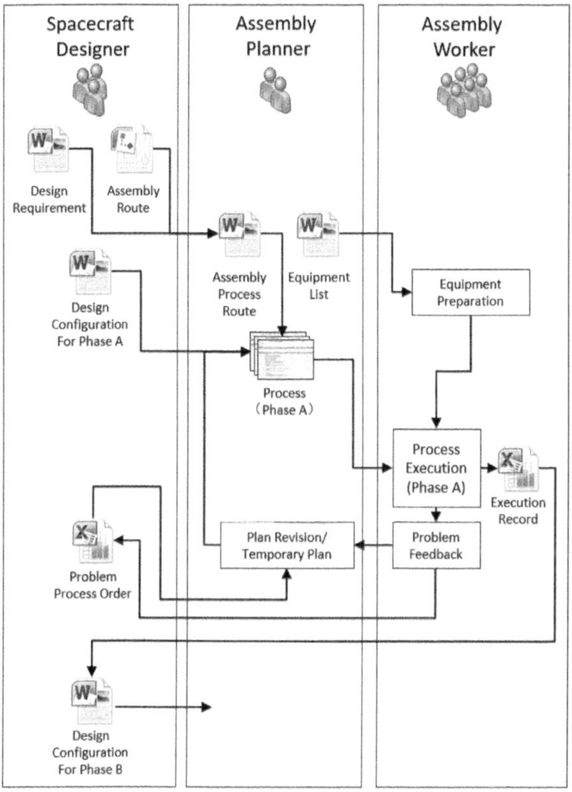

Fig. 1. Current pattern of configuration information generation and modification.

2.1 View of Assembly Design Configuration (ADC)

ADC is proposed for spacecraft designer. The product designer is responsible for making the requirements of spacecraft configuration during the development phase. By monitoring the problems arising in the assembly process, the designer should judge actual accessibility of the configuration requirements in current phase and make the revision and release of the requirements in time.

ADC includes the configuration requirements of spacecraft designers for all parts of the product at each stage.

2.2 View of Assembly Plan Configuration (APC)

APC is proposed for assembly planner. The assembly planner is responsible for the assembly process plan according to the phase configuration requirements defined by the spacecraft designer. By monitoring the problems arising in the process of the process, the planner should modify process plan in time to ensure that the implementation results conform to the requirements of the phase configuration.

APC includes parts and equipment configuration requirements defined by the process planner. All the requirements are indexed by three-layer structured assembly process document. When an APC corresponding to specific process is looked up, it adds up all the requirements before this process and tells what spacecraft looks like when the process is done.

2.3 View of Assembly Implementation Configuration (AIC)

AIC is proposed for assembly worker. The assembly workers are responsible for the implementation of the assembly according to the assembly process defined by the assembly planner. They record actual configuration parameters of the spacecraft and feedback the problems in real time.

AIC includes configuration parameters of all parts and related equipment at any time when assembly workers are fed back at any time.

2.4 Relationship Among Views

Figure 2 illustrates the relationship among the three views with the change of a configuration parameter in different assembly process. The ADC point is the requirements of spacecraft configuration in each phase. The points of APC are on the path of realize points of ADC. The points of AIC are values of configuration parameter in real processes. Their relationship can be explained by three kinds of event as following:

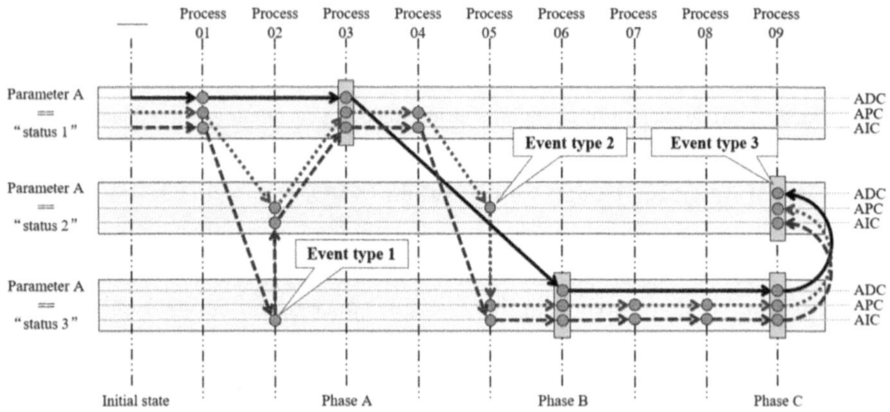

Fig. 2. Demo example of ADC, APC, AIC.

Event type 1: when the AIC point deviates from corresponded APC point (for example, the position interference occurs during the installation process.), the assemble planner takes advantage of the means of adding some temporary process (for example: adding extra repair) to make AIC point detour to the APC point.

Event type 2: if the problem is not solved through the plan revision (for example, the equipment planned to be installed cannot be supplied in time). the APC point needs to be revised to match the corresponded AIC point.

Event type 3: if there is a modification of ADC point, APC and AIC point should be modified in correspondence.

It is easy to realize six kinds of comparisons between different views, which can effectively support awareness of different role involved in quick-change assembly process: ADC-ADC, ADC-APC, APC-APC, ADC-AIC, APC-AIC, AIC-AIC. These comparisons concrete a new collaborative pattern for spacecraft assembly as next chapter proposed.

3 View-Based Collaborative Pattern

A view-based collaborative pattern is proposed as Fig. 3 shown. It consists of three modules (ADC definition, APC definition, and AIC process), which are responsible for spacecraft designer, assembly planner, and assembly worker. Benefitted by comparisons between different views, it is easy for all roles to aware of changes and situation of spacecraft assembly configuration in time. And a kind of collaboration can be concrete to promote assembly efficiency.

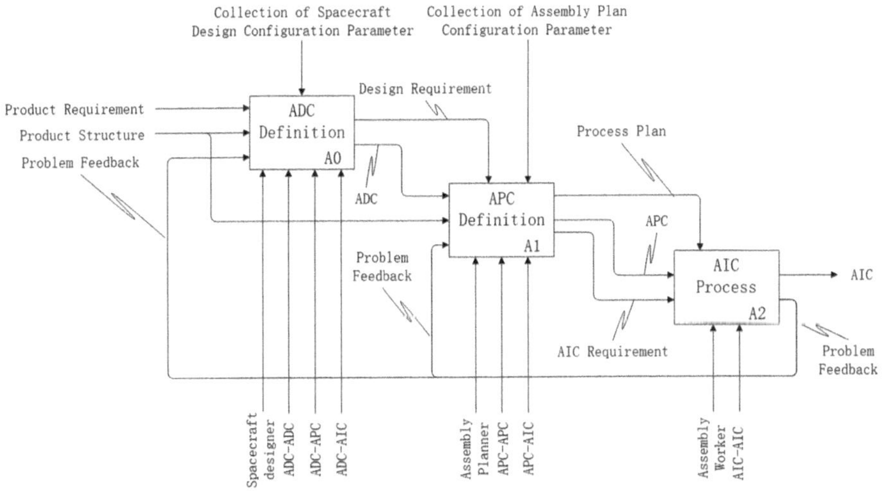

Fig. 3. View-based collaborative pattern of spacecraft assembly

3.1 ADC Definition

According to product requirements, product structure, and feedback from the assembly implementation unit, the spacecraft designer puts forward the ADC (requirements for the design of the assembly) definition. in this stage, ADC-ADC is used to clarify the progress plan of the previous stage, and to trace the change of the state of the design elements in each phase; ADC-APC is used to coordinate the correctness and completeness of the process with the assembly planer; ADC-AIC is used to analyze the on-site problems of the feedback from assembly workers.

3.2 APC Definition

According to ADC definition of each phase, assembly planner proposes process route and divide the process execution into process volumes, processes, and operation steps. Synchronously, APC (requirements for the assembly process execution) of each process is defined. In this stage, APC-APC is used to analyze and understand operations between different part of plan; APC-AIC is used to analyze the on-site problems of the feedback from assembly workers.

3.3 AIC Process

According to APC definition of each process, assembly worker executes process and concretes AIC (record of configuration parameter in reality) synchronously. When problem happens, it will be feedback to assembly planner or spacecraft designer, who amend the corresponding ADC or APC. In this stage, AIC-AIC is used to trace progress of the assembly process.

4 Conclusion

This paper analyzed requirements for configuration control of spacecraft assembly and proposed three kinds of spacecraft assembly configuration views: ADC, APC, and AIC. Based on these views, this paper proposed a collaborative pattern for spacecraft assembly. Benefit of comparisons between ADC, APC, AIC, changes and situation of spacecraft assembly configuration is easier to perceive.

References

1. Liu, G., Xiang, S., Yi, W., et al.: To carry forward the future, pioneering and innovative, and strive to create a world-class spacecraft AIT center. Spacecr. Environ. Eng. **02**, 135–146 (2015)
2. Fischer, P.M., Lüdtke, D., Lange, C., et al.: Im-plementing model-based system engineering for the whole lifecycle of a spacecraft. Ceas Space J. **9**(3), 351–365 (2017)
3. Wang, X., Liu, M., Ge, M., et al.: Research on assembly quality adaptive control system for complex mechanical products assembly process under uncertainty. Comput. Ind. **74**(9), 43–57 (2015)
4. Lan, W., Wei, Y., Ning, Z.: Application of spacecraft configuration baseline and file. Spacecr. Eng. **01**, 133–139 (2015)
5. Xin, G.: Spacecraft assembly process quality control method. Spacecr. Environ. Eng. **03**, 332–336 (2014)
6. Xue, H., Zheng, X.: Independently developed systems engineering of China's aerospace industry. In: Incose International Symposium, vol. 26, no. 1, pp. 1139–1388 (2016)

Multi-view Representation Learning via Canonical Correlation Analysis for Dysarthric Speech Recognition

Myungjong Kim[1], Beiming Cao[1], and Jun Wang[1,2(✉)]

[1] Speech Disorders and Technology Lab, Department of Bioengineering, University of Texas at Dallas, Richardson, TX, USA
wangjun@utdallas.edu

[2] Callier Center for Communication Disorders, University of Texas at Dallas, Richardson, TX, USA

Abstract. Although automatic speech recognition (ASR) has been commercially used for general public, it still does not perform sufficiently well for people with speech disorders (e.g., dysarthria). Multimodal ASR, which involves multiple sources of signals, has recently shown potential to improve the performance of dysarthric speech recognition. When multiple views (sources) of data (e.g., acoustic and articulatory) are available for training while only one view (e.g., acoustic) is available for testing, a better representation can be learned by simultaneously analyzing multiple sources of data. Although multi-view representation learning has recently used in normal speech recognition, it has rarely been studied in dysarthric speech recognition. In this paper, we investigate the effectiveness of multi-view representation learning via canonical correlation analysis (CCA) for dysarthric speech recognition. A representation of acoustic data is learned using CCA from the multi-view data (acoustic and articulatory). The articulatory data was simultaneously recorded with acoustic data using electromagnetic articulograph. Experimental evaluation on a database collected from nine patients with dysarthria due to Lou Gehrig's disease demonstrated the effectiveness of the multi-view representation learning via CCA on deep neural network-based speech recognition systems.

Keywords: Dysarthria · Canonical correlation analysis · Speech recognition
Deep neuron network · Multi-view representation learning

1 Introduction

Dysarthria is a motor speech disorder that caused by neurological injury of the motor speech system, which affects speech production and swallowing [1]. Dysarthric speech is generally characterized by breathy voice, poor articulation, and monotonic intonation [1]; thus, there is typically a reduced speech intelligibility in proportion to the severity of dysarthria. As advances in automatic speech recognition (ASR) techniques, it can be used in human-robot interface for the general public. However, current ASR systems are not well-suited to dysarthric speech, which limits the use of ASR for people with speech and voice disorders.

K. Deng et al. (Eds.): ICMIR 2018, AISC 856, pp. 1085–1095, 2019.
https://doi.org/10.1007/978-3-030-00214-5_133

Related work on the recognition of dysarthric speech has mostly focused on acoustic modeling to capture the acoustic cues of dysarthric speech. A variety of acoustic models such as Gaussian mixture model (GMM)-hidden Markov models (HMMs), support vector machines (SVM), and artificial neural networks (ANNs) were studied [2, 3]. Also, acoustic models with deep architectures including deep neural networks (DNNs) [4, 5], convolutional neural networks (CNNs) [6], and long short-term memory recurrent neural networks (LSTM-RNNs) [7], were recently applied to dysarthric speech recognition.

Articulatory movement information has been demonstrated effective in a variety of ASR applications such as normal speech recognition [8], whispered speech recognition [9], and dysarthric speech recognition [3, 10]. In [10], articulatory movement data such as the tongue and lips measured by electromagnetic articulograph (EMA) were used as additional features besides acoustic features. They reported that adding articulatory movement data produced better ASR performance than using only acoustic features for dysarthric speech recognition. EMA accurately captures flesh-point movement on the tongue and lips but is not practical because its tracking method is based on wired sensors attached on the articulators. In addition, EMA is cost-prohibitive and not easy to use due to complex setup and operation [11]. Therefore, indirect methods to incorporate articulatory information is highly needed.

When multiple modalities of data (e.g., acoustic and articulatory) are available for learning (training) while only one view (e.g., acoustic) is available for testing, we can possibly learn better representation by simultaneously analyzing multiple sources of data based on multi-view representation learning. Multi-view representation learning [12], an unsupervised learning/training technique, is typically based on learning transformed features from the primary view (available for both training and testing) to capture useful information from the secondary view (available for training only) using a paired two-view training set. Therefore, the learned features (transformed features) can capture primary as well as second view information, which is more informative for unimodal system. Multi-view representation learning has been recently successfully used in normal speech recognition [13]. However, it has rarely been studied in dysarthric speech recognition.

In this project, we investigate the effectiveness of multi-view representation learning via canonical correlation analysis (CCA) [14], which is to learn features in multi-views that are maximally correlated, for speaker-independent dysarthric speech recognition. Acoustic data were the primary view and articulatory data were the second view. Therefore, we expected that the transformed features from CCA are able to capture useful information from articulatory view, which then potentially improves the phonetic recognition accuracy.

2 Dysarthric Speech Dataset

2.1 Participants and Speech Tasks

Nine patients with amyotrophic lateral sclerosis (ALS) (6 females and 3 males) participated in the data collection. The participants were all American English speakers.

ALS is also known as Lou Gehrig's disease, which is one of the most common motor neuron diseases [15], resulting in progressive degeneration of both upper and lower motor neurons [15]. Four of the patients participated in more than one data collection session. These consecutive sessions within each patient were averagely six months apart. In total, eighteen sessions of speech acoustic and articulatory data were collected from nine patients. The average age at their patients' first visit was 62.8 ± 8.8 years old. Speech intelligibility in sessions of those patients with ALS varied from normal (100%) to severely unintelligible speech (0%). Speech intelligibility is the percentage of words that could be understood, which was measured by a certified speech-language pathologist. We roughly categorized sessions into three levels per their speech intelligibility scores: high (above 90%, eleven sessions), middle (65–90%, four sessions), and low (below 65%, three sessions).

Each participant produced about four repetitions of twenty unique phrases at their habitual speaking rate and loudness in each recording session. These phrases were selected from lists in people's daily conversations (e.g., Good afternoon.) or from those that are related to patient's use (e.g., I need to see a doctor.). There were 1,289 phrase utterances were collected in total, which consisted of 17,712 phoneme production samples of 39 unique phonemes, which covers most phonemes in English.

In addition, normal speech data from seven American English speakers (4 females and 3 males) were collected. The average age of the normal participants was 25.4 ± 3.6 years. No history of speech, language, or cognitive problems from any participants was reported. Each healthy subject participated in one session only, where he or she repeated a list of 132 phrases twice at their habitual speaking rate. The phrases were selected from [17], which are commonly spoken in people daily life (e.g., how are you doing?). There were 14 overlapped phrases from the datasets from the patients with ALS and the healthy participants. We used the normal speech data for acoustic model training.

2.2 Acoustic and Articulatory Data

We used the Wave Speech Research system (Northern Digital Inc., Waterloo, Canada) for data collection for all participants. Wave is one of the two commercially available electromagnetic tongue and lip motion tracking devices (See Fig. 1a). Safe, dental glue (PeriAcryl 90, GluStitch) or tape were used to attach four small sensors to the surface of each articulator using including Tongue Tip (TT), Tongue Body Back (TB), Upper Lip (UL), and Lower Lip (LL). An additional sensor was attached to the middle of forehead (Head Center, HC) for head correction. We placed the head sensor on the goggle rather than on skin directly to avoid skin motion artifact. Figure 1b shows the locations of the sensors. The four-sensor set (TT, TB, UL, and LL) was identified as an optimal set for this application in our prior work [17]. During the participants' speech production, three-dimensional movement data (x, y, z coordinates) of the sensors on the articulators were recorded. We set the system with a sampling rate 100 Hz. Wave system has a spatial resolution 0.5 mm in the center of the magnetic field [18].

Before data analysis, head-independent articulatory data was obtained by removing the translation and rotation of the head sensor from the raw articulatory data. This procedure was performed by the Wave system automatically. Figure 1b illustrates the

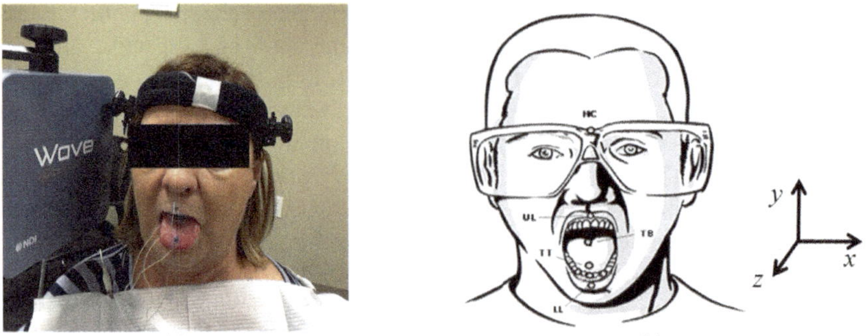

(a) (b)

Fig. 1. Speech and articulatory data collection: (a) the NDI Wave system and (b) sensor locations and labels: head center (HC), tongue tip (TT), tongue body back (TB), upper lip (UL), and lower lip (LL). In this coordinate system, x is horizontal or left-right, y is vertical or up-down, and z is frontier-posterior. Figure 1b was adapted from our prior work [20].

derived Cartesian coordinates system in Wave, where x is horizontal or left-right; y is vertical or up-down; and z is anterior-posterior. Lateral (x) movement of articulator is not expected significant during speech production [19], thus we used only y and z coordinates of the tongue and lip movement for analysis. The Wave system collected acoustic data synchronously with the articulatory data using a built-in microphone.

Figure 2 gives an example data when a subject produced "Call me back when you can". The top panel is acoustic waveform; The bottom panel is the articulatory movement time-series data (x, y, z coordinates) of the four sensors (TT, TB, UL, and LL) that were attached to the tongue and lips.

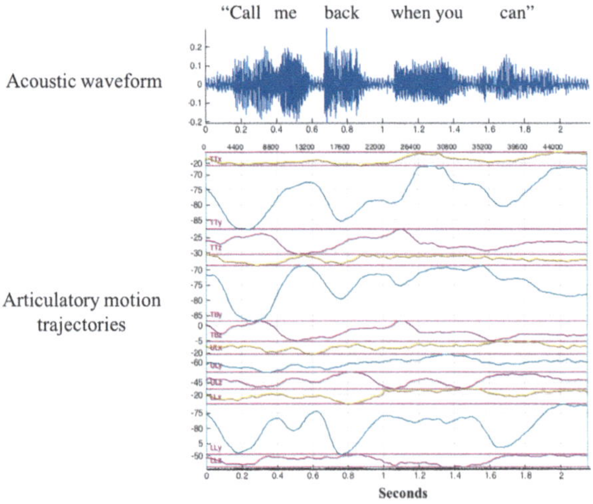

Fig. 2. Example of acoustic data (top) and time-series articulatory (sensor) movement (x, y, z coordinate) data (bottom). Four sensors are attached on tongue and lips (TT, TB, UL, and LL).

3 Multi-view Representation Learning via CCA

3.1 Canonical Correlation Analysis (CCA)

In multi-view representation learning, more than one type of information of the same underlying goal/task are accessible. Specifically, in this dysarthric speech recognition project, acoustic and articulatory data are the available two types of information. In this work, the training data set includes paired-observations $\{(\mathbf{x}_i, \mathbf{y}_i)\}_{i=1}^{N}$, where $\mathbf{x}_i \in \mathbb{R}^{D_x}$ denotes the ith input acoustic features, and $\mathbf{y}_i \in \mathbb{R}^{D_y}$ denote the ith input articulatory features recorded simultaneously.

CCA is a technique for learning a compact representation from multi-view data [13, 14]. CCA is to find linear projection matrices $\mathbf{U} \in \mathbb{R}^{D_x \times L}$ and $\mathbf{V} \in \mathbb{R}^{D_y \times L}$, where $L \leq min(D_x, D_y)$, from multi-view data that maximally correlate with each other whereas the dimensions are not correlated. We describe CCA as below:

$$\begin{aligned} &\max_{\mathrm{U,V}} \frac{1}{N} tr\left(\mathrm{U}^T \mathrm{X} \mathrm{Y}^T \mathrm{V}\right) \\ &\text{s.t.} \quad \mathrm{U}^T \left(\frac{\mathrm{XX}^T}{N} + r_x I\right) \mathrm{U} = \mathrm{V}^T \left(\frac{\mathrm{YY}^T}{N} + r_y I\right) \mathrm{V} = I \end{aligned} \tag{1}$$

where $\mathbf{X} = [\mathbf{x}_1, \ldots, \mathbf{x}_N] \in \mathbb{R}^{D_x \times N}$, $\mathbf{Y} = [\mathbf{y}_1, \ldots, \mathbf{y}_N] \in \mathbb{R}^{D_y \times N}$, and $(\boldsymbol{r}_x, \boldsymbol{r}_y) \geq 0$ are parameters of regularization. Let $\mathbf{\Sigma}_{12} = \frac{1}{N}\mathbf{X}\mathbf{Y}^T$ be the cross-covariance matrix and let $\mathbf{\Sigma}_{11} = \frac{1}{N}\mathbf{X}\mathbf{X}^T + \boldsymbol{r}_x \boldsymbol{I}$, and $\mathbf{\Sigma}_{22} = \frac{1}{N}\mathbf{Y}\mathbf{Y}^T + \boldsymbol{r}_y \boldsymbol{I}$ be the regularized auto-covariance matrices in acoustic and articulatory views of data. The subscripts 1 and 2 in the equations above denote the 1st (X) and 2nd (Y) view in the data which are acoustic and articulatory respectively. A singular value decomposition (SVD) of the matrix $\boldsymbol{T} = \mathbf{\Sigma}_{11}^{-1/2}\mathbf{\Sigma}_{12}\mathbf{\Sigma}_{22}^{-1/2}$ can be used to obtain the solution. The features (projections): $\tilde{\mathbf{x}} = \mathbf{U}^T\mathbf{x}$ for the primary (acoustic) view and $\tilde{\mathbf{y}} = \mathbf{V}^T\mathbf{y}$ for the secondary (articulatory) view will be the final CCA features. Finally, the CCA features for the acoustic view ($\tilde{\mathbf{x}}$) were used for the recognition of dysarthric speech.

3.2 Application to Dysarthric Speech Recognition

To obtain better articulatory representations of multiple speakers, we first applied Procrustes matching for across-speaker articulatory normalization. Procrustes matching was a statistical technique for shape analysis where shapes are represented as a set of landmarks on the surface of an object. In this work, the time-series data points of sensors attached on the tongue and lips were considered as articulatory shapes. These articulatory shapes from different speakers were translated and rotated to reduce the interspeaker physiological differences [16]. More specifically, two-dimensional (i.e., y and z coordinates) movement data of the four sensors (i.e., TT, TB, UL, and LL) were considered as a shape, which was then translated and rotated to a normalized shape, which has a centroid at the origin (0,0) and the centroids of the UL and LL formed a vertical line (to make the articulatory shapes the same angle relative to the head). Then,

CCA transformation matrix was estimated from acoustic features and Procrustes matched articulatory data. Next, the final CCA features were appended to the acoustic features as the final input features [13, 20].

Figure 3 provides examples how articulatory data looks like and the inconsistency of articulatory "shape" angle between two patients. Figure 3a is a scatter plot of the four sensors of a patient producing a short phrase "Can that wait until another time?". Figure 3b shows the plot of the same phrase produced by another patient. As shown in Fig. 3, different speakers have different angle of the articulatory "shape". A normalized step is to rotate them to a reference. For example, we can rotate the whole shapes such that the UL and LL form a vertical line. Figure 3c demonstrates how a data sample (from Fig. 3b) is translated and rotated in Procrustes matching.

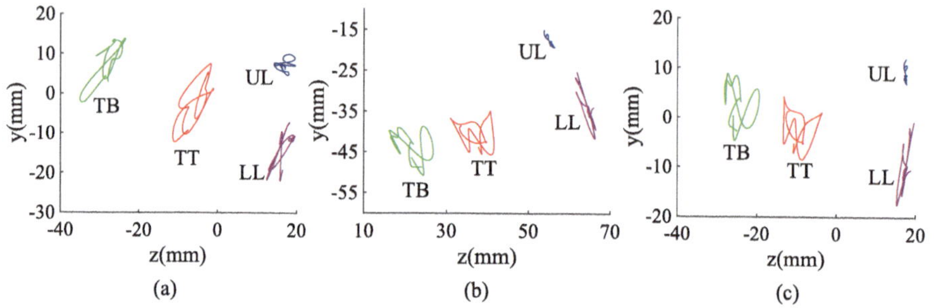

Fig. 3. Example of shapes (scatter plots of motion trajectories of the four sensors; TT, TB, UL, and LL) when producing phrase "Call me back when you can". (a) is a shape example of a patient with dysarthria due to ALS; (b) is a shape example of another patient; (c) is the shape in (b) after Procrustes matching.

3.3 Speech Recognizers

The target of a speech recognizer is to find appropriate phoneme/word sequences based on speech (and/or articulatory motion) time-series signals. Current ASR methods typically include an acoustic model (AM)/feature model (e.g., Gaussian mixture model) and a pattern recognizer (e.g., hidden Markov model, HMM).

An acoustic model integrates acoustic and phonetic knowledge and takes acoustic features (e.g. Mel-frequency cepstral coefficients, MFCCs) as input and generate acoustic scores for frames/segments of the variable-length data sequence for the pattern recognizer [21]. Gaussian mixture model (GMM) is a typical acoustic model which use a number of mixture-of-Gaussian random variables to fit speech features [21]. GMM performs well in modeling the probability density functions (PDFs) of speech feature but it is not able to model sequence information of speech. Therefore, a hidden Markov model (HMM) [22] is adopted to address the sequence modeling and variable-length feature problem. HMM captures sequential information of acoustic features with states, then given fixed states of HMM, models like Gaussian mixture model(GMM) could model the PDFs of features allocated to the states.

GMM-HMM was the long-standing ASR acoustic model in the field. Until recently, discriminative hierarchical models like deep neural network (DNN) and recurrent neural network (RNN) became feasible in ASR and reduced error rate significantly [21, 23]. A DNN is an artificial neural network (ANN) with more than two hidden layers, which is a conventional multilayer perception (MLP). The RNNs is a class of neural network where connections between neurons form a directed cycle. These cycles enable RNN to model the states of neurons along time sequence. Especially, long short-term memory (LSTM)-RNN performed extremely well in both small- and large-scale ASR [22]. LSTM-RNN impose a memory structure in RNN so that it is able to look far back to previous states rather than using recent state only in traditional RNN [24, 25]. Therefore, LSTM-HMM outperforms DNN-HMM and conventional RNN-HMM in many applications like language processing and ASR.

3.4 Experimental Setup

The baseline input features to a speech recognizer were 39 dimensional MFCCs, which consists of 13 static and their 1st and 2nd derivatives that were extract from frames of 25 ms and a shift size of 10 ms, which is typical setup used in ASR [2, 4, 6, 9, 10, 21, 23]. Articulatory data are 24 dimensional coordinates, which consisted of 8 static data (2 dimensions $\times$ 4 sensors) and their 1st and 2nd order derivatives, respectively. The shift size is also 10 ms. In addition to Procrustes matching, mean normalization along each dimension was used for both acoustic and articulatory data. In multi-view feature learning setting, the inputs are acoustic and articulatory features of the current frame concatenated with 8 context frames around each frame (4 frames before and 4 frames after the current frame), resulting in 351 (39 MFCCs $\times$ 9 frames) dimensional acoustic $(D_x = 351)$ and 216 (24 articulatory features $\times$ 9 frames) dimensional articulatory inputs $\left(D_y = 216\right)$ for the CCA models. After training a CCA feature transformation matrix, we applied feature transformation to the acoustic features. Finally, we appended the L-dimensional CCA features to D_x acoustic features, which was the final feature vector for each frame for testing.

The three speech recognizers that were described in the previous section were used and compared in this project, including the long-standing GMM-HMM, recently available DNN-HMM, and LSTM-HMM. Although it has been shown DNN-HMM and LSTM-HMM outperformed long-standing GMM-HMM, we still included GMM-HMM as a base-line approach because the rapid implementation of GMM-HMM [10, 16]. Each recognizer consisted of 719 tied-state (senone) left-to-right, three-state triphone HMMs. 719 was the average number of triphone combinations found in the training set in the 6-fold cross validations. The GMM included 8–10 Gaussian distributions, which was trained using standard maximum likelihood estimation. DNN was trained using 40 log filterbank energies within the 9-frame context window.

To obtain the best structure of the DNN, we conducted a preliminary experiment to test 1 to 6 layers with different numbers of hidden nodes in each layer: 256, 512, and 1,024. The best result was obtained with 3 hidden layers with 512 hidden units. Therefore, the DNN had 3 hidden layers with 512 hidden units in each layer and a 719-dimensional softmax output layer, corresponding to the number of senones of the

GMM-HMM system. We used the parametric rectified linear unit (PReLU) as the activation function and backpropagation to train the network. The LSTM network had 2 hidden layers, where each layer had 320 LSTM cells (nodes) plus 200 recurrent units [25]. The output layer had 719-dimensional softmax nodes. The network parameters were trained using backpropagation through time. The IRSTLM toolkit [26] was used to train bigram phoneme language models. Kaldi speech recognition toolkit [27] was used to perform the model training and decoding (testing).

3.5 Performance Measure

Phoneme error rate (PER), a standard measure in ASR, was used as the performance measure in this dysarthric speech recognition experiment. PER (in percentage) is all prediction errors over the total number of samples. Prediction errors include substitution (wrong prediction), deletion (missing a phoneme), and addition/insertion (prediction of a phoneme that does not exist).

As mentioned previously, we performed leave-one-subject-out (6-fold) cross validation in the speaker-independent phoneme recognition experiment. Data sessions with low speech intelligibility (below 65%) were excluded during training because adding these data degraded the performance. Thus, only the data sessions with high and mid speech intelligibility from each speaker as well as normal speech data were used as a training set. The small amount of training data for dysarthric speech is usually available. Thus, a larger training set by adding a portion of normal speech data benefits the model training [2, 28, 29]. The averaged performance of all validations was considered as the final performance.

4 Results and Discussion

Table 1 shows the PERs on GMM-HMM based ASR systems to show the effectiveness of the CCA. For the comparison of individual acoustic and articulatory features, the performance of acoustic features (a PER of 47.1%) was much better than articulatory features (a PER of 72.1%). When we append the articulatory features to acoustic features, the performance was improved compared to acoustic features, giving a PER of 45.2%. Next, we compared the performance of CCA with the number of CCA dimensions. When we used 50 dimensional CCA features alone (CCA50), the performance was comparable to the performance of acoustic features. Finally, the best performance was obtained when we combined the acoustic features and 50 dimensional CCA features (Acoust. + CCA50), producing a PER of 42.1%.

Our experimental results implied that by incorporating articulatory information in transforming features using CCA is effective for recognizing dysarthric speech. The performance improvement could be up to 12.3% PER reduction, if we compare the best approach (CCA + LSTM) with the baseline approach (GMM without CCA).

In addition, Fig. 4 shows the results using CCA with different speech recognizers including GMM-HMM, DNN-HMM, and LSTM-HMM. First, all of the speech recognizers + CCA demonstrated an improved performance. Second, Fig. 4 also demonstrated LSTM-HMM outperformed the other standard recognizers GMM-HMM and DNN-HMM, which is consistent with the literature (e.g., [16]). CCAs in Fig. 4 are 50-dimensional (CCA50) since Table 1 shows it outperform CCA30 and CCA70.

Table 1. PERs (%) obtained using GMM-HMM with different CCA dimensions

Features	PER (%)
Acoust.	47.1
Articul.	72.1
Acoust. + Articul.	45.2
CCA30	50.5
CCA50	47.9
CCA70	47.9
Acoust. + CCA30	43.3
Acoust. + CCA50	**42.1**
Acoust. + CCA70	43.0

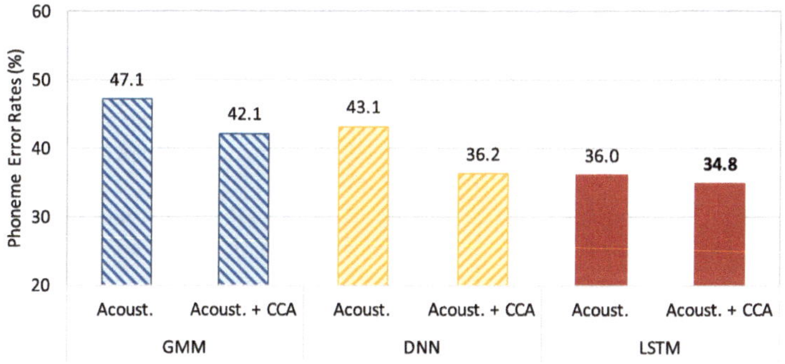

Fig. 4. PERs obtained by CCA50 with GMM-HMM, DNN-HMM, and LSTM-RNN-HMM.

Although this data set was relatively small, we believe that the recognition performance may improve if using a larger training dataset from more patients with ALS. We are actively recruiting more normal subjects and patients with ALS and collecting both acoustic and articulatory movement data. A further study with a larger data set collected from more patients will be conducted to verify these findings.

5 Conclusions

This paper explored the effectiveness of canonical correlation analysis (CCA) in multi-view representation to improve the speaker-independent dysarthric speech recognition accuracy. In this experiment with multi-view representation learning via CCA acoustic data were used as the primary view and articulatory movement data were used as the secondary view. Three speech recognizers including GMM-HMM, DNN-HMM, and LSTM-HMM, were used. Experimental results showed that the CCA-based multi-view representation learning approach provides significant improvement over acoustic features alone. Future directions of this work may include a more comprehensive evaluation of this approach using a larger data set collected from more patients with ALS or other neurological diseases.

Acknowledgments. This work was supported by the National Institutes of Health of the United States through grants R03DC013990 and R01DC013547 and by the American Speech-Language-Hearing Foundation through a New Century Scholar Research Grant. We would like to thank Dr. Jordan R. Green, Dr. Thomas F. Campbell, and Dr. Yana Yunusova, Jennifer McGlothlin, Kristin Teplansky, and the volunteering participants.

References

1. Duffy, J.R.: Motor Speech Disorders-E-Book: Substrates, Differential Diagnosis, and Management. Elsevier, New York City (2013)
2. Kim, M., Yoo, J., Kim, H.: Dysarthric speech recognition using dysarthria-severity-dependent and speaker-adaptive models. In: Proceeding of Interspeech, pp. 3622–3626 (2013)
3. Rudzicz, F.: Articulatory knowledge in the recognition of dysarthric speech. IEEE Trans. Audio Speech Lang. Process. **19**(4), 947–960 (2011)
4. Kim, M., Wang, J., Kim, H.: Dysarthric speech recognition using Kullback–Leibler divergence-based hidden markov model. In: Proceeding of Interspeech, pp. 2671–2675 (2016)
5. Yilmaz, E., Ganzeboom, M.S., Cucchiarini, C., Strik, H.: Multi-stage DNN training for automatic recognition of dysarthric speech. In: Proceeding of Interspeech, pp. 2685–2689 (2017)
6. Nakashika, T., Yoshioka, T., Takiguchi, T., Ariki, Y., Duffner, S., Garcia, C.: Dysarthric speech recognition using a convolutive bottleneck network. In: Proceeding of 12th IEEE International Conference on Signal Processing (ICSP), pp. 505–509 (2014)
7. Espana-Bonet, C., Fonollosa, J.A.: Automatic speech recognition with deep neural networks for impaired speech. In: Proceeding of Third International Conference on Advances in Speech and Language Technologies for Iberian Languages, pp. 97–107. Springer, Heidelberg (2016)
8. Kirchhoff, K., Fink, G.A., Sagerer, G.: Combining acoustic and articulatory feature information for robust speech recognition. Speech Commun. **37**(3), 303–319 (2002)
9. Cao, B., Kim, M., Mau, T., Wang, J.: Recognizing whispered speech produced by an individual with surgically reconstructed larynx using articulatory movement data. In: Proceeding of ACL/ISCA Workshop Speech Language Processing Assistive Technology, pp. 80–86 (2016)
10. Hahm, S., Heitzman, D., Wang, J.: Recognizing dysarthric speech due to amyotrophic lateral sclerosis with across-speaker articulatory normalization. In: Proceeding of the ACL/ISCA Workshop on Speech and Language Processing for Assistive Technologies, pp. 47–54 (2015)
11. Wang, J., Samal, A., Green, J.: Preliminary test of a real-time, interactive silent speech interface based on electromagnetic articulograph. In: Proceeding of ACL/ISCA Workshop on Speech and Language Processing for Assistive Technologies, Baltimore, USA, pp. 38–45 (2014)

12. Wang, W., Arora, R., Livescu, K., Bilmes, J.A.: On deep multiview representation learning. In: ICML, pp. 1083–1092 (2015)
13. Wang, W., Arora, R., Livescu, K., Bilmes, J.A.: Unsupervised learning of acoustic features via deep canonical correlation analysis. In: Proceeding of ICASSP, pp. 4590–4594 (2015)
14. Borga, M.: Canonical correlation: a tutorial. Online tutorial (2001)
15. Green, J.R., Yunusova, Y., Kuruvilla, M.S., Wang, J., Pattee, G.L., Synhorst, L., Zinman, L., Berry, J.D.: Bulbar and speech motor assessment in ALS: challenges and future directions. Amyotroph. Lateral Scler. Frontotemporal Degener. **14**(7–8), 494–500 (2013)
16. Kim, M., Cao, B., Mau, T., Wang, J.: Speaker-independent silent speech recognition from flesh-point articulatory movements using an LSTM neural network. In: IEEE/ACM Transactions on Audio, Speech, and Language Processing, vol. 25, no. 12, 2323–2336 (2017)
17. Wang, J., Samal, A., Rong, P., Green, J.R.: An optimal set of flesh points on tongue and lips for speech-movement classification. J. Speech Lang. Hear. Res. **59**, 15–26 (2016)
18. Berry Jeffery, J.: Accuracy of the NDI wave speech research system. J. Speech Lang. Hear. Res. **54**(5), 295–301 (2011)
19. Wang, J., Green, J.R., Samal, A., Yunusova, Y.: Articulatory distinctiveness of vowels and consonants: a data-driven approach. J. Speech Lang. Hear. Res. **56**(5), 1539–1551 (2013)
20. Kim, M., Cao, B, Mau, T, Wang, J.: Multiview representation learning via deep CCA for silent speech recognition. In: Proceeding of Interspeech, pp. 2769–2773 (2017)
21. Deng, L., Yu, D.: Deep learning: methods and applications. Found. Trends Signal Process. **3–4**, 197–387 (2014)
22. Rabiner, L., Juang, B.H.: An introduction to hidden Markov models. IEEE ASSP Mag. **3**(1), 4–16 (1986)
23. Hinton, G., Deng, L., Yu, D., Dahl, G.E., Mohamed, A.R., Jaitly, N., Senior, A., Vanhoucke, V., Nguyen, P., Sainath, T.N., et al.: Deep neural networks for acoustic modeling in speech recognition: the shared views of four research groups. IEEE Signal Process. Mag. **29**(6), 82–97 (2012)
24. Hochreiter, S., Schmidhuber, J.: Long short-term memory. Neural Comput. **9**(8), 1735–1780 (1997)
25. Sak, H., Senior, A., Beaufays, F.: Long short-term memory recurrent neural network architectures for large scale acoustic modeling. In: Fifteenth Annual Conference of the International Speech Communication Association, pp. 338–342 (2014)
26. Federico, M., Bertoldi, N., Cettolo, M.: IRSTLM: an open source toolkit for handling large scale language models. In: Proceeding of Interspeech, pp. 1618–1621 (2008)
27. Povey, D., Ghoshal, A., Boulianne, G., Burget, L., Glembek, O., Goel, N., Hannemann, M., Motlicek, P., Qian, Y., Schwarz, P., Silovsky, J.: The Kaldi speech recognition toolkit. In: Proceeding of IEEE Workshop on Automatic Speech Recognition and Understanding, Waikoloa, USA, pp. 1–4 (2011)
28. Christensen, H., Aniol, M.B., Bell, P., Green, P.D., Hain, T., King, S., Swietojanski, P.: Combining in-domain and out-of-domain speech data for automatic recognition of disordered speech. In: Proceeding of Interspeech, pp. 3642–3645 (2013)
29. Kim, M., Kim, Y., Yoo, J., Wang, J., Kim, H.: Regularized speaker adaptation of KL-HMM for dysarthric speech recognition. IEEE Trans. Neural Syst. Rehabil. Eng. **25**(9), 1581–1591 (2017)

Image Inpainting Based on Image Structure and Texture Decomposition

Zhenping Qiang[1], Hui Liu[2], and Zhenhong Shang[2(✉)]

[1] College of Big Data and Intelligent Engineering, Southwest Forestry University, Kunming, China
[2] Faculty of Information Engineering and Automation, Kunming University of Science and Technology, Kunming, China
szh@kmust.edu.cn

Abstract. Image inpainting is a major research content in the field of computer vision and is intended to solve the problem of high-quality repairing of the damaged images automatically. In this paper, considering the image containing structural component and texture component generally, an Adaptive Local Laplacian Filter (ALLF) is first proposed to decompose the image, and the different components obtained through ALLF decomposition are further repaired by different methods. The restoration results are added together to achieve the completion of the whole image. Experiments on synthetic and natural images show the advantages of our approach, especially, in small scratch repair and text removal task.

Keywords: Image inpainting · Image decomposition · Image completion
Adaptive local Laplacian filter

1 Introduction

Image completion, which is also known as "image inpainting", involves the issue of filling missing information in images. This is a challenging task in computer vision [1]. First of all, the filling algorithm should be able to successfully complete complex natural images. Secondly, it should be able to handle incomplete images with large missing parts. Finally, the algorithm's execution should be in a fully automatic manner, i.e. without intervention from the user.

So far, many inpainting algorithms have been proposed to resolve these issues. According to the different types of problem solving methods, the inpainting algorithms can be summed up into four types, namely, partial differential and variational based method [2–5], exemplar-based method [1, 6–10], transform domain based method [11–13] and deep learning based method [14–17]. The methods based on partial differential or total variation, and the methods based on transform domain can achieve good inpainting effect in the repair of small size damaged images; the exemplar-based methods can obtain relatively good results in the image restoration of the large missing region; the deep learning based methods are still in the initial stage, and most of the successful methods are to repair a specific type of small scale image. There are some problems to be solved in each type of methods, and this study is based on that the

K. Deng et al. (Eds.): ICMIR 2018, AISC 856, pp. 1096–1106, 2019.
https://doi.org/10.1007/978-3-030-00214-5_134

natural image itself including the structural component and the texture component, so the image can be decomposed first, and further to repair different components by different methods, in order to improve the quality of image restoration.

One of the simplest and effective method of image decomposition is accomplished by user interaction. In paper [18], the authors first described the lost structure information in the missing regions by the users. Then, the proposed inpainting method propagate the structure of the known patches along with the direction of these artificial curves to achieve structural propagation. Finally, they use the texture synthesis method which proposed in paper [19] to estimate the other unknown pixels value, and completes the whole image inpainting.

The texture and structural components of the image can be decomposed through automatic method. For example, in paper [20], the authors decomposed the image into two functions, one is to represent the image structural component, the other is used to capture the image texture. The structural component and texture component of the image can be obtained by solving the joint function. Further, the texture layer image is repaired by texture synthesis method and the structural layer is repaired using the PDE method [2]. Finally, the two parts are fused together to generate the result.

In this paper, an image decomposition method based on Adaptive Local Laplacian Filter (ALLF) is proposed, and the proposed decomposition method is used to decompose image. Further, texture synthesis technology is used to repair the decomposed texture component, and the partial differential method is applied to reconstruct the structural component. Experimental results showed that the proposed method is effective in small scratch image restoration and image text removal application.

2 Related Work

The first problem to be solved in layered image inpainting method is the decomposition of the image, and the method of image decomposition is varied too. Considering the multi-scale property of the image itself, that is, the image has different features at different scales. Therefore, image decomposition based on multi-scale as a very effective method has been widely studied.

Using multi-scale methods to decompose the image are to consider the different features such as edge and texture in different scales, and there are certain relationships of them in the spatial domain of the image. In this way, different image components can be judged by many aspects. In this section, we will mainly introduce the basic principle of multi-scale image decomposition based on Pyramid.

As early as 1987, Burt et al. proposed an image decomposition algorithm based on Laplacian Pyramid. For the image I, the initial image represent A_0 in Pyramid is the I itself (A_0 = I), and the approximate image of other levels in Pyramid can be obtained by filtering the approximate image of the upper level in Pyramid:

$$A_{k+1}(m,n) = \sum_{i=-N}^{N} \sum_{j=-N}^{N} w(i,j) A_k(2m+i, 2n+j) \tag{1}$$

where $w(i,j)$ is the filter coefficient, and different filters can be designed to obtain different filtering effects, such as using the Gauss filter, and the Gauss Pyramid representation of the image can be established according to the formula (1).

The process of image decomposition based on Pyramid is as follows: first, the cartoon component A'_k of k layer is obtained by using the approximately predicting based on the $k+1$ layer Pyramid representation image A'_{k+1}, and then the detail image D_k of the k layer is obtained by the calculating the difference between the Pyramid representation image A_k and the predicted image A'_k:

$$A'_k(m,n) = 4\sum_{i=-N}^{i=N}\sum_{j=-N}^{N} w(i,j)A_{k+1}\left(\frac{m+i}{2},\frac{n+j}{2}\right) \tag{2}$$

The detail image D_k is obtained by the following formula:

$$D_k(m,n) = A_k(m,n) - A'_k(m,n) \quad 0 \le k < K \tag{3}$$

As showed in Fig. 1, the first row is the image representation by Gauss Pyramid, the second row is the predicted images based on formula (2), and the third row is the

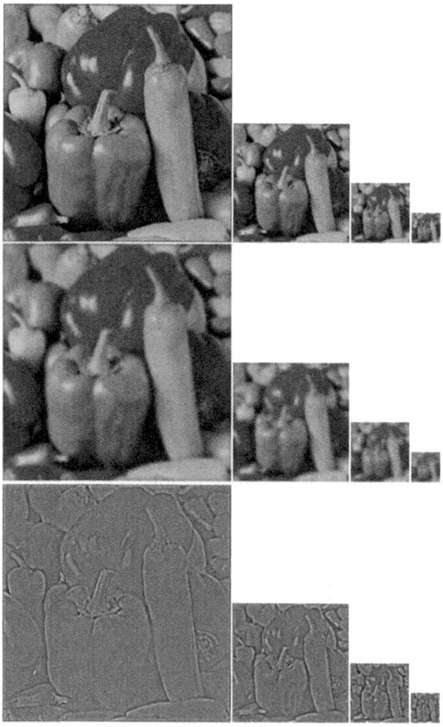

Fig. 1. Image decomposition process based on Pyramid

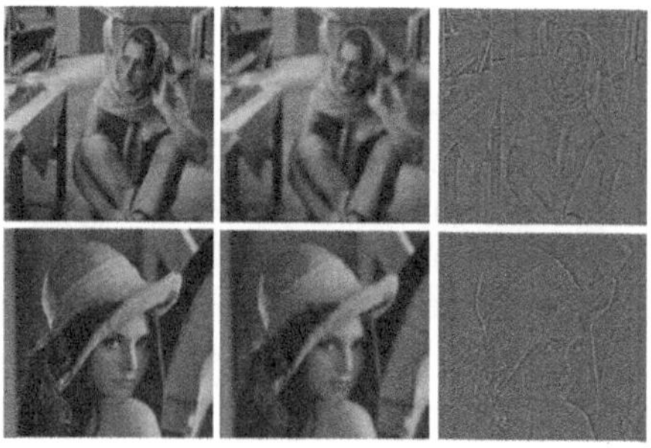

Fig. 2. Image decomposition examples based on Pyramid

detailed images calculated according to formula (3). As showed in Fig. 2 is two examples of image decomposition based on Gauss Pyramid.

In this paper, an adaptive fast local Laplacian filter method is proposed to decompose the image which to avoid the blur in the predicted image in Fig. 2, and further repairs the whole image with different restoration methods for the different components obtained.

3 Our Method

3.1 Image Decomposition

In this section, we will propose an Adaptive Local Laplacian Filter (ALLF), which adaptively sets the filter parameters according to the representation of the local change of the process image. The structural components of the image decomposition can be obtained by the image edge-preserving smooth based on ALLF, and the texture components cab be obtained by the calculating the difference between the original image and the image structural components.

3.1.1 Local Laplacian Filter

The Local Laplacian Filter (LLF) uses different mapping functions to process the original image according to the different application purposes, and calculates the Laplacian Pyramid coefficient based on the processed images, and finally collapses the Laplacian Pyramid to reconstruct the result image.

In this way, the applications based on LLF include three steps. The first step is to transform the input image by point-wise nonlinear mapping, and the remapping function is:

$$r(i) = i + (i - g)m_{factor}G_{\sigma_r}(i - g) \tag{4}$$

where i is the pixel value of the original image, g is the corresponding coefficient of the Gauss Pyramid, σ_r is the threshold of the contour differentiation, $G_{\sigma_r}(i - g)$ is standard Gauss function, and m_{factor} is a amplitude factor of the remapping function. For different values of g, a set of remapping images can be obtained.

The second step is to calculate the coefficient of the output Laplacian Pyramid based on the images obtained in the first step, and the last step is to get the final result image through collapsing the Laplacian Pyramid. In the paper [21], the authors perform filtering processing by setting fixed parameters σ_r and m_{factor}. In order to extract structural components more effectively, such as setting small σ_r and m_{factor} in the strong edge regions of the image to avoid smooth the edge contour and to guarantee the pixels' value in the value range after the remapping; and the larger σ_r can be used to smooth the homogeneous regions meanwhile. Therefore, the parameters of LLF should be set based on the local variation of the image adaptively.

3.1.2 The Representation of Image Local Variation

Considering the discrete Laplace operator is defined as the divergence of gradient, which describes the degree of the variation of the corresponding pixels. And the Gauss kernel weighted method can smooth the influence of the noise. Therefore, we used an improved Gauss kernel weighted extended discrete Laplace operator to extract the representation of image local variation. The formula is:

$$V_\sigma(p_{x,y}) = \sum_{i=x-k}^{x+k} \sum_{j=y-k}^{y+k} \left| G'_{k,\sigma_g} \odot L_{2k+1} \odot \Psi_{p_{x,y},2k+1} \right|_1 \tag{5}$$

where $\odot$ means the pixel-wise multiply of the each element in matrix, $\Psi_{p_{x,y}2k+1}$ is an image patch which the center is $p_{x,y}$, width and height is $2k+1$. G'_{k,σ_g} is the improved Gauss kernel (the value of the center point is 1, and the sum of the other elements is 1). k is the radius of the improved Gauss kernel, L_{2k+1} is an extended matrix form of the discrete Laplace operator:

$$L_{2k+1}(i,j) = \begin{cases} -1, & i = k+1, j = k+1; \\ 1, & \text{otherwise.} \end{cases} \tag{6}$$

The representation of image local variation obtained by formula (5) can not only avoid noise influence, but also represent the local variation of image well. Figure 3 gives two examples of local variation representations calculated based on formula (5).

3.1.3 Image Decomposition

The structural components of the image can be obtained by using the ALLF to smooth the image details with the edge-preserving, and the texture components of the image can be obtained by calculating the difference between the original image and the structural components.

Such as the 3.1.1 section analysis, the adaptive setting of parameter σ_r should be inversely proportional to the local variation V_σ obtained by formula (3), and the parameter $m_{factor} \in [-1,0]$ to avoid contrast enhancement in the result image. At the same time, the absolute value of m_{factor} should be proportional to V_σ. Therefore, we can get the adaptive parameter σ_r and the adaptive parameter m_{factor} through the following formula:

$$\sigma_{r,s}(x,y) = c(\sigma_{\max} - V_\sigma(x,y)) + \varepsilon_\sigma \tag{7}$$

$$m_{factor,s}(x,y) = -(a_s V_\sigma(x,y) + b_s) \tag{8}$$

where $\sigma_{\max}$ is the maximum value of V_σ, parameter c is a scale factor to avoid the element value in $\sigma_{r,s}$ is too large and resulting in the reduction the contrast of the whole image after filtering, ε_σ is a very small constant to avoid the element value in $\sigma_{r,s}$ is 0, a_s and b_s are normalized linear scaling factors to ensure $m_{factor} \in [-1,0]$.

Using the formula (7) and (8), we can calculate the adaptive parameters $\sigma_{r,s}$ and m_{factor}, and bring them into the formula (4) can smooth the image details with edge-preserving, Fig. 4 showed two examples of the filtering results.

a. input b. local variation representation

Fig. 3. Examples of local variation representaion.

a. input b. structural components of input

Fig. 4. Examples of smooth the image details (to get structural components).

3.2 Image Inpainting

After obtaining the image structural component, we can further calculate the texture component of the image based on the original image, and Fig. 5 shows an example of image decomposition result, (a) is the original image, (b) is the decomposed structural component, (c) is the decomposed texture component, and in order to display clearly, (e) is the result of the contrast extension of (c).

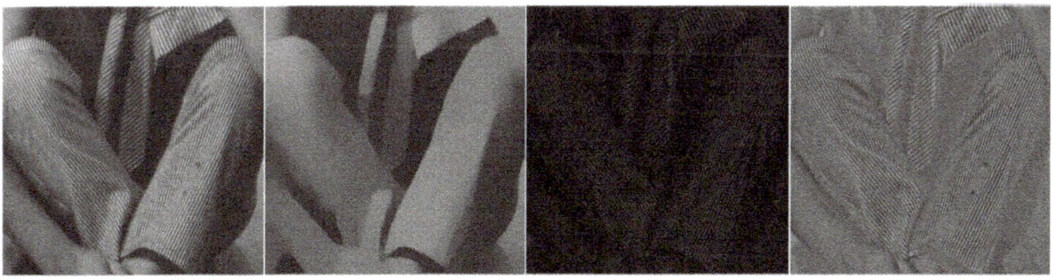

(a) (b) (c) (d)

Fig. 5. Image decomposition based on ALLF

After image decomposition, the different inpainting methods are used to repair the different components, and then the results of each component can be used to integrate the restoration of the whole image. Based on the image decomposition method we proposed in Sect. 3.1, the texture component is repaired by texture synthesis technology. If the missing area is smaller, partial differential method can be used to repair the structural component, if the missing region is large and the partial differential or total variational methods will cause fuzzy inevitably, so the exemplar-based inpainting methods will be a feasible solution strategy when the missing region is large.

4 Results and Compare

Figure 6 is two inpainting examples in accordance with the method describe in the last section. The texture component is repaired by using the method proposed in paper [19], and the structure component is repaired by the method proposed in paper [2]. The missing area in the first row is smaller (the scratch line added by the paint pen with 5 pixels), and the scratch in the second row is larger (added by paint pen with 13 pixels). In the result of the fist row, there is no repaired traces basically. However, the repaired traces can be seen obviously in the second row, which is mainly due to the fact that with the increase of the missing areas, the PDE-based methods will bring obvious fuzzy traces in the repair of the structure component.

Figure 7 is the results of the texture components still being repaired by using the texture synthesis method proposed in paper [19] and the structure components are repaired by using the Gibbs sampling method. Although the repaired results for large scratch are improved visually compared to Fig. 6, there are still significant ambiguities in the repaired regions with strong contour.

(a) (b) (c) (d)

Fig. 6. Image inpainting based on image decomposition

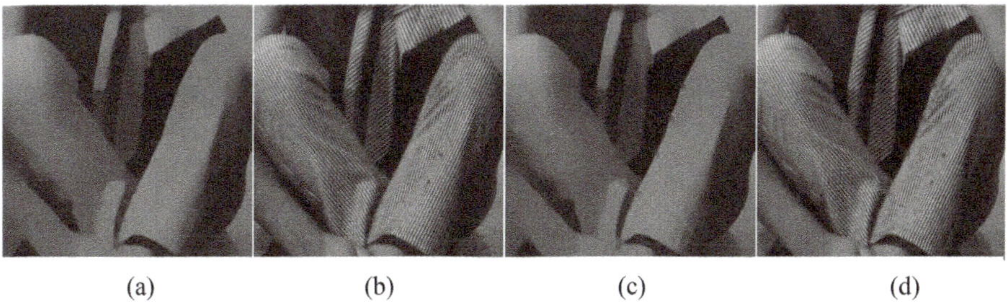

Fig. 7. The repaired results of Gibbs sampling method used for structure component

For image inpainting of image with small damaged areas or scratches, it is almost impossible to see the repaired traces from the results of Figs. 6 and 7. In order to further illustrate the effectiveness of the inpainting method we proposed in this paper, in Figs. 8 and 9 we showed the comparison of our method with the other successful methods on the synthesized image and true color image.

Because the repair area is very small, we use a quantitative method to analyze the results. In Tables 1 and 2, the PSNR and SSIM values of the repaired results for each method are given.

As shown in Table 1, the PSNR values of the inpainted results based on the texture synthesis method and we proposed method on the regular texture image is much larger than the methods based on diffusion, and the PSNR values of the repaired results of various methods are not have big difference.

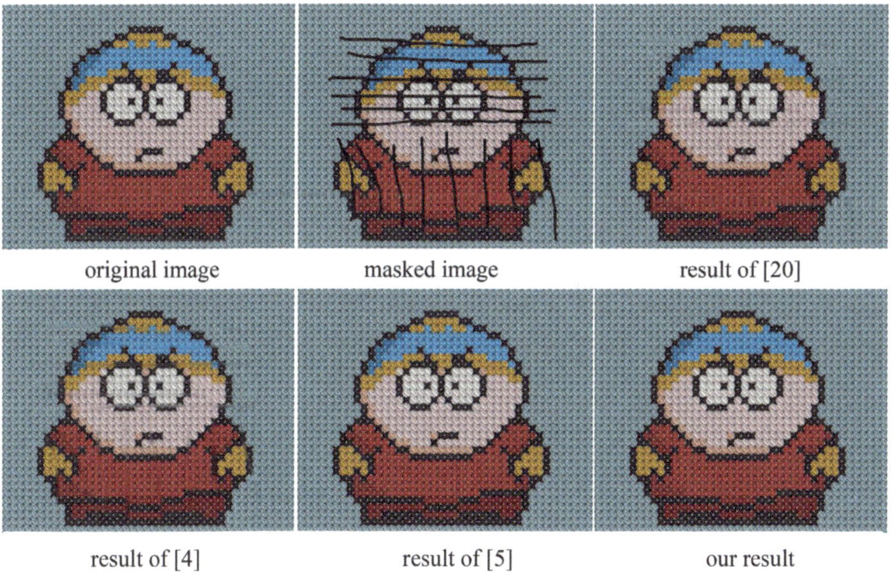

Fig. 8. Image inpainting results comparison on the synthesized image

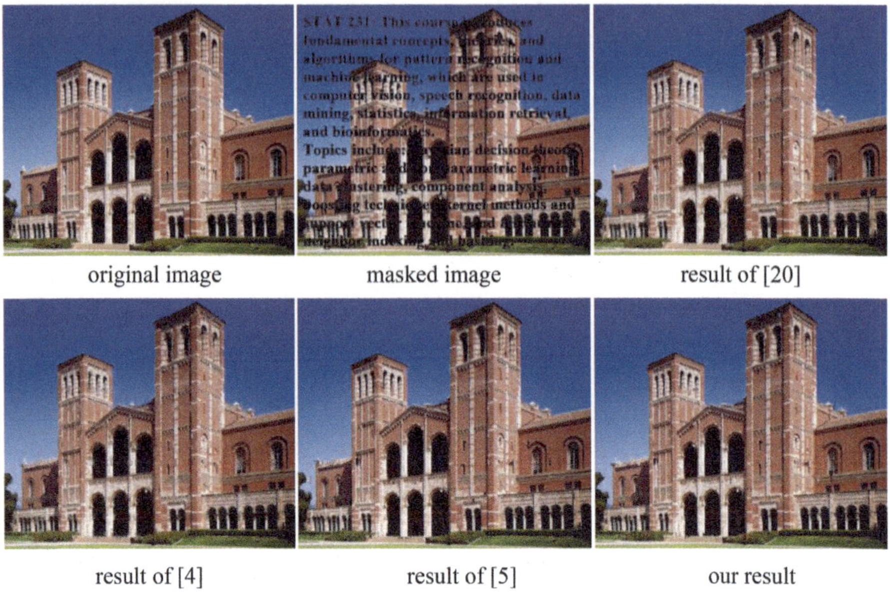

Fig. 9. Image inpainting results comparison on the true color image

Table 1. Comparison for PSNR of the inpainted results.

	Navier-Stokes method	Lexandru Telea method	Texture synthesis method	Our method
Results in Fig. 8	26.9434	26.6830	38.4387	38.7938
Results in Fig. 9	26.4704	25.3787	25.6571	25.8243

As shown in Table 2, for the synthetic image with regular texture components, the SSIM values displayed that the good results achieved by the texture synthesis method and we proposed method, and for the application of texture removal of true color image, the methods proposed in this paper and Navier-Stokes method have achieved good results.

Table 2. Comparison for SSIM of the inpainted results.

	Navier-Stokes method	Lexandru Telea method	Texture synthesis method	Our method
Results in Fig. 8	0.9651	0.9628	0.9981	0.9984
Results in Fig. 9	0.9618	0.9481	0.9578	0.9628

5 Conclusion

In this paper, we proposed an Adaptive Local Laplacian Filter (ALLF) for image processing. Based on this method, the image can be decomposed into structural component and texture component, and an image inpainting method based on image decomposition is proposed too. That is, the texture synthesis method is used to repair the texture component of the image, and the inpainting methods based on diffusion are used to repair the structural component of the image. Finally, the repaired results of different components are adding together to achieve the restoration of the entire image. Qualitative and quantitative experimental analysis has verified the effectiveness of the proposed image inpainting method in image small scratch repair and text removal application.

Acknowledgments. This work is supported by the projects of National Natural Science Foundation of China (11603016, 61462052) and project of Research Center of Kunming Forestry Information Engineering Technology (2015FBI06).

References

1. Komodakis, N., Tziritas, G.: Image completion using efficient belief propagation via priority scheduling and dynamic pruning. IEEE Trans. Image Process. **16**(11), 2649–2661 (2007)
2. Bertalmio, M., Sapiro, G., Caselles, V., et al.: Image inpainting. In: Proceedings of the 27th Annual Conference on Computer Graphics and Interactive Techniques. ACM Press, pp. 417–424 (2000). Addison-Wesley Publishing Co.
3. Chan, T., Shen, J.: Local inpainting models and TV in-painting. SIAM J. Appl. Math. **62**(3), 1019–1043 (2001)
4. Bertalmio, M., Bertozzi, A.L., Sapiro, G.: Navier–Stokes, fluid dynamics, and image and video inpainting. In: IEEE Computer Society Conference on Computer Vision and Pattern Recognition, vol. 1, no. 1, pp. 355–362 (2001)
5. Telea, A.: An image inpainting technique based on the fast marching method. J. Graph. Tools **9**(1), 23–34 (2004)
6. He, K., Sun, J.: Image completion approaches using the statistics of similar patches. IEEE Trans. Pattern Anal. Mach. Intell. **36**(12), 2423–2435 (2014)
7. Criminisi, A., Perez, P., Toyama, K.: Object removal by exemplar-based inpainting. In: IEEE Computer Society Conference on Computer Vision and Pattern Recognition, pp. 721–728 (2003)
8. Xu, Z., Sun, J.: Image inpainting by patch propagation using patch sparsity. IEEE Trans. Image Process. **19**(5), 1153–1165 (2010)
9. Meur, O.L., Gautier, J., Guillemot, C.: Examplar-based inpainting based on local geometry. In: IEEE International Conference on Image Processing, pp. 3401–3404. IEEE (2011)
10. Qiang, Z., He, L., Xu, D.: Exemplar-based pixel by pixel inpainting based on patch shift. In: CCF Chinese Conference on Computer Vision, pp. 370–382. Springer, Singapore (2017)
11. Fadili, M.J., Starck, J.L.: EM algorithm for sparse representation-based image inpainting. In: IEEE International Conference on Image Processing, vol. 2, pp. 61–64. IEEE (2005)
12. Fadili, M.J., et al.: MCALab: reproducible research in signal and image decomposition and inpainting. Comput. Sci. Eng. **12**(1), 44–63 (2009)

13. Fadili, M.J., Starck, J.L., Murtagh, F.: Inpainting and zooming using sparse representations. Comput. J. **52**(1), 64–79 (2009)
14. Yang, C., Lu, X., Lin, Z., et al.: High-resolution image inpainting using multi-scale neural patch synthesis. In: IEEE Conference on Computer Vision and Pattern Recognition, vol. 1, no. 2, pp. 6721–6729 (2017)
15. Demir, U., Ünal, G.: Inpainting by deep autoencoders using an advisor network. In: Signal Processing and Communications Applications Conference, pp. 1–4. IEEE (2017)
16. Yeh, R.A., Chen, C., Lim, T.Y., et al.: Semantic image inpainting with deep generative models. In: Proceedings of the IEEE Conference on Computer Vision and Pattern Recognition, pp. 5485–5493 (2017)
17. Ishikawa, H., Ishikawa, H., Ishikawa, H.: Globally and locally consistent image completion. ACM Trans. Graph. **36**(4), 1–14 (2017)
18. Sun, J., Yuan, L., Jia, J., et al.: Image completion with structure propagation. ACM Trans. Graph. **24**(3), 861–868 (2005)
19. Ashikhmin, M.: Synthesizing natural textures. In: Symposium on Interactive 3D Graphics, pp. 217–226. ACM (2001)
20. Bertalmio, M., Vese, L., Sapiro, G., et al.: Simultaneous structure and texture image inpaintin. IEEE Trans. Image Process. **12**(8), 882–889 (2003)
21. Paris, S., Hasinoff, S.W., Kautz, J.: Local Laplacian filters: edge-aware image processing with a Laplacian pyramid. ACM Trans. Graph. **30**(4), 1–12 (2011)

To Assess the Accuracy of Color Doppler Ultrasound in Diagnosis of Renal Artery Stenosis

Yang Bai[1], Guang-Hong Han[2], and Ying Xu[3](✉)

[1] Department of Ultrasound, The First Hospital of Jilin University, Changchun 130021, Jilin, People's Republic of China
[2] Department of Oral Geriatrics, School and Hospital of Stomatology, Jilin University, Changchun 130021, People's Republic of China
[3] Department of Nephrology, The First Hospital of Jilin University, Changchun 130021, Jilin, People's Republic of China
jlujdyy@126.com

Abstract. Objective: To evaluate the accuracy and clinical application value of color Doppler ultrasound in the diagnosis of accessory renal artery (ARA) stenosis. Methods: a retrospective analysis on 80 patients from January 2014 to November 2017, due to hypertension in our hospital. All of them accepted ultrasound examination, preliminary diagnosis with suspected accessory renal artery or accessory renal artery stenosis and underwent CTA scan within 2 weeks. We record the number of accessory renal artery, origin, course, stenosis. Results: 80 cases of patients with 86 accessory renal arteries. Taking CTA as the reference standard, in the 54 right accessory renal arteries, 36 with stenosis confirmed by CTA, ultrasound confirmed 34(34/36), the diagnosis rate was 94.4%, 32 left accessory renal arteries, 20 with stenosis confirmed by CTA, ultrasound diagnosed 12(12/20) with diagnosis rate 60%. Conclusion: color Doppler ultrasound can be used as the main means of screening accessory renal artery and accessory renal artery stenosis, but in ectopic origin of accessory renal artery, with relatively higher missed diagnosis rate, if necessary with CTA or DSA.

Keywords: Accessory renal artery · Hypertension
Color Doppler ultrasonography

1 Introduction

Accessory renal artery (ARA) is an extra kidney artery except the main renal artery, it is the most common anatomic variation of renal artery [1]. Accessory renal artery and main renal artery branch provide blood supply to the corresponding region of the kidney. Due to intersegmental separation between segments of the kidney and there was no cross branch arterial anastomosis, once ARA stenosis occurred, blood supply disorder of the corresponding region would happen, then stimulate the juxtaglomerular receptor, activate the RAS system, induced hypertension [2]. At present, the research of ARA is mainly based on morphological diagnosis, such as DSA, CTA, US and so on.

K. Deng et al. (Eds.): ICMIR 2018, AISC 856, pp. 1107–1111, 2019.
https://doi.org/10.1007/978-3-030-00214-5_135

Although the diagnostic accuracy of DSA is high, it is an invasive inspection project, and it is expensive, so it is not used as a screening test. Although CTA is noninvasive examination, but the use of contrast agents will cause certain effect on renal function, and Inspection cost is expensive for Chinese people, especially for those from countryside. Although Chinese economy develop rapidly recent years, the agricultural population is still more than 60% of the total population, per capita annual income of only 9000 yuan. DSA or CTA examination fees are still burden for part of the patients.

In view of this, the clinical doctor need a relatively reliable screening methods for accessory renal artery, so the ultrasound has become the most commonly used means of screening, however, because of the instrument conditions, the condition of the patients, complexity of disease, operator experience and other factors, the accessory renal artery detection rate between different Research Center variate greatly, so comprehend the function of ultrasound in the diagnosis of ARA and the stenosis of ARA, become our concern.

2 Experimental and Computational Details

2.1 Research Objects

There were 80 patients from September 2015 to December 2017 year, due to hypertension in our hospital and received ultrasound examination, preliminary diagnosis of ARA or ARA stenosis, and receiving CTA examination confirmed within 2 weeks. There were 80 cases enroll, male 43, female 37, age 17–78 years old, the average age of 53 ± 10 years old, the patients had more than 6 months history of high blood pressure with the diastolic blood pressure 100–160 mmHg. 21 cases of them with renal dysfunction.

2.2 Instruments and Methods

The instrument used for ultrasound examination is a Logic E9 color Doppler ultrasonic diagnostic instrument with a convex probem (2.8–5.0 MHz). Fasting after 8 p.m. before the check, the time scheduled for t 8 to 11 am, the subjects supine rest for 5 min, check on the subjects of abdominal aorta and the main branches, determine the number, position, origin position of and hemodynamic parameters of ARA. Such as peak velocity, PSV, resistance index (RI) and Pulsation index (PI), early systolic acceleration time, the level of renal artery abdominal aortic peak velocity and spectrum shape, and calculate the renal artery and abdominal aorta flow rate than RAR.

2.3 Statistical Method

Spss17.0 software was used for statistical analysis. All the measurement data were checked by normal distribution. The data of normal distribution were expressed by (x ± SD). The stenosis rate of >50% was defined as morphological stenosis.

3 Results and Discussion

In the 80 cases with ARA, there were a total of 86 branches, right 54, left 32, 74 cases and 6 cases with unilateral and bilateral ARA respectively. No variation of renal artery 74 cases Renal artery stenosis is one of the most common causes of secondary hypertension. It is not easy to control by drug, but in recent years it is found that ARA can also cause hypertension [3]. And the change of ARA pressure gradient also can not accurately get through the noninvasive measurement method [4]. For patients with refractory hypertension, it is difficult to determine the cause only through laboratory indicators. So judging whether there is an accessory renal artery or renal accessory artery stenosis is of profound significance for the diagnosis of hypertension (Tables 1 and 2).

Table 1. Right ARA examination

	US	CTA
Stenosis	32	36
Non stenosis	20	18

Table 2. Left ARA examination

	US	CTA
Stenosis	12	20
Non stenosis	20	12

There are many variations in the renal artery, the most common variation is renal accessory arteries which means more than one renal artery to the renal for blood supply. We regard the grossest renal artery from the kidney door into the kidney as main renal artery, other artery is called accessory renal artery (ARA). ARA origin varied, most of accessory renal artery originated from the abdominal aorta wall, and running parallel to the main renal artery. A few were from the superior mesenteric artery, the inferior mesenteric artery, the adrenal artery, the inferior phrenic artery, and the lumbar artery, and also reported from the testicular artery.

Some scholars have pointed out that patients with refractory hypertension with accessory renal artery variation rate is as high as 59%. Some scholars believe that when the accessory renal artery diameter >2.0 mm, renal sympathetic nerve transmission in the pipe wall will have function, renal blood flow and perfusion mismatch, can lead to renin angiotensin II increased, resulting in renal vascular hypertension [5]. Accessory renal artery diameter is generally thinner than main renal artery, easily lead to low perfusion and high resistance. MARSHALL raised the accessory renal artery can cause the kidney area for reducing blood flow and microcirculation disorder, affect the juxtaglomerular apparatus function, leading to the synthesis and release of renin, activate the renin angiotensin aldosterone system, finally leading to vasoconstriction and sodium retention, increasing blood pressure.

In the 80 cases with ARA, there were a total of 86 branches, right 54, left 32, 74 cases and 6 cases with unilateral and bilateral ARA respectively. No variation of renal artery 74 cases. The accuracy of Color Doppler ultrasound in the diagnosis of accessory renal artery stenosis has significant difference between both sides, which may be related to renal artery anatomy affects accuracy of ultrasound scanning. Right ARA origin from the right anterior wall of aorta, relatively easy to check, and the left ARA origin from the left posterior wall of aorta with deep location, gastrointestinal gas interference, showed relatively more difficult. The diameter of ARA stenosis position is more slender than normal renal artery, not easy to display, so the left accessory renal artery stenosis diagnosis rate decreased. In contrast, the CTA is not affected by the above factors, not only for morphology change of accessory renal artery stenosis, but also for the origin of ectopic adrenal arteries. The accuracy of CTA diagnosis is second only to DSA, and is considered as the most accurate imaging examination except DSA in clinical work. CTA can clearly display the number, diameter and position of ARA through three-dimensional reconstruction. It has a accurate diagnostic rate for renal artery anatomic variation and stenosis, and to a certain extent, it has made up for the lack of DSA which lack of Stereoscopic sense. But CTA has x ray radiation damage, and it is contraindicated with contrast agent allergy and renal insufficiency. Therefore, it can't be used as a clinical screening method for suspected renal hypertension.

US belongs to tomography, controlled by the operator, were more flexible. While due to restrictive conditions including operator experience, instrument conditions etc., imaging quality is often affected, especially for the patients with ectopic origin. It is difficult to find ectopic origin of small branches. Color doppler ultrasound can show the normal ARA lumen with full uniform slender, smooth wall, color flow filling good, no filling defect. The Doppler spectrum shape and parameter values in the normal range. If there are multiple blood supply arteries, the diameter of the artery will be smaller than that of the single blood supply artery. Ultrasonography in patients with hypertension, such as bilateral renal size had little difference, but one side renal artery diameter is slender than contralateral, color flow Doppler spectrum shape and parameter values in the normal range, we should consider the possible existence of accessory renal artery [6].

4 Conclusions

ARA is a kind of congenital variation of renal artery, ultrasound doctors should understand and be familiar with the anatomical location and its relationship with the main renal artery, to avoid misdiagnosis and missed diagnosis. Color Doppler ultrasound is a simple and economical imaging method, can be used as the preferred method of screening for ARA.

Acknowledgments. This work was supported by Health Department of Jilin Province (No. 2009S011) and Finance Department of Jilin Province [2015].

References

1. Marshall, A.G.: Aberrant renal arteries and hypertension. Lancet **2**(6686), 701–705 (1951)
2. Derrick, J.R., Tyson, D.R.: The association of aberrant renal arteries and systemic hypertension. Surgery **48**, 907–912 (1960)
3. Derrick, J.R., Tyson, K.R.: The surgical significance of aberrant renal arteries in relation to systemic hypertension. Circulation **24**, 1192–1196 (1961)
4. De Bruyne, B., Manoharan, G., Pijls, N.H., et al.: Assessment of renal artery stenosis severity by pressure gradient measurements. J. Am. Coll. Cardiol. **48**(9), 1851–1855 (2006)
5. Cuxart, M., Picazo, M., Matas, M., Canalias, J., Nadal, C., Falcó, J.: Arterial hypertension and stenosis of the accessory renal artery. Nefrologia **27**(4), 509–510 (2007)
6. Hall, N.J., Thorpe, R.J., MacKechnie, S.G.: Stenosis of the accessory renal artery: Doppler ultrasound findings. Australas. Radiol. **39**(1), 73–77 (1995)

Computer Aided Diagnosis for Cerebral Venous Sinus Thrombosis

Lianghui Fan[1(✉)], Jungang Han[1], Jiangang Duan[2], and Chen Zhao[1]

[1] College of Computer Science and Technology, Xi'an University of Posts and Telecommunications, Xi'an, China
fh592556555@gmail.com

[2] Neurology Department, Xuanwu Hospital Capital Medical University, Beijing, China

Abstract. Cerebral Venous Sinus Thrombosis (CVST) is a rare disease which accounts for about 0.5% to 1% of all strokes. Due to the clinical symptoms lack of specificity, it is easy to be missed and misdiagnose. In order to assist doctors with less experiences, especially doctors in small cities and rural area, with diagnosing the disease as soon as possible, we study the use of Computer Aided Diagnosis (CAD) for CVST. Firstly, according to the various symptoms of CVST, combined with decision tree and expert diagnostic experience, we summarized the knowledge and obtained the 179 rules. To apply these rules in correct order, decision tree is employed to sort the rules into appropriate order according to the information gain rate. Rete algorithm is used to speed up the rule matching process. The CAD system has been tested in the dataset of case from Xuanwu Hospital, Capital, Medical University and can be used in the Web.

Keywords: Cerebral Venous Sinus Thrombosis · Computer Aided Diagnosis
Decision tree · Rete algorithm

1 Introduction

Cerebral Venous Sinus Thrombosis (CVST) is one of the rare diseases of the neurology, and its clinical manifestations are complex, with high rate of disability and mortality [1]. The reason is that few physicians have experienced the disease and thus patients would not get correct treatment in time [2–4]. According to authoritative experts, early detection and early treatment can effectively reduce mortality.

During the recent decades, expert systems has been developed in the field of medicine [7–10], and in recent years, machine learning as an important branch of artificial intelligence, can learn features through many cases, which reflects the knowledge and experience of domain experts. Expert systems and machine learning have achieved outstanding results in medical diagnosis [11–15].

There are many induces factors for CVST, including genetic hypercoagulability, high coagulation state, infection, hematopathy, autoimmune diseases, drugs and other physical factors [5, 6]. And its clinical manifestations are complex and diverse. Thrombus may occur 14 locations in Cerebral veins, different location may cause different symptom. For example, the superior sagittal sinus thrombosis is characterized

K. Deng et al. (Eds.): ICMIR 2018, AISC 856, pp. 1112–1119, 2019.
https://doi.org/10.1007/978-3-030-00214-5_136

by increased intracranial pressure, scalp venous irritation, and psychotic symptoms, whereas the sinus sigmoideus thrombosis is characterized by somnolence, diplopia, and hemiplegia. At the same time, the inducing factors are correlated, and the complex relationship between various inducing factors needs to be summarized. However, it is unclear in pathological and difficult to exactly acquire pathogenic factors because of the small number of cases in the world. The source of our knowledge is the summary of experience of diagnosis and treatment in Xuanwu hospital for many years. The hospital has the largest neurology department in China.

In order to help physicians with diagnosing CVST, based on decision tree algorithm and Rete [16] algorithm. this paper studies related algorithm and CAD system for CVST. The diagnosis knowledges are from the expert experience of Xuanwu hospital of capital medical university and other authoritative guidelines [1–6]. The data set of case is from the actual case provided by Neurology department of Xuanwu Hospital which has largest neurology department in China.

Decision tree algorithm of machine learning can identify the most discriminating properties and divide the data into multiple subsets. The C4.5 algorithm [17] of decision tree is used to obtain information gain rate [16] of the attributes of the disease.

The Rete algorithm, one of the most efficient algorithms for inference, can greatly speed up the reasoning process by sharing and saving rule conditions and the temporary matching results [11–13].

Based on the patient's medical history, clinical symptoms and all kinds of inspection results, including CT, MRI and other imaging, reasoning is carried out using the above rule matching algorithm and the probability of suffering CVST can be predicted.

2 The Framework of CAD System

The overall structure of the CAD system in this paper is shown in Fig. 1, which consists of decision system module, data module, knowledge management module and diagnostic module. The data module defines the data structure in the system and unified data transmission. The decision module is the core of the system, which is responsible for the reasoning logic of diagnosis. Knowledge management module maintains system knowledge base. The diagnostic module connects with other modules to realize the function of diagnosis. Other modules provide basic database operating functions, such as management and maintaining diagnostic history.

2.1 The Knowledge Base

The bases of Knowledge include a fact library and a rule library. The fact library contains the basic facts and the intermediate facts produced in the process of reasoning. The basic facts come from the basic information. The basic information packet contains information about the patient's medical history, drug history, symptoms, routine examination results and imaging findings. For example, cranial pressure, clotting examination, and the history of taking hormone drugs are closely associated with venous sinus thrombosis. The information is stored in knowledge base in the form of a fact as follow:

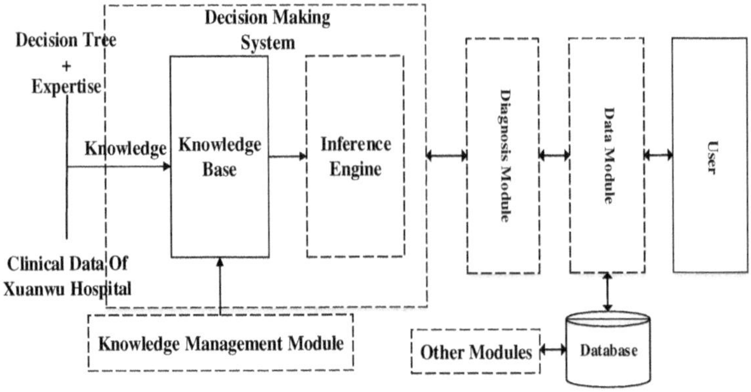

Fig. 1. CAD system framework

Symptom (symptom_name, attribute)

The intermediate facts contain the underlying facts that are obtained through the information of the disease and its combinatorial reasoning, which is used to guide the reasoning proceed smoothly. The rules library contains rules for decision making based on expert's diagnostic experience, diagnostic guidelines, and auxiliary rules for controlling reasoning. And the rules incorporate gold standard of imaging for diagnosing the disease, which ensures the accuracy of reasoning.

2.2 The Fact Base

The main facts stored in the fact library are the patient's clinical manifestation, which constitute the source of characteristic analysis of venous sinus thrombosis. The basic facts contain the patient's basic information and the symptoms.

According to the main pathogenic factors of CVST, the 64 factors and related information are shown in Table 1. A symptom template for these 64 etiologies is established including defining the basic attributes, names, and impact factors.

In the process of reasoning, auxiliary facts need to be added to control the direction of reasoning. In the reasoning process, the control algorithm is designed to control the normal reasoning of the system. Auxiliary facts include the number of rules to be executed, the label of rule with next highest priority, and the probability of suffering from CVST. Each reasoning produces a added fact, which indicates the state of reasoning and points the direction of the next reasoning.

2.3 The Rule Library

The objective of the reasoning algorithm is to deduce the patient's possibility of suffering from CVST by automatic inferencing. When the facts are insufficient, the algorithm gives a reasonable suggestion. After determining the impact factors of the

Table 1. Basic information and pathogenic factors

Basic information			
Sex	Age	Abortion	Hemopathy
Smoke	Drink	Fetation	Tumor
Anemia	Hyperthyroidism	Trauma	Glycuresis
Facial infection	Menstrual disorders	Dietary habit	Predisposing cause
Hypertension	Tympanitis	Mastoiditis	Meningitis
Antiphospholipid antibody syndrome	Antithrombin III lack of syndrome	Systemic lupus erythematosus	History of hormonal application
Hypertriglyceridemia	Protein C, S deficiency	Nasal sinusitis	Contraceptive
Symptoms			
Headache	Vomit	Vision	Epilepsy
Temperature	Consciousness	Facioplegia	Tinnitus
Neck discomfort	Optic disk edema	Dyskinesia	Papilledema
Routine inspection			
Fibrinogen	Red-cell count	Hb	HDL
RDW	Leukocyte count	D-dimer	triglyceride
LDL	Platelet count	Cholesterin	CSF
CSF glucose	Lumbar puncture pressure	CSF chloride	CSF protein
Imaging results			
Jugular ultrasound	CT direct sign	CT indirect sign	CTV sign
MRI	MRV	DSA	MRBTI

disease, two aspects should be considered according to expert knowledge: (1) Because the correlation between symptoms, the rules of logic decision should consider causal relationship among all kinds of symptoms, such as high cranial pressure will cause headache, but headache is not necessarily the result of high intracranial pressure. (2) Due to the strong diversity of clinical manifestations of CVST, it is necessary to consider the influence of the combination of various symptom. In our CAD system. All together 179 rules are summarized. Here are some simple rules:

1. IF dyskinesia AND/OR slurred speech THEN focal nerve function defect.
2. IF headache AND vomiting, AND optic disk edema, THEN high intracranial pressure.
3. IF headache AND vomiting BUT NO optic disk edema THEN high intracranial pressure in a certain probability.
4. IF focal neurological defect AND high intracranial pressure, THEN high probability of CVST, AND Continue for other factors.
5. IF defect of the focal nerve, BUT intracranial pressure normal, THEN a small probability of CVST, AND other rules are considered.

6. IF high intracranial pressure in a certain BUT NO focal neurological deficit, THEN there is a certain probability of CVST, AND continue to consider other rule.
7. IF partial or systemic seizures, THEN a certain probability of CVST, AND continue to consider another rule.

In the reasoning process, a patient's symptoms will trigger the corresponding rule. If the rule 1 deduced focal neurologic deficits, and rules 2 or 3 deduced the probability of increased intracranial pressure then rule 4 can obtain the probability of CVST. Rule 2 with rule 3, rule 4 with rule 5 and rule 4 with rule 6 are mutually exclusive respectively, meaning they are not happen at the same time. Rule 7 occurs simultaneously with rule 4 and increases the probability of CVST. The probability of each rule is calculated by data statistics and expert guideline and corrected in the process of test, resulting in the optimal value.

3 Key Algorithm

3.1 Rete Algorithm

In rule-based languages, the process of pattern matching is iterative. The fact table is modified for each execution, i.e. adding new fact tables or deleting old facts. These changes may satisfy the previously unmet rules, thus producing new facts. In each iteration, as the facts are added and removed, the set of rules that meet the criteria must be maintained and updated. The Rete algorithm is used to maintain the set of rules. For details of Rete algorithm, please refer to [16].

3.2 C4.5 Algorithm of Decision Tree

C4.5 is an improvement of ID3 algorithm, which adopts information gain rate to classify objects, and the information gain is normalized by "split information" value. C4.5 searches for the optimal classification in all possible decision-making spaces in the top-down manner. It has high accuracy and adaptability [15].

The information entropy of the classification of attribute A is defined as follows:

$$SplitInfo_A(D) = -\sum_{j=1}^{v} \frac{|D_j|}{|D|} \times log_2\left(\frac{|D_j|}{|D|}\right)$$

Where the training data set D corresponding to 148 actual cases, SplitInfoA(D) represents the information entropy generated by partition the training data set D into v output according to attribute A. In this study, attribute A is one of the 64 kinds of risk factors.

Information gain of the sample set after the partition of attribute A:

$$InfoGain(D, A) = E(D) - E_A(D)$$

where E(D) represents the amount of information before the partition, and EA(D) represents the amount of information after the partition by attribute A. The information gain rate is defined as follows:

$$GainRatio(A) = \frac{Gain(D, A)}{SplitInfo_A(D)}$$

In the iterative process, the information gain rate is selected as the attribute of partition. Applying the rules according to the information gain rate, the data can be categorized faster and more accurately.

3.3 Optimization Rules

In this paper, the information gain rate of 64 kinds of factors was calculated by using C4.5 algorithm. According to the descending order of information gain rate, The higher the information gain rate of an attribute, The greater the effect of the attribute on the diagnosis. The rule that contains facts with higher information gain rate will be apply first and will have a greater impact on the diagnosis of CVST. The top ten attributes with higher information gain rate are shown in Table 2. Note that attributes of imaging check are not included here.

Table 2. The top 10 information gain rate

Attribute	Information gain rate
Cranial pressure	0.3681
Headache	0.3228
Vision	0.1183
Epilepsy	0.1034
Predisposing cause	0.0731
Facioplegia	0.0556
Consciousness	0.0535
Facial infection	0.0534
Tinnitus	0.0402
Neck discomfort	0.0375

The information gain rate listed in Table 1 have also been validated by domain experts. In our system 179 rules are ordered according to information gain rate and given priorities accordingly thus the process of matches and reasoning are speed up.

4 Experiments and Results

In order to verify the accuracy of the system diagnosis, the actual cases from Xuanwu hospital were used. Most of the 148 cases have already been definitively diagnosed by experts. The set of 148 cases including CVST cases and also cases of intracranial hypertension, venous sinus stenosis and meningitis. All of these diseases are quite similar in clinical manifestation. The experimental results are shown in Table 3. Note that the all the test case did not use the medical image as evidence but rely solely on patients' symptoms and other test results.

The 5 cases of the 148 test cases are discarded due to incompleteness of data. The results of 20 cases are false positive. Of these 20 cases, even for human experts it is very difficult to diagnose without advanced imaging. The 123 cases in Table 3 are correctly diagnosed.

Table 3. Test results

Total cases	Correct diagnosis	False positive diagnosis
143	123	20

In the 123 cases, there are 5 cases with a less than 50% probability of diagnosis of CVST, but the results were more than 60% after referenced the indirect sign of CT. This means that CT can be used for diagnosis without MRI image. There 16 cases which are correctly eliminated possibility of CVST, but are diagnosed as anxiety, cerebral infarction ultimately or unknown.

In the diagnosis process, it is very easy to obtain the diagnosis results if the patient has MRI or DSA imaging results. The significance of our CAD system is that in the absence of imaging examination, the system is able to diagnose the possibility of CVST based on the patient's symptoms and other test results. The system has auxiliary functions for primary hospitals where no MRI imaging equipment available.

Our CAD system is developed on the platform of MyEclipse 2015, MySQL database is used to build the library of rule and facts, and the decision-making system is embedded in the platform in a modular way.

5 Conclusions

A computer Aided diagnosis system for CVST is implemented by using machine learning and expert knowledge. The system can correctly recognize the possibility of patient suffering CVST which is a very rare and very difficult to diagnosed disease for many physicians especially in small hospitals and rural area. Our system can recognize the possibility of CVST without MRI images or with incomplete imaging examination. The experimental results show that the system can not only improve the efficiency of doctors, but also can guide the diagnosis to less experienced doctor and thus effectively reduce missed diagnosis rate of CVST.

Acknowledgments. The authors like to thank the support of Xuanwu Hospital, Capital Medical University. This work is partially supported by the Graduate Innovation Foundation in Xi`an University of Posts and Telecommunications under Grant (CXJJ2017018).

References

1. Neurology Branch of Chinese Medical Association, Chinese Society of Neurology: Guideline for diagnosis and treatment of cerebral venous sinus thrombosis in China 2015. Chin. J. Neurol. **48**(10), 819–829 (2015). https://doi.org/10.3760/cma.j.issn.1006-7876.2015.10.002. (in Chinese)
2. Einhupl, K., Bousser, M.G., de Bruijn, S.F.T.M., Ferro, J.M., Martinelli, I., Masuhr, F., Stam, J., Dong, R., Sun, J., Li, X.: EFNS guideline on the treatment of cerebral venous and sinus thrombosis. Int. J. Cerebrovasc. Dis. **15**(10), 721–726 (2007). (in Chinese)
3. Sudhaker, B., Dnyaneshwar, M.P., Jaidip, C.R., et al.: Cerebral Venous Sinus Thrombosis (CVST) secondary to varicella induced hypercoagulable state in a adult. Intern. Med. Inside **2**(1), 1 (2014)
4. Karaca, M., Hismi, B., Ozgul, R.K., et al.: High prevalence of Cerebral Venous Sinus Thrombosis (CVST) as presentation of cystathionine beta-synthase deficiency in childhood: molecular and clinical findings of Turkish probands. Gene **534**(2), 197 (2014)
5. Saposnik, G., Barinagarrementeria, F., Brown, R.D., et al.: Diagnosis and management of cerebral venous thrombosis a statement for healthcare professionals from the American Heart Association/American Stroke Association. Stroke J. Cereb. Circu. **42**(4), 1158–1192 (2011)
6. Saposnik, G., Barinagarrementeria, F., Brown, R.D.: Diagnosis and management of cerebral venous thrombosis a statement for healthcare professionals from the American Heart Association/American Stroke Association. Int. J. Cerebrovasc. Dis. **19**(10), 724–757 (2011)
7. Li, J., Wang, Z., Lu, Z., Luo, Y., Zhou, J., Pang, W., Xu, Y.: Clinical features of cerebral venous sinus thrombosis in pregnancy or puerperium. J. Clin. Neurol. **26**(05), 375–377 (2013). (in Chinese)
8. Huang, H., Ren, S.: The study and application of rule-based expert system in heart rate automatic analysis. Outlook Sci. Technol. **27**(07), 93–95 (2017). (in Chinese)
9. Hàmbali M. Rule-based expert system for disease diagnosis. In: isteams nexus (2015)
10. Singla, J., Grover, D., Bhandari, A.: Medical expert systems for diagnosis of various diseases. Int. J. Comput. Appl. **93**(7), 36–43 (2014)
11. Aishwarya, S., Anto, S.: A medical expert system based on genetic algorithm and extreme learning machine for diabetes disease diagnosis. Int. J. Sci. Eng. Technol. Res. (IJSETR) **3**(5), 1375 (2014)
12. Zeki, T.S., Malakooti, M.V., Ataeipoor, Y., et al.: An expert system for diabetes diagnosis. Am. Acad. Sch. Res. J. **4**(5), 1 (2012)
13. Patra, P.S.K., Sahu, D.P., Indrajit, M.: An expert system for diagnosis of human diseases. Int. J. Comput. Appl. **1**(13), 70–73 (2010)
14. Liao, S.H.: Expert system methodologies and applications—a decade review from 1995 to 2004. Expert Syst. Appl. **28**(1), 93–103 (2005)
15. Nakai, K., Kanehisa, M.: Expert system for predicting protein localization sites in gram-negative bacteria. Proteins Struct. Funct. Bioinform. **11**, 95–110 (1991)
16. Forgy C.L.: Rete: a fast algorithm for the many pattern/many object pattern match problem. Expert Syst., 547–559 (1991). IEEE Computer Society Press
17. Quinlan, J.R.: C4.5: Programs for Machine Learning, vol. **1** (1993)

Machine Learning Soft Computing

Study on the Fuzzy Control Algorithm for Knee Exoskeleton Based on Co-simulation

Lei Fan, Lei Yan, and Jiang Xiao(✉)

Beijing Forestry University, Beijing, China
xiaojiang56@126.com

Abstract. A new type of knee joint exoskeleton, which consists of mechanical structure and a hydraulic damper, is described in this paper, the knee joint rotation is driven by the hydraulic damper. Firstly, on the basis of the exoskeleton mechanical structure, the working principle of exoskeleton were introduced, and the spring parameters of hydraulic dampers were selected by ADAMS software. Secondly, the assemble of exoskeleton and human model is imported into the ADAMS software for simulation. After building the control model in MATLAB software, the co-simulation of MATLAB and ADAMS software is carried out. In order to verify the effect of the fuzzy PID algorithm, the fuzzy PID algorithm is compared with the PID. By modifying the algorithm parameters, it is find that the fuzzy control algorithm has better following effect, which can make the wearer's gait characteristics closer to the standard gait. The simulation results show that the fuzzy PID algorithm has a better control effect in this kind of exoskeleton.

Keywords: Knee exoskeleton system · Co-simulation technology
Fuzzy PID control

1 Introduction

An exoskeleton attached to the leg is used to increase the leg muscle strength and to help those bedridden aged persons and patients with gait disorders to walk again like a normal person [1, 2]. Some lower extremity exoskeletons were also developed to help the paraplegia patients to complete a certain degree of leg movement and some of them were successfully applied to clinical gait corrections [3, 4]. As a result, rehabilitation robots are becoming more and more common [5].

Most of the knee joint exoskeletons are driven by a motor, some use spring auxiliary motor to store energy. Compared with motor drive, hydraulic equipment is simple and compact [6, 7]. Thus, a new type of hydraulic damping cylinder design applied in the exoskeleton, The spring in the damper is used as the drive, it absorbs the energy of the walking process and releases it at the right time.

One of the major tasks of the knee exoskeleton is to select the parameters of the hydraulic damper. Spring as the main part of the damper, it affects the whole equipment. Therefore, the parameters were continually adjusted through the ADAMS simulation analysis until the suitable spring parameters were selected.

K. Deng et al. (Eds.): ICMIR 2018, AISC 856, pp. 1123–1129, 2019.
https://doi.org/10.1007/978-3-030-00214-5_137

2 Structure Design and Working Principle

The knee exoskeleton has a degree of freedom, where the connection between the ankle and the calf uses a flexible carbon fiber material which has good flexibility property. A spring in a hydraulic damper is used to drive the extension of the knee joint. The hydraulic damper is a light weighted design, which has good driving capability and is easy to control while providing smooth driving. The use of hydraulic damper can significantly reduce the quantity of structural components and meet the patients' requirement - outdoor walking.

A gait classification of the walking cycle is shown in Fig. 1, in which the exoskeleton is equipped on patient's right leg. In this classification, the stand phase consists of five phases and the swing period consists of three phases. In the middle of the stance phase (3rd phase in Fig. 1), the right leg is in the upright position and at that moment, the signal is detected by the knee joint angle sensor and the servo valve is closed. When the right heel is off the ground in 4th phase in Fig. 1, the calf is equivalent to a simple pendulum and the knee bends in the process of moving. At this moment, the signal is detected by pressure sensors and the servo valve make the left and right oil cavity open. When it reach initial swing phase (6th phase in Fig. 1), right toes are completely off the ground, at that moment, spring compression reaches the maximum, the damper state as is shown in Fig. 2. When the right leg goes forward, the spring is stretched, which drives the piston to push the piston rod out. In order to control human walking gait in a more natural way, the precise control of the flow rate in the servo value is the key in the process of walking. In other words, the oil mouth size in the servo valve needs to be precisely control in the walking process so that damping can be easily adjusted and ultimately the rate of pistod extending and retracting is controlled.

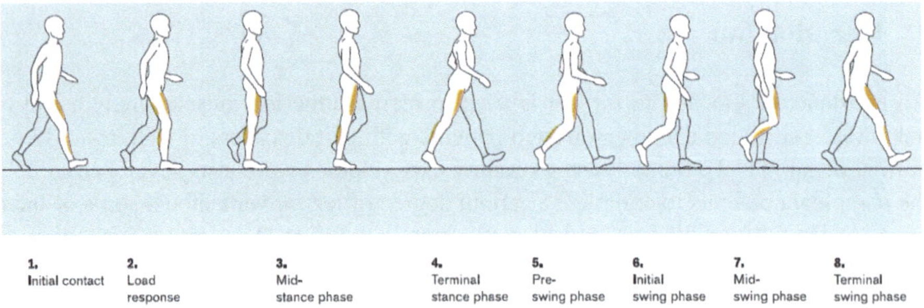

Fig. 1. The 8 phases of walking

The internal structure of the hydraulic damper is shown in Fig. 2. Three-dimensional mechanical model of the knee exoskeleton is shown in Fig. 3.

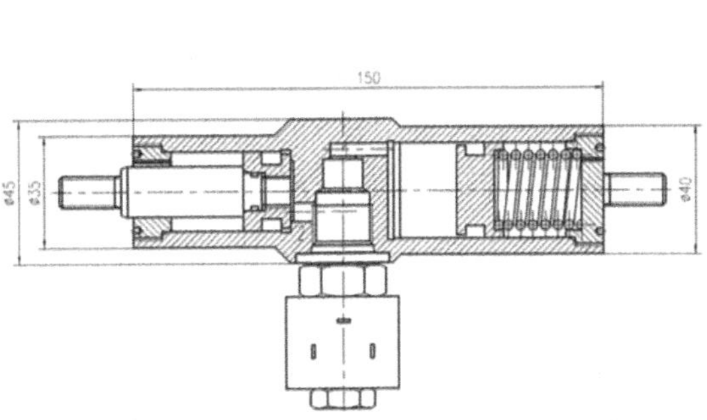

Fig. 2. Hydraulic damper (Unit: mm)

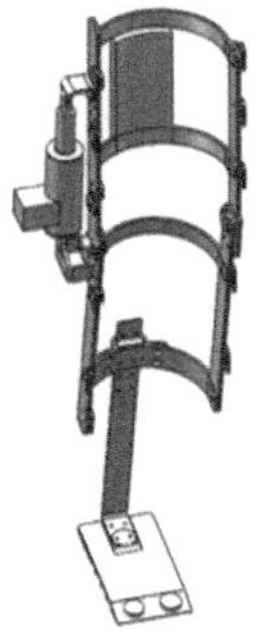

Fig. 3. Mechanical structure

3 Fuzzy PID Control

This control algorithm has been widely used in many fields, such as natural sciences and social sciences since it was proposed by L.A. Zadeh, a control expert at the University of California, USA in 1965. It expresses the control experience and knowledge of the expert as the control Rules, Which Language variables can describe, and then use these rules to control the system. Fuzzy control is especially suitable for the control of complex nonlinear systems with unknown mathematical models. It describes people's manual control experience and forms a series of control rules, and then uses fuzzy theory, fuzzy linguistic variables and fuzzy logic reasoning to raise the fuzzy control rules to numerical operations, let the computer use programs to implement these control rules so that computer simulation can automatically control the object [8, 9].

The specific process of converting input non-fuzzy quantities into fuzzy sub-sets and membership functions is called fuzzification. Fuzzy control process is divided into fuzzification, fuzzy logic reasoning and fuzzy solution judgment, which are respectively completed by fuzzy controller, fuzzy inference engine and defuzzifier. The performance of fuzzy system depends on the structure of fuzzy controller, fuzzy rules, reasoning algorithms, decision-making methods and other factors [10].

The basic principle of fuzzy control shown in Fig. 4, the two-dimensional fuzzy controller designed in this paper use knee angle deviation(e) and angle deviation rate

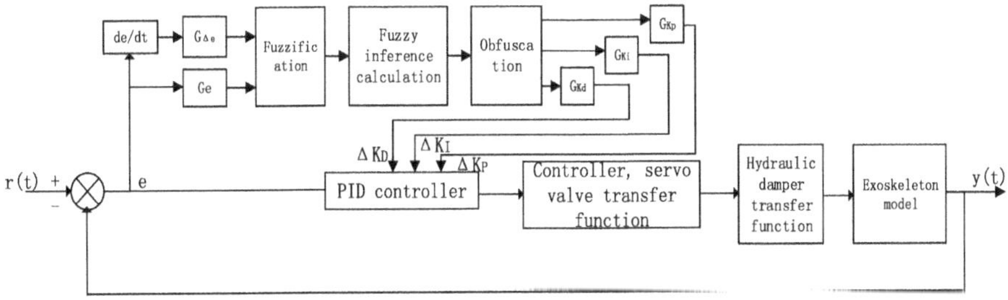

Fig. 4. Schematic diagram of fuzzy PID control system

(ec) as the input, Δk_P, Δk_I, Δk_D as the output to achieve real-time adjustment of parameters. For the sake of fuzzy control, their values need to be transformed into fuzzy values on the fuzzy domain.

Among them, the transfer function of the controller servo valve is established as: $\frac{0.056}{\frac{S^2}{500^2}+\frac{2\times 0.7}{500}S+1}=\frac{28}{S^2+700S+500^2}$, The hydraulic damper's transfer function is: $\frac{\frac{1}{5.26\times 10^{-4}}}{S\left(\frac{S^2}{156^2}+\frac{0.4}{156}S+1\right)}=\frac{46266159.7}{S^3+62.4S^2+24336S}$.

The basic domain of deviation is set to [−3, +3]. The basic universe of deviation rate of change is [−3, +3]. Define fuzzy subsets of input variables e and ec: {NB, NM, NS, ZO, PS, PM, PB}, the universe of deviation variables is {−3, −2, −1, 0, 1, 2, 3}, The k_P, k_I, k_D fuzzy rules used in Table 1.

Table 1. Fuzzy rules table about k_P, k_I, k_D

e \ ec	NB	NM	NS	ZO	PS	PM	PB
NB	PB, NB, PS	PB, NB, NS	PM, NM, NB	PM, NM, NB	PS, NS, NB	ZO, ZO, NM	ZO, ZO, NS
NM	PB, NB, PS	PB, NB, NS	PM, NM, NB	PS, NS, NM	PS, NS, NM	ZO, ZO, NS	NS, ZO, ZO
NS	PM, NB, ZO	PM, NM, NS	PM, NS, NM	NS, NS, NM	ZO, ZO, NS	NS, PS, NS	NS, PS, ZO
ZO	PM, NM, ZO	PM, NM, NS	PS, NS, NS	ZO, ZO, NS	NS, PS, NS	NM, PM, NS	NM, PM, ZO
PS	PS, NM, ZO	PS, NS, ZO	ZO, ZO, ZO	NS, PS, ZO	NS, PS, ZO	NM, PM, ZO	NM, PM, ZO
PM	PS, ZO, PB	ZO, ZO, PS	NS, PS, PS	NM, PS, PS	NM, PM, PS	NM, PB, PS	NB, PB, PB
PB	ZO, ZO, PB	ZO, ZO, PM	NM, PS, PM	NM, PM, PM	NM, PM, PS	NB, PB, PS	NB, PB, PB

Membership functions can be the same or different in the same fuzzy set. Among them, triangular, Gaussian and bell-shaped membership functions are often used. Here the triangular membership function is selected.

4 ADAMS Control Model

The knee angle and angular velocity as ADAMS output to the controller, which calculated by the controller, fed back to ADAMS, the feedback control the damping force of hydraulic damper. After drawing the 3D model of the human body, add the knee exoskeleton device, the quality of each part, the movement restraint and the driving force, among them, the standard human gait data is used to control the rest of the joint movement angle. The model is shown in Fig. 5.

After constantly changing the spring parameters to observe the walking posture of the human body, the appropriate spring parameters are finally selected as: the stiffness coefficient is 15 N/mm and the preload is 100 N.

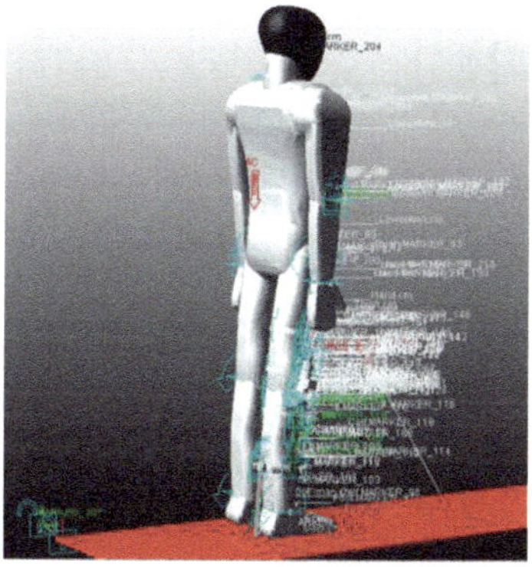

Fig. 5. ADAMS assembly effect

The final co-simulation is to be done in MATLAB, and the model made by ADAMS need to be exported into MATLAB. The exoskeleton system built by ADAMS is imported into MATLAB software to ensure the same path of ADAMS and MATLAB.

5 Simulation Test

Co-simulation using MATLAB and Adams, first, the exoskeleton model made by ADAMS is imported into MATLAB to verify the control effect. The PID control model is shown in Fig. 6.

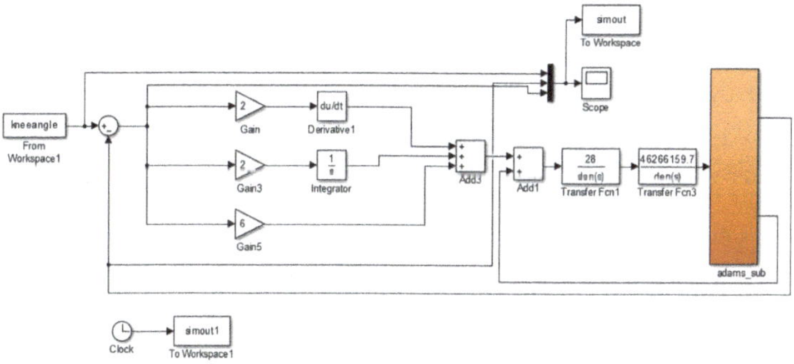

Fig. 6. PID control model

After adding ADAMS module, fuzzy PID control model is shown in Fig. 7.

When the transfer switch is set to 0, the knee on the right, which is equivalent to the human body model, is moved only by the internal spring force. When the transfer switch is adjusted to non-zero, the fuzzy PID control algorithm accesses the exoskeleton control model. Three types of simulation curves were draw as shown in Fig. 8.

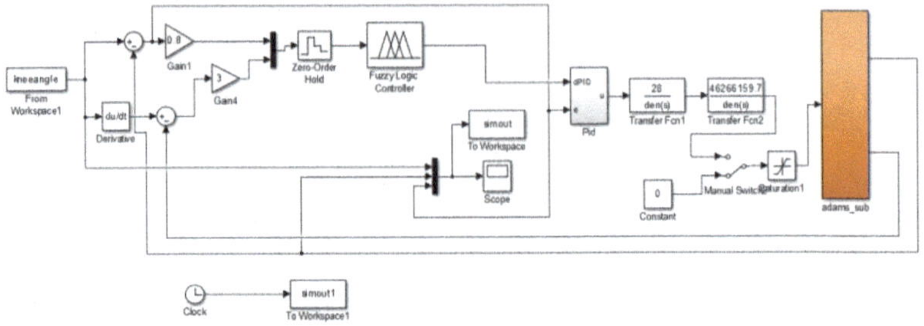

Fig. 7. Fuzzy PID control model

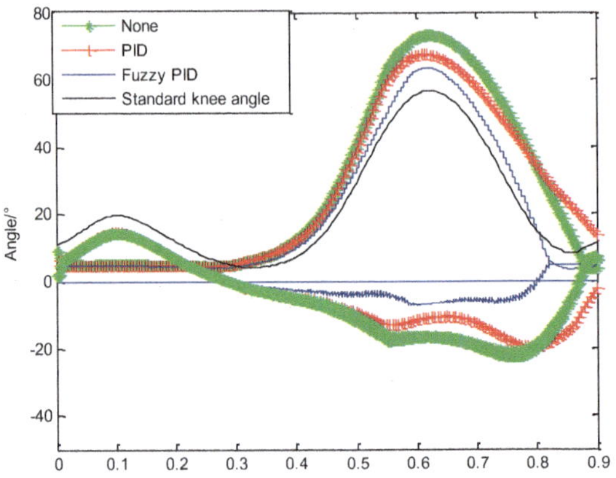

Fig. 8. The contrast of knee angle and deviation

In the absence of any control of the knee joint, compared with the standard human gait curve (black curve), the average deviation was 11.5909°. Compared with the standard human gait curve, the average deviation of the PID control algorithm is 9.3530°, and the average deviation of the fuzzy PID control algorithm is 5.5626°. Compared with the PID control algorithm, fuzzy PID control algorithm can greatly reduce the knee angle deviation, make the human body walk closer to the real gait.

6 Conclusion

Single limb knee exoskeleton has been an important branch of exoskeleton boosting system, the previous experience of knee exoskeleton at home and abroad is summarized, then developed this new knee exoskeleton.

1. The structure and working principle of knee exoskeleton are introduced. The structure mainly includes the structure of exoskeleton and hydraulic damper. The hydraulic damper is used as the actuator to promote the knee joint extension.

2. In the ADAMS, the spring parameters are constantly modified, and finally the appropriate spring parameters are selected.
3. In order to verify the selected fuzzy control algorithm, using MATLAB and Adams, fuzzy PID control, PID control and no control are compared, the simulation shows that the fuzzy PID algorithm in this paper can effectively control the knee flexion, make the model walk close to standard gait as soon as possible.

Acknowledgments. This research was financially supported by the Fundamental Research Funds for the Central Universities (Grant No. 2015ZCQ-GX-03).

References

1. Erhan, A., Mehmet, A.A.: The design and control of a therapeutic exercise robot for lower limb rehabilitation: physiotherabot. Mechatronics **21**(3), 509–522 (2011)
2. Lu, R., Li, Z., Su, C.Y., Xue, A.: Development and learning control of a human limb with a rehabilitation exoskeleton. IEEE Trans. Ind. Electron. **61**, 3776–3785 (2014)
3. Herr, H.: Exoskeletons and orthoses: classification, design challenges and future directions. J. Neuroeng. Rehabil. **6**, 21 (2009)
4. Banala, S.K., Kim, S.K., Agrawal, S.K., Scholz, J.P.: Robot assisted gait training with active leg exoskeleton (ALEX). IEEE Trans. Neural Syst. Rehabil. Eng. **17**(1), 2–8 (2009)
5. Varol, H.A., Sup, F., Goldfarb, M.: Real-time gait mode intent recognition of a powered knee and ankle prosthesis for standing and walking. In: 2nd IEEE RAS & EMBS International Conference on Biomedical Robotics and Biomechatronic, Biorob 2008, pp. 66–72, 19–22 October 2008
6. Pahl, E., Auberger, R.: Knee ankle foot orthosis with controlled swing and stance phase. In: Orthopadie-technik, January 2013
7. Dollar, M., Herr, H.: Design of a quasi-passive knee exoskeleton to assist running. In: 2008 IEEE/RSJ International Conference on Intelligent Robots and Systems, Nice, pp. 747–754 (2008)
8. Uçak, K., Günel, G.Ö.: Fuzzy PID type STR based on SVR for nonlinear systems. In: 2017 10th International Conference on Electrical and Electronics Engineering (ELECO), Bursa, Turkey, pp. 764–768 (2017)
9. Zuo, D., Qian, L., Yang, T., Cui, X., Luo, Q.: Coupling leveling control based on fuzzy PID for synchronous loading system of load-bearing test bed. Chin. J. Electron. **26**(6), 1206–1212 (2017)
10. Sharma, K., Palwalia, D.K.: A modified PID control with adaptive fuzzy controller applied to DC motor. In: 2017 International Conference on Information, Communication, Instrumentation and Control (ICICIC), Indore, India, pp. 1–6 (2017)
11. Fu, X.D., Cao, G.Z., Huang, S.D., Wu, C.: Fuzzy PID control of the planar switched reluctance motor for precision positioning. In: 2017 7th International Conference on Power Electronics Systems and Applications - Smart Mobility, Power Transfer & Security (PESA), Hong Kong, Hong Kong, pp. 1–4 (2017)
12. Jingjie, L., Dongmei, Y., Ruiyi, Z., Hui, Z.: A humidity control system based on T&H-decoupling and PID self-tuning fuzzy algorithm. In: 2017 13th IEEE International Conference on Electronic Measurement & Instruments (ICEMI), Yangzhou (2017)

A Parallel Framework for Fuzzy Membrane Clustering Based on P Systems and Improved PSO

Chengfang Zhang[1], Zhen Yue[2], Jie Jin[3], and Dan Yan[1](✉)

[1] Sichuan Police College, Luzhou 646000, Sichuan, China
619134000@qq.com
[2] Information Security Center, Beijing University of Posts and Telecommunications, Beijing 100876, China
[3] Chengdu Company, Space Star Technology Co., LTD., Chengdu 610100, Sichuan, China

Abstract. The traditional Fuzzy c-means (FCM) clustering algorithm is sensitive to initial seeds and noise. A novel fuzzy clustering algorithm is proposed by Peng Hong, which is named as Fuzzy-MC. However, due to the limitation of the serial architecture of current computer, the parallel and distributed computing characteristics of P system was not able to be exhibited. Therefore, Fuzzy-MC algorithm increases its computing time. To reduce the computing time, this paper proposes a GPU-based parallel Fuzzy-MC algorithm. In the parallel algorithm, block layer in CUDA programming model is used to represent the cells, while threads are used to realize the evolution and communication of the objects. Two artificial data sets and four real-life data sets from the UCI data sets are chosen to compare parallel version and the corresponding serial version. The computing time and clustering performance are used to demonstrate the effectiveness of the proposed algorithm. Compared with the Fuzzy-MC algorithm, parallel Fuzzy-MC algorithm not only ensures the clustering performance but also reduce its computing time.

Keywords: Fuzzy-MC · Improved PSO · GPU · P systems · CUDA

1 Introduction

Clustering analysis, as an important technology of data processing, has received extensive attention and application. Fuzzy c-means (FCM) is a typical method for fuzzy clustering algorithm [1]. The "hard" division of the K-Means clustering algorithm is transformed into "soft" division, which is more applicable in reality. FCM clustering algorithm has been research focus in clustering field [2, 3].

Although FCM clustering algorithm solves the defects of traditional hard clustering, the algorithm is sensitive to initial value and noise data. In the past few years, various improved approaches having been proposed to overcome the defects of FCM clustering algorithm at home and abroad. Generally, research progress can be grouped into three categories: (1) Other clustering algorithms are proposed based on the FCM algorithm, such as the fuzzy C line (FCL) algorithm [4]. (2) Fuzzy clustering algorithm

K. Deng et al. (Eds.): ICMIR 2018, AISC 856, pp. 1130–1137, 2019.
https://doi.org/10.1007/978-3-030-00214-5_138

combined with natural evolutionary algorithm [5–7]. (3) Clustering algorithm under computing framework [8, 9].

Membrane computing (MC) has been successfully employed in fuzzy clustering applications, namely, Fuzzy-MC [10]. Experimental results demonstrate Fuzzy-MC can clearly outperform other evolutionary clustering algorithms and classical FCM. The study in [10, 11] shows that the computational cost will increase a lot when number of cells becomes larger. To overcome the above shortcomings of conventional Fuzzy-MC-based clustering methods, we propose a novel parallel framework for fuzzy membrane clustering in this paper.

This paper is structured as follows. Section 2 introduces the designed P system. The serial algorithm and parallel algorithm of Fuzzy-MC is introduced in Sect. 3. The experimental results are given in Sect. 4. Finally, Sect. 5 concludes the paper.

2 P System Framework for Fuzzy Clustering

2.1 Membrane Structure

In this Section, the designed membrane structure is shown in Fig. 1. It can be seen from Fig. 1 that our system contains q cells, labeled with $1, 2, \ldots, q$, respectively. The detail tissue-like catalytic P system refers to [11].

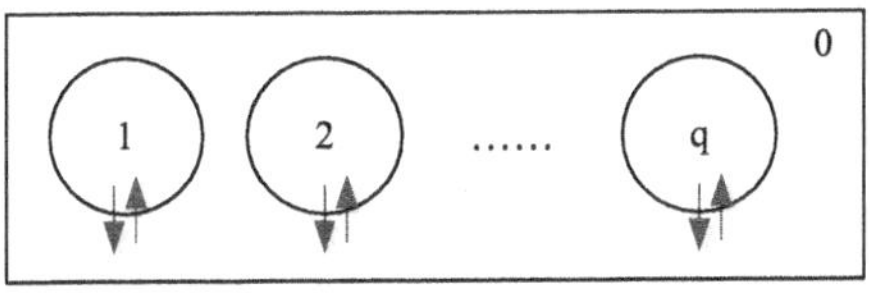

Fig. 1. The membrane structure of tissue-like P Systems

A (tissue-like) catalytic P system is given in the form

$$\prod = (O_1, w_1, w_2, \ldots, w_q, R_1, R_2, \ldots, R_q, R', i_0) \quad (1)$$

where O_1 is the alphabet of all objects, $w_i (1 \leq i \leq q)$ are multi-sets of objects present in the i cell (the number of objects in each cell is set to n). $R_i (1 \leq i \leq q)$ are finite sets of evolution rules in the i cell. R' represents transport rules set, it can be the following form: $< i, Z/Z', O >$, $i = 1, 2, \ldots, q$, among which Z and Z' are objects. i_0 is output area.

2.2 Object

In the D dimension space, the FCM fuzzy clustering divides the data set $D = \{X_1, X_2, \ldots, X_n\}$ into K clusters $\{C_1, C_2, \ldots, C_K\}$. $z_1, z_2, \ldots, z_k$ are corresponding fuzzy clustering centers. In order to deal with clustering problem, the object in the system is considered as $K \times d$ dimensional vector. The object representation is shown in Fig. 2.

$$Z = (z_1, z_2, \ldots, z_k) = (z_{11}, z_{12}, \ldots, z_{1d}; z_{21}, z_{22}, \ldots z_{2d}; , \ldots z_{k1}, z_{k2}, \ldots, z_{kd}) \tag{2}$$

where, $z_i = (z_{i1}, z_{i2}, \ldots, z_{id})$ is *ith* clustering center.

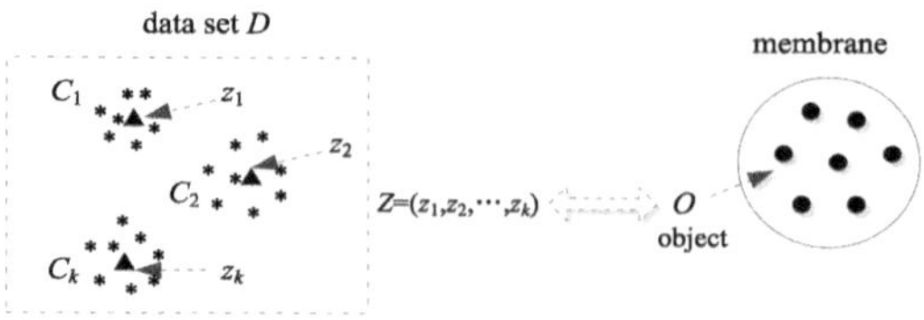

Fig. 2. The representation of objects in membranes

2.3 Improved PSO-Based Evolution Rule

The conventional position-velocity model of the evolution rule is as follows.

$$\begin{cases} O_j^i = \omega V_j^i + c_1 r_1 \left(O_{jbest}^i - O_j^i\right) + c_2 r_2 \left(O_{ibest}^{n-1} - O_j^i\right) + c_3 r_3 \left(O_{ebest}^{q-1} - O_j^i\right) \\ O_j^i = O_j^i + V_j^i \end{cases} \tag{3}$$

where ω was termed the 'intertia weight'. c_1, c_2, c_3 represent constant learning factor. r_1, r_2, r_3 were three random numbers between 0 and 1.

An improved position-velocity model is applied to obtain better object in our paper. First, objects of each membrane can remember its optimal position (O_{jbest}^i) when executing evolutionary computation. Second, optimal value (O_{ibest}^{n-1}) from other $N-1$ objects in the membrane was found. Finally, the optimal object (O_{ebest}^{q-1}) can be obtained from the other $q-1$ membranes with the help of P system from Fig. 1.

2.4 Evaluation Function

For the object evaluation in the framework of P system, the evaluation function in FCM is used to complete the evaluation of objects in our paper.

$$J_{FCM}(U, c_1, \ldots c_c) = \sum_{i=1}^{c} J_i = \sum_{i=1}^{c} \sum_{j}^{n} u_{ij}^m d_{ij}^2 \tag{4}$$

$$u_{ij} = \frac{1}{\sum_{k=1}^{c} \left(\frac{d_{ij}}{d_{kj}}\right)^{2/(m-1)}} \tag{5}$$

where, c_i is *ith* clustering center. $d_{ij} = \|c_i - x_j\|$ is Euclidean distance between the *ith* clustering center and the *jth* data point, $m \in [1, \infty)$ is weighted index to control flexibility. Generally, the smaller the value of J is, the better the object is.

3 Parallel Fuzzy-MC Algorithm Based on GPU

3.1 Serial Fuzzy-MC Algorithm

A novel fuzzy clustering algorithm called Fuzzy-MC is proposed in [11]. Their experimental results show that he Fuzzy-MC owns clear advantages over traditional FCM-based methods and biological evolutionary algorithms (PSO, GA) for cluster analysis, and can lead to state-of-the-art results.

3.2 Parallel Fuzzy-MC Algorithm

We can see from literature [11] that GPU-based is better than CPU-based in terms of time performance. In this paper, an improved PSO is used as the object evolution rule based on the ring tissue-like P system. The detailed flow of parallel computation is showed in Table 1.

4 Experiments

4.1 Experimental Setups

In this experiment, n is set as 1024; we also set the weighted exponent is 2 in FCM algorithm; parameters of PSO are set as follows: $\omega = 1.0, c_1 = c_2 = c_3 = 1.2$; The maximum number of iterations was set to 200. When testing the influence of the number of membranes on the performance, q will be set to 4, 8, 16, 32, 64, 128 and 256, respectively.

4.2 Data Sets

As shown in Table 2, six data are employed to verify the effectiveness of the proposed clustering framework (Data set is available on ***UCI***).

4.3 Index Analysis

(1) Clustering Performance

f is used to verify the performance of parallel fuzzy clustering algorithm. If f is too bad, the proposed parallel algorithm can be invalid. The comparison results of clustering performances for the parallel and serial version of fuzzy membrane clustering algorithm is given in Table 3. Four data sets (Wine, Square4, Iris, New Thyroid Disease) are selected in the Fig. 3. The following conclusions can be obtained from Table 3 and Fig. 3.

(1) By comparing the Iris and New Thyroid Disease, we can see that the performance of serial and parallel clustering is basically no difference from the beginning of q = 16.

Table 1. The parallel of Fuzzy-MC algorithm on the GPU

Input:
Data-set: X
Parameters: K; q; n; T
Improving PSO parameters: ω , c_1, c_2 , c_3
Output:
Optimal object { O^i_{jbest} , O^{n-1}_{ibest} , O^{q-1}_{ebest} }
Beginning:
Step1:/*(CPU). Initialization */
For i =1 to q
for j =1 to n
in ith membrane, jth object O^i_j is obtained;
end for
O^i_j is conveyed to GPU;
end for
Set initial iteration times t=0;
Step2:/* (GPU). Object evolution */
for collateral execution $i(1,2,...,q)th$ membranes.
for collateral execution $j(1,2,...,q)th$ objects.
Evolving object O^i_j based on improved PSO rule model;
Synchronous operation;
Evaluating object O^i_j based on evaluation function J_{FCM} ;
Synchronous operation;
end for
end for
Step 3:/* (GPU). Object communication */
for collateral execution $i(1,2,...,q)th$ membranes.
for collateral execution $j(1,2,...,q)th$ objects.
Updating J_{FCM} according to O^i_{jbest} ;
O^{n-1}_{ibest} is updated according to inter membrane communication rules;
O^{q-1}_{ebest} is updated according to intra membrane communication rules;
end for
Step 4:/* Stopping rule*/
if $t < T$
$t = t + 1$;
Turning Step 2;
end if
Output optimal object,{ O^i_{jbest} , O^{n-1}_{ibest} , O^{q-1}_{ebest} };
Computing membership matrix U^i_j ;
The fuzzy partition from data points to K clusters is completed according to the membership matrix U^i_j ;
END

(2) It can been seen that the local fluctuation of computing performance in GPU at q = 16 for Iris, and the fluctuation of computing performance in CPU at q = 32, but eventually tends to be stable and convergent.

(2) Time Consuming

Table 2. The description of Data_set

Data_set	No. of data	Dimension	No. of clusters	Description
New Thyroid Disease	215	5	3	Real-life
Wine	178	13	3	Real-life
Live Disorder	345	6	2	Real-life
Iris	150	4	3	Real-life
AD_5_2	250	2	5	Artificial-life
Square_4	1000	2	4	Artificial-life

Table 3. The comparison results of clustering performances for the parallel and serial version of fuzzy membrane clustering algorithm on the six data sets (f value)

Data Sets	Version	The number of cells						
		4	8	16	32	64	128	256
New Thyroid	Parallel	**1992.32**	**1976.67**	**1979.01**	1973.65	1973.16	1972.09	1972.15
	Serial	1977.95	1978.22	1977.01	**1975.23**	**1972.56**	**1970.44**	**1971.41**
Live Disorder	Parallel	**9858.10**	**9856.03**	**9854.21**	**9851.98**	**9851.76**	**9851.69**	**9851.72**
	Serial	9860.32	9863.56	9856.32	9853.27	9852.42	9852.00	9851.73
Wine	Parallel	17302.21	17100.32	**16295.62**	**16294.2**	16311.97	16298.1	**16291.96**
	Serial	**17298.31**	**17000.11**	16800.54	16305.01	**16300.26**	**16293.01**	16292.62
Iris	Parallel	110.02	104.66	**98.56**	**97.30**	**97.19**	97.28	97.13
	Serial	**108.36**	**98.39**	100.1	97.52	97.68	**96.99**	**96.91**
Square4	Parallel	2385.69	2382.68	2380.76	2382.92	2380.74	2380.11	2380.20
	Serial	**2382.50**	**2381.57**	**2379.94**	**2380.56**	**2379.89**	**2380.02**	**2379.77**
AD_5_2	Parallel	330.01	**328.92**	**328.20**	327.48	326.51	326.50	326.47
	Serial	**329.25**	329.10	328.57	**326.42**	**326.44**	326.50	**326.44**

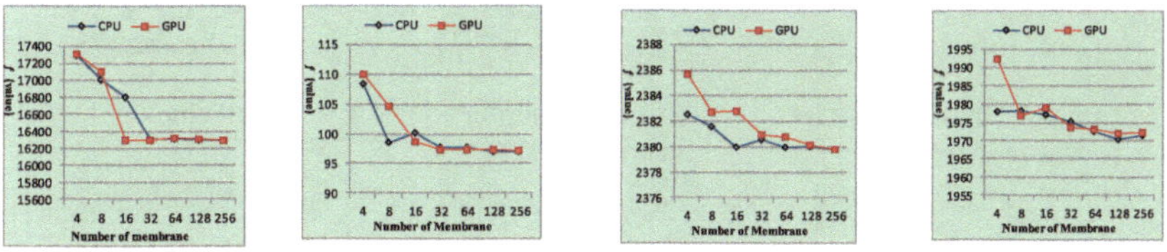

Fig. 3. The comparison of the parallel and serial versions of Fuzzy-MC algorithm in terms of computing performance. From left to right: Wine, Square4, Iris, New –Thyroid Disease.

The computing times of the parallel and serial versions is given in Table 4. The following conclusions can be obtained from Table 4. The strip comparative analysis diagrams of time consuming in CPU and GPU are shown in Fig. 4. The following conclusions can be obtained from Table 4 and Fig. 4.

(1) It can be seen from Fig. 4 that time consuming is almost doubly increasing with the number of membrane on CPU. As shown in the parallel version of (a) (b) (d) (f), the computing time is almost unchanged or the change is small with the increase of the amount of tasks, when the number of membrane is 4 and 8 in the initial stage.
(2) When the number of membrane increases to a certain value, the computational cores of GPU and CPU have been fully utilized, and the computational workload has almost become the only factor restricting the time consuming.

Table 4. The computing times of the parallel and serial versions of fuzzy membrane clustering algorithm on the six data sets

Data Sets	Version	The number of cells						
		4	8	16	32	64	128	256
New Thyroid	Parallel	**26.15**	**27.95**	**40.26**	**71.62**	**125.21**	**223.38**	**411.72**
	Serial	28.66	48.43	90.08	199.36	460.52	1005.25	2256.39
Live Disorder	Parallel	**32.68**	**34.61**	**61.20**	**104.04**	**217.33**	**585.07**	**1296.34**
	Serial	46.29	88.06	203.43	498.04	1244.52	3250.59	8509.15
Wine	Parallel	**992.06**	**1349.12**	**2428.41**	**4516.85**	**7782.02**	**14217.14**	**30523.88**
	Serial	1260.21	2873.27	6608.54	14803.13	35527.51	85216.51	204510.45
Iris	Parallel	**52.24**	**61.32**	**104.62**	**184.46**	**230.83**	**399.17**	**694.89**
	Serial	96.90	209.75	384.35	911.55	1782.53	4006.32	9726.64
Square4	Parallel	**113.28**	**200.27**	**384.76**	**767.63**	**1251.11**	**2432.02**	**4608.58**
	Serial	295.10	570.55	969.53	1884.11	4055.56	8420.29	19836.56
AD_5_2	Parallel	**51.35**	**53.65**	**120.31**	**263.83**	**510.67**	**983.06**	**2272.93**
	Serial	118.23	230.78	598.47	1334.60	2456.25	5656.35	12901.74

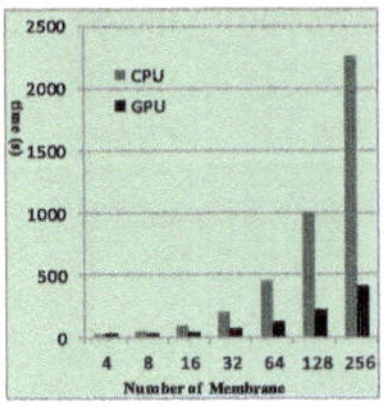

(a)New Thyroid Desease

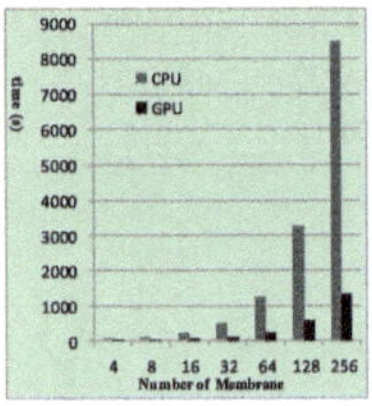

(b)Live Disorder

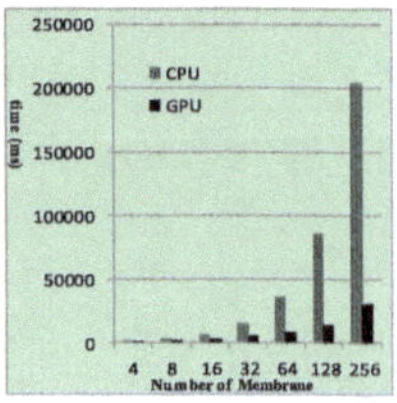

(c)Wine

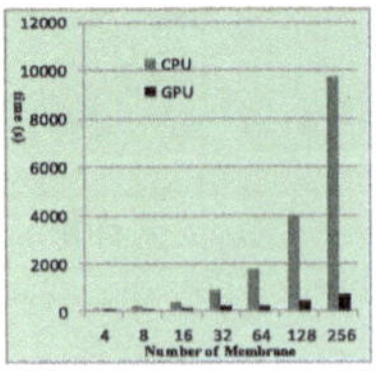

(d)Iris

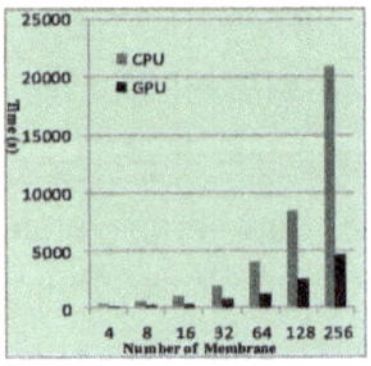

(e)Square4

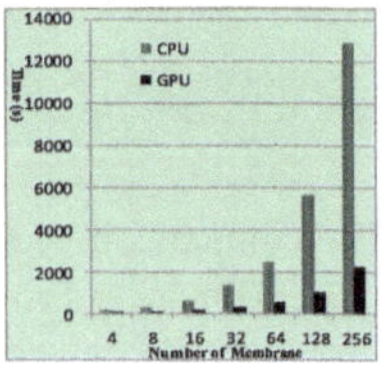

(f)AD_5_2

Fig. 4. The comparison of the parallel and serial versions of Fuzzy-MC algorithm in terms of computing times

5 Conclusion

In this paper we have developed a new parallel framework that allows us to apply the fuzzy membrane clustering to take full advantage of parallel characteristics of P system. In our experiments, parallel Fuzzy-MC algorithm and serial Fuzzy-MC algorithm are employed for clustering of four real-life data and two artificial-life data use, respectively. The experimental results confirm that the proposed framework exhibits better results than serial Fuzzy-MC algorithm both in clustering performance and time consuming.

Acknowledgments. The authors first sincerely thank the editors and anonymous reviewers for their constructive comments and suggestions. The authors would also like to thank Prof. H Peng from XiHua University (China). This work is supported by the Key research base of philosophy and Social Sciences in Sichuan province and key research base of Humanities and Social Sciences in Sichuan colleges and Universities (No. SHZLQN1701).

References

1. Sun, J.-G.: Clustering algorithms research. J. Softw. **19**(1), 48–61 (2008)
2. Li, Y.-C., Yin, P.-P.: A FCM-based thread partitioning algorithm for speculative multithreading. Chin. J. Comput. **37**(3), 580–592 (2014)
3. Guo, H., Liu, L., Zhao, Y., Xu, F.: Sonar image segmentation based on MAR and FCM clustering. Chin. J. Sci. Instrum. **10**(34), 2322–2327 (2013)
4. Bezdek, J.C.: Detection and characterization of cluster substructure 1. Linear structure fuzzy c-lines. SIAM J. Appl. Math. **40**(2), 339–357 (1981)
5. Bhuyan, J.N.: Genetic algorithm for clustering with an ordered representation. In: ICGA, pp. 408–415 (1991)
6. Niknam, T., Olamaie, J., Amiri, B.: A hybrid evolutionary algorithm based on ACO and SA for cluster analysis. J. Appl. Sci. **8**(15), 1695–2702 (2008)
7. Omran, M.G., Engelbrecht, A.P.: Self-adaptive differential evolution methods for unsupervised image classification. In: Cybernetics and Intelligent Systems, pp. 1–6 (2006)
8. Hebboul, A., Hachouf, F., Boulemnadjel, A.: A new incremental neural network for simultaneous clustering and classification. Neurocomputing **169**(2), 89–99 (2015)
9. Peng, H., Zhang, J., Jiang, Y.: DE-MC: a membrane clustering algorithm based on differential evolution mechanism. Rom. J. Inf. Sci. Technol. **17**(1), 76–88 (2014)
10. Peng, H., Wang, J., Pérez-Jiménez, M.J., Riscos-Núñez, A.: An unsupervised learning algorithm for membrane computing. Inf. Sci. **304**(C), 80–91 (2015)
11. Jin, J., Liu, H., Wang, F., Peng, H.: Parallel implementation of P systems for data clustering on GPU. Bio Inspired Comput. Theor. Appl. **562**, 200–211 (2015)

Intelligent Aided Design of Textile Patterns Based on Deep Learning

Ying Wang[1(✉)] and Zhengdong Liu[2]

[1] Basic Teaching Department, Beijing Institute of Fashion Technology, Beijing, China
jsjwy@bift.edu.cn
[2] School of Fashion, Beijing Institute of Fashion Technology, Beijing, China

Abstract. Deep learning technology has developed significantly in the field of pattern recognition in recent years. As an important research achievement in the field of artificial intelligence, multi-layer artificial neural network has achieved impressive practical results in visual processing. In particular, the style migration algorithm based on the convolution neural network CNN provides a new opportunity for the design of textile patterns by mixing visual quality from multiple source images to create new output images. This paper combines the fractal pattern with the style migration algorithm based on deep learning, and uses the deep learning framework Tensorflow to realize the intelligent aided design system of textile pattern based on deep learning. The integrity of the section is more in line with the textile printing requirements.

Keywords: Deep learning · Fractal pattern · Style migration · Textile pattern

The textile pattern is a product that integrates artistry and practicality. However, in the field of design of textile patterns, it is a very tedious process of creative design and modification. Whether traditional hand drawing or computer rendering, it still displays in graphic fabric drawings or other media. Therefore, it is necessary to expand the design vision and introduce a new design method.

This paper introduces the automatic generation technology of textile patterns based on deep learning. The use of deep learning architecture and technology to generate textile patterns (and other artistic contents) is a field of art and science developed in recent years with the help of scientific visualization. Its main characteristics lie in the fact that it not only allows the production of automated works of art, but also extends the field of creative expression. Therefore, it has opened up a new world for pattern designers.

1 The Formation of Fractal Pattern

The fractal graph is based on the mathematical theory, and it is programmed by adjusting the data of iterative function, change rule and color parameter to realize the visualization of the program code, and transform the boring data into gorgeous graphics or images. Due to the generation principle of fractal map, it has a very fine structure. If there is no source code of fractal graph, it will not be easy to copy. This fractal graph

K. Deng et al. (Eds.): ICMIR 2018, AISC 856, pp. 1138–1145, 2019.
https://doi.org/10.1007/978-3-030-00214-5_139

not only realizes the perfect combination of fractal geometry and computer graphics, but also the integration of science and art. Besides, the beauty of fractal maps is immeasurable, and adds a new element [1] to the design of textile patterns. This paper mainly uses Tensorflow to calculate the fractal graph on the complex plane. By changing the form and parameters of the iterative function, the fractal graph is generated, and it is saved in the picture form, which is convenient for the style design of the textile pattern in the later period.

Mandelbrot set is the most classical example of fractals. Consideration of iterative formula:

$$z\_\{n+1\} = z\_\{n\}^2 + c \tag{1}$$

Both Z and C are plural. When z_0 is 0, the value obtained can form a sequence of numbers c, $c^2 + c$, $(c^2 + c)^2 + c$……. When the series diverges to infinity, the corresponding points belong to the Mandelbrot set.

In TensorFlow, we define the following calculation steps:

xs = tf.constant(Z.astype(np.complex64))zs = tf.Variable(xs)ns = tf.Variable(tf.zeros_like(xs, tf.float32)) with tf.Session(): tf.global_variables_initializer().run() zs_ = tf.where(tf.abs(zs) < R, zs**2 + xs, zs) not_diverged = tf.abs(zs_) < R step = tf.group (zs.assign(zs_), ns.assign_add(tf.cast(not_diverged, tf.float32)) for i in range(ITER_-NUM): step.run() final_step = ns.eval() final_z = zs_.eval()

ZS corresponds to the Z of our previous iteration formula, while XS corresponds to C in the iteration formula. For convenience, as long as the absolute value of the numerical value is greater than a predetermined value R, it is considered to be divergent. For each calculation, tf.where is used to calculate the values that are not yet divergent. Combining ns and zs_, we can calculate the color and get the classic Mandelbrot image.

2 Real Time Style of Fractal Pattern

2.1 Deep Learning Network Structure

The deep learning network architecture designed in this paper consists of two parts (Fig. 1). One is the image transformation network F_W, that learns the transform of the style picture to the content picture, and the W (weights) is the optimal weight of all the neurons of the F_W; the two is the convolution neural network structure VGG16 developed by the Visual Geometry Group (VGG) of the Oxford visual geometry group. The VGG16 network model is used to calculate the loss function of loss network φ. The network structure extracts convolution features (Feature Map) of different layers through the convolution layer, which can express human visual perception of images. The convolution features are used to calculate the visual perception comparison loss of fractal pattern and style picture, and a series of loss functions, L_1, L_2, … L_k can be defined according to the Feature Map of different convolution layers. Assuming a fractal pattern X is calculated by deep convolution neural network F_W to get the creative pattern Y, then $y = F_W(x)$ is defined, and each loss function calculates a scalar value L_i (y, y_i) to measure the gap between the output y and the target image y_i, and the loss function can be considered in many aspects, such as texture, light, histogram and so on. Therefore, the solution process of F_W is transformed into formula (2) for gradient descent to get the optimal solution, where λ_i is the weight proportion of each loss function i.

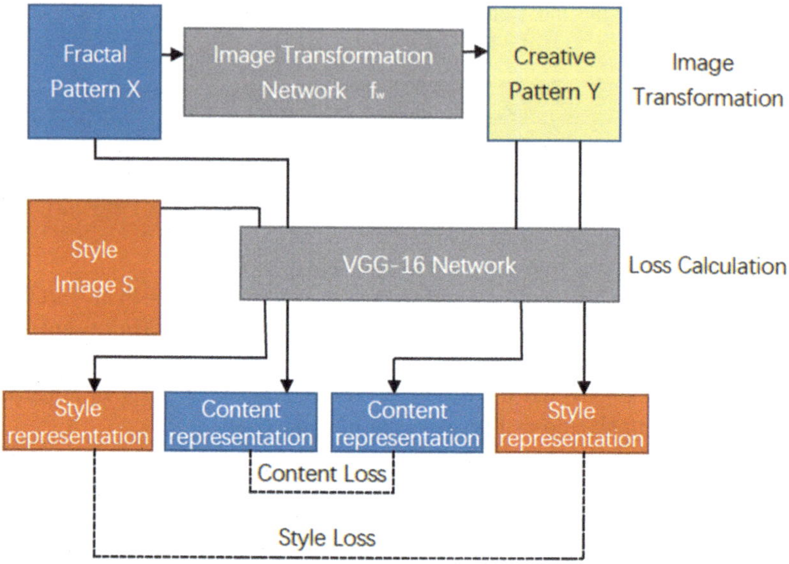

Fig. 1. Deep learning network structure

2.2 Image Transformation Network

The structure of the image transformation network is shown in Table 1. Three convolution layers are followed by 5 residual blocks, then the two unpooling layers, and the last convolution layer. The convolution core of the first and the last layers are all 9 × 9, and the rest are all 3 × 3. Each of the residual blocks contains two layers of convolution. We replace the Batch-normalization with instance-normalization [2] in our network structure, and all of the batch-normalization in the generator network are replaced. Doing so can prevent the instance-specific mean and covariance from simplifying the learning process, increasing the style migration speed and making the resulting image better. At the same time, deconvolution (transposed convolution) is also replaced by resize-convolution, which prevents the occurrence of tessellation.

$$W^* = \arg\min_W E_{x,\{y_i\}}\left[\sum_{i=1} \lambda_i l_i(f_W(x), y_i)\right] \quad (2)$$

Table 1. Image transformation network structure

Layer	Activation size	Nonlinearity
Input	3 × 256 × 256	
Convolution	32 × 256 × 256	ReLU
Convolution	64 × 128 × 128	ReLU
Convolution	128 × 64 × 64	ReLU
5 Residual block	128 × 64 × 64	
Unpooling	64 × 128 × 128	
Unpooling	32 × 256 × 256	
Convolution	3 × 256 × 256	Sigmoid

Furthermore, in order to be more suitable for the requirements of the high boundary of the textile pattern, the usual zero padding is replaced by reflect padding, which reduces the border effect.

2.3 Perceptual Loss Function

Because the generated fractal patterns are mostly pure color background images, if they are directly stylized, the edges of the stylized image will be extended to the color of the object, the boundary will become unevenness. In order to create the creative pattern, it will not lose the details of the fractal pattern itself, so that it is more in line with the spinning. There are two strategies in the definition of loss function. Firstly, the fractal pattern is segmented by semantic [3], and the segmented template is added to the neural network as the augmentation of the fractal pattern, so that the content that we are interested in is only processed in the optimization process. Secondly, a local affine transformation [4] in color space is established between the fractal pattern and the creative pattern of the final output image, that is, each pixel value in each color channel of the creative pattern can be obtained by the linear combination of the local region pixels of the fractal pattern. By establishing this relationship, we can achieve the goal of constrain the creative pattern by fractal pattern, so that we can not lose the details of the fractal pattern.

The specific mathematical forms are as follows:

$$L_l^1 = \frac{1}{2N_l D_l} \sum\nolimits_{ij} (F_l[O] - F_l[I])_{ij}^2 \tag{3}$$

Among them, L is the total number of convolution layers, and l is the L layer of the deep convolution neural network. Each layer has Nl filters each of which is a vector feature map of D size. $F_l[\cdot]$ is a characteristic matrix.

$$L_S^l = \sum_{c=1}^{C} \frac{1}{2N_{l,l}^2} \sum_{ij} \left(G_{l,l}[O] - G_{l,l}[S]\right)_{ij}^2 \tag{4a}$$

$$F_{l,l}[O] = F_l[O]M_{l,l}[I] \quad F_{l,l}[S] = F_l[S]M_{l,l}[S] \tag{4b}$$

C is the number of channels in the semantic partition mask, and $M_{l,I}[\cdot]$ is the channel C for the semantic segmentation mask of the L layer, and $G_{l,I}[\cdot]$ is the Gram Matrix [5] corresponding to $F_{l,I}[\cdot]$. In each layer of the convolution neural network, the mask is sampled to match the size of the feature map space.

$$L_m = \sum_{c=1}^{3} V_c[O]^T M_I V_c[O] \tag{5}$$

Among them, MI is the Matting Laplacian of input image I, and $V_c[O]$ is the quantized representation of C channel of output image O. Finally, by combining these functions, the final optimization objective function is obtained.

$$L_{total} = \sum_{l=1}^{L} \alpha_l L_l^l + \Gamma \sum_{l=1}^{L} \beta_l L_s^l + \lambda L_m \tag{6}$$

Where λ is the weight of the Photorealism regular term. The rest of the parameters are described in the previous section. In the same way, the optimization method is gradient descent.

3 Experiment and Analysis

In this paper, a pre-trained VGG-16 is used as a feature extractor. Select conv3_3 as content representation, conv1_2, conv2_2, conv3_3, conv4_3 as style representation. In view of the particularity of texture and edge requirements of textile patterns, we compare the generated fractal patterns with different style pictures, and compare the results as shown in Table 2. We see a slight distortion of the boundary effect of the textile pattern generated by a rich painting and color picture, such as the effect of 2 rows and 3 columns in the table and the 3 row of 3 rows. If the color of the style picture is single or similar to the fractal pattern, the texture and boundary of the textile pattern are very clear, which is basically in line with the printing requirement of the textile pattern.

In addition, we also compare the speed of the trained model with the other mature neural style migration algorithms in the size of 256 * 256 pixels, and the results are shown in Table 3. It can be seen that if we train the model well, the speed of adopting this method is acceptable.

Table 2. Comparison of stylization of creative patterns

Fractal Pattern	Style Image	Creative Pattern
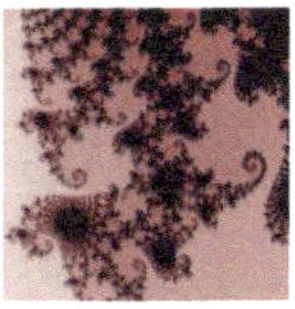		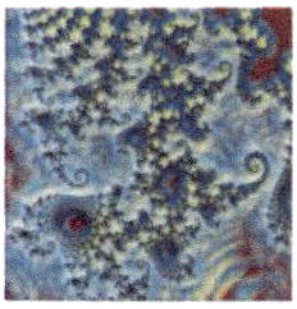
		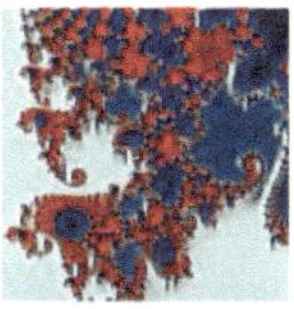
		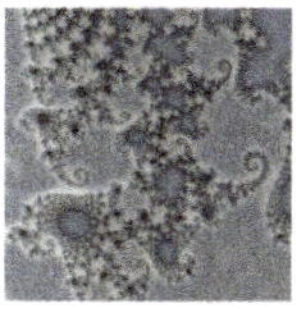
		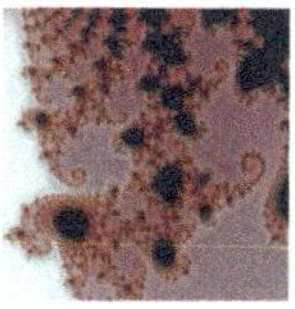

Table 3. Speed comparison of different neural style migration algorithms

Algorithm	Time
Gatys et al. [6]	14.34
Ulyanov et al. [7]	0.025
Johnson et al. [8, 9]	0.013
This paper	0.145

4 System Implementation

When creating creative patterns, designers need the integration of man-machine and integrated design. This process requires not only dialogue but also exchange of information. The process of design is not a simple input procedure – data processing output results, but a process of human-machine interaction. The designer puts forward the plan and the design parameters – the computer carries on the data analysis and completes the display result. The specific plan is as follows: as shown in Fig. 2, the

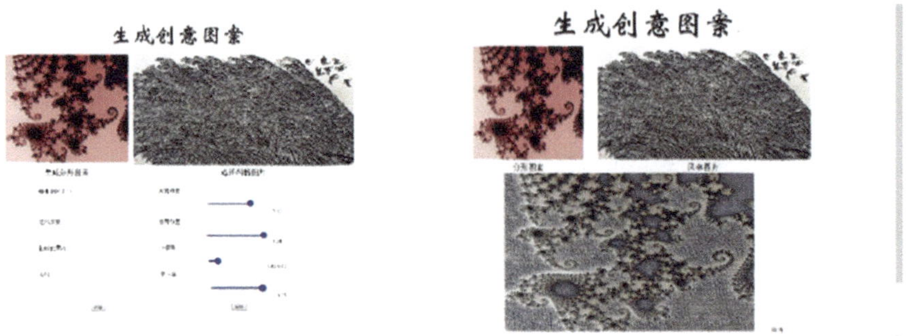

Fig. 2. The main browser interface of textile pattern intelligent aided design system

designer generates fractal images on the main page according to the needs of customers. After that, load a picture card on the main page. In the picture card, users can combine the current trend of the year to determine the creative elements and choose the style pictures. And can set the size of the output picture, the number of iterations, initialization of the picture, content weight, style weight, TV weight, learning rate, when choosing a good picture, you can choose to start, or reset the reselection.

In Fig. 2, whenever a task of creating creative patterns is created, an output picture card is generated. However, only 1 tasks are in operation, other tasks are waiting, and the Cancel button is applied to cancel the task. In the output picture card, the fractal image and the style picture are displayed on the top, and the middle result and the final generation diagram are displayed below. When you click the Cancel button, the system will cancel the task of ID.

Acknowledgments. This work was funded by Beijing Science and Technology Program (Z171100005017004).

References

1. Wang, S., Yang, X.: Generation of fractal image on complex plane and its application in textiles. J. Silk **54**(8), 56–61 (2017)
2. Ulyanov, D., Vedaldi, A., Lempitsky, V.: Instance normalization: the missing ingredient for fast stylization. arXiv preprint arXiv:1607.08022v3 [cs.CV], 6 November 2017
3. Chen, L.-C., Papandreou, G., Kokkinos, I., Murphy, K., Yuille, A.L.: Deeplab: semantic image segmentation with deep convolutional nets, atrous convolution, and fully connected CRFs (2016). arXiv preprint arXiv:1606.00915
4. Luan, F.J., Paris, S., Shechtman, E., Bala, K.: Deep photo style transfer. arXiv preprint arXiv: 1703.07511v3 [cs.CV], 11 April 2017
5. Weisstein, E.W.: Gram Matrix. MathWorld–A WolframWeb Resource. http://mathworld.wolfram.com/GramMatrix.html
6. Gatys, L.A., Ecker, A.S., Bethge, M.: Image style transfer using convolutional neural networks. In: Proceedings of the IEEE Conference on Computer Vision and Pattern Recognition, pp. 2414–2423 (2016)

7. Ulyanov, D., Lebedev, V., Vedaldi, A., Lempitsky, V.: Texture networks: feed-forward synthesis of textures and stylized images. In: Proceedings of the 33nd International Conference on Machine Learning, ICML 2016, New York City, NY, USA, 19–24 June 2016, pp 1349–1357
8. Johnson, J., Alahi, A., Fei, L.F.: Perceptual losses for real-time style transfer and super-resolution. In: Leibe, B., Matas, J., Sebe, N., et al. (eds.) Computer Vision-ECCV 2016. Springer, Cham (2016)
9. Johnson, J.: Neural-style (2015). https://github.com/jcjohnson/neural-style

Function Optimization Based on an Improved Adaptive Genetic Algorithm

Dake Zhang, Qian Qian(✉), and Feng Wang

Yunnan Key Laboratory of Computer Technology Applications, Kunming University of Science and Technology, Kunming, China
qianqian_yn@126.com

Abstract. Traditional genetic algorithms have some disadvantages in the one-dimensional multi-peaked function optimization problems. For example, it is easy to fall into local extremes, the rate of convergence is low and the stability is not ideal. This paper proposes a new adaptive genetic algorithm. Nonlinear functions are used to measure the probabilities of crossover and mutation operators. In addition, a modified algorithm flow based on the adaptive differential evolution algorithm and an elitist strategy are applied. Compared with other existing algorithms, the results of experiments indicate that the proposed algorithm has a higher convergence probability, a smaller number of iterations and a more reliable stability.

Keywords: Genetic algorithm · Adaptive · Function optimization Mutation probability · Crossover probability

1 Introduction

As one kind of intelligent algorithms, genetic algorithms are based on Darwin's theory of evolution and Mendel's theory of genetic variation [1–4]. As a random and global searching algorithm, the genetic algorithm has strong robustness and universal optimization ability [5]. Traditional genetic algorithms have a population that represents the potential solutions to the optimization problem. Each individual in the population evolves under the selection, crossover and mutation operators, so as to achieve the optimal solution of the problem. However, traditional genetic algorithms have some disadvantages in solving the problem of function optimization. Therefore, Srinivas and Pantaikl [6] proposed Adaptive Genetic Algorithms (AGA).

In AGA, the two control parameters in crossover and mutation operators are the key factors to affect the performance of the algorithm [7]. Crossover and mutation operators with adaptive regulation function can adjust the crossover probability and the mutation probability according to the concentration and dispersion of the current population during the process of evolution. The size of the crossover probability determines the abundance of the population. The larger the value of the crossover probability, the population will become richer, but the outstanding individuals in the population will be broken more easily. The probability of mutation determines how many new individuals are generated. The higher the mutation probability is, the more new individuals are generated, and the algorithm will be easier to jump out of the local extreme, so as to

K. Deng et al. (Eds.): ICMIR 2018, AISC 856, pp. 1146–1151, 2019.
https://doi.org/10.1007/978-3-030-00214-5_140

find the global optimal solution. However, if the mutation probability is too large, the algorithm will become a random search algorithm. As a result, AGA adaptively adjust these two genetic parameters. It can effectively improve the convergence accuracy and speed of genetic algorithms. AGA ensure both the diversity of the population and the convergence of the algorithm, leading the algorithm to the optimal solution [8].

The AGA based on the linear change of crossover and mutation probabilities cannot ensure to avoid falling into the local extreme and to find the global optimal individuals. Therefore, Shi and Li [9] proposed cosine adaptive genetic algorithms (CAGA). The function optimization problem is processed by means of non-linear transformation formula of crossover probability and mutation probability with the average, the maximum, and the individual fitness values of the population.

This paper proposes a new adaptive genetic algorithm (Improved Adaptive Genetic Algorithms, IAGA). A new transformation formula from adaptive differential evolution algorithms is proposed to calculate the crossover and mutation probabilities. Moreover, the processing sequence of selection, crossover, and mutation operators in classic genetic algorithms is also changed into sequence of mutation, crossover, and selection operators. The proposed algorithm can take into account both the global search performance and the convergence stability in the optimization problem of one-dimensional multi-peaked functions. The experimental results showed that the IAGA are better than GA, AGA and CAGA in function optimization with higher convergence degrees, smaller number of iterations, and better optimal solutions.

2 The Improved Adaptive Genetic Algorithm

2.1 The Basic Steps of the Improved Adaptive Genetic Algorithm

Step 1: Determining the encoding strategy, such as the binary encoding.
Step 2: Initializing the population and setting convergence conditions.
Step 3: Calculating the fitness values of each individual.
Step 4: Checking whether the function $\arctan(f_{avg}/f_{max}) \leq \pi/3$ is satisfied. A positive answer indicates that the average and the maximum fitness values of the population are relatively close. The smaller the difference of the two fitness values, the more the fitness values of the population are concentrated in a certain region. Then go to step 5. If the function is violated, go to step 6.
Step 5: In accordance with adaptive differential evolution algorithm flow, the implementation of mutation, crossover, selection operations will be done sequentially.
Step 6: In accordance with the traditional genetic algorithm flow, the implementation of selection, crossover, mutation operations will be done sequentially.
Step 7: An elitist strategy is performed to keep the best individual.
Step 8: Determining whether the convergence conditions are satisfied, and a positive answer leads to the next operation, otherwise, back to step 3.
Step 9: Calculating the fitness values of each individual, outputting the optimal solution.

2.2 Improved Crossover and Mutation Operators of Adaptive Genetic Algorithm

In the adaptive differential evolution algorithm, formula (1) is often used to construct the mutation operator. The operator can maintain the diversity of the individuals at the initial stage of the evolution. In addition, good individuals can be retained in the latter stage of the evolution.

$$\lambda = e^{1} - \frac{G_m}{G_m + 1 - G}, \quad F = F_0 \cdot 2^{\lambda} \tag{1}$$

In this paper, the IAGA crossover and mutation probabilities, as shown in formula (2), are designed based on the formula (1) to solve the one-dimensional multi-peaked function optimization problem.

$$P = \begin{cases} \frac{P_1 + P_2}{2} + \frac{P_1 - P_2}{2} \cdot 2^{e^{\beta}}, \quad \beta = \dfrac{1 - \left(\frac{f - f_{avg}}{f_{max} - f_{avg}}\right)}{\left(\frac{f - f_{avg}}{f_{max} - f_{avg}}\right) + f} & \arctan\left(\frac{f_{avg}}{f_{max}}\right) \le \frac{\pi}{3} \\ P_1 & \arctan\left(\frac{f_{avg}}{f_{max}}\right) > \frac{\pi}{3} \end{cases} \tag{2}$$

In formula (2), f_{max}: the maximum fitness of the population, f_{avg}: the average fitness of the population, f: Current individual fitness, P_1: Maximum probability of crossover (0.9) or mutation (0.2) operator, P_2: Minimum probability of crossover (0.1) or mutation (0.001) operator.

From formula (2), it can be seen that the probabilities of the improved crossover and mutation operators will not equal to zero, no matter the current fitness becomes larger or smaller. When the current individual is close to the optimal individual, the crossover and mutation probabilities are gradually reduced, and when the current individual is far away from the optimal individual, the crossover probability and the mutation probability are gradually increased. As a result, the search speed to reach the optimal solution is accelerated and the chance to fall into the local optimal solution is reduced.

2.3 Elitist Strategy

In order to ensure that the best individuals in each generation are not destroyed, it is very important to adopt an elitist strategy. The elitist strategy is that if the fitness values of the best individual in the current generation is less than the best individual of the previous generation, then replacing the worst individual in the current generation with the best individual of the previous generation to construct the population of the next generation. The elitist strategy is an important method for keeping the convergence of the IAGA.

3 Experimental Tests and Performance Analysis

3.1 Test Function

One-dimensional multi-peaked functions are usually used to test the actual efficiency of the genetic algorithms. This paper selected two representative test functions.

(1) $f_1(x) = 5\exp(-0.5x)\sin(30x) + \exp(0.2x)\ \sin(20x)\ + 6 \quad x \in [0.8]$

The optimal theoretical solution of the function $f_1(x)$ in the $x \in [0, 8]$ is 0.0540202. The optimal value is 11.7524517.

(2) $f_2(x) = x\sin(10\pi \cdot x)\ + 2 \quad x \in [-1.2]$

The maximum theoretical value of the function $f_2(x)$ in the $x \in [-1, 2]$ is slightly larger than 3.85, and the convergence value is 3.85.

3.2 Performance Evaluation

In this paper, IAGA and GA, AGA and CAGA are compared using two functions (1) and (2). Each function is calculated 50 times for each algorithm. If the optimal solution of each iteration (accurate to 10 decimal places) is not changed, then the convergence condition is met. In order to get accurate performance analysis results, the selection, crossover and mutation operators used in this paper are all the most basic methods. Specifically, the classic roulette method is adopted for selection. Single point crossover and single point mutation methods are used for crossover and mutation operations. Other parameters are as follows: population size (100), evolutionary generation (500), crossover probability (0.6), and mutation probability (0.01).

Combining the experimental results of Table 1, we can get the following conclusions. The proposed IAGA in this paper operated 50 times respectively in two test functions, and the convergence probabilities of each function were both over 0.8. Especially for the second test function, the convergence probability reached 0.92. The probabilities were 0.2 higher than convergence probability of GA and AGA, and 0.1 higher than that of CAGA. For the number of iterations, IAGA achieved the optimal value about 100 generations quicker than that of GA and AGA, and about 50 generations quicker than that of CAGA. Therefore, the proposed IAGA improved in the global convergence and the number of iterations than other algorithms.

Figure 1 show the results of each iteration for function (1). GA in Fig. 1 are obviously trapped into a local extreme. AGA are hovering around the maximum, which is not an ideal result for population means. CAGA fall into an unstable state in the early stages of evolution. In contrast, IAGA represent a stable iteration state. Because of the limitation of length, the figure results of function (2) are not shown, but similar tendencies are got. In all, the results show that the new algorithm has a higher convergence probability, a smaller number of iterations and a more reliable stability.

Table 1. Function of $f_1(x)$ and $f_2(x)$

Algorithm	Convergence times	Convergence probability	Iterations	Optimal solution
GA	27 and 31	0.54 and 0.62	389 and 367	11.691549005393387 and 3.849332211080026
AGA	31 and 37	0.62 and 0.74	356 and 318	11.739322652766329 and 3.849332183432195
CAGA	35 and 40	0.7 and 0.8	337 and 312	11.752451623765989 and 3.850073766214082
IAGA	42 and 46	0.84 and 0.92	265 and 279	11.752451667263683 and 3.850273695203874

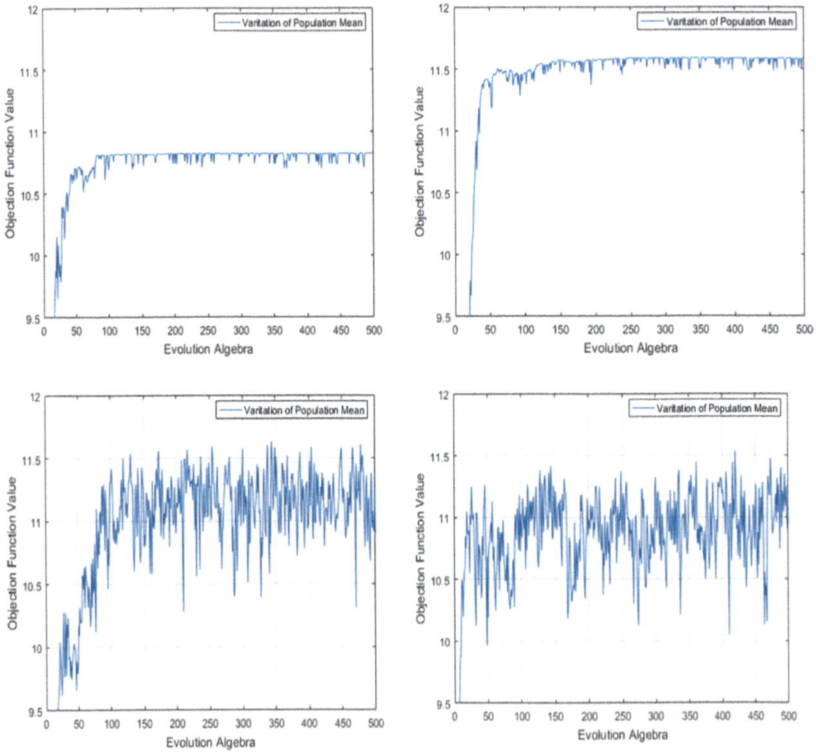

Fig. 1. Function $f_1(x)$ iteration result. (Top left: GA, Lower left: CAGA, Top right: AGA, Lower right: IAGA)

4 Conclusion

For the problem of crossover and mutation operators, this paper pointed out the deficiencies of GA, AGA and CAGA in function optimization. Along with the sequential modification of genetic operators, an improved adaptive genetic algorithm was proposed. The proposed IAGA was compared with GA, AGA and CAGA in optimization problem of two one-dimensional multi-peaked functions. As a result, IAGA has a higher convergence probability, a smaller number of iterations, and more reliable stability. Therefore, IAGA is an effective algorithm in one-dimensional multi-peaked function optimization.

Acknowledgments. This research is supported by the National Science Foundation of China (No. 31300938 and No. 61462053), and the Natural Science Foundation of Yunnan Province of China (No. 2016FB107).

References

1. DeJong, K.A.: The Analysis of the Behavior of a Class of Genetic Adaptive Systems. University of Michigan Press, Ann Arbor (1975)
2. Xi, Y., Chai, T.: Overview of genetic algorithms. Control Theory Appl. **13**(6), 697–708 (1996)
3. Ji, G.: Review of genetic algorithms. Comput. Appl. Softw. **21**(2), 69–73 (2004)
4. Ge, J., Qiu, Y.: Review of genetic algorithms. Comput. Appl. Res. **25**(10), 2911–2916 (2008)
5. Ou, Y., Wang, J.: A new improved genetic algorithm and its application. J. Syst. Simul. **15**(8), 1066–1068 (2003)
6. Srinivas, M., Pantaikl, M.: Adaptive probabilities of crossover and mutation ingenetic algorithms. IEEE Trans. Syst. Man Cybern. **24**(4), 656–667 (1994)
7. Wang, S.: A New Study on the Self-adaptive Genetic Algorithm. North China Electric Power University, Beijing (2009)
8. Yang, C., Qian, Q.: An improved adaptive genetic algorithm for function optimization. In: IEEE International Conference on Information and Automation, pp. 675–680 (2017)
9. Shi, S., Li, Q.: Optimization design of brushless DC motor based on adaptive genetic algorithm. J. Xi'an Jiaotong Univ. **36**(12), 1215–1280 (2002)

An Improved Pareto Ant Colony Algorithm for Interference Vehicle Scheduling Problem

Wang Rui[1(✉)], Luo Yumei[1], and Guo Ning[2]

[1] Faculty of Computer Information, Kunming Metallurgy College, Kunming 650033, China
164551048@qq.com
[2] Faculty of Information Engineering and Automation, Kunming University of Science and Technology, Kunming 650500, China

Abstract. In order to solve the vehicle routing problem with time windows of interference in the process of logistics distribution such as time window, demand changes and so on, an improved Pareto ant colony algorithm is proposed to minimize the length in this paper. Firstly, according to the customer time window changes and the characteristics of this kind of interference, two kinds of jamming events disturbance recovery strategies are put forward. Then the ant transition probability is defined based on the principles of saving, and the Pareto optimal solution set of global pheromone updating rule is used to improve the performance of the standard ant colony algorithm. Through the experiment, the simulation result shows that the algorithm improves the global search ability of the algorithm which effectively avoids the algorithm falling into local optimum and reduces the total distribution.

Keywords: Disturbance recovery strategy · Ant colony algorithm
Pareto optimal

1 Introduction

Vehicle routing problem was first proposed by Dantizig and Ramser in 1959 [1], which has important theoretical and practical significance in the field of logistics. VRPTW is a type of problem derived from VRP. As a typical NP problem, with the increase of the scale of the problem, the complexity of the problem will increase dramatically, it is difficult to get the exact solution of the problem. Therefore, it becomes an important research direction to solve the VRPTW by using the intelligent optimization algorithm [2].

Literature [3] uses a tabu algorithm (TS) of the vehicle scheduling problem with time window constraints. Literature [4] uses the genetic algorithm (GA) to solve the minimum number of vehicles and the minimum cost of vehicle routing problem with time Windows. But the genetic algorithm is easy to premature, the disadvantage of local search ability is not strong, tabu algorithm (TS) of the initial solution has the defects of high dependency. Ant colony optimization algorithm, a heuristic simulation algorithm, was first proposed by the first Italian scholars Dorigo [5]. As a general stochastic optimization method, solving results in a series of difficult combinatorial optimization problem are achieved by the exchange of information between simulated

K. Deng et al. (Eds.): ICMIR 2018, AISC 856, pp. 1152–1158, 2019.
https://doi.org/10.1007/978-3-030-00214-5_141

ants and positive feedback mechanism. But the standard ant colony algorithm has certain defects, such as slow convergence speed, long operation time, easy to fall into local optimal solution, etc [6]. Literature [7] introduces the searching hotspots strategy which improves the searching efficiency and prevents the ant colony trapping in local optimal solutions.

2 Description of Vehicle Routing Problem

Interference management is the solution of the original model on the basis of VRPTW. According to the original distribution scheme of VRPTW, judgment interference events affect the scheme and the quantification for some data in the mathematical model of the target. According to the constraint conditions of VRPTW, objective function and solution, the customer time window changes or changes in demand for VRPTW disturbance events can mainly quantify the influence of three factors: time, path and the cost. In this paper, the method of new customer point is adopted to deal with these three factors which affect the original distribution plan through the analysis of the new customer point. For example, changing a customer point b with some jamming events to the customer point c, then the number of new customer c is analyzed.

For the customer point c, firstly we analyze from the time. About the interference events with the customer time window changes, the deviation of time should be considered when vehicles in the original distribution plan cannot arrived during the window time after the c changes. This is the primary factors influencing the customer satisfaction degree of logistics enterprises. Secondly in the homogeneous goods, the migration which is relative to the path of the original plan caused by the customer time window changes or changes in customer demand point, results in the change of the cost. According to the literature [8], three factors which affect the interference treatment named the time deviation, the path deviation, the cost change are described as follows:

(1) Time deviation

The ∂r_i stands for the time when the vehicle arrives the customer point, s_c is the service time at c, and $[Te_c, Tl_c]$ is the customer time window, then the time deviation at point c can be represented as $\lambda_1(Te_c - \partial r_c) + \lambda_2(\partial r_c - Tl_c) + \lambda_3(\partial r_c + Te_c - Tl_c)$, among them $\lambda_1, \lambda_2, \lambda_3 \in \{0, 1\}$, when $\partial r_c \leq Te_c$, $\lambda_1 = 1$; when $\partial r_c \geq Tl_c$, $\lambda_2 = 1$; when $\partial r_c + Te_c \geq Tl_c$, $\lambda_3 = 1$. The rest of the cases, $\lambda_1, \lambda_2, \lambda_3$, take (System overall time off for $\sum_{i=1}^{N'} (\lambda_1(Te_c - \partial r_c) + \lambda_2(\partial r_c - Tl_c) + \lambda_3(\partial r_c + Te_c - Tl_c))$.

(2) Path deviation

There are two attributes in the path of logistics distribution vehicle scheduling system. One is the length of the distance between customers and clients, the other is the straight line connected to the clients called edge numbers. To rescue a new customer, there must be the edge of distribution which increases or decreases, and the total distribution path length will also be changed accordingly. Path deviation corresponds distribution vehicles, which can be represented as $\sum_{i=0}^{N_0'} \sum_{j=0}^{N_0'} C_{ij}\left(x_{ij}' - x_{ij}\right) + u\sum_{i=0}^{N_0'} \sum_{j=0}^{N_0'} C_{ij}\left(x_{ij}' - x_{ij}\right)$, among them, u means to increase or decrease the edge of a penalty coefficient, x_{ij}, $x_{ij}' \in \{0, 1\}$,when an edge existing between customer point i and customer point j, $x_{ij} = 1$, $x_{ij}' = 1$, otherwise $x_{ij} = 0$, $x_{ij}' = 0$.

(3) Cost changes

The corresponding transport costs will increase while adding a car for distribution. After the incident of disturbance, the path deviation is inevitably produced when re-planning the original scheme. Thus the distribution distance increases, and the cost of transportation also increases. New customer points must be considered in the enterprise.

The cost deviation between the interfere rescue scheme and the original one can be expressed as: $\sum_{k=1}^{K} \sum_{i=0}^{N_0'} \sum_{i=0,j\neq i}^{N_0'} C_{ij}x_{ijk}' - \sum_{k=1}^{K} \sum_{i=0}^{N_0'} \sum_{i=0,j\neq i}^{N_0'} C_{ij}x_{ijk} + \sum_{k=1}^{k} C_k\left(V_k' - V_k\right)$. In which $\sum_{k=1}^{K} \sum_{i=0}^{N_0'} \sum_{i=0,j\neq i}^{N_0'} C_{ij}x_{ijk}'$ represents the gross vehicle moving path in the rescue plan, $\sum_{k=1}^{K} \sum_{i=0}^{N_0'} \sum_{i=0,j\neq i}^{N_0'} C_{ij}x_{ijk}$ represents the gross vehicle driving path in the original plan, and C_k represents the fixed costs per car. Let $\sum_{K=1}^{K} C_k\left(V_k' - V_k\right)$ is the increased vehicle fixed cost of the rescue plan, in which V_k', $V_k \in \{0, 1\}$ $V_k' = 1$ represents the participation of vehicle k for distribution in the rescue plan, $V_k' = 0$ represents the non-participation of vehicle k for distribution in the rescue plan. v_k is similar with V_k', but its role is in the original distribution plan. Therefore, the cost factor should be considered in the final interference disturbance recovery scheme to solve VRPTW. That is to say, the transportation cost of logistics system should be controlled as far as possible, in order to make the cost changes as small as possible.

3 Ant Colony Algorithm with Interference Treatment

3.1 Disturbance Recovery Strategy for VRPTW Interference Event

According to the VRPTW disturbance recovery demand analysis of the customer time window changes and demand changes, and the characteristics of this kind of interference, this paper puts forward the following two disturbance recovery strategies:

(1) Increase vehicles to reinforce: In the case of a free vehicle in the distribution center, if there is a disturbance event, it may be considered to use the addition of vehicle to increase the aid. For example, a distribution center have M cars, and the number of cars which participate in the task of distribution is m(m < M), then the number of free vehicle is M − m. When t = thd, reinforcement strategy which does not affect the original delivery plan can be directly to deal with disruptive event after interference occurred. However, when the following two situations occur, the method of directly adding the vehicle to increase the assistance will not solve the disturbance to the logistics system:

- The number of free vehicles in the distribution center is less than the number of new customers if the interference event occurs at the same time as the customer point of s, namely, the new customer point s, and s > m.
- Reinforcement of the vehicle can't reach you within the customer time window starts at the customer service.

At this point, it is necessary to further consider how to insert new distribution vehicles and user points with minor disturbance in the original scheme, therefore, this paper puts forward a second rescue package - hybrid reinforcements.

(2) Hybrid reinforcements: For extra buses reinforcements can't solve the interference of events, this paper puts forward a hybrid based on extra buses reinforcement, the reinforcement method was introduced to the extra buses reinforcements on the basis of adjacent relief thoughts. Due to the delivery of the goods is the same, in the process of distribution, for extra buses reinforcements can not solve the interference of the incident point a customer, using the ideas of adjacent relief according to the original distribution scheme to determine the current vehicle position, adding the demand changes of point a customer in existing load in the car, if overload the weed out the corresponding vehicle, if do not overload, sets the corresponding position of the vehicle to the virtual customer, the customer point coordinates are the coordinates of the current position, demand is zero, no time windows restriction. Then, according to the matching degree of the next several factors, the distribution of vehicles is selected:

- The distance between customer point a and the imaginary customer point of each new increase represents the vehicle.
- The matching degree of the time from each virtual customer point to the customer point a and the time window of the customer point a.
- Distribution of customer point a, and path deviation from the original path.

Once you select a car which in the original distribution delivery to a customer, then the original distribution scheme will no longer applicable, at this point, according to the current path condition, The route of the vehicle is reprogrammed according to the time window of each customer, and decide whether to send reinforcements vehicle.

3.2 Algorithm Steps

Based on the improved Pareto ant colony algorithm, according to the characteristics of the artificial ants and interference management, this paper puts forward a improved Pareto ant colony algorithm which based on management thought t(Disruption Management − Improved Pareto Ant Colony Algorithm, DMIPACA). The disturbance recovery of VRPTW was performed with DMIPACA which through the following steps:

step 1: The optimal vehicle based on ant colony algorithm for the initial distribution plan.

step 2: In times of t_{dm}, interference event happened at x customer point, and x client all can be converted to new customers, then comprehensively determine the

perturbation recovery strategy which is based on the time deviation, path deviation and the cost changes brought by the new customer points.
step 3: According to the selected disturbance strategy, from the potential benefits and long-term perspective, make the completion of the customer service as a primary goal with a small amount of time deviation, on the basis of ensure the first goal, The path deviation is minimized and the cost is minimized as the second and third targets respectively, Update the current data after the new customer point is converted, using this data, DMIPACA is used to generate the new scheme.

4 Simulation Experiments and Result Analysis

Using literature data from [9], There are 7 vehicles can be set up distribution centers to participate in the distribution, the maximum load for each vehicle are for 5 × 103. Assume that the vehicle speed is RMB 1 unit, ignore unloading time at customer points. The data of yard and customer points is shown in Table 1. Respectively using virtual bike field interference in literature [10] management of genetic algorithm and new customers based on the virtual point method for interference management improvement of Pareto ant colony algorithm for VRPTW model and interfere with the management, the distribution problem of the model are to verify the effectiveness of the algorithm.

Table 1. Yard point data table with the customers

Custerm	x_i	y_i	d_i	Te_i	Tl_i
0	50	50	0	0	99,999
1	19	0	1.00×10^3	74	144
2	33	3	1.80×10^3	58	128
3	35	21	1.10×10^3	15	85
4	53	19	0.60×10^3	96	166
5	70	94	1.90×10^3	47	117
6	27	44	1.40×10^3	85	155
7	10	69	1.20×10^3	21	91
8	56	4	0.20×10^3	9	79
9	16	81	1.70×10^3	37	107
10	68	76	0.80×10^3	21	121
11	41	10	0.90×10^3	74	174
12	83	43	0.80×10^3	58	158
13	25	91	1.90×10^3	15	125
14	73	29	1.60×10^3	56	156
15	70	18	0.90×10^3	87	187

The initial approximate optimal solution of VRPTW vehicle distribution path obtained by literature [10]. Literature [10] with the improved genetic algorithm (ga) after 10 experiments for VRPTW approximate optimal moving path is as follows:

$$\begin{array}{l} 1 \text{ vehicle path}: 0 \rightarrow 3 \rightarrow 11 \rightarrow 1 \rightarrow 2 \rightarrow 8 \rightarrow 0; \\ 2 \text{ vehicle path}: 0 \rightarrow 4 \rightarrow 15 \rightarrow 14 \rightarrow 12 \rightarrow 0; \\ 3 \text{ vehicle path}: 0 \rightarrow 10 \rightarrow 5 \rightarrow 13 \rightarrow 0; \\ 4 \text{ vehicle path}: 0 \rightarrow 9 \rightarrow 7 \rightarrow 6 \rightarrow 0; \end{array}$$

The initial scheduling results shows that distribution center has four car can be used to deal with unexpected events, when $t_{dm} = 32.65$, customer point 4 and 14 require change of time window, customer point 8 and 11 require change in demand, and added customer point 16 which requires the delivery of goods, the specific data changed is shown in Table 2.

Table 2. The customer changes

	4	8	11	14	16
Original coordinate	(53, 19)	(56, 4)	(41, 10)	(73, 29)	No
Original time window	[96, 156]	[9, 79]	[74, 174]	[56, 156]	No
Original demand	0.6	0.2	0.9	1.8	No
New coordinates	(52, 29)	Don't change	Don't change	Don't change	(55, 60)
New time window	[10, 54]	Don't change	Don't change	[20, 70]	[30, 75]
New demand	1.6	1.4	1.9	Don't change	2.0

The rescheduling method, interference management method based on improved genetic algorithm and the interference of the Pareto ant colony algorithm based on improved management (DMIPACA) are adopted after the interference occurs, the mileage charge, path deviation, customer event time window deviation and the number of vehicles used are shown in Table 3.

Table 3. Solve VRPTW events happened with comparing the results of three methods

Methods	Rescheduling	GA	DMIPACA
The mileage charge	840.76	841.69	655.47
Path deviation	16	9	16
The service time window deviation	210.75	37.03	43.25
Number of vehicles	7	6	6

From Table 3, for the customer time window changes in VRPTW interference events, the path deviation amount of DMIPACA which based on the strategy of hybrid reinforcements is higher; Customer service time window deviations from below the rescheduling method, higher than the genetic algorithm from the literature [10]; But get mileage costs far less than the other two algorithms.

5 Conclusions

This paper proposes an improved Pareto ant colony algorithm to solve the vehicle routing problem with time windows events in the process of the interference. First of all, according to the customer time window changes and the disturbances restore demand analysis of changing demand for VRPTW as well as the characteristics of this kind of interference, the paper presents two disturbance recovery strategies of the interference event. Secondly, in view of the ant colony algorithm is easily plunged into local optimum in solving VRPTW problem, this paper adopted the Pareto optimal solution set of global pheromone to update rules. Finally, by comparing and analyzing the simulation experiments, results show that the improved algorithm can enhance the global search ability of ant colony algorithm, effectively avoid the algorithm falling into the local optimum, and reduce the total distribution costs. Thus the results also verifies the effectiveness and superiority of the improved algorithm.

References

1. Dantizig, G., Ramser, J.: The truck dispatching problem. Manage. Sci. **6**(1), 80–91 (1959)
2. Gong, G.: Optimization study of logistics distribution vehicle routing problem. Eng. Comput. Sci. **5**(33), 106–111 (2011)
3. Zhang, J.: The tabu search algorithm of distribution vehicle scheduling problem with time windows. J. North. Jiaotong Univ. **2**(28), 103–106 (2004)
4. Xie, B.L.: Full of vehicle scheduling problem with time windows genetic algorithm. J. Syst. Eng. **3**(15), 290–294 (2000)
5. Dorifo, M., Maniezzo, V., Colorni, A.: Ant system: optimization by a colony of cooperating agents. IEEE Trans. Syst. Man Cybern. **1**(26), 29–41 (1996)
6. Yang, Z.Q.: Improved ant colony algorithm for the shortest path problem in transportation system study. Sci. Comput. Domain Inf. Process. **8**(32), 76–78 (2009)
7. Chen, Y.X.: Vehicle routing optimization problem based on improved ant colony algorithm research. Comput. Appl. Res. **6**(29), 2031–2034 (2012)
8. Liang, A.M.: Nested Segmentation Algorithm to Solve VRPTW Mixed Interference Problem. Dalian University of Technology, Dalian (2010)
9. Wang, X.P.: Have the customer time window and shipments changes vehicle scheduling management research. Manage. Sci. **5**, 111–120 (2008)
10. Xu, C.L.: Logistics Distribution Changes the Customer Time Window Interference Management Research. Dalian University of Technology Electronic Commerce and Logistics Management, Dalian (2009)

A Privacy Protection Method for Learning Artificial Neural Network on Vertically Distributed Data

Guang Li[1(✉)], Xiaohong Su[2], and Yadong Wang[2]

[1] School of Electronic and Control Engineering, Chang'an University, Xi'an 710064, China
hit6006@126.com
[2] School of Computer Science and Technology, Harbin Institute of Technology, Harbin 150001, China

Abstract. For mining privacy data that can not be seen directly, the privacy preserving data mining (PPDM) is needed. As far as we know, for neural network learning on vertically distributed databases, there is no good enough PPDM method. For solving it, a privacy preserving method for learning neural networks on vertically distributed data is proposed by this paper. This method designs protocols to exchange essential information for learning neural networks without opening private data. The learning results with this proposed method are the same as the results with the original BP algorithm without considering privacy preservation. And, in the learning process, every node cannot get the details of other nodes' data.

Keywords: Vertically distributed databases · Privacy preserving Neural network · Secure multi-party computation

1 Introduction

Privacy preserving data mining (PPDM) is an important research area [1, 2]. Many researchers study this problem for distributed databases. It means the whole data set is decentralized storage on multiple nodes. It wants to mine on the whole data set but does not want these nodes to open their data. Now, many classifications can be trained privately on distributed databases [3, 4]. However, to the best of our knowledge [5–10], they do not contain a practical feedforward neural network (FNN).

Secretan [5] studies on the PNN (probabilistic neural network), not FNN. Barni [6] and Orlandi [7] mainly study how to use FNN, but not how to learn it. Yancheng [8] assumes all the data saved in one node. Saeed [9] and Li [10] both make simplifying assumptions. Saeed [9] assumes a linear activation function is used in the network. Li [10] assumes that the network only has one level.

In summary, the FNN can not be trained by using distributed privacy data. Considering it is a very important data mining and artificial intelligence method [11, 12], it is valuable to solve this problem. This paper studies this problem and gives a privacy preserving FNN learning algorithm on vertical partitioned data, which is based on BP

K. Deng et al. (Eds.): ICMIR 2018, AISC 856, pp. 1159–1167, 2019.
https://doi.org/10.1007/978-3-030-00214-5_142

(Back Propagation) algorithm. For vertically partitioned data, every node knows all tuples. But these tuples only includes part of the attributes. There is no node which has complete tuples.

2 Privacy Preserving Neural Network Learning Algorithm

This paper assumes there are n nodes: P_1, P_2, …, P_n. For every sample α, P_i has α_i, which is part of α. $\alpha_1 \cup \alpha_2 \cup \ldots \cup \alpha_n = \alpha$ and $\alpha_i \cap \alpha_j = \Phi$. These nodes want to learn a neural network without opening their data.

The method proposed by this paper uses secure multi-party computation [13–15] to change information. The secure product and secure sum protocol are mainly used in it.

Our algorithm evolves from BP algorithm [11], which assumes the network's structure and the weights' initial value are known. The gradient descent iterative method is used to optimize weights.

Like most of the similar methods, our algorithm supposes every node do not collude and there is an auxiliary node as semi-trusted third party. It trains d networks at the same time. Data nodes arrange the encryption E and the key K. The semi-trusted third party does not know K. M is a message, and $E(M)$ is its cipher text. $X = (x_1, x_2, \ldots, x_m)$ is a network, where x_i represents a weight. X's update is $Y = (y_1, y_2, \ldots, y_m)$, where y_i is x_i's update. Our algorithm has two parts: initialization and optimization.

2.1 Initialization

Customarily, network's input units map the data's attributes one-to-one, so we can divide X to X_1, …, X_n and X'', where X_i includes weights linked to the input units mapping to the attributes of data nodes P_i and $X'' = X - X_1 - X_2 - \ldots - X_n$.

In initialization, each node generates random vectors A, B and C. X'', A and B have same dimensions. Let $X'' = X'' + A - B$. C is an n-dimensional vector, let $X_i = C_i X_i$, where C_i is the i-th dimension of C. Each node gives its X, $E(A)$, $E(B)$ and $E(C)$ to the third party Q. Q stores only one set of them randomly.

Then a data node is selected randomly, and transmits $E(J)$ to Q, where J means the error for testing sets and is set to be big enough.

Repeating this process d times, Q will get d networks. Every network is expressed as (X, $E(A)$, $E(B)$, $E(C)$, t) and (X, $E(A)$, $E(B)$, $E(C)$, $E(J)$, l), where t is the node's symbol, which transmits X, $E(A)$, $E(B)$ and $E(C)$ to Q, and l is the node's symbol, which transmits $E(J)$ to Q.

2.2 Optimization

Two problems have to be solved in optimization process. The fist one is that the nodes can build equations from the detail training process. And these equations may be used to get the information of data on other nodes. We should not give the detailed training process to any nodes to eliminate this privacy threat. To solve this problem, more than one network is trained simultaneously.

At beginning, the third party Q has d networks. One of them is selected by Q randomly and is transmitted to data nodes. Then data nodes calculate the error and the weights' update for this network, and give them to Q after randomization. After that, Q updates the records to finish this iteration. The cipher text of random numbers used by data nodes is also transmitted to Q.

Problem 2: How can we calculate the error and the weights' update when each data node cannot see the samples on other nodes directly? We solve this problem by dividing the network and using the below two properties of neural networks.

Property 1: If a sample is α, and its component on data node P_i is α_i. Vector X_i includes weights linked to the input units mapping to the attributes of P_i, and vector X' includes all the weights linked to the input layer. Then, $X'\alpha$ is the vector on the network's second layer for α. It is easy to find $X'\alpha = X_1\alpha_1 + X_2\alpha_2 + \ldots + X_n\alpha_n$.

Property 2: If the training sample set is $D = \{\alpha_1, \alpha_2, \ldots, \alpha_{|D|}\}$, Y_i and Y are the update for network weights corresponding to sampleα_i and the whole training samples set D respectively. Obviously, $Y = Y_1 + Y_2 + \ldots + Y_{|D|}$.

The method of calculating the error and the weights' update is as follows. At the beginning, $\{P_1, P_2, \ldots, P_n\} - \{P_i\}$ arrange random number C_i, $i = 1, 2, \ldots, n$. Then, P_i gets X_i (the vector of weights linked to input units mapping to the attributes of P_i) and $X'' = X - X_1 - X_2 - \ldots - X_n$. Then for every sample$\alpha$, node $R = P_r$ is selected randomly as the master node. P_i calculates $X_i\alpha_i$, and R gets $X'\alpha = X_1\alpha_1 + X_2\alpha_2 + \ldots + X_n\alpha_n$ by using a secure sum protocol, where X' is the vector including all the weights linked to the input layer. After that, R can calculate the error and Y'' (the update of X'') using $X'\alpha$ and X''. Then, R calculates B_p (the back propagation of Y'') and Z_r (the update of X_r). R transmits C_iB_p to P_i to calculate C_iY_i, where Y_i is the updates of X_i.

D is the training sample set. D_{1i} is the set of training samples processed when P_i is the master node. $D_{2i} = D - D_{1i}$. After D has been processed, P_i can get three kinds of data. They are Y'' (the update of X'' for D_{1i}), Z_i (the update of X_i for D_{1i}) and C_iY_i, where Y_i is the update of X_i for D_{2i} and C_i is a random number. The update of X'' for D can be gotten though a secure sum protocol for the update of X'' for $D_{11}, D_{12}, \ldots, D_{1n}$. Then X'' can be updated. After that, a node R is selected randomly from $\{P_1, P_2, \ldots, P_n\} - \{P_i\}$. Because R has C_i, P_i has Z_i and X_i, so Q can get C_iZ_i and C_iX_i though a secure product protocol running by Q, R and P_i. Then P_i gives C_iY_i to Q. X_i can be updated by Q using $C_iZ_i + C_iY_i$ and C_iX_i.

2.3 Optimization of Termination and Result Selection

There are two sample sets, D and T. All networks are optimized using D to iterate enough. The network with the smallest error on T is the final result.

Our algorithm calculated error as calculating the weights' update. After that, data node P_i has J_i, which is the network's error for D_{1i} (D_{1i} is the set of samples processed when P_i is the master node). Then, Q selects P_r from $\{P_1, P_2, \ldots, P_n\} - \{P_l\}$ randomly (P_l is the node transmitting the training network's historical smallest error to Q), and transmits the training network's historical smallest error's cipher text $E(J')$ to P_r. P_r gets the new error $J = J_1 + J_2 + \ldots + J_n$ through a secure sum protocol, compares J with J', and transmits the comparison result and $E(J)$ to Q. Then Q will update the record.

When the optimization has finished, the third party Q has d network. The cipher texts of their errors are $E(J_1)$, $E(J_2)$, …, $E(J_d)$. The network with the smallest error will be selected as the final result. First, a data node $R = P_r$ is selected randomly. Then each data node in $\{P_1, P_2, \ldots, P_n\} - \{P_r\}$ generates a random number and gives its cipher text to Q. We call them $E(J_{d+1})$, $E(J_{d+2})$, …, $E(J_{d+n-1})$. If h is a random permutation, $(E(J'_1), E(J'_2), \ldots, E(J'_{d+n-1})) = h(E(J_1), E(J_2), \ldots, E(J_{d+n-1}))$ is gives to R by Q. R decrypts and sorts them. R gives Q the result by a permutation of 1, 2, …, $d + n - 1$ (if $J'_i < J'_j$, i is in front of j). Q gets the true index by h^{-1} and finds the network having the smallest error.

2.4 The Detailed Process of the Iterative Optimization

On each iteration, the third party Q selects a network randomly. First, we calculate the network's error for testing data. If it is less than the historical smallest error, update the historical smallest error and the corresponding network. After that, update the selected network using the training set. The iteration is done many times. Final result is the network with the smallest error. The detailed process is as follows.

Q generates $NUM = \{0, 0, \ldots, 0\}$ and a threshold M. M is the biggest iteration times for each network. NUM is a vector to record the number of iterations. NUM has d-dimensional.

while Q does not give a stop signal
  if for every $i \in \{1,2,\ldots, d\}$, $NUM[i] \geq M$
    SelectResult ( ) // selects the network with smallest error as the final result
    Q gives a stop signal
  else
    SelectOneNetwork ( ) // selects a network randomly
    CalculateError ( ) // calculates the error of this network on testing set
    UpdateRecord ( ) // if the selected network's error is less than its historical smallest error, update its historical smallest error and the corresponding network record
    UpdateWeights ( ) // update the selected network
  end if
end while

SelectResult () function does the work of comparing errors of different networks and selects the network with the smallest error as the final result.

SelectResult ( ) {

Q selects data nodes $R=P_r$ randomly. $\{P_1, P_2,\ldots, P_n\}-\{P_r\}$ respectively gives a random number's cipher text to Q. These cipher texts are $E(J_{d+1}), E(J_{d+2}), \ldots, E(J_{d+n-1})$.

If $E(J_1), E(J_2), \ldots, E(J_d)$ are the cipher texts of d network's error, and h is a random permutation, Q gives$(E(J'_1), E(J'_2), \ldots, E(J'_{d+n-1}))=h(E(J_1), E(J_2), \ldots, E(J_{d+n-1}))$ to R.

R sorts the plaintext of $E(J'_1), E(J'_2), \ldots, E(J'_{d+n-1})$, gives Q the result as a permutation of index $1,2,\ldots, d+n-1$. If $J'_i<J'_j$, i is in front of j.

Q uses h^{-1} to get the true index, and selects the network with the smallest error as the final result. }

SelectOneNetwork () function selects a neural network randomly and revert it to the real weight value.

SelectOneNetwork ( ) {

Q selects a network randomly, which is recorded as $V=(X_v, E(A_v), E(B_v), E(C_v), E(J_v), l)$ and $U=(X_u, E(A_u), E(B_u), E(C_u), t)$.

Q divided X_u and $E(C_u)$ to $X_{u1}, X_{u2}, \ldots, X_{un}, X_u''$, and $E(C_{u1}), E(C_{u2}), \ldots, E(C_{un})$. X_{ui} is the vector of weights that are linked to the network's input units mapping to P_i's data. X_u'' is the vector of weights that are not linked to the input layer. C_{ui} is the i-th dimension component of C_u.

Q selects P' randomly from $\{P_1, P_2,\ldots, P_n\}-\{P_t\}$. P_t is the node giving $E(A_u), E(B_u)$ to Q on this network's last iteration.

Q gives X_u'', $E(A_u)$ and $E(B_u)$ to P'. Q gives X_{ui} and $E(C_{ui})$ to P_i. $i=1,2,\ldots,n$.

P' decrypts $E(A_u)$ and $E(B_u)$, calculates $X''=X_u''-A_u+B_u$, and gives it to other data nodes. P_i decrypts $E(C_{ui})$, calculates $X_i=X_{ui}/C_{ui}$.

}

CalculateError () function is used to get the error of the network for testing samples. After that, data node P_i has J_i, which is the network's error for D_{1i} (D_{1i} is the set of samples processed when P_i is the master node), $i = 1, 2, \ldots, n$.

CalculateError ( ) {

$T=\{\beta_1, \beta_2, \ldots, \beta_{|T|}\}$ is the testing set, and β_j's component on P_i is β_{ji}, $i=1,2,\ldots,n$.

P_i set up $J_i=0$.

for $k=1$ **to** $|T|$

Data node $R=P_r$ is selected randomly as the master node.

P_i calculates $Net_i=X_i\beta_{ki}$. X_i is the vector of weights that are linked to the network's input units mapping to P_i's data.

R gets $Net=Net_1+Net_2+\ldots+Net_n$ using a security sum protocol.

R calculates J', which is the trained network's error for β_k.

$J_r=J_r+J'$.

end for }

UpdateRecord () does the work of error comparison and updates the historical smallest error and the corresponding network. Before using this function, the CalculateError () function is used. Data node P_i has J_i, which is the network's error for D_{1i} (D_{1i} is the set of samples processed when P_i is the master node), $i = 1, 2, \ldots, n$.

UpdateRecord ( ) {

Q selects R=P_r from {P_1, P_2,…, P_n}-{P_l} randomly. P_l is the node transmitting the training network's historical smallest error to Q. Q gives $E(J_v)$ to R, and records r as R's symbol. J_v is the trained network's historical smallest error.

R calculates the security sum $J=J_1+J_2+\ldots+J_n$, gives $E(J)$ to Q and decrypts $E(J_v)$.

If $J<J_v$, R gives $BOOL$=1 to Q, else, R gives $BOOL$=0 to Q.

if $BOOL$=1

Q uses (X_u, $E(A_u)$, $E(B_u)$, $E(C_u)$, $E(J)$, r) to replace V. V is the historical smallest error's corresponding network, and r is R's symbol. U=(X_u, $E(A_u)$, $E(B_u)$, $E(C_u)$, t) is the trained network's current weights.

end if }

UpdateWeights () function does the work of calculating the update of the network's weights. The network X is divided to X_1, …, X_n and X'', where X_i includes weights linked to the input units mapping to data node P_i's data and $X'' = X - X_1 - X_2 - \ldots - X_n$. P_i has X'' and X_i.

UpdateWeights ( ) {

D={ α_1, α_2, …, $\alpha_{|D|}$} is the training set. α_j's component on P_i is α_{ji}, i=1,2,…,n.

{P_1, P_2,…, P_n}-{P_i} arranges the random number C_i. P_i set up Y_i'', Y_i and Z_i all are 0.

for k=1 **to** $|D|$

R=P_r is selected randomly as the master node.

P_i calculates $Net_i = X_i \alpha_{ki}$. X_i is the vector of weights that are linked to the network's input units mapping to P_i's data.

R gets $Net=Net_1+Net_2+\ldots+Net_n$ using a security sum protocol.

R calculates Y^* (the update of X'' for α_k), and lets $Y_r''= Y_r''+ Y^*$.

R calculates Δ (the back propagation of Y^*), and gives $C_i\Delta$ to P_i, where $i \neq r$.

R calculates Z_r' (the update of X_r for α_k), and lets $Z_r = Z_r + Z_r'$.

P_i calculates Y_i' (the product of C_i and the update of X_i for α_k), and lets $Y_i=Y_i+Y_i'$.

end for

P_i generates random vectors A_i and B_i. They both have the same dimension as X''.

P_i calculates $X_i''=X''+A_i$, $Y_i''=Y_i''+B_i$.

Q update X'' to be $(X_1''+X_2''+\ldots+X_n'')/n$ and calculate $Y''=Y_1''+ Y_2''+\ldots+ Y_n''$ by the secure sum protocol.

P_t, which is selected randomly, calculates security sum $A=(A_1+ A_2+\ldots+A_n)/n$, $B=B_1+ B_2+\ldots+ B_n$, and gives $E(A)$, $E(B)$ and its symbol t to Q.

for k=1 **to** n

Q selects $R \neq P_k$ randomly. A security product protocol is run among Q, R and P_k to let Q get $Z_k'=C_kZ_k$ and $X_k'=C_kX_k$.

R gives $E(C_k)$ to Q. P_k gives Y_k to Q. Q calculates $Y_k'=Y_k+Z_k'$.

end for

Q respectively links X_i' and X'', Y_i' and Y'', $E(C_i)$ to be X', Y' and $E(C)$. i=1,2, …, n.

Q calculates $X=X'-Y'$, replaces U with (X, $E(A)$, $E(B)$, $E(C)$, t). U is the trained network's latest record. P_t is the node giving $E(A)$, $E(B)$ to Q.

$NUM[c_1]= NUM[c_1]+1$. c_1-th network is the trained network on this iteration. }

3 Algorithm Analysis and Experiment

Security, accuracy and efficiency are the three mainly requests of PPDM method. Security means after running the algorithm, the private information is not opened. Accuracy requests the method will get good results. Efficiency requests the method take little cost of communication and computation. Our method has no problem for accuracy because it is equivalent to BP algorithm. So we only introduce the efficiency and security.

3.1 Efficiency

It assumes that there are k samples, d training networks, n data nodes and every network have m weights. $O(G)$ and $O(E)$ are assumed to be the time need by decrypting and encrypting a number. Running BP algorithm on global data requires $O(F)$ operations. If there are $n + 1$ nodes, calculating a security sum need $O(S_1)$ time and $O(W_1)$ communication cost. That parameters for calculating a security product are $O(S_2)$ and $O(W_2)$. The greatest number of iteration time of each network is M. The expected whole computation and communication cost is $O(dMm(G + kS_1 + nS_2 + E) + dF)$ and $O(dMm(n + kW_1 + nW_2))$ respectively. The simulation results are shown in Fig. 1.

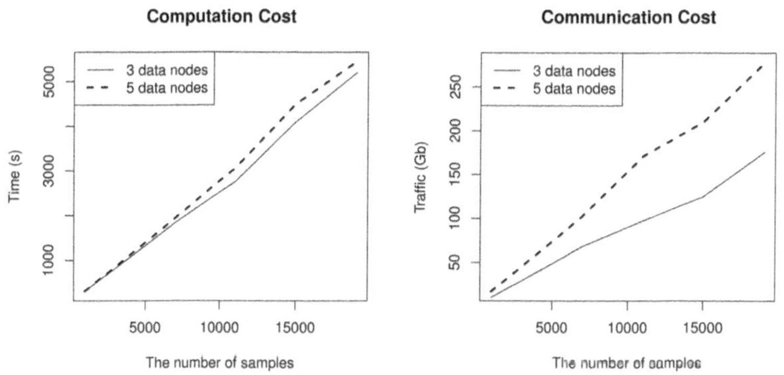

Fig. 1. The simulation computation and communication cost of out method.

3.2 Security

For the third part, it can get three kinds of data, the cipher text of random numbers and network's error on testing set, the randomized weights and weights' update, and the symbol of the data nodes that transmit data to the third part. The first two kinds of data are encrypted or randomized. The symbol of the nodes does not contain any information of the samples. So the third part cannot get privacy data.

For the data nodes, in each iteration, they can not find which network is training. If they cannot get what they want in one iteration, they cannot get that in the whole algorithm. So we only need to analyze one iteration.

When calculating the error, if P is the master node, a sample's component on other nodes is α, and the weights that are linked to the network's input units mapping non-master nodes form A, P will get $A\alpha$. On each iteration, P will get $A\alpha_1, A\alpha_2, \ldots, A\alpha_m$ for all the m testing samples. An equation set $\sum$ can be set up by using these components. But because the $k\alpha_i$ and the $(1/k)A$, where k is an arbitrary number, also meet$\sum$, $\sum$ has infinite solutions. So P cannot get data on other nodes from$\sum$.

When calculating the weights' update, in addition to the above data which is no use for leaking privacy, non-master nodes will get the randomized back propagation of weights' update. It is also no use for leaking privacy because it is randomized.

Calculating the error and the weights updated uses different samples. The information gotten in one process is no use for another. So for data nodes, our method has good performance in security. Each node cannot get other nodes' data.

4 Conclusion

This paper gives a privacy protection method for learning artificial neural network on vertically distributed data. This method makes using BP algorithm to learn a neural network privately on vertically distributed database possible. In this algorithm's learning process, every node cannot get the data on other nodes. It provides a solution to use neural network on distributed privacy data.

Acknowledgments. The Project Supported by Natural Science Basic Research Plan in Shaanxi Province of China (Program No. 2016JQ6078), the Fundamental Research Funds for the Central Universities of Chang'an University (300102328107, 0009-2014G6114024).

References

1. Khan, S., Gorhe, T., Vig, R., Patil, B.: Enabling Multi-level trust in privacy preserving data mining. In: Proceedings of 2015 ICGCIoT, pp. 1369–1372 (2015)
2. Bonawitz, K., Ivanov, V., Kreuter, B., Marcedone, A., McMahan, H., Patel, S., Ramage, D., Segal, A., Seth, K.: Practical secure aggregation for privacy-preserving machine learning. In: Proceedings of 2017 ACM SIGSAC CCS, pp. 1175–1191 (2017)
3. Gambs, S., Kegl, B., Aimeur, E.: Privacy-preserving boosting. Data Min. Knowl. Discov. **14**(1), 131–170 (2007)
4. Emekci, F., Sahin, O.D., Agrawal, D., El Abbadi, A.: Privacy preserving decision tree learning over multiple parties. Data Knowl. Eng. **63**(2), 348–361 (2007)
5. Secretan, J., Georgiopoulos, M., Castro, J.: A privacy preserving probabilistic neural network for horizontally partitioned databases. In: Proceeding of IJCNN 2007, pp. 1554–1559 (2007)
6. Barni, M., Orlandi, C., Piva, A.: A Privacy-preserving protocol for neural-network-based computation. In: Proceedings of 8th Workshop on Multimedia and Security, pp. 146–151 (2006)
7. Orlandi, C., Piva, A., Barni, M.: Oblivious neural network computing via homomorphic encryption. EURASIP J. Inf. Secur. **18**, 1–18 (2007)
8. Chang, Y.-C., Chi-Jen, L.: Oblivious polynomial evaluation and oblivious neural learning. Theor. Comput. Sci. **341**(1–3), 39–54 (2005)

9. Samet, S., Miri, A.: Privacy-preserving protocols for perceptron learning algorithm in neural networks. In: Proceedings of 4th IEEE Conference on Intelligent Systems, pp. 1065–1070 (2008)
10. Wan, L., Keong Ng, W., Han, S., Lee, V.: Privacy-preservation for gradient descent methods. In: Proceedings of 13th ACM SIGKDD, pp. 775–783 (2007)
11. Russell, S.: Artificial Intelligence: A Modern Approach. Pearson, London (2015)
12. Goodfellow, I., Bengio, Y., Courville, A.: Deep Learning (Adaptive Computation and Machine Learning Series). The MIT Press, Cambridge (2016)
13. Clifton, C., Kantarcioglou, M., Lin, X., Zhu, M.: Tools for privacy preserving distributed data mining. SIGKDD Explor. **4**(2), 28–34 (2002)
14. Prabhakaran, M., Sahai, A.: Secure Multi-Party Computation Cryptology and Information Security). IOS Press, Amsterdam (2013)
15. Frikken, K.: Secure Multiparty Computation. CRC Press, Boca Raton (2010)

Research of Simulated Annealing for BP Network and Its Application

Ming-liang Yang(✉), Hua Long, Yu-bin Shao, and Qing-zhi Du

Faculty of Information Engineering and Automation, Kunming University of Science and Technology, Kunming 650000, Yunnan, China
1520890081@qq.com

Abstract. In order to avoid overfitting and fall into local minima, a hybrid algorithm combining simulated annealing algorithm and BP algorithm is used to train neural network. The strong global optimization ability of simulated annealing algorithm and the fast local optimization ability of BP algorithm are combined to ensure that the global minimum point can be found more accurately without reducing the learning rate of the network. The model is applied to Taobao e-commerce sales forecast and the results are satisfactory.

Keywords: BP algorithm · Simulated annealing algorithm · Hybrid algorithm Neural network

1 Introduction

BP neural network is not only a functional approximator with universal applicability but also the neural network is also a nonlinear method to realize the nonlinear mapping [1, 3, 4] of the whole network from the input state to the output state due to its strong adaptive characteristics and excellent generalization ability. But the BP neural network also has some defects: for example the initial value is difficult, The convergent speed is slow and the BP training process is easy to fall into the local minimum point [2, 5–7].

This thesis will introduce some idea of simulated annealing algorithm [8–11] to solve the above problems. The simulated annealing algorithm has strong global optimization ability but not dependence on initial value deeply, however the local optimization ability is weak [12]. So we combine it with BP neural network to find the best point more accurately without reducing the learning rate.

2 A Algorithm Based on Simulated Annealing and BP Network

2.1 Definition of Simulated Annealing Algorithm

In the solution space [13, 14] of simulated annealing algorithm. When the system changes from one energy state to another, the corresponding energy also changes from E_1 to E_2, and the probability is as follows. Where the Boltzmann constant is define K, T is the object's temperature in the current state.

K. Deng et al. (Eds.): ICMIR 2018, AISC 856, pp. 1168–1174, 2019.
https://doi.org/10.1007/978-3-030-00214-5_143

$$p(\Delta E) = \exp\left(-\frac{E_2 - E_1}{KT}\right) \tag{1}$$

At a given temperature, the probability of the state satisfies the Poltzmann distribution. When T is high enough, the search probability in all directions in space is approximately equal. When T tends to 0, the minimum energy state is obtained with a high probability [15, 16]. The probability of state 2 being accepted by

$$P(1 \rightarrow 2) = \begin{cases} 1 & E_2 < E_1 \\ \exp\left(-\frac{E_2 - E_1}{T}\right) & E_2 \geq E_1 \end{cases} \tag{2}$$

If the new state energy is small, it is accepted (local optimum); if the global search is large, the probability is accepted by Eq. (2).

2.2 Definition of the BP Neural Network Algorithm

The main idea of BP neural network algorithm is that $x = (x_1, x_2, \ldots, x_i)^T$ is the input sample about learning,the corresponding output is $y = (y_1, y_2, \ldots, y_j)^T$, the output layer whose output is $z = (z_1, z_2, \ldots, z_k)^T$ and $t = (t_1, t_2, \ldots, t_k)^T$ is the target vecto. The network connection weight and threshold are modified by the error between the actual output of the network and the target vector. Make the output and expectation as close as possible. Three layers of neural networks can achieve the approximation of any non-linear function. Its model looks like the following (Fig. 1).

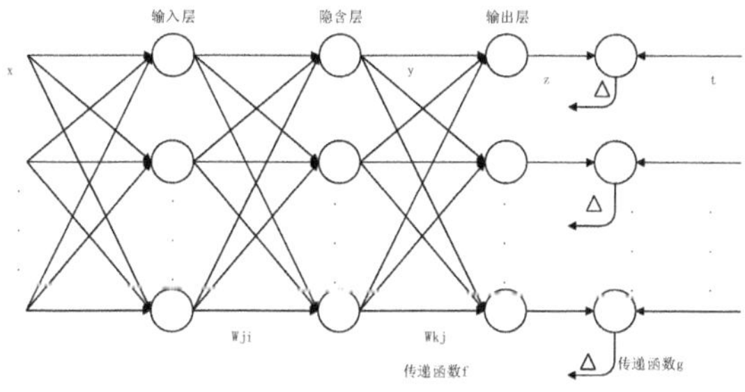

Fig. 1. Three-layer BP neural network structure

Defind $y_j = f(\sum_{i=1}^{n} w_{ji}x_i + b_{j1})$ as output of the jth neuron in hidden layers, b_j defined as the adjusted offsets at the jth neuron. $z_k = g(\sum_{j=1}^{h} w_{jk}y_j + bb_{k1})$ defined as the kth neuron output. bb_k is the threshold value of the kth neuron in the output layer. $\varepsilon = \frac{1}{2}\sum_{k=1}^{m}(t_k - z_k)^2$ indicated as the sum of square errors. The weight adjustment formula is

$$\Delta w_{pq} = -\lambda \frac{\partial \varepsilon}{\partial w_{pq}} \tag{3}$$

λ defined as learning rate, in reverse propagation, the weights of the hidden layer to the output layer are first adjusted. The input of the kth neuron in the output layer is v_k. To calculate the gradient of the sum of the squared errors. In turn, the weight and threshold adjustment from the hidden layer to the output layer can be obtained as follows:

$$\frac{\partial \varepsilon}{\partial w_{kj}} = \frac{\frac{1}{2}\sum_{k=1}^{m}(t_k - z_k)^2}{\partial z_k} \frac{\partial z_k}{\partial v_k}\frac{\partial v_k}{\partial w_{kj}} = -(t_k - z_k)g'(v_k)y_i = -\delta_k y_i \tag{4}$$

$$w_{kj}(t+1) = w_{kj}(t) + \lambda\delta_k y_i \tag{5}$$

$$bb_{k1}(t+1) = bb_{k1}(t) + \lambda\delta_k \tag{6}$$

Defined the input of the jth neuron of the hidden layer as u_j. To calculate the gradient of w_{ji} of the output layer to the hidden layer on the sum of the square of the error. Therefore, the weight and threshold adjustment iteration formula from the input layer to the hidden layer is

$$\frac{\partial \varepsilon}{\partial w_{ji}} = -\sum_{k=0}^{m}[(t_k - z_k)f'(u_k)w_{jk}]f'(u_j)x_i = -\delta_j x_i \tag{7}$$

$$w_{ji}(t+1) = w_{ji}(t) + \lambda\delta_j x_i \tag{8}$$

$$b_{ji}(t+1) = b_{ji}(t) + \lambda\delta_j \tag{9}$$

2.3 A New Algorithm BP of Network Based on Simulated Annealing

First, the approximate convergence direction of connection weight is determined by simulated annealing algorithm, that is the global optimum. The change of connection weight of neural network can be synthesized by backpropagation error and random number, if the sum of square of output error is reduced, then the connection weight is accepted, if the sum of output error squared increases, then the adjustment of the connection weight is accepted with a certain probability of Eq. (2). As the number of iterations increases, the probability of acceptance decreases until it is 0. The steps of a new BP neural network algorithm based on simulated annealing are as follows.

First step is to add input and expected output and initialize the weight matrix and threshold with random number and Initializing temperature $T = T_0$. Initialize iterations record f = 0; Second step is: when f < L (maximum number of iterations), continue to iterate and define f = f + 1. The training sample input model calculates the weight adjustment and modifies the weight and threshold of network connection according to the formula of (3)–(9) and accepts the adjustment if the sum of square error is reduced.

Otherwise, the adjustment is accepted or discarded with the random probability of (2): Third step: the optimal value obtained by simulated annealing algorithm back to BP neural network for further iterative evaluation to make its optimal value more accurate: Fourth step: repeat the above training calculation with the sample data, and the cooling function adopts the exponential temperature, $T(f+1) = K * T(f)$, where $0 < K < 1$ is a constant very close to 1.

In order to make the optimal value more quickly and accurately, the iterative step size is adaptively optimized here. So, the iterative step size is adjusted by output error, defined $e = \varepsilon(f) - \varepsilon(f + 1)$ as the error difference between the first iteration and the last iteration I in some cases, Step micro adjustment with output error difference. In other cases, a fixed iteration step is used.

3 Model Application

3.1 Experimental Data Source

In This article selects a Taobao store for the past four years of monthly (M) sales of Y. Use historical data to predict the last three months (Table 1).

Table 1. Merchandise sales of a Taobao store, 2014–2017 (piece)

Y\M	1	2	3	4	5	6	7	8	9	10	11	12
2014	9	54	57	43	73	79	60	106	259	135	111	110
2015	108	92	170	208	211	191	217	300	854	728	653	512
2016	406	264	596	654	996	949	591	865	1818	1089	1179	1038
2017	1089	989	1734	1765	2184	2732	1416	1529	2726	2475	2303	2169

3.2 Data Preprocessing

The data are normalized to the [0 1] interval before training.

$$X = \frac{x - x_{\min}}{x_{\max} - x_{\min}} \tag{10}$$

3.3 Model Building

The model of this simulation and he BP network model has three layers: the input layer node is 4, the hidden layer node number is 5, the output layer node number is 1, and the activation function of the hidden layer and output layer is

$$f(x) = \frac{1}{1 + e^x} \quad (0 < f(x) < 1) \tag{11}$$

The mean square is calculated by absolute error and mean square error (U are error) as the standard of model prediction and evaluation accuracy, The value z_i denied as output of the model, the value t_i defined as the expected.

$$e_{ae} = \frac{|z_i - t_i| \times 100}{t_i}\% \tag{12}$$

$$e_{mse} = \frac{1}{n}\sqrt{\sum_{i=1}^{n}\left[(t_i - z_i)^2 / t_i\right]} \tag{13}$$

The follow model of table are gray/BP network/this paper's model. Through the table below, it can be seen that the prediction accuracy of the proposed model has been improved obviously. Moreover, it also has certain ability to predict the fluctuation of data, and the accuracy is up to the expected requirement (Table 2).

Table 2. Comparison of prediction accuracy between three models

Mouth	Predict (piece)	Model one		Model two		Model three	
10	2575	2480	3.69	2335	9.32	2721	5.68
11	2303	2641	14.67	1956	15.08	2552	10.81
12	2169	2813	29.69	1869	13.80	2089	3.66
mse		11.11		7.49		4.25	

The initial temperature is 1000 °C, the termination temperature is 10^{-7} °C, the step length of the iteration temperature is $K = 0.998$, the maximum number of iterations are 1000, respectively. The average solution time of the iteration is taken as the convergence speed. The mean minimum and variance as fitness are compared with the basic BP Network algorithm as shown in Table 3 (Fig. 2).

Table 3. Comparison of convergence of two model algorithms

Algorithm	Convergence time/second	Sufficiency
BP neural network	8.25	22.46
Model algorithm	3.75	12.75

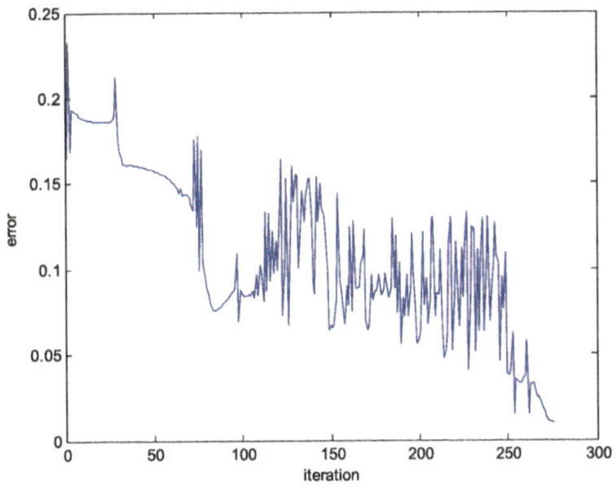

Fig. 2. Iterative variation diagram of model error

4 Conclusion

By comparing the results of the three models in Table 1, we can see that the prediction accuracy of the proposed model is obviously higher than other two models. In Table 2, we can also see that this paper has obvious advantages. The BP neural network prediction model, which jumps out of the local optimum and looks for the global optimum, it is true and effective. However, improving the accuracy and searching efficiency of the model are the main directions of the future research based on the neural network model.

References

1. Bao, Z., Yu, J.: Intelligent optimization algorithm and its MATLAB example, pp. 125–136. Electronic Industry Press (2016)
2. Chen, M.: MATLAB neural network principle and examples, pp. 183–194. Tsinghua University Press, Beijing (2017)
3. Guo, B., Zhang, P., Shen, Y., Huang, Y., Cao, Y.: Research on energy saving technology of processor based on BP neural network. Eng. Sci. Technol. (EST) **50**(1), 107–112 (2018)
4. Qin, S., Liu, C.: Huang bid for the Olympic Games. Identification of rice varieties based on K-means clustering algorithm and BP neural network. Chin. J. Cerea. Oils **2**, 128–131 (2018)
5. Yam, J.Y.F., Chow, T.W.S.: A weight initialization method for improving training speed infeed forward neural network. Neurocomputing **1**(30), 219–232 (2000)
6. Liping, L., Zhigang, L., Sunjinsheng, L.: Prediction and realization of DO in sewage treatment based on machine vision and BP netural network. Telkomnika **12**(4), 890–896 (2014)
7. Zhao, J.L., Sun, Z.J., Yang, N.D.: A feeding system control of NC honing machines based on fuzzy neural network. Adv. Mater. Res. **411**, 169–173 (2011)
8. Fang, L., Zuo, H., Yang, Z., Zhang, X., Du, J., Pang, L.: Binary wavefront optimization using a simulated annealing algorithm. Appl. Opt. **57**(8), 1744–1751 (2018)

9. Bouttier, C., Gavra, I.: Convergence rate of a simulated annealing algorithm with noisy observations. Statistics (2017)
10. Dong, Y., Wang, J., Chen, F., Hu, Y., Deng, Y.: Location of facility based on simulated annealing and "ZKW" algorithms. Math. Prob. Eng. **2017**, 9 pages (2017)
11. Siva, C., Meetei, T.S., Shiva, P., Narayanan, B., Arvind, G., Boomadevi, S., Pandiyan, K.: Engineering phase shifter domains for multiple QPM using simulated annealing algorithm. J. Opt. **19**(10), 105–507 (2017)
12. Qing, H., Yile, W., Tongwei, X.: Application of improved genetic simulated annealing algorithm in TSP optimization. Control Decis. Mak. **2**, 219–225 (2018)
13. Alcántar, V., Ledesma, S., Aceves, S.M., Ledesma, E., Saldaña, A.: Optimization of type III pressure vessels using genetic algorithm and simulated annealing. Int. J. Hydrog. Energy **42** (31), 20125–20132 (2017)
14. Mahmoodpour, S., Masihi, M.: An improved simulated annealing algorithm in fracture network modeling. J. Nat. Gas Sci. Eng. **33**, 538–550 (2016)
15. Wenyuan, F., Chaodon, L.: Brownian motion simulated annealing algorithm. J. Comput. **6**, 1301–1308 (2014)
16. Javad, M.O.M., Karimi, B.: A simulated annealing algorithm for solving multi-depot location routing problem with backhaul. Int. J. Ind. Syst. Eng. **25**(4), 460–477 (2017)

Research on Equipment Fault Prediction Method Based on Adaptive Genetic Neural Network

Hongtu Cai(✉), Yuwen Liu, Tingchao Kang, Shidong Fang, and Hanyu Ma

Army Artillery and Air Defense College, Hefei 230031, China
sypycht@126.com

Abstract. In order to effectively improve the capability of equipment fault prediction and realize state maintenance, this paper presents an improved hybrid model fault prediction method based on adaptive genetic neural network. It mainly uses genetic algorithm to improve the BP neural network, and iterates out the optimized initial weight gradually, introduces the strategy of adaptive adjustment of cross probability and mutation probability, and adjusts the probability of crossing or mutation according to the fitness of the individual to be crossed or changed. The example shows that the model of improved adaptive genetic neural network has better prediction effect, faster convergence speed and more generalization ability than simple BP neural network, and the performance of fault prediction is also enhanced greatly.

Keywords: Fault prediction · Neural network · Genetic algorithm

1 Introduction

Condition Based Maintenance (CBM) is a kind of maintenance method to solve "insufficient maintenance" or "excess maintenance" effectively. It has become a hot spot in the field of equipment maintenance theory research, however, the implementation of state maintenance must be based on the effective prediction of equipment failure. Back propagation (BP) neural network is one of the most commonly used prediction methods, It has good characteristics in terms of adaptation, robustness, etc. So far it is the most widely and successfully used neural network. But BP also has some shortcomings, such as slow convergence speed, only solve small scale problems, difficult to realize the global optimal solution. Genetic Algorithms (GA) is a global search method by natural selection randomness, which can realize the search of the best individual in the target group; At the same time, the search mode and structure used will not be affected by any group gradient change. These advantages can make up for the shortcomings of BP. But in the later stage of genetic algorithm search, the convergence of individual population and the reduction of genetic diversity are prone to lead to premature convergence. In this case, this paper combined GA and BP, and introduced the strategy of cross probability and mutation probability of adaptive adjustment,which can avoid the premature convergence to a certain extent by adjusting the probability of

K. Deng et al. (Eds.): ICMIR 2018, AISC 856, pp. 1175–1182, 2019.
https://doi.org/10.1007/978-3-030-00214-5_144

crossing or variation according to the fitness of the crossed or mutated individual. The example shows that the improved hybrid model fault prediction method based on adaptive genetic neural network not only effectively guarantees the global optimality of the training results, but also improves the convergence speed greatly.

2 Basic Model of Genetic Algorithm

2.1 GA Principle

GA is also called evolutionary algorithm, simulates the natural evolution process of the biological world, and realizes the optimization of complex problems through selection, crossover and mutation operation. It is a heuristic algorithm for solving optimization problems. It has inherent implicit parallelism and better global search capability. It uses probabilistic optimization method to automatically obtain the spatial information of search, and adjust the search direction adaptively. The basic genetic algorithm consists of 4 components: chromosome coding, individual fitness, genetic operator and operating parameter, and uses 3 genetic operators: selection, cross and mutation.

2.2 Genetic Algorithm Implementation Steps

(1) code the problem, that is, expressing the problem as chromosome.
(2) establish initial group.
(3) calculate the fitness of individuals as the basis of genetic operation.
(4) conduct the selection, crossover and mutation on groups to produce new population.

3 Basic Model of BP

3.1 BP Principle

BP is a multilayer feed-forward artificial neural net which is studied according to the error backpropagation, its basic principle of learning algorithm is the gradient descent method, the central idea is to minimize the network error by adjusting the weights and thresholds. The learning process of BP includes two stages: forward and backward. In the forward stage, the inputs is processed from the first layer through the middle layer to the last layer, and the outputs of current layer affects the next outputs only. If the value of the last layer doesn't agree with the expected value, then the signal enters backward transmission stage, and the error signal is returned along the former path, and the error is gradually reduced by adjusting the weights and thresholds of each layer, until the precision is met.

3.2 BP Algorithm

(1) initialize the connection weights: generate a relatively small number of non-zero values, and assign them to each neuron randomly as the initial weight.
(2) determine the topological parameters of BP and the training samples.
(3) determine the sample pairs of input training according to the set of samples provided.
(4) the information is transmitted forward, and the error information is obtained according to the information of the network.
(5) error signal back propagation. According to the error signal, update the threshold and weight of each layer until the error meets the accuracy requirement.

4 GA-BP Model

4.1 GA-BP Principle

GA is a robust global optimization algorithm based on random search and natural selection evolution. It can find the better connection weight before the neural network training, avoid the local optimal and improve the prediction accuracy [1]. Therefore, we combine genetic algorithm with BP net and determine the weight of network connection by GA. The basic idea is shown in Fig. 1.

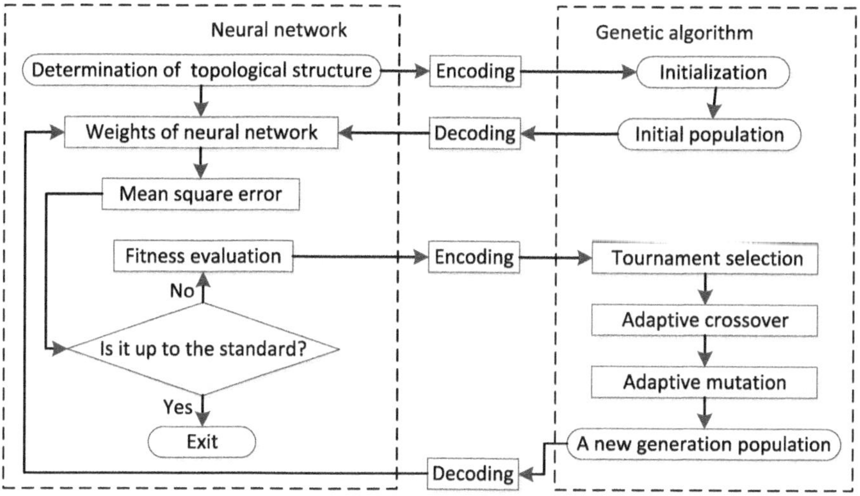

Fig. 1. Principles of genetic neural networks

4.2 Genetic Neural Network Algorithm Implementation

(1) Selection of coding methods

The studying of BP net weights is a complex continuous process of optimizing parameter. If the binary code is used, the independent variable will be much more, and the solution space becomes larger and is not conducive to the fast search. And the binary encoding will result in the change of the weight value in the decoding process, thus affecting the learning accuracy of the network, so the real number coding is adopted. The weights of neural network are cascaded into a long string in order, sequentially arranged from left to right, from top to bottom.

(2) The generation of the initial group

To make the genetic algorithm search all feasible solution spaces, the initial chromosome population is randomly generated, its expression is

$$P_{initial} = \pm \exp(-|\gamma|), |\gamma| < 4 \tag{1}$$

(3) The determination of the evaluation function

The weight values expressed on the chromosomes are assigned to the specified network in corresponding ways, the training set sample is input and output, and the mean square error (e_{MSE}) is returned after operation. Then the evaluation function of the chromosome is

$$f = 1/(1 + e_{MSE}) \tag{2}$$

(4) The determination of the selection operator

The selection of genetic operators is adopted in league matches, and the scale of tournament selection is 2.

(5) The determination of the crossover operator of weight value

For different application problems, the form of crossover operator is different. Effective linear crossover operation is adopted here:

$$\begin{cases} P_1^{g+1} = \alpha P_1^g + (1-\alpha)P_2^g \\ P_2^{g+1} = \alpha P_2^g + (1-\alpha)P_1^g \end{cases}, \quad \alpha \in (1.2, 1.5) \tag{3}$$

where P_1^g, P_2^g is two individuals of the previous generation; P_1^{g+1}, P_2^{g+1} is two new individuals after cross operation: α is constant.

(6) Determination of weight variation operator

Gauss mutation method is used to perform mutation operation with formula (5).

$$P^{g+1} = P^{g} + \mu(0,1)\beta\sqrt{f} \tag{4}$$

where $\mu(0, 1)$ is Gauss operator; f is corresponding fitness; β is constant coefficient.

(7) Adaptive cross and mutation rate

The value of p_c and p_m has a great influence on the performance of GA. In order to effectively improve the speed of convergence and avoid the occurrence of premature, the adaptive p_c and p_m can be used. For a solution with high fitness value, a lower p_c and p_m is taken to increase the opportunity to enter the next generation, and a higher p_c and p_m is taken for a solution with a lower fitness value, which increases the probability of the solution being eliminated. The formula for calculating the adaptive change of p_c and p_m with the change of fitness value is

$$p_c = \begin{cases} k_c \cdot (f_{\max} - f_c)/(f_{\max} - \bar{f}) & (f_c \geq \bar{f}) \\ k_c & (f_c < \bar{f}) \end{cases} \tag{5}$$

$$p_m = \begin{cases} k_m \cdot (f_{\max} - f_m)/(f_{\max} - \bar{f}) & (f_m \geq \bar{f}) \\ k_m & (f_m < \bar{f}) \end{cases} \tag{6}$$

where f_c is the bigger value among the two individuals who want to carry out cross operation; f_m is the bigger value among the two individuals who want to carry out the mutation operation; $f_{\max}$, $\bar{f}$ is the max and average value in the group; $f_{\max} - \bar{f}$ shows the convergence degree of the group.

4.3 Model Performance Evaluation Index

To evaluate the performance of the model better, the average absolute error, the mean square error, the percentage of mean absolute error, the percentage of mean square error should be measured. Set t_i as the actual measured value, t_i' as the prediction value, $\bar{t}$ as the real sequence mean and $\overline{t'}$ as the prediction sequence mean, then the calculation formula of each evaluation index is as follows:

(1) the average absolute error

$$e_{MAE} = \frac{1}{k}\sum_{i=1}^{k} \left| t_i' - t_i \right| \tag{7}$$

(2) the percentage of mean absolute error

$$e_{MAPE} = \frac{1}{k}\sum_{i=1}^{k} \left(\left| t_i' - t_i \right| / t_i \right) \tag{8}$$

(3) the mean square error

$$e_{MSE} = \frac{1}{k}\sqrt{\sum_{i=1}^{k} (t_i' - t_i)^2} \tag{9}$$

(4) the percentage of mean square error

$$e_{MSPE} = \frac{1}{k}\sqrt{\sum_{i=1}^{k} ((t_i' - t_i)/t_i)^2} \tag{10}$$

5 Application Example Analysis

The operating voltage range of a certain type of information system is 27 V to 29 V, with the increase of working time, its performance gradually attenuates and the voltage increases gradually, failure occurs when it is beyond the specified range. According to the actual working condition of the system, the data of the system's characteristic parameters changing with accumulative working time are monitored, as shown in Table 1.

Table 1. Cumulative working time data under normal condition

Voltage/V	27.00	27.10	27.20	27.30	27.40	27.50	27.60
Time/h	0	577	866	1123	1221	1290	1433
Voltage/V	27.80	28.00	28.20	28.40	28.60	28.80	29.00
Time/h	1597	1677	1780	1881	1967	2045	2096

(1) Structural design of BP net

According to the Kolmogorov theorem [2], if the unit numbers in the hidden layer are enough, the 3 layer BP can almost accurately approximate any complex nonlinear function. Therefore, we select the 3 layer BP, the input layer is 4 units t_{i-3}, t_{i-2}, t_{i-1} and t_i, the output layer is 2 units t_{i+1} and t_{i+2}, and the unit numbers of hidden layer is gradually adjusted, that is 9, 10, 11, 12, 13 respectively.

(2) Parameter setting of genetic algorithm

The crossover operator of genetic algorithm is $\alpha = 1.3$, the parameter of mutation operator is $\beta = 0.7$, the basic crossing rate is $k_c = 0.5$, and the basic mutation rate is $k_m = 0.06$.

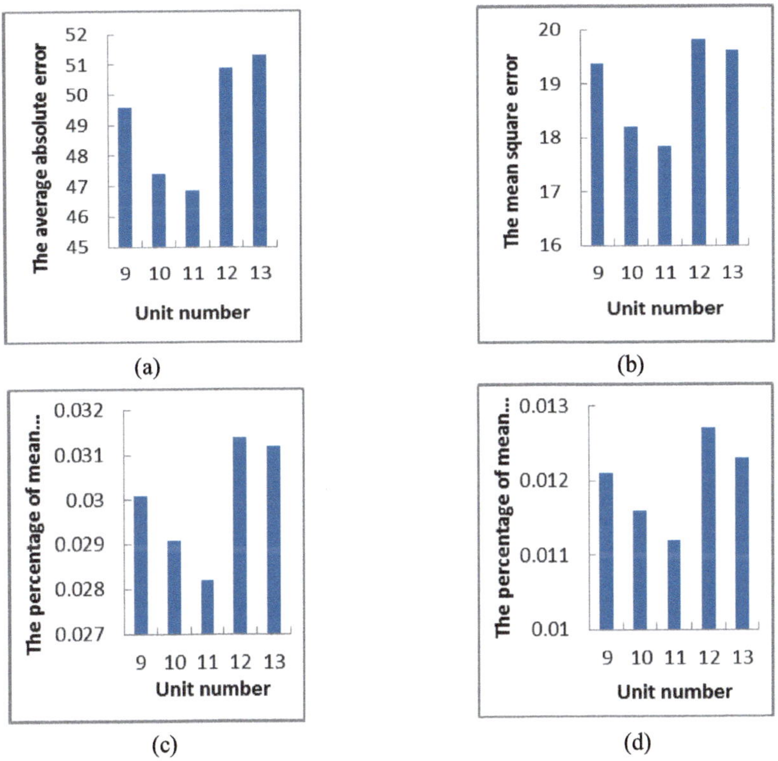

Fig. 2. Analysis of one step prediction results of the different hidden layer unit number

(3) Genetic neural network training

Select the data in Table 1 to train genetic neural network, and adjust the connection weights of BP.

(4) Determination of hidden layer units number

The trained genetic neural network is used to predict the system failure and get the evaluation indicators of the prediction, as shown in Fig. 2.

From Fig. 2, we can see that when hidden units is 11, the performance of one step prediction is the best. According to the characteristics of one step prediction value is more paid attention to the prediction of faults, and determine that hidden units is 11.

(5) Prediction of genetic neural network

The data in Table 1 are predicted by the established GA-BP model, and compared with the traditional BP prediction algorithm, the prediction are shown in Fig. 3.

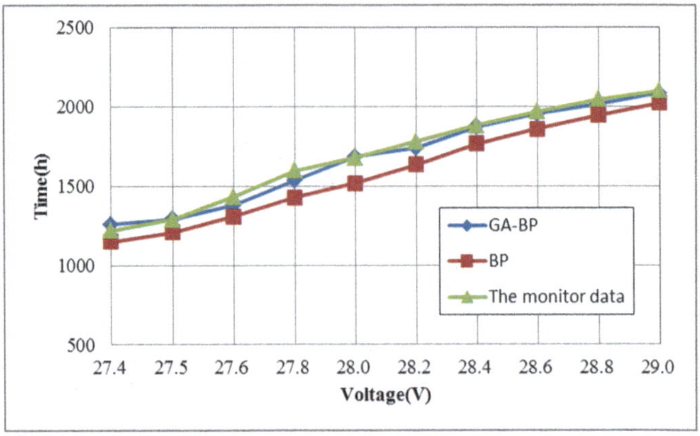

Fig. 3. Comparison of predictive performance of different algorithms

6 Conclusion

This paper proposes an improved hybrid model fault prediction method based on back propagation net optimized by GA; the BP net is improved through GA, the optimized initial weight is iterated out gradually, the network training time is reduced, and the neural network is effectively avoided into the local extremum. By means of the strategy of introducing adaptive crossover probability and mutation probability, premature convergence is avoided to a certain extent. It is proved by the example that the model established in this paper has better prediction effect and stronger generalization ability, which provides a new idea for equipment fault prediction.

References

1. Huang, Y.W., Yu, F., Hang, C., et al.: Securing Web application code by static analysis and runtime protection. In: Proceedings of the International Conference on World Wide Web (WWW 2004), pp. 40–52, New York, USA (2015)
2. Wang, S.X., Zhang, N., Wu, L., et al.: Wind speed forecasting based on the hybrid ensemble empirical mode decomposition and GA-BP neural network method. Renew. Energy **94**, 629–636 (2016)
3. Hao, J., Li, H.: Application of BP neural network to short-term-ahead generating power forecasting for PV system. Adv. Mater. Res. **609**, 128–131 (2013)
4. Pan, X.T., Qu, K.Q.: Wind power output prediction with BP neural network combining genetic algorithms. Adv. Mater. Res. **860–863**, 2526–2529 (2014)
5. Paoli, C., Notton, G., Nivet, M.L., et al.: A neural network model forecasting for prediction of hourly ozone concentration in corsica. In: 2011 10th International Conference on Environment and Electrical Engineering (EEEIC) (2011)
6. Hassan- Esfahani, L., Torres- Rua, A., McKee, M.: Assessment of optimal irrigation water allocation for pressurized irrigation system using water balance approach, learning machines, and remotely sensed data. Agric. Water Manag. **153**, 42–50 (2015)

A Review of Classical First Order Optimization Methods in Machine Learning

Zhang Menghan(✉)

Army Academy of Artillery and Air Defense, Hefei 230031, Anhui, China
13998826728@163.com

Abstract. In the development of machine learning, the first-order classical gradient optimization method plays a crucial role in optimizing the field, and many efficient algorithms are improved on this basis. Under the condition of the regularization and loss function as a whole, this paper mainly introduces three black-box methods that use first-order step information: Projection Sub-gradient Descent Method PGD; Mirror Descent Method MD; Dual Averaging Method DA. This paper summarizes the development course and the improvement method, analyzing and comparing the convergence rate and bound of different conditions. Since the solution of almost every model problem is confronted with large-scale data, considering the special meaning of the regularization and loss function is given to the machine learning model. In order to ensure the structure of the machine learning problem solution, we summed up the conditions of solving the regularization problem and some methods of structural optimization and improvement methods, refining the stochastic optimization methods in proving some of the math skills and the part of the study classified and summarized, analyzing the significance of the era for the current big data. In addition, because the noise interference gradient unbiased hypothesis is often not established in the real world, we also put forward some problems worthy of further study under the condition that the gradient information is biased.

Keywords: Machine learning · First-order gradient optimization
Individual convergence · Stochastic optimization

1 Research Background and Current Situation of Machine Learning and Optimization Problem

In computer systems, the usual experience is stored in the form of data, and machine learning (Machine Learning) is an algorithm that generates a model from the data for a given task, i.e., a learning algorithm to continue predicting data or classification. With the arrival of the era of big data and the promotion of machine learning algorithms, we have found that machine learning is everywhere: Google's algorithms can almost determine your hobbies, Amazon shopping can know your shopping habits, and so on. In fact, machine learning technology not only plays a role in computer science such as computer vision, multimedia, graphics, natural language processing, and software engineering, but has even become an important supporting technology in many cross-disciplinary fields such as bioinformatics and computational finance, such as: Biometrics, DNA sequence

K. Deng et al. (Eds.): ICMIR 2018, AISC 856, pp. 1183–1190, 2019.
https://doi.org/10.1007/978-3-030-00214-5_145

testing, market analysis of securities and detection of credit card fraud, etc. The development of machine learning technology has attracted much attention. Many developed countries have considered machine learning techniques from the perspective of national security.

Machine learning is strictly a subfield of artificial intelligence and includes many research directions: pattern recognition, statistical modeling, and data mining. Machine learning algorithms are essentially algorithms that learn prior experience from known data, use experience to predict or classify unknown data, and are classified into supervised learning, unsupervised learning, and semi-supervised learning according to the given types of label items. Due to a large number of statistical theoretical foundations, it is often referred to as statistical learning theory. In the process of optimizing the machine learning model, a large part also uses the convex optimization method [1] to solve the optimal value and the convergence bound.

When a convex optimization problem is determined, the study of optimization theory and optimization methods becomes an important part of machine learning. At present, many optimization algorithms have evolved by combining optimization methods with different principles. For example, in the classical first-order method, the most popular methods are Stochastic Gradient Descent (SGD) [2], Mirror Descent (MD) [3], and Dual Average Method [4], Alternating Direction Method of Multipliers (ADMM), etc. When the Bregman function of MD takes a special Euclidean distance, MD and SGD are equivalent [5], because of the two The methods are simple to operate and the theoretical analysis is relatively complete, so they are highly praised by the majority of scholars. Among them, the MD algorithm was once considered by the majority of optimization scholars as the optimal one-step optimization method. When the information of a level of gradient is not clear, the second-order method will be considered, and the second-order method will use the Newton method (Newton method) and the quasi-Newton method. (Quasi-Newton method), the step size for each step of iterative optimization is more accurate and reasonable than the first-order method, and the faster convergence rate can be obtained when processing small-scale databases, but The data today seems to require too much storage space and low efficiency in processing big data. This article is an overview of a gradient optimization method, so it does not provide a detailed comparison of the second-order methods.

2 Convergence Rate of Blackbox Method Optimized by First Order Gradient

From the perspective of optimization, the regularization term and the loss function are both convex functions, and the result of the combination can still be solved by general convex optimization methods. The early learning algorithm considers the objective function consisting of regularization and loss functions as a whole without considering the details of each part. This method is called the black box method. The common black-box gradient optimization methods mentioned in the first chapter have the most classic steepest descent method (also known as gradient descent method), and the two-step optimization method Newton method, and there are also between first-order and second-order. There are many conjugate gradient methods between the orders and so on [6]. However, these classical methods have more or less obvious flaws in some aspects of optimization algorithm metrics. Gradient descent method produces sawtooth phenomenon, its convergence speed is relatively slow, the CPU time to calculate the value of one objective function is almost equivalent to the total time of the random solution algorithm vector iteration, so the algorithm is used to stop the algorithm by comparing the difference of the objective function value. It is simply not practical. The relative Newton method using the two-step information Hesse matrix has a faster convergence speed, but even if the sample size is assumed to be the number of samples 10^7, the sample dimension 10^9, the required storage space for the Hessian matrix is 4×10^9 GB, the current 64-bit computer The storage limit is about 16 GB, and even if it is used to store the Hessian matrix, its capacity is still far away. Overall, the defects of the traditional optimization methods are still quite obvious, so many scholars have made many minor improvements. In this chapter, the article summarizes three more practical and efficient optimization methods.

2.1 Projection Sub-gradient Descent

Sub-gradients are a special extension of gradients. When the function is continuously irreducible, the gradient loses its own meaning due to the presence of non-conducting points. We note that the set $\boldsymbol{x} \in S$ of all subgradients at the function f is $\partial f(\boldsymbol{x})$. Its special case is that when the function is subtle, there is only one subgradient, i.e. $\partial f(\boldsymbol{x}) = \nabla f(\boldsymbol{x})$. The subgradient method is actually the general generalization of the steepest descent method [3, 4]. The continuous convex optimization problem can be solved by a method similar to gradient descent. The significance of the projection operator is that in the process of solving the convex feasible problem, the iteration often needs to calculate the projection onto the closed convex set. In practical applications, the cost of the projection on the closed convex set is large, and the projection operator exists. Let us only calculate the projection of a half-space containing a set of constraints, which greatly optimizes the space-time complexity.

The iterative formula for optimizing the model problem described in formula (1) is expressed as:

$$\boldsymbol{w}_{k+1} = \prod\nolimits_P (\boldsymbol{w}_k - \eta_k g(\boldsymbol{x}_i, \boldsymbol{w}_k)) \tag{1}$$

The initial step size (i.e. learning rate) η_1 and the number of iterations of the algorithm T are set and the weight vector $\boldsymbol{w}_1$ is initialized. During the first k iteration, calculate $f(\boldsymbol{w}_k)$ and set the step size η_k, calculate the second-order gradient $g(\boldsymbol{x}_i, \boldsymbol{w}_k)$ at the point $\boldsymbol{w}_k$, the algorithm needs to input the projection threshold P, and the initialization $\boldsymbol{w}_1$ should meet the conditions $||\boldsymbol{w}_1|| \in P$.

Convergence is the main criterion for measuring the pros and cons of an algorithm. For the projected subgradient method, many scholars obtained the optimal convergence rate by means of multiple proof techniques for the weighted average of all iterations of the projected subgradient method, Yu. Nesterov, 2015 The annual literature [7] summarizes the convergence bounds of some of the most commonly used classical one-degree-degree methods.

$$\sum_{k=0}^{t} a_k f(\boldsymbol{x_k}) - \sum_{k=0}^{t} a_k f(\boldsymbol{x_*}) \leq \left[\frac{1}{2} \|x_0 - \boldsymbol{x_*}\|_2^2 - \frac{1}{2} \sum_{k=0}^{t} a_k^2 \nabla f(\boldsymbol{x_k})_2^2 \right] \tag{2}$$

Where a_k is the step size and is always greater than zero, the initial value is $\boldsymbol{x}_0$ the optimal solution $\boldsymbol{x}_*$, the subgradient $\nabla f(\boldsymbol{x_k})$ at the point $\boldsymbol{x_k}$, and t is the number of iterations $A_t = \sum_{k=0}^{t} a_k$. This bound can take the precision of O($\frac{\sqrt{T}}{T}$) when taking a proper step, which can achieve the best. by slightly modifying its internal iterative method through linear interpolation techniques, the convergence rate of the individual is optimal under the general convex condition, which is not a small breakthrough.

From a historical point of view, the projected subgradient method originated from the study of the non-smooth function that Shor NZ published in 1985. In recent years, Bertrakes NZ published the convex analysis in 2003. In the optimization paper [1], after a long improvement, many kinds of improved optimization have appeared successively on the projected subgradient method, such as Mirror Descent [3] and Dual Averaging [4] mentioned earlier.

2.2 Mirror Descent

In the following, we introduce a gradient method that was once considered optimal by related scholars. If the sample size is large enough, the MD (Mirror Descent) [3] method is considered to be the optimal first-order method. Actually, SGD [8] is a special case of MD expansion. The model problem described in Eq. (1) is also optimized. The definition of all variables is the same as the projected subgradient method above. The iteration form of the MD algorithm is:

$$\boldsymbol{w}_{t+1} = \arg\min_{w \in \Omega} B_\psi(\boldsymbol{w}, \boldsymbol{w}_t) + \eta < f_t'(\boldsymbol{w}_t) + r_t'(\boldsymbol{w}_t), \boldsymbol{w} > \tag{3}$$

The initial step size η (i.e., the learning rate) and the number of iterations of the algorithm T are set, and the initialization weight vector $\boldsymbol{w}_1$, B_ψ represents the Bregman divergence. In the first k iteration, the sub-gradient $f_t'(\boldsymbol{w}_t) + r_t'(\boldsymbol{w}_t)$ is newly defined as the entire objective function and its result is calculated, and the step length η_k is set. The initialization $\boldsymbol{w}_1$ should satisfy the condition $||\boldsymbol{w}_1|| \in P$. It is not difficult to see that the MD method is equivalent to the GD method when taking B_ψ the European squared distance. This again shows that the GD method is a special case of the MD method.

Similarly, we do the convergence analysis of the MD algorithm. In 2015, an article [7] of Yu. Nesterov explicitly stated the bounded output of the weighted average output of the dual-form MD algorithm:

$$\sum_{k=0}^{t} a_k f(\boldsymbol{x_k}) - \sum_{k=0}^{t} a_k f(\boldsymbol{x_*}) \leq \sum_{k=0}^{t} \frac{1}{2} a_k^2 \nabla f^2(\boldsymbol{x_k}) + d(\boldsymbol{x_*}) \tag{4}$$

Its iteration method under dual conditions is:

$$\mathbf{x}_{t+1} = argmin\left\{\sum_{k=0}^{t}(a_k\nabla f(\boldsymbol{x}_k), \boldsymbol{x}) + d(\boldsymbol{x})\right\} \tag{5}$$

The definition of variables is basically the same as that of primal [3]. In contrast, a Proximal strong convex function d(x) is newly defined. Like the projected subgradient method above, under the general convex and strong convex conditions, the convergence rate can reach O($\frac{\sqrt{T}}{T}$) and O($\frac{1}{T}$) optimally by taking appropriate steps respectively.

The MD algorithm was first started in 2002 by a collaboration between Amir Beck and Marc Teboulle [3]. It is also one of the most classical methods for optimizing a gradient. After several improvements, many excellent algorithms with high efficiency in space and time and high efficiency in space and time have been derived. The most typical one is the COMID algorithm based on the regularization optimization problem. This algorithm was proposed by Duchi et al. in 2010. Its main improvement content is to process the regularization and loss functions in the optimized objective function model respectively. For example, only the loss function is linearly approximated, and the regularization term is not processed. At this time, the subproblem has an analytical solution, and the algorithm can not only converge quickly but also guarantee various properties of the regularization term. However, when $r(\boldsymbol{w})$ is a regularization term L_1, a sparse solution can be obtained.

2.3 Dual Average Method

The dual-average method first appeared in Nesterov.Y's paper [4] in 2009. Compared with the first two methods, the dual-average method is more flexible in gradient weight and step length, and can be obtained in a variety of cases without synchronization. The rate of convergence is excellent, but so far it has only been iterating over the weighted average output method in the general convex case to obtain the theoretical optimum. Its iteration method is also similar to the dual-form MD algorithm iterative method. The iteration method of the DA algorithm is given below.

The same optimization of the model problem described in Eq. (1), the iterative formula becomes:

$$\boldsymbol{x}_{t+1} = argmin\left\{\sum_{k=0}^{t}(a_k\nabla f(\boldsymbol{x}_k), \boldsymbol{x}) + \gamma_t d(\boldsymbol{x})\right\} \tag{6}$$

Compared with the dual-form MD algorithm, the variable definition is the same. Based on this, a step length factor γ_t is added to regulate the step size, so that it is more flexible in selecting the step size, and the optimal convergence rate can be achieved in more steps. This is a significant advantage of DA, but it is not perfect. It can only get the optimal solution of the weighted average output in the general convex case. In 2015, Nesterov and Shikhman [7] embedded a linear interpolation technique in the iteration of the dual average method, which proved that the iterative method of dual average can achieve the optimal individual convergence rate in the general convex case, and gave this kind of Application of convergence results in real-time system stabilization. In contrast, I embed the same linear interpolation technique in the case of strong convexity in the research process, γ_t and a_k then take another appropriate and obtained the approximate optimal solution $O(\frac{\log T}{T})$ of the individual convergence rate under strong convexity, which is not theoretical. The strong convex case has the optimal solution $O(\frac{1}{T})$, so the individual convergence rate for the strong convex case of the DA algorithm needs further study.

The above three classical one-step optimization methods are relatively basic classic algorithms. After they have been continuously improved, many efficient algorithms emerged. However, among the mass data walks and machine learning big data craze, the traditional batch The processing method is also the same as the above method that treats the regularization term and the loss function as an integrated objective function (we call it the black box method). Not enough to solve the real problem, can not achieve the bounds in the reality, sparseness and enough storage space, so we will briefly introduce the structure optimization method and random optimization method.

3 Stochastic Optimization

3.1 Stochastic Optimization and Batch Processing

The regularized loss function optimization problem has the form of the sum of the subfunctions, and the subfunctions have a definite meaning that each subfunction represents the regularization loss caused by a certain sample. When we randomly select a subfunction instead of sequentially optimizing At this time, the incremental gradient descent method is called Stochastic Gradient Descent (SGD) [8]. Since each iteration calculation is only a loss caused by optimizing one sample point, i.e., only the gradient information of the loss function of a single sample is needed without traversing all the samples, the memory space is largely saved compared to the batch processing algorithm, but At the same time, since the optimization is the loss of a single sample, the convergence rate will be affected. For an ideal case of an independent and identically distributed sample set, redundancy and generalization make the required sample points only a small part of the large sample set, so the results of the optimized learning accuracy will eventually converge to a stable value, so the stochastic optimization method is particularly suitable for solving large-scale machine learning problems. Although the convergence speed is not affected to a great extent, it does not mean that SGD is applicable under all conditions to obtain the theoretical optimal convergence rate. When the loss function is smooth and strongly convex, in the mathematical proof, SGD scales the unbiased estimate of the gradient information of the single sample loss function, and the more expected terms will affect the convergence of the bound, and thus cannot be optimal.

3.2 Random Optimization and Online Convex Optimization

In 2003, Zinkevich proposed an online learning algorithm. The samples processed by this method are generated with time series, that is, one sample is processed by one sample, and each step iterative process only deals with a single sample causing loss suboptimization problems. Compared with the processing method, the memory overhead and the computational cost are effectively saved. At the same time, this method gives a theoretical analysis tool that can quantitatively describe the solutions of the solution and batch processing methods in the iterative process of the algorithm—regist bound, and proves that the $O(\sqrt{T})$ is obtained for general convex problems. A unified theoretical framework [6]. Hazan in 2006 gave the algorithm bound under the strong convex condition, and obtained a log bound form that is closer to the exact solution than the root number form. In 2007, Shalev-Shwartz et al. used the stochastic projection subgradient method to solve large-scale SVMs based on two studies of general convexity and strong convexity, and used "(online-to-batch)". The conversion technique is to further develop the online algorithm into a stochastic optimization algorithm, the Pegasos algorithm. The success of Pegasos lies not only in the author's proposed random algorithm based on online algorithms, but more importantly in the corresponding verification experiments.

Although formally, the difference between online and random optimizations seems to be very small, it seems that the differences between sampling methods are only drawn, but in fact, the goal of online and stochastic optimization algorithm optimization is different. The goal pursued by the online optimization algorithm is generally to make the trend tend to zero, and the random optimization is to minimize the empirical risk of the objective function. It is precisely because of this that the regret bound may not be optimal, and the random optimization is different. The difference between the sampling methods of the two determines that the meaning of the entire sample selection and iterative calculation is not the same. Random optimization is to extract a single sample as an unbiased estimate, while the loss function of the online algorithm is given as a sequence over time. The calculation of the registry bound is more similar to the pipeline operation. The two methods are different. However, for the mathematic expressions of the bound and the registry, the two are only one T on the score line, so there is an online-to-batch conversion technique. Also talked about by many scholars.

4 Conclusion

In this paper, the history of the first-order classical gradient optimization method is combed in detail, the development and application of rapid machine learning are introduced, and the general framework of the optimized regularization plus loss function is described in detail. And introduced the three most classic and basic one-level optimization methods and focused on their differences between the links, the development process of their own thoughts were summed up, and analyze their bound convergence and its direction of development and improvement. In the last two chapters, from the perspective of reality in the era of big data, two kinds of more efficient and adaptive optimization methods, namely stochastic optimization and black-box optimization, are compared and summarized, and the development directions of different methods are also proposed.

However, there are still many deficiencies in the actual work: for example, due to the presence of noise interference in the actual problem or if the prior samples are not completely independent and identically distributed, the objective function gradient estimation bias tends to occur. Most of the optimization methods commonly used gradients Unbiased assumptions are often untenable, and deviations in the convergence of accelerating methods will accumulate with iterations in the biased case, and thus cannot be applied. Therefore, attention is paid to the field of machine learning. This is also the direction the author took to focus on the need to understand the learning process. And not all methods are suitable and the linear interpolation technique can get the optimal convergence rate under the corresponding situation. The proof of the individual convergence rate of some methods is still worth our continued study.

References

1. Bertsekas, D.P., Nedić, A., Ozdaglar, A.E.: Convex Analysis and Optimization. Athena Scientific, Belmont (2003)
2. Robbins, H., Monro, S.: A stochastic approximation method. Ann. Math. Stat. **22**(3), 400–407 (1951)
3. Beck, A., Teboulle, M.: Mirror descent and nonlinear projected sub-gradient methods for convex optimization. Oper. Res. Lett. **31**, 167–175 (2003)
4. Nesterov, Y.: Primal-dual subgradient methods for convex problems. Math. Program. **120**, 261–283 (2009)
5. Nesterov, Y.: A method of solving a convex programming problem with convergence rate O (1/k2). Soviet Math. Dokl. **27**(2), 372–376 (1983)
6. Duchi, J., Singer, Y.: Efficient online and batch learning using forward backward splitting. J. Mach. Learn. Res. **10**, 2899–2934 (2009)
7. Nesterov, Y., Shikhman, V.: Quasi-monotone subgradient methods for nonsmooth convex minimization. J. Optim. Theory Appl. **165**(3), 917–940 (2015)
8. Zhang, T.: Solving large scale linear prediction problems using stochastic gradient descent algorithms. In: Proceedings of the International Conference on Machine Leaning, New York, pp. 919–926. ACM (2004)

Research on Products Safety Evaluation Methodology of Spacecraft Assembly Process Based on Fuzzy Evaluation and AHP

Zhibin Liu[1,2(✉)], Gang Sun[1,2], Yidan Zhang[1,2], Yi Lu[1,2], Xiaohui Song[1,2], Zhe Liu[1,2], and Zheng Liu[1]

[1] Beijing Institute of Spacecraft Environment Engineering, Beijing 100094, China
raindown@126.com

[2] Beijing Engineering Research Center of the Intelligent Assembly Technology and Equipment for Aerospace Product, Beijing 100094, China

Abstract. The problems of quality and safety in AIT (assembly Integration and test), especially the damage of the crucial equipment or key parts, will have a great impact on quality, schedule, and cost of the spacecraft development process. Therefore, an engineering method of risk identification and prevention to evaluate the safety of key process in AIT is needed in order to improve the safety risk control ability for spacecraft. In this paper, we use the fuzzy comprehensive evaluation and analytic hierarchy process (AHP) method to construct the spacecraft assembly safety evaluation model. Based on the expert scoring, combining the multilevel fuzzy comprehensive evaluation method and directly scoring method, the weights of each evaluation index are analyzed and determined. After calculation, the risk value of the key activities for the spacecraft assembly process is obtained. The verification work on a certain type of spacecraft shows good effect that the assessment results accord with experience well and have high degree of distinction.

Keywords: Spacecraft assembly process · AHP (Analytical Hierarchy Process) Fussy evaluation · Safety evaluation

1 Introduction

Spacecraft assembly is a system engineering process which is multiple processed, interchange, fly-over and continuous. The quality safety problems occurred in the assembly stage, especially the damage of the crucial equipment or key parts, will have an unpredictable impact on the spacecraft development process [1]. Through the analysis of quality and safety problems occurred in recent years in the process of spacecraft assembly, lack of the assembly security, scratch knocking, crush, sparking brings on product safety problems is account for 41.94%. An engineering method of risk identification and prevention to evaluate the safety of key process in AIT is needed in order to improve the safety risk control ability for spacecraft has become one of the key problems to be solved. By constructing the spacecraft assembly safety evaluation model to evaluate the risk level of spacecraft assembly and provide reference for risk

K. Deng et al. (Eds.): ICMIR 2018, AISC 856, pp. 1191–1198, 2019.
https://doi.org/10.1007/978-3-030-00214-5_146

management and decision making. At the same time, with the risk assessment results we can identify the weak links in the assembly process and provide quantitative basis for the improvement of assembly process design. The weak links include risk level of operating activities and flows are not acceptable and the reason which occurs the risk overrun [2, 3].

2 Research on the Methods of Safety Risk Assessment

There are many techniques and methods of safety risk assessment safety risk. According to the spacecraft assembly, the evaluation method shall satisfy the following requirements:

(1) The data of experts can be preset in the model and the evaluation result is standard;
(2) The evaluation results are more objectively to decrease the errors of different evaluation due to experience difference;
(3) Evaluation results can not only be qualitative, but also can quantified the risk value;

After the comparative and analysis, analytic hierarchy process (quantitative method) and fuzzy evaluation (non-quantitative method) are selected for the safety evaluation method work.

3 Build a Safety Model of Spacecraft Assembly

3.1 Analysis of Risk Factors for Spacecraft Assembly Process

Satellite product in assembly process is likely to suffer damage. The reason can be summarized into two aspects: one is the vulnerability characteristic of satellite products; the two is the improper implementation of assembly activities. The former is the prerequisite and reason of assembly accident, the latter is the trigger condition of assembly accident.

The two factors play a role in the results which make the accident happen.

Figure 1 is the schematic diagram for analysis framework. It identifies the potential risk factors in the spacecraft assembly process which includes staff, products, environment, tooling and the type of operation and so on including vulnerability characteristic lists of satellite products (14 categories) and risk factors lists of satellite products assembly in test activities (a total of 3 categories and 34 small categories).

3.2 Establishment of Safety Evaluation Parameter System

According to the traditional theory of quality management, 5 major factors including people, machine, material, method and ring affect the field quality of production.

In order to get a comprehensive evaluation for the safe level of spacecraft assembly process, the 5 factors possibly effecting the quality of spacecraft assembly process needs to be considered comprehensively. By evaluating operational request which can be implemented and the actual situation of the project combined with risk, we establish

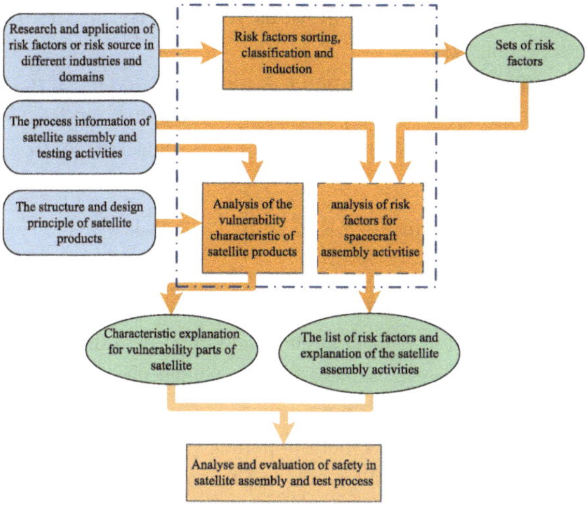

Fig. 1. Analysis framework of accident causes in the spacecraft assembly.

the safety evaluation parameter system including the top-level parameters and second parameters (as shown in Fig. 2) to provide the necessary basis and precondition for semi quantitative risk assessment.

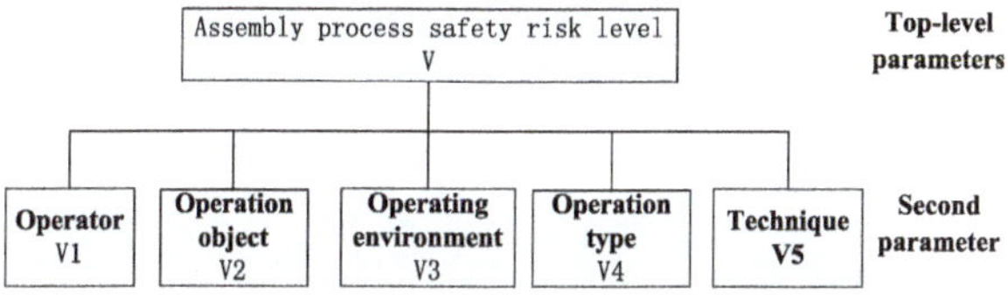

Fig. 2. Safety evaluation parameter system of spacecraft assembly product.

Second parameter includes 5 dimensions: operator, operation object, operating environment, operation type and technique.

4 Application of Safety Evaluation Method

4.1 Analytical Hierarchy Process

The basic idea of analytic hierarchy process is determining the relative importance of various parameters in different hierarchy through multiple comparisons by according to the evaluation parameters system which has been established as a hierarchical structure [7, 8].

Then we use the method through calculating eigenvector by matrix judgment to get the relative weight of parameters for each level compared with the anterior level. It can achieve comprehensive security risk assessment by using fuzzy comprehensive evaluation method. Evaluation steps show as follows: By two-two comparison, constructing a judgment matrix.

We suppose the next layer of parameter A includes u_1, u_2, …, u_n, and identify a_{ij} as a scale of importance relative to the parameters between u_i and u_j. We use the "1 ~ 9 scale" values to evaluate a_{ij}. The meaning of 1 ~ 9 scale shows in Table 1.

Table 1. Meaning of 1 ~ 9 scale.

Scale	Meaning
1	Two parameters have the same importance
3	The former was slightly important than the latter
5	The former is more important than the latter obviously
7	The former is more important than the latter strongly
9	The former is more important than the latter extremely
2, 4, 68	The intermediate value of the adjacent judge
Reciprocal	$u_i/u_j = a_{ij}$, $u_j/u_i = a_{ji} = 1/a_{ij}$

For the target A, comparing with the relative importance of n elements can get a two-two comparison judgment matrix $A = (a_{ij})_{n \times n}$.

$$A = \begin{vmatrix} 1 & a_{12} & \cdots & a_{1n} \\ a_{21} & 1 & \cdots & a_{2n} \\ \vdots & \vdots & \vdots & \vdots \\ \frac{1}{a_{1n}} & \frac{1}{a_{2n}} & \cdots & 1 \end{vmatrix} \quad (1)$$

4.2 Fussy Evaluation

Due to the security state of spacecraft assembly and expression of safety evaluation results reflect a certain ambiguity, such as boundary of the security level is not clear. Therefore, the fuzzy evaluation method is used to compose safety evaluation indexes. Evaluation of the basic steps show as follows [9]:

(1) Determining the factor levels: Set the level factor which is evaluated for $U = \{u_1, u_2, \ldots u_i, \ldots, u_m\}$, $i = 1, 2, \ldots, m$. u_i is the i factor in the first level. It is determined by the n factor of the second level. $U_i = \{u_{i1}, u_{i2}, \ldots u_{ij}, \ldots, u_{in}\}$, $j = 1, 2, \ldots, n$.
(2) To build weight: According to the importance degree of every factor in each level, we give corresponding weight to each factor, the weights show as follows: the first level, $A = \{a_1, a_2, \ldots a_i, \ldots, a_m\}$, $i = 1, 2, \ldots, m$; the second level, $A = \{a_{i1}, a_{i2}, \ldots a_{ij}, \ldots, a_{in}\}$, $j = 1, 2, \ldots, n$.
(3) Determine the alternative set V: The alternative set is composed of variety evaluations and results which evaluated by valuator for evaluation objects. No matter the quantity of the evaluation levels, the quantity of alternative set is only 1. Alternative set can be expressed as, $V = \{v_1, v_2, \ldots, v_p\}$, $p = 1, 2, \ldots, n$. From the technology perspective, the number of comment grade p is generally greater than 4

and less than 9. On the one hand, if the value of p is too big to overrun one's semantic distinguish ability, it is not easy to judge the object grade belonging. On the other hand, if the value of p is too small, it can't meet the quality of fuzzy comprehensive evaluation. Therefore, the moderate value of p is appropriate. This research adopts 5 levels of expression such as high risk, higher risk, medium risk, lower risk and low risk.

(4) Fuzzy comprehensive evaluation: Because of every factor of each level is determined by several factors of the lower level, the single factor evaluation of s level is determined by several factors of t level which is lower than s lever. We set multi factor evaluation matrix to Ri. Combined with the weight of factors in t level, we can obtain fuzzy comprehensive evaluation set of a factor in s level Bi. Ai is a vector representation of weight in t level.

$$R_i = \begin{bmatrix} r_{i11} & a_{i12} & \cdots & r_{i1p} \\ r_{i21} & r_{i22} & \cdots & r_{i2p} \\ \vdots & \vdots & \vdots & \vdots \\ r_{in1} & r_{in2} & \cdots & r_{inp} \end{bmatrix}$$

$$B_i = A_i \times R_i = [a_{i1}, a_{i2}, \ldots a_{in}] \times \begin{bmatrix} r_{i11} & a_{i12} & \cdots & r_{i1p} \\ r_{i21} & r_{i22} & \cdots & r_{i2p} \\ \vdots & \vdots & \vdots & \vdots \\ r_{in1} & r_{in2} & \cdots & r_{inp} \end{bmatrix} \tag{2}$$

Because of the 5 parameters of the second level in spacecraft assembly risk evaluation parameter system are different essentially. We should use different methods to obtain the membership degree. The membership degree of V1 and V3's determination method is similar as the conventional method. The parameters of V2, V4 and V5 in lower level change with the object and the parameters which are dynamic need to deal with improved algorithm. As shown in Table 2. V1: Operator; V2: Operation object; V3: Operating environment; V4: Operating environment; V5: Technique. The second parameter matrix of V1 ~ V5 judgment for spacecraft assembly process shows in Table 2.

Table 2. The parameter V1 ~ V5 judgment matrix.

	V1	V2	V3	V4	V5
V1	1	1	1	1	1
V2	–	1	1	1	1
V3	–	–	1	1	1
V4	–	–	–	1	1
V5	–	–	–	–	1

4.3 The Results of Comprehensive Evaluation Based on Membership Degree

Because of many parameters based on spacecraft assembly safety evaluation, this paper only shows the parameter calculation of the operation object (V1) to represent each parameter [10].

(1) We use AHP and expert scoring method to determine the weight coefficient. With the weight coefficient we get a matrix W_{CI} with 1 row and 4 columns consisting of the third level parameters (V11, V12, V13 and V14).
(2) To determine the criterion from alternative set in 5 grades of parameters (V11, V12, V13 and V14), as shown in Table 3. V11: number of people; V12: continuous operation time; V13: operation posture; V14: cooperative mode.

Table 3. The parameter V11 ~ V14 criterion.

Parameter	Criterion (5 grades)					
	Input condition	Judgment result				
		High risk	Higher risk	Medium risk	Lower risk	Low risk
V11	1					√
	2 ~ 3				√	
	4 ~ 5		√			
	More than 5	√				
V12	>2	√				
	1 ~ ≤ 2		√			
	0.5 ~ < 1			√		
	<0.5				√	
V13	Low comfort	√				
	Lower comfort		√			
	Medium comfort			√		
	Higher comfort				√	
	High comfort					√
V14	Multi department		√			
	One department				√	

(3) We get information of the second parameter V1 by questionnaire. Compared with criterion in Table 3 we respectively calculate the membership degree of parameters (V11, V12, V13 and V14) in each risk level and get a membership degree matrix R_{CI} with 4 rows and 5 columns.
(4) We calculate $R_{BI}= W_{CI} \cdot R_{CI}$ as the membership degree in each risk level of the second parameter V1. And in the same way, we calculate the membership degree in each risk level of parameters (V1, V2, V3, V4 and V5).

Using above calculation methods, we get the membership degree in each risk level of the second parameters (V1, V2, V3, V4 and V5) and a membership degree matrix $R_B = [R_{B1}, R_{B2}, R_{B3}, R_{B4}, R_{B5}]^T$ with 5 rows and 5 columns.

We use AHP and expert scoring method to determine the weight coefficient of the second parameters (V1, V2, V3, V4 and V5). We use normalizations to deal with the weight coefficient to get a matrix W_B with 1 row and 5 columns.

The weight coefficient matrix multiplies the membership degree matrix and we can get the membership degree of top-level parameters in risk level show as $R_A = W_B \times R_B$.

According to the principle of maximum membership degree, membership degrees of top-level parameters contain 5 numerical values. The maximum expresses the risk level for the result of comprehensive risk assessment.

With assembly safety evaluation model, we can analyze each single factor of the spacecraft assembly process and get the operation object which has high risk. In order to reduce and control the risk, we take control measures for the high-risk object to improve safety of assembly process.

5 Verification of the Safety Assessment Method

An assembly process of a spacecraft contains a total of 138 projects. With the assembly safety evaluation model, we get the risk of all projects. Statistical results show in Table 4.

Table 4. Safety evaluation results in an assembly process of a spacecraft.

No.	Risk value	Amount	Proportion (%)
1	High risk	20	14.49
2	Higher risk	24	17.39
3	Medium risk	28	20.29
4	Lower risk	11	7.97
5	Low risk	55	39.86

The result is accurate and effective. Risk is identified comparatively mainly reflected in the following two aspects:

(1) The tendency of risk values in process accord with experience well:
   - (a) Joint operation, large hoisting and large test which hold high severe degrees need complex operation and their risk value is high.
   - (b) Cable laying, cable lashing, antenna and solar wing reflecting and initiating explosive device need less complex operation and their risk value is medium.
   - (c) Installing parts directly under the spacecraft, measuring the torque, fasteners glue-sealing, installing the deck, installing equipment and precision measurement need simple operation and their risk value is low.

(2) Different projects with the same risk value have a high degree of distinction. We choose high risk value process to make an example. The risk value of the key process is greater than the non-key process.

6 Conclusions

This paper builds the spacecraft assembly safety evaluation model based on the analysis of risk factors. With further research related to safety evaluation method, structure and calculation of evaluation methods are completed to ensure that the results of evaluation are accurate and objective. Finally, with the verification work in a spacecraft, the evaluation results and experience are correlated well. And they have a high degree of distinction and achieve good results. Based on the evaluation results, technicians can carry out the assembly process simulation for ergonomics by selecting higher risk work. They can also analyse the core cause and classify the risk. If the risk can't be directly removed, technicians can take the management measures of key processes for controlling the implementation process. Technicians establish the typical process protection scheme for eliminating the risk. By taking the corresponding control measures of risk, the damage of key products will be reduced or avoided. The occurrence probability of safety and quality problems in spacecraft assembly process will be greatly reduced. The quality and progress of spacecraft assembly process will be guaranteed well.

References

1. Xu, F., Lin, B., Hou, S.: Satellite Engineering Conspectus, pp. 135–142. Springer, Beijing (2003)
2. Swain, A.D., Guttmann, H.E.: Handbook of human reliability analysis with emphasis on nuclear power applications. NUREG/CR-1278 (2005)
3. Huang, J., Shen, X.: The process risk analysis and the control strategy researches of satellite integration. Qual. Reliab. **3**, 112–116 (2009)
4. Nuclear Energy Agency: Research strategies for human performance. NEA/CSNI/R (97)24 (2008)
5. Oliver Strater: Evaluation of human reliability on the basis of operational experience. GRS 170 (2010)
6. Lumsden, J.M.: Operations safety risk management: managing integration test safety risk. In: IEEE Jet Propulsion Laboratory. California Institute of Technology, Under a Contract with the National Aeronautics and Space Administration, pp. 12–16 (2001)
7. Ericson, C.: Hazard Analysis Technique for System Safety. Wiley, New York (2005)
8. Michael, W.J.: Consequence and likelihood in risk estimation: a matter of balance in UK health and safety risk assessment practice. Saf. Sci. **43**(8), 345–353 (2005)
9. Relax Software Co & Intellect: Reliability: A Practitioner's Guide. Beihang University Press, Beijing (2005)
10. Leveson, N.G.: Applying systems thinking to analyze and learn from events. Saf. Sci. **49**(5), 55–64 (2011)

Parallel Strategy Based on Parameter Selection of Machine Learning Model

Yue Qin(✉) and Zhenzhou Ji

School of Computer Science and Technology,
Harbin Institute of Technology Weihai, Weihai, China
QYZW1027@163.com

Abstract. With the improvement of hardware level and the development of artificial intelligence rapidly, machine learning algorithms are commonly used in the field of intelligent products and intelligent analysis, and have greatly improved our quality of life. Machine learning model parameters include ordinary parameters and hyper-parameters. Hyper-parameters are external configuration of the model. The values of the parameters must be manually set. The optimization of hyper-parameters is often based on a large number of parameter iterations and empirical methods. In addition, the selection range of hyper-parameters is relatively large, and it takes a lot of time to select hyper-parameter, resulting in waste of resources and expenses. This paper designs a general parallelization strategy based on hyper-parameters selection of machine learning model and implements the process of parallel hyper-parameters selection by multi-process, which can greatly reduce the running time of the program, save time and improve the development efficiency.

Keywords: Hyper-parameters selection · Parallelization strategy
Multi-process · Multiprocessing module

1 Introduction

In the field of machine learning, deep learning plays an important role. At present, deep learning has its complete framework and many open source libraries can be used, such as Google Open Source Library TensorFlow, Open Source Library Caffe, Berkeley, California, Montreal Institute of Technology Open Source Theano (including Blocks and Keras) and more. For developers who want to incorporate deep learning capabilities in their applications, there are actually a lot of good open source projects on GitHub that deserved attention, and what people need to do is design deep neural networks like stacked wood. Due to the flexibility of the structural design in neural network framework, the range of hyper-parameters selection is relatively large. At the current hardware level, optimizing a very large neural network may run for several days. Therefore, for deep learning programmers, how to perform efficient model hyper-parameters selection becomes the most important issue.

As a popular language, Python has modules that can implement multi-process parallelism easily. This paper proposes a parallel strategy based on the hyper-parameters selection of machine learning model, and experiments are performed on the

K. Deng et al. (Eds.): ICMIR 2018, AISC 856, pp. 1199–1206, 2019.
https://doi.org/10.1007/978-3-030-00214-5_147

Iris data set to calculate the parallel speed up ratio and accuracy. Experimental results show that the parallel strategy can ensure that the selected model reach more than 96% accuracy, and accelerate the selection of hyper-parameters, improve the development efficiency greatly.

2 Theory Overview

2.1 Hyper-parameters of Model

The main ideas of Naive Bayes Model is the Bayesian formula, which is used to calculate the posterior probability, and the class with the maximum posterior probability is selected as final class which the object belongs to. Adjustable parameters are shown in Table 1.

Table 1. Hyper-parameters of Naive Bayes Model and SVM

Model	Parameters to optimize	Range of values
BernoulliNB	alpha	Float
	binarize	Float/None
	fit_prior	True/False
	class_prior	array
MultinomialNB	alpha	Float
	fit_prior	True/False
	class_prior	array
GaussianNB	None	None
SVM	C	Float
	kernel	linear/PRS......
	degree	Integer (ploy)
	gamma	Float (PRS) or auto
	coef0	(poly/sigmoid)
	tol	a Threshold and Float

The support vector machine model is to find the hyperplane with the largest interval in the feature space, which is abbreviated as SVM. Adjustable parameters are shown in Table 1, among them, (PRS) refers it is effective when kernel function is poly, rbf, sigmoid.

K-Nearest Neighbor is abbreviated as KNN. KNN is to select K ($K \geq 1$) data points, and calculate the distance between new data and feature values of K data point, then statistics max number class of data near the new data point, as the final class of the new data point, the classification task is completed. Adjustable parameters of KNN are shown in Table 2.

Gradient Boosting decision trees, which is abbreviated as GBDT is based on the gradient tree model. Gradient Boosting algorithms use the negative gradient of the loss function to fit the value of the current model as an approximation of the residuals, then train a new decision tree. Adjustable parameters are shown in Table 2. And Other machine learning models parameters can be know by help function in python.

Table 2. Hyper-parameters of KNN and GBDT

Model	Parameters to optimize	Range of values
KNN	n_neighbors	Integer (≥ 1)
	weights	uniform/distance
	algorithm	ball_tree/kd_tree/.....
	p	1 or 2
Gradient boosting	loss	Deviance/exponential
	n_estimators	Integer
	learning_rate	Float
	subsample	Float (0–1)

2.2 Hyper-parameters Selection Strategy

There are many methods for searching hyper-parameter. Gradient-based optimization means that for a particular learning algorithm, gradients relative to hyper-parameters can be calculated, and then gradients are used to optimize hyper-parameters. The first use of these techniques focused on neural networks [1]. Since then, these methods have been extended to other models, such as support vector machines [2] or logistic regression [3]. In general, Grid Search is used to perform brute force search on various hyper-parameters combined spaces [4]. [5] shows that the random selection test is more effective than the grid test for hyper-parameters optimization in aspects of experience and theory. James Bergstra etc. use nonparametric learning in other fields into deep learning filed, which is called sequential optimization [6]. Snoek et al. developed a software widely used by related scholars, called spearmint.

2.3 Multi-process Parallel Modules

At present, computers generally enter the multi-core era. When we need to do some computation intensive, using parallel computing to achieve CPU performance can also save us a lot of time. Due to the existence of GIL, multithreading in Python is not really multi-threaded. If you want to make full use of the resources of multi-core CPUs, most of the situations in python require the use of multiple processes. Python provides a very easy-to-use multi-process package multiprocessing. The main module in multiple processes is process and pool.

Pool module is used to create a management process pool. This module can be used when there are many child processes and the number of child processes needs to be controlled. The Pool.apply method can only allow one process to enter the pool. After one process is over, another process can enter the pool. The number of processes is best equal to the number of computer cores. Above is the execution process of the multi-processing module.

3 Parallel Strategy Based on Hyper-parameters Selection

3.1 Analysis of Parallelism

The process of model hyper-parameters selection is to set various machine learning models according to given training data set, form a hyper-parameters set of different models, and iterate over the hyper-parameters one by one. Finally, select the appropriate hyper-parameters according to the accuracy level. We use the maximum time among the parallelized child processes as the total time-consuming of classification. In this sequential serial strategy, because the combination of hyper-parameters is a very large collection, although an experienced scientist, it will also waste a lot of time in hyper-parameters selection iterations.

Through the above analysis, we found that the hyper-parameters selection is a process that can be parallelized and has a parallel nature. The parallel strategy of hyper-parameters selection by multi-process is an effective method to reduce the selection time.

3.2 Algorithm Implementation

Python uses a global interpreter lock (GIL), it serializes the threads in the process and can not use multi-threading to achieve performance improvements, can not actually achieve the goal of increasing speed in parallel. However, using multiple processes is not limited, this article uses multiprocessing module in python to implement multiple processes, the steps for parallel strategy designed by this paper are as follows and the detailed process of implementation is shown in Fig. 1.

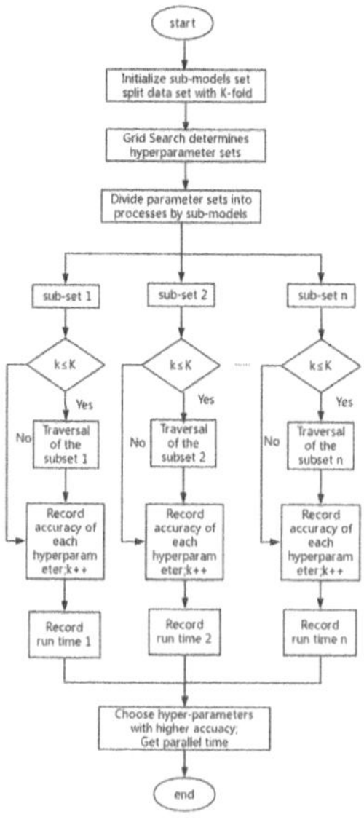

Fig. 1. Parallel strategy of hyper-parameters selection

1. Define the set of machine learning models manually. The sub-models used in this paper include: Support Vector Machine Model (SVM), Decision Tree Model, Naive Bayes Model, Gradient Boosting Tree Model (GBDT), AdaBoost integrated Learning Model, KNN. Of course, this method is also applicable to deep neural network, recurrent neural networks, convolutional neural networks, even designed networks by yourself.
2. Define the hyper-parameters set under each sub-models according to the grid search. The theoretical basis in second part details the hyper-parameters that can be adjusted in each sub-model. We can design the hyper-parameters set according to the requirements and the range of the hyper-parameters.
3. The hyper-parameters set is divided according to the type of the sub-models, the hyper-parameters under the same sub-model are divided into the same sub-set.
4. Adjust the number of processes, and then iterate over the hyper-parameters subsets and train models on each process independently. Use the cross-validation method to select the optimal hyper-parameters in the hyper-parameters subset. In this way, obtain an sub-model with optimal hyper-parameters. Record running time of each process and hyper-parameters accuracy of the sub-models.
5. Collect the sub-models obtained by each process, compare the accuracy, and select the model hyper-parameters with higher accuracy as the optimal model under the data set.

4 Performance Analysis

4.1 Experimental Preparation

The experiment is conducted on a PC with CPU i7-7700 2.80 GHz has 8 cores and 8 GB memory. Default compilation environment is anaconda integrated development environment based on Windows 10 system. The development language is python.

In the experiment, the hyper-parameters sets of the six sub-models are used to complete the classification tasks on the iris data set, including: Naive Bayes Model, SVM, KNN, AdaBoost, GradientBoosting, and Decision Tree. These sub-models form the basic hyper-parametric set space, and is marked as Base. N*Base indicates that the current size of the hyper-parameters subset is N times of the basic hyper-parametric set. The hyper-parameters subset size is 6*base, 30*base, 60*base, 100*base, 150*base in the experiment, we adjust the range of hyper-parameters to achieve the purpose of expanding the size of hyper-parameters subset.

This iris data set commonly used for data mining and contains information on 150 species of irises. Each 50 species is derived from one of three iris species, five attributes (Sepal.Length, Sepal.Width, Petal.Length, Petal.Width, and Species). The distribution of the data set in the two-dimensional plane is shown in Fig. 2. Different colors represent different species of data points.

Performance metrics are measured using run time, speed up ratio, and models accuracy. Because this article proposes a parallel strategy, it pays more attention to the value of the speed up ratio. The accuracy rate of models is defined as formula (1). In the two-classification problem, TP is a positive sample divided into positive class samples, and TN is a negative example divided into negative class samples, FP is the number of positive samples in the negative case, and FN is the number of negative samples in the negative case.

$$accuracy = \frac{TP + FN}{(TP + TN + FN + FP)} \tag{1}$$

Use the maximum time among the parallelized child processes as the total time-consuming of classification. And the definition of speed up ratio as formula (2), where T_S is running time of the serial approach and T_P is running time of the parallel strategy, S_n is speed up radio.

$$S_{\mathrm{n}} = \frac{T_S}{T_P} \tag{2}$$

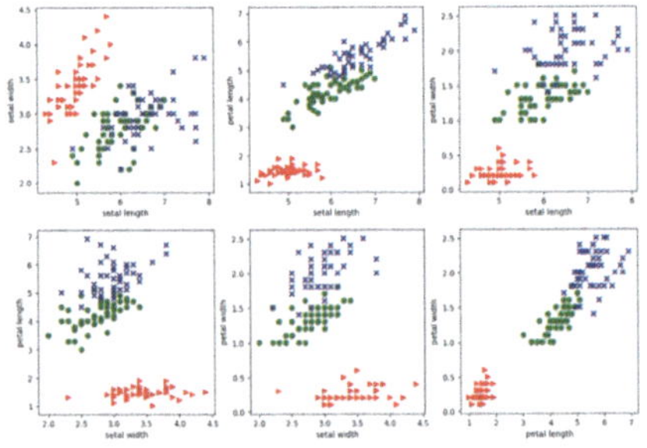

Fig. 2. The distribution of the Iris data set in the two-dimensional plane

4.2 Experimental Analysis

We conducted experiments on two variables: different processes numbers and different hyper-parameter subset sizes. The results are shown in Fig. 3. The curves of different colors in the graph represent different processes numbers, and each curve shows when the number of processes fixed, the running time gradually increases as the subset of hyper-parameters increases. By comparing the variation of the running time each curve, under the same hyper-parameters subset, it can be found that from a single process to double processes, the running time is reduced most obviously, and prove the multiple process parallel strategy does achieve the goal of reducing the hyper-parameters selection time.

The max number of fixed processes is 4, and we adjust the size of the hyper-parameters subset. The experimental results are shown in Fig. 4. The each curve in the figure shows when the number of processes fixed, the speed up ratio gradually increases as the subset of hyper-parameters increases, and eventually it tends to be gentle or slightly declining.

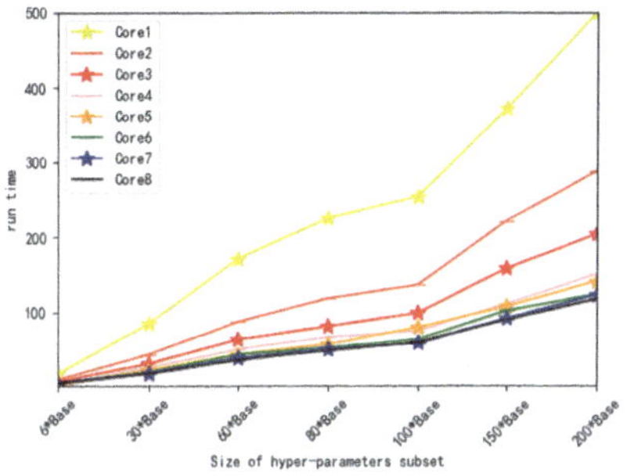

Fig. 3. Speed up radio of different processes when different hyper-parameters subsets

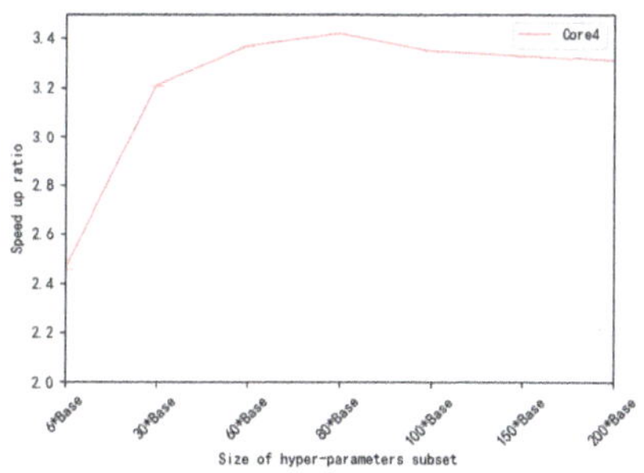

Fig. 4. The relationship between hyper-parameters subset size and speed up ratio when max process number is 4

Figure 5 shows in detail the speedup ratio of different hyper-parameters in the case of single process to eight processes. The change trend of these subsets: firstly increase gradually, and finally tend to be flat or slightly lower.

To fix the size of the hyper-parameters subset, we adjust the number of processes. The experimental results are shown in Fig. 6. Each curve in the graph shows that as the number of processes increases, the speed up ratio gradually increases, and eventually it tends to be flattened or decreased. This shows that for a fixed amount of tasks, an appropriate number of processes must be properly selected, too many processes can also cause waste of resources.

The accuracy of all models in the experiment is above 96%, and the speed up ratio under the different experimental conditions is at least 1.74. Now summarize the following conclusions.

1. When a large set of hyper-parameters, proper multiple processes do significantly reduce development time, double processes are the most efficient.

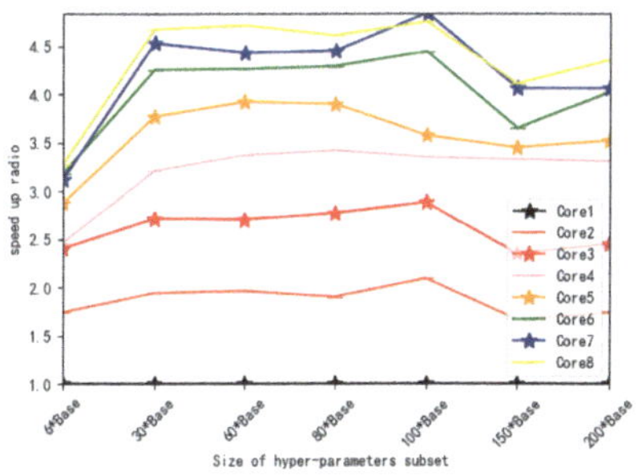

Fig. 5. The relationship between size of hyper-parameters subset and speed up ratio when different processes

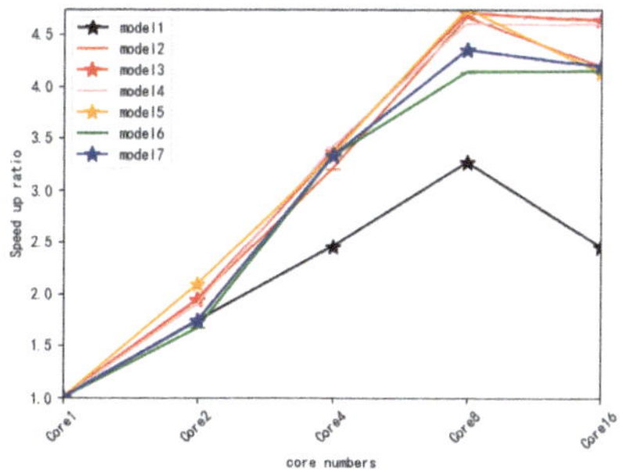

Fig. 6. The relationship between different processes and the speed up ratio when different size of hyper-parameter subsets

2. Python can introduce multiple process calculations with simple code, greatly reducing the time consumed by intensive computations.
3. When the set of hyper-parameters is small, the more processes, the more time is consumed. When select the number of processes, we must choose according to the actual amount of data. Otherwise, we will not only reduce the time, but also cause waste of resources.

5 Conclusions

The hyper-parameters optimization efficiency of machine learning model is a common problem in the field of artificial intelligence. Although there is empirical knowledge as a benchmark, the traditional sequential serialization method will still spend a lot of time iterating over parameter sets. With the gradual development of multi-cores and the development of system languages, parallelized algorithm can be easily and conveniently implemented. This paper proposes a parallel strategy for hyper-parameters selection based on machine learning model, also can be used in various structures of neural networks. Experiment shows that the proposed method can reduce the time and improve the development efficiency.

References

1. Larsen, J., Hansen, L.K., Svarer, C., Ohlsson, M.: Design and regularization of neural networks: the optimal use of a validation set. In: Neural Networks for Signal Processing, vol. 17, pp. 62–71. IEEE (1996)
2. Chapelle, O., Vapnik, V., Bousquet, O., Mukherjee, S.: Choosing multiple parameters for support vector machines. Mach. Learn. **46**(1–3), 131–159 (2002)
3. Do, C.B., Foo, C.S., Ng, A.Y.: Efficient multiple hyperparameter learning for log-linear models. In: International Conference on Neural Information Processing Systems, pp. 377–384. Curran Associates Inc (2007)
4. Hsu, C.W.: A practical guide to support vector classification. **67**(5) (2003)
5. Bergstra, J., Bengio, Y.: Random search for hyper-parameter optimization. J. Mach. Learn. Res. **13**(1), 281–305 (2012)
6. Author list, paper title, journal name, vol. no. pages-, year
7. Bergstra, J., Bengio, Y.: Algorithms for hyper-parameter optimization. In: International Conference on Neural Information Processing Systems, vol. 24, pp. 2546–2554. Curran Associates Inc (2011)

Semi-Supervised Representation Learning for Infants Biliary Atresia Screening Using Deep CNN-Based Variational Autoencoder

Zhiyong Wan[1], Wenjian Qin[1], Kuiliang Song[2], Bin Wang[3], Dahao Zhang[3], and Ling Li[1](✉)

[1] Shenzhen Institutes of Advanced Technology, Chinese Academy of Sciences, Beijing, China
912645099@qq.com
[2] Shandong Zhongbaokang Medical Implements Co. LTD, Zibo, China
[3] ShenZhen Children's Hospital China, Shenzhen, China

Abstract. The challenge of convolutional networks (CNNs) for medical imaging analysis is to train the network model with limited well labeled dataset. Since a variational autoencoder (VAE) is able to learn the probability distribution on data for describing an observation in terms of its latent attributes by unsupervised manner, it has emerged as one of the most popular unsupervised learning technology in computer vision applications. In this paper, we presented a semi-supervised representation learning approach for screening infant's biliary atresia using convolutional variational autoencoder (CVAE). Firstly, we leveraged a smartphone's camera for infant's stool images collection. Secondly, a pre-trained deep convolutional variation autoencoder was used to train the feature extractor for the infant's stool image-features extraction, and finally we fine-tuned the last classification layers for identifying acholic from normal stools. We compared our screening approach with "tradition stool color card" method; the results demonstrated CVAE model has a higher accuracy rate 92.16%. Universal screening biliary atresia by semi-supervised representation learning may be a valuable technology to help parents identify acholic stools in the perinatal period, which may ultimately lead to improved native liver survival probabilities.

Keywords: Variational autoencoder · Biliary atresia · CNN

1 Introduction

Biliary atresia is one of the most common causes of [4] cholestasis in infants <3 month of age; it is also the first common indication of pediatric liver. Transplantation[4] Patients with biliary atresia have 3 main clinical characteristic: Pale-pigmented stools, prolonged jaundice, and dark urine [3]. Pale-pigmented stools appears within the first month after birth for most patients. The Kasai operation is used as a first-line treatment for all types of biliary atresia. The treated time is of great concern, the success rate of surgery 60 days before birth is higher than that of 60 days [4]. To improve the prognosis of infants with BA, early identification is very important.

K. Deng et al. (Eds.): ICMIR 2018, AISC 856, pp. 1207–1212, 2019.
https://doi.org/10.1007/978-3-030-00214-5_148

Biliary atresia accounted for half of the cases of neonatal chronic obstructive jaundice, with an incidence of 1:8000–1:14000 births, but there was a great difference between the region and the race. The number of cases reported in Asia is 4–5. The incidence of Oriental races is about 1 times higher than that of the number of cases reported in Asia is 4–5. The incidence of Oriental races is about 1 times higher than that of men. The ratio of men to women is 1:2. Congenital biliary atresia is a kind of obstruction of intrahepatic and extra hepatic bile ducts, which can lead to congenital biliary atresia of cholesterol cirrhosis and eventually liver failure. It is one of the most important digestive diseases in the field of pediatric surgery, and is also the most common indication in children's liver transplantation [5].

The previous ways to screening biliary atresia was based on the infant stool card, which consist of 6 numbers of different stool pictures form infants. This design was inspired from Matsui et al. Picture (Nos. 1–3) on the stool card are labeled as abnormal, the others (Nos. 4–6) are normal. Currently, the infant stool card 4 was used for preliminary screening biliary atresia. If the stool color resembles the numbers 1–3, the possibility on your baby suffering from biliary atresia is higher. So the baby will be able to get the earliest treatment. And the success rate of the operation will be greatly improved. However, it is difficult to extend for universal screening and follow-up, as well as early recognition and diagnosis is not easy during the neonatal period due to parents' awareness and diverse diagnosis criteria.

Recently, variational autoencoder (VAE) has become a popular generative model. In general, pixel-by-pixel measurement like L2 loss and logistic regression loss is used to [1] measure the difference between the original and the reconstructed images [1]. So variational autoencoder (VAE) has become a popular simple and practical model. In general, pixel-by-pixel measurement like L2 loss and logistic regression loss is used to measure the difference between the original and the reconstructed images [1]. Such measurements are universal implemented for deep neural network training. Meanwhile, the generated images tend to be very blurry when compared to original images [1]. We developed a new variational autoencoder to analyze stool image (Fig. 1).

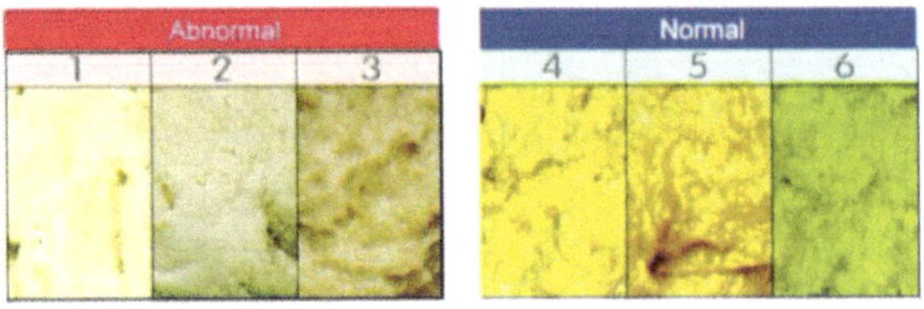

Fig. 1. Infant stool color card

2 Materials and Methods

The overview of the proposed framework is depicted in Fig. 2. In this paper, the VAE model is applied to recognize infant stool images. Supervised learning through deep learning, constantly earning the characteristics of abnormal and normal images, so as to achieve a good training accuracy.

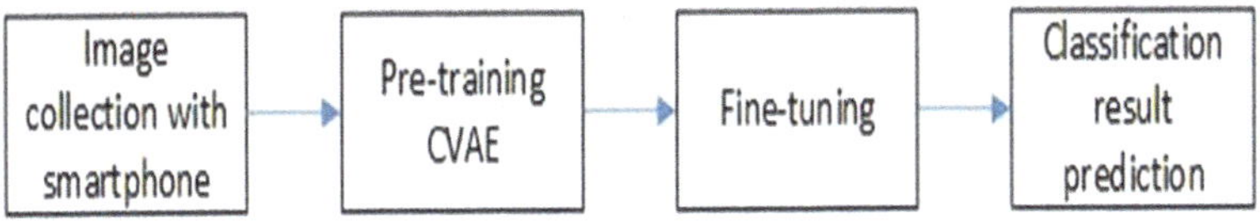

Fig. 2. Experimental flow chart

2.1 Datasets and Preprocessing

The subjects were all newborns from Shenzhen children's hospital. Thirty patients were diagnosed with biliary atresia from May 2017 to December 2017. Stools were collected in the morning, and were photographed with smartphone camera. The stool images were reviewed by two experienced doctors in biliary atresia and were marked as "normal" and "abnormal" independently. Only the stool color was agreed on by the two doctors were defined as "normal" and "abnormal".

In our experiments, we have collected 254 cases, 108 for normal, and 146 for the abnormal. In train model, we use 4 solutions for data augmentation and model performance. (1) Create a batch of new data from existing data by means of panning, flipping, and adding noise. (2) Regularization, by adding regular entries after Loss Function, we can restrain the generation of over fitting. (3) Dropout is also a kind of regularization method. But it is different from above by random output neurons to zero. (4) Unsupervised pre-training, we use auto-Encoder to do unsupervised pre training one level at a time, and finally add the classification layer to do the supervised Fine-Tuning.

2.2 Variational Autoencoder Image Model

Variational Autoencoder (VAE) can be used to model a priori data distribution. It consists of two parts: encoder and decoder. The encoder maps the high-level features of the data distribution to low-level representations of the data [2]. The low-level representations are called latent vectors. The decoder absorbs the low-level representation of the data and then outputs a high-level representation of the same data. We can do two things with VAE model. Firstly it allows us to encoder an image x to a latent vector z, with an encoder network, and then a decoder network is used to decode the latent vector z back to an image that will be as similar as the original image x^1 [1]. This means that, we need to maximize the marginal logarithmic likelihood of every observation in x, and the loss of VAE reconstruction is the expected logarithmic probability of observations in x [1]. Another important feature of VAE is the ability to control the distribution of latent vector z, which has characteristic of Gauss random variables with independent units, $z \sim N(0, I)$. In addition, the difference between the distribution of $q(z|x)$ and the Gaussian distribution can be quantified and minimized by gradient descent algorithm. Therefore, we can optimize the reconstruction loss and KL divergence loss to train VAE model through gradient descent.

$$L_{rec} = -E_{q(z|x)}[\log(X|Z)] \tag{1}$$

$$L_{kl} = D_{kl}(q(z|x)|p(z)) \tag{2}$$

$$L_{vae} = L_{rec} + L_{kl} \tag{3}$$

2.3 Image Encoder: Deep CNN

The input images to be encoder into a vector as latent representation, its shape are (64, 64, 64, 1). We apply 4 convolutional layers in the encoder network with 5 × 5 kernels, and the stride is fixed to be 2 to achieve spatial sampling instead of using deterministic spatial functions such as maxpooling [1]. After each layer of convolutional is a batch normalization layer. After convolutional and batch normalization, we use flattenLayer, denseLayer, and batchNormLayer to get latent vector z. The shape of z is (64, 128).

2.4 Image Decoder: Deep Convolutional Generative Model

The input of image decoder is the latent representation from encoder, which size is (64, 128). Firstly, we input the latent vector into decoder network, with denseLayer, reshapeLayer, and batchnormlayer the shape of the latent vector z has become (64, 8, 8, 256). We apply 4 deconvolution layers in the decoder network with 5 × 5kernels, and the stride is fixed to be 2. The first 3 Deconvolutional layer is followed by a batch normalization layer. The last Deconvolutional is not followed by a batch normalization layer. At last, the Decoder out shape is (64, 64, 64, 1), as the same as the encoder input image x.

2.5 Train Infant Stool Image

Reference was made to the relevant methods for early screening of biliary atresia disease. The infant stool card is a particularly effective method. It records the color of normal and abnormal feces in a card. The abnormal cases include white to light yellow; Normal includes yellow to green. The usual practice is to attach a card to the neonatal health handbook to the parents. If abnormal color of stool is found, contact with medical institutions should be made as soon as possible to further diagnose the disease. With the development of big data and artificial intelligence in recent years, image recognition has gained more and more attention. In this paper, the VAE model is applied to recognize infant stool images. Supervised learning through deep learning, constantly earning the characteristics of abnormal images and normal images, so as to achieve a good training accuracy.

3 Results and Discussion

The architecture of models and initialization of model parameters are provided in the tensorflow platform [2]. The Adam algorithm with learning rate 0.001. We set mini-batches of size 64. In the experiment, we divide the data into a training data and a test set. In the end of the experiment, we use this model to get a 98.56% accuracy rate on the training set, and 92.16% accuracy rate on testing set. The results are compared with previous studies, as shown in Table 1. The classification results obtained in this paper are the best.

Table 1. Comparison with previous research results

Methods	Recognition rate
Infant stool color card	60%–80%
Saturation of stool color in HSV color model	70%–90%
VAE (ours)	92.16%

4 Conclusions

We propose a method of deep learning image recognition to classify infant stool image. Compared with infant stool color card, this method has a higher accuracy rate. However, there are still many areas to be improved and inadequate, (1) the lack of training data may affect the accuracy of our experiments. (2) We need to try different neural networks to train the data. In a word, we propose a new method for screening neonatal biliary atresia based on deep learning. In the following work. Our job is to combine this approach with the application; it automatically recognizes the risk of biliary atresia.

Acknowledgments. This study was financed partially by the National Natural Science Foundation of China (NSFC: 61501444), Shenzhen Technology Development Project Fund under Grant JSGG20160429192140681.

Guangdong province science and technology plan projects (Grant No.2015B020233004), Shenzhen basic technology research project (Grant No. JCYJ20160429174611494, JCYJ20170818160306270).

References

1. Hou, X., Shen, L., Sun, K., et al.: Deep feature consistent variational autoencoder. In: Applications of Computer Vision, pp. 1133–1141. IEEE (2017)
2. Pu, Y., Gan, Z., Henao, R., et al.: Variational autoencoder for deep learning of images, labels and captions (2016)
3. Hsiao, C.H., Chang, M.H., Chen, H.L., et al.: Universal screening for biliary atresia using an infant stool color card in Taiwan. Hepatology **47**(4), 1233–1240 (2008)
4. Chen, S.M., Chang, M.H., Du, J.C., et al.: Screening for biliary atresia by infant stool color card in Taiwan. Pediatrics **117**(4), 1147 (2006)

5. Logan, S., Stanton, A.: Screening for biliary atresia. Lancet **342**(8874), 1–9 (1993)
6. Dilokthanakul, N., Mediano, P.A.M., Garnelo, M., et al.: Deep unsupervised clustering with gaussian mixture variational autoencoders (2016)
7. Santara, A., Maji, D., Tejas, D.P., et al.: Faster learning of deep stacked autoencoders on multi-core systems using synchronized layer-wise pre-training (2016)
8. Semeniuta, S., Severyn, A., Barth, E.: A hybrid convolutional variational autoencoder for text generation (2017)

Inverse Solution of Planar Redundant Manipulator Based on Cuckoo Search Algorithm with Dynamic Step Size Regulating

Changsheng Zhang[1], Wei Li[1(✉)], Biaofa Chen[2], Bin Qian[1], Rong Hu[1], and Chuan Li[1]

[1] Faculty of Information Engineering and Automation, Kunming University of Science and Technology, Kunming 650500, China
ttztty@sina.com

[2] Guangzhou Kossi Intelligent Technology Co. Ltd., Guangzhou 510000, China

Abstract. Aiming at the complex problem of solving the planar redundant manipulator inverse solution, the method of optimizing the inverse solution based on the cuckoo search algorithm (CS) is studied. To improve the poor adaptability of the step size update method of the traditional CS, a cuckoo search algorithm with dynamic step size regulating (DRCS) is presented, in which the updating step size is adjusted dynamically according to the difference between the previous generation optimal solution and the current updated position and that between the previous generation and the current optimal solutions. The stability and accuracy of the proposed theory are verified through the test function simulation. In the application of the inverse solution of the manipulator, the smaller rotation angle and less kinetic energy are obtained. Experimental results show that the rotation angle of manipulator can be reduced by 5.1 degrees averagely in once location shift and kinetic energy can be saved about 3.73% by using DRCS strategy.

Keywords: Redundant robot manipulators · Inverse solution optimization Cuckoo search algorithm · Dynamic step length regulating

1 Introduction

The planar redundant manipulator is widely used in various complex working environment due to its high flexibility. However, because the redundant characteristics may cause excessive degrees of freedom (DOF) and the complicated process of solving, the traditional inverse kinematic analysis algorithm can not meet the requirements of calculation precision and speed [1, 2]. Cuckoo search (CS) algorithm has been applied well in optimization calculating field because of its merits such as understanding and operating easily as well as excellent random search path [3].

Rodrigues D et al. presented a binary cuckoo search algorithm for feature selection in the context of theft detection in power distribution system [4]. M. Balasubbareddy proposed a non-dominated sorting hybrid cuckoo search theory combinated with genetic algorithm to solve multi-objective optimal power flow problem [5]. Nguyen applied the cuckoo search algorithm for the reconfiguration of the distribution network

K. Deng et al. (Eds.): ICMIR 2018, AISC 856, pp. 1213–1222, 2019.
https://doi.org/10.1007/978-3-030-00214-5_149

to minimize active power loss and the maximize voltage magnitude [6]. Yu Zhenzhong gave an inverse kinematics solution based on the combination of genetic trust region algorithm and analytical solution, which can quickly find accurate and optimized computation result for redundant mobile manipulator [7]. Singh found out that all possible solutions of a seven-DOF robot are on a torus curve, and then the inverse kinematics solution was obtained according to the determined position [8]. Ayusawa got the inverse kinematic solution of a fast multi-DOF robot through the Newton–Euler iterative algorithm [9]. Luo Tianhong optimized the redundant mechanical arm inverse solution by improving the glowworm swarm algorithm and achieved higher accuracy and stability [10]. Ou Qunwen put forward a method based on RBF neural network to solve the inverse solution of redundant manipulators with high precision and fast convergence speed [11].

In this paper, a cuckoo search algorithm with dynamic step size regulating (DRCS) is proposed for solving the inverse solution problem of planar redundant manipulator. According to the difference between the previous generation optimal solution and the current bird nest position and that between the previous generation and the current optimal solutions, the classical CS method is promoted by improving adaptability of nest position updating, and its validity is verified by the test function. Then the inverse solution of the three-axis planar redundant manipulator is optimized by DRCS compared with CS, which results in the manipulator rotation at a minimum angle to reach the target position, and the testing experiment is realized in co-simulation environment of MATLAB and ADAMS.

2 Cuckoo Search Algorithm Based on Dynamic Step Size Regulating

2.1 Basic Cuckoo Search Algorithm and Its Defects

To simulate the reproductive strategy of cuckoo parasitic behavior, Yang and Deb focused on three idealized rules when presenting the CS algorithm [12–14]:

(1) Each cuckoo lays one egg at a time, and dumps its egg in a randomly chosen nest;
(2) The best nests with high quality of eggs will carry over to the next generation;
(3) The number of available hosts nests is fixed, and the egg laid by a cuckoo is discovered by the host bird with a probability $p_a \in [0\ 1]$.

Based on the above three rules, according to the Lévy flight random walk to update the nest location which can be written as:

$$x_i^{k+1} = x_i^k + \alpha \otimes \mathrm{Levy}(\beta) \tag{1}$$

where, x_i^k denotes the position of the ith bird nest in the kth generation, α is the step size factor, $\otimes$ is entry wise multiplication, and the Lévy flight equation is:

$$\mathrm{Levy}(\beta) \sim \mu = t^{-\beta}, \text{and } 1 < \beta \le 3 \tag{2}$$

Levy(β) is defined by the following formula:

$$\mathrm{Levy}(\beta) \sim \frac{\phi \times \mu}{|v|^{1/\beta}} \tag{3}$$

where, $\mu \sim N(0,1)$, $v \sim N(0,1)$, and ϕ is given by:

$$\phi = \left(\frac{\Gamma(1+\beta) \times sin(\pi \times \beta/2)}{\Gamma\left(\left(\frac{1+\beta}{2}\right) \times \beta \times 2^{(\beta-1)/2}\right)} \right)^{1/\beta} \tag{4}$$

After giving up some nests according to p_a, the new solution is regenerated as:

$$x_i^{k+1} = x_i^k + \gamma(x_j^k - x_e^k) \tag{5}$$

where, $\gamma \sim N(0,1)$, x_j^k and x_e^k are two random solutions of kth generation.

In the Eq. (1) and the specific flight path shown in Fig. 1, it clearly indicates that the optimization route of the original CS algorithm is based on the Lévy flight random search strategy with uncertain step size. The search speed is improved by a larger step length in the optimization process so as to makes it easier to realize the global optimum, meanwhile, the local search accuracy is reduced. On the contrary, the search speed is decreased by a smaller step length, which will increase the accuracy of the optimization, but the global optimization ability will be poor and the solution easily falls into the local optimal.

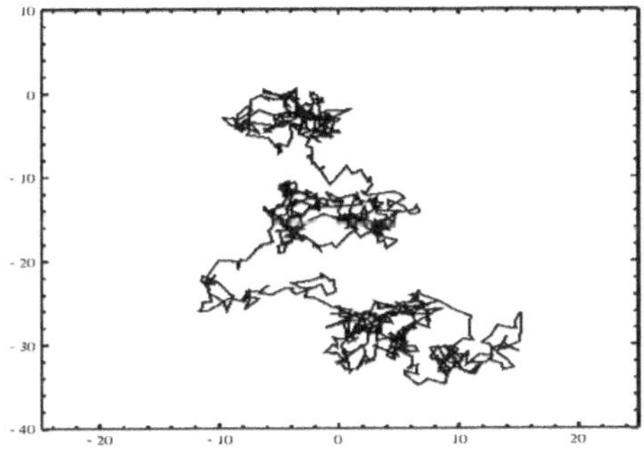

Fig. 1. Lévy flight search path

2.2 The Improvement of Cuckoo Search Algorithm

A kind of DRCS algorithm below is presented to make up for the defect mentioned above of the original CS, in which the bird nest position updating can be affected by the present and previous generation optimal solutions, thus the step length can be regulated dynamically. The new update of the nest location is given by the following formula:

$$x_i^{k+1} = x_i^k + \alpha \otimes \frac{\left|x_{best}^{k-1} - x_i^k\right|}{\left|x_{best}^{k-1} - x_{best}\right|} \otimes \mathrm{Levy}(\lambda) \tag{6}$$

where, x_{best}^{k-1} is the optimal solution of the (k−1)th generation, x_{best} is the current global optimal solution. When x_{best}^{k-1} is close to x_{best}, the step length is increased to avoid falling into local optimization. When x_{best}^{k-1} is close to x_i^k, it indicates that the result approaches to the global optimal solution, that is, the step length is reduced and the precision of the optimization result is improved at the same time.

To test the optimized search performance of DRCS method, six test functions are used for analysis, which global optimal value is 0 respectively, the description of test functions is given in Table 1.

Table 1. Description of test function

Test function	Objective function	Search range
Sphere	$f_1(x) = \sum_{i=1}^{D} x_i^2$	[−30, 30]
Rastrigin	$f_2(x) = \sum_{i=1}^{D} [x_i^2 - 10\cos(2\pi x_i) + 10]$	[−5.12, 5.12]
Ackly	$f_3(x) = -20\exp\left(-0.2\sqrt{\frac{1}{D}\sum_{i=1}^{D} x_i^2}\right) - exp\left(\frac{1}{D}\sum_{i=1}^{D}\cos(2\pi x_i)\right) + 20 + e$	[−30, 30]
Schwefel	$f_4(x) = \sum_{i=1}^{D} \left[x_i \sin\left(\sqrt{\lvert x_i \rvert}\right)\right]$	[−500, 500]
Zakharov	$f_5(x) = \sum_{i=1}^{D} x_i^2 + \left(\sum_{i=1}^{D} 0.5ix_i\right)^2 + \left(\sum_{i=1}^{D} 0.5ix_i\right)^4$	[−5, 10]
Griewank	$f_6(x) = \frac{1}{4000}\sum_{i=1}^{D} x_i^2 - \prod_{i=1}^{D}\left(\cos\left(\frac{x_i}{\sqrt{i}}\right)\right) + 1$	[−600, 600]

Both population size and iteration times of CS and DRCS are set at 200, and 0.25 for p_a. The experimental hardware configuration is Intel Core i5-4590 CPU @ 3.30 GHz and 4 GB memory, and the software environment is Windows 7 (64 bits) and MATLAB R2016b.

The 200 iteration times computation results with the same parameters setting for six test functions are shown in Fig. 2. When the fitness function converges, the iteration times of DRCS is obviously less than that of CS algorithm, meanwhile, the former's global optimization ability and convergence performance are better. As presented in Table 2, the average optimal value of DRCS is smaller than that of CS algorithm, which solution accuracy is higher and time property is sacrificed at the same time. In general, the convergence, stability and accuracy of the DRCS strategy are improved effectively.

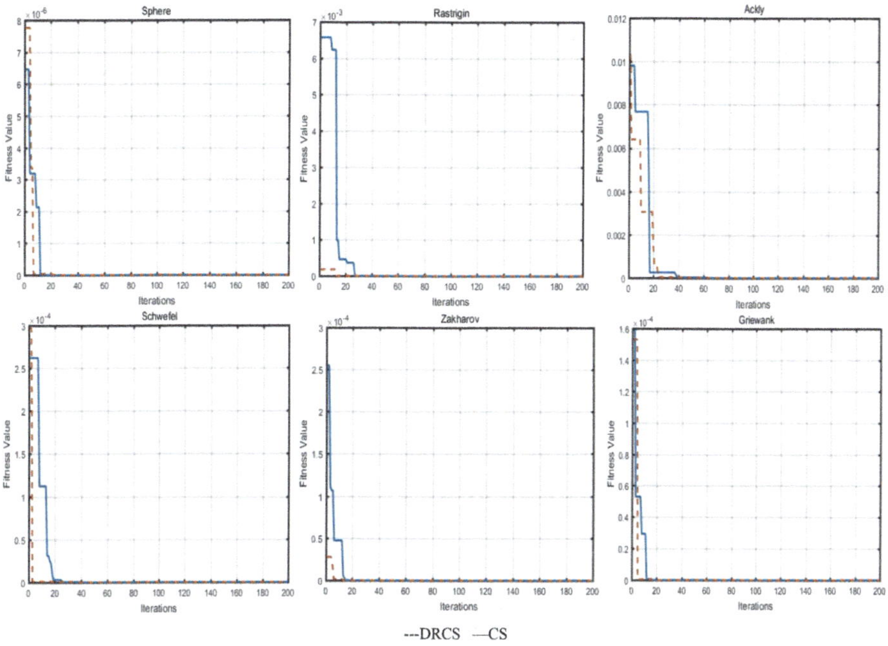

Fig. 2. Convergence curves of six test functions for CS and DRCS

Table 2. Test results

Test function	Algorithm	Average optimal value	Average time (s)
Sphere	CS	8.07E−21	0.28
	DRCS	3.92E−21	0.57
Rastrigin	CS	8.53E−14	3.36
	DRCS	1.78E−15	3.66
Ackly	CS	8.26E−11	0.30
	DRCS	4.29E−12	0.60
Schwefel	CS	2.99E−16	0.38
	DRCS	9.89E−17	0.51
Zakharov	CS	1.69E−22	0.39
	DRCS	6.62E−23	0.51
Griewank	CS	3.33E−16	0.32
	DRCS	1.11E−16	0.39

3 Inverse Solution of Planar Redundant Manipulator

The planar redundant manipulator is shown in Figs. 3 and 4, and its topology and the coordinate layout are given in Fig. 5 [15, 16].

Fig. 3. Redundant manipulator

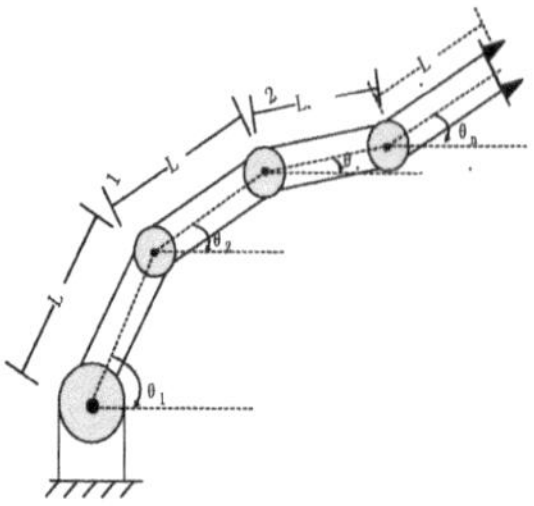

Fig. 4. Mechanism diagram

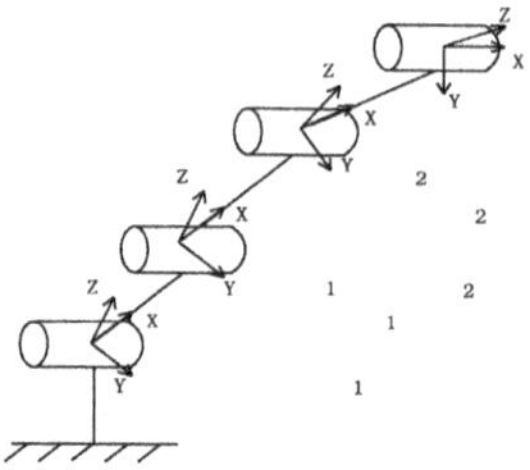

Fig. 5. Topology and coordinate layout

The manipulator target position (x_g, y_g) can be calculated by:

$$\begin{cases} x_g = \sum_{i=0}^{n} L_i \cos(\theta_i) \\ y_g = \sum_{i=0}^{n} L_i \sin(\theta_i) \end{cases} \tag{7}$$

The sum of the rotation angles of all joints is given below:

$$\Delta\theta = \sum_{i=0}^{n} (\theta_i - \theta_{ini}) \tag{8}$$

where, L_i is the length of the ith robot arm, θ_i is the turning angle of the ith robot joint, and θ_{ini} is the initial angle of the ith manipulator joint.

Thus, when solving the inverse solution of the planar redundant manipulator without special constraint, there will be numerous solutions. To reduce energy consumption, the turning angle of the robot arm should be minimized, which is the multi-objective function optimization problem. The initial position and target position of the manipulator are known, $\Delta\theta$ is the objective function, and the DRCS algorithm is used to find the optimization. Only the requirements of the precision and low energy consumption of the manipulator, rather than the flexibility and safety of the robot arm, are considered in this paper.

4 Experiments and Results

In order to make the experiment simple, without loss of generality, three-axis redundant manipulator is chosen as the object of the experiment below. L = [0.8 m, 0.8 m, 1 m], and the initial location is P_0(2.6 m, 0 m). The experimental hardware environment is same as above. The DRCS compared with CS algorithm is used to calculate the inverse solution when the end of manipulator moves to eight positions of P_1(0.5 m, 0.6 m), P_2(0.2 m, 0.4 m), P_3(1.5 m, 1.2 m), P_4(1.8 m, 0.7 m), P_5(0.8 m, 1.7 m), P_6(1.0 m, 0.3 m), P_7(1.6 m, 1.9 m) and P_8(0.9 m, 2.3 m) in turn. After 100 iteration times, both the minimum value of $\Delta\theta$ and the turning angle θ of each axis could be got, which are shown in Table 3. Aiming at verify the validity of the theories, a redundant manipulator model is established in Adams software, and rotation pairs and constraints should be created for the joint position. The manipulator is driven by the S function (in the combined simulation environment of MATLAB) based on θ. These positions of the end of robot arm are recorded in Fig. 6, and their trajectories are shown in Fig. 7.

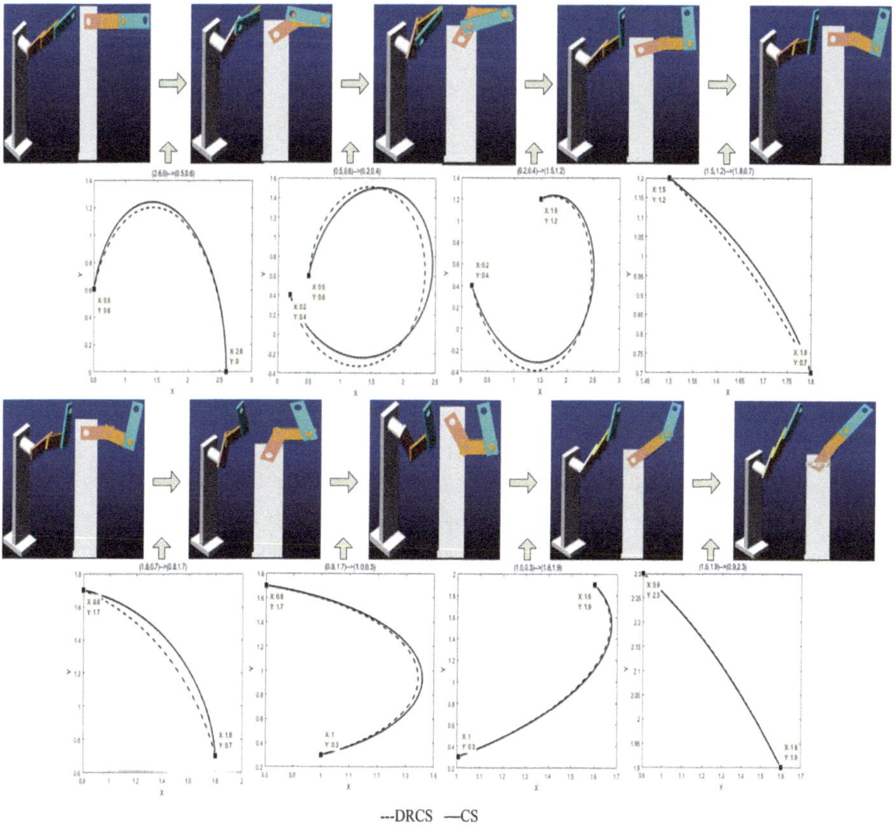

Fig. 6. Single step movement results

As illustrated in Table 3, the DRCS algorithm is obviously better than the CS for the calculation of redundant manipulator inverse solution, that is, Δθ calculated by DRCS algorithm is smaller, which can make the manipulator turning angle averagely decrease 5.1 degrees in once location shift, and the kinetic energy is saved about 3.73%. We can state that the trajectory obtained by the DRCS algorithm is significantly shorter than that by the CS, and the two trajectories shown in Figs. 6 and 7 for both methods are same smooth in ADAMS, which verifies the validity and application prospect of the DRCS strategy.

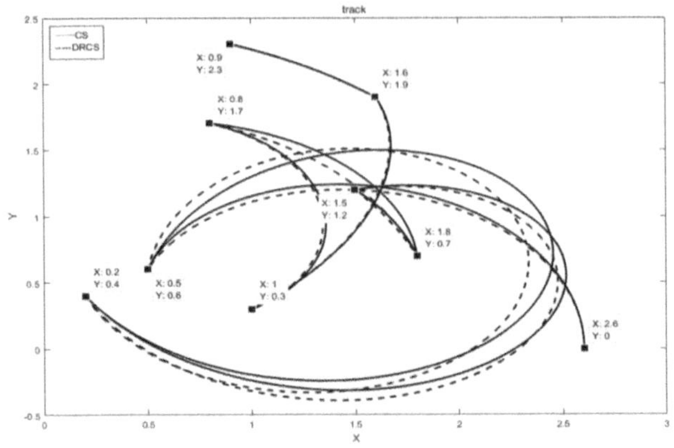

Fig. 7. Track chart of the end of manipulator

Table 3. The result of optimization

Location	Algorithm	$\theta_1/°$	$\theta_2/°$	$\theta_3/°$	$\Delta\theta/°$
P_1(0.5 m, 0.6 m)	CS	−7.07	33	164.76	204.83
	DRCS	31.38	0	169.43	200.81
P_2(0.2 m, 0.4 m)	CS	66.12	4	−157.21	26.52
	DRCS	64.28	0	−161.29	24.76
P_3(1.5 m, 1.2 m)	CS	0.67	14	94.37	333.27
	DRCS	14.68	0	94.24	305.13
P_4(1.8 m, 0.7 m)	CS	1	−20.63	75.44	53.11
	DRCS	0	−19.66	75.72	52.86
P_5(0.8 m, 1.7 m)	CS	−0.07	68.56	107	119.57
	DRCS	68.46	0	107.08	119.48
P_6(1.0 m, 0.3 m)	CS	−60.21	1	101.38	135.37
	DRCS	−57.19	0	103.5	129.23
P_7(1.6 m, 1.9 m)	CS	39.01	34	71.63	162.07
	DRCS	35.87	37	71.77	161.79
P_8(0.9 m, 2.3 m)	CS	53	55.79	91.79	55.94
	DRCS	58	51.02	91.55	55.93

5 Conclusion

In order to solve the problem of the planar redundant manipulator inverse solution, the DRCS algorithm is proposed in this paper. The distance between the previous generation optimal solution and the current position is divided by that between the previous generation optimal solution and the current optimal value to obtain the time-varying factor, which can be used to improve Lévy flight search strategy. When the previous optimal solution approaches the current optimal value, the step size is increased to enhance the global optimization ability. On the other hand, when the current generation optimal solution is close to the current position, the step size is reduced to improve the optimization accuracy. The dynamic adjustment method of updating step size increases the adaptability of CS, and the test functions are also used to verify that DRCS has better convergence and solution accuracy. The improved algorithm is applied in the inverse solution calculation for the 3-axis planar redundant manipulator, and the results of the co-simulation with MATLAB and ADAMS show that the rotation angle of manipulator can be reduced by 5.1 degrees averagely in once location shift and kinetic energy can be saved about 3.73%, which provides an effective method for inverse solution of planar redundant manipulators.

Acknowledgments. This work has been supported under R&D projects (Kunming university of science and technology: 2017FW73, 2017FW74) by Guangzhou Kossi intelligent technology Co. Ltd., China.

References

1. Artemiadis, P.K., Katsiaris, P.T., Kyriakopoulos, K.J.: A biomimetic approach to inverse kinematics for a redundant robot arm. Auton. Robots **29**(3–4), 293–308 (2010)
2. Martin, P.: Closed form inverse kinematics solution for a redundant anthropomorphic robot arm. Comput. Aided Geom. Des. **47**, 163–171 (2016)
3. Huang, J.D.: Research on Cuckoo Search and Its Applications. Huazhong University of Science and Technology, Wuhan (2014)
4. Rodrigues, D., Pereira, L.A.M., Almeida, T.N.S., et al.: BCS: a binary cuckoo search algorithm for feature selection. In: 2013 IEEE International Symposium on Circuits and Systems (ISCAS), pp. 465–468. IEEE (2013)
5. Balasubbareddy, M., Sivanagaraju, S., Suresh, C.V.: Multi-objective optimization in the presence of practical constraints using non-dominated sorting hybrid cuckoo search algorithm. Eng. Sci. Technol. Int. J. **18**(4), 603–615 (2015)
6. Nguyen, T.T., Truong, A.V.: Distribution network reconfiguration for power loss minimization and voltage profile improvement using cuckoo search algorithm. Int. J. Electr. Power Energy Syst. **68**, 233–242 (2015)
7. Yu, Z.Z., Yan, J.H., Zhao, J., et al.: A rapid inverse kinematics solution for super-redundant mobile manipulator. J. Beijing Univ. Posts Telecommun. **32**(5), 51–55 (2009)
8. Singh, G.K., Claassens, J.: An analytical solution for the inverse kinematics of a redundant 7DoF manipulator with link offsets. In: IEEE/RSJ International Conference on Intelligent Robots and Systems, pp. 2976–2982 (2010)
9. Ayusawa, K., Nakamura, Y.: Fast inverse kinematics algorithm for the large DOF system with decomposed gradient computation based on recursive formulation of equilibrium. In: IEEE/RSJ International Conference on Intelligent Robots and Systems, pp. 3447–3452 (2012)
10. Luo, T.H., Chen, C., Li, F.Y.: Inverse solution of redundant robot arm based on glowworm swarm optimization algorithm of time varying. Comput. Integr. Manuf. Syst. **22**(2), 576–582 (2016)
11. Ou, Q.W., Yun, C., Yang, X.B., et al.: Algorithm based on neural network for inverse kinematics of redundant manipulator. J. Mech. Electr. Eng. **33**(6), 663–667 (2016)

12. Yang, X.S., Deb, S.: Cuckoo search via levy flights. In: Proceeding of World Congress on Nature and Biologically Inspired Computing, World Congress, pp. 210–214. IEEE Publication (2009)
13. Yang, X.S., Deb, S.: Engineering optimization by cuckoo search. Int. J. Math. Model. Numer. Optim. **1**(4), 330–343 (2010)
14. Yang, W.H., Liu, J.R., Zhang, Y.: A new local-enhanced cuckoo search algorithm. Int. J. Comput. Sci. Math. **8**(2), 175–182 (2017)
15. Dong, H., Du, Z.J.: Inverse kinematics of planar redundant manipulators based on workspace density function. J. Mech. Eng. **51**(17), 8–14 (2015)
16. Craig, J.J.: Introduction to Robotics: Mechanics and Control, 4th edn. Pearson Education, New York (2017)

Prediction of Skin Cancer Based on Convolutional Neural Network

Jianfeng He, Qingqing Dong, and Sanli Yi(✉)

School of Information Engineering and Automation, Kunming University of Science and Technology, Kunming 650500, China
120112624@qq.com

Abstract. Skin cancer is one of the most lethal cancers in the world, which is a common type of cancer. And the early diagnosis of it is quite important. Aiming to improve the classification of skin cancer, a modified convolutional neural network model is proposed. Firstly, a convolutional layer and a pooling layer is added on the original network, constructed an 8-layer convolutional neural network. And then the network output value and the data label by the backward propagation adjust the network parameters. Lastly, by constantly debugging the network, the best model for the recognition network is selected. The proposed network applied to the dataset consisting of 900 cases of melanoma with the malignancy status pathologically confirmed. The results of the revised network prediction accuracy got 91.92% on training set and 89.5% on test set. The results obtained demonstrate that the proposed method is performant and can be used to predict whether the skin cancer is benign or malignant. With the development of deep learning and the clinical application of medical images, this method is expected to become a diagnostic tool for doctors to diagnose skin cancer.

Keywords: Deep learning · Melanoma · Image classification
Convolutional neural network

1 Introduction

Melanoma occurs on the surface of the skin, and is the most malignant tumor in the skin. At present, clinical diagnosis depends mainly on the visual diagnosis, its disadvantage is subject to the individual subjective influence of the doctor [1]. Therefore, here is a lack of high-precision and non-invasive method to diagnose benign and malignant skin cancer.

Image classification is an image processing process which distinguishes different categories of features [2]. At present, the common image classification method is mainly deep neural network, such as back propagation (BP). The BP model, which is modified weights of multilayer neural networks, which can process input in parallel, effectively solving the nonlinear problem. Compared with other methods, convolutional neural networks (CNN) is proposed by Lecun [3], because of introducing the weight sharing mechanism, which greatly reduces the number of parameters. Esteva [4] and others used a large amount of clinical data to study the transfer learning of GoogleNet Inception v3 network, the result of CNN training is comparable to that of

K. Deng et al. (Eds.): ICMIR 2018, AISC 856, pp. 1223–1229, 2019.
https://doi.org/10.1007/978-3-030-00214-5_150

dermatologist's diagnosis. The above algorithm obtains better classification effect in image classification task. CNN greatly reduces the number of parameters for the introduction of weight sharing mechanism, but the algorithm has a slow convergence speed and the problem of overfitting [5].

In this paper, in order to improve the image classification effect, we applied a CNN model, which a convolution layer and a pooling layer are added between the last pooling layer and the output layer of the convolution neural network to the classification of skin cancer malignancy, where the malignancy status is pathologically confirmed. Relu is selected as the activation function and SGD as the optimization method.

2 Proposed Method

Convolution neural network is widely used in many computer vision tasks such as image classification, face recognition and so on, it has the characteristics of local connection, weight sharing. The algorithm has the problems of slow convergence speed and overfitting in image classification [6]. According to the characteristics of melanoma data, an improved convolution neural network algorithm is proposed to establish a benign and malignant skin cancer prediction model, which is better in predicting malignant melanoma and quicker in convergence rate.

2.1 Pre-processing

Deep learning models perform better when there is large datasets, the common way to make our datasets bigger is data augmentation [7], to perform data augmentation, the simplest way is to apply geometric transformation to existing data. Applying transformations to images of lesions makes sense since this kind of data is very likely to be found in different sizes and orientations. Hence, all the transformations would boost the models to learn better.

Normalization, is a important process in image classification. It attempts to deal with external sources like data variation and how it can affect the pixel values. Normalization computes the mean of intensities for all image, and then subtracts it from each pixel of the image.

2.2 Convolution Neural Network Algorithm

Convolutional neural network is a multi-layer neural network composed of feature extraction of convolutional layers and feature processing of sub-sampling layers [8]. The bottom layer of the network inputs data and the top layer outputs its recognition result.

For the convolution layer, each element in the convolution kernel is taken as the weight parameter, multiplied by the pixel value in the input image. The jth feature graph matrix X_j^l of the L-layer is obtained by convolution of several feature graphs of the previous layer and is expressed as:

$$X_j^l = f\left(\sum\nolimits_{i \in N_j} X_i^{l-1} * k_{ij}^l + b_j^i\right) \quad (1)$$

f(·) is a neuron activation function, Nj represents a combination of input feature maps, * represents a convolution operation, kij^l is a convolution kernel matrix, and b_j^l is a bias matrix.

For the sampling layer, the role is based on the local correlation principle to pool sampling, so as to reduce the amount of data while retaining useful information. The sampling process can be expressed as:

$$X_j^l = f(down(X_j^{l-1})) \quad (2)$$

In which, down (•) represents the down-sampling function, commonly used by the maximum sampling function and the mean sampling function.

The output layer of the model is a logistic regression layer, each node output by the convolutional neural network represents the probability that the input picture belongs to a certain class i:

$$P(Y = i|x, w, b) = soft\max(wx + b) = \frac{\exp(w_i x + b_i)}{\sum_j \exp(w_j x + b_i)} \quad (3)$$

Where w is the weight parameter of the last layer and b is the corresponding bias parameter.

2.3 Improved CNN

There is a slow computational speed and the prediction of low accuracy in the construction of benign and malignant skin cancer prediction model. To solve these problems, we add a convolution layer and a pooling layer between the last pooling layer and the output layer of the convolution neural network, by increasing the depth of the network, it can be better to solve the problem that the classification effect is limited in the output layer. We experimentally determined the best Relu activation function, which greatly reduces the computational cost and the problem of overfitting.

This paper constructs the convolution neural network algorithm structure as shown in Fig. 1, the network structure includes the data layer, the convolution layer, the pooling layer, the full connection etc.

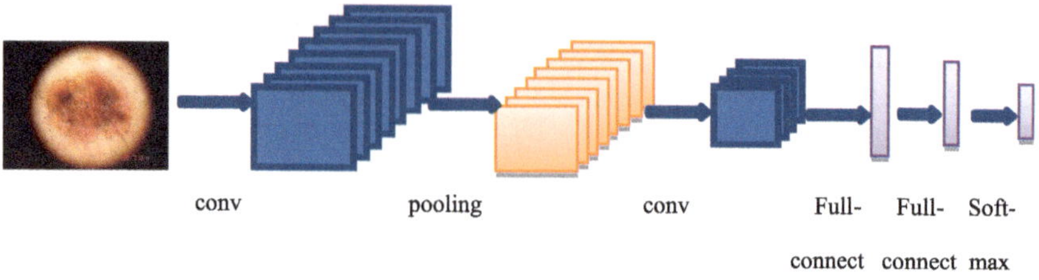

Fig. 1. Forecast model structure char

The activation functions commonly used in CNN network are Relu, Absval, Tanh, in which the linear activation function Relu is the most frequently used. The expression as follows:

$$\max(0,x)=\begin{cases}0, x\leq 0\\ x, x>0\end{cases} \tag{4}$$

where x represents the input value, the Relu function is a piecewise function, and if x is greater than 0, the output holds the original value, otherwise the output is 0. This method of forcing part of the data to be 0 makes the network sparse, reducing the likelihood of overfitting, and improving the convolution efficiency.

Three different activation functions are used in the Mnist data, trained 1000 iterations, the loss value and computation time of three activation functions are shown in Table 1. It can be seen that the loss value obtained by using the Relu function is the smallest, and the operation time is shortest, and the network first achieves convergence.

Table 1. Comparison of different activation function loss values

Activation function	Tanh	AbsVal	ReLU
Loss value	0.1113	0.0230	0.0131
Operation time/s	75.28	74.60	72.05

3 Experiment and Analysis

This paper is based on the Caffe deep learning platform [9] to build the network structure and train the model. The platform is developed by C++ to build a modular framework with Python, Matlab and command line interface, taking into account the acceleration and switching between GPU and CPU. This paper uses 900 cases of annotated melanoma data provided by the open data set ISBI International Symposium on Biomedical Imaging. According to the csv file provided, the data are divided into benign and malignant categories, of which benign are 724 cases, malignant are 176 cases.

3.1 Parameter Settings

An improved convolution neural network method is used to set the parameters of the network for melanoma data, and the data sets are trained and tested. The weights in the conv layers are initialized using random Gaussian distributions with a standard deviation of 0.001, SGD [10] with a momentum of 0.9 is used for the training, and the weights in the FC layer is initialized according to the Xavier algorithm. The network is trained with the batch size of 64 skin images.

3.2 Improved CNN Method Experiment

In this paper, we set the network 60,000 iterations, the number of training set is 5600, and the test set is 1000, experiment results as shown in Table 2, we can see that based on the improved CNN model of benign and malignant skin cancer, it's recognition rate of training set is up to 91.92%, the test set identification rate is 89.5%. The Loss and Accuracy curves during the CNN and the modified CNN iterative training process are shown in Fig. 2, where the blue curve represents the change in accuracy and the red curve represents the change in the loss value.

Table 2. The recognition rate of CNN prediction on skin cancer of benign and malignant

Number of training set samples	Training set recognition rate	Number of test set numbers	Test set recognition rate
5600	91.92%	1000	89.50%

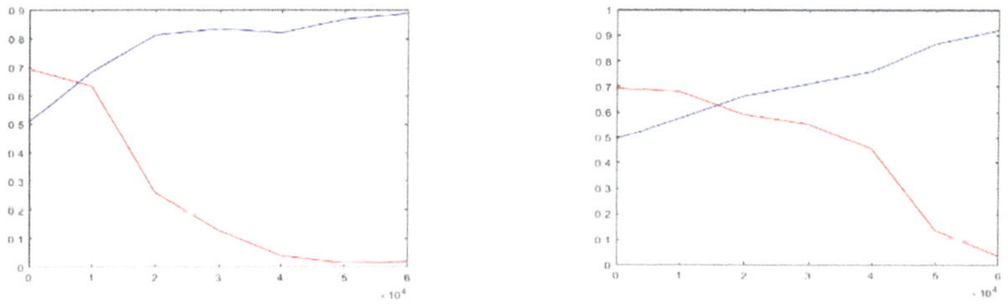

(a) (b)

Fig. 2. CNN Loss with Improved CNN Loss and Accuracy Rise Process. (a) CNN Loss and Accuracy curves; (b) Improved CNN Loss and Accuracy curves

From Fig. 2, the comparison between the improved CNN and the original CNN's prediction accuracy and the loss value in the data set shows that the improved accuracy of CNN prediction increases steadily with the increasing of iterative times, while the original CNN is iteratively with the number of iterative times increases, the accuracy gradually becomes stable.

3.3 Other Methods of Experiment

In order to verify the effectiveness of improved CNN for melanoma data recognition, this method was compared with KNN, SVM and other methods. As can be seen from Table 3, the improved CNN recognition rate of benign and malignant skin cancer in the training set and test set is higher than KNN and SVM. Therefore, improved CNN has better predictive effect on benign and malignant skin cancer.

Table 3. Improved CNN and KNN, SVM comparison results

Different methods	KNN recognition rate	SVM recognition rate	Improved CNN recognition rate
Number of training set (5600)	68.25%	61.25%	91.92%
Number of test set (1000)	64.34%	60.02%	89.5%

3.4 Conclusions

Convolution neural network is characterized by automatic learning features, if there are enough samples for training, the network can learn a good feature to classify. In this paper, the predictive model of benign and malignant skin cancer based on convolution neural network is helpful to the diagnosis of clinicians. Through the amplification to melanoma data, the structure of convolution neural network is deepened, the Relu is selected as activating function, the SGD is used to train the network, and the deep learning network with higher classification precision and better convergence effect is realized. In this paper, the accuracy training set of malignant melanoma was 91.92%, the test set reached 89.5%, and the result was better than other methods.

In the follow-up work, it will be more closely combined with the relevant knowledge of clinical medicine, for the establishment of benign and malignant skin cancer prediction model, then to find more general network structure design method.

References

1. Zhang, L., Liu, X., Wang, J., et al.: A method for predicting benign and malignant lung neoplasms based on quantitative image histology. Acta Autom. Sin. **43**(12), 2109–2114 (2017)
2. Luo, J., Wu, J.: Review of fine grained image classification based on deep convolution. Acta Autom. Sin. **43**(08), 1306–1318 (2017)
3. Wan, L., Zeiler, M., Zhang, S., Le Cun, Y., Fergus, R.: Regularization of neural networks using dropconnect. In: Proceedings of the International Conference on Machine Learning, Atlanta, USA, pp. 1058–1066 (2013)
4. Esteva, A., Kuprel, B., Novoa, R.A., et al.: Dermatologist-level classification of skin cancer with deep neural networks. Nature **542**(7639), 115–118 (2017)

5. Goodfellow, I.J., Warde-Farley, D., Mirza, M., Courville, A.C., Bengio, Y.: Maxout networks. In: Proceedings of the International Conference on Machine Learning, Atlanta, USA, pp. 1319–1327 (2013)
6. Vasconcelos, C.N., Vasconcelos, B.N.: Experiments using deep learning for dermoscopy image analysis. Pattern Recognit. Lett. (2017). ISSN: 0167-8655
7. Chougrad, H., Zouaki, H., Alheyane, O.: Deep Convolutional Neural Networks for breast cancer screening. Comput. Methods Programs Biomed. **157**, 19–30 (2018)
8. Yang, T., Huang, S.: Examples of application of improved convolutional neural networks in classification and recommendation. Appl. Comput. Res. (04) 2018
9. Jia, Y., Shelhamer, E., Donahue, J., Karayev, S., Long, J., Girshick, R., Guadarrama, S., Darrell, T.: Caffe: convolutional architecture for fast feature embedding. In: ACM International Conference on Multimedia, pp. 675–678 (2014)
10. Kingma, D.P., Ba, J.: Adam: a method for stochastic optimization. Comput. Sci. (2014)

Sentiment Classification Method for Chinese and Vietnamese Bilingual News Sentence Based on Convolution Neural Network

Xiaoxu He[1,2], Shengxiang Gao[1,2(✉)], Zhengtao Yu[1,2], Shulong Liu[1,2], and Xudong Hong[1,2]

[1] Faculty of Information Engineering and Automation, Kunming University of Science and Technology, Kunming 650500, China
gaoshengxiang.yn@foxmail.com
[2] Key Lab of Computer Technologies Application of Yunnan Province, Kunming University of Science and Technology, Kunming 650500, China

Abstract. Sentiment classification of the Chinese and Vietnamese news is important to analysis public opinion of Chinese and Vietnamese. For cross-lingual sentiment analysis, this paper studies the method of sentiment analysis for Chinese and Vietnamese bilingual news sentence based on convolution neural network (CNN). Firstly, collect Chinese and Vietnamese news data from Internet, use the data collected to train word vector. Secondly, employ the pre-trained word vector and bilingual lexicon to project the Chinese news labeled data to bilingual word vector space, the problem of the lack of annotated corpus in Vietnamese is solved in a certain level. Finally, using convolution neural network to make cross-lingual sentiment classification. Experiment result shows that the F score of Vietnamese news classification has reached 86%. Compared with the traditional methods, the proposed method does not need to construct sentiment lexicon and perform feature extraction, and the performance of cross-lingual sentiment classification is also remarkable.

Keywords: Sentiment analysis · Chinese news · Vietnamese news
Convolution neural network

1 Introduction

With the rapid developing of Internet and the rising of Web2.0, the text on the network is increasing. It is very important to realize the analysis and utilization of the Internet text. Opinion Mining [1] is generated in this context, the purpose is to automate the analysis of text views or emotional information [2]. In the early years, sentiment analysis for critical texts are mainly studied in this domain. For example, Liu divided sentiment analysis into three classifications: document level, sentence level and factor level [3], Sentence level sentiment classification assumes that the emotion expressed by a sentence is consistent, and that the object of the emotion is only one (sentence level sentiment classification can not achieve fine-grained recognition). The comment text obviously does not satisfy the above hypothesis, such as hotel comment sentence "the location of the hotel is far away, but the environment and the service is very good."

K. Deng et al. (Eds.): ICMIR 2018, AISC 856, pp. 1230–1239, 2019.
https://doi.org/10.1007/978-3-030-00214-5_151

The sentiment of this sentence is positive, but the evaluation of the position is negative. The news text differs from the comment text that a piece of news usually only discusses an event, and the emotion expressed in the news is usually fixed, and there are few different emotions at the same time in a sentence. Based on the above considerations, it is reasonable to analyze the sentimental orientation of news texts. The purpose of this paper is to analyze the sentimental orientation of Chinese and Vietnamese news sentences, and find whether the sentence is positive or negative (two classification).

At present, in the research of sentiment classification, the following main algorithms are proposed, which are dictionary based method [4–6], machine learning based method [7, 8] and depth learning based method [9–13]. In the dictionary based method, we first need to construct an emotion dictionary and classify the sentences according to the emotion dictionary. For example, Mihalcea uses two types of bilingual dictionaries to construct subjective word dictionary [4] with bilingual alignment, and then constructs a rule-based sentiment classifier based on the bilingual subjective lexicon. The main idea of machine learning based method is to extract emotion features, and then classify them by classification algorithms. For example, Pang extracts the unigram of text as a feature, and uses SVM algorithm and Naive Bayes Algorithm to divide the text into two classifications: positive and negative [7]. In recent years, deep learning has become a hot topic in the field of artificial intelligence. Compared with traditional machine learning algorithms, deep learning has the main advantage that it does not require tedious feature extraction steps. In the early days, deep learning achieved better results in image processing. With further exploration, deep learning has been well applied in the field of Natural Language Processing. For example, Kim [11] proposed the convolutional neural network applied to sentence classification task, the experimental results show that this method achieves better effect compared with the traditional machine learning algorithms, and the convolutional neural network classifier does not need artificial feature extraction.

On the basis of previous studies, this paper proposes a sentiment classification method based on the convolution neural network for Chinese and Vietnamese Bilingual News sentence. This method uses the Chinese Vietnamese bilingual dictionary to realize cross language sentiment analysis, and on this basis, uses the convolution neural network to realize the sentiment classification of Chinese and Vietnamese news sentences. In addition, this paper compares the difference between the use of word ID as input (one-hot vector) and the use of word vector (word2vec) as input, and experiments show that using word vector as input has achieved good results.

2 Sentiment Classification Algorithm for Chinese and Vietnamese Bilingual News Sentences Based on CNN

In order to classify the Chinese and Vietnamese Bilingual News sentences, a kind of cross language sentiment classification algorithm based on convolutional neural network is studied in this paper, as shown in Fig. 1. First, we collect Chinese and Vietnamese news texts, use participle tools to segment Chinese and Vietnamese news texts, and use word2vec tools to train Chinese word vectors and Vietnamese word vectors. Then, according to the word vectors obtained by training and bilingual dictionaries, the

Chinese News sentence is represented as a bilingual vector matrix, which is used as the input of the convolution neural network model. Finally, the convolution neural network classification model is constructed, and the labeled Chinese sentences are used as training sets to train the model. The algorithm only needs to annotate Chinese News sentences, and uses the Chinese Vietnamese bilingual dictionary to realize the sentiment classification of Vietnamese news sentences.

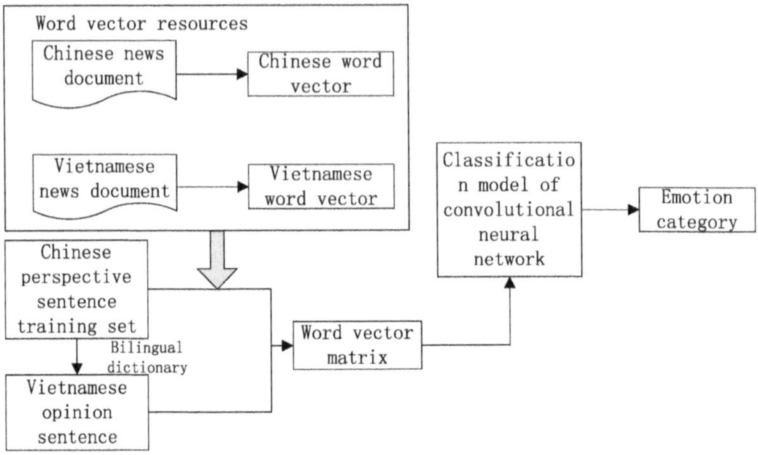

Fig. 1. The frame word of Chinese and Vietnamese news sentence sentiment classification based on convolution neural network

3 A Classification Model Based on Convolution Neural Network

In order to realize the sentiment classification of Chinese and Vietnamese Bilingual News sentences, this paper studies a Chinese and Vietnamese bilingual sentiment classification algorithm based on the convolution neural network. The algorithm model is shown in Fig. 2. The sentiment classification model based on convolution neural network consists of four layers, which are input layer, convolution layer, max-pooling layer and output layer. The functions of each layer are explained as follows.

First layer: the input layer. the main purpose of this layer is to represent the Chinese News sentence as the bilingual word vector that the computer can recognize and process. Due to the lack of Vietnamese corpus, different input strategies are used in model training and model testing. In the model training stage, according to the bilingual dictionary, the Chinese News sentence that has marked the sentiment classification (positive sign is 1, negative mark is 0) is mapped to Vietnamese, and the Vietnamese expression of Chinese News sentence is obtained. In the model testing phase, the Vietnamese unclassified news sentence is mapped to Chinese for processing. We use the above strategy to ensure that each language has two perspectives.

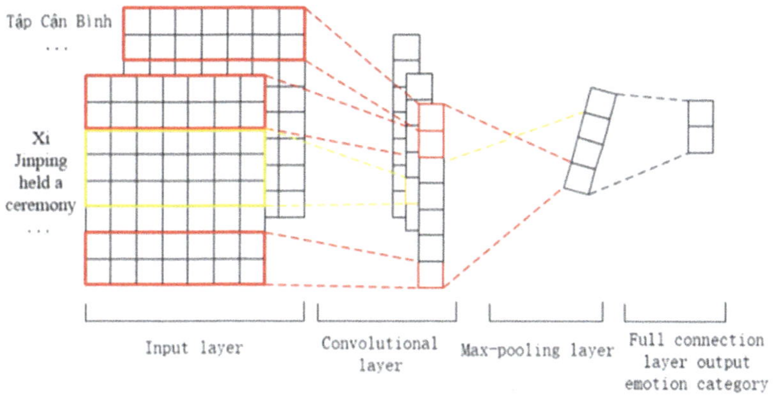

Fig. 2. Chinese and Vietnamese news sentence sentiment classification based on convolution neural network

The sentence "Xi Jinping held a ceremony to welcome the Vietnamese Communist Party General Secretary Ruan Fuzhong's visit to China " is shown as an example in the Fig. 2. First, the Chinese News sentence is mapped to Vietnamese using a bilingual dictionary. Then, we use Chinese segmentation tool and Vietnamese word segmentation tool to segment Chinese sentence and Vietnamese sentence. Finally, according to the pre trained word vectors, the sentences after segmentation are represented as matrices, and each row of the matrix represents the word vector of a word. Suppose that the sentence contains M words, the word vector length is n, and the size of the matrix is m * n.

Second layer: the convolution layer. The main job of this layer is to extract features from the word vector representation of sentences using convolution operations. $x_i \in R^N$ represent an n-dimensional word vector of the i-th word in Chinese sentences or Vietnamese sentences. And a sentence containing m words can be expressed as:

$$x_{1:m} = x_1 \oplus x_2 \oplus \ldots \oplus x_m \quad (1)$$

Wherein, ⊕ represents connection operations, such as (1,1,1,1) ⊕ (2,2,2,2) = (1,1, 1,1,2,2,2,2).Convolution operation refers to the sliding window with length h, from which the $x_{1:m}$ intercept vector is used as the input of the convolution layer activation function. After the activation function is calculated, the output is the feature to be extracted. The following is shown:

$$c_i = f(W \cdot x_{i:i+h-1} + b) \quad (2)$$

Function *f* represents activation function, and ReLU function is chosen as activation function in this paper. *W* and *b* represent the weight matrix and the offset unit, which are the parameters of the neural network, and can be trained by the neural network. C_i represents the characteristic calculated according to the activation function *f* and the input vector $x_{i:i+h-1}$.

The above formula describes the length of sliding window h at position $x_{i:i+h-1}$ on the operation, when the sliding window from the beginning of the sentence slide to the end, can obtain a set of input, denoted as ($x_{1:h}$, $x_{2:1+h}$,..., $x_{m-h+1:m}$), then we can get a group according to the input and output, said (C_1, C_2,..., C_{m-h+1}), a group of output is called feature mapping. It should be noted that the input layer of the model contains two channels, and each sliding window needs to glide through two channels when calculating the feature mapping.

Third layer: max-pooling layer, the main purpose of this layer is to process the feature mapping of the upper layer, and obtain the final feature vector. Specifically, the maximum value of the feature map is extracted, and the idea is that the maximum in each feature map represents the most important feature of the mapping. In addition, a major advantage of the max-pooling layer is that a fixed length feature vector can be obtained with the aid of this layer without requiring attention to the length of the original input sentence.

The fourth layer: the output layer. the main task of this layer is to calculate the emotion category based on the feature vector of the upper layer. This layer is the full connection layer, using the sigmoid function as the activation function, and the result is processed by softmax to obtain the probability. Let K dimension be characterized as X_k, then:

$$y_i = \frac{e^{(W^T \cdot X_k + b)}}{1 + e^{(W^T \cdot X_k + b)}} \tag{3}$$

Among them, W^T represents the transpose of vector W, and W and b represent the parameters of the fully connected layer network, which represents the output value of the output layer neuron. The output layer of the model contains only two neurons. In order to get the probability that the sentence is positive or negative (positive by 1 and negative by 0), further processing is needed, as shown in the following formula.

$$\hat{P}(y = 1|s_{cn}, s_{vn}) = \frac{e^{y_1}}{\sum_{i=1}^{2} e^{y_i}} \tag{4}$$

$$\hat{P}(y = 0|s_{cn}, s_{vn}) = \frac{e^{y_0}}{\sum_{i=1}^{2} e^{y_i}} \tag{5}$$

$S_{cn,}$ S_{vn} expresses Chinese sentences and Vietnamese sentences respectively.

4 Experiment and Result Analysis

4.1 Experimental Design

Using the Chinese Vietnamese news sentence as the experimental data set, in order to test the proposed model, the sentiment polarity of a part of Chinese News sentence and

Vietnamese news sentence is marked artificially. For example, the sentence "Xi Jinping held a ceremony to welcome the Vietnamese Communist Party General Secretary Ruan Fuzhong's visit to China. " to the positive polarity, the label is 1, annotation data are shown in Table 1. In addition, the Chinese Vietnamese bilingual dictionary consists of 256752 word pairs.

Table 1. Information of experiment data

Language	Number	Average length	Emotional polarity (positive)	Emotional polarity (negative)
Chinese	3000	33	1430	1570
Vietnamese	3000	45	1320	1680

4.2 Evaluation of Method

The accuracy, recall rate and F value were used to evaluate the experiment. Let PR denote positive prediction for positive numbers, PN expressed negative positive predictive data, MR said negative predictive negative number, MN expressed negative predictive positive number, while the accuracy rate, recall rate and F-measure calculation formula is as follows:

$$precise = \frac{PR}{PR + MN} \tag{6}$$

$$recall = \frac{PR}{PR + PN} \tag{7}$$

$$F_1 = 2 \cdot \frac{precise \cdot recall}{precise + recall} \tag{8}$$

4.3 Experiment Results and Analysis

In order to obtain the word vector in Chinese and Vietnamese, first use crawlers collected a large number of Chinese and Vietnamese unlabeled corpus, including about 1000000 Chinese sentences and 500000 Vietnamese sentences, from Chinese and Vietnamese major news sites. After word segmentation, gensim is used to train the word vector, and the word vector dimension is set to 100. In order to evaluate the effect of word vectors, the first five most similar words are selected for testing, and the results are shown in Table 2. It can be observed from the table that the training of Chinese word vector and Vietnamese word vector has good effect on word similarity calculation. In the experiment, it is necessary to pay attention to the Chinese News annotation corpus used in the training phase of the model. At the model testing stage, we used Vietnamese unlabeled corpus.

In order to compare the effects of different word vectors on the results, the one-hot word vectors and the word2vector word vectors are used as input respectively, and the

Table 2. Chinese word vector and Vietnamese word vector result

Chinese words	The 5 most similar words	Vietnamese words	The 5 most similar words
Jinping Xi	Ministry of Foreign Affairs, China, Yi Wang, Keqiang Li, Summit	NguyễnPhúTrọng (Zhongfu Yuan)	NguyễnChíVịnh (Nguyen Chi Vinh), PhùngQuangThanh (PhungQuangThanh), VIệt (Vietnam), BiểnĐông (east sea), PhạmBình Minh (Pham Minh Minh)
China	Nation, America, foreign, country, inland	Việt Nam (Vietnam)	Việt (Vietnam), HàNội (Hanoi), Mỹ (Inch), HồChí Minh (Ho Chi Ming), LiênHiệpQuốc (United Nation)
economy	Build, militar, market, medical care, investment	kinhtế (economy)	văn hóa (civilization), quân đội (military), cải cách (reform), buôn bán (business), thị trường (market)

word vectors are shown in Fig. 3. Since the parameters of the neural network are numerous, it is necessary to set these parameters according to the task before training the convolution neural network model. The parameters of this paper are set as follows.

$$onehot(\text{经济}) = [0,0,0,0,0,0,0,0,0,0,\ldots,0,0,0,1,000,0,\ldots,0,0,0,0,0,0,0,0,0,0,0,0]_{1\times|D|}$$

$$word2vector(\text{经济}) = [\ 1.67981553,\text{-}0.82601517,\ldots,1.64033353,\ldots,\text{-}0.15920518\ ]_{1\times100}$$

Fig. 3. The comparison of one-hot word vector and word2vector word vector

Sliding window: the sliding window is generally set to 1–5. If the parameter is too small, the input of the model will lack the semantic information. Or if the parameter is too large, the model will be too complex and difficult to train. Because the text is processed by word segmentation in the input layer, using word vectors instead of character vectors and the unlabeled corpus is widely used in the process of training vector in the word, the unlabeled corpora contain semantic information and reflect semantic information into the word vector. Therefore, the sliding window is set to 1, 2, and 3, and different sliding windows are configured with 100 different feature mappings (corresponding to different neurons in the neural network).

Mini-batch size: the traditional gradient descent algorithm is used to train the model.

Each iteration of the parameter needs to observe all the training samples, and the overhead is great. Using Mini-batch strategy for training, training process faster, better results. According to the size of the training set, this article sets mini-batch size to 30.

Loss function: cross entropy loss function.

Activation function: in this paper, the ReLU function is used as the activation function in the convolution layer, and the output layer uses the sigmoid function as the activation function.

The Dropout: Dropout parameter is used to prevent the overfitting of the model. The dropout layer is added after the input layer and the max-pooling layer, and the dropout value is set to 0.2.

Using the above parameters to train the model, in addition, to illustrate the effectiveness of the proposed method, this paper compares it with the dictionary based approach [4] and the support vector machine [7]. Method for word vector one-hot is marked M1 and method for word2vector word vector is marked as M2 in the input layer, The dictionary based method is marked as C1, and the SVM based method is marked as C2 (select unigram as feature). To achieve the above method, half off cross validation is carried out to obtain the accuracy, recall and F values of different methods, as shown in the following Figs. 4, 5, and 6.

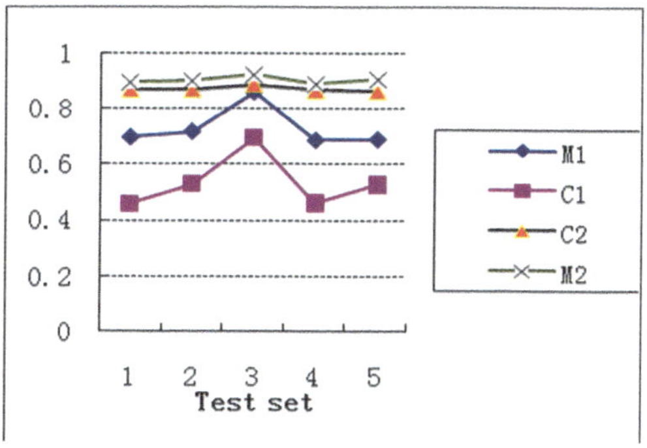

Fig. 4. Precision

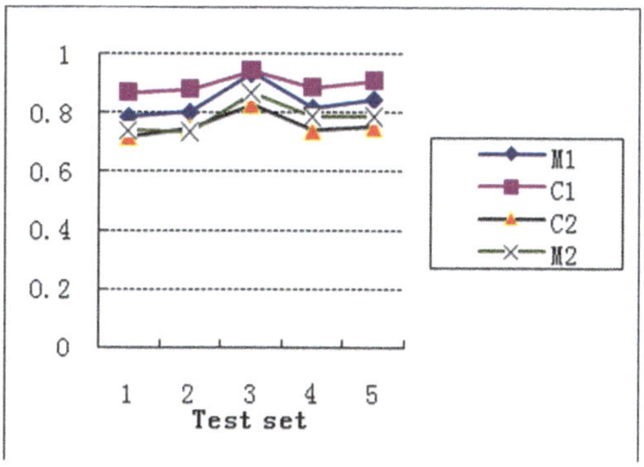

Fig. 5. Recall

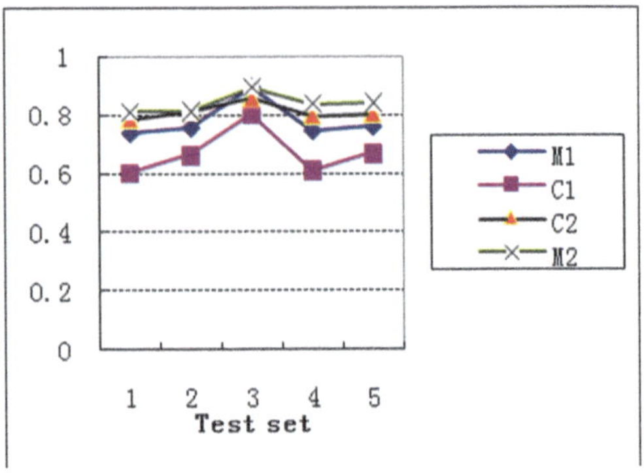

Fig. 6. F score

As can be seen from the diagram, using the pre trained word2vector word vector as input, the model achieves better results than using the one-hot word vector as input. This is because the pre - trained word vectors contain large amounts of semantic information in unlabeled corpora, whereas one-hot, a high-dimensional sparse word vector, does not have this function. In addition, the method of using word2vector word vector is better than dictionary based method on the whole. Although the dictionary based method achieves higher recall rate, the single rate of accuracy is poor. It is proved that the method presented in this paper not only has the function of cross linguistic analysis, but also achieves good results in various evaluation indexes.

5 Conclusions

In order to achieve Chinese and Vietnamese Bilingual News sentence sentiment classification, this paper proposes a cross language classification algorithm based on convolution neural network. The method uses a bilingual dictionary as a cross language bridge, and uses the word vector tool to map the natural language text into the vector space as the input layer of the convolution neural network. First of all, in order to get better word vectors, a large number of Chinese and Vietnamese news corpora are collected as word vectors for training corpus. Secondly, the Chinese Vietnamese bilingual dictionary is used to map Chinese into Vietnamese, and get the two perspectives of the text. It is trained as the two channel of the neural network. With the help of bilingual dictionaries, the model can not only classify the sentiment in Chinese, but also classify the sentiment in Vietnamese. The experimental results show that the model has achieved good results.

Acknowledgements. This work was supported by National Nature Science Foundation (Grant Nos. 61472168, 61672271, 61732005), and Science and Technology Innovation Talents Fund Project of Ministry of Science and Technology (Grant No. 2014HE001),Innovation Team Project of Yunnan Province (Grant No. 2014HC012).

References

1. Dave, K., Lawrence, S., Pennock, D.M.: Mining the peanut gallery: opinion extraction and semantic classification of product reviews. In: International Conference on World Wide Web, pp. 519–528. ACM (2003)
2. Pang, B., Lee, L.: Opinion mining and sentiment analysis. Found. Trends® Inf. Retrieval **2** (1–2), 1–135 (2008)
3. Liu, B.: Sentiment analysis: Mining opinions, sentiments, and emotions. Comput. Linguist. **42**(3), 1–4 (2016)
4. Mihalcea, R., Banea, C., Wiebe, J.M.: Learning multilingual subjective language via cross-lingual projections. In: Meeting of the Association of Computational Linguistics (2007)
5. Maks, I., Vossen, P.: A lexicon model for deep sentiment analysis and opinion mining applications. Decis. Support Syst. **53**(4), 680–688 (2012)
6. Taboada, M., Brooke, J., Tofiloski, M., et al.: Lexicon-based methods for sentiment analysis. Comput. Linguist. **37**(2), 267–307 (2011)
7. Pang, B., Lee, L., Vaithyanathan, S.: Thumbs up? Sentiment classification using machine learning techniques. In: Proceedings of Emnlp, pp. 79–86 (2002)
8. Mullen, T., Collier, N.: Sentiment analysis using support vector machines with diverse information sources. In: Conference on Empirical Methods in Natural Language Processing, EMNLP 2004, A Meeting of Sigdat, A Special Interest Group of the Acl, Held in Conjunction with ACL 2004, 25–26 July 2004, Barcelona, Spain, pp. 412–418. DBLP (2004)
9. Poria, S., Cambria, E., Gelbukh, A.: Aspect extraction for opinion mining with a deep convolutional neural network. Knowl. Based Syst. **108**, 42–49 (2016)
10. Gao, Y., Rong, W., Shen, Y., et al.: Convolutional Neural Network based sentiment analysis using Adaboost combination. In: International Joint Conference on Neural Networks, pp. 1333–1338 (2016)
11. Kim, Y.: Convolutional neural networks for sentence classification. In: Proceedings of the 2014 Conference on Empirical Methods in Natural Language Processing (EMNLP), pp. 1746–1751 (2014)
12. Santos, C.N.D., Gattit, M.: Deep convolutional neural networks for sentiment analysis of short texts. In: International Conference on Computational Linguistics (2014)
13. Severyn, A., Moschitti, A.: Twitter sentiment analysis with deep convolutional neural networks. In: Proceedings of the 38th International ACM SIGIR Conference on Research and Development in Information Retrieval, pp. 959–962. ACM (2015)

A Neural Network Face Recognition Algorithm Based on Surface-Simplex Swarm Evolution

Zhe Lin[1], Haiyan Quan[1(✉)], and Xinbao Yi[2]

[1] College of Information Engineering and Automation, Kunming University of Science and Technology, Kunming, China
quanhaiyan@163.com

[2] Xishuangbanna Vocational and Technical College, Jinghong, China

Abstract. In the BP neural network training algorithm, the optimization of weights is easy to fall into the local extreme points and the convergence rate is slow. Many researches introduce the intelligent optimization algorithm to improve them, but the traditional intelligent optimization algorithm usually has many Control parameters, if not correctly selected parameters, or did not properly select the initial point position, it is difficult to search for the optimal neural network weights. In this paper, a BP neural network learning algorithm based on surface-simplex swarm evolution is proposed. It reduces the control parameters of the algorithm by all random search. The diversity of the particles is maintained by the polygonal state of the population, and the algorithm is prevented from falling into the local extreme points. Dependence on initial values. In the application, the algorithm is applied to the training algorithm of face recognition neural network. The experimental results show that the trained neural network effectively improves the recognition rate.

Keywords: Neural network · Intelligent optimization
Surface-simplex swarm evolution · Face recognition

1 Introduction

With the continuous development of artificial intelligence, artificial neural network [1], as a mathematical operation model, successfully solves practical problems that many computers cannot complete in the fields of pattern recognition, prediction and estimation, and automatic control. Among them, BP neural network with supervised learning is the most widely used.

The traditional BP neural network learning algorithm uses the principle of gradient descent, which has the disadvantages of slow convergence and easy to fall into local extremes [2]. In recent years, many studies have introduced intelligent optimization algorithms to improve the learning phase [3–7]. However, these algorithms also have some disadvantages. Traditional intelligent optimization algorithms usually have multiple control parameters. If the parameters cannot be correctly selected or the initial point position is not properly selected, it is difficult to search for optimal neural network weights. At the same time, the improved algorithms for these algorithms [8]

K. Deng et al. (Eds.): ICMIR 2018, AISC 856, pp. 1240–1246, 2019.
https://doi.org/10.1007/978-3-030-00214-5_152

mostly increase the complexity of the algorithm and introduce more control parameters to achieve better algorithm performance.

This paper proposes a BP neural network learning algorithm based on simplex evolution. Its advantage lies in that the simplex evolutionary learning algorithm reduces the control parameters of the algorithm through a full random search, and maintains the diversity of the particles through the multi-role state of the population, avoiding the algorithm being trapped. The local extreme points reduce the dependence on the initial value. In this paper, the algorithm is applied to the training of face recognition neural network. Experimental results show that the BP neural network trained by the learning algorithm can effectively improve the recognition rate and convergence speed compared with the standard BP algorithm.

2 BP Neural Network Learning Algorithm Based on Surface Simplex Swarm Evolution

2.1 Surface-Simplex Swarm Evolution Algorithm

The Surface-Simplex Swarm Evolution Algorithm [9] is a new type of swarm intelligence optimization algorithm proposed by Quan et al. The principle of the simplex evolution algorithm is as follows:

First, in the search space R^n, two dimensions p, q are randomly selected in a uniform distribution manner to construct a search subspace R^2. In this search subspace, the search for each particle in the population is defined as follows:

$$W_{i1}^{p,q}(t+1) = r_{11} \times W_i^{p,q}(t) + r_{12} \times W_j^{p,q}(t) + (1 - r_{11} - r_{12}) \times W_o^{p,q}(t) \quad (1)$$

$$W_{i2}^{p,q}(t+1) = r_{21} \times W_i^{p,q}(t) + r_{22} \times \overline{W}_j^{p,q}(t) + (1 - r_{21} - r_{22}) \times W_o^{p,q}(t) \quad (2)$$

$$W_{i3}^{p,q}(t+1) = r_{31} \times W_i^{p,q}(t) + r_{32} \times \overline{W}_j^{p,q}(t) + (1 - r_{31} - r_{32}) \times \overline{W}_o^{p,q}(t) \quad (3)$$

$$W_{i4}^{p,q}(t+1) = r_{41} \times W_i^{p,q}(t) + r_{42} \times W_j^{p,q}(t) + (1 - r_{41} - r_{42}) \times \overline{W}_o^{p,q}(t) \quad (4)$$

Among them, $W_{i1}^{p,q}(t+1)$, $W_{i2}^{p,q}(t+1)$, $W_{i3}^{p,q}(t+1)$, $W_{i4}^{p,q}(t+1)$ are the four new positions searched for in the search subspace by the particle i in the t + 1th iteration; $W_i^{p,q}(t)$ is the original search of the search subspace in the t iteration of the particle i. Position; $W_j^{p,q}(t)$ is the original position of particle j in the search subspace in the t iteration; $W_o^{p,q}(t)$ is the optimal position of the optimal particle o in the search in the search subspace in the t iteration. $\overline{W}_j^{p,q}(t)$ is the symmetrical position of position $W_i^{p,q}(t)$ as the center, position $W_j^{p,q}(t)$; $\overline{W}_o^{p,q}(t)$ is the symmetrical position of position $W_i^{p,q}(t)$ as the center, position $W_o^{p,q}(t)$. r_{11}, r_{12}, r_{21}, r_{22}, r_{31}, r_{32}, r_{41} and r_{42} are 8 random numbers generated in a uniform distribution over the interval [0, 1].

Each particle then uses the simplex neighborhood search operator to search for new locations and update the four new central role locations on each particle on R^n:

$$W_{ic1}(t+1) = [W_{ic1}^1(t), W_{ic1}^2(t), \ldots, W_{ic1}^p(t+1), \ldots, W_{ic1}^q(t+1), \ldots, W_{ic1}^{n-1}(t), \ldots W_{ic1}^n(t)] \quad (5)$$

$$W_{ic2}(t+1) = [W_{ic2}^1(t), W_{ic2}^2(t), \ldots, W_{ic2}^p(t+1), \ldots, W_{ic2}^q(t+1), \ldots, W_{ic2}^{n-1}(t), \ldots W_{ic2}^n(t)] \quad (6)$$

$$W_{ic3}(t+1) = [W_{ic3}^1(t), W_{ic3}^2(t), \ldots, W_{ic3}^p(t+1), \ldots, W_{ic3}^q(t+1), \ldots, W_{ic3}^{n-1}(t), \ldots W_{ic3}^n(t)] \quad (7)$$

$$W_{ic4}(t+1) = [W_{ic4}^1(t), W_{ic4}^2(t), \ldots, W_{ic4}^p(t+1), \ldots, W_{ic4}^q(t+1), \ldots, W_{ic4}^{n-1}(t), \ldots W_{ic4}^n(t)] \quad (8)$$

Each particle in the defined population has three role states:

The central role state - defined by the principle of greediness, is the optimal position searched for by each particle: $W_{ic}(t+1)$.

Mining the role of state - based on the principle of recentness, define the latest position searched for each particle, namely:

$$W_{il}(t+1) = \{W_{ic1}(t+1), W_{ic2}(t+1), W_{ic3}(t+1), W_{ic4}(t+1)\} \quad (9)$$

The exploration of the role of state - in the principle of uniform distribution of randomness, is defined as each particle is uniformly distributed in the search space randomly positioned: $W_{ig}(t+1)$.

Evaluate the merits of the new position of the particle according to the error function, determine the position of each particle's three roles; finally record the optimal particle and its optimal position in this iteration cycle, end the search cycle of this iteration, and start the next iteration search cycle until The particles converge to the global best position.

2.2 BP Neural Network Learning Algorithm Based on Surface-Simplex Swarm Evolution

It has been proved theoretically that a three-layer BP neural network can achieve arbitrary precision and approximate any continuous function [10]. The specific training process is as follows:

Let a three-layer BP neural network have input layer m, hidden layer l and output layer n. Let a nonlinear system have an input of $X_i = (x_{i,1}, x_{i,2}, \ldots, x_{i,m})$, and an output of $Y_i = (y_{i,1}, y_{i,2}, \ldots, y_{i,n})$, where i is the sample category. The neuron's excitation function selects the symbolic function:$f(x) = sign(x)$.

The hidden layer output is:

$$h_{i,l} = f(\sum_{p=1}^{m} W_{i,l}^{(1)} x_{i,p} + b_{i,l}^{(1)}) \quad p = 1, 2, \ldots, m \quad (10)$$

The output layer output is:

$$y_{i,n} = f(\sum_{q=1}^{l} W_{i,n}^{(2)} h_{i,q} + b_{i,n}^{(2)}) \quad q = 1, 2, \ldots, l \tag{11}$$

In the formula, $h_{i,l}$ is the output of the hidden layer of the l neuron of the i sample, f is the hidden layer excitation function; $W_{i,l}^{(1)}$ is the input layer and the hidden layer of the l neuron of the i sample. Connection weights, $W_{i,n}^{(2)}$ is the connection weight between the hidden layer and the output layer of the n neuron of the i sample, $b_{i,l}^{(1)}$ is the hidden layer threshold of the l neuron of the i sample, and $b_{i,n}^{(2)}$ is the i. The sample output threshold for the n neuron.

The goal of the standard BP algorithm learning is to optimize the error function J. The BP neural network error function is defined as:

$$J = \frac{1}{m} \sum_{p=1}^{m} \sum_{r=1}^{n} (y_{i,n}^{d} - y_{i,n})^2 \tag{12}$$

In the formula, m represents the number of input samples of the training set, n represents the number of output neurons ($r = 1, 2, \ldots, n$); $y_{i,n}^{d}$ represents the expected output of the n neuron of the i sample, and $y_{i,n}$ represents the n neuron of the i sample The actual output.

The steps of the simplex evolutionary learning algorithm applied to the BP neural network are as follows:

(1) The m particles are initialized randomly based on the uniform distribution in the search space:

$$W_{i,l}(0) = \underline{W} + rand(0, 1) \times (\overline{W} - \underline{W}) \tag{13}$$

Where $W_{i,l}(0)$ is the position of the i-th particle in the kth dimension of the Rn search subspace. $\overline{W}$ and $\underline{W}$ are the upper and lower bounds of the search subspace in the k-th dimension, and rand (0, 1) is a random number uniformly distributed in the interval [0, 1].

(2) For each particle i in the population, use the 4 new central role positions searched in the search subspace: $W_{i,c1}^{p,q}(N+1)$, $W_{i,c2}^{p,q}(N+1)$, $W_{i,c3}^{p,q}(N+1)$, $W_{i,c4}^{p,q}(N+1)$ and keep their positions in other dimensions constant, updating the 4 new ones on each particle. Central role position;

(3) Evaluate the pros and cons of each particle based on the error function J and determine the three character states of each particle;

(4) Record the position of the optimal particle in the population: $W_{oc}(N+1)$, Start the next search cycle until the particles in the population converge to the optimal position, i.e. the position of the optimal particle in the population is stable to the given accuracy;

(5) Taking the optimal position of the particles as the weights and thresholds of the BP neural network, input the feature values of the training samples to train and learn the BP neural network, and classify the output results.

3 Experiments and Analysis

Using ORL face database for face training test set. The convergence performances of the standard BP algorithm and the BP learning algorithm based on single evolution are compared in terms of convergence accuracy and convergence speed (Figs. 1 and 2).

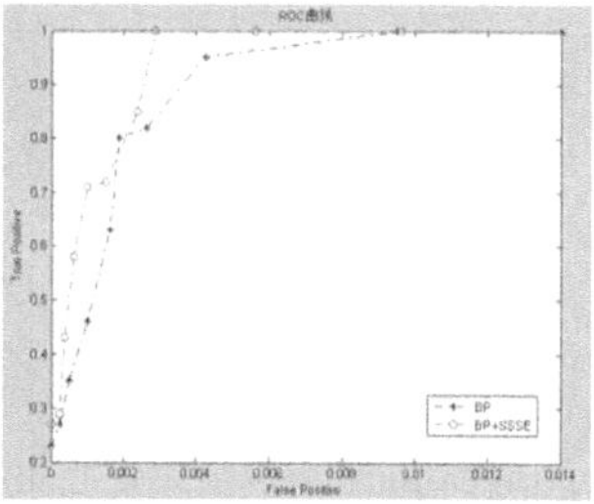

Fig. 1. ROC curve

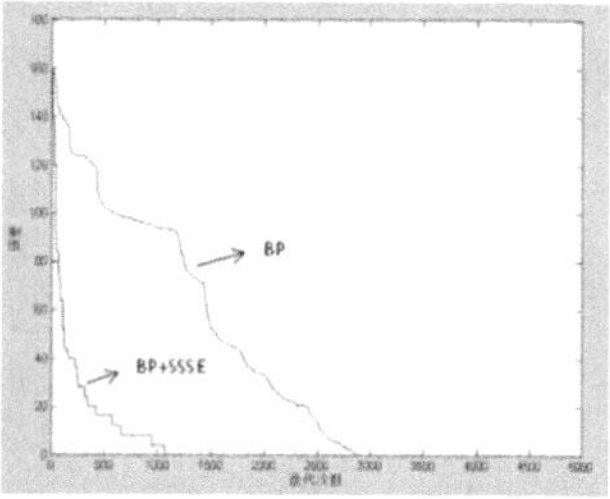

Fig. 2. Convergence curves

After 100 repeated experiments, the statistical output classification average recognition accuracy rate is based on the recognition result. The experimental results show the correctness of face recognition obtained by using the standard BP neural network [10] and the BP neural network optimized by the simplex evolution algorithm respectively. Among them, the recognition accuracy is the average value obtained from 100 experiments (Tables 1, 2 and 3).

After data comparison, the following conclusions are drawn: Compared with the standard BP algorithm, the BP neural network optimized by the simplex evolution algorithm proposed in this paper has a significant improvement in the recognition rate; at the same time, the convergence speed is faster in the experimental process.

Table 1. .

Feature number	BP	BP + SSSE
5	78%	83%
10	89%	90%
15	91%	93%
20	92%	95%
25	93%	96%
30	95%	98%

Table 2. .

Number of training	BP	BP + SSSE
5	84%	88%
6	88%	91%
7	92%	95%
8	95%	97%
9	96%	98%

Table 3. .

Threshold	BP	BP + SSSE
0.0	112/8000	77/8000
0.1	76/8000	45/8000
0.2	34/8000	23/8000
0.3	21/8000	19/8000
0.4	15/8000	12/8000
0.5	13/8000	8/8000
0.6	8/8000	5/8000
0.7	4/8000	3/8000
0.8	2/8000	2/8000

4 Conclusion

This paper presents a BP neural network learning algorithm based on simplex evolution for face recognition. The algorithm reduces the control parameters of the algorithm through a full random search, and uses the multi-character state of the population to maintain the diversity of the group and avoids the algorithm falling into a local extreme point. Experimental results show that the BP neural network trained by the simplex evolutionary learning algorithm effectively converges to the optimal value of the error function, and has a good convergence speed, which effectively improves the performance of the traditional BP algorithm.

In the experiment of BP neural network training and recognition for face recognition, the comparison with the traditional BP algorithm shows that the learning algorithm can make the neural network converge to the error curve quickly and consistently without relying on parameter selection. The minimum point improves face recognition accuracy and learning efficiency, and verifies the feasibility and effectiveness of the algorithm in practical applications.

Acknowledgements. This work is supported by National Natural Science Foundation of China (No. 41364002), The authors are graceful for the support of the grants.

References

1. Yuan, Z.: Artificial Neural Network and Its Application. Guangxi Science and Technology Press, Zhengzhou (1999)
2. Hu, J., Zeng, X.: A fast learning algorithm of global convergence for BP-neural network. J. Syst. Sci. Math. Sci. **30**(05), 604–610 (2010)
3. Li, S., Luo, Y., Zhang, M.: Prediction method for chaotic time series of optimized BP neural network based on genetic algorithm. Comput. Eng. Appl. **47**(29), 52–55 (2011)
4. Chen, J., Wu, Z., Zhao, D., Zhou, J., Li, S., Shang, J.: Back propagation neural network classification of sediment seabed acoustic sonar images based on particle swarm optimization algorithms. Acta Oceanol. Sin. **39**(09), 51–57 (2017)
5. Zhang, Q.-Y., Liang, L.: Facial recognition based on Gabor filter and ant colony optimizing BP neural network. J. Hunan Univ. Sci. Technol. **27**(04), 69–72 (2012)
6. Wang, L., Peng, L., Xia, D., Zeng, Y.: BP neural network incorporating self-adaptive differential evolution algorithm for time series forecasting. Comput. Eng. Sci. **37**(12), 2270–2275 (2015)
7. Wang, G., Guo, L., Duan, H., Liu, L., Wang, H.: Target threat assessment using glowworm swarm optimization and BP neural network. J. Jilin Univ. **43**(04), 1064–1069 (2013)
8. Leema, N., Khanna Nehemiah, H., Kannan, A.: Neural network classifier optimization using differential evolution with global information and back propagation algorithm for clinical datasets. Appl. Soft Comput. (2016)
9. Quan, H., Shi, X.: A surface-simplex swarm evolution algorithm. Wuhan Univ. J. Nat. Sci. **22**(01), 38–50 (2017)
10. Li, K., Li, K., Zhang, W.: PCA face recognition algorithm based on improved BP neural network. Comput. Appl. Softw. **31**(01), 158–161 (2014)

Malware Detection via Graph Based Access Behavioral Description and Semi-supervised Learning

Zhihui Zhao[1(✉)], Minle Wang[1], Weixuan Mao[2], and Subing Liu[1]

[1] Xi'an Research Institute of Hi-Tech, Xi'an, Shaanxi, China
zzhjcb706@163.com
[2] Xi'an Jiaotong University, Xi'an, Shaanxi, China

Abstract. Malicious code is the root cause of many security incidents and still the major threats for the Internet. Understandings on the access behaviors of programs provide ways of malware detection. In this paper, we propose a graph based representations for access behaviors of programs. With similarity metrics on the access behavior graph, we employ a semi-supervised learning algorithm to infer the intent of the programs. The promising result, 98.8% true positive rate at 0.5% false positive rate, shows the ability of our technique on malware detection and the benefits of the access behavior graph. Ability of our technique on malware detection and the benefits of the access behavior graph.

Keywords: Malware detection · Semi-supervised learning

1 Introduction

Malicious code is the root cause of many security incidents and still the major threats for the Internet. Detecting and defending against malware, which stands for malicious software, provide effective ways of protecting computers from attacks.

Nowadays, the techniques of malware detection can be summarized as two main steps. (1) unveiling the features of the program; (2) analyzing whether it is malicious or not. The techniques in the first step are usually divided as two categories, i.e., dynamic analysis and static analysis, according to whether the program will be executed or not. The dynamic analysis is conducted to execute the program and record its behaviors with help of sand-box technique. The static analysis focuses on the binary file of the program, reverses the features, e.g., opcode, control-flow, etc., of the program. Both of these two techniques aims to discover the essential features indicating whether the program is malicious or not with help of the second step [3, 5, 8]. The second step has been more and more rely on the statistics and the machine learning technique [4, 6, 9]. It discriminates the malicious programs from the benign programs based on modeling the features of programs.

With these guidances, we propose a graph based description for the access behaviors of programs. With similarity metrics on graphs, we apply semi-supervised learnings to determine the label of the undetermined programs, i.e., malicious or benign. More specifically: the access behaviors of programs lead to the data flow between system

K. Deng et al. (Eds.): ICMIR 2018, AISC 856, pp. 1247–1253, 2019.
https://doi.org/10.1007/978-3-030-00214-5_153

objects, e.g., processes, files and registries, which cause the exchange of information among system objects to guarantee the operation of the system. Since the data can flow into or flow out of the system objects, we model the system objects as a directed graph whose edges represent the direction of data flows between system objects. In this access behavior graph, we can measure the similarity of pairs of nodes, e.g., the program, based on the structural information. These similarities offer us evidences to infer the label of the undetermined programs given other known programs, since the more similarity of two programs in behaviors, the more similarity of their labels.

Moreover, we conduct the inference with help of a semi-supervised learning technique [12]. Semi-supervised learning problem is: given a data set consisting of both labeled and unlabeled points, predict the labels of the unlabeled points with modeling the relationships between data points. The assumption of consistency is the key to the semi-supervised learning problem: (1) nearby points are likely to have the same label; (2) points on the same structure (a cluster, a manifold or a graph) are likely to have the same label [11]. We construct the affinity matrix for programs with the behavioral similarities between programs which are obtained from the directed access behavior graph. If we treat this affinity graph as a system of programs, we can estimate the stablest state (label) for the programs with some whether they are malicious or not, requires much human labor, whereas, semi-supervised learning for malware detection is practicable in face of the rapid growth of both benign and malicious programs.

To summarize, the contributions of this paper are listed as follows:

- We propose a access behavior graph to describe the features of programs. This graph help us unify the access behaviors of programs into the same framework. And, it offers insights the similarities between the programs in terms of a graph framework.
- We apply a semi-supervised learning technique based on the similarity metric on the access behavior graph to discriminate malicious programs from the benign programs. the promising result, 98.8% true positive rate at 0.5% false positive rate, indicate the feasibility of our model in malware detection.

2 Access Behavior Graph

Access behaviors are characterized by programs' externally observable events/operations, e.g., system call events, I/O requests, etc. In this paper, the access behaviors is referred to system calls related to the access of system resources. In this paper, the system resource objects are refer to either file objects or registry objects. Meanwhile, these behaviors result in data flow between system objects.

To analyze the access behaviors of programs, we first provide a unified view of them in terms of the directions of the data/information flow: (1) when a subject a read an object b, we say the data/information of b flows to a; (2) when a subject b writes and object a, we say the data/information of a flows to b. All observed access behaviors in terms of system calls, e.g., *CreateFile*, *QueryInformation*, *ReadFile*, *RegQueryValue*, etc., can be treated as either class. Based on these observations, we can encode all externally observable access behaviors into the directions of data/information flow.

Then, we have the definition of our access behavior graph describing the access behaviors of programs.

Definition 0.1 (Access Behavior Graph). *A directed graph G(V, E):*

- *V* is a set of process objects distinguished by program name.
- *U* is a set of system resource objects distinguished by path.
- $E = \{e(n_1, n_2)|n_1 \in V, n_2 \in U, \text{or}, n_1 \in U, n_2 \in V\}$, $e(n_1, n_2)$ denotes an access behavior in terms of information flow $n_1 \Rightarrow n_2$, the information flows from n_1 to n_2.

3 Link Based Similarity Metrics

The above access behavior graph not only unifies the representation of access behaviors for programs, but also offers ways of evaluating the similarity between programs. Under the graph structure, we can measure the similarity between nodes according to their links, their neighbors, the links of their neighbors, the neighbors of their neighbors and so forth [7]. Obviously, the simplest and most efficient metric is the "one-hop" only information on the graph, i.e., their neighbors. Here, we list three metrics that we study in this paper as follows. In following metrics, $\Gamma(x)$ indicates the set of neighbors for x in the graph, and |S| means the number of elements in the set *S*.

Common Neighbors. The most direct metric of the similarity is to define $s_c(x, y) = |\Gamma(x) \cap \Gamma(y)|$. The similarity in this metric is the number of neighbors that *x* and *y* have in common.

Adamic/Adar. The metric of Admic/Adar is proposed in [1] to refine the simply counting of common neighbors by weighting rarer neighbors more heavily and more frequent neighbors less heavily. The weight is defined as the logarithm of inverse frequency, then the score of similarity in this metric is

$$s_a(x, y) = \sum_{z \in \{\Gamma(x) \cap \Gamma(y)\}} \frac{1}{\log|\Gamma(z)|} \tag{1}$$

Jaccard Index. The Jaccard coefficient is a commonly used similarity metric in information retrieval. This metric measures the probability that a neighbor of either *x* or *y* is a common neighbor of both *x* and *y*, i.e.,

$$s_j(x, y) = \frac{|\Gamma(x) \cup \Gamma(y)|}{|\Gamma(x) \cap \Gamma(y)|} \tag{2}$$

It is worth mentioning that there are two kinds of neighbors in our directed access behavior graph, in-neighbors and out-neighbors. Hence, we extend the metrics with respect to different neighbors. Generally, we can write the extended metrics as

$$s(x, y) = \alpha s^{(in)}(x, y) + (1 + \alpha)s^{(out)}, \quad 0 \leq \alpha \leq 1 \tag{3}$$

where α balance the weights between the metrics on in-neighbors and those on out-neighbors. In this paper, we set $\alpha = 0.5$ which treat equal weights for in- and out-neighbors. In our access behavior graph, the in-neighbors correspond to the read objects, while the out-neighbors correspond to the written objects.

4 Semi-supervised Learning Based Malware Detection

With the similarity metrics shown in previous section, we can construct the affinity matrix describe the similarities between programs in terms of a matrix. The affinity matrix M is a square matrix with each row i representing a program i and each entry e_{ij} representing the similarity between program i and j. In our case, the affinity matrix is nonnegative symmetric. This matrix specifies that manifold of programs under the similarity metric.

Consider the label of each program f_i as its state in a system of the manifold. Then, the energy function of the system under different states of programs can be defined as

$$E(\mathbf{f}) = \frac{1}{2}\sum_{i,j} e_{ij}(f(i) - f(j))^2 \tag{4}$$

The stablest states of all programs can be derived by minimizing the energy. In matrix notation, we can represent Eq. as $\frac{1}{2}\mathbf{f}^T\Delta\mathbf{f}$, where Δ is the normalized graph Laplacian defined as

$$\Delta = I - D^{-1/2}MD^{-1/2}, \quad D_{ij} = \begin{cases} \sum_k e_{ik} & \text{if } i = j, \\ 0 & \text{otherwise} \end{cases} \tag{5}$$

If we separate all programs into labeled set L and unlabeled set U, i.e., $\mathbf{f}^T = [\mathbf{f}_L^T, \mathbf{f}_U^T]$. Then Eq. turns to be $E(\mathbf{f}) = \frac{1}{2}[f_L^T\Delta_{LL}f_L^T + f_U^T\Delta_{UL}f_L^T + f_L^T\Delta_{LU}f_U^T + f_U^T\Delta_{UU}f_U^T]$. We will get $f_U^* = -(\Delta_{UU})^{-1}\Delta_{UL}f_L$ when $\frac{\partial L}{\partial f_U} = 0$, where f_U^* is the predicted states/labels of the undetermined programs. With these derived states/labels of programs in set U, we can determine whether the program is malicious or not according to the label.

5 Experiments

Data Collection. We collect process behavior data on one user's machine for 14 days, without interfering with their daily use. The machine was running in Microsoft Windows XP SP3. Finally we get 9,913 access traces as benign instances in our malware detection. For malicious programs, we focus on 32-bit portable executable (PE) samples obtained from VxHeaven [10], which is a well known malware repository. We collected 5,402 malware samples consisting of Trojan, Worms and Virus. We run the sample on Windows XP sp3 in the virtual machine. After each run, we revert the virtual machine to a clean snapshot so that malware samples will not interfere with each other.

Evaluation. We evaluate the performances with the ROC curve and area under the ROC curve (AUC). ROC curves allow us to visualize the performance of the model on malware detection by exhibiting false positive rates and true positive rates, while AUCs summarize the performance. The true positive rate is the rate that we truly detect the malware, while the false positive rate is the rate that we mistreat benign programs as malware.

Detection Results. We parse all access behaviors, in terms of system calls, of both benign and malicious programs, and construct the access behavior graph. Then, we calculate the similarities between programs according to the similarity metrics, which provide the affinity graph for the malware detection. To evaluate the performance, we randomly select a portion of instances/programs as known which is the training set, while the others are set to be unknown which is the testing set. Then, we conduct the semi-supervised learning algorithm to label the unknown in the testing set. The performance will be derived when we compare the labels with the ground truth.

Comparisons of Detection Results. Our technique focus on either file objects or registry objects, or both. We explore results on all these three conditions. Meanwhile, we propose three similarity metrics in previous section, i.e., Common neighbor, Adamic/Adar, Jaccard. Here, we compare the detection results of these three metrics. Figure 1 illustrates the results under different experimental settings when the fraction of training set is 80%. As we can see, the better result is obtained when we use Jaccard metric on both file and registry objects if we are not tolerant to the false positives. It is 98.8% true positive rate at 0.5% false positive rate. In average, examining file objects provides better detection results than registry objects. Combining both of these two objects does not always achieve better performance. And, the Jaccard metric offers better performances than the other two metrics, since Jaccard metric involve more information than the other two, i.e., the union of neighbors, which is more suitable in our application.

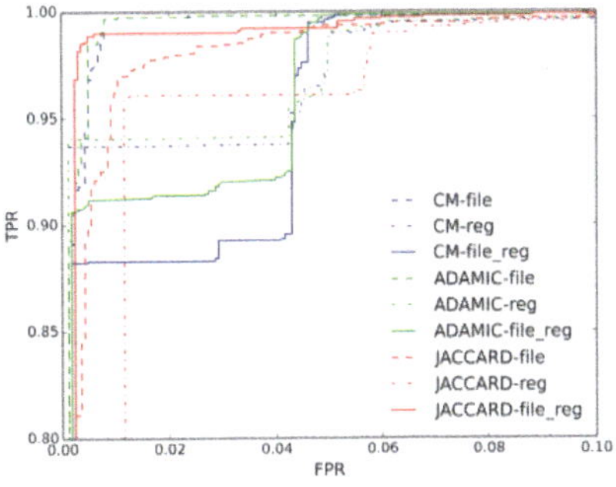

Fig. 1. Detection results with 80% training instances

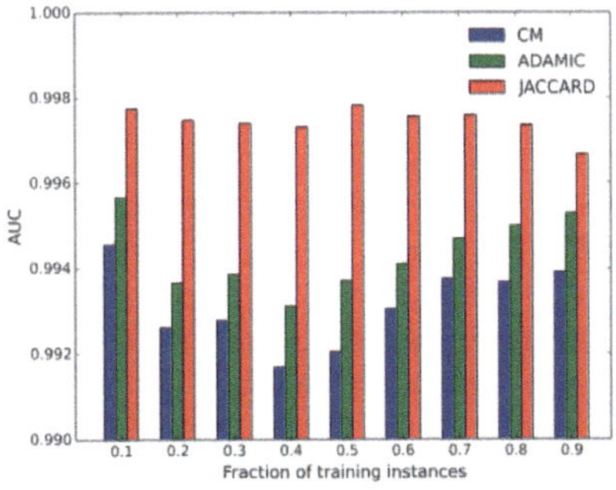

Fig. 2. Detection results on both file and registry objects with different portions of training instances

Different Sizes of Training Sets. We also explore the differences result from the different portions of training sets. Figure 2 illustrates the AUCs under different sizes of training sets examining both file and registry objects. From this figure, the best performances are achieved under Jaccard metrics, and the Adamic/Adar metric performs better than the common neighbor metric. Under Jaccard metric, similar results are obtained under different sizes of training sets, which shows the robust and practicality of this metric. Noticeable, there is a decline when we use 90% training instances. This decline results from the over fitting with such large portion of training instances.

6 Conclusion and Future Work

In this paper, we propose a way of understanding the access behaviors of programs and a semi-supervised learning based malware detection. We propose an access behavior graph to describe the access behaviors of programs from a system-centric perspective. The access behavior graph is motivated by the information flows between objects in access behaviors. Meanwhile, encoded into the form of graph, network science theory provides ways of measuring the similarity of programs. Applying a semi-supervised learning technique with the similarity metrics, a technique for malware detection is derived in our paper. The experimental results on the real data exhibit the feasibility of our technique. Graph based models on access behaviors offer a way of program analysis from a system wide perspective. With the help of network science theory, more work is worth to do under this framework, e.g., security assessment and vulnerability analysis.

References

1. Adamic, L.A., Adar, E.: Friends and neighbors on the web. Soc. Netw. **25**(3), 211–230 (2003)
2. Altman, N.S.: An introduction to kernel and nearest-neighbor nonparametric regression. Am. Stat. **46**(3), 175–185 (1992)
3. Apap, F., Honig, A., Hershkop, S., Eskin, E., Stolfo, S.: Detecting malicious software by monitoring anomalous windows registry accesses. In: Proceedings of the 5th International Conference on Recent advances in Intrusion Detection, RAID 2002, pp. 36–53. Springer, Berlin (2002)
4. Fredrikson, M., Jha, S., Christodorescu, M., Sailer, R., Yan, X.: Synthesizing near-optimal malware specifications from suspicious behaviors. In: IEEE Symposium on Security and Privacy (S&P), pp .45–60 (2010)
5. Heller, K.A., Svore, K.M., Keromytis, A.D., Stolfo, S.J.: One class support vector machines for detecting anomalous windows registry accesses. In: Proceedings of the Workshop on Data Mining for Computer Security (2003)
6. Kruegel, C., Mutz, D., Valeur, F., Vigna, G.: On the detection of anomalous system call arguments. In: Computer Security–ESORICS 2003, pp. 326–343 (2003)
7. Liben-Nowell, D., Kleinberg, J.: The link-prediction problem for social networks. J. Am. Soc. Inf. Sci. Technol. **58**(7), 1019–1031 (2007)

8. Mao, W., Cai, Z., Guan, X., Towsley, D.: Centrality metrics of importance in access behaviors and malware detections. In: Proceedings of the 30th Annual Computer Security Applications Conference. ACM (2014)
9. Martignoni, L., Stinson, E., Fredrikson, M., Jha, S., Mitchell, J.: A layered architecture for detecting malicious behaviors. In: Recent Advances in Intrusion Detection (RAID), vol. 5230, pp. 78–97. Springer, Berlin (2008)
10. VXHeaven (2010). http://vx.netlux.org/
11. Zhou, D., Bousquet, O., Lal, T.N., Weston, J., Schölkopf, B.: Learning with local and global consistency. Adv. Neural. Inf. Process. Syst. **16**(16), 321–328 (2004)
12. Zhu, X., Ghahramani, Z., Lafferty, J., et al.: Semi-supervised learning using gaussian fields and harmonic functions. In: ICML, vol. 3, pp. 912–919 (2003)

An Adaptive Genetic Algorithm with Dual Adjustment Strategies and Its Performance Analysis

Yuqiang Zhao, Qian Qian(✉), and Feng Wang

Yunnan Key Laboratory of Computer Technology Applications,
Kunming University of Science and Technology, Kunming 650500, China
qianqian_yn@126.com

Abstract. Genetic algorithm is an effective global optimization method and it has been widely used in many fields. In order to solve some problems of standard genetic algorithms, such as slow convergence and easy to fall into local extremum, an improved adaptive genetic algorithm is presented. Specifically, in crossover and mutation operations, the algorithm considers both the influence of the generation number of the evolution and the effect of the different adaptive individuals in the contemporary population. In addition, a new parameter (i.e., the generation number by which the same optimal solution is maintained) is introduced to adjust the crossover and mutation probabilities, so as to improve the exploration ability of the algorithm. With the above two adjustment strategies, the performance of the adaptive algorithm can be greatly improved. The experimental results of four functions showed that the proposed algorithm has a good improvement in aspects of convergent ability and the ability to jump out of local extremum.

Keywords: Genetic algorithms · Adaptation · Dual adjustment
Function optimization · Convergence

1 Introduction

Genetic algorithm (GA) [1] is a kind of global random searching algorithms that simulates the evolution of biological population. In order to find the optimal solution, GA abstracts the solution set of the optimization problem into a population, and then evolves under the guidance of genetic manipulations, such as selection, crossover and mutation operators. Genetic algorithms are widely used in industrial production, artificial intelligence, systems engineering and other fields [2], because the optimization problems under GA are not required to be continuous and differentiable and the algorithms have strong global optimization ability and robustness.

Genetic operators play an important role in GA. The selection operation is able to excavate the active individuals of the population, and the crossover and mutation operations can obtain new patterns through continuous exploration of the solution space. During the crossover and mutation operations, the crossover and mutation probabilities are the key to do the exploration and to remain the balance [3, 4]. Traditional genetic algorithms (Standard GA, SGA) use fixed crossover and mutation

K. Deng et al. (Eds.): ICMIR 2018, AISC 856, pp. 1254–1261, 2019.
https://doi.org/10.1007/978-3-030-00214-5_154

probabilities, however, it is difficult to choose the adequate probability values. If the probability values are large, the steadiness of the population will be destroyed. As a result, the algorithm can become a random searching without guidance; conversely, if the probability values are small, the algorithm will be easy to fall into the local extreme values [5]. Many researchers have proposed their own adaptive mechanisms to make the algorithm automatically adjust the required parameters in crossover and mutation operations [6, 7]. But there is still room for improvement. Therefore, this paper proposes a new adaptive genetic algorithm with dual adjustment strategies to improve the algorithm's searching ability.

2 Adaptive Genetic Algorithm for Dual Adjustment Strategies

2.1 Introduction of Existing Adaptive Algorithms

A typical adaptive genetic algorithm (SAGA) for adjusting crossover and mutation probabilities is proposed by Srinivas [6], and the adjustment rules are as follows:

$$Pc = \begin{cases} k_1 \frac{F_{\max} - F}{F_{\max} - F_{\text{avg}}}, & F > F_{\text{avg}} \\ \mathrm{k}_3, & \mathrm{F} < \mathrm{F}_{\text{avg}} \end{cases} \tag{1}$$

$$Pm = \begin{cases} k_2 \frac{F_{\max} - F'}{F_{\max} - F_{avg}}, & F' > F_{avg} \\ k_4, & F' < F_{avg} \end{cases} \tag{2}$$

In formula (1) and (2): $F_{\max}$, F_{avg}, $F_{\min}$ are the maximum fitness value, the average fitness value, and the minimum fitness value of the current population, respectively.

According to the formulas, the closer the average fitness is to the maximum fitness, the greater the Pc and Pm *will be*, so as to enrich the population diversity; and the individuals that have large fitness values will lead to small Pc and Pm, by which the good individuals can be preserved. However, in this algorithm, super individuals with Pc and Pm equal to 0 in the early stage could appear, leading to the "premature" phenomenon and making the algorithm not be able to evolve anymore. Therefore, a new adaptive Genetic Algorithm (WAGA) is proposed by Ren [7], the algorithm avoids the situation in which Pc and Pm equal to 0 by adding a numerical value, so that the algorithm is no longer easy to fall into local extremum. But the method is still a kind of linear adjustment, that means Pc and Pm will change quickly when the individual fitness is greater than the average fitness values. Therefore, Pc and Pm will have a big difference in the vicinity of the average fitness value, which is unfavorable to the stability of the algorithm. In document [8], the Nonlinear Adaptive Genetic Algorithm (CAGA) is constructed by using the characteristics of cosine function. The most important improvement of the study is that Pc and Pm are non-linear, so as to avoid the premature situation. However, as mentioned in document [9], the CAGA degrades to linear transformation when the average fitness and maximum fitness are close, so that it is hard to converge to the global optimal solution.

To sum up, the existing algorithms have improved the adaptive genetic algorithm, but there is still room for improvement. In the present paper, an adaptive genetic algorithm with double adjustment strategies (Double Adjustment Adaptive Genetic Algorithm, called DAAGA) is proposed and presented. The algorithm introduces a new parameter (i.e., the number of the generations by which the same optimal solution does not change, called G_f) to adjust Pc and Pm. In combination with the adjustment strategy of the average fitness structure, the double adaptive adjustment of Pm and Pc is implemented to ensure the good excavation ability and exploration ability of GA.

2.2 Adaptive Crossover and Mutation Operations

The important function of crossover operator is to find new solutions and to compose new individuals. From the perspective of the whole evolution, the algorithm should gradually reduce the exploration frequency of the new solution and ensure the convergence of the algorithm. From the perspective of one generation, because better solutions may also be distributed around "inferior" individuals, the same crossover probability should be used in among the whole population to ensure that the new individual will not miss the "good" solution, this viewpoint is also in line with the law of natural evolution.

Mutation operators play an important role in finding new genes and jumping out of local extremum. It also can restore the outstanding genes that lost in genetic manipulation. Because the initial population of the algorithm is limited to a part of the solution space, the algorithm should apply a large mutation probability to enlarge the range of the solution in the early stage, and the mutation probability should be gradually reduced to improve the convergence ability of the algorithm. In addition, from the internal view of a certain generation of population, the individuals with high adaptability should be retained by reducing the mutation probability, and the mutation probability should be increased for the individuals with low adaptability.

The above strategies consider the different effects of crossover and mutation operations on genetic algorithms. Furthermore, we also introduce a control parameter G_f (the number of the generations by which the same optimal solution does not change), which is the important symbol of whether the evolution falls into a stagnation state. The greater the value of the G_f, the more likely the algorithm is stagnated. Pc and Pm are important indexes to affect the performance, and they should increase with the increment of the G_f, so as to improve the capability of the algorithm to jump out of the local optimal solution during the stagnation state. Therefore, the following nonlinear adjustment formulas are designed:

$$Pc = Pc_2 + \frac{(Pc_1 - Pc_2)}{1 + \exp\left[A\left(2\frac{g}{G}\right) - 1\right]} + \beta\frac{G_f - G_t}{G_f} \tag{3}$$

$$Pm = \begin{cases} \alpha\frac{-\lg\sqrt[G]{g}}{1 + \exp[A(2\frac{f'}{f_{\max}} - 1)]} + \beta\frac{G_f - G_t}{G_f}, & F' > F_{avg} \\ Pm_1, & F' < F_{avg} \end{cases} \tag{4}$$

$$\beta = \begin{cases} 0.08, & G_f > G_t \\ 0, & G_f < G_t \end{cases} \tag{5}$$

In the formulas: g is the number of the current generations, G is the total number of generations, α, A and β are control parameters, and G_t is the generation threshold.

From formula (3) and formula (4), the transformation of Pc and Pm considers both the influence of the operators to the algorithm and the exploratory ability of the operators. Under the dual adjustments, the searching ability of the algorithm and the capability of jumping out of local extremum are improved.

The proposed algorithm also adjusts the processing sequence of the operators. As shown in Fig. 1, when G_f is greater than the given threshold G_t, it is believed that the algorithm is in the later stage or falls into the local optimal solution, so mutation operations will be performed first to find new genes; when G_f is smaller than the given threshold G_t, the crossover operation will be performed first.

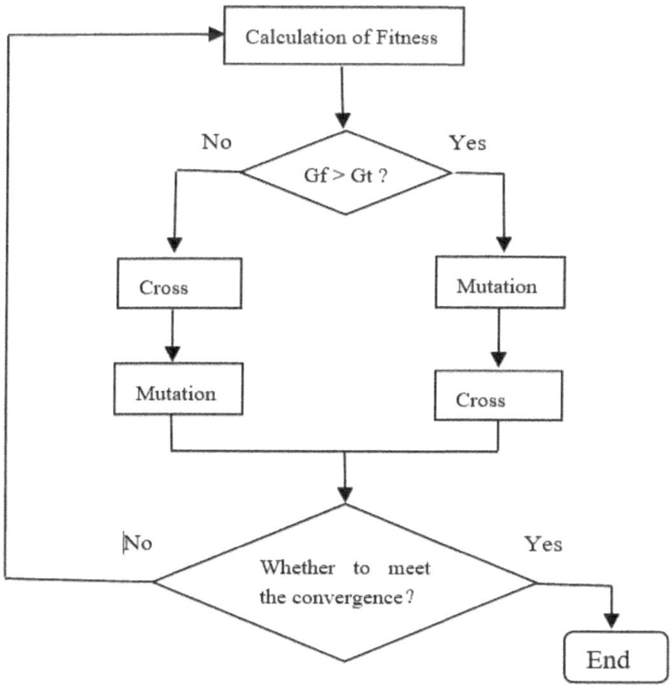

Fig. 1. Flow chart of improved algorithm

2.3 Optimal Individual Preservation Strategy

In the complex problem, the population's good genes are easy to lost prematurely, and the mutation operation cannot always effectively recover these good genes, leading to a slow convergence speed [10]. The article [11] proves that the optimal individual preservation strategy can lead to global convergence. Therefore, the optimal individual preservation strategy is also used in the present algorithm [12, 13]: if the optimal solution of the next-generation is worse than the optimal individual of the current population, the worst adaptive individuals in the next population will be replaced by the best individual of the present population.

3 Simulation Test and Result Analysis

3.1 Test Functions

In this paper, four complex two-dimensional functions are selected to test SAGA, WAGA, CAGA and DAAGA:

$$\begin{aligned} &\min:\ F_1(x,y) = 100(x^2+y)^2 + (1-x)^2, \\ &x, y \in (-2.048, 2.048) \end{aligned} \tag{6}$$

$$\begin{aligned} &\min:\ F_2(x,y) = \left(4 - 2.1x^2 + \frac{x^4}{3}\right)x^2 + xy + (-4+4y^2)y^2, \\ &x, y \in (-10, 10) \end{aligned} \tag{7}$$

$$\begin{aligned} &\max:\ F_3(x,y) = \left[\frac{3.0}{0.05 + (x^2+y^2)}\right]^2 + (x^2+y^2), \\ &x, y \in (-5.12, 5.12) \end{aligned} \tag{8}$$

$$\begin{aligned} &\max:\ F_4(x,y) = -[x^2 + 2y^2 - 0.4\cos(3\pi x) - 0.6\cos(4\pi y)], \\ &x, y \in (-10, 10) \end{aligned} \tag{9}$$

3.2 Result Analysis

SAGA, WAGA, CAGA and OAAGA are all binary coded, adaptive proportional selection, two-point crossover and multiple-gene mutation. Genetic algorithm parameters are as follows: chromosome length (40), population size (80). In addition, in SAGA, k_1 and k_2 are both 1.0, k_2 and k_4 are both 0.5; in WAGA, Pc_1 is 0.9, Pc_2 is 0.6, Pm_1 is 0.1, Pm_2 is 0.01; in CAGA, $Pc_{\max}$ is 0.9, $Pc_{\min}$ is 0.6, $Pm_{\max}$ is 0.1, $Pm_{\min}$ is 0.02; in DAAGA, A is 10, α is 10, G_t is 20, Pc_1 is 0.9, Pc_2 is 0.6, Pm_1 is 0.1.

Figure 2 shows a performance comparison of the maximum function values for SAGA, WAGA, CAGA, and DAAGA. Although the WAGA algorithm converges to the global optimal solution, the curve is easy to fall into stagnation, so that the convergent speed is slow. According to Fig. 2(c) and (d), the curve of SAGA algorithm is easy to show a flat change trend when approaching the global optimal solution, which indicates that the algorithm is often immersed in local optimal solution, so that it cannot converge to the global optimal solution. The CAGA algorithm lacks the searching ability and the ability to jump out of the local best, because the evolutionary curve of

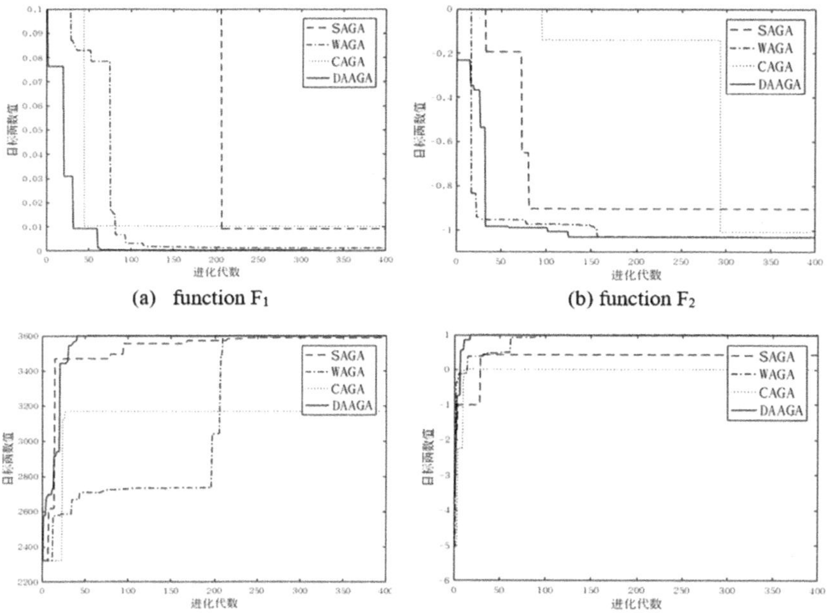

(a) function F_1 (b) function F_2

(b) function F_3 (d) function F_4

Fig. 2. Optimization results

Table 1. Comparison of the results of 4 test function algorithms

Function	Algorithm	Convergent times	Average convergent algebra	Average convergence value
F_1	SAGA	11	370	0.0089
	WAGA	21	250	0.0018
	CAGA	10	389	0.0266
	DAAGA	27	101	0.0005
F_2	SAGA	15	212	−0.9021
	WAGA	28	78	−1.0914
	CAGA	19	209	−0.8730
	DAAGA	30	41	−1.0316
F_3	SAGA	20	129	3441
	WAGA	22	118	3578
	CAGA	17	198	3398
	DAAGA	30	20	3600
F_4	SAGA	20	237	0.7990
	WAGA	23	169	0.8537
	CAGA	17	252	0.6011
	DAAGA	29	101	0.9972

the algorithm maintains a very small frequency of changes. The DAAGA algorithm has a short stagnation period during the overall evolution, and can converge to the global optimal solution with a small number of generations.

Table 1 shows the optimized test results of 4 test functions that have been run 30 times each. Compared with SAGA, WAGA and CAGA, the average convergence value of the function calculated by the DAAGA algorithm is closest to the global optimal value of the function, and the number of convergence to the global optimal solution is closest to the experimental number, so DAAGA's stability is better than the other three algorithms.

4 Conclusion

A new and original adaptive genetic algorithm with dual adjustments is proposed. Specifically, the crossover probability decreases nonlinearly along with the increment of evolution iterations, and it is considered that the individuals within the same generation should have the same possibility to join crossover operations. As for the mutation operation, the mutation probability decreases with evolution, and the influence of the fitness variation of individuals in the same generation is also considered. Furthermore, in order to enhance the exploration ability of crossover and mutation operators, a parameter (G_f) is introduced to adjust the crossover mutation probabilities. The algorithm also uses G_f to control the processing sequence of crossover and mutation operations. The experiment results show that the dual adjustment adaptive genetic algorithm presented in this paper can effectively improve the ability of searching and convergence, and it also has a good performance in the ability to jump out of the local optimal solution.

Acknowledgements. This research is supported by the National Science Foundation of China (No. 31300938 and No. 61462053), and the Natural Science Foundation of Yunnan Province of China (No. 2016FB107).

References

1. Holland, J.H.: Adaptation in Natural and Artificial Systems, vol. 6, no. 2, pp. 126–137. Michigan University Press, Ann Arbor (1975)
2. Ma, Y., Yun, W.: Advances in genetic algorithms. Comput. Appl. Res. **29**(4), 1201–1206 (2012)
3. Jiang, R., Luo, Y., Hu, D.: An adaptive genetic algorithm based on population entropy estimation. J. Tsinghua Univ. (Nat. Sci. Ed.) **42**(3), 358–361 (2002)
4. Huang, J., Fu, Z.: Optimization and simulation of function based on adaptive genetic algorithm. Comput. Simul. **28**(5), 237–240 (2011)
5. Yang, C., Qian, Q., Wang, F., et al.: Application of improved adaptive genetic algorithm in function optimization. Comput. Appl. Res. **35**(4), 1042–1045 (2018)
6. Srinivas, M., Patnaik, L.M.: Adaptive probabilities of crossover and mutation in genetic algorithm. IEEE Trans. Syst. Man Cybernet. **24**(4), 656–667 (1994)

7. Rin, Z., San, Z.: Improvement of adaptive genetic algorithm and its application in system identification. J. Syst. Simul. **18**(1), 41–66 (2006)
8. Shi, S., Li, Q., Wang, X.: Optimal design of brushless DC motor based on Adaptive genetic algorithm. J. Xian Jiaotong Univ. **36**(12), 1215–1218 (2002)
9. Kuang, H., Jin, J., Su, Y.: Improvement of crossover mutation operator by adaptive genetic algorithm. Comput. Eng. Appl. **42**(12), 93–99 (2006)
10. Rudolph, G.: Convergence properties of canonical genetic algorithms. IEEE Trans. Neural Netw. **5**(1), 96–101 (1994)
11. Yun, W., Xi, Y.: Global convergence and computational efficiency of genetic algorithms. Control Theory Appl. **13**(4), 455–460 (1996)
12. Lin, M., Li, M., Zhou, L.: An adaptive genetic algorithm based on evolution. Comput. Eng. **36**(20), 173–175 (2010)
13. Xu, C., Lin, W., Hong, X.: An adaptive genetic algorithm based on weighted Hamming distance. J. South China Normal Univ. (Nat. Sci. Ed.) **47**(6), 121–127 (2015)

Pulmonary Vascular Segmentation Algorithm Based on Fractional Differential Enhancement

Yuqing Xiao[1], Wenjun Tan[1,2](✉), Qing Zhou[1], and Yu Ji[1]

[1] College of Computer Science and Engineering, Northeastern University, Shenyang, China
tanwenjun@cse.neu.edu.cn

[2] Key Laboratory of Medical Image Computing of Ministry of Education, Northeastern University, Shenyang, China

Abstract. With one of the highest incidence rates, Lung cancer accounts for 26.8% of cancer deaths. Due to the pulmonary segment having a separate pulmonary artery and pulmonary vein, the division of the pulmonary artery and pulmonary vein is critical for path planning in surgery of segmental resection of the lung. As a result of this previous research, CT images provided are used as data sources. Combined with the anatomical structure characteristics of pulmonary vessels, the following segmentation methods of the pulmonary artery and pulmonary vein are studied in this thesis. (a) image segmentation and center line extraction is introduced to lung parenchyma segmentation, based on the 3D growth region method. (b) Morphology used to fill the holes in the segmented pulmonary parenchyma, operations made between filled images and the original image to obtain the lungs filled with pulmonary vessels, fractional differential operator designed to enhance the gray level of the small blood vessels, and complete pulmonary blood vessel segmented by local optimal threshold. (c) Algorithm for extracting the centerline of lung blood vessels based on two distance field is researched and arteriovenous conglutination part of the pulmonary centerline detected and removed automatically to obtain the vascular subtree. Finally, Using topological anatomy of the pulmonary artery and vein, the vascular tree classification algorithm based on the subtree leaf node matching is studied along with the pulmonary blood vessels which are grown by region growth method through the center line. The pulmonary artery and vein are then extracted from the chest CT image. The experimental results show that this proposed pulmonary artery and vein segmentation method has high accuracy and short operation time, and has good clinical application value. This newly developed algorithm and implementation will have important applications for treatment planning, dose calculations and treatment validation during cancer radiation treatment.

Keywords: CT image · Pulmonary vascular segmentation
Fractional differentiation · Centerline · Subtree leaf node matching

K. Deng et al. (Eds.): ICMIR 2018, AISC 856, pp. 1262–1274, 2019.
https://doi.org/10.1007/978-3-030-00214-5_155

1 Introduction

1.1 Background

The United States National Cancer Institute predicted that in 2016, there would be 1 685 210 new cases of cancer diagnosis and 595 690 people would die of cancer nationally. About 60% of cancer patients receive radiation therapy, and for early stage non-small cell lung cancer, radiation therapy is recommended as a component of standard treatment care. Depending on the location of the tumor in the lung whether or not it is attached to other tissue structures, a tumor may be highly mobile and also able to change shape during a patient's breathing cycle. The primary goal of cancer radiation treatment is to ensure the accurate delivery of the dose of radiation to the tumor and simultaneously minimizing the side-effects for the surrounding tissue and critical structures. However, these previous tracking approaches are not achievable for use with three-dimensional (3D) tumor shapes in real time, a potentially important factor for accurate dose delivery. Tumor respiratory motion also induces tumor deformations. A facet of this personalization of radiation treatment of lung cancer is assessing creating parameters the motion trajectory of a patient's mobile lung tumor during the course of their breathing cycle. This knowledge is then used to either apply radiation to the tumor during a small interval of that trajectory, or control the treatment delivery so that radiation is continuously applied to the tumor by directing radiation to the full motion trajectory of the tumor while the patient is breathing freely.

1.2 Research Meaning

In recent times, lung disease, especially lung cancer, has become one of the main threats to human life and consequently causes the highest number of deaths in all types of cancer. Lobectomy is still the primary and main choice for the treatment of lung cancer as doctors are confident on the position of lung tissue and surgery is more likely to be successful. Therefore, it is vital we find a reasonable method for the segmentation of the pulmonary vessels imminently, and are able to make a reasonable scheme for the resection of the lesion. Compared to an entire lung resection, resection of one or several sections of the pulmonary system of patients has little impact on their health. Due to the pulmonary segments having a separate pulmonary artery and vein, it has been found that the pulmonary artery and pulmonary vein are an important aspect needed to find lung segments, and facilitate a more precise surgery scheme which can greatly improve the success rate of the operation and therefore, reduce the mortality of lung cancer.

Medical image segmentation is an important part of the medical image processing, which plays a very important role in clinical medical treatment, [1]. Image segmentation is used to divide the image into a series of areas that do not coincide with each other, and then extract the target area of interest. Segmentation of pulmonary vessels is the extraction of pulmonary vessels from the lungs, which in turn plays an important role in the successful removal of lesions. Pulmonary and pulmonary veins extracted from pulmonary vessels have significant practical and social value for early screening prevention and early treatment of lung cancer.

Pulmonary CT images not only have poor contrast, background noise and small pulmonary vascular structure, but for people with unusual organ shape or other characteristics, they pose difficulties. Scholars all over the word have studied the segmentation of blood vessels for many years. Several approaches of segmenting the pulmonary vessels are relatively well researched. Compared with the segmentation of the pulmonary vessels, extraction of pulmonary arteries and pulmonary veins is a lot more complicated, but some advances have been achieved. Yamaguchi [3] used CT images of the key layer of the stable position of the pulmonary artery along with the two characteristics of the bronchus and extracted them using the region growing algorithm, which requires manual interaction.

When the pulmonary artery and pulmonary veins make contact with each other, it can lead to unreliable results in segmentation. Park and his colleague [5] used an algorithm which is based on the minimum spanning tree to extract the pulmonary artery and pulmonary vein. This method processes the fine peripheral branches of the tree to achieve high segmentation accuracy. However, this method is used for non-angiographic imaging and requires manual intervention. To fix this problem, Gao [6] uses the ambiguous distance transform and morphological operations to correctly extract the pulmonary and pulmonary veins under manual interaction and achieves a higher accuracy. In addition to this, Bülow T and his colleague [7] proposed direct separation of the pulmonary and pulmonary veins by exploiting the spatial distribution of the pulmonary artery with tracheal tree growth, which meant there was no need to separate the vascular tree first. Sebbe [8] uses a fast-moving algorithm which spreads in the direction of minimum cost to extract pulmonary arteries. The disadvantage of this method is that it is dramatically affected by the resolution and noise. To find a more suitable treatment process, Ebrahimdoost [9] used a three-dimensional level set algorithm to extract the pulmonary artery. Although this method is performed on angiographic images, it can only be used for pulmonary artery segmentation so it needs to be modified. Mekada [10] catches the spatial distribution characteristics of pulmonary arteries and pulmonary veins to extract them resulted in high accuracy of this method at 80–95% and allowed automatic segmentation to be performed on angiographic images. Unfortunately, this method still requires improvements such as the use of Pulmonary artery and pulmonary vein characteristics to induce pulmonary vascular segmentation accuracy improvement. Due to the resolution of the CT scan and high pulmonary vascular spatial density, pulmonary arteries and pulmonary veins in many parts of the CT image showed some adhesion and part of the volume effect exacerbates this point by reducing the spatial resolution. A number of papers have found that pulmonary and parts of the pulmonary venous adhesions complicate the automatic separation of the pulmonary and pulmonary veins [4–6]. In light of this, many people are already studying the segmentation of the pulmonary vessels, but only a few are attempting automated separation of pulmonary and pulmonary veins [4–7].

2 Overall Flow

2.1 Expansion

Let A and B be a set of two-dimensional integer controls Z2, where B is the structural element and A is the target element (Fig. 1).

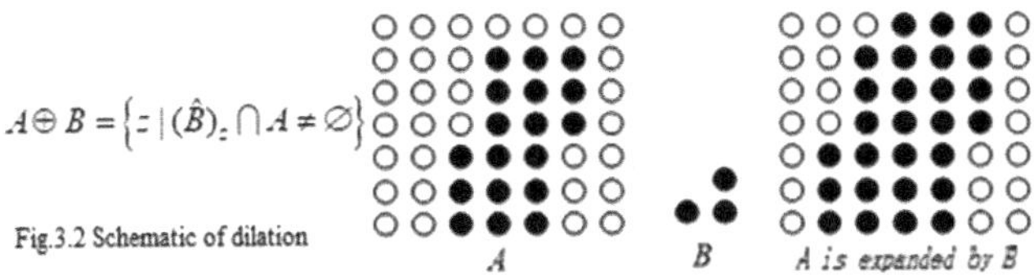

Fig. 1. Expansion process

2.2 Erosion

The process of erosion is shown in Fig. 2. Fractional differential enhancement method principle can improve the signal quality. Applying fractional differential to image enhancement[30] can also make the details of the image clearer and the edge information more prominent without affecting the smooth area in the image.

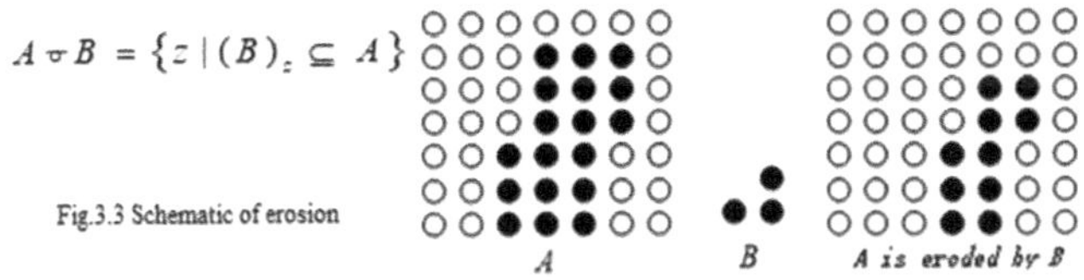

Fig. 2. A eroded by B

In this paper, Grümwald-Letnikov (G-L) is used to define the numerical method of fractional differential in an image. The period of the unary signal w(t) is t $\in [a, t]$ and the signal period [a, t] equally divided by a time interval (such as h = 1, 2, …). From this, with regard to CT images, the differential expression of the fractional differential and the separability of the two-dimensional Fourier transform and the differential expression of the partial fractional differential, can be obtained as follows:

$$\frac{dw(t)}{dt} \approx w(t) + (-v)w(t-1) + \frac{-v(-v+1)}{2}w(t-2) + \cdots + \frac{\Gamma(-v+1)}{n!(-v+n+n)}w(t-n) \tag{1}$$

$$\frac{\partial w(x,y)}{\partial x} \approx w(x,y) + (-v)w(x-1,y) + \frac{-v(-v+1)}{2}w(x-2,y) + \cdots + \frac{\Gamma(-v+1)}{(n-1)!\Gamma(-v+n+n)}w(x-n+1,y) \tag{2}$$

$$\frac{\partial w(x,y)}{\partial x} \approx w(x,y) + (-v)w(x,y-1) + \frac{-v(-v+1)}{2}w(x,y-2) + \cdots + \frac{\Gamma(-v+1)}{(n-1)!\Gamma(-v+n+n)}w(x,y-n) \tag{3}$$

Using the above differential expressions to construct differential operators such as 3 * 3, 5 * 5 to process the image, greatly improves the detail provided by the image in (1)–(3).

3 Design and Implementation of Pulmonary Vascular Division

The pulmonary vessels are connected to the heart and their grayscales in the CT image are basically the same. Therefore, in order to completely separate the pulmonary vessels from other tissues such as the heart, the lungs must be extracted separately, then segmenting pulmonary vascular in the pulmonary parenchyma can occur. In this paper, Fig. 3, we design an overall flow chart of segmented pulmonary vessels.

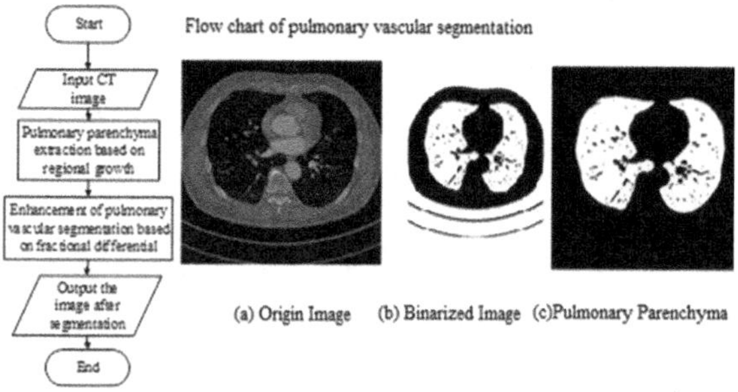

Fig. 3. Segmentation of lung parenchyma

4 Pulmonary Parenchyma Segmentation Method Based on Adaptive Threshold 3D Region Growth

Since the distribution of lung parenchyma density is relatively uniform, the grayscale value of pulmonary parenchyma on the CT image is also relatively uniform. As a result of this, the optimal segmentation threshold T of the intermediate layer CT image can be obtained by the maximum inter-class variance method. This can replace the whole threshold lung parenchymal segmentation, reducing calculation time and improving operational efficiency to meet actual needs. The threshold T is used as the binary form of the segmentation threshold for CT images. The binary CT images also carry other tissues. In order to only extract the lung parenchyma, a pixel with a grayscale value of 255 was taken as the initial seed from the middle layer CT image by 3D region growing method. The growth rule is threshold T = 255. As shown in Fig. 3, the image (b) can be obtained by obtaining the binary form of the original image (a) which is to be segmented, and finally the image (c) will be obtained after the region growth.

As the grayscale value of the pulmonary blood vessels is significantly higher than the grayscale value of the lung parenchyma, there are a number of holes in the extracted lung parenchyma. These holes are most likely to be pulmonary nodules and high-density blood vessels. In order to segment this part of the pulmonary blood vessels, these holes need to be filled by the closed-form operation of morphological pre-swelling corrosion, as shown in Fig. 3. The images are, in turn, CT images after segmentation of lung parenchyma, morphologically processed CT images and the original CT image of pulmonary parenchymal.

From the above analysis, we can obtain the flow chart of pulmonary parenchyma segmentation, as shown in Fig. 4.

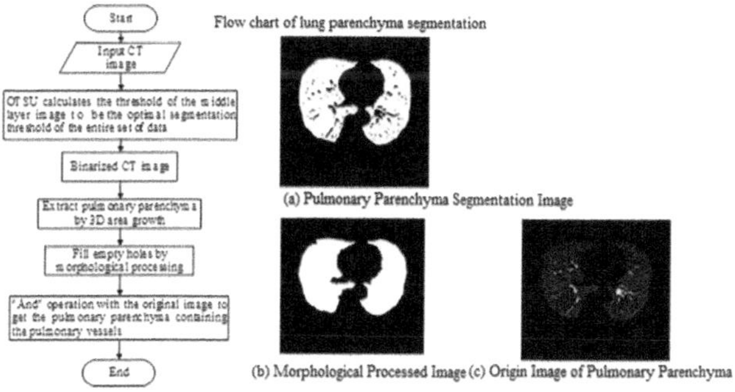

Fig. 4. Flow chart of pulmonary parenchyma segmentation

5 Pulmonary Vascular Segmentation Method Based on Fractional Differential Enhancement

As the differential operator is isotropic, small blood vessels with characteristic direction do not emerge. Therefore, the fractional differential operator template can be constructed according to (1)–(3). The operator has the following functions: little influence on the surrounding smooth region with small changes of grayscale value and enhancing the pixels with larger grayscale value changes around. By analyzing the characteristics of the blood vessels and their distribution in the CT images of the lungs, the pulmonary blood vessels appear cylindrical, and most of them are in four diagonal directions. A 3 * 3 differential operator template is constructed as shown in Fig. 6 (Fig. 5).

Due to the operator not having symmetry, the following method should be used in the calculation: The coordinate of "1" in the operator template should coincide with the coordinate of the pixel to be subjected to fractional differential operation. The templates in the four directions are respectively convoluted with the CT images of the lungs, then the convolution results in all directions are weighted, and the results of the four

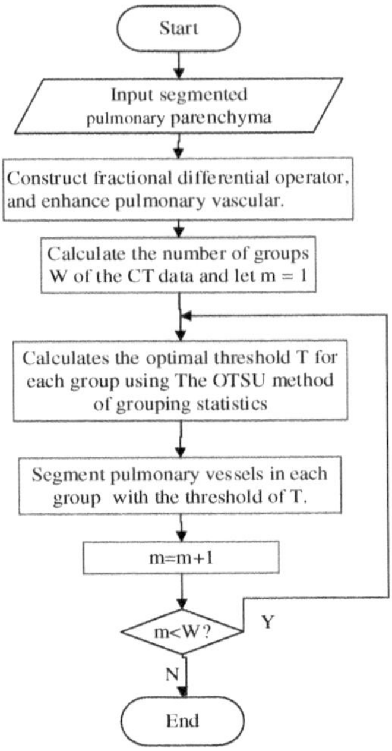

Fig. 5. Differential operator template

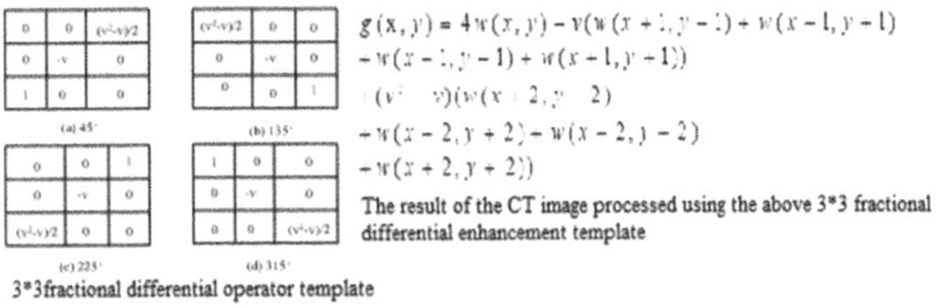

Fig. 6. Fractional differential

directions are summed. The pulmonary vessels enhanced in the CT images can be represented by the Fig. 3.

Figure 7 demonstrates using the fractional differential enhance operator to process pulmonary parenchyma and the fine branches of the pulmonary vessels in the image are highlighted. Pulmonary vascular structures are and its grayscale distribution is uneven. Using global optimal threshold for pulmonary vascular segmentation [36] will result in increased error. Due to the size and shape of those pulmonary vessels, layers which are far away from each other are different. Due to this, the segmentation results are satisfactory when using the optimal segmentation threshold of the middle layer instead of the optimal segmentation threshold of the whole set of data. To segment the pulmonary vessels from enhanced the CT images, we propose a new segmentation method - Grouping Statistics to Calculate the Optimal Threshold of Segmentation. The specific idea of this method is as follows: Each W layer of CT image is divided into a group, the

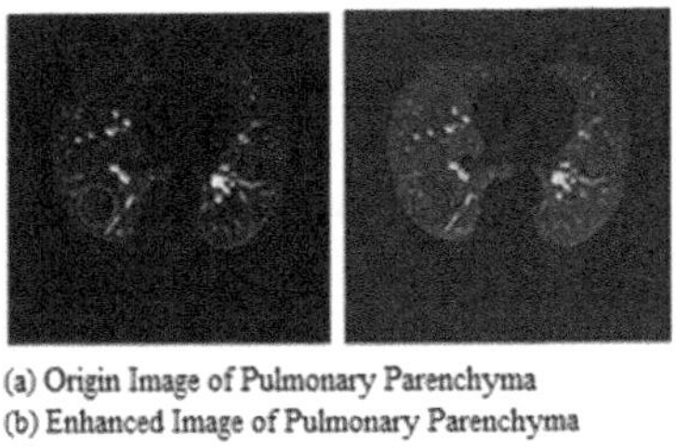

Fig. 7. Pulmonary parenchyma enhancement Schematic

segmentation threshold of this group is replaced by the optimal segmentation threshold of the middle layer of this W layers, and the segmentation threshold of each group is deduced in turn.

Using the analysis above, a pulmonary vascular segmentation flow chart is as shown in Fig. 8.

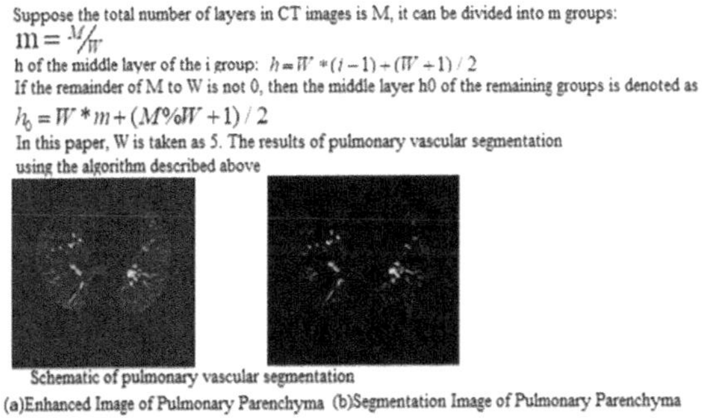

Fig. 8. Flow chart of pulmonary vascular segmentation

6 Experimental Results and Analysis

6.1 Experimental Data

In order to validate the algorithm proposed in this paper to segment the pulmonary vessels, we used 10 sets of CT image data provided by a tertiary referral hospital. Resolution of data is 512 * 512 and layers of data are between 368 and 436.

Experiment Environment

Hardware and software environment the experiment based on is as follows.

Hardware Environment:

CPU: Inter(R) Core(TM) i7-6700 CPU @ 3.40 GHz

RAM: 16.00 GB

HDD: Western Digital, 1 TB

Software Environment:
OS: Windows10, 64bit
Development Language: Matlab, C++
Programming environment and compiler: Visual Studio 2010, Matlab 2016a
Other Software Tools: Microsoft Office Visio

In Experiment 1, CT images were processed with different differential orders. It can be seen in Fig. 9, that the change of pulmonary vascular enhancement in CT images was observed as the differential order was increased. By comparison and analysis, it is obvious that with the increase of differential order, the detail of blood vessel in CT image is augmented, and the effect of fine blood vessel with the order of 0.1 to 0.2 is found to be the best. When the order is greater than 0.2, although the detail of the blood vessels are enhanced, it also suppresses the effect of large blood vessels and reduces the contrast between the thick blood vessel and other tissues. In Experiment 2, data showing pulmonary vascular segmentation from 10 different sets of CT images was analyzed qualitatively and quantitatively on two-dimensional and three-dimensional scale. Firstly, the methods used for global optimum and the grouping statistics were used on the segmentation results of the enhanced pulmonary parenchyma to find the results of the thresholding split. Figure 10 compares these results to the data set of 512 * 512 * 378, the segmentation results of the 150th, the 250th and the 350th were accepted.

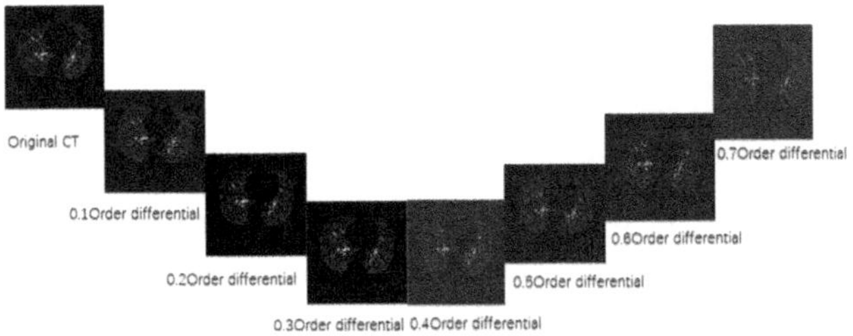

Fig. 9. Different order of CT

The first line in Fig. 10 is an enhanced CT image. The second line is the pulmonary vascular segmentation processed by the global optimal threshold. The third line is the result of the segmentation using the optimal threshold method of the stratified statistical calculation. It can be observed that, the statistical calculation of the optimal threshold for the pulmonary vascular segmentation method can also be divided into smaller blood vessels, not only large ones, thereby enhancing the accuracy of vascular segmentation.

In Experiment 2, we also conducted an experiment to trial methods of obtaining the optimal segmentation threshold by using the stratified statistical calculation for the original CT image and the enhanced CT image. Taking the data of 512 * 512 * 403 as an example, as shown in Fig. 10, the 150th, 250th and 350th layer segmentation

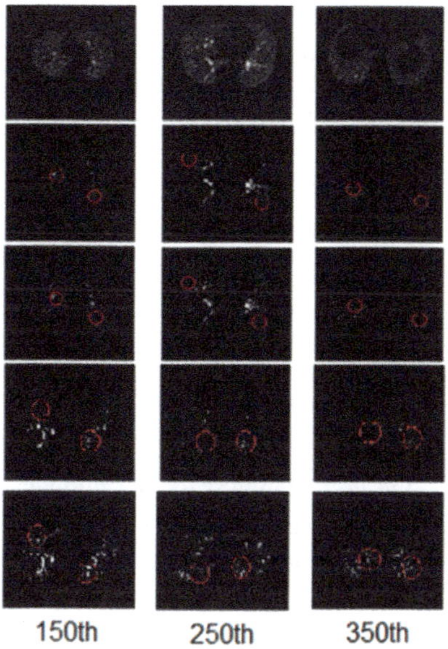

Fig. 10. CT images of different order differential operator treatment

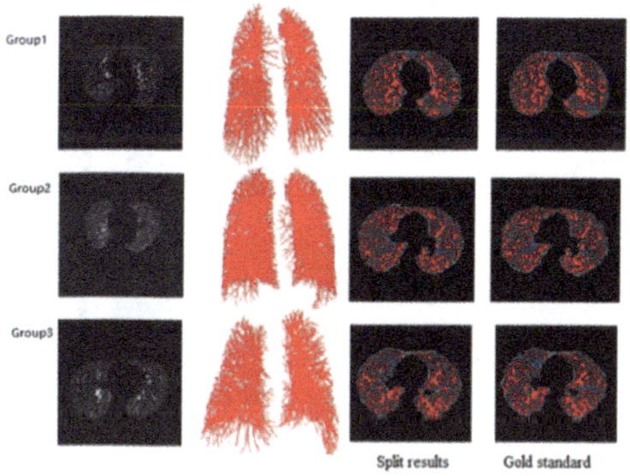

Fig. 11. Aschematic of pulmonary vascular segmentation of this algorithm

results. On the first line of Fig. 10, the pulmonary blood vessel segmentation was performed using the original thresholder tomography under the optimal threshold. The second line refers to the fractional differential enhanced CT image, which was used to calculate the optimal threshold for pulmonary vascular segmentation, using stratified statistics. It can be seen from the figure after the enhancement, by retaining more small blood vessels, the segmentation accuracy is improved.

The above two groups of experiments will be qualitatively verified by the observation of the superiority of the algorithm in this paper, and then we will quantitatively analyze the accuracy of the algorithm to segment the pulmonary vessels. By comparing the segmentation results using this algorithm with the golden standard image, we calculated the leak rate and the over-segmentation rate of this algorithm to objectively evaluate the segmentation algorithm in this paper. The gold standard image is the result of a hand-drawn medical certificate by a professional on the original CT image. Pulmonary vascular segmentation assessment criteria are as follows:

$$\text{Accuracy} = \frac{S \cap T}{T} \tag{4}$$

$$\text{Over rate} = \frac{S - S \cap T}{T} \tag{5}$$

$$\text{Leakage rate} = \frac{T - S \cap T}{T} \tag{6}$$

The above formula is the gold standard for the number of pixels for a pulmonary vascular image This algorithm is divided into pulmonary vascular pixels. The third experiment is to divide the pulmonary blood vessels by this method and conduct quantitative analysis. Taking the CT data of the three groups as an example, Fig. 12 shows the result of the 3D segmentation. In order to see the result of the segmentation and the gold standard image Fig. 11 shows the comparison between the two-dimensional display of the algorithm segmentation results and the golden standard image.

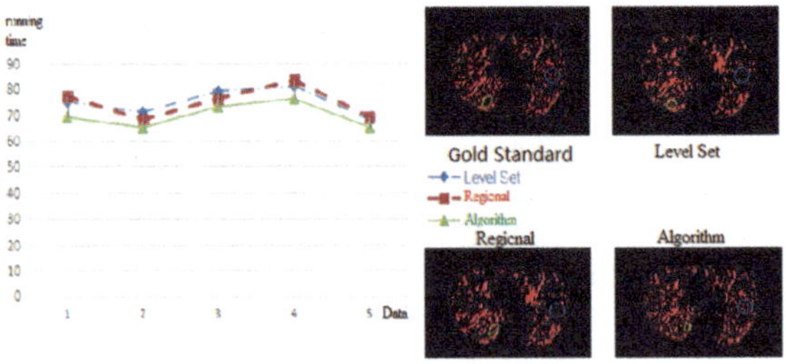

Fig. 12. Running time contrast chart of three methods

The algorithm proposed in this paper is used to calculate the correctness rate, unwanted segmentation rate and missing segmentation rate of the three groups of experimental results using the calculation formula of evaluation criteria for the pulmonary vascular segmentation results. The algorithm is evaluated under these three aspects of evaluation criteria. The experimental results and the evaluation results are shown in Table 1 which shows the statistical data of the results of pulmonary vascular segmentation obtained by the proposed algorithm. By analyzing the three evaluation criteria against the correctness rate, leakage rate and over-segmentation rate of the three experimental results, it can be concluded that the algorithm proposed in this paper is reliable for pulmonary vascular segmentation and can accurately extract pulmonary blood vessels in the lung parenchyma.

Table 1. Pulmonary vascular segment data algorithm

Data	Level	T	S	Accuracy	Leakage	Over seg
1	378	11207111	10952069	90.31%	9.69%	7.41%
2	403	15425248	15112697	89.86%	11.66%	8.11%
3	386	16103221	15667602	91.51%	11.49%	5.78%

Experiment 4 compares the algorithm proposed in this paper with the commonly used methods of pulmonary vascular segmentation, such as regional growth method and level set method. Comparing it with the gold standard image, the correctness, miss rate and over segmentation rate were then calculated. Figure 12 is a comparison of the results of the three methods using the gold standard image. Figure 12(a) is the golden standard image, Fig. 12(b) is the result of level set segmentation, Fig. 12(c) is the result of segmentation by region growing, and Fig. 12(d) is the segmentation result of this algorithm. As can be seen from the results in the figure, the three methods are more than acceptable for the segmentation of the rough blood vessels. However, when comparing the marked part of the blue circle in the figure, we can see that the algorithm can segment small blood vessels. This is finding is reinforced by the smallest mark in the yellow circle in the comparison chart and we can say that the over-segmentation in this paper is the smallest.

Due to the resolution of the CT data being 512 * 512, the difference is that each layer provides a different number of data. Table 2 is the algorithm of this paper, the level set method and the regional growth method of three methods of segmentation results statistics.

Table 2 compares and analyzes three methods using the three evaluation criteria. It can be seen from the statistical results that the segmentation method proposed in this paper is superior to the other two methods both in correctness rate and in the missed segmentation rate and the unwanted-segmentation rate. We are able to say that the accuracy level has reached that of clinical needs.

Table 2. Segmentation result statistics table based on three segmentation methods

Data	F	Segmentation method								
		Level set			Regional growth			Algorithm		
		Accuracy	Leakage	Over-Seg	Accuracy	Leakage	Over-Seg	Accuracy	Leakage	Over-Seg
1	378	90.42%	9.58%	4.43%	89.53%	10.47%	4.11%	91.87%	8.13%	3.15%
2	386	88.51%	11.49%	3.78%	88.17%	11.83%	4.06%	90.53%	9.47%	3.42%
3	375	90.36%	9.64%	3.61%	89.24%	10.76%	3.56%	92.45%	7.55%	3.71%
4	426	90.43%	9.57%	3.55%	90.35%	9.65%	3.63%	92.70%	7.30%	3.11%
5	421	89.58%	10.42%	3.67%	88.91%	11.09%	4.02%	91.36%	8.64%	3.24%
Average	89.86%	10.14%	3.81%	89.24%	10.76%	3.88%	91.78%	8.22%	3.33%	

Figure 12 shows the running time of the three methods. It can be seen from the figure that the running time of the method in this paper is less than 2 min. Compared with the method of pulmonary blood vessel segmentation and based on level set or region growth, the method introduced in this paper has advantages in time performance.

References

1. Wang, Y.: Biomedical Image and Image Processing. Tsinghua University Press, Beijing
2. Zhang, J., Zhan, T., Chen, Y., et al.: Hierarchical segmentation model of improved multi-objective image. J. Image Graphics **15**(4), 617–623 (2010)
3. Yamaguchi, T., Kitasaka, T., Mori, K., et al.: A preliminary study for automated recognition of branches of pulmonary artery and vein using anatomical positional relations from a 3-D chest X-ray CT image. Computer Assisted Radiology and Surgery, Berlin (2002)
4. Saha, P.K., Gao, Z., Alford, S.K., et al.: Topomorphologic separation of fused isointensity objects via multiscale opening: separating arteries and veins in 3-D pulmonary CT. IEEE Trans. Med. Imaging **29**(3), 840–851 (2010)
5. Park, S., Min Lee, S., Kim, N., et al.: Automatic reconstruction of the arterial and venous trees on volumetric chest CT. Med. Phys. **40**(7), 071906 (2013)
6. Gao, Z., Grout, R.W., Holtze, C., et al.: A new paradigm of interactive artery/vein separation in noncontrast pulmonary CT imaging using multiscale topomorphologic opening. IEEE Trans. Biomed. Eng. **59**(11), 3016–3027 (2012)
7. Bülow, T., Wiemker, R., Blaffert, T., et al.: Automatic extraction of the pulmonary artery tree from multi-slice CT data. In: Proceedings of SPIE, vol. 5746, p. 731 (2005)
8. Sebbe, R., Gosselin, B., Coche, E., et al.: Pulmonary arteries segmentation and feature extraction through slice marching. In: ProRISC Workshop on Circuits, Systems and Signal Processing, pp. 1–5 (2003)
9. Ebrahimdoost, Y., Qanadli, S.D., Nikravanshalmani, A., et al.: Automatic segmentation of pulmonary artery (PA) in 3D pulmonary CTA images. In: 2011 17th International Conference on IEEE, pp. 1–5 (2011)
10. Mekada, Y., Nakamura, S., Ide, I., et al.: Pulmonary artery and vein classification using spatial arrangement features from X-ray CT images. In: 7th Asia-pacific Conference on Control and Measurement, pp. 232–235 (2006)
11. Geng, H., Tan, W., Yang, J.: A review of lung tissue segmentation based on CT images. Appl. Res. Comput. **33**(7), 1929–1935 (2016)

Author Index

K. Deng et al. (Eds.): ICMIR 2018, AISC 856, pp. 1275–1280, 2019.
https://doi.org/10.1007/978-3-030-00214-5

GPSR Compliance

The European Union's (EU) General Product Safety Regulation (GPSR) is a set of rules that requires consumer products to be safe and our obligations to ensure this.

If you have any concerns about our products, you can contact us on ProductSafety@springernature.com

In case Publisher is established outside the EU, the EU authorized representative is:

Springer Nature Customer Service Center GmbH
Europaplatz 3
69115 Heidelberg, Germany

Batch number: 10761599

Printed by Printforce, the Netherlands